T0205330

Lecture Notes in Computer Science 14079

The series Lecture Notes in Computer Science (LNCS), including its subseries Lecture Notes in Artificial Intelligence (LNAI) and Lecture Notes in Bioinformatics (LNBI), has established itself as a medium for the publication of new developments in computer science and information technology research, teaching, and education.

LNCS enjoys close cooperation with the computer science R & D community, the series counts many renowned academics among its volume editors and paper authors, and collaborates with prestigious societies. Its mission is to serve this international community by providing an invaluable service, mainly focused on the publication of conference and workshop proceedings and postproceedings. LNCS commenced publication in 1973.

Pat Morin · Subhash Suri
Editors

Algorithms and Data Structures

18th International Symposium, WADS 2023
Montreal, QC, Canada, July 31 – August 2, 2023
Proceedings

Editors
Pat Morin
Carleton University
Ottawa, ON, Canada

Subhash Suri
University of California
Santa Barbara, CA, USA

ISSN 0302-9743 ISSN 1611-3349 (electronic)
Lecture Notes in Computer Science
ISBN 978-3-031-38905-4 ISBN 978-3-031-38906-1 (eBook)
https://doi.org/10.1007/978-3-031-38906-1

This Springer imprint is published by the registered company Springer Nature Switzerland AG
The registered company address is: Gewerbestrasse 11, 6330 Cham, Switzerland

Preface

This volume contains the papers presented at WADS 2023: 18th Algorithms and Data Structures Symposium, held on July 31–August 2, 2023 in Montreal.

WADS, which alternates with the Scandinavian Symposium and Workshops on Algorithm Theory (SWAT), is a venue for researchers in the area of design and analysis of algorithms and data structures to present their work. This year, there were 92 submissions. Each submission was refereed by an average of 3.1 program committee members. The committee decided to accept 47 papers.

The program also includes three invited talks by Valerie King (University of Victoria), Joe Mitchell (Stony Brook University), and Tselis Schramm (Stanford University). This year, WADS took place back to back with the 35th Canadian Conference on Computational Geometry (CCCG 2023) at Concordia University, in Montreal.

Special issues of papers selected from WADS 2023 are planned for two journals, Algorithmica, and Computational Geometry: Theory and Applications.

The Alejandro López-Ortiz Best Paper Award for WADS 2023 was awarded to the paper *Zip-zip Trees: Making Zip Trees More Balanced, Biased, Compact, or Persistent* by Ofek Gila, Michael Goodrich, and Robert Tarjan.

We thank the Program Committee for their hard work and good judgement and thank all the subreviewers who contributed to the reviewing process. We thank Denis Pankratov at Concordia Unversity for taking on the job of local organizer for both WADS and CCCG.

We gratefully acknowledge the WADS 2023 sponsors: Concordia University Gina Cody School of Engineering and Computer Science, Springer, Elsevier, and the Fields Institute.

June 2023

Pat Morin
Subhash Suri

Preface

This volume contains the papers presented at WADS 2023: 18th Algorithms and Data Structures Symposium, held July 31–August 2, 2023 in Montreal.

WADS, which alternates with the Scandinavian Symposium and Workshops on Algorithm Theory (SWAT), is a venue for researchers in the area of design and analysis of algorithms and data structures to present their work. This year, there were 92 submissions. Each submission was refereed by an average of [illegible] program committee members. The committee decided to accept [illegible] papers.

The program also includes three invited talks by Valerie King (University of Victoria), [illegible] Mitchell (Stony Brook University), and [illegible] Schramm (Stanford University). This year WADS took place back to back with the 35th Canadian Conference on Computational Geometry (CCCG 2023) at Concordia University in Montreal.

Special issues of papers selected from WADS 2023 are planned for two journals: Algorithmica and Computational Geometry: Theory and Applications.

The [illegible] Best Paper Award for WADS 2023 was awarded to the paper [illegible] by [illegible] Gibb, Michael [illegible] and [illegible].

We thank the Program Committee for their hard work and good judgement and all the subreviewers who contributed to the reviewing process. We thank Denis [illegible] (Concordia University) for [illegible] WADS and CCCG.

We gratefully acknowledge the WADS 2023 sponsors: Concordia University Gina Cody School of Engineering and Computer Science, Springer, Elsevier, and the Fields Institute.

[illegible] 2023 Pat Morin

Subhash Suri

Preface

This volume contains the papers presented at WADS 2023: 18th Algorithms and Data Structures Symposium, held on July 31–August 2, 2023 in Montreal.

WADS, which alternates with the Scandinavian Symposium and Workshops on Algorithm Theory (SWAT), is a venue for researchers in the area of design and analysis of algorithms and data structures to present their work. This year, there were 92 submissions. Each submission was refereed by an average of 3.1 program committee members. The committee decided to accept 47 papers.

The program also includes three invited talks by Valerie King (University of Victoria), Joe Mitchell (Stony Brook University), and Tselis Schramm (Stanford University). This year, WADS took place back to back with the 35th Canadian Conference on Computational Geometry (CCCG 2023) at Concordia University, in Montreal.

Special issues of papers selected from WADS 2023 are planned for two journals, Algorithmica, and Computational Geometry: Theory and Applications.

The Alejandro López-Ortiz Best Paper Award for WADS 2023 was awarded to the paper *Zip-zip Trees: Making Zip Trees More Balanced, Biased, Compact, or Persistent* by Ofek Gila, Michael Goodrich, and Robert Tarjan.

We thank the Program Committee for their hard work and good judgement and thank all the subreviewers who contributed to the reviewing process. We thank Denis Pankratov at Concordia Unversity for taking on the job of local organizer for both WADS and CCCG.

We gratefully acknowledge the WADS 2023 sponsors: Concordia University Gina Cody School of Engineering and Computer Science, Springer, Elsevier, and the Fields Institute.

June 2023

Pat Morin
Subhash Suri

Organization

Program Committee

Pankaj Agarwal	Duke University, USA
Therese Biedl	University of Waterloo, Canada
Prosenjit Bose	Carleton University, Canada
Maike Buchin	Ruhr Universität Bochum, Germany
Sergio Cabello	University of Ljubljana, Slovenia
Siu-Wing Cheng	Hong Kong University of Science & Technology, China
Maria Chudnovsky	Princeton University, USA
Leah Epstein	University of Haifa, Israel
Omrit Filtser	Open University of Israel, Israel
Pawel Gawrychowski	University of Wroclaw, Poland
Daniel Lokshtanov	University of California, Santa Barbara, USA
Michael Mitzenmacher	Harvard University, USA
Pat Morin	Carleton University, Canada
David Mount	University of Maryland, USA
Lata Narayanan	Concordia University, Canada
Yakov Nekrich	Michigan Technological University, USA
Michał Pilipczuk	University of Warsaw, Poland
Manish Purohit	Google, USA
Benjamin Raichel	University of Texas at Dallas, USA
Jared Saia	University of New Mexico, USA
Saket Saurabh	Institute of Mathematical Sciences, India
Rodrigo Silveira	Universitat Politècnica de Catalunya, Spain
Tatiana Starikovskaya	École Normale Supérieure, France
Subhash Suri	University of California, Santa Barbara, USA
Csaba Toth	California State University Northridge, USA
Birgit Vogtenhuber	Graz University of Technology, Austria
Haitao Wang	University of Utah, USA
Jie Xue	NYU Shanghai, China
Meirav Zehavi	Ben-Gurion University, Israel

Contents

Geometric Spanning Trees Minimizing the Wiener Index

A. Karim Abu-Affash[1(✉)], Paz Carmi[2], Ori Luwisch[2], and Joseph S. B. Mitchell[3]

[1] Software Engineering Department, Shamoon College of Engineering, 84100 Beer-Sheva, Israel
abuaa1@sce.ac.il

[2] Department of Computer Science, Ben-Gurion University, 84105 Beer-Sheva, Israel
carmip@cs.bgu.ac.il, orilu@post.bgu.ac.il

[3] Department of Applied Mathematics and Statistics, Stony Brook University, Stony Brook, NY, USA
joseph.mitchell@stonybrook.edu

Abstract. The Wiener index of a network, introduced by the chemist Harry Wiener [30], is the sum of distances between all pairs of nodes in the network. This index, originally used in chemical graph representations of the non-hydrogen atoms of a molecule, is considered to be a fundamental and useful network descriptor. We study the problem of constructing geometric networks on point sets in Euclidean space that minimize the Wiener index: given a set P of n points in $\mathbb{R}^d$, the goal is to construct a network, spanning P and satisfying certain constraints, that minimizes the Wiener index among the allowable class of spanning networks.

In this work, we focus mainly on spanning networks that are trees and we focus on problems in the plane ($d = 2$). We show that any spanning tree that minimizes the Wiener index has non-crossing edges in the plane. Then, we use this fact to devise an $O(n^4)$-time algorithm that constructs a spanning tree of minimum Wiener index for points in convex position. We also prove that the problem of computing a spanning tree on P whose Wiener index is at most W, while having total (Euclidean) weight at most B, is NP-hard.

Computing a tree that minimizes the Wiener index has been studied in the area of communication networks, where it is known as the *optimum communication spanning tree problem.*

Keywords: Wiener Index · Optimum communication spanning tree · Minimum routing cost spanning tree

This work was partially supported by Grant 2016116 from the United States – Israel Binational Science Foundation. Work by P. Carmi and O. Luwisch was partially supported by the Lynn and William Frankel Center for Computer Sciences. Work by J. Mitchell was partially supported by NSF (CCF-2007275) and by DARPA (Lagrange).

P. Morin and S. Suri (Eds.): WADS 2023, LNCS 14079, pp. 1–14, 2023.
https://doi.org/10.1007/978-3-031-38906-1_1

1 Introduction

The *Wiener index* of a weighted graph $G = (V, E)$ is the sum, $\sum_{u,v \in V} \delta_G(u, v)$, of the shortest path lengths in the graph between every pair of vertices, where $\delta_G(u, v)$ is the weight of the shortest (minimum-weight) path between u and v in G. The Wiener index was introduced by the chemist Harry Wiener in 1947 [30]. The Wiener index and its several variations have found applications in chemistry, e.g., in predicting the antibacterial activity of drugs and modeling crystalline phenomena. It has also has been used to give insight into various chemical and physical properties of molecules [28] and to correlate the structure of molecules with their biological activity [20]. The Wiener index has become part of the general scientific culture, and it is still the subject of intensive research [2,10,12, 32]. In its applications in chemistry, the Wiener index is most often studied in the context of unweighted graphs. The study of minimizing the sum of interpoint distances also arises naturally in the network design field, where the problem of computing a spanning tree of minimum Wiener index is known as the *Optimum Communication Spanning Tree* (OCST) problem [15,18].

Given an undirected graph $G = (V, E)$ and a (nonnegative) weight function on the edges of G, representing the delay on each edge, the routing cost $c(T)$ of a spanning tree T of G is the sum of the weights (delays) of the paths in T between every pair of vertices: $c(T) = \sum_{u,v \in V} \delta_T(u, v)$, where $\delta_T(u, v)$ is the weight of the (unique) path between u and v in T. The OCST problem aims to find a minimum routing cost spanning tree of a given weighted undirected graph G, thereby seeking to minimize the expected cost of a path within the tree between two randomly chosen vertices. The OCST was originally introduced by Hu [18] and is known to be NP-complete in graphs, even if all edge weights are 1 [19]. Wu et al. [31] presented a polynomial time approximation scheme (PTAS) for the OCST problem. Specifically, they showed that the best k-star (a tree with at most k internal vertices) yields a $(\frac{k+3}{k+1})$-approximation for the problem, resulting in a $(1 + \varepsilon)$-approximation algorithm of running time $O(n^{2\lceil \frac{2}{\varepsilon} \rceil - 2})$.

While there is an abundance of research related to the Wiener index, e.g., computing and bounding the Wiener indexes of specific graphs or classes of graphs [16,17,24] and explicit formulas for the Wiener index for special classes of graphs [3,23,26,29,30], to the best of our knowledge, the Wiener index has not received much attention in geometric settings. In this work, we study the Wiener index and the optimum communication spanning tree problem in selected geometric settings, hoping to bring this important and highly applicable index to the attention of computational geometry researchers.

Our Contributions and Overview. Let P be a set of n points in the plane. We study the problem of computing a spanning tree on P that minimizes the Wiener index when the underlying graph is the complete graph over P, with edge weights given by their Euclidean lengths. In Sect. 2, we prove that the optimal tree (that minimizes the Wiener index) has no crossing edges. As our main algorithmic result, in Sect. 3, we give a polynomial-time algorithm to solve the problem when the points P are in convex position; this result strongly utilizes the structural result that the edges of an optimal tree do not cross, which enables us to devise a dynamic programming algorithm to optimize. Then, in Sect. 4, we prove that the "Euclidean Wiener Index Tree Problem", in which we seek a spanning tree on P whose Wiener index is at most W, while having total (Euclidean) weight at most B, is (weakly) NP-hard. Finally, in Sect. 5, we discuss the problem of finding a minimum Wiener index *path* spanning P.

Related Work. A problem related to ours is the minimum latency problem, also known as the traveling repairman problem TRP: Compute a path, starting at point s, that visits all points, while minimizing the sum of the distances (the "latencies") along the path from s to every other point (versus between *all* pairs of points, as in the Wiener index). There is a PTAS for TRP (and the k-TRP, with k repairmen) in the Euclidean plane and in weighted planar graphs [27].

Wiener index optimization also arises in the context of computing a noncontracting embedding of one metric space into another (e.g., a line metric or a tree metric) in order to minimize the average distortion of the embedding (defined to be the sum of all pairs distances in the new space, divided by the sum of all pairs distances in the original space). It is NP-hard to minimize average distortion when embedding a tree metric into a line metric; there is a constant-factor approximation (based on the k-TRP) for minimizing the average distortion in embedding a metric onto a line (i.e., finding a spanning path of minimum Wiener index) [11], which, using [27], gives a $(2 + \varepsilon)$-approximation in the Euclidean plane.

A related problem that has recently been examined in a geometric setting is the computation of the Beer index of a polygon P, defined to be the probability that two randomly (uniformly) distributed points in P are visible to each other [1]; the same paper also studies the problem of computing the expected distance between two random points in a polygon, which is, like the Wiener index, based on computing the sum of distances (evaluated as an integral in the continuum) between all pairs of points.

Another area of research that is related to the Wiener index is that of *spanners*: Given a weighted graph G and a real number $t > 1$, a *t-spanner* of G is a spanning sub-graph G^* of G, such that $\delta_{G^*}(u, v) \leq t \cdot \delta_G(u, v)$, for every two vertices u and v in G. Thus, the shortest path distances in G^* approximate the shortest path distances in the underlying graph G, and the parameter t represents the approximation ratio. The smallest t for which G^* is a t-spanner of G is known as the *stretch factor*. There is a vast literature on spanners, especially in geometry (see, e.g., [4–8,13,22,25]). In a geometric graph, G, the *stretch factor* between two vertices, u and v, is the ratio between the length of the shortest path from u to v in G and the Euclidean distance between u and v. The *average stretch factor* of G is the average stretch factor taken over all pairs of vertices in G. For a given weighted connected graph $G = (V, E)$ with positive edge weights and a positive value W, the *average stretch factor spanning tree* problem seeks a spanning tree T of G such that the average stretch factor (over $\binom{n}{2}$ pairs of vertices) is bounded by W. For points in the Euclidean plane, one can construct in polynomial time a spanning tree with constant average stretch factor [9].

2 Preliminaries

Let P be a set of n points in the plane and let $G = (P, E)$ be the complete graph over P. For each edge $(p, q) \in E$, let $w(p, q) = |pq|$ denote the weight of (p, q), given by the Euclidean distance, $|pq|$, between p and q. Let T be a spanning tree of P. For points $p, q \in P$, let $\delta_T(p, q)$ denote the weight of the (unique) path between p and q in T. Let $W(T) = \sum_{p,q \in P} \delta_T(p, q)$ denote the Wiener index of T, given by the sum of the weights of the paths in T between every pair of points. Finally, for a point $p \in P$, let $\delta_p(T) = \sum_{q \in P} \delta_T(p, q)$ denote the total weight of the paths in T from p to every point of P.

Theorem 1. *Let T be a spanning tree of P that minimizes the Wiener index. Then, T is planar.*

Proof. Assume towards a contradiction that there are two edges (a, c) and (b, d) in T that cross each other. Let F be the forest obtained by removing the edges (a, c) and (b, d) from T. Thus F contains three sub-trees. Assume, w.l.o.g., that a and b are in the same sub-tree T_{ab}, and c and d are in separated sub-trees T_c and T_d, respectively; see Fig. 1. Let n_{ab}, n_c, and n_d be the number of points in T_{ab}, T_c, and T_d, respectively. Thus,

$$\begin{aligned} W(T) = W(T_{ab}) &+ n_c \cdot \delta_a(T_{ab}) + n_d \cdot \delta_b(T_{ab}) \\ &+ W(T_c) + (n_{ab} + n_d) \cdot \delta_c(T_c) + n_c(n_{ab} + n_d) \cdot |ac| \\ &+ W(T_d) + (n_{ab} + n_c) \cdot \delta_d(T_d) + n_d(n_{ab} + n_c) \cdot |bd| \\ &+ n_c \cdot n_d \cdot \delta_T(a, b) . \end{aligned}$$

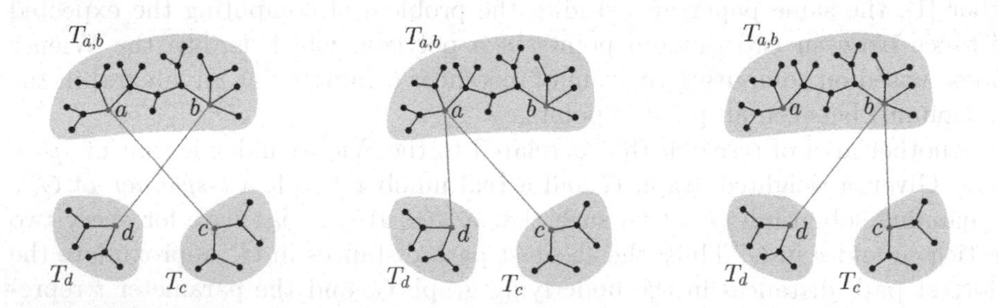

Fig. 1. The trees T, T', and T'' (from left to right).

Let T' be the spanning tree of P obtained from T by replacing the edge (b, d) by the edge (a, d). Similarly, let T'' be the spanning tree of P obtained from T by replacing the edge (a, c) by the edge (b, c). Thus,

$$\begin{aligned} W(T') = W(T_{ab}) &+ (n_c + n_d) \cdot \delta_a(T_{ab}) \\ &+ W(T_c) + (n_{ab} + n_d) \cdot \delta_c(T_c) + n_c(n_{ab} + n_d) \cdot |ac| \\ &+ W(T_d) + (n_{ab} + n_c) \cdot \delta_d(T_d) + n_d(n_{ab} + n_c) \cdot |ad| , \end{aligned}$$

and

$$\begin{aligned}W(T'') = W(T_{ab}) &+ (n_c + n_d) \cdot \delta_b(T_{ab}) \\ &+ W(T_c) + (n_{ab} + n_d) \cdot \delta_c(T_c) + n_c(n_{ab} + n_d) \cdot |bc| \\ &+ W(T_d) + (n_{ab} + n_c) \cdot \delta_d(T_d) + n_d(n_{ab} + n_c) \cdot |bd| \,.\end{aligned}$$

Therefore,

$$\begin{aligned}W(T) - W(T') = n_d\big(\delta_b(T_{ab}) - \delta_a(T_{ab})\big) + n_d(n_{ab} + n_c)\big(|bd| - |ad|\big) \\ + n_c \cdot n_d \cdot \delta_T(a, b) \,,\end{aligned}$$

and

$$\begin{aligned}W(T) - W(T'') = n_c\big(\delta_a(T_{ab}) - \delta_b(T_{ab})\big) + n_c(n_{ab} + n_d)\big(|ac| - |bc|\big) \\ + n_c \cdot n_d \cdot \delta_T(a, b) \,.\end{aligned}$$

If $W(T) - W(T') > 0$ or $W(T) - W(T'') > 0$, then this contradicts the minimality of T, and we are done.

Assume that $W(T) - W(T') \le 0$ and $W(T) - W(T'') \le 0$. Since $n_c > 0$ and $n_d > 0$, we have

$$\delta_b(T_{ab}) - \delta_a(T_{ab}) + (n_{ab} + n_c)\big(|bd| - |ad|\big) + n_c \cdot \delta_T(a, b) \le 0 \,,$$

and

$$\delta_a(T_{ab}) - \delta_b(T_{ab}) + (n_{ab} + n_d)\big(|ac| - |bc|\big) + n_d \cdot \delta_T(a, b) \le 0 \,.$$

Thus, by summing these inequalities, we have

$$(n_{ab} + n_c)\big(|bd| - |ad|\big) + (n_{ab} + n_d)\big(|ac| - |bc|\big) + (n_c + n_d) \cdot \delta_T(a, b) \le 0 \,.$$

That is,

$$\begin{aligned}n_{ab}\big(|bd| + |ac| - |ad| - |bc|\big) + n_c\big(|bd| + \delta_T(a, b) - |ad|\big) \\ + n_d\big(|ac| + \delta_T(a, b) - |bc|\big) \le 0 \,.\end{aligned}$$

Since $n_{ab}, n_c, n_d > 0$, and, by the triangle inequality, $|bd| + |ac| - |ad| - |bc| > 0$, $|bd| + \delta_T(a, b) - |ad| > 0$, and $|ac| + \delta_T(a, b) - |bc| > 0$, this is a contradiction. □

3 An Exact Algorithm for Points in Convex Position

Let $P = \{p_1, p_2, \ldots, p_n\}$ be a set of n points in convex position in the plane, ordered in clockwise-order with an arbitrary first point p_1; see Fig. 2. For simplicity of presentation, we assume that all indices are taken modulo n. For each $1 \le i \le j \le n$, let $P[i, j] \subseteq P$ be the set $\{p_i, p_{i+1}, \ldots, p_j\}$. Let $T_{i,j}$ be a spanning tree of $P[i, j]$, and let $W(T_{i,j})$ denote its Wiener index. For a point $x \in \{i, j\}$,

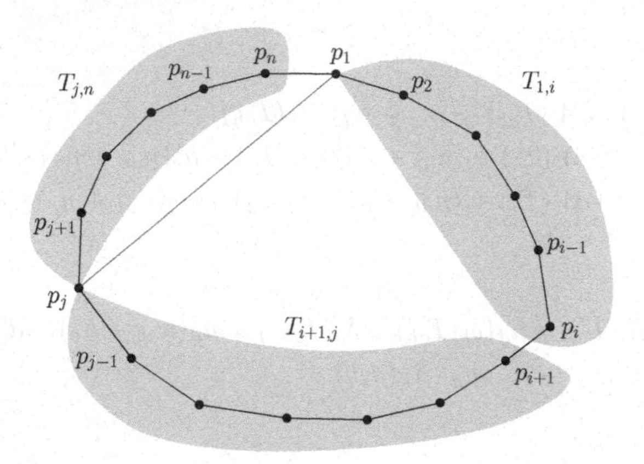

Fig. 2. The convex polygon that is obtained from P. p_1 is connected to p_j in T.

let $\delta_x(T_{i,j})$ be the total weight of the shortest paths from p_x to every point of $P[i,j]$ in $T_{i,j}$. That is $\delta_x(T_{i,j}) = \sum_{p \in P[i,j]} \delta_{T_{i,j}}(p_x, p)$.

Let T be a minimum Wiener index spanning tree of P and let W^* be its Wiener index, i.e., $W^* = W(T)$. Notice that, for any $1 \le i < j \le n$, the points in $P[i,j]$ are in convex position, since the points in P are in convex position. Since T is a spanning tree, each point, particularly p_1, is adjacent to at least one edge in T. Let p_j be the point with maximum index j that is connected to p_1 in T. Moreover, by Theorem 1, T is planar. Thus, there exists an index $1 \le i \le j$ such that all the points in $P[1,i]$ are closer to p_1 than to p_j in T, and all the points in $P[i+1,j]$ are closer to p_j than to p_1 in T. Let $T_{1,i}$, $T_{i+1,j}$, and $T_{j,n}$ be the sub-trees of T containing the points in $P[1,i]$, $P[i+1,j]$, and $P[j,n]$, respectively; see Fig. 2. Hence,

$$\begin{aligned} W^* &= W(T_{1,i}) + (n-i) \cdot \delta_1(T_{1,i}) \\ &\quad + W(T_{i+1,j}) + (n-j+i) \cdot \delta_j(T_{i+1,j}) \\ &\quad + W(T_{j,n}) + (j-1) \cdot \delta_j(T_{j,n}) \\ &\quad + i(n-i) \cdot |p_1 p_j|. \end{aligned}$$

For $1 \le i < j \le n$, let $W_j[i,j] = W(T_{i,j}) + (n-j+i-1) \cdot \delta_j(T_{i,j})$ be the minimum value obtained by a spanning tree $T_{i,j}$ of $P[i,j]$ rooted at p_j. Similarly, let $W_i[i,j] = W(T_{i,j}) + (n-j+i-1) \cdot \delta_i(T_{i,j})$ be the minimum value obtained by a spanning tree $T_{i,j}$ of $P[i,j]$ rooted at p_i. Thus, we can write W^* as

$$W^* = W_1[1,n] = W_1[1,i] + W_j[i+1,j] + W_j[j,n] + i(n-i) \cdot |p_1 p_j|.$$

Therefore, in order to compute W^*, we compute $W_1[1,i]$, $W_j[i+1,j]$, $W_j[j,n]$, and $i(n-i) \cdot |p_1 p_j|$ for each j between 2 and n and for each i between 1 and j, and take the minimum over the sum of these values. In general, for every $1 \le i < j \le n$, we compute $W_j[i,j]$ and $W_i[i,j]$ recursively using the following formulas; see also Fig. 3.

$$W_j[i,j] = \min_{\substack{i \le k < j \\ k \le l < j}} \{W_k[i,k] + W_k[k,l] + W_j[l+1,j] + (l-i+1)(n-l+i-1) \cdot |p_k p_j|\},$$

and

$$W_i[i,j] = \min_{\substack{i < k \le j \\ i \le l < k}} \{W_i[i,l] + W_k[l+1,k] + W_k[k,j] + (j-l)(n-j+l) \cdot |p_i p_k|\}.$$

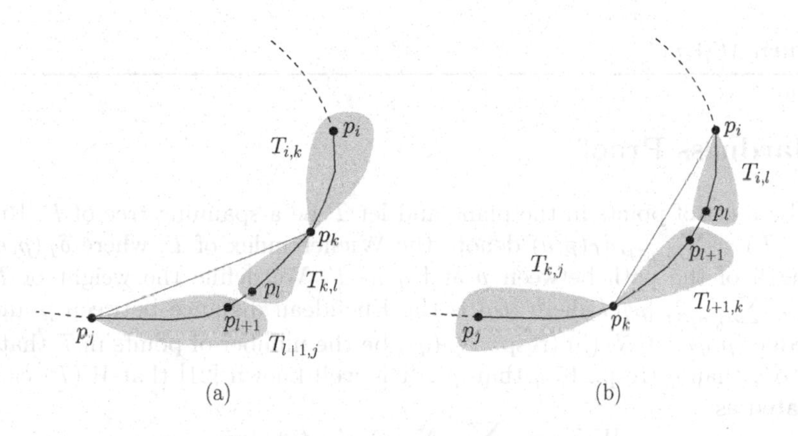

Fig. 3. A sub-problem defined by $P[i,j]$. (a) Computing $W_j[i,j]$. (b) Computing $W_i[i,j]$.

We compute $W_j[i,j]$ and $W_i[i,j]$, for each $1 \le i < j \le n$, using dynamic programming as follows. We maintain two tables $\overrightarrow{M}$ and $\overleftarrow{M}$ each of size $n \times n$, such that $\overrightarrow{M}[i,j] = W_j[i,j]$ and $\overleftarrow{M}[i,j] = W_i[i,j]$, for each $1 \le i < j \le n$. We fill in the tables using Algorithm 1.

Notice that when we fill the cell $\overrightarrow{M}[i,j]$, all the cells $\overrightarrow{M}[i,k]$, $\overleftarrow{M}[k,l]$, and $\overrightarrow{M}[l+1,j]$, for each $i \le k < j$ and for each $k \le l < j$, are already computed, and when we fill the cell $\overleftarrow{M}[i,j]$, all the cells $\overleftarrow{M}[i,l]$, $\overrightarrow{M}[l+1,k]$, and $\overleftarrow{M}[k,j]$, for each $i < k \le j$ and for each $i \le l < k$, are already computed. Thus, each cell in the table is computed in $O(n^2)$ time, and the whole table is computed in $O(n^4)$ time. Therefore, $W^* = W_1[1,n] = \overleftarrow{M}[1,n]$ can be computed in $O(n^4)$ time.

The following theorem summarizes the result of this section.

Theorem 2. *Let P be a set of n points in convex position. Then, a spanning tree of P of minimum Wiener index can be computed in $O(n^4)$ time.*

Algorithm 1. *ComputeOptimal*(P)

1: $n \leftarrow |P|$
2: **for** each $i \leftarrow 1$ to n **do**
  $\overrightarrow{M}[i,i] \leftarrow 0$
  $\overleftarrow{M}[i,i] \leftarrow 0$
3: **for** each $j \leftarrow n$ to 1 **do**
  for each $i \leftarrow j$ to n **do**
   $\overrightarrow{M}[i,j] \leftarrow \min_{\substack{i\le k<j \\ k\le l<j}} \{\overrightarrow{M}[i,k] + \overleftarrow{M}[k,l] + \overrightarrow{M}[l+1,j] + (l-i+1)(n-l+i-1)\cdot|p_k p_j|\}$
   $\overleftarrow{M}[i,j] \leftarrow \min_{\substack{i< k\le j \\ i\le l<k}} \{\overleftarrow{M}[i,l] + \overrightarrow{M}[l+1,k] + \overleftarrow{M}[k,j] + (j-l)(n-j+l)\cdot|p_i p_k|\}$
4: **return** $\overleftarrow{M}[1,n]$

4 Hardness Proof

Let P be a set of points in the plane and let T be a spanning tree of P. Recall that $W(T) = \sum_{p,q\in P} \delta_T(p,q)$ denote the Wiener index of T, where $\delta_T(p,q)$ is the length of the path between p and q in T. We define the weight of T as $wt(T) = \sum_{(p,q)\in T} |pq|$, where $|pq|$ is the Euclidean distance between p and q. For a edge (p,q), let $N_T(p)$ (resp., $N_T(q)$) be the number of points in T that are closer to q than q (resp., to q than p). It is well known [21] that $W(T)$ can be formulated as:

$$W(T) = \sum_{(p,q)\in T} N_T(p)\cdot N_T(q)\cdot |pq|.$$

In this section, we prove that the following problem is NP-hard.

Euclidean Wiener Index Tree Problem: Given a set P of points in the plane, a cost W, and a budget B, decide whether there exists a spanning tree T of P, such that $W(T) \le W$ and $wt(T) \le B$.

Theorem 3. *The Euclidean Wiener Index Tree Problem is weakly NP-hard.*

Proof. Inspired by Carmi and Chaitman-Yerushalmi [8], We reduce from the Partition problem, which is known to be NP-hard [14], to the Euclidean Wiener Index Tree Problem. In the Partition problem, we are given a set $X = \{x_1, x_2, \ldots, x_n\}$ of n positive integers with even $R = \sum_{i=1}^{n} x_i$, and the goal is to decide whether there is a subset $S \subseteq X$, such that $\sum_{x_i\in S} x_i = \frac{1}{2}R$.

Given an instance $X = \{x_1, x_2, \ldots, x_n\}$ of the Partition problem, where x_i's are integers, we construct a set P of $m = n^3 + 3n$ points as follows. The set P consists of n points $p_1, p_2, \ldots, p_n$ located equally spaced on a circle of radius nR, a cluster C of n^3 points located on the center of the circle. Moreover, for each $1 \le i \le n$, we locate two points l_i and r_i both at distance x_i from p_i and the distance between them is $\frac{1}{2}x_i$; see Fig. 4. Finally, we set

$$B = \left(n^2 + \frac{7}{4}\right)R, \text{ and}$$
$$W = 3n^2(m-3)R + \left(\frac{9}{4}m - \frac{13}{4}\right)R$$
$$= 3n^5R + \frac{45}{4}n^3R - 9n^2R + \frac{27}{4}nR - \frac{13}{4}R.$$

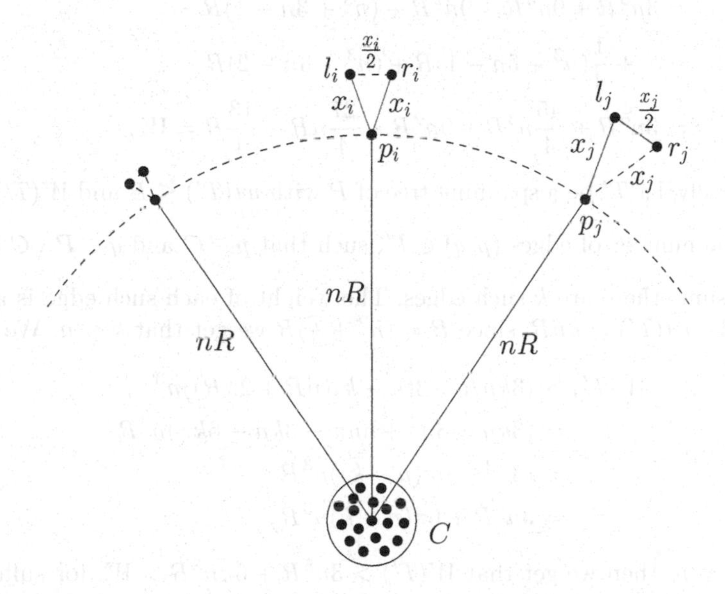

Fig. 4. The set P produced by the reduction. Connecting the points l_j, r_j, and p_j for $x_j \in S$ (blue) and connecting the points l_i, r_i, and p_i for $x_i \in X \setminus S$ (red). (Color figure online)

Assume that there exists a set $S \subseteq X$, such that $\sum_{x_i \in S} x_i = \frac{1}{2}R$. We construct a spanning tree T for the points in P as follows:

- Select an arbitrary point $s \in C$ and connect it to all the points in $C \cup \{p_1, p_2, \ldots, p_n\}$ as a star centered at s.
- For each $1 \le i \le n$, connect the points p_i and l_i.
- For each $x_i \in S$, connect the points p_i and r_i.
- For each $x_i \in X \setminus S$, connect the points r_i and l_i; see Fig. 4.

It is easy to see that $wt(T) = n^2R + R + \frac{3}{4}R = (n^2 + \frac{7}{4})R = B$. Moreover, the Wiener index of T is:

$$\begin{aligned}
W(T) &= \sum_{(p,q)\in T} N_T(p)\cdot N_T(q)\cdot |pq| \\
&= 3(n^3+3n-3)n^2R + \sum_{x_i\in S'} 2(n^3+3n-1)x_i \\
&\quad + \sum_{x_i\notin S'} \left((n^3+3n-1)\frac{1}{2}x_i\right) + \sum_{x_i\notin S'} \left(2(n^3+3n-2)x_i\right) \\
&= 3n^5R + 9n^3R - 9n^2R + (n^3+3n-1)R \\
&\quad + \frac{1}{4}(n^3+3n-1)R + (n^3+3n-2)R \\
&= 3n^5R + \frac{45}{4}n^3R - 9n^2R + \frac{27}{4}nR - \frac{13}{4}R = W\,.
\end{aligned}$$

Conversely, let T' be a spanning tree of P with $wt(T') \le B$ and $W(T') \le W$.

Claim. The number of edges $(p,q)\in T'$, such that $p\in C$ and $q\in P\setminus C$ is n.

Proof. Assume there are k such edges. The weight of each such edge is at least nR thus the $wt(T') \ge knR$, since $B = (n^2+\frac{7}{4})R$ we get that $k \le n$. We have

$$\begin{aligned}
W(T') &> (3knR + 3(n-k)(nR+2\pi R))n^3 \\
&= (3kn + 3n^2 + 6n\pi - 3kn - 6k\pi)n^3R \\
&= (3n^2 + 6\pi(n-k))n^3R \\
&= 3n^5R + 6\pi(n-k)n^3R\,.
\end{aligned}$$

Thus, if $k < n$, then we get that $W(T') > 3n^5R + 6\pi n^3R > W$, for sufficiently large n. □

Let $P_i = \{p_i, l_i, r_i\}$, for every $1 \le i \le n$. From the proof of Claim 4, it follows that for every $1 \le i \le n$, there is an exactly one edge (p,q) in T', where $q \in P_i$ and $p \in C$. Moreover, it is easy to see that $q = p_i$. Thus, in every P_i, we have $(p_i,l_i)\in T'$ or $(p_i,r_i)\in T'$. Assume w.l.o.g., that $(p_i,l_i)\in T'$. Therefore, either $(p_i,r_i)\in T'$ or $(l_i,r_i)\in T'$. Let $S'\subseteq X$, such that $x_i\in S'$ if and only if $(p_i,r_i)\in T'$, and let $R' = \sum_{x_i\in S'} x_i$.

Thus, to finish the proof we show that if $R' \ne \frac{1}{2}R$, then either $wt(T') > B$ or $W(T) > W$.

Case 1: $R' > \frac{1}{2}R$. In this case, we have

$$\begin{aligned}
wt(T') &\ge n^2R + \sum_{x_i\in S'} 2x_i + \sum_{x_i\notin S'} \frac{3}{2}x_i \;=\; n^2R + 2R' + \frac{3}{2}(R-R') \\
&= n^2R + \frac{1}{2}R' + \frac{3}{2}R \;>\; n^2R + \frac{1}{4}R + \frac{3}{2}R \;=\; (n^2+\frac{7}{4})R \;= B\,.
\end{aligned}$$

Therefore, $wt(T') > B$.

Case 2: $R' < \frac{1}{2}R$. In this case, we have

$$
\begin{aligned}
W(T) &= \sum_{(p,q)\in T} N_T(p)\cdot N_T(q)\cdot |pq| \\
&= 3(n^3+3n-3)n^2R + \sum_{x_i\in S'} 2(n^3+3n-1)x_i \\
&\quad + \sum_{x_i\notin S'} \Big((n^3+3n-1)\frac{1}{2}x_i\Big) + \sum_{x_i\notin S'} \Big(2(n^3+3n-2)x_i\Big) \\
&= 3n^5R + 9n^3R - 9n^2R + 2(n^3+3n-1)R' \\
&\quad + \frac{1}{2}\Big(n^3+3n-1\Big)(R-R') + 2(n^3+3n-2)(R-R') \\
&= 3n^5R + 9n^3R - 9n^2R + 2(n^3+3n-2)R \\
&\quad - \Big(\frac{1}{2}\Big(n^3+3n-1\Big)-2\Big)R' + \frac{1}{2}\Big(n^3+3n-1\Big)R \\
&\quad - \Big(\frac{1}{2}\Big(n^3+3n-1\Big)-2\Big)R' + \frac{1}{2}\Big(n^3+3n-1\Big)R \\
&> 3n^5R + 9n^3R - 9n^2R + 2(n^3+3n-2)R \\
&\quad - \frac{1}{2}\Big(\frac{1}{2}\Big(n^3+3n-1\Big)-2\Big)R + \frac{1}{2}\Big(n^3+3n-1\Big)R \\
&= 3n^5R + \frac{45}{4}n^3R - 9n^2R + \frac{27}{4}nR - \frac{13}{4}R = W\,.
\end{aligned}
$$

□

5 Paths that Optimize Wiener Index

We consider now the case of spanning paths that optimize the Wiener index.

Theorem 4. *Let P be a set of n points. The path that minimizes the Wiener index among all Hamiltonian paths of P is not necessarily planar.*

Proof. Consider the set P of $n = 2m + 2$ points in convex position as shown in Fig. 5. The set P consists of two clusters P_l and P_r and two points p and q,

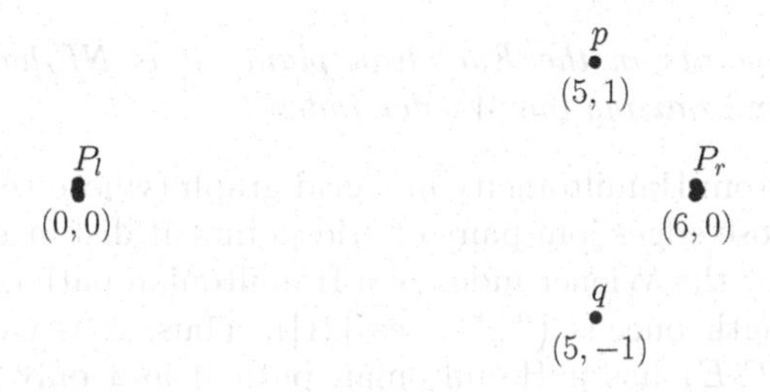

Fig. 5. A set P of $n = 2m + 2$ points in a convex position.

where $|P_l| = |P_r| = m$. The points in cluster P_l are arbitrarily close to the origin $(0,0)$, and the points in cluster P_r are arbitrarily close to coordinate $(6,0)$. The points p and q are located on coordinates $(5,1)$ and $(5,-1)$, respectively.

Since the points in P_l are arbitrarily close to the origin $(0,0)$, any path connecting these points has a Wiener index zero. Thus, any Hamiltonian path Π of P that aims to minimize the Wiener index will connect the points in P_l by a path. Similarly, any Hamiltonian path Π of P that aims to minimize the Wiener index will connect the points in P_r by a path. Therefore, it is sufficient to consider the 12 possible Hamiltonian paths defined on points $(0,0)$, $(6,0)$, p, and q, while treating each one of the points $(0,0)$ and $(6,0)$ as a path (of Wiener index zero) containing m points starting and ending at this point. We computed the Wiener index of these Hamiltonian paths, and this computation shows that the Hamiltonian path that minimizes the Wiener index is not planar (for sufficiently large n); see Fig. 6. □

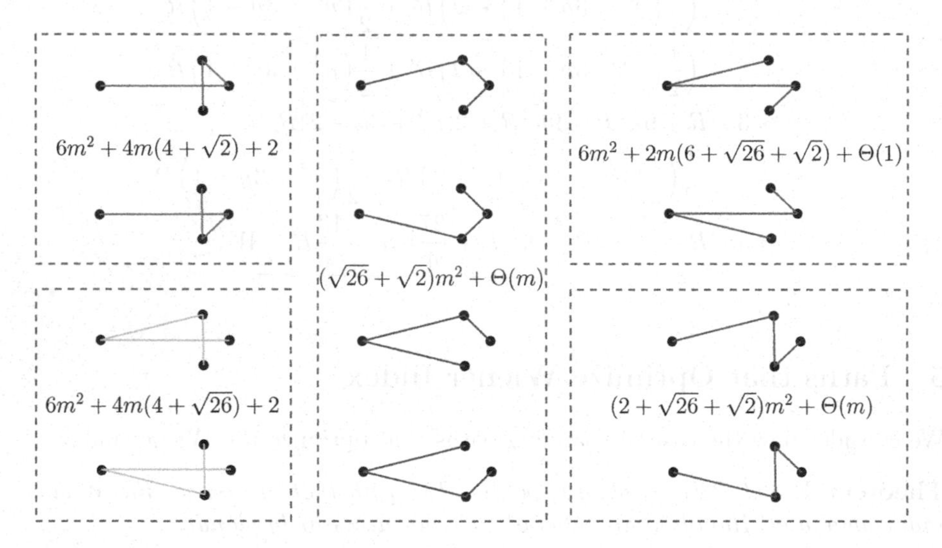

Fig. 6. The 12 possible Hamiltonian paths that are defined on points $(0,0)$, $(6,0)$, p, and q, and their Wiener index.

Theorem 5. *For points in the Euclidean plane, it is NP-hard to compute a Hamiltonian path minimizing the Wiener index.*

Proof. We reduce from Hamiltonicity in a grid graph (whose vertices are integer grid points and whose edges join pairs of grid points at distance one; see Fig. 7). It is well known that the Wiener index of a Hamiltonian path of n points, where each edge is of length one, is $\binom{n+1}{3}$ (see [21]). Thus, it is easy to see that a grid graph $G = (P, E)$ has a Hamiltonian path if and only if there exists a Hamiltonian path in the complete graph over P of Wiener index $\binom{n+1}{3}$. □

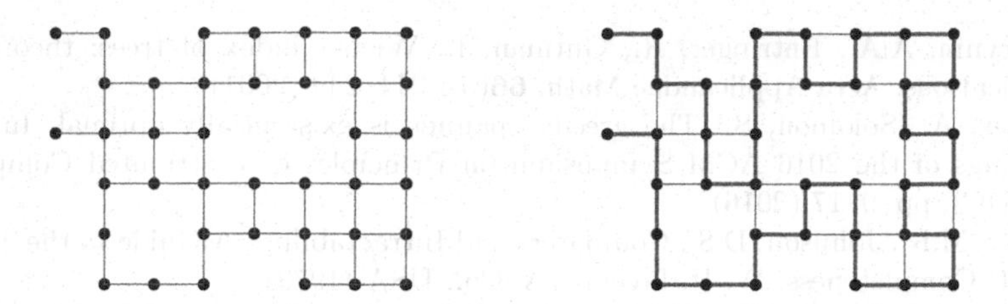

Fig. 7. A grid graph G and a Hamiltonian path with Wiener index $\binom{n+1}{3}$ in G.

Theorem 6. *There exists a set P of n points in the plane, such that the Wiener index of any Hamiltonian path is at least $\Theta(\sqrt{n})$ times the Wiener index of the complete Euclidean graph over P*

Proof. Let P be a set of n points located on a $\sqrt{n} \times \sqrt{n}$ integer grid. The Wiener index of any Hamiltonian path of P is at least $\binom{n+1}{3}$, which is the Wiener index of a Hamiltonian path whose all its edges are of length one. Thus, the Wiener index of any Hamiltonian path of P is $\Theta(n^3)$. On the other hand, since the distance between every two points in P is at most $\sqrt{2n}$ and there are $\binom{n}{2}$ pairs of points, the Wiener index of the complete graph over P is $O(n^{2.5})$. □

References

1. Abrahamsen, M., Fredslund-Hansen, V.: Degree of convexity and expected distances in polygons. arXiv preprint arXiv:2208.07106 (2022)
2. Bonchev, D.: The Wiener number: some applications and new developments. In: Topology in Chemistry, pp. 58–88 (2002)
3. Bonchev, D., Trinajstić, N.: Information theory, distance matrix, and molecular branching. J. Chem. Phys. **67**(10), 4517–4533 (1977)
4. Bose, P., Carmi, P., Chaitman-Yerushalmi, L.: On bounded degree plane strong geometric spanners. J. Discrete Algorithms **15**, 16–31 (2012)
5. Bose, P., Gudmundsson, J., Smid, M.: Constructing plane spanners of bounded degree and low weight. In: Möhring, R., Raman, R. (eds.) ESA 2002. LNCS, vol. 2461, pp. 234–246. Springer, Heidelberg (2002). https://doi.org/10.1007/3-540-45749-6_24
6. Bose, P., Hill, D., Smid, M.H.M.: Improved spanning ratio for low degree plane spanners. Algorithmica **80**(3), 935–976 (2018)
7. Cardinal, J., Collette, S., Langerman, S.: Local properties of geometric graphs. Comput. Geom. **39**(1), 55–64 (2008)
8. Carmi, P., Chaitman-Yerushalmi, L.: Minimum weight Euclidean t-spanner is np-hard. J. Discrete Algorithms **22**, 30–42 (2013)
9. Cheng, S.-W., Knauer, C., Langerman, S., Smid, M.H.M.: Approximating the average stretch factor of geometric graphs. J. Comput. Geom. **3**(1), 132–153 (2012)
10. Das, K.C., Gutman, I., Furtula, B.: Survey on geometric-arithmetic indices of graphs. Match (Mulheim an der Ruhr, Germany) **65**, 595–644 (2011)
11. Dhamdhere, K., Gupta, A., Ravi, R.: Approximation algorithms for minimizing average distortion. Theory Comput. Syst. **39**(1), 93–111 (2006)

12. Dobrynin, A.A., Entringer, R., Gutman, I.: Wiener index of trees: theory and applications. Acta Applicandae Math. **66**(3), 211–249 (2001)
13. Filtser, A., Solomon, S.: The greedy spanner is existentially optimal. In: Proceedings of the 2016 ACM Symposium on Principles of Distributed Computing (PODC), pp. 9–17 (2016)
14. Garey, M.R., Johnson, D.S.: Computers and Intractability: A Guide to the Theory of NP-Completeness. W. H. Freeman & Co., USA (1979)
15. Gonzalez, T.F.: Handbook of Approximation Algorithms and Metaheuristics, Chapter 59. Chapman & Hall/CRC, Boca Raton (2007)
16. Graovac, A., Pisanski, T.: On the wiener index of a graph. J. Math. Chem. **8**, 53–62 (1991)
17. Harary, F.: Graph Theory. Addison-Wesley, Boston (1969)
18. Hu, T.C.: Optimum communication spanning trees. SIAM J. Comput. **3**(3), 188–195 (1974)
19. Johnson, D.S., Lenstra, J.K., Kan, A.R.: The complexity of the network design problem. Networks **8**, 279–285 (1978)
20. Kier, L.: Molecular Connectivity in Chemistry and Drug Research. Elsevier, Amsterdam (1976)
21. Knor, M., Škrekovski, R., Tepeh, A.: Mathematical aspects of Wiener index. Ars Math. Contemp. **11**, 327–352 (2015)
22. Li, X.Y., Wang, Y.: Efficient construction of low weighted bounded degree planar spanner. Int. J. Comput. Geom. Appl. **14**(1–2), 69–84 (2004)
23. Mekenyan, O., Bonchev, D., Trinajstić, N.: Structural complexity and molecular properties of cyclic systems with acyclic branches. Croat. Chem. Acta **56**(2), 237–261 (1983)
24. Mohar, B., Pisanski, T.: How to compute the wiener index of a graph. J. Math. Chem. **2**(3), 267 (1988)
25. Narasimhan, G., Smid, M.: Geometric Spanner Networks. Cambridge University Press, Cambridge (2007)
26. Ronghua, S.: The average distance of trees. J. Syst. Sci. Complex. **6**, 18–24 (1993)
27. Sitters, R.: Polynomial time approximation schemes for the traveling repairman and other minimum latency problems. SIAM J. Comput. **50**(5), 1580–1602 (2021)
28. Trinajstić, N.: Mathematical and Computational Concepts in Chemistry. Ellis Horwood, Chichester (1986)
29. Weiszfeld, E., Plastria, F.: On the point for which the sum of the distances to n given points is minimum. Ann. Oper. Res. **167**(1), 7–41 (2009)
30. Wiener, H.: Structural determination of paraffin boiling points. J. Am. Chem. Soc. **69**(1), 17–20 (1947)
31. Wu, B.Y., Lancia, G., Bafna, V., Chao, K.M., Ravi, R., Tang, C.Y.: A polynomial-time approximation scheme for minimum routing cost spanning trees. SIAM J. Comput. **29**(3), 761–778 (2000)
32. Xu, K., Liu, M., Gutman, I., Furtula, B.: A survey on graphs extremal with respect to distance-based topological indices. Match (Mulheim an der Ruhr, Germany) **71**, 461–508 (2014)

The Mutual Visibility Problem for Fat Robots

Rusul J. Alsaedi, Joachim Gudmundsson, and André van Renssen(✉)

University of Sydney, Sydney, Australia
rals2984@uni.sydney.edu.au,
{joachim.gudmundsson,andre.vanrenssen}@sydney.edu.au

Abstract. Given a set of $n \geq 1$ unit disk robots in the Euclidean plane, we consider the fundamental problem of providing mutual visibility to them: the robots must reposition themselves to reach a configuration where they all see each other. This problem arises under obstructed visibility, where a robot cannot see another robot if there is a third robot on the straight line segment between them. This problem was solved by Sharma *et al.* [18] in the luminous robots model, where each robot is equipped with an externally visible light that can assume colors from a fixed set of colors, using 9 colors and $O(n)$ rounds. In this work, we present an algorithm that requires only 2 colors and $O(n)$ rounds. The number of colors is optimal since at least two colors are required for point robots [10].

Keywords: Mutual visibility · Fat robots · Obstructed visibility · Collision avoidance · Robots with lights

1 Introduction

We consider a set of n unit disk robots in $\mathbb{R}^2$ and aim to position these robots in such a way that each pair of robots can see each other (see Fig. 1 for an example initial configuration where not all robots can see each other and an end configuration where they can). This problem is fundamental in that it is typically the first step in solving more complex problems. We consider the problem under the classical oblivious robots model [15], where robots are autonomous (no external control), anonymous (no unique identifiers), indistinguishable (no external markers), history-oblivious (no memory of activities done in the past), silent (no means of direct communication), and possibly disoriented (no agreement on their coordination systems). We consider this problem under the fully synchronous model, where in every *round* all robots are activated. All robots execute the same algorithm, following Look-Compute-Move (LCM) cycles [9] (i.e., when a robot becomes active, it uses its vision to get a snapshot of its surroundings (Look), computes a destination point based on the snapshot (Compute), and finally moves towards the computed destination (Move)).

This classical robot model has a long history and has many applications including coverage, exploration, intruder detection, data delivery, and symmetry

P. Morin and S. Suri (Eds.): WADS 2023, LNCS 14079, pp. 15–28, 2023.
https://doi.org/10.1007/978-3-031-38906-1_2

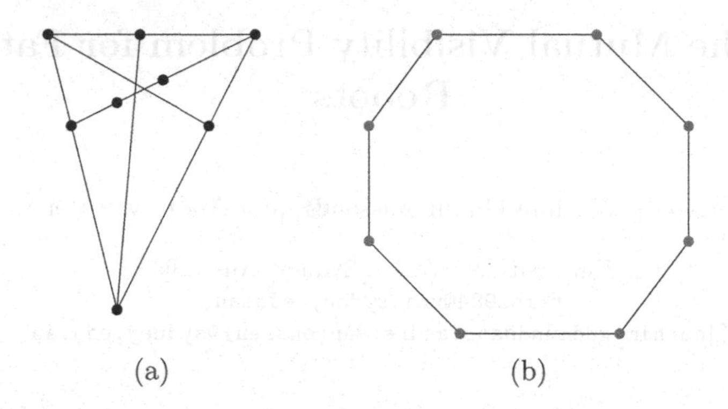

(a) (b)

Fig. 1. An example of an initial instance (a) and an end configuration (b).

breaking [5]. Unfortunately, most of the previous work considered the robots to be dimensionless point robots which do not occupy any space.

The classical model also makes the important assumption of unobstructed visibility, i.e., any three collinear robots are mutually visible to each other. This assumption, however, does not make sense for the unit disk robots we consider. To remove this assumption, robots under obstructed visibility have been the subject of recent research [1,3,4,6–8,10–14,19,20,22,24]. Under obstructed visibility, robot r_i can see robot r_j if and only if there is no third robot on the straight line segment connecting r_i and r_j.

Additionally, a variation on this model received significant attention: the luminous robots model (or robots with lights model) [10–12,16,19,20,24]. In this model, robots are equipped with an externally visible light which can assume colors from a fixed set. The lights are persistent, i.e., the color of the light is not erased at the end of the LCM cycle. When the number of colors in the set is 1, this model corresponds to the classical oblivious robots model [10,15].

Being the first step in a number of other problems, including the Gathering and Circle Formation problems [18], the Mutual Visibility problem received significant attention in this new robots with lights model. When robots are dimensionless points, the Mutual Visibility problem was solved in a series of papers [10–12,19,20,24]. Unfortunately, the techniques developed for point robots do not apply directly to the unit disk robots, due to the lack of collision avoidance. For unit disk robots, much progress has been made in solving the Mutual Visibility problem [1,3,4,7,8,13,17,18,21], however these approaches either require additional assumptions such as chirality (the robots agree on the orientation of the axes, i.e., on the meaning of clockwise), knowledge of n, or without avoiding collisions. Additionally, some approaches require a large number of colors and not all approaches bound the number of rounds needed.

1.1 Related Work

Most of the existing work in the robots with lights model considers point robots [10,11,20,24]. Di Luna *et al.* [10] solved the Mutual Visibility problem

lem for those robots with obstructed visibility in the lights model, using 2 and 3 colors under semi-synchronous and asynchronous computation, respectively. Sharma *et al.* [20] provided a solution for point robots that requires only 2 colors, which is optimal since at least two colors are needed [10]. Unfortunately, the required number of rounds is not analyzed. Sharma *et al.* [23] also considered point robots in the robots with lights model. In the asynchronous setting, they provide an $O(1)$ time and $O(1)$ colors solution using their Beacon-Directed Curve Positioning technique to move the robots.

Mutual visibility has also been studied for fat robots. Agathangelou *et al.* [1] studied it in the fat robots model of Czyzowicz *et al.* [8], where robots are not equipped with lights. Their approach allows for collisions, assumes chirality, and the robots need to know n, making it unsuited for our setting. Sharma *et al.* [21] developed an algorithm that solves coordination problems for fat robots in $O(n)$ rounds in the classical oblivious model, assuming n is known to the robots.

Poudel *et al.* [17] studied the MUTUAL VISIBILITY problem for fat robots on an infinite grid graph G and the robots have to reposition themselves on the vertices of the graph G. They provided two algorithms; the first one solves the MUTUAL VISIBILITY problem in $O(\sqrt{n})$ time under a centralized scheduler. The second one solves the same problem in $\Theta(\sqrt{n})$ time under a distributed scheduler, but only for some special instances.

When considering both fat robots and the robots with lights model, the main result is by Sharma *et al.* [18]. Their solution uses 9 colors and solves the MUTUAL VISIBILITY problem in $O(n)$ rounds.

1.2 Contributions

We consider $n \geq 1$ unit disk robots in the plane and study the problem of providing mutual visibility to them. We address this problem in the lights model. In particular, we present an algorithm that solves the problem in $O(n)$ rounds using only 2 colors while avoiding collisions. The number of colors is optimal since at least two colors are needed for point robots [10]. Our results improve on previous work in terms of the number of colors used [18] and by using fat robots and having a linear number of rounds [10,20].

Our algorithm works under fully synchronous computation, where all robots are activated in each round and they perform their LCM cycles simultaneously in synchronized rounds. The moves of the robots are rigid, i.e., an adversary does not have the power to stop a moving robot before reaching its computed destination [15].

2 Preliminaries

Consider a set of $n \geq 1$ anonymous robots $\mathcal{R} = \{r_1, r_2, \ldots, r_n\}$ operating in the Euclidean plane. During the entire execution of the algorithm, we assume that n is *not* known to the robots. Each robot $r_i \in \mathcal{R}$ is a non-transparent disk with diameter 1, sometimes referred to as a fat robot. The center of the robot r_i is denoted by c_i and the position of c_i is also said to be the position of r_i. We denote by $\text{dist}(r_i, r_j)$ the Euclidean distance between the two robots, i.e., the distance from c_i to c_j. To avoid collisions among robots, we have to ensure that $\text{dist}(r_i, r_j) \geq 1$ between any two robots r_i and r_j at all times. Each robot r_i has its own coordinate system centered at itself, and it knows its position with respect to its coordinate system. Robots may not agree on the orientation of their coordinate systems, i.e., there is no common notion of direction. Since all the robots are of unit size, they agree implicitly on the unit of measure of other robots. The robots have a camera to take a snapshot, and the visibility of the camera is unlimited provided that there are no obstacles (i.e., other robots) [1].

Following the fat robot model [1,8], we assume that a robot r_i can see another robot r_j if there is at least one point on the bounding circle of r_j that is visible from r_i. Similarly, we say that a point p in the plane is visible by a robot r_i if there is a point p_i in the bounding circle of r_i such that the straight line segment $\overline{p_i p}$ does not intersect any other robot. We say that robot r_i fulfills the mutual visibility property if r_i can see all other robots in $\mathcal{R}$. Two robots r_i and r_j are said to *collide* at time t if the bounding circles of r_i and r_j share a common point at time t. For simplicity, we use r_i to denote both the robot r_i and the position of its center c_i.

Each robot r_i is equipped with an externally visible light that can assume any color from a fixed set $\mathcal{C}$ of colors. The set $\mathcal{C}$ is the same for all robots in $\mathcal{R}$. The color of the light of robot r at time t can be seen by all robots that are visible to r at time t.

A *configuration* $\mathbb{C}$ is a set of n tuples in $\mathcal{C} \times \mathbb{R}^2$ which define the colors and positions of the robots. Let $\mathbb{C}_t$ denote the configuration at time t. Let $\mathbb{C}_t(r_i)$ denote the configuration $\mathbb{C}_t$ for robot r_i, i.e., the set of tuples in $\mathcal{C} \times \mathbb{R}^2$ of the robots visible to r_i. A configuration $\mathbb{C}$ is *obstruction-free* if for all $r_i \in \mathcal{R}$, we have that $|\mathbb{C}(r_i)| = n$. In other words, when all robots can see each other.

Let $\mathbb{H}_t$ denote the convex hull formed by $\mathbb{C}_t$. Let $\partial\mathbb{H}_t = \mathcal{V}_t \cup \mathcal{S}_t$ denote the set of robots on the boundary of $\mathbb{H}_t$, where $\mathcal{V}_t \subseteq \mathcal{R}$ is the set of corner robots lying on the corners of $\mathbb{H}_t$ and $\mathcal{S}_t \subseteq \mathcal{R}$ is the set of robots lying in the interior of the edges of $\mathbb{H}_t$. The robots in the set $\mathcal{V}_t$ are called *corner robots* and those in the set $\mathcal{S}_t$ are called *side robots*. The robots in the set $\mathcal{I}_t = \mathbb{H}_t \backslash \partial\mathbb{H}_t$ are called *interior robots*. Given a robot $r_i \in \mathcal{R}$, we denote by $\mathbb{H}_t(r_i)$ the convex hull of $\mathbb{C}_t(r_i)$. Note that $\mathbb{H}_t(r_i)$ can differ from $\mathbb{H}_t$ if r_i does not see all robots on the convex hull.

Given two points $a, b \in \mathbb{R}^2$, we denote by $|\overline{ab}|$ the length of the straightline segment $\overline{ab}$ connecting them. Given $a, b, d \in \mathbb{R}^2$, we use $\angle abd$ to denote the clockwise angle at point b between ab and bd.

At any time t, a robot $r_i \in \mathcal{R}$ is either active or inactive. When active, r_i performs a sequence of *Look-Compute-Move* (LCM) operations:

- *Look:* a robot takes a snapshot of the positions of the robots visible to it in its own coordinate system;
- *Compute:* executes its algorithm using the snapshot. This returns a destination point $x \in \mathbb{R}^2$ and a color $c \in \mathcal{C}$; and

– *Move:* moves to the computed destination $x \in \mathbb{R}^2$ (if x is different than its current position) and sets its own light to color c.

We assume that the execution starts at time 0. Therefore, at time $t = 0$, the robots start in an arbitrary configuration $\mathbb{C}_0$ with $\mathrm{dist}(r_i, r_j) \geq 1$ for any two robots $r_i, r_j \in \mathbb{R}^2$, and the color of the light of each robot is set to *Off*.

Formally, the MUTUAL VISIBILITY problem is defined as follows: Given any $\mathbb{C}_0$, in finite time, reach an obstruction-free configuration without having any collisions in the process. An algorithm is said to solve the MUTUAL VISIBILITY problem if it always achieves an obstruction-free configuration from any arbitrary initial configuration. Each robot executes the same algorithm locally every time it is activated. We measure the quality of the algorithm both in terms of the number of colors and the number of rounds needed to solve the MUTUAL VISIBILITY problem.

Finally, we need the following definitions to present our MUTUAL VISIBILITY algorithm. Let $e = \overline{v_1v_2}$ be a line segment connecting two corner robots v_1 and v_2 of $\mathbb{H}_t$. Following Di Luna *et al.* [11], we define the *safe zone* $S(e)$ as a portion of the plane outside $\mathbb{H}_t$ such that the corner robots v_1 and v_2 of $\mathbb{H}_t$ remain corner robots when a side robot is moved into this area: for all points $x \in S(e)$, we ensure that $\angle xv_1v_2 \leq \frac{180-\angle v'v_1v_2}{4}$ and $\angle v_1v_2x \leq \frac{180-\angle v_1v_2v_3}{4}$, where v', v_1, v_2, and v_3 are consecutive vertices of the convex hull of $\mathbb{H}_t$ (see Fig. 2(a)). Side robots may not always be able to compute $S(e)$ exactly due to obstructions of visibility leading to different local views. However, if there is only one side robot on e, then it can compute $S(e)$ exactly. When there is more than one robot on e, $S'(e)$ is the safe region computed by a robot based on its local view. It is guaranteed that $S'(e) \subseteq S(e)$ (see Fig. 2(b) for the safe zone of robot r_2).

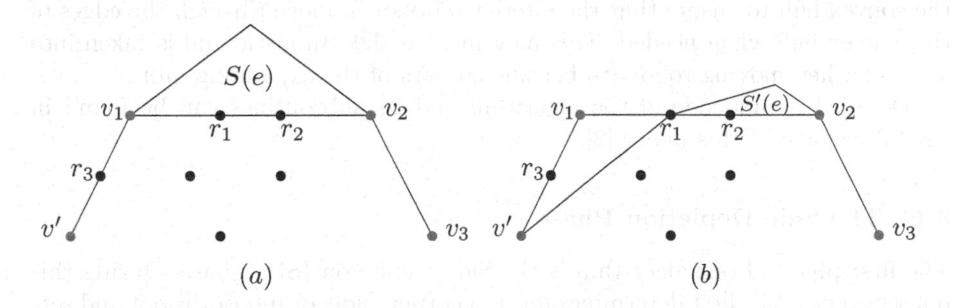

Fig. 2. (a) The safe zone of $e = \overline{v_1v_2}$. (b) The safe zone of a side robot r_2 on e.

3 The Mutual Visibility Algorithm

In this section, we present an algorithm that solves the MUTUAL VISIBILITY problem for $n \geq 1$ unit disk robots under rigid movement in the robots with lights model. Our algorithm assumes the fully synchronous setting of robots. The algorithm needs two colors: $\mathcal{C} = \{Off, Red\}$. A red robot represents a corner robot. A robot whose light is off represents any other robot. See Fig. 3 for an example. Initially, the lights of all robots are off.

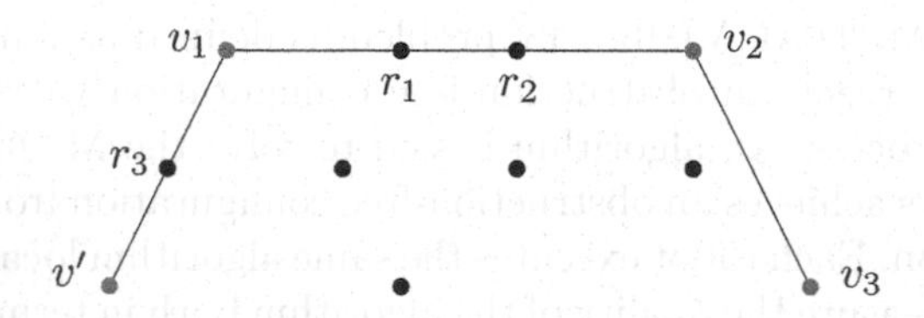

Fig. 3. The different colors of the robots: corner robots (red), side robots (off), and interior robots (off). (Color figure online)

It has been shown that positioning the robots in the corners (i.e., vertices) of an n-vertex convex hull provides a solution to the Mutual Visibility problem [10,11,14,19,20,24]. Hence, our algorithm also ensures that the robots eventually position themselves in this way.

Conceptually, our general strategy consists of two phases, though the robots themselves do not explicitly discern between them. In the *Side Depletion* phase, some side robots move to become corner robots, ensuring that there are only corner and interior robots left. In the *Interior Depletion* phase, interior robots move and become corner robots of the convex hull. The move-algorithm checks if the robot's path shares any point with any other robots, ensuring that no collision occur. Throughout both phases, corner robots slowly move to expand the convex hull to ensure that the interior robots can move through the edges of the convex hull when needed. This movement is deterministic and is taken into account when moving robots to become corners of the expanding hull.

Detailed pseudocode of the algorithm and its subroutines can be found in the full version of this paper [2].

3.1 The Side Depletion Phase

The first phase of our algorithm is the Side Depletion (SD) phase. During this phase, every robot first determines if it is a corner, side, or interior robot and sets their light accordingly. Note that robots can make this distinction themselves, by checking what angle between consecutive robots it sees: if some angle is larger than 180° it is a corner robot, if the angle is exactly 180° it is a side robot, and otherwise it is an interior robot.

All corner robots move a distance of 1 along the angle bisector determined by its neighbors in the direction that does not intersect the interior of the convex hull. In other words, the corner robots move to expand the size of the convex hull. We note that since all corner robots move this way, they all stay corner robots throughout this process.

Side robots that see at least one corner robot move to become new corner robots of $\mathbb{H}$ (using the safe zone described earlier and taking the above movement of corner robots into account) and change their light to red. Side robots that do not see a corner robot on their convex hull edge do not move and will become interior robots in the next round (due to the change to the convex hull), while keeping their light off.

More precisely, a side robot r on edge $e = \overline{v_1 v_2}$ of $\mathbb{H}_k$ moves as follows: If at least one of its neighbors on $\overline{v_1 v_2}$ is a corner robot, r moves to a point in the safe zone $S(e)$. There are at most two such robots r_1 and r_2 on each edge $\overline{v_1 v_2}$ (see Fig. 4 and 5). Sharma *et al.* [18] showed that these can move simultaneously to the safe zone outside the hull. Both r_1 and r_2 become new corners of $\mathbb{H}$ and change their lights to red (see Fig. 5).

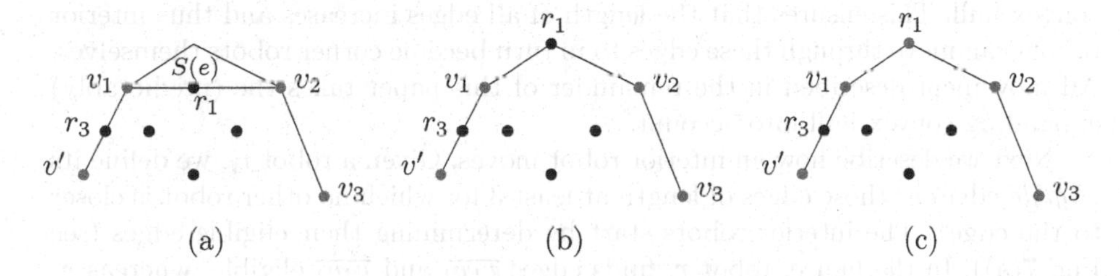

Fig. 4. One side robot r_1 on an edge $e = \overline{v_1 v_2}$ moves to become a corner robot.

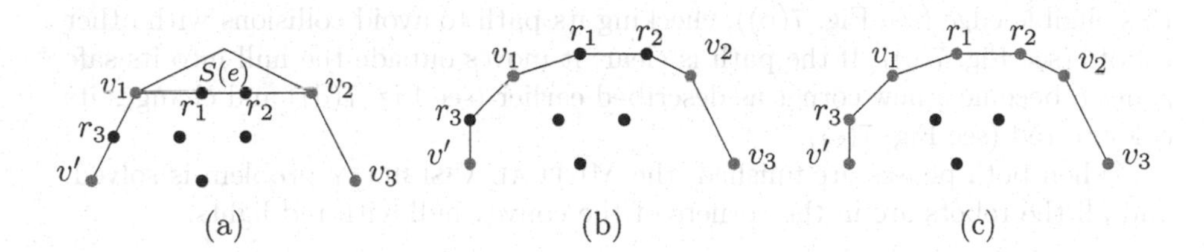

Fig. 5. Two side robots r_1 and r_2 on an edge $e = \overline{v_1 v_2}$ move to become corner robots.

If both of its neighbors on $\overline{v_1 v_2}$ are not corners (see Fig. 6), robot r does not move and stay in its place, and it will become an interior robot in the next round.

We only execute this phase once, at the start of our algorithm and only move each robot once.

3.2 The Interior Depletion Phase

Once the SD phase finishes, the Interior Depletion (ID) phase starts. During this phase the robots in the interior of the hull move such that they become new vertices of the hull.

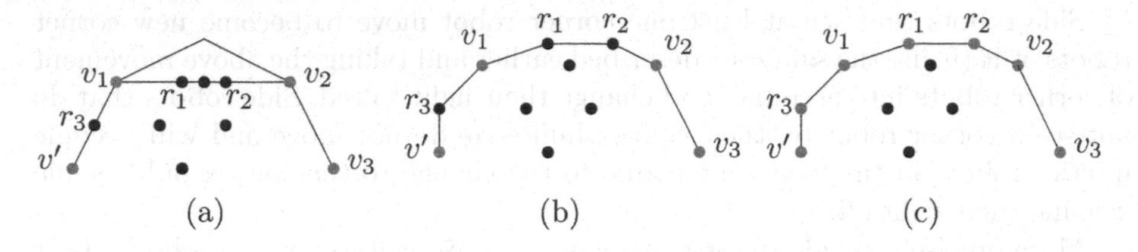

Fig. 6. When there are more than two side robots on an edge of the convex hull, only two side robots on the edge move to become corner robots. These are the clockwise and the counterclockwise extreme side robots. In this case, robots r_1 and r_2 move to become corner robots.

In every round, all corner robots move as in the SD phase, expanding the convex hull. This ensures that the length of all edges increases and thus interior robots can move through these edges to in turn become corner robots themselves. All movement described in the remainder of this paper takes the (predictably) expanding convex hull into account.

Next we describe how an interior robot moves. Given a robot r_i, we define its *eligible* edges as those edges of length at least 3 for which no other robot is closer to the edge[1]. The interior robots start by determining their eligible edges (see Fig. 7(a)). In the figure, robot r_i finds edges $\overline{v_1v_2}$ and $\overline{v_2v_3}$ eligible, whereas r_j finds $\overline{v_2v_3}$, and r_l finds $\overline{v_3v_4}$ eligible. However, the robots between r_i, r_j find no edge eligible. Let Q denote the set of edges that are eligible to an interior robot r_i. Every interior robot that has an eligible edge moves perpendicular to some eligible edge towards this edge to become a corner robot by moving through this eligible edge (see Fig. 7(b)), checking its path to avoid collisions with other robots (see Fig. 7(c)). If the path is clear, it moves outside the hull into its safe zone to become a new corner as described earlier (see Fig. 7(d)) and changes its color to red (see Fig. 7(e)).

When both phases are finished, the Mutual Visibility problem is solved, and all the robots are in the corners of the convex hull with red lights.

3.3 Special Cases

There are two special cases to consider: $n = 1$, and the case where the initial configuration is a line. The case $n = 1$ can be easily recognized by the only robot, since it does not see any other robot and thus it can terminate.

If in the initial configuration all robots lie on a single line, we differentiate between the robots that see only one other robot and the robots that see two other robots. If a robot r_i sees only one other robot r_j, when r_i is activated for the very first time it sets its light to red and moves orthogonal to the line $\overline{r_ir_j}$

[1] The length of 3 is used to ensure that two robots can move through the same edge without colliding with each other (requiring a length of 2) while ensuring that they also do not collide with the corner robots on the edge (adding a length of 0.5 per corner robot).

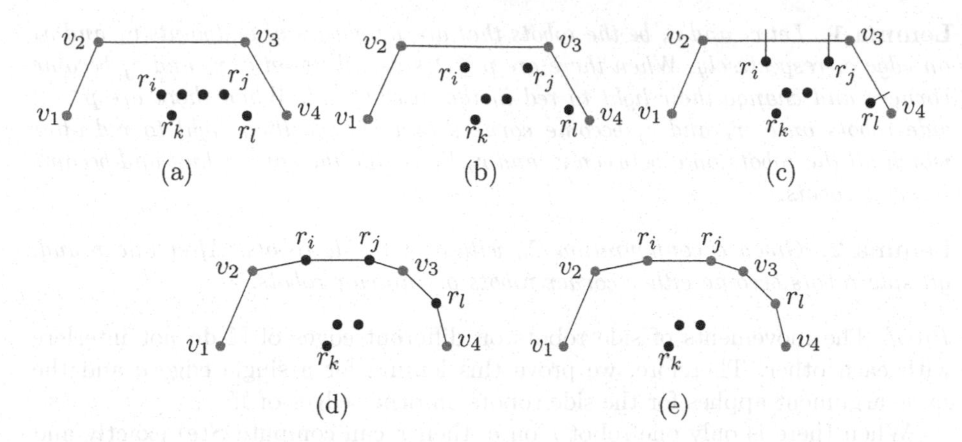

Fig. 7. (a) The eligible edge computation. The robot r_i finds edges $\overline{v_1 v_2}$ and $\overline{v_2 v_3}$ eligible, whereas r_j finds $\overline{v_2 v_3}$, and r_l finds $\overline{v_3 v_4}$ eligible. The robots between r_i, r_j find no eligible edges. (b) The intermediate movements of the interior robots r_i, r_j and r_l. (c) The interior robots r_i, r_j and r_l check their paths to avoid collisions with other robots while moving outside the hull to become corners. (d) After the interior robots r_i, r_j and r_l move, they become corners. (e) Robots r_i, r_j and r_l change their lights to red.

for some arbitrary positive distance. When r_i is activated in future rounds and $\mathbb{H}_k(r_i)$ is still a line segment, it can conclude that there are only two robots and it does nothing until it sees r_j set its light to red. Once r_j sets its light to red, r_i terminates.

If a robot r_i sees two other robots r_j and r_l, robot r_i will be able to tell if $\mathbb{H}_k(r_i)$ is a line segment as follows. Robot r_i will move orthogonal to line $\overline{r_j r_l}$ and set its light to red if and only if it sees that the lights of r_j and r_l are set to red, as this indicates that both other robots see only a single other robot, i.e., $n = 3$. Otherwise, moving the two extremal robots of the initial configuration as described above ensures that the configuration is no longer a line segment, allowing the SD and ID phase to solve the problem.

As these special cases add only a constant number of rounds to the running time and do not influence the number of colors, we focus on the general case in the remainder of this paper.

4 Analysis

We proceed to prove that our algorithm solves the Mutual Visibility problem in a linear number of rounds, using only two colors and while avoiding collisions between the robots. Due to space constraints, supporting lemmas and additional proofs can be found in the full version of this paper [2].

We start with some properties of the Side Depletion phase.

Lemma 1. *Let r_i and r_j be the robots that are neighbors of endpoints v_1 and v_2 on edge e, respectively. When there are $p \leq 2$ side robots on e, r_i and r_j become corners and change their light to red in the next round. When there are $p > 2$ side robots on e, r_i and r_j become corners and change their light to red after which all the robots on e between r_i and r_j lie inside the convex hull and become interior robots.*

Lemma 2. *Given a configuration $\mathbb{C}_0$ with $q \geq 1$ side robots. After one round, all side robots become either corner robots or interior robots.*

Proof. The movements of side robots on different edges of $\mathbb{H}$ do not interfere with each other. Therefore, we prove this lemma for a single edge e and the same argument applies for the side robots on other edges of $\mathbb{H}$.

When there is only one robot r on e, then r can compute $S(e)$ exactly and move to a point $x \in S(e)$ as soon as it is activated. When there are two or more robots on e, two side robots (the extreme ones on this edge) become corners in one round by Lemma 1. This causes the other robots on e to become interior robots.

Since the robots on different edges do not influence each other and the moves on any edge end in one round, this phase ends in one round. □

Now that there are no more side robots, we argue that the interior robots also eventually become corners. We first show that the interior robots can determine whether the SD phase has finished.

Lemma 3. *Given a configuration $\mathbb{C}_k$ and an edge $e = \overline{v_1v_2}$ of $\mathbb{H}_k$, no robot in the interior of $\mathbb{H}_k$ moves to $S(e)$ if there is a side robot on e.*

Proof. If there are side robots in $\mathbb{H}_k$, it is easy to see that every corner robot of $\mathbb{H}_k$ on an edge that contains side robots sees at least one side robot. Similarly, when there are side robots, interior robots can easily infer that the SD phase is not finished, and hence they do not move to their respective $S(e)$. □

Next, we argue that every interior robot will eventually become a corner robot during the ID phase.

Lemma 4. *Let I_k be the number of interior robots in round k. In each round $k \in n^+$ until $I_k = \emptyset$, if there is an edge of length at least 3, there is at least one robot in I_k for which the set of line segments Q is not empty.*

Proof. We note that every edge of the convex hull of the corner robots $\mathbb{H}_k$ is closest to some interior robot(s). In particular, this holds for any edge of length at least 3. We note that this set of interior robots forms a line, as they all have the same closest distance to the edge. Out of these robots, by definition, the left and right extreme ones have the edge in their Q. □

Lemma 5. *Given any initial configuration $\mathbb{C}_0$, there exists a $k \in n^+$ such that $I_k = \emptyset$ in $\mathbb{C}_k$ and the corner robots do not move in any round $k' > k$.*

Theorem 1. *Given any initial configuration $\mathbb{C}_0$, there is some round $k \in n^+$ such that all robots lie on $\mathbb{H}_k$ and have their lights set to red.*

Proof. Lemma 5 shows that there exists a round k such that there are no interior robots left. Interior robots that moved to become corner robots changed their lights to red as soon as they reached their corner positions. Furthermore, the interior robots move to the safe zone where they by definition become corners. Since Lemma 5 guarantees that there are no collisions, the robots occupy different positions of $\mathbb{H}_k$ and all their lights will be red. □

We are now ready to prove that the Mutual Visibility problem is solvable using only two colors. Let $\mathbb{C}_{ID}$ denote the configuration of robots after the ID phase is finished and let $\mathbb{H}_{ID}$ be the convex hull created by $\mathbb{C}_{ID}$.

Lemma 6. *Given any initial configuration $\mathbb{C}_0$, no collisions of robots occur until $I_k = \emptyset$.*

Lemma 7. *Given a robot $r_i \in \mathcal{R}$ with its light set to red and a round $k \in n^+$, if all robots in $\mathbb{C}_k(r_i)$ have their light set to red, and no robot is in the interior of $\mathbb{H}_k(r_i)$, then $\mathbb{C}_k$ does not contain interior robots.*

Theorem 2. *The Mutual Visibility problem is solvable without collisions for unit disk robots in the fully synchronous setting using two colors in the robots with lights model.*

Proof. We have from Lemma 2 that from any initial non-collinear $\mathbb{C}_0$, we reach a configuration $\mathbb{C}_{SD}$ without side robots after one round, some becoming corner robots and some becoming interior robots. Once the SD phase is over, Theorem 1 shows that the ID phase moves all interior robots to become corner robots. We have from Lemma 7 that robots can locally detect whether the ID phase is over and configuration $\mathbb{C}_{ID}$ is reached. By Lemma 6, no collisions occur in the SD and ID phases.

Therefore, starting from any non-collinear configuration $\mathbb{C}_0$, all robots eventually become corners of the convex hull, solving the Mutual Visibility problem without collisions.

It remains to show that starting from any initial collinear configuration $\mathbb{C}_0$ the robots correctly evolve into some non-collinear configuration from which we can apply the above analysis. If $n \leq 3$, this can be shown through a simple case analysis: For $n = 1$, when the only robot becomes active, it sees no other robot, changes its color to red and immediately terminates. For $n = 2$, robot r_i changes its color to red when it becomes active for the first time and moves orthogonal to line $\overline{r_i r_j}$ that connects it to the only other robot r_j it sees in $\mathbb{C}(r_i)$. When r_i later realizes that $|\mathbb{C}(r_i)|$ is still 2 and $r_j.light = red$, it simply terminates. For $n = 3$, when r_i realizes that both of its neighbors in $\mathbb{C}(r_i)$ have light set to red and are collinear with it, it moves orthogonal to that line and sets its light to red. The next time it becomes active, it finds itself at one of the corners and simply terminates as it sees all the other robots in the corners of the hull with light set to red.

For $n \geq 4$, let a and b be the two robots that occupy the corners of the line segment $\mathbb{H}_0$ (i.e. the endpoint robots of $\mathbb{H}_0$). Nothing happens until a or b is activated, setting its light to red, and moving orthogonal to $\mathbb{H}_0$. After a or b moves, when another robot becomes active, it realizes that the configuration is not a line anymore and enters the normal execution of our algorithm. It is easy to see that after the line segment $\mathbb{H}_0$ evolves into a polygonal shape, it never reverts to being a line.

Finally, since our algorithm uses only two colors, the theorem follows. □

Finally, we analyze the number of rounds needed by our algorithm.

Lemma 8. *After $O(n)$ rounds, the convex hull has grown enough in size to allow all n robots to become corners.*

Proof. Since in every round all corner robots move a distance of 1 along the bisector of their exterior angle, the length of the convex hull grows by at least 1 in every round. Note that when a robot becomes a corner, it moves outside the current convex hull and thus, by triangle inequality, extends the hull that way as well.

Hence, after at most $4n$ rounds the convex hull is long enough to ensure that there is space for all interior robots: there are at most n edges of the convex hull and for each of them to *not* be long enough, their total length is strictly less than $3n$. Hence, by expanding the convex hull by a total of $4n$, we ensure that there is enough space for each of the less than n interior robots of diameter 1. Expanding the convex hull a total of $4n$ takes $O(n)$ rounds, completing the proof. □

We note that for the above lemma the corner robots do not need to know n, as they can simply keep moving until the algorithm finishes.

Lemma 9. *The Interior Depletion phase of the mutual visibility algorithm finishes in $O(n)$ rounds.*

Proof. When an interior robot can move outside the convex hull to become a corner robot, it needs at most a constant rounds to do so. During those rounds the robot becomes active, checks its path while moving to the safe zone to become a corner robot, and changes its light to red. There are fewer than n interior robots and by Lemma 4 at least one robot can move when there is an edge of length at least 3. By Lemma 8 in $O(n)$ rounds there are sufficient long edges to allow the less than n interior robots to move through them. Therefore, the Interior Depletion phase of the mutual visibility algorithm finishes in $O(n)$ rounds. □

We now have the following theorem bounding the running time of our algorithm using Lemmas 2 and 9 and Theorem 2.

Theorem 3. *Our algorithm solves the* Mutual Visibility *problem for unit disk robots in $O(n)$ rounds without collisions in the fully synchronous setting using two colors.*

References

1. Agathangelou, C., Georgiou, C., Mavronicolas, M.: A distributed algorithm for gathering many fat mobile robots in the plane. In: Proceedings of the 2013 ACM Symposium on Principles of Distributed Computing, pp. 250–259 (2013)
2. Alsaedi, R.J., Gudmundsson, J., van Renssen, A.: The mutual visibility problem for fat robots. ArXiv e-prints (2022). http://arxiv.org/abs/2206.14423
3. Bolla, K., Kovacs, T., Fazekas, G.: Gathering of fat robots with limited visibility and without global navigation. In: Proceedings of the 2012 International Symposium on Swarm Intelligence and Differential Evolution, pp. 30–38 (2012)
4. Chaudhuri, S.G., Mukhopadhyaya, K.: Leader election and gathering for asynchronous fat robots without common chirality. J. Discrete Algorithms **33**, 171–192 (2015)
5. Cieliebak, M., Flocchini, P., Prencipe, G., Santoro, N.: Distributed computing by mobile robots: gathering. SIAM J. Comput. **41**(4), 829–879 (2012)
6. Cohen, R., Peleg, D.: Local spreading algorithms for autonomous robot systems. Theor. Comput. Sci. **399**(1–2), 71–82 (2008)
7. Cord-Landwehr, A., et al.: Collisionless gathering of robots with an extent. In: Černá, I., et al. (eds.) SOFSEM 2011. LNCS, vol. 6543, pp. 178–189. Springer, Heidelberg (2011). https://doi.org/10.1007/978-3-642-18381-2_15
8. Czyzowicz, J., Gasieniec, L., Pelc, A.: Gathering few fat mobile robots in the plane. Theor. Comput. Sci. **410**(6–7), 481–499 (2009)
9. Das, S., Flocchini, P., Prencipe, G., Santoro, N., Yamashita, M.: Autonomous mobile robots with lights. Theor. Comput. Sci. **609**, 171–184 (2016)
10. Di Luna, G.A., Flocchini, P., Chaudhuri, S.G., Poloni, F., Santoro, N., Viglietta, G.: Mutual visibility by luminous robots without collisions. Inf. Comput. **254**, 392–418 (2017)
11. Di Luna, G.A., Flocchini, P., Gan Chaudhuri, S., Santoro, N., Viglietta, G.: Robots with lights: overcoming obstructed visibility without colliding. In: Felber, P., Garg, V. (eds.) SSS 2014. LNCS, vol. 8756, pp. 150–164. Springer, Cham (2014). https://doi.org/10.1007/978-3-319-11764-5_11
12. Di Luna, G.A., Flocchini, P., Poloni, F., Santoro, N., Viglietta, G.: The mutual visibility problem for oblivious robots. In: Proceedings of the 26th Canadian Conference on Computational Geometry (2014)
13. Dutta, A., Gan Chaudhuri, S., Datta, S., Mukhopadhyaya, K.: Circle formation by asynchronous fat robots with limited visibility. In: Ramanujam, R., Ramaswamy, S. (eds.) ICDCIT 2012. LNCS, vol. 7154, pp. 83–93. Springer, Heidelberg (2012). https://doi.org/10.1007/978-3-642-28073-3_8
14. Flocchini, P.: Computations by luminous robots. In: Papavassiliou, S., Ruehrup, S. (eds.) ADHOC-NOW 2015. LNCS, vol. 9143, pp. 238–252. Springer, Cham (2015). https://doi.org/10.1007/978-3-319-19662-6_17
15. Flocchini, P., Prencipe, G., Santoro, N.: Distributed computing by oblivious mobile robots. Synth. Lect. Distrib. Comput. Theory **3**(2), 1–185 (2012)
16. Peleg, D.: Distributed coordination algorithms for mobile robot swarms: new directions and challenges. In: Pal, A., Kshemkalyani, A.D., Kumar, R., Gupta, A. (eds.) IWDC 2005. LNCS, vol. 3741, pp. 1–12. Springer, Heidelberg (2005). https://doi.org/10.1007/11603771_1
17. Poudel, P., Sharma, G., Aljohani, A.: Sublinear-time mutual visibility for fat oblivious robots. In: Proceedings of the 20th International Conference on Distributed Computing and Networking, pp. 238–247 (2019)

18. Sharma, G., Alsaedi, R., Busch, C., Mukhopadhyay, S.: The complete visibility problem for fat robots with lights. In: Proceedings of the 19th International Conference on Distributed Computing and Networking, pp. 1–4 (2018)
19. Sharma, G., Busch, C., Mukhopadhyay, S.: Bounds on mutual visibility algorithms. In: Proceedings of the 27th Canadian Conference on Computational Geometry, pp. 268–274 (2015)
20. Sharma, G., Busch, C., Mukhopadhyay, S.: Mutual visibility with an optimal number of colors. In: Bose, P., Gąsieniec, L.A., Römer, K., Wattenhofer, R. (eds.) ALGOSENSORS 2015. LNCS, vol. 9536, pp. 196–210. Springer, Cham (2015). https://doi.org/10.1007/978-3-319-28472-9_15
21. Sharma, G., Busch, C., Mukhopadhyay, S.: How to make fat autonomous robots see all others fast? In: Proceedings of the 2018 IEEE International Conference on Robotics and Automation, pp. 3730–3735 (2018)
22. Sharma, G., Busch, C., Mukhopadhyay, S., Malveaux, C.: Tight analysis of a collisionless robot gathering algorithm. ACM Trans. Auton. Adapt. Syst. **12**, 1–20 (2017)
23. Sharma, G., Vaidyanathan, R., Trahan, J.L.: Constant-time complete visibility for asynchronous robots with lights. In: Spirakis, P., Tsigas, P. (eds.) SSS 2017. LNCS, vol. 10616, pp. 265–281. Springer, Cham (2017). https://doi.org/10.1007/978-3-319-69084-1_18
24. Vaidyanathan, R., Busch, C., Trahan, J.L., Sharma, G., Rai, S.: Logarithmic-time complete visibility for robots with lights. In: Proceedings of the 2015 IEEE International Parallel and Distributed Processing Symposium, pp. 375–384 (2015)

Faster Algorithms for Cycle Hitting Problems on Disk Graphs

Shinwoo An(✉), Kyungjin Cho, and Eunjin Oh(✉)

POSTECH, Pohang, Korea
{shinwooan,kyungjincho,eunjin.oh}@postech.ac.kr

Abstract. In this paper, we consider three hitting problems on a disk intersection graph: Triangle Hitting Set, Feedback Vertex Set, and Odd Cycle Transversal. Given a disk intersection graph G, our goal is to compute a set of vertices hitting all triangles, all cycles, or all odd cycles, respectively. Our algorithms run in time $2^{\tilde{O}(k^{4/5})}n^{O(1)}$, $2^{\tilde{O}(k^{9/10})}n^{O(1)}$, and $2^{\tilde{O}(k^{19/20})}n^{O(1)}$, respectively, where n denotes the number of vertices of G. These do not require a geometric representation of a disk graph. If a geometric representation of a disk graph is given as input, we can solve these problems more efficiently. In this way, we improve the algorithms for those three problem by Lokshtanov et al. [SODA 2022].

Keywords: Disk graphs · feedback vertex set · triangle hitting set

1 Introduction

In this paper, we present subexponential parameterized algorithms for the following three well-known parameterized problems on disk graphs: Triangle Hitting Set, Feedback Vertex Set, and Odd Cycle Transversal. Given a graph $G = (V, E)$ and an integer k, these problems ask for finding a subset of V of size k that hits all triangles, cycles, and odd-cycles, respectively. On general graphs, the best-known algorithms for Triangle Hitting Set, Feedback Vertex Set, and Odd Cycle Transversal run in $2.1^k n^{O(1)}$, $2.7^k n^{O(1)}$, and $2.32^k n^{O(1)}$ time, respectively, where n denotes the number of vertices of a graph [14,15,19]. All these problems are NP-complete, and moreover, no $2^{o(k)}n^{O(1)}$ algorithm exists for these problems on general graphs unless ETH fails [9]. This motivates the study of these problems on special graph classes such as planar graphs and geometric intersection graphs.

Although these problems are NP-complete even for planar graphs, much faster algorithms exist for planar graphs. More specifically, they can be solved in time $2^{O(\sqrt{k})}n^{O(1)}$, $2^{O(\sqrt{k})}n^{O(1)}$, and $2^{O(\sqrt{k}\log k)}n^{O(1)}$, respectively, on planar graphs [10,17]. Moreover, they are optimal unless ETH fails. The $2^{O(\sqrt{k})}n^{O(1)}$-time algorithms for Triangle Hitting Set and Feedback Vertex Set follow from the *bidimensionality theory* of Demaine at al [10]. A planar graph either

This work was supported by the National Research Foundation of Korea(NRF) grant funded by the Korea government(MSIT) (No.RS-2023-00209069).

P. Morin and S. Suri (Eds.): WADS 2023, LNCS 14079, pp. 29–42, 2023.
https://doi.org/10.1007/978-3-031-38906-1_3

has an $r \times r$ grid graph as a minor, or its treewidth[1] is $O(r)$. This implies that for a YES-instance (G, k) of Feedback Vertex Set, the treewidth of G is $O(\sqrt{k})$. Then a standard dynamic programming approach gives a $2^{O(\sqrt{k})}n^{O(1)}$-time algorithm for Feedback Vertex Set. This approach can be generalized for H-minor free graphs and bounded-genus graphs.

Recently, several subexponential-time algorithms for cycle hitting problems have been studied on *geometric intersection graphs* [2,5,6,11–13]. Let $\mathcal{D}$ be a set of geometric objects such as disks or polygons. The intersection graph $G = (V, E)$ of $\mathcal{D}$ is the graph where a vertex of V corresponds to an element of $\mathcal{D}$, and two vertices of V are connected by an edge in E if and only if their corresponding elements in $\mathcal{D}$ intersect. In particular, the intersection graph of (unit) disks is called a *(unit) disk graph*, and the intersection graph of interior-disjoint polygons is called a *map graph*. Unlike planar graphs, map graphs and unit disk graphs can have large cliques. For unit disk graphs, Feedback Vertex Set can be solved in $2^{O(\sqrt{k})}n^{O(1)}$ time [2], and Odd Cycle Transversal can be solved in $2^{O(\sqrt{k}\log k)}n^{O(1)}$ time [4]. For map graphs, Feedback Vertex Set can be solved in $2^{O(\sqrt{k}\log k)}n^{O(1)}$ time [11]. These are almost tight in the sense that no $2^{O(\sqrt{k})}n^{O(1)}$-time algorithm exists unless ETH fails [11,13].

However, until very recently, little has been known for disk graphs, a broad class of graphs that generalizes planar graphs and unit disk graphs. Very recently, Lokshtanov et al. [16] presented the first subexponential-time algorithms for cycle hitting problems on disks graphs. The explicit running times of the algorithms are summarized in the first row of Table 1. On the other hand, the best-known lower bound on the computation time for these problems is $2^{\Omega(k^{1/2})}n^{O(1)}$ assuming ETH. There is a huge gap between the upper and lower bounds.

Our Results. In this paper, we make progress on the study of subexponential-time parameterized algorithms on disk graphs by presenting faster algorithms for the cycle hitting problems. We say an algorithm is *robust* if this algorithm does not requires a geometric representation of a disk graph. We present $2^{\tilde{O}(k^{4/5})}n^{O(1)}$-time, $2^{\tilde{O}(k^{9/10})}n^{O(1)}$-time, and $2^{\tilde{O}(k^{19/20})}n^{O(1)}$-time robust algorithms for Triangle Hitting Set, Feedback Vertex Set, and Odd Cycle Transversal, respectively. Furthermore, we present $2^{\tilde{O}(k^{2/3})}n^{O(1)}$-time, $2^{\tilde{O}(k^{7/8})}n^{O(1)}$-time, and $2^{\tilde{O}(k^{15/16})}n^{O(1)}$-time algorithms for Triangle Hitting Set, Feedback Vertex Set, and Odd Cycle Transversal which are not robust. These results are summarized in the second and third rows of Table 1.

For Triangle Hitting Set, we devised a kernelization algorithm based on a crown decomposition by modifying the the algorithm in [1]. After a branching process, we obtain a set of $O((k/p)\log k)$ instances (G', k') such that G' has a set of $O(kp)$ triangles, which we call a *core*, for a value p. Every triangle of G' shares at least two vertices with at least one triangle of a core. This allows us to remove several vertices from G' so that the number of vertices of G' is $O(kp)$.

[1] The definition of the treewidth can be found in Sect. 2.

Table 1. Comparison of the running times of our algorithms and the algorithm by Lokshtanov et al. [16]. All algorithms for ODD CYCLE TRANSVERSAL are randomized.

TRIANGLE HITTING	FVS	OCT	robust?	
$2^{O(k^{9/10}\log k)}n^{O(1)}$	$2^{O(k^{13/14}\log k)}n^{O(1)}$	$2^{O(k^{27/28}\log k)}n^{O(1)}$	yes	[16]
$2^{O(k^{4/5}\log k)}n^{O(1)}$	$2^{O(k^{9/10}\log k)}n^{O(1)}$	$2^{O(k^{19/20}\log k)}n^{O(1)}$	yes	this paper
$2^{O(k^{2/3}\log k)}n^{O(1)}$	$2^{O(k^{7/8}\log k)}n^{O(1)}$	$2^{O(k^{15/16}\log k)}n^{O(1)}$	no	this paper

Then we can obtain a tree decomposition of small treewidth, and then we apply dynamic programming on the tree decomposition.

For FEEDBACK VERTEX SET and ODD CYCLE TRANSVERSAL, we give an improved analysis of the algorithms in [16] by presenting improved bounds on one of the two main combinatorial results presented in [16] concerning the treewidth of disk graphs (Theorem 2.1). More specifically, Theorem 1 of [16] says that for any subset M of V such that $N(v)\cap M \neq N(u)\cap M$ for any two vertices u and v not in M, the size of U is $O(|M|\cdot p^6)$, where p is the ply of the disks represented by the vertices of G. We improve an improved bound of $O(|M|\cdot p^2)$. To obtain an improved bound, we classify the vertices of $V-M$ into three classes in a more sophisticated way, and then use the concept of the additively weighted higher-order Voronoi diagram. We believe the additively weighted higher-order Voronoi diagram will be useful in designing optimal algorithms for TRIANGLE HITTING SET, FEEDBACK VERTEX SET, and ODD CYCLE TRANSVERSAL. Indeed, the Voronoi diagram (or Delaunay triangulation) is a main tool for obtaining an optimal algorithm for FEEDBACK VERTEX SET on unit disk graphs [2].

2 Preliminaries

For a graph G, we let $V(G)$ and $E(G)$ be the sets of vertices and edges of G, respectively. For a subset U of G, we use $G[U]$ to denote the subgraph of G induced by U. Also, for a subset $U \subset V$ of V, we simply denote $G[V\backslash U]$ by $G-U$. For a vertex v of G, let $N(v)$ be the set of the vertices of G adjacent to v. We call it the *neighborhood* of v.

A triangle of G is a cycle of G consisting of three vertices. We denote a triangle consisting of three vertices x, y and z by $\{x,y,z\}$. We sometimes consider it as the set $\{x,y,z\}$ of vertices. For instance, the union of triangles is the set of all vertices of the triangles. A subset F of V is called a *triangle hitting set* of G if $G[V\backslash F]$ has no triangle. Also, F is called a *feedback vertex set* if $G[V\backslash F]$ has no cycle (and thus it is forest). Finally, F is called a *odd cycle transversal* if $G[V\backslash F]$ has no odd cycle. In other words, a triangle hitting set, a feedback vertex set, and a odd cycle transversal hit all triangles, cycles and odd cycles, respectively. Notice that a feedback vertex set of G is also a triangle hitting set.

Disk Graphs. Let $G = (V, E)$ be a disk graph defined by a set $\mathcal{D}$ of disks. In this case, we say that $\mathcal{D}$ is a *geometric representation* of G. For a vertex v of G, we let $D(v)$ denote the disk of $\mathcal{D}$ represented by v. The *ply* of $\mathcal{D}$ is defined as the maximum number of disks of $\mathcal{D}$ containing a common point. Note that the disk graph of a set of disks of ply p has a clique of size at least p. If it is clear from the context, we say that the ply of G is p. The *arrangement* of $\mathcal{D}$ is the subdivision of the plane formed by the boundaries of the disks of $\mathcal{D}$ that consists of vertices, edges and faces. The *arrangement graph* of G, denoted by $\mathcal{A}(G)$, is the plane graph where every face of the arrangement of $\mathcal{D}$ contained in at least one disk is represented by a vertex, and vertices are adjacent if the faces they represent share an edge. Given a disk graph G, it is NP-hard to compute its geometric representation [8]. We say an algorithm is *robust* if this algorithm does not requires a geometric representation of a disk graph.

Tree Decomposition. A *tree decomposition* of an undirected graph $G = (V, E)$ is defined as a pair (T, β), where T is a tree and β is a mapping from nodes of T to subsets of V (called bags) with the following conditions. Let $\mathcal{B} := \{\beta(t) : t \in V(T)\}$ be the set of bags of T.

- For $\forall u \in V$, there is at least one bag in $\mathcal{B}$ which contains u.
- For $\forall (u, v) \in E$, there is at least one bag in $\mathcal{B}$ which contains both u and v.
- For $\forall u \in V$, the nodes of T containing u in their bags are connected in T.

The *width* of a tree decomposition is defined as the size of its largest bag minus one, and the *treewidth* of G is the minimum width of a tree decomposition of G. The treewidth of a disk graph G is $O(p\sqrt{|V(G)|})$, where p is the ply of G [16].

Higher-Order Voronoi Diagram. To analyze the treewidth of G of bounded ply, we use the concept of the *higher-order Voronoi diagram.* Given a set of n weighted sites (points), its order-k Voronoi diagram is defined as the subdivision of $\mathbb{R}^2$ into maximal regions such that all points within a given region have the same k nearest sites. Here, the distance between a site s with weight w and a point x in the plane is defined as $d(s, x) - w$, where $d(s, x)$ is the Euclidean distance between s and x.

Lemma 1 (Theorem 4 in [18]). *The complexity of the addictively weighted order-k Voronoi diagram of m point sites in the plane is $O(mk)$.*

Due to page limit, some proofs and details are omitted. In particular, we omit the details of the algorithms that require a geometric representation. Since the concept of the weighted treewidth is used only for those algorithms, we also omit the definition of the weighted treewidth. All missing proofs and details can be found in the full version of this paper.

3 Triangle Hitting Set

In this section, we present a robust $2^{O(k^{4/5} \log k)} n^{O(1)}$-time algorithm for Triangle Hitting Set, which improves the $2^{O(k^{9/10} \log k)} n^{O(1)}$-time algorithm of [16].

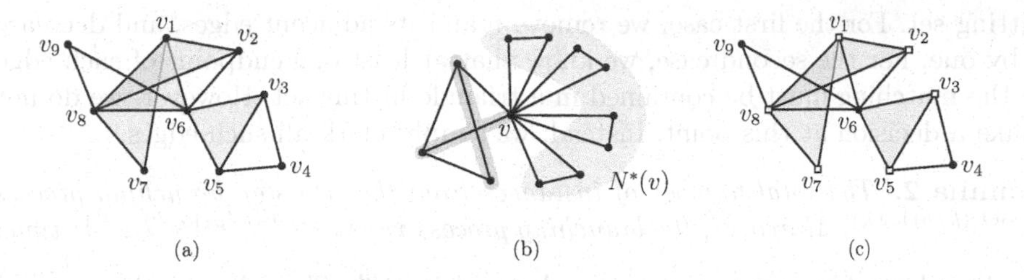

Fig. 1. (a) The four gray triangles, $v_1v_2v_8, v_1v_6v_8, v_1v_7v_8$ and $v_3v_5v_6$, form a core. A triangle of G not in the core shares two vertices with at least one gray triangle. (b) The marked edges are colored gray. Then $N^*(v)$ has a matching of size four. (c) The vertices in the initial set of F are $v_1, v_2, v_3, v_5, v_7, v_8$. We add two three gray triangles $v_1v_2v_8, v_1v_7v_8, v_3v_5v_6$ to W. Note that W is not yet a core because of $v_6v_8v_9$. (Color figure online)

3.1 Two-Step Branching Process

We first apply a branching process as follows to obtain $2^{O((k/p)\log k)}$ instances one of which is a YES-instance (G', k') of TRIANGLE HITTING SET where every clique of G' has size $O(p)$. For a clique C of G and a triangle hitting set F, all except for at most two vertices of C are contained in F. In particular, if G has a triangle hitting set of size k, any clique of G has size at most $k+2$. For a clique of size at least p, we branch on which vertices of the clique are not included in a triangle hitting set. After this, the solution size k decrease by at least $p-2$. We repeat this until every clique has size $O(p)$. In the resulting branching tree, every node has at most $O(k^2)$ children since any clique of G has size at most $k+2$. And the branching tree has height $O(k/p)$. In this way, we can obtain $2^{O((k/p)\log k)}$ instances of TRIANGLE HITTING SET one of which is a YES-instance (G', k') of TRIANGLE HITTING SET where G' has a geometric representation of ply at most p. Moreover, G' is an induced subgraph of G, and $k' \leq k$. Using the EPTAS for computing a maximum clique in a disk graph [7], we can complete the branching step in $2^{O((k/p)\log k)}n^{O(1)}$ time. Note that the algorithm in [7] does not require a geometric representation of a graph.

For each instance (G, k) obtained from the first branching, we apply the second branching process to obtain $2^{O(k/p)}$ instances one of which is a YES-instance having a *core* of size $O(pk)$. A set W of triangles of G is called a *core* if for a triangle of G, a triangle of W shares at least two vertices with the triangle. See Fig. 1(a). During the branching process, we mark an edge to remember that one of its endpoints must be added to a triangle hitting set. The marking process has an invariant that no two marked edges share a common endpoint, and thus the number of marked edges is at most k. The marks will be considered in the dynamic programming procedure. Initially, all edges are unmarked.

Let v be a vertex such that $N^*(v)$ has a matching of size at least p, where $N^*(v)$ be the set of neighbors of v not incident to any marked edge. See Fig. 1(b). In this case, a triangle hitting set contains either v or at least one endpoint of each edge in the matching. We branch on whether or not v is added to a triangle hitting set. For the first case, we remove v and its adjacent edges, and decrease k by one. For the second case, we know that at least one endpoint of each edge in the matching must be contained in a triangle hitting set. However, we do not make a decision at this point. Instead, we simply mark all such edges.

Lemma 2. *The total number of instances from the two-step branching process is $2^{O((k/p)\log k)}$. Moreover, the branching process runs in $2^{O((k/p)\log k)}n^{O(1)}$ time.*

This branching process was already used in [16]. The following lemma is a key for our improvement over [16].

Lemma 3. *Let (G, k) be a YES-instance obtained from the two-step branching. Then G has a core of size $O(pk)$, and we can compute one in polynomial time.*

Proof. We construct a core W of G as follows. Let F_0 be the the union of a triangle hitting set of size at most $3k$ and the set of all endpoints of the marked edges, which can be computed in polynomial time.[2] Note that, the size of F_0 is at most $O(k)$. Then let W be the set of triangles constructed as follows: for each edge xy of $G[F_0]$, we add an arbitrary triangle of G formed by x, y and $v \in V\backslash F_0$ to W, if it exists. The number of triangles in W is $O(pk)$ by Lemma 4.

At this moment, W is not necessarily a core. See Fig. 1(c). Thus we add several triangles further to W to compute a core of G. By the branching process, for every vertex v of F_0, $N(v)\backslash F_0$ has a maximum matching of size at most p. We add the triangles formed by v and the edges of the maximum matching to W for every vertex v of F_0. Note that we do not update F_0 during this phase, and thus triangles added to W due to two different vertices might intersect. This algorithm clearly runs in polynomial time. Moreover, since the size of F_0 is $O(k)$, we add at most $O(pk)$ triangles to W, and thus the size of W is $O(pk)$.

We claim that W is a core of G. Let $\{x, y, z\}$ be a triangle of G not in W. Since F_0 contains a triangle hitting set of size at most $3k$, it must contain at least one of x, y and z. If at least two of them, say x and y, are contained in F_0, then there exists a triangle having edge xy in W by construction. Thus, $\{x, y, z\}$ shares two vertices with such a triangle. The remaining case is that exactly one of them, say x, is contained in F_0. In this case, we have considered x and a maximum matching of $N(x)\backslash F_0$. The only reason why $\{x, y, z\}$ is not added to W is that the edge yz is adjacent to another edge $y'z'$ for some other triangle $\{x, y', z'\}$. Then $\{x, y', z'\}$ is added to W. Note that $\{x, y, z\}$ and $\{x, y', z'\}$ share at least two vertices, and thus $\{x, y, z\}$ satisfies the condition for W being a core. ◻

Lemma 4. *For a subset F of V of size $O(k)$, $G[F]$ has $O(pk)$ edges.*

The following observation will be used in the correctness proof of the kernelization step in Sect. 3.2. The observation holds because G' does not have any triangle not appearing in G.

[2] We can find a hitting set of size at most $3k$ as follows: Find a triangle, and add all its vertices to a triangle hitting set. Then remove all its vertices from the graph.

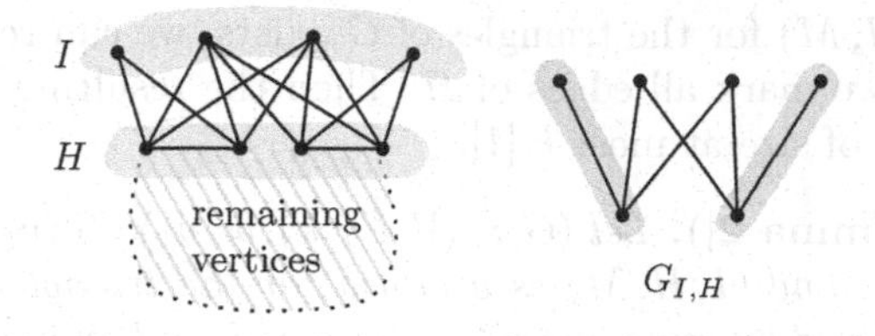

Fig. 2. The edges in the matching M in $G_{I,H}$ are colored gray.

Observation 1. *Let G' be an induced subgraph of G such that $V(G')$ contains all vertices of a core W of G. Then W is a core of G'.*

For each instance (G, k) obtained from the branching process, we apply the cleaning step that removes all vertices not hitting any triangle of G from G. Specifically, we remove a vertex of degree less than two. Also, we remove a vertex whose neighbors are independent. Whenever we remove a vertex, we also remove its adjacent edges.

3.2 Kernelization Using Crown Decomposition

Let (G, k) be a YES-instance we obtained from the branching process. We show that if the number of vertices of G contained in the triangles of G is at least pk, we can construct a *crown* for the triangles of G. Then using this, we can produce a YES-instance (G', k) of TRIANGLE HITTING SET in polynomial time where G' is a proper induced subgraph of G. Thus by repeatedly applying this process (at most n^2 times) and then by removing all vertices not contained in any triangle of G, we can obtain a YES-instance (G', k) where G' is a disk graph of complexity $O(pk)$.

We first define a crown decomposition of a graph for triangles. For illustration, see Fig. 2. A crown decomposition of a graph was initially introduced to construct a linear kernel for VERTEX COVER. Later, Abu-Khzam [1] generalized this concept to hypergraphs. Using this, he showed that a triangle hitting set admits a quadratic kernel for a general graph. In our case, we will show that the size of a kernel is indeed $O(pk)$ due to the branching process. More specifically, it is due to the existence of a core of size $O(pk)$.

Definition 1 ([1]). *A* crown *for the triangles of G is a triple (I, H, M) with a subset I of $V(G)$, a subset H of $E(G)$, and a matching M of $G_{I,H}$ s.t.*

- *no two vertices of I are contained in the same triangle of G,*
- *each edge of H forms a triangle with some vertex of I, and*
- *every edge (vertex in $G_{I,H}$) of H is matched under M,*

where $G_{I,H}$ is the bipartite graph with vertex set $I \cup H$ such that $x \in I$ and $uv \in H$ are connected by an edge if and only if u, v and x form a triangle.

If a crown (I, H, M) for the triangles of G exists, we can remove all vertices in I, but instead, we mark all edges of H. Then the resulting graph also has a triangle hitting set of size at most k [1].

Lemma 5 ([1, Lemma 2]). *Let $(G = (V, E), k)$ be a YES-instance of* Triangle Hitting Set, *and (I, H, M) is a crown for the triangles of G. Then the subgraph of G obtained by removing all vertices of I and by marking all edges of H has a trianlge hitting set of size at most k.*

Now we show that (G, k) has a crown for its triangles if cpk vertices are contained in the triangles of G, where c is a sufficiently large constant. Let W be a core of G of size $O(pk)$, which exists due to Lemma 3 and Observation 1. Let I be the set of vertices of G not contained in any triangle of W but contained in some triangle of G. If cpk vertices are contained in the triangles of G, the size of I is $c'pk$ for a sufficiently large constant c' since $|W| = O(pk)$. Then let H be the set of all edges of G which form triangles together with the vertices of I. Note that for every edge of H, its endpoints are contained in the same triangle of W by the definition of the core. Thus the size of H is at most $3 \cdot |W| = O(pk)$. Therefore, if more than cpk vertices are contained in the triangles of G, $|I| > |H|$.

Lemma 6. *If $|I| > |H|$, there are two subsets $H' \subseteq H$ and $I' \subseteq I$ such that (I', H', M') is a crown for the triangles of G.*

By Lemma 5 and Lemma 6, we can obtain an instance (G', k') equivalent to (G, k) such that the union of the triangles has complexity $O(pk)$ for each instance (G, k) obtained from Sect. 3.1 in polynomial time.

Theorem 1. *Given a disk graph G with its geometric representation, we can find a triangle hitting set of G of size k in $2^{O(k^{2/3} \log k)} n^{O(1)}$ time, if it exists. Without a geometric representation, we can do this in $2^{O(k^{4/5} \log k)} n^{O(1)}$ time.*

Proof. In this proof, we only show how to solve the triangle hitting set problem in $2^{O(k^{4/5} \log k)} n^{O(1)}$ time without using a geometric representation of G. After the branching and kernelization processes, we have $2^{O((k/p) \log k)}$ instances one of which is a YES-instance such that the size of the union of the triangles of G' is at most $O(pk)$. For each instance (G', k'), we remove all vertices not contained in any triangle of G'. Then the resulting graph G' has $O(pk)$ vertices. Then we compute a tree decomposition (T, β) of G' of treewidth $O(p\sqrt{pk})$. Then the total running time is $2^{O(p^{3/2}\sqrt{k})} \cdot 2^{O((k/p) \log k)}$. By letting $p = k^{1/5}$, we have $2^{O(k^{4/5} \log k)} n^{O(1)}$-time. $\square$

4 Feedback Vertex Set and Odd Cycle Transversal

In this section, we show that the algorithms in [16] for Feedback Vertex Set and Odd Cycle Transversal indeed take $2^{\tilde{O}(k^{9/10})} n^{O(1)}$ time and

$2^{\tilde{O}(k^{19/20})}n^{O(1)}$ time algorithms respectively. It is shown in [16] that they take $2^{O(k^{13/14})}n^{O(1)}$ time and $2^{O(k^{27/28})}n^{O(1)}$ time, respectively, but we give a better analysis. We can obtain non-robust algorithms for these problems with better running times, but we omit the description of them.

We obtained better time bounds by classifying the vertices in a kernel used in [16] into two types, and by using the higher order additively weighted Voronoi diagrams. To make the paper self-contained, we present the algorithms in [16] here. The following lemma is a main observation of [16]. Indeed, the statement of the lemma given by [16] is stronger than this, but the following statement is sufficient for our purpose. We say a vertex v is *deep* for a subset F of V if all neighbors of v in G are contained in F. For a subset Q of $V(G)$, let G/Q be the graph obtained from G by contracting each connected component of $G[Q]$ into a single vertex.

Lemma 7 (Theorem 1.2 of [16]). *Let G be a disk graph that has a realization of ply p. For a subset F of V, let F^* be the union of F and the set of all deep vertices for F. If F contains a core of G, then the arrangement graph of G/Q has treewidth $O(\max\{\sqrt{|F^*| \cdot w} \cdot p^{1.5}, w\})$ for a set $Q \subseteq V \backslash F^*$, where w is the treewidth of $(G - F^*)/Q$.*

In Sect. 4.1, we show that the size of F^* is $O(p^2|F|)$. This is our main contribution in this section. Then we solve the two cycle hitting problems using dynamic programming on a tree decomposition of bounded treewidth.

Branching and Cleaning Process. We first apply the two-step branching process as we did for Triangle Hitting Set. Note that if a vertex set F is a feedback vertex set or odd cycle transversal, then F is a triangle hitting set.

We apply the cleaning process for each instance (G, k) we obtained from the earlier branching process. First, we remove all vertices of degree one. Then we keep $O(1)$ vertices from each class of *false twins* as follows. A set of vertices is called a *false twin* if they are pairwise non-adjacent, and they have the same neighborhood. As observed in [16], for each class of false twins, every minimal feedback vertex set either contains the entire class except for at most one vertex, or none of the vertices in that class. Thus we keep only one vertex (an arbitrary one) from each class of *false twins* if (G, k) is an instance of Feedback Vertex Set. Similarly, for each class of false twins, every odd cycle transversal contains the entire class except for at most two vertices, or none of the vertices in that class. Therefore, when we deal with Odd Cycle Transversal, we keep only two vertices from each class of *false twins*. We remember how many vertices each kept vertex represents and make use of this information in DP.

We have $2^{O((k/p)\log k)}$ instances one of which is a YES-instance (G, k) where G has a core of size $O(pk)$ and a geometric representation of ply p. Moreover, at most two vertices are in each class of false twins.

4.1 The Number of Deep Vertices

Let (G, k) be an YES-instance obtained from the branching and cleaning process. Let F be a subset of $V(G)$ containing a core of G. The disks of $G - F$ have ply at most two. In this section, we give an upper bound on the number of *deep vertices* for F. For this, we classify the vertices of $V - F$ in two types: regular and irregular vertices. See Fig. 3(a).

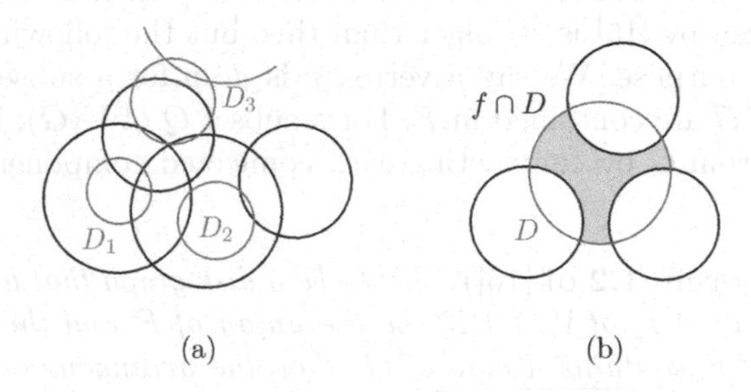

Fig. 3. (a) The vertices of F are colored black, and the vertices of $V - F$ are colored red. D_1 is deep and irregular, D_2 is deep and regular, and D_3 is shallow and irregular. (b) D is irregular due to the fourth case of Definition 2 (Color figure online).

Definition 2. *A vertex v of $V - F$ is said to be* irregular *if v belongs to one of the following types:*

- *$D(v)$ contains a vertex of $\mathcal{A}$,*
- *$D(v)$ is contained in a face of $\mathcal{A}$,*
- *$D(v)$ contains a disk represented by a vertex of F, or*
- *$D(v)$ intersects three edges of $\mathcal{A}$ incident to the same face of $\mathcal{A}$,*

where $\mathcal{A}$ denotes the arrangement of the disks represented by F. If v does not belong to any of the types, we say v is regular.

Lemma 8. *The number of deep and regular vertices is $O(p^2|F|)$.*

Proof. If v is regular, the neighbors of v in G form two cliques. To see this, observe that the part of $\mathcal{A}$ restricted to $D(v)$ consists of *parallel* arcs. That is, the arcs can be sorted in a way that any two consecutive arcs come from the same face of $\mathcal{A}$. Also, any two arcs which are not consecutive in the sorted list are not incident to a common face of $\mathcal{A}$. In this case, there are two points x and y on the boundary of $D(v)$ such that the line segment xy intersects all disks represented by F intersecting $D(v)$. Then the disks of F intersecting $D(v)$ contains either x or y. Therefore, the neighbors of v form at most two cliques.

Since each clique has size at most p, a deep and regular vertex has at most $2p$ neighbors in G. Let v be a deep and regular vertex having exactly r neighbors. Consider the additively weighted order-r Voronoi diagram rVD of F. Here, the distance between a disk D' of F and a point x in the plane is defined as

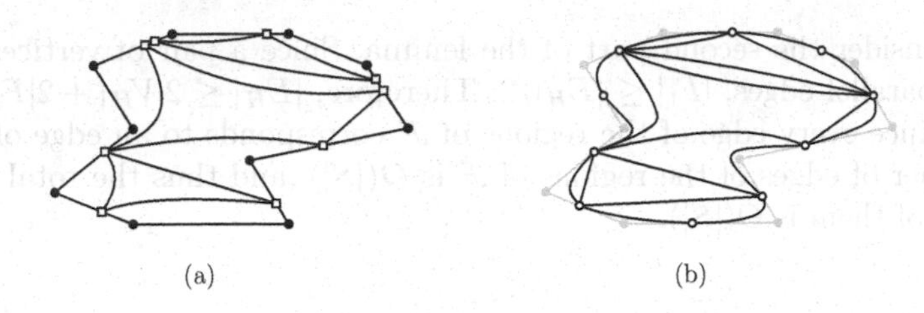

(a) (b)

Fig. 4. (a) S is the polygon with vertices marked with black disks, and each region of $\mathcal{F}$ has vertices marked with white boxes. (b) The vertices of H are marked with white.

$d(c, x) - w$, where c is the center of D' and w is the radius of D'. Then the center of $D(v)$ is contained in a Voronoi region of the order-r Voronoi diagram whose sites are exactly the neighbors of v. Since no two deep vertices have the same neighborhood in F, each Voronoi region contains at most one deep vertex. Therefore, the number of deep and regular vertices having exactly r neighbors is linear in the complexity of rVD for $r \leq 2p$. By Lemma 1, its complexity is $O(r|F|)$. Thus, the number of deep and regular vertices of G is $O(p^2|F|)$. ◻

For irregular vertices, we make use of the fact that the ply of the disks represented by those vertices is at most two. It is not difficult to see that the number of irregular vertices of the first three types is $O(p|F|)$. This is because the complexity of $\mathcal{A}$ is $O(p|F|)$. To analyze the number of irregular vertices of the fourth type, we use the following lemma.

Lemma 9. *Let S be a connected region with $|S|$ edges in the plane, and $\mathcal{F}$ be a set of interior-disjoint regions contained in S such that each region intersects at least three edges of S. Then the number of regions of $\mathcal{F}$ is $O(|S|)$. Also, the total number of vertices of the regions of $\mathcal{F}$ is $O(|S|)$.*

Proof. We consider the following planar graph H: each vertex of H corresponds to an edge of S. Also, for each region of $\mathcal{F}$ with vertices $x_1, x_2, \ldots, x_t$ in order, we add edges connecting the vertices corresponding to the edges of S containing x_i and x_{i+1} for all indices $1 \leq i < t$. In addition to this, for each vertex v of S, we add an edge connecting the vertices of H corresponding the two edges of S incident to v. See Fig. 4. There might be parallel edges in H, but a pair of vertices has at most two parallel edges. Let F_1 be the set of faces bounded by a pair of parallel edges, and let F_2 be the set of the other faces of H. Then each region of $\mathcal{F}$ corresponds to a face of F_2. To prove the first part of the lemma, it suffices to show that the size of F_2 is linear in the number of edges of S.

By the Euler's formula, $|E_H| \leq |V_H|+|F_1|+|F_2|-2$, where V_H and E_H denote the set of vertices and edges of H, respectively. Note that no vertex of H has degree at most one. Since every face of F_2 is incident to at least three edges of H, and every edge of E_H is incident to exactly two faces of H, $2|E_H| \geq 3|F_2|+2|F_1|$. Thus we have $2|F_1|+3|F_2| \leq 2|V_H|+2|F_1|+2|F_2|-4$. That is, $|F_2| \leq 2|V_H|-4$. This implies the first part of the lemma.

Now consider the second part of the lemma. Since a pair of vertices has at most two parallel edges, $|F_1| \leq |E_H|/2$. Therefore, $|E_H| \leq 2|V_H| + 2|F_2| - 4 = O(|V_H|)$. Since every edge of the regions of $\mathcal{F}$ corresponds to an edge of H, the total number of edges of the regions of $\mathcal{F}$ is $O(|S|)$, and thus the total number of vertices of them is $O(|S|)$. □

Lemma 10. *The number of irregular vertices of* V *is* $O(p|F|)$.

4.2 Feedback Vertex Set

In this section, we show how to compute a feedback vertex set of size k in $2^{\tilde{O}(k^{9/10})}n^{O(1)}$ time without a geometric representation of G. Let (G, k) be an YES-instance of Feedback Vertex Set from the branching and cleaning processes. Then we have a core W of G of size $O(pk)$. Let F be the union of a feedback vertex set of size $2k$ and the vertex set of all triangles of core W of G, and let F^* be the union of F and all deep vertices for F. The size of F^* is $O(p^3k)$ by Lemmas 8 and 10. Since F^* contains a feedback vertex set, $G - F^*$ is a forest. Therefore, the treewidth of G is $O(p^4\sqrt{k})$ by Lemma 7. Then we compute a tree decomposition of treewidth $O(p^4\sqrt{k})$, and then apply a dynamic programming on the tree decomposition to solve the problem in $2^{O(p^4\sqrt{k})}$ time. The total running time is $2^{O(k^{9/10}\log k)}n^{O(1)}$ by setting $p = k^{1/10}$.

Theorem 2. *Given a disk graph* G*, we can find a feedback vertex set of size* k *in* $2^{O(k^{9/10}\log k)}n^{O(1)}$ *time, if it exists. Given a disk graph* G *together with its geometric representation, we can do this in in* $2^{O(k^{7/8}\log k)}n^{O(1)}$ *time.*

4.3 Odd Cycle Transversal

In this section, we present an $2^{O(k^{19/20}\log k)}n^{O(1)}$-time randomised algorithm for Odd Cycle Transversal when a geometric representation of G is given. Here, we give a sketch of the algorithm.

Let (G, k) be an YES-instance of Odd Cycle Transversal obtained from the branching and cleaning processes. We have a core W of G of size $O(pk)$. Let F be the union of the triangles of W. Let F^* be the union of F and the set of all deep vertices for F of size $O(p^3k)$ by Lemmas 8 and 10. Let $G^* = G - F^*$. Since G^* does not have a triangle, it is planar. We compute disjoint sets $Z_1, \ldots, Z_p \subseteq V(G)$ such that for every $i \in [\sqrt{k}]$ and every $Z' \subseteq Z_i$, the treewidth of $G/(Z_i \backslash Z')$ is $O(\sqrt{k} + |Z'|)$. This can be done in polynomial time [3].

Then there is an index i such that for a fixed odd cycle transversal S, $S \cap Z_i$ has size $O(\sqrt{k})$. We iterate over every choice of i, and some choice of $Z' = S \cap Z_i$ of size at most $O(\sqrt{k})$. For each set Z_i, we can compute a candidate set for Z' of size $2^{O(\sqrt{k}\log k)}$ with high probability [17]. Therefore, the number of choices of Z' is $2^{O(\sqrt{k}\log k)}$ for each Z_i. For each iteration, we contract $Z_i \backslash Z'$ and then apply a dynamic programming on a tree decomposition of $G/(Z_i \backslash Z')$. The number of iterations is $2^{O(\sqrt{k}\log k)}$, and $G^*/(Z_i \backslash Z')$ has treewidth $O(\sqrt{k})$. Therefore, by Lemma 7, the treewidth of $G/(Z_i \backslash Z')$ is $O(p^4 k^{3/4})$, and we can obtain the desired running time by setting $p = k^{1/20}$ using Proposition 6.10 in [16].

Theorem 3. *Given a disk graph G, we can find a odd cycle transversal of size k in $2^{O(k^{19/20}\log k)} n^{O(1)}$ time* w.h.p. *Given a disk graph G together with its geometric representation, we can do this in in $2^{O(k^{15/16}\log k)} n^{O(1)}$ time* w.h.p.

References

1. Abu-Khzam, F.N.: A kernelization algorithm for *d*-hitting set. J. Comput. Syst. Sci. **76**(7), 524–531 (2010)
2. An, S., Oh, E.: Feedback vertex set on geometric intersection graphs. In: Proceedings of the 32nd International Symposium on Algorithms and Computation (ISAAC 2021), pp. 47:1–47:12 (2021)
3. Bandyapadhyay, S., Lochet, W., Lokshtanov, D., Saurabh, S., Xue, J.: Subexponential parameterized algorithms for cut and cycle hitting problems on H-minor-free graphs⋆. In: Proceedings of the 2022 Annual ACM-SIAM Symposium on Discrete Algorithms (SODA), pp. 2063–2084. SIAM (2022)
4. Bandyapadhyay, S., Lochet, W., Lokshtanov, D., Saurabh, S., Xue, J.: True contraction decomposition and almost eth-tight bipartization for unit-disk graphs. In: 38th International Symposium on Computational Geometry (SoCG 2022), vol. 224, pp. 11:1–11:16 (2022)
5. de Berg, M., Bodlaender, H.L., Kisfaludi-Bak, S., Marx, D., Van Der Zanden, T.C.: A framework for exponential-time-hypothesis-tight algorithms and lower bounds in geometric intersection graphs. SIAM J. Comput. **49**(6), 1291–1331 (2020)
6. de Berg, M., Kisfaludi-Bak, S., Monemizadeh, M., Theocharous, L.: Clique-based separators for geometric intersection graphs. In: 32nd International Symposium on Algorithms and Computation (ISAAC 2021), pp. 22:1–22:15 (2021)
7. Bonamy, M., et al.: EPTAS and subexponential algorithm for maximum clique on disk and unit ball graphs. J. ACM **68**(2), 1–32 (2021)
8. Breu, H., Kirkpatrick, D.G.: Unit disk graph recognition is NP-hard. Comput. Geom. **9**(1–2), 3–24 (1998)
9. Cygan, M.: Parameterized Algorithms. Springer, Switzerland (2015)
10. Demaine, E.D., Fomin, F.V., Hajiaghayi, M., Thilikos, D.M.: Subexponential parameterized algorithms on bounded-genus graphs and H-minor-free graphs. J. ACM (JACM) **52**(6), 866–893 (2005)
11. Fomin, F.V., Lokshtanov, D., Panolan, F., Saurabh, S., Zehavi, M.: Decomposition of map graphs with applications. In: Proceedings of the 46th International Colloquium on Automata, Languages, and Programming (ICALP 2019), pp. 60:1–60:15 (2019)
12. Fomin, F.V., Lokshtanov, D., Panolan, F., Saurabh, S., Zehavi, M.: Finding, hitting and packing cycles in subexponential time on unit disk graphs. Discrete Comput. Geom. **62**(4), 879–911 (2019)
13. Fomin, F.V., Lokshtanov, D., Saurabh, S.: Bidimensionality and geometric graphs. In: Proceedings of the Twenty-Third Annual ACM-SIAM Symposium on Discrete Algorithms (SODA 2012), pp. 1563–1575 (2012)

14. Li, J., Nederlof, J.: Detecting feedback vertex sets of size k in $O^*(2.7k)$ time. In: Proceedings of the Fourteenth Annual ACM-SIAM Symposium on Discrete Algorithms (SODA 2022), pp. 971–989 (2020)
15. Lokshtanov, D., Narayanaswamy, N., Raman, V., Ramanujan, M., Saurabh, S.: Faster parameterized algorithms using linear programming. ACM Trans. Algorithms (TALG) **11**(2), 1–31 (2014)
16. Lokshtanov, D., Panolan, F., Saurabh, S., Xue, J., Zehavi, M.: Subexponential parameterized algorithms on disk graphs (extended abstract). In: Proceedings of the 2022 Annual ACM-SIAM Symposium on Discrete Algorithms (SODA 2022), pp. 2005–2031
17. Lokshtanov, D., Saurabh, S., Wahlström, M.: Subexponential parameterized odd cycle transversal on planar graphs. In: IARCS Annual Conference on Foundations of Software Technology and Theoretical Computer Science (FSTTCS 2012). Schloss Dagstuhl-Leibniz-Zentrum fuer Informatik (2012)
18. Rosenberger, H.: Order-k voronoi diagrams of sites with additive weights in the plane. Algorithmica **6**(1), 490–521 (1991)
19. Wahlström, M.: Algorithms, measures and upper bounds for satisfiability and related problems. Ph.D. thesis, Department of Computer and Information Science, Linköpings universitet (2007)

Tight Analysis of the Lazy Algorithm for Open Online Dial-a-Ride

Júlia Baligács(✉), Yann Disser(✉), Farehe Soheil(✉), and David Weckbecker(✉)

TU Darmstadt, Darmstadt, Germany
{baligacs,disser,soheil,weckbecker}@mathematik.tu-darmstadt.de

Abstract. In the open online dial-a-ride problem, a single server has to deliver transportation requests appearing over time in some metric space, subject to minimizing the completion time. We improve on the best known upper bounds on the competitive ratio on general metric spaces and on the half-line, for both the preemptive and non-preemptive version of the problem. We achieve this by revisiting the algorithm LAZY recently suggested in [WAOA, 2022] and giving an improved and tight analysis. More precisely, we show that it has competitive ratio 2.457 on general metric spaces and 2.366 on the half-line. This is the first upper bound that beats known lower bounds of 2.5 for schedule-based algorithms as well as the natural REPLAN algorithm.

Keywords: online algorithms · dial-a-ride · competitive analysis

1 Introduction

In the open online dial-a-ride problem, we are given a metric space (M, d) and have control of a server that can move at unit speed. Over time, *requests* of the form $(a, b; t)$ arrive. Here, $a \in M$ is the *starting position* of the request, $b \in M$ is its *destination*, and $t \in \mathbb{R}_{\geq 0}$ is the *release time* of the request. We consider the online variant of the problem, meaning that the server does not get to know all requests at time 0, but rather at the respective release times. Our task is to control the server such that it serves all requests, i.e., we have to move the server to position a, load the request $(a, b; t)$ there after its release time t, and then move to position b where we unload the request. The objective is to minimize the *completion time*, i.e., the time when all requests are served.

We assume that the server always starts at time 0 in some fixed point, which we call the *origin* $O \in M$. The server has a *capacity* $c \in (\mathbb{N} \cup \{\infty\})$ and is not allowed to load more than c requests at the same time. Furthermore, we consider the *non-preemptive* version of the problem, that is, the server may not unload a request preemptively at a point that is not the request's destination. In the dial-a-ride problem, a distinction is made between the open and the closed

Supported by DFG grant DI 2041/2.

P. Morin and S. Suri (Eds.): WADS 2023, LNCS 14079, pp. 43–64, 2023.
https://doi.org/10.1007/978-3-031-38906-1_4

Table 1. State of the art of the open online dial-a-ride problem and overview of our results: Bold bounds are original results, other bounds are inherited.

metric space		old bounds		new bounds upper
		lower	upper	
general	non-preemptive	2.05	**2.618** [5]	**2.457** (Thm 1)
	preemptive	2.04	2.618	2.457
line	non-preemptive	**2.05** [10]	2.618	2.457
	preemptive	**2.04** [11]	**2.41** [11]	—
half-line	non-preemptive	**1.9** [25]	2.618	**2.366** (Thm 2)
	preemptive	**1.62** [25]	2.41	2.366

variant. In the *closed* dial-a-ride problem, the server has to return to the origin after serving all requests. By contrast, in the *open* dial-a-ride problem, the server may finish anywhere in the metric space. In this work, we only consider the open variant of the problem. By letting $a = b$ for all requests $(a, b; t)$, we obtain the *online travelling salesperson problem (TSP)* as a special case of the dial-a-ride problem.

In this work, we only consider deterministic algorithms for the online dial-a-ride problem. As usual in competitive analysis, we measure the quality of a deterministic algorithm by comparing it to an optimum offline algorithm. The measure we apply is the completion time of a solution. For a given sequence of requests σ and an algorithm ALG, we denote by $\text{ALG}(\sigma)$ the completion time of the algorithm for request sequence σ. Analogously, we denote by $\text{OPT}(\sigma)$ the completion time of an optimal offline algorithm. For some $\rho \geq 1$, we say that an algorithm ALG is *ρ-competitive* if, for all request sequences σ, we have $\text{ALG}(\sigma) \leq \rho \cdot \text{OPT}(\sigma)$. The *competitive ratio* of ALG is defined as $\inf\{\rho \geq 1 \mid \text{ALG is } \rho\text{-competitive}\}$. The *competitive ratio of a problem* is defined as $\inf\{\rho \geq 1 \mid \text{there is some } \rho\text{-competitive algorithm}\}$

Our Results. We consider the parametrized algorithm $\text{LAZY}(\alpha)$ that was presented in [5] and prove the following results (see Table 1).

Our main result is an improved general upper bound for the open online dial-a-ride problem.

Theorem 1. *For $\alpha = \frac{1}{2} + \sqrt{11/12}$, $\text{LAZY}(\alpha)$ has a competitive ratio of $\alpha + 1 \approx 2.457$ for open online dial-a-ride on general metric spaces for every capacity $c \in \mathbb{N} \cup \{\infty\}$.*

Prior to our work, the best known general upper bound of $\varphi + 1 \approx 2.618$ on the competitive ratio for the open online dial-a-ride problem was achieved by $\text{LAZY}(\varphi)$ and it was shown that $\text{LAZY}(\alpha)$ has competitive ratio at least $\frac{3}{2} + \sqrt{11/12} \approx 2.457$ for any choice of α, even on the line [5]. This means that we give a conclusive analysis of LAZY(α) by achieving an improved upper bound that tightly matches the previously known lower bound. In particular, $\alpha = 1/2 + \sqrt{11/12}$ is the (unique) best possible waiting parameter for LAZY(α), even on the line. The best known general lower bound remains 2.05 [10].

Crucially, our upper bound beats, for the first time, a known lower bound of 2.5 for the class of *schedule-based* algorithms [8], i.e., algorithms that divide the execution into subschedules that are never interrupted. Historically, all upper bounds, prior to those via LAZY, were based on schedule-based algorithms [9,10]. Our result means that online algorithms cannot afford to irrevocably commit to serving some subset of requests if they hope to attain the best possible competitive ratio.

Secondly, our upper bound also beats the same lower bound of 2.5 for the REOPT (or REPLAN) algorithm [3], which simply reoptimizes its solution whenever new requests appear. While this algorithm is very natural and may be the first algorithm studied for the online dial-a-ride problem, it has eluded tight analysis up to this day. So far, it has been a canonical candidate for a best-possible algorithm. We finally rule it out.

In addition to the general bound above, we analyze LAZY(α) for open online dial-a-ride on the half-line, i.e., where $M = \mathbb{R}_{\geq 0}$, and show that, in this metric space, even better bounds on the competitive ratio are possible for different values of α. More precisely, we show the following.

Theorem 2. *For* $\alpha = \frac{1+\sqrt{3}}{2}$, LAZY($\alpha$) *has a competitive ratio of* $\alpha + 1 \approx 2.366$ *for open online dial-a-ride on the half-line for every capacity* $c \in \mathbb{N} \cup \{\infty\}$.

This further improves on the previous best known upper bound of 2.618 [5]. The best known lower bound is 1.9 [25].

We go on to show that the bound in Theorem 2 is best-possible for LAZY(α) over all parameter choices $\alpha \geq 0$.

Theorem 3. *For all* $\alpha \geq 0$, LAZY(α) *has a competitive ratio of at least* $\alpha + 1 \approx 2.366$ *for open online dial-a-ride on the half-line for every capacity* $c \in \mathbb{N} \cup \{\infty\}$.

In the preemptive version of the online dial-a-ride problem, the server is allowed to unload requests anywhere and pick them up later again. In this version, prior to our work, the best known upper bound on general metric spaces was 2.618 [5] and the best known upper bound on the line and the half-line was 2.41 [11]. Obviously, every non-preemptive algorithm can also be applied in the preemptive setting, however, its competitive ratio may degrade since the optimum might have to use preemption. Our algorithm LAZY repeatedly executes optimal solutions for subsets of requests and can be turned preemptive by using preemptive solutions. With this change, our analysis of LAZY still carries through in the preemptive case and improves the state of the art for general metric spaces and the half-line, but not the line. The best known lower bound in the preemptive version on general metric spaces is 2.04 [11] and the best known lower bound on the half-line is 1.62 [25].

Corollary 1. *The competitive ratio of the open preemptive online dial-a-ride problem with any capacity $c \in \mathbb{N} \cup \{\infty\}$ is upper bounded by*

a) $\frac{3}{2} + \sqrt{\frac{11}{12}} \approx 2.457$ *and this bound is achieved by* $\text{LAZY}\left(\frac{1}{2} + \sqrt{\frac{11}{12}}\right)$,

b) $1 + \frac{1+\sqrt{3}}{2} \approx 2.366$ *on the half-line and this bound is achieved by* $\text{LAZY}(\frac{1+\sqrt{3}}{2})$.

Related Work. Two of the most natural algorithms for the online dial-a-ride problem are IGNORE and REPLAN. The basic idea of IGNORE is to repeatedly follow an optimum schedule over the currently unserved requests and ignoring all requests released during its execution. The competitive ratio of this algorithm is known to be exactly 4 [8,21]. By contrast, the main idea of REPLAN is to start a new schedule over all unserved requests whenever a new request is released. While this algorithm has turned out to be notoriously difficult to analyze, it is known that its competitive ratio is at least 2.5 [3] and at most 4 [8]. Several variants of these algorithms have been proposed such as SMARTSTART [21], SMARTERSTART [10] or WAITORIGNORE [25], which lead to improvements on the best known bounds on the competitive ratio of the dial-a-ride problem. In this work, we study the recently suggested algorithm LAZY [5], which is known to achieve a competitive ratio of $\varphi + 1$ for the open online dial-a-ride problem, where $\varphi = \frac{\sqrt{5}+1}{2} \approx 1.618$ denotes the golden ratio.

For the preemptive version of the open dial-a-ride problem, the best known upper bound on general metric spaces is $\varphi + 1 \approx 2.618$ [5]. Bjelde et al. [11] proved a stronger upper bound of $1 + \sqrt{2} \approx 2.41$ for when the metric space is the line.

In terms of lower bounds, Birx et al. [10] were able to prove that every algorithm for the open online dial-a-ride problem has a competitive ratio of at least 2.05, even if the metric space is the line. This separates dial-a-ride from online TSP on the line, where it is known that the competitive ratio is exactly 2.04 [11]. For open online dial-a-ride on the half-line, Lipmann [25] established a lower bound of 1.9 for the non-preemptive version and a lower bound of 1.62 for the preemptive version.

For the closed variant of the online dial-a-ride problem, the competitive ratio is known to be exactly 2 on general metric spaces [1,3,14] and between 1.76 and 2 on the line [8,11]. On the half-line the best known lower bound is 1.71 [1] and the best known upper bound is 2 [1,14]. The TSP variant of the closed dial-a-ride problem is tightly analyzed with a competitive ratio of 2 on general metric spaces [1,3,14], of 1.64 on the line [3,11], and of 1.5 on the half-line [12].

Other variants of the problem have been studied in the literature. This includes settings where the request sequence has to fulfill some reasonable additional properties [12,16,22], where the server is presented with additional [4] or less [26] information, where the server has some additional abilities [13,19], where the server has to handle requests in a given order [15,19], or where we consider different objectives than the completion time [2,6,7,16–18,22–24]. Other examples include the study of randomized algorithms [21], or other metric spaces, such as a circle [20]. Moreover, it has been studied whether some natural classes of algorithms can have good competitive ratios. For example, *zealous* algorithms always have to move towards an unserved request or the origin [12]. A *schedule-based* algorithm operates in schedules that are not allowed to be interrupted. Birx [8] showed that all such algorithms have a competitive ratio of at least 2.5. Together with our results, this implies that schedule-based algorithms algorithms cannot be best-possible.

2 Algorithm Description and Notation

In this section, we define the algorithm LAZY introduced in [5]. The rough idea of the algorithm is to wait until several requests are revealed and then start a schedule serving them. Whenever a new request arrives, we check whether we can deliver all currently loaded requests and return to the origin in a reasonable time. If this is possible, we do so and begin a new schedule including the new requests starting from the origin. If this is not possible, we keep following the current schedule and consider the new request later.

More formally, a *schedule* is a sequence of actions specifying the server's behaviour, including its movement and where requests are loaded or unloaded. By OPT$[t]$, we denote an optimal schedule beginning in O at time 0 and serving all requests that are released not later than time t. By OPT(t), we denote its completion time. Given a set of requests R and some point $x \in M$, we denote by $S(R, x)$ a shortest schedule serving all requests in R beginning from point x at some time after all requests in R are released. In other words, we can ignore the release times of the requests when computing $S(R, x)$. As waiting is not beneficial for the server if there are no release times, the *length of the schedule*, i.e., the distance the server travels, is the same as the time needed to complete it and we denote this by $|S(R, x)|$.

Now that we have established the notation needed, we can describe the algorithm (cf. Algorithm 1). By t, we denote the current time. By p_t, we denote the position of the server at time t, and by R_t, we denote the set of requests that have been released but not served until time t. The variable i is a counter over the schedules started by the algorithm. The waiting parameter $\alpha \geq 1$ specifies how long we wait before starting a schedule. The algorithm uses the following commands: DELIVER_AND_RETURN orders the server to finish serving all currently loaded requests and return to O in the fastest possible way, WAIT_UNTIL(t) orders the server to remain at its current location until time t, and FOLLOW_SCHEDULE(S) orders the server to execute the actions defined by schedule S. When any of these commands is invoked, the server aborts what it is doing and executes the new command. Whenever the server has completed a command, we say that it becomes *idle*.

We make a few comments for illustration of the algorithm. If the server returns to the origin upon receiving a request, we say that the schedule it was currently following is *interrupted*. Observe that, due to interruption, the sets $R^{(i)}$ are not necessarily disjoint. Also, observe that $p^{(1)} = O$, and if schedule $S^{(i)}$ was interrupted, we have $p^{(i+1)} = O$ and $t^{(i+1)} = \alpha \cdot \text{OPT}(t^{(i+1)})$. If $S^{(i)}$ was not interrupted, $p^{(i+1)}$ is the ending position of $S^{(i)}$.

Algorithm 1: $\text{Lazy}(\alpha)$

initialize: $i \leftarrow 0$

upon receiving a request:

if *server can serve all loaded requests and return to O until time* $\alpha \cdot \text{Opt}(t)$ **then**
  execute DELIVER_AND_RETURN

upon becoming idle:

if $t < \alpha \cdot \text{Opt}(t)$ **then**
  execute WAIT_UNTIL($\alpha \cdot \text{Opt}(t)$)
else if $R_t \neq \emptyset$ **then**
  $i \leftarrow i+1$, $R^{(i)} \leftarrow R_t$, $t^{(i)} \leftarrow t$, $p^{(i)} \leftarrow p_t$
  $S^{(i)} \leftarrow S(R^{(i)}, p^{(i)})$
  execute FOLLOW_SCHEDULE($S^{(i)}$)

The following observations were already noted in [5] and follow directly from the definitions above and the fact that requests in $R^{(i)} \backslash R^{(i-1)}$ were released after time $t^{(i-1)}$.

Observation 1 ([5]). *For every request sequence, the following hold.*

a) For every $i > 1$, $\text{Opt}(t^{(i)}) \geq t^{(i-1)} \geq \alpha \cdot \text{Opt}(t^{(i-1)})$.

b) For every $x, y \in M$ *and every subset of requests* R, *we have* $|S(R,x)| \leq d(x,y) + |S(R,y)|$.

c) Let $i > 1$ *and assume that* $S^{(i-1)}$ *was not interrupted. Let* a *be the starting position of the request in* $R^{(i)}$ *that is picked up first by* $\text{Opt}(t^{(i)})$. *Then,*

$$\text{Opt}(t^{(i)}) \geq t^{(i-1)} + |S(R^{(i)}, a)| \geq \alpha \cdot \text{Opt}(t^{(i-1)}) + |S(R^{(i)}, a)|.$$

3 Factor-Revealing Approach

The results in this paper were informed by a factor-revealing technique, inspired by a similar approach of Bienkowski et al. [6], to analyze a specific algorithm Alg, in our case Lazy. The technique is based on a formulation of the adversary problem, i.e., the problem of finding an instance that maximizes the competitive ratio, as an optimization problem of the form

$$\max\left\{ \frac{\text{Alg}(x)}{\text{Opt}(x)} \,\middle|\, x \text{ describes a dial-a-ride instance} \right\}. \quad (1)$$

An optimum solution to this problem immediately yields the competitive ratio of Alg. Of course, we cannot hope to solve this optimization problem or even describe it with a finite number of variables. The factor-revealing approach consists in relaxing (1) to a practically solvable problem over a finite number of variables.

The key is to select a set of variables that captures the structure of the problem well enough to allow for meaningful bounds. In our case, we can, for example, introduce variables for the starting position and duration of the second-to-last as well as for the last schedule. We then need to relate those variables via constraints that ensure that an optimum solution to the relaxed problem actually has a realization as a dial-a-ride instance. For example, we might add the constraint that the distance between the starting positions of the last two schedules is upper bounded by the duration of the second-to-last schedule.

The power of the factor-revealing approach is that it allows to follow an iterative process for deriving structurally crucial inequalities: When solving the relaxed optimization problem, we generally have to expect an optimum solution that is not realizable and overestimates the competitive ratio. We can then focus our efforts on understanding why the corresponding variable assignment cannot be realized by a dial-a-ride instance. Then, we can introduce additional variables and constraints to exclude such solutions. In this way, the unrealizable solutions inform our analysis in the sense that we obtain bounds on the competitive ratio that can be proven analytically by only using the current set of variables and inequalities. Once we obtain a realizable lower bound, we thus have found the exact competitive ratio of the algorithm under investigation.

In order to practically solve the relaxed optimization problems, we limit ourselves to linear programs (LPs). Note that the objective of (1) is linear if we normalize to $\text{OPT}(x) = 1$. We can do this, since the competitive ratio is invariant with respect to rescaling the metric space and release times of requests. Another advantage of using linear programs is that we immediately obtain a formal proof of the optimum solution from an optimum solution to the LP dual. Of course, the correctness of the involved inequalities still needs to be established.

In the remainder of this paper, we present a purely analytic proof of our results. Many of the inequalities we derive in lemmas were informed by a factor-revealing approach via a linear program with a small number of binary variables. This means that we additionally need to branch on all binary variables in order to obtain a formal proof via LP duality. We refer to Appendix A for more details of the binary program that informed our results for the half-line.

4 Analysis on General Metric Spaces

This section is concerned with the proof of Theorem 1. For the remainder of this section, let $(r_1, \dots, r_n)$ be some fixed request sequence. Let k be the number of schedules started by $\text{LAZY}(\alpha)$, and let $S^{(i)}$, $t^{(i)}$, $p^{(i)}$, $R^{(i)}$ $(1 \leq i \leq k)$ be defined as in the algorithm. Note that we slightly abuse notation here because k, $S^{(i)}$, $t^{(i)}$, $p^{(i)}$, and $R^{(i)}$ depend on α. As it will always be clear from the context what α is, we allow this implicit dependency in the notation.

As it will be crucial for the proof in which order OPT and LAZY serve requests, we introduce the following notation. Let

- $r_{f,\text{OPT}}^{(i)} = (a_{f,\text{OPT}}^{(i)}, b_{f,\text{OPT}}^{(i)}; t_{f,\text{OPT}}^{(i)})$ be the first request in $R^{(i)}$ picked up by $\text{OPT}[t^{(i)}]$,
- $r_{l,\text{OPT}}^{(i)} = (a_{l,\text{OPT}}^{(i)}, b_{l,\text{OPT}}^{(i)}; t_{l,\text{OPT}}^{(i)})$ be the last request in $R^{(i)}$ delivered by $\text{OPT}[t^{(i+1)}]$,

- $r^{(i)}_{f,\textsc{Lazy}} = (a^{(i)}_{f,\textsc{Lazy}}, b^{(i)}_{f,\textsc{Lazy}}; t^{(i)}_{f,\textsc{Lazy}})$ be the first request in $R^{(i)}$ picked up by $\textsc{Lazy}(\alpha)$,
- $r^{(i)}_{l,\textsc{Lazy}} = (a^{(i)}_{l,\textsc{Lazy}}, p^{(i+1)}; t^{(i)}_{l,\textsc{Lazy}})$ be the last request in $R^{(i)}$ delivered by $\textsc{Lazy}(\alpha)$.

Definition 1. *We say that the i-th schedule is* α-good *if*

a) $|S^{(i)}| \leq \textsc{Opt}(t^{(i)})$ *and*
b) $t^{(i)} + |S^{(i)}| \leq (1+\alpha) \cdot \textsc{Opt}(t^{(i)})$.

In this section, we prove by induction on i that, for $\alpha \geq \frac{1}{2} + \sqrt{11/12}$, every schedule is α-good. Note that this immediately implies Theorem 1.

As our work builds on [5], the first few steps of our proof are the same as in [5]. For better understandability and reading flow, we repeat the proofs of some important but simple steps and mark the results with appropriate citations. The results starting with Lemma 2 are new and improve on the analysis in [5].

We begin with proving the base case.

Observation 2 (Base case, [5]). *For every* $\alpha \geq 1$*, the first schedule is* α*-good.*

Proof. Recall that $S^{(1)}$ begins in O and is the shortest tour serving all requests in $R^{(1)}$. $\textsc{Opt}[t^{(1)}]$ begins in O and serves all requests in $R^{(1)}$, too, which yields $|S^{(1)}| \leq \textsc{Opt}(t^{(1)})$. The fact that we have $t^{(1)} = \alpha \cdot \textsc{Opt}(t^{(1)})$ implies
$t^{(1)} + |S^{(1)}| \leq (1+\alpha) \cdot \textsc{Opt}(t^{(1)})$. □

Next, we observe briefly that the induction step is not too difficult when the last schedule was interrupted.

Observation 3 (Interruption case, [5]). *Let* $\alpha \geq 1$*. Assume that schedule* $S^{(i)}$ *was interrupted. Then,* $S^{(i+1)}$ *is* α*-good.*

Proof. If schedule $S^{(i)}$ was interrupted, we have $p^{(i+1)} = O$ and $t^{(i+1)} = \alpha \cdot \textsc{Opt}(t^{(i+1)})$. Therefore, $|S^{(i+1)}| = |S(R^{(i+1)}, O)| \leq \textsc{Opt}(t^{(i+1)})$ and $t^{(i+1)} + |S^{(i+1)}| \leq (1+\alpha) \cdot \textsc{Opt}(t^{(i+1)})$. □

For this reason, we will assume in many of the following statements that the schedule $S^{(i)}$ was not interrupted.

By careful observation of the proof in [5], one can see that the following fact already holds for smaller α. For convenience, we repeat the proof of the following Lemma with an adapted value of α.

Lemma 1 ([5]). *Let* $\alpha \geq \frac{1+\sqrt{17}}{4} \approx 1.281$ *and* $i \in \{1, \ldots, k-1\}$*. If* $S^{(i)}$ *is* α*-good, then* $|S^{(i+1)}| \leq \textsc{Opt}(t^{(i+1)})$.

Proof. First, observe that if $S^{(i)}$ was interrupted, we have $p^{(i+1)} = O$. Note that $\textsc{Opt}(t^{(i+1)})$ begins in O and serves all requests in $R^{(i+1)}$ so that we have

$$|S^{(i+1)}| = |S(R^{(i+1)}, O)| \leq \textsc{Opt}(t^{(i+1)}).$$

Therefore, assume from now on that $S^{(i)}$ was not interrupted. Also, if $\textsc{Opt}[t^{(i+1)}]$ serves $r^{(i)}_{l,\textsc{Lazy}}$ at $p^{(i+1)}$ before collecting any request from $R^{(i+1)}$, we trivially have

$$|S^{(i+1)}| = |S(R^{(i+1)}, p^{(i+1)})| \leq \textsc{Opt}(t^{(i+1)}).$$

Therefore, assume additionally that $\textsc{Opt}[t^{(i+1)}]$ collects $r^{(i+1)}_{f,\textsc{Opt}}$ before serving $r^{(i)}_{l,\textsc{Lazy}}$. Next, we prove the following assertion.

Claim: In the setting described above, we have

$$d(a^{(i+1)}_{f,\textsc{Opt}}, p^{(i+1)}) \leq \left(1 + \frac{2}{\alpha} - \alpha\right) \textsc{Opt}(t^{(i)}). \tag{2}$$

To prove the claim, note that $r^{(i+1)}_{f,\textsc{Opt}}$ is released not earlier than $\alpha \cdot \textsc{Opt}(t^{(i)})$. Since we assume that $\textsc{Opt}(t^{(i+1)})$ collects $r^{(i+1)}_{f,\textsc{Opt}}$ before serving $r^{(i)}_{l,\textsc{Lazy}}$ at $p^{(i+1)}$, we obtain

$$\textsc{Opt}(t^{(i+1)}) \geq \alpha \cdot \textsc{Opt}(t^{(i)}) + d(a^{(i+1)}_{f,\textsc{Opt}}, p^{(i+1)}). \tag{3}$$

Upon the arrival of the last request in $R^{(i)}$, we have $\textsc{Opt}(t) = \textsc{Opt}(t^{(i+1)})$ and the server can finish its current schedule and return to the origin in time $t^{(i)} + |S^{(i)}| + d(p^{(i+1)}, O)$. As we assume that $S^{(i)}$ was not interrupted, this yields

$$t^{(i)} + |S^{(i)}| + d(p^{(i+1)}, O) > \alpha \cdot \textsc{Opt}(t^{(i+1)}). \tag{4}$$

Combined, we obtain that

$$\begin{aligned}
d(a^{(i+1)}_{f,\textsc{Opt}}, p^{(i+1)}) &\overset{(3)}{\leq} \textsc{Opt}(t^{(i+1)}) - \alpha \cdot \textsc{Opt}(t^{(i)}) \\
&\overset{(4)}{\leq} \frac{1}{\alpha} \cdot \left(t^{(i)} + |S^{(i)}| + d(p^{(i+1)}, O)\right) - \alpha \cdot \textsc{Opt}(t^{(i)}) \\
&\overset{S^{(i)}\,\alpha\text{-good}}{\leq} \frac{1}{\alpha} \cdot \left((1+\alpha) \cdot \textsc{Opt}(t^{(i)}) + d(p^{(i+1)}, O)\right) - \alpha \cdot \textsc{Opt}(t^{(i)}) \\
&\leq \left(1 + \frac{2}{\alpha} - \alpha\right) \textsc{Opt}(t^{(i)}),
\end{aligned}$$

where we have used in the last inequality that $d(p^{(i+1)}, O) \leq \textsc{Opt}(t^{(i)})$ because $\textsc{Opt}(t^{(i)})$ begins in O and has to serve $r^{(i)}_{l,\textsc{Lazy}}$ at $p^{(i+1)}$. This completes the proof of the claim.

Now, we turn back to proving Lemma 1. We obtain

$$\begin{aligned}
|S^{(i+1)}| &\leq d(p^{(i+1)}, a^{(i+1)}_{f,\textsc{Opt}}) + |S(R^{(i+1)}, a^{(i+1)}_{f,\textsc{Opt}})| \\
&\overset{\text{Obs 1c)}}{\leq} d(p^{(i+1)}, a^{(i+1)}_{f,\textsc{Opt}}) + \textsc{Opt}(t^{(i+1)}) - \alpha \cdot \textsc{Opt}(t^{(i)}) \\
&\overset{(2)}{\leq} \left(1 + \frac{2}{\alpha} - 2\alpha\right) \textsc{Opt}(t^{(i)}) + \textsc{Opt}(t^{(i+1)}) \\
&\leq \textsc{Opt}(t^{(i+1)}),
\end{aligned}$$

where the last inequality follows from the fact that $1 + \frac{2}{\alpha} - 2\alpha \leq 0$ if and only if $\alpha \geq \frac{1+\sqrt{17}}{4} \approx 1.2808$. □

Recall that the goal of this section is to prove that every schedule is α-good. So far, we have proven the base case (cf. Observation 2) and $|S^{(i+1)}| \leq \text{OPT}(t^{(i+1)})$ (Lemma 1) in the induction step. It remains to show that $t^{(i+1)} + |S^{(i+1)}| \leq (1+\alpha) \cdot \text{OPT}(t^{(i+1)})$ assuming $S^{(1)}, \dots, S^{(i)}$ are α-good. In Observation 3, we have already seen that this holds if $S^{(i)}$ was interrupted. To show that the induction step also holds if $S^{(i)}$ was not interrupted, we distinguish several cases for the order in which OPT serves the requests. We begin with the case that $\text{OPT}[t^{(i+1)}]$ picks up some request in $R^{(i+1)}$ before serving $r^{(i)}_{l,\text{LAZY}}$, i.e., that $\text{OPT}[t^{(i+1)}]$ does not follow the order of the $S^{(i)}$.

Lemma 2. *Let $\alpha \geq 1$. Assume that $S^{(i)}$ is α-good and was not interrupted, and that $\text{OPT}[t^{(i+1)}]$ picks up $r^{(i+1)}_{f,\text{OPT}}$ before serving $r^{(i)}_{l,\text{LAZY}}$. Then, $t^{(i+1)} + |S^{(i+1)}| \leq (1+\alpha) \cdot \text{OPT}(t^{(i+1)})$.*

Proof. Using the order in which OPT handles the requests, we obtain the following. After picking up $r^{(i+1)}_{f,\text{OPT}}$ at $a^{(i+1)}_{f,\text{OPT}}$ after time $t^{(i)}$, $\text{OPT}[t^{(i+1)}]$ has to serve $r^{(i)}_{l,\text{LAZY}}$ at $p^{(i+1)}$ so that

$$\text{OPT}(t^{(i+1)}) \geq t^{(i)} + d(p^{(i+1)}, a^{(i+1)}_{f,\text{OPT}}). \tag{5}$$

After finishing schedule $S^{(i)}$, the server either waits until time $\alpha \cdot \text{OPT}(t^{(i+1)})$ or immediately starts the next schedule, i.e., we have

$$t^{(i+1)} = \max\{\alpha \cdot \text{OPT}(t^{(i+1)}), t^{(i)} + |S^{(i)}|\}.$$

If $t^{(i+1)} = \alpha \cdot \text{OPT}(t^{(i+1)})$, the assertion follows immediately from Lemma 1. Thus, assume $t^{(i+1)} = t^{(i)} + |S^{(i)}|$. This yields

$$\begin{aligned}
t^{(i+1)} + |S^{(i+1)}| &\overset{\text{Obs 1b)}}{\leq} t^{(i)} + |S^{(i)}| + d(p^{(i+1)}, a^{(i+1)}_{f,\text{OPT}}) + |S(R^{(i+1)}, a^{(i+1)}_{f,\text{OPT}})| \\
&\overset{S^{(i)}\ \alpha\text{-good}}{\leq} (1+\alpha) \cdot \text{OPT}(t^{(i)}) + d(p^{(i+1)}, a^{(i+1)}_{f,\text{OPT}}) \\
&\qquad + |S(R^{(i+1)}, a^{(i+1)}_{f,\text{OPT}})| \\
&\overset{\text{Obs 1a)}}{\leq} \frac{1+\alpha}{\alpha} t^{(i)} + d(p^{(i+1)}, a^{(i+1)}_{f,\text{OPT}}) + |S(R^{(i+1)}, a^{(i+1)}_{f,\text{OPT}})| \\
&\overset{\text{Obs 1c)}}{\leq} \frac{1}{\alpha} t^{(i)} + d(p^{(i+1)}, a^{(i+1)}_{f,\text{OPT}}) + \text{OPT}(t^{(i+1)}) \\
&\overset{(5)}{\leq} \frac{1}{\alpha} \left(\text{OPT}(t^{(i+1)}) - d(p^{(i+1)}, a^{(i+1)}_{f,\text{OPT}})\right) + d(p^{(i+1)}, a^{(i+1)}_{f,\text{OPT}}) \\
&\qquad + \text{OPT}(t^{(i+1)}) \\
&= \left(1 + \frac{1}{\alpha}\right) \text{OPT}(t^{(i+1)}) + \left(1 - \frac{1}{\alpha}\right) d(p^{(i+1)}, a^{(i+1)}_{f,\text{OPT}}) \\
&\leq 2 \cdot \text{OPT}(t^{(i+1)}),
\end{aligned}$$

where we have used in the last inequality that $d(p^{(i+1)}, a^{(i+1)}_{f,\text{OPT}}) \leq \text{OPT}(t^{(i+1)})$ as $\text{OPT}[t^{(i+1)}]$ has to visit both points. □

Next, we consider the case where OPT handles $r_{l,\text{LAZY}}^{(i)}$ and $r_{f,\text{OPT}}^{(i+1)}$ in the same order as LAZY.

Lemma 3. *Let $\alpha \geq 1$. Assume that schedules $S^{(1)}, \ldots, S^{(i)}$ are α-good, $S^{(i)}$ was not interrupted, and $\text{OPT}[t^{(i+1)}]$ serves $r_{l,\text{LAZY}}^{(i)}$ before collecting $r_{f,\text{OPT}}^{(i+1)}$. If we have $d(p^{(i+1)}, a_{f,\text{OPT}}^{(i+1)}) + \text{OPT}(t^{(i)}) \leq \alpha \cdot \text{OPT}(t^{(i+1)})$, then $t^{(i+1)} + |S^{(i+1)}| \leq (1+\alpha) \cdot \text{OPT}(t^{(i+1)})$.*

Proof. Similarly as in the proof of Lemma 2, we can assume

$$t^{(i+1)} = t^{(i)} + |S^{(i)}|. \tag{6}$$

We have

$$\begin{aligned}
t^{(i+1)} + |S^{(i+1)}| &\overset{\text{Obs 1b)}}{\leq} t^{(i)} + |S^{(i)}| + d(p^{(i+1)}, a_{f,\text{OPT}}^{(i+1)}) + |S(R^{(i+1)}, a_{f,\text{OPT}}^{(i+1)})| \\
&\overset{S^{(i)}\,\alpha\text{-good}}{\leq} (1+\alpha) \cdot \text{OPT}(t^{(i)}) + d(p^{(i+1)}, a_{f,\text{OPT}}^{(i+1)}) \\
&\quad + |S(R^{(i+1)}, a_{f,\text{OPT}}^{(i+1)})| \\
&\overset{\text{Obs 1c)}}{\leq} (1+\alpha) \cdot \text{OPT}(t^{(i)}) + d(p^{(i+1)}, a_{f,\text{OPT}}^{(i+1)}) \\
&\quad + \text{OPT}(t^{(i+1)}) - \alpha \cdot \text{OPT}(t^{(i)}) \\
&= \text{OPT}(t^{(i+1)}) + d(p^{(i+1)}, a_{f,\text{OPT}}^{(i+1)}) + \text{OPT}(t^{(i)}) \\
&\leq (1+\alpha) \cdot \text{OPT}(t^{(i+1)}),
\end{aligned}$$

where the last inequality follows from the assumption that $d(p^{(i+1)}, a_{f,\text{OPT}}^{(i+1)}) + \text{OPT}(t^{(i)}) \leq \alpha \cdot \text{OPT}(t^{(i+1)})$. □

Now that we have proven the case described in Lemma 3, we will assume in the following that

$$d(p^{(i+1)}, a_{f,\text{OPT}}^{(i+1)}) > \alpha \cdot \text{OPT}(t^{(i+1)}) - \text{OPT}(t^{(i)}). \tag{7}$$

The following lemma states that, in this case, the $(i-1)$-th schedule (if it exists) was interrupted, i.e., the i-th schedule starts in the origin at time $\alpha \cdot \text{OPT}(t^{(i)})$.

Lemma 4. *Let $\alpha \geq \frac{1+\sqrt{3}}{2} \approx 1.366$. Assume that the i-th schedule is α-good and was not interrupted, and $\text{OPT}[t^{(i+1)}]$ serves $r_{l,\text{LAZY}}^{(i)}$ before collecting $r_{f,\text{OPT}}^{(i+1)}$. If (7) holds, then $p^{(i)} = O$ and $t^{(i)} = \alpha \cdot \text{OPT}(t^{(i)})$.*

Proof. If $i = 1$, we obviously have $p^{(i)} = O$ and $t^{(i)} = \alpha \cdot \text{OPT}(t^{(i)})$. Thus, assume that $i \geq 2$. If $r_{l,\text{LAZY}}^{(i)} \in (R^{(i-1)} \cap R^{(i)})$, schedule $S^{(i-1)}$ was interrupted and, thus, the statement holds. Otherwise, request $r_{l,\text{LAZY}}^{(i)}$ is released while schedule $S^{(i-1)}$

is running, i.e., $t_{l,\text{LAZY}}^{(i)} \geq t^{(i-1)} \geq \alpha \cdot \text{OPT}(t^{(i-1)})$. Combining this with the assumption that $\text{OPT}[t^{(i+1)}]$ serves $r_{l,\text{LAZY}}^{(i)}$ before collecting $r_{f,\text{OPT}}^{(i+1)}$, we obtain

$$\text{OPT}(t^{(i+1)}) \geq t_{l,\text{LAZY}}^{(i)} + d(p^{(i+1)}, a_{f,\text{OPT}}^{(i+1)}) \geq \alpha \cdot \text{OPT}(t^{(i-1)}) + d(p^{(i+1)}, a_{f,\text{OPT}}^{(i+1)}). \quad (8)$$

Rearranging yields

$$\begin{aligned} \alpha \cdot \text{OPT}(t^{(i-1)}) &\overset{(8)}{\leq} \text{OPT}(t^{(i+1)}) - d(p^{(i+1)}, a_{f,\text{OPT}}^{(i+1)}) \\ &\overset{(7)}{<} \text{OPT}(t^{(i)}) - (\alpha - 1)\text{OPT}(t^{(i+1)}) \\ &\overset{\text{Obs 1a)}}{\leq} (1 + \alpha - \alpha^2)\text{OPT}(t^{(i)}), \end{aligned}$$

which is equivalent to

$$\text{OPT}(t^{(i-1)}) < \left(1 + \frac{1}{\alpha} - \alpha\right)\text{OPT}(t^{(i)}). \quad (9)$$

By the assumption that $S^{(i-1)}$ is α-good, the server finishes schedule $S^{(i-1)}$ not later than time $(\alpha + 1) \cdot \text{OPT}(t^{(i-1)})$. Thus, at the time where request $r_{l,\text{LAZY}}^{(i)}$ is released, the server can serve all loaded requests and return to the origin by time

$$\max\{(\alpha + 1)\text{OPT}(t^{(i-1)}), t_{l,\text{LAZY}}^{(i)}\} + \text{OPT}(t^{(i-1)}).$$

We have

$$\begin{aligned} (\alpha + 2) \cdot \text{OPT}(t^{(i-1)}) &\overset{(9)}{<} (\alpha + 2)\left(1 + \frac{1}{\alpha} - \alpha\right)\text{OPT}(t^{(i)}) \\ &= \left(3 + \frac{2}{\alpha} - \alpha^2 - \alpha\right)\text{OPT}(t^{(i)}) \\ &\leq \alpha \cdot \text{OPT}(t^{(i)}), \end{aligned}$$

where the last inequality holds for $\alpha \geq 1.343$. Furthermore, as $r_{l,\text{LAZY}}^{(i)} \in R^{(i)}$, it holds that

$$\begin{aligned} t_{l,\text{LAZY}}^{(i)} + \text{OPT}(t^{(i-1)}) &\leq \text{OPT}(t^{(i)}) + \text{OPT}(t^{(i-1)}) \\ &\overset{(9)}{\leq} \left(2 + \frac{1}{\alpha} - \alpha\right)\text{OPT}(t^{(i)}) \leq \alpha \cdot \text{OPT}(t^{(i)}), \end{aligned}$$

where the last inequality holds for $\alpha \geq \frac{1+\sqrt{3}}{2} \approx 1.366$. This implies that the server can return to the origin by time $\alpha \cdot \text{OPT}(t^{(i)})$, i.e., we have $p^{(i)} = O$. □

We now come to the technically most involved case.

Lemma 5. *Let* $\alpha \geq \frac{1}{2} + \sqrt{11/12} \approx 1.457$. *Assume that the i-th schedule is* α*-good and was not interrupted, and* $\text{OPT}[t^{(i+1)}]$ *serves* $r^{(i)}_{l,\text{LAZY}}$ *before collecting* $r^{(i+1)}_{f,\text{OPT}}$. *If* (7) *holds, then* $t^{(i+1)} + |S^{(i+1)}| \leq (1+\alpha) \cdot \text{OPT}(t^{(i+1)})$.

Proof. We begin by proving the following assertion.
Claim: $\text{OPT}[t^{(i+1)}]$ *serves all requests in* $R^{(i)}$ *before picking up* $r^{(i+1)}_{f,\text{OPT}}$ *in* $a^{(i+1)}_{f,\text{OPT}}$.
To prove the claim, assume otherwise, i.e., that $\text{OPT}[t^{(i+1)}]$ serves $r^{(i)}_{l,\text{OPT}}$ after collecting $r^{(i+1)}_{f,\text{OPT}}$. The request $r^{(i+1)}_{f,\text{OPT}}$ is released after schedule $S^{(i)}$ is started, i.e., after time $\alpha \cdot \text{OPT}(t^{(i)})$. Thus,

$$\begin{aligned}\text{OPT}(t^{(i+1)}) &\geq \alpha \cdot \text{OPT}(t^{(i)}) + d(a^{(i+1)}_{f,\text{OPT}}, b^{(i)}_{l,\text{OPT}}) \\ &\overset{\Delta\text{-ineq}}{\geq} \alpha \cdot \text{OPT}(t^{(i)}) + d(a^{(i+1)}_{f,\text{OPT}}, p^{(i+1)}) - d(b^{(i)}_{l,\text{OPT}}, O) - d(O, p^{(i+1)}). \quad (10)\end{aligned}$$

Since $S^{(i)}$ starts in O, ends in $p^{(i+1)}$ and serves $r^{(i)}_{l,\text{OPT}}$, we obtain

$$d(O, b^{(i)}_{l,\text{OPT}}) + d(b^{(i)}_{l,\text{OPT}}, O) \leq |S^{(i)}| + d(p^{(i+1)}, O) \overset{\text{Lem 1}}{\leq} \text{OPT}(t^{(i)}) + d(p^{(i+1)}, O). \quad (11)$$

Furthermore, because $\text{OPT}[t^{(i+1)}]$ serves $r^{(i)}_{l,\text{LAZY}}$ at $p^{(i+1)}$ before picking up $r^{(i+1)}_{f,\text{OPT}}$ at $a^{(i+1)}_{f,\text{OPT}}$, we have

$$\begin{aligned}\text{OPT}(t^{(i+1)}) &\geq d(O, p^{(i+1)}) + d(p^{(i+1)}, a^{(i+1)}_{f,\text{OPT}}) \\ &\overset{(7)}{>} d(O, p^{(i+1)}) + \alpha \cdot \text{OPT}(t^{(i+1)}) - \text{OPT}(t^{(i)}). \quad (12)\end{aligned}$$

Combining all of the above yields

$$\begin{aligned}\text{OPT}(t^{(i+1)}) &\overset{(10),(11)}{\geq} \alpha \cdot \text{OPT}(t^{(i)}) + d(a^{(i+1)}_{f,\text{OPT}}, p^{(i+1)}) \\ &\quad - \frac{\text{OPT}(t^{(i)}) + d(p^{(i+1)}, O)}{2} - d(O, p^{(i+1)}) \\ &\overset{(12)}{>} \alpha \cdot \text{OPT}(t^{(i)}) + d(a^{(i+1)}_{f,\text{OPT}}, p^{(i+1)}) - \frac{\text{OPT}(t^{(i)})}{2} \\ &\quad - \frac{3}{2}\left(\text{OPT}(t^{(i)}) - (\alpha - 1)\text{OPT}(t^{(i+1)})\right) \\ &= \left(\frac{3}{2}\alpha - \frac{3}{2}\right)\text{OPT}(t^{(i+1)}) - (2 - \alpha)\text{OPT}(t^{(i)}) + d(a^{(i+1)}_{f,\text{OPT}}, p^{(i+1)}) \\ &\overset{(7)}{>} \left(\frac{3}{2}\alpha - \frac{3}{2}\right)\text{OPT}(t^{(i+1)}) - (2 - \alpha)\text{OPT}(t^{(i)}) \\ &\quad + \alpha \cdot \text{OPT}(t^{(i+1)}) - \text{OPT}(t^{(i)}) \\ &= \left(\frac{5}{2}\alpha - \frac{3}{2}\right)\text{OPT}(t^{(i+1)}) - (3 - \alpha)\text{OPT}(t^{(i)})\end{aligned}$$

$$\begin{aligned}
&\overset{\text{Obs 1a)}}{\geq} \left(\frac{5}{2}\alpha - \frac{3}{2}\right) \textsc{Opt}(t^{(i+1)}) - \left(\frac{3}{\alpha} - 1\right) \textsc{Opt}(t^{(i+1)}) \\
&= \left(\frac{5}{2}\alpha - \frac{1}{2} - \frac{3}{\alpha}\right) \textsc{Opt}(t^{(i+1)}) \\
&\geq \textsc{Opt}(t^{(i+1)})
\end{aligned}$$

where the last inequality holds if and only if $\alpha \geq \frac{1}{10} \cdot (3 + \sqrt{129}) \approx 1.436$. As this is a contradiction, we have that $\textsc{Opt}[t^{(i+1)}]$ serves all requests in $R^{(i)}$ before picking up $r^{(i+1)}_{f,\textsc{Opt}}$ in $a^{(i+1)}_{f,\textsc{Opt}}$. This completes the proof of the claim.

Now that we have established the claim, we turn back to the proof of Lemma 5. Let $T \geq 0$ denote the time it takes $\textsc{Opt}[t^{(i+1)}]$ until it has served $r^{(i)}_{l,\textsc{Opt}}$, i.e., all requests from $R^{(i)}$. First, observe that

$$T \geq \textsc{Opt}(t^{(i)}). \tag{13}$$

By the claim, we have

$$\textsc{Opt}(t^{(i+1)}) \geq T + d(b^{(i)}_{l,\textsc{Opt}}, a^{(i+1)}_{f,\textsc{Opt}}) + |S(R^{(i+1)}, a^{(i+1)}_{f,\textsc{Opt}})|. \tag{14}$$

The algorithm $\textsc{Lazy}(\alpha)$ finishes $R^{(i+1)}$ by time

$$\begin{aligned}
t^{(i+1)} + S^{(i+1)} &\overset{\text{Lem 4}}{=} \alpha \cdot \textsc{Opt}(t^{(i)}) + |S^{(i)}| + |S^{(i+1)}| \\
&\leq \alpha \cdot \textsc{Opt}(t^{(i)}) + |S^{(i)}| + d(p^{(i+1)}, a^{(i+1)}_{f,\textsc{Opt}}) + |S(R^{(i+1)}, a^{(i+1)}_{f,\textsc{Opt}})| \\
&\leq \alpha \cdot \textsc{Opt}(t^{(i)}) + |S^{(i)}| + d(p^{(i+1)}, b^{(i)}_{l,\textsc{Opt}}) + d(b^{(i)}_{l,\textsc{Opt}}, a^{(i+1)}_{f,\textsc{Opt}}) \\
&\quad + |S(R^{(i+1)}, a^{(i+1)}_{f,\textsc{Opt}})| \\
&\overset{(14)}{\leq} \alpha \cdot \textsc{Opt}(t^{(i)}) + |S^{(i)}| + d(p^{(i+1)}, b^{(i)}_{l,\textsc{Opt}}) + \textsc{Opt}(t^{(i+1)}) - T.
\end{aligned} \tag{15}$$

As $S^{(i)}$ visits $b^{(i)}_{l,\textsc{Opt}}$ before $p^{(i+1)}$ and $\textsc{Opt}[t^{(i+1)}]$ visits $p^{(i+1)}$ before $b^{(i)}_{l,\textsc{Opt}}$,

$$\begin{aligned}
|S^{(i)}| + T &\geq \left(d(O, b^{(i)}_{l,\textsc{Opt}}) + d(b^{(i)}_{l,\textsc{Opt}}, p^{(i+1)})\right) \\
&\quad + \left(d(O, p^{(i+1)}) + d(p^{(i+1)}, b^{(i)}_{l,\textsc{Opt}})\right) \\
&= 2 \cdot d(p^{(i+1)}, b^{(i)}_{l,\textsc{Opt}}) + d(O, b^{(i)}_{l,\textsc{Opt}}) + d(O, p^{(i+1)}) \\
&\geq 3 \cdot d(p^{(i+1)}, b^{(i)}_{l,\textsc{Opt}}).
\end{aligned} \tag{16}$$

Combined, we obtain that the algorithm finishes not later than

$$\begin{aligned}
t^{(i+1)} + |S^{(i+1)}| &\overset{(15)}{\leq} \alpha \cdot \textsc{Opt}(t^{(i)}) + |S^{(i)}| + d(p^{(i+1)}, b^{(i)}_{l,\textsc{Opt}}) + \textsc{Opt}(t^{(i+1)}) - T \\
&\overset{(16)}{\leq} \alpha \cdot \textsc{Opt}(t^{(i)}) + |S^{(i)}| + \frac{|S^{(i)}| + T}{3} + \textsc{Opt}(t^{(i+1)}) - T
\end{aligned}$$

$$\overset{\text{Obs 1a)}}{\leq} 2 \cdot \textsc{Opt}(t^{(i+1)}) + \frac{4}{3}|S^{(i)}| - \frac{2}{3}T$$
$$\overset{\text{Lem 1, (13)}}{\leq} 2 \cdot \textsc{Opt}(t^{(i+1)}) + \frac{2}{3}\textsc{Opt}(t^{(i)})$$
$$\overset{\text{Obs 1a)}}{\leq} \left(2 + \frac{2}{3\alpha}\right) \cdot \textsc{Opt}(t^{(i+1)})$$
$$\leq (1+\alpha) \cdot \textsc{Opt}(t^{(i+1)})$$

where the last inequality holds if and only if $\alpha \geq \frac{1}{2} + \sqrt{11/12}$. □

The above results enable us to prove Theorem 1.

Proof (of Theorem 1). Our goal was to prove by induction that every schedule is α-good for $\alpha \geq \frac{1}{2} + \sqrt{11/12}$. In Observation 2, we have proven the base case. In the induction step, we have distinguished several cases. First, we have seen in Observation 3 that the induction step holds if the previous schedule was interrupted. Next, we have seen in Lemma 1 that the induction hypothesis implies $|S^{(i+1)}| \leq \textsc{Opt}(t^{(i+1)})$. If the previous schedule was not interrupted, we have first seen in Lemma 2 that the induction step holds if $\textsc{Opt}[t^{(i+1)}]$ loads $r_{f,\textsc{Opt}}^{(i+1)}$ before serving $r_{l,\textsc{Opt}}^{(i)}$. If $\textsc{Opt}[t^{(i+1)}]$ serves $r_{l,\textsc{Opt}}^{(i)}$ before loading $r_{f,\textsc{Opt}}^{(i+1)}$, the induction step holds by Lemma 3 and Lemma 5. □

5 Analysis on the Half-Line

In this section, we prove that Lazy is even better if the metric space considered is the half-line. In particular, we prove Theorem 2. To do this, we begin by showing that $\alpha + 1$ is an upper bound on the competitive ratio of Lazy(α) for $\alpha = \frac{1+\sqrt{3}}{2} \approx 1.366$. Later, we complement this upper bound with a lower bound construction and show that, for all $\alpha \geq 0$, Lazy(α) has a competitive ratio of at least $\frac{3+\sqrt{3}}{2} \approx 2.366$.

Since, for all $\alpha \geq \frac{1+\sqrt{3}}{2} \approx 1.366$, Observations 2 and 3, as well as Lemmas 1-4 hold, it remains to show a counterpart to Lemma 5 for $\alpha \geq \frac{1+\sqrt{3}}{2}$ on the half-line. Similarly to the proof of Theorem 1, combining Observations 2 and 3 and Lemmas 1–4 with Lemma 6 then yields Theorem 2.

Lemma 6. *Let $\frac{1+\sqrt{3}}{2} \leq \alpha \leq 2$, and let $M = \mathbb{R}_{\geq 0}$. Assume that the i-th schedule is α-good and was not interrupted, and that $\textsc{Opt}[t^{(i+1)}]$ serves $r_{l,\textsc{Lazy}}^{(i)}$ before collecting $r_{f,\textsc{Opt}}^{(i+1)}$. If (7) holds, then $t^{(i+1)} + |S^{(i+1)}| \leq (1+\alpha) \cdot \textsc{Opt}(t^{(i+1)})$.*

Proof. First, observe that, in Lemma 5, we have the same assumptions except that we worked on general metric spaces. Therefore, all the inequalities shown in Lemma 5 hold in this setting, too, so that we can use them for our proof. Next, note that on the half-line, we have for any $x, y \in M$

$$d(x,y) \leq \max\{d(x,O), d(y,O)\}. \tag{17}$$

We show that this implies that a similar claim as in Lemma 5 holds.
Claim: $\text{OPT}[t^{(i+1)}]$ *serves all requests in* $R^{(i)}$ *before picking up* $r^{(i+1)}_{f,\text{OPT}}$ *in* $a^{(i+1)}_{f,\text{OPT}}$.
To prove the claim, assume otherwise, i.e., that $\text{OPT}[t^{(i+1)}]$ serves $r^{(i)}_{l,\text{OPT}}$ after collecting $r^{(i+1)}_{f,\text{OPT}}$. The request $r^{(i+1)}_{f,\text{OPT}}$ is released after schedule $S^{(i)}$ is started, i.e., after time $\alpha \cdot \text{OPT}(t^{(i)})$. Thus,

$$\begin{aligned}\text{OPT}(t^{(i+1)}) &\geq \alpha \cdot \text{OPT}(t^{(i)}) + d(a^{(i+1)}_{f,\text{OPT}}, b^{(i)}_{l,\text{OPT}}) \\ &\geq \alpha \cdot \text{OPT}(t^{(i)}) + d(a^{(i+1)}_{f,\text{OPT}}, p^{(i+1)}) - d(b^{(i)}_{l,\text{OPT}}, p^{(i+1)}) \\ &\overset{(17)}{\geq} \alpha \cdot \text{OPT}(t^{(i)}) + d(a^{(i+1)}_{f,\text{OPT}}, p^{(i+1)}) \\ &\quad - \max\{d(b^{(i)}_{l,\text{OPT}}, O), d(O, p^{(i+1)})\}. \end{aligned} \tag{18}$$

Combining the above with the results from the proof of Lemma 5 yields

$$\begin{aligned}\text{OPT}(t^{(i+1)}) &\overset{(18),(11)}{\geq} \alpha \cdot \text{OPT}(t^{(i)}) + d(a^{(i+1)}_{f,\text{OPT}}, p^{(i+1)}) \\ &\quad - \max\Big\{\frac{\text{OPT}(t^{(i)}) + d(p^{(i+1)}, O)}{2}, d(O, p^{(i+1)})\Big\} \\ &\overset{(12)}{>} \alpha \cdot \text{OPT}(t^{(i)}) + d(a^{(i+1)}_{f,\text{OPT}}, p^{(i+1)}) \\ &\quad - \max\Big\{\frac{2\text{OPT}(t^{(i)}) - (\alpha-1)\text{OPT}(t^{(i+1)})}{2}, \\ &\qquad \text{OPT}(t^{(i)}) - (\alpha-1)\text{OPT}(t^{(i+1)})\Big\} \\ &= \frac{\alpha-1}{2}\text{OPT}(t^{(i+1)}) + (\alpha-1)\text{OPT}(t^{(i)}) + d(a^{(i+1)}_{f,\text{OPT}}, p^{(i+1)}) \\ &\overset{(7)}{>} \Big(\frac{\alpha-1}{2} + \alpha\Big)\text{OPT}(t^{(i+1)}) + (\alpha-2)\text{OPT}(t^{(i)}) \\ &\overset{\text{Obs 1a)}}{\geq} \Big(\frac{\alpha-1}{2} + \alpha + \frac{\alpha-2}{\alpha}\Big)\text{OPT}(t^{(i+1)}) \\ &\geq \text{OPT}(t^{(i+1)})\end{aligned}$$

where the last inequality holds for all $\alpha \geq \frac{4}{3}$. As this is a contradiction, we have that $\text{OPT}[t^{(i+1)}]$ serves all requests in $R^{(i)}$ before picking up $r^{(i+1)}_{f,\text{OPT}}$ in $a^{(i+1)}_{f,\text{OPT}}$. This completes the proof of the claim.

Now that we have established the claim, we turn back to the proof of Lemma 6. Let $T \geq 0$ denote the time it takes $\text{OPT}[t^{(i+1)}]$ until it has served $r^{(i)}_{l,\text{OPT}}$, i.e., all requests from $R^{(i)}$. First, observe that

$$T \geq \text{OPT}(t^{(i)}). \tag{19}$$

If $p^{(i+1)} \geq b^{(i)}_{l,\text{OPT}}$, as $\text{OPT}[t^{(i+1)}]$ visits $p^{(i+1)}$ before $b^{(i)}_{l,\text{OPT}}$, we have

$$T \geq d(O, p^{(i+1)}) + d(p^{(i+1)}, b^{(i)}_{l,\text{OPT}}) \overset{(17)}{\geq} 2 \cdot d(p^{(i+1)}, b^{(i)}_{l,\text{OPT}}).$$

Otherwise, if $p^{(i+1)} < b^{(i)}_{l,\mathrm{OPT}}$, as $S^{(i)}$ visits $b^{(i)}_{l,\mathrm{OPT}}$ before $p^{(i+1)}$, we have

$$T \overset{(19)}{\geq} \mathrm{OPT}(t^{(i)}) \geq |S^{(i)}| \geq d(O, b^{(i)}_{l,\mathrm{OPT}}) + d(b^{(i)}_{l,\mathrm{OPT}}, p^{(i+1)}) \overset{(17)}{\geq} 2 \cdot d(b^{(i)}_{l,\mathrm{OPT}}, p^{(i+1)}).$$

Thus, in either case, we have

$$d(b^{(i)}_{l,\mathrm{OPT}}, p^{(i+1)}) \leq \frac{T}{2}. \tag{20}$$

Combined, we obtain that the algorithm finishes not later than

$$\begin{aligned} t^{(i+1)} + |S^{(i+1)}| &\overset{(15)}{\leq} \alpha \cdot \mathrm{OPT}(t^{(i)}) + |S^{(i)}| + d(p^{(i+1)}, b^{(i)}_{l,\mathrm{OPT}}) + \mathrm{OPT}(t^{(i+1)}) - T \\ &\overset{(20)}{\leq} \alpha \cdot \mathrm{OPT}(t^{(i)}) + |S^{(i)}| + \frac{T}{2} + \mathrm{OPT}(t^{(i+1)}) - T \\ &\overset{\text{Obs 1a)}}{\leq} 2 \cdot \mathrm{OPT}(t^{(i+1)}) + |S^{(i)}| - \frac{T}{2} \\ &\overset{\text{Lem 1, (19)}}{\leq} 2 \cdot \mathrm{OPT}(t^{(i+1)}) + \frac{1}{2}\mathrm{OPT}(t^{(i)}) \\ &\overset{\text{Obs 1a)}}{\leq} \left(2 + \frac{1}{2\alpha}\right) \cdot \mathrm{OPT}(t^{(i+1)}) \\ &\leq (1 + \alpha) \cdot \mathrm{OPT}(t^{(i+1)}) \end{aligned}$$

where the last inequality holds if and only if $\alpha \geq \frac{1+\sqrt{3}}{2} \approx 1.366$. □

5.1 Lower Bound on the Half-Line

In this section, we prove Theorem 3, i.e., we show that $\mathrm{LAZY}(\alpha)$ has a competitive ratio of at least $\frac{3+\sqrt{3}}{2} \approx 2.366$ for every choice of the parameter α, even when the metric space is the half line. Our proof builds on some lower bound constructions given in [5]. We begin by restating needed results.

Lemma 7 ([5]). *$\mathrm{LAZY}(\alpha)$ has competitive ratio at least $1+\alpha$ for the open online dial-a-ride problem on the half-line for all $\alpha \geq 0$ and every capacity $c \in \mathbb{N} \cup \{\infty\}$.*

The following bound holds, since [5, Proposition 2] only uses the half line.

Lemma 8 ([5]). *For every $\alpha \in [0, 1)$, $\mathrm{LAZY}(\alpha)$ has a competitive ratio of at least $1 + \frac{3}{\alpha+1} > 2.5$ for the open online dial-a-ride problem on the half-line for every capacity $c \in \mathbb{N} \cup \{\infty\}$.*

Combining these two results gives that, for $\alpha \in [0, 1) \cup [\frac{1+\sqrt{3}}{2}, \infty)$, the competitive ratio of $\mathrm{LAZY}(\alpha)$ is at least $\frac{3+\sqrt{3}}{2} \approx 2.366$.

The next proposition closes the gap between $\alpha < 1$ and $\alpha \geq 1.366$ and completes the proof of Theorem 3. An overview of the lower bounds for different domains of α can be found in Fig. 2.

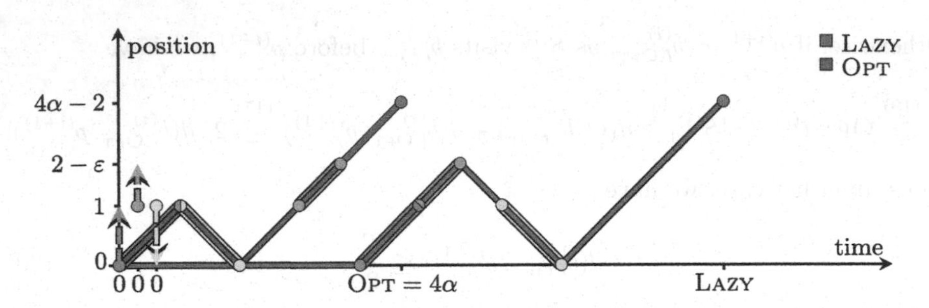

Fig. 1. Instance of the open online dial-a-ride problem on the half-line where LAZY(α) has a competitive ratio of at least $2 + \frac{1}{2\alpha}$ for all $\alpha \in [1, 1.366)$.

Proposition 1. *For every $\alpha \in [1, 1.366)$, the algorithm* LAZY(α) *has a competitive ratio of at least $2 + \frac{1}{2\alpha}$ for the open online dial-a-ride problem on the half-line for every capacity $c \in \mathbb{N} \cup \{\infty\}$.*

Proof. Let $\alpha \in [1, 1.366)$ and let $\varepsilon > 0$ be sufficiently small. We define an instance (cf. Fig. 1) by giving the request sequence

$$r_1 = (0, 1; 0), \, r_2 = (1, 0; 0), \, r_3 = (1, 2 - \varepsilon; 0), \\ \text{and } r_4 = (4\alpha - 2, 4\alpha - 2; 4\alpha).$$

The offline optimum delivers the requests in the order (r_1, r_2, r_3, r_4) with no waiting times. This takes 4α time units.

On the other hand because OPT$(0) = 4 - 2\varepsilon$, LAZY(α) waits in the origin until time $\alpha(4 - 2\varepsilon)$ and starts serving requests r_1, r_2, r_3 in the order (r_1, r_3, r_2). At time 4α, request r_4 is released. At this time, serving the loaded request r_1 and returning to the origin takes time

$$\alpha(4 - 2\varepsilon) + 2 = 4\alpha + 2 - 2\alpha\varepsilon \overset{\alpha \in [1, 1.366), \varepsilon \ll 1}{>} 4\alpha^2 = \alpha \cdot \text{OPT}(4\alpha).$$

Thus, LAZY(α) continues its schedule and afterwards serves r_4. Overall, this takes time (at least) $\alpha(4 - 2\varepsilon) + (4 - 2\varepsilon) + 4\alpha - 2 = 8\alpha + 2 - (2\alpha + 2)\varepsilon$. Letting $\varepsilon \to 0$, we obtain that the competitive ratio of LAZY(α) is at least

$$\lim_{\varepsilon \to 0} \frac{8\alpha + 2 - (2\alpha + 2)\varepsilon}{4\alpha} = 2 + \frac{1}{2\alpha}.$$

□

A Factor-Revealing Approach for the Half-Line

We show how to use the factor revealing approach from Sect. 3 for the dial-a-ride problem on the half-line. Consider the following variables (recall that $k \in \mathbb{N}$ is the number of schedules started by LAZY(α)).

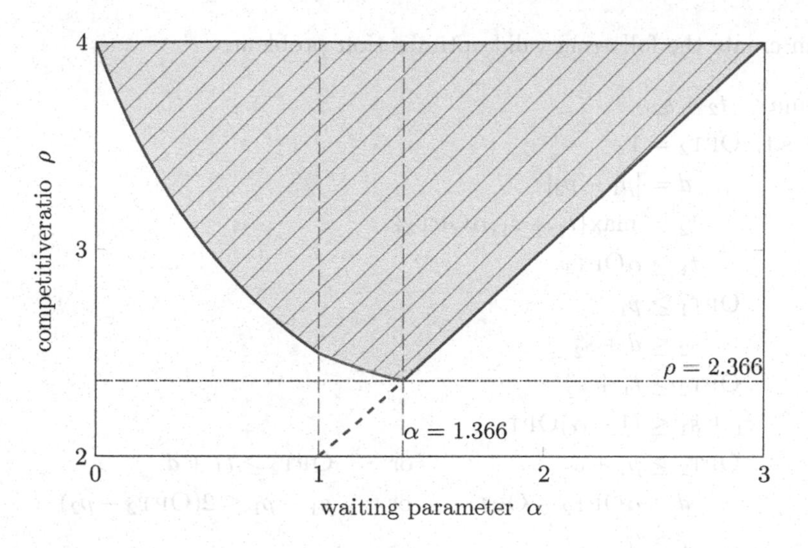

Fig. 2. Lower bounds on the competitive ratio of LAZY(α) depending on α. The lower bound of Lemma 7 is depicted in red, the lower bound of Lemma 8 in blue, and the lower bound of Proposition 1 in green. (Color figure online)

- $t_1 = t^{(k-1)}$, the start time of the second to last schedule
- $t_2 = r^{(k)}$, the start time of the last schedule
- $s_1 = |S^{(k-1)}|$, the duration of the second to last schedule
- $s_2 = |S^{(k)}|$, the duration of the last schedule
- $\text{OPT}_1 = \text{OPT}(t^{(k-1)})$, duration of the optimal tour serving requests released until $t^{(k-1)}$
- $\text{OPT}_2 = \text{OPT}(t^{(k)})$, duration of the optimal tour
- $p_1 = p^{(k)}$, the position where LAZY(α) ends the second to last schedule
- $p_2 = a^{(k)}_{f,\text{OPT}}$, the position of the first request in $R^{(k)}$ picked up first by the optimal tour
- $s_2^a = |S(R^{(k)}, a^{(k)}_{f,\text{OPT}})|$, duration of the schedule serving $R^{(k)}$ starting in p_2
- $d = d(p^{(k)}, a^{(k)}_{f,\text{OPT}})$, the distance between p_1 and p_2

With these variables

$$x = (t_1, t_2, s_1, s_2, \text{OPT}_1, \text{OPT}_2, p_1, p_2, s_2^a, d),$$

we can create the following valid optimization problem.

$$
\begin{aligned}
\max\quad & t_2 + s_2 && \\
\text{s.t.}\quad & \textsc{Opt}_2 = 1 && (21)\\
& d = |p_1 - p_2| && (22)\\
& t_2 = \max\{t_1 + s_1, \alpha\textsc{Opt}_2\} && (23)\\
& t_1 \geq \alpha\textsc{Opt}_1 && (24)\\
& \textsc{Opt}_1 \geq p_1 && (25)\\
& s_2 \leq d + s_2^a && (26)\\
& \textsc{Opt}_2 \geq t_1 + s_2^a && (27)\\
& t_1 + s_1 \leq (1+\alpha)\textsc{Opt}_1 && (28)\\
& \textsc{Opt}_2 \geq p_1 + d \quad \text{or} \quad \textsc{Opt}_2 \geq t_1 + d && (29)\\
& d \geq \alpha\textsc{Opt}_2 - \textsc{Opt}_1 \quad \text{or} \quad s_1 - p_1 \leq 2(\textsc{Opt}_2 - p_2) && (30)\\
& x \geq 0 && (31)
\end{aligned}
$$

Note that in (29) and (30), at least one of the two inequalities has to be satisfied in each case. In order to obtain an MILP, one can introduce four binary variables $b_1, \ldots, b_4$ to model constraints (22), (23), (29), and (30).

With $M > 0$ being a large enough constant, equality (22) can be replaced by the inequalities

$$
\begin{aligned}
d &\geq p_1 - p_2,\\
d &\geq p_2 - p_1,\\
d &\leq p_1 - p_2 + b_1 \cdot M,\\
d &\leq p_2 - p_1 + (1 - b_1) \cdot M.
\end{aligned}
$$

Equality (23) can be replaced by the inequalities

$$
\begin{aligned}
t_2 &\geq t_1 + s_1,\\
t_2 &\geq \alpha\textsc{Opt}_2,\\
t_2 &\leq t_1 + s_1 + b_2 \cdot M,\\
t_2 &\leq \alpha\textsc{Opt}_2 + (1 - b_2) \cdot M.
\end{aligned}
$$

Constraint (29) can be replaced by the inequalities

$$
\begin{aligned}
\textsc{Opt}_2 &\geq p_1 + d - b_3 \cdot M,\\
\textsc{Opt}_2 &\geq t_1 + d - (1 - b_3) \cdot M,
\end{aligned}
$$

and, likewise, (30) by the inequalities

$$
\begin{aligned}
d &\geq \alpha\textsc{Opt}_2 - \textsc{Opt}_1 - b_4 \cdot M,\\
s_1 - p_1 &\leq 2(\textsc{Opt}_2 - p_2) + (1 - b_4) \cdot M.
\end{aligned}
$$

The resulting MILP has the optimal solution

$$(t_1, t_2, s_1, s_2, \text{OPT}_1, \text{OPT}_2, p_1, p_2, s_2^a, d, b_1, b_2, b_3, b_4)$$
$$= \left(1, \frac{\alpha+1}{\alpha}, \frac{1}{\alpha}, 2-\alpha, \frac{1}{\alpha}, 1, 0, 2-\alpha, 0, 2-\alpha, 0, 1, 1, 1\right)$$

and optimal value $\max\{3 + \frac{1}{\alpha} - \alpha, 1 + \alpha\}$. For $\alpha = \frac{1+\sqrt{3}}{2} > 1.366$, this expression is minimized.

References

1. Ascheuer, N., Krumke, S.O., Rambau, J.: Online dial-a-ride problems: minimizing the completion time. In: Reichel, H., Tison, S. (eds.) STACS 2000. LNCS, vol. 1770, pp. 639–650. Springer, Heidelberg (2000). https://doi.org/10.1007/3-540-46541-3_53
2. Ausiello, G., Demange, M., Laura, L., Paschos, V.: Algorithms for the on-line quota traveling salesman problem. Inf. Process. Lett. **92**(2), 89–94 (2004). https://doi.org/10.1007/978-3-540-27798-9_32
3. Ausiello, G., Feuerstein, E., Leonardi, S., Stougie, L., Talamo, M.: Algorithms for the on-line travelling salesman. Algorithmica **29**(4), 560–581 (2001)
4. Ausiello, G., Allulli, L., Bonifaci, V., Laura, L.: On-line algorithms, real time, the virtue of laziness, and the power of clairvoyance. In: Cai, J.-Y., Cooper, S.B., Li, A. (eds.) TAMC 2006. LNCS, vol. 3959, pp. 1–20. Springer, Heidelberg (2006). https://doi.org/10.1007/11750321_1
5. Baligács, J., Disser, Y., Mosis, N., Weckbecker, D.: An improved algorithm for open online dial-a-ride. In: Chalermsook, P., Laekhanukit, B. (eds.) WAOA 2022. Lecture Notes in Computer Science, vol. 13538, pp. 154–171. Springer, Cham (2022). https://doi.org/10.1007/978-3-031-18367-6_8
6. Bienkowski, M., Kraska, A., Liu, H.: Traveling repairperson, unrelated machines, and other stories about average completion times. In: Proceedings of the 48th International Colloquium on Automata, Languages, and Programming (ICALP), pp. 28:1–28:20 (2021)
7. Bienkowski, M., Liu, H.: An improved online algorithm for the traveling repairperson problem on a line. In: Proceedings of the 44th International Symposium on Mathematical Foundations of Computer Science (MFCS), pp. 6:1–6:12 (2019)
8. Birx, A.: Competitive analysis of the online dial-a-ride problem. Ph.D. thesis, TU Darmstadt (2020)
9. Birx, A., Disser, Y.: Tight analysis of the smartstart algorithm for online dial-a-ride on the line. SIAM J. Discrete Math. **34**(2), 1409–1443 (2020)
10. Birx, A., Disser, Y., Schewior, K.: Improved bounds for open online dial-a-ride on the line. Algorithmica **85**(5), 1372–1414 (2022)
11. Bjelde, A., et al.: Tight bounds for online TSP on the line. ACM Trans. Algorithms **17**(1), 1–58 (2020)
12. Blom, M., Krumke, S.O., de Paepe, W.E., Stougie, L.: The online TSP against fair adversaries. INFORMS J. Comput. **13**(2), 138–148 (2001)
13. Bonifaci, V., Stougie, L.: Online k-server routing problems. Theory Comput. Syst. **45**(3), 470–485 (2008)
14. Feuerstein, E., Stougie, L.: On-line single-server dial-a-ride problems. Theor. Comput. Sci. **268**(1), 91–105 (2001)

15. Hauptmeier, D., Krumke, S., Rambau, J., Wirth, H.C.: Euler is standing in line dial-a-ride problems with precedence-constraints. Discrete Appl. Math. **113**(1), 87–107 (2001)
16. Hauptmeier, D., Krumke, S.O., Rambau, J.: The online dial-a-ride problem under reasonable load. In: Bongiovanni, G., Petreschi, R., Gambosi, G. (eds.) CIAC 2000. LNCS, vol. 1767, pp. 125–136. Springer, Heidelberg (2000). https://doi.org/10.1007/3-540-46521-9_11
17. Jaillet, P., Lu, X.: Online traveling salesman problems with service flexibility. Networks **58**(2), 137–146 (2011)
18. Jaillet, P., Lu, X.: Online traveling salesman problems with rejection options. Networks **64**(2), 84–95 (2014)
19. Jaillet, P., Wagner, M.R.: Generalized online routing: new competitive ratios, resource augmentation, and asymptotic analyses. Oper. Res. **56**(3), 745–757 (2008)
20. Jawgal, V.A., Muralidhara, V.N., Srinivasan, P.S.: Online travelling salesman problem on a circle. In: Gopal, T.V., Watada, J. (eds.) TAMC 2019. LNCS, vol. 11436, pp. 325–336. Springer, Cham (2019). https://doi.org/10.1007/978-3-030-14812-6_20
21. Krumke, S.O.: Online optimization competitive analysis and beyond. Habilitation thesis, Zuse Institute Berlin (2001)
22. Krumke, S.O., et al.: Non-abusiveness helps: An O(1)-competitive algorithm for minimizing the maximum flow time in the online traveling salesman problem. In: Jansen, K., Leonardi, S., Vazirani, V. (eds.) APPROX 2002. LNCS, vol. 2462, pp. 200–214. Springer, Heidelberg (2002). https://doi.org/10.1007/3-540-45753-4_18
23. Krumke, S.O., de Paepe, W.E., Poensgen, D., Lipmann, M., Marchetti-Spaccamela, A., Stougie, L.: On minimizing the maximum flow time in the online dial-a-ride problem. In: Erlebach, T., Persinao, G. (eds.) WAOA 2005. LNCS, vol. 3879, pp. 258–269. Springer, Heidelberg (2006). https://doi.org/10.1007/11671411_20
24. Krumke, S.O., de Paepe, W.E., Poensgen, D., Stougie, L.: News from the online traveling repairman. Theor. Comput. Sci. **295**(1–3), 279–294 (2003)
25. Lipmann, M.: On-line routing. Ph.D. thesis, Technische Universiteit Eindhoven (2003)
26. Lipmann, M., Lu, X., de Paepe, W.E., Sitters, R.A., Stougie, L.: On-line dial-a-ride problems under a restricted information model. Algorithmica **40**(4), 319–329 (2004)

Online TSP with Known Locations

Evripidis Bampis[1], Bruno Escoffier[1,2](✉), Niklas Hahn[1], and Michalis Xefteris[1]

[1] Sorbonne Université, CNRS, LIP6, 75005 Paris, France
bruno.escoffier@lip6.fr
[2] Institut Universitaire de France, Paris, France

Abstract. In this paper, we consider the Online Traveling Salesperson Problem (OLTSP) where the locations of the requests are known in advance, but not their arrival times. We study both the open variant, in which the algorithm is not required to return to the origin when all the requests are served, as well as the closed variant, in which the algorithm has to return to the origin after serving all the requests. Our aim is to measure the impact of the extra knowledge of the locations on the competitiveness of the problem. We present an online 3/2-competitive algorithm for the general case and a matching lower bound for both the open and the closed variant. Then, we focus on some interesting metric spaces (ring, star, semi-line), providing both lower bounds and polynomial time online algorithms for the problem.

Keywords: TSP · Online algorithms · Competitive analysis

1 Introduction

In the classical Traveling Salesperson Problem (TSP), we are given a set of locations in a metric space. The objective is to find a tour visiting all the locations minimizing the total traveled time [15]. Ausiello et al. [5] introduced the Online Traveling Salesperson Problem (OLTSP) where the input arrives over time, i.e., during the travel new requests (locations) appear that have to be visited by the algorithm. The time in which a request is communicated to the traveler (we will refer to them as the server) is called release time (or release date). Since the article by Ausiello et al. [5], a series of papers considered many versions of OLTSP in various metric spaces (general metric space [5], the line [5,8,11], or the semi-line [4,7,9]). Several applications can be modeled as variants of OLTSP, e.g., applications in logistics and robotics [2,16].

The performance of the proposed algorithms for OLTSP has been evaluated in the framework of competitive analysis, by establishing upper and lower bounds of the competitive ratio. This is defined as the maximum ratio between the cost of the online algorithm and the cost of an optimal offline algorithm over all input instances. However, it is admitted that the competitive analysis approach can be very pessimistic as it gives a lot of power to the adversary. Hence, many papers try to limit the power of the adversary [9], or give extra knowledge and hence more power to the online algorithm [1,13]. More recently, OLTSP has also been studied in the framework of *Learning-Augmented Algorithms* [7,11,12].

P. Morin and S. Suri (Eds.): WADS 2023, LNCS 14079, pp. 65–78, 2023.
https://doi.org/10.1007/978-3-031-38906-1_5

Here, we study another natural way of offering more power to the online algorithm by considering that the location of each request is known in advance, but its release date arrives over time. The release date, in this context, is just the time after which a request can be served. This is the case for many applications where the delivery/collection locations are known (parcel collection from fixed storage facilities, cargo collection on a harbour etc.). Think for example of a courier that has to deliver packets to a fixed number of customers. These packets may have to be delivered in person, so the customers should be at home to receive them. Each customer can inform the courier through an app when he/she returns home and is ready to receive the packet. We refer to this problem as the online Traveling Salesperson Problem with known locations (OLTSP-L) and consider two variants: in the *closed* variant the server is required to return to the origin after serving all requests, while in the *open* one the server does not have to return to the origin after serving the last request.

Previous Results. The offline version of the open variant of the problem in which the requests are known in advance can be solved in quadratic time when the metric space is a line [8,16]. For the online problem, Ausiello et al. [5] showed a lower bound of 2 for the open version and a lower bound of 1.64 for the closed version of OLTSP, even when the metric space is the line. They also proposed an optimal 2-competitive algorithm for the closed version and a 2.5-competitive algorithm for the open version of OLTSP in general metric spaces. For the line, they presented a (7/3)-competitive algorithm for the open case and a 1.75-competitive algorithm for the closed one. More recently, Bjelde et al., in [8], proposed a 1.64-competitive algorithm for the closed case (on the line) matching the lower bound of [5]. They also provided a lower bound of 2.04 for the open case on the line, as well as an online algorithm matching this bound. In [9], Blom et al. proposed a best possible 1.5-competitive algorithm for the closed case when the metric space is a semi-line. Chen et al., in [10], presented lower and upper bounds of randomized algorithms for closed OLTSP on the line.

Hu et al. [12] and Bernardini et al. [7] introduced learning-augmented algorithms for OLTSP. They considered various prediction models and error measures. The prediction models where each request is associated to a prediction for both its release time and location are related to our work, however no direct comparison can be made with our results. The most related work is the one by Gouleakis et al. [11]. They define a prediction model in which the predictions correspond to the locations of the requests. They establish lower bounds by assuming that the predictions are perfect, i.e., the locations are given, and the adversary can only control the release times of the requests, as well as upper bounds as a function of the error value. Notice that when the error is equal to 0, their model coincides with ours (see Table 1 for a comparison with our results).

Our Contribution. In this work we study both the closed (closed OLTSP-L) and the open case of OLTSP-L (open OLTSP-L) and present several lower and upper bounds for the problem. In Table 1, we give an overview of our results and the state of the art.

Table 1. Results for the open and closed variants of OLTSP-L. Polynomial time algorithms are denoted by * and tight results in bold.

	Open OLTSP-L		Closed OLTSP-L	
	Lower Bound	Upper Bound	Lower Bound	Upper Bound
Semi-line	4/3 Thm. 4	13/9* Thm. 5	**1**	**1*** Prop. 2
Line	13/9 [11, Thm. 4]	3/2 Thm. 1; 5/3* [11, Thm. 3]	**3/2** [11, Thm. 2]	**3/2*** [11, Thm. 1]
Star	13/9 [11, Thm. 4]	3/2 Thm. 1	**3/2** [11, Thm. 2]	**3/2** Thm. 1; $(7/4+\epsilon)^*$ Thm. 3
Ring	**3/2** Prop. 1	**3/2** Thm. 1	**3/2** [11, Thm. 2]	**3/2** Thm. 1; 5/3* Thm. 2
General	**3/2** Prop. 1	**3/2** Thm. 1	**3/2** [11, Thm. 2]	**3/2** Thm. 1

We first consider general metric spaces (dealt with in Sect. 2). For the closed version, a lower bound of 3/2 has been shown in [11] (valid in the case of a line). We show that a lower bound of 3/2 holds also for the open case (on rings). We then provide a 3/2-competitive online algorithm for both variants, thus matching the lower bounds. Although our algorithm does not run in polynomial time, the online nature of the problem is a source of difficulty independent of its computational complexity. Thus, such algorithms are of interest even if their running time is not polynomially bounded.

However, it is natural to also focus on polynomial time algorithms. We provide several bounds in specific metric spaces. In Sect. 3, we focus on rings and present a polytime 5/3-competitive algorithm for the closed OLTSP-L. Next, we give a polytime $(7/4 + \epsilon)$-competitive algorithm, for any constant $\epsilon > 0$, in the closed case on stars (Sect. 4). In Sect. 5, we study the problem on the semi-line. We present a simple polytime 1-competitive algorithm for the closed variant and, for the open case, we give a lower bound of 4/3 and an upper bound (with a polytime algorithm) of 13/9.

To measure the gain of knowing the locations of the requests, we also provide some lower bounds on the case where the locations are unknown, and more precisely on the case where the number of locations is known but not the locations themselves.

In the open case of the problem, there is a lower bound of 2 in [5] on the line that holds even when the number of requests is known. Hence, the same lower bound holds for the star and for the ring. We provide a lower bound of 3/2 for the semi-line. For the closed case, the existing lower bound of 2 for classic OLTSP without knowledge of the number of requests for the ring [5] and the star [13] can easily be extended to the scenario augmented by the knowledge of the number of requests. We also present a lower bound of 4/3 for the semi-line. For the line, it is easy to see (with a trivial modification in the proof) that the lower bound of 1.64 on the line [5] still holds in this model where we know the number of locations. These lower bounds show that knowing only the number of requests is not sufficient to get better competitive ratios in most cases, and thus it is meaningful to consider the more powerful model with known locations.

Due to space constraints, we omit a detailed discussion of these lower bounds and some of our proofs. The full version of this paper [6] includes all missing proofs as well as additional details.

Preliminaries. The input of OLTSP-L consists of a metric space M with a distinguished point O (the origin), and a set $Q = \{q_1, ..., q_n\}$ of n requests. Every request q_i is a pair (t_i, p_i), where p_i is a point of M (which is known at $t = 0$) and $t_i \geq 0$ is a real number. The number t_i represents the release time of q_i. A server located at the origin at time 0, which can move with at most unit speed, must serve all the requests after their release times with the goal of minimizing the total completion time (makespan).

When we refer to general metric spaces or general metrics, we mean all metric spaces that are *continuous*, in the sense that the shortest path between $x \in M$ and $y \in M$ is continuous in M and has length $d(x, y)$, as defined in [5].[1]

For the rest of the paper, we denote the total completion time of an online algorithm ALG by $|ALG|$ and that of an optimal (offline) solution OPT by $|OPT|$. We recall that an algorithm ALG is r-competitive if on all instances we have $|ALG| \leq r \cdot |OPT|$.

2 General Metrics

As mentioned earlier, a lower bound of 3/2 for closed OLTSP-L, even in the case of a line, was shown in [11]. In this section, we first show that the same lower bound of 3/2 holds for open OLTSP-L (in the case of a ring). We then devise a 3/2-competitive algorithm, for general metrics, both in the open and in the closed cases, thus matching the lower bounds in both cases.

Proposition 1. *For any $\epsilon > 0$, there is no $(3/2 - \epsilon)$-competitive algorithm for open OLTSP-L on the ring.*

Proof. Consider a ring with circumference 1 with 2 requests A and B, with a distance of 1/3 from O and from each other. At $t = 1/3$, w.l.o.g. due to symmetry, we can assume that the algorithm is in the segment (arc of the ring) $[OA]$ (including both O and A). Then, B is released. Request A is released at $t = 2/3$. Hence, OPT can visit A and B in time 2/3, whereas the online algorithm cannot finish before $t = 1$. It cannot serve the first request before $t = 2/3$ and will have to go a distance of 1/3 to serve the second request, as well. □

[1] We note that our 3/2-competitive algorithm for general metric spaces also works for discrete metric spaces (where you do not continuously travel from one point to another, but travel in a discrete manner from point x at time t to point y at time $t + d(x, y)$).

Now, let us present the 3/2-competitive algorithm. Roughly speaking, the principle of the algorithm is:

- first, to wait (at O) a well chosen amount of time T. This time T depends both on the requests' locations and on their release times. It is chosen so that (1) OPT cannot have already served the requests on a large part of its route (tour/path for the closed/open case, resp.) and (2) there is a route for which a large part is fully released.
- then, to choose an order of serving requests that optimizes some criterion mixing the length of the corresponding route and the fraction of it which is released at time T, and to follow this route, waiting at unreleased requests.

More formally, we consider Algorithm 1, which uses the following notation (see Example 1 for an illustration of the notation and the execution of the algorithm). For a given order σ_i on the requests (where $\sigma_i(1)$ denotes the first request in the order, $\sigma_i(2)$ the second request, ...), we denote:

- by ℓ_i the length of the route associated to σ_i (starting at O), i.e., $\ell_i = d(O, \sigma_i(1)) + \sum_{j=1}^{n-1} d(\sigma_i(j), \sigma_i(j+1))$ in the open case, $\ell_i = d(O, \sigma_i(1)) + \sum_{j=1}^{n-1} d(\sigma_i(j), \sigma_i(j+1)) + d(\sigma_i(n), O)$ in the closed case;
- by $\alpha_i(t)$ the fraction of the length of the route associated to σ_i, starting at O, which is fully released at time t. More formally, if requests $\sigma_i(1), \ldots, \sigma_i(k-1)$ are released at t but $\sigma_i(k)$ is not, then the route is fully released up to $\sigma_i(k)$, and $\alpha_i(t) = (d(O, \sigma_i(1)) + \sum_{j=1}^{k-1} d(\sigma_i(j), \sigma_i(j+1)))/\ell_i$. If all the requests are released, then $\alpha_i(t) = 1$.

Algorithm 1: Algorithm for closed and open OLTSP-L

Input: Offline: Request locations $p_1, \ldots, p_n$

Online: Release times $t_1, \ldots, t_n$

1 Let $\sigma_1, \ldots, \sigma_{n!}$ be all possible orders of requests.
2 For all $i \leq n!$, compute ℓ_i the length of the tour/path associated to σ_i.
3 Wait at O until time T defined as the first time t such that there exists an order σ_{i_0} with (1) $t \geq \ell_{i_0}/2$ and (2) $\alpha_{i_0}(t) \geq 1/2$.
4 At time T:
  - Compute an order σ_{i_1} which minimizes, over all orders σ_i $(1 \leq i \leq n!)$, $(1-\beta_i)\ell_i$, where $\beta_i = \min\{\alpha_i(T), 1/2\}$.
  - Follow (starting at time T) the tour/path associated to σ_{i_1}, by serving the requests in this order, waiting at each request location for its release.

Example 1. Let us consider the following example for the closed case. There are three requests q_1, q_2, q_3, released at time 2, 6 and 8, respectively, with $d(O, q_1) = d(q_1, q_2) = d(q_2, q_3) = d(q_3, O) = 3$, $d(q_1, q_3) = 1$, and $d(O, q_2) = 2$.

Let us consider the order $\sigma_0 = (q_1, q_2, q_3)$, with $\ell_0 = 12$. Then for $0 \leq t < 2$ we have $\alpha_0(t) = 1/4$, for $2 \leq t < 6$ we have $\alpha_0(t) = 1/2$, for $6 \leq t < 8$ we have $\alpha_0(t) = 3/4$, and for $t \geq 8$ we have $\alpha_0(t) = 1$.

Now, let us look at the algorithm, and first the determination of T (line 3 of the algorithm). At time 6, we have $6 \geq \ell_0/2$ and $\alpha_0(6) \geq 1/2$, so $T \leq 6$. At any time $t < 6$: on the one hand $t < \ell_0/2$, and on the other hand neither q_2 nor q_3 is released, and one can see that no tour $\sigma_i \neq \sigma_0$ has $\alpha_i(t) \geq 1/2$. This means that $T = 6$, and that $i_0 = 0$. So the algorithm starts moving at $T = 6$. The order minimizing $(1 - \beta_i)\ell_i$ is $\sigma_{i_1} = (q_2, q_1, q_3)$ (with $\ell_{i_1} = 9$, $\alpha_{i_1}(T) = 6/9$, so $\beta_{i_1} = 1/2$ and $(1 - \beta_{i_1})\ell_1 = 9/2$). The algorithm will follow this tour, serving q_2 at $6 + 2 = 8$, q_1 at 11, q_3 at 12, and be back in O at 15.

Theorem 1. *Algorithm 1 is 3/2-competitive both for closed and open OLTSP-L.*

Proof. At time T, there exists σ_{i_0} with $T \geq \ell_{i_0}/2$ and $\alpha_{i_0}(T) \geq 1/2$. Then $\beta_{i_0} = 1/2$, and $(1-\beta_{i_0})\ell_{i_0} = \ell_{i_0}/2 \leq T$. By definition of σ_{i_1}, we have $(1-\beta_{i_1})\ell_{i_1} \leq (1 - \beta_{i_0})\ell_{i_0}$. So we get

$$(1 - \beta_{i_1})\ell_{i_1} \leq T \ . \tag{1}$$

Now, let us consider an optimal solution OPT, and denote by σ_{i^*} the order in which OPT serves the requests. We assume w.l.o.g. that OPT follows the tour/path σ_{i^*}, waiting only at requests' positions: it goes from O to (the position of) $\sigma_{i^*}(1)$ in time $d(O, \sigma_{i^*}(1))$, waits at (the position of) $\sigma_{i^*}(1)$ if the request is not released, then from $\sigma_{i^*}(1)$ to $\sigma_{i^*}(2)$ in time $d(\sigma_{i^*}(1), \sigma_{i^*}(2))$, ...[2]

- If $\alpha_{i^*}(T) \leq 1/2$, then $\beta_{i^*} = \alpha_{i^*}(T)$, and $|OPT| \geq T + (1 - \alpha_{i^*}(T))\ell_{i^*} = T + (1 - \beta_{i^*})\ell_{i^*}$.
- If $\alpha_{i^*}(T) > 1/2$, then $\beta_{i^*} = 1/2$. We look at the position of OPT at time T in the tour/path associated to σ_{i^*}. Suppose that it is (strictly) on the second half of this tour/path. Then, at $T - \epsilon$ (for a sufficiently small $\epsilon > 0$), it was already on the second part. But then $T - \epsilon \geq \ell_{i^*}/2$ (as OPT has already visited half of the tour/path), and $\alpha_{i^*}(T - \epsilon) \geq 1/2$ (for the same reason). Then $T - \epsilon$ would fulfill the two conditions in Line 3 of the algorithm, a contradiction with the definition of T. Consequently, at T, OPT is in the first half of its tour/path. So $|OPT| \geq T + \ell_{i^*}/2 = T + (1 - \beta_{i^*})\ell_{i^*}$.

Then, in both cases we have $|OPT| \geq T + (1 - \beta_{i^*})\ell_{i^*}$. By definition of σ_{i_1} we deduce that

$$T + (1 - \beta_{i_1})\ell_{i_1} \leq |OPT|. \tag{2}$$

Now we look at the value $|ALG|$ of the solution ALG output by the algorithm.

- If ALG never waits after T, we get that $|ALG| = T+\ell_{i_1}$. As by definition $\beta_{i_1} \leq 1/2$, $|ALG| \leq T + 2(1-\beta_{i_1})\ell_{i_1}$. But adding Eqs. (1) and (2) with coefficients 1/2 and 3/2, we get $T+2(1-\beta_{i_1})\ell_{i_1} \leq 3|OPT|/2$. Hence, $|ALG| \leq 3|OPT|/2$.

[2] Indeed, if OPT does not do this we can easily transform it into another optimal solution (with the same order of serving requests) that acts like this.

– Otherwise, ALG waits after T for some request to be released. Let t^* be the last time ALG waits. As a fraction $\alpha_{i_1}(T)$ of σ_{i_1} is completely released at T (i.e., when ALG starts), ALG has distance at most $(1-\alpha_{i_1}(T))\ell_{i_1}$ to perform after t^*. Since, by definition, $\beta_{i_1} \leq \alpha_{i_1}(T)$, we get

$$|ALG| \leq t^* + (1 - \alpha_{i_1}(T))\ell_{i_1} \leq t^* + (1 - \beta_{i_1})\ell_{i_1}. \tag{3}$$

We have $t^* \leq |OPT|$, as a request is released at t^*. Adding Eqs. (1) and (2) gives $2(1 - \beta_{i_1})\ell_{i_1} \leq |OPT|$. Putting these two inequalities in Equation (3) gives $|ALG| \leq 3|OPT|/2$.

□

Note that Algorithm 1 also solves the more general Online Asymmetric TSP with known locations and achieves again a ratio of 3/2 for both variants. For the problem without the knowledge of the locations, there is an optimal 2.62-competitive algorithm for the closed variant, and it has been proved that there is no constant competitive algorithm for the open variant [3].

3 Ring

In this section, we discuss the problem on the metric space induced by the border of a ring. W.l.o.g. we assume that the ring is a circle with a circumference of 1. The distance $d(x, y)$ between points x, y on the border of the circle is the smaller arc length of the segment between x and y (or y and x). We denote the location of points on the ring by their clockwise distance from the origin. Further, we assume that the requests are ordered such that p_i is located next to p_{i+1}, and p_1 and p_n are the closest requests to the origin on their respective sides.

For the closed case, using the location data of the requests, we describe an algorithm beating the competitive ratio of 2 in polynomial time. For a simpler exposition, we assume that the shortest time to visit all request locations and return to the origin when ignoring release times is 1. This means that the shortest way to visit all requests is a round trip on the ring. Otherwise, the instance can be interpreted as an instance on the line.

The algorithm (see Fig. 1 for an illustration) first deals with a specific case, when there is a large interval of size at least 1/3 without any request inside (so $p_{i+1} - p_i > 1/3$ for some i), see left part of the figure. In this case the server goes to p_i or p_{i+1} - the farthest from O (arc 1 in the figure). Afterwards, it moves back to O (arc 2), serving requests along the way and waiting for their release if necessary. Then, it goes to the other extremity (arc 3). There, it turns around, moving back to O (arc 4), again serving the requests on the way, waiting for their release if necessary.

For the other case (right part of the figure), when there is no such large empty interval, the server acts as follows: First, it waits at the origin to see whether to take a clockwise or a counter-clockwise tour around the ring. In order to decide this, the server waits for a contiguous part of at least 1/3 of the ring to be released in either half of the ring. More precisely, all consecutive requests inside a segment of length at least 1/3 are released and the whole segment lies completely in $(0, 1/2]$ or in $[1/2, 1)$. We denote this segment by s and the time that this occurs by $t^{(1)}$. The server then starts moving on the shortest path to segment s (arc 1 in the figure), serving released requests it encounters on the way. It then continues in the same direction and serves segment s (arc 2), which has been completely released. Having visited s, the server continues until O (arc 3), waiting at unreleased requests if any. Once the server reaches O, there might be some requests which have not been served because they were unreleased when the server passed their locations. To complete the TSP tour, the server will therefore go to the furthest unserved request (arc 4, that may be shorter than on the figure) and back to the origin (arc 5), waiting at any unreleased request if necessary.

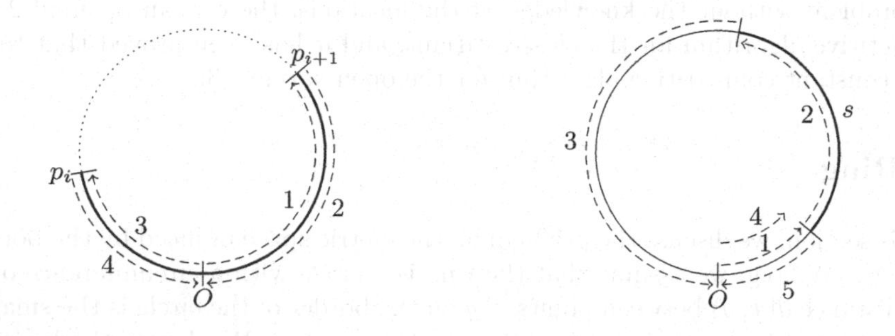

Fig. 1. The two main cases of the ring algorithm.

For pseudocode, see Algorithm 2. The algorithm guarantees a competitive ratio of 5/3, which we formalize in Theorem 2. The full proof can be found in the full version of this paper [6].

Theorem 2. *Algorithm 2 has a competitive ratio of* 5/3 *for closed OLTSP-L on the ring.*

Sketch of Proof. We only consider in this sketch of proof the second case (right part of the figure).

We first look at the case where ALG waits at some request after step (arc on the figure) 2. Let $t^{(0)}$ denote the last time that ALG has to wait at an unreleased request p_{j^*} and let x be the distance between O and p_{j^*}. Then, $|OPT| \geq t^{(0)} + x$. If $x \geq 1/3$, then $|ALG| \leq t^{(0)} + 1 - x + 2 \cdot 1/6 \leq t^{(0)} + 1$, and $|OPT| \geq t^{(0)} + 1/3$, so $|ALG| \leq |OPT| + 2/3 \leq 5|OPT|/3$. Otherwise ($x < 1/3$) $|ALG| \leq t^{(0)} + x + 2 \cdot 1/3$, and again $|ALG| \leq |OPT| + 2/3 \leq 5|OPT|/3$.

Now, we consider that ALG does not wait after step/arc 2. Then, after leaving O at $t^{(1)}$, ALG follows the five arcs, of total length at most $1 + 2 \cdot 1/6$, so $|ALG| \leq t^{(1)} + 4/3$.

Algorithm 2: Algorithm for closed OLTSP-L on the ring

Input: Offline: Request locations $p_1, \ldots, p_n$, numbered clockwise from O

Online: Release times $t_1, \ldots, t_n$

1 **if** *there exists a segment* $s = [p_i, p_{i+1}]$ *(with* $1 \leq i \leq n-1$*) of size at least 1/3* **then**

2 Go to the farmost request from O between p_i and p_{i+1}, say p_{i+1}.

3 Move from p_{i+1} to O clockwise. Serve the requests on the way. If a request is unreleased, wait until it is released.

4 Move from O to p_i (clockwise). Then go back to O, serving the requests on the way. If a request is unreleased, wait until it is released.

5 **else**

6 Wait at the origin until $t^{(1)}$, when a segment $s \subseteq (0, 1/2]$ or $s \subseteq [1/2, 1)$ of length $|s| \geq 1/3$ is completely released, i.e., $\forall i \in [n]$ s.t. $p_i \in s$ it holds that $t_i \leq t^{(1)}$. If $s \subseteq (0, 1/2]$, move clockwise, else, move counter clockwise, till the farmost (from O) extremity of s is reached, serving requests along the way without waiting.

7 Continue in the same direction back to the origin, serving requests along the way, waiting if necessary. Having reached O, continue in the same direction and stop at the unserved request which is furthest from the origin. Wait until it can be served, turn around and move back to the origin, on the way serving the remaining requests, waiting if necessary.

8 **end**

As at $t^{(1)} - \epsilon$ there is one unreleased request in $[1/6, 1/2]$ and one in $[1/2, 5/6]$, it is easy to see that $OPT \geq t^{(1)} + 1/2$. Actually, at $t^{(1)} - \epsilon$, on the interval $[0, 1/2]$ (and symmetrically in the other half $[1/2, 1]$) either there is some unreleased request on $[1/6, 1/3]$, or two unreleased requests: one in $[0, 1/6]$, and one in $[1/3, 1/2]$. Then we can show that in all cases either $|OPT| \geq t^{(1)} + 2/3$ (and then $|ALG| \leq t^{(1)} + 4/3 \leq |OPT| + 2/3 \leq 5|OPT|/3$), or $|OPT| \geq 4/3$ (and then $|ALG| \leq t^{(1)} + 4/3 \leq |OPT| + (4/3 - 1/2) = |OPT| + 5/6 \leq 5|OPT|/3$, as $5/6 \leq 2/3 \cdot 4/3 \leq 2|OPT|/3$). □

4 Stars

We consider in this section the case of stars. There is a star, centered at O (the initial position), with k branches or rays. If two requests q_i and q_j, located at p_i and p_j, are not in the same branch, then the distance between them is $d(p_i, p_j) = d(p_i, O) + d(O, p_j)$.

We focus on the closed case, and devise a polytime algorithm which is $(7/4 + \epsilon)$-competitive (on stars). Recall that the lower bound on stars when we know the number of requests but not their locations is 2 (proof omitted).

Let us define the released segment of a ray j (at some time t) to be the whole ray if all its requests are released, otherwise the segment from the extremity of the ray to its farthest (from O) unreleased request. The idea of the algorithm is the following. If a long ray (i.e., a ray of length at least 1/4 of the overall length of all rays) exists, it already gives a good competitive ratio to first serve this ray completely before going over to the remaining rays. Otherwise, if all rays are short, the algorithm waits in the origin until time t, which is exactly the combined length of all rays. At this time, it identifies a set R of rays maximizing the combined length of the released segments in R under the constraint that the set R can be traversed completely, including going back to the origin, in time t. Having identified the set R, our algorithm then serves the released requests in the set R and waits at the origin until all requests are released. Afterwards, it serves the unserved requests in an optimal manner, e.g., serving the rays in an arbitrary order by going to the furthest unserved request in each ray, and serving the remaining requests on that ray on the way back to the origin without waiting.

Algorithm 3: Algorithm for closed OLTSP-L on the star

Input: Offline: Request locations $p_1, \ldots, p_n$

Online: Release times $t_1, \ldots, t_n$

1 Let $R_1, \ldots, R_k$ be the rays of the star and $r_1, \ldots, r_k$ the lengths of the rays, i.e., the distance from the origin to the respective extremity of the ray.
2 **if** *there exists* $r_j, j \in [k]$, *with* $r_j \geq 1/4 \sum_{j'=1}^{k} r_{j'}$ **then**
3 | Instantly traverse ray R_j, turning once reaching the extremity and start to serve the requests. If a request is unreleased, wait at this location until it is released before continuing.
4 **else**
5 | Wait in the origin until $t = \sum_{j'=1}^{k} r_{j'}$.
6 | Let R be the set of indices $R \subseteq [k]$ maximizing the combined lengths of released segments s.t. $\sum_{j' \in R} r_{j'} \leq 1/2 \sum_{j'=1}^{k} r_{j'}$.
7 | Traverse the rays of set R, serving the released requests and return to O.
8 **end**
9 Wait in O until all requests are released, serve the unserved requests in an optimal manner and return back to the origin.

Theorem 3. *For any constant* $\epsilon > 0$, *Algorithm 3 achieves a competitive ratio of* $7/4 + \epsilon$ *for closed OLTSP-L on the star in polynomial time.*

Proof. We assume without loss of generality that $\sum_{j=1}^{k} r_j = 1$. This means that $|OPT| \geq 2$, since every ray has to be fully traversed and the server has to come back to O. Further, for the analysis we assume w.l.o.g. that requests are served on the way back to the origin, and that a request is served only if all requests which are farther from the origin on the same ray have already been served.

For the first case, assume that there is a ray r of length at least 1/4. Then, the algorithm instantly goes to the far end of the ray and serves its requests on the way back, waiting at each until it is released. Having returned to the origin at time t^*, the algorithm can serve the remaining requests in time 3/2 once everything is released. It further holds that $|OPT| \geq t^*$, i.e., OPT cannot serve all requests and return to the origin before ALG has served the ray r and returned to the origin. Since ALG starts serving the remaining requests when all requests are released and OPT cannot finish before that time, we get $|ALG| \leq |OPT| + 3/2$. This means $\rho \leq 1 + 3/4 = 7/4$.

Otherwise, if all rays have a length of less than 1/4, the algorithm waits until time $t = 1$. At this time, it identifies the set R and serves the released requests in R. For a simpler exposition, in the following, we assume that an optimal set R is found. Since the combined length of rays in R is at most 1/2, ALG will be back at the origin by time $t = 2$. Denote the combined length of the released segments in the rays with index in R by ℓ. ALG then serves the requests and returns to the origin, waiting there until all remaining requests are released, which is at most $t = |OPT|$. For the remaining requests, ALG needs time at most $2 - 2\ell$, so we get $|ALG| \leq |OPT| + 2 - 2\ell$. Clearly, if $\ell \geq 1/4$, this means that $|ALG| \leq |OPT| + 3/2 \leq 7/4 \cdot |OPT|$ as $|OPT| \geq 2$.

Otherwise, if $\ell < 1/4$, we first observe that even though at most ℓ can be covered by time $t = 1$, this does not mean that OPT can have served only released segments of length at most ℓ. It could be that OPT is already on some ray R_j, having visited some requests, but not able to return to the origin by time $t = 1$. However, by the definition of R, OPT can have visited at most ℓ before traversing R_j (the ray on which it is at $t = 1$). Denote by ℓ' the length of the segment of requests visited by OPT on R_j at time 1. It must hold that $\ell' \leq \ell$. Thus, OPT must serve $r_j - \ell'$ on R_j and at least $1 - r_j - \ell$ on the remaining rays. Since $r_j \leq 1/4$, this implies $|OPT| \geq 1 + r_j - \ell' + 2(1 - r_j - \ell) \geq 3 - r_j - 3\ell \geq 11/4 - 3\ell$. Since $|ALG| \leq |OPT| + 2 - 2\ell$, we have $\rho \leq 1 + \frac{2-2\ell}{11/4-3\ell} \leq 1 + \frac{3}{4}$ as $\frac{2-2\ell}{11/4-3\ell} \leq \frac{3}{4}$ for $0 \leq \ell \leq 1/4$.

Using a knapsack FPTAS (see e.g., [14, Section 6.2]), a $(1-\epsilon)$-approximation to R can be found in polynomial time, for any constant $\epsilon > 0$. This increases the competitive ratio by at most ϵ. □

5 Semi-Line

In this section, we study the case of a semi-line starting at the origin O. We denote the location of requests by their distance from the origin, the distance of the furthest request by L and say that the semi-line has length L.

For the open version of the problem, we give a lower bound of 4/3 on the semi-line (Theorem 4, proof omitted). We then describe an algorithm that is 13/9-competitive (Theorem 5), thus improving on the lower bound of 3/2 when only the number of requests is known (proof omitted). We finally show that for the closed case competitive ratio 1 is achievable.

Theorem 4. *For any $\epsilon > 0$, there is no $(4/3 - \epsilon)$-competitive algorithm for open OLTSP-L on the semi-line.*

Now, let us describe an algorithm that is 13/9-competitive for open OLTSP-L on the semi-line. The principle of this algorithm (see Algorithm 4 for a formal description) is to reach position $L/2$ and wait until segment $[0, L/4]$ or $[3L/4, L]$ is completely released. More precisely, it waits until all consecutive requests inside the segment $[0, L/4]$ or $[3L/4, L]$ are released. After that, the algorithm moves to position 0 or L and serves all requests on the semi-line without turning back again. To keep the competitive ratio small, we have to be careful with how we move away from the origin to reach $L/2$.

- At first, ALG moves from the origin only if it does not leave any unreleased requests behind. It follows the same strategy until $t = \min\{L/2 + x, 3L/4\}$, where x is the furthest request position such that by time t, the interval $[0, x)$ is completely released.[3] By construction, the contiguous segment starting from 0 which is served by ALG is at least the size of that of OPT.
- If $x \geq L/4$, then ALG continues the same strategy until it has served all requests. Otherwise, if $x < L/4$, ALG is at position x at $t = L/2 + x$ and abandons that strategy to reach position $L/2$ at exactly $t = L$, where it waits for $[0, L/4]$ or $[3L/4, L]$ to be completely released as described above.

Algorithm 4: Algorithm for open OLTSP-L on the semi-line

Input: Offline: Request locations $p_1, \ldots, p_n$, where $p_1 \leq p_2 \leq \ldots \leq p_n$

Online: Release times $t_1, \ldots, t_n$

1 $L \leftarrow d(O, p_n)$
2 Starting from the origin, move in the direction of p_n and serve all released requests that are encountered. Stop at the first unreleased request. Wait until it can be served, serve it and continue this procedure until $t = \min\{L/2 + x, 3L/4\}$, where x is the furthest position such that $[0, x)$ is completely released by time t.
3 **if** $x < L/4$ **then**
4   Start moving to position $p = L/2$.
5   **if** $[0, L/4]$ *is completely released before* $t = L$ **then**
6   return to position x and serve the requests in $[x, L/4]$.
7   **else**
8   Wait at $L/2$ until $[0, L/4]$ (or $[3L/4, L]$) is completely released and serve it moving from x to $L/4$ (or from L to $3L/4$).
9   **end**
10 Continue in the same direction, serving requests on the way, waiting if necessary, until all requests are served.

Theorem 5. *Algorithm 4 has a competitive ratio of* 13/9 *for open OLTSP-L on the semi-line.*

---

[3] Strictly speaking, if $x \geq L$, all requests and thus $[0, x]$ would be released. We will ignore this and only use half-open intervals to avoid technicalities.

Let $t^{(1)}$ be the first time when $[0, L/4]$ is completely released (when $t^{(1)} < L$) or the first time when $[0, L/4]$ or $[3L/4, L]$ is completely released (when $t^{(1)} \geq L$).

Next, we present some observations on which the proof is based. The full proof is omitted.

Observation 1. *If OPT turns around at least once, then, without loss of generality, the first time it turns is at L.*

Indeed, if OPT does not do this we can easily transform it into another optimal solution that acts like this.

Observation 2. *Assume that ALG has already served* $[0, x_{ALG})$ *at* $t = t^{(1)}$. *If* $t^{(1)} \leq L/2 + x$ *and* $x < L/4$, *then OPT as served at most* $[0, x_{OPT})$ *at* $t = t^{(1)}$, *where* $x_{OPT} \leq x_{ALG}$.

The above observation is true by construction. ALG has served the maximal continuous segment $[0, a)$ that any algorithm can serve, for some $a \geq 0$, until $t^{(1)} \leq L/2 + x$ (when $x < L/4$).

Observation 3. *When* $x < L/4$ *and* $t = L/2 + x$, *ALG is at position* x.

It is easy to show that by construction ALG will always reach position x by time $t = L/2 + x$ when $x < L/4$.

Roughly speaking, in the proof we distinguish several cases based on the value of $t^{(1)}$ and OPT's behavior. If $x \geq L/4$, then OPT has to turn around at least once to get a competitive ratio greater than 1 (Observation 2), and, in that case, the ratio would be at most 10/7. On the other hand, if $x < L/4$ and OPT does not turn around at any time, the ratio would be 1 when $t^{(1)} \leq L/2 + x$ (from Observation 2) and at most 13/9 otherwise (as OPT cannot have served a longer continuous segment starting from 0 than ALG until time $t = L/2 + x$). In the case that OPT turns around at least once, $|OPT|$ is large enough to show that the competitive ratio can be bounded from above by 13/9.

Finally, for the closed variant of the problem, we show that there exists an algorithm with competitive ratio 1 (the details can be found in the full version [6] of this paper).

Proposition 2. *For closed OLTSP-L, there is a polytime algorithm with competitive ratio* 1 *on the semi-line.*

Acknowledgments. This work was partially funded by the grant ANR-19-CE48-0016 from the French National Research Agency (ANR).

References

1. Allulli, L., Ausiello, G., Laura, L.: On the power of lookahead in on-line vehicle routing problems. In: Wang, L. (ed.) COCOON 2005. LNCS, vol. 3595, pp. 728–736. Springer, Heidelberg (2005). https://doi.org/10.1007/11533719_74

2. Ascheuer, N., Grötschel, M., Krumke, S.O., Rambau, J.: Combinatorial online optimization. In: Kall, P., Lüthi, H.J. (eds.) Operations Research Proceedings 1998, pp. 21–37. Springer, Heidelberg (1999). https://doi.org/10.1007/978-3-642-58409-1_2
3. Ausiello, G., Bonifaci, V., Laura, L.: The on-line asymmetric traveling salesman problem. J. Discrete Algorithms **6**(2), 290–298 (2008)
4. Ausiello, G., Demange, M., Laura, L., Paschos, V.T.: Algorithms for the on-line quota traveling salesman problem. Inf. Process. Lett. **92**(2), 89–94 (2004)
5. Ausiello, G., Feuerstein, E., Leonardi, S., Stougie, L., Talamo, M.: Algorithms for the on-line travelling salesman. Algorithmica **29**, 560–581 (2001)
6. Bampis, E., Escoffier, B., Hahn, N., Xefteris, M.: Online TSP with known locations. CoRR abs/2210.14722 (2022)
7. Bernardini, G., Lindermayr, A., Marchetti-Spaccamela, A., Megow, N., Stougie, L., Sweering, M.: A universal error measure for input predictions applied to online graph problems. In: Oh, A.H., Agarwal, A., Belgrave, D., Cho, K. (eds.) Advances in Neural Information Processing Systems (2022)
8. Bjelde, A., et al.: Tight bounds for online TSP on the line. ACM Trans. Algorithms **17**(1), 3:1–3:58 (2021)
9. Blom, M., Krumke, S.O., de Paepe, W., Stougie, L.: The online TSP against fair adversaries. INFORMS J. Comput. **13**(2), 138–148 (2001)
10. Chen, P., Demaine, E.D., Liao, C., Wei, H.: Waiting is not easy but worth it: the online TSP on the line revisited. CoRR abs/1907.00317 (2019)
11. Gouleakis, T., Lakis, K., Shahkarami, G.: Learning-augmented algorithms for online TSP on the line. In: Proceedings of the 37th AAAI Conference on Artificial Intelligence (2023, to appear). CoRR abs/2206.00655
12. Hu, H., Wei, H., Li, M., Chung, K., Liao, C.: Online TSP with predictions. CoRR abs/2206.15364 (2022)
13. Jaillet, P., Wagner, M.: Online routing problems: value of advanced information as improved competitive ratios. Transp. Sci. **40**, 200–210 (2006)
14. Kellerer, H., Pferschy, U., Pisinger, D.: Knapsack Problems. Springer, Heidelberg (2004). https://doi.org/10.1007/978-3-540-24777-7
15. Lawler, E.L., Lenstra, J.K., Kan, A.H.G.R., Shmoys, D.B.: The Traveling Salesman Problem: A Guided Tour of Combinatorial Optimization. Wiley Series in Discrete Mathematics & Optimization (1991)
16. Psaraftis, H.N., Solomon, M.M., Magnanti, T.L., Kim, T.U.: Routing and scheduling on a shoreline with release times. Manage. Sci. **36**(2), 212–223 (1990)

Socially Fair Matching: Exact and Approximation Algorithms

Sayan Bandyapadhyay[1], Fedor Fomin[2], Tanmay Inamdar[2](✉), Fahad Panolan[3], and Kirill Simonov[4]

[1] Portland State University, Portland, USA
sayanb@pdx.edu
[2] University of Bergen, Bergen, Norway
{Tanmay.Inamdar,Fedor.Fomin}@uib.no
[3] Indian Institute of Technology, Hyderabad, India
fahad@cse.iith.ac.in
[4] Hasso Plattner Institute, University of Potsdam, Potsdam, Germany
Kirill.Simonov@hpi.de

Abstract. Matching problems are some of the most well-studied problems in graph theory and combinatorial optimization, with a variety of theoretical as well as practical motivations. However, in many applications of optimization problems, a "solution" corresponds to real-life decisions that have major impact on humans belonging to diverse groups defined by attributes such as gender, race, or ethnicity. Due to this motivation, the notion of *algorithmic fairness* has recently emerged to prominence. Depending on specific application, researchers have introduced several notions of fairness.

In this paper, we study a problem called Socially Fair Matching, which combines the traditional Minimum Weight Perfect Matching problem with the notion of *social fairness* that has been studied in clustering literature [Abbasi et al., and Ghadiri et al., FAccT, 2021]. In our problem, the input is an edge-weighted complete bipartite graph, where the bipartition represent two groups of entities. The goal is to find a perfect matching as well as an assignment that assigns the cost of each matched edge to one of its endpoints, such that the maximum of the total cost assigned to either of the two groups is minimized.

Unlike Minimum Weight Perfect Matching, we show that Socially Fair Matching is weakly NP-hard. On the positive side, we design a *deterministic* PTAS for the problem when the edge weights are arbitrary. Furthermore, if the weights are integers and polynomial in the number of vertices, then we give a randomized polynomial-time algorithm that solves the problem exactly. Next, we show that this algorithm can be used to obtain a *randomized* FPTAS when the weights are arbitrary.

Keywords: Fairness · Matching · Approximation Algorithms

The research leading to these results has received funding from the Research Council of Norway via the project BWCA (grant no. 314528), and the European Research Council (ERC) via grant LOPPRE, reference 819416.

P. Morin and S. Suri (Eds.): WADS 2023, LNCS 14079, pp. 79–92, 2023.
https://doi.org/10.1007/978-3-031-38906-1_6

1 Introduction

Matching is a ubiquitous problem in computer science, since many optimization problems in practice can be interpreted as *assignment* problems, and matchings in (bipartite) graphs are a natural candidate for modeling such problems. The polynomial-time *Blossom* algorithm of Edmonds [9] for computing a maximum matching is one of the cornerstones of algorithmic graph theory and combinatorial optimization. Traditionally, optimization problems have focused on optimizing a single objective function subject to certain constraints based on the problem. In this viewpoint, all the different aspects of a solution are condensed into a single number, called the *cost* of the solution. This model allows for a clean abstraction of the problem, which is useful for studying the problem from theoretical point of view.

In many real-world applications, however, an optimization problem inherently involves different tradeoffs. For example, suppose there are two possible locations for building a new school in a community – one location is cheap, but the location is extremely inconvenient for students of one demographic group in the community over the other; on the other hand the second location is relatively expensive, but is easily accessible to students of all demographic groups. In such a situation, it is vastly preferable to choose the second location for the school. Researchers have considered several approaches to alleviate this issue. One approach is the problem of multi-objective optimization (see [3,21]), which adds several objective functions to an optimization problem. Another related approach is to model the *fairness* aspect directly into the problem, where the notion of fairness may be specific to the problem at hand. In this paper, we follow the second approach.

We study SOCIALLY FAIR MATCHING, which introduces a notion of fairness called *social fairness* in the classical MINIMUM WEIGHT PERFECT MATCHING problem. Social fairness was introduced very recently in the context of clustering problems [1,12,20], which is useful to balance the total clustering cost over all groups. In this work, we introduce the SOCIALLY FAIR MATCHING problem. To put it in the context, let us consider the classical edge-weighted bipartite matching problem, where the goal is to find a minimum cost perfect matching in a complete bipartite graph between two groups R and B each containing n vertices. Now, for any matched edge (u, v) with $u \in R$ and $v \in B$, depending on the application, the cost might be paid by either u or v. Thus, the total cost for the two groups R and B might not be well-balanced. To address a similar issue in the context of clustering problems, Abbasi et al. [1], and Ghadiri et al. [12] proposed the notion of *social fairness.* We adopt this notion in the context of matching. Thus, in SOCIALLY FAIR MATCHING, the goal is to find a perfect matching in a complete bipartite graph as well as an assignment that assigns the cost of each matched edge to one of its endpoints, such that the maximum of the total cost assigned to either of the two groups is minimized.

Twinning of cities is a legal or social agreement between two geographically and politically distinct cities to promote cultural and commercial ties [23]. This phenomenon goes back centuries, but in modern history, cities in two different countries are twinned as an alternative channel for diplomacy. Consider the situation where $2n$ cities from two different countries desire to be twinned with each other, such that each city of one country is twinned with a city of another country. For each pair of twin cities, the headquarters may be located in one of the two cities, which must bear the administrative cost that depend on the specific parameters for twinning the specific pair of cities. In this application, it might be desirable that we come up with a pairing, such that the total expenses born by cities of each country is minimized. Note that this can be modeled as an instance of SOCIALLY FAIR MATCHING, where the weight of each edge represents the administrative cost of twinning two cities.

Our Results and Contributions. We first observe that SOCIALLY FAIR MATCHING is weakly NP-hard, when the edge weights are arbitrary integers. The reduction is via the well-known PARTITION problem, which asks whether it is possible to partition a given set of integers into two parts with equal weights. In contrast, we show when the edge weights are integers and polynomial in n, the problem can be solved exactly in polynomial time using a randomized algorithm. For this result, we reduce SOCIALLY FAIR MATCHING to the problem of polynomial identity testing, which can be solved in randomized polynomial time via an application of the Schwartz-Zippel Lemma. For the case of general weights, we show how to obtain a $(1+\epsilon)$-approximation using two different approaches. First, we show that the case of general weights can be reduced to that with polynomial integer weights at a small loss in the approximation guarantee. Thus, we can obtain a *randomized* FPTAS[1] via the previous result. In a different direction, we show that one can also obtain a *deterministic* PTAS[2] in this case. Despite having a worse running time as compared to the previous FPTAS, we believe that this result is interesting for a few reasons. First, the PTAS is deterministic, unlike the inherently randomized nature of the previous FPTAS due to its reliance on the Schwartz-Zippel lemma. Another reason is that the PTAS is entirely combinatorial – we first guess (i.e., enumerate) a subset of *heavy* edges of an optimal solution, and essentially reduce the problem to classical MINIMUM WEIGHT PERFECT MATCHING. Thus, another advantage is that it does not rely on any sophisticated algebraic machinery such as polynomial identity testing.

Related Work. So-called PARTITIONED MIN-MAX WEIGHTED MATCHING (PMMWM), has been studied in the operations research literature [19]. In this problem, we are given an edge weighted bipartite graph $G = (R \uplus B, E)$, and an integer $m \le |R|$. The goal is to find a partition $R_1, R_2, \dots, R_m$ of R, and a matching M saturating R, such that the maximum total weight of matched edges (i.e., edges of M) incident to all vertices in R_i is minimized. Furthermore, it is required that for every $1 \le i \le m$, $|R_i| \le u$, where u is an upper bound given in the input. Kress et al. [19] establish hardness results and approximation algorithms

[1] FPTAS stands for *Fully Polynomial-Time Approximation Scheme*, i.e., for any $\epsilon > 0$, an algorithm that returns a $(1+\epsilon)$-approximation in time $(n/\epsilon)^{O(1)}$.

[2] PTAS stands for *Polynomial-Time Approximation Scheme*, i.e., for any $\epsilon > 0$, an algorithm that returns a $(1+\epsilon)$-approximation in time $n^{f(1/\epsilon)}$.

for PMMWM in general case. Even though seemingly unrelated to Socially Fair Matching, note that if for the input to PMMWM, it holds that (i) G is a complete bipartite graph with $|R| = |B| = n$, (ii) m, the number of parts of the partition is equal to 2, and (iii) the upper bound u is equal to $|R|$ (i.e., there is no upper bound on the size of any part), then the resulting problem is equivalent to Socially Fair Matching. To the best of our knowledge, for this special case of PMMWM, no improved approximation results are known. Another problem related to PMMWM is the so-called Min-Max Weighted Matching (MMWM) problem [2,6], where the only difference from PMMWM is that partition of R into $R_1, R_2, \ldots, R_m$ is given in the input, G is complete, and $n = |R| = |B|$. Duginov [6] establishes several results for this problem. Among these, they observe that MMWM is related to the well-known Exact Perfect Matching problem, which has a randomized polynomial-time algorithm, but obtaining a deterministic one is a long-standing open problem. Although the setting of MMWM appears more similar to Socially Fair Matching as compared to PMMWM, the fact that the partition $R_1, R_2, \ldots, R_m$ is given in the input in MMWM, makes the two problems quite different.

Further Related Work on Fairness and Matching. In recent years, researchers have introduced and studied several different notions of fairness, e.g., disparate impact [10], statistical parity [15,22], individual fairness [7] and group fairness [8]. Kleinberg et al. [18] formulated three notions of fairness and showed that it is theoretically impossible to satisfy them simultaneously. See also [4,5] for similar exposures.

Several different fair matching problems have been studied in the literature. Huang et al. [13] studied fair b-matching, where matching preferences for each vertex are given as ranks, and the goal is to avoid assigning vertices to high ranked preferences as much as possible. Fair-by-design-matching is studied by Garcia-Soriano and Bonchi [11], where instead of a single matching, a probability distribution over all feasible matchings is computed which guarantees individual fairness. See also [14,17].

Organization. In Sect. 2, we formally define our problem. The deterministic PTAS is discussed in Sect. 3. Section 4 contains the randomized polynomial-time algorithm, the FPTAS, and the hardness result. Finally, in Sect. 5, we conclude with some interesting open questions.

2 Preliminaries

For an integer $\ell \geq 1$, we use the notation $[\ell] := \{1, 2, \ldots, \ell\}$.

Socially Fair Matching. In Socially Fair Matching, the input is a complete bipartite graph $G = (R \uplus B, E)$, where $R \uplus B$ is a bipartiton of $V(G)$, with $|R| = |B| = n$. We will often refer to the vertices of R and B as red and blue respectively. Each edge in $E(G)$ has a non-negative weight. The goal is to compute a perfect matching $M \subseteq E(G)$, and an assignment $f : M \rightarrow \{\mathsf{red}, \mathsf{blue}\}$, such that $\max\left\{\sum_{e\in M(\mathsf{red})} w(e), \sum_{e\in M(\mathsf{blue})} w(e)\right\}$ is minimized, where $M(\mathsf{red})$ is the set of edges in M such that $f(e) = \mathsf{red}$, and $M(\mathsf{blue})$ is defined analogously.

Fields, Polynomials, Vectors and Matrices

Here, we review some definitions from linear algebra. We refer to any graduate textbook on algebra for more details. For a finite field $\mathbb{F}$ and a set of variables $X = \{x_1, \ldots, x_n\}$, $\mathbb{F}[X]$ denotes the ring of polynomials in X over $\mathbb{F}$. The *characteristic* of a field is defined as least positive integer m such that $\sum_{i=1}^{m} 1 = 0$.

A vector v over a field $\mathbb{F}$ is an array of values from $\mathbb{F}$. The matrix is said to have dimension $n \times m$ if it has n rows and m columns. For a vector v, we denote its *transpose* by v^T. The the rank of a matrix is the maximum number k such that there is a $k \times k$ submatrix whose determinant is non-zero.

3 A Deterministic PTAS for Arbitrary Weights

Let M be a perfect matching, and $f : M \to \{\mathsf{red}, \mathsf{blue}\}$ be an arbitrary assignment. Then, for a vertex $r \in R$, we define

$$\mu_{M,f}(r) := \begin{cases} 0 & \text{if } e = \{r, b\} \in M \text{ with } f(e) = \mathsf{red} \\ w(e) & \text{if } e = \{r, b\} \in M \text{ with } f(e) = \mathsf{blue} \end{cases}$$

and for $b' \in B$, we define

$$\mu_{M,f}(b') := \begin{cases} 0 & \text{if } e = \{r', b'\} \in M \text{ with } f(e) = \mathsf{blue} \\ w(e) & \text{if } e = \{r', b'\} \in M \text{ with } f(e) = \mathsf{red} \end{cases}.$$

For a subset $R' \subseteq R$ (resp. $B' \subseteq B$), we define $\mu_{M,f}(R') := \sum_{r\in R'} \mu_{M,f}(r)$ (resp. $\sum_{b\in B'} \mu_{M,f}(b)$).

Fix an optimal solution $M^* \subseteq E(G)$, and the corresponding assignment $f^* : M^* \to \{\mathsf{red}, \mathsf{blue}\}$. Define $M^*(\mathsf{red})$ and $M^*(\mathsf{blue})$ as the sets of edges assigned red and blue by f^* respectively. For $v \in V(G)$, we use the shorthand $\mu^*(v) := \mu^*_{M^*,f^*}(v)$. Note that $OPT = \max\{\mu^*(R), \mu^*(B)\}$.

Let $t = 1 + \lceil 1/\epsilon \rceil$. Let $R_1 \subseteq R$ be the set of vertices incident to the heaviest $\max\{t, |M^*(\mathsf{red})|\}$ edges in $M^*(\mathsf{red})$. Similarly, let $B_1 \subseteq B$ be the set of vertices incident on the heaviest $\max\{t, |M^*(\mathsf{blue})|\}$ edges in $M^*(\mathsf{blue})$. Here, we assume that the ties are broken arbitrarily in the previous definitions.

Let R_1' denote the matched endpoints of vertices in B_1, i.e., $R_1' := \{r \in R : \exists b \in B_1 \text{ such that } \{r, b\} \in M^*\}$. Similarly, define $B_1' := \{b \in B : \exists r \in R_1 \text{ such that } \{r, b\} \in M^*\}$. Note that $|R_1| + |R_1'| = |B_1| + |B_1'| \leq 2t$. Now, define $R' = R \setminus (R_1 \cup R_1')$, and $B' = B \setminus (B_1 \cup B_1')$.

The following observation is easy to follow.

Observation 1. *For any $r \in R'$, let $\{r, b\} \in M^*$ be the matched edge incident on r. Then, $b \in B'$. For any $b' \in B'$, let $\{r', b'\} \in M^*$ be the matched edge incident on b'. Then, $r \in R'$.*

We also have the following observation.

Observation 2. *For any* $r \in R'$, $\mu^*(r) \leq \epsilon \cdot \mu^*(R) \leq \epsilon \cdot OPT$. *For any* $b \in B'$, $\mu^*(b) \leq \epsilon \cdot \mu^*(B) \leq \epsilon \cdot OPT$.

Proof. Suppose for contradiction that there is some $r' \in R' \neq \emptyset$ (wlog) such that $\mu^*(r') > \epsilon \cdot \mu^*(R)$. Then, by the definition of R_1, for all $r \in R_1$, $\mu^*(r) \geq \mu^*(r') > \epsilon \cdot \mu^*(R)$. This implies that $\mu^*(R_1) > \lceil 1/\epsilon \rceil \cdot \epsilon \cdot \mu^*(R) \geq OPT$, which is a contradiction. The proof is exactly the same for any $b' \in B'$.

Let $\tilde{M}^* \subseteq M^*$ be the subset of edges of an optimal solution that are incident on $V' := R_1 \cup R_1' \cup B_1 \cup B_1'$. The first step is to guess $\tilde{M}^*$ and the partial optimal assignment $f^* : \tilde{M}^* \to \{\text{red}, \text{blue}\}$. Note that since $|\tilde{M}^*| \leq 2t = O(1/\epsilon)$, we can enumerate all $n^{O(t)} = n^{O(1/\epsilon)}$ possible choices. We are left with a smaller instance $R' \cup B'$, such that $|R'| = |B'| = n - 2t$, where $R' = R \setminus (R_1 \cup R_1')$ and $B' = B \setminus (B_1 \cup B_1')$. Assuming we are working with the correct guess, we also have an upper bound of $U \leq \epsilon \cdot OPT$, which can be inferred from the smallest distance in the partial solution already guessed.

Let $OPT' = \max\{\mu^*(R'), \mu^*(B')\}$ denote the optimal assignment cost in the remaining instance. Henceforth, wlog assume that $OPT' = \mu^*(R') = \frac{1}{\delta}\mu^*(B')$ for some $\delta \in [0, 1]$. The other case is symmetric, and we can run the algorithm by exchanging the roles of red and blue points, and select the solution with smaller cost.

Let $G' = (R' \cup B', E')$ be the subgraph of $G[R' \cup B']$, where we delete all edges with weight greater than U. Let M denote a minimum weight perfect matching in G'. We know such a perfect matching exists, because we are working with the correct guess, which implies that the assignment corresponding to OPT' is a perfect matching such that the weight of any edge is at most $U \leq \epsilon \cdot OPT$.

Lemma 1. $OPT' \leq w(M) \leq \mu^*(R') + \mu^*(B') = (1 + \delta) \cdot OPT'$. *Furthermore, there is an assignment* $f' : M \to \{\text{red}, \text{blue}\}$ *such that* $\max\{\mu_{M,f'}(R'), \mu_{M,f'}(B')\} \leq (1 + \delta) \cdot OPT'$.

Proof. Let M' be a matching corresponding to the optimal solution OPT'. Then,

$$w(M) \leq w(M') = \mu^*(R') + \mu^*(B') = (1 + \delta) \cdot \mu^*(R')$$

Now, we construct an arbitrary assignment $f' : M \to \{\text{red}, \text{blue}\}$, and note that

$$\max\{\mu_{M,f'}(R'), \mu_{M,f'}(B')\} \leq w(M) = (1 + \delta) \cdot \mu^*(R') = (1 + \delta) \cdot OPT'.$$

Finally, since this is a valid solution of cost at most $w(M)$, the cost of optimal solution OPT' must be at most $w(M)$.

As a first step, we try to obtain an assignment $f : M \to \{\text{red}, \text{blue}\}$ that achieves the red to blue cost ratio approximately equal to $1/\delta$. Since we do not know the exact value of δ, we will try the ratios $1/(1 + \epsilon)^s$, for $s = 0, 1, \ldots, q$. In order to upper bound q, let us consider a simpler case when δ is very small.

Case 1: $\delta < 2\epsilon$**.** If $\mu^*(B') < 2\epsilon \cdot OPT$, then for every edge $e \in M$, we define the assignment $f'(e) = B$. We construct the assignment $f : M \cup \tilde{M}^* \to \{\mathsf{red}, \mathsf{blue}\}$ by defining $f(e) = f'(e)$ if $e \in M$; and $f(e) = f^*(e)$ otherwise. Note that,

$$\begin{aligned}\text{Blue cost} &= \sum_{b \in B'} \mu_{M,f'}(b) + \sum_{b \in B_1} \mu^*(b) \\ &= 0 + \mu^*(B_1) \leq \mu^*(B) \leq OPT,\end{aligned}$$

And,

$$\begin{aligned}\text{Red cost} &= \sum_{r \in R'} \mu_{M,f'}(r) + \sum_{r \in R_1} \mu^*(r) \qquad \text{(From Lemma 1)} \\ &= w(M) + \mu^*(R_1) \\ &\leq \mu^*(R') + \mu^*(B') + \mu^*(R_1) \\ &\leq \mu^*(R) + 2\epsilon \cdot OPT \\ &\leq (1+2\epsilon) \cdot OPT.\end{aligned}$$

Case 2: $2\epsilon \leq \delta \leq 1$**.** In this case, have that $\mu^*(B') = \delta OPT' \geq 2\epsilon OPT \geq 2\epsilon OPT'$. By trying $s = 1, 2, \ldots, q$, we want to find a value s such that $(1+\epsilon)^{-(s+1)} \leq \frac{\delta}{1+\delta} < (1+\epsilon)^{-s}$. Note that $\frac{\delta}{1+\delta} \geq \frac{2\epsilon}{2} = \epsilon$, since $2\epsilon \leq \delta \leq 1$. This implies that $q \leq -\lceil \log_{1+\epsilon}\left(\frac{\delta}{1+\delta}\right) \rceil \leq -\lceil \frac{\log \epsilon}{\log(1+\epsilon)} \rceil = g(\epsilon)$ for some g. Furthermore, this implies that the weight of any non-removed edge is at most $\epsilon \cdot OPT \leq (1+\epsilon)^{-s} \cdot OPT$ for any value of s we will consider.

Lemma 2. *When* $2\epsilon \leq \delta \leq 1$*, we can obtain an assignment* $f : M_c \to \{\mathsf{red}, \mathsf{blue}\}$*, where* $M_c = M \cup M^*$ *such that,*

$$\max\left\{\sum_{r \in R} \mu_{M_c,f}(r), \sum_{b \in B} \mu_{M_c,f}(b)\right\} \leq (1+2\epsilon) \cdot OPT.$$

Proof. Order the edges in the matching M in an arbitrary order, and let the weight of edge e_i be w_i. Let j be the index such that $\sum_{i=1}^{j} w_i < (1+\epsilon)^{-s} \cdot w(M)$, but $\sum_{i=1}^{j+1} w_i \geq (1+\epsilon)^{-s} \cdot w(M)$.

Note that such an index i always exists, since the weight of any non-removed edge – in particular that of e_1 – is at most $\epsilon \cdot OPT \leq (1+\epsilon)^{-q} OPT$, which corresponds to the largest value of s we consider.

Now, let R_C denote the red endpoints of edges $e_1, e_2, \ldots, e_{j+1}$, and let B_C denote the blue endpoints of edges $e_{j+2}, e_{j+3}, \ldots, e_{k'}$. We construct the assignment $f : M \to \{\mathsf{red}, \mathsf{blue}\}$ as follows. For each edge $e \in \{e_1, \ldots, e_{j+1}\}$, we let $f(e) = \mathsf{red}$, and for each edge $e' \in \{e_{j+2}, \ldots, ek'\}$, we let $f(e') = \mathsf{blue}$. Finally, we extend this assignment to $\tilde{M}^*$ by assigning for each edge $e \in \tilde{M}^*$ as $f(e) = f^*(e)$. Thus, now f is an assignment from $M \cup \tilde{M}^*$ to $\{\mathsf{red}, \mathsf{blue}\}$. Now we analyze the cost of this assignment.

Consider the following,

$$
\begin{aligned}
\sum_{i=1}^{j} w_i &\leq (1+\epsilon)^{-s} w(M) \\
&\leq \frac{(\mu^*(R') + \mu^*(B'))}{(1+\epsilon)^s} \\
&= (1+\epsilon) \cdot (1+\epsilon)^{-(s+1)} \cdot \left(\frac{\delta+1}{\delta}\right) \cdot \mu^*(B') \\
&\leq (1+\epsilon) \cdot \mu^*(B') \qquad (1)
\end{aligned}
$$

$$
\begin{aligned}
\sum_{i=j+2}^{k'} w_i &= w(M) - \sum_{i=1}^{j+1} w_i \\
&\leq w(M) - \frac{w(M)}{(1+\epsilon)^s} \\
&\leq \frac{\mu^*(R') + \delta \mu^*(R')}{\delta+1} \\
&\leq \mu^*(R'). \qquad (2)
\end{aligned}
$$

Here, the last inequality in (1) and the second-last inequality in (2) follow from the definition of s, i.e., $(1+\epsilon)^{-(s+1)} \leq \frac{\delta}{1+\delta} < (1+\epsilon)^{-s}$. Therefore,

$$
\begin{aligned}
\text{Blue cost} &= \sum_{i=1}^{j+1} w_i + \mu^*(B_1) \\
&\leq (1+\epsilon) \cdot \mu^*(B') + \mu(B_1) + w_{j+1} \\
&\leq (1+\epsilon) \cdot \mu^*(B) + \epsilon \cdot OPT
\end{aligned}
$$

$$
\begin{aligned}
\text{Red cost} &= \sum_{i=1}^{j+2} w_i + \mu^*(R_1) \\
&\leq \mu^*(R') + \mu^*(R_1) \\
&= \mu^*(R).
\end{aligned}
$$

Since the cost of the solution returned is the maximum of red cost and the blue cost, it is easy to show that the cost of our solution is upper bounded by $(1+2\epsilon) \cdot OPT$.

Theorem 1. *There exists a deterministic PTAS for* Socially Fair Matching *for arbitrary weights. In other words, for any $\epsilon > 0$, there exists a deterministic algorithm that returns a $(1+\epsilon)$-approximation for* Socially Fair Matching *in time $n^{O(1/\epsilon)}$.*

4 Randomized Polynomial Time Algorithm for Polynomial Weights

First, we assume that the weights are all integers in the range $[0, N]$, where N is an integer that is at least n (if not, a simple scaling ensures this property). We will describe an exact randomized algorithm that runs in time polynomial in n and N in this case.

Let $\mathbb{F}$ be a field of characteristic 2 containing at least $4(N+1)^2 \geq n^2$ distinct elements. If $X = \{x_1, x_2, \ldots, x_t\}$ is a set of t variables, then we use $\mathbb{F}[X]$ to denote the ring of polynomials in X.

Let $Z = \{z_{ij} : 1 \leq i, j \leq n\}$ be a set of n^2 variables, and let $X = \{x, y\} \cup Z$, where a variable z_{ij} corresponds to the edge $e_{ij} = \{i, j\} \in E(G)$. We define a matrix $A = (A_{ij})$, where

$$A_{ij} = \begin{cases} 0 & \text{if } \{i,j\} \notin E(G) \\ (x^{w_{ij}} + y^{w_{ij}}) \cdot z_{ij} & \text{if } \{i,j\} \in E(G) \text{ with weight } w_{ij} \end{cases}$$

First, we observe that the permanent of the matrix A computed in $\mathbb{F}[X]$ is equal to the determinant of A, which is a polynomial in X. Let Π be the set of permutations of n. Then, we have the following equality:

$$Q = \det(A) = \sum_{\sigma \in \Pi} \prod_{q=1}^{n} A_{q,\sigma(q)} = \sum_{i=0}^{N} x^i P_i(y, Z) = \sum_{i=0}^{N} x^i \sum_{j=0}^{N} y^j \cdot P_{i,j}(Z)$$

where $Q = Q(x, y, Z)$ is a polynomial in variables x, y and Z, each $P_i(y, Z)$ is a polynomial in y and Z, and $P_{i,j}(Z)$ is a polynomial in variables Z. Note that the degree of each polynomial $P_{i,j}(Z)$, which is equal to the maximum degree of any of its monomials, is at most n.

Observation 3. *There exists a perfect matching M and an assignment $f : M \rightarrow \{\mathsf{red}, \mathsf{blue}\}$ such that $\mu_{M,f}(R) = w_r$ and $\mu_{M,f}(B) = w_b$ iff the polynomial $P_{w_r,w_b}(z)$ is not identically equal to zero.*

Next, using the definition of polynomial Q, the following equalities are easy to see:

$$\left(1\; x \ldots x^N\right) \cdot \left(P_0(y, Z)\; P_1(y, Z) \ldots P_N(y, Z)\right)^\top = Q(x, y, Z) \tag{3}$$

and for each $0 \leq i \leq N$, we have that

$$\left(1\; y \ldots y^N\right) \cdot \left(P_{i,0}(Z)\; P_{i,1}(y, Z) \ldots P_{i,N}(Z)\right)^\top = P_i(y, Z) \tag{4}$$

Computing Polynomials at Specified Values. For a set $P = \{p_1, p_2, \ldots, p_{k+1}\} \subseteq \mathbb{F}$ of size $k+1$, let $V(P) \in \mathbb{F}^{(k+1)\times(k+1)}$ be the *Vandermonde matrix*, whose entries are given by $V(P)_{ij} = (p_i)^{j-1}$. That is, $V(P)$ looks as follows:

$$V(P) = \begin{pmatrix} p_1^0 & p_1^1 & p_1^2 & \cdots & p_1^k \\ p_2^0 & p_2^1 & p_2^2 & \cdots & p_2^k \\ \vdots & & & \ddots & \vdots \\ p_{k+1}^0 & p_{k+1}^1 & p_{k+1}^2 & \cdots & p_{k+1}^k \end{pmatrix}$$

Note that if P consists of $k+1$ distinct non-zero elements of $\mathbb{F}$, then $V(P)$ is invertible over $\mathbb{F}$. In this case, we let $W(P) = V^{-1}(P)$ be its inverse.

Next, we observe the following:

Let $T = \{y_1, y_2, \ldots, y_{N+1}\}$ be a set of distinct non-zero values of $\mathbb{F}$, then (4) implies that for any $0 \leq i \leq N$, the following holds:

$$V(T) \cdot \begin{pmatrix} P_{i,0}(Z) \\ P_{i,1}(Z) \\ \vdots \\ P_{i,N}(y, Z) \end{pmatrix} = \begin{pmatrix} P_i(y_1, Z) \\ P_i(y_2, Z) \\ \vdots \\ P_i(y_N, Z) \end{pmatrix} \tag{5}$$

which implies that

$$\begin{pmatrix} P_{i,0}(Z) \\ P_{i,1}(Z) \\ \vdots \\ P_{i,N}(y, Z) \end{pmatrix} = W(T) \begin{pmatrix} P_i(y_1, Z) \\ P_i(y_2, Z) \\ \vdots \\ P_i(y_N, Z) \end{pmatrix} \tag{6}$$

In particular, the polynomials $P_{i,j}(Z)$ at the given values $Z \leftarrow Z'$ can be evaluated in time polynomial in N using the computation above, assuming we can evaluate the polynomial $P_i(y, Z)$ at values $y \leftarrow y'$, and $Z \leftarrow Z'$. Next, we show how to do this computation.

From (3), we get that if $S = \{x_1, x_2, \ldots, x_{N+1}\}$ are distinct non-zero values of $\mathbb{F}$, then:

$$V(S) \cdot \begin{pmatrix} P_0(y, Z) \\ P_1(y, Z) \\ \vdots \\ P_N(y, Z) \end{pmatrix} = \begin{pmatrix} Q(x_1, y, Z) \\ Q(x_2, y, Z) \\ \vdots \\ Q(x_N, y, Z) \end{pmatrix} \implies \begin{pmatrix} P_0(y, Z) \\ P_1(y, Z) \\ \vdots \\ P_N(y, Z) \end{pmatrix} = W(S) \cdot \begin{pmatrix} Q(x_1, y, Z) \\ Q(x_2, y, Z) \\ \vdots \\ Q(x_N, y, Z) \end{pmatrix}$$

In particular, given the values $y \leftarrow y'$, and $z_{ij} \leftarrow z'_{ij}$, where $y', z'_{ij} \in \mathbb{F}$, the polynomials $P_i(y', Z')$ can be evaluated in time polynomial in N, assuming the polynomial $Q(x, y, Z)$ can be evaluated at the specified values $x \leftarrow x'$, $y \leftarrow y'$ and $Z \leftarrow Z'$. However, note that the polynomial Q is equal to the determinant of the matrix A. Thus, this can be implemented in polynomial time.

Recall that we want to determine whether the polynomial $P_{w_r,w_b}(Z)$ is identically equal to zero (cf. Proposition 3). To this end, we sample the values $Z' = \{z'_{ij}\}$ from $\mathbb{F}$ – note that $\mathbb{F}$ contains at least $4(N+1)^2 \geq n^2$ distinct elements, and the degree of the polynomial P_{w_r,w_b} is at most n. Therefore, by Schwartz-Zippel lemma, the probability that the polynomial is non-zero, when evaluated at Z' is equal to zero is at most $n/(N+1) \leq 1/n$. Thus, we obtain the following theorem.

Theorem 2. *There exists a randomized algorithm that, given an* Socially Fair Matching *instance on n vertices, and where all edge weights are integers in range $[0, N]$, with $N \geq n$, runs in time $(n+N)^{O(1)}$, and finds an optimal solution with probability at least $1 - 1/n$.*

4.1 FPTAS for General Weights via Reduction to Polynomial Integer Weights

Let $0 < \epsilon \leq 1$ be a fixed constant. By appropriately scaling, we assume that the smallest positive weight is at least $3/\epsilon$. Then, we round all weights of all the edges up to the nearest integer. Note that the weight of any edge is increased by strictly smaller than 1, which is at most $\epsilon/3$ factor of its original weight. Thus, assume that all weights are non-negative integers. Say, this is preprocessing step A.

By iterating over all edges, we "guess" the largest weight of an edge (after rounding up) that is part of an optimal solution. Let L denote a guess for the largest weight, and note that L is an integer. Then, we delete all the edges with weight larger than L. Suppose $L \leq 2n/\epsilon$. Then, we skip the following preprocessing step B, and directly use Theorem 2 as described subsequently.

Now, suppose that $L > 2n/\epsilon$. Then, for each edge with weight (after preprocessing step A) w, we define its weight to be $\lceil \frac{w}{L/(2n/\epsilon)} \rceil \cdot \frac{L}{2n/\epsilon}$. We say that this is preprocessing step B.

Claim. Suppose we guess the maximum weight L of an edge in an optimal solution correctly. Then, after preprocessing step A, and step B in the iteration corresponding to L, the optimal solution w.r.t. new weights is at most $1 + \epsilon$ times the original optimal weight.

Proof. As argued previously, step A incurs at most an $(1+\epsilon/3)$ factor increase in the cost of any solution. Consider the iteration corresponding to L, the maximum weight of an edge in some optimal solution $F \subseteq E$. By removing edges with weight larger than L, we do not delete any edge of an optimal solution. Note that the total increase in the weight of any edge due to step B is at most $\frac{L}{2n/\epsilon}$. Thus, for any set of edges of size at most n, the total increase in the weight is at most $\frac{\epsilon L}{2n} \cdot n \leq \frac{\epsilon L}{2} \leq \frac{\epsilon \cdot OPT}{2}$. Thus, the total increase in the weight due to preprocessing steps A and B can be upper bounded by $(1+\epsilon/3) \cdot (1+\epsilon/2)OPT \leq (1+\epsilon) \cdot OPT$.

After preprocessing step B, the weights are of the form $t \cdot \frac{L}{2n/\epsilon}$, where t is an integer in the range $[0, \lceil n/\epsilon \rceil]$. By dividing each weight by a factor of $L/(2n/\epsilon)$, we obtain an instance where all the weights are integers in the range $[0, \lceil n/\epsilon \rceil]$, i.e., $N = \lceil n/\epsilon \rceil \geq n$. Then, the algorithm from Theorem 2 can be used to find an optimal solution in time $(n/\epsilon)^{O(1)}$, with probability at least $1 - 1/n$. Therefore, we obtain the following theorem.

Theorem 3. *There exists a randomized FPTAS for* SOCIALLY FAIR MATCHING *for arbitrary weights. In other words, for any* $\epsilon > 0$, *there exists a randomized algorithm that returns a* $(1+\epsilon)$*-approximation for* SOCIALLY FAIR MATCHING *in time* $(n/\epsilon)^{O(1)}$, *with probability at least* $1 - 1/n$.

4.2 NP-Hardness of SOCIALLY FAIR MATCHING

The reduction is from a variant of PARTITION. The input is a set $A = \{a_1, a_2, \ldots, a_n\}$ of n positive integers, and an integer k. The problem asks whether it is possible to partition A into two sets A_1 and A_2 such that the sum of the integers in A_1 and A_2 are equal. It is known that PARTITION is weakly NP-hard, i.e., if the integers in A are given in binary [16].

We reduce this to SOCIALLY FAIR MATCHING as follows. First, let $R = \{r_1, r_2, \dots, r_n\}$, and $B = \{b_1, b_2, \dots, b_n\}$ be two disjoint sets of $2n$ vertices. Let $G = (R \cup B)$ be a *complete* bipartite graph, i.e., there is an edge between every r_i and b_j, $1 \leq i, j \leq n$. Now we define the weights on the edges. For $1 \leq i \leq n$, set $w(r_i, b_i) = a_i$. For $1 \leq i \neq j \leq n$, let $w(r_i, b_j) = n \cdot L$, where $L = \sum_{i=1}^{n} a_i$.

Note that any solution of cost at most L must output the matching $M = \{\{r_i, b_i\} : 1 \leq i \leq n\}$. Restricting our attention to such a solution, now the task is to find an assignment $f : M \to \{\mathsf{red}, \mathsf{blue}\}$. It is easy to see that there is a bijection between an assignment f, and a partition $\{A_1, A_2\}$ of the integers A in the given PARTITION instance. In particular, deciding whether there exists an assignment $f : M \to \{\mathsf{red}, \mathsf{blue}\}$, such that $\mu_{M,f}(R) = \mu_{M,f}(B) = \frac{L}{2}$ is equivalent to determining that the input A of PARTITION can be partitioned into two sets with equal sum. Therefore, finding an optimal solution to SOCIALLY FAIR MATCHING is weakly NP-hard.

5 Conclusions

In this work, we introduce a well-motivated matching problem, namely SOCIALLY FAIR MATCHING, and systemically study the complexity of the problem in terms of exact and approximate computation. Our results draw a nearly complete picture of the computational complexity of the problem. On the one hand, we show that the problem is weakly NP-hard when the edge weights are arbitrary integers. On the other hand, we obtain a randomized polynomial-time algorithm when the weights are polynomially bounded. The latter result leads to a randomized FPTAS for the general problem. We also obtain a deterministic PTAS in the general case, which is a simple, combinatorial algorithm.

Our work leads to several interesting open questions. An obvious question is to obtain a deterministic FPTAS for the problem. Also, it would be interesting to see for which subclasses of graphs our problem admits polynomial-time algorithms. Finally, one might be interested in suitably extending our model to multiple groups.

References

1. Abbasi, M., Bhaskara, A., Venkatasubramanian, S.: Fair clustering via equitable group representations. In: Elish, M.C., Isaac, W., Zemel, R.S. (eds.) FAccT 2021: 2021 ACM Conference on Fairness, Accountability, and Transparency, Virtual Event/Toronto, Canada, 3–10 March 2021, pp. 504–514. ACM (2021)
2. Barketau, M., Pesch, E., Shafransky, Y.: Minimizing maximum weight of subsets of a maximum matching in a bipartite graph. Discrete Appl. Math. **196**, 4–19 (2015)
3. Berger, A., Bonifaci, V., Grandoni, F., Schäfer, G.: Budgeted matching and budgeted matroid intersection via the gasoline puzzle. Math. Program. **128**(1–2), 355–372 (2011). https://doi.org/10.1007/s10107-009-0307-4

4. Chouldechova, A.: Fair prediction with disparate impact: a study of bias in recidivism prediction instruments. Big Data **5**(2), 153–163 (2017)
5. Corbett-Davies, S., Pierson, E., Feller, A., Goel, S., Huq, A.: Algorithmic decision making and the cost of fairness. In: Proceedings of the 23rd ACM SIGKDD International Conference on Knowledge Discovery and Data Mining, Halifax, NS, Canada, 13–17 August 2017, pp. 797–806. ACM (2017)
6. Duginov, O.: Weighted perfect matching with constraints on the total weight of its parts. J. Appl. Ind. Math. **15**(3), 393–412 (2021)
7. Dwork, C., Hardt, M., Pitassi, T., Reingold, O., Zemel, R.: Fairness through awareness. In: Proceedings of the 3rd Innovations in Theoretical Computer Science Conference, pp. 214–226 (2012)
8. Dwork, C., Ilvento, C.: Group fairness under composition. In: Proceedings of the 2018 Conference on Fairness, Accountability, and Transparency (FAT* 2018) (2018)
9. Edmonds, J.: Paths, trees, and flowers. Can. J. Math. **17**, 449–467 (1965)
10. Feldman, M., Friedler, S.A., Moeller, J., Scheidegger, C., Venkatasubramanian, S.: Certifying and removing disparate impact. In: Proceedings of the 21th ACM SIGKDD International Conference on Knowledge Discovery and Data Mining, pp. 259–268 (2015)
11. García-Soriano, D., Bonchi, F.: Fair-by-design matching. Data Min. Knowl. Disc. **34**(5), 1291–1335 (2020). https://doi.org/10.1007/s10618-020-00675-y
12. Ghadiri, M., Samadi, S., Vempala, S.S.: Socially fair k-means clustering. In: Elish, M.C., Isaac, W., Zemel, R.S. (eds.) FAccT 2021: 2021 ACM Conference on Fairness, Accountability, and Transparency, Virtual Event/Toronto, Canada, 3–10 March 2021, pp. 438–448. ACM (2021)
13. Huang, C., Kavitha, T., Mehlhorn, K., Michail, D.: Fair matchings and related problems. Algorithmica **74**(3), 1184–1203 (2016)
14. Kamada, Y., Kojima, F.: Fair matching under constraints: Theory and applications (2020)
15. Kamishima, T., Akaho, S., Sakuma, J.: Fairness-aware learning through regularization approach. In: Spiliopoulou, M., et al. (eds.) Data Mining Workshops (ICDMW), 2011 IEEE 11th International Conference on, Vancouver, BC, Canada, 11 December 2011, pp. 643–650. IEEE Computer Society (2011)
16. Karp, R.M.: Reducibility among combinatorial problems. In: Miller, R.E., Thatcher, J.W., Bohlinger, J.D. (eds.) Complexity of Computer Computations. The IBM Research Symposia Series, pp. 85–103. Springer, Boston (1972). https://doi.org/10.1007/978-1-4684-2001-2_9
17. Klaus, B., Klijn, F.: Procedurally fair and stable matching. Econ. Theory **27**(2), 431–447 (2006)
18. Kleinberg, J., Mullainathan, S., Raghavan, M.: Inherent trade-offs in the fair determination of risk scores. In: 8th Innovations in Theoretical Computer Science Conference (ITCS 2017). Schloss Dagstuhl-Leibniz-Zentrum fuer Informatik (2017)
19. Kress, D., Meiswinkel, S., Pesch, E.: The partitioning min-max weighted matching problem. Eur. J. Oper. Res. **247**(3), 745–754 (2015)
20. Makarychev, Y., Vakilian, A.: Approximation algorithms for socially fair clustering. CoRR abs/2103.02512 (2021)
21. Papadimitriou, C.H., Yannakakis, M.: On the approximability of trade-offs and optimal access of web sources. In: 41st Annual Symposium on Foundations of Computer Science, FOCS 2000, 12–14 November 2000, Redondo Beach, California, USA, pp. 86–92. IEEE Computer Society (2000). https://doi.org/10.1109/SFCS.2000.892068

22. Thanh, B.L., Ruggieri, S., Turini, F.: k-NN as an implementation of situation testing for discrimination discovery and prevention. In: Apté, C., Ghosh, J., Smyth, P. (eds.) Proceedings of the 17th ACM SIGKDD International Conference on Knowledge Discovery and Data Mining, San Diego, CA, USA, 21–24 August 2011, pp. 502–510. ACM (2011)
23. Wikipedia contributors: Sister city — Wikipedia, the free encyclopedia (2022). https://en.wikipedia.org/w/index.php?title=Sister_city&oldid=1107517947. Accessed 25 Sept 2022

A Parameterized Approximation Scheme for Generalized Partial Vertex Cover

Sayan Bandyapadhyay[1], Zachary Friggstad[2], and Ramin Mousavi[2](✉)

[1] Department of Computer Science, Portland State University, Portland, USA
sayanb@pdx.edu

[2] Department of Computing Science, University of Alberta, Edmonton, Canada
{zacharyf,mousavih}@ualberta.ca

Abstract. Partial Vertex Cover is a well-studied generalization of the classic Vertex Cover problem, where we are given a graph $G = (V, E)$ along with a non-negative integer k, and the goal is to cover the maximum number of edges possible by picking exactly k vertices. In this paper, we study a natural extension of Partial Vertex Cover to multiple color classes of the edges. In our problem, we are additionally given a partition of E into m color classes $E_1, E_2, ..., E_m$ and coverage requirements $c_i \geq 1$ for all $1 \leq i \leq m$. The goal is to find a subset of vertices of size k that covers at least $\beta \cdot c_i$ edges from each E_i and the contraction factor $\beta \leq 1$ is maximized.

As we prove in our paper, the multi-colored extension becomes very difficult to approximate in polynomial time to any reasonable factor. Consequently, we study the parameterized complexity of approximating this problem in terms of various parameters such as k and m. Our main result is a $(1-\epsilon)$-approximation for the problem that runs in time $f(k, m, \epsilon) \cdot \text{poly}(|V|)$ for some computable function f. As we argue, our result is tight, in the sense that it is not possible to remove the dependence on k or m from the running time of such a $(1-\epsilon)$-approximation.

Keywords: Partial vertex cover · Fixed parameter tractable · Approximation algorithms

1 Introduction

In the classic VERTEX COVER problem, we are given a graph and the goal is to select as few nodes as possible such that all edges of the graph are covered, where we say an edge is covered if at least one of its endpoints is selected. This problem is well-studied in both approximation algorithms and parameterized complexity literature. There are multiple folklore 2-approximations for this problem, and currently the best approximation ratio in polynomial time is $2 - \frac{1}{O(\sqrt{\log n})}$ [25]. On the hardness side, the problem is NP-hard to approximate within a factor better than 1.36 [16]. Also assuming UNIQUE GAMES CONJECTURE (UGC), there is no polynomial time α-approximation for the problem with $\alpha < 2$ [26].

P. Morin and S. Suri (Eds.): WADS 2023, LNCS 14079, pp. 93–105, 2023.
https://doi.org/10.1007/978-3-031-38906-1_7

Another way of coping with NP-hard problems is to study their *parameterized complexity*, where the running time is analyzed in terms of the input size as well as a parameter of the problem [15]. In parameterized complexity, *fixed-parameter tractability* (FPT) is a popular notion, which allows running time to be expressed as a function of a parameter. More precisely, an algorithm has FPT running time parameterized by p if it runs in time $f(p) \cdot |x|^{O(1)}$, where f is a function that solely depends on p, and $|x|$ is the size of the input.

In the parameterized version of VERTEX COVER, we are given a graph G with parameter k, and the goal is to decide whether G has a vertex cover of size at most k or not. Since the first FPT algorithm for this problem [9], several improved FPT algorithms have been designed culminating in the current fastest $(1.286^k + k \cdot n)$ time algorithm [14].

Several variants of VERTEX COVER have been studied in the literature, e.g., CAPACITATED VERTEX COVER [17,21] and CONNECTED VERTEX COVER [21, 29]. In this paper, we focus on another variant called PARTIAL VERTEX COVER. In an instance of PARTIAL VERTEX COVER, denoted by $(G = (V, E), k, c)$, we are given a graph $G = (V, E)$ with two non-negative integers k and c. In the decision version, the goal is to determine if there is a subset of vertices of size at most k that covers at least c many edges. Interestingly, two different optimization versions of PARTIAL VERTEX COVER have been studied. One is a minimization version where the coverage requirement c is treated as a hard constraint, which must be satisfied, and we would like to minimize the size of the subset of vertices that cover the c edges. The other is a maximization version where instead the size requirement k of the subset is a hard constraint, and we would like to maximize the number of edges covered by the subset. To distinguish between these two versions we use a suitable Min/Max prefix to the problem name.

MIN PARTIAL VERTEX COVER: This version admits polynomial-time 2-approximation [8], i.e., there is an algorithm that returns a subset of vertices of size at most $2 \cdot \text{opt}$ and covers at least c edges. Here opt is the minimum size of any subset of vertices that covers at least c edges. Note this problem is a generalization of vertex cover, where $c = |E|$. Hence, the hardness results for vertex cover carry over for this problem as well. It follows that the polynomial-time 2-approximation is asymptotically tight under UGC.

MAX PARTIAL VERTEX COVER: In contrast to the minimization version, this version is less understood. There are a simple greedy $(1 - \frac{1}{e})$-approximation [22] and an LP rounding based $\frac{3}{4}$-approximation [1]. In fact, the best-known approximation factor is a slightly larger $\frac{3}{4} + \epsilon$ for some absolute constant $\epsilon > 0$ [19]. On the other hand, the problem is NP-hard to approximate within a factor of $(1 - \delta)$ for some small constant $\delta > 0$ [30]. Furthermore, it is UGC-hard to approximate within a factor of 0.944, see Appendix A in [27]. Note that even under UGC, there is a large gap between the best-known lower and upper bounds.

PARTIAL VERTEX COVER has also been studied in the parameterized complexity literature. It is known that the problem is FPT in parameter c [7] and $W[1]$-hard when k is the parameter [20]. Circumventing this hardness in a seminal work, Marx [28] designed an *FPT approximation scheme* (FPT-AS) for Max Partial Vertex Cover, which runs in time $(\frac{k}{\epsilon})^{O(\frac{k^3}{\epsilon})} \cdot \text{poly}(n)$. That is the algorithm finds a subset of vertices of size at most k that covers at least $(1-\epsilon)\cdot$ opt many edges, where opt is the maximum coverage possible with k vertices. Recently, this has been improved to a faster FPT-AS with running time $O((\frac{1}{\epsilon})^k)\cdot \text{poly}(n)$ [27].

In this paper, we study a natural generalization of Partial Vertex Cover with multiple coverage requirements, which we refer to as Generalized Partial Vertex Cover. In an instance $(G=(V,E),k,c_1,...,c_m)$ of this problem, we are given a graph $G=(V,E)$, and the edge set E is partitioned into m color classes $E_1, E_2, ..., E_m$. Each E_i has an integer coverage requirement $c_i \geq 1$. The goal is to find a subset $S \subseteq V$ of size at most k such that for each $1 \leq i \leq m$, at least c_i edges from E_i are covered by S.

Generalized Partial Vertex Cover is a natural model for geometric covering in the presence of colorful points. Consider the situation where we have a collection of geometric objects in the plane. The collection can have either axis-parallel lines, line segments, or circular arcs where a pair of intersecting objects intersect only at a single point and no three objects intersect at a common point (see Fig. 1). We are also given points of m colors. The goal is to cover a target number of points from each class by selecting a set of k objects. It is possible to model this problem using Generalized Partial Vertex Cover by taking one vertex for each object and a colored edge for each pair of objects whose intersection point is a colored input point.

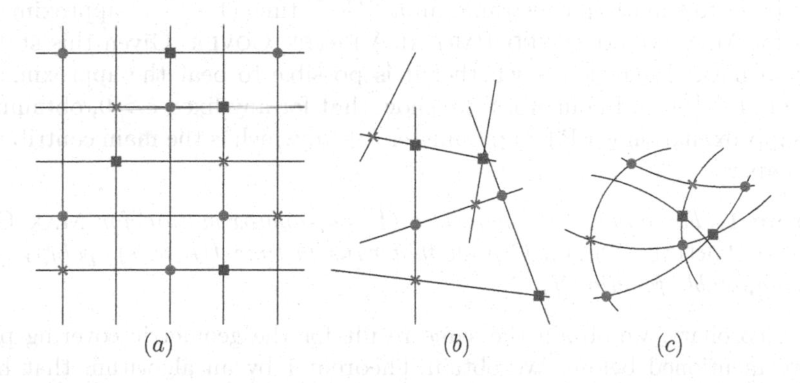

Fig. 1. Geometric covering of colorful points with (a) lines, (b) segments, and (c) arcs.

Covering problems with multiple coverage requirements have been studied before in the literature, e.g., colorful k-center [2,4,24] and generalized set cover [11,12,23]. Note that similar to Partial Vertex Cover, one can also consider two optimization versions of Generalized Partial Vertex Cover. Indeed, the minimization version, where we want to minimize the size of the vertex cover while satisfying all the coverage requirements, has been studied before from the perspective of approximation. Bera et al. [6] designed an $O(\log m)$-approximation (recall m is the number of color classes) for this version and proved the approximation factor to be asymptotically tight unless P = NP. Furthermore, there is a $(2+\epsilon)$-approximation with $O(n^{\frac{m}{\epsilon}})$ running time [3].

In this paper, we initiate the study of the maximization version of GENERALIZED PARTIAL VERTEX COVER. We treat the size of the solution set of vertices as a hard constraint, i.e., we need to output a subset of vertices of size at most k. Recall that for MAX PARTIAL VERTEX COVER, we would like to maximize the number of edges covered. However, here we have multiple color classes. Consequently, we define the problem in the following natural way. We assume that we are given the values c_i for $1 \leq i \leq m$ and there is a feasible subset of vertices in G of size at most k that covers at least c_i edges from each E_i. The goal is to find a subset of vertices of size at most k that covers at least $\beta \cdot c_i$ edges from each E_i and the contraction factor $\beta \leq 1$ is maximized.

Note that MAX PARTIAL VERTEX COVER is not exactly the same as MAX GENERALIZED PARTIAL VERTEX COVER with $m = 1$, as the parameter c_1 is not present in the former. However, MAX GENERALIZED PARTIAL VERTEX COVER with $m = 1$ is a strict generalization of MAX PARTIAL VERTEX COVER, in the sense that if we can solve the former, we can also solve the latter by trying out different values of c_1. This also holds in terms of approximate solutions. Hence, similar to MAX PARTIAL VERTEX COVER, our problem must also be APX-hard. Moreover, as we show in Sect. 3, for an unbounded number of colors, it is not possible to obtain any approximation in polynomial time for this generalization. Due to the work of Chekuri et al. [13] on maximizing monotone submodular functions subject to a matroid constraint, an $n^{O(m/\epsilon^3)}$ time $(1-\frac{1}{e}-\epsilon)$-approximation follows for MAX GENERALIZED PARTIAL VERTEX COVER. Given this state of the art, a natural question is whether it is possible to beat the approximation bound of $(1-\frac{1}{e}-\epsilon)$. In our work, we show that for any fixed $\epsilon > 0$, obtaining a $(1-\epsilon)$-approximation is FPT in parameter $k+m$, which is the main contribution of our paper.

Theorem 1. *For any $\epsilon > 0$, there is a $(1-\epsilon)$-approximation for* MAX GENERALIZED PARTIAL VERTEX COVER *that runs in time $f(k, m, \epsilon) \cdot poly(|V|)$ for some computable function f.*

As a corollary, we obtain the same result for the geometric covering problems we mentioned before. We obtain Theorem 1 by an algorithm that helps us approximately decide the decision version of MAX GENERALIZED PARTIAL VERTEX COVER. Here, we are simply to decide if there is a GENERALIZED PARTIAL VERTEX COVER solution that covers at least c_i edges from each color class i.

Theorem 2. *For any $\epsilon > 0$, there is an $2^{O(\frac{m \cdot k^2 \cdot \log k}{\epsilon})} \cdot n^{O(1)}$ time algorithm for* GENERALIZED PARTIAL VERTEX COVER *that either returns `FAIL` or a subset $S \subseteq V$ of k nodes that covers $\geq (1-\epsilon) \cdot c_i$ edges for each color class i. If there is a feasible* GENERALIZED PARTIAL VERTEX COVER *solution, the algorithm does not return `FAIL`.*

Theorem 1 simply follows by using a binary search to compute the maximum β such that the algorithm from Theorem 2 does not return FAIL on the instance with coverage requirements $\lceil \beta \cdot c_i \rceil$ for each color class i.

The proof of Theorem 2 borrows many ideas from the special case of $m = 1$ from [28]. However, there is a simple greedy step in [28] that does not generalize to our setting, we replace this step with a dynamic programming algorithm that computes approximate coverages.

Intuitively, if c_i is small (i.e., at most $\binom{k}{2}/\epsilon$) for each color class i then a random labeling scheme can be used just like in [28]. If c_i is large for each color class i, we compute a set of nodes S that has a total degree of almost c_i in each color class i. Since c_i is large and the graph is simple, having a total degree of almost c_i means almost c_i edges are in fact covered since the double-counting for edges with both endpoints in S is limited. We note that if we only have big color classes then one can view this problem as a restricted version of the Multi-Objective Knapsack problem, which admits FPTASes [5,18]. Multi-Objective Knapsack is an extension of 0–1 Knapsack where instead of one, m profit functions are given for each item. There are two obstacles in applying their FPTASes as a black-box to our problem: (i) we need to incorporate the small color classes as well, and (ii) the running time of these algorithms are XP in m.

We use dynamic programming to address the above issues. However running time of a straightforward dynamic programming algorithm would involve the term $|E|^m$, for example if the DP table parameters guess how much of the total degree coverage in each color class comes from the first j nodes. We consider an alternative dynamic programming algorithm that only tracks the total degree in each large color class (i.e., $c_i \geq \binom{k}{2}/\epsilon$) approximately in order to get the desired running time. We also show how one can combine our dynamic programming approach for the large color classes with the approach for small color classes to get a single algorithm that works for instances that have a mix of large and small color classes.

Note that as Max Partial Vertex Cover is APX-hard and it is a special case of Max Generalized Partial Vertex Cover with $m = 1$, the time bound of the FPT-AS in Theorem 1 cannot be improved to $f(m, \epsilon) \cdot \text{poly}(|V|)$ by removing the dependence on k. Moreover, in Sect. 3, we prove that for any $\gamma \leq 1$, there is no γ-approximation for Max Generalized Partial Vertex Cover that runs in $f(k) \cdot |V|^{o(k)}$ time, assuming the Exponential Time Hypothesis (ETH). Hence, the time bound of the FPT-AS in Theorem 1 cannot be improved to $f(k, \epsilon) \cdot \text{poly}(|V|)$ either, by removing the dependence on m. Moreover, Generalized Partial Vertex Cover is W[1]-hard with parameter $k + m$, as Partial Vertex Cover is W[1]-hard in k. Hence, there is no exact algorithm for these multi-coverage extensions in $f(k, m) \cdot \text{poly}(|V|)$ time, and our algorithmic result can be considered to be tight. See Table 1 for a comprehensive summary of our results.

Table 1. A summary of results on MAX GENERALIZED PARTIAL VERTEX COVER in terms of different parameters of the problem.

Parameter	Result
k	$\forall \gamma \leq 1$, γ-inapproximable in time $f(k) \cdot n^{o(k)}$ (Theorem 3)
	Exact in time $n^{O(k)}$ (Trivial)
m	APX-hard for $m = 1$
	$(1 - \frac{1}{e} - \epsilon)$-approximation in time $n^{O(m/\epsilon^3)}$ [13]
$k + m$	No exact algorithm in time $f(k, m) \cdot \text{poly}(\lvert V \rvert)$
	$(1 - \epsilon)$-approximation in time $f(k, m, \epsilon) \cdot \text{poly}(\lvert V \rvert)$ (Theorem 1)

2 Approximation Scheme for GENERALIZED PARTIAL VERTEX COVER

2.1 Overview and Preliminaries

Let $G = (V, E)$ be a simple, undirected graph whose edges are partitioned into m color classes $E_1, E_2, \ldots, E_m$. Each E_i has a coverage requirement of $c_i \geq 1$. Finally, let k be the bound on the size of a GENERALIZED PARTIAL VERTEX COVER solution. The goal is to find a subset $S \subseteq V, |S| \leq k$ such that for each color class i, the number of edges in E_i with at least one endpoint in S is at least c_i. That is, S *covers* at least c_i edges from E_i for each $1 \leq i \leq m$.

The algorithm in [28] for the case $m = 1$ essentially does one of two things. If c_1 is sufficiently large, i.e. $c_1 \geq \binom{k}{2}/\epsilon$ then it simply picks the k highest-degree vertices. The total degree of these vertices is $\geq c_1$ so, recalling the graph is simple, the number of edges covered is at least $c_1 - \binom{k}{2} \geq (1 - \epsilon) \cdot c_1$.

Otherwise, if $c_1 < \binom{k}{2}$ is small then the approach in [28] solves the problem exactly with a randomized algorithm (that is easy to derandomize) via random label coding[1]: each edge in E is labelled randomly with a label from $\{1, 2, \ldots, c_1\}$. Supposing there is a partial vertex cover S covering some set of edges F with $|F| = c_1$, the probability all edges in F receive different labels is bounded by a function of k and ϵ. Then each vertex in the optimal solution S sees a particular subset of labels. If one guesses these subsets of labels for each of the k vertices in S, it is simply a matter of checking if, for each subset A of these labels in our guess, the number of vertices in V that sees A (plus, perhaps, more labels) is at least the number of times we used subset A in our guess.

Call a color class i *large* if $c_i \geq \binom{k}{2}/\epsilon$, and *small* otherwise. Many ideas from [28] extend to our setting with one notable exception: we cannot just greedily pick the highest-degree vertices when there are multiple color classes. We still have that for large color class i, it suffices to ensure that the set S we pick has

[1] We use the term label coding rather than the traditional term "color coding" so that there is no confusion between this coding step and the colors of the edges in the problem GENERALIZED PARTIAL VERTEX COVER we will address.

total degree at least c_i so that the number of edges that are double-counted is at most $\epsilon \cdot c_i$.

If one views each vertex v as having an m-dimensional vector $x^v \in \mathbb{Z}^m$ where $x_i^v = |\delta(v) \cap E_i|$, then if $c_i \geq \binom{k}{2}/\epsilon$ for each color class i we just need to pick k vectors of this form whose sum is $\geq (c_1, c_2, \ldots, c_m)$ component-wise. This can be viewed as the MULTI-OBJECTIVE KNAPSACK problem; however, the known algorithms for MULTI-OBJECTIVE KNAPSACK involves $O(|V|^m)$ term in the running time which is not FPT in m. So we cannot use these algorithms as black-box.

As we mentioned before, a natural DP algorithm for our problem is also too slow. Our main idea is a variation of the standard DP algorithm: the entries of the DP table, which are meant to track how many edges from each color class have been covered by the subproblem, are only tracked approximately for large color classes and exactly for small color classes.

In order to track the coverage values approximately, for each v and each large color class i, we round $|\delta(v) \cap E_i|$ to the nearest integer multiple of $\frac{\epsilon \cdot c_i}{k}$. Then, for a vertex, the number of different possible coverage values needed to be tracked, reduces from c_i to $\frac{k}{\epsilon}$. The rounding error for large color class i and a fixed vertex is at most $\frac{\epsilon \cdot c_i}{k}$; together with the fact that we only choose k vertices, the total error for color class i is at most $\epsilon \cdot c_i$, as desired.

2.2 The Algorithm

Assume, without loss of generality, $c_1 \geq c_2 \geq \ldots \geq c_m$. Recall we say a color class i is *large* if $c_i > \binom{k}{2}/\epsilon$, otherwise color class i is *small.* Let $0 \leq \ell \leq m$ be the number of large color classes, so $c_1, \ldots, c_\ell \geq \binom{k}{2}/\epsilon$ and $c_{\ell+1}, \ldots, c_m < \binom{k}{2}/\epsilon$. Throughout, we let $\delta(v)$ be the set of edges in E having v as an endpoint. Our final algorithm will return one of two things:

- a declaration that there is no set $S \subseteq V, |S| \leq k$ such that $|\{(u,v) \in E_i : S \cap \{u,v\} \neq \emptyset\}| \geq c_i$ for each $1 \leq i \leq m$, or
- a set $S \subseteq V, |S| \leq k$ such that $|\{(u,v) \in E_i : S \cap \{u,v\} \neq \emptyset\}| \geq (1-\epsilon) \cdot c_i$ for large color classes i and $|\{(u,v) \in E_i : S \cap \{u,v\} \neq \emptyset\}| \geq c_i$ for small color classes i.

In our analysis, we will assume there is indeed a set $S^* \subseteq V$ with $|S^*| = k$ that covers at least c_i edges from each color class E_i. We fix one such subset and, further, order its nodes arbitrarily so $S^* = \{v_1^*, v_2^*, \ldots, v_k^*\}$. For each $1 \leq i \leq m$, let F_i be any set of exactly c_i edges covered by S^*. A concise summary of our algorithm is found at the end of this section in Algorithm 1.

Label Coding. We take care of the standard steps first: random label coding and guessing the subsets of labels seen by the vertices in S^*.

Step 1 - Label Coding
For each small color class i and each $e \in E_i$, sample a label $\chi_i(e) \in \{(i,1),(i,2),\ldots,(i,c_i)\}$ uniformly at random for e. Do this independently for each edge in a small color class. Note, we are using a different set of labels for each color class.

Lemma 1. *The probability each* $e \in \cup_{i=\ell+1}^m F_i$ *receives a different label is at least* $e^{-m\cdot k^2/\epsilon}$.

Proof. Out of the $c_i^{c_i}$ labellings for F_i, exactly $c_i!$ of them assign different labels to each edge in F_i. A weak form of Stirling's approximation shows $c_i! > (c_i/e)^{c_i}$. So the probability each edge of F_i receives a different label is at least $(c_i/e)^{c_i}/(c_i^{c_i}) = e^{-c_i} \geq e^{-k^2/\epsilon}$. By independence when we assigned labels, the probability that all edges in F_i receive different labels for all small color classes i is at least $e^{-m\cdot k^2/\epsilon}$.

This is the only randomized step in our entire algorithm. If we repeat our algorithm $\log n \cdot e^{m\cdot k^2/\epsilon}$ times, then with probability $\geq 1 - 1/n$ at least one iteration will have each $e \in \cup_{i=\ell+1}^m F_i$ receiving different colors. When this happens, our algorithm is guaranteed to succeed. Alternatively, this step can be entirely derandomized using families of perfect hash functions. That is, for each small color class i we consider a $(|V|, c_i)$-perfect hash family $\mathcal{H}_i$. By trying all $(m-\ell)$-tuples of functions from $\mathcal{H}_{\ell+1} \times \cdots \times \mathcal{H}_m$, we can construct a family of perfect hash functions. The number of such tuples is bounded by $O(f(k,m,\epsilon)\cdot \log^m |V|)$ for some function $f()$ which can be seen to be FPT in k, m, ϵ (see the running time discussion at the end of this section). See Chap. 5.6 of [15] for more information on perfect hash families in the context of FPT algorithms.

Step 2 - Guessing Subsets of Labels
Let $\Lambda = \cup_{i=\ell+1}^m \{(i,1),(i,2),\ldots,(i,c_i)\}$ be the set of all possible labels used so far. Let $L_v = \cup_{i=\ell+1}^m \{\chi_i(e) : e \in E_i \cap \delta(v)\}$ be the set of labels of edges (from small color classes) that are incident to v. If we are able to find k nodes $S \subseteq V$ such that $\cup_{v\in S} L_v = \Lambda$, then S will have covered at least c_i edges from each small color class i.

Now, $|\Lambda| \leq m \cdot \binom{k}{2}/\epsilon$. So we "guess" label subsets $L^1, L^2, \ldots, L^k \subseteq \Lambda$ for the k nodes in the optimum, i.e. in a correct guess L^j will be the set of different edge labels for edges incident to $v_j^* \in S$. There are at most $k^{|\Lambda|}$ ways to partition Λ into at most k sets. We reject a guess if $\cup_{j=1}^k L^j \neq \Lambda$. From this point on, we assume the guess for these L^j labels is consistent with the optimum solution. That is, $L^j = L_{v_j^*}$.

2.3 Approximate Dynamic Programming

We assume Step 2 above (guessing subsets of labels for small color classes) succeeded and, further, that $L^j, j \in \{1,\ldots,k\}$ are correct guesses for the labels of edges in small color classes covered by the nodes in a feasible partial vertex cover solution.

Below, we present a dynamic programming algorithm that runs in FPT time parameterized by k, m and $1/\epsilon$.

For each $v \in V$ and large color class $1 \leq i \leq \ell$, define $d'_{v,i} = \left\lceil \frac{k}{\epsilon} \cdot \frac{\min\{|\delta(v) \cap E_i|, c_i\}}{c_i} \right\rceil$. That is, $d'_{v,i}$ is a scaled version of the degree of v in color class i so that it is an integer at most $\lceil k/\epsilon \rceil$.

Lemma 2. *If there is a feasible partial vertex cover solution S, then for each $1 \leq i \leq \ell$ we have $\sum_{v \in S} d'_{v,i} \geq \lceil \frac{k}{\epsilon} \rceil$. Conversely, if there is a subset $S \subseteq V$ with $|S| = k$ such that for each $1 \leq i \leq \ell$ we have $\sum_{v \in S} d'_{v,i} \geq \lceil k/\epsilon \rceil$ and also that each $L \subseteq \Lambda$ appears as many times in the form $L_v, v \in S$ as it does of the form $L^j, j \in \{1, \ldots, k\}$, then S covers at least $(1 - 2 \cdot \epsilon) \cdot c_i$ edges for each large color class $1 \leq i \leq \ell$ and at least c_i edges for small color class $\ell < i \leq m$.*

Proof. For the first statement, let S be a feasible partial vertex cover solution. Then for each $1 \leq i \leq \ell$ we have $\sum_{v \in S} d'_{v,i} \geq \frac{k}{\epsilon \cdot c_i} \sum_{v \in S} |\delta(v) \cap E_i| \geq \frac{k}{\epsilon}$, with the last bound following from the fact that S covers at least c_i edges in E_i, so the total degree of its nodes in color class i should also be at least c_i. Since the left-hand side of the expression is an integer, then it is in fact at least $\lceil k/\epsilon \rceil$, as required.

For the second part, consider such a set S. For each $1 \leq i \leq \ell$ we note $\frac{k}{\epsilon} < \sum_{v \in S} d'_{v,i} \leq k + \sum_{v \in S} \frac{k}{\epsilon} \cdot \frac{|\delta(v) \cap E_i|}{c_i} - k + \frac{k}{\epsilon \cdot c_i} \sum_{v \in S} |\delta(v) \cap E_i|$. Rearranging shows $c_i \cdot (1 - \epsilon) \leq \sum_{v \in S} |\delta(v) \cap E_i|$. As noted in [28], the actual number of edges covered by color class i is at least $\sum_{v \in S} |\delta(v) \cap E_i| - \binom{k}{2}$. That is, since G is a simple graph and $|S| = k$, then there are at most $\binom{k}{2}$ edges double-counted by $\sum_{v \in S} |\delta(v) \cap E_i|$. So S covers at least $c_i \cdot (1 - \epsilon) - \binom{k}{2}$ edges in E_i, which is at least $c_i \cdot (1 - 2 \cdot \epsilon)$ because this is a large color class, i.e. $c_i \geq \binom{k}{2}/\epsilon$.

Also, since each subset of labels $L \subseteq \Lambda$ appears as many times of the form $L_v, v \in S$ as it does of the form $L^j, j \in \{1, \ldots, k\}$. As $L^1, L^2, \ldots, L^k$ are presumed to be correct guesses for the subsets of labels covered by the k vertices in a feasible solution, we have that S also covers c_i edges from each small color class $\ell < i \leq m$.

We now present our recurrence. Order the nodes arbitrarily as $v_1, v_2, \ldots, v_n$. For each $0 \leq k' \leq k$, each $0 \leq j \leq n$, each $T \subseteq \{1, \ldots, k\}$ and each $c' \in \{0, 1, \ldots, \lceil k/\epsilon \rceil\}^\ell$ we let $f(k', j, T, c')$ be a boolean value that is true if and only if there is some $S' \subseteq \{v_1, v_2, \ldots, v_j\}$ with $|S'| = k'$ such that $\sum_{v \in S'} d'_{v,i} \geq c'_i$ for each $1 \leq i \leq \ell$ and each $L \in \Lambda$ appears as many times of the form $L_v, v \in S'$ as it does $L^j, j \in T$. The base cases are when $j = 0$ (i.e. we may not select any vertices), in which case $f(k', 0, T, c')$ will be `True` if $k' = 0, T = \emptyset$ and $c' = (0, 0, \ldots, 0)$, and `False` otherwise.

For $1 \leq j \leq n$, the choice is whether v_j should be included or not. If either $k' = 0$ or if L_{v_j} is not of the form $L^{j'}$ for any $j' \in T$, then we do not consider it and simply have:

$$f(k', j, T, c') = f(k', j - 1, T, c')$$

Otherwise, pick any $j' \in T$ such that $L_v = L^{j'}$. In this case, we have:

$$f(k', j, T, c') = f(k', j-1, T, c') \bigvee f(k'-1, j-1, T-\{j'\}, (\max\{c'_i - d'_{v_j,i}, 0\})_{1 \le i \le \ell}).$$

A straightforward proof by induction shows that this calculation is correct, i.e., $f(k', j, T, c')$ is True if and only if at least one of the values on the right-hand side of the recurrence is True.

The number of different parameters is $O(k \cdot n \cdot 2^k \cdot (k/\epsilon)^\ell)$, each step only makes at most two recursive calls, and each step can be executed in polynomial time (eg. checking if L_{v_j} is among $\{L^{j'}, j' \in T\}$, updating the c' for the recurrence, etc.). The corresponding sets S witnessing that $f(k', j, T, c') =$ True can also be computed along with the recurrence in a straightforward manner.

2.4 Putting It All Together

Algorithm 1 summarizes the steps in the randomized version of our algorithm. The value $f()$ referenced in the algorithm is described in Sect. 2.3 and should be computed using dynamic programming.

Algorithm 1 Randomized GENERALIZED PARTIAL VERTEX COVER Approximation Scheme

Randomly sample $\chi_i(e) \in \{(i,1), \ldots, (i, c_i)\}$ for each $\ell+1 \le i \le m$ and each $e \in E_i$
$\Lambda \leftarrow \cup_{i=\ell+1}^{m} \{(i,1), \ldots, (i, c_i)\}$
for each k-tuple of subsets $L^1, L^2, \ldots, L^k \subseteq \Lambda$ with $\cup_{j=1}^{k} L^j = \Lambda$ **do**
  if $f(k, n, \{1, \ldots, k\}, (\lceil k/\epsilon \rceil)_{1 \le i \le \ell}) =$ TRUE and the corresponding set S covers at least $(1 - 2 \cdot \epsilon) \cdot c_i$ edges from E_i for each $1 \le i \le \ell$ and c_i edges from E_i for each $\ell < i \le m$ **then**
    return S
return FAIL

From Lemma 1, we can repeat the algorithm $\log n \cdot e^{m \cdot k^2/\epsilon}$ times to get a high probability of success. Putting this all together, we have a randomized algorithm that runs in time $O\big(\log n \cdot e^{\frac{m \cdot k^2}{\epsilon}} \cdot k^{\frac{m \cdot k^2}{\epsilon}} \cdot k \cdot n \cdot 2^k \cdot (\frac{k}{\epsilon})^m \cdot n^{O(1)}\big)$ which can be simplified to $2^{O(\frac{m \cdot k^2 \cdot \log k}{\epsilon})} \cdot n^{O(1)}$.

With high probability in the **yes** case, it returns a set $S \subseteq V, |S| \le k$ that covers at least $(1 - 2 \cdot \epsilon) \cdot c_i$ edges from each large color class i and covers at least c_i edges from each small color class. The algorithm can be derandomized using a $|\Lambda|$-wise independent family of random variables, completing the proof of Theorem 2.

3 Hardness of Approximation

In this section, we prove the following theorem.

Theorem 3. *For any $\gamma \leq 1$, there is no γ-approximation for* Max Generalized Partial Vertex Cover *that runs in $f(k) \cdot n^{o(k)}$ time, assuming the Exponential Time Hypothesis (ETH).*

Proof. To prove this theorem, we show a gap reduction from Set Cover to Generalized Partial Vertex Cover. In Set Cover, we are given a ground set A of elements and a family $\mathcal{F}$ of subsets of the ground set. We are also given an integer $k > 0$. The goal is to decide whether there are k subsets in $\mathcal{F}$ whose union contains all the elements in A. Assuming the Exponential Time Hypothesis (ETH), Set Cover cannot be solved in $f(k) \cdot n^{o(k)}$ time, where $n = |\mathcal{F}| + |A|$ [10].

We construct an instance $I' = (G = (V, E), k', c_1, c_2, ..., c_m)$ of Generalized Partial Vertex Cover from the given instance I of Set Cover. V contains a vertex for every subset in $\mathcal{F}$ and a vertex for every element in A. A subset is connected to an element in G with an edge if and only if it contains the element. Note that G is a bipartite graph with bi-partition $(\mathcal{F}, A)$. Moreover, we set $k' = k$, $m = |A|$ and $c_i = 1$ for all i. Finally, for each $1 \leq i \leq |A|$, the edges incident on the i-th element form the edge set E_i.

Now, if there is a Set Cover of size k, we pick the corresponding k subset vertices. As each element is covered by the Set Cover, there must be an edge in G that connects each element to one of these k subset vertices. It follows that the coverage requirements are satisfied exactly in this case.

On the other hand, suppose there is no Set Cover of size k. We claim that for any $\gamma \leq 1$, there do not exist k vertices in G which cover at least $\gamma \cdot c_i$ edges of E_i for all i. The claim proves an inapproximability gap of γ. Suppose the claim is not true. Then there exist a set S of k vertices in G which cover at least $\gamma \cdot c_i$ edges of E_i. As $c_i = 1$ and $\gamma \leq 1$, at least $\lceil \gamma \rceil = 1$ edge must be covered from each color class. Also, wlog we can assume that S contains only subset vertices. If it contains an element vertex, we can replace it with a subset vertex without decreasing the coverage. But, this gives us k subsets that cover all the elements in A contradicting our assumption. Hence, the claim is true.

Now, as Set Cover cannot be solved in $f(k) \cdot n^{o(k)}$ time and $k' = k$, the theorem follows.

Acknowledgment. We thank anonymous reviewers for pointing out a simpler dynamic programming approach.

References

1. Ageev, A.A., Sviridenko, M.I.: Approximation algorithms for maximum coverage and max cut with given sizes of parts. In: Cornuéjols, G., Burkard, R.E., Woeginger, G.J. (eds.) IPCO 1999. LNCS, vol. 1610, pp. 17–30. Springer, Heidelberg (1999). https://doi.org/10.1007/3-540-48777-8_2
2. Anegg, G., Angelidakis, H., Kurpisz, A., Zenklusen, R.: A technique for obtaining true approximations for k-center with covering constraints. In: Bienstock, D., Zambelli, G. (eds.) IPCO 2020. LNCS, vol. 12125, pp. 52–65. Springer, Cham (2020). https://doi.org/10.1007/978-3-030-45771-6_5

3. Bandyapadhyay, S., Banik, A., Bhore, S.: On fair covering and hitting problems. In: Kowalik, Ł, Pilipczuk, M., Rzążewski, P. (eds.) WG 2021. LNCS, vol. 12911, pp. 39–51. Springer, Cham (2021). https://doi.org/10.1007/978-3-030-86838-3_4
4. Bandyapadhyay, S., Inamdar, T., Pai, S., Varadarajan, K.: A constant approximation for colorful k-center. In: 27th Annual European Symposium on Algorithms (ESA 2019). Schloss Dagstuhl-Leibniz-Zentrum fuer Informatik (2019)
5. Bazgan, C., Hugot, H., Vanderpooten, D.: Implementing an efficient fptas for the 0–1 multi-objective knapsack problem. Eur. J. Oper. Res. **198**(1), 47–56 (2009)
6. Bera, S.K., Gupta, S., Kumar, A., Roy, S.: Approximation algorithms for the partition vertex cover problem. Theor. Comput. Sci. **555**, 2–8 (2014)
7. Bläser, M.: Computing small partial coverings. Inf. Process. Lett. **85**(6), 327–331 (2003)
8. Bshouty, N.H., Burroughs, L.: Massaging a linear programming solution to give a 2-approximation for a generalization of the vertex cover problem. In: Morvan, M., Meinel, C., Krob, D. (eds.) STACS 1998. LNCS, vol. 1373, pp. 298–308. Springer, Heidelberg (1998). https://doi.org/10.1007/BFb0028569
9. Buss, J.F., Goldsmith, J.: Nondeterminism within p*. SIAM J. Comput. **22**(3), 560–572 (1993)
10. Karthik, C.S., Laekhanukit, B., Manurangsi, P.: On the parameterized complexity of approximating dominating set. J. ACM **66**(5), 33:1–33:38 (2019). https://doi.org/10.1145/3325116
11. Chekuri, C., Inamdar, T., Quanrud, K., Varadarajan, K., Zhang, Z.: Algorithms for covering multiple submodular constraints and applications. J. Comb. Optim. **44**(2), 979–1010 (2022)
12. Chekuri, C., Quanrud, K., Zhang, Z.: On approximating partial set cover and generalizations. arXiv preprint arXiv:1907.04413 (2019)
13. Chekuri, C., Vondrák, J., Zenklusen, R.: Dependent randomized rounding for matroid polytopes and applications. arXiv preprint arXiv:0909.4348 (2009)
14. Chen, J., Kanj, I.A., Xia, G.: Simplicity is beauty: improved upper bounds for vertex cover. Manuscript communicated by email (2005)
15. Cygan, M., et al.: Parameterized Algorithms. Springer, Cham (2015). https://doi.org/10.1007/978-3-319-21275-3
16. Dinur, I., Safra, S.: The importance of being biased. In: Proceedings of the Thiry-fourth Annual ACM Symposium on Theory of Computing, pp. 33–42 (2002)
17. Dom, M., Lokshtanov, D., Saurabh, S., Villanger, Y.: Capacitated domination and covering: a parameterized perspective. In: Grohe, M., Niedermeier, R. (eds.) IWPEC 2008. LNCS, vol. 5018, pp. 78–90. Springer, Heidelberg (2008). https://doi.org/10.1007/978-3-540-79723-4_9
18. Erlebach, T., Kellerer, H., Pferschy, U.: Approximating multiobjective knapsack problems. Manage. Sci. **48**(12), 1603–1612 (2002)
19. Feige, U., Langberg, M.: Approximation algorithms for maximization problems arising in graph partitioning. J. Algorithms **41**(2), 174–211 (2001)
20. Guo, J., Niedermeier, R., Wernicke, S.: Parameterized complexity of generalized vertex cover problems. In: Dehne, F., López-Ortiz, A., Sack, J.-R. (eds.) WADS 2005. LNCS, vol. 3608, pp. 36–48. Springer, Heidelberg (2005). https://doi.org/10.1007/11534273_5
21. Guo, J., Niedermeier, R., Wernicke, S.: Parameterized complexity of vertex cover variants. Theory Comput. Syst. **41**(3), 501–520 (2007)
22. Hochba, D.S.: Approximation algorithms for np-hard problems. ACM Sigact News **28**(2), 40–52 (1997)

23. Inamdar, T., Varadarajan, K.: On the partition set cover problem. arXiv preprint arXiv:1809.06506 (2018)
24. Jia, X., Sheth, K., Svensson, O.: Fair colorful k-center clustering. Math. Program. **192**, 339–360 (2022)
25. Karakostas, G.: A better approximation ratio for the vertex cover problem. ACM Trans. Algorithms (TALG) **5**(4), 1–8 (2009)
26. Khot, S., Regev, O.: Vertex cover might be hard to approximate to within 2- ε. J. Comput. Syst. Sci. **74**(3), 335–349 (2008)
27. Manurangsi, P.: A note on max k-vertex cover: faster FPT-AS, smaller approximate kernel and improved approximation. arXiv preprint arXiv:1810.03792 (2018)
28. Marx, D.: Parameterized complexity and approximation algorithms. Comput. J. **51**(1), 60–78 (2008)
29. Mölle, D., Richter, S., Rossmanith, P.: Enumerate and expand: improved algorithms for connected vertex cover and tree cover. Theory Comput. Syst. **43**(2), 234–253 (2008)
30. Petrank, E.: The hardness of approximation: gap location. Comput. Complex. **4**(2), 133–157 (1994)

Dominator Coloring and CD Coloring in Almost Cluster Graphs

Aritra Banik[1(✉)], Prahlad Narasimhan Kasthurirangan[2(✉)], and Venkatesh Raman[3(✉)]

[1] National Institute of Science Education and Research, HBNI, Bhubaneswar, India
arita@niser.ac.in

[2] Department of Applied Mathematics and Statistics, Stony Brook University, Stony Brook, NY, USA
prahladnarasim.kasthurirangan@stonybrook.edu

[3] The Institute of Mathematical Sciences, HBNI, Chennai, India
vraman@imsc.res.in

Abstract. In this paper, we study two popular variants of Graph Coloring – Dominator Coloring and Class Domination Coloring. In both problems, we are given a graph G and a $\ell \in \mathbb{N}$ as input and the goal is to properly color the vertices with at most ℓ colors with specific constraints. In Dominator Coloring, we require for each $v \in V(G)$, a color c such that v dominates all vertices colored c. In Class Domination Coloring, we require for each color c, a $v \in V(G)$ which dominates all vertices colored c. These problems, defined due to their applications in social and genetic networks, have been studied extensively in the last 15 years. While it is known that both problems are fixed-parameter tractable (FPT) when parameterized by (t, ℓ) where t is the treewidth of G, we consider strictly structural parameterizations which naturally arise out of the problems' applications.

We prove that Dominator Coloring is FPT when parameterized by the size of a graph's *cluster vertex deletion* (CVD) set and that Class Domination Coloring is FPT parameterized by CVD set size plus the number of remaining cliques. En route, we design a simpler and faster FPT algorithms when the problems are parameterized by the size of a graph's *twin cover*, a special CVD set. When the parameter is the size of a graph's *clique modulator*, we design a randomized single-exponential time algorithm for the problems. These algorithms use an inclusion-exclusion based polynomial sieving technique and add to the growing number of applications using this powerful algebraic technique.

1 Introduction

Graphs motivated by applications in bio-informatics, social networks, and machine learning regularly define edges between data points based on some notion of similarity. As a consequence, we are often interested in how "close" a given graph is to a (special type of) *cluster graph* – a graph where every component is a clique. A popular measure of this "closeness" is the *cluster editing distance*. A graph G has cluster-editing distance k if it is the smallest number

P. Morin and S. Suri (Eds.): WADS 2023, LNCS 14079, pp. 106–119, 2023.
https://doi.org/10.1007/978-3-031-38906-1_8

such that there exists a set of k edges whose addition to or deletion from G results in a cluster graph. As an introduction to the extensive literature surrounding this parameter, we refer the reader to [6,7,19,25].

Another popular parameter of this type is the *cluster vertex deletion set size* (CVD set size) [16,24,28,32]. A CVD set in a graph is a subset of vertices whose deletion leaves a cluster graph. The CVD set size of a graph is the size of a smallest sized CVD set. Note that a graph with cluster-editing distance k has a CVD set of size $2k$. Thus, CVD set size is a smaller parameter than cluster-editing distance.

In this paper we use CVD set size to study the (parameterized) complexity of two variants GRAPH COLORING – DOMINATOR COLORING and CLASS DOMINATION COLORING. A *coloring* of a graph G is a function $\chi\colon V(G) \rightarrow C$, where C is a set of *colors*. A *proper coloring* of G is a coloring of G such that $\chi(u) \neq \chi(v)$ for all $(u, v) \in E(G)$. The set of all vertices which are colored c, for a $c \in C$, is called the *color class* c. We sometimes refer to the color c itself as a color class. We let $|\chi|$ denote $|Im(\chi)|$, the size of the image of χ. A vertex $v \in V(G)$ *dominates* $S \subseteq V(G)$ if $S \subseteq N_G[v]$. A *dominator coloring* χ of G is a proper coloring of the graph such that for all $v \in V(G)$, v dominates a color class $c \in Im(\chi)$. A *CD coloring* χ of G is a proper coloring of the graph such that for all $c \in Im(\chi)$ there exist a $v \in V(G)$ such that v dominates all vertices in the color class c. We are now ready to define our problems of interest.

DOMINATOR COLORING (DOMCOL)
Input: A graph G; an integer ℓ
Question: Does there exist a dominator coloring χ of G with $|\chi| \leq \ell$?

CLASS DOMINATION COLORING (CD COLORING)
Input: A graph G; an integer ℓ
Question: Does there exist a CD coloring χ of G with $|\chi| \leq \ell$?

We use (G, ℓ) to denote an instance of both these problems since it will be clear from context which problem we are referring to. While both problems have a rich theoretical history (see Sect. 1.1), CD COLORING has garnered renewed interest due to its practical applications in social networks and genetic networks – the problem is equivalent to finding the minimum number of (i) *stranger groups* with a common *friend* in social network graphs [11]; and (ii) *gene groups* that do not directly regulate each other but are regulated by a common gene in genetic networks [30]. Most notations used in this paper are standard and are referred from textbooks such as [14,15,20]. The full version [3] contains formal definitions.

1.1 Related Work

DOMCOL was introduced by Gera *et al.* in 2006 [23] while CD COLORING was introduced by Merouane *et al.* in 2012 [34] (the problem was termed DOMINATED

Coloring here). These papers proved that DomCol (CD Coloring) is NP-hard (even for a fixed $\ell \geq 4$). Unlike Graph Coloring, both DomCol and CD Coloring can be solved in polynomial time when $\ell = 3$ [8,34]. These problems, which marry two of the most well-studied problems in graph theory – Graph Coloring and Dominating Set, have been studied in several papers in the last 15 years. Results in these papers can be broadly categorized into two.

First, there have been several crucial results which establish lower and upper bounds on the size of an optimal dominator coloring (CD coloring) of graphs belonging to special graph classes. For example, refer papers [1,2,10,22,27,35] for results on DomCol and [2,11,30,34] for results on CD Coloring. The second (seemingly more sparse) are algorithmic results on these two problems. Even for simple graph classes such as trees, algorithmic results have been hard to obtain – indeed, after Gera *et al.* showed that DomCol can be solved in constant-time for paths in [23], it took close to a decade and incremental works [8,9] before a polynomial time algorithm was developed for trees in [33]! It is still unknown if DomCol restricted to forests is polynomial time solvable. While DomCol and CD Coloring seem extremely similar on the surface, we note a striking dichotomy in complexity results involving the two problems – DomCol is NP-hard restricted to *claw-free graphs* while CD Coloring is polynomial time solvable for the same graph class [2].

The parameterized complexity of DomCol and CD Coloring were first explored by Arumugam *et al.* in 2011 [1] and by Krithika *et al.* in 2021 [31] respectively. The authors expressed the problems in *Monodic Second-Order Logic* (MSOL) and used a theorem due to Courcelle and Mosbah [12] to prove that DomCol and CD Coloring parameterized by (t, ℓ), where t is the *treewidth* of the input graph, is FPT. Their expression of these problems in MSOL immediately also shows (by [13]) that DomCol and CD Coloring parameterized by (w, ℓ), where w is the *clique-width* of the input graph, is FPT. However, both problems have remained unexplored when viewed through the lens of other *structural parameters* that measure the distance (commonly vertex deletion) from a tractable graph class. They have become increasingly popular in the world of parameterized algorithms since they are usually small in practice. We refer the interested reader to the following survey by Fellows *et al.* for an overview of structural parameterization [17] and [24,29] for its use in studying Graph Coloring and Dominating Set. Our paper initiates the study of such structural parameterizations of DomCol and CD Coloring.

1.2 Our Results, Techniques, and Organization of the Paper

As a graph with bounded vertex cover has bounded treewidth, using results from [1,31], it is easy to show that the DomCol and CD Coloring are FPT parameterized by a graph's *vertex cover*. We give details in Sect. 2 of [3]. Due to its general nature, the algorithm has a large runtime. We design faster algorithms for more natural (and larger) parameters. Our main results are tabulated in Table 1. Our overarching result is the following: DomCol parameterized by CVD set size is FPT. This is shown through an involved branching algorithm. We also show that CD COLORING parameterized by (k, q), where q the number of cliques that remain on deleting a CVD (which is of size k) is FPT. We design much faster algorithms for larger parameters (i.e., special CVD sets).

We consider two well-studied parameters of this type – the size of a *clique modulator* (CLQ) and that of a *twin cover* (TC). Since optimal CVD sets, TCs, and CLQs can be found quickly [21,26,28], we implicitly assume that these sets are also given as input. In Sect. 3, we design randomized algorithms for the two problems which run in $\mathcal{O}^*(c^k)$-time for a small constant c. We show that our algorithm for CD COLORING is optimal unless the *Exponential-Time Hypothesis* fails. These algorithms use an inclusion-exclusion based polynomial sieving technique in addition to an exact single-exponential algorithm to solve DOMCOL that we develop in Sect. 2. We believe that this algebraic method holds great potential for use in other GRAPH COLORING variants.

Table 1. A summary of results. Cells marked ♣ are proved in this paper.

	Exact	CLQ	TC	CVD Set
DOMCOL	$\tilde{\mathcal{O}}(4^n)$ ♣	$\mathcal{O}^*(16^k)$ ♣	$\mathcal{O}^*(2^{\mathcal{O}(k \log k)})$ ♣	$\mathcal{O}^*(2^{\mathcal{O}(2^k)})$ ♣
CD COLORING	$\tilde{\mathcal{O}}(2^n)$ [31]	$\mathcal{O}^*(2^k)$ ♣	$\mathcal{O}^*(2^{\mathcal{O}(k \log k)})$ ♣	$\mathcal{O}^*(2^{\mathcal{O}(2^k kq \log q)})$ ♣

We show that DOMCOL and CD COLORING admit $\mathcal{O}^*(2^{\mathcal{O}(k \log k)})$-time algorithms when k is the size of a twin cover in Sect. 4. For this purpose, we introduce the notion of a partial dominator coloring and a partial CD coloring and show that their corresponding extension problems (similar to PRE-COLORING EXTENSION) can be solved quickly. We prove that the extension problem involving DOMCOL can be solved using a relationship between DOMCOL and LIST COLORING that we establish. On the other hand, we formulate the CD COLORING extension problem as an *Integer Linear Program* which, in turn, can be solved using well known methods. We then show that an optimal-sized dominator coloring (CD coloring) can be obtained as an extension of a small number of partial dominator colorings (CD colorings). Section 6 establishes some lower bounds for DOMCOL and CD COLORING with respect to these parameters.

2 Exact Algorithm for DomCol

We present an inclusion-exclusion based algorithm to solve DOMCOL. First, we require some definitions. Let $\mathcal{U}$ be a finite set of elements, $\mathscr{P}(\mathcal{U})$ its power set, and $\mathcal{S} \subseteq \mathscr{P}(\mathcal{U})$. An *$\ell$-partization* of the system $(\mathcal{U}, \mathcal{S})$ is a set $\{S_1, S_2 \dots S_\ell\} \subseteq \mathcal{S}$ such that (i) $\cup_{i=1}^{\ell} S_i = \mathcal{U}$ and (ii) $S_i \cap S_j = \emptyset$ for $i \neq j$. The following theorem, which forms the bedrock of popular inclusion-exclusion based algorithms, was proved in [5]. We refer an interested reader to Chap. 4 of [18] for a discussion on this algorithmic technique. We use $\tilde{\mathcal{O}}(\cdot)$ to hide polynomial terms.

Theorem 2.1. *One can decide, in $\tilde{\mathcal{O}}(2^{|\mathcal{U}|})$-time and exponential space, if there exists an ℓ-partization of $(\mathcal{U}, \mathcal{S})$.*

Given an instance (G, ℓ) of DomCol, we use Theorem 2.1 to decide if this instance is a YES in $\tilde{\mathcal{O}}(4^n)$-time. To the best of our knowledge, this is the first algorithm described for the problem that betters the trivial $\tilde{\mathcal{O}}(n^n)$-time algorithm. Consider an instance (G, ℓ) of DomCol. Let $V(G) = \{v_1, v_2 \dots v_n\}$ where the vertices are ordered arbitrarily. Let $\mathcal{I}_G \subseteq \mathscr{P}(V(G))$ denote the collection of independent sets of G and $V'(G) = \{v'_i\}_{i=1}^n$ be a copy of $V(G)$. Define $\mathcal{U}_G = V(G) \cup V'(G)$ and $\mathcal{S}_G = \{I \cup \Delta(I) \mid I \in \mathcal{I}_G \text{ and } \Delta(I) \subseteq \{v'_i \in V'(G) \mid N_G[v_i] \supseteq I\}\}$. That is, $\mathcal{S}_G$ consists of all sets where the first component is an independent set of G and the second component is a (possibly empty) set of vertices which dominate all the vertices in the first component. Note that $\mathcal{I}_G$ can be constructed in $\tilde{\mathcal{O}}(2^n)$-time, and therefore $\mathcal{S}_G$ in $\tilde{\mathcal{O}}(4^n)$-time.

$\star$ **Observation 2.2.** *(G, ℓ) is a* YES *instance of* DomCol *if, and only if, there exists an ℓ-partization of $(\mathcal{U}_G, \mathcal{S}_G)$.*

From Theorem 2.1 and Observation 2.2, we have the following theorem.

Theorem 2.3. *There exists an algorithm running in $\tilde{\mathcal{O}}(4^n)$-time and exponential space to solve* DomCol.

3 Parameterized by Clique Modulator Size

Consider a graph G. A subset M of $V(G)$ is a *clique modulator* if $G - M$ is a clique. We use Q to denote this clique and let $k = |M|$. For convenience, we sometimes use Q in place of $V(Q)$. Our goal for this section is to prove the following: DomCol and CD Coloring can be solved in $\mathcal{O}^*(16^k)$ and $\mathcal{O}^*(2^k)$-time respectively. Under the Exponential-Time Hypothesis, neither of these problems can be solved in $2^{o(k)}$-time (Lemma 6.2).

Our algorithms follow a similar randomized strategy using polynomial sieving and *Schwartz-Zippel Lemma*, as that of [26] for List Coloring parameterized by clique modulator. The overall idea of this technique is as follows: we design a *weighted Edmonds matrix* A whose entries are polynomials over $\mathbb{R}$ such that G admits a dominator coloring (CD coloring) of size ℓ if, and only if, a specified monomial containing $2k$-many (k-many) variables divides a term T of $\det A$. We use Theorem 3.1, Lemma 2.5 in [26], (with $|J| = 2k$ for DomCol and k for CD Coloring) followed by Theorem 3.2 [36,37] to determine whether such a T exists in $\det A$.

Theorem 3.1. *Let $J \subseteq \{x_1, x_2 \dots x_n\}$ and $P(x_1, x_2 \dots x_n)$ be a polynomial over $\mathbb{R}$. Then, there is a polynomial $Q(x_1, x_2 \dots x_n)$ whose evaluation takes $\mathcal{O}(2^{|J|})$-time such that $Q \not\equiv 0$ if, and only if, P contains a term divisible by $\prod_{x_j \in J} x_j$.*

Results preceded by a "$\star$" are proved in the full version of the paper [3]

Theorem 3.2. *Let $P(x_1, x_2 \dots x_n)$ be a non-zero polynomial over a field $\mathbb{F}$ with maximum degree d. Then,* $\Pr[P(r_1, r_2 \dots r_n) = 0] \leq \frac{d}{|\mathbb{F}|}$ *if $\{r_1, r_2 \dots r_n\}$ is picked randomly from $\mathbb{F}$.*

Since the result for DomCol involves several key observations in addition to a non-trivial adaptation of [26], we focus on its algorithm in the next subsection. Our $\mathcal{O}^*(2^k)$-time algorithm for CD Coloring is in Sect. 4.2 of [3].

3.1 DomCol Parameterized by Clique Modulator Size

Assume that we have an instance (G, ℓ) of DomCol where the size of Q is at most k. Then, $|V(G)| \leq 2k$ and therefore, by the proof of Theorem 2.3, (G, ℓ) can be solved in $\mathcal{O}(16^k)$-time. Now, consider the case where the size of clique is at least $k + 1$. Then, in any proper coloring of G, there exists a color which is used (exactly once) in Q but not in M. An important observation ensues.

Observation 3.3. *If $|Q| > k$, every vertex in Q dominates a color class in any proper coloring of G.*

Our main theorem in this subsection is Theorem 3.4. We prove this by designing a weighted Edmonds matrix A whose elements are polynomials containing the variables $X = \{x_v \mid v \in M\}$ and $Y = \{y_v \mid v \in M\}$ with the following property: the existence of x_v in a term T of $\det A$ will mean that v has been properly colored and y_v would mean that v dominates a color class. We will therefore need to look for a term of $\det A$ which is divisible by $\prod_{v \in M} x_v y_v$.

Theorem 3.4. DomCol *can be solved in $\mathcal{O}^*(16^k)$-time where k is the size of a clique modulator of the input graph.*

Let $C = \{c_1, c_2 \dots c_\ell\}$ denote a set of ℓ-many colors. Moreover, let $C' = \{c'_v \mid v \in M\}$ be a set of k-many artificial colors. If $|V(G)| > |C \cup C'|$, this is a No instance of DomCol as vertices in Q must get different colors. Pad $V(G)$ with $(|C \cup C'| - |V(G)|)$-many artificial vertices. Let this set be $V'(G)$. We construct a balanced bipartite graph B with bipartition $(V(G) \cup V'(G), C \cup C')$ by defining its edges as follows. Every vertex in $V(G)$ is connected all vertices in C. In addition, $v \in M$ is also connected to the artificial color $c'_v \in C'$ corresponding to it. Finally, every vertex in $V'(G)$ is connected to all vertices in $C \cup C'$. Each edge $(v, c) \in E(B)$ is associated with a $\mathcal{S}_{(v,c)} \subseteq \mathcal{P}(M)^2$ by the following relation:

- If $(v, c) \in Q \times C$, $\mathcal{S}_{(v,c)}$ is the collection of sets $S = (S_1, S_2) \in \mathcal{P}(M)^2$ where $S_1 \cup \{v\}$ is an independent set of G **and** $S_2 = \{u \in M \mid N_G[u] \supseteq S_1 \cup \{v\}\}$.
- If $(v, c) \in M \times C$, $\mathcal{S}_{(v,c)}$ is the collection of sets $S = (S_1, S_2) \in \mathcal{P}(M)^2$ where S_1 is an independent set of G with $v \in S_1$ **and** $S_2 = \{u \in M \mid N_G[u] \supseteq S_1\}$.
- If v or c is an artificial vertex or color, $\mathcal{S}_{(v,c)} = \{\emptyset\}$.

We now define our matrix A with dimensions $|V(G) \cup V'(G)| \times |C \cup C'|$. Its rows labeled are by $V(G) \cup V'(G)$ and columns by $C \cup C'$ and its entries are polynomials over $\mathbb{R}$. In addition to the sets of variables X and Y, let $Z =$

$\{z_{(v,c)} \mid (v,c) \in E(B)\}$ be a set of variables indexed by edges in $E(B)$. For each $(v,c) \in E(B)$, we define:

$$P(v,c) = \sum_{S \in \mathcal{S}_{(v,c)}} \left(\prod_{s \in S_1} x_s \cdot \prod_{s \in S_2} y_s \right) \text{ and } A(v,c) = z_{(v,c)} \cdot P(v,c)$$

Here, we assume that the empty product equals 1. All other entries of A are 0. The proof of the following crucial theorem has a similar flavor as Lemma 3.2 of [26] and can be found as Theorem 4.5 in [3].

⋆ **Theorem 3.5.** *(G,ℓ) is a* YES *instance of* DOMCOL *if, and only if,* det A *contains a monomial divisible by $\prod_{x \in X} x \cdot \prod_{y \in Y} y$.*

We apply Theorem 3.2 to the polynomial Q obtained when Theorem 3.1 is applied to $\det A$ and $J = X \cup Y$. Since $|J| = 2k$, we have a randomized algorithm (whose correctness follows from Theorem 3.5) which runs in $\mathcal{O}^*(16^k)$-time to solve DOMCOL when restricted to instances where the size of the clique is greater than k. With a more complicated polynomial sieving method as in [26], we can improve this to an algorithm which runs in $\mathcal{O}^*(4^k)$-time. Combining this result with the discussion preceding Observation 3.3 gives us the proof of Theorem 3.4. Note, therefore, that the bottleneck in the running time of Theorem 3.4 is due to Theorem 2.3.

4 Parameterized by Twin Cover Size

Consider a graph G. A subset M of $V(G)$ is a *twin cover* if for all $(u,v) \in E(G)$ either (i) $u \in M$ or $v \in M$ **or** (ii) $N_G[u] = N_G[v]$ (note that (ii) can be rephrased as "u and v are *true twins*"). This parameter was introduced by Ganian in 2015 [21] and has been used extensively in the world of parameterized complexity since. As before, we let k denote the size of a twin cover throughout this section. We use the following observation due to Ganian – Observation 1 in [21].

Observation 4.1. *Let G be a graph. $M \subseteq V(G)$ is a twin cover of G if, and only if, $G - M$ is a cluster graph and, if $\mathcal{Q} = \{Q_i\}_{i=1}^q$ are its connected components, for any $Q_i \in \mathcal{Q}$ and any $u, v \in Q_i$, $N_G[u] = N_G[v]$.*

For a $Q \in \mathcal{Q}$ and an arbitrary $v \in Q$, we let $N_G[Q] = N_G[v]$ and say that $u \in V(G)$ is dominated by Q if v dominates it. These notions are well defined as a consequence of Observation 4.1. We prove that DOMCOL and CD COLORING are FPT parameterized by the size of a twin cover – specifically, that they both admit algorithms running in $\mathcal{O}^*(2^{\mathcal{O}(k \log k)})$-time.

A *partial coloring* $\chi^S : S \to C$ is a proper coloring of $G[S]$ (note that the superscript of χ specifies the domain of the function). A proper coloring of G is an *extension* of χ^S if it agrees with χ^S on S. In the well-studied PRE-COLORING EXTENSION problem (first introduced in [4]), we are given a partial coloring χ^S

of a graph G and an $\ell \in \mathbb{N}$, and ask if there exists an extension $\mathcal{X}$ of $\mathcal{X}^S$ with $|\mathcal{X}| \leq \ell$. It is known that PRE-COLORING EXTENSION is FPT parameterized by the size of a twin cover [21].

We introduce a similar notion of a *partial CD coloring* and a *partial dominator coloring* and prove that their corresponding extension problems can be solved quickly – in FPT time in case of CD COLORING and in polynomial time in case of DOMCOL. We will then prove that an optimal CD coloring (dominator coloring) can be constructed as an extension of an $\mathcal{O}(2^{\mathcal{O}(k \log k)})$-sized set of partial CD colorings (partial dominator colorings). Furthermore, we show that such a set can be be constructed in FPT time. This, indeed, proves the running time of the aforementioned FPT algorithms for the two problems. Although our proof techniques for the two algorithms are appreciably different, space constrains force us to move the entire section focusing on DOMCOL to the full version (Sect. 5.1 in [3]).

4.1 CD Coloring Parameterized by Twin Cover Size

We say that a clique $Q \in \mathcal{Q}$ is *isolated* if $N_G[Q_i] \cap M = \emptyset$. We have the following simple observations regarding isolated cliques.

Observation 4.2. *Consider an isolated clique $Q \in \mathcal{Q}$. Then, in any CD coloring of G, vertices in Q must get unique colors.*

Corollary 4.3. *If $\mathcal{Q}' = \{Q \in \mathcal{Q} \mid Q$ is isolated$\}$, $(G, \ell) \equiv (G - \mathcal{Q}', \ell - |V(\mathcal{Q}')|)$.*

We therefore have a reduction rule – given a CD COLORING instance, we can consider the equivalent instance which has no isolated cliques. *We assume that graphs in the rest of this subsection do not have any isolated cliques.* We make a few small observations before defining a partial CD coloring.

Observation 4.4. *Let $\mathcal{X}$ be an arbitrary CD coloring of G and let $c \in Im(\mathcal{X})$ denote a color used by $\mathcal{X}$ outside M. Then, c is dominated by a vertex in M.*

Proof. Clearly, if c is used in two cliques in $\mathcal{Q}$, it can only be dominated by a vertex in M. If c is used in exactly one clique, then there must exist a vertex in M which dominates it – G has no isolated cliques! Now, assume that c is used both in M and in exactly one clique $Q \in \mathcal{Q}$. Let $\mathcal{X}(u) = \mathcal{X}(v) = c$ where $u \in Q$ and $v \in M$. Assume that a $w \in V(\mathcal{Q})$ dominates c. Then $w \in Q$ and $w \in N_G(v)$. However, this implies that $u \in N_G(v)$ (by Observation 4.1), a contradiction. Thus, only a vertex in M can dominate c. □

Observation 4.5. *Let $\mathcal{X}$ be an arbitrary CD coloring of G and let $u, v \in V(G)$ such that $N_G[u] = N_G[v]$. Then, $\mathcal{X}'$, the mapping which switches the colors assigned to u and v by $\mathcal{X}$ and retains the coloring of the rest of the vertices, is a CD coloring of G with $|\mathcal{X}| = |\mathcal{X}'|$.*

Definition 4.6. *Let $S \subseteq V(G)$. A partial CD coloring is a mapping $\mathcal{X}^S : S \to C$ with the following properties: (i) it is a partial coloring of G, and (ii) every $c \in Im(\mathcal{X}^S)$ is dominated by some vertex in $V(G)$.*

Definition 4.7. *A CD coloring $\mathcal{X}$ is an extension of a partial CD coloring $\mathcal{X}^S$ if it agrees with $\mathcal{X}^S$ on S. An extension $\mathcal{X}$ of $\mathcal{X}^S$ is disjoint if no vertex outside S is colored using a $c \in Im(\mathcal{X}^S)$.*

We will prove the following theorem, which establishes that an optimal disjoint extension of a partial CD coloring can be found quickly.

Theorem 4.8. *Let $\mathcal{X}^S$ be a partial CD coloring of G where $M \subseteq S$. Then, given an $\ell \in \mathbb{N}$, we can decide in $\mathcal{O}^*(2^{\mathcal{O}(k \log k)})$-time if there exists a disjoint extension $\mathcal{X}$ of $\mathcal{X}^S$ with $|\mathcal{X}| \leq \ell$.*

For each $Q_i \in \mathcal{Q}$, let b_i denote the number of vertices in Q_i that are uncolored by $\mathcal{X}^S$. Let $\mathbf{b} = (b_1, b_2 \ldots b_q)$ be a column vector and let $\{v_1, v_2 \cdots v_k\}$ be an arbitrary ordering of M. Consider the matrix $A \in M(q, k)$ with $A(i, j) = 1$ if $v_j \in N_G[Q_i]$ and is 0 otherwise (similar to an adjacency matrix). Let $\mathbf{1}_k$ denote the vector in $\mathbb{R}^k$ containing only 1s. Consider the following Integer Linear Program (ILP):

$$\text{Minimize } \{\mathbf{1}_k \cdot \mathbf{x} \mid A\mathbf{x} \geq \mathbf{b},\ \mathbf{x} \in \mathbb{Z}^k,\ \mathbf{x} \geq 0\}$$

We refer the interested reader to Chapter 6 of [14] for a discussion on the use of ILPs in the world of parameterized algorithms. We describe the intuition behind designing this ILP. The ith component of $\mathbf{x}$, denoted by x_i, represents the number of color classes outside $Im(\mathcal{X}^S)$ that $v_i \in M$ is required to dominate. Consider an arbitrary disjoint extension $\mathcal{X}$ of $\mathcal{X}^S$. By Observation 4.4, every $c \in Im(\mathcal{X}) \backslash Im(\mathcal{X}^S)$ is dominated by some vertex in M. Since no two vertices in a clique can get the same color, for each $Q_i \in \mathcal{Q}$ the total sum of colors that the vertices in $N_G[Q_i] \cap M$ dominate must be at least b_i (the number of uncolored vertices in Q_i). This is exactly the ith constraint in the linear program. We now formalize this notion below.

Note that the above program is feasible since $n \cdot \mathbf{1}$ is a feasible solution. Let ℓ^* denote the optimal value of the above program and $\ell' = |\mathcal{X}^S|$. The following lemma is proved as Lemma 5.18 in [3].

$\star$ **Lemma 4.9.** *There exists a disjoint extension $\mathcal{X}$ of $\mathcal{X}^S$ with $|\mathcal{X}| \leq \ell$ if, and only if, $\ell^* \leq \ell - \ell'$.*

Since our ILP has k variables, it can be solved in $\mathcal{O}^*(2^{\mathcal{O}(k \log k)})$-time [14]. Thus, Theorem 4.8 follows as a corollary of Lemma 4.9.

Consider Γ, a collection of partial CD colorings, with the following properties:

(i) For all $\mathcal{X}^S \in \Gamma$, $M \subseteq S$.
(ii) Given a CD coloring $\mathcal{X}'$ of G, there exists an disjoint extension $\mathcal{X}$ of a $\mathcal{X}^S \in \Gamma$ with $|\mathcal{X}| \leq |\mathcal{X}'|$.
(iii) $|\Gamma| = \mathcal{O}(2^{\mathcal{O}(k \log k)})$ and Γ can be constructed in $\mathcal{O}^*(2^{\mathcal{O}(k \log k)})$-time.

Theorem 4.10. *Assume that a Γ satisfying (i), (ii), and (iii) exists. Then, there exists an algorithm to solve* CD Coloring *in $\mathcal{O}^*(2^{\mathcal{O}(k \log k)})$-time, where k is the size of a twin cover of the input graph.*

Proof. Consider a collection of partial CD colorings Γ with the three properties elucidated above. By Theorem 4.8, for each $\chi^S \in \Gamma$ we can decide in $\mathcal{O}^*(2^{\mathcal{O}(k \log k)})$-time if there exists a disjoint extension χ of χ^S with $|\chi| \leq \ell$. By the third property of Γ, we can decide in $\mathcal{O}^*(2^{\mathcal{O}(k \log k)})$-time if there exists a disjoint extension χ of one of the partial CD colorings in Γ. By its second property, (G, ℓ) is a YES instance if, and only if, such a disjoint extension exists. □

All that is left to do is to prove that such a collection exists. We do so below.

Constructing a Γ. We construct this collection of partial CD colorings in two steps. Let $\mathscr{P}$ denote the set of partitions of M. For each partition $\mathcal{P} = \{P_1, P_2 \dots P_\kappa\} \in \mathscr{P}$ (κ is at most k), check if each part $P_i \in \mathcal{P}$ is independent and if there exists a $v \in V(G)$ which dominates P_i. If so, color all vertices in each P_i by c_i. If not, reject this partition of M. Note that we have now obtained a collection Γ' of $\mathcal{O}(2^{k \log k})$-many partial CD colorings of G. Moreover, constructing Γ' required $\mathcal{O}^*(2^{k \log k})$-time. Note that $S = M$ for all $\chi^S \in \Gamma'$ by construction.

Consider a $\chi^M \in \Gamma'$. Let $\mathscr{P}'$ be the set of "partitions" of $Im(\chi^M)$ into two sets. Partitions is in quotations here since we allow for $Im(\chi^M)$ to be split into $\{Im(\chi^M), \emptyset\}$. Consider a $\mathcal{P}' = \{P'_0, P'_1\} \in \mathscr{P}'$. We interpret colors in P'_0 as those that are only used in M and those in P'_1 as those that can also be used in $G - M$. Define $\Delta = \{\delta \colon P'_1 \to M \mid \delta(c) \text{ dominates } c \text{ for all } c \in P'_1\}$.

Note that $|\Delta| = \mathcal{O}(2^{k \log k})$. Consider a $\delta \in \Delta$. For each $c \in P'_1$, let $\mathcal{Q}_c$ denote the collection of cliques which are adjacent to $\delta(c)$ but not adjacent to vertices colored c (refer Fig. 1). That is, $\mathcal{Q}_c = \{Q \in \mathcal{Q} \mid \delta(c) \in N_G[Q], N_G[Q] \cap {\chi^M}^{-1}(c) = \emptyset\}$.

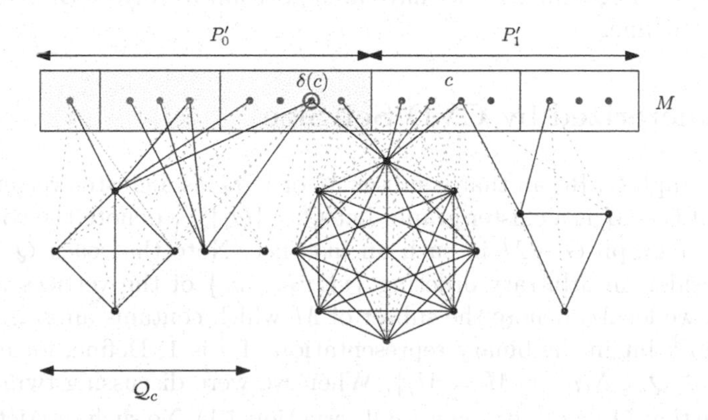

Fig. 1. Given a $\chi^M \in \Gamma'$, a $\{P'_0, P'_1\} \in \mathcal{P}'$, $\delta \in \Delta$, and a $c \in P'_1$ (the green vertices), we define $\mathcal{Q}_c$ to be the collection of cliques which are adjacent to $\delta(c)$ but not a vertex colored c.

Now, for each $c \in P'_1$, color an arbitrary uncolored vertex (if such a vertex exists) of each $Q \in \mathcal{Q}_c$. Note that this indeed produces a partial CD coloring of G. Let Γ be the collection of partial CD colorings obtained by varying over $\mathcal{X}^S \in \Gamma'$, $\mathcal{P}' \in \mathscr{P}'$, and $\delta \in \Delta$.

***Γ* Satisfies the Three Properties.** Since $|\mathscr{P}'| \leq 2^k$ and $|\Delta| \leq 2^{k \log k}$, $|\Gamma| \leq |\Gamma'| \cdot 2^k \cdot 2^{k \log k}$. Thus, $|\Gamma| \in \mathcal{O}(2^{\mathcal{O}(k \log k)})$. Moreover, Γ took $\mathcal{O}^*(2^{\mathcal{O}(k \log k)})$-time to construct. Note that $M \subseteq S$ for all $\mathcal{X}^S \in \Gamma$ by construction. To complete our result, we must prove property (ii) for Γ - given a CD coloring $\mathcal{X}'$ of G, there exists an disjoint extension $\mathcal{X}$ of a $\mathcal{X}^S \in \Gamma$ with $|\mathcal{X}| \leq |\mathcal{X}'|$.

Consider a CD coloring $\mathcal{X}'$ of G. Let $\mathcal{P} = \{P_1, P_2 \dots P_\kappa\}$ be the partition of M such that for two vertices u and v, $\mathcal{X}'(u) = \mathcal{X}'(v)$ if, and only if, $u, v \in P_i$ for some $P_i \in \mathcal{P}$. We can, without loss in generality, assume that the vertices in P_i are colored c_i. Let $\mathcal{X}^M \in \Gamma'$ be this partial CD coloring of G. Let P_1' denote the set of colors used in both M and $\mathcal{Q}$ by $\mathcal{X}'$. Let P_0' be the rest of the colors. For each $c \in P_1'$, let $\delta(c) \in M$ denote an arbitrary vertex in M that dominates c. Note that such a vertex exists due to Observation 4.4. Consider the $\mathcal{X}^S \in \Gamma$ constructed due to the selection of the triplet $(\mathcal{X}^M, (P_0', P_1'), \delta)$.

Assume there exists a clique $Q \in \mathcal{Q}$ such that $\mathcal{X}'(v) \in P_1'$ for all $v \in Q$. Then, by construction of $\mathcal{X}^S$, every vertex of Q will be colored. Now, consider a clique $Q \in \mathcal{Q}$ such that $\mathcal{X}'(v) \in P_1'$ for all $v \in Q'$ where $\emptyset \subset Q' \subseteq Q$. Then, $\mathcal{X}^S$ colors at least $|Q'|$-many vertices of Q. By repeated application of Observation 4.5, we can assume that $\mathcal{X}^S$ colors a superset of Q'. Thus, we can assume that $\mathcal{X}'$ does not use colors of P_1' outside S. Let $\mathcal{X}$ be a mapping which agrees with $\mathcal{X}^S$ on S and with $\mathcal{X}'$ on $V(G) \backslash S$. This is clearly a proper coloring of G. Moreover, a $c \in Im(\mathcal{X}) \subseteq Im(\mathcal{X}')$ is dominated by the vertex that dominates c in $\mathcal{X}'$. Thus, $\mathcal{X}$ is a disjoint extension of $\mathcal{X}^S$ where $|\mathcal{X}| \leq |\mathcal{X}'|$. Since $\mathcal{X}'$ was arbitrary, we have proved that Γ satisfies (ii), the last remaining property that we were looking to prove. Thus, by Theorem 4.10, we have an algorithm to solve CD COLORING in $\mathcal{O}^*(2^{\mathcal{O}(k \log k)})$-time.

5 Parameterized by CVD Set Size

Consider a graph G. Recall that a subset M of $V(G)$ is a cluster vertex deletion (CVD) set if $G - M$ is a cluster graph. Let $\mathcal{Q} = \{Q_i\}_{i=1}^q$ denote the components of the cluster graph $G - M$ in such an instance. Note that each $Q_i \in \mathcal{Q}$ is a clique. Consider an arbitrary ordering $\{v_1, v_2 \dots v_k\}$ of the vertices in M. For $0 \leq j < 2^k$, we let M_j denote the subset of M which contains an $v_i \in M$ if, and only if, the ith bit in the binary representation of j is 1. Define, for each i and j, $Q_i^j = \{v \in Q_i \mid N_G[v] \cap M = M_j\}$. When we were discussing twin covers in Sect. 4, note that $Q_i = Q_i^j$ for some j (Observation 4.1). No such restriction exists for CVD sets. We will exploit this partitioning of $\mathcal{Q}$ to show that there exists an $\mathcal{O}^*(2^{\mathcal{O}(2^k)})$ and $\mathcal{O}^*(2^{\mathcal{O}(2^k kq \log q)})$-time algorithm which solves DOMCOL and CD COLORING respectively. Due to page constraints, our complicated branching algorithm for DOMCOL is deferred to Sect. 6.1 of [3].

We discuss our much simpler algorithm for CD COLORING here. By Observation 4.2 and Corollary 4.3, we can assume that there exists a $v \in Q_i$, for arbitrary $Q_i \in \mathcal{Q}$, with $N_G[v] \cap M \neq \emptyset$. Pick one arbitrary vertex from each non-empty Q_i^j and call this collection M'. Note that $|M \cup M'| \in 2^{\mathcal{O}(2^k q)}$. We have the following observation with the same flavor as Observation 4.4.

Observation 5.1. *Let $\mathcal{X}$ be an arbitrary CD coloring of G and let $c \in Im(\mathcal{X})$ denote a color used by $\mathcal{X}$ outside M. Then, c is dominated by a vertex in $M \cup M'$.*

Note that the proof of Theorem 4.8 and the construction of partial CD colorings that followed hinged entirely on Observation 4.4 and not the fact that M was a twin cover. Thus, using Observation 5.1, along with the ideas in Sect. 4.1, we have the following theorem.

Theorem 5.2. *There exists an algorithm to solve* Class Domination Coloring *in $\mathcal{O}^*(2^{\mathcal{O}(2^k kq \log q)})$-time, where $k = |M|$, where M is a CVD set of the input graph and q is number of connected components in $G - M$.*

6 Lower Bounds

In this section, we establish some lower bounds for DomCol and CD Coloring with the parameters discussed in the previous sections. Consider a graph G. Construct a graph G' from G as follows: $V(G') = V(G) \cup \{v_0\}$ for a new vertex v_0 and $E(G') = E(G) \cup \{(v_0, v) \mid v \in V(G)\}$. By [23,34] we have the following.

Observation 6.1. *G has a proper coloring using ℓ-many colors if, and only if, G' has a dominator coloring (CD coloring) using $(\ell + 1)$-many colors.*

Note that $V(G)$ is a clique modulator and a twin cover in G'. Therefore, if DomCol and CD Coloring can be solved in $\mathcal{O}^*(f(k))$-time, where k is the size of the modulator and f is some computable function, then Graph Coloring can be solved in $\mathcal{O}(f(n))$-time where n is the number of vertices in an instance of this problem. It is known, under the Exponential-Time Hypothesis (ETH), that Graph Coloring cannot be solved in $2^{o(n)}$-time (refer [14,18]). We state this discussion below as a lemma.

Lemma 6.2. DomCol *and* CD Coloring *do not admit $\mathcal{O}^*(2^{o(k)})$-time algorithms unless ETH fails. Here, k is the size of a clique modulator or a twin cover of the input graph.*

The proofs of the following results utilize a relationship between Hitting Set and DomCol that we establish and can be found in Sect. 7 of [3].

⋆ **Theorem 6.3.** *Unless the Set Cover Conjecture is false,* DomCol *does not admit a $\mathcal{O}^*((2-\epsilon)^k)$-time algorithm where k is the size of a CVD set.*

⋆ **Theorem 6.4.** *Unless NP $\subseteq$ coNP/poly* DomCol, *parameterized by the size of a CVD set, does not admit a polynomial kernel.*

References

1. Arumugam, S., Chandrasekar, K.R., Misra, N., Philip, G., Saurabh, S.: Algorithmic aspects of dominator colorings in graphs. In: Iliopoulos, C.S., Smyth, W.F. (eds.) IWOCA 2011. LNCS, vol. 7056, pp. 19–30. Springer, Heidelberg (2011). https://doi.org/10.1007/978-3-642-25011-8_2
2. Bagan, G., Merouane, H.B., Haddad, M., Kheddouci, H.: On some domination colorings of graphs. Discret. Appl. Math. **230**, 34–50 (2017)
3. Dominator coloring and CD coloring in almost cluster graphs. arXiv arXiv:2210.17321 (2023)
4. Biró, M., Hujter, M., Tuza, Z.: Precoloring extension. i. interval graphs. Discret. Math. **100**(1-3), 267–279 (1992)
5. Björklund, A., Husfeldt, T., Koivisto, M.: Set partitioning via inclusion-exclusion. SIAM J. Comput. **39**(2), 546–563 (2009)
6. Böcker, S., Briesemeister, S., Bui, Q.B.A., Truß, A.: Going weighted: parameterized algorithms for cluster editing. Theor. Comput. Sci. **410**(52), 5467–5480 (2009)
7. Böcker, S., Briesemeister, S., Klau, G.W.: Exact algorithms for cluster editing: evaluation and experiments. Algorithmica **60**(2), 316–334 (2011)
8. Chellali, M., Maffray, F.: Dominator colorings in some classes of graphs. Graphs Comb. **28**(1), 97–107 (2012)
9. Chellali, M., Merouane, H.B.: On the dominator colorings in trees. Discuss. Math. Graph Theory **32**(4), 677–683 (2012)
10. Chen, Q., Zhao, C., Zhao, M.: Dominator colorings of certain cartesian products of paths and cycles. Graphs Comb. **33**, 73–83 (2017)
11. Chen, Y.H.: The dominated coloring problem and its application. In: Murgante, B., et al. (eds.) ICCSA 2014. LNCS, vol. 8584, pp. 132–145. Springer, Cham (2014). https://doi.org/10.1007/978-3-319-09153-2_10
12. Courcelle, B., Mosbah, M.: Monadic second-order evaluations on tree-decomposable graphs. Theor. Comput. Sci. **109**(1&2), 49–82 (1993)
13. Courcelle, B., Olariu, S.: Upper bounds to the clique width of graphs. Discret. Appl. Math. **101**(1–3), 77–114 (2000)
14. Cygan, M., Fomin, F.V., Kowalik, L., Lokshtanov, D., Marx, D., Pilipczuk, M., Pilipczuk, M., Saurabh, S.: Parameterized Algorithms. Springer, Switzerland (2015). https://doi.org/10.1007/978-3-319-21275-3
15. Diestel, R.: Graph Theory, Graduate Texts in Mathematics, vol. 173, 4th edn. Springer, Heidelberg (2012)
16. Doucha, M., Kratochvíl, J.: Cluster vertex deletion: a parameterization between vertex cover and clique-width. In: Rovan, B., Sassone, V., Widmayer, P. (eds.) MFCS 2012. LNCS, vol. 7464, pp. 348–359. Springer, Heidelberg (2012). https://doi.org/10.1007/978-3-642-32589-2_32
17. Fellows, M.R., Jansen, B.M.P., Rosamond, F.A.: Towards fully multivariate algorithmics: Parameter ecology and the deconstruction of computational complexity. Eur. J. Comb. **34**(3), 541–566 (2013)
18. Fomin, F.V., Kratsch, D.: Exact Exponential Algorithms. Texts in Theoretical Computer Science. An EATCS Series, Springer, Heidelberg (2010). https://doi.org/10.1007/978-3-642-16533-7
19. Fomin, F.V., Kratsch, S., Pilipczuk, M., Pilipczuk, M., Villanger, Y.: Tight bounds for parameterized complexity of cluster editing with a small number of clusters. J. Comput. Syst. Sci. **80**(7), 1430–1447 (2014)

20. Fomin, F.V., Lokshtanov, D., Saurabh, S., Zehavi, M.: Kernelization: Theory of Parameterized Preprocessing. Cambridge University Press, Cambridge (2019)
21. Ganian, R.: Improving vertex cover as a graph parameter. Discrete Math. Theor. Comput. Sci. **17** (2015)
22. Gera, R.: On the dominator colorings in bipartite graphs (2007)
23. Gera, R., Rasmussen, C., Horton, S.: Dominator colorings and safe clique partitions. Congr. Numer. **181**, 19–32 (2006)
24. Goyal, D., Jacob, A., Kumar, K., Majumdar, D., Raman, V.: Structural parameterizations of dominating set variants. In: Fomin, F.V., Podolskii, V.V. (eds.) CSR 2018. LNCS, vol. 10846, pp. 157–168. Springer, Cham (2018). https://doi.org/10.1007/978-3-319-90530-3_14
25. Guo, J.: A more effective linear kernelization for cluster editing. Theor. Comput. Sci. **410**(8–10), 718–726 (2009)
26. Gutin, G.Z., Majumdar, D., Ordyniak, S., Wahlström, M.: Parameterized precoloring extension and list coloring problems. SIAM J. Discret. Math. **35**(1), 575–596 (2021)
27. Henning, M.A.: Total dominator colorings and total domination in graphs. Graphs Comb. **31**(4), 953–974 (2015)
28. Hüffner, F., Komusiewicz, C., Moser, H., Niedermeier, R.: Fixed-parameter algorithms for cluster vertex deletion. Theory Comput. Syst. **47**(1), 196–217 (2010)
29. Jansen, B.M.P., Kratsch, S.: Data reduction for graph coloring problems. Inf. Comput. **231**, 70–88 (2013)
30. Klavzar, S., Tavakoli, M.: Dominated and dominator colorings over (edge) corona and hierarchical products. Appl. Math. Comput. **390**, 125647 (2021)
31. Krithika, R., Rai, A., Saurabh, S., Tale, P.: Parameterized and exact algorithms for class domination coloring. Discrete Appl. Math. **291**, 86–299 (2021)
32. Majumdar, D., Raman, V.: FPT algorithms for FVS parameterized by split and cluster vertex deletion sets and other parameters. In: Xiao, M., Rosamond, F. (eds.) FAW 2017. LNCS, vol. 10336, pp. 209–220. Springer, Cham (2017). https://doi.org/10.1007/978-3-319-59605-1_19
33. Merouane, H.B., Chellali, M.: An algorithm for the dominator chromatic number of a tree. J. Comb. Optim. **30**, 27–33 (2015)
34. Merouane, H.B., Haddad, M., Chellali, M., Kheddouci, H.: Dominated colorings of graphs. Graphs Comb. **31**(3), 713–727 (2015)
35. Panda, B.S., Pandey, A.: On the dominator coloring in proper interval graphs and block graphs. Discrete Math. Algorithms Appl. **7**, 1550043 (2015)
36. Schwartz, J.T.: Fast probabilistic algorithms for verification of polynomial identities. J. ACM **27**(4), 701–717 (1980)
37. Zippel, R.: Probabilistic algorithms for sparse polynomials. In: Ng, E.W. (ed.) Symbolic and Algebraic Computation. LNCS, vol. 72, pp. 216–226. Springer, Heidelberg (1979). https://doi.org/10.1007/3-540-09519-5_73

Tight Approximation Algorithms for Ordered Covering

Jatin Batra[1(✉)], Syamantak Das[2], and Agastya Vibhuti Jha[3]

[1] Tata Institute of Fundamental Research, Mumbai, India
jatinbatra50@gmail.com
[2] Indraprastha Institute of Information Technology, Delhi, India
syamantak@iiitd.ac.in
[3] Ecole Polytechnique Fédérale de Lausanne, Lausanne, Switzerland
agastya.jha@epfl.ch

Abstract. The classical unweighted set cover problem aims to pick a minimum number of subsets from a given family of subsets whose union would cover the universe of elements. The vertex cover problem is an important special case where the subsets and elements are respectively vertices and edges of a given undirected graph. On the other hand, the *min-sum* versions of both problems offer a different perspective - the objective is to find a linear ordering of the covering entities so that the sum of cover times of the elements is minimized. In this paper, we study common generalizations of these two classical problems. In particular, we study the Top-ℓ-norm problem for set cover and the even more general ordered norm (a linear combination of Top-ℓ norms) for vertex cover.

1. We give a polynomial time randomized rounding based $(2+\varepsilon)$-approx. for ordered norm vertex cover.
2. We show that a natural greedy algorithm gives $\mathcal{O}(\log(n/\ell))$-approx. for Top-$\ell$ set cover and show a matching hardness up to constants.

At a technical level, we exploit a certain known progress property of the set cover greedy algorithm for the Top-ℓ set cover problem. For the ordered norm vertex cover problem, we employ a major shift from the standard independent rounding techniques used for the min-sum vertex cover problem. We instead use a slightly weaker version of dependent rounding on a carefully constructed bipartite assignment instance.

Keywords: Scheduling · Randomized algorithms · Approximation algorithms

1 Introduction

The *set cover* problem forms a cornerstone of combinatorial optimization. In this problem, given a set of elements E and a family V of subsets whose union is E, the goal is to pick the minimum number of subsets from V such that their union is E. It is well-known that a natural greedy algorithm is an $\mathcal{O}(\ln n)$-approximation (where $n = |E|$) to this problem [Joh73,Sla96] and this bound is tight unless **NP**$\subseteq$ **DTIME**$(n^{\mathcal{O}(\log\log n)})$ [Fei98].

P. Morin and S. Suri (Eds.): WADS 2023, LNCS 14079, pp. 120–135, 2023.
https://doi.org/10.1007/978-3-031-38906-1_9

An important special case of set cover which is fundamental in its own right is the *vertex cover* problem where the underlying structure is a graph $G = (V, E)$ and the objective is to pick the fewest possible vertices to cover every edge in E. There are several 2-approximation algorithms known for the vertex cover problem, both purely combinatorial as well as based on rounding linear programs (see [Hoc96] and the references therein). Although not conclusively proved, there are strong evidences which suggest that a 2-approximation is possibly tight [KR08].

Min sum set cover (MSSC) and *min sum vertex cover* (MSVC) introduced by Feige, Lovasz and Tetali [FLT04] offer an ordering perspective to both these problems. In MSSC (and analogously in MSVC), given the set of elements E and a family $\mathcal{F}$ of subsets V, we are required to find a linear ordering of the subsets in $\mathcal{F}$. The *cover time* of an element $e \in E$ can be thought of as the position of the first subset in the above ordering that contains (or covers) e. The objective is to determine the ordering that minimizes the *sum of cover times of elements in* E. It is worth noting that, from the linear ordering perspective, the classical set cover and vertex cover problem can be viewed as finding the ordering of the subsets that minimizes the *maximum cover time* of any element. A natural greedy algorithm provides a 4-approximation to MSSC [FLT04] while MSVC admits a $16/9 \simeq 1.778$-approximation [BBFT21].

It turns out that the techniques to solve the classical versions of the above covering problems are quite different from those used for studying the min-sum versions. Hence, a natural question is - *can we unify the above variants through a common algorithmic framework?* One such generalization that has garnered significant attention is the *ordered optimization* framework [Aou19, BSS18, CSa, CS19, CSb]. This framework has been successful in unifying variants of k-clustering and scheduling, and has been also been successfully extended to stochastic settings. In this paper, we study ordered minimization versions of set cover and vertex cover.

Ordered and Top-ℓ Covering. Given a linear ordering σ of the subsets, let $\mathbf{Cov}_\sigma \in \mathbb{R}_+^{|E|}$ be the vector induced by the cover times of elements written in non-increasing order. Further, we are given a *weight* vector $\mathbf{w} \in \mathbb{R}_+^{|E|}$, such that w_i are non-increasing in i. In the *ordered min-sum set cover* (OMSSC) problem, we define the cost of the ordering σ as $\langle \mathbf{w}, \mathbf{Cov}_\sigma \rangle$.

An important and non-trivial special case of ordered covering is the setting when $\mathbf{w} \in \{0,1\}^{|E|}$ - we refer to this problem as Top-ℓ Set Cover , where ℓ is the number of ones in $\mathbf{w}$. This problem, in fact, is a unification of classical set cover ($\ell = 1$) and MSSC ($\ell = |E|$) (and analogously for the vertex cover variants).

Our Results. In this paper, we give a $(2 + \varepsilon)$-approximation algorithm running in time $n^{O(1/\varepsilon)}$ for ordered min-sum vertex cover (OMSVC) for arbitrary weight vectors w. This result is tight assuming popular conjectures [KR08] since this captures the classical vertex cover problem as a special case.

As our second main result, we give a $\mathcal{O}(\log(n/\ell))$-approximation algorithm for Top-ℓ Set Cover . In fact, we show that the natural greedy algorithm (which does not require explicit knowledge of the value of ℓ) admits the above approximation ratio. The key takeaway from this result is that the natural greedy ordering admits an approximation ratio that smoothly transitions with ℓ between the two extremes of Top-ℓ Set Cover , that is $\mathcal{O}(\log n)$ for $\ell = 1$ (set cover) and $\mathcal{O}(1)$ for $\ell = n$ (MSSC). We complement this with a matching hardness (up to constant factor) result for *any given* ℓ*, possibly growing with* n.

1.1 Related Work

Ordered optimization for combinatorial optimization problems have received substantial attention in the recent years, primarily triggered by the work of Byrka et al. [BSS18] on the ordered k-median problem and the subsequent body of work by Chakrabarty and Swamy [CS19,CSa,CSb]. These last series of papers consider both k-median clustering and load balancing and generalize the result to *any symmetric norm* of the induced cost vector. The above results have been extended to the stochastic settings [IS20]. There has also been recent developments in coreset construction for ordered clustering problems [BJKW19] which has found interest in the machine learning community.

In contrast, the progress on set cover or vertex cover has been sparse. To the best of our knowledge, the only work related to our settings is due to [GGKT08]. Among other results, they claim a $\mathcal{O}(\log n)$-approximation for the All-Norm set cover problem where one minimizes any symmetric norm of the cost vector induced by the linear ordering of sets and cover time of elements while the natural greedy algorithm gives a tight $p + \ln p + \mathcal{O}(1)$-approximation for minimizing the ℓ_p-norm of the cost vector. For vertex cover, they give a 8-approximation for the All-Norm setting. We would like to note here that the ordered vertex cover problem we consider is indeed a special case of the above All-Norm problem. However, we are able to exploit several structural insights of the ordered setting and hence are able to obtain a 2-approximation. Similarly, for the Top-ℓ Set Cover problem, we are able to prove that the natural greedy algorithm gives a tight (up to constants) approximation ratio for any $\ell, 1 \leq \ell \leq n$, without an explicit knowledge of ℓ.

1.2 Overview of Techniques

Ordered Min-Sum Vertex Cover. The main challenge in ordered optimization (even the Top-ℓ special case) is that the weights for the cover times depend on the ordering of the cover times in the schedule. Roughly speaking, the existing line of work [BSS18,CSa,CSb,CS19] shows that it suffices to minimize the alternate objective (which refer to as the Discounted Min-Sum Vertex Cover (DMSVC) objective) : $\sum_{e \in E}(\max\{\mathbf{Cov}_\sigma(e) - \lambda_\ell), 0\}$, where $\lambda_\ell \geq 0$ is an additional input parameter. The reader may think of λ_ℓ as a discount i.e. only the part of the cover time over λ_ℓ counts towards the cost. Crucially, the DMSVC objective can be written without reference to the ordering of the cover times.

Our technique is to solve the natural LP for DMSVC (see Sect. 2.2 for notation):

$$\text{Minimize} \quad \sum_{t \geq \lambda_\ell} \sum_{e \in E} u_{e,t}, \quad s.t.$$

$$u_{e,t} + \sum_{t'<t} (x_{u,t'} + x_{v,t'}) \geq 1, \quad \forall\, e = (u,v), t = 1,2,\ldots \tag{1}$$

$$\sum_{v \in V} x_{v,t} \leq 1, \quad \forall\, t = 1,2,\ldots \tag{2}$$

$$u, x \geq 0.$$

Note that the above LP closely resembles that of MSVC (for $\lambda_\ell = 0$). Hence, it is natural to try to extend the rounding techniques for MSVC [FLT04, BBFT21].

Prior Approaches for MSVC: The randomized rounding algorithms of [FLT04, BBFT21] work by first applying a linear transformation (called a *kernel*) $z = Kx$ to an LP solution x. The actual rounding of Kx happens in further two steps. First, they construct a *tentative* schedule τ from z by standard randomized rounding (called α-point rounding) that rounds each vertex independently. Importantly, the number of vertices landing in a slot in τ (*its load*) can be arbitrarily greater than 1 (the *expected* loads are 2 and 4/3 in [FLT04] and [BBFT21] respectively), which necessitates a second phase where ties are (randomly) broken among vertices in the same slot in τ.

However, the second phase in the approach above is untenable for DMSVC. Consider the following example - suppose an edge $e = (u, v)$ is covered by time λ_ℓ by the LP solution x, say both u, v have been picked to an extent $1/2$ by time λ_ℓ in x. Then, to obtain an $O(1)$-approximation, our solution must pick at least one of u, v by time λ_ℓ *with probability 1*. However, the inherent tie-breaking in the algorithms of [FLT04, BBFT21] leads to a non-zero probability that e will not even be covered by time $\beta\lambda_\ell$ for any constant β.

Our Technique. We replace α-point *independent* rounding of [FLT04, BBFT21] with a more sophisticated *dependent* rounding that ensures that each time-slot is assigned at most 2 vertices (with probability 1) in the intermediate schedule τ. This guarantees that if the LP solution covers an edge e by time t, our schedule covers e by time $2t$, which can be shown to be sufficient for our purposes. In particular, we use dependent rounding [GKPS06] to construct τ. Following [GKPS06], we construct a bipartite graph instance as base for the rounding, with vertices on one side, and time slots on other. The edges are assigned probabilities (via a kernel $z = 2x$ similar to [FLT04]). Crucially, we do not use the negative-correlation property of [GKPS06] and instead exploit the special property of our bipartite graph that the total weighted degree of any vertex $v \in V$ is at most 1.

Top-ℓ Set Cover. We analyze the natural greedy algorithm, inspired by [INZ16]. Note that the objective function can be equivalently written as $\sum_t \min(|R_t|, \ell)$ where R_t is the set of elements uncovered at time t. The key technical idea is a progress lemma that shows that the number of sets needed to be picked by the greedy algorithm to halve the number of uncovered elements from $2p$ to p can be upper bounded by the number of sets needed by the optimal schedule to cover all but $p/2$ elements.

2 Ordered Vertex Cover

In this section, we give a polynomial time $(2+\varepsilon)$-approx. algorithm for OMSVC. Let us first recall some notation.

Notation. Under a schedule σ for vertices, let the edges be ordered $e_1, e_2, \ldots e_n$ in non-increasing order of cover times. $\mathbf{Cov}_\sigma$ denotes the ordered vector of cover times of edges. We will slightly abuse notation and let $Cov_\sigma(e)$ also denote the cover time of edge e under σ.

We divide this section into two parts:

1. In the first part, we sketch a reduction from OMSVC to (a generalization of) DMSVC following previous works [Aou19,BSS18,CSa,CS19,CSb].
2. In the second part, we give an LP-rounding based $(2,2)$-bicriteria approx. algorithm for generalized DMSVC. Combined with the reduction of the first part, this implies a polynomial time $(2+\varepsilon)$-approx. algorithm for OMSVC (for any $\varepsilon > 0$).

Our key technical contribution is a $(2,2)$-bicriteria approx. algorithm for the discounted version. Previous randomized rounding approaches such as [FLT04,BBFT21] fail to provide even an $(O(1), O(1))$-approximation when there are discounts.

2.1 Reduction Sketch

As a warm up, let us consider a special case of the above problem when $\mathbf{w} = \mathbb{1}_\ell$, corresponding to Top-$\ell$ Vertex Cover . Note that Top-ℓ Vertex Cover interpolates between the Min-Sum Vertex Cover and the Vertex Cover problem.

The main technical challenge is that to write the objective $\langle \mathbb{1}_\ell, \mathbf{Cov}_\sigma \rangle$, we need to know the ordering of cover times of edges beforehand. To get around this, it is helpful to come up with an upper bound for the objective that can be written as a sum over edges e of terms that depend only on the cover times of the respective edges. Suppose we knew the ℓth-largest cover time λ_ℓ in the optimum schedule. Then, it is not hard to see that $\ell\lambda_\ell + \sum_{e\in E}(Cov_\sigma(e) - \lambda_\ell)_+$ is an upper bound on $\langle \mathbb{1}_\ell, \mathbf{Cov}_\sigma \rangle$, that is tight for the optimum schedule. Hence with a correct guess of λ_ℓ, it suffices to minimize the sum of the cover times *above the discount* λ_ℓ.

Motivated by the above, we introduce the following problem:

Definition 1. ***Discounted Min-Sum Vertex Cover (DMSVC):*** *Given a graph $G = (V, E)$ and a discount λ_ℓ, output a schedule σ of vertices that minimizes $\sum_{e \in E}(Cov_\sigma(e) - \lambda_\ell)_+$. A schedule σ is said to be (α, β)-approx. if it satisfies:*

$$\sum_{e \in E}(Cov_\sigma(e) - \beta\lambda_\ell)_+ \leq \sum_{e \in E}\alpha(Cov_{\sigma^*}(e) - \lambda_\ell)_+$$

for some $\alpha, \beta \geq 1$ and any feasible schedule σ^*.

It can be shown that obtaining an (α, β)-approx. schedule for DMSVC suffices to get a max (α, β)-approx. for Top-ℓ Vertex Cover .

Theorem 1 [Aou19,CS19]. *Suppose there exists a polynomial time* $\max((\alpha, \beta))$-*approx. algorithm for DMSVC, then there exists an* $\max((\alpha, \beta))(1 + O(\varepsilon))$-*approx. algorithm for Top-ℓ Vertex Cover with running time $n^{O(1/\varepsilon)}$.*

For completeness, we will provide a proof-sketch of a generalization of 1 in Appendix A. The following observation helps us in extending the above intuition to OMSVC:

Observation 2 *The objective function of OMSVC can be expressed in the following way: $\sum_{\ell \in E}(w_\ell - w_{\ell+1})\langle \mathbb{1}_\ell, \mathbf{Cov}_\sigma \rangle$ where $w_{n+1} = 0$.*

Note that since $\mathbf{w}$ is a non-negative vector, the above linear combination is in fact a conic combination of $\langle \mathbb{1}_\ell, \mathbf{Cov}_\sigma \rangle$. Plugging in the discounted cost for $\langle \mathbb{1}_\ell, \mathbf{Cov}_\sigma \rangle$ for each ℓ such that $w_\ell - w_{\ell+1} > 0$ gives us a discounted objective function for the ordered case. This motivates the following problem:

Definition 2. ***Generalized Discounted Min-Sum Vertex Cover (GDMSVC):*** *Given a graph $G = (V, E)$, a set T of indices in $|E|$, non-negative weights $(\Delta_\ell)_{\ell \in T}$ and a vector of discounts $\lambda \in \mathbb{R}^{|T|}$, output a schedule σ of vertices that minimizes $\sum_{\ell \in T}\Delta_\ell \sum_{e \in E}(Cov_\sigma(e) - \lambda_\ell)_+$. A schedule σ is said to be an (α, β)-approx. if it satisfies:*

$$\sum_{\ell \in T}\Delta_\ell \sum_{e \in E}(Cov_\sigma(e) - \beta\lambda_\ell)_+ \leq \alpha \sum_{\ell \in T}\Delta_\ell \sum_{e \in E}(Cov_{\sigma^*}(e) - \lambda_\ell)_+$$

for some $\alpha, \beta > 0$ and any feasible schedule σ^*.

Intuitively, the set T corresponds to indices such that $w_\ell - w_{\ell+1} > 0$, and $\Delta_\ell = w_\ell - w_{\ell+1}$. The following result by [CS19] shows that it suffices to solve GDMSVC to get an approximation algorithm for OMSVC. We give a proof-sketch in Appendix A.

Theorem 3 [CS19]. *Given a polynomial time (α, β)-approx. algorithm for GDMSVC, there exists a* $\max(\alpha, \beta)(1 + O(\varepsilon))$-*approx. algorithm for OMSVC running in time $n^{O(1/\varepsilon)}$.*

2.2 Generalized Discounted Min-Sum Vertex Cover

In the following, we give an LP-based polynomial time (2,2)-approx. for GDMSVC. Coupled with Theorem 3, this will imply a $(2+\varepsilon)$-approx. algorithm for OMSVC running in time $n^{O(1/\varepsilon)}$.

Before describing the algorithm, we lay down some notation: For a vector $\mathbf{a} \in \mathbb{R}^n$, let $\mathbf{a}_{\leq k}, \mathbf{a}_{<k}$ and $\mathbf{a}_{[k,\ell]}$ denote $\sum_{i=1}^{k} a_i, \sum_{i=1}^{k-1} a_i$ and $\sum_{i=k}^{\ell} a_i$ respectively. Also for $x \in \mathbb{R}$, let x_+ denote $\max(x, 0)$.

LP Relaxation for GDMSVC. Let time be divided into time-slots $1, 2, \ldots$ where time-slot t denotes the interval $[t-1, t)$. Let $u_{e,t}$ and $x_{v,t}$ indicate whether e is uncovered at the start of time-slot t and whether v is scheduled in time-slot t respectively. Then we have the following natural LP relaxation of GDMSVC:

$$\text{Minimize} \quad \sum_{\ell \in T} \Delta_\ell \sum_{t \geq \lambda_\ell} \sum_{e \in E} u_{e,t}, \quad s.t.$$

$$u_{e,t} + x_{u,<t} + x_{v,<t} \geq 1, \quad \forall\, e = (u,v), t = 1, 2, \ldots \tag{3}$$

$$\sum_{v \in V} x_{v,t} \leq 1, \quad \forall\, t = 1, 2, \ldots \tag{4}$$

$$u, x \geq 0.$$

Constraint (3) says that edge e is uncovered at the start of time-slot t if none of its two endpoints have been scheduled before t. Constraint (4) says that only one vertex may be scheduled in any time-slot.

Since our algorithm uses key elements from the dependent rounding framework of Gandhi, Khuller, Parthasarathy and Srinivasan [GKPS06], we give an overview of this result first.

Overview of Dependent Rounding

In a series of works, [AS, Sri01, GKPS06] gave efficient algorithms for randomized rounding satisfying cardinality constraints and negative correlation. We will use the following special case of their results.

Theorem 4 (Dependent rounding [AS, Sri01, GKPS06]**).** *There exists a polynomial time algorithm for the following sampling problem. We are given a set U and a number $p_i \in [0,1]$ for each element $i \in U$, such that $\sum_{i \in U} p_i = k$ for some integer k. Sample $S \subseteq U$ satisfying the following properties:*

1. Marginal distribution. *For any element $i \in U$, $\Pr(i \in S) = p_i$.*
2. Cardinality preservation. $|S| = k$.
3. Negative correlation. *For any pair of elements $i, j \in U$, $\Pr(i \in S \cap j \in S) \leq \Pr(i \in S)\Pr(j \in S)$.*

Remark 1. We only use properties (1) and (2) of Theorem 4 in the analysis of our algorithm.

2.3 Overview of the Algorithm

Given an optimum fractional solution x of the LP relaxation for GDMSVC, the algorithm rounds x as follows:

1. Construct a bipartite graph $D = (V \cup [|V|],\ E')$. Define a weight function on edges $\varphi : E' \to [0,1]$ using x. Intuitively, the set $[|V|]$ denotes the set of time slots and φ denotes the probabilities used by the dependent rounding algorithm.
2. Carry out dependent rounding on (D, φ). This will assign vertices to time slots. However multiple vertices could be assigned to the same time slot. Denote this tentative schedule by τ.
3. Schedule all the vertices assigned to the same time slot in an arbitrary order. Denote the final schedule with σ (Fig. 1).

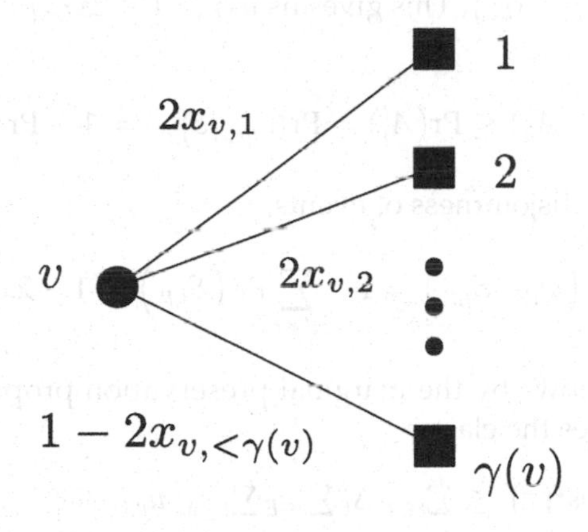

Fig. 1. Bipartite graph for dependent rounding

Construction of D. D is a bipartite graph defined on $2|V|$ vertices $V \cup [|V|]$. For each vertex $v \in V$, let $\gamma(v) : V \to [|V|]$ denote the first time instant t such that $2\sum_{t'<t} x_{v,t} \geq 1$. We will let $E' = \{(v,t) \mid v \in V,\ 1 \leq t' \leq \gamma(v)\}$. Define $\varphi : E' \to [0,1]$ as follows:

$$\varphi(e) = \begin{cases} 2x_{v,t} & 1 \leq t < \gamma(v) \\ 1 - 2x_{v,<t} & t = \gamma(v) \end{cases}.$$

We let $\mathcal{N}(v)$ denote the neighbouring vertices of $v \in V$ in D.

Dependent Rounding Phase. We will carry out dependent rounding on $(D = (V \cup [|V|], E'),\ \varphi)$. This will return a subset of edges $M \subseteq E'$ with the following properties:

1. Let $\mathcal{E}_{v,j}$ denote the event that vertex v was assigned to time slot j for $j \in \mathcal{N}(v)$ in set E'. Then $\Pr(\mathcal{E}_{v,j}) = \varphi((v,j))$ and $\{\mathcal{E}_{v,j}\}_{j \in \mathcal{N}(v)}$ are disjoint events. The second property holds due to the following reason: since for all $v \in V$, $\sum_{e \in \delta(v)} \varphi(e) \leq 1$, (where $\delta(v)$ denotes the set of edges incident on v in E') at most one edge of M is incident on v by property (2) of Theorem 4.
2. For any $t \in [|V|]$, $\sum_{e \in \delta(t)} \varphi_e \leq 2 \sum_{v \in V} x_{v,t} \leq 2$ by the LP constraint.

Analysis. Let $p_t(e)$ denote the probability that edge $e \in E$ has not been covered till the beginning of time-slot t. For $u \in V$, define A_u^t to be the event that u has not been scheduled in any time-slots before t. Note that, $A_u^t = \cap_{j<t} \mathcal{E}_{u,j}^c$.

Lemma 1. *For* $e = (j,k) \in E$, $p_t(e) \leq u_{e,t}$.

Proof. By the LP constraint 4, we have that $1 - x_{j,<t} - x_{k,<t} \leq u_{e,t}$. Without loss of generality, let $x_{j,<t} \geq x_{k<t}$. This gives us $u_{e,t} \geq 1 - 2x_{j,<t}$.

Now,

$$p_t(e) = \Pr(A_j^t \cap A_k^t) \leq \Pr(A_j^t) = \Pr(\cap_{t'<t} \mathcal{E}_{j,t'}^c) = 1 - \Pr\left(\cup_{t'<t} \mathcal{E}_{j,t'}\right).$$

Further note that by disjointness of events,

$$1 - \Pr\left(\cup_{t'<t} \mathcal{E}_{j,t'}\right) = 1 - \sum_{t'<t} \Pr\left(\mathcal{E}_{j,t'}\right) = 1 - 2x_{j,<t}.$$

The last equality follows by the marginal preservation property of dependent rounding. This proves the claim.

Observation 5. $\mathbb{E}[cost(\tau)] \leq \sum_{\ell \in T} \Delta_\ell \sum_{e \in E} \sum_{t \geq \lambda_\ell} u_{e,t}$.

Proof. Fix $\ell \in T$. Define $\text{cost}_\ell(\tau) = \Delta_\ell \sum_{e \in E} (cov_\tau(e) - \lambda_\ell)_+$.

Now,

$$\mathbb{E}[cost_\tau(\ell)] = \sum_{e \in E} \Delta_\ell \mathbb{E}\left[(cov_\tau(e) - \lambda_\ell)_+\right] = \Delta_\ell \sum_{e \in E} \sum_{t \geq \lambda_\ell} p_t(e) \leq \Delta_\ell \sum_{e \in E} \sum_{t \geq \lambda_\ell} u_{e,t}.$$

Note that by linearity of expectation, $\mathbb{E}[cost(\tau)] = \sum_{\ell \in T} \mathbb{E}[cost_\tau(\ell)]$. This completes the proof.

Observation 6 *For any edge* $e \in E$ *and for all* $\ell \in T$, $(cov_\sigma(e) - 2\lambda_\ell)_+ \leq 2 \cdot (cov_\tau(e) - \lambda_\ell)_+$.

Proof. By property 2 in 2.3, at most 2 vertices are assigned to any time slot. Thus breaking ties arbitrarily can cause the cost to go up by a factor at most 2.

Theorem 7. *The algorithm described above is a* (2,2)*-approx. to GDMSVC.*

Remark 2. Note that the rounding algorithm itself is independent of **w**.

3 Top-ℓ Set Cover

In this section we show that the greedy algorithm achieves an $O(\log(n/\ell))$-approximation ratio for the Top-ℓ Set Cover problem and that this is tight up to constant factors unless $P = NP$. Recall that the greedy algorithm picks the set which covers the maximum number of uncovered elements in each iteration.

The idea of the analysis of the upper bound is to break up the schedule into two phases: the time t_ℓ until which all but at most ℓ elements have been covered and the additional time to cover the remaining at most ℓ elements. The contribution of the first phase towards the cover times of the Top-ℓ elements is $\ell \cdot t_\ell$ which we upper bound by $O(\log(n/\ell))$OPT analogous to the Set Cover problem. We bound the overall contribution of the second phase by $O(1)$OPT analogous to the Min Sum Set Cover problem.

3.1 Notation for Top-ℓ Set Cover

We will abuse notation slightly and use GREEDY (resp. OPT) to denote both the schedule returned by the greedy (resp. optimal) algorithm and its cost. R_i denotes the set of elements uncovered after GREEDY has scheduled i sets for all $i \geq 1$. t_x (resp. t_x^*) denotes the first time when at most x elements are uncovered in GREEDY (resp. OPT).

3.2 Upper Bound for the Approximation Ratio of GREEDY

Begin by expressing GREEDY in terms of t_ℓ and $\langle |R_i| \rangle_{i \geq t_\ell}$ (such an expression holds for any schedule).

Proposition 1. *GREEDY* $= \ell \cdot t_\ell + \sum_{i=t_\ell}^{n} |R_i|$.

Proof. For any time i, let $\hat{R}_i$ be R_i restricted to the elements with the largest ℓ cover times in τ. Clearly, $|\hat{R}_i| = \ell$ for $i < t_\ell$ and $\hat{R}_i = R_i$ otherwise. Hence, GREEDY $= \sum_{i \geq 0} |\hat{R}_i| = \sum_{i=0}^{t_\ell - 1} \ell + \sum_{i=t_\ell}^{n} |R_i|$.

We now compare the progress of GREEDY versus OPT (similar to [INZ16]). The following lemma shows that the time it takes for GREEDY to reduce the number of uncovered elements from $2p$ to p can be bounded by $2t_{p/2}^*$.

Lemma 2. *For any* p, $t_p - t_{2p} \leq 2t_{p/2}^*$.

Proof. At $t_{p/2}^*$, at most $p/2$ elements are uncovered in OPT. Hence, there is a collection C of $t_{p/2}^*$ sets which covers all but $p/2$ elements. Therefore, for any $t \in [t_{2p}, t_p)$, C covers at least $|R_t| - p/2 \geq p/2$ elements of R_t. Since there exists some set in C which covers at least $p/(2|C|)$ elements of R_t, GREEDY also covers at least $p/\left(2 \cdot t_{p/2}^*\right)$ additional elements at time t. The lemma now follows as GREEDY can only cover p additional elements in $[t_{2p}, t_p)$.

Let us prove a lower bound on OPT.

Proposition 2. $OPT \geq (1/2) \sum_{j=0}^{\infty} \ell \cdot 2^{-j} t^*_{\ell \cdot 2^{-j}}$.

Proof. At least $\ell \cdot 2^{-(j+1)}$ elements are covered by OPT in $\mathcal{I}^*_j := \left[t^*_{\ell \cdot 2^{-j}}, t^*_{\ell \cdot 2^{-(j+1)}}\right)$, each having a cover time at least $t^*_{\ell \cdot 2^{-j}}$. Since $\mathcal{I}^*_j$ are disjoint for $j \geq 0$,

$$\text{OPT} \geq \sum_{j=0}^{\infty} \left(\ell \cdot 2^{-(j+1)}\right) t^*_{\ell \cdot 2^{-j}} = (1/2) \sum_{j=0}^{\infty} \left(\ell \cdot 2^{-j}\right) t^*_{\ell \cdot 2^{-j}}.$$

Now employ Propositions 1 and 2 to upper bound GREEDY.

Claim 1. $\sum_{i=t_\ell}^{\infty} |R_i| \leq 16 OPT$.

Proof. At any time in the interval $\mathcal{I}_j : \left[t_{\ell \cdot 2^{-j}}, t_{\ell \cdot 2^{-(j+1)}}\right)$, at most $\ell \cdot 2^{-j}$ elements are uncovered. Since $\mathcal{I}_j$ cover $[t_\ell, \infty)$ for $j \geq 0$,

$$\sum_{i=t_\ell}^{\infty} |R_i| \leq \sum_{j=0}^{\infty} \ell \cdot 2^{-j} \left(t_{\ell \cdot 2^{-(j+1)}} - t_{\ell \cdot 2^{-j}}\right) \leq \sum_{j=0}^{\infty} \ell \cdot 2^{-j} \cdot 2 t^*_{\ell \cdot 2^{-(j+2)}} \leq 16 \text{OPT}.$$

The second to last and last inequalities follow from Lemma 2 and Proposition 2 respectively.

Claim 2. $\ell \cdot t_\ell \leq 8 \log_2 \left(\frac{n}{\ell}\right) OPT$.

Proof. By Lemma 2 for $p = \ell \cdot 2^j$ and all $j \geq 0$, $t_{\ell \cdot 2^j} \leq t_{\ell \cdot 2^{j+1}} + 2 t^*_{\ell \cdot 2^{j-1}}$. Adding these inequalities together for all $j \geq 0$ and using that t^*_x is non-increasing in x, we get that $t_\ell \leq 2 \lceil \log_2 (n/\ell) \rceil t^*_{\ell/2}$.

Now bound $\ell \cdot t_\ell$ as

$$\ell \cdot t_\ell \leq 8 \left\lceil \log_2 \left(\frac{n}{\ell}\right) \right\rceil \left(\frac{1}{2} \cdot \frac{\ell}{2} \cdot t^*_{\ell/2}\right) \leq 8 \left\lceil \log_2 \left(\frac{n}{\ell}\right) \right\rceil \text{OPT},$$

where the last inequality follows from Proposition 2.

Proposition 1 together with Claims 1 and 2 immediately gives the following theorem.

Theorem 8.

$$GREEDY \leq \left(8 \log_2 \left(\frac{n}{\ell}\right) + 16\right) OPT.$$

3.3 Hardness

In this section, we prove that any polynomial time approximation algorithm must have an approximation factor of $\Omega\left(\log \frac{n}{\ell}\right)$. The starting point is the following result of [FLT04].

Theorem 9 [FLT04]. *For all $c_0, \varepsilon > 0$, there are instances of set cover where all the sets have the same size such that it is* **NP**-*hard to distinguish between the following two cases:*

1. *There are t disjoint sets which cover all the elements.*
2. *For all integers x s.t. $1 \leq x \leq c_0 t$, every collection of x subsets leaves at least a fraction of $\left(1 - \frac{1}{t}\right)^x - \varepsilon$ elements uncovered.*

The hardness of Top-ℓ set cover directly follows from Theorem 9.

Theorem 10. *There are instances of the Top-ℓ Set Cover problem such that it is* **NP**-*hard to distinguish between the following two cases:*

1. *There is a schedule which incurs cost at most $\ell \cdot t$.*
2. *Every schedule incurs cost at least $\Omega(\ell \cdot t \log(n/\ell))$ (Fig. 2).*

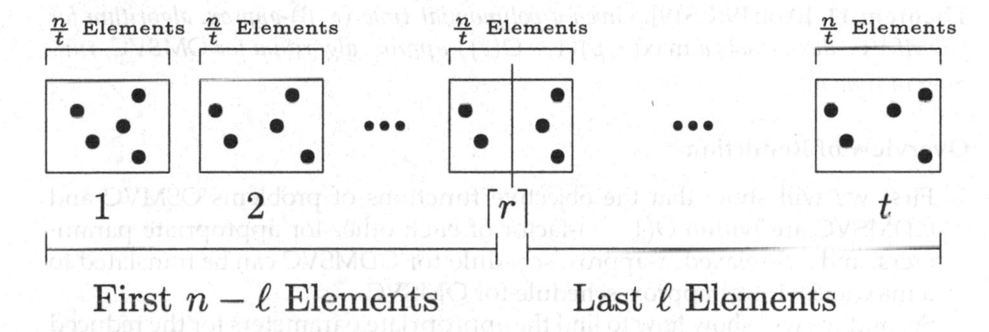

Fig. 2. Theorem 10 Case I

Proof. We reduce the problem in Theorem 9 to our problem. We are given a set cover instance as in Theorem 9. We will slightly abuse notation t_ℓ, R_i from Subsect. 3.1. The schedule to which the notation applies will be clear from the context.

Case I: There are t disjoint subsets which cover all the elements. Since all the sets have same size, each set has size n/t. Note that $t_\ell = \lfloor t(n-\ell)/n + 1 \rfloor$.

The cost incurred is at most $(n/t)\sum_{j=t_\ell}^{t} j$. This cost can be upper bounded in the following way:

$$\frac{n}{t}\sum_{j=t_\ell}^{t} j \leq \frac{n}{t}\left(\sum_{j=1}^{t} j - \sum_{j=1}^{t_\ell - 1} j\right) \leq \frac{3}{2}(\ell \cdot t) - \frac{\ell^2 t}{n}.$$

Case II: By Proposition 1, the cost of any schedule is at least $\ell \cdot t_\ell + \sum_{i=t_\ell}^{c_0 t} |R_i|$. Since every collection of x subsets leaves at least a fraction of $(1 - 1/t)^x - \epsilon$ elements uncovered, $|R_i| \approx n \cdot e^{-i/t}$ (provided $t = \omega(1)$). Thus,

$$\sum_{i=t_\ell}^{c_0 t} |R_i| \approx n \int_{t_\ell}^{c_0 t} \left(e^{-x/t} - \varepsilon\right) dx = n \cdot \left(te^{-t_\ell/t} - te^{-c_0} - \varepsilon\left(c_0 t - t_\ell\right)\right) \approx n \cdot te^{-t_\ell/t}/3.$$

Hence the, total cost of any schedule is lower bounded by $(\ell \cdot t_\ell + n \cdot te^{-t_\ell/t})/3$ for $c_0 \geq ln(3)$ and $\varepsilon \approx 10^{-2}$. Varying over t_ℓ, this quantity is minimized at $t_\ell = t \cdot ln\,(n/\ell)$ where it is equal to $\ell \cdot t\,(1 + ln\,(n/\ell))\,/3$.

Any approximation algorithm which gets an approximation factor better than $\max\{\Omega\,(1)\,, \Omega\,(\log\,(n/\ell))\}$ will help us distinguish between the two cases. Thus any such algorithm can not run in polynomial time unless $\mathbf{P} = \mathbf{NP}$.

A Appendix

In this section, we provide the reduction from OMSVC to GDMSVC. We point the interested reader to [Aou19] and [CS19] for a more general version of this reduction.

The main result of this section is as follows:

Theorem 11 [Aou19,CS19]. *Given a polynomial time (α, β)-approx. algorithm for GDMSVC, there exists a $\max(\alpha, \beta)(1 + O(\varepsilon))$-approx. algorithm for OMSVC, running in time $n^{O(1/\varepsilon)}$.*

Overview of Reduction

1. First we will show that the objective functions of problems OSMVC and GDMSVC are within $O(1 + \varepsilon)$-factor of each other for appropriate parameters, and a β-relaxed, α-approx. schedule for GDMSVC can be translated to a $\max(\alpha, \beta)(1 + \varepsilon)$-approx. schedule for OMSVC.
2. Second we will show how to find the appropriate parameters for the reduced instance in polynomial time with at most $O(1 + \varepsilon)$-loss in approximation ratio.

Step 1: The following observation helps us simplify the objective function of GDMSVC:

Observation 12. *The objective function of GDMSVC can be expressed in the following way: $\sum_{\ell \in E}(w_\ell - w_{\ell+1})\,\langle \mathbb{1}_\ell, \mathbf{Cov}_\sigma \rangle$ where $\mathbf{w}_{n+1} = 0$.*

Intuitively, appropriate discounts λ_ℓ in GDMSVC preserve the Top-ℓ Vertex Cover cost upto ε-factors. Formally:

Lemma 3 [Aou19,CS19]. *Let $\ell \in [n]$ and $\varepsilon > 0$. Then,*

$$\langle \mathbb{1}_\ell, \mathbf{Cov}_\sigma \rangle \leq \ell \cdot \lambda_\ell + \sum_{e \in E} (Cov_\sigma(e) - \lambda_\ell)_+ \leq (1 + \varepsilon)\langle \mathbb{1}_\ell, \mathbf{Cov}_\sigma \rangle$$

if $Cov_\sigma(\ell) \leq \lambda_\ell \leq (1 + \varepsilon) Cov_\sigma(\ell)$.

We can set $\Delta_\ell = w_\ell - w_{\ell+1}$ and guess the discounts λ for all $\ell \in |E|$ such that $w_\ell - w_{\ell+1} > 0$ (We call such indices as break-points). In this case T exactly corresponds to the set of break-points.

The following lemma in [Aou19] and [CS19] completes the first step in the reduction:

Lemma 4 [Aou19,CS19]. *Given $\alpha, \beta \geq 1$, $Cov_\sigma, Cov_{\sigma^*}$ for schedules σ, σ^*, and a discount vector λ satisfying the conditions of lemma 3 and*

$$\sum_{\ell \in T} \Delta_\ell \sum_{e \in E} (Cov_\sigma(e) - \beta\lambda_\ell)_+ \leq \alpha \sum_{\ell \in T} \Delta_\ell \sum_{e \in E} (Cov_{\sigma^*}(e) - \lambda_\ell)_+ \tag{5}$$

then $\langle \mathbf{w}, \mathbf{Cov}_\sigma \rangle \leq \max(\alpha, \beta)(1 + O(\varepsilon))\langle \mathbf{w}, \mathbf{Cov}_{\sigma^*} \rangle$.

To summarize, it is possible to come up with a $\max(\alpha, \beta)(1 + O(\varepsilon))$-approx. schedule for OMSVC, given an appropriate discount vector λ and an (α, β)-approx. solution for GDMSVC.

However the caveat in the above procedure is the following: Δ_ℓ might be positive for all $\ell \in [n]$. Therefore, guessing textbf$\lambda \in \mathbb{R}^{|E|}$ will take exponential time in general. To make the above procedure run in polynomial time, [Aou19] and [CS19] give the following procedure:

Step 2: Define a new weight-vector $\widetilde{\mathbf{w}}$ with at most $O(\delta^{-1} \log n)$ (for arbitrarily small $\delta > 0$) break-points.

Formally, for any fixed $\delta > 0$, let $B_{n,\delta} = \{\min\{\lceil (1+\delta)^s \rceil, n\} \mid s \geq 0\}$. We will denote $B_{n,\delta}$ by B for simplicity. $\widetilde{\mathbf{w}}$ is defined as follows:

$$\widetilde{w}_\iota = \begin{cases} w_i & i \in B \\ w_{next(i)} & \text{o.w.} \end{cases},$$

where $next(i)$ is the smallest index $j \in B$ bigger than i. $\widetilde{\mathbf{w}}$ will be referred as the δ-sparsification of $\mathbf{w}$.

The following lemma shows that $\widetilde{\mathbf{w}}$ preserves the inner product $\langle \mathbf{w}, \mathbf{Cov}_\sigma \rangle$ upto δ-factors.

Lemma 5 [Aou19,CS19].

$$\langle \widetilde{\mathbf{w}}, \mathbf{Cov}_\sigma \rangle \leq \langle \mathbf{w}, \mathbf{Cov}_\sigma \rangle \leq (1+\delta)\langle \widetilde{\mathbf{w}}, \mathbf{Cov}_\sigma \rangle.$$

Since there are at most $O(\delta^{-1} \log n)$ break-points in $\widetilde{\mathbf{w}}$, guessing the discounts naively takes quasi-polynomial time. The following lemma from [CS19] provides a polynomial time approach to guess the right discounts. Intuitively, it uses the fact that the discounts λ_ℓ are monotone in ℓ. In the following, we will set $\delta = \varepsilon$.

Lemma 6 [Aou19,CS19]. *Suppose we can guess the threshold λ_1 in $n^{O(1)}$ time, then for any ϵ-sparsified $\widetilde{\mathbf{w}}$, it is possible to come up with a discount vector $\lambda \in \mathbb{R}^{|B|}$ which satisfying the conditions of Lemma 3 in time $n^{O(1/\varepsilon)}$.*

Guessing λ_1 in our case takes $O(n)$ time in the worst case, and thus by setting $\delta = \varepsilon$, $T = B$, $\Delta_\ell = (\widetilde{w}_\ell - \widetilde{w}_{next(\ell)})$ (where $next(\ell)$ is the smallest index in B bigger than ℓ), we produce an instance of GDMSVC such that the objective functions are preserved upto $O(\varepsilon)$-factors. The time taken to produce such an instance is $n^{O(1/\varepsilon)}$.

Putting Everything Together

Instantiating Lemma 4 with ε-sparsified $\widetilde{\mathbf{w}}$, and then applying Lemma 5 twice shows the following inequality:

$$\langle \mathbf{w}, \mathbf{Cov}_{\sigma} \rangle (1 - O(\varepsilon)) \leq \max(\alpha, \beta)(1 + O(\varepsilon)) \langle \mathbf{w}, \mathbf{Cov}_{\sigma} \rangle.$$

Multiplying by $(1 - O(\varepsilon))^{-1}$ on both sides completes the proof.

References

[Aou19] Aouad, A., Segev, D.: The ordered k-median problem: surrogate models and approximation algorithms. Math. Program. **177**(4), 55–83 (2019)

[AS] Ageev, A.A., Sviridenko, M.I.: Pipage rounding: a new method of constructing algorithms with proven performance guarantee. J. Comb. Optim. **8**, 307–328 (2004)

[BBFT21] Bansal, N., Batra, J., Farhadi, M., Tetali, P.: Improved approximations for min sum vertex cover and generalized min sum set cover. In: SODA 2021, pp. 986–1005. Society for Industrial and Applied Mathematics (2021)

[BJKW19] Braverman, V., Jiang, S.H.-C., Krauthgamer, R., Wu, X.: Coresets for ordered weighted clustering. In: Chaudhuri, K., Salakhutdinov, R. (eds.) Proceedings of the 36th International Conference on Machine Learning. Proceedings of Machine Learning Research, vol. 97, pp. 744–753. PMLR (2019)

[BSS18] Byrka, J., Sornat, K., Spoerhase, J.: Constant-factor approximation for ordered k-median. In: Proceedings of the 50th Annual ACM SIGACT Symposium on Theory of Computing, STOC 2018, New York, NY, USA, pp. 620–631. Association for Computing Machinery (2018). https://doi.org/10.1145/3188745.3188930

[CSa] Chakrabarty, D., Swamy, C.: Interpolating between k-median and k-center: approximation algorithms for ordered k-median. In: 45th International Colloquium on Automata, Languages, and Programming, ICALP 2018 (2018)

[CSb] Chakrabarty, D., Swamy, C.: Simpler and better algorithms for minimum-norm load balancing. In: 27th Annual European Symposium on Algorithms. LIPIcs, vol. 144, pp. 27:1–27:12 (2019)

[CS19] Chakrabarty, D., Swamy, C.: Approximation algorithms for minimum norm and ordered optimization problems. In: STOC 2019, New York, NY, USA. Association for Computing Machinery (2019)

[Fei98] Feige, U.: A threshold of ln n for approximating set cover. J. ACM **45**(4), 634–652 (1998). https://doi.org/10.1145/285055.285059

[FLT04] Feige, U., Lovász, L., Tetali, P.: Approximating min sum set cover. Algorithmica **40**(4), 219–234 (2004)

[GGKT08] Golovin, D., Gupta, A., Kumar, A., Tangwongsan, K.: All-norms and all-l_p-norms approximation algorithms. In: Leibniz International Proceedings in Informatics. LIPIcs, vol. 2 (2008)

[GKPS06] Gandhi, R., Khuller, S., Parthasarathy, S., Srinivasan, A.: Dependent rounding and its applications to approximation algorithms. J. ACM **53**(3), 324–360 (2006)

[Hoc96] Hochbaum, D.S.: Approximating Covering and Packing Problems: Set Cover, Vertex Cover, Independent Set, and Related Problems, pp. 94–143. PWS Publishing Co., USA (1996)

[INZ16] Im, S., Nagarajan, V., Van Der Zwaan, R.: Minimum latency submodular cover. ACM Trans. Algorithms **13**, 1–28 (2016)

[IS20] Ibrahimpur, S., Swamy, C.: Approximation algorithms for stochastic minimum-norm combinatorial optimization. In: 2020 IEEE 61st Annual Symposium on Foundations of Computer Science (FOCS), pp. 966–977 (2020). https://doi.org/10.1109/FOCS46700.2020.00094

[Joh73] Johnson, D.S.: Approximation algorithms for combinatorial problems. In: Proceedings of the Fifth Annual ACM Symposium on Theory of Computing, STOC 1973, New York, NY, USA, pp. 38–49. Association for Computing Machinery (1973). https://doi.org/10.1145/800125.804034

[KR08] Khot, S., Regev, O.: Vertex cover might be hard to approximate to within $2-\varepsilon$. J. Comput. Syst. Sci. **74**(3), 335–349 (2008). Computational Complexity 2003

[Sla96] Slavík, P.: A tight analysis of the greedy algorithm for set cover. In: Proceedings of the Twenty-Eighth Annual ACM Symposium on Theory of Computing, STOC 1996, New York, NY, USA, pp. 435–441. Association for Computing Machinery (1996). https://doi.org/10.1145/237814.237991

[Sri01] Srinivasan, A.: Distributions on level-sets with applications to approximation algorithms. In: Proceedings 2001 IEEE International Conference on Cluster Computing, pp. 588–597 (2001)

Online Minimum Spanning Trees with Weight Predictions

Magnus Berg, Joan Boyar, Lene M. Favrholdt, and Kim S. Larsen(✉)

University of Southern Denmark, Odense, Denmark
{magbp,joan,lenem,kslarsen}@imada.sdu.dk
https://imada.sdu.dk/u/magbp, https://imada.sdu.dk/u/joan,
https://imada.sdu.dk/u/lenem, https://imada.sdu.dk/u/kslarsen/

Abstract. We consider the minimum spanning tree problem with predictions, using the weight-arrival model, i.e., the graph is given, together with predictions for the weights of all edges. Then the actual weights arrive one at a time and an irrevocable decision must be made regarding whether or not the edge should be included into the spanning tree. In order to assess the quality of our algorithms, we define an appropriate error measure and analyze the performance of the algorithms as a function of the error. We prove that, according to competitive analysis, the simplest algorithm, Follow-the-Predictions, is optimal. However, intuitively, one should be able to do better, and we present a greedy variant of Follow-the-Predictions. In analyzing that algorithm, we believe we present the first random order analysis of a non-trivial online algorithm with predictions, by which we obtain an algorithmic separation. This may be useful for distinguishing between algorithms for other problems when Follow-the-Predictions is optimal according to competitive analysis.

Keywords: Online Algorithms · Predictions · Random Order Analysis · Minimum Spanning Tree

1 Introduction

The *Minimum Spanning Tree* (MST) problem is one of the classical graph algorithms problems, where one must select edges from a weighted graph such that these constitute a spanning tree of minimal weight. We consider an online version of this problem in the relatively new context of predictions, a direction that emerged following the successes of machine learning that has provided more accessible and reliable predictions.

In the area of online algorithms, we consider problems, many of which have offline counterparts, where input is presented to an algorithm in a piece-wise fashion (often referred to as *requests*), and irrevocable decisions must be made

Supported in part by the Independent Research Fund Denmark, Natural Sciences, grant DFF-0135-00018B and in part by the Innovation Fund Denmark, grant 9142-00001B, Digital Research Centre Denmark, project P40: Online Algorithms with Predictions.

P. Morin and S. Suri (Eds.): WADS 2023, LNCS 14079, pp. 136–148, 2023.
https://doi.org/10.1007/978-3-031-38906-1_10

when each item is presented. The quality of an online algorithm is often assessed using competitive analysis, which essentially focuses on the worst-case ratio of the cost of the online algorithm to the cost of an optimal, offline algorithm, OPT.

When considering graph problems, various models, inspired by different application scenarios, exist. In the vertex-arrival model, the requests are the vertices of the graph, arriving together with the subset of its incident edges that connect to vertices that have already arrived. In the edge-arrival model, requests are the edges, identified by their two endpoints. For weighted graphs, there is also the *weight-arrival model*, where the graph is known, and the weights arrive online. In the vertex-arrival and edge-arrival models, there is only one possible online algorithm, the one that accepts every edge that does not create a cycle, since otherwise the algorithm's output might not span the entire graph. Even in the weight-arrival model, no deterministic algorithm for online MST can be competitive [15]. This makes the problem hard, but interesting in the context of advice or predictions.

Partially in an attempt to measure how much information about the future is needed for various online problems, online algorithms with advice were introduced [4,7,9,11]. In the model used most often, it is an information-theoretical game of how few oracle-produced bits in total are needed to obtain a particular competitive ratio or optimality. Obviously, the connection here is that oracle-based advice can be considered infallible predictions. The MST problem has been considered by Bianchi et al. in this model [3]. They obtain results for various arrival models and restricted graph classes, including the weight-arrival model, but with only two different weights allowed.

The seminal paper by Lykouris and Vassilvitskii [16], introducing machine-learned advice, which is now more often referred to as predictions, has inspired rapidly growing [1] efforts in the area [17]. In this context, ideally we want algorithms to use the predictions and perform optimally when predictions are correct (referred to as *consistency*), perform as well as a good online algorithm when predictions are all wrong (*robustness*), and degrade gracefully from one to the other as the predictions become increasingly erroneous (*smoothness*). The ideal situation described above can of course often not be reached, so one proves upper and lower bounds, as is customary in the field. Discussing smoothness requires a definition of error. This is problem-dependent and requires some thought. We want to distinguish between good and bad algorithms, and defining error measures that exaggerate or underestimate the importance of errors leads to unreliable results.

For the online MST problem with predictions, there are natural error measures. We arrive at an error measure, defined as the sum of differences between the predicted and actual values of the $n-1$ edges (the number of edges in a spanning tree) with the largest discrepancies; a measure with desirable properties.

We focus on the MST problem with predictions in the weight-arrival model. Our first somewhat surprising result is that with this error measure (or any of some reasonable alternatives), competitive analysis [13,18] cannot distinguish between different, correct algorithms. This means that the most naïve algorithm, Follow-the-Predictions (FTP), is optimal under this measure, with a competitive ratio of $1 + 2\varepsilon$, where ε is the error, normalized by the value of OPT. Of course, this also means that the perhaps more reasonable algorithm, we call Greedy Follow-the-Predictions (GFTP), that switches to another edge when a revealed actual weight matches or does better than the predicted weight of an edge it could replace, is indistinguishable from FTP under competitive analysis.

In online algorithms, there are other performance measures one can turn to when competitive analysis is insufficient, as discussed in [5,6,8]. One of the most well accepted is Random Order Analysis [14], also called the Random Order Model; a chapter in [10] discusses some results. Note that the problem from [10] of finding a maximum forest is not very similar to our problem, since the forest is not required to be spanning. The random order analysis technique reduces the power of the adversary, compared to competitive analysis. In competitive analysis, the adversary chooses the requests and the order in which they a presented, while in random order analysis, the adversary chooses the requests, but those requests are presented to the algorithm uniformly at random. Using random order analysis, we establish a separation between FTP and GFTP. We believe this is the first time random order analysis has been applied in the context of predictions.

Omitted details and proofs can be found in the full paper [2].

2 Preliminaries

Given an online algorithm ALG for an online minimization problem Π, and an instance I of Π, we let $\text{ALG}[I]$ denote ALG's solution on instance I, and $\text{ALG}(I)$ denote the cost of $\text{ALG}[I]$. Then, the *competitive ratio* of ALG is

$$\text{CR}_{\text{ALG}} = \inf\{c \mid \exists b \colon \forall I \colon \text{ALG}(I) \leqslant c\text{OPT}(I) + b\}.$$

When online algorithms have access to a predictor, a further parameter is introduced into the problem, namely the accuracy of that predictor. Throughout this paper, we let η be the error measure that computes the quality of the predictions, and we let $\varepsilon = \frac{\eta}{\text{OPT}}$ be the normalized error measure. Our error measure is defined later (see Definition 1).

Given an online algorithm with predictions, ALG, we express the competitive ratio of ALG as a function of ε, and evaluate it based on the three criteria: *consistency*, *robustness*, and *smoothness*. Following [16], we define consistency as ALG's competitive ratio, when the prediction error is 0. ALG is *α-consistent* if there exists a constant, α, such that $\text{CR}_{\text{ALG}}(0) = \alpha$.

As ε grows, the competitive ratio of ALG will decay as a function of ε. For a function, β, we say that ALG is *β-smooth*, if $\text{CR}_{\text{ALG}}(\varepsilon) \leqslant \beta(\varepsilon)$, for all $\varepsilon \geqslant 0$.

An algorithm, ALG, is said to be *γ-robust*, if there exists some constant γ such that $\text{CR}_{\text{ALG}}(\varepsilon) \leqslant \gamma$, for all $\varepsilon \geqslant 0$. Since no online algorithm for the WMST problem can be competitive, the robustness of any deterministic online algorithm with predictions cannot be worse than the competitive ratio of any online algorithm.

2.1 Random Order Analysis

Given an online algorithm, ALG, for a problem, Π, and an instance of Π with request sequence $I = \langle i_1, i_2, \ldots, i_n \rangle$, a permutation σ of I is chosen uniformly at random, and $\sigma(I)$ is presented to ALG. The *random order ratio* of ALG is defined as

$$\text{ROR}_{\text{ALG}} = \inf\{c \mid \exists b\colon \forall I\colon \mathbb{E}_\sigma[\text{ALG}(\sigma(I))] \leqslant c\text{OPT}(I) + b\},$$

As with the competitive ratio, we express the random order ratio of algorithms with predictions as a function of ε.

2.2 Weight-Arrival MST Problem

The offline MST problem is a thoroughly studied problem, for which efficient optimal algorithms are known, for example Kruskal's and Prim's Algorithms. Given a graph $G = (V, E, w)$, the task is to find a spanning tree T for G that minimizes the objective function $c(T) = \sum_{e \in E(T)} w(e)$. For the MST problem in the weight-arrival model (WMST), online algorithms are initially provided with the underlying graph $G = (V, E)$, and then the weights of the edges in G arrive online. At the time the true weight of an edge e arrives, the online algorithm has to irrevocably accept or reject e for its final tree. We focus on the WMST problem where we assume that an online algorithm has access to predicted weights for all edges in G before the online computation is initiated.

2.3 Notation and Nomenclature

We use the notation $\mathbb{R}^+$ and $\mathbb{Z}^+$ to denote the positive real numbers and the positive integers, respectively. Graphs, in the following, are weighted, simple, connected and undirected, with weights in $\mathbb{R}^+$. Given a graph G, $n = |V(G)|$ and $m = |E(G)|$. For any clarification on graph theory, we refer to [19]. Further, we define a *WMST-instance* to be a triple $(G, \hat{w}, w)$ consisting of a graph G, and two maps $\hat{w}\colon E(G) \to \mathbb{R}^+$ and $w\colon E(G) \to \mathbb{R}^+$, defining for each edge $e \in E(G)$, a predicted weight $\hat{w}(e)$ and a true weight $w(e)$. Given a graph G and a tree $T \subset G$, when writing T, we implicitly refer to $E(T)$.

Given an algorithm with predictions, ALG, for the WMST problem, and a WMST-instance $(G, \hat{w}, w)$, we let $\text{ALG}[\hat{w}(E(G)), w(E(G))]$ denote the tree that ALG outputs. When G is clear from the context, we write $\text{ALG}[\hat{w}, w]$ and let $\text{ALG}(\hat{w}, w)$ denote the cost of $\text{ALG}[\hat{w}, w]$. We let $\text{OPT}[w]$ be an optimal MST of G, and $\text{OPT}[\hat{w}]$ be an optimal MST of G with respect to $\hat{w}$.

2.4 Error Measure

We use the following measure, denoted by η, selected due to its desirable properties and its ability to distinguish between algorithms under random order analysis. In the full paper [2], we show that intuitive alternatives have flaws.

Definition 1. *Let* $(G, \hat{w}, w)$ *be any WMST-instance,* $e_1, e_2, \ldots, e_m$ *be any ordering of* $E(G)$*, and* $\{p_i\}_i$ *be the sequence where* $p_i := |w(e_i) - \hat{w}(e_i)|$*. Further, let* $\{p_{i_j}\}_j$ *be the sequence* $\{p_i\}_i$*, sorted such that* $p_{i_1} \geqslant p_{i_2} \geqslant \cdots \geqslant p_{i_m}$*. The* error*,* η*, is given by* $\eta(\hat{w}, w) = \sum_{j=1}^{n-1} p_{i_j}$*. When* $(G, \hat{w}, w)$ *is clear from the context, we write* η *for* $\eta(\hat{w}, w)$*. The* normalized error *is* $\varepsilon = \frac{\eta}{\text{OPT}}$*.*

Note that $n - 1$ is the number of edges in a spanning tree. Thus, the risk of unreasonably large prediction errors for dense graphs as with other possible error measures has been eliminated (see the full paper [2]). This measure also satisfies the monotonicity and Lipschitzness properties from [12].

3 Optimal Algorithms Under Competitive Analysis

We prove that our two algorithms FTP and GFTP, defined in Algorithms 1 and 2, respectively, are 1-consistent and $(1 + 2\varepsilon)$-smooth algorithms and that this is best possible. First, we focus on the simplest algorithm, called Follow-the-Predictions (FTP), defined in Algorithm 1.

Algorithm 1 FTP

```
1: Input: A WMST-instance (G, ŵ, w)
2: Let T be a MST of G w.r.t. ŵ
3: while receiving inputs (w(e_i), e_i) do
4:     if e_i ∈ T then
5:         Accept e_i                              ▷ Add e_i to the solution
```

Theorem 1. $\text{CR}_{\text{FTP}}(\varepsilon) \leqslant 1 + 2\varepsilon$.

We also present a non-trivial algorithm, called Greedy-FTP (GFTP) that starts by producing the tree that FTP outputs. Whenever the true weight of an edge, e, that is not contained in GFTP's current tree is revealed, the algorithm checks whether e can replace an edge in its current tree. It does so by comparing the predicted weights of a subset of edges in its current tree by the newly revealed true weight. We formalize the strategy of GFTP in Algorithm 2.

Theorem 2. $\text{CR}_{\text{GFTP}}(\varepsilon) \leqslant 1 + 2\varepsilon$.

Theorem 3. *For any* $\varepsilon \in [0, 1/2)$*, any deterministic online algorithm,* ALG*, for the WMST problem with weight predictions has* $\text{CR}_{\text{ALG}}(\varepsilon) \geqslant 1 + 2\varepsilon$*.*

Proof. We only sketch the proof. The adversary uses the following graph, where $k = \left\lceil \frac{5}{2-r} \right\rceil$ and $\ell = k^2$.

- $V = \{v_1, v_2, \ldots, v_{2k}\} \cup \{z_j \mid 1 \leqslant j \leqslant \ell\}$,
- $E = I \cup \bigcup_{j=1}^{\ell} E_j$, where

Algorithm 2 GFTP

1: **Input:** A WMST-instance $(G, \hat{w}, w)$
2: Let T be a MST of G w.r.t. $\hat{w}$
3: $S = \emptyset$ ▷ S contains the *seen* edges
4: **while** receiving inputs $(w(e_i), e_i)$ **do**
5: $S = S \cup \{e_i\}$
6: **if** $e_i \in T$ **then**
7: Accept e_i ▷ Add e_i to the solution
8: **else** ▷ $e_i \notin T$
9: C is the cycle e_i introduces in T
10: $C' = C \setminus S$
11: **if** $C' \neq \emptyset$ **then**
12: $e_{\max} = \arg\max_{e_j \in C'} \{\hat{w}(e_j)\}$
13: **if** $w(e_i) \leqslant \hat{w}(e_{\max})$ **then**
14: $T = (T \setminus \{e_{\max}\}) \cup \{e_i\}$ ▷ Update T
15: Accept e_i ▷ Add e_i to the solution

- $I = \bigcup_{i=1}^{2k-1} \{(v_i, v_{i+1})\}$ and
- $E_j = \{(z_j, v_i) \mid 1 \leqslant i \leqslant 2k\}$, for all $j = 1, 2, \ldots, \ell$,

- $\hat{w}_{k,\ell}(e) = w_{k,\ell}(e) = 1$, for all $e \in I$,
- $\hat{w}_{k,\ell}((z_j, v_i)) = k + i - 1$, for all $(z_j, v_i) \in E_j$, and
- For $1 \leq j \leq \ell$, if ALG accepts an edge (z_j, v_i), $1 \leq i \leq 2k - 1$, then $w_{k,\ell}((z_j, v_{i+1})) = i$. Otherwise, $w_{k,\ell}((z_j, v_{2k})) = 4k$.

To see that the result holds for any $\varepsilon < 1/2$, note that adding some real number to all predicted and true weights changes OPT, but ALG − OPT as well as η remain unchanged. □

In the full paper, we prove that, for FTP and GFTP, the lower bound of $1 + 2\varepsilon$ holds for any $\varepsilon \in \mathbb{R}^+$. Thus, we obtain:

Corollary 1. $\text{CR}_{\text{FTP}}(\varepsilon) = \text{CR}_{\text{GFTP}}(\varepsilon) = 1 + 2\varepsilon$, *and this is optimal among deterministic algorithms.*

4 Separation by Random Order Analysis

We show that GFTP has a better random order ratio than FTP, separating the two algorithms. Throughout, we set $T_{\text{FTP}} = \text{OPT}[\hat{w}]$, $T_{\text{OPT}} = \text{OPT}[w]$, and $T_{\text{GFTP}} = \text{GFTP}[\hat{w}, w]$. Further, we denote by T the tree that GFTP makes online changes to. Note that initially $T = T_{\text{FTP}}$, and after GFTP has processed the full input sequence, $T = T_{\text{GFTP}}$. Finally, we denote by S the collection of *seen* edges in $E(G)$, i.e., those edges whose true weight has been revealed.

Theorem 4. $\text{ROR}_{\text{FTP}}(\varepsilon) = 1 + 2\varepsilon$.

Proof. Since FTP does not make online changes to T_{FTP}, the competitive analysis of FTP translates directly to a random order analysis of FTP. Hence, the result follows from Corollary 1. □

We start with the following lower bound on GFTP.

Theorem 5. $\text{ROR}_{\text{GFTP}}(\varepsilon) \geqslant 1 + \varepsilon$.

We now turn to proving an upper bound of $1+(1+\ln(2))\varepsilon \approx 1+1.69\,\varepsilon$ on the random order ratio of GFTP (Theorem 6). To this end, we apply the following lemmas.

Lemma 1. *Let G be a graph, and let T_1 and T_2 be two spanning trees of G. Then, for any edge $e_1 \in T_1 \setminus T_2$, there exists an edge $e_2 \in T_2 \setminus T_1$ such that e_2 introduces a cycle into T_1 that contains e_1, and e_1 introduces a cycle into T_2 that contains e_2.*

Lemma 2. *Let $e \in T_{\text{FTP}} \setminus S$. If, at any point, an edge e' introduces a cycle in T that contains e, then $\hat{w}(e) \leqslant \hat{w}(e')$.*

Lemma 3. *For all integers $n \geqslant 2$, we have that*

$$\frac{1}{n-1}\sum_{i=0}^{n-2}\left(1+\frac{n-1}{2n-2-i}\right) \leqslant 1+\ln(2)\,.$$

Lemma 4. *Suppose that* GFTP *has just rejected $e' \notin T$. Then, at any future point, any unseen edge e that is contained in the cycle that e' introduces into T, at that point, satisfies that $\hat{w}(e) < w(e')$.*

Theorem 6. $\text{ROR}_{\text{GFTP}}(\varepsilon) \leqslant 1 + (1+\ln(2))\varepsilon$.

Proof. Given a WMST-instance $(G, \hat{w}, w)$, we let $T_{\text{GFTP},\sigma}$ denote the output tree that GFTP constructs when run on $(G, \hat{w}, w)$, where the order in which the weights arrive has been permuted according to a uniformly randomly chosen permutation σ of $\{1, 2, \ldots, m\}$. Further, we denote by $\text{GFTP}(\hat{w}, w, \sigma)$ the cost of $T_{\text{GFTP},\sigma}$.

The idea towards a random order ratio upper bound for GFTP is to prove the existence of a subset $\text{E}_{\text{BLAME}} \subset T_{\text{OPT}} \cup T_{\text{GFTP},\sigma}$ such that

$$\mathbb{E}_\sigma[\text{GFTP}(\hat{w}, w, \sigma)] - \text{OPT}(w) \leqslant \sum_{e\in \text{E}_{\text{BLAME}}} |\hat{w}(e) - w(e)|\,.$$

and

$$\mathbb{E}_\sigma[|\text{E}_{\text{BLAME}}|] \leqslant (n-1)(1+\ln(2))\,.$$

More specifically, we define a function $f\colon T_{\text{GFTP},\sigma} \to T_{\text{OPT}}$ and prove that f is bijective, implying that

$$\mathbb{E}_\sigma[\text{GFTP}(\hat{w}, w, \sigma)] - \text{OPT}(w) = \sum_{e\in T_{\text{GFTP},\sigma}} (w(e) - w(f(e)))\,,$$

and then, for each $e \in T_{\text{GFTP},\sigma}$, argue that $w(e) - w(f(e))$ is upper bounded by the prediction error of either e or $f(e)$, or the sum of the two. Then, we show that the expected number of edges for which the upper bound is the error of both e and $f(e)$ is upper bounded by $(n-1)\ln(2)$. We also show that the edges whose errors are used as upper bounds are all distinct.

Recall that throughout the execution of GFTP, its current tree is called T. For the remainder of this proof, we denote by T' a spanning tree of G that is initially set to T_{OPT}. We use T' to keep track of which edges in T_{OPT} have been associated with an edge in $T_{\text{GFTP},\sigma}$ under f. Any time GFTP accepts an edge e, we associate e with an edge $e' \in T'$ under f. We consider two cases:

(a) If $e \in T'$, we set $f(e) = e$ and leave T' unchanged.
(b) If $e \notin T'$, then Lemma 1 implies that there exists an edge $e' \in T' \setminus T$ such that e' introduces a cycle into T that contains e, and e introduces a cycle into T' that contains e'. We select such an edge e', set $f(e) = e'$, and replace e' by e in T'.

We repeat this process every time GFTP accepts an edge. This, however, requires T' to remain a spanning tree at all times. To see that T' remains a spanning tree, we note that in case (a), T' remains unchanged and is therefore still a spanning tree. In case (b), we replace e' with e in T'. Since e introduces a cycle into T' that contains e', it follows that T' remains acyclic after the replacement, and so T' is still a spanning tree.

Towards Bijectivity of f: In case (a), $f(e) = e$, and so $e \in (T \cap T' \cap S)$. Since $e \notin ((T \cap T') \setminus S) \cup (T' \setminus T)$, we never map to e again later. In case (b), $f(e) = e'$, and after replacing e' by e in T', we find that $e \in (T \cap T' \cap S)$, and $e' \in \overline{T \cup T'}$. Hence, as neither e nor e' is contained in $(T \cap T' \setminus S) \cup (T' \setminus T)$, we never map to either again later. Hence f is injective, and since $|T_{\text{GFTP},\sigma}| = |T_{\text{OPT}}|$, f is bijective.

We now describe how the set E_{BLAME} is constructed. For each edge e accepted by GFTP, we do the following, based on the situation at the time when $w(e)$ has been revealed, but e has not yet been handled by GFTP.

- If $e \in T'$, no edge is added to E_{BLAME}.
- If $e \in \overline{T \cup T'}$, we add $f(e)$ to E_{BLAME}.
- If $e \in T \setminus T'$, then
  - if $w(f(e))$ arrives before $w(e)$, we add e to E_{BLAME},
  - otherwise, we add both e and $f(e)$ to E_{BLAME}.

For each edge e accepted by GFTP, let E_e be the set of edges added to E_{BLAME} because of e by the above scheme. By the following case analysis, we prove that

$$w(e) - w(f(e)) \leq \sum_{e' \in E_e} |\hat{w}(e') - w(e')| .$$

Whenever an edge, e_{next}, has its weight revealed, we consider the following cases.

Case $e_{\text{next}} \in T \cap T'$: GFTP accepts e_{next}. By (a), $f(e_{\text{next}}) = e_{\text{next}}$, implying that $w(e_{\text{next}}) - w(f(e_{\text{next}})) = 0$.

Case $e_{\text{next}} \in \overline{T \cup T'}$: If GFTP accepts $e_{\text{next}} \in \overline{T \cup T'}$, it does so due to e_{next} replacing some edge $e \in T$ that is contained in the cycle that e_{next} introduces into T. We let e_{OPT} denote $f(e_{\text{next}})$ and argue that

$$w(e_{\text{next}}) - w(e_{\text{OPT}}) \leqslant |\hat{w}(e_{\text{OPT}}) - w(e_{\text{OPT}})|\,.$$

Note that since GFTP swapped out e for e_{next}, we have that $w(e_{\text{next}}) \leqslant \hat{w}(e)$. Further, we can argue that $\hat{w}(e) \leqslant \hat{w}(e_{\text{OPT}})$. Indeed, if $e = e_{\text{OPT}}$, this is trivial. If $e \neq e_{\text{OPT}}$, then, since e_{OPT} introduces a cycle that contains e_{next}, it follows that before swapping out e for e_{next}, e_{OPT} would introduce a cycle into T containing e, and so $\hat{w}(e) \leqslant \hat{w}(e_{\text{OPT}})$, by Lemma 2. Hence,

$$\begin{aligned} w(e_{\text{next}}) - w(e_{\text{OPT}}) &\leqslant \hat{w}(e) - w(e_{\text{OPT}}) \\ &\leqslant \hat{w}(e_{\text{OPT}}) - w(e_{\text{OPT}}) \\ &\leqslant |\hat{w}(e_{\text{OPT}}) - w(e_{\text{OPT}})|\,. \end{aligned}$$

Case $e_{\text{next}} \in T' \setminus T$: In this case, e_{next} introduces a cycle C in T. Denote by e an edge in $C \setminus S$ for which $e = \arg\max_{e_i \in C \setminus S}\{\hat{w}(e_i)\}$. We split the remaining analysis of this case into two subcases.

Subcase (accept): If $w(e_{\text{next}}) \leqslant \hat{w}(e)$, then GFTP accepts e_{next} and removes e from its tree. Then, by (a), $f(e_{\text{next}}) = e_{\text{next}}$ and so $w(e_{\text{next}}) - w(f(e_{\text{next}})) = 0$.

Subcase (reject): If $\hat{w}(e) < w(e_{\text{next}})$, then GFTP rejects e_{next}. Since each edge in T' will, at some point, be associated with an edge in $T_{\text{GFTP},\sigma}$, by the bijectivity of f, it follows that GFTP will later accept some edge that will be associated with e_{next} under f. Denote this edge by e_{future}, such that $f(e_{\text{future}}) = e_{\text{next}}$.

Note that GFTP can accept e_{future} either due to a swap, or because $w(e_{\text{future}})$ was revealed while contained in T. In the latter case, at the time where GFTP accepts e_{future}, we find that e_{future} is contained in the cycle that e_{next} introduces into T, and so, by Lemma 4, $\hat{w}(e_{\text{future}}) < w(e_{\text{next}})$, implying that $w(e_{\text{future}}) - w(e_{\text{next}}) < w(e_{\text{future}}) - \hat{w}(e_{\text{future}})$.

Case $e_{\text{next}} \in T \setminus T'$: In this case, GFTP accepts e_{next}. By (b), there exists an edge $e_{\text{OPT}} \in T' \setminus T$ such that $f(e_{\text{next}}) = e_{\text{OPT}}$. Since T remains unchanged when $w(e_{\text{next}})$ is revealed, it follows that e_{OPT} would introduce a cycle in T containing e_{next} before $w(e_{\text{next}})$ was revealed. Hence, by Lemma 2, we find that $\hat{w}(e_{\text{next}}) \leqslant \hat{w}(e_{\text{OPT}})$, and so

$$\begin{aligned} w(e_{\text{next}}) - w(e_{\text{OPT}}) &= w(e_{\text{next}}) - \hat{w}(e_{\text{next}}) + \hat{w}(e_{\text{next}}) - w(e_{\text{OPT}}) \\ &\leqslant w(e_{\text{next}}) - \hat{w}(e_{\text{next}}) + \hat{w}(e_{\text{OPT}}) - w(e_{\text{OPT}}) \\ &\leqslant |w(e_{\text{next}}) - \hat{w}(e_{\text{next}})| + |\hat{w}(e_{\text{OPT}}) - w(e_{\text{OPT}})|\,. \end{aligned}$$

If $w(f(e_{\text{next}}))$ is revealed before $w(e_{\text{next}})$, we obtain a stronger upper bound of $|w(e_{\text{next}}) - \hat{w}(e_{\text{next}})|$, as shown in Case $e_{\text{next}} \in T' \setminus T$ Subcase (reject).

This ends the case analysis.

Next, we show that all edges in E_{BLAME} are distinct. To this end, let e be an edge that GFTP has just accepted. By construction of E_{BLAME}, we either add e, $f(e)$, or both to E_{BLAME}. After e has been accepted, $e \in T \cap T'$, and can therefore never be hit under f. Hence, e will not be added to E_{BLAME} again later. Similarly, if $f(e) = e$, $f(e)$ will not be added to E_{BLAME} again later. On the other hand, if $f(e) \neq e$, then after replacing $f(e)$ with e in T', $f(e) \in \overline{T \cup T'}$, and $f(e)$ may therefore be accepted later due to a swap. In this case, by construction of E_{BLAME}, $f(f(e)) \neq f(e)$ is added to E_{BLAME}, and so we do not add $f(e)$ twice.

Now, all that remains is to show that $\mathbb{E}_\sigma[|E_{BLAME}|] \leqslant (n-1)(1+\ln(2))$.

For the remainder of this proof, let i be the random variable that counts the number of edges that have either been accepted by GFTP (now in $T' \cap T \cap S$), or belong to $T' \setminus T$ and have been rejected (now in $(T' \setminus T) \cap S$). Thus, $i = |T' \cap T \cap S| + |(T' \setminus T) \cap S| = |T' \cap S|$. One may observe that i remains unchanged when $e_{\text{next}} \in \overline{T \cup T'}$ and GFTP rejects e_{next}. In all other of the above cases, i is incremented.

We prove the following invariants.

Invariant (i): Any edge in $T \setminus T'$ is unseen.

Proof of (i): Initially, all edges are unseen. Whenever the weight of an edge $e \in T \setminus T'$ is revealed, we replace $f(e)$ with e in T', so now $e \in T \cap T'$. Hence, after replacing $f(e)$ with e in T', all edges in $T \setminus T'$ are again unseen.

Invariant (ii): For any $0 \leqslant i \leqslant n-2$, the probability that the next edge is contained in $T \setminus T'$, denoted p_i, satisfies

$$p_i = \frac{|(T \setminus T') \setminus S|}{|E(G) \setminus S|} \leqslant \frac{n-1}{2n-2-i},$$

Proof of (ii): We let j count the number of seen edges. At any point in time, $i \leq j$. Now, observe that

$$p_i = \frac{|(T \setminus T') \setminus S|}{|E(G) \setminus S|} \leqslant \frac{|(T \setminus T') \setminus S|}{|(T \cup T') \setminus S|}$$

From Invariant (i), it follows that $|(T \setminus T') \setminus S| = n-1-a_j-x_j$, where a_j is the number of edges that have been accepted after j edges have had their weights revealed, i.e., the number of edges in $T \cap T' \cap S$, and x_j is the number of edges in $T \cap T' \setminus S$. Then,

$$p_i \leqslant \frac{n-1-x_j-a_j}{|(T \cup T') \setminus S|}.$$

Now,

$$|(T \cup T') \setminus S| = |T \cup T'| - |(T \cup T') \cap S|.$$

For any $0 \leqslant i \leqslant n-2$,

$$\begin{aligned} &|T \cup T'| = |T| + |T'| - |T \cap T'| = 2n-2-x_j-a_j, \\ \text{and} \quad &|(T \cup T') \cap S| = |(T \setminus T') \cap S| + |T' \cap S| = |T' \cap S| = i. \end{aligned}$$

Here the second to last equality follows from Invariant (i), and the last equality follows from the definition of i. Hence,

$$p_i \leqslant \frac{n-1-x_j-a_j}{2n-2-i-x_j-a_j}.$$

Using that $a_j + x_j \geqslant 0$ and $i \leqslant n-1$, it follows that

$$p_i \leqslant \frac{n-1}{2n-2-i}.$$

The only time we add two edges to E_{BLAME} is when $e_{\text{next}} \in T \setminus T'$ and $f(e_{\text{next}}) \notin (T' \setminus T) \cap S$. For each $i = 0, 1, \ldots, n-2$, the probability that e_{next} is in $T \setminus T'$ is p_i. In any other case, we add at most one edge to E_{BLAME}.

Note that i is the size of the set of edges, e, for which either $w(e)$ has been revealed or $w(f(e))$ but not $w(e)$ has been revealed. If the weight of $f(e)$ but not that of e has been revealed, then $w(e) - w(f(e))$ is accounted for in Case $e_{\text{next}} \in T' \setminus T$, Subcase (reject), where $e_{\text{next}} = f(e)$. Therefore, when $i = n-1$, all edges in $T_{\text{GFTP},\sigma}$ have been accounted for, and so we can compute the size of E_{BLAME}. We obtain

$$\mathbb{E}_\sigma[|\text{E}_{\text{BLAME}}|] \leqslant \sum_{i=0}^{n-2}(1+p_i) \leqslant \sum_{i=0}^{n-2}\left(1 + \frac{n-1}{2n-2-i}\right) \leqslant (n-1)(1+\ln(2)),$$

where the last inequality follows from Lemma 3.

Since the prediction error of any edge is only used to upper bound incurred cost once, and since the average of the $n-1$ largest prediction errors upper bound the average prediction error of any set of $(n-1)(1+\ln(2))$ edges, it follows that

$$\mathbb{E}_\sigma[\text{GFTP}(\hat{w}, w, \sigma) - \text{OPT}(w)] \leqslant (1+\ln(2))\eta,$$

so

$$\frac{\mathbb{E}_\sigma[\text{GFTP}(\hat{w}, w, \sigma)]}{\text{OPT}(w)} \leqslant 1 + (1+\ln(2))\varepsilon,$$

and, hence, $\text{ROR}_{\text{GFTP}}(\varepsilon) \leqslant 1 + (1+\ln(2))\varepsilon$. □

5 Open Problems

An obvious open problem is to determine the exact random order ratio of GFTP, in the range $1+\varepsilon$ to $1+\ln(2)\varepsilon$.

GFTP can be seen as an improvement of FTP, and we are interested in what we believe could be a further improvement: In addition to accepting some edges that are not in the chosen minimum spanning tree based on predictions, also reject *some* that *are* in that tree, if the actual weight is higher than the predicted. The obvious approach gives an algorithm with a worse competitive ratio than FTP's, but restricting which edges the algorithm can accept after such a rejection gives rise to another optimal algorithm under competitive analysis. It would be interesting to apply random order analysis to such an algorithm as well.

More generically, it would be interesting to apply random order analysis to other online problems with predictions, as well as to consider error measures similar to ours for other problems.

References

1. Algorithms with Predictions. https://algorithms-with-predictions.github.io/. Accessed 14 Feb 2023
2. Berg, M., Boyar, J., Favrholdt, L.M., Larsen, K.S.: Online minimum spanning trees with weight predictions (2023). arXiv:2302.12029
3. Bianchi, M.P., Böckenhauer, H., Brülisauer, T., Komm, D., Palano, B.: Online minimum spanning tree with advice. Int. J. Found. Comput. Sci. **29**(4), 505–527 (2018)
4. Boyar, J., Favrholdt, L.M., Kudahl, C., Larsen, K.S., Mikkelsen, J.W.: Online algorithms with advice: a survey. ACM Comput. Surv. **50**(2), 1–34 (2017). Article No. 19
5. Boyar, J., Favrholdt, L.M., Larsen, K.S.: Relative worst-order analysis: a survey. ACM Computing Surv. **54**(1), 1–21 (2020). Article No. 8
6. Boyar, J., Irani, S., Larsen, K.S.: A comparison of performance measures for online algorithms. Algorithmica **72**(4), 969–994 (2015)
7. Dobrev, S., Královič, R., Pardubská, D.: Measuring the problem-relevant information in input. RAIRO - Theor. Inform. Appl. **43**(3), 585–613 (2009)
8. Dorrigiv, R., López-Ortiz, A.: A survey of performance measures for on-line algorithms. SIGACT News **36**(3), 67–81 (2005)
9. Emek, Y., Fraigniaud, P., Korman, A., Rosén, A.: Online computation with advice. Theor. Comput. Sci. **412**(24), 2642–2656 (2011)
10. Gupta, A., Singla, S.: Random-order models. In: Roughgarden, T. (ed.) Beyond the Worst-Case Analysis of Algorithms, pp. 234–258. Columbia University, New York (2020)
11. Hromkovič, J., Královič, R., Královič, R.: Information complexity of online problems. In: Hliněný, P., Kučera, A. (eds.) MFCS 2010. LNCS, vol. 6281, pp. 24–36. Springer, Heidelberg (2010). https://doi.org/10.1007/978-3-642-15155-2_3
12. Im, S., Kumar, R., Qaem, M.M., Purohit, M.: Non-clairvoyant scheduling with predictions. In: 33rd ACM Symposium on Parallelism in Algorithms and Architectures (SPAA), pp. 285–294. ACM (2021)
13. Karlin, A.R., Manasse, M.S., Rudolph, L., Sleator, D.D.: Competitive snoopy caching. Algorithmica **3**, 77–119 (1988)
14. Kenyon, C.: Best-fit bin-packing with random order. In: 7th Annual ACM-SIAM Symposium on Discrete Algorithms (SODA), pp. 359–364. SIAM (1996)
15. Komm, D.: An Introduction to Online Computation: Determinism, Randomization, Advice. Springer, Cham (2016). https://doi.org/10.1007/978-3-319-42749-2
16. Lykouris, T., Vassilvitskii, S.: Competitive caching with machine learned advice. J. ACM **68**(4), 1–25 (2021)
17. Mitzenmacher, M., Vassilvitskii, S.: Algorithms with predictions. Commun. ACM **65**(7), 33–35 (2022)

18. Sleator, D.D., Tarjan, R.E.: Amortized efficiency of list update and paging rules. Commun. ACM **28**(2), 202–208 (1985)
19. West, D.B.: Introduction to Graph Theory. Featured Titles for Graph Theory, 2nd edn. Prentice Hall, Hoboken (2001)

Compact Distance Oracles with Large Sensitivity and Low Stretch

Davide Bilò[1], Keerti Choudhary[2], Sarel Cohen[3], Tobias Friedrich[4], Simon Krogmann[4], and Martin Schirneck[5](✉)

[1] Department of Information Engineering, Computer Science and Mathematics, University of L'Aquila, L'Aquila, Italy
davide.bilo@univaq.it

[2] Department of Computer Science and Engineering, Indian Institute of Technology Delhi, Delhi, India
keerti@iitd.ac.in

[3] School of Computer Science, The Academic College of Tel Aviv-Yaffo, Tel Aviv-Yafo, Israel
sarelco@mta.ac.il

[4] Hasso Plattner Institute, University of Potsdam, Potsdam, Germany
{tobias.friedrich,simon.krogmann}@hpi.de

[5] Faculty of Computer Science, University of Vienna, Vienna, Austria
martin.schirneck@univie.ac.at

Abstract. An *f-edge fault-tolerant distance sensitive oracle* (f-DSO) with stretch $\sigma \geqslant 1$ is a data structure that preprocesses an input graph G. When queried with the triple (s, t, F), where $s, t \in V$ and $F \subseteq E$ contains at most f edges of G, the oracle returns an estimate $\widehat{d}_{G-F}(s,t)$ of the distance $d_{G-F}(s,t)$ between s and t in the graph $G - F$ such that $d_{G-F}(s,t) \leqslant \widehat{d}_{G-F}(s,t) \leqslant \sigma \cdot d_{G-F}(s,t)$.

For any positive integer $k \geqslant 2$ and any $0 < \alpha < 1$, we present an f-DSO with sensitivity $f = o(\log n/ \log\log n)$, stretch $2k-1$, space $O(n^{1+\frac{1}{k}+\alpha+o(1)})$, and an $\widetilde{O}(n^{1+\frac{1}{k}-\frac{\alpha}{k(f+1)}})$ query time.

Prior to our work, there were only three known f-DSOs with subquadratic space. The first one by Chechik et al. [Algorithmica 2012] has a stretch of $(8k-2)(f+1)$, depending on f. Another approach is storing an *f-edge fault-tolerant* $(2k-1)$-*spanner* of G. The bottleneck is the large query time due to the size of any such spanner, which is $\Omega(n^{1+1/k})$ under the Erdős girth conjecture. Bilò et al. [STOC 2023] gave a solution with stretch $3+\varepsilon$, query time $O(n^\alpha)$ but space $O(n^{2-\frac{\alpha}{f+1}})$, approaching the quadratic barrier for large sensitivity.

In the realm of subquadratic space, our f-DSOs are the first ones that guarantee, at the same time, large sensitivity, low stretch, and non-trivial query time. To obtain our results, we use the approximate distance oracles of Thorup and Zwick [JACM 2005], and the derandomization of the f-DSO of Weimann and Yuster [TALG 2013] that was recently given by Karthik and Parter [SODA 2021].

Keywords: Approximate shortest paths · Distance sensitivity oracle · Fault-tolerant data structure · Spanner · Subquadratic space

P. Morin and S. Suri (Eds.): WADS 2023, LNCS 14079, pp. 149–163, 2023.
https://doi.org/10.1007/978-3-031-38906-1_11

1 Introduction

There are applications, like routing on edge devices, where we want to quickly find out the distances between pairs of vertices, but we cannot store the entire graph topology due to memory restrictions. This problem is solved by a class of data structures called *distance oracles* (DO). Typically, not a single structure serves every use case and constructions need to provide reasonable trade-offs between the space requirement, query time, and stretch, that is, the quality of the estimated distance. We are interested in the fault-tolerant setting. Here, the data structure must additionally be able to tolerate multiple edge failures in the underlying graph. An *f-edge fault-tolerant distance sensitivity oracles* (*f*-DSO) for a graph $G = (V, E)$ is able to report, for any two $s, t \in V$ and set $F \subseteq E$ of at most f failing edges, an estimate $\widehat{d}_{G-F}(s,t)$ of the *replacement distance* $d_{G-F}(s,t)$ in the graph $G-F$. The parameter f is the *sensitivity* of the oracle. We say the *stretch* of the data structure is σ if $d_{G-F}(s,t) \leqslant \widehat{d}_{G-F}(s,t) \leqslant \sigma \cdot d_{G-F}(s,t)$, for any admissible query (s, t, F).

Several f-DSOs with different space-stretch-time trade-offs have been designed in the last decades, most of which can only handle a very small number $f \leqslant 2$ of failures [3–5,9,15,19–21,24,25,32]. We highlight those with sensitivity $f \geqslant 3$. The f-DSO of Duan and Ren [22] requires $O(fn^4)$ space,[1] returns exact distances in $f^{O(f)}$ query time, but the preprocessing algorithm that builds it requires exponential-in-f time $n^{\Omega(f)}$. The data structure by Chechik, Cohen, Fiat, and Kaplan [16] is more compact, requiring $O(n^{2+o(1)} \log W)$ space, and can be preprocessed in time[2] $\widetilde{O}(n^{5+o(1)} \log W)$, where W is the weight of the heaviest edge of G. The oracle has stretch $1 + \varepsilon$, for any constant $\varepsilon > 0$, with an $O(f^5 \log n \log\log W)$ query time, and handles up to $f = o(\log n / \log\log n)$ failures. Finally, the f-DSO of Chechik, Langberg, Peleg, and Roditty [18] requires a subquadratic space of $O(fkn^{1+1/k} \log(nW))$, where $k \geqslant 1$ is an integer parameter, and has a fast query time of $\widetilde{O}(f \log\log d_{G-F}(s,t))$, but guarantees only a stretch of $(8k-2)(f+1)$ that depends on the number f of failures.

Another common way to provide approximate replacement distances in the presence of transient edge failures are fault-tolerant spanners [30]. An *f-edge fault-tolerant spanner with stretch σ (fault-tolerant σ-spanner)* is a subgraph H of G such that $d_{H-F}(s,t) \leqslant \sigma \cdot d_{G-F}(s,t)$, for every suitable triple (s, t, F), with $|F| \leqslant f$. For any positive integer k, Chechik, Langberg, Peleg, and Roditty [17] gave an algorithm computing a fault-tolerant $(2k-1)$-spanner with $O(fn^{1+1/k})$ edges. This was recently improved by Bodwin, Dinitz, and Robelle by reducing the size to $O(f^{1-1/k} n^{1+1/k})$ [10] and eventually to $f^{1/2} n^{1+1/k} \cdot \mathsf{poly}(k)$ for any even k and $f^{1/2-1/(2k)} n^{1+1/k} \cdot \mathsf{poly}(k)$ for odd k [11]. The authors of the last work also show almost matching lower bounds of $\Omega(f^{1/2-1/(2k)} n^{1+1/k} + fn)$ for general $k > 2$, and $\Omega(f^{1/2} n^{3/2})$ for $k = 2$ assuming the Erdős girth conjecture [23].

The main problem with the spanner approach is the high query time. In fact, to retrieve the approximate distance between a given pair of vertices, one

[1] The space is measured in the number of machine words on $O(\log n)$ bits.

[2] For a non-negative function $g(n)$, we use $\widetilde{O}(g)$ to denote $O(g \cdot \mathsf{polylog}(n))$.

has to compute the single-source distance from one of them, say with Dijkstra's algorithm, in time that is at least linear in the size of the spanner. Therefore, an important problem in the field of fault-tolerant data structures to design f-DSOs with subquadratic space, that simultaneously guarantee a non-trivial $o(n^{1+1/k})$ query time, a low stretch of $2k-1$, and a large sensitivity f.[3]

Very recently, Bilò, Chechik, Choudhary, Cohen, Friedrich, Krogmann, and Schirneck [6] addressed the same problem. They presented, for all $\varepsilon > 0$, and constants $1/2 > \alpha > 0$, and f, a $(3+\varepsilon)$-approximate f-DSO for unweighted graphs taking space $\widetilde{O}_\varepsilon(n^{2-\frac{\alpha}{f+1}}(\log n)^{f+1})$ and has a query time of $\widetilde{O}_\varepsilon(n^\alpha)$. While their query time is sub-linear, their space converges to quadratic for large sensitivity.

In contrast, we design a deterministic oracle for weighted graphs that can handle up to $f = o(\log n/\log\log n)$ edge failures and provides a trade-off between stretch, space, and query time. Namely, for any positive integer $k \geqslant 2$ and constant $1-\frac{1}{k} > \alpha > 0$, our data structure has stretch $2k-1$, requires $kn^{1+\alpha+\frac{1}{k}+o(1)}$ space, and can be queried in time $\widetilde{O}(n^{1+\frac{1}{k}-\frac{\alpha}{k(f+1)}})$. The query time improves to $\mathsf{poly}(D, f, \log n)$ for graphs in which the pair-wise hop distances are bounded by D. If, for example, D is polylogarithmic, the query time is as well. We note that the query time of our f-DSO for general graphs is $\Omega(n)$ for all choices of α.

Both [6] and this work approach the problem by handling hop-short and hop-long paths separately, as is common in the area [24,36], and use the distance oracle of Thorup and Zwick [35] on the lowest level. Apart from that, the techniques are different. We highlight ours below.

Our Techniques. Our f-DSO for bounded hop diameter is the result of combining the approximate distance oracles of Thorup and Zwick [35] with randomized replacement path covering (RPC), a collection of certain subgraphs of G, introduced by Weimann and Yuster [36]. Such coverings are very large, even larger than the underlying graph itself. They are thus unusable when emphasizing subquadratic space, barring additional processing. The main issue when compressing an RPC is retaining the information which subgraph is relevant for which query. We provide two different ways to solve this. One is based on the idea of using sparse spanners as proxies for the subgraphs in the covering, and the other one uses the recent derandomization technique of Karthik and Parter [27]. To lift this to an arbitrary hop diameter, we borrow from fault-tolerant spanners. There, a single graph is constructed up front to handle all queries. To achieve a compact oracle with a query time better than any spanner, we instead turn this process around and use the hop-short f-DSO to combine only the subgraphs we need.

Other Related Work. Demetrescu and Thorup [19] designed the first exact 1-DSO for directed edge-weighted graphs with $O(n^2 \log n)$ space and $O(\log n)$ query time. Demetrescu, Thorup, Chowdhury, and Ramachandran [20] improved the query time to $O(1)$ and generalized the oracle to handle also a single vertex failure. Later, in two consecutive papers, Bernstein and Karger improved the preprocessing time from $\widetilde{O}(mn^2)$ to $\widetilde{O}(mn)$ [4,5]. Khanna and Baswana [28]

[3] Subquadratic space f-DSOs with stretch $2k-1$ can only exist for $k \geqslant 2$. There is an $\Omega(n^2)$-bit lower bound for exact f-DSOs, regardless of the query time [35].

designed 1-DSO for unweighted graphs having size $O(k^5 n^{1+1/k} \frac{\log^3 n}{\varepsilon^4})$, a stretch of $(2k-1)(1+\varepsilon)$, and $O(k)$ query time. The problem of 1-DSO was also studied with a special focus on the preprocessing time [15,24,25,32,36].

For the case of multiple failures, other than the results we explicitly mentioned in the introduction [16,18,22], it is worth mentioning the 2-DSO of Duan and Pettie [21] with $O(n^2 \log^3 n)$ size and $O(\log n)$ query time and the work by van den Brand and Saranurak [12].

Outline. Section 2 provides an overview of our approach and presents our results. The preliminaries and notation needed to follow the technical part are given in Sect. 3. In Sect. 4, we first describe the randomized subquadratic-space f-DSO for short hop distances and then derandomize it, not only to obtain a deterministic construction but also to accelerate the query time to $\mathsf{poly}(D, f, \log n)$. Section 5 then describes how to use this to develop a deterministic subquadratic-space f-DSO also for hop-long replacement paths.

Some of the proofs are deferred to the full version [7] due to space reasons.

2 Overview

Our first goal is to develop an f-DSO whose space is subquadratic in n, provided that the hop diameter[4] D and the sensitivity f are not too large. One of the first DSOs was given by Weimann and Yuster [36]. It reports exact distances but, on graphs with a large hop diameter, it is too large and too slow. We first give an overview of their techniques and then describe the steps we take to reduce the space as well as the query time using approximation.

Given the graph $G = (V, E)$ as well as positive integers L and f, the DSO in [36] samples a family $\mathcal{G}$ of $\widetilde{O}(fL^f)$ random spanning subgraphs of G, that is, all the subgraphs have the same vertex set V. Each graph $G_i \in \mathcal{G}$ is generated by removing each edge of G with probability $1/L$. With high probability,[5] for all vertices $s, t \in V$ and sets $F \subseteq E$ of at most f edge failures, if there is a replacement path from s to t that has at most L edges and none of them is in F, then there exists a subgraph $G_i \in \mathcal{G}$ that does not contain any edge of F but such an replacement path. Let $\mathcal{G}_F \subseteq \mathcal{G}$ be the subfamily of all the G_i in which at least all of F was removed. In other words, if s and t have a *hop-short* shortest path in $G - F$, at least one of their replacement paths survives in a graph in $\mathcal{G}_F$.

To handle hop-short replacement paths, it is enough to go over the subgraphs and report the minimum distance $d_{G_i}(s,t)$ over all $G_i \in \mathcal{G}_F$. For the *hop-long* replacement paths on more than L edges, a random subset $B \subseteq V$ of $\widetilde{O}(fn/L)$ of *pivots* is sampled. This way any hop-long replacement path decomposes into short subpaths such that both endpoints are in B. To answer a hop-long query (s, t, F), a dense weighted graph H^F is created on the vertex set

[4] The *hop diameter* of a weighted graph is the minimum integer D such that all shortest paths between pairs of vertices have at most D edges.

[5] An event occurs *with high probability* (w.h.p.) if it has success probability at least $1 - n^{-c}$ for some constant $c > 0$. In fact, c can always be made arbitrarily large without affecting the asymptotics.

$V(H^F) = B \cup \{s,t\}$ such that for any two $u, v \in V(H^F)$ the edge $\{u,v\}$ has weight $\min_{G_i \in \mathcal{G}_F} d_{G_i}(u,v)$. Those edges represent the subpaths. The oracle's eventual answer to the query is the distance $d_{H^F}(s,t)$ in H^F.

The replacement distances reported by the DSO are exact w.h.p. However, this approach has several drawbacks. The most important one for us is that each of the graphs G_i has $\Omega(m)$ edges, raising the space to store them all to $\Omega(fL^f m)$, which is super-quadratic in n for dense graphs G. Also, the query time is rather high, the bottleneck is computing the weight of the $O(|B|^2) = \widetilde{O}(f^2 n^2/L^2)$ edges of H^F for the hop-long paths.

The key observation for improving this result in graphs with a small hop diameter is that there *all* replacement paths are hop-short. Afek, Bremler-Barr, Kaplan, Cohen, and Merritt [1] showed that for undirected, weights graphs G and failure sets $F \subseteq E$ with $|F| \leqslant f$, every shortest path in $G-F$ is a concatenation of at most $f+1$ shortest paths in G interleaved with at most f edges. So if D is a bound on the hop diameter of G, the hop diameter of $G - F$ is at most $L = (f+1)D + f$. With this definition of L, we can safely ignore hop-long replacement paths. Note that the assumption of G being undirected is essential here: The Afek et al. [1] result fails in directed graphs. Moreover, there is no hope for a subquadratic DSO in that case. Thorup and Zwick [35] showed that every data structure reporting pairwise distances in a directed graph must take $\Omega(n^2)$ space. This holds even if the data structure does not support a single edge failure and only provides an arbitrary finite approximation of the distance.

Nevertheless, we can use approximation in order to reduce the space of the DSO for undirected graphs. Instead of storing the subgraphs G_i, we replace them by the *distance oracle* (DO) of Thorup and Zwick [35]. For any positive integer k and G_i, we get a DO of size $O(kn^{1+1/k})$ that, when queried with two vertices s, t, reports the distance $d_{G_i}(s,t)$ but with a *stretch* of $2k-1$. That means, the returned value $\widehat{d}(s,t)$ satisfies $d_{G_i}(s,t) \leqslant \widehat{d}(s,t) \leqslant (2k-1) \cdot d_{G_i}(s,t)$. While the use of more efficient data structures reduces the space of our DSO to $\widetilde{O}(fL^f n^{1+1/k})$, discarding the actual subgraphs G_i makes it impossible to recover the information which edges have been removed in which graph, that is, to compute the subfamily $\mathcal{G}_F$. We provide two different ways to solve this. The first one is to use spanners. The DO in [35] is accompanied by a spanner of the same size and we show that if the spanner associated with G_i does not contain an edge of F then it is safe to rely on G_i for the replacement distances, even if the graph itself has some failing edges from F.

Interestingly, the other solution comes from derandomization. Karthik and Parter [27] showed how to make the subgraph creation deterministic, albeit now with $O((cfL\log m)^{f+1})$ such graphs for some constant $c > 0$. This makes the resulting DSO less compact and also increases the preprocessing time. However, they presented a way to compute the now deterministic family $\mathcal{G}_F$ using error-correcting codes. This allows us to significantly improve the query time if the diameter is small. For this, we show how to implement the encoding procedure without using additional storage space.

We present our results in the following setting. We consider graphs with *polynomial edge weights*, meaning that they are edge-weighted by positive reals from a range of size $\mathsf{poly}(n)$, where n is the number of vertices. While the weights themselves may have arbitrary precision, the number of values that can be written as sums of at most n weights is again polynomial. Therefore, we can encode any graph distance in a constant number of $O(\log n)$-bit machine words. The restriction on the range is justified as follows. Let $W = \max_{e \in E} w(e) / \min_{e \in E} w(e)$ be the ratio between the maximum and minimum weight. Chechik et al. [16, Lemma 4.1] gave a reduction from approximate DSOs for general weighted graphs to approximate DSOs for graphs with polynomial weights that increases the space and preprocessing time only by a factor $O(\frac{\log W}{\log n})$, the query time by a factor $O(\log \log W)$, and the stretch by a factor $1 + \frac{1}{n}$.

In the statements below, k controls the stretch vs. space trade-off is an arbitrary positive integer, possibly even depending on the number of vertices n. However, there are only space improvements to be had for values $k = O(\log n)$.

Theorem 1. *Let $G = (V, E)$ be an undirected graph with polynomial edge weights, and hop diameter D. For all positive integers k and $f = o(\log n / \log \log n)$, there is an f-DSO for G that has stretch $2k - 1$ and satisfies the following properties.*

1. *(Randomized.) The DSO takes space $D^f k n^{1+\frac{1}{k}+o(1)}$, has a preprocessing time of $D^f kmn^{\frac{1}{k}+o(1)}$, and answers queries correctly w.h.p. in time $D^f n^{o(1)}$.*
2. *(Deterministic.) The DSO takes space $D^{f+1} k n^{1+\frac{1}{k}+o(1)}$, has preprocessing time of $D^{f+1} kmn^{\frac{1}{k}+o(1)}$, and query time $O(f^3 D^{\frac{\log n \log \log n}{\log D}})$.*

Corollary 1. *If G has a polylogarithmic hop diameter, then there is an f-DSO for G with stretch $2k - 1$ that takes $kn^{1+\frac{1}{k}+o(1)}$ space, has a preprocessing time of $kmn^{\frac{1}{k}+o(1)}$, and $\widetilde{O}(1)$ query time.*

We also devise a solution for graphs with an arbitrarily large hop diameter. To do so, we have to compute the correct distances for hop-long replacement paths. In [36], this was the role of the dense subgraph H^F on the pivots in B. Imagine we would sparsify it using the spanner construction above. This would significantly reduce the number of edges we need and stretch the distance $d_{H^F}(s,t)$ to at most $2k-1$ times the correct replacement distance. But computing first the graph and then the distance would still take a lot of time. Instead, the idea of our solution is to prepare a spanner on vertex set B for each subgraph and to combine only those we need for the result. This way, we achieve both low memory and $o(n^{1+1/k})$ query time, as stated in the following theorem.

We remark again that Bilò et al. [6, Theorem 1.1] gave an oracle for unweighted graphs whose query time is sublinear, at the expense of the space being only marginally subquadratic.

Theorem 2. *Let $G = (V, E)$ be an undirected graph polynomial edge weights. For all positive integers k and $f = o(\log n / \log \log n)$, and every $0 < \alpha < 1$, there is an f-DSO for G with stretch $2k - 1$, space $kn^{1+\alpha+\frac{1}{k}+o(1)}$, preprocessing time $kmn^{1+\alpha+\frac{1}{k}+o(1)}$, and query time $\widetilde{O}(n^{1+\frac{1}{k}-\frac{\alpha}{k(f+1)}})$.*

3 Preliminaries

Shortest Paths and Hop Diameter. We let $G = (V, E)$ denote the undirected base graph with n vertices and m edges, edge-weighted by a function $w\colon E \to \mathcal{W}$, where the set of admissible weights $\mathcal{W} \subseteq \mathbb{R}^+$ is of size $|\mathcal{W}| = \mathsf{poly}(n)$. We tacitly assume $m = \Omega(n)$. For any undirected graph H (that may differ from the input G) we denote by $V(H)$ and $E(H)$ the set of its vertices and edges, respectively. Let P be a path in H from a vertex $s \in V(H)$ to $t \in V(H)$, we say that P is an *s-t-path* in H. We denote by $|P| = \sum_{e \in E(P)} w(e)$ the *length* of P, that is, its total weight. For vertices $u, v \in V(P)$, we let $P[u..v]$ denote the subpath of P from u to v. For two paths P, Q in H that share an endpoint, we use $P \circ Q$ for their concatenation. For $s, t \in V(H)$, the *distance* $d_H(s,t)$ is the minimum length of any s-t-path in H; if s and t are disconnected, we set $d_H(s,t) = +\infty$. When talking about the base graph G, we drop the subscripts if this does not create any ambiguities. The *hop diameter* of H is the maximum number of edges of any shortest path between pairs of vertices in $V(H)$.

Spanners and Distance Sensitivity Oracles. A *spanner of stretch* $\sigma \geqslant 1$, or σ*-spanner*, for H is a subgraph $S \subseteq H$ such that for any two vertices $s, t \in V(S) = V(H)$, the inequality $d_H(s,t) \leqslant d_S(s,t) \leqslant \sigma \cdot d_H(s,t)$ holds. For a set $F \subseteq E$ of edges, let $G{-}F$ be the graph obtained from G by removing all edges in F. For any two $s, t \in V$, a *replacement path* $P(s,t,F)$ is a shortest path from s to t in $G{-}F$. Its length $d(s,t,F) = d_{G-F}(s,t)$ is the *replacement distance*. For a positive integer f, an f*-distance sensitivity oracle* (DSO) reports, upon query (s,t,F) with $|F| \leqslant f$, the replacement distance $d(s,t,F)$. It has *stretch* $\sigma \geqslant 1$, or is σ*-approximate*, if the reported value $\widehat{d}(s,t,F)$ satisfies $d(s,t,F) \leqslant \widehat{d}(s,t,F) \leqslant \sigma \cdot d(s,t,F)$ for any admissible query. We measure the space complexity of a data structure in the number of $O(\log n)$-bit machine words. The size of the input G does not count against the space unless it is stored explicitly.

Error-Correcting Codes. For a positive integer h, we set $[h] = \{0, 1, \dots, h{-}1\}$. For positive integers q, p, and ℓ with $p \leqslant \ell$, a *code with alphabet size* q*, message length* p*, and block length* ℓ is a set $C \subseteq [q]^\ell$ such that $|C| \geqslant q^p$. An *encoding* for C is a computable injective mapping $[q]^p \to C$. Two codewords $x, y \in C$ have *(relative) distance* $\Delta(x,y) = |\{j \in [\ell] \mid x_j \neq y_j\}|/\ell$. For a positive real $\delta > 0$, code C is *error-correcting with (relative) distance* δ, if for any two $x, y \in C$, $\Delta(x,y) \geqslant \delta$. In this case, we say C is a $[p, \ell, \delta]_q$-code. It will be sufficient to focus on Reed-Solomon codes, which are $[p, q, 1 - \frac{p-1}{q}]_q$-codes for any $p \leqslant q$. When choosing q (and therefore $\ell = q$) as a power of 2 and $p < q$, there is an encoding algorithm for Reed-Solomon codes that takes $O(\ell \log p)$ time and $O(\ell)$ space using fast Fourier transform [31].

4 Small Hop Diameter

We first describe the simpler randomized version of our distance sensitivity oracle for graphs with small hop diameter. Afterwards, we derandomize it using more involved techniques like error-correcting codes. Throughout, we assume that the base graph G has edge weights from a polynomial-sized range.

4.1 Preprocessing

In the setting of Theorem 1, all shortest paths have at most D edges. Let $f = o(\log n/\log\log n)$ be the sensitivity of the oracle and $L \geqslant \max(f, 2)$ be an integer parameter which will be fixed later (depending on D). An *(L, f)-replacement path covering* (RPC) [27] is a family $\mathcal{G}$ of spanning subgraphs of G such that for any set $F \subseteq E$, $|F| \leqslant f$, and pair of vertices $s, t \in V$ such that s and t have a shortest path in $G - F$ on at most L edges, there exists a subgraph $G_i \in \mathcal{G}$ that does not contain any edge of F but an s-t-path of length $d(s, t, F)$. That means, some replacement path $P(s, t, F)$ from $G - F$ also exists in G_i. Let $\mathcal{G}_F \subseteq \mathcal{G}$ be the subfamily of all graphs that do not contain an edge of F. The definition of an RPC implies that if s and t have a replacement path w.r.t. F on at most L edges, then $\min_{G_i \in \mathcal{G}_F} d_{G_i}(s, t) = d(s, t, F)$.

To build the DSO, we first construct an (L, f)-RPC. This can be done by generating $|\mathcal{G}| = cfL^f \ln n$ random subgraphs for a sufficiently large constant $c > 0$. Each graph G_i is obtained from G by deleting any edge with probability $1/L$ independently of all other choices. As shown in [36], the family $\mathcal{G} = \{G_i\}_i$ is an (L, f)-RPC with high probability It is also easy to see using Chernoff bounds[6] that for any failure set F, $|\mathcal{G}_F| = O(|\mathcal{G}|/L^{|F|}) = \widetilde{O}(fL^{f-|F|})$.

We do not allow ourselves the space to store all subgraphs. We therefore replace each G_i by a distance oracle D_i, a data structure that reports, for any two $s, t \in V$, (an approximation of) the distance $d_{G_i}(s, t)$. For any positive integer k, Thorup and Zwick [35] devised a DO that is computable in time $\widetilde{O}(kmn^{1/k})$, has size $O(kn^{1+1/k})$, query time $O(k)$, and a stretch of $2k - 1$. Roddity, Thorup, and Zwick [33] derandomized the oracle, and Chechik [13,14] reduced the query time to $O(1)$ and the space to $O(n^{1+1/k})$. Additionally, we store, for each G_i, a spanner S_i. The same work [35] contains a spanner construction with stretch $2k - 1$ that is compatible with the oracle, meaning that the oracle D_i reports exactly the value $d_{S_i}(s, t)$. The spanner is computable in time $\widetilde{O}(kmn^{1/k})$ and has $O(kn^{1+1/k})$ edges. We store it as a set of edges. There are static dictionary data structures known that achieve this in $O(kn^{1+1/k})$ space such that we can check in $O(1)$ worst-case time whether an edge is present or retrieve an edge. They can be constructed in time $\widetilde{O}(kn^{1+1/k})$ [26]. The total preprocessing time of the distance sensitivity oracle is $\widetilde{O}(|\mathcal{G}|m + |\mathcal{G}|kmn^{1/k}) = \widetilde{O}(fL^f kmn^{1/k})$ and it takes $\widetilde{O}(fL^f kn^{1+1/k})$ space.

4.2 Query Algorithm

Assume for now that the only allowed queries to the DSO are triples (s, t, F) of vertices $s, t \in V$ and a set $F \subseteq E$ of at most f edges such that any shortest path

[6] There is a slight omission in [36, Lemma 3.1] for $|\mathcal{G}_F|$ is only calculated for $|F| = f$.

from s to t in $G - F$ has at most L edges. We will justify this assumption later with the right choice of L. The oracle has to report the replacement distance $d(s,t,F)$. Recall that $\mathcal{G}_F$ is the family of all graphs in $\mathcal{G}$ that have at least all edges of F removed. Since $\mathcal{G}$ is an (L,f)-RPC, all we have to do is compute (a superset of) $\mathcal{G}_F$ and retrieve (an approximation of) $\min_{G_i \in \mathcal{G}_F} d_{G_i}(s,t)$. The issue is that we do not have access to the graphs G_i directly.

The idea is to use the spanners S_i as proxies. This is justified by the next lemma that follows from a connection between the spanners and oracles presented in [35]. Let $D_i(s,t)$ denote the answer of the distance oracles D_i.

Lemma 1. *Let $G_i \in \mathcal{G}$ be a subgraph, S_i its associated spanner, and D_i its $(2k-1)$-approximate distance oracle. For any two vertices $s,t \in V$ and set $F \subseteq E$ with $|F| \leqslant f$, if $F \cap E(S_i) = \emptyset$, then $d(s,t,F) \leqslant D_i(s,t) \leqslant (2k-1)\, d_{G_i}(s,t)$.*

A proof is available in the full version of this paper [7].

Let $\mathcal{G}^S = \{S_i\}_{i \in [r]}$ be the collection of spanners for all $G_i \in \mathcal{G}$, and $\mathcal{G}_F^S \subset \mathcal{G}^S$ those that do not contain an edge of F. Below, we hardly distinguish between a set of spanners (or subgraphs) and their indices, thus e.g. $S_i \in \mathcal{G}_F^S$ is abbreviated as $i \in \mathcal{G}_F^S$. Since $E(S_i) \subseteq E(G_i)$ and using the convention, we get $\mathcal{G}_F^S \supseteq \mathcal{G}_F$.[7] To compute $\mathcal{G}_F^S$, we cycle through all of $\mathcal{G}^S$ and probe each dictionary with the edges in F, this takes $O(f|\mathcal{G}|) = \widetilde{O}(f^2 L^f)$ time and dominates the query time. If $i \in \mathcal{G}_F^S$, then we query the distance oracle D_i with the pair (s,t) in constant time. As answer to the query (s,t,F), we return $\min_{i \in \mathcal{G}_F^S} D_i(s,t)$. By Lemma 1, the answer is at least as large as the sought replacement distance and, since there is a graph $G_i \in \mathcal{G}_F \subseteq \mathcal{G}_F^S$ with $d_{G_i}(s,t) = d(s,t,F)$, it is at most $(2k-1)\, d(s,t,F)$.

Let D be an upper bound on the hop diameter of G. As mentioned above, Afek et al. [1, Theorem 2] showed that the maximum hop diameter of all graphs $G - F$ for $|F| \leqslant f$ is bounded by $(f+1)D + f$. Using this value for L implies that indeed all queries admit a replacement path on at most L edges. For the DSO, it implies a preprocessing time of $\widetilde{O}(fL^f kmn^{1/k}) = \widetilde{O}(f^{f+1} D^f kmn^{1/k})$, which for $f = o(\log n / \log\log n)$ is $D^f kmn^{1/k+o(1)}$. The space requirement is $\widetilde{O}(fL^f kn^{1+1/k}) = \widetilde{O}(f^{f+1} D^f kn^{1+1/k}) = D^f kn^{1+1/k+o(1)}$, and the query time $\widetilde{O}(f^2 L^f) = \widetilde{O}(f^{f+2} D^f) = D^f n^{o(1)}$. This proves the first part of Theorem 1.

4.3 Derandomization

We now make the DSO deterministic via a technique by Karthik and Parter [27]. The derandomization will allow us to find the relevant subgraphs faster, so we do not need the spanners anymore. Recall that the distance oracles D_i were already derandomized in [33]. The only randomness left is the generation of the subgraphs G_i. Getting a deterministic construction offers an alternative approach to dealing with the issue that discarding the subgraphs for space reasons

[7] We do mean here that $\mathcal{G}_F^S$ is a superset of $\mathcal{G}_F$. Since the spanner contain fewer edges than the graphs, F may be missing from $E(S_i)$ even though $F \cap E(G_i) \neq \emptyset$. This is fine as long as we take the *minimum* distance over all spanners from $\mathcal{G}_F^S$.

deprives us of the information which edges have been removed. Intuitively, we can now reiterate this process at query time to find the family $\mathcal{G}_F$. Below we implement this idea in a space-efficient manner.

We identify the edge set $E = \{e_0, e_1, \dots, e_{m-1}\}$ with $[m]$. Let q be a positive integer. Assume that $p = \log_q m$ is integral, otherwise one can replace $\log_q m$ with $\lceil \log_q m \rceil$ without any changes. We interpret any edge $e_i \in E$ as a base-q number $(c_0, c_1, \dots, c_{p-1}) \in [q]^p$ by requiring $i = \sum_{j=0}^{p-1} c_j q^j$. Consider an error-correcting $[p, \ell, \delta]$-code with distance $\delta > 1 - \frac{1}{fL}$ and (slightly abusing notation) let C be the $(m \times \ell)$-matrix with entries in $[q]$ whose i-th row is the codeword encoding the message $e_i = (c_0, c_1, \dots, c_{p-1})$. The key contribution of the work by Karthik and Parter [27] is the observation that the *columns* of C form a family of hash functions $\{h_j : E \to [q]\}_{j \in [\ell]}$ such that for any pair of disjoint sets $P, F \subseteq E$ with $|P| \leqslant L$ and $|F| \leqslant f$ there exists an index $j \in [\ell]$ with $\forall x \in P, y \in F : h_j(x) \neq h_j(y)$.

An (L, f)-replacement path covering can be constructed from this as follows. Choose q as the smallest power of 2 greater[8] than $fL \log_L m$. Note that $q \leqslant \frac{2fL \log_2 m}{\log_2 L} \leqslant \frac{4fL \log_2 n}{\log_2 L}$. A Reed-Solomon code with alphabet size q, message length $p = \log_q m$, and block length $\ell = q$ has distance greater than $1 - \frac{1}{fL}$ [27, Corollary 18]. The resulting covering $\mathcal{G}$ consists of $|\mathcal{G}| = O(\ell \cdot q^f) = O(q^{f+1})$ subgraphs, each one indexed by a pair (j, S) where $j \in [\ell]$ and $S \subseteq [q]$ is a set with $|S| \leqslant f$. In the subgraph $G_{(j,S)}$, an edge e_i is removed if and only if $h_j(e_i) \in S$. It is verified in [27] that the family $\mathcal{G} = \{G_{(j,S)}\}_{j,S}$ is indeed an (L, f)-RPC. Moreover, for a fixed set $F = \{e_{i_1}, \dots, e_{i_{|F|}}\}$ of edge failures, define $\mathcal{G}_F \subseteq \mathcal{G}$ to be the subfamily consisting of the graphs indexed by $(j, \{h_j(e_{i_1}), \dots, h_j(e_{i_{|F|}})\})$ for each $j \in [\ell]$. Then, the construction ensures that no graph in $\mathcal{G}_F$ contains any edge of F and, for each pair of vertices $s, t \in V$ with an replacement path (w.r.t. F) on at most L edges, there is graph in $\mathcal{G}_F$ in which s and t are joined by a path of length $d(s, t, F)$. The ℓ graphs in $\mathcal{G}_F$ contain all the information we need for the short replacement distances with respect to the failure set F. The number of subgraphs in the covering is $O((4fL \log_L n)^{f+1})$, this is a factor $O((4f \log_L n)^f L)$ larger than what we had for the randomized variant.

In turn, we can make use of the extreme locality of the indexing scheme for $\mathcal{G}$. Since we chose q as a power of 2, we get the letters of the message $e_i = (c_0, c_1, \dots, c_{p-1})$ by reading off blocks of $\log_2 q$ bits of the binary representation of i. The codeword of C corresponding to e_i is computable in time $O(p + \ell \log p) = O(fL(\log_L n) \log \log n)$ and space $O(p + \ell) = O(fL \log_L n)$ with the encoding algorithm of Lin, Al-Naffouri, Han, and Chung [31]. The whole matrix C and from it the family $\mathcal{G}$ can be generated in time $O(fLm(\log_L n) \log \log n + |\mathcal{G}|m) = O((4fL \log_L n)^{f+1} m)$. Note that the codeword of e_i is $(h_1(e_i), h_2(e_i), \dots, h_\ell(e_i))$. So even after discarding C and the subgraphs, we can find the indices of graphs in $\mathcal{G}_F$ by encoding the edges of F in time $O(|F|\ell \log p) = O(f^2 L(\log_L n) \log \log n)$

[8] The original construction in [27] sets q as a prime number. We use a power of 2 instead to utilize the encoding algorithm in [31]. All statements hold verbatim for both cases.

and rearranging the values into the ℓ sets $\{h_j(e_{i_1}), \dots, h_j(e_{i_{|F|}})\}$. In particular, using the algorithm in [31], we do not need to store the generator matrix of the code C.

The remaining preprocessing is similar as in Sect. 4.1, but we need neither the spanners nor the dictionaries anymore. We set $L = O(fD)$ again and, for each subgraph $G_{(j,S)}$, we only build the distance oracle $D_{(j,S)}$. This dominates the preprocessing time $\widetilde{O}(|\mathcal{G}|kmn^{1/k}) = \widetilde{O}(4^{f+1}f^{2f+2}D^{f+1}(\log_D n)^{f+1}kmn^{1/k}) = D^{f+1}kmn^{1/k+o(1)}$. The total size is now $\widetilde{O}(|\mathcal{G}|kn^{1+1/k}) = D^{f+1}kn^{1+1/k+o(1)}$. However, due to the derandomization, the query time is now much faster than before, in particular, polynomial in f, D and $\log n$. We do not have to cycle through all spanners anymore and instead compute $\mathcal{G}_F$ and query the DO only for those ℓ graphs. As a result, the time to report the replacement distance is $O(f^2L(\log_L n)\log\log n + \ell) = O(f^3D\,(\log_D n)\log\log n)$, completing Theorem 1.

5 Large Hop Diameter

We also devise a distance sensitivity oracle for graphs with an arbitrary hop diameter while maintaining a small memory footprint. For this, we have to handle hop-long replacement paths, that is, those that have more than L edges. We obtain a subquadratic-space distance sensitivity oracle with the same stretch of $2k-1$ but an $o(n^{1+1/k})$ query time. This is faster than computing the distance in any possible spanner.

5.1 Deterministic Pivot Selection

We say a query with vertices $s, t \in V$ and set $F \subseteq E$, $|F| \leqslant f$ has long replacement paths if every $P(s,t,F)$ has at least L edges. Those need to be handled in general DSOs. This is usually done by drawing a random subset $B \subset V$ of $\widetilde{O}(fn/L)$ pivots, as in [36], or essentially equivalent sampling every vertex independently with probability $\widetilde{O}(f/L)$ [24,34]. With high probability, B hits one path for every query with long replacement paths.

There are different approaches known to derandomize this depending on the setting [2,5,8,9,29]. In our case, we can simply resort to the replacement path covering to obtain $\mathcal{P}$ since we have to preprocess it anyway. We prove the following lemma for the more general class of arbitrary positive weights. Note that the key properties of an (L,f)-RPC remain in place as all definitions are with respect to the number of edges on the replacement paths. We make it so that B hits the slightly shorter paths with $L/2$ edges (instead of L). We are going to use this stronger requirement in Lemma 4.

Lemma 2. *Let $G = (V,E)$ be an undirected graph with positive edge weights. Let Q be the set of all queries (s,t,F), with $s,t \in V$ and $F \subseteq E$, $|F| \leqslant f$, for which every s-t-replacement path w.r.t. F has at least $L/2$ edges. Given an (L,f)-replacement path covering $\mathcal{G}$ for G, there is a deterministic algorithm that computes in time $\widetilde{O}(|\mathcal{G}|(mn+Ln^2/f))$ a set $B \subseteq V$ of size $\widetilde{O}(fn/L)$ such that, for* *all* $(s,t,F) \in Q$, *there is a replacement path* $P = P(s,t,F)$ *with* $B \cap V(P) \neq \emptyset$. *At the same time, one can build a data structure of size* $O(|\mathcal{G}||B|^2)$ *that reports, for every* $G_i \in \mathcal{G}$ *and* $x,y \in B$, *the distance* $d_{G_i}(x,y)$ *in constant time.*

A proof is available in the full version of this paper [7].

5.2 Preprocessing

Our solution for large hop diameter builds on the deterministic DSO in Sect. 4.3. As for the case of a small hop diameter, we construct an (L,f)-replacement path covering $\mathcal{G}$ and, for each $G_{(i,S)} \in \mathcal{G}$, the distance oracle $D_{(i,S)}$. Recall that this part takes time $\widetilde{O}(|\mathcal{G}|kmn^{1/k})$ and $O(|\mathcal{G}|kn^{1+1/k})$ space.

We invoke Lemma 2 to obtain the set B. Additionally, for each subgraph, we build a complete weighted graph $H_{(i,S)}$ on the vertex set B where the weight of edge $\{x,y\}$ is $d_{G_{(i,S)}}(x,y)$, which we retrieve from the data structure mentioned in Lemma 2. We then compute a $(2k-1)$-spanner $T_{(i,S)}$ for $H_{(i,S)}$ with $O(k|B|^{1+1/k})$ edges via the same deterministic algorithm by Roditty, Thorup, and Zwick [33]. We store the new spanners for our DSO. The time to compute them is $\widetilde{O}(|\mathcal{G}|k|B|^{2+1/k})$ and, since $|B| = \widetilde{O}(fn/L)$, the preprocessing time is

$$\widetilde{O}(|\mathcal{G}|(kmn^{1/k} + |B|m + k|B|^{2+1/k})) = \widetilde{O}(|\mathcal{G}|(mn + k|B|^{2+1/k})) \leqslant L^{f+1}mn^{1+o(1)} + L^{f-1-1/k}kn^{2+1/k+o(1)} \leqslant L^{f+1}kmn^{1+1/k+o(1)}.$$

To obtain the bounds of Theorem 2, we set $L = n^{\frac{\alpha}{f+1}}$. Parameter $0 < \alpha < 1$ allows us to balance the space and query time. With this, we get a preprocessing time of $kmn^{1+\alpha+1/k+o(1)}$. and a space of $O(|\mathcal{G}|(kn^{1+1/k} + n + k|B|^{1+1/k}) = \widetilde{O}(|\mathcal{G}|kn^{1+1/k}) = L^{f+1}kn^{1+1/k+o(1)} = kn^{1+\alpha+1/k+o(1)}$.

5.3 Updated Query Algorithm

The algorithm to answer a query (s,t,F) starts similarly as before. We use the error-correcting codes to compute the subfamily $\mathcal{G}_F$ and the estimate $\widehat{d}_1(s,t,F) = \min_{(i,S)\in\mathcal{G}_F} D_{(i,S)}(s,t)$ is retrieved. However, this is no longer guaranteed to be an $(2k-1)$-approximation if the query is hop-long, i.e., if every shortest s-t-path in $G - F$ has at least L edges. It could be that no replacement paths survive and, in the extreme case, s and t are disconnected in each $G_{(i,S)} \in \mathcal{G}_F$, while they still have a finite distance in $G - F$. To account for long queries, we join all the spanners T_i for $i \in \mathcal{G}_F$. In more detail, we build a multigraph[9] H^F on the vertex set $V(H^F) = B \cup \{s,t\}$ whose edge set (restricted to pairs of pivots) is the disjoint union of all the sets $\{E(T_i)\}_{i\in\mathcal{G}_F}$ and, for each subgraph $(i,S) \in \mathcal{G}_F$ and pivot $x \in B$ contains the edges $\{s,x\}$ and $\{x,t\}$ with respective weights $D_{(i,S)}(s,x)$ and $D_{(i,S)}(x,t)$, where $D_{(i,S)}$ is the corresponding DO. The oracle then computes the second estimate $\widehat{d}_2(s,t,F) = d_{H^F}(s,t)$ and returns $\widehat{d}(s,t,F) = \min\{\widehat{d}_1(s,t,F),\ \widehat{d}_2(s,t,F)\}$.

[9] The multigraph is only used to ease notation.

Lemma 3. *The distance sensitivity oracle has stretch* $2k-1$ *and the query takes time* $\widetilde{O}(\frac{n^{1+1/k}}{L^{1/k}}) = \widetilde{O}(n^{1+\frac{1}{k}-\frac{\alpha}{k(f+1)}})$.

Proving this lemma is enough to complete Theorem 2. In order to do so, we first establish the fact that $d_{H^F}(s,t)$ is a $(2k-1)$-approximation for $d(s,t,F)$ in case of a long query.

Lemma 4. *Let* $s,t \in V$ *be two vertices and* $F \subseteq E$ *a set of edges with* $|F| \leqslant f$. *It holds that* $d(s,t,F) \leqslant d_{H^F}(s,t)$. *If additionally every shortest s-t-path in* $G-F$ *has more than L edges, then we have* $d_{H^F}(s,t) \leqslant (2k-1)\,d(s,t,F)$.

Proofs of these lemmas are available in the full version of this paper [7].

Acknowledgement. The authors thank Merav Parter for raising the question of designing distance sensitivity oracles that require only subquadratic space.

This project received funding from the European Research Council (ERC) under the European Union's Horizon 2020 research and innovation program (Grant agreement No. 101019564 "The Design of Modern Fully Dynamic Data Structures (MoDynStruct)").

References

1. Afek, Y., Bremler-Barr, A., Kaplan, H., Cohen, E., Merritt, M.: Restoration by path concatenation: fast recovery of MPLS paths. Distrib. Comput. **15**, 273–283 (2002). https://doi.org/10.1007/s00446-002-0080-6
2. Alon, N., Chechik, S., Cohen, S.: Deterministic combinatorial replacement paths and distance sensitivity oracles. In: Proceedings of the 46th International Colloquium on Automata, Languages, and Programming, (ICALP), pp. 12:1 12:14 (2019). https://doi.org/10.4230/LIPIcs.ICALP.2019.12
3. Baswana, S., Khanna, N.: Approximate shortest paths avoiding a failed vertex: near optimal data structures for undirected unweighted graphs. Algorithmica **66**, 18–50 (2013). https://doi.org/10.1007/s00453-012-9621-y
4. Bernstein, A., Karger, D.R.: Improved distance sensitivity oracles via random sampling. In: Proceedings of the 19th Symposium on Discrete Algorithms (SODA), pp. 34–43 (2008). https://dl.acm.org/doi/abs/10.5555/1347082.1347087
5. Bernstein, A., Karger, D.R.: A nearly optimal oracle for avoiding failed vertices and edges. In: Proceedings of the 41st Symposium on Theory of Computing (STOC), pp. 101–110 (2009). https://doi.org/10.1145/1536414.1536431
6. Bilò, D., et al.: Approximate distance sensitivity oracles in subquadratic space. In: Proceedings of the 55th Symposium on Theory of Computing (STOC), pp. 1396–1409 (2023). https://doi.org/10.1145/3564246.3585251
7. Bilò, D., Choudhary, K., Cohen, S., Friedrich, T., Krogmann, S., Schirneck, M.: Compact distance oracles with large sensitivity and low stretch (2023). https://arxiv.org/abs/2304.14184
8. Bilò, D., Choudhary, K., Cohen, S., Friedrich, T., Schirneck, M.: Deterministic sensitivity oracles for diameter, eccentricities and all pairs distances. In: Proceedings of the 49th International Colloquium on Automata, Languages, and Programming (ICALP), pp. 22:1–22:19 (2022). https://doi.org/10.4230/LIPIcs.ICALP.2022.22

9. Bilò, D., Cohen, S., Friedrich, T., Schirneck, M.: Near-optimal deterministic single-source distance sensitivity oracles. In: Proceedings of the 29th European Symposium on Algorithms (ESA), pp. 18:1–18:17 (2021). https://doi.org/10.4230/LIPIcs.ESA.2021.18
10. Bodwin, G., Dinitz, M., Robelle, C.: Optimal vertex fault-tolerant spanners in polynomial time. In: Proceedings of the 32nd Symposium on Discrete Algorithms (SODA), pp. 2924–2938 (2021). https://doi.org/10.1137/1.9781611976465.174
11. Bodwin, G., Dinitz, M., Robelle, C.: Partially optimal edge fault-tolerant spanners. In: Proceedings of the 33rd Symposium on Discrete Algorithms (SODA), pp. 3272–3286 (2022). https://doi.org/10.1137/1.9781611977073.129
12. van den Brand, J., Saranurak, T.: Sensitive distance and reachability oracles for large batch updates. In: Proceedings of the 60th Symposium on Foundations of Computer Science (FOCS), pp. 424–435 (2019). https://doi.org/10.1109/FOCS.2019.00034
13. Chechik, S.: Approximate distance oracles with constant query time. In: Proceedings of the 46th Symposium on Theory of Computing (STOC), pp. 654–663 (2014). https://doi.org/10.1145/2591796.2591801
14. Chechik, S.: Approximate distance oracles with improved bounds. In: Proceedings of the 47th Symposium on Theory of Computing (STOC), pp. 1–10 (2015). https://doi.org/10.1145/2746539.2746562
15. Chechik, S., Cohen, S.: Distance sensitivity oracles with subcubic preprocessing time and fast query time. In: Proceedings of the 52nd Symposium on Theory of Computing (STOC), pp. 1375–1388 (2020). https://doi.org/10.1145/3357713.3384253
16. Chechik, S., Cohen, S., Fiat, A., Kaplan, H.: $(1+\varepsilon)$-approximate f-sensitive distance oracles. In: Proceedings of the 28th Symposium on Discrete Algorithms (SODA), pp. 1479–1496 (2017). https://doi.org/10.1137/1.9781611974782.96
17. Chechik, S., Langberg, M., Peleg, D., Roditty, L.: Fault tolerant spanners for general graphs. SIAM J. Comput. **39**, 3403–3423 (2010). https://doi.org/10.1137/090758039
18. Chechik, S., Langberg, M., Peleg, D., Roditty, L.: f-sensitivity distance oracles and routing schemes. Algorithmica **63**, 861–882 (2012). https://doi.org/10.1007/s00453-011-9543-0
19. Demetrescu, C., Thorup, M.: Oracles for distances avoiding a link-failure. In: Proceedings of the 13th Symposium on Discrete Algorithms (SODA), pp. 838–843 (2002). https://dl.acm.org/doi/10.5555/545381.545490
20. Demetrescu, C., Thorup, M., Chowdhury, R.A., Ramachandran, V.: Oracles for distances avoiding a failed node or link. SIAM J. Comput. **37**, 1299–1318 (2008). https://doi.org/10.1137/S0097539705429847
21. Duan, R., Pettie, S.: Dual-failure distance and connectivity oracles. In: Proceedings of the 20th Symposium on Discrete Algorithms (SODA), pp. 506–515 (2009). https://dl.acm.org/doi/10.5555/545381.545490
22. Duan, R., Ren, H.: Maintaining exact distances under multiple edge failures. In: Proceedings of the 54th Symposium on Theory of Computing (STOC), pp. 1093–1101 (2022). https://doi.org/10.1145/3519935.3520002
23. Erdős, P.: Extremal Problems in Graph Theory. Theory of Graphs and its Applications, pp. 29–36 (1964)
24. Grandoni, F., Vassilevska Williams, V.: Faster replacement paths and distance sensitivity oracles. ACM Trans. Algorithms **16**, 15:1–15:25 (2020). https://doi.org/10.1145/3365835

25. Gu, Y., Ren, H.: Constructing a distance sensitivity oracle in $O(n^{2.5794}M)$ time. In: Proceedings of the 48th International Colloquium on Automata, Languages, and Programming (ICALP), pp. 76:1–76:20 (2021). https://doi.org/10.4230/LIPIcs.ICALP.2021.76
26. Hagerup, T., Miltersen, P.B., Pagh, R.: Deterministic dictionaries. J. Algorithms **41**, 69–85 (2001). https://doi.org/10.1006/jagm.2001.1171
27. Karthik, C.S., Parter, M.: Deterministic replacement path covering. In: Proceedings of the 32nd Symposium on Discrete Algorithms (SODA), pp. 704–723 (2021). https://doi.org/10.1137/1.9781611976465.44
28. Khanna, N., Baswana, S.: Approximate shortest paths avoiding a failed vertex: optimal size data structures for unweighted graphs. In: Proceedings of the 27th Symposium on Theoretical Aspects of Computer Science (STACS), pp. 513–524 (2010). https://doi.org/10.4230/LIPIcs.STACS.2010.2481
29. King, V.: Fully dynamic algorithms for maintaining all-pairs shortest paths and transitive closure in digraphs. In: Proceedings of the 40th Symposium on Foundations of Computer Science (FOCS), pp. 81–91 (1999). https://doi.org/10.1109/SFFCS.1999.814580
30. Levcopoulos, C., Narasimhan, G., Smid, M.H.M.: Efficient algorithms for constructing fault-tolerant geometric spanners. In: Proceedings of the 30th Symposium on Theory of Computing (STOC), pp. 186–195 (1998). https://doi.org/10.1145/276698.276734
31. Lin, S., Al-Naffouri, T.Y., Han, Y.S., Chung, W.: Novel polynomial basis with fast Fourier transform and its application to Reed-Solomon erasure codes. IEEE Trans. Inf. Theory **62**, 6284–6299 (2016). https://doi.org/10.1109/TIT.2016.2608892
32. Ren, H.: Improved distance sensitivity oracles with subcubic preprocessing time. J. Comput. Syst. Sci. **123**, 159–170 (2022). https://doi.org/10.1016/j.jcss.2021.08.005
33. Roditty, L., Thorup, M., Zwick, U.: Deterministic constructions of approximate distance oracles and spanners. In: Caires, L., Italiano, G.F., Monteiro, L., Palamidessi, C., Yung, M. (eds.) ICALP 2005. LNCS, vol. 3580, pp. 261–272. Springer, Heidelberg (2005). https://doi.org/10.1007/11523468_22
34. Roditty, L., Zwick, U.: Replacement paths and k simple shortest paths in unweighted directed graphs. ACM Trans. Algorithms **8**, 33:1–33:11 (2012). https://doi.org/10.1145/2344422.2344423
35. Thorup, M., Zwick, U.: Approximate distance oracles. J. ACM **52**, 1–24 (2005). https://doi.org/10.1145/1044731.1044732
36. Weimann, O., Yuster, R.: Replacement paths and distance sensitivity oracles via fast matrix multiplication. ACM Trans. Algorithms **9**, 14:1–14:13 (2013). https://doi.org/10.1145/2438645.2438646

Finding Diameter-Reducing Shortcuts in Trees

Davide Bilò[1], Luciano Gualà[2(✉)], Stefano Leucci[1], and Luca Pepè Sciarria[2]

[1] Department of Information Engineering, Computer Science and Mathematics, University of L'Aquila, L'Aquila, Italy
{davide.bilo,stefano.leucci}@univaq.it
[2] Department of Enterprise Engineering, University of Rome "Tor Vergata", Rome, Italy
guala@mat.uniroma2.it

Abstract. In the *k-Diameter-Optimally Augmenting Tree Problem* we are given a tree T of n vertices as input. The tree is embedded in an unknown *metric* space and we have unlimited access to an oracle that, given two distinct vertices u and v of T, can answer queries reporting the cost of the edge (u, v) in constant time. We want to augment T with k shortcuts in order to minimize the diameter of the resulting graph.

For $k = 1$, $O(n \log n)$ time algorithms are known both for paths [Wang, CG 2018] and trees [Bilò, TCS 2022]. In this paper we investigate the case of multiple shortcuts. We show that no algorithm that performs $o(n^2)$ queries can provide a better than 10/9-approximate solution for trees for $k \geq 3$. For any constant $\varepsilon > 0$, we instead design a linear-time $(1+\varepsilon)$-approximation algorithm for paths and $k = o(\sqrt{\log n})$, thus establishing a dichotomy between paths and trees for $k \geq 3$. We achieve the claimed running time by designing an ad-hoc data structure, which also serves as a key component to provide a linear-time 4-approximation algorithm for trees, and to compute the diameter of graphs with $n + k - 1$ edges in time $O(nk \log n)$ even for non-metric graphs. Our data structure and the latter result are of independent interest.

Keywords: Tree diameter augmentation · Fast diameter computation · Approximation algorithms · Time-efficient algorithms

1 Introduction

The *k-Diameter-Optimally Augmenting Tree Problem* (k-DOAT) is defined as follows. The input consists of a tree T of n vertices that is embedded in an *unknown* space c. The space c associates a non-negative cost $c(u, v)$ to each pair of vertices $u, v \in V(T)$, with $u \neq v$. The goal is to quickly compute a set S of k *shortcuts* whose addition to T minimizes the *diameter* of the resulting graph $T + S$. The diameter of a graph G, with a cost of $c(u, v)$ associated with each edge (u, v) of G, is defined as $\text{diam}(G) := \max_{u,v \in V(G)} d_G(u, v)$, where $d_G(u, v)$

P. Morin and S. Suri (Eds.): WADS 2023, LNCS 14079, pp. 164–178, 2023.
https://doi.org/10.1007/978-3-031-38906-1_12

is the distance in G between u and v that is measured w.r.t. the edge costs. We assume to have access to an oracle that answers a query about the cost $c(u, v)$ of any tree-edge/shortcut (u, v) in constant time.

When c satisfies the triangle inequality, i.e., $c(u, v) \le c(u, w) + c(w, v)$ for every three distinct vertices $u, v, w \in V(T)$, we say that T is embedded in a *metric space*, and we refer to the problem as *metric* k-DOAT.

k-DOAT and metric k-DOAT have been extensively studied for $k = 1$. k-DOAT has a trivial lower bound of $\Omega(n^2)$ on the number of queries needed to find S even if one is interested in any finite approximation and the input tree is actually a path.[1] On the positive side, it can be solved in $O(n^2)$ time and $O(n \log n)$ space [2], or in $O(n^2 \log n)$ time and $O(n)$ space [20]. Interestingly enough, this second algorithm uses, as a subroutine, a linear time algorithm to compute the diameter of a unicycle graph, i.e., a connected graph with n edges.

Metric k-DOAT has been introduced in [11] where an $O(n \log^3 n)$-time algorithm is provided for the special case in which the input is a path. In the same paper the authors design a less efficient algorithm for trees that runs in $O(n^2 \log n)$ time. The upper bound for the path case has been then improved to $O(n \log n)$ in [19], while in [2] it is shown that the same asymptotic upper bound can be also achieved for trees. Moreover, the latter work also gives a $(1+\varepsilon)$-approximation algorithm for trees with a running time of $O(n + \frac{1}{\varepsilon} \log \frac{1}{\varepsilon})$.

Our Results. In this work we focus on metric k-DOAT for $k > 1$. In such a case one might hope that $o(n^2)$ queries are enough for an exact algorithm. Unfortunately, we show that this is not the case for $3 \le k = o(n)$, even if one is only searching for a σ-approximate solution with $\sigma < \frac{10}{9}$. Our lower bound is unconditional, holds for trees (with many leaves), and trivially implies an analogous lower bound on the time complexity of any σ-approximation algorithm.

Motivated by the above lower bound, we focus on approximate solutions and we show two linear-time algorithms with approximation ratios of 4 and $1 + \varepsilon$, for any constant $\varepsilon > 0$. The latter algorithm only works for trees with few leaves and $k = o(\sqrt{\log n})$. This establishes a dichotomy between paths and trees for $k \ge 3$: paths can be approximated within a factor of $1 + \varepsilon$ in linear-time, while trees (with many leaves) require $\Omega(n^2)$ queries (and hence time) to achieve a better than 10/9 approximation. Notice that this is not the case for $k = 1$. We leave open the problems of understanding whether exact algorithms using $o(n^2)$ queries can be designed for 2-DOAT on trees and k-DOAT on paths.

To achieve the claimed linear-time complexities of our approximation algorithms, we develop an ad-hoc data structure which allows us to quickly compute a small set of well-spread vertices with large pairwise distances. These vertices are

[1] As an example, consider an instance I consisting of two subpaths P_1, P_2 of $\Theta(n)$ edges each and cost 0. The subpaths are joined by an edge of cost 1, and the cost of all shortcuts is 1. Any algorithm that does not examine the cost of some shortcut (u, v) with one endpoint in P_1 and the other in endpoint in P_2 cannot distinguish between I and the instance I' obtained from I by setting $c(u, v) = 0$. The claim follows by noticing that there are $\Theta(n^2)$ such shortcuts (u, v) and that the optimal diameters of I and I' are 1 and 0, respectively.

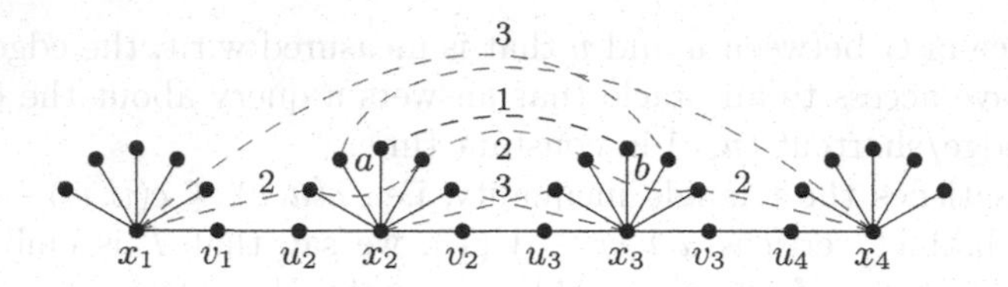

Fig. 1. The graph G of the lower bound construction. The edges of the tree T are solid and have cost 2; the non-tree edges are dashed and their colors reflect the different types of augmenting edges as defined in the proof of Lemma 1. To reduce clutter, only some of the augmenting edges are shown.

used as potential endvertices for the shortcuts. Interestingly, our data structure can also be used to compute the diameter of a *non-metric* graph with $n + k - 1$ edges in $O(nk \log n)$ time. For $k = O(1)$, this extends the $O(n)$-time algorithm in [20] for computing the diameter of a unicycle graph, with only a logarithmic-factor slowdown. We deem this result of independent interest as it could serve as a tool to design efficient algorithms for k-DOAT, as shown for $k = 1$ in [20].

Other Related Work. The problem of minimizing the diameter of a graph via the addition of k shortcuts has been extensively studied in the classical setting of optimization problems. This problem is shown to be NP-hard [18], not approximable within logarithmic factors unless P = NP [3], and some of its variants – parameterized w.r.t. the overall cost of the added shortcuts and w.r.t. the resulting diameter – are even W[2]-hard [8,9]. As a consequence, the literature has focused on providing polynomial time approximation algorithms for all these variants [1,3,6–8,16]. This differs from k-DOAT, where the emphasis is on $o(n^2)$-time algorithms.

Finally, variants of 1-DOAT in which one wants to minimize either the radius or the continuous diameter of the augmented tree have been studied [4,5,12,13,15,17].

Paper Organization. In Sect. 2 we present our non-conditional lower bound. Section 3 is devoted to the algorithm for computing the diameter of a graph with $n + k - 1$ edges in time $O(nk \log n)$. This relies on our data structure, which is described in Sect. 3.2. Finally, in Sect. 4 we provide our linear-time approximation algorithms for metric k-DOAT that use our data structure. The implementation of our data structure and its analysis, along with some proofs throughout the paper, can be found in the full version of the paper.

2 Lower Bound

This section is devoted to proving our lower bound on the number of queries needed to solve k-DOAT on trees for $k \geq 3$. We start by considering the case $k = 3$.

Lemma 1. *For any sufficiently large n, there is a class $\mathcal{G}$ of instances of metric* 3-DOAT *satisfying the following conditions:*

(i) In each instance $\langle T, c\rangle \in \mathcal{G}$, T is a tree with $\Theta(n)$ vertices and all tree-edge/shortcut costs assigned by c are positive integers;
(ii) No algorithm can decide whether an input instance $\langle T, c\rangle$ from $\mathcal{G}$ admits a solution S such that $\text{diam}(T + S) \leq 9$ *using $o(n^2)$ queries.*

Proof. We first describe the 3-DOAT instances. All instances $\langle T, c\rangle \in \mathcal{G}$ share the same tree T, and only differ in the cost function c. The tree T is defined as follows: consider 4 identical stars, each having n vertices and denote by x_i the center of the i-th star, by L_i the set of its leaves, and let $X_i = L_i \cup \{x_i\}$. The tree T is obtained by connecting each pair x_i, x_{i+1} of centers with a path of three edges $(x_i, v_i), (v_i, u_{i+1})$, and (u_{i+1}, x_{i+1}), as in Fig. 1. All the tree edges (u, v) have the same cost $c(u, v) = 2$.

The class $\mathcal{G}$ contains an instance $I_{a,b}$ for each pair $(a, b) \in L_2 \times L_3$, and an additional instance I. Fix $(a, b) \in L_2 \times L_3$. The costs of the shortcuts in $I_{a,b}$ are defined w.r.t. a graph $G_{a,b}$ obtained by augmenting T with the following edges:

- All edges (x_1, y) with $y \in L_2$, and all edges (x_4, y) with $y \in L_3$ with cost 2.
- The edges (y, z) for every $y \in L_2$ and every $z \in L_3$. The cost (a, b) is 1, while the cost of all other edges is 2;
- The edges (y, z) for every distinct pair of vertices y, z that are both in L_2 or both in L_3. The cost of all such edges is 3.
- All edges (x_2, y) with $y \in L_3$, and all edges (x_3, y) with $y \in L_2$ with cost 3;
- All edges (x_1, y) with $y \in L_3$, and all edges (x_4, y) with $y \in L_2$ with cost 3.

We define the cost $c(u, v)$ of all remaining shortcuts (u, v) as $d_{G_{a,b}}(u, v)$.

We now argue that c satisfies the triangle inequality. Consider any triangle in $G_{a,b}$ having vertices u, v, and w. We show that the triangle inequality holds for the generic edge (u, v). As the costs of all edges of $G_{a,b}$, except for the shortcut (a, b), are either 2 or 3 and since $c(a, b) = 1$, we have that $c(u, v) \leq 3 \leq c(u, w) + c(w, v)$. Any other triangle clearly satisfies the triangle inequality as it contains one or more edges that are not in $G_{a,b}$ and whose costs are computed using distances in $G_{a,b}$.

To define the cost function c of the remaining instance I of $\mathcal{G}$, choose any $(a, b) \in L_2 \times L_3$, and let G be the graph obtained from $G_{a,b}$ by changing the cost of (a, b) from 1 to 2. We define $c(u, v) = d_G(u, v)$. Notice that, in G, all edges $(u, v) \in L_2 \times L_3$ have cost 2 and that the above arguments also show that c still satisfies the triangle inequality.

Since our choice of $I_{a,b}$ and I trivially satisfies (i), we now focus on proving (ii).

We start by showing the following facts: (1) each instance $I_{a,b}$ admits a solution S such that $\text{diam}(T + S) \leq 9$; (2) all solutions S to I are such that $\text{diam}(T + S) \geq 10$; (3) if $u \neq a$ or $v \neq b$ then $d_G(u, v) = d_{G_{a,b}}(u, v)$.

To see (1), consider the set $S = \{(x_1, a), (a, b), (b, x_4)\}$ of 3 shortcuts. We can observe that $\text{diam}(T+S) \leq 9$. This is because $d_{T+S}(x_i, x_j) \leq 5$ for every two star centers x_i and x_j (see also Fig. 1). Moreover, each vertex $u \in L_i$ is at a distance of at most 2 from x_i. Therefore, for every two vertices $u \in X_i$ and $v \in X_j$, we have that $d_{T+S}(u,v) \leq d_T(u,x_i) + d_{T+S}(x_i,x_j) + d_T(x_j,v) \leq 2+5+2 = 9$.

Concerning (2), let us consider any solution S of 3 shortcuts and define B_i as the set of vertices X_i plus the vertices u_i and v_i, if they exist. We show that if there is no edge in S between B_i and B_{i+1} for some $i = 1,\ldots,3$, then $\text{diam}(T+S) \geq 10$. To this aim, suppose that this is the case, and let $u \in L_i$ and $v \in L_{i+1}$ be two vertices that are not incident to any shortcut in S. Notice that the shortest path in $T+S$ between u and v traverses x_i and x_{i+1}, and hence $d_{T+S}(u,v) = d_{T+S}(u,x_i)+d_{T+S}(x_i,x_{i+1})+d_{T+S}(x_{i+1},v) = 4+d_{T+S}(x_i,x_{i+1})$. We now argue that $d_{T+S}(x_i,x_{i+1}) \geq 6$. Indeed, we have that all edges in $T+S$ cost at least 2, therefore $d_{T+S}(x_i,x_{i+1}) < 6$ would imply that the shortest path π from x_i to x_{i+1} in $T+S$ traverses a single intermediate vertex z. By assumption, z must belong to some B_j for $j \notin \{i, i+1\}$. However, by construction of G, we have $c(x_i,z) \geq 3$ and $c(x_{i+1},z) \geq 3$ for every such $z \in B_j$.

Hence, we can assume that we have a single shortcut edge between B_i and B_{i+1}, for $i = 1,\ldots,3$. Let $x \in L_1$ (resp. $y \in L_4$) such that no shortcut in S is incident to x (resp. y). Notice that every path in $T+S$ from x to y traverses at least 5 edges and, since all edges cost at least 2, we have $\text{diam}(T+S) \geq d_{T+S}(x,y) \geq 10$.

We now prove (3). Let u,v be two vertices such that there is a shortest path π in $G_{a,b}$ from u to v traversing the edge (a,b). We show that there is another shortest path from u to v in $G_{a,b}$ that avoids edge (a,b). Consider a subpath π' of π consisting of two edges one of which is (a,b). Let $w \neq a,b$ be one of the endvertices of π' and let $w' \in \{a,b\}$ be the other endvertex of π'. Observe that edges (a,b), (a,w), and (w,b) forms a triangle in $G_{a,b}$ and $c(a,w), c(w,b) \in \{2,3\}$. Since $c(a,b) = 1$, we have $c(w,a) \leq c(w,b) + c(b,a)$ and $c(w,b) \leq c(w,a) + c(a,b)$. This implies that we can shortcut π' with the edge (w,w') thus obtaining another shortest path from u to v that does not use the edge (a,b).

We are finally ready to prove (ii). We suppose towards a contradiction that some algorithm $\mathcal{A}$ requires $o(n^2)$ queries and, given any instance $\langle T,c\rangle \in \mathcal{G}$, decides whether $\text{diam}(T+S) \leq 9$ for some set S of 3 shortcuts.[2] By (2), $\mathcal{A}$ with input I must report that there is no feasible set of shortcuts that achieves diameter at most 9. Since $\mathcal{A}$ performs $o(n^2)$ queries, for all sufficiently large values of n, there must be an edge (a,b) with $a \in L_2$ and $b \in L_3$ whose cost $c(a,b)$ is not inspected by $\mathcal{A}$. By (3), the costs of all the edges (u,v) with $(u,v) \neq (a,b)$ are the same in the two instances I and $I_{a,b}$ and hence $\mathcal{A}$ must report that $I_{a,b}$ admits no set of shortcuts S such that $\text{diam}(T+S) \leq 9$. This contradicts (1). □

With some additional technicalities we can generalize Lemma 1 to $3 \leq k = o(n)$, which immediately implies a lower bound on the number of queries needed by any σ-approximation algorithm with $\sigma < 10/9$.

[2] For the sake of simplicity, we consider deterministic algorithms only. However, standard arguments can be used to prove a similar claim also for randomized algorithms.

Lemma 2. *For any sufficiently large n and $3 \leq k = o(n)$, there is a class $\mathcal{G}$ of instances of metric k-*DOAT *satisfying the following conditions:*

(i) In each instance $\langle T, c\rangle \in \mathcal{G}$, T is a tree with $\Theta(n)$ vertices and all tree-edge/shortcut costs assigned by c are positive integers;
(ii) No algorithm can decide whether an input instance $\langle T, c\rangle$ from $\mathcal{G}$ admits a solution S such that $\text{diam}(T + S) \leq 9$ using $o(n^2)$ queries.

Theorem 1. *There is no $o(n^2)$-query σ-approximation algorithm for metric k-*DOAT *with $\sigma < 10/9$ and $3 \leq k = o(n)$.*

3 Fast Diameter Computation

In this section we describe an algorithm that computes the diameter of a graph on n vertices and $n + k - 1$ edges, with non-negative edge costs, in $O(nk \log n)$ time. Before describing the general solution, we consider the case in which the graph is obtained by augmenting a path P of n vertices with k edges as a warm-up.

3.1 Warm-Up: Diameter on Augmented Paths

Given a path P and a set S of k shortcuts we show how to compute the diameter of $P+S$. We do that by computing the *eccentricity* $\mathcal{E}_s := \max_{v \in V(P)} d_{P+S}(s, v)$ of each vertex s. Clearly, the diameter of $P+S$ is given by the maximum value chosen among the computed vertex eccentricities, i.e., $\text{diam}(P+S) = \max_{s \in V(P)} \mathcal{E}_s$. Given a subset of vertices $X \subseteq V(P)$, define $\mathcal{E}_s(X) := \max_{v \in X} d_{P+S}(s, v)$, i.e., $\mathcal{E}_s(X)$ is the eccentricity of s restricted to X.

In the rest of the section we focus on an fixed vertex s. We begin by computing a condensed weighted (multi-)graph G'. To this aim, we say that a vertex v is *marked as terminal* if it satisfies at least one of the following conditions: (i) $v = s$, (ii) v is an endvertex of some shortcut in S, or (iii) v is an endpoint of the path. Traverse P from one endpoint to the other and let $v_1, \dots, v_h$ be the marked vertices, in order of traversal. We set the vertex set of G' as the set M of all vertices marked as terminals while the edge set of G' contains (i) all edges $e_i = (v_i, v_{i+1})$ for $i = 1, \dots, h-1$, where the cost of edge e_i is $d_P(v_i, v_{i+1})$, and (ii) all edges in S, with their respective costs. The graph G' has $O(k)$ vertices and edges, and it can be built in $O(k)$ time after a one-time preprocessing of P which requires $O(n)$ time. See Fig. 2 for an example.

We now compute all the distances from s in G' in $O(k \log k)$ time by running Dijkstra's algorithm. Since our construction of G' ensures that $d_{P+S}(s, v) = d_{G'}(s, v)$ for every terminal v, we now know all the distances $\alpha_i := d_{P+S}(s, v_i)$ with $v_i \in M$.

For $i = 1, \dots, h-1$, define P_i as the subpath of P between v_i and v_{i+1}. To find $\mathcal{E}_s(V(P))$, we will separately compute the quantities $\mathcal{E}_s(P_1), \dots, \mathcal{E}_s(P_{h-1})$.[3]

[3] With a slight abuse of notation we use P_i to refer both to the subpath of P between v_i and v_{i+1} and to the set of vertices therein.

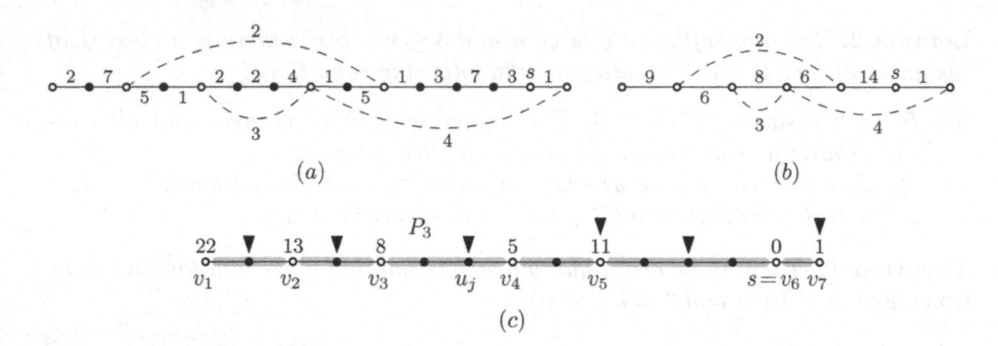

Fig. 2. An example showing how to compute the eccentricity of a vertex s in $P+S$. (a) Edges of P are solid, while edges in S are dashed. Vertices marked as terminals are white. (b) The condensed graph G'. (c) Every terminal v_i is labeled with its corresponding α_i and the arrows point to the vertices u_j in each (shaded) path P_i such that $j \geq 2$ is the smallest index with $\ell(j) \geq r(j)$.

Fix an index i and let u be a vertex in P_i. Consider a shortest path π from s to u in $P+S$ and let x be the last marked vertex traversed by π (this vertex always exists since s is marked). We can decompose π into two subpaths: a shortest path π_x from s to x, and a shortest path π_u from x to u. By the choice of x, x is the only marked vertex in π_u. This means that x is either v_i or v_{i+1} and, in particular:

$$d_{P+S}(s,u) = \min\{\alpha_i + d_{P_i}(v_i,u), \alpha_{i+1} + d_{P_i}(v_{i+1},u)\}.$$

Hence, the farthest vertex u from s among those in P_i is the one that maximizes the right-hand side of the above formula, i.e.:

$$\mathcal{E}_s(P_i) = \max_{u \in P_i} \min\{\alpha_i + d_{P_i}(v_i,u), \alpha_{i+1} + d_{P_i}(v_{i+1},u)\}.$$

We now describe how such a maximum can be computed efficiently. Let u_j denote the j-th vertex encountered when P_i is traversed from v_i to v_{i+1}. The key observation is that the quantity $\ell(j) = \alpha_i + d_{P_i}(v_i,u_j)$ is monotonically non-decreasing w.r.t. j, while the quantity $r(j) = \alpha_{i+1} + d_{P_i}(v_{i+1},u_j)$ is monotonically non-increasing w.r.t. j. Since both $\ell(j)$ and $r(j)$ can be evaluated in constant time once α_i and α_{i+1} are known, we can binary search for the smallest index $j \geq 2$ such that $\ell(j) \geq r(j)$ (see Fig. 2(c)). Notice that index j always exists since the condition is satisfied for $u_j = v_{i+1}$.

This requires $O(\log|P_i|)$ time and allows us to return $\mathcal{E}_s(P_i) = \max\{\ell(j-1), r(j)\}$.

After the linear-time preprocessing, the time needed to compute the eccentricity of a single vertex s is then $O(k \log k + \sum_{i=1}^{h-1} \log|P_i|)$. Since $\sum_{i=1}^{h-1} |P_i| < n+h$ and $h = O(k)$, this can be upper bounded by $O(k\log k + k \log \frac{n}{k}) = O(k \log \max\{k, \frac{n}{k}\})$. Repeating the above procedure for all vertices s, and accounting for the preprocessing time, we can compute the diameter of $P+S$ in time $O(nk \log \max\{k, \frac{n}{k}\})$.

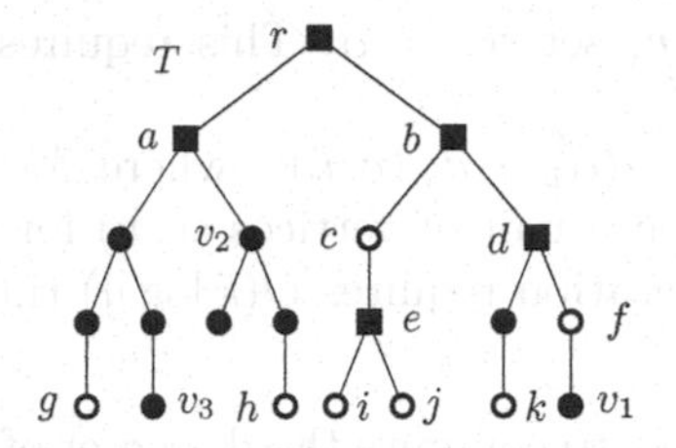

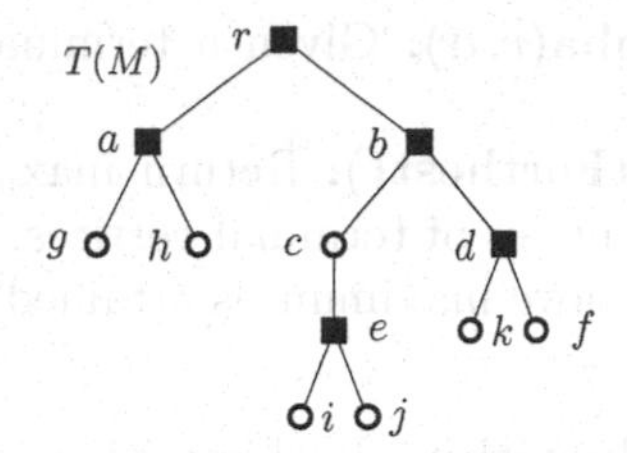

Fig. 3. The rooted binary tree T with root r is depicted on the left side. The set of terminal vertices is $M = \{c, f, g, h, i, j, k\}$ and is represented using white vertices. The square vertices are the Steiner vertices. The corresponding shrunk tree $T(M)$ is depicted on the right side.

3.2 Diameter on Augmented Trees

We are now ready to describe the algorithm that computes the diameter of a tree T augmented with a set S of k shortcuts. The key idea is to use the same framework used for the path. Roughly speaking, we define a condensed graph G' over a set of $O(k)$ marked vertices, as we did for the path. Next, we use G' to compute in $O(k \log k)$ time the distance $\alpha_v := d_{T+S}(s, v)$ for every marked vertex v, and then we use these distances to compute the eccentricity $\mathcal{E}_s$ of s in $T + S$. This last step is the tricky one. In order to efficiently manage it, we design an ad-hoc data structure which will be able to compute $\mathcal{E}_s$ in $O(k \log n)$ time once all values α_v are known. In the following we provide a description of the operations supported by our data structure, and we show how they can be used to compute the diameter of $T + S$. The full version of the paper contains a description of how such a data structure can be implemented.

An Auxiliary Data Structure: Description. Given a tree T that is rooted in some vertex r and two vertices u, v of T, we denote by $\text{LCA}_T(u, v)$ their *lowest common ancestor* in T, i.e., the deepest node (w.r.t. the hop-distance) that is an ancestor of both u and v in T. Moreover, given a subset of vertices M, we define the *shrunk* version of T w.r.t. M as the tree $T(M)$ whose vertex set consists of r along with all vertices in $\{\text{LCA}_T(u, v) \mid u, v \in M\}$ (notice that this includes all vertices $v \in M$ since $\text{LCA}(v, v) = v$), and whose edge set contains an edge (u, v) for each pair of distinct vertices u, v such that the only vertices in the unique path between u and v in T that belong to $T(M)$ are the endvertices u and v. We call the vertices in M *terminal* vertices, while we refer to the vertices that are in $T(M)$ but not in M as *Steiner* vertices. See Fig. 3 for an example.

Our data structure can be built in $O(n)$ time and supports the following operations, where k refers to the current number of terminal vertices:

MakeTerminal(v): Mark a given vertex v of T as a terminal vertex. Set $\alpha_v = 0$. This operation requires $O(\log n)$ time.
Shrink(): Report the shrunk version of T w.r.t. the set M of terminal vertices, i.e., report $T(M)$. This operation requires $O(k)$ time.

SetAlpha(v, α): Given a terminal vertex v, set $\alpha_v = \alpha$. This requires $O(1)$ time.

ReportFarthest(): Return $\max_{u \in T} \min_{v \in M}(\alpha_v + d_T(v, u))$, where M is the current set of terminal vertices, along with a pair of vertices (v, u) for which the above maximum is attained. This operation requires $O(k \log n)$ time.

Our Algorithm. In this section we show how to compute the diameter of $T+S$ in time $O(nk \log n)$. We can assume w.l.o.g. that the input tree T is binary since, if this is not the case, we can transform T into a binary tree having same diameter once augmented with S, and asymptotically the same number of vertices as T. This is a standard transformation that requires linear time and is described in the full version of the paper for completeness. Moreover, we perform a linear-time preprocessing in order to be able to compute the distance $d_T(u, v)$ between any pair of vertices in constant time.[4]

We use the data structure $\mathcal{D}$ of Sect. 3.2, initialized with the binary tree T. Similarly to our algorithm on paths, we compute the diameter of T by finding the eccentricity $\mathcal{E}_s$ in $T + S$ of each vertex s. In the rest of this section we fix s and show how to compute $\mathcal{E}_s$.

We start by considering all vertices x such that either $x = s$ or x is an endvertex of some edge in S, and we mark all such vertices as terminals in $\mathcal{D}$ (we recall that these vertices also form the set M of terminals). This requires time $O(k \log n)$. Next, we compute the distances from s in the (multi-)graph G' defined on the shrunk tree $T(M)$ by (i) assigning weight $d_T(u, v)$ to each edge (u, v) in $T(M)$, and (ii) adding all edges in S with weight equal to their cost. This can be done in $O(k \log k)$ time using Dijkstra's algorithm. We let α_v denote the computed distance from s to terminal vertex v.

We can now find $\mathcal{E}_s$ by assigning cost α_v to each terminal v in $\mathcal{D}$ (using the SetAlpha(v, α_v) operation), and performing a ReportFarthest() query. This requires $O(k + k \log n) = O(k \log n)$ time. Finally, we revert $\mathcal{D}$ to the initial state before moving on to the next vertex s.[5]

Theorem 2. *Given a graph G on n vertices and $n+k-1$ edges with non-negative edge costs and $k \geq 1$, we can compute the diameter of G in time $O(nk \log n)$.*

Since there are $\eta = \binom{n}{2} - (n-1) = \frac{(n-1)(n-2)}{2}$ possible shortcut edges, we can solve the k-DOAT problem by computing the diameter of $T + S$ for each of the $\binom{\eta}{k} = O(n^{2k})$ possible sets S of k shortcuts using the above algorithm. This yields the following result:

[4] This can be done by rooting T in an arbitrary vertex, noticing that $d_T(u, v) = d_T(u, \mathrm{LCA}_T(u, v)) + d_T(\mathrm{LCA}_T(u, v), v)$, and using an *oracle* that can report $\mathrm{LCA}(u, v)$ in constant time after a $O(n)$-time preprocessing [14].

[5] This can be done in $O(k \log n)$ time by keeping track of all the memory words modified as a result of the operations on $\mathcal{D}$ performed while processing s, along with their initial contents. To revert $\mathcal{D}$ to its initial state it suffices to rollback these modifications.

Corollary 1. *The* k-DOAT *problem on trees can be solved in time* $O(n^{2k+1} \cdot k \log n)$.

4 Linear-Time Approximation Algorithms

In this section we describe two $O(n)$-time approximation algorithms for metric k-DOAT. The first algorithm guarantees an approximation factor of 4 and runs in linear time when $k = O\left(\sqrt{\frac{n}{\log n}}\right)$. The second algorithm computes a $(1+\varepsilon)$-approximate solution for constant $\varepsilon > 0$, but its running time depends on the number λ of leaves of the input tree T and it is linear when $\lambda = O\left(n^{\frac{1}{(2k+2)^2}}\right)$ and $k = o(\sqrt{\log n})$.

Both algorithms use the data structure introduced in Sect. 3.2 and are based on the famous idea introduced by Gonzalez to design approximation algorithms for graph clustering [10]. Gonzalez' idea, to whom we will refer to as Gonzalez algorithm in the following, is to compute h suitable vertices $x_1, \ldots, x_h$ of an input graph G with non-negative edge costs in a simple iterative fashion.[6] The first vertex x_1 has no constraint and can be any vertex of G. Given the vertices $x_1, \ldots, x_i$, with $i < h$, the vertex x_{i+1} is selected to maximize its distance towards $x_1, \ldots, x_i$. More precisely, $x_{i+1} \in \arg\max_{v \in V(G)} \min_{1 \leq j \leq i} d_G(v, x_j)$. We now state two useful lemmas, the first of which is proved in [10].

Lemma 3. *Let* G *be a graph with non-negative edge costs and let* $x_1, \ldots, x_h$ *be the vertices computed by Gonzalez algorithm on input* G *and* h. *Let* $D = \min_{1 \leq i < j \leq h} d_G(x_i, x_j)$. *Then, for every vertex* v *of* G, *there exists* $1 \leq i \leq h$ *such that* $d_G(v, x_i) \leq D$.

Lemma 4. *Given as input a graph* G *that is a tree and a positive integer* h, *Gonzalez algorithm can compute the vertices* $x_1, \ldots, x_h$ *in* $O(n + h^2 \log n)$ *time.*

Proof. We can implement Gonzalez algorithm in time $O(n + h^2 \log n)$ by constructing the data structure $\mathcal{D}$ described in Sect. 3.2 and use it as follows. We iteratively (i) mark the vertex x_i as a terminal (in $O(\log n)$ time), and (ii) query $\mathcal{D}$ for the vertex x_{i+1} that maximizes the distance from all terminal vertices (in $O(h \log n)$ time). □

In the remainder of this section let S^* be an optimal solution for the k-DOAT instance consisting of the tree T embedded in a metric space c, and let $D^* = \text{diam}(T + S^*)$.

The 4-Approximation Algorithm. The 4-approximation algorithm we describe has been proposed and analyzed by Li *et al.* [16] for the variant of k-DOAT in which we are given a graph G as input and edge/shortcut costs are

[6] In the original algorithm by Gonzalez, these h vertices are used to define the centers of the h clusters.

uniform. Li *et al.* [16] proved that the algorithm guarantees an approximation factor of $(4 + \frac{2}{D^*})$; the analysis has been subsequently improved to $(2 + \frac{2}{D^*})$ in [3]. We show that the algorithm guarantees an approximation factor of 4 for the k-DOAT problem when c satisfies the triangle inequality.

The algorithm works as follows. We use Gonzalez algorithm on the input $G = T$ and $h = k + 1$ to compute $k + 1$ vertices $x_1, \dots, x_{k+1}$. The set S of k shortcuts is given by the star centered at x_1 and having $x_2, \dots, x_{k+1}$ as its leaves, i.e., $S = \{(x_1, x_i) \mid 2 \le i \le k+1\}$. The following lemma is crucial to prove the correctness of our algorithm.

Lemma 5 **([8, Lemma 3]).** *Given $G = T$ and $h = k+1$, Gonzalez algorithm computes a sequence of vertices $x_1, \dots, x_{k+1}$ with $\min_{1 \le i \le k+1} d_T(x_i, v) \le D^*$ for every vertex v of G.*

Theorem 3. *In a k-DOAT input instance in which T is embedded in a metric space and $k = O\left(\sqrt{\frac{n}{\log n}}\right)$, the algorithm computes a 4-approximate solution in $O(n)$ time.*

The $(1+\varepsilon)$-Approximation Algorithm. We now describe an algorithm that, for any constant $\varepsilon > 0$, computes a $(1+\varepsilon)$-approximate solution for metric k-DOAT. The running time of the algorithm is guaranteed to be linear when $k = o(\sqrt{\log n})$ and T has $\lambda = O\left(n^{\frac{1}{(2k+2)^2}}\right)$ leaves.

As usual for polynomial-time approximation schemes, we will consider only large enough instances, i.e., we will assume $n \ge n_0$ for some constant n_0 depending only on ε. We can solve the instances with $n < n_0$ in constant time using, e.g., the exact algorithm of Sect. 3.

In particular, we will assume that

$$n > \left(\frac{12\lambda(k+2)^2}{\varepsilon}\right)^{2k+2}. \tag{1}$$

Notice that, for any constant $\varepsilon > 0$, when $k = o(\sqrt{\log n})$ and $\lambda = O\left(n^{\frac{1}{(2k+2)^2}}\right)$, the right-hand side of (1) is in $o(n)$. As a consequence, it is always possible to choose a constant n_0 such that (1) is satisfied by all $n \ge n_0$.

The main idea is an extension of the similar result proved in [2] for the special case $k = 1$. However, we also benefit from the fast implementation we provided for Gonzalez algorithm to obtain a linear running time.

The idea borrowed from [2] is that of reducing the problem instance into a smaller instance formed by a tree T' induced by few vertices of T and by a suitable cost function c'. Next, we use the exact algorithm of Corollary 1 to compute an optimal solution S' for the reduced instance. Finally, we show that S' is a $(1+\varepsilon)$-approximate solution for the original instance. The quasi-optimality of the computed solution comes from the fact that the reduced instance is formed by a suitably selected subset of vertices that are close to the unselected ones.

The reduced instance is not exactly a k-DOAT problem instance, but an instance of a variant of the k-DOAT problem in which each edge (u, v) of the tree T' has a known non-negative cost associated with it, we have a shortcut for each pair of distinct vertices, and the function c' determines the cost of each shortcut (u, v) that can be added to the tree. Therefore, in our variant of k-DOAT we are allowed to add the shortcut (u, v) of cost $c'(u, v)$ even if (u, v) is an edge of T' (i.e., the cost of the shortcut (u, v) may be different from the cost of the edge (u, v) of T'). We observe that all the results discussed in the previous sections hold even for this generalized version of k-DOAT.[7]

Let B be the set of *branch* vertices of T, i.e., the internal vertices of T having a degree greater than or equal to 3. It is a folklore result that a tree with λ leaves contains at most $\lambda - 1$ branch vertices. Therefore, we have $|B| \leq \lambda - 1$.

The reduced instance T' has a set V' of $\eta = 2n^{\frac{1}{2k+2}}$ vertices defined as follows. V' contains all the branch vertices B plus $\eta - |B|$ vertices $x_1, \ldots, x_{\eta-|B|}$ of T that are computed using Gonzalez algorithm on input $G = T$ and $h = \eta - |B|$. By Lemma 4, the vertices $x_1, \ldots, x_{\eta-|B|}$ can be computed in $O(n + (\eta - |B|)^2 \log n) = O(n)$ time. As $|B| \leq \lambda - 1 = O\left(n^{\frac{1}{(2k+2)^2}}\right)$, it follows that $\eta - |B| > n^{\frac{1}{2k+2}}$. The edges of T' are defined as follows. There is an edge between two vertices $u, v \in V'$ iff the path P in T from u to v contains no vertex of V' other than u and v, i.e., $V(P) \cap V' = \{u, v\}$. The cost of an edge (u, v) of T' is equal to $d_T(u, v)$. Then, the cost function c' of T' is defined for every pair of vertices $u, v \in V'$, with $u \neq v$, and is equal to $c'(u, v) = c(u, v)$.

Given the vertices V', the $\eta - 1$ costs of the edges (u, v) of T', that are equal to the values $d_T(u, v)$, can be computed in $O(n)$ time using a depth-first traversal of T (from an arbitrary vertex of T').

We use the exact algorithm of Corollary 1 to compute an optimal solution S' for the reduced instance in time $O(\eta^{2k+1} \cdot k \log \eta) = O\left(2^{2k+2} \cdot n^{1-\frac{1}{2k+2}} \cdot k \log n\right) = O(n)$. The algorithm returns S' as a solution for the original problem instance. We observe that the algorithm runs in $O(n)$ time. In order to prove that the algorithm computes a $(1 + \varepsilon)$-approximate solution, we first give a preliminary lemma showing that each vertex of T is not too far from at least one of the vertices in $\{x_1, \ldots, x_{\eta-|B|}\}$.

Lemma 6. *If T has $\lambda = O\left(n^{\frac{1}{(2k+2)^2}}\right)$ leaves, then, for every $v \in V(T)$, there exists $i \in \{1, \ldots, \eta - |B|\}$ such that $d_T(x_i, v) \leq \frac{\varepsilon}{4(k+2)} D^*$.*

Theorem 4. *Let $\varepsilon > 0$ be a constant. Given a metric k-DOAT instance with $k = o(\sqrt{\log n})$ and such that T is a tree with $\lambda = O(n^{\frac{1}{(2k+2)^2}})$ leaves, the algorithm computes a $(1 + \varepsilon)$-approximate solution in $O(n)$ time.*

---

[7] This is because we can reduce the generalized k-DOAT problem instance into a k-DOAT instance in linear time by splitting each edge of T' of cost χ into two edges, one of cost 0 and the other one of cost χ, to avoid the presence of shortcuts that are parallel to the tree edges. All the shortcuts that are incident to the added vertex used to split an edge of T' have a sufficiently large cost which renders them useless.

Proof. We already proved through the section that the algorithm runs in $O(n)$ time. So, it only remains to prove the approximation factor guarantee.

We define a function $\phi : V(T) \to \{x_1, \ldots, x_{\eta-|B|}\}$ that maps each vertex $v \in V(T)$ to its closest vertex x_i, with $i \in \{1, \ldots, \eta - |B|\}$ w.r.t. the distances in T, i.e., $d_T(v, \phi(v)) = \min_{1 \le i \le \eta-|B|} d_T(v, x_i)$.

We now show that there exists a set S of at most k shortcuts such that (i) each edge $e \in S$ is between two vertices in $\{x_1, \ldots, x_{\eta-|B|}\}$ and (ii) $\text{diam}(T' + S) \le (1 + \frac{k\varepsilon}{k+2})D^*$. The set S is defined by mapping each shortcut $e = (u, v) \in S^*$ of an optimal solution for the original k-DOAT instance to the shortcut $(\phi(u), \phi(v))$ (self-loops are discarded). Clearly, (i) holds. To prove (ii), fix any two vertices u and v of T'. We first show that $d_{T+S}(u, v) \le (1 + \frac{k\varepsilon}{k+2})D^*$ and then prove that $d_{T'+S}(u, v) \le (1 + \frac{k\varepsilon}{k+2})D^*$.

Let P be a shortest path in $T + S^*$ between u and v and assume that P uses the shortcuts $e_1 = (u_1, v_1), \ldots, e_t = (u_t, v_t) \in S^*$, with $t \le k$. Consider the (not necessarily simple) path P' in $T + S$ that is obtained from P by replacing each shortcut e_i with a *detour* obtained by concatenating the following three paths: (i) the path in T from u_i to $\phi(u_i)$; (ii) the shortcut $(\phi(u_i), \phi(v_i))$; (iii) the path in T from $\phi(v_i)$ to v_i.

The overall cost of the detour that replaces the shortcut e_i in P' is at most $c(e_i) + \frac{\varepsilon}{k+2}D^*$. Indeed, using that $c(\phi(u_i), \phi(v_i)) \le d_T(u_i, \phi(u_i)) + c(e_i) + d_T(\phi(v_i), v_i)$ together with Lemma 6 that implies $d_T(u_i, \phi(u_i)), d_T(\phi(v_i), v_i) \le \frac{\varepsilon}{4(k+2)}D^*$, we obtain

$$
\begin{aligned}
&d_T(u_i, \phi(u_i)) + c(\phi(u_i), \phi(v_i)) + d_T(\phi(v_i), v_i) \\
&\qquad \le 2d_T(u_i, \phi(u_i)) + c(e_i) + 2d_T(\phi(v_i), v_i) \le c(e_i) + \frac{\varepsilon}{k+2}D^*.
\end{aligned}
$$

As a consequence, $c(P') \le c(P) + \frac{t\varepsilon}{k+2}D^* \le c(P) + \frac{k\varepsilon}{k+2}D^*$ and, since $c(P) \le \text{diam}(T + S^*) \le D^*$, we obtain $c(P') \le (1 + \frac{k\varepsilon}{k+2})D^*$.

To show that $d_{T'+S}(u, v) \le (1 + \frac{k\varepsilon}{k+2})D^*$, i.e., that $\text{diam}(T' + S) \le (1 + \frac{k\varepsilon}{k+2})D^*$, it is enough to observe that P' can be converted into a path P'' in $T'+S$ of cost that is upper bounded by $c(P')$. More precisely, we partition the edges of P' except those that are in S into subpaths, each of which has two vertices in $\{x_1, \ldots, x_{\eta-|B|}\}$ as its two endvertices and no vertex in $\{x_1, \ldots, x_{\eta-|B|}\}$ as one of its internal vertices. The path P'' in $T' + S$ is defined by replacing each subpath with the edge of T' between its two endvertices. Clearly, the cost of this edge, being equal to the distance in T between the two endivertices, is at most the cost of the subpath. Therefore, the cost of P'' in $T' + S$ is at most the cost of P' in $T + S$; hence $d_{T'+S}(u, v) \le (1 + \frac{k\varepsilon}{k+2})D^*$.

We conclude the proof by showing that the solution S' computed by the algorithm satisfies $\text{diam}(T + S') \le (1 + \varepsilon)D^*$. The solution S' is an optimal solution for the reduced instance. As a consequence, $\text{diam}(T' + S') \le \text{diam}(T' + S) \le (1 + \frac{k\varepsilon}{k+2})D^*$. Let u and v be any two vertices of T. We have that $d_{T'+S'}(\phi(u), \phi(v)) \le (1 + \frac{k\varepsilon}{k+2})D^*$. Moreover, by Lemma 6, we have that $d_T(u, \phi(u)), d_T(v, \phi(v)) \le \frac{\varepsilon}{4(k+2)}D^*$. Therefore,

$$\begin{aligned} d_{T+S'}(u,v) &\leq d_T(u,\phi(u)) + d_{T+S'}(\phi(u),\phi(v)) + d_T(v,\phi(v)) \\ &\leq \frac{\varepsilon}{4(k+2)}D^* + d_{T'+S'}(\phi(u),\phi(v)) + \frac{\varepsilon}{4(k+2)}D^* \\ &\leq \frac{2\varepsilon}{4(k+2)}D^* + \left(1+\frac{k\varepsilon}{k+2}\right)D^* < (1+\varepsilon)D^*. \end{aligned}$$

Hence $\mathrm{diam}(T+S') \leq (1+\varepsilon)D^*$. This completes the proof.

References

1. Adriaens, F., Gionis, A.: Diameter minimization by shortcutting with degree constraints. CoRR abs/2209.00370 (2022). https://doi.org/10.48550/arXiv.2209.00370. Accepted at the IEEE ICDM 2022 conference
2. Bilò, D.: Almost optimal algorithms for diameter-optimally augmenting trees. Theor. Comput. Sci. **931**, 31–48 (2022). https://doi.org/10.1016/j.tcs.2022.07.028
3. Bilò, D., Gualà, L., Proietti, G.: Improved approximability and non-approximability results for graph diameter decreasing problems. Theor. Comput. Sci. **417**, 12–22 (2012)
4. Carufel, J.D., Grimm, C., Maheshwari, A., Schirra, S., Smid, M.H.M.: Minimizing the continuous diameter when augmenting a geometric tree with a shortcut. Comput. Geom. **89**, 101631 (2020). https://doi.org/10.1016/j.comgeo.2020.101631
5. Carufel, J.D., Grimm, C., Maheshwari, A., Smid, M.H.M.: Minimizing the continuous diameter when augmenting paths and cycles with shortcuts. In: Pagh, R. (ed.) 15th Scandinavian Symposium and Workshops on Algorithm Theory, SWAT 2016. LIPIcs, vol. 53, pp. 27:1–27:14. Schloss Dagstuhl - Leibniz-Zentrum fuer Informatik (2016)
6. Chepoi, V., Vaxès, Y.: Augmenting trees to meet biconnectivity and diameter constraints. Algorithmica **33**(2), 243–262 (2002)
7. Demaine, E.D., Zadimoghaddam, M.: Minimizing the diameter of a network using shortcut edges. In: Kaplan, H. (ed.) SWAT 2010. LNCS, vol. 6139, pp. 420–431. Springer, Heidelberg (2010). https://doi.org/10.1007/978-3-642-13731-0_39
8. Frati, F., Gaspers, S., Gudmundsson, J., Mathieson, L.: Augmenting graphs to minimize the diameter. Algorithmica **72**(4), 995–1010 (2015)
9. Gao, Y., Hare, D.R., Nastos, J.: The parametric complexity of graph diameter augmentation. Discret. Appl. Math. **161**(10–11), 1626–1631 (2013)
10. Gonzalez, T.F.: Clustering to minimize the maximum intercluster distance. Theor. Comput. Sci. **38**, 293–306 (1985). https://doi.org/10.1016/0304-3975(85)90224-5
11. Große, U., Knauer, C., Stehn, F., Gudmundsson, J., Smid, M.H.M.: Fast algorithms for diameter-optimally augmenting paths and trees. Int. J. Found. Comput. Sci. **30**(2), 293–313 (2019). https://doi.org/10.1142/S0129054119500060
12. Gudmundsson, J., Sha, Y.: Algorithms for radius-optimally augmenting trees in a metric space. In: Lubiw, A., Salavatipour, M. (eds.) WADS 2021. LNCS, vol. 12808, pp. 457–470. Springer, Cham (2021). https://doi.org/10.1007/978-3-030-83508-8_33
13. Gudmundsson, J., Sha, Y., Yao, F.: Augmenting graphs to minimize the radius. In: Ahn, H., Sadakane, K. (eds.) 32nd International Symposium on Algorithms and Computation, ISAAC 2021, 6–8 December 2021, Fukuoka, Japan. LIPIcs, vol. 212, pp. 45:1–45:20. Schloss Dagstuhl - Leibniz-Zentrum für Informatik (2021). https://doi.org/10.4230/LIPIcs.ISAAC.2021.45

14. Harel, D., Tarjan., R.E.: Fast algorithms for finding nearest common ancestors. SIAM J. Comput. **13**(2), 338–355 (1984). https://doi.org/10.1137/0213024
15. Johnson, C., Wang, H.: A linear-time algorithm for radius-optimally augmenting paths in a metric space. Comput. Geom. **96**, 101759 (2021). https://doi.org/10.1016/j.comgeo.2021.101759
16. Li, C., McCormick, S.T., Simchi-Levi, D.: On the minimum-cardinality-bounded-diameter and the bounded-cardinality-minimum-diameter edge addition problems. Oper. Res. Lett. **11**(5), 303–308 (1992). https://doi.org/10.1016/0167-6377(92)90007-P
17. Oh, E., Ahn, H.: A near-optimal algorithm for finding an optimal shortcut of a tree. In: Hong, S. (ed.) 27th International Symposium on Algorithms and Computation, ISAAC 2016, 12–14 December 2016, Sydney, Australia. LIPIcs, vol. 64, pp. 59:1–59:12. Schloss Dagstuhl - Leibniz-Zentrum fuer Informatik (2016)
18. Schoone, A.A., Bodlaender, H.L., van Leeuwen, J.: Diameter increase caused by edge deletion. J. Graph Theory **11**(3), 409–427 (1987)
19. Wang, H.: An improved algorithm for diameter-optimally augmenting paths in a metric space. Comput. Geom. **75**, 11–21 (2018). https://doi.org/10.1016/j.comgeo.2018.06.004
20. Wang, H., Zhao, Y.: Algorithms for diameters of unicycle graphs and diameter-optimally augmenting trees. Theor. Comput. Sci. **890**, 192–209 (2021). https://doi.org/10.1016/j.tcs.2021.09.014

Approximating the Smallest k-Enclosing Geodesic Disc in a Simple Polygon

Prosenjit Bose[1], Anthony D'Angelo[2(✉)], and Stephane Durocher[3]

[1] School of Computer Science, Carleton University, Ottawa, Canada
jit@scs.carleton.ca
[2] Carleton University, Ottawa, Canada
anthony.dangelo@carleton.ca
[3] University of Manitoba, Winnipeg, Canada
stephane.durocher@umanitoba.ca

Abstract. We consider the problem of finding a geodesic disc of smallest radius containing at least k points from a set of n points in a simple polygon that has m vertices, r of which are reflex vertices. We refer to such a disc as a SKEG disc. We present an algorithm to compute a SKEG disc using higher-order geodesic Voronoi diagrams with worst-case time $O(k^2 n + k^2 r + \min(kr, r(n-k)) + m)$ ignoring polylogarithmic factors.

We then present a 2-approximation algorithm that finds a geodesic disc containing at least k points whose radius is at most twice that of a SKEG disc. Our algorithm runs in $O(n \log^2 n \log r + m)$ expected time using $O(n+m)$ expected space if $k \in O(n/\log n)$; if $k \in \omega(n/\log n)$, the algorithm computes a 2-approximation solution with high probability in $O(n \log^2 n \log r + m)$ worst-case time with $O(n+m)$ space.

Keywords: Minimum/smallest enclosing circle/disc · Geodesic · Simple polygon · 2-approximation · Computational geometry

1 Introduction

The *smallest enclosing disc* problem[1] takes as input a set S of n points in the plane and returns the smallest Euclidean disc that contains S. This can be solved in $O(n)$ expected time [47] and $O(n)$ worst-case time [36]. The *smallest k-enclosing disc* problem is a generalization that asks for a smallest disc that contains at least $k \leq |S|$ points[2] of S, for any given k, and has been well-studied [1,19,22,23,27,34,35]. It is conjectured that an exact algorithm that computes the smallest k-enclosing disc in the plane requires $\Omega(nk)$ time [26, §1.5].

---

[1] Also known as the *minimum enclosing disc* problem.

[2] In this paper, we use the notation $|Z|$ to denote the number of points in Z if Z is a point set, or the number of vertices of Z if Z is a face or a polygon.

Supported by the Natural Sciences and Engineering Research Council of Canada (NSERC).

P. Morin and S. Suri (Eds.): WADS 2023, LNCS 14079, pp. 179–192, 2023.
https://doi.org/10.1007/978-3-031-38906-1_13

Matoušek [34] presented an algorithm that first computes a constant-factor approximation[3] in $O(n \log n)$ time and $O(n)$ space (recently improved to $O(n)$ expected-time for a 2-approximation that uses $O(n)$ expected space [27]), and then uses that approximation to seed an algorithm for solving the problem exactly in $O(n \log n + nk)$ expected time using $O(nk)$ space or $O(n \log n + nk \log k)$ expected time using $O(n)$ space (recently improved to $O(nk)$ expected time using $O(n + k^2)$ expected space [19,27]). Matoušek [35] also presented an algorithm for computing the smallest disc that contains all but at most q of n points (i.e., the minimum $(n - q)$-enclosing disc) in $O(n \log n + q^3 n^\epsilon)$ time for ϵ "a positive constant that can be made arbitrarily small by adjusting the parameters of the algorithms; multiplicative constants in the $O()$ notation may depend on ϵ" [35].

In this paper we generalize the smallest k-enclosing disc problem to simple polygons using the *geodesic metric* meaning that the distance $d_g(a, b)$ between two points a and b is the length of the shortest path $\Pi(a, b)$ between a and b that lies completely inside the simple polygon P. A *geodesic disc* $D(c, \rho)$ of radius ρ centred at $c \in P$ is the set of all points in P whose geodesic distance to c is at most ρ. The main focus of our article is the following problem.

Smallest k-Enclosing Geodesic (*SKEG*) Disc Problem

Consider a simple polygon P_{in} defined by a sequence of m vertices in $\mathbb{R}^2$, r of which are reflex vertices, and a set S of n points of $\mathbb{R}^2$ contained in P_{in}.[4] Find a *SKEG disc*, i.e., a geodesic disc of minimum radius ρ^* in P_{in} that contains at least k points of S.

We make the general position assumptions that no two points of S are equidistant to a vertex of P_{in}, and no four points of S are geodesically co-circular. Under these assumptions, a SKEG disc contains exactly k points. Let $D(c^*, \rho^*)$ be a SKEG disc for the points of S in P_{in}. For convenience, at times we will refer to this as simply D^*. A k-enclosing geodesic disc (*KEG* disc) is a geodesic disc in P_{in} that contains exactly k points of S. A 2-SKEG disc is a KEG disc with radius at most $2\rho^*$ (i.e., it is a 2-approximation). The main result of our article is the following theorem.

Theorem 4. *If $k \in O(n/\log n)$, **Main-Algo** computes a 2-SKEG disc in expected time $O(n \log^2 n \log r + m)$ and expected space $O(n + m)$; if $k \in \omega(n/\log n)$, **Main-Algo** computes a 2-SKEG disc with high probability*[5] *in $O(n \log^2 n \log r + m)$ deterministic time with $O(n + m)$ space.*

---

[3] A β-approximation means that the disc returned has a radius at most β times the radius of an optimal solution.

[4] When we refer to a point p being in a polygon P, we mean that p is in the interior of P or on the boundary, ∂P.

[5] We say an event happens with high probability if the probability is at least $1 - n^{-\lambda}$ for some constant λ.

1.1 Related Work

A region Q is *geodesically convex* relative to P if for all points $u, v \in Q$, the geodesic shortest path from u to v in P is in Q. The *geodesic convex hull* CH_g of a set of points S in a polygon P is the intersection of all geodesically convex regions in P that contain S. The geodesic convex hull of n points in a simple m-gon can be computed in $O(n \log n + m)$ time using $O(n + m)$ space [24,43].

The *geodesic centre* problem asks for a smallest enclosing geodesic disc that lies in the polygon and encloses all vertices of the polygon (stated another way, a point that minimizes the distance to the farthest point). This problem is well-studied [2,7,12,41,43] and can be solved in $O(m)$ time and space [2]. The geodesic centre problem has been generalized to finding the geodesic centre of a set of points S inside a simple polygon. The problem can be solved by finding the geodesic centre of the weakly simple polygon formed by the geodesic convex hull of the point set S [6], and thus runs in $O(n \log n + m)$ time. For this approach, computing the solution is dominated by the time to compute CH_g.

We are not aware of other work tackling the subject of this paper, but there has been other work done with geodesic discs in polygons. Generalized versions of the geodesic centre for polygons [8,39,40,44,45]; packing and covering [42,44]; and clustering [11] have all been studied. Dynamic k-nearest neighbour queries were studied by de Berg and Staals [20]. They presented a static data structure for geodesic k-nearest neighbour queries for n sites in a simple m-gon that is built in $O(n(\log n \log^2 m + \log^3 m))$ expected time using $O(n \log n \log m + m)$ expected space and answers queries in $O(\log(n + m) \log m + k \log m)$ expected time.[6]

If P_{in} has no reflex vertices, it is a convex polygon and the SKEG disc problem is solved by the algorithm for planar instances which uses a grid-refinement strategy. This works in the plane because $\mathbb{R}^2$ with the Euclidean metric is a doubling metric space, meaning that for any disc of radius $\rho > 0$ in $\mathbb{R}^2$ it can be covered by $O(1)$ discs of radius $\rho/2$ [28]. Geodesic discs do not have this property; it may take $\Theta(r)$ smaller discs to cover the larger one. Another difficulty of the geodesic metric is that for two points u and v of S on opposite sides of a given chord, their geodesic bisector (formed by concatenating their bisector and hyperbolic arcs) can cross the chord $\Omega(r)$ times.

1.2 An Exact Algorithm

In this section we present an exact algorithm for the SKEG disc problem that uses higher-order geodesic Voronoi diagrams to find the exact solution.

Rather than working with our m-gon, we use the polygon simplification algorithm of Aichholzer et al. [3] to transform P_{in} into a simple polygon $P \supseteq P_{in}$ of size $O(r)$ in $O(m)$ time and space such that P preserves the visibility of points

[6] We note that depending on the relations of the values m, r, and n to each other, there may be situations in which our algorithms may be improved by polylogarithmic factors by using the k-nearest neighbour query data structure.

in P_{in}, as well as their shortest paths. The reflex vertices in P_{in} also appear in P. We again assume that S is in general position with respect to P. The polygon simplification allows the running time of the algorithm to depend on the number of *reflex* vertices of P_{in} rather than the total number of vertices of P_{in}. Thus, the algorithm runs faster on polygons with fewer reflex vertices.

For $k \leq n-1$ we use the shortest-path data structure of Guibas and Hershberger [24,29]. This shortest-path data structure can be built in $O(r)$ time and space and, given any two query points in the polygon, returns a tree of $O(\log r)$ height in $O(\log r)$ time representing the shortest path between the two query points as well as the length of this path. The in-order traversal of this tree gives the shortest path. The data structure uses additional $O(\log r)$ space to build the result of the query (i.e., the tree) by linking together precomputed structures. Given the result of a query, we can perform a search through this tree in $O(\log r)$ time to find the midpoint of the shortest path [46, Lemma 3], or the first or last edge of the path.

Higher-order Voronoi diagrams have been considered to solve the smallest k-enclosing disc problem in the plane [1,22]. This approach can be generalized to a point set S contained in simple polygons, but it requires computing the order-k geodesic Voronoi diagram (OKGVD),[7] or the order-$(k-1)$ and order-$(k-2)$ diagrams. Using the current results for OKGVDs [5,6,9,32,37,38,46], we have the following theorem.

Theorem 1. *Given an integer $1 < k \leq n$, a SKEG disc of the points of S can be computed in*

- $O(n \log n + m)$ *time and* $\Theta(n+m)$ *space for* $k = n$ *(in which case it has become the geodesic centre problem for points in a simple polygon);*
- $O(nr \log n \log^2 r + m)$ *time and* $\Theta(n+m)$ *space for* $k = n-1$ *and* $r/\log^2 r \in \Omega(k \log k)$*;*
- $O(n^2 \log n + n^2 \log r + nr \log n + nr \log r + r^2 + m)$ *time and* $\Theta(n+m)$ *space for* $k = n-1$ *and* $r/\log^2 r \in o(k \log k)$*;*
- $O(k^2 n \log n \log^2 r + \min(rk, r(n-k)) \log r + m)$ *time and* $\Omega(k(n-k) + \min(rk, r(n-k)) + m)$ *space for* $n \log n \in o(r/\log r)$ *and* $k < n-1$*;*
- $O(k^2 n \log n + k^2 r \log r + k(n-k) \log r + \min(kr, r(n-k)) \log r + m)$ *time and* $\Omega(k(n-k) + \min(rk, r(n-k)) + m)$ *space for* $n \log n \in \Omega(r/\log r)$ *and* $k < n-1$*;*

Ignoring polylogarithmic factors, the worst-case runtime for $k = n$ *is* $O(n+m)$*; for* $k = n-1$ *and* $r/\log^2 r \in \Omega(k \log k)$ *is* $O(nr+m)$*; for* $k = n-1$ *and* $r/\log^2 r \in o(k \log k)$ *is* $O(n^2 + nr + r^2 + m)$*; for* $k < n-1$ *and for* $n \log n \in o(r/\log r)$ *is* $O(k^2 n + \min(rk, r(n-k)) + m)$*; and* $O(k^2 n + k^2 r + \min(kr, r(n-k)) + m)$ *otherwise.*

[7] Stated briefly, the order-k Voronoi diagram is a generalization of the Voronoi diagram such that each face is the locus of points whose k nearest neighbours are the k points of S associated with (i.e., that define) the face.

The main result of our paper is an improvement in the runtime for computing a SKEG disc (Theorem 1), but it comes at the expense of a 2-approximation. This is summarized in Theorem 4. The runtime of Theorem 4 is derived by balancing the runtimes of two algorithms: RS-Algo (a random sampling algorithm described in Sect. 2) and DI-Algo (a Divide-and-Conquer algorithm described in Sect. 3).

Ignoring polylogarithmic factors, the expected runtime of the approximation algorithm of Theorem 4 is $O(n + m)$. Thus, the approximation algorithm of Theorem 4 is roughly expected to be faster by a factor of k or r for $k = n - 1$, and a factor of k^2 otherwise. For example, if $k \in \Omega(n)$ and $k < n - 1$, compare $O(n+m)$ to: $O(n^3+r+m)$ for k close to n, or $O(n^3+nr+m)$ for k a fraction of n; and $O(n^3+n^2r+m)$.

Section 4 shows how to use a randomized iterative search as the merge step in Sect. 3 to compute a 2-approximation. We summarize our results in Sect. 5.

2 Random Sampling Algorithm

In this section we present a random sampling algorithm to compute a 2-SKEG disc. Preprocessing for our first algorithm, RS-Algo, includes converting P_{in} into a simplified polygon P, and building the shortest-path data structure, all in $O(m)$ time and space. **RS-Algo** proceeds as follows:

1. Compute a random sample of $(n/k)\ln(n)$ points of S.
2. For each point c in the random sample, find its $(k-1)^{\text{st}}$-closest point in S using the geodesic distance.
3. Return the KEG disc of minimum computed radius.

Theorem 2. ***RS-Algo*** *computes a 2-SKEG disc with high probability in deterministic time $O((n^2/k)\log n \log r + m)$ using $O(n+m)$ space.*

Remark. There is a simple deterministic algorithm that runs in $O(n^2 \log r+m)$ time by computing the $(k-1)^{\text{st}}$-closest neighbour for each point. However, we design a randomized algorithm, RS-Algo, that is faster when k is $\omega(\log n)$. Having a k term in the denominator of the runtime allows us to improve the runtime of our algorithm as k approaches n. When combining RS-Algo and DI-Algo and ignoring polylogarithmic factors, the runtime is at least a factor of $\Theta(n)$ faster than this simple approach.

3 Divide-and-Conquer Algorithm

In this section we describe DI-Algo, a Divide-and-Conquer algorithm to compute a 2-SKEG disc.

Let D^* be a SKEG disc for the points of S in a polygon (the polygon being referred to will be clear from the context). In each recursive call, we split the current polygon by a diagonal into two subpolygons of roughly equal size and recursively compute a 2-SKEG disc for each of the two subpolygons. The merge step involves computing a 2-approximation to the optimal disc that contains k points under the assumption that it intersects the diagonal used to generate the recursive calls. We delay discussion of the merge step until Sect. 4. DI-Algo requires the following preprocessing that takes $O(n \log r + m)$ time and $O(n+m)$ space.

Simplification and Shortest-Path Data Structure Refer to Sect. 2.

Balanced Hierarchical Polygon Decomposition We compute a balanced hierarchical polygon decomposition tree T_B [16,24,25]. (Note that T_B is built by the shortest-path data structure.) Until all faces are triangles, each polygonal face f is split into two subpolygons of between $|f|/3$ and $2|f|/3$ vertices by a diagonal of f. Decomposing our simplified polygon, this takes $O(r)$ time and $O(r)$ space [14,25]. We end up with a decomposition tree, based on a triangulation of our polygon, of $O(\log r)$ height whose leaves store the triangles. The internal nodes store the diagonals of the triangulation. A diagonal's position in the tree corresponds to when it was inserted. Conceptually, we associate with each internal node the subpolygon split by the diagonal contained in the node (though we do not store this subpolygon in the node).

Build Point-Location Data Structure Recall that the leaves of T_B represent triangles in a triangulation of P. We build Kirkpatrick's [15,30] $O(\log r)$ query-time point-location data structure on these triangles in $O(r)$ time with $O(r)$ space.

Augment T_B for Point Location We augment T_B to store which points of S are in which subpolygon represented by the internal nodes of the tree.

Lemma 1. *In $O(n \log r + r)$ time and $O(n + r)$ space, we augment the decomposition tree T_B to know which points of S are in which subpolygon.*

3.1 Algorithm Description: DI-Algo

Let τ denote the node of T_B associated with the current recursive call, let P_τ denote the current polygon split by the diagonal ℓ stored in τ, and let S_τ be the set of points of S in P_τ. For ease of discussion, we abuse notation and say P_τ is stored in τ. Let $|P_\tau| = r'$ and $|S_\tau| = n'$. Let P_1 (resp., P_2) be the subpolygon associated with the left (resp., right) child of τ on which we recursed that contains the points $S_1 \subseteq S$ (resp., $S_2 \subseteq S$). We also use the notation $P_1 \cup P_2$ to refer to P_τ.

The merge step (described in Sect. 4.1) involves computing $D(c, \rho)$, a KEG disc for $S_1 \cup S_2$ where the centre $c \in \ell$. If $D^* \cap S_1 \neq \emptyset$ and $D^* \cap S_2 \neq \emptyset$, then $D^* \cap \ell \neq \emptyset$ and either the specially-chosen centre c is inside $D^* \cap \ell$ (implying a 2-approximation), or $D(c, \rho)$ has a radius smaller than a disc centred on such a point (also implying a 2-approximation). If either $D^* \cap S_1 = \emptyset$ or $D^* \cap S_2 = \emptyset$, then the optimal disc is centred in one of the subpolygons and contains only points of S in that subpolygon, thus we have a 2-approximation by recursion. Therefore, the smallest disc computed among the three discs (i.e., the two discs returned by the recursion and the disc computed in the merge step) is a 2-approximation to D^*. For convenience, in the sequel we refer to the disc $D(c, \rho)$ whose centre is on the chord ℓ as a *merge disc.*

Lemma 2. *At the end of the merge step, the smallest of the three candidate discs is a 2-SKEG disc for the subpolygon under consideration (i.e., $P_1 \cup P_2$).*

DI-Algo proceeds as follows.

1. (Base case) When the recursive step reaches a triangle,[8] we use the planar 2-approximation algorithm which runs in expected time linear in the number of points of S in the subpolygon under consideration [26,27].
2. We recurse on the subpolygons P_1 and P_2 stored in the left and right child of τ respectively. Note that a recursive call is not necessary if a subpolygon contains fewer than k points of S in it. Let the returned disc with the smaller radius be the current solution.
3. (Merge Step) Consider the diagonal ℓ stored in τ. Compute a merge disc centred on ℓ for the points of $S_1 \cup S_2$.
4. Return the smallest of the three discs.

Theorem 3. ***DI-Algo*** *computes a 2-SKEG disc in $O(n \log^2 n \log r + m)$ expected time and $O(n + m)$ expected space.*

Proof (Sketch).
The recurrence tree of the Divide-and-Conquer algorithm mimics T_B and has $O(\log r)$ depth. By Lemma 2, the result of the merge step is a 2-approximation for P_τ. Thus, when we finish at the root, we have a 2-approximation to D^*.

The base case of the recursion is triggered when we reach a leaf of T_B. Here P_τ is a triangle. When the base case is reached, DI-Algo runs the planar 2-approximation algorithm in expected time and expected space linear in the number of points of S_τ (i.e., $O(n')$) [26,27].

We assume for the moment that the merge step runs in expected time $O(n' \log n' \log r + n' \log^2 n')$ and uses $O(n' + r')$ space (we show this in Sect. 4.1). Each node in a given level of T_B has a different value for n', but across the whole level their sum is n. This implies the expected runtime across a level of T_B is $O(n \log n \log r + n \log^2 n)$.

The $O(\log r)$ height of the recurrence tree implies that (including preprocessing) the algorithm runs in $O(n \log n \log^2 r + n \log^2 n \log r + m)$ expected time. We can simplify this expression to $O(n \log^2 n \log r + m)$ by starting with the assumption that the dominant term is $n \log n \log^2 r$ and then arriving at the contradiction that $m \in \omega(n \log n \log^2 r)$.

The space bound follows from the space for preprocessing and the space in the merge step which is released after the merge. □

[8] We could identify convex subpolygons before we get to the triangles, but it would not improve the asymptotic runtime.

We can balance this expected runtime against the runtime of Theorem 2. When $k \in O(n/\log n)$ the expected runtime of DI-Algo is faster by Theorem 3; when $k \in \omega(n/\log n)$ the runtime of RS-Algo is faster by polylogarithmic factors by Theorem 2. Although the runtimes are asymptotically identical when $k \in \Theta(n/\log n)$, RS-Algo gets faster as k gets larger. For example, if $k \in \Omega(n)$ RS-Algo runs in time $O(n \log n \log r + m)$ and finds a 2-approximation with high probability, whereas Theorem 3 finds a 2-approximation in expected time $O(n \log^2 n \log r + m)$.

This leads to our main theorem and our main algorithm, **Main-Algo**, which first performs the mentioned preprocessing, then runs RS-Algo if $k \in \omega(n/\log n)$; otherwise, it runs DI-Algo.

Theorem 4. *If $k \in O(n/\log n)$, **Main-Algo** computes a 2-SKEG disc in expected time $O(n \log^2 n \log r + m)$ and expected space $O(n + m)$; if $k \in \omega(n/\log n)$, **Main-Algo** computes a 2-SKEG disc with high probability[9] in $O(n \log^2 n \log r + m)$ deterministic time with $O(n + m)$ space.*

4 Merge

In this section we describe how to perform the merge step of our Divide-and-Conquer algorithm. First we point out that it is not clear whether it is possible to apply the recursive random sampling technique of Chan [13]. His approach requires one to partition the points of S into a constant number of fractional-sized subsets such that the overall solution is the best of the solutions of each of the subsets. It is not clear how to partition the points of S to allow for an efficient merge step. In fact, this is an issue that Chan [13] points out in his paper for a related problem of finding the smallest square containing k points before showing how to circumvent this issue in an orthogonal setting.

Assume a SKEG disc in P_τ for k points of S_τ intersects ℓ and contains at least one point of S_1 and one point of S_2. In this case our merge step either returns: 1) a KEG disc centred on a special point of ℓ that is guaranteed to be inside a SKEG disc; or 2) a disc centred on ℓ whose radius is smaller than that of such a KEG disc.

4.1 Merge Algorithm

For each $u \in S_\tau$, let u_c be the closest point of the chord ℓ to u. Let the set of all such closest points be S_τ^c. We refer to the elements of S_τ^c as *projections* of the elements of S_τ onto ℓ. For any $u \in S_\tau$ and radius ρ, let the interval $I(u, \rho) \subseteq \ell$ be $D(u, \rho) \cap \ell$ (i.e., the set of points on ℓ within geodesic distance ρ of u). $I(u, \rho)$ is empty if $\rho < d_g(u, u_c)$. For $u_c \in S_\tau^c$ and radius ρ, we say the *depth* of u_c is the number of intervals from points of S_τ that contain u_c using the same distance ρ (i.e., the number of discs of radius ρ centred on points of S_τ that contain u_c).

[9] We say an event happens with high probability if the probability is at least $1 - n^{-\lambda}$ for some constant λ.

Merge-Algo proceeds as follows.

1. Compute $\mathbb{C} = S_\tau^c$.
2. Initialize set $\mathbb{S} = S_\tau$.
3. While $|\mathbb{C}| > 0$:
   - Pick a point $u_c \in \mathbb{C}$ uniformly at random.
   - Find the point $z \in \mathbb{S}$ that is the kth-closest neighbour of u_c.
   - Let $\rho = d_g(u_c, z)$.
   - If $|\mathbb{C}| = 1$, break the while-loop.
   - For each $v \in \mathbb{S}$, compute $I(v, \rho)$. If $I(v, \rho)$ is empty, remove v from $\mathbb{S}$.
   - For each $w \in \mathbb{C}$, compute the depth of w. If the depth of w is less than k, remove w from $\mathbb{C}$.
   - Remove u_c from $\mathbb{C}$.
4. Return u_c and ρ.

Observation 5. *If the SKEG disc D^* contains at least one point of S_1 and at least one point of S_2, then $D^* \cap \ell$ is a continuous non-empty interval of ℓ due to geodesic convexity.*

Lemma 3. *If a SKEG disc D^* contains at least one point of S_τ from each side of ℓ, then for some $u \in S_\tau$ that is in D^*, $D^* \cap \ell$ contains $u_c \in S_\tau^c$.*

Proof (Sketch). Follows from applying Pollack et al. [41, Corollary 2] to the projections u_c and the closest point on ℓ to the centre c^* of D^* assuming, without loss of generality, that u is below ℓ and c^* is to the right of u and on or above ℓ. □

Lemma 4. ***Merge-Algo*** *runs in $O(n' \log n' \log r + n' \log^2 n')$ time with high probability and $O(n' + r')$ space and produces a 2-approximation if $D^* \cap S_1 \neq \emptyset$ and $D^* \cap S_2 \neq \emptyset$.*

Proof (Sketch). If $D^* \cap S_1 \neq \emptyset$ and $D^* \cap S_2 \neq \emptyset$, then we either return a disc centred on a projection of some point of S_τ inside D^*, or a disc centred on a projection whose radius is less than that of such a disc. Thus, the result of the algorithm is a 2-approximation by Lemma 3.

We assume for the moment that we can compute S_τ^c in $O(n' \log r)$ time and $O(n' + r')$ space (we will show this in Sect. 4.2). Given the n' elements of S_τ^c or S_τ: we build sets $\mathbb{C}$ and $\mathbb{S}$ in $O(n')$ time and space; we can delete an identified element from the sets in constant time and space; we can iterate through a set in constant time and space per element; and we can pick a point $u_c \in \mathbb{C}$ uniformly at random in $O(1)$ time and space.

Using the shortest-path data structure, we can find the kth-closest neighbour of u_c in $O(n' \log r)$ time and $O(n' + r')$ space by first computing the distance of everyone in $\mathbb{S}$ to u_c in $O(\log r)$ time and $O(r')$ space each (the space is re-used for the next query), and then using a linear-time rank-finding algorithm to find the kth-ranking distance in $O(n')$ time and space [4,10,18]. Let ρ be this k^{th}-ranking distance. We assume for the moment that we can compute, for all $O(n')$ elements $v \in \mathbb{S}$, the intervals $I(v, \rho)$ in $O(n' \log r)$ time and $O(n' + r')$

space (we will show this in Sect. 4.3). Given the intervals we also assume for the moment that we can compute, for all $O(n')$ elements $w \in \mathbb{C}$, the depth of w in overall $O((|\mathbb{S}| + |\mathbb{C}|)\log(|\mathbb{S}| + |\mathbb{C}|)) = O(n' \log n')$ time and $O(|\mathbb{S}| + |\mathbb{C}|) = O(n')$ space (we will show this in Sect. 4.3). In $O(n')$ time and space we can remove from $\mathbb{S}$ and $\mathbb{C}$ elements whose interval I is empty or whose depth is less than k, respectively. Elements of $\mathbb{S}$ whose intervals are empty are too far away from ℓ to be contained in the geodesic disc of any element of $\mathbb{C}$ for any radius of value ρ or less, and since ρ is non-increasing in each iteration, we can remove these elements of $\mathbb{S}$ from consideration.[10] Elements of $\mathbb{C}$ whose depth is less than k will contain fewer than k elements of $\mathbb{S}$ in their geodesic discs for any radius of value ρ or less. Since we seek to minimize the value ρ, we can remove these elements of $\mathbb{C}$ from consideration. The removal of these elements ensures ρ is non-increasing in each iteration. Computing the kth-nearest neighbour of a projection, the intervals I, and the depths of elements of $\mathbb{C}$ dominates the complexity of the while-loop.

We associate with each projection in $\mathbb{C}$ the distance associated with its kth-nearest neighbour (and thus the disc's radius) and assume these distances are unique. By considering a random permutation of these distances, we can use Devroye's "Theory of Records" [21] to show that we require $O(\log n')$ iterations of the while-loop with high probability.

The work done in one iteration given sets $\mathbb{S}$ and $\mathbb{C}$ with $|\mathbb{S}| = n'$ and $|\mathbb{C}| \leq n'$ is $O(n' \log r + n' \log n')$ using $O(n' + r')$ space. With high probability we have $O(\log n')$ iterations, thus $O(n' \log n' \log r + n' \log^2 n')$ time with high probability. □

4.2 Computing Projections

Funnel Now let us briefly review the notion of a *funnel* [24,31]. The vertices of the geodesic shortest path between two points a and b, $\Pi(a, b)$, consists of the vertices a, b, and a subset of the vertices of the polygon P forming a polygonal chain [17,33]. Consider a diagonal ℓ of P, its two endpoints ℓ_1 and ℓ_2, and a point $p \in \{P \cup S\}$. The union of the three paths $\Pi(p, \ell_1)$, $\Pi(p, \ell_2)$, and ℓ form what is called a funnel. As Guibas and Hershberger [24] point out, this funnel represents the shortest paths from p to the points on ℓ in that their union is the funnel. Starting at p, the paths $\Pi(p, \ell_1)$ and $\Pi(p, \ell_2)$ may overlap during a subpath, but there is a unique vertex p_a (which is the farthest vertex on their common subpath from p) where the two paths diverge. After they diverge, the two paths never meet again. This vertex p_a is called the *apex*[11] of the funnel. The path from the apex to an endpoint of ℓ forms an *inward-convex* polygonal chain (i.e., a convex path through vertices of P with the bend protruding into the interior of P).

[10] At the moment, removing points from $\mathbb{S}$ is a practical consideration; no lower bound is clear on the number of elements from $\mathbb{S}$ that can be discarded in each iteration, though if one were to find a constant fraction lower bound, one could shave a logarithmic factor off the runtime.

[11] Sometimes, as in [31], this is called a *cusp*. In [31] the funnel is defined as beginning at the cusp and ending at the diagonal.

Before the main while-loop of Merge-Algo begins, we precompute S_τ^c, i.e., the closest point of ℓ to u for each point $u \in S_\tau$. As before, let u_c be the closest point of ℓ to u.

Lemma 5. *In $O(n' \log r)$ time and $O(n'+r')$ space we compute the closest point of ℓ to each $u \in S_\tau$.*

Proof (Sketch). We can compute the funnel of a point $u \in S_\tau$ to ℓ in $O(\log r)$ time and $O(r')$ space. Given this funnel, we can perform a binary search along ℓ in $O(\log r)$ time and $O(r')$ space (using extensions of the funnel edges) to find u_c. We know from Pollack et al. [41, Corollary 2] that this point is where the last edge connecting u to ℓ is closest to $\pi/2$. □

4.3 Intersecting the Chord with Discs

Let $\partial D(u, \rho)$ be the boundary of the disc $D(u, \rho)$. Consider the funnel of $u \in S_\tau$ and ℓ. Beginning at the apex of the funnel, for each vertex w on the convex chains ending at the endpoints of ℓ, there is a *domain* along ℓ in which w is the last vertex on the path from u to ℓ. We refer to the extremal points of w's domain as *domain markers*.

Since domain markers are points along ℓ, they provide distances against which to compare. We can use these distances when searching for points on ℓ that are a specified distance away from the point u (which is the point used to define the funnel and hence the markers). For any point $u \in S_\tau$, the distance to ℓ increases monotonically as we move from its projection u_c to the endpoints of ℓ [41]. Thus if we are given a radius ρ, we can use two funnels to perform two binary searches among these markers between u_c and the endpoints of ℓ to find the at most two domains which contain $\partial D(u, \rho) \cap \ell$. If we have a particular radius in mind, it takes constant time and space to test to which side of a domain marker there might exist a point whose distance to u is our radius (since we know u_c, and that the distance from u to ℓ between u_c and the endpoints of ℓ increases monotonically).

As a subroutine for Merge-Algo, for a given radius ρ, for each u in $\mathbb{S}$ we compute the interval $I(u, \rho) = D(u, \rho) \cap \ell$.

Lemma 6. *Given a radius ρ and the set $\mathbb{S}$, for each $u \in \mathbb{S}$ we compute $I(u, \rho)$ in $O(|\mathbb{S}| \log r)$ time and $O(|\mathbb{S}| + r')$ space.*

Proof. Consider the geodesic disc of radius ρ, $D(u, \rho)$. Since the disc is geodesically convex, if the chord intersects the disc in only one point, it will be at the projection u_c. If it does not intersect the disc, then at u_c the distance from u to ℓ will be larger than ρ. Otherwise, if the chord intersects the disc in two points, u_c splits ℓ up into two intervals, each with one intersection point (i.e., each one contains a point of $\partial D(u, \rho) \cap \ell$). If u_c is an endpoint of ℓ, assuming ℓ has positive length, one of these intervals may degenerate into a point, making u_c coincide with one of the intersection points.

These two intervals to either side of u_c have the property that on one side of the intersection point contained within, the distance from u to ℓ is larger than ρ, and on the other side, the distance is less than ρ. Therefore, if $\partial D(u, \rho)$ does intersect ℓ in two points we can proceed as in the proof of Lemma 5: in $O(\log r)$ time and $O(r')$ space we can build the two funnels of u between u_c and the endpoints of ℓ (truncated at the apices) and then perform a binary search in each to locate the domain in which a point at distance ρ lies. We find the subinterval delimited by the domain markers of the reflex vertices wherein the distance from u to ℓ changes from being more (less) than ρ to being less (more) than ρ. Once we find this domain, we can compute $\partial D(u, \rho) \cap \ell$ in $O(1)$ time and space. □

Once the intervals $\{I(u, \rho) \mid u \in \mathbb{S}\}$ are computed, we compute the overlay of these intervals as well as with $\mathbb{C}$ and compute the depth of the elements of $\mathbb{C}$ (i.e., how many intervals contain the element in question) in $O((|\mathbb{S}|+|\mathbb{C}|) \log(|\mathbb{S}|+|\mathbb{C}|))$ time and $O(|\mathbb{S}| + |\mathbb{C}|)$ space using a plane sweep.

5 Concluding Remarks

In this paper we have described two methods for computing a 2-SKEG disc using randomized algorithms. With DI-Algo, we find a 2-SKEG disc using $O(n + m)$ expected space in $O(n \log^2 n \log r + m)$ expected time; and using RS-Algo, in deterministic time that is faster by polylogarithmic factors when $k \in \omega(n/\log n)$, we find a 2-approximation with high probability using worst-case deterministic space $O(n+m)$. We leave as an open problem to solve the 2-SKEG disc problem in $O(n \log r + m)$ time.

Acknowledgements. The authors thank Pat Morin, Jean-Lou de Carufel, Michiel Smid, and Sasanka Roy for helpful discussions as well as anonymous reviewers.

References

1. Aggarwal, A., Imai, H., Katoh, N., Suri, S.: Finding k points with minimum diameter and related problems. J. Algorithms **12**(1), 38–56 (1991)
2. Ahn, H., Barba, L., Bose, P., De Carufel, J.L., Korman, M., Oh, E.: A linear-time algorithm for the geodesic center of a simple polygon. Discrete Comput. Geom. **56**(4), 836–859 (2016)
3. Aichholzer, O., Hackl, T., Korman, M., Pilz, A., Vogtenhuber, B.: Geodesic-preserving polygon simplification. Int. J. Comput. Geom. Appl. **24**(4), 307–324 (2014)
4. Alexandrescu, A.: Fast deterministic selection. In: SEA. LIPIcs, vol. 75, pp. 24:1–24:19. Schloss Dagstuhl - Leibniz-Zentrum für Informatik (2017)
5. Aronov, B.: On the geodesic Voronoi diagram of point sites in a simple polygon. Algorithmica **4**(1), 109–140 (1989)
6. Aronov, B., Fortune, S., Wilfong, G.T.: The furthest-site geodesic Voronoi diagram. Discrete Comput. Geom. **9**, 217–255 (1993)

7. Asano, T., Toussaint, G.: Computing the geodesic center of a simple polygon. In: Discrete Algorithms and Complexity, pp. 65–79. Elsevier (1987)
8. Bae, S.W., Korman, M., Okamoto, Y.: Computing the geodesic centers of a polygonal domain. Comput. Geom. **77**, 3–9 (2019)
9. Barba, L.: Optimal algorithm for geodesic farthest-point Voronoi diagrams. In: SoCG. LIPIcs, vol. 129, pp. 12:1–12:14. Schloss Dagstuhl - Leibniz-Zentrum für Informatik (2019)
10. Blum, M., Floyd, R.W., Pratt, V.R., Rivest, R.L., Tarjan, R.E.: Time bounds for selection. J. Comput. Syst. Sci. **7**(4), 448–461 (1973)
11. Borgelt, M.G., van Kreveld, M.J., Luo, J.: Geodesic disks and clustering in a simple polygon. Int. J. Comput. Geom. Appl. **21**(6), 595–608 (2011)
12. Bose, P., Toussaint, G.T.: Computing the constrained Euclidean geodesic and link center of a simple polygon with application. In: Computer Graphics International, pp. 102–110. IEEE Computer Society (1996)
13. Chan, T.M.: Geometric applications of a randomized optimization technique. Discret. Comput. Geom. **22**(4), 547–567 (1999)
14. Chazelle, B.: A theorem on polygon cutting with applications. In: FOCS, pp. 339–349. IEEE Computer Society (1982)
15. Chazelle, B.: Triangulating a simple polygon in linear time. Discrete Comput. Geom. **6**(3), 485–524 (1991). https://doi.org/10.1007/BF02574703
16. Chazelle, B., Guibas, L.J.: Visibility and intersection problems in plane geometry. Discrete Comput. Geom. **4**(6), 551–581 (1989). https://doi.org/10.1007/BF02187747
17. Chein, O., Steinberg, L.: Routing past unions of disjoint linear barriers. Networks **13**(3), 389–398 (1983)
18. Cormen, T.H., Leiserson, C.E., Rivest, R.L., Stein, C.: Introduction to Algorithms, 3rd edn. MIT Press, Cambridge (2009)
19. Datta, A., Lenhof, H., Schwarz, C., Smid, M.H.M.: Static and dynamic algorithms for k-point clustering problems. J. Algorithms **19**(3), 474–503 (1995)
20. de Berg, S., Staals, F.: Dynamic data structures for k-nearest neighbor queries. Comput. Geom. **111**, 101976 (2023)
21. Devroye, L.: Applications of the theory of records in the study of random trees. Acta Inform. **26**(1/2), 123–130 (1988)
22. Efrat, A., Sharir, M., Ziv, A.: Computing the smallest k-enclosing circle and related problems. Comput. Geom. **4**, 119–136 (1994)
23. Eppstein, D., Erickson, J.: Iterated nearest neighbors and finding minimal polytopes. Discrete Comput. Geom. **11**(3), 321–350 (1994). https://doi.org/10.1007/BF02574012
24. Guibas, L.J., Hershberger, J.: Optimal shortest path queries in a simple polygon. J. Comput. Syst. Sci. **39**(2), 126–152 (1989)
25. Guibas, L.J., Hershberger, J., Leven, D., Sharir, M., Tarjan, R.E.: Linear-time algorithms for visibility and shortest path problems inside triangulated simple polygons. Algorithmica **2**, 209–233 (1987)
26. Har-Peled, S.: Geometric Approximation Algorithms, vol. 173. American Mathematical Society (2011)
27. Har-Peled, S., Mazumdar, S.: Fast algorithms for computing the smallest k-enclosing circle. Algorithmica **41**(3), 147–157 (2005)
28. Heinonen, J.: Lectures on Analysis on Metric Spaces. Springer, New York (2001). https://doi.org/10.1007/978-1-4613-0131-8
29. Hershberger, J.: A new data structure for shortest path queries in a simple polygon. Inf. Process. Lett. **38**(5), 231–235 (1991)

30. Kirkpatrick, D.G.: Optimal search in planar subdivisions. SIAM J. Comput. **12**(1), 28–35 (1983)
31. Lee, D., Preparata, F.P.: Euclidean shortest paths in the presence of rectilinear barriers. Networks **14**(3), 393–410 (1984)
32. Liu, C., Lee, D.T.: Higher-order geodesic Voronoi diagrams in a polygonal domain with holes. In: SODA, pp. 1633–1645. SIAM (2013)
33. Lozano-Pérez, T., Wesley, M.A.: An algorithm for planning collision-free paths among polyhedral obstacles. Commun. ACM **22**(10), 560–570 (1979)
34. Matoušek, J.: On enclosing k points by a circle. Inf. Process. Lett. **53**(4), 217–221 (1995)
35. Matoušek, J.: On geometric optimization with few violated constraints. Discrete Comput. Geom. **14**(4), 365–384 (1995). https://doi.org/10.1007/BF02570713
36. Megiddo, N.: Linear-time algorithms for linear programming in $\mathbb{R}^3$ and related problems. SIAM J. Comput. **12**(4), 759–776 (1983)
37. Oh, E.: Optimal algorithm for geodesic nearest-point Voronoi diagrams in simple polygons. In: SODA, pp. 391–409. SIAM (2019)
38. Oh, E., Ahn, H.: Voronoi diagrams for a moderate-sized point-set in a simple polygon. Discrete Comput. Geom. **63**(2), 418–454 (2020)
39. Oh, E., Bae, S.W., Ahn, H.: Computing a geodesic two-center of points in a simple polygon. Comput. Geom. **82**, 45–59 (2019)
40. Oh, E., De Carufel, J.L., Ahn, H.: The geodesic 2-center problem in a simple polygon. Comput. Geom. **74**, 21–37 (2018)
41. Pollack, R., Sharir, M., Rote, G.: Computing the geodesic center of a simple polygon. Discrete Comput. Geom. **4**(6), 611–626 (1989). https://doi.org/10.1007/BF02187751
42. Rabanca, G., Vigan, I.: Covering the boundary of a simple polygon with geodesic unit disks. CoRR abs/1407.0614 (2014)
43. Toussaint, G.: Computing geodesic properties inside a simple polygon. Revue D'Intell. Artif. **3**(2), 9–42 (1989)
44. Vigan, I.: Packing and covering a polygon with geodesic disks. CoRR abs/1311.6033 (2013)
45. Wang, H.: On the geodesic centers of polygonal domains. JoCG **9**(1), 131–190 (2018)
46. Wang, H.: An optimal deterministic algorithm for geodesic farthest-point voronoi diagrams in simple polygons. In: SoCG. LIPIcs, vol. 189, pp. 59:1–59:15. Schloss Dagstuhl - Leibniz-Zentrum für Informatik (2021)
47. Welzl, E.: Smallest enclosing disks (balls and ellipsoids). In: Maurer, H. (ed.) New Results and New Trends in Computer Science. LNCS, vol. 555, pp. 359–370. Springer, Heidelberg (1991). https://doi.org/10.1007/BFb0038202

Online Interval Scheduling with Predictions

Joan Boyar[1], Lene M. Favrholdt[1], Shahin Kamali[2], and Kim S. Larsen[1](✉)

[1] University of Southern Denmark, Odense, Denmark
{joan,lenem,kslarsen}@imada.sdu.dk
[2] York University, Toronto, Canada
kamalis@yorku.ca
https://imada.sdu.dk/u/joan, https://imada.sdu.dk/u/lenem,
https://www.eecs.yorku.ca/~kamalis/, https://imada.sdu.dk/u/kslarsen

Abstract. In online interval scheduling, the input is an online sequence of intervals, and the goal is to accept a maximum number of non-overlapping intervals. In the more general disjoint path allocation problem, the input is a sequence of requests, each involving a pair of vertices of a known graph, and the goal is to accept a maximum number of requests forming edge-disjoint paths between accepted pairs. These problems have been studied under extreme settings without information about the input or with error-free advice. We study an intermediate setting with a potentially erroneous prediction that specifies the set of intervals/requests forming the input sequence. For both problems, we provide tight upper and lower bounds on the competitive ratios of online algorithms as a function of the prediction error. For disjoint path allocation, our results rule out the possibility of obtaining a better competitive ratio than that of a simple algorithm that fully trusts predictions, whereas, for interval scheduling, we develop a superior algorithm. We also present asymptotically tight trade-offs between consistency (competitive ratio with error-free predictions) and robustness (competitive ratio with adversarial predictions) of interval scheduling algorithms. Finally, we provide experimental results on real-world scheduling workloads that confirm our theoretical analysis.

Keywords: Online interval scheduling · Algorithms with prediction · Competitive analysis · Disjoint paths

The first, second, and fourth authors were supported in part by the Danish Council for Independent Research grant DFF-0135-00018B and in part by the Innovation Fund Denmark, grant 9142-00001B, Digital Research Centre Denmark, project P40: Online Algorithms with Predictions.
The third author was supported in part by the Natural Sciences and Engineering Research Council of Canada (NSERC).

P. Morin and S. Suri (Eds.): WADS 2023, LNCS 14079, pp. 193–207, 2023.
https://doi.org/10.1007/978-3-031-38906-1_14

1 Introduction

In the interval scheduling problem, the input is a set of intervals with integral endpoints, each representing timesteps at which a process starts and ends. A scheduler's task is to decide whether to accept or reject each job so that the intervals of accepted jobs do not overlap except possibly at one of their endpoints. The objective is to maximize the number of accepted intervals, referred to as the *profit* of the scheduler. This problem is also known as *fixed job scheduling* and *k-track assignment* [35].

Interval scheduling is a special case of the *disjoint path allocation problem*, where the input is a graph G and a set of n *requests*, each defined by a pair of vertices in G. An algorithm can accept or reject each pair, given that it can form edge-disjoint paths between vertices of accepted pairs. Interval scheduling is the particular case when G is a path graph. The disjoint path allocation problem can be solved in polynomial time for trees [30] and outerplanar graphs by a combination of [26,29,34], but the problem is NP-complete for general graphs [28], and even on quite restricted graphs such as series-parallel graphs [42]. The disjoint path problem is the same as call control/call allocation with all bandwidths (both of the calls and the edges they would be routed on) being equal to 1 and as the maximum multi-commodity integral flow problem with edges having unit capacity.

In this work, we focus on the online variant of the problem, in which the set of requests is not known in advance but is revealed in the form of a request sequence, I. A new request must either be irrevocably accepted or rejected, subject to maintaining disjoint paths between accepted requests. We analyze an online algorithm via a comparison with an optimal offline algorithm, OPT. The *competitive ratio* of an online algorithm ALG is defined as $\inf_I \{\text{ALG}(I)/\text{OPT}(I)\}$, where ALG$(I)$ and OPT(I), respectively, denote the profit of ALG and OPT on I (for randomized algorithms, ALG(I) is the expected profit of ALG). Since we consider a maximization problem, our ratios are between zero and one.

For interval scheduling on a path graph with m edges, the competitive ratios of the best deterministic and randomized algorithms are respectively $\frac{1}{m}$ and $\frac{1}{\lceil \log m \rceil}$ [20]. These results suggest that the constraints on online algorithms must be relaxed to compete with OPT. Specifically, the problem has been considered in the *advice complexity model* for path graphs [16,31], trees [18], and grid graphs [19]. Under the advice model, the online algorithm can access error-free information on the input called advice. The objective is to quantify the trade-offs between the competitive ratio and the size of the advice.

In recent years, there has been an increasing interest in improving the performance of online algorithms via the notion of *prediction*. Here, it is assumed that the algorithm has access to machine-learned information in the form of a prediction. Unlike the advice model, the prediction may be erroneous and is quantified by an *error measure* η. The objective is to design algorithms whose competitive ratio degrades gently as a function of η. Several online optimization problems have been studied under the prediction model, including non-clairvoyant scheduling [43,45], makespan scheduling [36], contract scheduling [6,7], and other variants of scheduling problems [11,13,14,39].

Other online problems studied under the prediction model include bin packing [3,4], knapsack [21,33,47], caching [8,40,44,46], matching problems [9,37,38], time series search [5], and various graph problems [12,15,24,25,27]. See also the survey by Mitzenmacher and Vassilvitskii [41] and the collection at [1].

1.1 Contributions

We study the disjoint path allocation problem under a setting where the scheduler is provided with a set $\hat{I}$ of requests predicted to form the input sequence I. Given the erroneous nature of the prediction, some requests in $\hat{I}$ may be incorrectly predicted to be in I (false positives), and some requests in I may not be included in $\hat{I}$ (false negatives). We let the *error set* be the set of requests that are false positives or false negatives and define the error parameter $\eta(\hat{I}, I)$ to be the cardinality of the largest set of requests in the error set that can be accepted. For interval scheduling, this is the largest set of non-overlapping intervals in the error set. Thus, $\eta(\hat{I}, I) = \text{OPT}(\text{FP} \cup \text{FN})$. We explain later that this definition of η satisfies specific desired properties for the prediction error (Proposition 1). In the following, we use $\text{ALG}(\hat{I}, I)$ to denote the profit of an algorithm ALG for prediction $\hat{I}$ and input I. We also define $\gamma(\hat{I}, I) = \eta(\hat{I}, I)/\text{OPT}(I)$; this *normalized error* measure is helpful in describing our results because the point of reference in the competitive analysis is $\text{OPT}(I)$. Our first result concerns general graphs:

- **Disjoint-Path Allocation:** We study a simple algorithm TRUST, which accepts a request only if it belongs to the set of predictions in a given optimal solution for $\hat{I}$. We show that, for any graph G, any input sequence I, and any prediction $\hat{I}$, $\text{TRUST}(\hat{I}, I) \geq (1 - 2\gamma(\hat{I}, I))\text{OPT}(I)$ (Theorem 1). Furthermore, for any algorithm ALG and any positive integer p, there are worst-case input sequence I_w and prediction set $\hat{I}_w$ over a star graph, S_{8p}, with $8p$ leaves, such that $\eta(\hat{I}_w, I_w) = p$ and $\text{ALG}(\hat{I}_w, I_w) \leq (1 - 2\gamma(\hat{I}_w, I_w))\text{OPT}(I_w)$ (Theorem 2). Thus, TRUST achieves an optimal competitive ratio in any graph class that contains S_8.

The above result demonstrates that even for trees, the problem is so hard that no algorithm can do better than the trivial TRUST. Therefore, our main results concern the more interesting case of path graphs, that is, interval scheduling:

- **Interval scheduling:** We first show a negative result for deterministic interval scheduling algorithms. Given any deterministic algorithm ALG and integer p, we show there are worst-case instances I_w and predictions $\hat{I}_w$ such that $\eta(\hat{I}_w, I_w) = p$ and $\text{ALG}(\hat{I}_w, I_w) \leq (1 - \gamma(\hat{I}_w, I_w))\text{OPT}(I_w)$ (Theorem 3, setting $c = 2$).

  Next, we present a negative result for TRUST. For any positive integer, p, we show there are worst-case instances I_w and predictions $\hat{I}_w$ such that $\eta(\hat{I}_w, I_w) = p$ and $\text{TRUST}(\hat{I}_w, I_w) = (1 - 2\gamma(\hat{I}_w, I_w))\text{OPT}(I_w)$. (Theorem 4).

This suggests that there is room for improvement over TRUST.
Finally, we introduce our main technical result, a deterministic algorithm TRUSTGREEDY that achieves an optimal competitive ratio for interval scheduling. TRUSTGREEDY is similar to TRUST in that it maintains an optimal solution for $\hat{I}$, but unlike TRUST, it updates its planned solution to accept requests greedily when it is possible without a decrease in the profit of the maintained solution. For any input I and prediction $\hat{I}$, we show that $\text{TRUSTGREEDY}(\hat{I}, I) \geq (1 - \gamma(\hat{I}, I))\text{OPT}(I)$ (Theorem 5), which proves optimality of TRUSTGREEDY in the light of Theorem 3.

- **Consistency-Robustness Trade-off:** We study the trade-off between *consistency* and *robustness*, which measure an algorithm's competitive ratios in the extreme cases of error-free prediction (consistency) and adversarial prediction (robustness) [40]. We focus on randomized algorithms because a non-trivial trade-off is infeasible for deterministic algorithms (Proposition 2). Suppose that for any input I, an algorithm ALG guarantees a consistency of $\alpha < 1$ and robustness of $\beta \leq \frac{1}{\lceil \log m \rceil}$. We show $\alpha \leq 1 - \frac{\lfloor \log m \rfloor - 1}{2}\beta$ and $\beta \leq \frac{2}{\lfloor \log m \rfloor - 1} \cdot (1 - \alpha)$ (Theorem 6). For example, to guarantee a robustness of $\frac{1}{10\lceil \log m \rceil}$, the consistency must be at most 19/20, and to guarantee a consistency of $\frac{2}{3}$, the robustness must be at most $\frac{2}{3}\frac{1}{\lfloor \log m \rfloor - 1}$. We also present a family of randomized algorithms that provides an almost *Pareto-optimal* trade-off between consistency and robustness (Theorem 7).
- **Experiments on Real-World Data:** We compare our algorithms with the online GREEDY algorithm (which accepts an interval if and only if it does not overlap previously accepted intervals), and OPT on real-world scheduling data from [23]. Our results are in line with our theoretical analysis: both TRUST and TRUSTGREEDY are close-to-optimal for small error values; TRUSTGREEDY is almost always better than GREEDY even for large values of error, while TRUST is better than GREEDY only for small error values.

Omitted details and all omitted proofs can be found in the full paper [17].

2 Model and Predictions

We assume that an oracle provides the online algorithm with a set $\hat{I}$ of requests predicted to form the input sequence I. Note that it is not interesting to consider alternative types of predictions that are very compact. For interval scheduling on a path with m edges, since the problem is AOC-complete, one cannot achieve a competitive ratio $c \leq 1$ with fewer than $cm/(e \ln 2)$ bits, even if all predictions are correct [22].

In what follows, true positive (respectively, negative) intervals are correctly predicted to appear (respectively, not to appear) in the request sequence. False positives and negatives are defined analogously as those incorrectly predicted to appear or not appear. We let TP, TN, FP, FN denote the four sets containing these different types of intervals. Thus, $I = \text{TP} \cup \text{FN}$ and $\hat{I} = \text{TP} \cup \text{FP}$. We use

$\eta(\hat{I}, I)$, to denote the error for the input formed by the set I, when the set of predictions is $\hat{I}$. When there is no risk of confusion, we use η instead of $\eta(\hat{I}, I)$.

The error measure we use here is $\eta = \text{OPT}(\text{FP} \cup \text{FN})$, and hence, the normalized error measure is $\gamma = \text{OPT}(\text{FP} \cup \text{FN})/\text{OPT}(I)$. Our error measure satisfies the following desirable properties, the first two of which were strongly recommended in Im, et al. [32]: First, the *monotonicity* property requires that increasing the number of true positives or true negatives does not increase the error. Second, the *Lipschitz* property ensures that $\eta(\hat{I}, I) \geq |\text{OPT}(I) - \text{OPT}(\hat{I})|$. Finally, *Lipschitz completeness* requires $\eta(I, \hat{I}) \leq \text{OPT}(\text{FP} \cup \text{FN})$. The Lipschitz and Lipschitz completeness properties enforce a range for the error to avoid situations where the error is too small or too large. We refer to the full paper [17] for details on these properties. There, we also discuss natural error models, such as Hamming distance between the request sequence and prediction, and explain why these measures do not satisfy our desired properties.

Proposition 1. *The error measure $\eta(\hat{I}, I) = \text{OPT}(\text{FP} \cup \text{FN})$ satisfies the properties of monotonicity, Lipschitz, and Lipschitz completeness.*

3 Disjoint-Path Allocation

In this section, we show that a simple algorithm TRUST for the disjoint path allocation problem has an optimal competitive ratio for any graph of maximal degree at least 8. TRUST simply relies on the predictions being correct. Specifically, it computes an optimal solution I^* in $\hat{I}$ before processing the first request. Then, it accepts any request in I^* that arrives and rejects all others.

We first establish that, on any graph, $\text{TRUST}(\hat{I}, I) \geq \text{OPT}(I) - 2\eta(\hat{I}, I) = (1 - 2\gamma(\hat{I}, I))\text{OPT}(I)$. The proof follows by observing that (i) false negatives cause a deficit of at most $\text{OPT}(\text{FN})$ in the schedule of TRUST compared to the optimal schedule for I^*, (ii) false positives cause a deficit of at most $\text{OPT}(\text{FP})$ in the optimal schedule of I^*, compared to the optimal schedule for I, and (iii) $\text{OPT}(\text{FP}) + \text{OPT}(\text{FN}) \leq 2\text{OPT}(\text{FP} \cup \text{FN}) = 2\eta$.

Theorem 1. *For any graph G, any prediction $\hat{I}$, and input sequence I, we have* $\text{TRUST}(\hat{I}, I) \geq (1 - 2\gamma(\hat{I}, I))\text{OPT}(I)$.

The following result shows that Theorem 1 is tight for star graphs of degree 8. One can conclude that TRUST is optimal for any graph class that contains stars of degree 8.

Theorem 2. *Let* ALG *be any deterministic algorithm and p be any positive integer. On the star graph S_{8p}, there exists a set of predicted requests $\hat{I}_w$ and a request sequence I_w such that $\eta(\hat{I}_w, I_w) = p$ and* $\text{ALG}(\hat{I}_w, I_w) \leq (1 - 2\gamma(\hat{I}, I))\text{OPT}(I_w)$.

Proof (sketch). We consider the non-center vertices of S_{8p} in p groups of eight, and handle them all identically, one group at a time, treating each group independently. The prediction is fixed, but the input sequence depends on the algorithm's actions. For each group, we show that the error in the prediction is

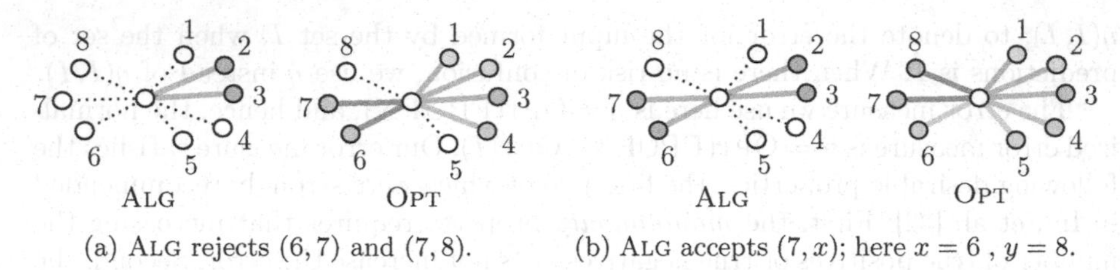

(a) ALG rejects (6, 7) and (7, 8). (b) ALG accepts $(7, x)$; here $x = 6$, $y = 8$.

Fig. 1. Illustration of the proof of Theorem 2 for the case that ALG accepts (2, 3). Highlighted edges indicate paths between accepted pairs.

1, and the profit of OPT is at least 2 units more than that of ALG. Given that groups do not share edges between themselves, the total error and algorithm's profits are summed over all groups. Hence, the total error will be equal to $\eta(\hat{I}_w, I_w) = p$, and we can write $\text{ALG}(I_w) \leq \text{OPT}(I_w) - 2\eta(\hat{I}_w, I_w)$, that is, $\text{ALG}(I_w) \leq (1 - 2\gamma(\hat{I}_w, I_w))\text{OPT}(I_w)$.

Next, we explain how an adversary defines the input for each group. For group $0 \leq i \leq s-1$, the non-center vertices are $8i + j$, where $1 \leq j \leq 8$, but we refer to these vertices by the value j. Let $\hat{I}_w = \{(1,2), (2,3), (3,4), (4,5), (6,7), (7,8)\}$ be the part of the prediction relevant for the current group of eight vertices. Both (6, 7) and (7, 8) are always included in the input sequence, with (6, 7) arriving immediately before (7, 8). ALG accepts at most one of them. This is discussed in the cases below. The first request in the input is always (2, 3). Here we discuss the case when ALG accepts (2, 3); the other case when it rejects (2, 3) follows with a similar case analysis (see the full paper [17]).

Case ALG accepts (2, 3): The next request to arrive is (6, 7). If ALG rejects this request, the next to arrive is (7, 8). If ALG also rejects this request, then the requests (1, 2) and (3, 4) also arrive, but (4, 5) is a false positive (see Fig. 1a). Then, OPT accepts $\{(1,2), (3,4), (6,7)\}$, ALG only accepts $\{(2,3)\}$, and $\text{OPT}(\text{FN} \cup \text{FP}) = 1$. Thus, we may assume that ALG accepts at least one of (6, 7) and (7, 8), which we call $(7, x)$ where $x \in \{6, 8\}$. We call the other of these two edges $(7, y)$. Then, the requests (1, 2) and (3, 4) also arrive, along with a false negative $(5, x)$. The request (4, 5) is a false positive and is not in the input (see Fig. 1b). Since (4, 5) and $(5, x)$ share an edge, $\text{OPT}(\text{FN} \cup \text{FP}) = 1$. ALG accepts $\{(2,3), (7,x)\}$, and OPT accepts $\{(1,2), (3,4), (5,x), (7,y)\}$. To conclude, the error increases by 1, and ALG's deficit to OPT increases by 2. □

4 Interval Scheduling

In this section, we show tight upper and lower bounds for the competitive ratio of a deterministic algorithm for interval scheduling. As an introduction to the difficulties in designing algorithms for the problem, we start by proving a general lower bound. We show that for any deterministic algorithm ALG, there exists an input sequence I_w and a set of predictions $\hat{I}_w$ such that $\text{ALG}(\hat{I}_w, I_w) = \text{OPT}(I_w) - \eta(\hat{I}_w, I_w)$, and that this can be established for any integer error of at least 2.

Theorem 3. *Let* ALG *be any deterministic algorithm. For any positive integers p and $c \in [2, m]$, there are instances I_w and predictions $\hat{I}_w$ such that $p \leq \eta(\hat{I}_w, I_w) \leq (c-1)p$ and* $\text{ALG}(\hat{I}_w, I_w) = (1 - \gamma(\hat{I}_w, I_w))\text{OPT}(I_w) \leq \frac{1}{c}\text{OPT}(I_w)$.

Proof. ALG will be presented with p intervals of length c, and the remainder of the sequence will depend on which of these it accepts. The prediction, however, will include the following $2p$ requests: $\hat{I} = \bigcup_{i=0}^{p-1} \{(ci, c(i+1)), (ci, ci+1)\}$.

The input I_w is formed by p phases, $i \in [0, p-1]$. The ith phase starts with the true positive $(ci, c(i+1))$. There are two cases to consider:

- If ALG accepts $(ci, c(i+1))$, then the phase continues with $\{(ci + j, ci + (j+1)) \mid 0 \leq j \leq c-1\}$. The first of these requests is a true positive, and the other $c-1$ are false negatives. Note that ALG cannot accept any of these c requests. The optimal algorithm rejects the original request $(ci, c(i+1))$ and accepts all of the c following unit-length requests.
- If ALG rejects $(ci, c(i+1))$, the phase ends with no further requests. In this case, $(ci, ci+1)$ is a false positive.

The contribution, η_i, of phase i to $|\text{FP} \cup \text{FN}|$ is $\eta_i = c-1$ in the first case and $\eta_i = 1$ in the second. Since the intervals in FP $\cup$ FN are disjoint, we can write $\text{OPT}(\text{FP} \cup \text{FN}) = \sum_{i=0}^{p-1} \eta_i$ and it follows that $p \leq \text{OPT}(\text{FP} \cup \text{FN}) \leq (c-1)p$. Moreover, the net advantage of OPT over ALG in phase i is at least η_i: in the first case, OPT accepts $\eta_i + 1$ and ALG accepts one request, and in the second case, OPT accepts $\eta_i = 1$ and ALG accepts no requests. Given that there are p phases, we can write $\text{ALG}(\hat{I}_w, I_w) \leq \text{OPT}(I_w) - \sum_{i=0}^{p-1} \eta_i = \text{OPT}(I_w) - \text{OPT}(\text{FP} \cup \text{FN}) = (1 - \gamma(\hat{I}_w, I_w))\text{OPT}(I_w)$.

In phases where ALG accepts the first request, OPT accepts c times as many requests as ALG. In phases where ALG rejects the first request, OPT accepts one interval, and ALG accepts no intervals. Thus, $\text{OPT}(I_w) \geq c \cdot \text{ALG}(\hat{I}_w, I_w)$. □

For $c = 2$, we get $\eta(\hat{I}_w, I) = p$ and $\text{ALG}(\hat{I}_w, I_w) = (1 - \gamma(\hat{I}_w, I_w))\text{OPT}(\hat{I}_w)$. The next theorem shows that the competitive ratio of TRUST compared to the lower bound of Theorem 3 is not tight. The proof follows from an adversarial sequence similar to that of Theorem 3 in which the profit of OPT and η grow in phases while the profit of TRUST stays 0.

Theorem 4. *For any integer $p \geq 1$, there exists a prediction $\hat{I}_w$ and an input sequence I_w so that $\eta(\hat{I}_w, I_w) = p$ and* $\text{TRUST}(\hat{I}_w, I_w) = (1 - 2\gamma(\hat{I}_w, I_w))\text{OPT}(I_w)$.

4.1 TrustGreedy

In this section, we introduce an algorithm TRUSTGREEDY, TG, which achieves an optimal competitive ratio for interval scheduling.

The Algorithm. TG starts by choosing an optimal solution, I^*, from the predictions in $\hat{I}$. This optimal offline solution is selected by repeatedly including an interval that ends earliest possible among those in $\hat{I}$ that do not overlap any already selected intervals. TG plans to accept those intervals in I^* and reject all others, and it just follows its plan, except possibly when the next request is in FN. During the online processing after this initialization, TG maintains an updated plan, A. Initially, A is I^*. When a request, r, is in FN, TG accepts if r overlaps no previously accepted intervals and can be accepted by replacing at most one other interval in A that ends no earlier than r. In that case, r is added to A, possibly replacing an overlapping interval to maintain the feasibility of A (no two intervals overlap). As a comment, only the first interval from FN that replaces an interval r in the current A is said to "replace" it. There may be other intervals from FN that overlap r and are accepted by TG, but they are not said to "replace" it. We let U denote the set of intervals in $I^* \cap \text{FP}$ that are not replaced during the execution of TG.

Analysis. Let TG denote the set of intervals chosen by TrustGreedy on input I and prediction $\hat{I}$, and Opt the intervals chosen by the optimal algorithm. We define the following subsets of TG and Opt:

- $\text{TG}^{\text{FN}} = \text{TG} \cap \text{FN}$ and $\text{OPT}^{\text{FN}} = \text{OPT} \cap \text{FN}$
- $\text{TG}^{\text{TP}} = \text{TG} \cap \hat{I} = \text{TG} \cap \text{TP}$ and $\text{OPT}^{\text{TP}} = \text{OPT} \cap \hat{I} = \text{OPT} \cap \text{TP}$

Lemma 1. *Each interval $i \in \text{OPT}^{\text{TP}}$ overlaps an interval in I^* extending no further to the right than i.*

Proof. Assume to the contrary that there is no interval in I^* that overlaps i and ends no later than i. If i does not overlap anything in I^*, we could have added i to I^* and have a feasible solution (non-overlapping intervals), contradicting the fact that I^* is optimal. Thus, i must overlap an interval r in I^*, which, by assumption, must end strictly later than i. This contradicts the construction of I^*, since i would have been in I^* instead of r. □

We define a set O^{FN} consisting of a copy of each interval in OPT^{FN} and let $\mathcal{F} = O^{\text{FN}} \cup U$. We define a mapping $f\colon \text{OPT} \rightarrow \text{TG} \cup \mathcal{F}$ as follows. For each $i \in \text{OPT}$:

1. If there is an interval in I^* that overlaps i and ends no later than i, then let r be the rightmost such interval.
   (a) If $r \in U \cup \text{TG}^{\text{TP}}$, then $f(i) = r$.
   (b) Otherwise, r has been replaced by some interval t. In this case, $f(i) = t$.
2. Otherwise, by Lemma 1, i belongs to OPT^{FN}.
   (a) If there is an interval in TG^{FN} that overlaps i and ends no later than i and an interval in U that overlaps i's right endpoint, let r be the rightmost interval in TG^{FN} that overlaps i and ends no later than i. In this case, $f(i) = r$.
   (b) Otherwise, let o_i be the copy of i in O^{FN}. In this case, $f(i) = o_i$.

We let F denote the subset of $\mathcal{F}$ mapped to by f and note that in step 1a, intervals are added to $F \cap U$ when $r \in U$. In step 2b, all intervals are added to $F \cap O^{\text{FN}}$.

Lemma 2. *The mapping f is an injection.*

Proof. Intervals in $U \cup \text{TG}^{\text{TP}}$ are only mapped to in step 1a. Note that U and TG are disjoint. If an interval $i \in \text{OPT}$ is mapped to an interval $r \in U \cup \text{TG}$ in this step, i overlaps the right endpoint of r. There can be only one interval in OPT overlapping the right endpoint of r, so this part of the mapping is injective. Intervals in TG^{FN} are only mapped to in steps 1b and 2a. In step 1b, only intervals that replace intervals in I^* are mapped to. Since each interval in TG^{FN} replaces at most one interval in I^* and the right endpoint of each interval in I^* overlaps at most one interval in OPT, no interval is mapped to twice in step 1b. If, in step 2a, an interval, i, is mapped to an interval, r, i overlaps the right endpoint of r. There can be only one interval in OPT overlapping the right endpoint of r, so no interval is mapped to twice in step 2a.

We now argue that no interval is mapped to in both steps 1b and 2a. Assume that an interval, i_1, is mapped to an interval, t, in step 1b. Then, there is an interval, r, such that r overlaps the right endpoint of t and i_1 overlaps the right endpoint of r. This means that the right endpoint of i_1 is no further to the left than the right endpoint of t. Assume for the sake of contradiction that an interval $i_2 \neq i_1$ is mapped to t in step 2a. Then, i_2 overlaps the right endpoint of t, and there is an interval, $u \in U$, overlapping the right endpoint of i_2. Since i_2 overlaps t, i_2 must be to the left of i_1. Since i_2 is mapped to t, t extends no further to the right than i_2. Thus, since r overlaps both t and i_1, r must overlap the right endpoint of i_2, and hence, r overlaps u. This is a contradiction since r and u are both in I^*.

Intervals in $F \cap O^{\text{FN}}$ are only mapped to in step 2b and no two intervals are mapped to the same interval in this step. □

Lemma 3. *The subset F of $\mathcal{F}$ mapped to by f is a feasible solution.*

Proof. We first note that $F \cap U$ is feasible since $F \cap U \subseteq U \subseteq I^*$ and I^* is feasible. Moreover, $F \cap O^{\text{FN}}$ is feasible since the intervals of $F \cap O^{\text{FN}}$ are identical to the corresponding subsets of OPT. Thus, we need to show that no interval in $F \cap U$ overlaps any interval in $F \cap O^{\text{FN}}$.

Consider an interval $u \in F \cap U$ mapped to from an interval $i \in \text{OPT}$. Since i is not mapped to its own copy in $\mathcal{F}$, its copy does not belong to F. Since $i \in \text{OPT}$, no interval in $F \cap O^{\text{FN}}$ overlaps i. Thus, we need to argue that $F \cap O^{\text{FN}}$ contains no interval strictly to the left of i overlapping u.

Assume for the sake of contradiction that there is an interval $\ell \in F \cap O^{\text{FN}}$ to the left of i overlapping u. Since ℓ ended up in F although its right endpoint is overlapped by an interval from U, there is no interval in I^* (because of step 1 in the mapping algorithm) or in TG^{FN} (because of step 2a in the mapping algorithm) overlapping ℓ and ending no later than ℓ. Thus, $I^* \cup \text{TG}^{\text{FN}}$ contains no interval strictly to the left of u overlapping ℓ. This contradicts the fact that u has not been replaced since the interval in OPT^{FN} corresponding to ℓ could have replaced it. □

The following theorem follows from Lemmas 2 and 3.

Theorem 5. *For any prediction $\hat{I}$ and any input sequence I, we have* $\textsc{TrustGreedy}(\hat{I}, I) \geq (1 - \gamma(\hat{I}, I))\textsc{Opt}(I)$.

5 Consistency-Robustness Trade-Off

We study the trade-off between the competitive ratio of the interval scheduling algorithm when predictions are error-free (consistency) and when predictions are adversarial (robustness). The following proposition shows an obvious trade-off between the consistency and robustness of deterministic algorithms.

Proposition 2. *If a deterministic algorithm has non-zero consistency, α, it has robustness $\beta \leq \frac{1}{m}$.*

The more interesting case is randomized algorithms. The proof of the following was inspired by the proof of Theorem 13.8 in [20] for the online case without predictions, and that $\Omega(\log m)$ result was originally proven in [10].

Theorem 6. *If a (possibly randomized) algorithm* $\textsc{Alg}$ *is both α-consistent and β-robust, then $\alpha \leq 1 - \frac{\lfloor \log m \rfloor - 1}{2}\beta$ and $\beta \leq \frac{2}{\lfloor \log m \rfloor - 1} \cdot (1 - \alpha)$.*

Proof. Let $r = \lfloor \log m \rfloor - 1$ and let $m' = 2^{r+1}$. Consider a prediction $\sigma = \langle \hat{I}_0, \hat{I}_1, \ldots, \hat{I}_r, \hat{I}' \rangle$, where $\hat{I}' = \langle (0,1), (1,2), \ldots, (m'-1, m') \rangle$ and, for $0 \leq i \leq r$, $\hat{I}_i = \langle (0, m'/2^i), (m'/2^i, 2m'/2^i), \ldots, (m' - m'/2^i, m') \rangle$. Note that $\hat{I}_i$ consists of 2^i disjoint intervals of length $m'/2^i$. For $0 \leq i \leq r$, let $\sigma_i = \langle \hat{I}_0, \hat{I}_1, \ldots, \hat{I}_i \rangle$.

In order to maximize the number of small intervals that can be accepted if they arrive, an algorithm would minimize the (expected) fraction of the line occupied by the larger intervals, to leave space for the small intervals, while maintaining β-robustness. Since $\textsc{Opt}(\sigma_0) = 1$ and $\textsc{Alg}$ is β-robust, $E[\textsc{Alg}(\sigma_0)] \geq \beta$. For σ_i with $i \geq 1$, $\textsc{Opt}$ accepts all intervals in $\hat{I}_i$, so $\textsc{Opt}(\sigma_i) = 2^i$. To be β-robust, the expected number of intervals of length at most $m'/2^i$ that $\textsc{Alg}$ accepts is at least $2^i\beta$. Inductively, for $i \geq 1$, by the linearity of expectations, this is at least $2^{i-1}\beta$ intervals of length $m'/2^i$, and these intervals have a total expected size of at least $2^{i-1}\beta \times m'/2^i = \frac{m'}{2}\beta$. Again, by the linearity of expectations, for σ_r, the expected sum of the lengths of the accepted intervals is at least $\sum_{i=0}^{r} \frac{m'}{2}\beta = \frac{m'(r+1)}{2}\beta$.

From σ_r, the expected number of intervals $\textsc{Alg}$ must have accepted is at least $2^r\beta$. If σ is the actual input sequence, then the predictions are correct, so for $\textsc{Alg}$ to be α-consistent, we must have $E[\textsc{Alg}(\sigma')] \geq m'\alpha$. Since also $2^r\beta + (m' - \frac{m'(r+1)}{2}\beta) \geq E[\textsc{Alg}(\sigma')]$, we can combine these two inequalities and obtain $\frac{2^r}{m'}\beta + 1 - \frac{r+1}{2}\beta \geq \alpha$. Since $\frac{2^r}{m'} = \frac{1}{2}$, this reduces to $\alpha \leq 1 - \frac{r}{2}\beta$. Solving for β, $\beta \leq \frac{2}{r}(1 - \alpha)$. □

Note that as α approaches 1 (optimal consistency), β goes to 0 (worst-case robustness) and vice-versa. Next, we present a family of algorithms, ROBUSTTRUST, which has a parameter $0 \leq \alpha \leq 1$ and works as follows. With a probability of α, ROBUSTTRUST applies TG. (Applying TRUST, instead of TG, gives the same consistency and robustness results.) With probability $1 - \alpha$, ROBUSTTRUST ignores the predictions, and applies the Classify-and-Randomly-Select (CRS) algorithm described in Theorem 13.7 in [20]. For completeness, we include the CRS algorithm in the full paper [17]. CRS is strictly $\lceil \log m \rceil$-competitive (they use ratios at least one). A similar algorithm was originally proven $O(\log m)$-competitive in [10].

When ROBUSTTRUST applies TG and the predictions are correct, it accepts exactly as many intervals as there are in the optimal solution. From these observations, we can get the following results.

Theorem 7. ROBUSTTRUST (RT) *with parameter α has consistency at least α and robustness at least $\frac{1-\alpha}{\lceil \log m \rceil}$.*

6 Experimental Results

We present an experimental evaluation of TRUST and TRUSTGREEDY in comparison with the GREEDY algorithm, which serves as a baseline online algorithm, and OPT, which serves as the performance upper bound. We evaluate our algorithms using real-world scheduling data for parallel machines [23]. Each benchmark from [23] specifies the start and finish times of tasks as scheduled on parallel machines with several processors (see also the full paper [17]). We use these tasks to generate inputs to the interval scheduling problem. For each benchmark with N tasks, we create an instance I of an interval scheduling problem by randomly selecting $n = \lfloor N/2 \rfloor$ tasks from the benchmark and randomly permuting them. This sequence serves as the input to all algorithms. To generate the prediction, we consider 1000 equally distanced values of $d \in [0, n]$. For each value of d, we initiate the prediction set $\hat{I}$ with the set of intervals in I, remove $|\text{FN}| = d$ randomly selected intervals from $\hat{I}$ and add to it $|\text{FP}| = d$ randomly selected intervals from the remaining $N - n$ tasks in the benchmark. The resulting set $\hat{I}$ is given to TRUST and TRUSTGREEDY as prediction $\hat{I}$. For each value of d, we compute the normalized error $\gamma(\hat{I}, I) = \frac{\text{OPT}(\text{FN} \cup \text{FP})}{\text{OPT}(I)}$, and report the profit of TRUST and TRUSTGREEDY as a function of γ.

Figure 2 shows the results for two representative benchmarks from [23], namely, LLNL (the workload of the BlueGene/L system installed at Lawrence Livermore National Lab) and SDSC (the workload log from San Diego Supercomputer Center). The results are aligned with our theoretical findings: TRUST quickly becomes worse than GREEDY as the error value increases, while TRUSTGREEDY degrades gently as a function of the prediction error. In particular, TRUSTGREEDY is better than GREEDY for almost all error values. We note that GREEDY performs better when there is less overlap between the input intervals, which is the case in LLNL compared to SDSC. In an extreme case,

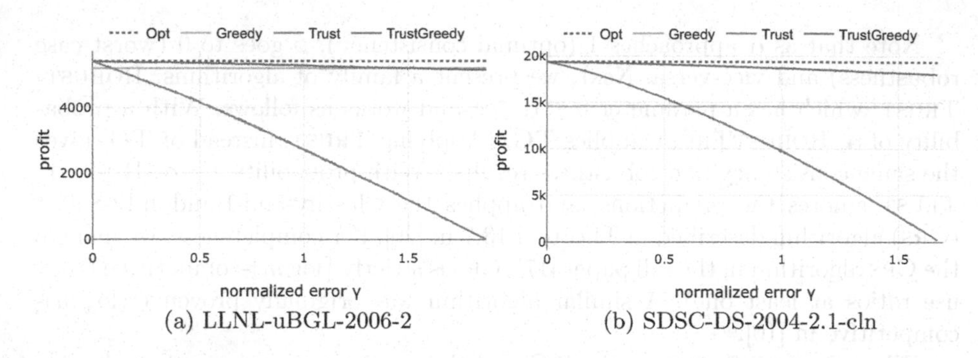

(a) LLNL-uBGL-2006-2 (b) SDSC-DS-2004-2.1-cln

Fig. 2. Profit as a function of normalized error value

when no two intervals overlap, GREEDY is trivially optimal. Nevertheless, even for LLNL, TRUSTGREEDY is not much worse than GREEDY for extreme values of error: the profit for the largest normalized error of $\gamma = 1.87$ was 5149 and 5198 for TRUSTGREEDY and GREEDY, respectively. Note that for SDSC, where there are more overlaps between intervals, TRUSTGREEDY is strictly better than GREEDY, even for the largest error values.

We present results for more benchmarks and situations where false negatives and false positives contribute differently to the error set in the full paper [17]. Our code and results are available at [2].

7 Concluding Remarks

In [30], the authors observe that finding disjoint paths on stars is equivalent to finding maximal matchings on general graphs, where each request in the input to the disjoint path selection bijects to an edge in the input graph for the matching problem. Therefore, we can extend the results of Sect. 3 to the following *online matching problem.* The input is a graph $G = (V, E)$, where V is known, and edges in E appear in an online manner; upon arrival of an edge, it must be added to the matching or rejected. The prediction is a set $\hat{E}$ that specifies edges in E. As before, we use FP and FN to indicate the set of false positives and false negatives and define $\gamma(\hat{E}, E) = \frac{\text{OPT}(\text{FP}\cup\text{FN})}{\text{OPT}(E)}$, where $\text{OPT}(S)$ indicates the size of the matching for graph $G = (V, S)$. The following is immediate from Theorems 1 and 2.

Proposition 3. *For any instance $G = (V, E)$ of the online matching problem under the edge-arrival model and a prediction set $\hat{E}$, there is an algorithm* TRUST *that matches at least $(1-2\gamma(\hat{E}, E))\text{OPT}(G)$ edges. Moreover, there are instances $G_w = (V, E_w)$ of the matching problem, along with predictions $\hat{E}_w$ for which any deterministic algorithm matches at most $(1 - 2\gamma(\hat{E}, E)_w)\text{OPT}(G_w)$ edges.*

Using the correspondence between matchings in a graph, G, and the independent set in the line graph of G, we can get a similar result for the independent set under the vertex-arrival model. We refer to the full paper [17] for details.

References

1. Algorithms with predictions. https://algorithms-with-predictions.github.io/. Accessed 19 Feb 2023
2. Interval scheduling with prediction. https://github.com/shahink84/IntervalSchedulingWithPrediction. Accessed 19 Feb 2023
3. Angelopoulos, S., Dürr, C., Jin, S., Kamali, S., Renault, M.: Online computation with untrusted advice. In: Proceedings of the ITCS. LIPIcs, vol. 151, pp. 52:1–52:15 (2020)
4. Angelopoulos, S., Kamali, S., Shadkami, K.: Online bin packing with predictions. In: Proceedings of the IJCAI, pp. 4574–4580 (2022)
5. Angelopoulos, S., Kamali, S., Zhang, D.: Online search with best-price and query-based predictions. In: Proceedings of the AAAI, pp. 9652–9660 (2023)
6. Angelopoulos, S., Kamali, S.: Contract scheduling with predictions. In: Proceedings of the AAAI, pp. 11726–11733. AAAI Press (2021)
7. Angelopoulos, S., Arsénio, D., Kamali, S.: Competitive sequencing with noisy advice. CoRR abs/2111.05281 (2021)
8. Antoniadis, A., et al.: Paging with succinct predictions. In: Proceedings of the ICML (2023, to appear)
9. Antoniadis, A., Gouleakis, T., Kleer, P., Kolev, P.: Secretary and online matching problems with machine learned advice. In: Proceedings of the NeurIPS (2020)
10. Awerbuch, B., Bartal, Y., Fiat, A., Rosén, A.: Competitive non-preemptive call control. In: Proceedings of the SODA, pp. 312–320 (1994)
11. Azar, Y., Leonardi, S., Touitou, N.: Flow time scheduling with uncertain processing time. In: Proceedings of the STOC, pp. 1070–1080 (2021)
12. Azar, Y., Panigrahi, D., Touitou, N.: Online graph algorithms with predictions. In: Proceedings of the SODA, pp. 35–66 (2022)
13. Balkanski, E., Gkatzelis, V., Tan, X.: Strategyproof scheduling with predictions. In: Proceedings of the ITCS. LIPIcs, vol. 251, pp. 11:1–11:22 (2023)
14. Bampis, E., Dogeas, K., Kononov, A.V., Lucarelli, G., Pascual, F.: Scheduling with untrusted predictions. In: Proceedings of the IJCAI, pp. 4581–4587 (2022)
15. Banerjee, S., Cohen-Addad, V., A., Li, Z.: Graph searching with predictions. In: Proceedings of the ITCS. LIPIcs, vol. 251, pp. 12:1–12:24 (2023)
16. Barhum, K., Böckenhauer, H.-J., Forišek, M., Gebauer, H., Hromkovič, J., Krug, S., Smula, J., Steffen, B.: On the power of advice and randomization for the disjoint path allocation problem. In: Geffert, V., Preneel, B., Rovan, B., Štuller, J., Tjoa, A.M. (eds.) SOFSEM 2014. LNCS, vol. 8327, pp. 89–101. Springer, Cham (2014). https://doi.org/10.1007/978-3-319-04298-5_9
17. Berg, M., Boyar, J., Favrholdt, L.M., Larsen, K.S.: Online interval scheduling with predictions. ArXiv (2023). arXiv:2302.13701. To appear in 18th WADS, 2023
18. Böckenhauer, H., Benz, N.C., Komm, D.: Call admission problems on trees. Theor. Comput. Sci. **922**, 410–423 (2022)
19. Böckenhauer, H., Komm, D., Wegner, R.: Call admission problems on grids with advice. Theor. Comput. Sci. **918**, 77–93 (2022)
20. Borodin, A., El-Yaniv, R.: Online Computation and Competitive Analysis. Cambridge University Press, Cambridge (1998)
21. Boyar, J., Favrholdt, L.M., Larsen, K.S.: Online unit profit knapsack with untrusted predictions. In: Proceedings of the SWAT. LIPIcs, vol. 227, pp. 20:1–20:17 (2022)

22. Boyar, J., Favrholdt, L.M., Kudahl, C., Mikkelsen, J.W.: The advice complexity of a class of hard online problems. Theory Comput. Syst. **61**(4), 1128–1177 (2017)
23. Chapin, S.J., et al.: Benchmarks and standards for the evaluation of parallel job schedulers. In: Feitelson, D.G., Rudolph, L. (eds.) JSSPP 1999. LNCS, vol. 1659, pp. 67–90. Springer, Heidelberg (1999). https://doi.org/10.1007/3-540-47954-6_4
24. Chen, J.Y., et al.: Triangle and four cycle counting with predictions in graph streams. In: Proceedings of the ICLR (2022)
25. Chen, J.Y., Silwal, S., Vakilian, A., Zhang, F.: Faster fundamental graph algorithms via learned predictions. In: Proceedings of the ICML. PLMR, vol. 162, pp. 3583–3602 (2022)
26. Wagner, D., Weihe, K.: A linear-time algorithm for edge-disjoint paths in planar graphs. Combinatorica **15**, 135–150 (1995)
27. Eberle, F., Lindermayr, A., Megow, N., Nölke, L., Schlöter, J.: Robustification of online graph exploration methods. In: Proceedings of the AAAI, pp. 9732–9740 (2022)
28. Even, S., Itai, A., Shamir, A.: On the complexity of timetable and multicommodity flow problems. SIAM J. Comput. **5**(4), 691–703 (1976)
29. Frank, A.: Edge-disjoint paths in planar graphs. J. Combin. Theory Ser. B **39**, 164–178 (1985)
30. Garg, N., Vazirani, V.V., Yannakakis, M.: Primal-dual approximation algorithms for integral flow and multicut in trees. Algorithmica **18**, 3–20 (1977)
31. Gebauer, H., Komm, D., Královič, R., Královič, R., Smula, J.: Disjoint path allocation with sublinear advice. In: Xu, D., Du, D., Du, D. (eds.) COCOON 2015. LNCS, vol. 9198, pp. 417–429. Springer, Cham (2015). https://doi.org/10.1007/978-3-319-21398-9_33
32. Im, S., Kumar, R., Qaem, M.M., Purohit, M.: Non-clairvoyant scheduling with predictions. In: Proceedings of the SPAA, pp. 285–294 (2021)
33. Im, S., Kumar, R., Qaem, M.M., Purohit, M.: Online knapsack with frequency predictions. In: Proceedings of the NeurIPS, pp. 2733–2743 (2021)
34. Matsumoto, K., Nishizeki, T., Saito, N.: An efficient algorithm for finding multicommodity flows in planar networks. SIAM J. Comput. **14**(2), 289–302 (1985)
35. Kolen, A.W., Lenstra, J.K., Papadimitriou, C.H., Spieksma, F.C.: Interval scheduling: a survey. Nav. Res. Logist. **54**(5), 530–543 (2007)
36. Lattanzi, S., Lavastida, T., Moseley, B., Vassilvitskii, S.: Online scheduling via learned weights. In: Proceedings of the SODA, pp. 1859–1877 (2020)
37. Lavastida, T., Moseley, B., Ravi, R., Xu, C.: Learnable and instance-robust predictions for online matching, flows and load balancing. In: Proceedings of the ESA. LIPIcs, vol. 204, pp. 59:1–59:17 (2021)
38. Lavastida, T., Moseley, B., Ravi, R., Xu, C.: Using predicted weights for ad delivery. In: Proceedings of the ACDA, pp. 21–31 (2021)
39. Lee, R., Maghakian, J., Hajiesmaili, M., Li, J., Sitaraman, R.K., Liu, Z.: Online peak-aware energy scheduling with untrusted advice. In: Proceedings of the e-Energy, pp. 107–123 (2021)
40. Lykouris, T., Vassilvitskii, S.: Competitive caching with machine learned advice. In: Proceedings of the ICML. PMLR, vol. 80, pp. 3302–3311 (2018)
41. Mitzenmacher, M., Vassilvitskii, S.: Algorithms with predictions. In: Roughgarden, T. (ed.) Beyond the Worst-Case Analysis of Algorithms, pp. 646–662. Cambridge University Press (2020)
42. Nishizeki, T., Vygen, J., Zhou, X.: The edge-disjoint paths problem is NP-complete for series-parallel graphs. Discret. Appl. Math. **115**(1–3), 177–186 (2001)

43. Purohit, M., Svitkina, Z., Kumar, R.: Improving online algorithms via ML predictions. In: Proceedings of the NeurIPS, pp. 9661–9670 (2018)
44. Rohatgi, D.: Near-optimal bounds for online caching with machine learned advice. In: Proceedings of the SODA, pp. 1834–1845 (2020)
45. Wei, A., Zhang, F.: Optimal robustness-consistency trade-offs for learning-augmented online algorithms. In: Proceedings of the NeurIPS (2020)
46. Wei, A.: Better and simpler learning-augmented online caching. In: Proceedings of the APPROX/RANDOM. LIPIcs, vol. 176, pp. 60:1–60:17 (2020)
47. Zeynali, A., Sun, B., Hajiesmaili, M., Wierman, A.: Data-driven competitive algorithms for online knapsack and set cover. In: Proceedings of the AAAI, pp. 10833–10841 (2021)

On Length-Sensitive Fréchet Similarity

Kevin Buchin[1], Brittany Terese Fasy[2], Erfan Hosseini Sereshgi[3](✉), and Carola Wenk[3]

[1] TU Dortmund, Dortmund, Germany
kevin.buchin@tu-dortmund.de
[2] Montana State University, Bozeman, USA
brittany.fasy@montana.edu
[3] Tulane University, New Orleans, USA
{shosseinisereshgi,cwenk}@tulane.edu

Abstract. Taking length into consideration while comparing 1D shapes is a challenging task. In particular, matching equal-length portions of such shapes regardless of their combinatorial features, and only based on proximity, is often required in biomedical and geospatial applications. In this work, we define the length-sensitive partial Fréchet similarity (LSFS) between curves (or graphs), which maximizes the length of matched portions that are close to each other and of equal length. We present an exact polynomial-time algorithm to compute LSFS between curves under L_1 and L_∞. For geometric graphs, we show that the decision problem is NP-hard even if one of the graphs consists of one edge.

Keywords: Fréchet distance · partial matching · curves · graphs in $\mathbb{R}^2$

1 Introduction

Measuring the similarity between geometric objects is a fundamental task with many applications. One way to measure similarity is to take the reciprocal of a distance measure, such as the Fréchet distance for curves [7]. While the Fréchet distance matches the entire curves to each other, many situations require only matching parts of the curves. For instance, if we want to evaluate whether one curve is a subset of the other, a perfect matching makes little sense. For such situations, partial Fréchet similarity has been proposed [11], which aims to match curves to each other so that the total length of the portions that are close to each other is maximized.

When maximizing lengths, the matching may be skewed towards noisier portions of a curve, as noise makes a curve longer. While this may be acceptable if the noise occurs on both curves, it should not be possible to increase the similarity score by adding noise on one of the curves, such as in the example of Fig. 1, where the partial Fréchet similarity matches the spiral of g to the middle of f. To mitigate this issue, we introduce the *length-sensitive Fréchet similarity (LSFS)*,

Supported by the National Science Foundation grant CCF 2046730.
Supported by the National Science Foundation grant CCF 2107434.

P. Morin and S. Suri (Eds.): WADS 2023, LNCS 14079, pp. 208–231, 2023.
https://doi.org/10.1007/978-3-031-38906-1_15

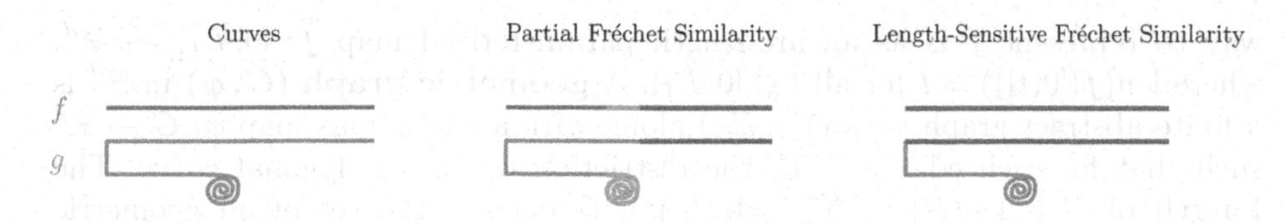

Fig. 1. Two curves f, g with different matchings, illustrated in blue. (Color figure online)

where only equal-length portions of curves that are close to each other may be matched as in Fig. 1 (right).

Non-partial length-sensitive Fréchet matchings have been considered previously. Buchin, Buchin, and Gudmundsson consider speed constraints on Fréchet matchings [10], and Maheshwari et al. study the Fréchet distance with *speed limits* [19]. Similar speeds on the curves result in matching similar lengths to each other. However, all of these approaches restrict to matching the whole curves within a given distance, which is very different from the setting of partial matchings, and makes the techniques non-applicable in our setting.

Our motivation for studying length-sensitive partial Fréchet matchings stems from the problem of comparing paths [18] and geometric networks [2,3,5,6,13], more specifically from the *map reconstruction problem.* To evaluate different reconstruction algorithms that construct road networks from movement trajectories (e.g., [4,9,12,15,16]), one needs to evaluate how similar the reconstructions are to the underlying network. Usually trajectories only cover portions of the network, and therefore a partial similarity is desired. However, to avoid favoring algorithms that build noisy reconstructions, length-sensitive matchings are desirable. In this context, length-sensitive Fréchet matchings can be seen as a continuous version of the commonly used graph-sampling technique [1,8].

Our Contributions. In this paper, we define length-sensitive partial Fréchet similarity (LSFS) for curves and geometric graphs. Specifically, the LSFS maximizes the length of matched portions that are close to each other and of equal length. In Sect. 3, we define LSFS for curves and graphs, providing a clean mathematical definition for this intuitive concept. For geometric graphs we show in Sect. 5 that LSFS is NP-hard even if one of the graphs consists of only one edge. In Sect. 4, we present a polynomial-time algorithm to compute LSFS for curves under L_1; the same approach generalizes to L_∞ as well.[1]

2 Preliminaries

A **polygonal curve** f in $\mathbb{R}^d$ is a finite sequence of line segments (or edges) in $\mathbb{R}^d$. Its **length**, $\text{len}(f) = L_f$, is the sum of the lengths of the edges. Another

[1] We do not consider L_2, since we expect to encounter the same algebraic obstacles that make partial Fréchet similarity unsolvable over the rational numbers [14].

way to represent f is as an arc-length parameterized map $f\colon [0, L_f] \to \mathbb{R}^d$, where $\mathrm{len}(f([0,t])) = t$ for all $t \in [0, L_f]$. A **geometric graph (G, ϕ)** in $\mathbb{R}^d$ is a finite abstract graph $G = (V_G, E_G)$ along with a continuous map $\phi\colon G \to \mathbb{R}^d$ such that for each edge $e \in E$, the restriction $\phi|_e$ is a polygonal curve. The **length** of G is $\mathrm{len}(G) = \sum_{e \in E} \mathrm{len}(\phi|_e)$. $\boldsymbol{\mathcal{G}}$ denotes the set of all geometric graphs in $\mathbb{R}^d$, up to reparameterization, i.e., (G, ϕ_G) and (H, ϕ_H) are equivalent in $\mathcal{G}$ if there exists a homeomorphism $h\colon G \to H$ such that $\phi_G = \phi_H \circ h$. Let $(G, \phi_G), (H, \phi_H) \in \mathcal{G}$ and let $h\colon H \to G$ be a function that is homeomorphic onto its image. If, for each path π in (H, ϕ_H), the path $h(\pi)$ in (G, ϕ_G) has the same (intrinsic) length, we say that h is **length-preserving**.

The **Fréchet distance** between two curves f and g in $\mathbb{R}^d$ is

$$\boldsymbol{d_F(f,g)} = \inf_{h:[0,L_f]\to[0,L_g]} \max_{t\in[0,L_f]} ||f(t) - g(h(t))||_p, \tag{1}$$

where h is an homeomorphism such that $h(0) = 0$, and $||.||_p$ is the p-norm. In this paper, we focus on the case when $p = 1$ or $p = \infty$.

For polygonal curves f and g and threshold $\varepsilon > 0$, Alt and Godau [7] provided a polynomial time dynamic programming algorithm based on the free-space diagram. The **free-space diagram $\mathcal{D}_\varepsilon(f, g)$** is a binary function defined over $[0, L_f] \times [0, L_g]$. For a point $(x, y) \in [0, L_f] \times [0, L_g]$, $\mathcal{D}_\varepsilon(f, g)$ is **free** (colored white) if $||f(x) - g(y)||_p \leq \varepsilon$; otherwise, it is **infeasible** (colored gray). The set of all free points is the **free-space**. See Fig. 2. If f, g are polygonal curves with m and n edges, respectively, then $[0, L_f] \times [0, L_g]$ can be decomposed into **cells** of an $m \times n$ grid, where the i-th column corresponds to the i-th edge of f and the j-th row corresponds to the j-th edge of g. By convexity of the L_p-norm, free space within a cell is convex; if $p = 2$ it is the intersection of an ellipse with the cell, if $p = 1$ or $p = \infty$ it is the intersection of a parallelogram with the cell. There exists a bi-monotone path in the free space from $(0, 0)$ to (L_f, L_g) if and only if $d_F(f, g) \leq \varepsilon$. The dynamic program marches through $\mathcal{D}_\varepsilon(f, g)$ cell-by-cell propagating reachability information from $(0, 0)$.

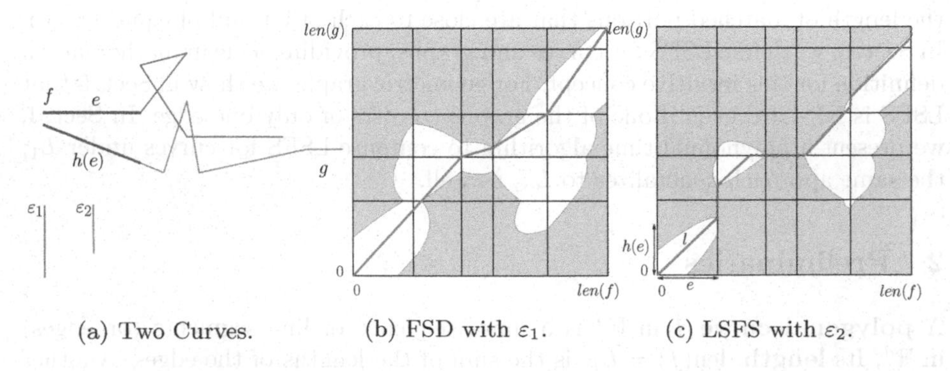

(a) Two Curves. (b) FSD with ε_1. (c) LSFS with ε_2.

Fig. 2. The free-space diagram $\mathcal{D}_\varepsilon(f, g)$, for $p = 2$, for two threshold values, ε_1 and ε_2. In (a), there is a path from (0,0) to (L_f, L_g), so $d_F(f, g) \leq \varepsilon_1$. In (b), the slope-one line segments (shown in blue) correspond to length-preserving matchings. e is a line segment on f that was matched with $h(e)$ on g. (Color figure online)

3 Defining Length-Sensitive Fréchet Similarity

Here, we introduce the length-sensitive Fréchet similarity (LSFS), a notion of similarity between curves (or between graphs).

3.1 LSFS for Curves

Let f, g be two curves and let $\varepsilon > 0$. To define length-sensitive Fréchet similarity, we consider the homeomorphisms h in Eq. (1), which correspond to paths γ_h in the free-space diagram $\mathcal{D}_\varepsilon(f,g)$ from $(0,0)$ to (m,n). We maximize the length of the portions of γ_h that are free and length-preserving. As before, free portions of γ_h are exactly the points such that $||f(x) - g(h(x))||_p \leq \varepsilon$. The length-preserving matchings between sub-curves of f and g correspond to slope-one segments of γ_h. Putting this together, we get $\boldsymbol{I_h^\varepsilon} =$

$$\{x \in [0, L_f] \mid (x, h(x)) \text{ is free and } \exists \delta > 0\colon h|_{(x-\delta, x+\delta)} \text{ is length-preserving.}\}$$

Here, I_h^ε is the portion of (the parameter space of) f that h maps to g in a length-preserving way while staying within distance ε. We quantify this by defining the **length-sensitive Fréchet similarity (LSFS)** as:

$$\mathbb{F}_\varepsilon(\boldsymbol{f}, \boldsymbol{g}) = \sup_{h:[0,L_f]\to[0,L_g]} \text{len}(I_h^\varepsilon),$$

where h ranges over all homeomorphisms such that $h(0) = 0$. We interpret I_h^ε as a set of (maximal) intervals, which means that $f(I_h^\varepsilon)$ is a set of subcurves in $\mathbb{R}^d$. Then $\text{len}(I_h^\varepsilon)$ measures the total length of these subcurves of f, since f is arc-length parameterized.[2] Note that the definition above only requires h to be *locally* length-preserving (i.e., only within δ of x). However, this is actually a global property, as proven in Corollary 1 in Appendix A.

We note that the condition of equal length is less harsh than it might seem at first: sub-curves that are close and have similar lengths can still be matched; the score is the length of the shorter portion. As an example of LSFS, consider Fig. 2. We are looking for a bi-monotone path from bottom-left to top-right that maximizes the total length of slope-one segments in the free (white) space. The line segment l, which has slope one, indicates a length-preserving matching because the corresponding matched line segments, e and $h(e)$ on the two curves.

3.2 LSFS for Graphs

Let $(G, \phi_G), (H, \phi_H) \in \mathcal{G}$ and $\varepsilon > 0$. We extend the definition of LSFS to a similarity measure between graphs. Let G be a graph, let C be a connected subgraph of G, and let $h\colon C \to H$ be a continuous map that is homeomorphic onto its image. We define $\boldsymbol{C_h^\varepsilon} =$

$$\{x \in C \mid ||\phi_G(x) - \phi_H(h(x))||_p \leq \varepsilon \text{ and } \exists \delta > 0\colon h|_{\mathbb{B}_p(x,\delta)} \text{ is length-preserving}\}.$$

[2] For LSFS it suffices to measure lengths of subcurves on f. Partial Fréchet similarity [11] measures lengths of subcurves on f and g, but is not length-preserving.

Here, $\mathbb{B}_p(x, \delta)$ is the open ball centered at x with radius $\delta \in \mathbb{R}_{\geq 0}$. Figure 3 shows an example. The set C_h^ε is a subgraph of (G, ϕ_G), and so $(C_h^\varepsilon, \phi_G|_{C_h^\varepsilon})$ is in $\mathcal{G}$. While C is a connected graph, we note that C_h^ε need not be connected.

The restriction to connected components of G is important. For example, consider Fig. 4, which demonstrates what can happen if we do not enforce this requirement.

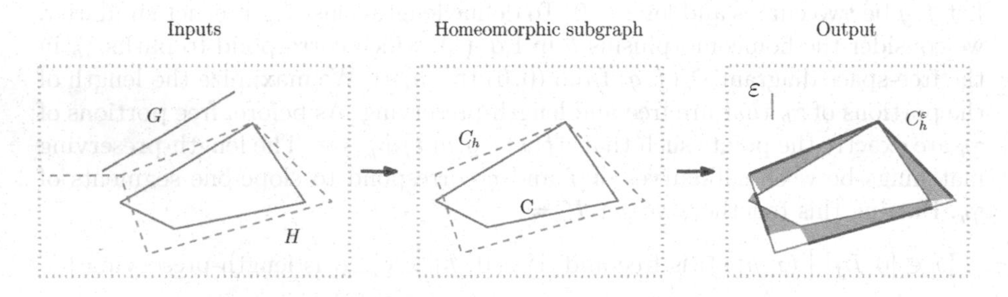

Fig. 3. Computing C_h^ε on two given graphs, G (in magenta) and H (orange). First, we take C, a connected subgraph of G. Then, we find $h\colon C \to H$, a map such that C is homeomorphic onto its image. For a given ε, dark blue segments are C_h^ε. The total length of those segments is the LSFS, $\mathbb{F}_\varepsilon(G, H)$.

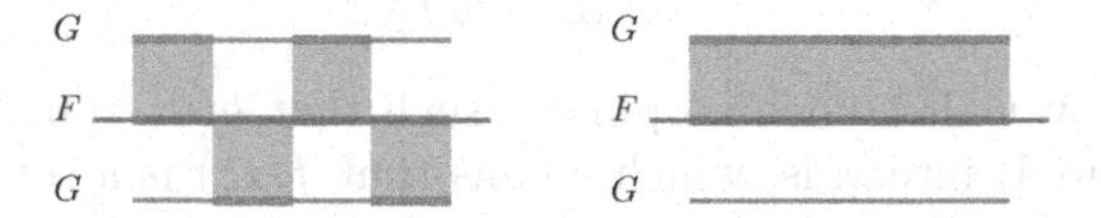

Fig. 4. Two graphs, F (in magenta) and G (orange). Blue represents the matched portions of the graphs. Although the total length matched is the same for both examples, the right is more preferable, as it matches one connected component of G to an entire interval instead of breaking it up into four connected matchings. (Color figure online)

Given this setup, the **length-sensitive Fréchet similarity** is the maximum matched length that can be obtained through such homeomorphisms:

$$\mathbb{F}_\varepsilon(G, H) = \sup_{C \subset G} \sup_{h: C \to H} \operatorname{len}(C_h^\varepsilon).$$

4 Computing LSFS for Curves

In this section, we present a polynomial-time dynamic programming algorithm for computing the LSFS for two curves in $\mathbb{R}^2$ under the L_1 and L_∞ norms. To enable efficient propagation of the score function (defined in Sect. 4.2), we first refine each cell of the free-space diagram (Sect. 4.1). We then concentrate on a single refined cell and show in Sect. 4.4 how to propagate the score function based on the lemmas and observations from Sect. 4.3. Moreover, we show the complexity of the score function in each refined cell. Finally, in Sect. 4.5, we describe the overall dynamic programming algorithm and its total complexity.

4.1 Refining the Free-Space Diagram

To compute the score function using dynamic programming, we refine the free-space diagram as follows: Consider $\mathcal{D}_\varepsilon(f,g)$, using $p = 1$ or $p = \infty$. Then the free space of a diagram grid cell $C' = \mathcal{D}_\varepsilon[i][j]$ is the intersection of a parallelogram with the cell. Thus, the free space in C' is defined by a polygon with up to eight vertices, and the infeasible space can have up to four connected components which may not be convex. For every vertex v of the free-space polygon in C', we extend a horizontal and a vertical line from v to the boundaries of the cell. This results in a subdivision of C' that splits all polygons in the cell into (simpler) triangles, quadrilaterals, and pentagons; see Fig. 5. In addition, each sub-cell now has at most two connected components of the infeasible space, each of which is a convex polygon defined by at most five vertices.

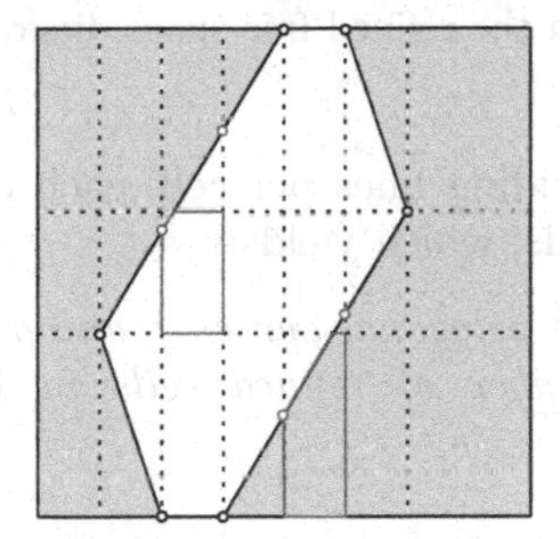
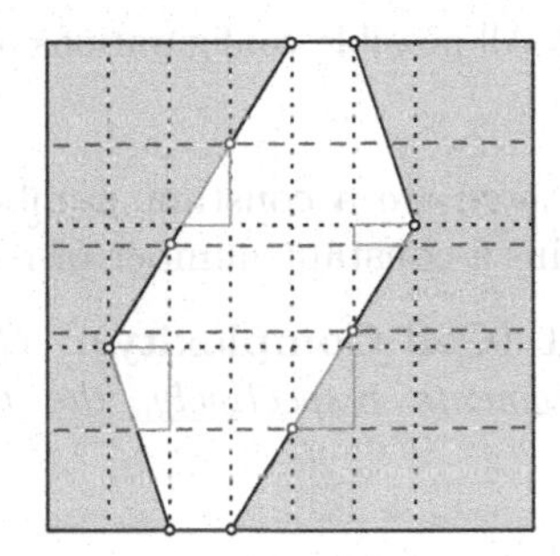

Fig. 5. Refining a cell C' in $\mathcal{D}_\varepsilon(f,g)$. The left shows C' after extending the horizontal and vertical lines through each vertex of the white polygon. The red pentagons (one in free space and one in infeasible space) result from these lines intersecting edges of the white polygon. We split these pentagons into two propagatable polygons by drawing the horizontal blue lines from those intersections (blue points). The resulting green polygons on the right are all propagatable.

All of the new polygons (in both the free and infeasible regions), except the pentagons, share a property that we call *propagatability.*[3] A propagatable polygon is a polygon with at most four edges, at least two of which are horizontal or vertical. For each (non-propagatable) pentagon we add an additional split by finding a vertex v that does not lie on a horizontal line, and then extend a horizontal line through v to the boundaries of C'. After this split, even the pentagons are now split into propagatable polygons. The arrangement of all horizontal and vertical lines subdivides C' into a set of *refined cells*, and every free-space polygon inside a refined cell is propagatable, see Fig. 6:

Observation 1 (Refined Cells Contain Propagatable Polygons). *The free-space polygons inside refined cells are propagatable. The set of possible configurations of the refined cells are:*

[3] We use the term *propagatable*, as these regions allow for easier propagation of a score function in the dynamic program, which we elaborate on in the next sub-sections.

a) All free. b) Free space on right; dividing line has positive slope. c) Free space on left; dividing line has positive slope. d) Free space on right; dividing line has negative slope. e) Free space on left; dividing line has negative slope. f) Two components of infeasible space; dividing lines have positive slopes. g) Two components of infeasible space; dividing lines have negative slopes. h) All infeasible.

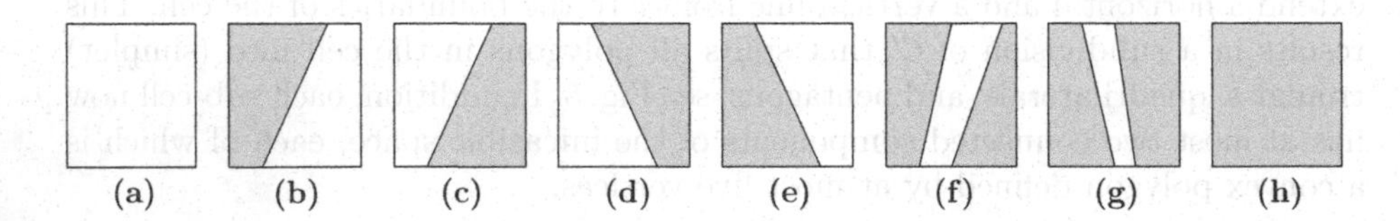

Fig. 6. All possible configurations of a cell in the refined free-space diagram.

Since there are a constant number of dividing lines per cell, each diagram cell contains a constant number of refined cells, which yields:

Lemma 1 (Cell Complexity). *Given two polygonal curves f and g with m and n segments respectively, the total number of refined cells in $\mathcal{D}_\varepsilon(f,g)$ is $\Theta(nm)$.*

4.2 Score Function in a Cell

To compute LSFS using dynamic programming in $\mathcal{D}_\varepsilon(f,g)$, we define a score function that maps $(x,y) \in \mathcal{D}_\varepsilon(f,g)$ to the LSFS defined for the corresponding prefixes of f and g. The score function $\mathcal{S} : [0, L_f] \times [0, L_g] \to \mathbb{R}$ is defined as

$$\mathcal{S}(x,y) = \mathbb{F}_\varepsilon(f|_{[0,x]}, g|_{[0,y]}) = \sup_{h:[0,x]\to[0,y]} \operatorname{len}(I_h^\varepsilon) \ . \quad (2)$$

The dynamic program in Sect. 4.5 computes the score function on the cell boundaries, by propagating from the left and bottom of a refined cell to the top and right of the cell. Let $\mathcal{C} = [x_L, x_L + x_R] \times [y_B, y_B + y_T] \subseteq [0, L_f] \times [0, L_g]$ be a refined cell, and denote with L, R, B, T the left, right, bottom, and top boundaries of $\mathcal{C}$, respectively. For ease of exposition we represent $\mathcal{C}$ using the local coordinate system $[0, x_R] \times [0, y_T]$ and use the following notation (see Fig. 7). The score functions restricted to L, R, B, T are:

$$\mathcal{S}_L(l) = \mathcal{S}(x_L, y_B + l) \qquad \mathcal{S}_R(r) = \mathcal{S}(x_R, y_B + r)$$
$$\mathcal{S}_B(b) = \mathcal{S}(x_L + b, y_B) \qquad \mathcal{S}_T(t) = \mathcal{S}(x_L + t, y_T)$$

Any bi-monotone path from $(0,0)$ in $\mathcal{D}_\varepsilon(f,g)$ to a point in $\mathcal{C}$ has to go through L or B. We can therefore express $\mathcal{S}_T$ and $\mathcal{S}_R$ as

$$\mathcal{S}_T(t) = \max(\mathcal{S}_{L\to T}(t), \mathcal{S}_{B\to T}(t)) \text{ and } \mathcal{S}_R(r) = \max(\mathcal{S}_{L\to R}(r), \mathcal{S}_{B\to R}(r)) \ . \quad (3)$$

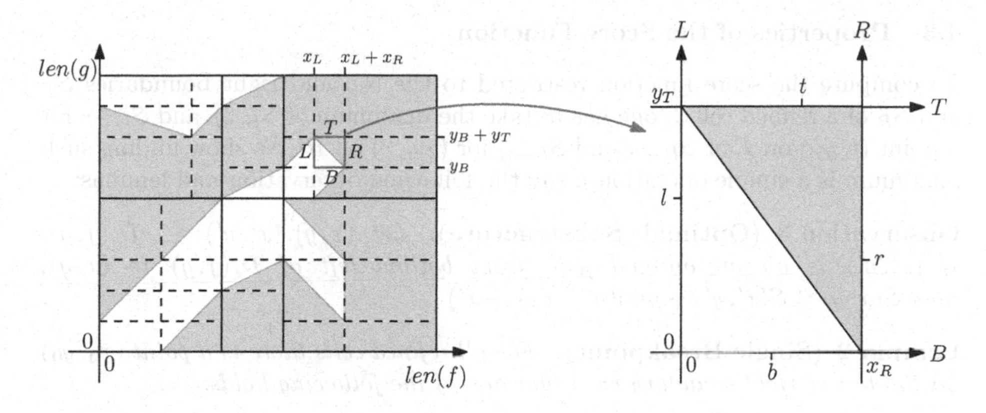

Fig. 7. A refined cell $\mathcal{C}$ shown in $\mathcal{D}_\varepsilon(f, g)$ (left) and in local coordinates (right).

Here, $\mathcal{S}_{L\to T}$ models the propagation from L to T. It is a restriction of LSFS on T that considers only those homeomorphisms h that pass through L. Such a homeomorphism corresponds to a bi-monotone path in $\mathcal{D}_\varepsilon(f, g)$ that is comprised of a path from $(0, 0)$ to a point on L concatenated with a path from this point to a point on T. $\mathcal{S}_{B\to T}, \mathcal{S}_{L\to R}, \mathcal{S}_{B\to R}$ model the remaining types of propagations from L or R to T or R and are defined as:

$$\mathcal{S}_{L\to T}(t) = \sup_{l\in[0,y_T]} \mathcal{S}_L(l) + \mathcal{L}((0,l),(t,y_T)) \quad \mathcal{S}_{B\to T}(t) = \sup_{b\in[0,x_R]} \mathcal{S}_B(b) + \mathcal{L}((b,0),(t,y_T))$$
$$\mathcal{S}_{L\to R}(r) = \sup_{l\in[0,y_T]} \mathcal{S}_L(l) + \mathcal{L}((0,l),(x_R,r)) \quad \mathcal{S}_{B\to R}(r) = \sup_{b\in[0,x_R]} \mathcal{S}_B(b) + \mathcal{L}((b,0),(x_R,r)),$$

where $\mathcal{L}((x_1, y_1), (x_2, y_2)) = \mathbb{F}_\varepsilon(f|_{[x_L+x_1,x_L+x_2]}, g|_{[y_B+y_1,y_B+y_2]})$ measures the LSFS between points $(x_1, y_1), (x_2, y_2)$ in the local coordinate system of $\mathcal{C}$. Note that a bi-monotone path from (x_1, y_1) to (x_2, y_2) is not necessarily unique, however $\mathcal{L}((x_1, y_1), (x_2, y_2))$ is, see Fig. 8.

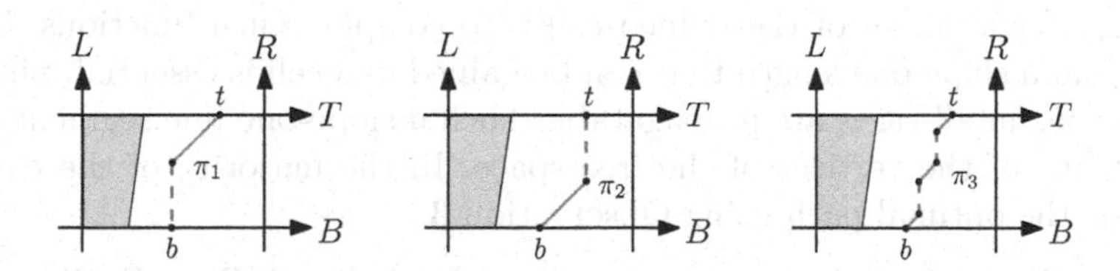

Fig. 8. π_1,π_2 and π_3 are different monotone paths from $(b, 0)$ to $(t, y_B + y_T)$ that have the same slope-one length, $\mathcal{L}((b, 0), (t, y_B + y_T))$ (blue). Other edges in these paths are horizontal or vertical. (Color figure online)

In the remainder, we extensively use the following observation:

Observation 2. *Let (x, y) and $(x + \Delta x, y + \Delta y)$ be points in $\mathcal{D}_\varepsilon(f, g)$. Then the maximum length-preserving portion that can be achieved between those two points is* $\min(\Delta x, \Delta y)$. *Therefore* $\mathcal{L}((x, y), (x + \Delta x, y + \Delta y)) \le \min(\Delta x, \Delta y)$.

4.3 Properties of the Score Function

To compute the score function restricted to the top and right boundaries $\mathcal{S}_T$ and $\mathcal{S}_R$ of a refined cell $\mathcal{C}$, one has to take the maximum of $\mathcal{S}_{B\to T}$ and $\mathcal{S}_{L\to T}$ for a point (t, y_T) on T or $\mathcal{S}_{B\to R}$ and $\mathcal{S}_{L\to R}$ for (x_R, r) on R. We show finding such maximum is a simple operation using the following observation and lemmas:

Observation 3 (Optimal Substructure). *Let $(x,y),(x',y') \in \mathcal{D}_\varepsilon(f,g)$. If (x',y') is on an optimal path from bottom left of $\mathcal{D}_\varepsilon(f,g)$ to (x,y), then $\mathcal{S}(x,y) \leq \mathcal{S}(x',y') + \min(y-y', x-x')$*

Lemma 2 (Single Breakpoint). *For all refined cells there is a point (x_0, y_0) on the top or right boundary such that one of the following holds:*

1. if $(x_0, y_0) \in T$:

$$\mathcal{S}_T(x) = \begin{cases} \mathcal{S}_{L\to T}(x) \text{ for } x < x_0 \\ \mathcal{S}_{B\to T}(x) \text{ for } x \geq x_0 \end{cases}, \text{ and } \mathcal{S}_R(y) = \mathcal{S}_{B\to R}(y) \quad (4)$$

2. if $(x_0, y_0) \in R$:

$$\mathcal{S}_T(x) = \mathcal{S}_{L\to T}(x)\ , \text{ and } \mathcal{S}_R(y) = \begin{cases} \mathcal{S}_{L\to R}(y) \text{ for } y > y_0 \\ \mathcal{S}_{B\to R}(y) \text{ for } y \leq y_0 \end{cases} \quad (5)$$

Lemma 3 (Slope Upper-Bound). *The score function on a refined cell boundary is piecewise lienar, with each piece of slope less than or equal to 1.*

We provide proofs for Lemmas 2 and 3 in Appendix B.

4.4 Score Function Propagation Within a Cell

In this section, we demonstrate how to compute the score function from a cell boundary to another. In particular, we seek the functions $\mathcal{S}_{L\to R}$, $\mathcal{S}_{B\to R}$, $\mathcal{S}_{L\to T}$, and $\mathcal{S}_{B\to T}$ for all cases of the refined cells. To compute such functions, finding the maximum slope-one length that can be gained in a cell is essential. Since the free polygons in all cases are propagatable, such a slope-one line segment has to intersect one of the vertices of the free space. In the majority of the cases we determine the optimal path using Observation 3.

Case (a): We explore the propagation from L to R and B to R. Since going from L and B to T are symmetrical to these, one can compute them in a similar manner. For every point (x_R, r) on R there is a path with maximum slope-one length that reaches it through L. We aim to find corresponding points $(0, l)$ on L for every point (x_R, r) on R such that $\mathcal{S}_{L\to R}(r) = \mathcal{S}_L(l) + \mathcal{L}((0,l),(x_R,r))$ is maximal. $\mathcal{S}_L(l)$ is a non-decreasing piecewise linear function and all the pieces have slopes less than or equal to one. Therefore, finding $(0,l)$ and (x_R, r) that result in a maximal $\mathcal{L}((0,l),(x_R,r))$ is crucial.

Consider cell $\mathcal{C}$ in Fig. 9, since $\mathcal{C}$ is free, $\mathcal{L}((0,l),(x_R,r)) = \min(y_T, x_R)$ and this value can be achieved by drawing slope-one line segment from $(0,0)$. Any

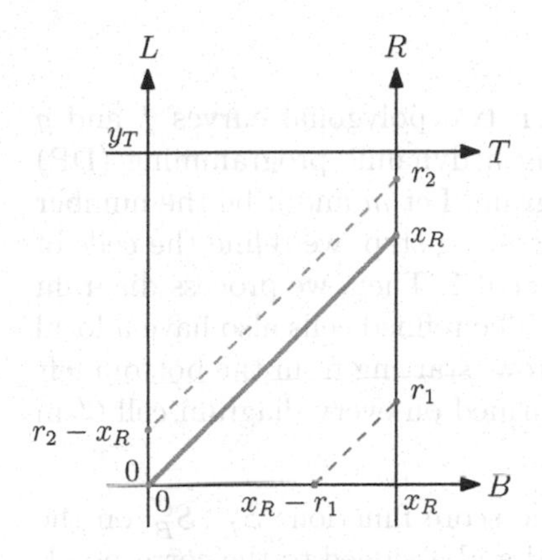

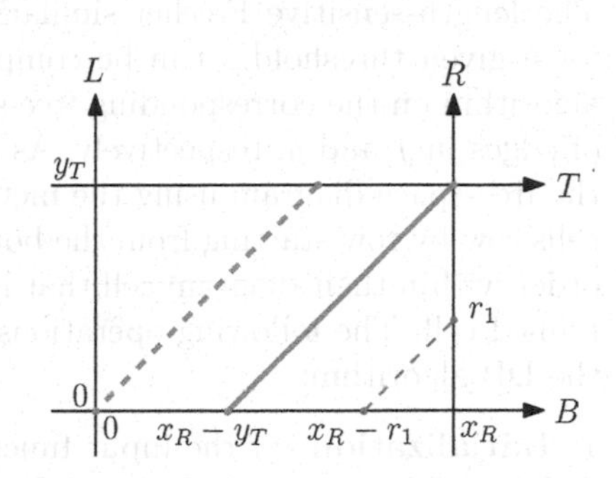

Fig. 9. Case (a), when $x_R < y_T$ on the left, and when $x_R > y_T$ on the right. The blue line shows the maximum slope-one length possible. Maximal slope-one segments to r_1 and r_2 are shown in magenta and orange, respectively. (Color figure online)

point (x_R, r_1) on R below this line segment can be reached from $(0,0)$ on L while $\mathcal{L}((0,0),(x_R,r_1)) = r_1$. Any point (x_R, r_2) above this line segment on R can be reached from $(0, r_2 - x_R)$ while achieving the maximum slope-one length. Note that if $x_R > y_T$, a slope-one line through $(0,0)$ does not intersect with R within C hence there is technically no point on R above such a line.

$$\mathcal{S}_{L \to R}(r) = \begin{cases} \mathcal{S}_L(0) + r & \text{for } r \leq x_R \\ \mathcal{S}_L(r - x_R) + x_R & \text{for } r \geq x_R \end{cases}$$

For constructing $\mathcal{S}_{B \to R}(r)$, note that $\mathcal{L}((b,0),(x_R,r))$ is maximal when a slope-one line segment going through (x_R, y_T) intersects B. If such line does not exist the maximal slope-one length is achieved by drawing a slope-one line from $(0,0)$. For any point (x_R, r_1) below this line we draw a slope-one line segment from (x_R, r_1) that cuts B on $(x_R - r_1, 0)$. For any point (x_R, r_2) above such a line (if only $x_R < y_T$), the optimal path from B starts at $(0,0)$.

$$\mathcal{S}_{B \to R}(r) = \begin{cases} \mathcal{S}_B(x_R - r) + r & \text{for } r \leq x_R \\ \mathcal{S}_B(0) + x_R & \text{for } r \geq x_R \end{cases}$$

Theorem 1. *The propagation within Case (a) adds $O(1)$ breakpoints/complexity.*

Proof. Assuming $\mathcal{S}_L$ has n breakpoints, we want to determine the number of breakpoints on $\mathcal{S}_{L \to R}$, which consists of two pieces. In the first piece, $\mathcal{S}_L(l)$ is added to an increasing linear function r so the resulting function has at most n breakpoints. The second piece of $\mathcal{S}_{L \to R}(r)$ is a constant function without breakpoints. Thus, $\mathcal{S}_{L \to R}(r)$ has at most $n+1$ breakpoints. Similarly, $\mathcal{S}_{B \to R}(r)$, $\mathcal{S}_{L \to T}(t)$, and $\mathcal{S}_{B \to T}(t)$ each also have at most $n+1$ breakpoints.

Case (c)–Case (h): We illustrate the rest of the cases in details in Appendix C.

4.5 Dynamic Programming Algorithm

The length-sensitive Fréchet similarity between two polygonal curves f and g for a given threshold ε can be computed using a dynamic programming (DP) algorithm on the corresponding free-space diagram. Let m and n be the number of edges in f and g, respectively. As a pre-processing step, we refine the cells of the free-space diagram using the method in Sect. 4.1. Then, we process diagram cells row by row starting from the bottom left. The refined cells also have a local order within their diagram cell that is row by row, starting from the bottom left refined cell. The following operations are performed on every diagram cell $\mathcal{C}'$ in the DP algorithm:

1. **Initialization** (of the input functions): The score functions $S_L^{\mathcal{C}'}, S_B^{\mathcal{C}'}$ on the left and bottom boundaries of $\mathcal{C}'$ are divided and assigned to the corresponding score functions $S_L^{\mathcal{C}}, S_B^{\mathcal{C}}$ on those refined cells $\mathcal{C}$ that lie on the left and bottom boundary of $\mathcal{C}'$.
2. **Propagation:** For each refined cell $\mathcal{C}$, in bottom up order within $\mathcal{C}'$, we:
   (a) Compute $\mathcal{S}_{L\to R}^{\mathcal{C}}, \mathcal{S}_{B\to R}^{\mathcal{C}}, \mathcal{S}_{B\to T}^{\mathcal{C}}$, and $\mathcal{S}_{L\to T}^{\mathcal{C}}$ using the methods in Sect. 4.4 and Appendix C
   (b) Compute $\mathcal{S}_R^{\mathcal{C}}$ and $\mathcal{S}_T^{\mathcal{C}}$ based on Eq. (3).
3. **Concatenation:** After all refined cells in the current diagram cell $\mathcal{C}'$ have been processed, we find the score function on the right boundary $S_R^{\mathcal{C}'}$ by concatenating the corresponding $\mathcal{S}_R$ functions on the refined cells. Similarly, we can compute the score function on the top boundary of the diagram cell.

As the DP algorithm finishes processing the last diagram cell, the *LSFS* between f and g can be found in the top right corner of the free-space diagram.

Theorem 2. *The Length-Sensitive Fréchet similarity between two curves f and g with m and n pieces, respectively, can be computed in $O(m^2n^2)$.*

Proof. The DP algorithm is nested and processes all refined cells within all diagram cells. The total number of refined cells $O(mn)$; see Lemma 1. The initialization and concatenation steps take time linear in the complexity of the involved score functions. As demonstrated in Theorems 1–10, computing $\mathcal{S}_{L\to R}^{\mathcal{C}}, \mathcal{S}_{B\to R}^{\mathcal{C}}, \mathcal{S}_{B\to T}^{\mathcal{C}}$, and $\mathcal{S}_{L\to T}^{\mathcal{C}}$ in each cell adds 1 breakpoint to $S_L^{\mathcal{C}}$ and $S_B^{\mathcal{C}}$. Constructing $S_R^{\mathcal{C}}$ and $S_T^{\mathcal{C}}$ adds at most 1 breakpoint to one of these functions according to Lemma 2. Thus, the number of breakpoints in cell $\mathcal{D}[i][j]$ is in $O(ij)$, i.e., linear in the number of previous cells. Hence, the complexity of the score functions on the top right cell is $O(mn)$ and the total runtime of the DP is $O(m^2n^2)$.

5 Hardness of LSFS for Graphs

Unfortunately, deciding whether an optimal Length-Sensitive Fréchet similarity measure is above a given threshold is NP-hard.

Theorem 3 (Maximum LSFS is NP-hard). *Deciding if $\mathbb{F}_\varepsilon(G, H) > L$ is NP-hard, even if G consists of only one edge and H is a plane graph.*

Proof. We reduce from the Hamiltonian path problem in grid graphs, which is known to be NP-hard [17], even for induced grid graphs of degree at most three [20]. Let $H' = (V', E')$ be a grid graph; that is, the vertex set is a finite subset of $\mathbb{Z}^2$ and there is an edge between two vertices u, v if and only if $\|u-v\| = 1$. We construct the graph H as follows: for every vertex, we add an edge to a new degree-one vertex at distance > 1; see Fig. 10. Formally, let $V'' = V' + (3/4, 3/4)$ be the set V' translated by $(3/4, 3/4)$ and $E'' = \{(v', v'') \in V' \times V'' \mid v'' = v' + (3/4, 3/4)\}$. The edges in E'' have length $\sqrt{2} \cdot 3/4 \approx 1.06$. We choose $H = (V' \cup V'', E' \cup E'')$. Without loss of generality, we assume that the coordinates of the vertices of H are between 0 and $n = |V'| > 1$, since we assume that H' is connected. Let G consist of only one edge $(0, n+1)$. We choose $\varepsilon = n + 1$. We claim that if H' has a Hamiltonian path then $\mathbb{F}_\varepsilon(G, H) = n + 1$, and otherwise $\mathbb{F}_\varepsilon(G, H) < n + 1/5$, as desired.

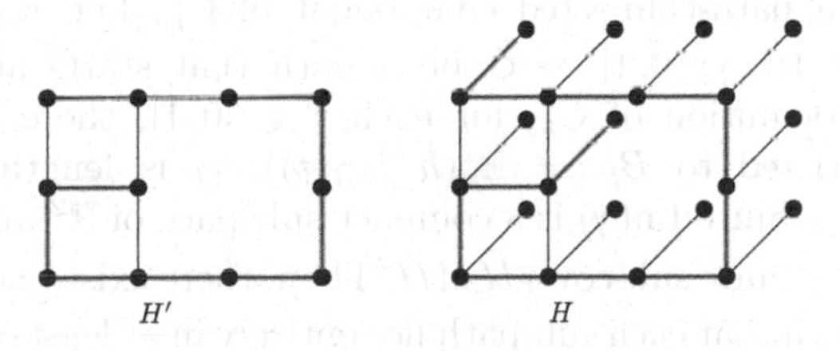

Fig. 10. A grid graph H' with Hamiltonian path in green. The graph H and the image of G corresponding to the Hamiltonian path. (Color figure online)

We have chosen ε sufficiently large such that h (defined in Sect. 3.2) maps G onto any simple path in H (not necessarily ending at vertices). If H' has a Hamiltonian path, then we map G length-preserving onto the corresponding path in H extended by parts (of length 1) of edges in E'' at the beginning and end. Thus, $\mathbb{F}_\varepsilon(G, H)$ is the length of the edge $(0, n+1)$, i.e., $n + 1$. If H' does not have a Hamiltonian path, then the longest simple path in H' that starts and ends at vertices has length at most $n - 2$. Thus, the longest simple path in H has length at most $n - 2 + 2\sqrt{2} \cdot 3/4 \approx n + 0.12 < n + 1/5$.

6 Conclusions and Discussion

We defined length-sensitive Fréchet similarity as a natural partial similarity measure for geometric graphs and curves in $\mathbb{R}^d$. We presented an efficient algorithm for computing it under the L_1 norm for curves, and showed that the corresponding decision problem for geometric graphs is NP-hard. However, there are several directions that can be explored in this area. Can our similarity measure be transformed into a distance measure that is a metric?

For curves in $\mathbb{R}^d$, we conjecture that the running time in Theorem 2 might not be optimal. In [11], a faster running time is achieved by making use of the fact that the score functions are piecewise concave, which is not the case for our functions. For geometric graphs, finding an approximation algorithm to compute LSFS could lead to a practical similarity measure between road networks. A first step might be to consider restricted graph classes. A natural next question is whether there is a polynomial time algorithm for trees.

A LSFS is Length-Preserving on Connected Components

By definition, C_h^ε only requires the restriction of h to small balls to be length-preserving, we show that indeed all connected components are length-preserving.

Lemma 4 (Path-Connected Components Are Length-Preserving). *Each restriction of h to the preimage of a path-connected component of C_h^ε is a length-preserving map.*

Proof. Let $\widetilde{C}$ be a path-connected component of C_h^ε. Let $x, y \in \widetilde{C}$. Since $\widetilde{C}$ is path-connected, let $\gamma\colon [0,1] \to \widetilde{C}$ be a path that starts at x and ends at y. Then, by the definition of C_h^ε, for each $t \in [0,1]$, there exists a $\delta_t > 0$ such that h restricted to $B_t := \mathbb{B}_p(h^{-1}(\gamma(t)), \delta_t)$ is length-preserving. Let $\mathcal{U} := \{\gamma(B_t)\}_{t\in[0,1]}$. Since $\mathrm{Im}(\gamma)$ is a compact subspace of $\mathbb{R}^2$ and since $\mathcal{U}$ covers $\mathrm{Im}(\gamma)$, there exists a finite sub-cover $\widehat{\mathcal{U}}$ of $\mathcal{U}$. Then, there exists a decomposition of γ into sub-paths such that each sub-path lies entirely in at least one open set in $\widehat{\mathcal{U}}$. Let $\{\gamma_i\}_{i=1}^n$ be one such decomposition. Then, we know that $\mathrm{len}(\gamma) = \sum_i \mathrm{len}(p_i)$.

Since each $\mathrm{Im}\, p_i$ is contained in some $U \in \widehat{\mathcal{U}}$ and since h restricted to U is length-preserving, we know that $\mathrm{len}(\gamma_i) = \mathrm{len}(h(\gamma_i))$. Taking the sum over all sub-paths, we find $\mathrm{len}(\gamma) = \mathrm{len}(h(\gamma))$. Thus, we have shown that h restricted to $\widetilde{C}$ is length-preserving.

Noting that a curve $f\colon [0, L_f] \to \mathbb{R}^d$ is simply a directed graph in $\mathbb{R}^d$, we obtain:

Corollary 1. *Each restriction of h to the preimage of a path-connected component of H^ε is a length-preserving map.*

B Proof of Lemmas 2 and 3

In this section we discuss the proofs for Lemmas 2 and 3.

B.1 Proof of Lemma 2

Proof. First we show that for any two points (x_1, y_1) and (x_2, y_2) on the top or right boundary, there exist optimal paths to (x_1, y_1) and (x_2, y_2) that might overlap but do not cross each other. Assuming that the optimal paths to (x_1, y_1) and (x_2, y_2) cross at a point a_0, there is another optimal path to (x_1, y_1) that goes through a_0 and overlaps with the optimal path to (x_2, y_2) before reaching a_0 (as shown in Fig. 11b).

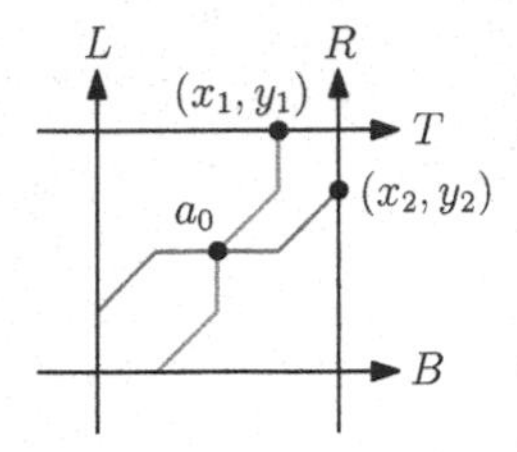

(a) Optimal paths intersect transversally.

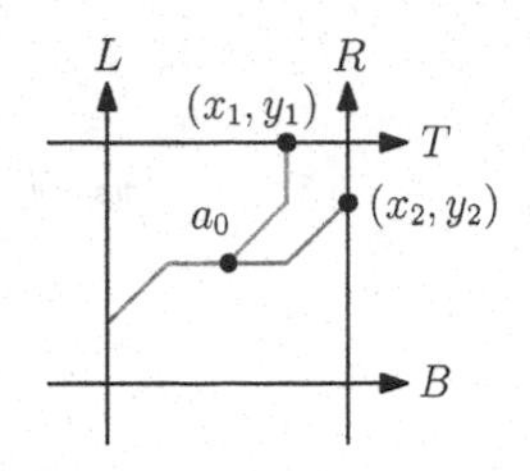

(b) Optimal paths have initial overlap.

Fig. 11. A refined cell $\mathcal{C}$ in the free-space diagram $\mathcal{D}_\varepsilon(f, g)$. In each sub-figure, we see two optimal paths to $(x_1, y_1) \in T$ and to $(x_2, y_2) \in R$. (a) The two paths intersect transversally, at a_0. (b) The two paths first overlap, then separate at a_0.

Therefore, the optimal path to any point on the top or right boundary of a cell cuts the cell into two parts such that for all points on the top or right boundary in one part there is an optimal path going through them within that part. Considering the example in Fig. 11b with the purple path being an optimal path to (x_2, y_2), without loss of generality, we can say any point (x, y) such that $y > y_2$ has an optimal path that comes from the left boundary (more specifically, above the purple path) which means all points on the top boundary have an optimal path coming from the left boundary, hence (4) applies. Similarly, consider (x_1, y_1) and the orange path in Fig. 11a to be optimal. In this case, all the points (x, y) such that $x > x_1$ have an optimal path that comes from the bottom boundary (the right side of the orange path) which results in (5).

B.2 Proof of Lemma 3

Proof. Let $\mathcal{C}$ be a refined cell in $\mathcal{D}_\varepsilon(f, g)$. Let $(x_1, y_B + y_T), (x_2, y_B + y_T)$ be two points on T of $\mathcal{C}$. We want to show: $\mathcal{S}_T(x_2) - \mathcal{S}_T(x_1) = \mathcal{S}(x_2, y_B + y_T) - \mathcal{S}(x_1, y_B + y_T) \leq x_2 - x_1$. Consider the vertical line $\ell = \{(x, y) \in D \mid x = x_1\}$. The optimal monotone path from $(0, 0)$ to $(x_2, y_B + y_T)$ has to cross ℓ at a point (x_1, y_0). Note that (x_1, y_0) does not have to be inside $\mathcal{C}$, see Fig. 12. We know $\mathcal{S}(x_1, y_B + y_T) \geq \mathcal{S}(x_1, y_0)$ and from Observation 3 follows that $\mathcal{S}(x_2, y_B + y_T) - \mathcal{S}(x_1, y_0) \leq \min(y_B + y_T - y_0, x_2 - x_1)$. Thus we can conclude: $\mathcal{S}_T(x_2) - \mathcal{S}_T(x_1) = \mathcal{S}(x_2, y_B + y_T) - \mathcal{S}(x_1, y_B + y_T) \leq \min(y_B + y_T - y_0, x_2 - x_1) \leq x_2 - x_1$. Similarly we can show the equation above for other boundaries of $\mathcal{C}$ as well.

C Score Function Propagation Within a Cell, Continued

We discuss the remainder of the cases and score functions. Due to symmetric nature of Case (b),Case (c), Case (d), and Case (e) with regard to $\mathcal{S}_{L \to T}$ and $\mathcal{S}_{B \to R}$ (likewise, for Case (f) and Case (g)), we do not explain $\mathcal{S}_{L \to T}$ for these cases to avoid repetition.

Case (b): Consider Fig. 13 and the cell $\mathcal{C}$ that is divided into two propagatable polygons by a line with a positive slope. The maximum value for $\mathcal{L}((0, l), (x_R, r))$

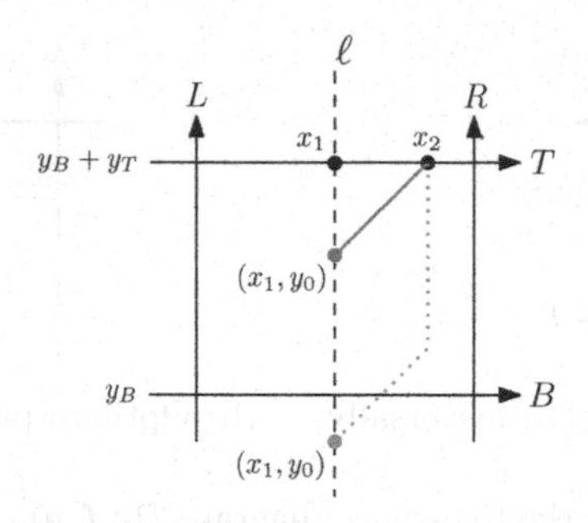

Fig. 12. A cell $\mathcal{C}$ in $\mathcal{D}_\varepsilon(f, g)$. (x_1, y_0) is on the optimal path to (x_2, y_T).

in this case is $x_R - j$ if $x_R < y_T$, and y_T if $x_R > y_T$. Such value can be achieved by drawing a slope-one line from $(j, 0)$ and if the divider's slope is less than one (see Fig. 13, right), by drawing a slope-one line from (x_R, y_T). For $\mathcal{S}_{L \to R}$, for any point (x_R, r_1) on R, below the maximum line, the optimal path goes from $(0, 0)$ to $(x_R - r_1, 0)$ horizontally and then to (x_R, r_1) via a slope-one segment (red dashed line).

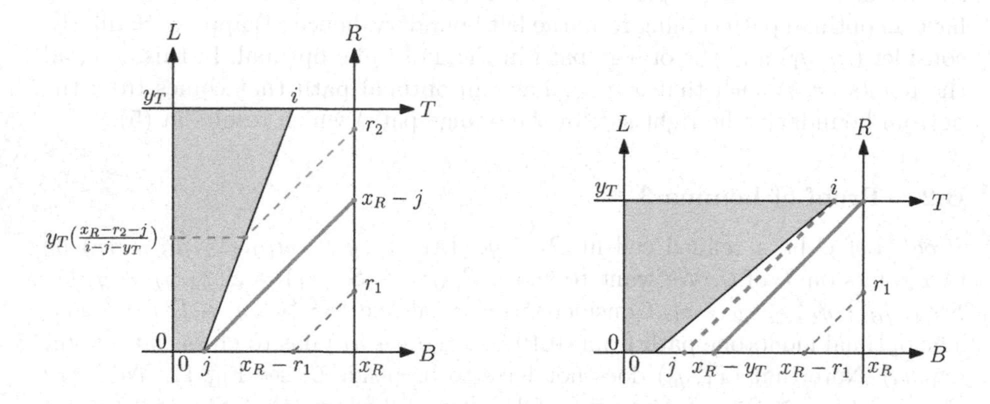

Fig. 13. Case (b) when $\frac{y_T}{i-j} > 1$ on the left, and when $\frac{y_T}{i-j} < 1$ on the right. The blue line shows the maximum slope-one length possible. Maximum slope-one segments to r_1 and r_2 are shown in magenta and orange, respectively. (Color figure online)

For all points (x_R, r_2) above the maximum, the start point of the optimal path depends on $\mathcal{S}_L(l)$. In other words, taking the longest slope-one path to (x_R, r_2) in $\mathcal{C}$ does not necessary yield the maximum score for (x_R, r_2) in this case. This is due to the fact that $\mathcal{L}((0, l), (x_R, r))$ over l (and r) has negative slope. In Theorem 4, we demonstrate how this decision is made and why the number of start-point candidates is at most the number of breakpoints on $\mathcal{S}_L(l)$.

$$\mathcal{S}_{L\to R}(r) = \begin{cases} \mathcal{S}_L(0) + r & \text{for } r \leq x_R - j \\ \max\limits_{l\in[0, y_T(\frac{x_R - r - j}{i-j-y_T})]} \mathcal{S}_L(l) + \mathcal{L}((\frac{i-j}{y_T}(l+j), l), (x_R, r)) & \text{for } r \geq x_R - j \end{cases}$$

Propagation from B to R can be done similar to Case (a), yielding:

$$\mathcal{S}_{B\to R}(r) = \begin{cases} \mathcal{S}_B(x_R - r) + r \text{ for } r \leq x_R - j \\ \mathcal{S}_B(j) + x_R - j \text{ for } r \geq x_R - j \end{cases}$$

$$\mathcal{S}_{B\to T}(t) = \begin{cases} \mathcal{S}_B(t) & \text{for } t \leq j \\ \mathcal{S}_B(j) + (t-j) & \text{for } t \geq j \end{cases}$$

Theorem 4. *The propagation within Case (b) adds $O(1)$ complexity.*

Proof. Assuming $\mathcal{S}_L$ has n breakpoints, we want to determine the number of breakpoints on $\mathcal{S}_{L\to R}$, which consists of two pieces. The first piece, $\mathcal{S}_L(0) + r$ is a linear function. The second piece, is the upper-envelope of the summation of a non-decreasing piecewise linear and a decreasing linear function. See Fig. 14 for an example. To compute $\mathcal{S}_{L\to R}$, one can find the summation on the breakpoints of $\mathcal{S}_L$ and connect them, which results in at most n breakpoints. Taking the upper-envelope makes the resulting function non-decreasing and can only reduce the number of breakpoints. Therefore, there are at most $n+1$ breakpoints on $\mathcal{S}_{L\to R}$, and the complexity of $\mathcal{S}_{L\to R}$ is equal to the complexity of $\mathcal{S}_L$. The complexity for $\mathcal{S}_{B\to R}$, $\mathcal{S}_{B\to T}$, and $\mathcal{S}_{L\to T}$ can be shown similar to Theorem 1.

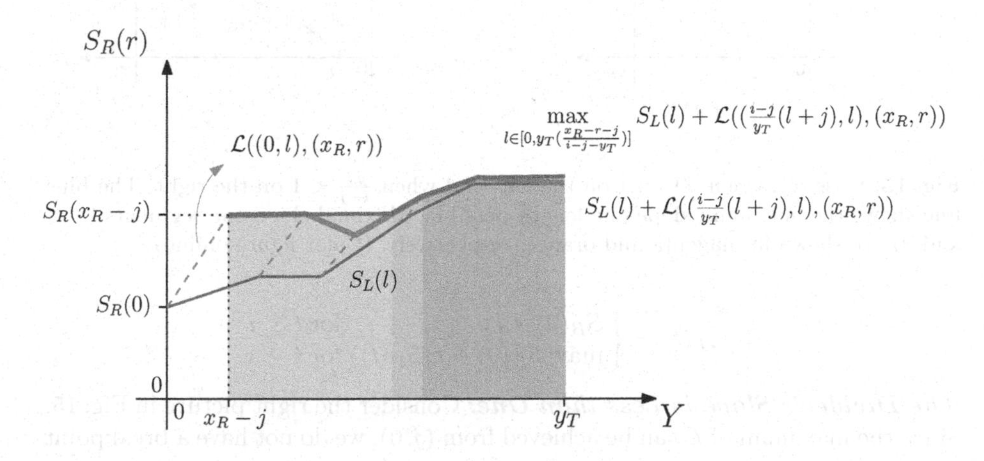

Fig. 14. The score function for $S_{L\to R}(r)$ when $r \geq x_R - j$.

Case (c): Case (c) represents a cell divided by a positive-slope line with free space on the left, therefore, the maximum length slope-one segment goes through (i, y_T), if the divider's slope is greater than one ($\frac{y_T}{i-j} > 1$). Otherwise the maximum segment intersects B at $(j, 0)$.

The Divider's Slope is Greater than One: Consider Fig. 15, we first explore the paths from L to R. We draw a slope-one segment starting at $(0,0)$ to find the score function's breakpoint. For any point below $(x_R, \frac{j}{\frac{y_T}{i-j}-1})$ the optimal path starts from $(0,0)$. For all points (x_R, r) above the breakpoint, the optimal path starts from $(0, (1-\frac{i-j}{y_T})r - j)$.

$$\mathcal{S}_{L\to R}(r) = \begin{cases} \mathcal{S}_L(0) + r & \text{for } r \le \frac{j}{\frac{y_T}{i-j}-1} \\ \mathcal{S}_L((1-\frac{i-j}{y_T})r - j) + \frac{i-j}{y_T}r + j & \text{for } r \ge \frac{j}{\frac{y_T}{i-j}-1} \end{cases}$$

Similarly for paths from B to R we have:

$$\mathcal{S}_{B\to R}(r) = \begin{cases} \mathcal{S}_B((\frac{i-j}{y_T}-1)r + j) + r & \text{for } r \le \frac{j}{\frac{y_T}{i-j}-1} \\ \mathcal{S}_B(0) + \frac{i-j}{y_T}r + j & \text{for } r \ge \frac{j}{\frac{y_T}{i-j}-1} \end{cases}$$

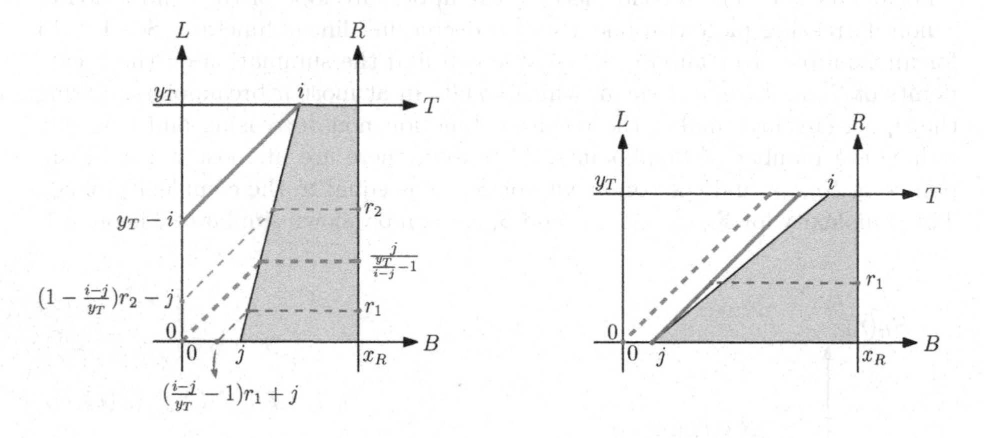

Fig. 15. Case (c), when $\frac{y_T}{i-j} > 1$ on the left, and when $\frac{y_T}{i-j} < 1$ on the right. The blue line shows the maximum slope-one length possible. Maximal slope-one segments to r_1 and r_2 are shown in magenta and orange, respectively. (Color figure online)

$$\mathcal{S}_{B\to T}(t) = \begin{cases} \mathcal{S}_B(0) + t & \text{for } t \le i \\ \max(\mathcal{S}_B(0) + t, \mathcal{S}_B(t)) & \text{for } t \ge i \end{cases}$$

The Divider's Slope is Less than One: Consider the right picture in Fig. 15. Since the maximum of $\mathcal{L}$ can be achieved from $(j,0)$, we do not have a breakpoint in the score functions:

$$\mathcal{S}_{L\to R}(r) = \mathcal{S}_L(0) + r \quad \text{and} \quad \mathcal{S}_{B\to R}(r) = \mathcal{S}_B(j) + r$$

However for $\mathcal{S}_{B\to T}(t)$ we have:

$$\mathcal{S}_{B\to T}(t) = \begin{cases} \mathcal{S}_B(0) + t & \text{for } t \le y_T \\ \mathcal{S}_B(t - y_T) + y_T & \text{for } y_T \le t \le j + y_T \\ \max(\mathcal{S}_B(t - y_T) + y_T, \mathcal{S}_B(t)) & \text{for } t \ge j + y_T \end{cases}$$

Theorem 5. *The propagation within Case (c) adds $O(1)$ complexity.*

Proof. The $O(1)$ complexity follows in a similar way to Theorem 1. Note that for $\mathcal{S}_{B\to T}$ the max function adds at most one breakpoint (Fig. 16).

Case (d):

$$\mathcal{S}_{L\to R}(r) = \begin{cases} \mathcal{S}_L(0) + r & \text{for } r \le \frac{j}{\frac{y_T}{i-j}-1} \\ \max\limits_{l\in[0,(1-\frac{i-j}{y_T})r-j]} \mathcal{S}_L(l) + \mathcal{L}((0,l),(\frac{i-j}{y_T}r+j,r)) & \text{for } r \ge \frac{j}{\frac{y_T}{i-j}-1} \end{cases}$$

$$\mathcal{S}_{B\to R}(r) = \begin{cases} \mathcal{S}_B((\frac{i-j}{y_T}-1)r+j) + r & \text{for } r \le \frac{j}{\frac{y_T}{i-j}-1} \\ \mathcal{S}_B(0) + \frac{j}{\frac{y_T}{i-j}-1} & \text{for } r \ge \frac{j}{\frac{y_T}{i-j}-1} \end{cases}$$

$$\mathcal{S}_{B\to T}(t) = \begin{cases} \mathcal{S}_B(0) + t & \text{for } t \le i \\ \max\limits_{b\in[0,t-(\frac{y_T}{i-j}(t-j))]} \mathcal{S}_B(b) + \mathcal{B}((b,0),(t,\frac{y_T}{i-j}(t-j))) & \text{for } t \ge i \end{cases}$$

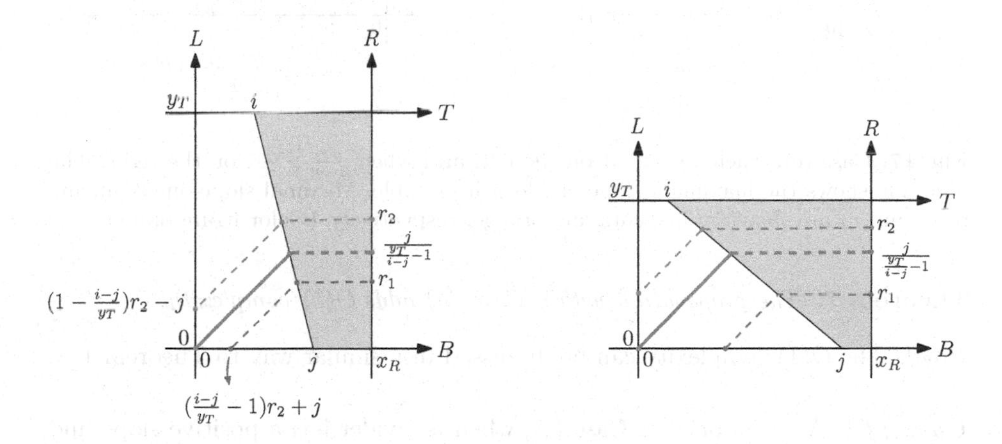

Fig. 16. Case (d), when $\frac{y_T}{i-j} < -1$ on the left, and when $\frac{y_T}{i-j} > -1$ on the right. The blue line shows the maximum slope-one length possible. Maximal slope-one segments to r_1 and r_2 are shown in magenta and orange, respectively. (Color figure online)

Theorem 6. *The propagation within Case (d) adds $O(1)$ complexity.*

Proof. The $O(1)$ complexity can be discussed in a similar way to Theorem 4 (Fig. 17)

Case (e):

$$\mathcal{S}_{L\to R}(r) = \begin{cases} \mathcal{S}_L(0) + r & \text{for } r \le x_R \\ \mathcal{S}_L(y_T(\frac{x_R-r-j}{i-j-y_T})) + r - y_T(\frac{x_R-r-j}{i-j-y_T}) & \text{for } r \ge x_R \end{cases}$$

$$\mathcal{S}_{B\to R}(r) = \begin{cases} \mathcal{S}_B(x_R - r) + r & \text{for } r \le x_R - j \\ \mathcal{S}_B\left(\frac{x_R - r - \frac{jy_T}{i-j}}{1-\frac{y_T}{i-j}}\right) + x_R - \frac{x_R - r - \frac{jy_T}{i-j}}{1-\frac{y_T}{i-j}} & \text{for } r \ge x_R - j \end{cases}$$

$$\mathcal{S}_{B\to T}(t) = \begin{cases} \mathcal{S}_B(t) & \text{for } r \le t \le i \\ \mathcal{S}_B\left(\frac{y_T}{i-j} + t - j\right) + t - \frac{y_T}{i-j} + t - j & \text{for } t \ge i \end{cases}$$

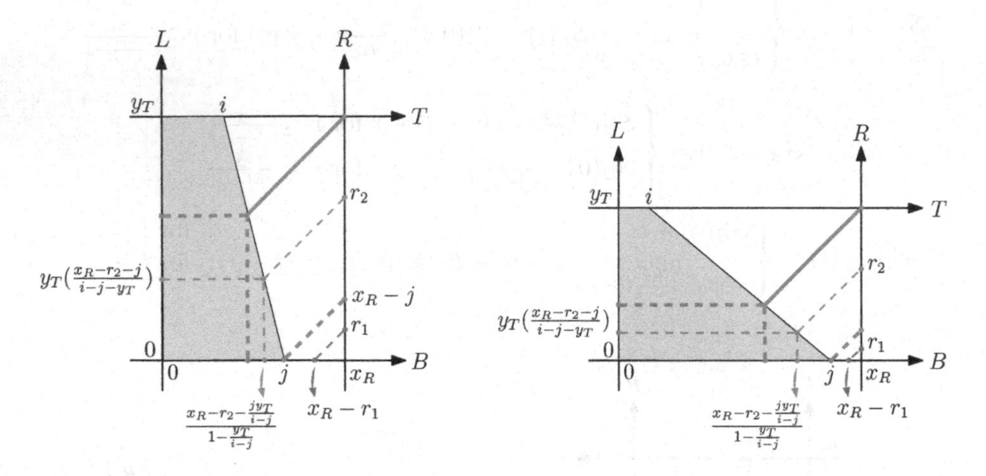

Fig. 17. Case (e), when $\frac{y_T}{i-j} < -1$ on the left, and when $\frac{y_T}{i-j} > -1$ on the right. The blue line shows the maximum slope-one length possible. Maximal slope-one segments to r_1 and r_2 are shown in magenta and orange, respectively. (Color figure online)

Theorem 7. *The propagation within Case (e) adds $O(1)$ complexity.*

Proof. The $O(1)$ complexity can be discussed in a similar way to Theorem 1

Case (f): As mentioned in Case (c), when a divider has a positive slope and is located on the right side of a free space, extra care is required. Suppose a divider's slope is less than one, a slope-one segment can be drawn above it, starting at $(k, 0)$. This segment can be a part of the optimal path from B to R. Therefore, we have four possibilities for this case:

Both Dividers' Slopes are Greater Than One: When both slopes are greater than one, the optimal path between two points does not necessarily have at most three segments. Consider Fig. 18 (right) and point (x_R, r_2'). The longest slope-one path to (x_R, r_2') is shown in orange, and has multiple slope-one segments. By Lemma 3, this path is always maximal when going through B. The same cannot be said for paths going through L in this case, since there are vertical segments in such a path. Therefore, for a point (x_R, r) above $(x_R, \frac{j+k}{1-\frac{w-k}{y_T}})$, $\mathcal{S}_{L\to R}(r)$ is the maximum over l of $\mathcal{S}_L(l)$ and the length of slope-one segments in the feasible space between l and r.

$$\mathcal{S}_{L\to R}(r) = \begin{cases} \mathcal{S}_L(0) + r & \text{for } r \le \frac{j+k}{1-\frac{w-k}{y_T}} \\ \max\limits_{l\in[0,\frac{(1-\frac{w-k}{y_T})r_2-k+j}{1-\frac{i-j}{y_T}}]} \mathcal{S}_L(l) + \mathcal{L}((\frac{(i-j)}{y_T}l + (i-j)j, l), (\frac{w-k}{y_T}r + k, r)) & \text{for } r \ge \frac{j+k}{1-\frac{w-k}{y_T}} \end{cases}$$

$$\mathcal{S}_{B\to R}(r) = \begin{cases} \mathcal{S}_B((\frac{i-j}{y_T} - 1)r + j) + r \text{ for } r \le \frac{j+k}{1-\frac{w-k}{y_T}} \\ \mathcal{S}_B(j) + \frac{w-k}{y_T}r + k - j \ \text{ for } r \ge \frac{j+k}{1-\frac{w-k}{y_T}} \end{cases}$$

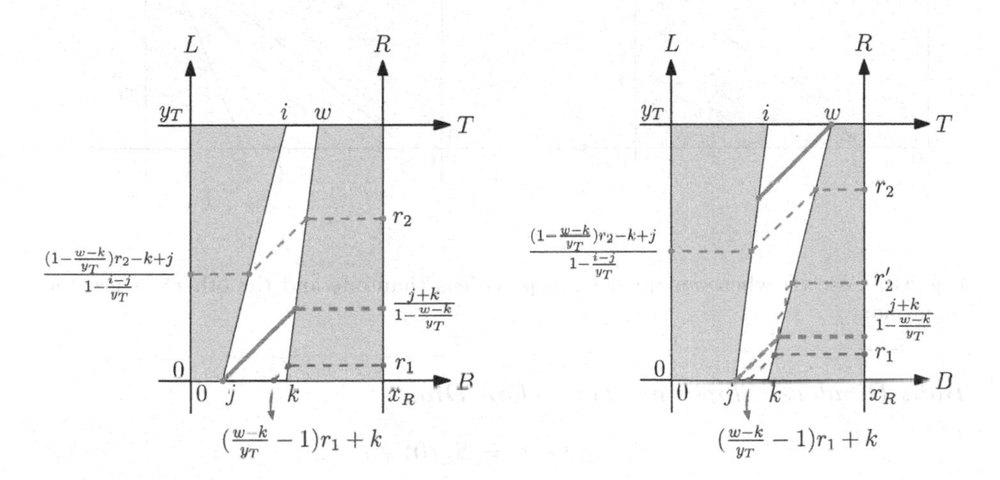

Fig. 18. Case (f) when both dividers have slope greater than one. The blue line shows the maximum slope-one length possible. Maximal slope-one segments to r_1, r_2 and r_2' are shown in magenta and orange. (Color figure online)

Theorem 8. *The propagation within Case (f) adds $O(1)$ complexity.*

The proof is omitted, as it is similar to Theorem 4 and Theorem 5.

Left Divider's Slope is Greater than One: Since a divider can only have slope less than one if $x_R > y_T$, the maximum length slope-one segment that can obtained in this case is y_T. This amount can be achieved by simply drawing a slope-one line from $(K, 0)$ (see Fig. 19) because the right divider's slope is less than one. We have:

$$\mathcal{S}_{L\to R}(r) = \mathcal{S}_L(0) + r$$
$$\mathcal{S}_{B\to R}(r) = \mathcal{S}_B(k) + r$$

Right divider's slope is greater than one: Similar to the previous sub-case, the maximum length slope-one segment is achievable. However, since the right divider's slope is greater than one, the maximum can be gained by drawing a slope-one line from (w, y_T) (Fig. 19).

$$\mathcal{S}_{L\to R}(r) = \mathcal{S}_L(0) + r$$

$$\mathcal{S}_{B\to R}(r) = \mathcal{S}_B((\frac{w-k}{y_T} - 1)r + k) + r$$

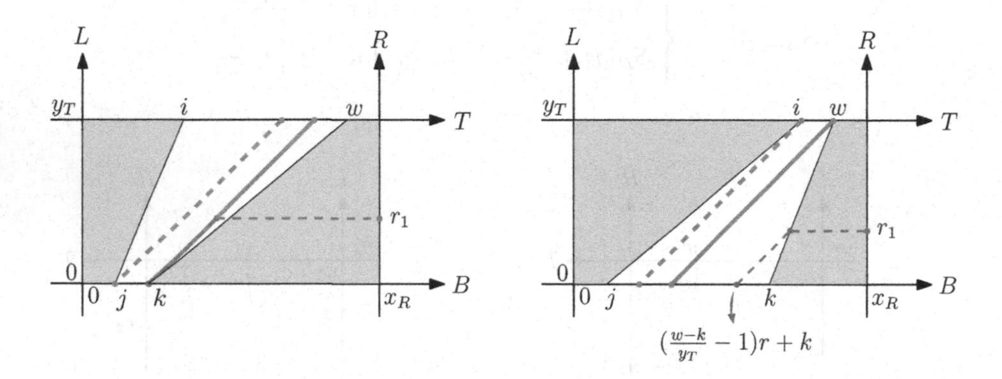

Fig. 19. Case (f) when one divider's slope is less than one and the other's is greater than one.

Both Dividers' Slopes are Less Than One:

$$\mathcal{S}_{L\to R}(x_R, r) = \mathcal{S}_L(0) + r$$

$$\mathcal{S}_{B\to R}(r) = \begin{cases} \mathcal{S}_B(k) + r & \text{for } r \leq \frac{(i-j)(j-k)}{(i-j)-y_T} \\ \max\limits_{b\in[k, \frac{(w-k)((1-\frac{i-j}{y_T})r + ky_T - j(i-j))}{y_T - w + k}]} \mathcal{S}_B(b) + \mathcal{L}((b, \frac{y_T}{w-k}b - ky_T), (\frac{i-j}{y_T}r + j(i-j), r)) & \text{for } r \geq \frac{(i-j)(j-k)}{(i-j)-y_T} \end{cases}$$

Case (g): Case (g) consists of two dividers with negative slope. Naturally, there are three possibilities for this case:

Two Parallel Dividers (II):

$$\mathcal{S}_{L\to R}(r) = \begin{cases} \mathcal{S}_L(0) + r & \text{for } r \leq \frac{j+k}{1-\frac{w-k}{y_T}} \\ \mathcal{S}_L(\frac{(1-\frac{w-k}{y_T})r - k + j}{1-\frac{i-j}{y_T}}) + \frac{(\frac{w-k-i+j}{y_T})r + k - j}{1-\frac{i-j}{y_T}} & \text{for } r \geq \frac{j+k}{1-\frac{w-k}{y_T}} \end{cases}$$

$$\mathcal{S}_{B\to R}(r) = \begin{cases} \mathcal{S}_B((\frac{w-k}{y_T})r + k) + r & \text{for } r \leq \frac{j+k}{1-\frac{w-k}{y_T}} \\ \mathcal{S}_B(j) + \frac{j+k}{1-\frac{w-k}{y_T}} & \text{for } r \geq \frac{j+k}{1-\frac{w-k}{y_T}} \end{cases}$$

Theorem 9. *The propagation within Case (g) adds $O(1)$ complexity.*

Proof. The $O(1)$ complexity for sub-case II follows in a way similar to Theorem 1. Likewise, for V and Λ we can use the same approach as in Theorem 4.

V-shaped and Λ-shaped cases share some functions and properties with the parallel case II. We discuss their corresponding functions below:

The Left Divider's Slope is Greater than the Right Divider's (V): The score function on R coming from L is identical to the parallel case (II), since $\mathcal{L}((0,l),(x_R,r))$ is non-decreasing over l and r (see Fig. 20). However, $\mathcal{L}((0,b),(x_R,r))$ is decreasing over b, so for $\mathcal{S}_{B\rightarrow R}(r)$ we have:

$$\mathcal{S}_{B\rightarrow R}(r)=\begin{cases}\mathcal{S}_B(j)+r & \text{for } r\leq\frac{j+k}{1-\frac{w-k}{y_T}}\\ \max\limits_{B\in[0,(\frac{w-k}{y_T}-1)r+k]}\mathcal{S}_B(b)+\mathcal{L}((b,\frac{y_T}{i-j}b-jy_T),(\frac{w-k}{y_T}r+k,r)) & \text{for } r\geq\frac{j+k}{1-\frac{w-k}{y_T}}\end{cases}$$

The Right Divider's Slope is Greater than the Left Divider's (Λ): The score function on R coming from B is identical to the parallel case (II) because $\mathcal{L}((0,b),(x_R,r))$ is non-decreasing over b.[4] However, that is not the case with $\mathcal{L}((0,l),(x_R,r))$:

$$\mathcal{S}_{L\rightarrow R}(r)=\begin{cases}\mathcal{S}_L(0)+r & \text{for } r\leq\frac{j+k}{1-\frac{w-k}{y_T}}\\ \max\limits_{l\in[0,\frac{(1-\frac{w-k}{y_T})r-k+j}{1-\frac{i-j}{y_T}}]}\mathcal{S}_L(l)+\mathcal{L}((\frac{(i-j)}{y_T}l+(i-j)j,l),(\frac{w-k}{y_T}r+k,r)) & \text{for } r\geq\frac{j+k}{1-\frac{w-k}{y_T}}\end{cases}$$

Case (h). Here, the cell is filled with infeasible space; therefore, for any pair of points p_1 and p_2 on the boundaries, $\mathcal{L}(p_1,p_2)=0$. Hence, going from L to R (or B to T) is simply: $\mathcal{S}_{L\rightarrow R}(r)=\mathcal{S}_L(r)$ (and $\mathcal{S}_{B\rightarrow T}(t)=\mathcal{S}_B(t)$). Since the score functions on the boundaries are non-decreasing (by Observation 3), we have: $\mathcal{S}_{B\rightarrow R}(r)=\mathcal{S}_B(x_R)$ (and $\mathcal{S}_{L\rightarrow T}(t)=\mathcal{S}_L(y_T)$).

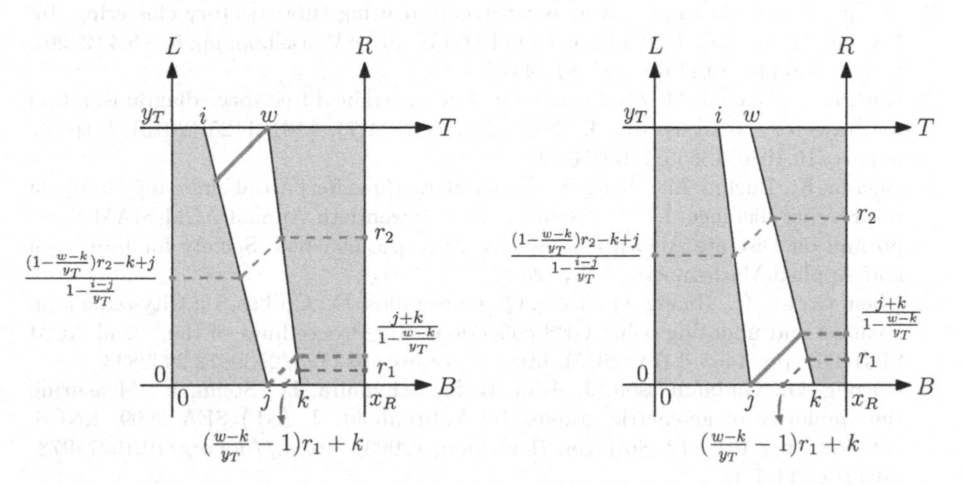

Fig. 20. Case (g) with V on the left and Λ on the right side. The blue line shows the maximum slope-one length possible. Maximal slope-one segments to r_1 and r_2 are shown in magenta and orange. (Color figure online)

[4] Note that in both sub-cases, if $\mathcal{S}_B(x_R)>\mathcal{S}_B(k)+\frac{j+k}{1-\frac{t-k}{y_T}}$, then $\mathcal{S}_R(r)=\mathcal{S}_B(x_R)$.

Theorem 10. *The propagation within Case (h) adds no complexity.*

Proof. Since $\mathcal{S}_{L\rightarrow R}(r) = \mathcal{S}_L(l)$ directly, there is no additional breaking point on $\mathcal{S}_{L\rightarrow R}$. Furthermore, $\mathcal{S}_{B\rightarrow R}$ is a constant function, which means it does not transfer any complexity from $\mathcal{S}_B$. We have similar explanations for $\mathcal{S}_{B\rightarrow T}$ and $\mathcal{S}_{L\rightarrow T}$.

References

1. Aguilar, J., Buchin, K., Buchin, M., Hosseini Sereshgi, E., Silveira, R.I., Wenk, C.: Graph sampling for map comparison. In: 3rd ACM SIGSPATIAL International Workshop on Spatial Gems (2021)
2. Ahmed, M., Fasy, B.T., Hickmann, K.S., Wenk, C.: Path-based distance for street map comparison. ACM Trans. Spat. Algorithms Syst. 28 (2015)
3. Ahmed, M., Fasy, B.T., Wenk, C.: Local persistent homology based distance between maps. In: Proceedings of the 22nd ACM SIGSPATIAL International Conference on Advances in Geographic Information Systems, pp. 43–52. ACM (2014)
4. Ahmed, M., Karagiorgou, S., Pfoser, D., Wenk, C.: Map Construction Algorithms. Springer, Heidelberg (2015). https://doi.org/10.1007/978-3-319-25166-0
5. Akitaya, H.A., Buchin, M., Kilgus, B., Sijben, S., Wenk, C.: Distance measures for embedded graphs. Comput. Geom.: Theory Appl. **95**, 101743 (2021)
6. Alt, H., Efrat, A., Rote, G., Wenk, C.: Matching planar maps. J. Algorithms **49**(2), 262–283 (2003). https://doi.org/10.1016/s0196-6774(03)00085-3
7. Alt, H., Godau, M.: Computing the Fréchet distance between two polygonal curves. IJCGA **5**(1–2), 75–91 (1995)
8. Biagioni, J., Eriksson, J.: Inferring road maps from global positioning system traces. Transp. Res. Rec.: J. Transp. Res. Board **2291**(1), 61–71 (2012)
9. Buchin, K., et al.: Improved map construction using subtrajectory clustering. In: Proceedings of the 4th ACM SIGSPATIAL LocalRec Workshop, pp. 5:1–5:4 (2020). https://doi.org/10.1145/3423334.3431451
10. Buchin, K., Buchin, M., Gudmundsson, J.: Constrained free space diagrams: a tool for trajectory analysis. Int. J. Geogr. Inf. Sci. **24**(7), 1101–1125 (2010). https://doi.org/10.1080/13658810903569598
11. Buchin, K., Buchin, M., Wang, Y.: Exact algorithms for partial curve matching via the Fréchet distance. In: Proceedings of the Twentieth Annual ACM-SIAM Symposium on Discrete Algorithms, SODA 2009, pp. 645–654. Society for Industrial and Applied Mathematics, USA (2009)
12. Chen, C., Lu, C., Huang, Q., Yang, Q., Gunopulos, D., Guibas, L.: City-scale map creation and updating using GPS collections. In: Proceedings of the/ 22nd ACM SIGKDD, pp. 1465–1474 (2016). https://doi.org/10.1145/2939672.2939833
13. Cheong, O., Gudmundsson, J., Kim, H.-S., Schymura, D., Stehn, F.: Measuring the similarity of geometric graphs. In: Vahrenhold, J. (ed.) SEA 2009. LNCS, vol. 5526, pp. 101–112. Springer, Heidelberg (2009). https://doi.org/10.1007/978-3-642-02011-7_11
14. De Carufel, J.L., Gheibi, A., Maheshwari, A., Sack, J.R., Scheffer, C.: Similarity of polygonal curves in the presence of outliers. Comput. Geom. **47**(5), 625–641 (2014). https://doi.org/10.1016/j.comgeo.2014.01.002
15. Duran, D., Sacristán, V., Silveira, R.I.: Map construction algorithms: a local evaluation through hiking data. GeoInformatica **24**(3), 633–681 (2020). https://doi.org/10.1007/s10707-019-00386-7

16. He, S., et al.: Roadrunner: improving the precision of road network inference from GPS trajectories. In: Proceedings of 26th ACM SIGSPATIAL GIS, pp. 3–12 (2018)
17. Itai, A., Papadimitriou, C.H., Szwarcfiter, J.L.: Hamilton paths in grid graphs. SIAM J. Comput. **11**(4), 676–686 (1982)
18. Koide, S., Xiao, C., Ishikawa, Y.: Fast subtrajectory similarity search in road networks under weighted edit distance constraints. Proc. VLDB Endow. **13**(12), 2188–2201 (2020). https://doi.org/10.14778/3407790.3407818 https://doi.org/10.14778/3407790.3407818
19. Maheshwari, A., Sack, J.R., Shahbaz, K., Zarrabi-Zadeh, H.: Fréchet distance with speed limits. Comput. Geom. **44**(2), 110–120 (2011). https://doi.org/10.1016/j.comgeo.2010.09.008. Special issue of selected papers from the 21st Annual Canadian Conference on Computational Geometry
20. Papadimitriou, C.H., Vazirani, U.V.: On two geometric problems related to the travelling salesman problem. J. Algorithms **5**(2), 231–246 (1984)

Hardness of Graph-Structured Algebraic and Symbolic Problems

Jingbang Chen[1(✉)], Yu Gao[2], Yufan Huang[3], Richard Peng[1], and Runze Wang[4]

[1] University of Waterloo, Waterloo, Canada
{j293chen,y5peng}@uwaterloo.ca
[2] Georgia Institute of Technology, Atlanta, USA
ygao380@gatech.edu
[3] Purdue University, West Lafayette, USA
huan1754@purdue.edu
[4] Carnegie Mellon University, Pittsburgh, USA
runzew@andrew.cmu.edu

Abstract. In this paper, we study the hardness of solving graph-structured linear systems with coefficients over a finite field $\mathbb{Z}_p$ and over a polynomial ring $\mathbb{F}[x_1, \ldots, x_t]$.

We reduce solving general linear systems in $\mathbb{Z}_p$ to solving unit-weight low-degree graph Laplacians over $\mathbb{Z}_p$ with a polylogarithmic overhead on the number of non-zeros. Given the hardness of solving general linear systems in $\mathbb{Z}_p$ [Casacuberta-Kyng 2022], this result shows that it is unlikely that we can generalize Laplacian solvers over $\mathbb{R}$, or finite-element based methods over $\mathbb{R}$ in general, to a finite-field setting. We also reduce solving general linear systems over $\mathbb{Z}_p$ to solving linear systems whose coefficient matrices are walk matrices (matrices with all ones on the diagonal) and normalized Laplacians (Laplacians that are also walk matrices) over $\mathbb{Z}_p$.

We often need to apply linear system solvers to random linear systems, in which case the worst case analysis above might be less relevant. For example, we often need to substitute variables in a symbolic matrix with random values. Here, a symbolic matrix is simply a matrix whose entries are in a polynomial ring $\mathbb{F}[x_1, \ldots, x_t]$. We formally define the reducibility between symbolic matrix classes, which are classified in terms of the degrees of the entries and the number of occurrences of the variables. We show that the determinant identity testing problem for symbolic matrices with polynomial degree 1 and variable multiplicity at most 3 is at least as hard as the same problem for general matrices over $\mathbb{R}$.

Keywords: Combinatorics and graph theory · Fine-grained complexity · Graph Laplacian · Algebraic computation

J. Chen, Y. Huang and R. Peng—Part of this work was done while at Georgia Institute of Technology.

R. Wang—Part of this work was done while at Swarthmore College.

P. Morin and S. Suri (Eds.): WADS 2023, LNCS 14079, pp. 232–246, 2023.
https://doi.org/10.1007/978-3-031-38906-1_16

1 Introduction

Algorithms for solving systems of linear equations have been long studied in mathematics, optimization, and more recently theoretical algorithms. Despite much attention on solvers for general systems, the current best bounds for them remain subquadratic. As a result, the pursuit of faster algorithms for solving linear systems, especially ones with subquadratic running times, increasingly focus on cases with additional structures.

Many, if not most, invocations of linear systems involve matrices with additional structures. There are two main approaches to leverage such structures for algorithmic efficiency. One is separator-based solvers such as nested dissection [9] and its generalizations [1,19]. Alon and Yuster [1] proved that if $\boldsymbol{A}$ is well-separable, i.e., the underlying graph of $\boldsymbol{A}$ is planar or avoids a fixed minor, then $\boldsymbol{Ax} = \boldsymbol{b}$ can be solved over an *arbitrary* field in subquadratic time [1]. Another approach is the finite-element based solvers where the structured linear system solvers utilize the combinatorial and algebraic structures of the coefficients to accelerate computation. An example is graph Laplacians and symmetric diagonally-dominant (SDD) matrices. After Spielman and Teng proposed the first near-linear time Laplacian solver [27], there has been a long line of research on further improving the runtime for solving Laplacian systems [6,12–14], leading to the current fastest running time of $O(m(\log\log n)^{O(1)}\log(1/\epsilon))$ [10] for Laplacians. Many generalizations of Laplacian linear systems also admit near-linear time or subquadratic solvers. [4,5,15,16].

A major limitation of these finite-element based methods is many of their building blocks only work over reals. This is acceptable, or even preferrable, for many of the original motivating applications such as optimization [18] and learning [31]. On the other hand, applications from cryptography [21], topology [33], as well as algorithms for producing exact fractional solutions [8,28] require solving systems over finite fields as intermediate steps. Such reliance on finite field based primitives is also the case in recent algorithms for solving general sparse systems over reals [20,22]. Most of these finite-field based primitives have running time more than quadratic, significantly more than the nearly-linear running times of graph-structured matrices over reals. As a result, it's natural to investigate whether ideas developed for nearly-linear time algorithms for graph-structured matrices can also provide gains in finite field settings, even in special cases of graph-structured matrices over finite fields.

In this paper, we show that generalizing graph-based methods over $\mathbb{R}$ to a finite-field setting is unlikely. Specifically, we provide hardness results showing that solving the following classes of linear systems over $\mathbb{Z}_p$ is as hard as solving general linear systems over $\mathbb{Z}_p$: *a)* Laplacians over $\mathbb{Z}_p$, *b)* unit-weight Laplacians over $\mathbb{Z}_p$, *c)* unit-weight Laplacians over $\mathbb{Z}_p$ with degrees at most $O(\log p)$, *d)* walk matrices over $\mathbb{Z}_p$, and *e)* normalized Laplacians over $\mathbb{Z}_p$.

Outline. In Sect. 2, we provide an overview of concepts required to state our main results. In Sect. 3, we state our hardness results and main technical lemmas.

In Sect. 4, we prove the hardness of unit-weight Laplacians systems and low-degree unit-weight Laplacian systems. We also show how it is possible to control magnitudes of diagonal entries of a Laplacian and make sure each row or column has $O(1)$ number of non-zeroes.

Due to space constraints, additional background and proofs can be found at [2]. Section 2 of [2] surveys current literature on solving linear systems in general and solving graph structured linear systems. Section 3 of [2] formally defines the notations and concepts relevant to the paper. Section 5 of [2] defines the Schur complements (SC) and proves lemmas about gadgets that is central to our reductions. Section 6 of [2] contains the proofs of the theorems on Laplacian system hardness. Section 7 of [2] contains the proofs about the hardness of solving walk matrix systems over $\mathbb{Z}_p$. Finally, Section 8 of [2] deals with symbolic graph-structured linear systems, which we find to be a potential field to work on about the hardness of structured linear systems in future.

2 Preliminaries

To formalize our reductions, we borrow the concepts of matrix classes and efficient f-reducibility from [17]. Additional background knowledge about notations and algebra can be found in Section 3 of [2].

A matrix class is an infinite set of matrices with a common structure. The class of general matrices over $\mathbb{Z}_p$ is denoted as $\mathcal{GS}_{\mathbb{Z}_p}$. Similarly, $\mathcal{GS}_{\mathbb{R}}$ denotes the class of matrices over $\mathbb{R}$. For formal definitions of the matrix classes used, see Section 3.2 of [2]. On a high level, our definition of Laplacian matrices over $\mathbb{Z}_p$ (Definition 5, [2]) simply means edge weights are now in $\mathbb{Z}_p - \{0\}$ instead of $\mathbb{R}^+$, and our definition of unit-weight Laplacian (Definition 7, [2]) means the off-diagonal entries are either -1 or 0. Unit-weight Laplacians are thus Laplacians for unit-weight (or unweighted) graphs. Our definition of combinatorial degree corresponds to the degree of the underlying graph of the given Laplacian. Finally, we refer to matrices of form $\boldsymbol{I}-\boldsymbol{A}$ where $\boldsymbol{A}$ is symmetric and has 0 on the diagonal as walk matrices, which has applications in graph *st*-connectivity [3].

Now we may formalize the notion of reducing from solving a linear system of given matrix class to solving a linear system of another matrix class.

Definition 1 (Solving Linear System over $\mathbb{Z}_p$, SLS). *Given a linear system* $(\boldsymbol{A}, \boldsymbol{b})$ *where* $\boldsymbol{A} \in \mathbb{Z}_p^{m\times n}, \boldsymbol{b} \in \mathbb{Z}_p^m$, *we define the SLS problem as finding a solution* $\boldsymbol{x}^* \in \mathbb{Z}_p^n$ *such that* $\boldsymbol{A}\boldsymbol{x}^* = \boldsymbol{b} \pmod p$.

In this paper, we care about exact solutions of linear systems over $\mathbb{Z}_p$, of which the running time mainly depends on the number of non-zero entries in $\boldsymbol{A}$, $\mathrm{nnz}(\boldsymbol{A})$. Since we are working in $\mathbb{Z}_p$, the bit complexity of arithmetic operations is less a concern. To measure the difficulty of solving a linear system over $\mathbb{Z}_p$, for any linear system $(\boldsymbol{A} \in \mathbb{Z}_p^{m\times n}, \boldsymbol{b} \in \mathbb{Z}_p^m)$, we denote the sparse complexity of it as

$$\mathcal{S}(\boldsymbol{A}, \boldsymbol{b}) = \mathrm{nnz}(\boldsymbol{A}). \tag{1}$$

Definition 2 (Efficient f-reducibility). *Suppose we have two matrix classes $\mathcal{M}^1$ and $\mathcal{M}^2$ and two algorithms $\mathcal{A}_{1\mapsto 2}$, $\mathcal{A}_{2\mapsto 1}$ such that given an SLS instance $(\boldsymbol{M}^1, \boldsymbol{c}^1)$ where $\boldsymbol{M}^1 \in \mathcal{M}^1$, the algorithm $\mathcal{A}_{1\mapsto 2}$ returns an SLS instance $(\boldsymbol{M}^2, \boldsymbol{c}^2) = \mathcal{A}_{1\mapsto 2}(\boldsymbol{M}^1, \boldsymbol{c}^1)$ such that $\boldsymbol{M}^2 \in \mathcal{M}^2$, and if $\boldsymbol{x}^2$ is a solution to the SLS instance $(\boldsymbol{M}^2, \boldsymbol{c}^2)$ then $\boldsymbol{x}^1 = \mathcal{A}_{2\mapsto 1}(\boldsymbol{M}^2, \boldsymbol{c}^2, \boldsymbol{x}^2)$ is a solution the SLS instance $(\boldsymbol{M}^1, \boldsymbol{c}^1)$.*

Consider a non-decreasing function $f : \mathbb{R}_+ \mapsto \mathbb{R}_+$. If we always have

$$\mathcal{S}(\boldsymbol{M}^2, \boldsymbol{c}^2) \leq f(\mathcal{S}(\boldsymbol{M}^1, \boldsymbol{c}^1)) \tag{2}$$

and the running times of $\mathcal{A}_{1\mapsto 2}$ and $\mathcal{A}_{2\mapsto 1}$ are both bounded by $\widetilde{O}(\text{nnz}(\boldsymbol{M}^1))$, we say $\mathcal{M}^1$ is efficiently f-reducible to $\mathcal{M}^2$, which we also writes

$$\mathcal{M}^1 \leq_f \mathcal{M}^2. \tag{3}$$

A weaker form of solving linear system is deciding whether a given linear system has a solution. The decision problem of linear system solving as well as decisional f-reducibility is treated in Section 3.3 and 3.4 of [2]. Our Laplacian and walk matrix reductions both work for the decisional f-reducibility and efficient f-reducibility.

3 Our Results

We now state and discuss our main results.

3.1 Hardness Result

Hardness of Laplacian Systems Over $\mathbb{Z}_p$. Our generalization of Laplacian to $\mathbb{Z}_p$ means a Laplacian over $\mathbb{Z}_p$ is simply a symmetric matrix with $\mathbf{1}$ in its kernel (see Definition 5, [2]). This makes Laplacians over $\mathbb{Z}_p$ more expressive than Laplacians over the reals. The theorem below says that for any prime p, solving an arbitrary linear system $\boldsymbol{A}$ over $\mathbb{Z}_p$ is as hard as solving a Laplacian linear system $\boldsymbol{L}$ where $\text{nnz}(\boldsymbol{L}) = O(\text{nnz}(\boldsymbol{A}))$. The proof can be strengthened to "over an arbitrary field, general linear systems are as hard as linear systems whose coefficient matrices are symmetric matrices with $\mathbf{1}$ in the kernels." The proof is in Section 6.1 of [2].

Theorem 1 (Hardness for Laplacian systems in $\mathbb{Z}_p$). *$\mathcal{GS}_{\mathbb{Z}_p} \leq_f \mathcal{LS}_{\mathbb{Z}_p}$ where $f(\text{nnz}) = O(\text{nnz})$.*

Hardness of Walk Matrix Systems Over $\mathbb{Z}_p$ and Extended Field. As an application of our hardness results for Laplacian systems over $\mathbb{Z}_p$, we show that solving a linear system whose coefficient matrix is a walk matrix (Definition 13, [2]) is as hard as solving general linear systems. We can use a similar construction as the reduction from general linear systems to Laplacian systems over $\mathbb{Z}_p$. This reduction involves constructing a 8×8 block matrix. The proof can be found in Section 7.1 of [2].

Theorem 2 (Hardness for Walk Matrix systems in $\mathbb{Z}_p$). *$\mathcal{GS}_{\mathbb{Z}_p} \leq_f \mathcal{WS}_{\mathbb{Z}_p}$ where $f(\text{nnz}) = O(\text{nnz})$.*

We can strengthen our hardness results on Laplacians and walk matrices by reducing arbitrary linear systems to normalized Laplacian systems (see Definition 15 of [2]). To further reduce a Laplacian $\boldsymbol{L} = \boldsymbol{D} - \boldsymbol{A}$ to a normalized Laplacian (and thus a walk matrix), we can simply normalize the Laplacian by multiplying $\boldsymbol{D}^{-1/2}$ on the left and on the right of $\boldsymbol{L}$. This normalization does not work directly over finite fields, since not all elements are quadratic residues and having a 0 on the diagonal does not imply the entire row is zero. Nevertheless, we are able to control the diagonal entries in the reduced Laplacian matrix so that we don't create zeros on the diagonal unless the entire row is zero using tools in Sect. 4.2. We can also work in a field extension so that all elements in $\mathbb{Z}_p$ can have a square root in the extension field. Finally, we can show it is easy to convert a solution in the extension to a solution in $\mathbb{Z}_p$ simply by keeping only the part in $\mathbb{Z}_p$. The proof can be found in Section 7.2 of [2].

Theorem 3 (Hardness for Normalized Laplacian systems in $\mathbb{Z}_p[\sqrt{t}]$). *If t is the primitive root of p, $\mathcal{LS}_{\mathbb{Z}_p} \leq_f \mathcal{NLS}_{\mathbb{Z}_p[\sqrt{t}]}$ for $f(\text{nnz}) = O(\text{nnz})$.*

Hardness of Unit-Weight Laplacian Systems Over $\mathbb{Z}_p$. Our definition of weighted Laplacian does not permit us to re-interpret a Laplacian in $\mathbb{Z}_p$ as a Laplacian in $\mathbb{R}$ simply by treating entries of $\mathbb{Z}_p$ as entries of $\mathbb{R}$. We are going to show that the Laplacians corresponding to simple, unweighted, undirected graphs are as hard as general linear systems over $\mathbb{Z}_p$. This is in contrast to the situation over $\mathbb{R}$, since we can solve these Laplacian systems efficiently.

One important tool of our reductions is the *Schur complement* (SC). In Lemma 5.6 of [2], we will show that Schur complement allows us to replace any edge in a graph (or in other words, a non-diagonal entry in our Laplacian $\boldsymbol{L}$) by a constructed gadget circuit, or vice versa, while preserving its effective resistance. This process will only slightly blow up the dimensions of the matrices.

We might treat a weighted edge of weight w_e as w_e parallel unit-weight edges. Lemma 5.6 of [2] can also be applied to eliminate parallel unit-weight edges. We can replace each edge with a circuit of weight 1 such that after replacement the parallel edges are no longer parallel, which means the resulting Laplacian have off-diagonal entries -1 or 0. This can be achieved by replacing an edge with two length-2 paths connected in parallel. In other words, we are replacing each of the parallel edges with a cycle consisting of four edges.

Using these tools, we show regardless of the original magnitudes of non-zero entries in the matrix, we can reduce solving a (sparse) general linear system over $\mathbb{Z}_p$ to solving a unit-weight Laplacian system with a blow-up factor of $O(\log^2 p/\log\log p)$ on the number of non-zero entries. This implies that solving unit-weight Laplacians are as hard as solving general linear systems over $\mathbb{Z}_p$, up to poly-logarithmic overhead in runtime.

Theorem 4 (Hardness for unit-weight Laplacian systems in $\mathbb{Z}_p$). *$\mathcal{LS}_{\mathbb{Z}_p} \leq_f \mathcal{UWLS}_{\mathbb{Z}_p}$ where $f(\text{nnz}) = O(\text{nnz} \cdot \log^2 p/\log\log p)$.*

Hardness of Low-Degree Unit-Weight Laplacian Systems Over $\mathbb{Z}_p$. Our results on the hardness of unit-weight Laplacians imply that given an arbitrary linear system $\boldsymbol{Ax} = \boldsymbol{b}$ over $\mathbb{Z}_p$, we can reduce it to a Laplacian system $\boldsymbol{Ly} = \boldsymbol{c}$ such that $\boldsymbol{L}$ is the Laplacian of an unweighted sparse graph. The reduction does not guarantee that the degrees of individual vertices in the graph corresponding to $\boldsymbol{L}$ are small. For example, this does not preclude we reduce $\boldsymbol{Ax} = \boldsymbol{b}$ to a unit-weight Laplacian $\boldsymbol{Ly} = \boldsymbol{c}$ where $\boldsymbol{L}$ corresponds to a star.

Our results on low-degree unit-weight Laplacians show that Laplacian systems corresponding to unit-weight graphs with maximum degree $O(\log p)$ are complete for all linear systems over $\mathbb{Z}_p$.

Theorem 5 (Hardness for Low-Degree Unit-weight Laplacian systems in $\mathbb{Z}_p$). $\mathcal{LS}_{\mathbb{Z}_p} \leq_f \mathcal{DLS}_{\log p, \mathbb{Z}_p}$ *where* $f(\mathrm{nnz}) = O(\mathrm{nnz} \cdot \log \mathrm{nnz} \cdot \log^2 p / \log \log p)$.

These results indicate that if one can develop efficient solvers of these special structured linear systems, all linear systems over $\mathbb{Z}_p$ can be solved equally efficiently. Moreover, if a fast solver exists, it might imply a non-trivially fast algorithm for solving linear systems exactly in $\mathbb{Q}$ since we may apply Dixon's scheme [8] to reconstruct rational solutions. This remains an interesting open problem.

3.2 Useful Tools

Construction of Arbitrary Resistances Modulo p. The key component to the reduction from Laplacian linear systems to unit-weight Laplacian is to construct circuits of arbitrary resistance modulo p. With the help of Lemma 5.6 of [2], we can then replace off-diagonal entries (or in other words, an edge in the graph corresponding to the Laplacian) with the corresponding circuit. Our results prove that we need only $O\left(\log^2 p / \log \log p\right)$ unit-weight edges to represent all resistance values $0, \ldots, p-1 \pmod{p}$. Furthermore, the maximum degree of the circuit is $O(\log p)$. We can construct the desired circuit in $O(\log^3 p)$ time. Notice this reduction only works for $\mathbb{Z}_p$. The proof is in Section 5.3 of [2].

Lemma 1 (Small Unit Circuits of Arbitrary Resistance over $\mathbb{Z}_p$). *For any prime* $p > 2$ *and any integer* r *in* $[1, p-1]$, *we can construct in* $O(\log^3 p)$ *time a network* $\boldsymbol{L}$ *of unit resistors (aka. an undirected unweighted graph) with two vertices* s *and* t *such that the effective resistance between* s *and* t *is* $r \pmod{p}$ *and that*

- *the number of non-zero entries of the circuit,* $\mathrm{nnz}(\boldsymbol{L})$, *is* $O\left(\log^2 p / \log \log p\right)$,
- *the maximum degree of the circuit,* $\Delta_{\boldsymbol{L}}^{w} = \Delta_{\boldsymbol{L}}$, *is* $O(\log p)$.

Combinatorial Degree Decrease of Laplacians Over $\mathbb{Z}_p$. The construction of arbitrary resistances modulo p proves we can replace undesired off-diagonal entries with -1. The following lemma proves we can also control the diagonal elements of the resulting Laplacian. In particular, we can guarantee that given a Laplacian linear system, we can reduce the Laplacian to another Laplacian where each row or column contains only $O(1)$ non-zero entries. Specifically, we achieve this by transforming it to a sparser graph. Our result shows that this only increases the number of non-zero entries by a factor of $O(\log \mathrm{nnz})$.

Lemma 2 ($O(1)$-Combinatorial-Degree Laplacian Construction over $\mathbb{Z}_p$). *For a Laplacian matrix $\boldsymbol{L} \in \mathbb{Z}_p^{n\times n}$ with an underlying graph $G = (V, E)$, we can construct a Laplacian matrix $\hat{\boldsymbol{L}} \in \mathbb{Z}_p^{k\times k}$ with an underlying graph $H = (V', E')$ such that $V \subseteq V'$ and $\mathbf{SC}(\hat{\boldsymbol{L}}, V) = \boldsymbol{L}$. We have $\Delta(H) = O(1)$ and $\mathrm{nnz}(\hat{\boldsymbol{L}}) = \Theta(|V'|) = \Theta(|E'|) = O(|E| \log |E|) = O(\mathrm{nnz}(\boldsymbol{L}) \log \mathrm{nnz}(\boldsymbol{L}))$.*

The above two tools directly help constructing the low-degree unit-weight Laplacian reduction. One can see that if we first reduce a Laplacian system to another Laplacian system whose combinatorial degree (or in other words, number of non-zero elements per row or column) is $O(1)$ by Lemma 2 and then replace every edge by the unit-weight gadget corresponding to its weight from Lemma 1, we will have an unit-weight Laplacian of degree $O(\log p)$.

3.3 Hardness of Symbolic Matrices

We now consider the decisional problem of linear system solving for $\boldsymbol{Ax} = \boldsymbol{b}$ where $\boldsymbol{A}, \boldsymbol{b}$ have entries in the polynomial ring $\mathbb{F}[x_1, \ldots, x_k]$ and we want to decide if there exists some $\boldsymbol{x} \in (F(x_1, \ldots, x_k))^n$ such that the linear system holds. A closely related problem is the determinantal polynomial identity testing (PIT) problem: deciding if $\det \boldsymbol{A} = 0$ where $\boldsymbol{A}$ has entries in a polynomial ring $\mathbb{F}[x_1, \ldots, x_k]$. The determinantal PIT problem, as well as PIT in general, has many direct connections with graph matching [29] and beyond. Saxena's surveys [23,24] and Chap. 4 of Shpilka and Yehudayoff's survey on arithmetic circuit complexity [26] are excellent sources for understanding the background and progresses about the PIT problem. While these problems are hard to solve deterministically [11,30], the Schwartz-Zippel lemma [7,25,32] justifies we can simply plug in random values to the variables. This implies that symbolic matrices enable us to study how linear systems behave when the coefficients are random scalars.

Edmonds matrices and Tutte matrices [29] are the two most important and relevant examples of graph-structured symbolic matrices. If each variable appears only once in symbolic matrix $\boldsymbol{A}$, then we may simply substitute entries of $\boldsymbol{A}$ with independent variables, resulting in an Edmonds matrix. If variables appear at most twice, then checking if $\det(\boldsymbol{A}) = 0$ is at least as hard as general graph matching. What if we allow variables to appear at least three times in $\boldsymbol{A}$? In this case, we can show checking if $\det(\boldsymbol{A}) = 0$ is at least as hard as checking if $\det(\boldsymbol{B}) = 0$ for a matrix $\boldsymbol{B} \in \mathcal{GS}_{\mathbb{R}}$:

Theorem 6 (Hardness of A Special Class of Symbolic Matrices). *Let $f(\mathrm{nnz}) = O(\mathrm{nnz})$, then ${\mathcal{GS}_{\mathbb{R}}}^{n\times n}$ is decisional f-reducible to a symbolic matrix class $\mathcal{SM}[\mathbb{R}[x_1, \ldots, x_{n^2}], 1, 3]$.*

4 Hardness of Solving Laplacian System over $\mathbb{Z}_p$

In this section, we present the proofs about the hardness of Laplacians over $\mathbb{Z}_p$. Some more elementary or technical proofs is provided in the full version of the paper: the hardness result for general Laplacians over $\mathbb{Z}_p$ is in Section 6.1 of [2].

4.1 Reduction to Unit-Weight Laplacian System

In this section, we show that the class of Laplacian matrices $\mathcal{LS}_{\mathbb{Z}_p}$ is efficiently f-reducible to the class of unit-weight Laplacians for some f that does not depend on the original size of the matrix. Specifically, the reduction is by replacing every edge in a Laplacian system in $\mathcal{LS}_{\mathbb{Z}_p}$ by a gadget with $O\left(\log^2 p/\log\log p\right)$ unit-weight edges.

Theorem 4 (Hardness for unit-weight Laplacian systems in $\mathbb{Z}_p$). $\mathcal{LS}_{\mathbb{Z}_p} \leq_f \mathcal{UWLS}_{\mathbb{Z}_p}$ *where* $f(\text{nnz}) = O(\text{nnz} \cdot \log^2 p/\log\log p)$.

Proof. We first use the circuit construction from Lemma 1 to show that given any Laplacian system $\boldsymbol{L} \in \mathbb{Z}_p^{n\times n}$, we can compute a unit-weight Laplacian system $\overline{\boldsymbol{L}} \in \mathbb{Z}_p^{k\times k}$ such that

1. $\text{nnz}(\overline{\boldsymbol{L}}) \leq O(\text{nnz}(\boldsymbol{L})\log^2 p/\log\log p)$.
2. For any linear system $\boldsymbol{L}\boldsymbol{x} = \boldsymbol{b}$, $\{\boldsymbol{x} \mid \boldsymbol{L}\boldsymbol{x} = \boldsymbol{b}\} = \left\{\boldsymbol{y}_{[n]} \mid \overline{\boldsymbol{L}}\boldsymbol{y} = \begin{bmatrix} \boldsymbol{b} \\ \boldsymbol{0} \end{bmatrix}\right\}$.

In addition, our construction also guarantees that

$$\text{maxdeg}(\overline{\boldsymbol{L}}) \leq O(\log p)\text{maxdeg}(\boldsymbol{L}). \tag{4}$$

Then we will let $\mathcal{A}_{1\mapsto 2}(\boldsymbol{L}, \boldsymbol{b}) = \left(\overline{\boldsymbol{L}}, \begin{bmatrix} \boldsymbol{b} \\ \boldsymbol{0} \end{bmatrix}\right)$, $\mathcal{A}_{2\mapsto 1}(\boldsymbol{Y}) = \{\boldsymbol{y}_{[n]} | y \in \boldsymbol{Y}\}$ and prove their running times.

Construction of $\overline{\boldsymbol{L}}$. We apply Lemma 5.6 of [2] repeatedly. We let $\boldsymbol{L}^{(0)}$ be $\boldsymbol{L}$. We fix n as the number of rows in $\boldsymbol{L}^{(0)}$. $\boldsymbol{L}^{(k)}$ is defined recursively:

1. If $\boldsymbol{L}_{i,j}^{(k-1)}$ is not 0 or $p-1$ for some $1 \leq i < j \leq n$ satisfying $\forall 1 \leq \ell < k, (i,j) \neq (i(\ell), j(\ell))$, apply Lemma 5.6 of [2] for $\boldsymbol{L} = \boldsymbol{L}^{(k-1)}$, $i_0 = i, j_0 = j$. Let Lemma 1 constructs the circuit $\boldsymbol{R}$ in Lemma 5.6 of [2] with size $O\left(\log^2 p/\log\log p\right)$ and resistance $-\boldsymbol{L}_{i,j}^{(k-1)}$. Let the resulting $\boldsymbol{U}$ be $\boldsymbol{L}^{(k)}$. We record the indices i and j as $i(k)$ and $j(k)$.
2. Otherwise, $\boldsymbol{L}^{(k)}$ is undefined.

Let $\overline{\boldsymbol{L}}$ be $\boldsymbol{L}^{(\overline{k})}$ where $\overline{k}$ is the largest k such that $\boldsymbol{L}^{(k)}$ is defined. $\{(i(1), j(1)), \ldots, (i(\overline{k}), j(\overline{k}))\}$ goes through every entry of $\boldsymbol{L}$ which is strictly above the diagonal and is not equal to 0 or $p-1$ exactly once. Thus, $\overline{k} \leq \text{nnz}(\boldsymbol{L})$.

By Item 1 of Lemma 5.6 of [2], we also have $\mathrm{nnz}(\boldsymbol{L}^{(i)}) \leq \mathrm{nnz}(\boldsymbol{L}^{(i-1)}) + O\left(\log^2 p/\log\log p\right)$ for $1 \leq i \leq \bar{k}$. Combining the two inequalities, we get

$$\mathrm{nnz}(\overline{\boldsymbol{L}}) \leq \mathrm{nnz}(\boldsymbol{L}) + \mathrm{nnz}(\boldsymbol{L})O\left(\log^2 p/\log\log p\right) = O\left(\mathrm{nnz}(\boldsymbol{L})\log^2 p/\log\log p\right). \tag{5}$$

Since each edge is either preserved or replaced by a circuit with maximum degree $O(\log p)$ constructed by Lemma 1, the maximum degree $\mathrm{maxdeg}(\overline{\boldsymbol{L}})$ is no more than $O(\log p)\mathrm{maxdeg}(\boldsymbol{L})$. This property is vital in the reduction to Low-Degree Laplacian System described in Sect. 4.3.

Next we prove that for any linear system $\boldsymbol{L}\boldsymbol{x} = \boldsymbol{b}$,

$$\{\boldsymbol{x} \mid \boldsymbol{L}\boldsymbol{x} = \boldsymbol{b}\} = \left\{\boldsymbol{y}_{[n]} \mid \overline{\boldsymbol{L}}\boldsymbol{y} = \begin{bmatrix}\boldsymbol{b}\\ \boldsymbol{0}\end{bmatrix}\right\} \tag{6}$$

by induction. Fix any vector $\boldsymbol{b}$. Suppose $\{\boldsymbol{x} \mid \boldsymbol{L}\boldsymbol{x} = \boldsymbol{b}\} = \left\{\boldsymbol{y}_{[n]} \mid \boldsymbol{L}^{(k)}\boldsymbol{y} = \begin{bmatrix}\boldsymbol{b}\\ \boldsymbol{0}\end{bmatrix}\right\}$.

By Lemma 5.6 of [2], $\left\{\boldsymbol{y} \mid \boldsymbol{L}^{(k)}\boldsymbol{y} = \begin{bmatrix}\boldsymbol{b}\\ \boldsymbol{0}\end{bmatrix}\right\} = \left\{\boldsymbol{z}_{[\|\boldsymbol{y}\|_0]} \mid \boldsymbol{L}^{(k+1)}\boldsymbol{z} = \begin{bmatrix}\boldsymbol{b}\\ \boldsymbol{0}\end{bmatrix}\right\}$ where $\|\boldsymbol{y}\|_0$ is the number of coordinates of $\boldsymbol{y}$. Thus,

$$\{\boldsymbol{x} \mid \boldsymbol{L}\boldsymbol{x} = \boldsymbol{b}\} = \left\{\boldsymbol{y}_{[n]} \mid \boldsymbol{L}^{(k)}\boldsymbol{y} = \begin{bmatrix}\boldsymbol{b}\\ \boldsymbol{0}\end{bmatrix}\right\} = \left\{\boldsymbol{z}_{[n]} \mid \boldsymbol{L}^{(k+1)}\boldsymbol{z} = \begin{bmatrix}\boldsymbol{b}\\ \boldsymbol{0}\end{bmatrix}\right\}. \tag{7}$$

By induction, we have

$$\begin{aligned}\{\boldsymbol{x} \mid \boldsymbol{L}\boldsymbol{x} = \boldsymbol{b}\} &= \left\{\boldsymbol{y}_{[n]} \mid \boldsymbol{L}^{(\bar{k})}\boldsymbol{y} = \begin{bmatrix}\boldsymbol{b}\\ \boldsymbol{0}\end{bmatrix}\right\} \\ &= \left\{\boldsymbol{y}_{[n]} \mid \overline{\boldsymbol{L}}\boldsymbol{y} = \begin{bmatrix}\boldsymbol{b}\\ \boldsymbol{0}\end{bmatrix}\right\}.\end{aligned} \tag{8}$$

Running Time of $\mathcal{A}_{1\mapsto 2}$ *and* $\mathcal{A}_{2\mapsto 1}$. $\mathcal{A}_{1\mapsto 2}$ simply replaces each non-diagonal non-zero entry by a matrix $\boldsymbol{U}$ with $\mathrm{nnz}(\boldsymbol{U}) = O(\log^2 p/\log\log p)$. This costs $O(\mathrm{nnz}(\overline{\boldsymbol{L}})) = O(\mathrm{nnz}(\boldsymbol{L})\log^2 p/\log\log p)$ time. We can implement $\mathcal{A}_{2\mapsto 1}$ as $\mathcal{A}_{2\mapsto 1}(\boldsymbol{y}^*) = (\boldsymbol{y}^*_{[n]})$. It costs nearly linear time as we only truncate vectors.

The proof of the decisional f-reducibility is provided in Corollary 6.5 of [2].

4.2 Controlling Laplacian Diagonals over $\mathbb{Z}_p$

In this section, we try controlling the diagonals of any Laplacian matrix $\boldsymbol{L} \in \mathbb{Z}_p^{n\times n}$ without changing the solution space of the linear system. We can control the range of off-diagonal entries by applying Lemma 1. Since a diagonal entry of $\boldsymbol{L}$ is the sum of off-diagonal entries in the row, we can apply Lemma 1 to indirectly control diagonal entries. In this section, we show that it is also possible to control diagonal entries directly. Specifically, we introduce the following three tools:

1. Adjusting Diagonals to be non-zero: In Section 7.2 of [2], we need to ensure all diagonal entries $\boldsymbol{L}_{i,i}$ are non-zero so that $\frac{1}{\boldsymbol{L}_{i,i}}$ is defined. The tool shows it is possible to achieve this guarantee.
2. Graph Stretching: By splitting an edge to two edges with an intermediate vertex, we are able to isolate the influence from both endpoints. This tool is used in Sect. 4.2 for decreasing combinatorial degree.
3. Combinatorial Degree Decrease: The diagonal value is influenced by both the density of the represented graph and the value of entries. This tool is able to decrease the graph density to $\Delta_{\boldsymbol{L}} = O(1)$ level. In Sect. 4.3, this serves as the first step of the reduction.

By Lemma 5.1 and Lemma 5.2 of [2], we know that we can modify the graph structure from $G(V, E)$ of $\boldsymbol{L}$ to $G'(V', E')$ of $\hat{\boldsymbol{L}}$ as long as $\mathbf{SC}(\hat{\boldsymbol{L}}, V) = \boldsymbol{L}$. Furthermore, by using Lemma 5.4 of [2], we can replace a subgraph with a new one if the Schur complement over the original vertices is equal to $\boldsymbol{L}$. We will apply Lemma 5.7 of [2] frequently to achieve our goal in this section.

Adjusting Diagonals to Be Non-Zero. Some of our reductions require the diagonal entries of the Laplacian must be non-zero. We show that it is always achievable by replacing some subgraphs. Each time, we select an arbitrary off-diagonal entry $\boldsymbol{L}_{i,j}$ with value $w \neq 0$ and try to split it to make the corresponding diagonal entry $\boldsymbol{L}_{i,i}$ and $\boldsymbol{L}_{j,j}$ non-zero. We can apply the process repeatedly until all diagonal entries of the Laplacian are non-zero. Let the resulting Laplacian be $\hat{\boldsymbol{L}}$. By Lemma 6.2 of [2], $\mathbf{SC}(\hat{\boldsymbol{L}}, V) = \boldsymbol{L}$, so the reduction preserves the solution space. There are at most n diagonal entries in $\boldsymbol{L}$, thus the reduction maintains $\mathrm{nnz}(\boldsymbol{L}) = O(\mathrm{nnz}(\boldsymbol{L}))$. The detail of the split process is discussed in Section 6.2.1 and Section 6.2.3 of [2].

Graph Stretching. The main intuition behind reducing combinatorial degree is by splitting high degree vertices to a few low degree vertices. Since this operation affects the degrees of the neighbors of the vertex being split, we need to isolate the influence by stretching each edge by an additional vertex. Specifically, for every edge $(x, y, w)(w \in [1, p-1])$, we remove it and add $(x, t, 2w)$ and $(t, y, 2w)$ where t is a new vertex for each edge. Since $w \in [1, p-1]$, we have $2w \not\equiv 0$. Since $\frac{1}{w} = \frac{1}{2w} + \frac{1}{2w}$, by Lemma 6.2 of [2], such stretching does not change the Schur complement of V. Note that the combinatorial degree of each t is $O(1)$. Since we do such split for every edge with non-zero weight, the number of vertices and edges are increased by $O(\mathrm{nnz}(\boldsymbol{L}))$, thus stays in $O(\mathrm{nnz}(\boldsymbol{L}))$.

Combinatorial Degree Decrease. With the following lemma, we can decrease the max combinatorial degree of any Laplacian matrix $\boldsymbol{L}$ but the weights of edges of the resulting graph are unbounded.

Lemma 2 (**$O(1)$-Combinatorial-Degree Laplacian Construction over $\mathbb{Z}_p$**). *For a Laplacian matrix $\boldsymbol{L} \in \mathbb{Z}_p^{n\times n}$ with an underlying graph $G = (V, E)$,* *we can construct a Laplacian matrix* $\hat{\boldsymbol{L}} \in \mathbb{Z}_p^{k\times k}$ *with an underlying graph* $H = (V', E')$ *such that* $V \subseteq V'$ *and* $\mathbf{SC}(\hat{\boldsymbol{L}}, V) = \boldsymbol{L}$*. We have* $\Delta(H) = O(1)$ *and* $\mathrm{nnz}(\hat{\boldsymbol{L}}) = \Theta(|V'|) = \Theta(|E'|) = O(|E|\log|E|) = O(\mathrm{nnz}(\boldsymbol{L})\log \mathrm{nnz}(\boldsymbol{L}))$*.*

To prove this lemma, the following lemma gives a construction to reduce any vertex with combinatorial degree > 2 by 2 with at most one extra edge and one extra vertex, while the Schur complement of V stays unchanged.

Lemma 3. *For the Laplacian* $\boldsymbol{L}$ *of any graph* G *with vertex set* $V = \{v_0, v_1, v_2, \ldots, v_k\}$ *and edge set* $E = \{(v_0, v_i, w_i)\}$ $(1 \le i \le k)$*, we construct the Laplacian* $\hat{\boldsymbol{L}}$ *of graph* G' *where* G' *is the union of the following parts:*

- H*:* $V_H = \{v_0, v_3, \ldots, v_k\}, E_H = \{(v_0, v_i, w)\}(3 \le i \le k)$*;*
- *an extra node* t *and three edges* (t, v_0, a)*,* (t, v_1, b)*,* (t, v_2, c)*;*
- *an extra edge* (v_1, v_2, d)*.*

If $w_1 + w_2 \not\equiv 0$*, we let* $a = 2(w_1+w_2), b = 2w_1, c = 2w_2, d = -\frac{w_1 w_2}{w_1+w_2}$*. Otherwise, we let* $a = 1, b = w_1, c = w_2, d = -w_1 w_2$*. With such a construction, we have* $\mathbf{SC}(\hat{\boldsymbol{L}}, V) = \boldsymbol{L}$*.*

Proof. See the proof of Lemma 6.4 in [2].

Proof (Proof of Lemma 2). The construction is processed by applying Lemma 3 repeatedly. We first apply the graph stretching described in Sect. 4.2 to G, denoted as G'. We also mark all original vertices before stretching as the internal vertices. We let $G^{(0)}$ be G', then $G^{(k)}$ is defined iteratively:

1. If $G^{(k-1)}$ has some internal vertex x with max combinatorial degree d larger than 2, we group the d edges out of x into $\lfloor \frac{d}{2} \rfloor$ pairs of edges $T_1, \ldots, T_{\lfloor \frac{d}{2} \rfloor}$. If there is an unpaired edge, the edge is ignored in this round. For each pair T_i, we view the subgraph formed by T_i as a star with center x and apply Lemma 3 to replace the star with the construction. Note that the new added vertex is not marked as an internal one. After processing every pair, x has a combinatorial degree of at most $\lfloor \frac{d}{2} \rfloor + 1$. Let the resulting graph be $G^{(k)}$. Let $\boldsymbol{L}^{(\overline{k})}$ be the Laplacian of G^k. By Lemma 3, we have $\mathbf{SC}(\boldsymbol{L}^{(\overline{k})}, V) = \boldsymbol{L}$.
2. Otherwise, we are done $G^{(k)}$ is undefined.

We let $\overline{\boldsymbol{L}}$ be $\boldsymbol{L}^{(\overline{k})}$ where $\overline{k}$ is the largest k such that $\boldsymbol{L}^{(k)}$ is defined. By Lemma 3, we have $\mathbf{SC}(\overline{\boldsymbol{L}}, V) = \boldsymbol{L}$.

In $G^{(\overline{k})}$, all internal (original) vertices have a combinatorial degree of at most 2 because of the graph stretching, and we only apply Lemma 3 with internal vertices. So only the degree of non-internal (newly added) vertices might be increasing. There are two types of newly added vertices. One type comes from the graph stretching in Sect. 4.2, the other type comes from applying Lemma 3. Now we analyze on these two types of newly added vertices.

Lemma 4. *In* $G^{(\overline{k})}$*, all newly added vertices' combinatorial degree is at most* 4*.*

Proof. See Section 6.4 of [2].

Since all vertices' combinatorial degree is not larger than 4, we prove that after such reduction, the max combinatorial degree is $O(1)$ level. Now we need to analyze the number of vertices and edges after the combinatorial degree reduction process. The graph stretching stage is adding $|E|$ edges and vertices, so both the number of vertices and the number of edges after stretching become $O(|E|)$. Since we only deal with the internal vertices, the number of invocations to Lemma 3 is $O(\sum_{i \in V} \deg_i \log \deg_i) = O(|E| \log |E|)$. Each time we replace a star, we add 2 more edges and 1 new node. Thus, we will add $O(|E| \log |E|)$ edges and nodes in the reduction phase. Therefore, $|V'|$ and $|E'|$ are in $O(|E| \log |E|)$.

4.3 Reduction to Low-Degree Laplacian System

In this section, we show that the class of Laplacian matrices $\mathcal{LS}_{\mathbb{Z}_p}$ is efficiently and decisional f-reducible to itself for some f that restrict the diagonal entries to $O(\log p)$, which we also define as the maxdeg of the matrix.

Theorem 5 (Hardness for Low-Degree Unit-weight Laplacian systems in $\mathbb{Z}_p$). $\mathcal{LS}_{\mathbb{Z}_p} \leq_f \mathcal{DLS}_{\log p, \mathbb{Z}_p}$ *where* $f(\text{nnz}) = O(\text{nnz} \cdot \log \text{nnz} \cdot \log^2 p / \log\log p)$.

Proof. We will prove that for any Laplacian system $\boldsymbol{L} \in \mathbb{Z}_p^{n \times n}$, we can compute a low degree Laplacian system $\overline{\boldsymbol{L}} \in \mathbb{Z}_p^{k \times k}$ such that

- $\text{nnz}(\overline{\boldsymbol{L}}) \leq O(\text{nnz}(\boldsymbol{L}) \cdot \log \text{nnz}(\boldsymbol{L}) \cdot \log^2 p / \log\log p)$.
- $\Delta^{w}_{\overline{\boldsymbol{L}}} = O(\log p)$.
- $\overline{\boldsymbol{L}}$ is also a unit-weight matrix.
- For any linear system $\boldsymbol{Lx} = \boldsymbol{b}$, $\{\boldsymbol{x} \mid \boldsymbol{Lx} = \boldsymbol{b}\} = \left\{ \boldsymbol{y}_{[n]} \mid \overline{\boldsymbol{L}}\boldsymbol{y} = \begin{bmatrix} \boldsymbol{b} \\ \boldsymbol{0} \end{bmatrix} \right\}$.

Construction of $\overline{\boldsymbol{L}}$. Initially, $\boldsymbol{L}$ has $\text{nnz}(\boldsymbol{L})$ entries. Therefore, there are $O(\text{nnz}(\boldsymbol{L}))$ edges, which means the maximum combinatorial degree is bounded above by $O(\text{nnz}(\boldsymbol{L}))$ and the maximum weighted degree is bounded above by $O(p \cdot \text{nnz}(\boldsymbol{L}))$. The unit-weight Laplacian reduction in Sect. 4.1 only guarantees the maximum weighted degree is bounded above by $O(\log p \cdot nnz(\boldsymbol{L}))$. Therefore, our main idea is to transform $\boldsymbol{L}$ to a Laplacian matrix $\hat{\boldsymbol{L}}$ such that $\Delta_{\hat{\boldsymbol{L}}} = O(1)$. Then we apply our unit-weight reduction and turn it to $\overline{\boldsymbol{L}}$.

We construct $\hat{\boldsymbol{L}}$ by applying Lemma 2. Therefore, $\text{nnz}(\hat{\boldsymbol{L}}) = O(\text{nnz}(\boldsymbol{L}) \log \text{nnz}(\boldsymbol{L}))$ and $\mathbf{SC}(\hat{\boldsymbol{L}}, [n]) = \boldsymbol{L}$. Most importantly, $\Delta_{\hat{\boldsymbol{L}}}$ is decreased to $O(1)$ level. Note that the edge weights can be $O(p)$ and thus the weighted degree of some nodes could be up to $O(p)$. Then we run the reduction from Lemma 4 on $\hat{\boldsymbol{L}}$, getting our $\overline{\boldsymbol{L}}$ eventually. Specifically, the reduction is done by replacing every edge to a gadget with $O(\log^2 p / \log\log p)$ unit-weight edges and $O(\log p)$ max degree. Therefore, every node in $\hat{\boldsymbol{L}}$'s degree is reduced to

$$O(1) \text{ (combinatorial degree)} \times O(\log p) \text{ (gadget max degree)} = O(\log p).$$

Since the extra vertices in the gadget also have $O(\log p)$ degree, $\Delta^{w}_{\overline{\boldsymbol{L}}} = O(\log p)$.

By Lemma 5.1 and 5.2 of [2], we know that $\mathbf{SC}(\hat{\boldsymbol{L}}, [n]) = \boldsymbol{L}$ preserves the solution space to arbitrary linear system $\boldsymbol{Lx} = \boldsymbol{b}$. Specifically, for any linear system $\boldsymbol{Lx} = \boldsymbol{b}$, $\{\boldsymbol{x} \mid \boldsymbol{Lx} = \boldsymbol{b}\} = \left\{\boldsymbol{y}_{[n]} \mid \overline{\boldsymbol{L}}\boldsymbol{y} = \begin{bmatrix}\boldsymbol{b}\\ \boldsymbol{0}\end{bmatrix}\right\}$. This is in the same form of our unit-weight Laplacian reduction, which also preserves the solution space. Therefore, our low degree reduction preserves the solution space after the two-step transform. Since $\mathrm{nnz}(\hat{\boldsymbol{L}}) = O(\mathrm{nnz}(\boldsymbol{L}) \log \mathrm{nnz}(\boldsymbol{L}))$, after the unit-weight gadget replacement (each gadget is $O(\log^2 p/ \log\log p)$, $\mathrm{nnz}(\overline{\boldsymbol{L}}) \leq O(\mathrm{nnz}(\boldsymbol{L}) \cdot \log \mathrm{nnz}(\boldsymbol{L}) \cdot \log^2 p/ \log\log p)$. Since every edge in the gadget is unit-weight, $\overline{\boldsymbol{L}}$ is also a unit-weight Laplacian by definition.

Running Time of $\mathcal{A}_{1\mapsto 2}$ *and* $\mathcal{A}_{2\mapsto 1}$. In $\mathcal{A}_{1\mapsto 2}$, the first step is do the combinatorial degree decrease, which costs $O(\mathrm{nnz}(\boldsymbol{L}) \log \mathrm{nnz}(\boldsymbol{L}))$ time. Then in the second step of applying the unit-weight Laplacian reduction, we simply replaces each non-diagonal non-zero entry by a matrix $\boldsymbol{U}$ with $\mathrm{nnz}(\boldsymbol{U}) = O(\log^2 p/ \log\log p)$. This costs $O(\mathrm{nnz}(\overline{\boldsymbol{L}})) = O(\mathrm{nnz}(\hat{\boldsymbol{L}}) \log^2 p/ \log\log p) = O(\mathrm{nnz}(\boldsymbol{L}) \log \mathrm{nnz}(\boldsymbol{L}) \cdot \log^2 p/ \log\log p)$. We can implement $\mathcal{A}_{2\mapsto 1}$ as $\mathcal{A}_{2\mapsto 1}(\boldsymbol{y}^*) = (\boldsymbol{y}^*_{[n]})$. It costs nearly linear time as we only truncate vectors.

The proof of decisional f-reducibility is in Corollary 6.6 of [2].

References

1. Alon, N., Yuster, R.: Solving linear systems through nested dissection. In: FOCS 2010, pp. 225–234. IEEE Computer Society (2010). https://doi.org/10.1109/FOCS.2010.28
2. Chen, J., Gao, Y., Huang, Y., Peng, R., Wang, R.: Hardness of graph-structured algebraic and symbolic problems (2022). https://arxiv.org/abs/2109.12736v3
3. Cheung, H.Y., Lau, L.C., Leung, K.M.: Graph connectivities, network coding, and expander graphs. In: 2011 IEEE 52nd Annual Symposium on Foundations of Computer Science, pp. 190–199. IEEE (2011)
4. Cohen, M.B., et al.: Solving directed Laplacian systems in nearly-linear time through sparse LU factorizations. In: 2018 IEEE 59th Annual Symposium on Foundations of Computer Science (FOCS), pp. 898–909. IEEE (2018)
5. Daitch, S.I., Spielman, D.A.: Support-graph preconditioners for 2-dimensional trusses. CoRR abs/cs/0703119 (2007). http://arxiv.org/abs/cs/0703119
6. Daitch, S.I., Spielman, D.A.: Faster approximate Lossy generalized flow via interior point algorithms. In: Proceedings of the Fortieth Annual ACM Symposium on Theory of Computing, STOC 2008, pp. 451–460. Association for Computing Machinery, New York (2008). https://doi.org/10.1145/1374376.1374441
7. Demillo, R.A., Lipton, R.J.: A probabilistic remark on algebraic program testing. Inf. Process. Lett. **7**(4), 193–195 (1978). https://doi.org/10.1016/0020-0190(78)90067-4. https://www.sciencedirect.com/science/article/pii/0020019078900674
8. Dixon, J.D.: Exact solution of linear equations using P-adic expansions. Numer. Math. **40**(1), 137–141 (1982)
9. George, A.: Nested dissection of a regular finite element mesh. SIAM J. Numer. Anal. **10**(2), 345–363 (1973). https://doi.org/10.1137/0710032

10. Jambulapati, A., Sidford, A.: Ultrasparse ultrasparsifiers and faster Laplacian system solvers. In: Proceedings of the Thirty-Second Annual ACM-SIAM Symposium on Discrete Algorithms, SODA 2021, pp. 540–559. Society for Industrial and Applied Mathematics, USA (2021)
11. Kabanets, V., Impagliazzo, R.: Derandomizing polynomial identity tests means proving circuit lower bounds. In: Proceedings of the Thirty-Fifth Annual ACM Symposium on Theory of Computing, STOC 2003, pp. 355–364. Association for Computing Machinery, New York (2003). https://doi.org/10.1145/780542.780595
12. Kelner, J.A., Orecchia, L., Sidford, A., Zhu, Z.A.: A simple, combinatorial algorithm for solving SDD systems in nearly-linear time. In: Proceedings of the Forty-Fifth Annual ACM Symposium on Theory of Computing, pp. 911–920 (2013)
13. Koutis, I., Miller, G.L., Peng, R.: A nearly-m log n time solver for SDD linear systems. In: 2011 IEEE 52nd Annual Symposium on Foundations of Computer Science, pp. 590–598 (2011). https://doi.org/10.1109/FOCS.2011.85
14. Koutis, I., Miller, G.L., Peng, R.: Approaching optimality for solving SDD linear systems. SIAM J. Comput. **43**(1), 337–354 (2014)
15. Kyng, R., Lee, Y.T., Peng, R., Sachdeva, S., Spielman, D.A.: Sparsified Cholesky and multigrid solvers for connection Laplacians. In: Proceedings of the Forty-Eighth Annual ACM Symposium on Theory of Computing, pp. 842–850 (2016)
16. Kyng, R., Peng, R., Schwieterman, R., Zhang, P.: Incomplete nested dissection. In: Proceedings of the 50th Annual ACM SIGACT Symposium on Theory of Computing, pp. 404–417 (2018)
17. Kyng, R., Zhang, P.: Hardness results for structured linear systems. SIAM J. Comput. **49**(4), FOCS17-280 (2020)
18. Lee, Y.T., Sidford, A.: Solving linear programs with Sqrt (rank) linear system solves. arXiv preprint arXiv:1910.08033 (2019)
19. Lipton, R.J., Rose, D.J., Tarjan, R.E.: Generalized nested dissection. SIAM J. Numer. Anal. **16**(2), 346–358 (1979). https://doi.org/10.1137/0716027
20. Nie, Z.: Matrix anti-concentration inequalities with applications. In: Proceedings of the 54th Annual ACM SIGACT Symposium on Theory of Computing, pp. 568–581 (2022)
21. Odlyzko, A.M.: Discrete logarithms in finite fields and their cryptographic significance. In: Beth, T., Cot, N., Ingemarsson, I. (eds.) EUROCRYPT 1984. LNCS, vol. 209, pp. 224–314. Springer, Heidelberg (1985). https://doi.org/10.1007/3-540-39757-4_20
22. Peng, R., Vempala, S.S.: Solving sparse linear systems faster than matrix multiplication. In: Proceedings of the 2021 ACM-SIAM Symposium on Discrete Algorithms (SODA), pp. 504–521. SIAM (2021)
23. Saxena, N.: Progress on polynomial identity testing. Bull. EATCS **99**, 49–79 (2009)
24. Saxena, N.: Progress on polynomial identity testing-II. In: Agrawal, M., Arvind, V. (eds.) Perspectives in Computational Complexity. PCSAL, vol. 26, pp. 131–146. Springer, Cham (2014). https://doi.org/10.1007/978-3-319-05446-9_7
25. Schwartz, J.T.: Fast probabilistic algorithms for verification of polynomial identities. J. ACM **27**(4), 701–717 (1980). https://doi.org/10.1145/322217.322225
26. Shpilka, A., Yehudayoff, A.: Arithmetic circuits: a survey of recent results and open questions. Found. Trends® Theor. Comput. Sci. **5**(3–4), 207–388 (2010)
27. Spielman, D.A., Teng, S.H.: Nearly-linear time algorithms for graph partitioning, graph sparsification, and solving linear systems. In: Proceedings of the Thirty-Sixth Annual ACM Symposium on Theory of Computing, pp. 81–90 (2004)

28. Storjohann, A.: The shifted number system for fast linear algebra on integer matrices. J. Complex. **21**(4), 609–650 (2005). https://doi.org/10.1016/j.jco.2005.04.002. https://www.sciencedirect.com/science/article/pii/S0885064X05000312. Festschrift for the 70th Birthday of Arnold Schonhage
29. Tutte, W.T.: The factorization of linear graphs. J. Lond. Math. Soc. **s1–22**(2), 107–111 (1947). https://doi.org/10.1112/jlms/s1-22.2.107
30. Valiant, L.G.: Completeness classes in algebra. In: Proceedings of the Eleventh Annual ACM Symposium on Theory of Computing, pp. 249–261 (1979)
31. Zhu, X., Ghahramani, Z., Lafferty, J.: Semi-supervised learning using gaussian fields and harmonic functions. In: Proceedings of the Twentieth International Conference on International Conference on Machine Learning, ICML 2003, pp. 912–919. AAAI Press (2003)
32. Zippel, R.: Probabilistic algorithms for sparse polynomials. In: Ng, E.W. (ed.) Symbolic and Algebraic Computation. LNCS, vol. 72, pp. 216–226. Springer, Heidelberg (1979). https://doi.org/10.1007/3-540-09519-5_73
33. Zomorodian, A., Carlsson, G.: Computing persistent homology. In: Proceedings of the Twentieth Annual Symposium on Computational Geometry, pp. 347–356 (2004)

Sublinear-Space Streaming Algorithms for Estimating Graph Parameters on Sparse Graphs

Xiuge Chen[1], Rajesh Chitnis[2], Patrick Eades[1], and Anthony Wirth[1(✉)]

[1] School of Computing and Information Systems, The University of Melbourne, Parkville, Victoria, Australia
xiugechen@gmail.com, patrick.f.eades@gmail.com, awirth@unimelb.edu.au
[2] School of Computer Science, University of Birmingham, Birmingham, UK
rajeshchitnis@gmail.com

Abstract. In this paper, we design sub-linear space streaming algorithms for estimating three fundamental parameters – maximum independent set, minimum dominating set and maximum matching – on sparse graph classes, i.e., graphs which satisfy $m = O(n)$ where m, n is the number of edges, vertices respectively. Each graph parameter we consider can have size $\Omega(n)$ even on sparse graph classes, and hence for sublinear-space algorithms we are restricted to parameter estimation instead of attempting to find a solution. We obtain these results:

- **Estimating Max Independent Set via the Caro-Wei bound**: Caro and Wei each showed $\lambda = \sum_v 1/(d(v)+1)$ is a lower bound on max independent set size, where vertex v has degree $d(v)$. If average degree, $\bar{d}$, is $\mathcal{O}(1)$, and max degree $\Delta = \mathcal{O}(\varepsilon^2 \bar{d}^{-3} n)$, our algorithms, with at least $1-\delta$ success probability:
  - In *online streaming*, return an actual independent set of size $1 \pm \varepsilon$ times λ. This improves on Halldórsson et al. [Algorithmica '16]: we have less working space, i.e., $\mathcal{O}(\log \varepsilon^{-1} \cdot \log n \cdot \log \delta^{-1})$, faster updates, i.e., $\mathcal{O}(\log \varepsilon^{-1})$, and bounded success probability.
  - In *insertion-only* streams, approximate λ within factor $1 \pm \varepsilon$, in one pass, in $\mathcal{O}(\bar{d}\varepsilon^{-2} \log n \cdot \log \delta^{-1})$ space. This aligns with the result of Cormode et al. [ISCO '18], though our method also works for *online streaming*. In a vertex-arrival and random-order stream, space reduces to $\mathcal{O}(\log(\bar{d}\varepsilon^{-1}))$. With extra space and post-processing step, we remove the max-degree constraint.
- **Sublinear-Space Algorithms on Forests**: On a forest, Esfandiari et al. [SODA '15, TALG '18] showed space lower bounds for 1-pass randomized algorithms that approximately estimate these graph parameters. We narrow the gap between upper and lower bounds:
  - Max independent set size within $3/2 \cdot (1 \pm \varepsilon)$ in one pass and in $\log^{\mathcal{O}(1)} n$ space, and within $4/3 \cdot (1 \pm \varepsilon)$ in two passes and in $\tilde{\mathcal{O}}(\sqrt{n})$ space; the lower bound is for approx. $\leq 4/3$.
  - Min dominating set size within $3 \cdot (1 \pm \varepsilon)$ in one pass and in $\log^{\mathcal{O}(1)} n$ space, and within $2 \cdot (1 \pm \varepsilon)$ in two passes and in $\tilde{\mathcal{O}}(\sqrt{n})$ space; the lower bound is for approx. $\leq 3/2$.

P. Morin and S. Suri (Eds.): WADS 2023, LNCS 14079, pp. 247–261, 2023.
https://doi.org/10.1007/978-3-031-38906-1_17

- Max matching size within $2 \cdot (1 \pm \varepsilon)$ in one pass and in $\log^{\mathcal{O}(1)} n$ space, and within $3/2 \cdot (1 \pm \varepsilon)$ in two passes and in $\tilde{\mathcal{O}}(\sqrt{n})$ space; the lower bound is for approx. $\leq 3/2$.

Keywords: Graph Algorithms · Data Streams Model · Caro-Wei Bound · Independent Set · Dominating Set · Matching

1 Introduction

Maximum independent set, minimum dominating set, and maximum matching are key graph problems. Independent set models, for example, optimization and scheduling problems where conflicts should be avoided [1,18,21,28], dominating set models guardian selection problems [32,33,35,36,40], while a matching models similarity [11,17,38] and inclusion dependencies [2,13]. When the input is presented as a data stream, we show new algorithms for computing the following parameters related to these problems on sparse graphs.

Parameters of Interest: Given an undirected graph G, comprising vertices V and edges E, let n and m be the sizes of V and E, respectively. A subset S of V is an *independent set* if and only if the subgraph induced by S contains no edges. Subset S is a *dominating set* if and only if every vertex in V is either in S or adjacent to some vertex in S. A subset M of E is a *matching* if and only if no pair of edges in M share a vertex. The three parameters of interest to us are the size of a maximum independent set, aka *independence number*, β; the size of a minimum dominating set, aka *domination number*, γ; and the size of a maximum matching, aka *matching number*, ϕ. It is well known that $n - \phi \geq \beta \geq n - 2\phi$.

Given some k, it is NP-complete [27] to decide whether $\beta \geq k$ and to decide whether $\gamma \leq k$. There are several *approximation algorithms* for these problems that run in polynomial time and return a solution within a guaranteed factor of optimum. For instance, a greedy algorithm for maximum matching outputs a 2-approximation. Similarly, there exist greedy algorithms that $\mathcal{O}(\Delta)$-approximate the maximum independent set [20] and approximate within $\mathcal{O}(\ln \Delta)$ the minimum dominating set [24,30], where Δ is the maximum degree. More promising results are obtained on sparse graphs, where $m \in \mathcal{O}(n)$, which we study in this paper. For example, the Caro-Wei bound [6,39] $\lambda = \sum_{v \in V(G)} (1 + \deg(v))^{-1}$ is a lower bound on β.

Data Streams: We focus on estimating these graph parameters in the data stream model. The *semi-streaming* model [16], with $\mathcal{O}(n \log n)$ bits of working space, is commonplace for general graphs. In this paper we consider *sparse* graphs: as $\mathcal{O}(n)$ bits would be sufficient to store the entire sparse graph, we restrict our space allowance to $o(n)$ bits. We *tune* our algorithms to specific stream formats: edge-arrival, vertex-arrival, insertion-only, turnstile, arbitrary, and random, as described fully in Sect. 2.1.

Halldórsson et al. [19] introduced the *online streaming* model, combining the data-stream and online models. In *online streaming*, after each stream item, on demand, the algorithm must efficiently report a valid solution. Typically, an online-streaming algorithm has a initial solution and modifies it element by element. Similar to the *online* model, each decision on the solution is irrevocable. We distinguish between *working space*, involved in computing the solution, and an additional *solution space* for storing or returning the solution. Solution space may be significantly larger than the working space, but is write only.

1.1 Previous Results

Estimating the size of a maximum matching in a sparse graph has been extensively studied in the streaming setting. When the input is a tree, several $(2+\varepsilon)$-approximation algorithms are known. Esfandiari et al. [15] designed an $\tilde{\mathcal{O}}(\sqrt{n})$-space algorithm for insertion-only streams, where $\tilde{\mathcal{O}}$ notation suppresses a polylogarithmic factor. The space was further reduced to $\log^{\mathcal{O}(1)} n$ by Cormode et al. [9], while Bury et al. [4] generalized it to turnstile streams.

In this scenario, there is relatively little research on independent set and dominating set. Halldórsson et al. [19] studied general hypergraphs and gave a one-pass insertion-only streaming algorithm in $\mathcal{O}(n)$ space, outputting, in expectation, an independent set with size at least the Caro-Wei bound, λ; their algorithm suits the *online streaming* model. Cormode et al. [7] designed an algorithm that $(1 \pm \varepsilon)$-approximates λ with constant success probability in $\mathcal{O}(\varepsilon^{-2}\bar{d}\log n)$ space. They also showed a nearly tight lower bound: every randomized one-pass algorithm with constant error probability requires $\Omega(\varepsilon^{-2}\bar{d})$ space to $(1 \pm \varepsilon)$-approximate λ. Meanwhile, Esfandiari et al. [15] established space lower bounds for estimating ϕ in a forest: every 1-pass randomized streaming approximation algorithm with factor better than 3/2 needs $\Omega(\sqrt{n})$ space. Adapting their approach gives other lower bounds: $\Omega(\sqrt{n})$ bits are required to approximate β better than 4/3, and γ better than 3/2.

In the full version of our paper[1], we summarize other results, for estimating the matching number and domination number in bounded arboricity or bounded-degree graphs that provide further context for our work.

1.2 Our Results

We design new sublinear-space streaming algorithms for estimating β, γ, ϕ in sparse graphs. With such little space, we focus on approximate parameter estimation, rather than finding solutions. Our results fall into two categories. First, approximating independent set via the Caro-Wei bound in bounded-average-degree graphs [2]. Second, approximating parameters on streamed forests. All results are summarized in tables below; due to space constraints we deferred some theorem descriptions to the full version of the paper (See footnote 1), henceforth "FV".

[1] Full version is at https://doi.org/10.48550/arXiv.2305.16815.

[2] This includes planar graphs, bounded treewidth, bounded genus, H-minor-free, etc.

Approximating the Caro-Wei Bound **(λ)** Boppana et al. [3] show that λ is at least the Turán Bound [37], i.e., $\beta \geq \lambda \geq n/(\bar{d}+1)$, indicating every $\mathcal{O}(1)$-estimate of λ is a $\mathcal{O}(\bar{d})$-approximation of β. We hence approximate β in bounded-average-degree graphs; we summarize results in Table 1.

Table 1. Streaming Caro-Wei Bound approximation. *Arrival* is *Edge* for edge-arrival or *Vertex* for vertex-arrival. *Type* is *Insert* for insertion-only or *Turns* for turnstile. *Order* is *Arb* for arbitrary or *Ran* for random. *Online* is "Yes" for *online streaming* algorithm. Theorems marked as '*' have the max-degree constraint, which can be removed if an *in-expectation guarantee* suffices. Constants $c \in (0,1)$ and $c' > 1$.

Arrival	Type	Order	Factor	(Work) Space	Online	Success Prob	Reference
Edge	Insert	Arb	1	$\mathcal{O}(n)$	Yes	*Expected*	[19]
Edge	Turns	Arb	$(1 \pm \varepsilon)$	$\mathcal{O}(\bar{d}\varepsilon^{-2} \log n)$	No	c	[7]
Edge	Insert	Arb	$(1 \pm \varepsilon)$	$\mathcal{O}(\bar{d}\varepsilon^{-2} \log n)$	No	c	Thm 6*
Edge	Insert	Arb	$(1 \pm \varepsilon)$	$\mathcal{O}(\log \varepsilon^{-1} \log n)$	Yes	c	FV (See footnote 1) *
Edge	Insert	Arb	$(1 \pm \varepsilon)$	$\mathcal{O}(\bar{d}^2\varepsilon^{-2} \log^2 n)$	No	$n^{-c'}$	FV (See footnote 1)
Vertex	Insert	Ran	$(1 \pm \varepsilon)$	$\mathcal{O}(\log(\bar{d}\varepsilon^{-1}))$	No	c	FV (See footnote 1) *
Either	Either	Arb	c	$\Omega(\bar{d}/c^2)$	-	c	[7]

With a random permutation of vertices, there is a known offline algorithm for estimating λ. Our algorithm, CARAWAY, simulates such a permutation via hash functions drawn uniformly at random from an ε-min-wise hash family. Let ε be the approximation error *rate* [3]. For graphs with average degree $\bar{d}$ and max degree $\Delta \in \mathcal{O}(\varepsilon^2 \bar{d}^{-3} n)$, CARAWAY $(1 \pm \varepsilon)$-approximates λ in an arbitrary-order edge-arrival stream. It fails with probability at most δ, and uses working space $\mathcal{O}(\bar{d}\varepsilon^{-2} \log n \cdot \log \delta^{-1})$. By allowing $\mathcal{O}(\log \varepsilon^{-1} \cdot \log n \cdot \log \delta^{-1})$ additional, write-only, *solution space* to store the output, the modified algorithm, CARAWAY$_1$, reports an actual solution set in the *online streaming* model with $\mathcal{O}(\log \varepsilon^{-1})$ update time. The max-degree constraint is required only to bound the failure probability, δ for these two algorithms; if an *in-expectation* bound suffices, this constraint can be disregarded.

Further variant CARAWAY$_2$ removes the max-degree constraint entirely, but requires more space. Failing with low probability, CARAWAY$_2$ returns a $(1 \pm \varepsilon)$-approximation using $\mathcal{O}(\bar{d}^2 \varepsilon^{-2} \log^2 n)$ space. Since post-processing is required, CARAWAY$_2$ works in the standard streaming model, but not in the *online streaming* model. Additionally, if the stream is vertex-arrival and random-order, CARAWAY can be modified to use only $\mathcal{O}(\log(\bar{d}\varepsilon^{-1}) \cdot \log \delta^{-1})$ space.

Bounding *rate* ε is non-trivial: there is positive correlation between vertices being in the sample whence we estimate λ. To bound ε, we start by constraining the max degree; we also bound ε in the offline algorithm and all its variants.

[3] The *relative error* between the estimate and the actual value.

Comparison with Cormode et al. [7] : Since Cormode et al. relied on estimating sampled vertices' degrees, we understand their methods report the parameter β, but not an actual independent set. In an insertion-only stream satisfying our Δ constraint, CARAWAY has asymptotically the same approximation ratio in the same space as Cormode et al. Importantly, in *online streaming*, CARAWAY$_1$, can output an actual independent set. Moreover, in vertex-arrival and random-order streams, CARAWAY requires comparatively less space.

Comparison with Halldórsson et al. [19] : The online streaming algorithm by Halldórsson et al. has $\mathcal{O}(\log n)$ update time and $\mathcal{O}(n)$ working space. CARAWAY$_1$ has faster update time, $\mathcal{O}(\log \varepsilon^{-1})$, and less working space, $\mathcal{O}(\log \varepsilon^{-1} \cdot \log n \cdot \log \delta^{-1})$. Besides, our estimate is within an error rate, ε, with a guaranteed constant probability; there is no such guarantee from Halldórsson et al.

Estimating β, γ and ϕ on Forests Table 2 has results on streamed forests. Esfandiari et al. [15] showed it is non-trivial to estimate fundamental graph parameters (e.g., β, γ, ϕ) in a one-pass streamed forest [4]). We trade off approximation ratio for space and number of passes, and obtain two results classes.

Table 2. Estimating γ, β, and ϕ in streamed forests. All succeed with high probability.

Problem	Arrival	Type	Order	Pass	Factor	Space	Reference
γ	Edge	Turns	Arb	1	$3(1 \pm \varepsilon)$	$\log^{\mathcal{O}(1)} n$	FV (See footnote 1)
γ	Edge	Turns	Arb	2	$2(1 \pm \varepsilon)$	$\tilde{\mathcal{O}}(\sqrt{n})$	FV (See footnote 1)
γ	Any	Any	Arb	1	$3/2 - \varepsilon$	$\Omega(\sqrt{n})$	[15]
β	Edge	Turns	Arb	1	$3/2(1 \pm \varepsilon)$	$\log^{\mathcal{O}(1)} n$	Thm 9
β	Edge	Turns	Arb	2	$4/3(1 \pm \varepsilon)$	$\tilde{\mathcal{O}}(\sqrt{n})$	Thm 10
β	Any	Any	Arb	1	$4/3 - \varepsilon$	$\Omega(\sqrt{n})$	[15]
ϕ	Edge	Insert	Arb	1	$2(1 \pm \varepsilon)$	$\tilde{\mathcal{O}}(\sqrt{n})$	[15]
ϕ	Edge	Insert	Arb	1	$2(1 \pm \varepsilon)$	$\log^{\mathcal{O}(1)} n$	[9]
ϕ	Edge	Turns	Arb	1	$2(1 \pm \varepsilon)$	$\log^{\mathcal{O}(1)} n$	[5], FV (See footnote 1)
ϕ	Edge	Turns	Arb	2	$3/2(1 \pm \varepsilon)$	$\tilde{\mathcal{O}}(\sqrt{n})$	FV (See footnote 1)
ϕ	Any	Any	Arb	1	$3/2 - \varepsilon$	$\Omega(\sqrt{n})$	[15]

First, one-pass $\log^{\mathcal{O}(1)} n$-space algorithms, which relying on approximating both the numbers of leaves and non-leaf vertices in a stream. We have

- a $3/2 \cdot (1 \pm \epsilon)$-approximation for β;
- a $3 \cdot (1 \pm \epsilon)$-approximation for γ;
- a $2 \cdot (1 \pm \epsilon)$-approximation for ϕ.

Second, two-pass $\tilde{\mathcal{O}}(\sqrt{n})$-space algorithms[5]: We further narrow the gap between the upper and lower bounds [15]:

[4] $\Omega(\sqrt{n})$ (or $\Omega(n)$) space is required for randomized (or deterministic) algorithms.
[5] $\tilde{\mathcal{O}}$-notation suppresses the poly-logarithmic factor in the bound.

- a $4/3 \cdot (1 \pm \epsilon)$-approximation algorithm for β;
- a $2 \cdot (1 \pm \epsilon)$-approximation algorithm for γ;
- a $3/2 \cdot (1 \pm \epsilon)$-approximation algorithm for ϕ.

Our innovation is introducing the notion of support vertices from structural graph theory [12]. A *support vertex* is one that is adjacent to one or more leaves: an approximate count could be a key component of streaming graph algorithms.

2 Preliminaries

Given a set of vertices in G, S, let $N(S)$ be the neighbors of S (excluding S), and $N[S]$ be $S \cup N(S)$. Let Δ be the max degree and $\bar{d}$ be the average degree of G, and $d(v)$ be the degree of vertex v. $Deg_i(G)$ is the set of vertices with $d(v) = i$, while $Deg_{\geq i}(G)$ is the set with $d(v) \geq i$. A support vertex [12] is a vertex adjacent to at least one leaf. We denote the set of support vertices by $Supp(G)$.

2.1 Stream Formats

A graph stream is *edge-arrival* if edges arrive one by one, sequentially, in arbitrary order. We call a graph stream *insertion-only* if there are no deletions. In a *turnstile* stream, an edge can be deleted, but only if its most recent operation was an insertion. It is *vertex-arrival* if it comprises a sequence of (vertex, vertex-list) pairs, (u_i, A_i), where the list A_i comprises the subset of the vertices $\{u_j\}_{j<i}$ that have occurred previously in the stream that are adjacent to u_i. Moreover, the order of vertex arrivals can be either *arbitrary* or (uniformly) *random*.

2.2 Fundamental Streaming Algorithms

We assume vertices are $[n] = \{1, 2, \ldots, n\}$, and that n is known in advance. Second, our results are developed for certain graph classes, hence we assume that at the *end* of the stream, the graph is guaranteed to be in the target class.

We employ several streaming primitives. Consider a vector $x \in \mathbb{R}^n$ whose coordinates are updated by a turnstile stream with each update in the range $[-M, M]$. Let x_i be the final value of coordinate i. For $p > 0$, the L_p norm of x is $\|x\|_p = (\sum_{i=1}^n |x_i|^p)^{1/p}$. The L_0 *norm* is $\|x\|_0 = (\sum_{i=1}^n |x_i|^0)$. Both *norms* can be $(1 \pm \varepsilon)$-approximated in streams in small space.

Theorem 1. [25] For $p \in (0, 2)$ and $\varepsilon, \delta \in (0, 1)$, there is an algorithm that $(1 \pm \varepsilon)$-approximates $\|x\|_p$ with probability at least $1 - \delta$ and $\mathcal{O}(\varepsilon^{-2} \log(mM) \log \delta^{-1})$ space usage. Both update and reporting time are in $\tilde{\mathcal{O}}(\varepsilon^{-2} \log \delta^{-1})$.

Theorem 2. [26] Given $\varepsilon, \delta \in (0, 1)$, there is an algorithm that, with probability at least $1 - \delta$, $(1 \pm \varepsilon)$-approximates $\|x\|_0$. It has an update and reporting time of $\mathcal{O}(\log \delta^{-1})$, and uses space $\mathcal{O}(\varepsilon^{-2} \log(\delta^{-1}) \log(n)(\log(1/\varepsilon) + \log\log(mM)))$.

Given $k \leq n$, x is *k-sparse* if it has at most k non-zero coordinates.

Theorem 3. [8] Given k and $\delta \in (0,1)$, there is an algorithm that recovers a k-sparse vector exactly with probability $1-\delta$ and space usage $\mathcal{O}(k \log n \cdot \log(k/\delta))$.

We focus on *heavy hitters* that are coordinates of vector x with $|x_i| \geq \varepsilon \|x\|_1$.

Theorem 4. [10] Given $\varepsilon, \tau \in (0,1)$, there is an algorithm that outputs all items with frequency at least $(\varepsilon+\tau)\|x\|_1$, and with probability $1-\delta$ outputs no items with frequency less than $\tau\|x\|_1$. It has update time $\mathcal{O}(\log n \cdot \log(2\log n/(\delta\tau)))$ and query time $\mathcal{O}(\varepsilon^{-1})$, and uses $\mathcal{O}(\varepsilon^{-1} \log n \cdot \log(2\log n/(\delta\tau)))$ space.

2.3 ε-min-wise Hash Functions

To simulate uniform-at-random selection, we invoke carefully chosen hash families. A family of hash functions, $\mathcal{H} = \{h : [n] \to [m]\}$, is "$\varepsilon$-min-wise" if for every $A \subset [n]$ and $x \in [n] \setminus A$, $\Pr_{h\in\mathcal{H}}[\forall a \in A, h(x) < h(a)] = (1 \pm \varepsilon)/(|A|+1)$. Indyk [22] showed that the ε-min-wise property can be achieved via a $\mathcal{O}(\log \varepsilon^{-1})$-wise hash family, under the constraints that $|A| \in \mathcal{O}(\varepsilon m)$. Its space usage is $\mathcal{O}(\log \varepsilon^{-1} \cdot \log m)$ and its computation time is $\mathcal{O}(\log \varepsilon^{-1})$.

2.4 Extension of Chernoff-Hoeffding Inequality

Panconesi and Srinivasan [34] extended the Chernoff-Hoeffding inequality to negatively correlated Boolean random variables as follows:

Theorem 5. [34] For negatively correlated Boolean random variables $X_1, \ldots, X_r$, let $X = \sum_{i=1}^{r} X_i$, $\mu = E[X]$ and $0 < \delta < 1$, then $\Pr\left[|X - \mu| \geq \delta\mu\right] \leq 2\exp(-\mu\delta^2/2)$.

3 Estimating β via the Caro-Wei Bound

In this section, we introduce efficient streaming algorithms to estimate the independence number, β in graphs with bounded average degree. A *folklore* offline greedy algorithm returns an independent set whose size is in expectation the Caro-Wei Bound, λ. Given a uniform-at-random permutation π of V, the greedy algorithm adds vertex v to the solution whenever v is *earlier* in π than all v's neighbors. Storing π explicitly requires $\Omega(n \log n)$ bits, not sublinear, so disallowed here.

We instead simulate a random permutation via function h drawn randomly from an ε-min-wise hash-family, $\mathcal{H}$. Vertex v is added whenever its $h(v)$ is less than all its neighbors' hash values. Family $\mathcal{H}$ ensures the probability v has smallest hash is approximately proportional to $\deg(v)$. As $\deg(v)$ could be $\Theta(n)$, we rely instead on standard sampling and recovery techniques in our algorithm, Caraway, see Algorithm 1. The co-domain of h in Caraway is $[n^3]$ for two reasons: (i) with high probability, there are no colliding pairs; (ii) we assume $\varepsilon \in \omega(n^{-2})$, so the ε-min-wise property applies to every subset of size $\geq n$.

Algorithm 1. CARAWAY, estimating Caro-Wei bound

1: **Input**: Average degree, $\bar{d}$, error rate, ε.
2: **Initialization**: $p \leftarrow 4(\bar{d}+1)/(\varepsilon^2 n)$
3: $S \leftarrow pn$ vertices sampled uniformly at random from $[n]$
4: Select uniformly at random $h \in \mathcal{H}$, ε-min-wise hash functions: $[n] \rightarrow [n^3]$
5: **for all** $e = (u, v)$ in the stream **do**
6: **if** $(u \in S) \wedge (h(u) \geq h(v))$ **then**
7: Remove u from S
8: **if** $(v \in S) \wedge (h(v) \geq h(u))$ **then**
9: Remove v from S
10: **return** $\widehat{\lambda} \leftarrow |S|/p$

Theorem 6. *Given* $\varepsilon \in (0,1)$, *if* $\Delta \leq \varepsilon^2 n/(3(\bar{d}+1)^3)$, *in an insertion-only stream,* CARAWAY $(1 \pm \varepsilon)$*-approximates* λ *w.p.* $2/3$ *in work space* $\mathcal{O}(\bar{d}\,\varepsilon^{-2}\log n)$.

To prove Theorem 6, we start with a result on the expectation of $\widehat{\lambda}$.

Lemma 1. [⋆][6] $E[\widehat{\lambda}] = (1 \pm \varepsilon)\,\lambda$.

We show that, with constant probability, estimate $\widehat{\lambda}$ is no more than a $1 \pm \varepsilon$ factor away from its mean, $E[\widehat{\lambda}]$. Let X_v be an indicator variable for the event that v is both sampled in, and not removed from, S. Let $X = \sum_{v \in V} X_v$. Since some pairs of X_v and X_u are positively correlated, to estimate X, we cannot directly apply a Chernoff(-like) bound. There are three cases to consider.

- Vertices u and v are adjacent: $u \in N[v]$ and $v \in N[u]$. If X_u is 1, wlog, we know that X_v is 0. Hence X_v and X_u are negatively correlated.
- Vertices u and v are not adjacent, but share at least one neighbor. Consider one such common neighbor, w, i.e., $w \in N[u] \cap N[v]$. Let $x < N(x)$ denote x having a smaller hash value than all elements in $N(x)$. Knowing $u < N(u)$ implies $h(v) < h(w)$ is more likely, and hence $v < N(v)$ is more likely, hence X_v and X_u are positively correlated, shown in Lemma 2.
- All other cases: X_v and X_u are independent (see Lemma 2).

Lemma 2. [⋆] *Let* x *and* y *be non-adjacent vertices with* $|N(x) \cap N(y)| = k$, $|N(x) \setminus N(y)| = l$, *and* $|N(y) \setminus N(x)| = r$. *From a uniform random permutation, we have* $\Pr[y < N(y) \mid x < N(x)] = \frac{(l+r+2k+2)}{(r+k+1)(l+k+r+2)}$.

Hence if $k = 0$, then $\Pr[y < N(y) \mid x < N(x)] = 1/(r+1) = \Pr[y < N(y)]$. However, if $k > 0$, then $\Pr[y < N(y) \mid x < N(x)] > 1/(r+k+1)$. So $x < N(x)$ and $y < N(y)$ are positively correlated iff x and y have some common neighbor.

In a permutation generated from a hash function from a ε-min-wise family, as shown in the full version (See footnote 1), Lemma 2 also holds within a factor $1 \pm \varepsilon$: it relies only the probability of an element being the smallest.

[6] Due to space constraints, each result labeled [⋆] has its proof in the full version (See footnote 1).

With a Chernoff-type bound unavailable, we bound the variance of $\widehat{\lambda}$, and apply Chebyshev's inequality: recall that $\text{Var}(\widehat{\lambda}) = \text{Var}(|S|/p) = \text{Var}(X)$.

Lemma 3. [$\star$] *If $\Delta \leq \varepsilon^2 n/(3(\bar{d}+1)^3)$ and $p \geq 4(\bar{d}+1)/(\varepsilon^2 n)$, then $Var(X) \leq \varepsilon^2 E^2[X]/3$.*

With these tools, we prove Theorem 6: details are in the full version (See footnote 1). Returning the median of several instances, the success probability of Algorithm 1 becomes $1-\delta$ in $\mathcal{O}(\bar{d}\varepsilon^{-2} \log \delta^{-1} \cdot \log n)$ total space.

3.1 Three Extensions of the CARAWAY algorithm

The CARAWAY algorithm (Algorithm 1) can be further adapted to suit the following different problem settings, with further details in the full version (See footnote 1).

- In the online streaming model, we are able to also output an actual independent set by allowing n bits of external, solution-space memory.
- The condition of having a bound on the maximum degree can be removed by excluding high-degree vertices, although this requires a post-processing stage and is thus not an online algorithm.
- In a vertex-arrival stream, with vertices in *random* order, the algorithm uses only $\mathcal{O}(\log(\bar{d}\varepsilon^{-2}))$ space.

4 Sublinear-Space Streaming Algorithms for Approximating Graph Parameters in Forests

We now describe our sublinear-space streaming algorithms for approximating graph parameters when the input graph is a forest[7]. Our algorithms work for turnstile edge-arrival streams. For each graph parameter, we design two algorithms: a 1-pass algorithm using $\log^{\mathcal{O}(1)} n$ space, and a 2-pass algorithm using $\tilde{\mathcal{O}}(\sqrt{n})$ space, with better approximation ratio (see Table 3).

For brevity, we describe in this section our algorithms for estimating β. Algorithms for γ and ϕ are in the full version (See footnote 1). There are two main steps:

- (Sect. 4.1) Obtain bounds on β in terms of other structural graph quantities such as number of leaves and number of support vertices,
- (Sect. 4.2) Estimate these structural graph quantities in sublinear-space.

[7] Unless otherwise stated, we assume that the input forest has no isolated vertices.

Table 3. Turnstile edge-arrival streaming algorithms for approximating γ, β, and ϕ in forests. Each succeeds with high probability.

Problem	Number of Passes	Approximation Factor	Space	Reference
β	1	$3/2 \cdot (1 \pm \varepsilon)$	$\log^{\mathcal{O}(1)} n$	Theorem 9
β	2	$4/3 \cdot (1 \pm \varepsilon)$	$\tilde{\mathcal{O}}(\sqrt{n})$	Theorem 10
γ	1	$3 \cdot (1 \pm \varepsilon)$	$\log^{\mathcal{O}(1)} n$	FV (See footnote 1)
γ	2	$2 \cdot (1 \pm \varepsilon)$	$\tilde{\mathcal{O}}(\sqrt{n})$	FV (See footnote 1)
ϕ	1	$2 \cdot (1 \pm \varepsilon)$	$\log^{\mathcal{O}(1)} n$	[5], FV (See footnote 1)
ϕ	2	$3/2 \cdot (1 \pm \varepsilon)$	$\tilde{\mathcal{O}}(\sqrt{n})$	FV (See footnote 1)

4.1 Structural Bounds on the Independence Number

Leaves and Non-leaf Vertices: Non-leaf vertices are used in 2-approximating the max matching [5,15] and 3-approximating the min domination set in forests [14,29,31]. They are helpful because one can always reason that some min dominating set must be a subset of them. Similarly, there exists a max independent set that includes all leaves. For a forest, F, let Deg_1 be the set of leaves, c be the number of tree components[8], and $Deg_{\geq 2}$ be the set of non-leaf vertices. We can both upper and lower bound β using only $|Deg_1|$ and c.

Theorem 7. [$\star$] *For every forest,* $\max\left\{\frac{n}{2}, |Deg_1| - c\right\} \leq \beta \leq \frac{1}{2}(n + |Deg_1|)$.

Support Vertices: We now demonstrate the power of support vertices: a vertex is a support vertex if it adjacent to at least one leaf. Each graph parameter can be found in linear space via standard dynamic programming and fixing the subset of leaves in the solution. Due to our (sublinear) space constraints, we cannot apply dynamic programming. However, going just one-level above the leaves – counting the support vertices – is sufficient to improve the approximations. Let *Supp* be the set of support vertices in F; for every F:

Theorem 8. [$\star$] $\frac{1}{2}(n + |Deg_1| - |Supp|) \leq \beta \leq \frac{2}{3}(n + |Deg_1| - |Supp|)$.

Furthermore, combining Theorem 8 with Theorem 7, we have:

Corollary 1. [$\star$] *If* $|Supp| \leq \frac{1}{2}|Deg_{\geq 2}|$, $\frac{3}{8}(n + |Deg_1|) \leq \beta \leq \frac{1}{2}(n + |Deg_1|)$.

4.2 One-Pass and Two-Pass Algorithms for Estimating β

We show how to estimate the structural graph quantities within factor $(1 \pm \varepsilon)$, in small space. Our algorithms rely on the concept of the degree vector $\mathcal{D}(F)$: the ith coordinate holds the degree of vertex i. Storing this vector would require $\Theta(n \log n)$ bits, exceeding our working-space bound. Hence we rely on sparse-recovery techniques to estimate quantities that are functions of $\mathcal{D}(F)$, supporting fast stream updates.

[8] The variable c of Table 1 is now out of scope.

Estimating Non-leaf Vertices Though β estimation does not use $|Deg_{\geq 2}|$, it helps to estimate γ and ϕ. Consider the vector $\mathcal{D}'(F) = \mathcal{D}(F) + \{-1\}^n$. Since we assume F has no isolated vertices, the number of non-zero entries in $\mathcal{D}'(F)$ is the number of vertices of degree at least 2, viz. $\|\mathcal{D}'(F)\|_0 = |Deg_{\geq 2}|$. Applying Theorem 2, we $(1 \pm \varepsilon)$-approximate $|Deg_{\geq 2}|$ in $\log^{\mathcal{O}(1)} n$ space with constant success probability. Since no existing L_0 sketch supports initialization with $\{-1\}^n$, degree decrements are post-processed. Running $\mathcal{O}(\log \delta^{-1})$ independent instances and returning the median, the failure probability is less than δ.

Lemma 4. *Given $\varepsilon, \delta \in (0,1)$, $|Deg_{\geq 2}|$ is $(1 \pm \varepsilon)$-approximated in a turnstile stream with probability $(1-\delta)$ and in $\mathcal{O}(\log^{\mathcal{O}(1)} n \cdot \log \delta^{-1})$ space.*

This allows us to $(2\pm\varepsilon)$-approximate ϕ and $(3\pm\varepsilon)$-approximate γ in streaming forests using only polylog space, details are deferred to the full version (See footnote 1).

Estimating Leaves We can estimate $|Deg_1|$ with $n - |Deg_{\geq 2}|$. But if $|Deg_1|$ is in $o(|Deg_{\geq 2}|)$, the relative error can be very large. We turn elsewhere.

Lemma 5. [$\star$] *For every forest, $|Deg_1| = 2c + \sum_{i=3}^{\Delta}(i-2) \cdot |Deg_i|$.*

Post-processing $\mathcal{D}$ with $\{-2\}^n$, to obtain $\mathcal{D}''$, we have $\|\mathcal{D}''(F)\|_1 = |Deg_1| + \sum_{i=3}^{\Delta}(i-2) \cdot |Deg_i|$. Folding in Lemma 5, we have $|Deg_1| = \|\mathcal{D}''(F)\|_1/2 + c$.

Lemma 6. *Given $\varepsilon, \delta \in (0,1)$, $|Deg_1|$ is $(1 \pm \varepsilon)$-approximated in a turnstile stream with probability $(1-\delta)$ in $\mathcal{O}(\log^{\mathcal{O}(1)} n \cdot \log \delta^{-1})$ space.*

Combining Lemma 6 with Theorem 7, we conclude:

Theorem 9. *Forest independence number β is $3/2 \cdot (1 \pm \varepsilon)$-approximated with probability $(1-\delta)$ on turnstile streams using $\mathcal{O}(\log^{\mathcal{O}(1)} n \cdot \log \delta^{-1})$ space.*

Estimating Support Vertices Degree-counting, above, does not suffice to estimate the number of support vertices. We first show, in Subroutine 2, given a turnstile streamed forest, how to output an $(1 \pm \varepsilon)$-estimate of $|Supp|$, when $|Supp|$ is at least a threshold K_1. When $|Supp(F)|$ is small, we can approximate β via either of the following methods:

- If few non-leaves, find $|Deg_{\geq 2}|$ and $|Supp(F)|$ exactly (Subroutine 3).
- If many non-leaves, Corollary 1 enables excluding $|Supp(F)|$.

Similar to Cormode et al. [9] and to Jayaram and Woodruff [23], we assume the number of deletions is $\mathcal{O}(n)$, preserving the analyses of Subroutines 2 and 3.

Lemma 7. [$\star$] *Given threshold K_1, constant c_1, and error rate $\varepsilon_1 \in (0,1)$, Subroutine 2 is a turnstile-stream algorithm that uses $\widetilde{\mathcal{O}}(n/(\varepsilon_1^2 K_1))$ space and: when $|Supp(F)| \geq K_1$, $(1 \pm \varepsilon_1)$-approximates $|Supp(F)|$ with probability $\geq 1 - 3e^{-\frac{c_1}{3}}$; when $|Supp(F)| < K_1$, returns an estimate at most $2K_1$ with probability $\geq 1 - 2e^{-\frac{c_1}{3}}$.*

Subroutine 2. Estimating $|Supp|$

1: **Input**: Size threshold K_1, constant c_1, and error rate $\varepsilon_1 \in (0,1)$
2: **Initialization:** $I \leftarrow \frac{c_1 n}{\varepsilon_1^2 K_1}$ vertices sampled uniformly at random from $[n]$
3: **for** $v \in I$ **do**
4: $l(v) \leftarrow \{\}$
5: $m, t \leftarrow 0$
6: **First Pass**:
7: **for all** $(e = (u, v), i)$ in the stream, with $i \in \pm 1$ **do**
8: $m \leftarrow m + i$
9: **if** $u \in I$ **then**
10: Toggle v's presence in $l(u)$
11: $t \leftarrow t + i$
12: Perform the same operation on v
13: Abort if $t \geq \frac{2m}{n} \frac{c_1 n}{\varepsilon_1^2 K_1} e^{\frac{c_1}{3}}$
14: **Second Pass**:
15: Count the degree of every vertex in $I \cup \left(\bigcup_{w \in I} l(w)\right)$
16: $C \leftarrow \{u \mid u \in I$ and there is some leaf $v \in l(u)\}$
17: **return** $|\widehat{Supp(F)}| \leftarrow |C| \times \frac{\varepsilon_1^2 K_1}{c_1}$

Sparse recovery on $\mathcal{D}'$ recovers all vertices in $Deg_{\geq 2}$, with probability $1 - \delta$, but no leaves. In a second pass, we verify whether each recovered vertex is a support vertex. For a length-two path (a P_2, i.e., an isolated edge), neither endpoint is recovered. Except in the case of a P_2 path, if some neighbour of vertex u is not recovered, then u is a support vertex. Since no leaf is recovered, Subroutine 3 shows how to verify $Deg_{\geq 2}$.

Lemma 8. [$\star$] *Given threshold K_2, constant c_2, Subroutine 3 in two passes returns both $|Supp(F)|$ and $|Deg_{\geq 2}(F)|$ in $\widetilde{\mathcal{O}}(K_2)$ space. When $|Deg_{\geq 2}(F)| \leq K_2$, the failure probability is at most $1/c_2$.*

Algorithm Summary: We run Lemma 6's algorithm, Subroutine 2, and Subroutine 3 concurrently, with $K_1 = K_2 = \mathcal{O}(\sqrt{n})$, $c_1 = \mathcal{O}(\ln \delta^{-1})$, and $c_2 = \mathcal{O}(\delta^{-1})$. Returning the minimum of $(3(n + |Deg_1(F)|))/8$ and $(n + |Deg_1(F)| - |Supp(F)|)/2$, we $4/3 \cdot (1 \pm \varepsilon)$-approximate β with probability $(1 - \delta)$ in $\widetilde{\mathcal{O}}(\sqrt{n})$ space. Details and the proof of Theorem 10 are in the full version (See footnote 1).

Theorem 10. [$\star$] *There is an algorithm that for every $\varepsilon, \delta \in (0,1)$, estimates the independence number $\beta(F)$ in a turnstile forest stream within factor $4/3 \cdot (1 \pm \varepsilon)$ with probability $(1 - \delta)$ in $\widetilde{\mathcal{O}}(\sqrt{n})$ space and two passes.*

5 Conclusion and Open Questions

We designed CARAWAY to estimate the Caro-Wei bound. It improves on Halldórsson et al. [19], to match Cormode et al. [7], also reporting a *solution*

Subroutine 3. Estimating $|Supp(F)|$ when $|Deg_{\geq 2}(F)| \leq K_2$

1: **Input**: A size threshold K_2, a large constant c_2
2: **Initialization**: Sparse recovery sketch $\mathcal{L}$ in [8] with $k = K_2$ and $\delta = 1/c_2$
3: **First Pass**:
4: **for all** $(e = (u, v), i)$ in the stream, $i \in \pm 1$ **do**
5: Update $\mathcal{L}$ with (e, i)
6: Decrement all coordinates in $\mathcal{L}$ by 1
7: Denote the result of sparse recovery from $\mathcal{L}$ by $\mathcal{R}$
8: $p, m \leftarrow 0$
9: **for** $v \in \mathcal{R}$ **do**
10: $d_v, l_v \leftarrow 0$ ▷ Degree and leaf counters
11: **Second Pass**:
12: **for all** $(e = (u, v), i)$ in the stream, $i \in \pm 1$ **do**
13: $m \leftarrow m + i$; $d_u \leftarrow d_u + i$; $d_v \leftarrow d_v + i$
14: **if** $u \notin \mathcal{R}$ and $v \notin \mathcal{R}$ **then**
15: $p \leftarrow p + i$
16: **if** $u \in \mathcal{R}$ and $v \notin \mathcal{R}$ **then**
17: $l_u \leftarrow l_u + i$
18: **if** $v \in \mathcal{R}$ and $u \notin \mathcal{R}$ **then**
19: $l_v \leftarrow l_v + i$
20: **if** $2m \neq n - |\mathcal{R}| + \sum_{v \in \mathcal{R}} d_v$ **then**
21: **Return** *FAIL*
22: **else**
23: $|\widehat{Supp(F)}| \leftarrow |\{u \mid u \in \mathcal{R} \text{ and } l(u) = 1\}| + 2p$
24: $|\widehat{Deg_{\geq 2}}(F)| \leftarrow |\mathcal{R}|$
25: **Return** $|\widehat{Supp(F)}|$ and $|\widehat{Deg_{\geq 2}}|$

in online streaming. A dedicated, write-only *solution space* admits computing $\mathcal{O}(n)$-size solutions, a likely fruitful future direction.

We then invoked the notion of *support vertices* from structural graph theory [12] in the streaming model. We hence approximated graph parameters, via sketches to estimate the number of leaves, non-leaves and support vertices. This progresses towards meeting the approximation ratios for the lower bounds of Esfandiari et al. [15]. Support vertices could also help estimate other streamed graph parameters in forests and other sparse graphs.

Acknowledgement. We would like to thank Robert Krauthgamer for helpful discussions.

Author note. Xiuge Chen is now with Optiver, Sydney. Patrick Eades is now with The University of Sydney.

References

1. Araujo, F., Farinha, J., Domingues, P., Silaghi, G.C., Kondo, D.: A maximum independent set approach for collusion detection in voting pools. J. Parallel Distrib. Comput. **71**(10), 1356–1366 (2011)
2. Bauckmann, J., Abedjan, Z., Leser, U., Müller, H., Naumann, F.: Discovering conditional inclusion dependencies. In: CIKM 2012, pp. 2094–2098 (2012)
3. Boppana, R.B., Halldórsson, M.M., Rawitz, D.: Simple and local independent set approximation. In: SIROCCO 2018, pp. 88–101 (2018)
4. Bury, M., et al.: Structural results on matching estimation with applications to streaming. Algorithmica **81**(1), 367–392 (2019)
5. Bury, M., Schwiegelshohn, C.: Sublinear estimation of weighted matchings in dynamic data streams. In: Bansal, N., Finocchi, I. (eds.) ESA 2015. LNCS, vol. 9294, pp. 263–274. Springer, Heidelberg (2015). https://doi.org/10.1007/978-3-662-48350-3_23
6. Caro, Y.: New results on the independence number. Technical report, Tel-Aviv University (1979)
7. Cormode, G., Dark, J., Konrad, C.: Approximating the Caro-Wei bound for independent sets in graph streams. In: Lee, J., Rinaldi, G., Mahjoub, A.R. (eds.) ISCO 2018. LNCS, vol. 10856, pp. 101–114. Springer, Cham (2018). https://doi.org/10.1007/978-3-319-96151-4_9
8. Cormode, G., Firmani, D.: A unifying framework for ℓ_0-sampling algorithms. Distrib. Parallel Databases **32**(3), 315–335 (2014)
9. Cormode, G., Jowhari, H., Monemizadeh, M., Muthukrishnan, S.: The sparse awakens: streaming algorithms for matching size estimation in sparse graphs. In: ESA 2017, pp. 29:1–29:15 (2017)
10. Cormode, G., Muthukrishnan, S.: An improved data stream summary: the count-min sketch and its applications. J. Algorithms **55**(1), 58–75 (2005)
11. Sarma, A.D., et al.: Finding related tables. In: SIGMOD 2012, pp. 817–828 (2012)
12. DeLaVina, E., Larson, C.E., Pepper, R., Waller, B., Favaron, O.: On total domination and support vertices of a tree. AKCE Int. J. Graphs Comb. **7**(1), 85–95 (2010)
13. Deng, D., et al.: The data civilizer system. In: CIDR (2017)
14. Eidenbenz, S.J.: Online dominating set and variations on restricted graph classes. Technical report/ETH Zurich, Department of Computer Science 380 (2002)
15. Esfandiari, H., Hajiaghayi, M.T., Liaghat, V., Monemizadeh, M., Onak, K.: Streaming algorithms for estimating the matching size in planar graphs and beyond. In: SODA 2015, pp. 1217–1233 (2015)
16. Feigenbaum, J., Kannan, S., McGregor, A., Suri, S., Zhang, J.: On graph problems in a semi-streaming model. Theor. Comp. Sci. **348**(2–3), 207–216 (2005)
17. Ganti, V., Sarma, A.D.: Data cleaning: a practical perspective. Synth. Lect. Data Manage. **5**(3), 1–85 (2013)
18. Gemsa, A., Nöllenburg, M., Rutter, I.: Evaluation of labeling strategies for rotating maps. J. Exp. Algorithmics (JEA) **21**, 1–21 (2016)
19. Halldórsson, B.V., Halldórsson, M.M., Losievskaja, E., Szegedy, M.: Streaming algorithms for independent sets in sparse hypergraphs. Algorithmica **76**(2), 490–501 (2016)
20. Halldórsson, M.M., Radhakrishnan, J.: Greed is good: approximating independent sets in sparse and bounded-degree graphs. Algorithmica **18**(1), 145–163 (1997)

21. Hossain, A.: Automated design of thousands of nonrepetitive parts for engineering stable genetic systems. Nat. Biotech. **38**(12), 1466–1475 (2020)
22. Indyk, P.: A small approximately min-wise independent family of hash functions. J. Algorithms **38**(1), 84–90 (2001)
23. Jayaram, R., Woodruff, D.P.: Data streams with bounded deletions. In: PODS 2018, pp. 341–354 (2018)
24. Johnson, D.S.: Approximation algorithms for combinatorial problems. J. Comput. Syst. Sci. **9**(3), 256–278 (1974)
25. Kane, D.M., Nelson, J., Woodruff, D.P.: On the exact space complexity of sketching and streaming small norms. In: SODA 2010, pp. 1161–1178 (2010)
26. Kane, D.M., Nelson, J., Woodruff, D.P.: An optimal algorithm for the distinct elements problem. In: PODS 2010, pp. 41–52 (2010)
27. Karp, R.M.: Reducibility among combinatorial problems. In: Miller, R.E., Thatcher, J.W., Bohlinger, J.D. (eds.) Complexity of Computer Computations. The IBM Research Symposia Series, pp. 85 103. Springer, Boston (1972). https://doi.org/10.1007/978-1-4684-2001-2_9
28. Kieritz, T., Luxen, D., Sanders, P., Vetter, C.: Distributed time dependent contraction hierarchies. In: SEA 2010, pp. 83 93 (2010)
29. Lemańska, M.: Lower bound on the domination number of a tree. Discussiones Math. Graph Theory **24**(2), 165–169 (2004)
30. Lovász, L.: On the ratio of optimal integral and fractional covers. Discrete Math. **13**(4), 383–390 (1975)
31. Meierling, D., Volkmann, L.: A lower bound for the distance k-domination number of trees. Results Math. **47**(3–4), 335–339 (2005)
32. Milenković, T., Memišević, V., Bonato, A., Pržulj, N.: Dominating biological networks. PLOS One **6**(8), 0023016 (2011)
33. Nacher, J.C., Akutsu, T.: Dominating scale-free networks with variable scaling exponent: heterogeneous networks are not difficult to control. New J. Phys. **14**(7), 073005 (2012)
34. Panconesi, A., Srinivasan, A.: Randomized distributed edge coloring via an extension of the Chernoff-Hoeffding bounds. SICOMP **26**(2), 350–368 (1997)
35. Pino, T., Choudhury, S., Al-Turjman, F.: Dominating set algorithms for wireless sensor networks survivability. IEEE Access **6**, 17527–17532 (2018)
36. Shen, C., Li, T.: Multi-document summarization via the minimum dominating set. In: COLING 2010, pp. 984–992 (2010)
37. Turán, P.: On an extremal problem in graph theory. Mat. Fiz. Lapok, 436–452 (1941)
38. Wang, J., Li, G., Fe, J.: Fast-join: an efficient method for fuzzy token matching based string similarity join. In: ICDE 2011, pp. 458–469 (2011)
39. Wei, V.: A lower bound on the stability number of a simple graph. Technical report, Bell Laboratories Technical Memorandum (1981)
40. Yu, J., Wang, N., Wang, G., Yu, D.: Connected dominating sets in wireless ad hoc and sensor networks-a comprehensive survey. Comput. Commun **36**, 121–134 (2013)

Efficient k-Center Algorithms for Planar Points in Convex Position

Jongmin Choi[1,2], Jaegun Lee[2], and Hee-Kap Ahn[3(✉)]

[1] CryptoLab, Seoul, Korea
icothos@cryptolab.co.kr

[2] Department of Computer Science and Engineering, Pohang University of Science and Technology, Pohang, Korea
jagunlee@postech.ac.kr

[3] Graduate School of Artificial Intelligence, Department of Computer Science and Engineering, Pohang University of Science and Technology, Pohang, Korea
heekap@postech.ac.kr

Abstract. We present an efficient algorithm for the planar k-center problem for points in convex position under the Euclidean distance. Given n points in convex position in the plane, our algorithm computes k congruent disks of minimum radius such that each input point is contained in one of the disks. Our algorithm runs in $O(n^2 \min\{k, \log n\} \log n + k^2 n \log n)$ time. This is the first polynomial-time algorithm for the k-center problem for points in convex position. For any fixed integer k, the running time is $O(n^2 \log n)$. Our algorithm works with little modification for the k-center problem of points in convex position under the Minkowski distance of order p for any fixed positive integer p.

1 Introduction

A finite set of points is said to be *in convex position* or *convex independent* if none of the points can be represented as a convex combination of the others. Thus, for points in convex positions, all of the points are the vertices of their convex hull. Some computationally hard problems can be solved trivially or efficiently under the assumption of convex position. For instance, the traveling salesman problem is NP-hard for arbitrary planar point sets, but it can be solved in polynomial time for points in convex position: the tour visiting the points in order along the convex hull of the points is an optimal tour. The minimum-weighted triangulation is NP-hard for arbitrary planar point sets, but it can be solve in polynomial time using dynamic programming for points in convex position. However, this is not always the case.

This research was supported by the Institute of Information & communications Technology Planning & Evaluation(IITP) grant funded by the Korea government(MSIT) (No. 2017-0-00905, Software Star Lab (Optimal Data Structure and Algorithmic Applications in Dynamic Geometric Environment)) and (No. 2019-0-01906, Artificial Intelligence Graduate School Program(POSTECH)).

P. Morin and S. Suri (Eds.): WADS 2023, LNCS 14079, pp. 262–274, 2023.
https://doi.org/10.1007/978-3-031-38906-1_18

In this paper, we consider the k-center problem for points in convex position. The k-center problem is defined as follows: Given a set of n points in space, cover the points using k balls so that the maximum radius of a ball is minimized. The k-center problem is a fundamental problem that has a long history. The best known algorithms for the Euclidean k-center for n arbitrary points take $O(n^{O(\sqrt{k})})$ time [20] in the plane, and $O(n^{O(k^{1-\frac{1}{d}})})$ time [1] in the d-dimensional Euclidean space. For $k = 2$, there were a fair amount of work on the planar k-center problem [2,7,14,15,17,19,22,23,26,27]. The problem was solved optimally in $O(n \log n)$ time [9,10] recently.

Much less is known about the k-center problem for points in convex position. Choi and Ahn gave a deterministic $O(n \log n)$-time algorithm for the two-center problem for n points in convex position in the plane [10]. Their algorithm is optimal as any deterministic algorithm for the two-center problem requires $\Omega(n \log n)$ time in the algebraic decision tree model of computation [15]. Very recently, an $O(n^2 \log^3 n)$-time algorithm is given for the planar Euclidean 3-center problem for points in convex position [4]. However, no efficient algorithm for the planar k-center problem with $k > 3$ is known for points in convex position.

There were a few polynomial-time algorithms known for points under certain restriction. For n weighted points in one-dimensional real line, the k-center problem can be solved in $O(n \log n)$ time [8]. The k-center problem for skylines was considered in the plane. A sequence of points are said to be in the skyline if their x-coordinates are increasing while their y-coordinates are decreasing in order along the sequence. The centers must be chosen among the input points in the discrete version while they can be arbitrary points in the plane in the continuous version. The best algorithm for the discrete Euclidean k-center problem for the skyline of n points in the plane takes $O(n \log k + n \log \log n)$ time [5]. If it is allowed to uncover m points among n points in the skyline, both the continuous and discrete version can be computed in $O(kn(1 + m) \log n)$ time [12].

Our result. We consider the planar k-center problem for points in convex position in the plane. Our main theoretical result is to show that the planar k-center problem can be solved in polynomial time. More specific, we show that the optimal radius and an optimal set of the centers of k disks covering n input points can be computed in $O(n^2 \min\{k, \log n\} \log n + k^2 n \log n)$ time under the Euclidean distance. This is the first polynomial-time algorithm for the k-center problem for points in convex position.

For any fixed integer k, the running time is $O(n^2 \log n)$. Our algorithm improves upon the $O(n^2 \log^3 n)$-time algorithm for $k = 3$ [4]. Our algorithm works with little modification for the k-center problem for points in convex position under the Minkowski distance of order p for any fixed positive integer p.

1.1 Sketch of Our Algorithm

Let r^* be the minimum radius of k congruent disks that cover an input sequence of n points in convex position. Then r^* belongs to the set of $O(n^3)$ radii, each by the circle determined by at most three input points. However, no efficient algorithm is known for deciding whether $r \geqslant r^*$ or not for a radius r.

First, we observe that for any fixed $r \geqslant r^*$, there are k disks with radius r satisfying the following condition: The input sequence can be partitioned into substrings such that each substring is assigned to exactly one disk covering the substring, at most two substrings are assigned to one disk, the two substrings assigned to a disk are nonconsecutive, and the substrings assigned to one disk can be separated from the substrings assigned to any other disk by a line.

Based on these observations, we give a decision algorithm that for a given r, determines whether $r \geqslant r^*$ or not. In specific, it computes the minimum number of disks with radius r that cover the input sequence. In doing so, it computes for each input point and for each $\ell = 1, \ldots, k$, the maximum length of the substring starting from the point for which there is a set of ℓ disks with radius r satisfying the condition. Then it determines whether any input point gets length value n.

We give two efficient methods that compute the lengths, one making use of a dynamic data structure for maintaining the intersections of disks and one pruning candidates using a lower bound. By combining these two methods, the decision algorithm takes $O(n^2 \min\{k, \log n\} + k^2 n)$ time.

Using the decision algorithm, we can compute r^* in $O(n^3 \log n)$ time by applying binary search over the set of $O(n^3)$ radii, each defined by two or three input points. To make the search efficiently without considering all the radii, we compute r^* in two stages. In the first stage, we compute $O(n)$ radii for each input point using $O(n)$ substrings. These substrings are defined by dividing a substring of length $n-1$ into halves recursively until it becomes a substring of length 1. In total, we compute $O(n^2)$ radii. Then we compute the minimal interval of the radii that contains r^* by applying binary search using the decision algorithm. In the second stage, we apply parametric search using the decision algorithm to find r^* in the interval. By combining sequential and parallel algorithms with Cole's parametric technique [11], we obtain $O(n^2)$ candidates for r^*. Then we apply binary search on the candidates and find r^* in $O(n^2 \min\{k, \log n\} \log n + k^2 n \log n)$ time. Once r^* is computed, our decision algorithm finds k disks with radius r^* covering P by bookkeeping.

The proofs of some lemmas and theorems are available in the full version of the paper.

2 Preliminaries

A finite set of points is said to be *in convex position* if none of the points can be represented as a convex combination of the others. Let $P = \{p_0, p_1, \ldots, p_{n-1}\}$ be a set of n points in the convex position in the plane. We assume that the points of P with their indices are ordered along their convex hull $\mathsf{CH}(P)$ in clockwise direction. We regard P as a *cyclic sequence* of points and denote it by $\langle p_0, p_1, \ldots, p_{n-1} \rangle$. We extend the indices to all integers so that they are understood modulo n; i.e. $p_i = p_j$ if and only if $i \equiv j \bmod n$.

A *subsequence* of a (cyclic) sequence is a sequence obtained from the original sequence by deleting zero or more elements without changing the order of the remaining elements. A *substring* of a (cyclic) sequence is a subsequence that consists of a consecutive run of elements from the original sequence. We say two substrings of a (cyclic) sequence are *consecutive* if their concatenation in some order is also a substring of the original sequence. We use $P(i,t)$ to denote the substring $\langle p_i, p_{i+1}, \ldots, p_{i+t-1}\rangle$ of t points of P in clockwise direction along P. We regard $P(i,t)$ of P as a *noncyclic* substring of P.

We say a disk D with radius r *covers* a point p if $p \in D$. We say a disk D *covers* a set S of points if $S \subset D$. A set D of disks covers a set of points if each point in the set is covered by one of the disks in D.

3 Properties

Let r^* be the minimum radius of k congruent disks that cover P. Then an optimal k-center of P consists of k disks with radius r^* such that each disk covers substrings of P. See Fig. 1 for an illustration for $k = 4$.

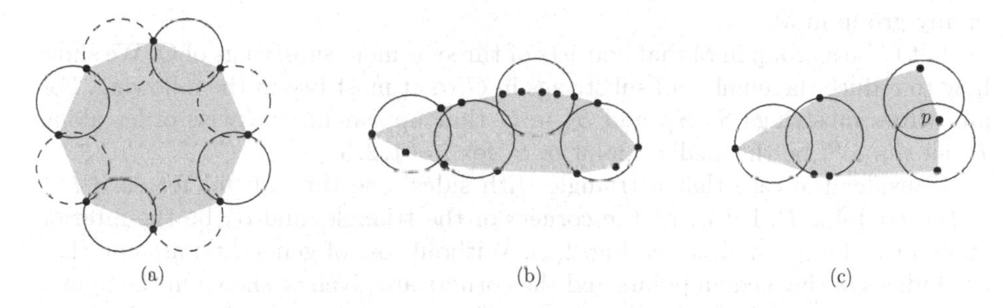

Fig. 1. Optimal solutions to the 4-center problem for point sets in convex position. (a) Two solutions, one consisting of all solid disks and one consisting of all dashed disks. Each disk covers a substring consisting of two consecutive points. (b) An optimal solution. A disk may cover two disjoint and nonconsecutive substrings. (c) Point p is covered by two disks in an optimal solution.

We provide a property of an optimal solution to the k-center problems for points in convex position. To ease the discussion, we define a few terms.

For a (cyclic) sequence $S = \langle s_1, s_2, \ldots, s_t\rangle$ of t points, consider a partition of S into nonempty substrings. We say that a set of disks *covers* the partition if each substring in the partition is covered by one disk of the set. Notice that a substring can be covered by one or more disks. A map for a partition Q is defined with a set D of disks that covers Q using $\mathsf{Q} \times \mathsf{D}$, where $\mathsf{Q} \times \mathsf{D}$ consists of all pairs (Q, D) with $Q \in \mathsf{Q}$ and $D \in \mathsf{D}$ such that the substrings in Q are covered by D. Thus, a map defines a collection of disjoint groups, one for each disk consisting of the substrings mapped to the disk. A map is called *balanced* if each group in the collection defined by the map has at most two nonconsecutive substrings. A map is called *line-separable* if every two groups can be separated by a line.

An (ℓ, r)*-partition* of S for a positive integer ℓ and a radius r is a partition of S into substrings such that there is a set of ℓ disks with radius r covering the partition. We say that S admits an (ℓ, r)*-cover* if there are an (ℓ, r)-partition Q for S and a balanced and line-separable map M for the partition. In this case, the (ℓ, r)-cover consists of Q and M. The following two lemmas provide a main property of an optimal solution to the k-center problem.

Lemma 1. *If P can be covered by ℓ disks with radius r, there is an (ℓ, r)-partition of P and a line-separable map for the partition.*

Lemma 2. *For any fixed $r \geqslant r^*$, P admits a (k, r)-cover.*

Proof. Suppose that for a fixed $r \geqslant r^*$, no (k, r)-partition of P has a balanced map. By Lemma 1, there is a (k, r)-partition and a map for the partition such that every two groups in the map can be separated by a line. Let Q be a (k, r)-partition and M be a map for Q such that every two groups in M can be separated by a line. If there are two consecutive substrings in a group, they can be concatenated into one. So we assume that there are no two consecutive substrings in any group in M.

Let G be a group in M that consists of three or more substrings of Q. We show how to reduce the number of substrings in G to at most two in the following. For any three substrings S_1, S_2, and S_3 in G that appear in clockwise order along P, let $m_i \in S_i$ be the median point of S_i for $i = 1, 2, 3$.

Consider the case that a triangle with sides, one through m_i for each $i = 1, 2, 3$, contains P. Let q_j be the corners of the triangle, and α_j be the interior angle at q_j for $j = 1, 2, 3$. See Fig. 2(a). Without loss of generality, assume that the indices of the median points and the corners are given as shown in the figure. For two points p and q in the plane, we use $\|p-q\|$ to denote the distance between p and q under the Euclidean distance.

If $\alpha_1 \geqslant \pi/2$, the substring $S = \langle m_1, \ldots, m_2 \rangle$ of P and q_1 can be covered by a disk with radius r because $\|m_1 - m_2\| \leqslant 2r$. See Fig. 2(b).

We modify Q and M as follows. We insert two substrings $S_1 \setminus S$ and $S_2 \setminus S$ to Q, and insert the group G' consisting of these two substrings and the substrings of G that share no point of S to M. We remove S_1 and S_2 from Q, and remove G from M. Each group in M, except G', belongs to one of the following two types: (A) it contains a substring that is contained in S, and (B) it contains no substring that is contained in S. Since the points in P are in convex position, G' and any group of type (A) can be separated by the line through m_1 and m_2. Let $\bar{G}$ be the group consisting of S. We remove all the groups of type (A) from M and the substrings of the groups from Q. Then we insert S to Q and $\bar{G}$ to M. We make no change to the groups of type (B) and their substrings.

Then the resulting partition is still a (k, r)-partition and the resulting map is line-separable. Moreover, the number of points contained in two substrings S_1 and S_2 of G gets halved by the modification in the new group G'. Each substring of G, except S_1 and S_2, does not belongs to G' if it is contained in S. It remains the same and belongs to G' if it is not contained in S. All groups of type (A) are merged to one group $\bar{G}$ consisting of a single substring S without changing any group of type (B) and its substrings. The case that $\alpha_2 \geqslant \pi/2$ or $\alpha_3 \geqslant \pi/2$ can be handled similarly.

Assume that $\alpha_i < \pi/2$ for all $i = 1, 2, 3$. Suppose that no triangle $\triangle m_i q_i m_{i+1}$ can be covered by a disk with radius r for all $i \in \{1, 2, 3\}$. Then $\|m_i - m_{i+1}\| > 2r \sin \alpha_i$ for $i = 1, 2, 3$. The disk D determined by m_1, m_2, and m_3 has the radius r' which is at most r. For the center c of D, we have $\cos(\angle c m_i m_{i+1}) > r \sin \alpha_i / r' \geqslant \sin \alpha_i$, so $\angle c m_i m_{i+1} < \pi/2 - \alpha_i$. Thus, the sum of angles at the corners of triangle $\triangle m_1 m_2 m_3$ is smaller than $3\pi - 2\sum_i \alpha_i = \pi$, which is a contradiction. See Fig. 2(c). Thus, $\triangle m_i q_i m_{i+1}$ can be covered by a disk with radius r for some $i \in \{1, 2, 3\}$, implying that substring $\langle m_i, \cdots, m_{i+1}\rangle$ can be covered by a disk with radius r. So we can update Q and M as in the previous paragraph.

Consider the case that no triangle with sides, one through m_i and for each $i = 1, 2, 3$, contains P. There always exists a wedge that contains P, touches m_i and m_{i+1} along its sides for some $i \in \{1, 2, 3\}$, and has interior angle at least $\pi/2$. So the substring $\langle m_i, \cdots, m_{i+1}\rangle$ can be covered by a disk with radius r. So we can update Q and M as in the former case.

By repeating the process for each group G in M that contains three or more substrings of Q, we can get an (k, r)-partition and a balanced and line-separable map for the partition in a finite number of iterations. Thus, P admits a (k, r)-cover. ◻

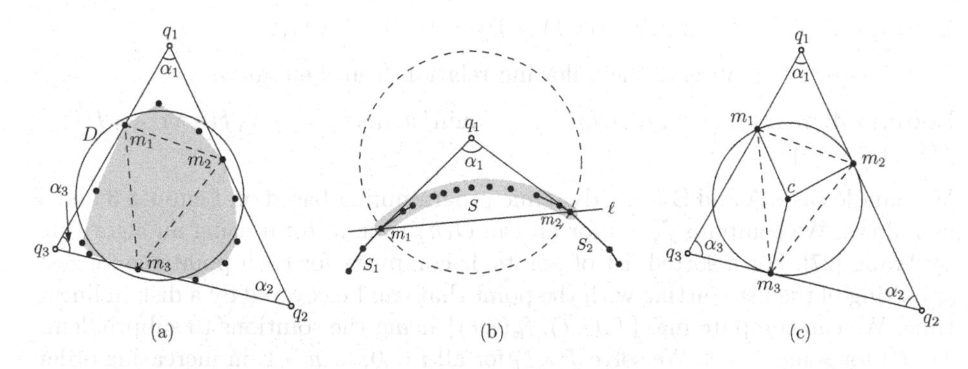

Fig. 2. (a) The triangle with sides tangent to CH(P) at m_1, m_2, and m_3. (b) $S = \langle m_1, \cdots, m_2\rangle$ is covered by a disk with radius r. (c) The circumcircle of m_1, m_2, m_3.

4 Decision Algorithm

We give an algorithm for the decision version of the problem: Given a set P of n points in convex position, a positive integer k, and a real value $r > 0$, determine whether P admits a (k, r)-cover. To ease the discussion, we assume that every disk has radius r unless otherwise stated.

For integers i with $0 \leqslant i \leqslant n-1$ and ℓ with $1 \leqslant \ell \leqslant k$, let $f(i,\ell)$ denote the length of the longest substring of P from p_i that admits an (ℓ,r)-cover. For a fixed ℓ, if $f(i,\ell)=n$ for some i, P can be covered by ℓ disks. If $f(i,\ell)<n$ for all i with $0 \leqslant i \leqslant n-1$, there is no way to cover P using ℓ disks.

Our algorithm works as follows. It computes $f(i,\ell)$ for all i and ℓ. Then it determines whether there are some integers i and ℓ with $f(i,\ell)=n$. Before describing the details of the algorithm, we give some observations and properties.

Observation 1 *The value $f(i,\ell)+i$ is nondecreasing while i or ℓ is increasing.*

Clearly, $f(i,\ell)+i$ is nondecreasing while ℓ increases. By definition, there are ℓ disks that together cover $P(i,f(i,\ell))$. Then $P(i+1,f(i,\ell)-1)$ can also be covered by the ℓ disks. Thus, we have $f(i,\ell)-1 \leqslant f(i+1,\ell)$, implying that $f(i,\ell)+i$ is nondecreasing while i is increasing.

Consider an (ℓ,r)-cover of $P(i,f(i,\ell))$ consisting of an (ℓ,r)-partition $\bar{\mathsf{Q}}$ and a map $\bar{\mathsf{M}}$. Let G be the group in $\bar{\mathsf{M}}$ that contains the first substring $S=P(i,t)$ for some integer t with $1 \leqslant t \leqslant f(i,1)$. Then there are three cases for G:

A. G consists of S.
B. G consists of S and another substring of $\bar{\mathsf{Q}}$ that is not the last one.
C. G consists of S and the last substring of $\bar{\mathsf{Q}}$.

Let $f_{\mathsf{A}}(i,\ell), f_{\mathsf{B}}(i,\ell)$ and $f_{\mathsf{C}}(i,\ell)$ denote the maximum $f(i,\ell)$ values for cases A, B, and C, respectively. Then $f(i,\ell)=\max\{f_{\mathsf{A}}(i,\ell), f_{\mathsf{B}}(i,\ell), f_{\mathsf{C}}(i,\ell)\}$. We have some properties and recurrence relations as in the followings.

Lemma 3. $f_{\mathsf{A}}(i,\ell)=\min\{n, f(i,1)+f(i+f(i,1),\ell-1)\}$.

For case B, we observe the following relation from Lemma 3.

Lemma 4. $\max\{f_{\mathsf{A}}(i,\ell), f_{\mathsf{B}}(i,\ell)\} = \min\{n, \max_{1\leqslant\gamma\leqslant\ell-1}\{f(i,\gamma)+f(i+f(i,\gamma),\ell-\gamma)\}\}$.

We handle cases A and B using dynamic programming based on Lemmas 3 and 4 as follows. We compute $f(i,1)$ for all i in $O(n)$ time in total using an algorithm by Wang [27]. For a sorted list of points, it computes for each point the longest substring of the list starting with the point that can be covered by a disk in linear time. We can compute $\max\{f_{\mathsf{A}}(i,\ell), f_{\mathsf{B}}(i,\ell)\}$ using the solutions to subproblems $f(\cdot,\ell')$ for some $\ell'<\ell$. We solve $f(i,\ell)$ for all $i=0,\dots n-1$, in increasing order of ℓ values, and compute $\max\{f_{\mathsf{A}}(i,\ell), f_{\mathsf{B}}(i,\ell)\}$ in $O(k^2n)$ time in total for all i and ℓ values.

Lemma 5. $\max\{f_{\mathsf{A}}(i,\ell), f_{\mathsf{B}}(i,\ell)\}$ *can be computed in* $O(k^2n)$ *time for all integers* $i=0,\dots,n-1$ *and* $\ell=1,\dots,k$.

For case C, G consists of S and the last substring of $\bar{\mathsf{Q}}$. To maximize $f_{\mathsf{C}}(i,\ell)$, we need to consider both substrings at the same time.

Observation 2 *$f_{\mathsf{C}}(i,\ell)$ is the largest possible integer β such that a disk covers both $P(i,t)$ and $P(i+t+f(i+t,\ell-1),\beta-t-f(i+t,\ell-1))$ for some integer t with $1 \leqslant t \leqslant f(i,1)$.*

To compute $f_{\mathsf{C}}(i,\ell)$ for fixed integers i and ℓ, there are $O(n)$ candidate values for each of t and β. We can determine whether n points in the plane can be covered by a disk with radius r in $O(n)$ time [25]. Thus, we can compute $f_{\mathsf{C}}(i,\ell)$ in $O(kn^4)$ time for all integers i and ℓ with $0 \leqslant i \leqslant n-1$ and $1 \leqslant \ell \leqslant k$. We present two efficient algorithms that compute $f_{\mathsf{C}}(i,\ell)$ for all i and ℓ, one in $O(kn^2)$ time and one in $O(n^2 \log n)$ time in total.

4.1 $O(kn^2)$-Time Algorithm for Case C

Let $f_{\mathsf{AC}}(i,\ell) = \max\{f_{\mathsf{A}}(i,\ell), f_{\mathsf{C}}(i,\ell)\}$. We present an efficient algorithm that computes f_{AC}. Since $f_{\mathsf{AC}}(i,\ell) \geqslant f_{\mathsf{A}}(i,\ell)$, we use $f_{\mathsf{A}}(i,\ell)$ as a lower bound in computing f_{AC}. We assume that $f_{\mathsf{A}}(i,\ell) < n$. If $f_{\mathsf{A}}(i,\ell) = n$, there is an (ℓ,r)-cover for P, and thus P admits a (k,r)-cover and the algorithm terminates.

By Observation 2, there are $O(n^2)$ candidates for t and $\beta = f_{\mathsf{C}}(i,\ell)$. We show how to compute the maximum possible value of $f_{\mathsf{C}}(i,\ell)$ by considering only $O(n)$ candidates in the following. From this, we construct a data structure determining whether a given substring can be covered by a disk with radius r or not in amortized $O(1)$ time.

Let $X_{\beta,t}$ be the Boolean variable that is set to true if $P(i,t)$ and $P(i+t+f(i+t,\ell-1), \beta-t-f(i+t,\ell-1))$ can be covered by a disk with radius r, and that is set to false if there is no such disk. We compute the largest β such that $X_{\beta,t} = \mathsf{true}$. Let $\mathsf{X} = \{X_{\beta,t}\}$ be the $n \times n$ matrix whose (β,t)-entry is $X_{\beta,t}$ for integers β and t with $1 \leqslant \beta \leqslant n$ and $1 \leqslant t \leqslant n$. Note that $X_{\beta',t} = \mathsf{false}$ for all $\beta' \geqslant \beta$ if $X_{\beta,t} = \mathsf{false}$. To find the largest β, we initially set $t=1$ and $\beta = f_{\mathsf{A}}(i,\ell) = f(i,1) + f(i+f(i,1),\ell-1)$. We traverse matrix X as follows; If $X_{\beta,t} = \mathsf{false}$, increase t by 1. If $X_{\beta,t} = \mathsf{true}$, increase β by 1. During the traversal, t is at most $f(i,1)$, and thus $i+t+f(i+t,\ell-1) \leqslant i+f(i,1)+f(i+f(i,1),\ell-1)$.

Let S_1 and S_2 be the two substrings of $P(i+t+f(i+t,\ell-1), \beta-t-f(i+t,\ell-1))$ such that S_1 is from the point with index $i+t+f(i+t,\ell-1)$ to the point with index $i+f(i,1)+f(i+f(i,1),\ell-1)$ and S_2 is from the point with index $i+f(i,1)+f(i+f(i,1),\ell-1)+1$ to the last point of $P(i+t+f(i+t,\ell-1), \beta-t-f(i+t,\ell-1))$ in clockwise direction. When t increases by one, a prefix of S_1 is removed from S_1. When β increases by one, the point of P next to S_2 is appended to S_2.

Our algorithm also maintains a data structure determining whether $X_{\beta,t}$ is true or false efficiently during the traversal. The data structure is based on the the dynamic data structure by Wang [27]. Wang's data structure maintains the intersection of disks centered at points in a sequence T sorted along the x-axis over insertions of points into T at one end of T and deletions of points from T at the other end of T. The data structure supports a sequence of n such insertions and deletions of points in $O(n)$ time in total for a sequence T of size $O(n)$. Since the data structure also works for a sequence of points in convex position [27], it can be extended to work with little modification for the concatenation of two disjoint substrings of P over such insertions and deletions of points.

Lemma 6 ([27]). *There is a data structure for two disjoint substrings S_1 and S_2 of P that maintains the intersection of disks with a fixed radius, each centered* *at a point in $S_1 \cup S_2$ over appending operations of the point in P next to S_2 to S_2 and deleting operations of the first point of S_1 from S_1. The data structure supports a sequence of n such appending and and deleting operations of points in $O(n)$ time in total.*

Our data structure is an extension of Wang's data structure in Lemma 6 that works on three disjoint substrings S_1, S_2, and S_3 of P. We can determine whether the circular hull of the points in $S_1 \cup S_2 \cup S_3$ is empty or not by maintaining Wang's data structures for some pairs of substrings and using the indices of the vertices of the circular hull.

Lemma 7. *We can construct a dynamic data structure in $O(m)$ time for three disjoint substrings S_1, S_2, and S_3 of P with $|S_1|+|S_2|+|S_3| = m$ that maintains the intersection of disks with a fixed radius, each centered at a point in $S_1 \cup S_2 \cup S_3$ over appending operations of the point of P next to S_1 (or S_3) to S_1 (or S_3) and deleting operations of the first point of S_2 from S_2 in amortized $O(1)$ time.*

While β increases, $i+t+f(i+t,\ell)$ is nondecreasing by Observation 1. Note that t is nondecreasing during our traversal in X. Thus we can find in $O(n)$ time $\max\{f_{\mathsf{A}}(i,\ell), f_{\mathsf{C}}(i,\ell)\}$ by Lemma 7. As there are $O(kn)$ pairs of i and ℓ values, it takes $O(kn^2)$ time to compute $\max\{f_{\mathsf{A}}(i,\ell), f_{\mathsf{C}}(i,\ell)\}$ for all i and ℓ values.

Lemma 8. *We can compute $f_{\mathsf{AC}}(i,\ell)$ for all integers $i = 0, \ldots n-1$ and $\ell = 1, \ldots k$ in $O(kn^2)$ time.*

4.2 $O(n^2 \log n)$-Time Algorithm for Case C

We present another efficient algorithm that computes $f_{\mathsf{AC}}(i,\ell)$.

Lemma 9. *We can compute $f_{\mathsf{AC}}(i,\ell)$ for all $i = 0, \ldots n-1$ and $\ell = 1, \ldots k$ in $O(n^2 \log n)$ time.*

By combining Lemmas 5, 8 and 9, we can determine whether $P(i,n)$ admits a (k,r)-cover or not for all $i = 0, \ldots, n-1$. Since this is equivalent to determining whether $r \geqslant r^*$ or not by Lemma 2, we have the following theorem.

Theorem 1. *Given a set of n points in convex position, an integer k, and a radius r, we can determine in $O(n^2 \min\{k, \log n\} + k^2 n)$ time whether the set admits a (k,r)-cover or not.*

The decision version of the k-center problem for planar points is closely related to covering the points using the minimum number of unit disks. Our decision algorithm for the k-center problem for points in convex position runs by increasing ℓ from 1 until $f(i,\ell) = n$ for some $i \in \{0, \ldots, n-1\}$. Thus, our decision algorithm works for the unit-disk covering problem for points in convex position. But to achieve $O(n^2 \min\{k, \log n\} + k^2 n)$ time, we use the $O(\ell n^2)$-time algorithm to compute case C for $\ell = O(\log n)$, and the $O(n^2 \log n)$-time algorithm for $\ell = \omega(\log n)$.

Corollary 1. *Given a set of n points in convex position in the plane, we can compute in $O(n^2 \min\{m, \log n\} + m^2 n)$ time a smallest set of unit disks that covers the points, where m is the number of unit disks in the set.*

5 Search Algorithm

We give an algorithm that computes the minimum radius r^* of k congruent disks covering P in $O(n^2 \min\{k, \log n\} \log n + k^2 n \log n)$ time. We first compute a coarse set of radii and the minimal interval of the radii that contains r^*. Then we find r^* by parametric search and binary search. Since r^* is the radius of a smallest disk D covering a group G in a (k, r^*)-cover, r^* is determined by a substring of G and at most one point of the other substring of G.

Observation 3 *The optimal radius r^* is the radius of the smallest disk that covers a substring S of P and contains at most one point of $P \backslash S$ on its boundary.*

To perform the search efficiently, we compute r^* in two stages.

5.1 Computing an Interval Containing r^*

We compute a coarse set of radii and the minimal interval of the radii that contains r^* using Observation 3. For a point $p_u \in P$, let P_u be the substring of P consisting of the points in $P \setminus \{p_u\}$. We define two substrings of P by dividing P_u into two halves. We do this recursively on each substring until it consists of one point. Let S be the set of these substrings. There are $O(n)$ substrings in S.

We compute the radius of the smallest disk that contains p_u on its boundary and covers a substring of S. Let R_u be the set of all radii, each defined for a substring of S. Instead of computing R_u, we compute a set $\bar{R}_u$ of $O(n)$ distances such that $R_u \subseteq \bar{R}_u$. For $S \in \mathsf{S}$, let $H(S) = \bigcap_{p \in S} H(p)$, where $H(p)$ is the closed half-plane bounded by the perpendicular bisector of p_u and p that does not contain p_u. As in the merge step of the mergesort, we take two halves $S_1, S_2 \in \mathsf{S}$ of S, compute $H(S)$ by using $H(S_1)$ and $H(S_2)$, and compute the distance between p_u and $H(S)$. In addition, we compute the distance between p_u and the intersection of the boundaries of $H(S_1)$ and $H(S_2)$. This can be done in a way similar to that of the data structure by Guibas and Hershberger [16].

Lemma 10 **([16]).** *Given n lines $l_0, \ldots, l_{n-1}$ sorted by slope, we can construct a data structure in $O(n)$ time using $O(n)$ space that given a query (i, j), returns a structure representing the upper envelope of $l_i, \ldots, l_j$ in $O(\log n)$ time.*

Let $\bar{R}_u$ be the set of those distances computed for p_u and the substrings of S. We show $R_u \subseteq \bar{R}_u$. Each radius $r \in R_u$ is the radius of the smallest disk D defined by p_u and a substring $S \in \mathsf{S}$. If D is determined by p_u and a point $p \in S$, r is the distance between p_u and $H(p)$, which is contained in $\bar{R}_u$. If D is determined by p_u and two points $p_1, p_2 \in S$, the intersection q of the boundaries of $H(p_1)$ and $H(p_2)$ is the center of D. If q does not appear as a vertex of $H(S)$, there is a point $p' \in S$ such that $q \notin H(p')$. Then $p' \notin D$, which is a contradiction. Therefore, q always appears as a vertex of $H(S)$, r is the distance between p_u and the vertex, and it is also contained $\bar{R}_u$. Therefore, $R_u \subseteq \bar{R}_u$.

We construct the data structure (a binary tree) in Lemma 10 for p_u, one for each $u = 0, \ldots, n-1$. So in total, we construct n data structures in $O(n^2)$

total time using $O(n^2)$ space. It takes $O(n^2 \log n)$ time to compute all $O(n^2)$ distances. We also compute $O(n^2)$ radii of disks that are determined by $P(i,t)$ for all pairs of indices i, t using Lemma 11.

Lemma 11. *For all pairs of indices i, t, the smallest disk enclosing $P(i,t)$ can be computed in $O(n^2 \log n)$ time.*

Let R be the set of the $O(n^2)$ distances and radii. We sort the values of R, apply binary search over the sorted list using the decision algorithm in Theorem 1, and find the minimal interval $(r_L, r_U]$ containing r^* in $O(T_S \log n)$ time, where $T_S = O(n^2 \min\{k, \log n\} + k^2 n)$ is the running time of the decision algorithm.

Lemma 12. *We can compute R and the minimal interval of the values in R that contains r^* in $O((n^2 \min\{k, \log n\} + k^2 n) \log n)$ time.*

5.2 Computing r^*

Now we have $(r_L, r_U]$ containing r^*. Based on Observation 3, we compute $O(n^2)$ candidates of r^* in $(r_L, r_U]$ by parametric search as follows.

Let $\mathcal{T}$ be the data structure in Lemma 10 corresponding to p_u that is constructed in Sect. 5.1. For a node w in $\mathcal{T}$, let $S(w)$ be the substring of P corresponding to w. Let $C_u(r)$ be the circle with radius r and centered at p_u. Consider the intersection $H(S(w)) \cap C_u(r)$, which is a circular arc on $C_u(r)$, possible empty. Each endpoint of the circular arc lies on a boundary segment of $H(S(w))$. Moreover, it remains to lie on the same boundary segment for all $r \in (r_L, r_U]$. Thus, the intersection can be represented by a linear function on $r \in (r_L, r_U]$. For each node w of $\mathcal{T}$, we store this function $\kappa(w, r)$ corresponding to w, which can be done in $O(n)$ time. We do this for all n data structures in $O(n^2)$ time.

For integers i, t with $0 \leqslant i \leqslant n-1$ and $1 \leqslant t \leqslant n$ such that $p_u \notin P(i,t)$, let $r_u(i,t)$ be the radius of the smallest disk covering $P(i,t)$ and p_u, and let $t_u(i,r) = \max\{t \mid r_u(i,t) \leqslant r\}$. Observe that $r_u(i,t)$ is nondecreasing while t increases, and $t_u(i,r)$ is nondecreasing while r increases.

Given $r \in (r_L, r_U]$ and two indices i and t, we want to determine whether $r \geqslant r_u(i,t)$ or not efficiently. $P(i,t)$ can be represented by a set W of $O(\log n)$ nodes of $\mathcal{T}$, at most two at each level of $\mathcal{T}$. By definition, $r \geqslant r_u(i,t)$ if and only if $H(P(i,t)) \cap C_u(r) \neq \emptyset$. This is equivalent to $\bigcap_{w \in W} \kappa(w,r) \neq \emptyset$.

Parametric Search. To apply parametric search, we run a parallel algorithm that given $r \in (r_L, r_U]$, computes $t_u(i,r)$ for all integer i with $0 \leqslant i \leqslant n-1$.

Lemma 13. *For any fixed integers u and i and any fixed $r \in (r_L, r_U]$, we can compute $t_u(i,r)$ in $O(\log n)$ time after $O(n)$-time preprocessing.*

By Lemma 13, we can design an algorithm for any fixed $r \in (r_L, r_U]$ that finds $t_u(i,r)$ for all integers u and i in $O(\log n)$ parallel time using $O(n^2)$ processors after $O(n^2)$-time preprocessing. By using Cole's technique [11], we can find such r as follows. Using a sequential decision algorithm with running time T_S and a parallel computation that can be done in T_P steps using N_P processors, each processor running independently, we have an $O((T_S + N_P)(T_P + \log N_P))$-time algorithm for finding such r.

Theorem 2. *Given n points in convex position in the plane and an integer k, we can find k congruent disks of minimum radius that cover the points in $O(n^2 \min\{k, \log n\} \log n + k^2 n \log n)$ time.*

Proof. We compute a coarse interval $(r_L, r_U]$ in $O(T_S \log n)$ time by Lemma 12, where $T_S = O(n^2 \min\{k, \log n\} + k^2 n)$. We set $T_P = O(\log n)$ and $N_P = O(n^2)$. By combining the sequential and parallel algorithms with Cole's parametric technique, we can compute $t_u(i, r^*)$ for all indices i and u in $O(T_S \log n)$ time. After computing $r_u(i, t_u(i, r^*))$ for all indices i and u in $O(n^2 \log n)$ time using Lemma 10, we can obtain r^* by applying binary search to the $O(n^2)$ values of $r_u(i, t_u(i, r^*))$ and the radii computed from Lemma 11, in $O(T_S \log n)$ time. Our decision algorithm gives k disks with radius r^* covering P by bookkeeping. ☐

We can compute k congruent disks of minimum radius that covers n points in convex position in the plane in $O(kn^2 \log n)$ time for $k = O(\log n)$, in $O(n^2 \log^2 n)$ time for $k = O(n^{\frac{1}{2}} \log^{\frac{1}{2}} n)$, and in $O(k^2 n \log n)$ time for $k = \omega(n^{\frac{1}{2}} \log^{\frac{1}{2}} n)$. If $k = O(1)$, the running time becomes $O(n^2 \log n)$.

Each comparison in the parallel algorithm is to find the circle center determined by two or three points, and thus it is not hard to implement the parametric search.

References

1. Agarwal, P.K., Procopiuc, C.M.: Exact and approximation algorithms for clustering. Algorithmica **33**(2), 201–226 (2002)
2. Agarwal, P.K., Sharir, M.: Planar geometric location problems. Algorithmica **11**(2), 185–195 (1994)
3. Aronov, B., et al.: Data structures for halfplane proximity queries and incremental Voronoi diagrams. Algorithmica **80**(11), 3316–3334 (2018)
4. Bian, D., Jiang, B., Cao, Z.: An efficient algorithm of the planar 3-center problem for a set of the convex position points. In: MATEC Web of Conferences, vol. 232, p. 03022. EDP Sciences (2018)
5. Cabello, S.: Faster distance-based representative skyline and k-center along pareto front in the plane. arXiv arXiv:2012.15381 (2020)
6. Capoyleas, V., Rote, G., Woeginger, G.: Geometric clusterings. J. Algorithms **12**(2), 341–356 (1991)
7. Chan, T.M.: More planar two-center algorithms. Comput. Geom. Theory Appl. **13**(3), 189–198 (1999)
8. Chen, D.Z., Li, J., Wang, H.: Efficient algorithms for the one-dimensional k-center problem. Theor. Comput. Sci. **592**, 135–142 (2015)

9. Cho, K., Oh, E.: Optimal algorithm for the planar two-center problem. arXiv:2007.08784 (2020)
10. Choi, J., Ahn, H.-K.: Efficient planar two-center algorithms. Comput. Geom. Theory Appl. **97**, 101768 (2021)
11. Cole, R.: Slowing down sorting networks to obtain faster sorting algorithms. J. ACM **34**(1), 200–208 (1987)
12. Dupin, N., Nielsen, F., Talbi, E.-G.: Unified polynomial dynamic programming algorithms for p-center variants in a 2d pareto front. Mathematics **9**(4), 453 (2021)
13. Edelsbrunner, H., Kirkpatrick, D., Seidel, R.: On the shape of a set of points in the plane. IEEE Trans. Inf. Theory **29**(4), 551–559 (1983)
14. Eppstein, D.: Dynamic three-dimensional linear programming. ORSA J. Comput. **4**(4), 360–368 (1992)
15. Eppstein, D.: Faster construction of planar two-centers. In: Proceedings of the 8th Annual ACM-SIAM Symposium on Discrete Algorithms (SODA 1997), pp. 131–138 (1997)
16. Guibas, L.J., Hershberger, J.: Optimal shortest path queries in a simple polygon. J. Comput. Syst. Sci. **39**(2), 126–152 (1989)
17. Hershberger, J.: A faster algorithm for the two-center decision problem. Inf. Process. Lett. **47**(1), 23–29 (1993)
18. Hershberger, J., Suri, S.: Finding tailored partitions. J. Algorithms **12**(3), 431–463 (1991)
19. Hershberger, J., Suri, S.: Off-line maintenance of planar configurations. J. Algorithms **21**(3), 453–475 (1996)
20. Hwang, R.Z., Lee, R.C.T., Chang, R.C.: The slab dividing approach to solve the Euclidean p-center problem. Algorithmica **9**(1), 1–22 (1993)
21. Jahn, T., Spirova, M.: On bisectors in normed planes. Contrib. Discrete Math. **10**(2), 1–9 (2015)
22. Jaromczyk, J.W., Kowaluk, M.: An efficient algorithm for the Euclidean two-center problem. In: Proceedings of the 10th Annual Symposium on Computational Geometry (SoCG 1994), pp. 303–311 (1994)
23. Katz, M.J., Sharir, M.: An expander-based approach to geometric optimization. In Proceedings of the 9th Annual Symposium on Computational Geometry (SoCG 1993), pp. 198–207 (1993)
24. Martín, P., Yáñez, D.: Geometric clustering in normed planes. Comput. Geom. **78**, 50–60 (2019)
25. Megiddo, N.: Linear-time algorithms for linear programming in R^3 and related problems. SIAM J. Comput. **12**(4), 759–776 (1983)
26. Sharir, M.: A near-linear algorithm for the planar 2-center problem. Discrete Comput. Geom. **18**(2), 125–134 (1997)
27. Wang, H.: On the planar two-center problem and circular hulls. Discr. Comput. Geom. **68**, 1175–1226 (2022)

Classification via Two-Way Comparisons (*Extended Abstract*)

Marek Chrobak(✉) and Neal E. Young

University of California, Riverside, USA
{marek.chrobak,neal.young}@ucr.edu

Abstract. Given a weighted, ordered query set Q and a partition of Q into classes, we study the problem of computing a minimum-cost decision tree that, given any query $q \in Q$, uses equality tests and less-than comparisons to determine the class to which q belongs. Such a tree can be much smaller than a lookup table, and much faster and smaller than a conventional search tree. We give the first polynomial-time algorithm for the problem. The algorithm extends naturally to the setting where each query has multiple allowed classes.

1 Introduction

Given a weighted, ordered *query* set Q partitioned into classes, we study the problem of computing a minimum-cost decision tree that uses equality tests (e.g., "$q = 4$?") and less-than tests (e.g., "$q < 7$?") to quickly determine the class of any given query $q \in Q$. (Here the cost of a tree is the weighted sum of the depths of all queries, where the depth of a given query $q \in Q$ is the number of tests the tree makes when given query q.) We call such a tree a *two-way-comparison decision tree (*2WCDT*)*. See Fig. 1.

A main use case for 2WCDTs is when the number of classes is small relative to the number of queries. In this case a 2WCDT can be significantly smaller than a lookup table, and, likewise, faster and smaller than a conventional search tree, because a search tree has to identify a given query q (or the inter-key interval that q lies in) whereas a decision tree only has to identify q's class. Because they can be faster and more compact, 2WCDTs are used in applications such as dispatch trees, which allow compilers and interpreters to quickly resolve method implementations for objects declared with type inheritance [2,3]. (Each type is assigned a numeric ID via a depth-first search of the inheritance digraph. For each method, a 2WCDT maps each ID to the appropriate method resolution.)

Chambers and Chen give a heuristic to construct low-cost 2WCDTs, but leave open whether minimum-cost 2WCDTs can be found in polynomial time [2,3]. We give the first polynomial-time algorithm to find minimum-cost 2WCDTs. The algorithm runs in time $O(n^4)$, where $n = |Q|$ is the number of distinct query values,

Most proofs from Sects. 3 and 4 are omitted. See [6] for full proofs.

M. Chrobak—Research partially supported by National Science Foundation grant CCF-2153723.

P. Morin and S. Suri (Eds.): WADS 2023, LNCS 14079, pp. 275–290, 2023.
https://doi.org/10.1007/978-3-031-38906-1_19

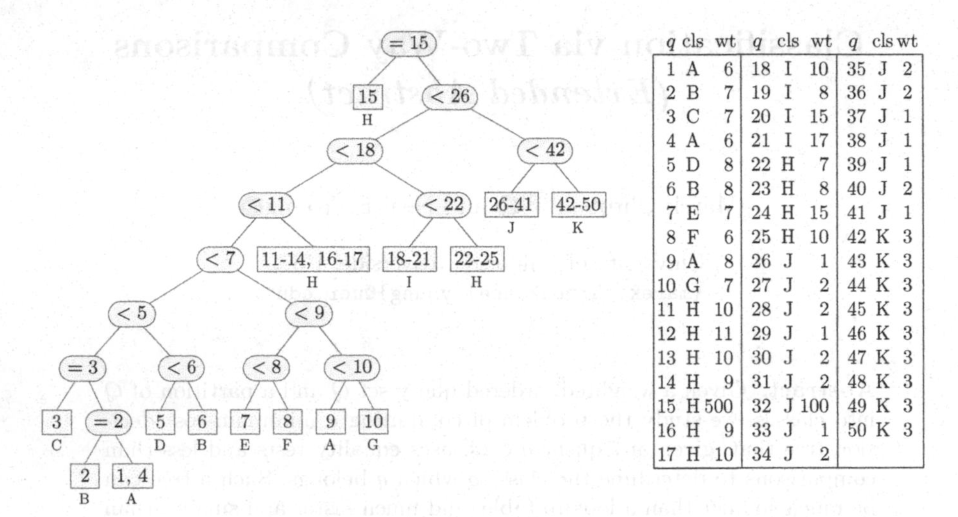

q	cls	wt	q	cls	wt	q	cls	wt
1	A	6	18	I	10	35	J	2
2	B	7	19	I	8	36	J	2
3	C	7	20	I	15	37	J	1
4	A	6	21	I	17	38	J	1
5	D	8	22	H	7	39	J	1
6	B	8	23	H	8	40	J	2
7	E	7	24	H	15	41	J	1
8	F	6	25	H	10	42	K	3
9	A	8	26	J	1	43	K	3
10	G	7	27	J	2	44	K	3
11	H	10	28	J	2	45	K	3
12	H	11	29	J	1	46	K	3
13	H	10	30	J	2	47	K	3
14	H	9	31	J	2	48	K	3
15	H	500	32	J	100	49	K	3
16	H	9	33	J	1	50	K	3
17	H	10	34	J	2			

Fig. 1. An optimal two-way-comparison decision tree (2WCDT) for the problem instance shown on the right. The instance (but not the tree) is from [2,3, Figure 6]. Each leaf (rectangle) is labeled with the queries that reach it, and below that with the class for the leaf. The table gives the class and weight of each query $q \in Q = [50] = \{1, 2, \ldots, 50\}$. The tree has cost 2055, about 11% cheaper than the tree from [2,3], of cost 2305.

matching the best time bound known for a special type of 2WCDTS called *two-way-comparison search trees (*2WCSTS*)*, discussed below. The algorithm extends naturally to the setting where each query can belong to multiple classes, any one of which is acceptable as an answer for the query. The extended algorithm runs in time $O(n^3 m)$, where m is the sum of the sizes of the classes.

Related Work. Various types of decision trees are ubiquitous in the areas of artificial intelligence, machine learning, and data mining, where they are used for data classification, clustering, and regression.

We study decision trees for one-dimensional data sets. Most work on such trees has focussed on *search trees*, that is, decision trees that must explicitly identify the query or the inter-key interval it lies in (essentially, each class is a singleton). Here is a brief summary of relevant work on such trees. One of our main contributions is to increase the understanding of trees based on two-way comparisons. These are not yet fully understood.

The tractability of finding a minimum-cost search tree depends heavily on the kind of tests that the tree can use. For some kinds of tests, the problem is NP-complete [13]. Early works considered trees in which each test compared the given query value q to some particular comparison key k, with *three* possible outcomes: the query value q is less than, equal to, or greater than k [7, §14.5], [15, §6.2.2] (See Fig. 2(a)). We call such trees *three-way-comparison search trees*, or 3WCSTS for short. In a 3WCST, the query values that reach any given node form an interval. This leads to a natural $O(n^3)$-time dynamic-programming algorithm

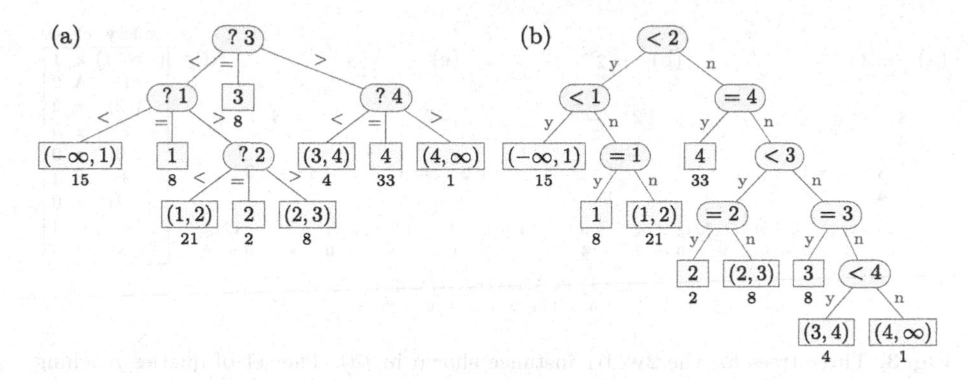

Fig. 2. Tree *(a)* is a three-way-comparison search tree (3WCST). Tree *(b)* is a two-way-comparison search tree (2WCST) for the same instance. The query (or interval of queries) reaching each (rectangular) leaf is within the leaf. The weight of the query (or interval) is below the leaf.

with $O(n^2)$ subproblems for finding minimum-cost 3WCSTs [9]. Knuth reduced the time to $O(n^2)$ [14].

In practice each three-way comparison is often implemented by doing a less-than test followed by an equality test. Knuth [15, §6.2.2, Example 33] proposed exploring binary search trees that use these two tests directly in any combination. Such trees are called *two-way-comparison search trees (*2WCST*s)* [1]. For the so-called *successful-queries* variant (defined later, after Theorem 2), assuming that the query weights are normalized to sum to 1, there is always a 2WCST whose cost exceeds the entropy of the weight distribution by at most 1 [8]. The entropy is a lower bound on the cost of any binary search tree that uses Boolean tests of any kind. This suggests that restricting to less-than and equality tests need not be too costly [8].

Stand-alone equality tests introduce a technical obstacle not encountered with 3WCSTs. Namely, while (analogously to 3WCSTs) each node of a 2WCST is naturally associated with an interval of queries, not all queries from this interval necessarily reach the node. For this reason the dynamic-programming approach for 3WCSTs does not extend easily to 2WCSTs. This led early works to focus on restricted classes of 2WCSTs, namely *median split trees* [17] and *binary split trees* [10,12,16]. These, by definition, constrain the use of equality tests so as to altogether sidestep the aforementioned technical obstacle. *Generalized binary split trees* are less restrictive, but the only algorithm proposed to find them [11] is incorrect [5]. Similarly, the first algorithms proposed to find minimum-cost 2WCSTs (without restrictions) were given without proofs of correctness [18,19], and the recurrence relations underlying some of those proposed algorithms turned out to be demonstrably wrong [5].

In 1994, Spuler made a conjecture that leads to a natural dynamic program for 2WCSTs. Namely, that every instance admits a minimum-cost 2WCST with the *heaviest-first* property: that is, at any equality-test node $\langle = h\rangle$, *the compari-*

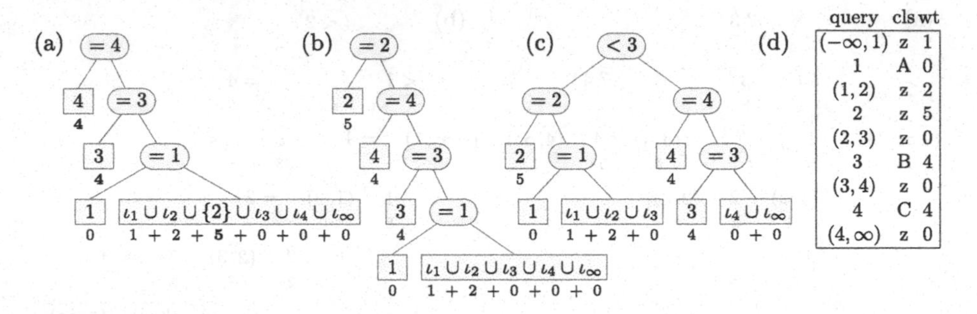

Fig. 3. Three trees for the 2WCDT instance shown in *(d)*. The set of queries reaching each (rectangular) leaf is shown within the leaf (to save space, there ι_i denotes the inter-key open interval with right boundary i, e.g. $\iota_1 = (-\infty, 1)$, $\iota_2 = (1, 2)$). The associated weights are below the leaf. The optimal tree *(a)* has cost 36 and is not heaviest-first. Each heaviest-first tree (e.g. *(b)* of cost 37 or *(c)* of cost 39) is not optimal. These properties also hold if each weight is perturbed to make the weights distinct.

son key h is heaviest among keys reaching the node [19]. In a breakthrough in 2002, Anderson et al. proved the conjecture for the aforementioned successful-queries variant, leading to an $O(n^4)$-time dynamic-programming algorithm to find minimum-cost 2WCSTs for that variant [1]. In 2021, Chrobak et al. simplified their result (in particular, the handling of keys of equal weights, as discussed later) obtaining an $O(n^4)$-time algorithm for finding minimum-cost 2WCSTs [4].

Our Contributions. Unfortunately these 2WCST algorithms don't extend easily to 2WCDTs. The main obstacle is that for some instances (e.g. in Fig. 3) no minimum-cost 2WCDT has the crucial heaviest-first property. To overcome this obstacle we introduce a *splitting* operation (Definition 7), a correctness-preserving local rearrangement of the tree that can be viewed as an extension of the well-studied rotation operation to a more general class of trees, specifically, to trees whose allowed tests induce a laminar set family (Property 1).

We use splitting to identify an appropriate relaxation of the heaviest-first property that we call being *admissible* (Definition 4). Most of the paper is devoted to proving the following theorem:

Theorem 1. *If the instance is feasible, then some optimal tree is admissible.*

Section 3 gives the proof. Along the way it establishes new structural properties of optimal 2WCSTs and 2WCDTs. Section 4 shows how Theorem 1 leads to a suitable dynamic program and our main result:

Theorem 2. *There is an $O(n^3m)$-time algorithm for finding a min-cost* 2WCDT.

Remarks. The presentation above glosses over a secondary technical obstacle for 2WCSTs. For 2WCST instances with *distinct* query weights, the heaviest-first property uniquely determines the key of each equality test, so that the subset of queries that reach any given node in a 2WCST with the heaviest-first property must be *one of* $O(n^4)$ *predetermined subsets.* This leads to a natural dynamic program with $O(n^4)$ subproblems (See Sect. 3). But this does not hold for instances with non-distinct weights. This obstacle turns out to be more challenging than one might expect. Indeed, there are instances for which, for each of the 2^n subsets S of Q, there is a minimum-cost 2WCST, having the heaviest-first property, with a node u such that the set of queries reaching u is S. It appears that one cannot just break ties arbitrarily: it can be that, for two maximum-weight keys h and h' reaching a given node u, there is an optimal subtree in which u does an equality-test to h, but none in which u does an equality-test to h' [4, Figure 3]. Similar issues arise in finding optimal *binary split trees*—these can be found in time $O(n^4)$ if the instance has distinct weights, while for arbitrary instances the best bound known is $O(n^5)$ [10].

Nonetheless, using a perturbation argument Chrobak et al. show that an arbitrary 2WCST instance can indeed be handled as if it is a distinct-weights instance just by breaking ties among equal weights in an appropriate way [4]. We use the same approach here for 2WCDTs.

Most search-tree problems come in two flavors: the *successful-queries* variant and the *standard* variant. In the former, the input is an ordered set K of weighted keys, each comparison must compare the given query value to a particular key in K, and each query must be a value in K. In the latter, the input is augmented with a weight for each open interval between successive keys. Queries (called *unsuccessful queries*) to values in these intervals are also allowed, and must be answered by returning the interval in which the query falls. Our formal definition of 2WCDTs generalizes both variants.

2 Definitions and Technical Overview

For the remainder of the paper, fix a 2WCDT instance $(Q, w, \mathcal{C}, K)$, where Q is a totally ordered finite set of *queries*, each with a weight $w(q) \geq 0$, the set $\mathcal{C} \subseteq 2^Q$ is a collection of *classes* of queries (where each class has a unique identifier), and $K \subseteq Q$ is the set of *keys*. Let $n = |Q|$ and $m = \sum_{c \in \mathcal{C}} |c|$. The problem is to compute a minimum-cost *two-way-comparison decision tree* (2WCDT, as defined below) for the instance.

To streamline presentation, throughout the paper we restrict attention to the model of decision trees that allows only less-than and equality tests. Our results extend naturally to decision trees that also use other inequality comparisons between queries and keys. See the end of Sect. 4 for details.

Definition 1 (2WCDT). *A* two-way-comparison decision tree (2WCDT) *is a rooted binary tree T where each non-leaf node is a* test *of the form* $\langle\, < k \rangle$ *for some* $k \in K \setminus \{\min Q\}$, *or of the form* $\langle\, = k \rangle$ *for some* $k \in K$, *and the two children of the node are labeled with the two possible test outcomes ("yes" or "no"). Each leaf node is labeled with the identifier of one class in $\mathcal{C}$. This class must contain every query $q \in Q$ whose search (as defined next) ends at the leaf.*

For each $q \in Q$, *the* search for q in T *starts at the root, then recursively searches for q in the root's yes-subtree if q satisfies the root's test, and otherwise* *in the no-subtree. The search stops at a leaf, called the* leaf for q. *The path from the root to this leaf is called q's* search path. *We say that q* reaches *each node on this path, and q's* depth *in T is defined as the length of this path (equivalently, the number of comparisons when searching for q). The* cost *of T is the weighted sum of the depths of all queries in Q.*

A tree T is called irreducible *if, for each node u in T, (i) at least one query in Q reaches u, and (ii) if some class $c \in \mathcal{C}$ contains all the queries that reach u, then u is a leaf.*

For any $\ell, r \in Q$, let $[\ell, r]_Q$ and $[\ell, r]_K$ denote the query interval $\{q \in Q : \ell \leq q \leq r\}$ *and* key interval $\{k \in K : \ell \leq k \leq r\} = K \cap [\ell, r]_Q$.

Allowing K and Q to be specified as we do captures both the successful-queries and standard variants. The successful-queries variant corresponds to the case when $K = Q$. The standard variant is modeled by having one non-key query between every pair of consecutive keys, and before the minimum key and after the maximum key (so $|Q \backslash K| = |K| + 1$). Each such non-key query represents an interval between keys.

For ease of exposition, assume without loss of generality that each query belongs to some class, so $m \geq |Q|$ and the input size is $\Theta(n + m) = \Theta(m)$. Note that the instance is not necessarily feasible, that is, it might not have a decision tree. (To be feasible, in addition to each query belonging to some class, each query interval that contains no keys must be contained in some class.) If the instance is feasible, some optimal tree is irreducible, so we generally restrict attention to irreducible trees. As we shall see, in an irreducible tree any given test is used in at most one node.

Definition 2 (ordering queries by weight). *For any query subset $R \subseteq Q$ and integer $i \geq 0$ define $\mathsf{heaviest}_i(R)$ to contain the i heaviest queries in R (or all of R if $i \geq |R|$). For $q \in Q$, define $\mathsf{heavier}(q)$ to contain the queries (in Q) that are heavier than q. Define $\mathsf{lighter}(q)$ to contain the queries (in Q) that are lighter than q. Break ties among query weights arbitrarily but consistently throughout.*

Formally, we use the following notation to implement the consistent tie-breaking mentioned above. Fix an ordering of Q by increasing weight, breaking ties arbitrarily. For $q \in Q$ let $\tilde{w}(q)$ denote the rank of q in the sorted order. Throughout, given distinct queries q and q', define q to be lighter than q' if $\tilde{w}(q) < \tilde{w}(q')$ and heavier otherwise ($\tilde{w}(q) > \tilde{w}(q')$). So, for example $\mathsf{heaviest}_i(R)$ contains the last i elements in the ordering of R by increasing $\tilde{w}(q)$. The symbol $\perp$ represents the undefined quantity $\arg\max \emptyset$. Define $\tilde{w}(\perp) = w(\perp) = -\infty$, $\mathsf{heavier}(\perp) = Q$, and $\mathsf{lighter}(\perp) = \emptyset$.

Definition 3 (intervals and holes). *Given any non-empty query subset $R \subseteq Q$, call $[\min R, \max R]_Q$ the* query interval *of R. Define $k^*(R)$ to be the heaviest key in R, if there is one (that is, $k^*(R) = \arg\max\{\tilde{w}(k) : k \in K \cap R\}$). Define also $\mathsf{holes}(R) = [\min R, \max R]_Q \backslash R$ to be the set of* holes *in R. We say that a hole $h \in \mathsf{holes}(R)$ is* light *if $\tilde{w}(h) < \tilde{w}(k^*(R))$, and otherwise* heavy.

The set of queries reaching a node u in a tree T is called u's query set, *denoted Q_u. The query interval, and light and heavy holes, for u are defined to be those for u's query set Q_u.*

Overview. Note that each hole $h \in \mathsf{holes}(Q_u)$ at a node u in a tree T must result from a failed equality test $\langle = h \rangle$ at a node v on the path from the root to u in T. In particular, $h \in K$. Further, if the hole is light, then h is not the heaviest key reaching v.

The problem has the following *optimal substructure* property. Any query subset $R \subseteq Q$ naturally defines the subproblem $\pi(R)$ induced by restricting the query set to R (that is, $\pi(R) = (R, w, \mathcal{C}_R, K)$ where $\mathcal{C}_R = \{c \cap R : c \in \mathcal{C}\}$). In any minimum-cost tree T for R, if T is not a leaf, then the yes-subtree and no-subtree of T must be minimum-cost subtrees for their respective subproblems.

Let $\mathsf{cost}(R)$ be the minimum cost of an irreducible tree for $\pi(R)$. (If R is empty, then $\mathsf{cost}(R) = \infty$, as no tree for R is irreducible.) Then the following recurrence holds:

Observation 1 (recurrence relation). *Fix any $R \subseteq Q$. If $R = \emptyset$, we have $\mathsf{cost}(R) = \infty$. Otherwise, if $(\exists c \in \mathcal{C})\, R \subseteq c$ (that is, R can be handled by a single leaf labeled c), then $\mathsf{cost}(R) = 0$. Otherwise, for any test u, let $(R_u^{\mathsf{yes}}, R_u^{\mathsf{no}})$ be the bipartition of R into those queries that satisfy u and those that don't. Then*

$$\mathsf{cost}(R) = w(R) + \min_u \left(\mathsf{cost}(R_u^{\mathsf{yes}}) + \mathsf{cost}(R_u^{\mathsf{no}}) \right),$$

where the variable u ranges over the allowed tests (per Definition 1) such that R_u^{yes} and R_u^{no} are non-empty. (If there are no such tests then $\mathsf{cost}(R) = \infty$.)

The goal is to compute $\mathsf{cost}(Q)$ using a dynamic program that applies the recurrence in Observation 1 recursively, memoizing results so that for each distinct query set R the subproblem for R is solved at most once. (The algorithm as presented computes only $\mathsf{cost}(Q)$. It can be extended in the standard way to also compute an optimal tree.) The obstacle is that exponentially many distinct subproblems can arise.

Identity Classification Without Equality Tests. For intuition, consider first the variant of our problem in which $\mathcal{C}$ is the *identity classification*, that is $\mathcal{C} = \{\{q\} : q \in Q\}$, and only less-than tests $\langle < k \rangle$ are allowed (equality tests are not). In the absence of equality tests, there are no holes. When applying the recurrence recursively to $\mathsf{cost}(Q)$, each query set R that arises is a query interval. There are $O(n^2)$ such query intervals, and for each the right-hand side of the recurrence can be evaluated in $O(n)$ time. This yields an $O(n^3)$-time algorithm. This approach mirrors a classical dynamic-programming algorithm for 3WCSTs [9], as discussed in the introduction.

The algorithm extends easily to arbitrary classifications $\mathcal{C}$. Recall that a given query set R can be handled by a leaf (at zero cost) if and only if $R \subseteq c$ for some $c \in \mathcal{C}$. This condition can be checked in constant time given (ℓ, r) such that $R = [\ell, r]_Q$ (after an appropriate precomputation, e.g., for each ℓ, precompute the maximum r for which the condition holds).

Identity Classification with Equality Tests Allowed. Next consider the variant when C is the identity classification but both kinds of tests are allowed. This is essentially the problem of computing a minimum-cost 2WCST. In this variant, each node u in a tree T has query set $Q_u = [\min Q_u, \max Q_u]_Q \setminus \mathsf{holes}(Q_u)$. Applying the recurrence naively to $\mathsf{cost}(Q)$ can yield exponentially many subproblems because $\mathsf{holes}(Q_u)$ can be almost any subset of $[\min Q_u, \max Q_u]_Q$. However, as mentioned in Sect. 1, it is known that some optimal tree T has the *heaviest-first* property [1,4]: for each node u in T that does an equality test $\langle = h \rangle$, the test key h is the heaviest key reaching u. (Our tie-breaking scheme makes h unique.) In such a tree there are no light holes. That is, the hole set of any given node u is the set of heavy holes at u:

$$\mathsf{holes}(Q_u) = [\min Q_u, \max Q_u]_K \cap \mathsf{heavier}(k^*(Q_u)).$$

(Note that, by the definition of $k^*(Q_u)$, no keys heavier than $k^*(Q_u)$ reach u, so the set $[\min Q_u, \max Q_u]_K \cap \mathsf{heavier}(k^*(Q_u))$ contains exactly the heavy holes at u.)

A non-empty query set R without light holes is determined by the triple $(\min R, \max R, k^*(R))$, so there are $O(n^3)$ query sets without light holes. This leads naturally to an $O(n^4)$-time algorithm for instances with distinct weights [1,4]. (Specifically, redefine $\mathsf{cost}(R)$ to be the minimum cost of any *heaviest-first*, irreducible tree for $\pi(R)$. Then $\mathsf{cost}(R) = \infty$ if R has at least one light hole. Add this case as a base case to the recurrence. Apply the modified recurrence recursively to calculate $\mathsf{cost}(Q)$. Then the number of distinct non-trivial subproblems that arise is $O(n^3)$, and each can be solved in $O(n)$ time.)

Allowing Equality Tests and An Arbitrary Classification. The existing results for 2WCSTs don't extend to 2WCDTs because, as shown in Fig. 3, there are 2WCDT instances with distinct weights for which no optimal tree is heaviest-first. But, in some sense, the example in Fig. 3 is as bad as it gets. There is an optimal tree in which an appropriate relaxation of the heaviest-first property holds, namely, that each node's query set is *admissible.* Roughly, this means that there are at most three light holes, and the light holes must be taken heaviest first from those keys that don't belong to some class $c \in C$ that contains k^* (the heaviest key reaching the node). Here's the formal definition:

Definition 4 (admissible). *Consider any query subset $R \subseteq Q$. The set R is called* admissible *if it is non-empty and the set of light holes in R is either empty or has the form*

$$\mathsf{heaviest}_b([\min R, \max R]_K \cap \mathsf{lighter}(k^*(R)) \setminus c)$$

for some $b \in [3]$ and $c \in C$ such that $k^*(R) \in c$. A tree (or subtree) T is called admissible *if the query set of each node in T is admissible.*

Above (and within any mathematical expression), for any integer i, the notation $[i]$ denotes the set $\{1, 2, \ldots, i\}$.

To gain some intuition note that, by definition, for any query set R its holes must be in $[\min R, \max R]_K$, and its light holes must be in $\mathsf{lighter}(k^*(R))$. For the algorithm, roughly, we redefine $\mathsf{cost}(R) = \infty$ if R is not admissible, add a corresponding base case to the recurrence and then recursively apply the modified recurrence to compute $\mathsf{cost}(Q)$. Each admissible query set R with no light holes is determined by the triple $(\min R, \max R, k^*(R))$. Per Definition 4, each admissible query set R with at least one light hole is determined by a triple $(\min R, \max R, k^*(R), b, c)$, where $(b, c) \in [3] \times \mathcal{C}$ with $k^*(R) \in c$. It follows that there are $O(n^3 + n^2 m) = O(n^2 m)$ admissible query subsets, so that, in the recursive evaluation of $\mathsf{cost}(Q)$, $O(n^2 m)$ distinct non-trivial subproblems arise. These are solvable in total time $O(n^3 m)$. Section 4 gives the detailed proof.

3 Some Optimal Tree is Admissible

This section proves Theorem 1: if the instance is feasible, then some optimal tree is admissible. Along the way we establish quite a bit more about the structure of optimal trees. Section 3.1 introduces the aforementioned *splitting* operation (Definition 7), a correctness-preserving local rearrangement of the tree that extends the well-studied rotation operation to trees whose allowed tests induce a laminar family (Property 1). (These trees subsume 2WCDTs and 2WCSTs.) Splitting is used to prove two weight bounds (Lemmas 3 and 4), which are used in turn to prove the main structural theorem (Theorem 3). The theorem says that if one child of a node u_1 is lighter than some descendant u_d of the other child, then the path from u_1 to u_d must have a highly restricted structure. (Roughly, for any two distinct nodes u_i and u_j along the path, the outcome from u_i that remains on the path is *consistent*, per Definition 5, with both outcomes leaving u_j.) Theorem 3 holds for any class of trees whose allowed tests induce a laminar family. Lemmas 5 and 6 use Theorem 3 to prove Theorem 1 for distinct-weights instances. A perturbation argument then extends the result to all instances.

We start with some general terminology for how pairs of tests can relate. Recall that $(Q, w, \mathcal{C}, K)$ is a problem instance with at least one correct tree. In any such tree, each edge $u \to v$ from a node to its child corresponds to one of the two possible outcomes of the test at u. We use $u \to v$ to denote both the edge and the associated outcome at u. For example, if u is $\langle < 3\rangle$, and v is the no-child of u, then outcome $u \to v$ means the queried value is at least 3.

Definition 5. *Two such outcomes $u \to v$ and $x \to y$ are called* consistent *if Q contains a query value that satisfies them both. Otherwise they are* inconsistent.

Two tests are said to be equivalent *if either for all $q \in Q$ the tests give the same outcome for q, or for all $q \in Q$ the tests give opposite outcomes for q.*

For example, assume $Q = [4]$. The yes-outcome of $\langle < 3\rangle$ is inconsistent with the yes-outcome of $\langle = 4\rangle$ and with the no-outcome of $\langle < 4\rangle$, but is consistent with both outcomes of $\langle = 2\rangle$, and with both outcomes of $\langle < 2\rangle$. The tests $\langle < 4\rangle$ and $\langle = 4\rangle$ are equivalent.

Most of the proof requires only the following property of tests:

Property 1 (laminarity). *Let u and x be test nodes. (i) If u and x do non-equivalent tests, then, among the four pairs of outcomes between the two nodes, exactly one pair is inconsistent, while the other three pairs are consistent. Formally, let $u \to v$, $u \to v'$, $x \to y$, and $x \to y'$ be the two outcomes from u and the two outcomes from x. Then exactly one pair in $\{u \to v, u \to v'\} \times \{x \to y, x \to y'\}$ is inconsistent. (ii) If u and x do equivalent tests, each outcome at u is consistent with a distinct outcome at x.*

Property 1 is easily verified. (Note that, by the definition of 2WCDTs in Sect. 2, and assuming there is more than one test, each outcome of each test is satisfied by at least one query in Q.) We call Property 1 *laminarity* because it is equivalent to the laminarity of the collection of sets that has, for each possible test, one set containing the queries that satisfy the test. In our case this laminar collection is

$$\{\{q \in Q : q < k\} : k \in K \setminus \{\min Q\}\} \cup \{\{q\} : q \in K\}.$$

As an example, consider the query set $Q = [4]$. Then the yes-outcome of $\langle < 3\rangle$ and the yes-outcome of $\langle = 4\rangle$ are inconsistent, while every other pair of outcomes is consistent; e.g., the yes-outcome of $\langle < 3\rangle$ and the no-outcome of $\langle = 4\rangle$ are consistent, as they are both satisfied by the query value 2.

Throughout most of the rest of this section (including Sects. 3.1 and 3.2), fix T to be an arbitrary irreducible tree.

Property 2. *(i) In T, if u is a proper ancestor of a test node v then the outcome of u leading to v is consistent with both outcomes at v, and the other outcome of u is consistent with exactly one outcome at v. (ii) No two nodes in T are equivalent.*

Proof. The irreducibility of T implies the first part of (i) (that the outcome of u leading to v is consistent with both outcomes at v). So Property 1(ii) implies that u and v are not equivalent. Then Property 1(i) implies the second part of (i) (that the outcome at u leading away from v is consistent with exactly one outcome at v.) To justify Property 2(ii), let x and y be two different test nodes in T. We have already established that if one of x, y is an ancestor of the other then they cannot be equivalent. In the remaining case, let u be the lowest common ancestor of x and y. Using (i) twice, the outcome at u leading to x is consistent with both outcomes at x but is consistent with exactly one of the outcomes at y. So x and y cannot be equivalent. □

3.1 Two Weight Bounds, via Splitting

This section defines splitting and proves the weight bounds (Lemmas 3 and 4).

Definition 6 (x-consistent path). *Let u be a node in T, T_u the subtree of T rooted at u, and x any allowed test (not necessarily in T). The* x-consistent path *from u is the maximal downward path from u in T_u such that each outcome along this path is consistent with both outcomes at x.*

The x-consistent path from u is unique because (by laminarity) at most one outcome out of any given node is consistent with both outcomes at x. In the case that T_u contains a node $\tilde{x}$ that is equivalent to x, the x-consistent path from u is the path from u to $\tilde{x}$ (using here the irreducibility of T and that neither outcome at $\tilde{x}$ is consistent with both outcomes at x). In the case that T_u contains no such node $\tilde{x}$, this x-consistent path from u ends at a leaf.

Fix a node u in T and a test node x, not necessarily in T. Informally, *splitting* T_u *around* x replaces subtree T_u of T by the subtree T'_x obtained by the following process: initialize T'_x to a subtree with root x, whose yes- and no-subtrees are each a copy of T_u, then splice out each redundant test (that is, each test w such that one of the outcomes at w is inconsistent with the outcome at x that leads to w, implying that every query reaching w satisfies the other outcome at w). The resulting subtree T'_x has a particular structure that we'll need to use. The formal definition, below, makes this structure explicit.

This construction and the proofs below use notation $u_1 \to u_2 \to \cdots \to u_j$ for a downward path in T, and use u'_i to denote the sibling of u_i, so each edge $u_i \to u'_{i+1}$ leaves the path.

Definition 7 (splitting). Splitting T_u around x *yields the subtree* T'_x *produced by the following process. Let* $u_1 \to u_2 \to \cdots \to u_d$ *be the* x*-consistent path from* $u = u_1$*, as defined in Definition 6. Initialize* T'_x *to have root* x*, with yes- and no-subtrees, denoted* T_u^{yes} *and* T_u^{no}*, each a copy of* T_u*.*

For each outcome $\alpha \in \{\mathsf{yes}, \mathsf{no}\}$ *at* x*, modify* T_u^α *within* T'_x *as follows. For each* $i \in [d-1]$*, if outcome* $u_i \to u'_{i+1}$ *is inconsistent with the* α*-outcome at* x*, within* T_u^α*, delete node* u_i *and the subtree* $T_{u'_{i+1}}$*, making* u_{i+1} *the child of the current parent of* u_i *in place of* u_i*. For* $i = d$*, if* u_d *is a leaf, stop. Otherwise (*u_d *is a test node), let* $u_d \to y'$ *be the outcome at* u_d *that is inconsistent with the* α*-outcome at* x*. Within* T_u^α*, delete node* u_d *and the subtree* $T_{y'}$*, making the other child* y *of* u_d *the child of the current parent of* u_d *in place of* u_d*.*

Note that, for each $\alpha \in \{\mathsf{yes}, \mathsf{no}\}$, by the definition of the x-consistent path from u and Property 1 (laminarity), outcome $u_i \to u'_{i+1}$ is inconsistent with exactly one outcome at x. Also, if u_d is a test node then it must be equivalent to x, so exactly one outcome at u_d is inconsistent with the α-outcome at x. (See Lemmas 1 and 2 in the full paper [6] for a more detailed characterization of the result of splitting.) Figs. 4 and 5 give examples of splitting. In Fig. 4, $d = 4$ and x is a test node in T_u (in fact $x = u_4$). In Fig. 5, x is a new node (not equivalent to any node in T_u, where $u = u_1$), $d = 5$ and u_5 is a leaf.

The proofs of the following weight bounds take advantage of laminarity. Specifically, as T is irreducible, Property 2(i) implies that if u_i is a proper ancestor of u_j then outcome $u_i \to u'_{i+1}$ is consistent with one outcome at u_j and inconsistent with the other.

Lemma 3. *Suppose* T *is optimal. Let* $u_1 \to \cdots \to u_{j+1}$ *be any downward path in* T*. For* $1 \le i \le j-1$*, let* δ_i *be the number of ancestors* u_s *of* u_i *on the path such that outcomes* $u_s \to u'_{s+1}$ *and* $u_i \to u'_{i+1}$ *are consistent with opposite outcomes*

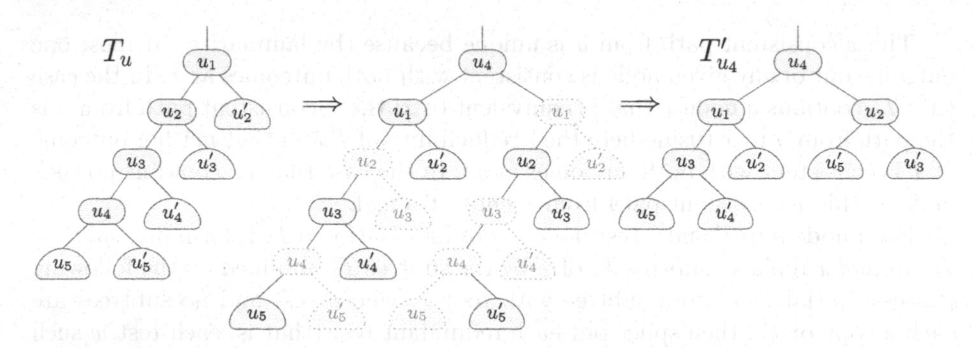

Fig. 4. Splitting a subtree T_u (where $u = u_1$) around descendant u_4. The figures in this section draw T_u by drawing u_i and u_i' as the left and right children of their parent u_{i-1}, so that the u_4-consistent path from u_1 is drawn as a prefix of the left spine. Each rounded half-circle represents a subtree, labeled with its root. Here outcomes $u_1 \to u_2'$ and $u_3 \to u_4'$ are consistent with the outcome $u_4 \to u_5$ at u_4 while outcome $u_2 \to u_3'$ is consistent with the other outcome $u_4 \to u_5'$. In the notation of Lemma 3 (taking $j = 4$) $\delta_2 = \delta_3 = 1$ and $\beta = 2$, and the lemma gives the bound $w(u_2') \geq w(u_5) + 2w(u_5')$.

at u_j. Let β be the number of ancestors u_s of u_{j-1} whose outcome $u_s \to u_{s+1}'$ is consistent with outcome $u_j \to u_{j+1}$ (so $0 \leq \beta \leq j-1$). Then

$$w(u_2') \geq (j-1-\beta)\, w(u_{j+1}) + \beta w(u_{j+1}') + \textstyle\sum_{i=3}^{j}(\delta_{i-1} - 1)w(u_i').$$

Lemma 4. *Suppose T is optimal. Let x be any test node, not necessarily in T. Let $u_1 \to \cdots \to u_{j+1}$ be a prefix of the x-consistent path from u_1. For $1 \leq i \leq j-1$, let δ_i be the number of ancestors u_s of u_i on the path such that outcomes $u_s \to u_{s+1}'$ and $u_i \to u_{i+1}'$ are consistent with opposite outcomes at u_j. Let β' be the number of ancestors u_s of u_j whose outcome $u_s \to u_{s+1}'$ is consistent with the yes-outcome of x (so $0 \leq \beta' \leq j$). Then*

$$w(u_2') \geq \min(j-1-\beta', \beta'-1)w(u_j) + \textstyle\sum_{i=3}^{j}(\delta_{i-1} - 1)w(u_i').$$

The proofs (in the full paper) show that if the claimed inequalities are not met, then splitting the subtree T_u around u_j (for Lemma 3) or x (for Lemma 4) would give a cheaper optimal tree.

3.2 Structural Theorem

As in the previous section, for any downward path $u_1 \to u_2 \to \cdots \to u_j$, the sibling of u_i is denoted u_i' (for $2 \leq i \leq j$).

Theorem 3. *Suppose T is optimal. Let $u_1 \to u_2 \to \cdots \to u_d$ be any downward path in T (not necessarily starting at the root) such that $w(u_2') < w(u_d)$. Then, for all different nodes u_i, u_j on the path, with $i, j < d$, both outcomes at u_i are consistent with outcome $u_j \to u_{j+1}$.*

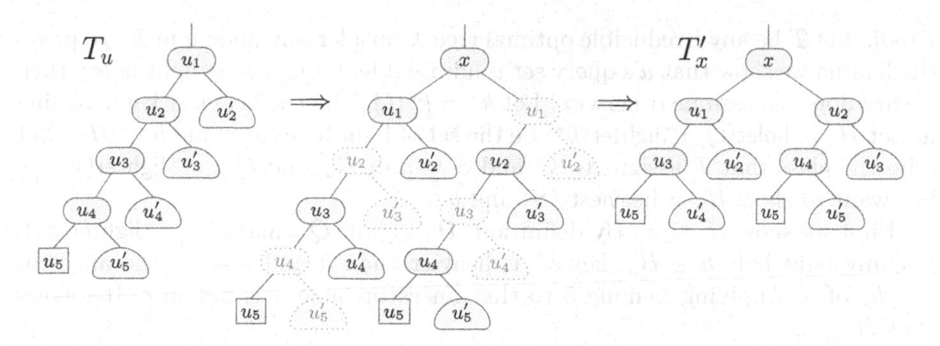

Fig. 5. Splitting a subtree T_u (where $u = u_1$) around a new node x (not equivalent to any node in T_u). The x-consistent path from u_1 is $u_1 \to \cdots \to u_5$. Here $u_1 \to u'_2$ and $u_3 \to u'_4$ are consistent with the left outcome at x, while $u_2 \to u'_3$ and $u_4 \to u'_5$ are consistent with the right outcome. In the notation of Lemma 4 (taking $j = 4$) $\delta_2 = \delta_3 = 1$ and $\beta' = 2$. The lemma gives the bound $w(u'_2) \geq w(u_4)$.

For intuition, suppose a node u in T does an equality test $\langle = h \rangle$, and, in the no-subtree of u, some node x has $w(x) > w(h)$. By the theorem, then, the query value $q = h$ satisfies all outcomes along the path from the no-child of u to x.

The only property of the admissible tests that Theorem 3 relies on is laminarity. The proof (in the full paper) uses Lemma 3 twice, and a careful induction.

3.3 Proof of Theorem 1 (Some Optimal Tree is Admissible)

The proofs above rely only on laminarity. The proofs below use the particular structure of less-than and equality tests, and the properties of u-consistent paths. In particular, when x is an equality test, say x is $\langle = h \rangle$, the x-consistent path from u is the path that a search for h would take if started at u.

Lemma 5. *Suppose the instance has distinct weights and T is optimal. Consider any equality-test node $\langle = h \rangle$ and a key k with $w(k) > w(h)$ reaching this node. Then a search for h from the no-child of $\langle = h \rangle$ would end at the leaf L_k for k, and the path from $\langle = h \rangle$ to L_k has at most four nodes (including $\langle = h \rangle$ and L_k). Also, h is not in the class that T assigns to k.*

The proof (in the full paper) has several parts. First it applies Theorem 3 to the path from $\langle = h \rangle$ to L_k to show that, for any two different test nodes u_i, u_j along the path, both outcomes at u_i are consistent with $u_j \to u_{j+1}$. This (with $i = 1$) implies that a search for h starting at u_2 ends at L_k. A local-exchange argument shows h is not in the class that T assigns to k. That the path has at most four nodes then follows from Lemma 4 and local-exchange arguments.

Lemma 6. *If the instance has distinct weights, every irreducible optimal tree is admissible.*

The lemma follows from a careful application of Lemma 5 and the definition of admissible trees. Here is the full proof.

Proof. Let T be any irreducible optimal tree. Consider any node u in T. To prove the lemma we show that u's query set is admissible. If Q_u has no light holes, then we are done, so assume otherwise. Let $k^* = k^*(Q_u)$ be the heaviest key reaching u. Let $H_u = \mathsf{holes}(Q_u) \cap \mathsf{lighter}(k^*)$ be the set of light holes at u and $b = |H_u|$. Let c be the class that T assigns to k^* and $S = [\min Q_u, \max Q_u]_K \cap \mathsf{lighter}(k^*) \setminus c$. We want to show $H_u = \mathsf{heaviest}_b(S)$ and $b \in [3]$.

First we show $H_u \subseteq S$. By definition, $H_u \subseteq [\min Q_u, \max Q_u]_K \cap \mathsf{lighter}(k^*)$. For any light hole $h \in H_u$, key k^* is heavier than h and reaches the ancestor $\langle = h \rangle$ of u. Applying Lemma 5 to that ancestor, hole h is not in c. It follows that $H_u \subseteq S$.

Next we show $H_u = \mathsf{heaviest}_b(S)$. Suppose otherwise for contradiction. That is, there are $k \in S \setminus H_u \subseteq Q_u$ and $h \in H_u$ such that k is heavier than h. Keys k^* and k reach the ancestor $\langle = h \rangle$ of u. Applying Lemma 5 (twice) to that ancestor, the search path for h starting from the no-child of $\langle = h \rangle$ ends both at L_{k^*} and at the leaf L_k for k. So $L_k = L_{k^*}$, which implies that k is in c, contradicting $k \in S$. Therefore $H_u = \mathsf{heaviest}_b(S)$.

Finally, we show that $b \leq 3$. Let $h \in H_u$ be the light hole whose test node $\langle = h \rangle$ is closest to the root. Key k^* reaches $\langle = h \rangle$ and weighs more than h. Applying Lemma 5 to $\langle = h \rangle$ and key k^*, the path from $\langle = h \rangle$ to L_{k^*} has at most four nodes (including the leaf). Each light hole has a unique equality-test node on that path. So (using that u is on this path) there are at most three light holes in Q_u. □

Finally we prove Theorem 1:

Theorem 1. *If the instance is feasible, then some optimal tree is admissible.*

The proof (in the full paper) is a somewhat subtle perturbation argument, showing (informally) that every instance I is "arbitrarily close" to a distinct-weights instance I^* that shares the same set of admissible trees, and such that any optimal tree for I^* is also optimal for I. By Lemma 6, I^* has an optimal tree that is admissible, which is therefore also optimal and admissible for I.

4 Algorithm

This section proves Theorem 2, that the problem admits an $O(n^3 m)$-time algorithm. The input is an arbitrary 2WCDT instance $(Q, w, \mathcal{C}, K)$. In this section, for any $R \subseteq Q$ redefine $\mathsf{cost}(R)$ to be the minimum cost of any *admissible* tree for the subproblem $\pi(R) = (R, w, \mathcal{C}, K)$ obtained by restricting the query set to R. (Take $\mathsf{cost}(R) = \infty$ if there is no admissible tree for $\pi(R)$.) The algorithm returns $\mathsf{cost}(Q)$, the minimum cost of any admissible tree for $(Q, w, \mathcal{C}, K)$. By Theorem 1, this equals the minimum cost of any tree.

The algorithm computes $\mathsf{cost}(Q)$ by using memoized recursion on the following recurrence relation:

Recurrence 1. *For any $R \subseteq Q$,*

$$\mathsf{cost}(R) = \begin{cases} \infty & (R \notin \mathcal{A}) \\ 0 & (R \in \mathcal{A} \wedge (\exists c \in \mathcal{C})\, R \subseteq c) \\ w(R) + \min_u \big(\mathsf{cost}(R_u^{\mathsf{yes}}) + \mathsf{cost}(R_u^{\mathsf{no}})\big), & (\textit{otherwise}) \end{cases}$$

where above $\mathcal{A}$ denotes the set of admissible query subsets of Q (per Definition 4), $(R_u^{\mathsf{yes}}, R_u^{\mathsf{no}})$ is the bipartition of R into those values that satisfy test u and those that don't, and u ranges over the allowed tests (per Definition 1) such that R_u^{yes} and R_u^{no} are admissible. (If there are no such tests then the minimum is infinite.)

There are $O(n^2m)$ admissible query sets. (Indeed, for any admissible set R, if R has no light holes it is determined by the triple $(\min R, \max R, k^*(R))$. Otherwise, per Definition 4, R is determined by a tuple $(\min R, \max R, k^*(R), b, c)$, where $(b, c) \in [3] \times \mathcal{C}$ with $k^*(R) \in c$.) So $O(n^2m)$ subproblems arise in recursively evaluating $\mathsf{cost}(Q)$. To finish we describe how to evaluate the right-hand side of Recurrence 1 for a given R in $O(n)$ amortized time.

Assume (by renaming elements in Q in a preprocessing step) that $Q = [n]$. Given a non-empty query set $R \subseteq Q$, define the *signature* of R to be

$$\tau(R) = (\min R, \max R, k^*(R), H(R)),$$

where $H(R) = \mathsf{holes}(R) \cap \mathsf{lighter}(k^*(R))$ is the set of light holes in R.

For any R, its signature is easily computable in $O(n)$ time (for example, bucket-sort R and then enumerate the hole set $[\ell, r]_Q \setminus R$ to find $H(R)$). Each signature is in the set

$$\mathcal{S} = Q \times Q \times (K \cup \{\bot\}) \times 2^Q$$

of *potential signatures.* Conversely, given any potential signature $t = (\ell, r, k, H') \in \mathcal{S}$, the set $\tau^{-1}(t)$ with signature t, if any, is unique and computable from t in $O(n)$ time. (Specifically, $\tau^{-1}(t)$ is $Q_t = [\ell, r]_Q \setminus ((K \cap \mathsf{heavier}(k)) \cup H')$ provided Q_t is non-empty and has signature t.)

Lemma 7. *After an $O(n^3m)$-time preprocessing step, given the signature $\tau(R)$ of $R \in \mathcal{A}$, the right-hand of Recurrence 1 is computable in amortized time $O(n)$.*

The proof of the lemma is in the full paper. Theorem 2 follows.

Extending the Algorithm to Other Inequality Tests. Our model considers decision trees that use less-than and equality tests. Allowing the negations of these tests is a trivial extension. (E.g., every greater-than-or-equal test $\langle \geq k \rangle$ is equivalent by swapping the children to the less-than test $\langle < k \rangle$.) We note without proof that our results also extend easily to the model that allows less-than-or-equal tests (of the form $\langle \leq k \rangle$). The proof of Theorem 3 requires only a minor adjustment: such tests need to be taken into account when proving the first claim in the proof of Lemma 5; the extended algorithm then allows such tests in Recurrence 1.

Acknowledgements. Thanks to Mordecai Golin and Ian Munro for introducing us to the problem and for useful discussions.

References

1. Anderson, R., Kannan, S., Karloff, H., Ladner, R.E.: Thresholds and optimal binary comparison search trees. J. Algorithms **44**, 338–358 (2002). https://doi.org/10.1016/S0196-6774(02)00203-1
2. Chambers, C., Chen, W.: Efficient multiple and predicated dispatching. In: Proceedings of the 1999 ACM SIGPLAN Conference on Object-Oriented Programming Systems, Languages & Applications (OOPSLA 1999), Denver, Colorado, USA, 1–5 November 1999, pp. 238–255 (1999)
3. Chambers, C., Chen, W.: Efficient multiple and predicated dispatching. SIGPLAN Not. **34**(10), 238–255 (1999). https://doi.org/10.1145/320385.320407
4. Chrobak, M., Golin, M., Munro, J.I., Young, N.E.: A simple algorithm for optimal search trees with two-way comparisons. ACM Trans. Algorithms **18**(1), 2:1–2:11 (2021). https://doi.org/10.1145/3477910
5. Chrobak, M., Golin, M., Munro, J.I., Young, N.E.: On Huang and Wong's algorithm for generalized binary split trees. Acta Informatica **59**(6), 687–708 (2022). https://doi.org/10.1007/s00236-021-00411-z
6. Chrobak, M., Young, N.E.: Classification via two-way comparisons (2023). https://doi.org/10.48550/ARXIV.2302.09692
7. Cormen, T.H., Leiserson, C.E., Rivest, R.L., Stein, C.: Introduction to Algorithms, 4th edn. The MIT Press, Cambridge (2022)
8. Dagan, Y., Filmus, Y., Gabizon, A., Moran, S.: Twenty (simple) questions. In: Proceedings of the 49th Annual ACM SIGACT Symposium on Theory of Computing, STOC 2017, Montreal, QC, Canada, 19–23 June 2017, pp. 9–21 (2017). https://doi.org/10.1145/3055399.3055422
9. Gilbert, E., Moore, E.: Variable-length binary encodings. Bell Syst. Tech. J. **38**(4), 933–967 (1959). https://doi.org/10.1002/j.1538-7305.1959.tb01583.x
10. Hester, J.H., Hirschberg, D.S., Huang, S.H., Wong, C.K.: Faster construction of optimal binary split trees. J. Algorithms **7**, 412–424 (1986). https://doi.org/10.1016/0196-6774(86)90031-3
11. Huang, S.H.S., Wong, C.K.: Generalized binary split trees. Acta Informatica **21**(1), 113–123 (1984). https://doi.org/10.1007/BF00289143
12. Huang, S.H.S., Wong, C.K.: Optimal binary split trees. J. Algorithms **5**, 69–79 (1984). https://doi.org/10.1016/0196-6774(84)90041-5
13. Hyafil, L., Rivest, R.L.: Constructing optimal binary decision trees is NP-complete. Inf. Process. Lett. **5**(1), 15–17 (1976). https://doi.org/10.1016/0020-0190(76)90095-8
14. Knuth, D.E.: Optimum binary search trees. Acta Informatica **1**, 14–25 (1971). https://doi.org/10.1007/BF00264289
15. Knuth, D.E.: The Art of Computer Programming, Volume 3: Sorting and Searching, 2nd edn. Addison-Wesley Publishing Company, Redwood (1998)
16. Perl, Y.: Optimum split trees. J. Algorithms **5**, 367–374 (1984). https://doi.org/10.1016/0196-6774(84)90017-8
17. Sheil, B.A.: Median split trees: a fast lookup technique for frequently occurring keys. Commun. ACM **21**, 947–958 (1978). https://doi.org/10.1145/359642.359653
18. Spuler, D.: Optimal search trees using two-way key comparisons. Acta Informatica **31**(8), 729–740 (1994). https://doi.org/10.1007/BF01178732
19. Spuler, D.A.: Optimal search trees using two-way key comparisons. Ph.D. thesis, James Cook University (1994)

Improved Bounds for Discrete Voronoi Games

Mark de Berg[1](✉) and Geert van Wordragen[2]

[1] Department of Mathematics and Computer Science, TU Eindhoven, Eindhoven, The Netherlands
M.T.d.Berg@tue.nl

[2] Department of Computer Science, Aalto University, Espoo, Finland
Geert.vanWordragen@aalto.fi

Abstract. In the planar one-round discrete Voronoi game, two players $\mathcal{P}$ and $\mathcal{Q}$ compete over a set V of n voters represented by points in $\mathbb{R}^2$. First, $\mathcal{P}$ places a set P of k points, then $\mathcal{Q}$ places a set Q of ℓ points, and then each voter $v \in V$ is won by the player who has placed a point closest to v. It is well known that if $k = \ell = 1$, then $\mathcal{P}$ can always win $n/3$ voters and that this is worst-case optimal. We study the setting where $k > 1$ and $\ell = 1$. We present lower bounds on the number of voters that $\mathcal{P}$ can always win, which improve the existing bounds for all $k \geqslant 4$. As a by-product, we obtain improved bounds on small ε-nets for convex ranges for even numbers of points in general position.

Keywords: Voronoi games · Competitive facility location · Spatial voting theory

1 Introduction

In the *discrete Voronoi game*, two players compete over a set V of n voters in $\mathbb{R}^d$. First, player $\mathcal{P}$ places a set P of k points, then player $\mathcal{Q}$ places a set Q of ℓ points disjoint from the points in P, and then each voter $v \in V$ is won by the player who has placed a point closest to v. In other words, each player wins the voters located in its Voronoi cells in the Voronoi diagram $\text{Vor}(P \cup Q)$. In case of ties, that is, when a voter v lies on the boundary between a Voronoi cell owned by $\mathcal{P}$ and a Voronoi cell owned by $\mathcal{Q}$, then v is won by player $\mathcal{P}$. Note that $\mathcal{P}$ first places all their k points and then $\mathcal{Q}$ places their ℓ points—hence, this is a *one-round Voronoi game*—and that k and ℓ need not be equal. The one-round discrete Voronoi game was introduced by Banik *et al.* [4].

There is also a version of the Voronoi game where the players compete over a continuous region [1,9,12]. For this version a multiple-round variant, where $k = \ell$ and the players place points alternatingly, has been studied as well. We will confine our discussion to the discrete one-round game.

MdB is supported by the Dutch Research Council (NWO) through Gravitation grant NETWORKS-024.002.003.

P. Morin and S. Suri (Eds.): WADS 2023, LNCS 14079, pp. 291–308, 2023.
https://doi.org/10.1007/978-3-031-38906-1_20

The discrete one-round Voronoi game for $k = \ell = 1$ is closely related to the concept of plurality points in spatial voting theory [14]. In this theory, there is a d-dimensional policy space, and voters are modelled as points indicating their preferred policies. A *plurality point* is then a proposed policy that would win at least $\lceil n/2 \rceil$ voters against any competing policy. Phrased in terms of Voronoi games, this means that $\mathcal{P}$ can place a single point that wins at least $\lceil n/2 \rceil$ voters against any single point placed by $\mathcal{Q}$. The discrete Voronoi game with $k > 1$ and $\ell = 1$ can be thought of as an election where a coalition of k parties is colluding against a single other party.

Another way to interpret Voronoi games is as a *competitive facility-location problem*, where two companies want to place facilities so as to attract as many customers as possible, where each customer will visit the nearest facility. Competitive facility location has not only been studied in a (discrete and continuous) spatial setting, but also in a graph-theoretic setting; see e.g. [3,13,16].

Previous Work. The one-round discrete Voronoi game leads to interesting algorithmic as well as combinatorial problems.

The algorithmic problem is to compute an optimal set of locations for the players. More precisely, for player $\mathcal{P}$ the goal is to compute, given a set V of n voters, a set P of k points that wins a maximum number of voters under the assumption that player $\mathcal{Q}$ responds optimally. For player $\mathcal{Q}$ the goal is to compute, given a voter set V and a set P of points placed by $\mathcal{P}$, a set Q of ℓ points that wins as many voters from V as possible. These problems were studied in $\mathbb{R}^1$ by Banik *et al.* [4] for the case $k = \ell$. They showed that an optimal set for $\mathcal{P}$ can be computed in $O(n^{k-\lambda_k})$ time, for some $0 < \lambda_k < 1$, and that an optimal set for $\mathcal{Q}$ can be computed in $O(n)$ if the voters are given in sorted order. The former result was improved by De Berg *et al.* [7], who presented an algorithm with $O(k^4 n)$ running time. They also showed that in $\mathbb{R}^2$ the problem for $\mathcal{P}$ is Σ_2^P-hard. The problem for $\mathcal{P}$ in the special case $k = \ell = 1$, is equivalent to finding the so-called Tukey median of V. This can be done in $O(n^{d-1} + n \log n)$ time, as shown by Chan [10].

The combinatorial problem is to prove worst-case bounds on the number of voters that player $\mathcal{P}$ can win, assuming player $\mathcal{Q}$ responds optimally. Tight bounds are only known for $k = \ell = 1$, where Chawla *et al.* [11] showed the following: for any set V of n voters in $\mathbb{R}^d$, player $\mathcal{P}$ can win at least $\lceil n/(d+1) \rceil$ voters and at most $\lceil n/2 \rceil$ voters, and these bounds are tight. Situations where $\mathcal{P}$ can win $\lceil n/2 \rceil$ voters are particularly interesting, as these correspond to the existence of a plurality point in voting theory. The bounds just mentioned imply that a plurality point does not always exist. In fact, a plurality point only exists for certain very symmetric point sets, as shown by Wu *et al.* [17]. De Berg *et al.* [6] showed how to test in $O(n \log n)$ time if a voter set admits a plurality point.

The combinatorial problem for $k > 1$ and $\ell = 1$ was studied by Banik *et al.* [5]. Here player $\mathcal{P}$ will never be able to win more than $\left(1 - \frac{1}{2k}\right) n$ voters, because player $\mathcal{Q}$ can always win at least half of the voters of the most crowded Voronoi cell in $\mathrm{Vor}(P)$. Banik *et al.* [5] present two methods to derive lower bounds on

Table 1. Lower bounds on the fraction of voters that $\mathcal{P}$ can win on any voter set in $\mathbb{R}^2$, when $\mathcal{P}$ has k points and $\mathcal{Q}$ has a single point. The stated fraction of our method for arbitrary k is for $n \to \infty$; the precise bound is $\left(1 - \frac{20\frac{5}{8}}{k}\right) n - 6$.

$k = 1$	$k = 2$	$k = 3$	$k = 4$	$k = 5$	arbitrary k	reference
1/3	3/7	7/15	15/31	21/41	$1 - \frac{42}{k}$	Banik *et al.* [5]
			1/2	11/21	$1 - \frac{20\frac{5}{8}}{k}$	this paper

the number of voters that $\mathcal{P}$ can always win. Below we discuss their results in $\mathbb{R}^2$, but we note that they generalize their methods to $\mathbb{R}^3$.

The first method uses a (weak) ε-net for convex ranges on the voter set V, that is, a point set N such that any convex range R containing at least εn voters, will also contain a point from N. Now, if $|Q| = 1$ then the voters won by $\mathcal{Q}$ lie in a single Voronoi cell in $\text{Vor}(P \cup Q)$. Since Voronoi cells are convex, this means that if we set $P := N$ then $\mathcal{P}$ wins at least $(1 - \varepsilon)n$ voters. Banik *et al.* use the ε-net construction for convex ranges by Mustafa and Ray [15]. There is no closed-form expression for the size of their ε-net, but the method can give a (4/7)-net of size 2, for instance, and an (8/15)-net of size 3. The smallest size for which they obtain an ε-net for some $\varepsilon \leqslant 1/2$, which corresponds to $\mathcal{P}$ winning at least half the voters, is $k = 5$. Banik *et al.* show that the ε-net of Mustafa and Ray can be constructed in $O(kn \log^4 n)$ time. The second method of Banik *et al.* uses an ε-net for disks, instead of convex sets. This is possible because one can show that a point $q \in Q$ that wins α voters, must have a disk around it that covers at least $\lfloor \alpha/6 \rfloor$ voters without containing a point from P. Banik *et al.* then present a $(7/k)$-net for disks of size k, which can be constructed in $O(n^2)$ time. This gives a method that ensures $\mathcal{P}$ wins at least $\left(1 - \frac{42}{k}\right) n$ voters, which is better than the first method when $k \geqslant 137$.

Our Results. We study the combinatorial question—how many voters can player $\mathcal{P}$ win from any voter set V of size n, under optimal play from $\mathcal{Q}$—in the planar setting, for $k > 1$ and $\ell = 1$. We obtain the following results, where we assume that V is in general position—no three voters are collinear—and that n is even.

In Sect. 2 we present an improvement[1] over the ε-net bounds by Mustafa and Ray [15] for convex ranges. This improves the results of Banik *et al.* [5] on the fraction of voters that $\mathcal{P}$ can win when $k \geqslant 4$ and k is relatively small. We do not have a closed-form expression for the size of our ε-net as function of ε. Theorem 2 gives a recurrence on these sizes, and Table 1 shows how our bounds compare to those of Banik *et al.* for $k = 4, 5$ (which follow from the bounds of Mustafa and Ray [15]). It is particularly interesting that our bounds improve the smallest k for which $\mathcal{P}$ can win at least half the voters, from $k = 5$ to $k = 4$.

[1] Our definition of ε-net is slightly weaker than usual, since a range missing the ε-net may contain up to $\lceil \varepsilon n \rceil$ points, instead of $\lfloor \varepsilon n \rfloor$ points.

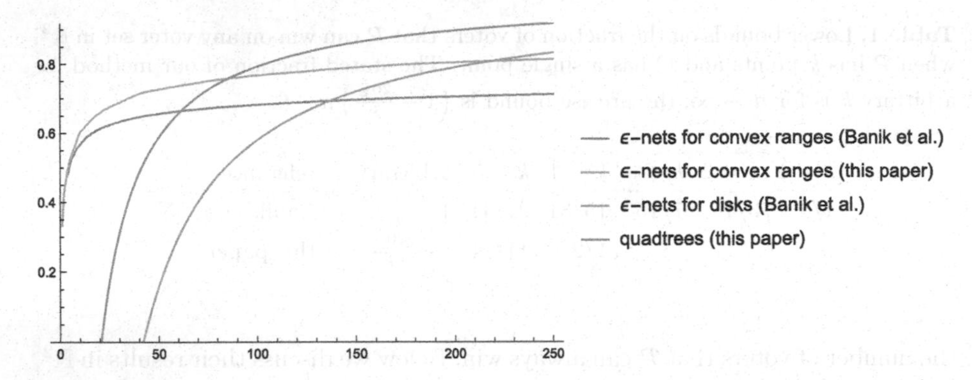

Fig. 1. Lower bounds on the fraction of voters that $\mathcal{P}$ can win as a function of k (the number of points of $\mathcal{P}$) when $\mathcal{Q}$ has a single point, for the L_2-metric. The red and green graphs do not intersect, so for large k the quadtree method gives the best solution. (Color figure online)

In Sect. 3 we present a new strategy for player $\mathcal{P}$. Unlike the strategies by Banik *et al.*, it is not based on ε-nets. Instead, it uses a quadtree-based approach. By combining this approach with several other ideas and using our ε-net method as a subroutine, we are able to show that there is a set P of k points that guarantees that $\mathcal{P}$ wins at least $\left(1 - \frac{20\frac{5}{8}}{k}\right) n - 6$ voters, which significantly improves the $\left(1 - \frac{42}{k}\right) n$ bound of Banik *et al.* Fig. 1 show the bounds obtained by the various methods in a graphical way.

We also study the discrete one-round Voronoi game in the L_1-metric, for $k > 1$ and $\ell = 1$. When $k = 1$, player $\mathcal{P}$ can win at least half the voters by placing a point on a multi-dimensional median, that is, a point whose x- and y-coordinate are medians among the x- and y-coordinates of the voter set V [6]. The case $k > 1$ and $\ell = 1$ has, as far as we know, not been studied so far. We first observe that for the L_1-metric, an ε-nets for axis-parallel rectangles can be used to obtain a good set of points for player $\mathcal{P}$. Using known results [2] this implies the results for $2 \leqslant k \leqslant 5$ in Table 1.

2 Better ε-Nets for Convex Ranges

Below we present a new method to construct an ε-net for convex ranges in the plane, which improves the results of Mustafa and Ray [15]. As mentioned in the introduction, this implies improved bounds on the number of voters $\mathcal{P}$ can win with k points when $\mathcal{Q}$ has a single point, for relatively small values of k.

Let L be a set of three concurrent lines and consider the six wedges defined by the lines. Bukh [8] proved that for any continuous measure there is a choice of L where each of the wedges has equal measure. Instead of a measure, we have V, a point set in general position in the plane, thus we need a generalisation where the wedges contain some specified number of points. In our generalisation, the *weight* of a wedge is given by the number of points from V assigned to it. If a point $v \in V$ lies in the interior of a wedge then we assign v to that wedge, and if v lies on the boundary of two or more wedges we assign v to one of them. (This assignment is not arbitrary, but we will do it is such a way as to obtain the desired number of points in each wedge.) We call this a *wedge assignment.*

Theorem 1. *Let V be a set of n points in general position in the plane, where $n \geqslant 8$ is even. For any given $\alpha, \beta, \gamma \in \mathbb{N}$ such that $2\alpha + 2\beta + 2\gamma = n$, we can find a set of three concurrent lines that partitions the plane into six wedges such that there is a wedge assignment resulting in wedges whose weights are $\alpha, \beta, \gamma, \alpha, \beta, \gamma$ in counterclockwise order.*

Proof (Sketch). Let $\ell(\theta)$ be the directed line making an angle θ with the positive x-axis and that has exactly weight $n/2$ on either side of it, for a suitable assignment of points to the half-planes on either side of $\ell(\theta)$. Consider the line $\ell(\theta)$ for $\theta = 0$. For some point $z = (x, 0) \in \ell(\theta)$, consider the rays $\rho_1, \ldots, \rho_4$ emanating from z, such that the six wedges defined by these rays and $\ell(0)$ have the desired number of voters; see Fig. 2. By varying θ and the point z, we can ensure that the rays $\rho_1, \ldots, \rho_4$ line up in such a way that, together with $\ell(\theta)$, they form three concurrent lines. □

We also need the following easy-to-prove observation.

Observation 1. Let L be a set of three lines intersecting in a common point p^*, and consider the six closed wedges defined by L. Any convex set S not containing p^* intersects at most four wedges, and the wedges intersected by S are consecutive in the clockwise order.

Notice that Observation 1 considers all *closed* wedges intersecting S. Thus, when S touches the boundary between two wedges, then both of them are taken into account. Hence, it will not be a problem that Theorem 1 could assign points on the boundary between two wedges to either one of them.

We now have all the tools to prove our new bounds on ε-nets for convex ranges. The guarantee they give is slightly weaker than usual: where ordinarily placing an ε-net for n points means a range not intersecting the ε-net can contain at most εn (and thus at most $\lfloor \varepsilon n \rfloor$) points, our *ceiling-based* ε-nets only guarantee that such a range contains at most $\lceil \varepsilon n \rceil$ points.

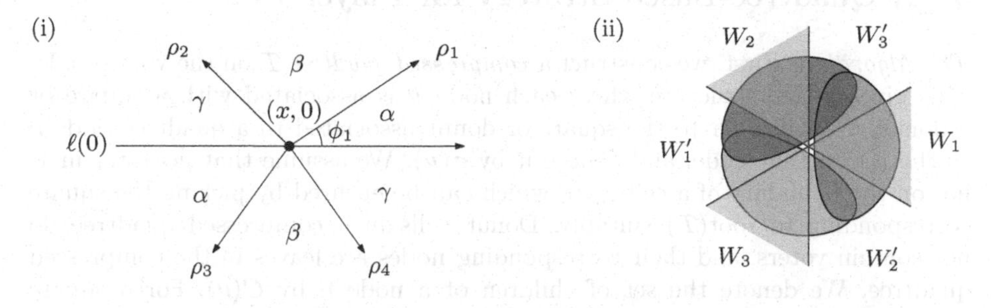

Fig. 2. Illustrations for the proofs of Theorem 1 and Theorem 2.

Theorem 2. *Let ε_k be the smallest value such that any finite point set in $\mathbb{R}^2$ admits a weak ε_k-net of size k for convex ranges. Then for any set V of $n \geqslant 8$ points in general position, with n even, and any $r_1, r_2, s \in \mathbb{N}_0$, we can make a ceiling-based ε-net for V with*

$$\varepsilon = \frac{1}{2}\left(\frac{1}{\varepsilon_{r_1}} + \frac{2}{\varepsilon_{r_2}}\right)^{-1} + \frac{1}{2}\varepsilon_s.$$

Proof (Sketch). Let $\mu := \frac{1}{2}\left(\frac{1}{\varepsilon_{r_1}} + \frac{2}{\varepsilon_{r_2}}\right)^{-1}$. We apply Theorem 1 with $\beta = \gamma = \left\lceil \frac{\mu}{\varepsilon_{r_2}} n \right\rceil$ and $\alpha = \frac{n}{2} - 2\beta$, which means $\alpha \leqslant \left\lfloor \frac{\mu}{\varepsilon_{r_1}} n \right\rfloor$, giving us a set L of three concurrent lines. To show that there exists a weak ceiling-based $(\varepsilon_{r_1+2r_2+3s+1})$-net N for V, number the wedges defined by L as $W_1, W_3', W_2, W_1', W_3, W_2'$ in clockwise order, as in Fig. 2(ii). Let $V_i \subset V$ and $V_i' \subset V$ be the subsets of points assigned to W_i and W_i', respectively. We can assume without loss of generality that $|V_1| = |V_1'| = \alpha$, and $|V_2| = |V_2'| = \beta$, and $|V_3| = |V_3'| = \gamma$. We add the following points to our net N: (i) the common intersection of the lines in L, denoted by p^*; (ii) an ε_{r_1}-net for V_1, an ε_{r_2}-net for V_2, and an ε_{r_2}-net for V_3; (iii) for each of the three collections of three consecutive wedges—these collections are indicated in red, green, and blue in Fig. 2(ii)—an ε_s-net. By construction, the size of our net N is $1 + r_1 + 2r_2 + 3s$. In the full paper we show that N is a ceiling-based $(\mu + \frac{1}{2}\varepsilon_s)$-net. □

Note that $\varepsilon_0 = 1$, since if the net is empty, a range can contain all n points from V. Moreover, $\varepsilon_1 = 2/3$, and $\varepsilon_2 = 4/7$, and $\varepsilon_3 \leqslant 8/15$ by the results of Mustafa and Ray [15]. Using Theorem 2 we can then set up a recursion to obtain ceiling-based ε-nets with $k \geqslant 4$ points, by finding the best choice of r_1, r_2, s such that $k = r_1 + 2r_2 + 3s + 1$. This gives $\varepsilon_4 \leqslant \frac{1}{2}$, by setting $r_1, r_2 = 0$ and $s = 1$. Hence, for even n, player $\mathcal{P}$ can always place four points to win at least as many voters as player $\mathcal{Q}$, as opposed to the five that were proven in earlier work. Note that this also holds for $n \leqslant 8$, since then player $\mathcal{P}$ can simply pick four points coinciding with four of the at most eight voters. A similar statement holds for larger k when $n \leqslant 8$.

3 A Quadtree-Based Strategy for Player $\mathcal{P}$

The Algorithm. First, we construct a *compressed quadtree* $\mathcal{T}$ on the voter set V. This gives a tree structure where each node ν is associated with a square or a donut. We will refer to the square or donut associated to a quadtree node ν as the *cell* of that node, and denote it by $\sigma(\nu)$. We assume that no voter in V lies on the boundary of a cell $\sigma(\nu)$, which can be ensured by picking the square corresponding to root($\mathcal{T}$) suitably. Donut cells in a compressed quadtree do not contain voters, and their corresponding nodes are leaves in the compressed quadtree. We denote the set of children of a node ν by $C(\nu)$. For a square quadtree cell σ, we denote its four quadrants by NE(σ), SE(σ), SW(σ), and NW(σ).

We define the *size* of a square σ, denoted by $size(\sigma)$, to be its edge length. Let $\text{dist}(\sigma_1, \sigma_2)$ denote the distance between the boundaries of two squares σ_1, σ_2. The distance between two quadtree cells satisfies the following property. Note that the property also holds when the cells are nested.

Observation 2. Let σ_1 and σ_2 be square cells corresponding to two nodes in $\mathcal{T}$ such that σ_1 and σ_2 are intersected by a common horizontal (or vertical) line. If $\text{dist}(\sigma_1, \sigma_2) > 0$ then $\text{dist}(\sigma_1, \sigma_2) \geqslant \min(size(\sigma_1), size(\sigma_2))$.

The idea of our algorithm to generate the k points played by player $\mathcal{P}$ is as follows. We pick a parameter m, which depends on k, and then we recursively traverse the tree $\mathcal{T}$ to generate a set $\mathcal{R}$ of regions, each containing between $m+1$ and $4m$ points. Each region $R(\nu) \in \mathcal{R}$ will be a quadtree cell $\sigma(\nu)$ minus the quadtree cells $\sigma(\mu)$ of certain nodes μ in the subtree rooted at ν. For each region $R \in \mathcal{R}$, we then generate a set of points that we put into P. The exact procedure to generate the set $\mathcal{R}$ of regions is described by Algorithm 1, which is called with $\nu = \text{root}(\mathcal{T})$.

Algorithm 1. *MakeRegions*(ν, m)

Input: A node ν in $\mathcal{T}$ and a parameter m

Output: A pair $(\mathcal{R}, V_{\text{free}})$, where $\mathcal{R}$ is a set of regions containing at least $m+1$ and at most $4m$ voters, and V_{free} contains the voters in the subtree rooted at ν that are not yet covered by a region in $\mathcal{R}$.

1: **if** ν is a leaf node **then**
2: Return $(\emptyset, \{v\})$ if ν contains a voter v, and return $(\emptyset, \emptyset)$ otherwise
3: **else**
4: Recursively call MakeRegions(μ, m) for all children $\mu \in C(\nu)$. Let $\mathcal{R}$ be the union of the returned sets of regions, and let V_{free} be the union of the sets of returned free voters.
5: **if** $|V_{\text{free}}| \leqslant m$ **then**
6: Return $(\mathcal{R}, V_{\text{free}})$
7: **else**
8: $R(\nu) \leftarrow \sigma(\nu) \setminus \bigcup_{R \in \mathcal{R}} R$ ▷ Note that $V_{\text{free}} = R(\nu) \cap V$.
9: Return $(\mathcal{R} \cup \{R(\nu)\}, \emptyset)$

We use the regions in $\mathcal{R}$ to place the points for player $\mathcal{P}$, as follows. For a region $R := R(\nu)$ in $\mathcal{R}$, define $\sigma(R) := \sigma(\nu)$ to be the cell of the node ν for which R was generated. For each $R \in \mathcal{R}$, player $\mathcal{P}$ will place a grid of 3×3 points inside $\sigma(R)$, plus four points outside $\sigma(R)$, as shown in Fig. 3(i). (Some points placed for R may coincide with points placed for some $R' \neq R$, but this will only help to reduce the number of points placed by $\mathcal{P}$).

Note that each $R \in \mathcal{R}$ contains more than m voters and the regions in $\mathcal{R}$ are disjoint. Hence, $|\mathcal{R}| < n/m$ and $|P| < 13n/m$. A compressed quadtree can be constructed in $\mathcal{O}(n \log n)$ time, and the rest of the construction takes $\mathcal{O}(n)$ time. The following lemma summarizes the construction.

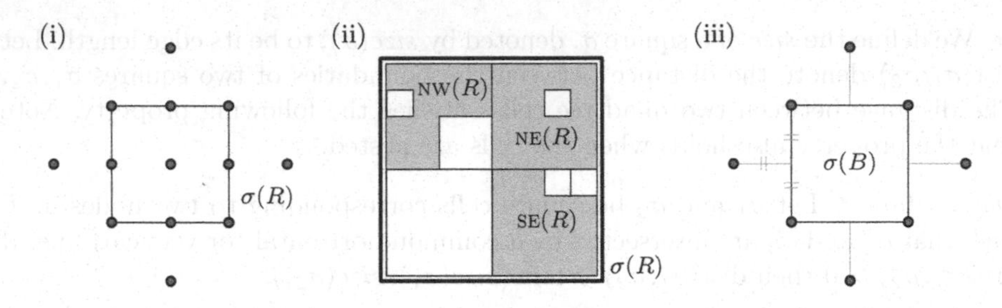

Fig. 3. (i) The 13 points (in red) placed in P for a region $R \in \mathcal{R}$. A block B such that $\sigma(B)$ is one of the quadrants of $\sigma(R)$ is called a type-I block. (ii) A region R (shown in green) and its blocks (that is, its child regions). The white area is covered by regions that have been created earlier. Since $\text{SW}(\sigma(R))$ has already been fully covered, $\text{SW}(R)$ does not exist. (iii) The eight points placed in P for a type-II block $B \in \mathcal{B}$. (Color figure online)

Lemma 1. *The quadtree-based strategy described above places fewer than* $13n/m$ *points for player* $\mathcal{P}$ *and runs in* $\mathcal{O}(n \log n)$ *time.*

An Analysis of the Number of Voters Player $\mathcal{Q}$ *Can Win.* To analyze the number of voters that $\mathcal{Q}$ can win, it will be convenient to look at the "child regions"of the regions in $\mathcal{R}$, as defined next. Recall that for a region $R := R(\nu)$ in $\mathcal{R}$, we defined $\sigma(R) := \sigma(\nu)$. Let $\text{NE}(R) := R \cap \text{NE}(\sigma(R))$ be the part of R in the NE-quadrant of $\sigma(R)$. We call $\text{NE}(R)$ a *child region* of R. The child regions $\text{SE}(R)$, $\text{SW}(R)$, and $\text{NW}(R)$ are defined similarly; see Fig. 3(ii) for an example.

Let $\mathcal{B}$ be the set of non-empty child regions of the regions in $\mathcal{R}$. From now on, we will refer to the child regions in $\mathcal{B}$ as *blocks*. Blocks are not necessarily rectangles, and they can contains holes and even be disconnected. For a block $B \in \mathcal{B}$, we denote its parent region in $\mathcal{R}$ by $\text{pa}(B)$, and we let $\sigma(B)$ denote the quadtree cell corresponding to B. For instance, if $B = \text{NE}(\text{pa}(B))$ then $\sigma(B) = \text{NE}(\sigma(\text{pa}(B)))$.

Note that at the end of Algorithm 1, the set V_{free} need not be empty. Thus the blocks in $\mathcal{B}$ may not cover all voters. Hence, we add a special *root block* B_0 to $\mathcal{B}$, with $\sigma(B_0) := \sigma(\text{root}(\mathcal{T}))$ and which consists of the part of $\sigma(\text{root}(\mathcal{T}))$ not covered by other blocks. Note that we do not add any points to P for B_0.

Because we will later refine our strategy for player $\mathcal{P}$, it will be convenient to analyze the number of voters that $\mathcal{Q}$ can win in an abstract setting. Our analysis requires the collection $\mathcal{B}$ of blocks and the set P of points played by $\mathcal{P}$ to have the following properties.

($\mathcal{B}$.1). The blocks in $\mathcal{B}$ are generated in a bottom-up manner using the compressed quadtree $\mathcal{T}$. More precisely, there is a collection $N(\mathcal{B})$ of nodes in $\mathcal{T}$ that is in one-to-one correspondence with the blocks in $\mathcal{B}$ such that the following holds:
Let $B(\nu)$ be the block corresponding to a node $\nu \in N(\mathcal{B})$. Then $B(\nu) = \sigma(\nu) \setminus \bigcup_{\mu} B(\mu)$, where the union is taken over all nodes $\mu \in N(\mathcal{B})$ that are a descendant of ν.

We also require that the blocks in $\mathcal{B}$ together cover all voters.

($\mathcal{B}$.2). For each block $B \in \mathcal{B}$, except possibly the root block B_0, the point set P includes the 13 points shown in Fig. 3(i) for the cell that is the parent of $\sigma(B)$, or it includes the eight points shown in Fig. 3(iii). In the former case we call B a *type-I block*, in the latter case we call B a *type-II block*. Note that in both cases P includes the four corners of $\sigma(B)$.

Observe that ($\mathcal{B}$.1) implies that the blocks $B \in \mathcal{B}$ are disjoint. Moreover, property ($\mathcal{B}$.2) implies the following. For a square σ, define $plus(\sigma)$ to be the plus-shaped region consisting of five equal-sized squares whose central square is σ.

Observation 3. Let q be a point played by player $\mathcal{Q}$ and let $B \in \mathcal{B}$ be a block. If q wins a voter v that lies in $\sigma(B)$, then $q \in plus(\sigma(B))$. Furthermore, if $q \in \sigma(B)$ then q can only win voters in $plus(\sigma(B))$.

It is easy to see that the sets $\mathcal{B}$ and P generated by the construction described above have properties ($\mathcal{B}$.1) and ($\mathcal{B}$.2). We proceed to analyze the number of blocks from which a point q played by $\mathcal{Q}$ can win voters, assuming the set $\mathcal{B}$ of blocks has the properties stated above.

We will need the following observation. It follows from ($\mathcal{B}$.1), which states that a block B completely covers the part of $\sigma(B)$ not covered by blocks that have been created earlier in the bottom-up process.

Observation 4. If $\sigma(B) \subset \sigma(B')$ for two blocks $B, B' \in \mathcal{B}$ then $B' \cap \sigma(B) = \emptyset$.

The following lemma states that the set P of points played by player $\mathcal{P}$ includes all vertices of the blocks in $\mathcal{B}$, except possibly the corners of the root block B_0.

Lemma 2. *Let p be a vertex of a block $B \in \mathcal{B}$. Then $p \in P$, except possibly when p is a corner of $\sigma(B_0)$.*

Proof. Property ($\mathcal{B}$.1) states that the blocks in $\mathcal{B}$ are created in a bottom-up order. We will prove the lemma by induction on this (partial) order.

Consider a block $B \in \mathcal{B}$ and let p be a vertex of B. Let s be a sufficiently small square centered at p and let s_1, s_2, s_3, s_4 be its quadrants. There are two cases; see Fig. 4.

If p is a reflex vertex of B, then B covers three of the four squares s_1, s_2, s_3, s_4. The remaining square must already have been covered by a region B' created before B, by Observation 4. By induction, we may conclude that $p \in P$.

If p is a convex vertex, then exactly one of the four squares s_1, s_2, s_3, s_4, say s_1, is contained in B. If p is a corner of $\sigma(B)$, then $p \in P$ by property ($\mathcal{B}$.2). Otherwise, at least one square $\sigma_i \neq \sigma_1$, say σ_2, is contained in $\sigma(B)$. We can now use the same argument as before: p is a vertex of a region B' created before B, and so $p \in P$ by induction. Note that this not only holds when p lies on an edge of $\sigma(B)$, as in Fig. 4, but also when p lies in the interior of $\sigma(B)$. □

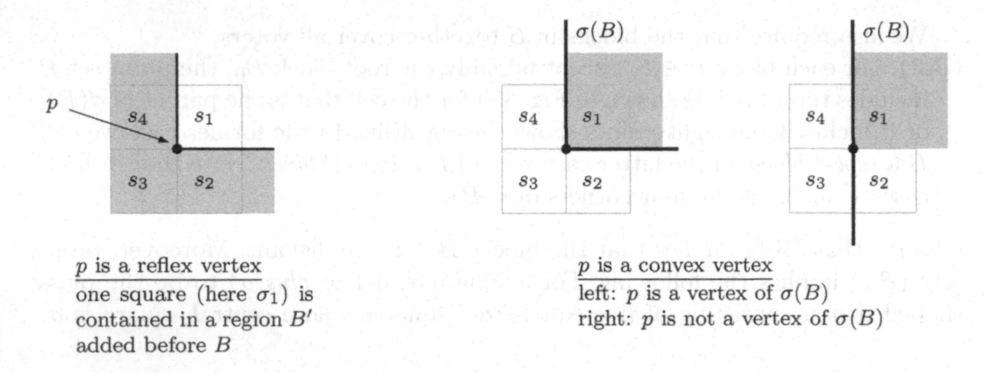

Fig. 4. Illustration for the proof of Lemma 2.

Now consider a point q played by player $\mathcal{Q}$, and assume without loss of generality that $q \in \sigma(B_0)$. We first show that q can win voters from at most five blocks $B \in \mathcal{B}$; later we will improve this to at most three blocks. We may assume that the horizontal and vertical lines through q do not pass through a vertex of any block $B \in \mathcal{B}$. This is without loss of generality, because an infinitesimal perturbation of q ensures this property, while such a perturbation does not change which voters are won by q. (The latter is true because voters at equal distance from q and P are won by player $\mathcal{P}$.)

Define $B(q) \in \mathcal{B}$ to be the block containing q. We start by looking more closely at which voters q might win from a block $B \neq B(q)$. Define $V(B) := V \cap B$ to be the voters lying in B. Let ρ_{left} be the axis-aligned ray emanating from q and going the left, and define $\rho_{\text{up}}, \rho_{\text{right}}, \rho_{\text{down}}$ similarly. Let e be the first edge of B that is hit by ρ_{right} and define

$$V_{\text{right}}(B) := \{v \in V(B) : v \text{ lies in the horizontal half-strip whose left edge is } e\}.$$

Define the sets $V_{\text{up}}(B)$, $V_{\text{down}}(B)$, and $V_{\text{left}}(B)$ similarly. See Fig. 5(i), where the voters from $V_{\text{right}}(B)$ are shown in dark green, the voters from $V_{\text{up}}(B)$ and $V_{\text{down}}(B)$ are shown in orange and blue, respectively, and $V_{\text{left}}(B) = \emptyset$. Because P contains all vertices of B by Lemma 2, the only voters from $V(B)$ that q can

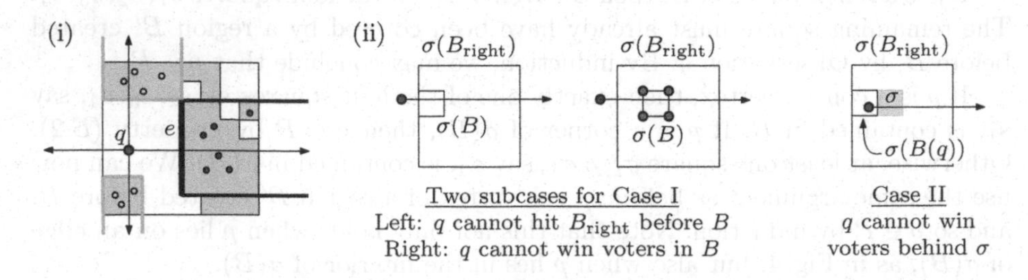

Fig. 5. (i) The sets of voters that q might be able to win in the green block B. (ii) Illustration for the proof of Lemma 3 (Color figure online).

possibly win are the voters in $V_{\text{left}}(B) \cup V_{\text{up}}(B) \cup V_{\text{right}}(B) \cup V_{\text{down}}(B)$. (In fact, we could restrict these four sets even a bit more, but this is not needed for our arguments.)

Let $B_{\text{right}} \neq B(q)$ be the first block in $\mathcal{B}$ hit by ρ_{right}, and define $B_{\text{left}}, B_{\text{up}}, B_{\text{down}}$ similarly for the rays $\rho_{\text{left}}, \rho_{\text{up}}, \rho_{\text{down}}$. The next lemma states that there is only one block $B \neq B(q)$ for which q might be able to win voters in $V_{\text{right}}(B)$, namely B_{right}. Similarly, q can only when voters from $V_{\text{left}}(B)$ for $B = B_{\text{left}}$, and so on.

Lemma 3. *If q wins voters from $V_{\text{right}}(B)$, where $B \neq B(q)$, then $B = B_{\text{right}}$.*

Proof. Suppose for a contradiction that q wins voters from $V_{\text{right}}(B)$ for some block $B \notin \{B(q), B_{\text{right}}\}$. We distinguish two cases.

Case I: $q \notin \sigma(B_{\text{right}})$
Since the corners of $\sigma(B_{\text{right}})$ are in P by ($\mathcal{B}$.2), the point q cannot win voters to the right of $\sigma(B_{\text{right}})$. Hence, if q wins voters from $V_{\text{right}}(B)$, then B must lie at least partially inside $\sigma(B_{\text{right}})$. Now consider $\sigma(B)$. We cannot have $\sigma(B_{\text{right}}) \subset \sigma(B)$ by Observation 4. Hence, $\sigma(B) \subset \sigma(B_{\text{right}})$ and so $B_{\text{right}} \cap \sigma(B) = \emptyset$. We now have two subcases, illustrated in Fig. 5(ii).

- If the left edge of $\sigma(B)$ is contained in the left edge of $\sigma(B_{\text{right}})$, then ρ_{right} would hit B before B_{right}, contradicting the definition of B_{right}.
- On the other hand, if the left edge of $\sigma(B)$ is not contained in the left edge of $\sigma(B_{\text{right}})$, then $\text{dist}(\sigma(B_{\text{right}}), \sigma(B)) \geqslant \textit{size}(\sigma(B))$ by Observation 2. Since P contains the four corners of $\sigma(B)$, this contradicts that q wins voters from $V_{\text{right}}(B)$.

Case II: $q \in \sigma(B_{\text{right}})$
We cannot have $\sigma(B_{\text{right}}) \subset \sigma(B(q))$, otherwise $B(q) \cap \sigma(B_{\text{right}}) = \emptyset$ by Observation 4, which contradicts $q \in B(q)$. Hence, $\sigma(B(q)) \subset \sigma(B_{\text{right}})$ and $B_{\text{right}} \cap \sigma(B(q)) = \emptyset$.

Consider the square σ with the same size of $\sigma(B(q))$ and immediately to the right of $\sigma(B(q))$; see the grey square in Fig. 5(ii). We must have $\sigma \subset \sigma(B_{\text{right}})$, otherwise the right edge of $\sigma(B(q))$ would be contained in the right edge of $\sigma(B_{\text{right}})$ and so ρ_{right} would exit $\sigma(B_{\text{right}})$ before it can hit B_{right}. By Observation 3, point q cannot win voters to the right of σ. Hence, $B \cap \sigma \neq \emptyset$. Now consider the relative position of $\sigma(B)$ and σ. There are two subcases.

- If $\sigma(B) \subset \sigma$, then either the distance from q to B is at least $\textit{size}(\sigma(B))$ by Observation 2, contradicting that q wins voters from $V_{\text{right}}(B)$; or $\sigma(B)$ lies immediately to the right of $\sigma(B(q))$, in which case ρ_{right} cannot hit B_{right} before B.
- Otherwise, $\sigma \subset \sigma(B)$. If $\sigma(B) \subset \sigma(B_{\text{right}})$, then $B_{\text{right}} \cap \sigma(B) = \emptyset$ by Observation 4, contradicting (since $\sigma \subset \sigma(B)$) that ρ_{right} hits B_{right} before B. Hence, $\sigma(B_{\text{right}}) \subset \sigma(B)$. But then $B \cap \sigma(B_{\text{right}}) = \emptyset$, which contradicts that $B \cap \sigma \neq \emptyset$.

□

Lemma 3 implies that q can only win voters from the five blocks $B(q)$, B_{left}, B_{right}, B_{up}, and B_{down}. The next lemma shows that q cannot win voters from all these blocks simultaneously.

Lemma 4. *Point q can win voters from at most three of the blocks $B(q)$, B_{left}, B_{right}, B_{up}, and B_{down}.*

Proof. First suppose that the size of $\sigma(B(q))$ is at most the size of any of the four cells $\sigma(B_{\text{left}}), \ldots, \sigma(B_{\text{down}})$ from which q wins voters. By Observation 4, this implies that all four blocks $B_{\text{left}}, B_{\text{right}}, B_{\text{up}}, B_{\text{down}}$ lie outside $\sigma(B(q))$. Then it is easy to see that q can win voters from at most two of the four blocks $B_{\text{left}}, B_{\text{right}}, B_{\text{up}}, B_{\text{down}}$, because all four corners of $\sigma(B(q))$ are in P by $(\mathcal{B}.2)$. For instance, if q lies in the NE-quadrant of $\sigma(B(q))$, then q can only win voters from B_{right} and B_{up}; the other cases are symmetrical.

Now suppose that $\sigma(B(q))$ is larger than $\sigma(B_{\text{right}})$, which we assume without loss of generality to be a smallest cell from which q wins voters among the four cells $\sigma(B_{\text{left}}), \ldots, \sigma(B_{\text{down}})$. We have two cases.

Case I: $q \notin \sigma(\text{pa}(B_{\text{right}}))$

Note that $\sigma(B_{\text{right}})$ must either be the NW- or SW-quadrant of $\sigma(\text{pa}(B_{\text{right}}))$, because otherwise $q \notin \textit{plus}(B_{\text{right}})$ and q cannot win voters from B_{right} by Observation 3. Assume without loss of generality that $\sigma(B_{\text{right}}) = \text{NW}(\sigma(\text{pa}(B_{\text{right}})))$. Then q must be located in the square of the same size as $\sigma(B_{\text{right}})$ and immediately to its left. In fact, q must lie in the right half of this square. We now define two blocks, σ and σ' that play a crucial role in the proof. Their definition depends on whether B_{right} is a type-I or a type-II block.

- If B_{right} is a type-I block, then we define $\sigma' := \sigma(B_{\text{right}})$ and we define σ to be the square of the same size as σ' and immediately to its left. Note that $q \in \sigma$, since q wins voters from $\sigma(B_{\text{right}})$. See Fig. 6(i).
- If B_{right} is a type-II block, then we define $\sigma' := \text{NW}(\sigma(B_{\text{right}}))$ or $\sigma' = \text{SW}(\sigma(B_{\text{right}}))$ and we define σ to be the square of the same size as σ' and immediately to its left. Whether we choose $\sigma' := \text{NW}(\sigma(B_{\text{right}}))$ or $\sigma' = \text{SW}(\sigma(B_{\text{right}}))$ depends on the position of q: the choice is made such that the square σ to the left of σ' contains q. See Fig. 6(ii) for an example.

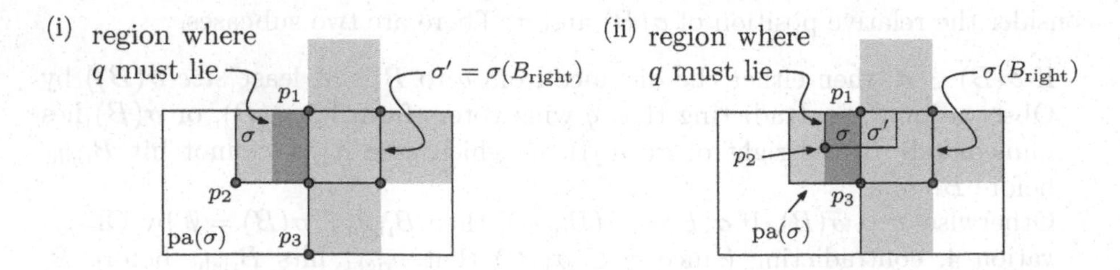

Fig. 6. Two cases for the definition of σ' and σ, (i) when B_{right} is a type-I block and (ii) when B_{right} is a type-II block. In the latter case σ' and σ' could also lie in the bottom half of their parent regions, depending on where q lies.

Since we will not use $\text{pa}(\text{pa}(\sigma'))$ in the proof, these two choices are symmetric as far as the proof is concerned—we only need to swap the up- and down-direction.

We now continue with the proof of Case I. All statements referring to σ and σ' will hold for both definitions just given.

Observe that $\sigma(B_{\text{down}}) \neq \sigma$, since otherwise $\sigma(B_{\text{down}}) \subset \sigma(B(q))$, contradicting by Observation 4 that $q \in B(q)$. We will now consider three subcases. In each subcase we argue that either we are done—we will have shown that q wins voters from at most one of the blocks B_{down}, B_{left}, and B_{up}—or q cannot win voters from B_{down}. After discussing the three subcases, we then continue the proof under the assumption that q does not win voters from B_{down}.

- *Subcase (i): $\sigma(B_{\text{down}}) = \text{SE}(\text{pa}(\sigma))$ and B_{down} is a type-I block.*
  In the case all four corners of σ are in P. If $q \in \text{NE}(\sigma)$ then q can only win voters from B_{up} and if $q \in \text{SE}(\sigma)$ then q can only win voters from B_{down} (this is in addition to voters won from $B(q)$ and B_{right}), and so we are done.
- *Subcase (ii): $\sigma(B_{\text{down}}) = \text{SE}(\text{pa}(\sigma))$ and B_{down} is a type-II block.*
  If $q \in \text{SE}(\sigma)$ then q cannot win voters from B_{up} or B_{left}, and so we are done. Otherwise q cannot win voters from B_{down}, as claimed.
- *Subcase (iii): $\sigma(B_{\text{down}}) \neq \text{SE}(\text{pa}(\sigma))$.*
  If $B_{\text{down}} \cap \text{pa}(\sigma) \neq \emptyset$ then both B_{down} and $B(q)$ intersect $\text{pa}(\sigma)$, and both $\sigma(B_{\text{down}})$ and $\sigma(B(q))$ contain $\text{pa}(\sigma)$. But this is impossible due to Observation 4. Hence, B_{down} must lie below $\text{pa}(\sigma)$. We claim that then q cannot win voters from B_{down}. The closest q can be to B_{down} is when it lies on the bottom line segment of σ. Hence, any voter in B_{down} won by q must be closer to that segment than to p_3, and also than to p_2. (See Fig. 6 for the locations of p_2 and p_3.) But this is clearly impossible. Hence, q cannot win voters from B_{down}.

Thus, in the remainder of the proof for Case I we can assume that q does not win voters in B_{down}. Hence, it suffices to show that q cannot win voters from B_{up} and B_{left} simultaneously. To this end, we assume q wins a voter v_{up} from B_{up} and a voter v_{left} from B_{left} and then derive a contradiction.

Let e_{up} be the first edge of B_{up} hit by ρ_{up} and let e_{left} be defined analogously; see Fig. 7. Note that e_{up} and e_{left} must lie outside $\text{pa}(\sigma)$, otherwise we obtain a contradiction with Observation 4.

Let $\ell_{\text{hor}}(q)$ be the horizontal line through q.

Claim. v_{left} must lie above $\ell_{\text{hor}}(q)$.

Proof of Claim. We need to show that the perpendicular bisector of q and p_2 will always intersect the left edge of $\text{pa}(\sigma)$ above $\ell_{\text{hor}}(q)$. For the situation in Fig. 6(i) this is relatively easy to see, since q lies relatively far to the left compared to p_2. For situation in Fig. 6(ii), it follows from the following argument. To win voters in B_{right}, the point q must lie inside the circle C through p_1, p_2, p_3. Now, suppose q actually lies on C and let $\alpha := \angle qzp_2$, where z is the center of C. Thus the bisector of q and p_3 has slope $-\tan(\alpha/2)$. The result then follows from the fact that $2\tan(\alpha/2) - \sin\alpha > 0$ for $0 < \alpha < \pi/2$. □

For v_{up} the situation is slightly different: q can, in fact, win voters to the right of $\ell_{vert}(q)$, the vertical line through q. In that case, however, it cannot win v_{left}.

Claim. If q wins a voter from B_{up} right of $\ell_{vert}(q)$, then q cannot win a voter from B_{left}.

Proof of Claim. It follows from Observation 2 that e_{up} must overlap with the top edge of σ. Because the edges e_{up} and e_{left} cannot intersect, one of them must end when or before the two meet.

If e_{left} ends before meeting (the extension of) e_{up}, then the top endpoint of e_{up}, which is in P by Lemma 2, prevents q from winning voters from B_{left}.

So now assume that e_{up} ends before meeting (the extension of) e_{left}. Then the left endpoint of e_{up}, which we denote by p_4, is in P. Without loss of generality, set $p_1 = (0, 1)$, $p_2 = (-1, 0)$ and $p_4 = (p_x, 1)$. Now, winning voters from B_{up} means q must lie inside the circle C_{up} with center on e_{up} that goes through p_1 and p_4. Thus it has center $c = (\frac{p_x}{2}, 0)$. It must also lie in the circle C through p_1, p_2, p_3 so it can win voters from B_{right}. The line through the circle centers makes an angle $\alpha := \arctan \frac{p_x}{2}$ with the line $x = 0$. The circles intersect at p_1, which lies on the line $x = 0$, so their other intersection point lies on the line ℓ that makes an angle 2α with $x = 0$. If q is to win voters from both B_{right} and B_{up} it must lie between ℓ and $x = 0$. Next we show that this implies that q cannot win voters from B_{left}.

We first show that if $q = \ell \cap C$, then p_4 prevents q from winning voters in B_{left}. Because then p_4 and q both lie on C_{up}, their perpendicular bisector $b(p_4, q)$ is the angular bisector of $\angle qcp_4$. Note that $\angle qcp_4 = 2\alpha$. Indeed, the line through the circle centers makes a right-angled triangle together with $y = 1$ and $x = 0$, so the angle at c must be $\frac{1}{2}\pi - \alpha$. Hence, $\angle qcp_1 = \pi - 2\alpha$, and so $\angle qcp_4 = 2\alpha$. Thus, $b(p_4, q)$ makes an angle α with $y = 1$ and so it intersects the line $x = -2$ at height $y = 1 - (2 + \frac{p_x}{2}) \tan\alpha$ which is $1 - (2 + \tan\alpha) \tan\alpha$. For $0 < 2\alpha < \pi/2$ this is below $\ell_{hor}(q)$ which lies at $y = \cos 2\alpha$. By the previous Claim, this means that q cannot win voters from B_{left}. Therefore, q cannot win voters from B_{left}.

To finish the proof, we must argue that q cannot win voters from B_{left} either when $q \neq \ell \cap C$. It clear that moving q to the left helps to win voters in B_{left}, so we can assume that $q \in C$. Then it is not hard to see (by following the calculations

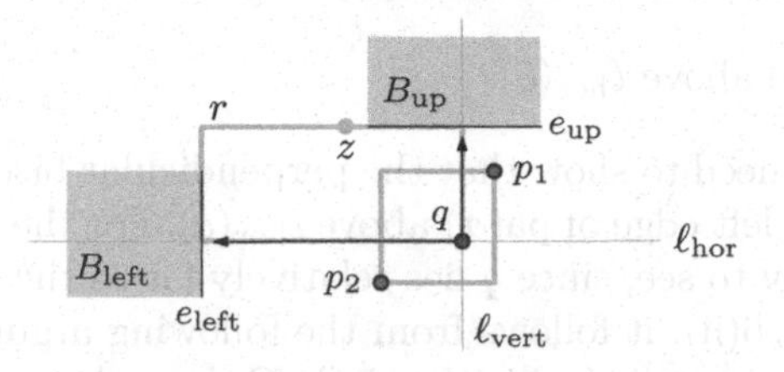

Fig. 7. Definition of e_{left} and e_{up}, and r and z.

above) that the best position for q is $\ell \cap C$, for which we just showed that q cannot win voters in B_{left}. This finishes the proof of the claim. □

We can now assume v_{left} lies above $\ell_{\text{vert}}(q)$ and v_{up} lies to the left of $\ell_{\text{vert}}(q)$. We will show that this leads to a contradiction. To this end, consider the rectangle r whose bottom-right corner is q, whose top edge overlaps e_{up} and whose left edge overlaps with e_{left}; see Fig. 7. Then the left edge of r contains the top endpoint of e_{left} and/or the top edge of r contains the left endpoint of e_{up}. By Lemma 2, we thus know that there is a point $p_4 \in P$ lying on the left or top edge of r. Now assume without loss of generality that the top edge of r is at least as long as its left edge, and let $z \in e_{\text{up}}$ be the point such that the $q_x - z_x = z_y - q_y$. Now, if p_4 lies on the left edge of r or to the left of z on the top edge, then p_4 prevents q from winning v_{left}. On the other hand, if p_4 lies to the right of z on the top edge of r, then p_4 prevents q from winning v_{up}. So in both cases we have a contradiction.

Case II: $q \in \sigma(\text{pa}(B_{\text{right}}))$
Assume without loss of generality that $\sigma(B_{\text{right}})$ is one of the two northern quadrants of $\text{pa}(\sigma(B_{\text{right}}))$. We cannot have $q \in \sigma(B_{\text{right}})$, since together with $\mathit{size}(\sigma(B_{\text{right}})) < \mathit{size}(\sigma(B(q)))$ this contradicts that $q \in B(q)$, by Observation 4. Hence, $q \in \text{NW}(\text{pa}(\sigma(B_{\text{right}})))$ and $\sigma(B_{\text{right}}) = \text{NE}(\text{pa}(\sigma(B_{\text{right}})))$.

If B_{right} is a type-I block then all corners of $\text{NW}(\text{pa}(\sigma(B_{\text{right}})))$ are in P, which (as we saw earlier) implies that q can win voters from at most three blocks. If B_{right} is a type-II block, then we can follow the proof of Case I. (For type-I blocks this is not true. The reason is that in the proof of the first Claim, we use that B_{left} does not lie immediately to the left of σ, which is not true for type-I blocks in Case 2. Note that this still is true for type-II blocks in Case 2.) □

By construction, each block contains at most $m < n/|\mathcal{R}|$ voters, where $\mathcal{R}$ is the set of regions created by Algorithm 1. Moreover, $\mathcal{P}$ places 13 points per region in $\mathcal{R}$, and so $k \leqslant 13|\mathcal{R}|$ points in total. Finally, Lemma 4 states that $\mathcal{Q}$ can win voters from at most three blocks. We can conclude the following.

Lemma 5. *Let V be a set of n voters in $\mathbb{R}^2$. For any given k, the quadtree-based strategy described above can guarantee that $\mathcal{P}$ wins at least $\left(1 - \frac{39}{k}\right) n$ voters by placing at most k points, against any single point placed by player $\mathcal{Q}$.*

A More Refined Strategy for Player $\mathcal{P}$. It can be shown that the analysis presented above is tight. Hence, to get a better bound we need a better strategy.

Recall that each region $R \in \mathcal{R}$ contains between $m + 1$ and $4m$ voters. Currently, we use the same 13 points for any R, regardless of the exact number of voters it contains and how they are distributed over the child regions of R. Our refined strategy takes this into account, and also incorporates the ε-nets developed in the previous section, as follows. Let n_R denote the number of voters in a region $R \in \mathcal{R}$. We consider two cases, with several subcases.

- *Case A:* $m < n_R \leqslant \frac{16}{11}m$. We place eight points in total for R, as in Fig. 3(iii). We also add between two and six extra points, depending on the subcase.

- If $m < n_R \leqslant \frac{7}{6}m$, we add two extra points, forming a $\frac{4}{7}$-net.
- If $\frac{7}{6}m < n_R \leqslant \frac{5}{4}m$, we add three extra points, forming a $\frac{8}{15}$-net.
- If $\frac{5}{4}m < n_R \leqslant \frac{4}{3}m$, we add four extra points, forming a $\frac{1}{2}$-net.
- If $\frac{4}{3}m < n_R \leqslant \frac{7}{5}m$, we add five extra points, forming a $\frac{10}{21}$-net.
- If $\frac{7}{5}m < n_R \leqslant \frac{16}{11}m$, we add six extra points, forming a $\frac{11}{24}$-net.

One can show that in each subcase above, player $\mathcal{Q}$ wins at most $\lceil 2m/3 \rceil + 1$ voters from inside R, due to the (ceiling-based) ε-nets. For example, in the first case $\mathcal{Q}$ wins at most $(7m/6) \cdot (4/7) = 2m/3$ voters, in the second subcase $\mathcal{Q}$ wins at most $(5m/4) \cdot (8/15) = 2m/3$ voters, etcetera. Furthermore, one easily verifies that in each subcase we have $\frac{\text{number of voters in } R}{\text{number of points placed}} > m/10$.

– *Case B:* $\frac{16}{11}m < n_R \leqslant 4m$. We first place the same set of 13 points as in our original strategy. We add two or four extra points, depending on the subcase.
  - If $\frac{16}{11}m < n_R \leqslant 2m$ we add two extra points, as follows. Consider the four child regions of R. Then we add a centerpoint—in other words, a $\frac{2}{3}$-net of size 1—for the voters in the two child regions with the largest number of voters.
  - If $2m < n_R \leqslant 4m$ we add four extra points, namely a centerpoint for each of the four child regions of R.

Note that in both subcases, $\mathcal{Q}$ wins at most $2m/3$ voters from any child region. For the child regions where we placed a centerpoint, this holds because a child region contains at most m voters by construction. For the two child regions where we did not place a centerpoint in the first subcase, this holds because these child regions contains at most $2m/3$ voters. Furthermore, in both subcases $\frac{\text{number of voters in } R}{\text{number of points placed}} > \frac{16}{165}m$.

Lemma 6. *Let V be a set of n voters in $\mathbb{R}^2$. For any given k, the refined quadtree-based strategy can guarantee that $\mathcal{P}$ wins at least $\left(1 - \frac{20\frac{5}{8}}{k}\right) n - 6$ voters by placing at most k points, against any single point placed by player $\mathcal{Q}$.*

Proof. The proof for the original quadtree-based strategy was based on two facts: First, player $\mathcal{Q}$ can win voters from at most three blocks $B \in \mathcal{B}$; see Lemma 4. Second, any block $B \in \mathcal{B}$ (which was a child region of some $R \in \mathcal{R}$) contains at most m voters.

In the refined strategy, we use a similar argument, but for a set $\mathcal{B}_{\text{new}}$ of blocks defined as follows. For the regions $R \in \mathcal{R}$ that fall into Case A, we put R itself (instead of its child regions) as a type-II block into $\mathcal{B}_{\text{new}}$. For the regions $R \in \mathcal{R}$ that fall into Case B, we put their child regions as type-I blocks into $\mathcal{B}_{\text{new}}$. By Lemma 4, $\mathcal{Q}$ can win voters from at most three blocks in $\mathcal{B}_{\text{new}}$. Moreover, our refined strategy ensures that $\mathcal{Q}$ wins at most $\lceil 2m/3 \rceil + 1$ voters from any $B \in \mathcal{B}_{\text{new}}$. Thus $\mathcal{Q}$ wins at most $2m + 6$ voters in total.

Finally, for each $R \in \mathcal{R}$ we have $\frac{\text{number of voters in } R}{\text{number of points placed}} > \frac{16}{165}m$. Hence, $m < \frac{165}{16k}n$ and so $\mathcal{Q}$ wins at most $\frac{165}{8k}n + 6 = \frac{20\frac{5}{8}}{k}n + 6$ voters. □

4 Conclusion

We studied the discrete one-round Voronoi game where player $\mathcal{P}$ can place $k > 1$ points and player $\mathcal{Q}$ can place a single point. We improved the existing bounds on the number of voters player $\mathcal{P}$ can win. For small k this was done by proving new bounds on ε-nets for convex ranges. For large k we used a quadtree-based approach, which uses the ε-nets as a subroutine. The main open problem is: Can player $\mathcal{P}$ always win at least half the voters in the L_2-metric by placing less than four points?

The discrete one-round Voronoi game can also be studied in the L_1-metric, instead of in the L_2-metric as we did here. Our quadtree-based strategy also works well in the version of the problem, as we show in the full version of our paper.

References

1. Ahn, H.-K., Cheng, S.-W., Cheong, O., Golin, M., van Oostrum, R.: Competitive facility location along a highway. In: Wang, J. (ed.) COCOON 2001. LNCS, vol. 2108, pp. 237–246. Springer, Heidelberg (2001). https://doi.org/10.1007/3-540-44679-6_26
2. Aronov, B., et al.: Small weak epsilon-nets. Comput. Geom. **42**(5), 455–462 (2009). https://doi.org/10.1016/j.comgeo.2008.02.005
3. Bandyapadhyay, S., Banik, A., Das, S., Sarkar, H.: Voronoi game on graphs. Theor. Comput. Sci. **562**, 270–282 (2015). https://doi.org/10.1016/j.tcs.2014.10.003
4. Banik, A., Bhattacharya, B.B., Das, S.: Optimal strategies for the one-round discrete Voronoi game on a line. J. Comb. Optim. **26**(4), 655–669 (2012). https://doi.org/10.1007/s10878-011-9447-6
5. Banik, A., Carufel, J.D., Maheshwari, A., Smid, M.H.M.: Discrete Voronoi games and ε-nets, in two and three dimensions. Comput. Geom. **55**, 41–58 (2016). https://doi.org/10.1016/j.comgeo.2016.02.002
6. de Berg, M., Gudmundsson, J., Mehr, M.: Faster algorithms for computing plurality points. ACM Trans. Algorithms **14**(3), 36:1–36:23 (2018). https://doi.org/10.1145/3186990
7. de Berg, M., Kisfaludi-Bak, S., Mehr, M.: On one-round discrete Voronoi games. In: Proceedings of the 30th International Symposium on Algorithms and Computation (ISAAC 2019). LIPIcs, vol. 149, pp. 37:1–37:17 (2019). https://doi.org/10.4230/LIPIcs.ISAAC.2019.37
8. Bukh, B.: A point in many triangles. Electron. J. Comb. **13**(1) (2006). http://www.combinatorics.org/Volume_13/Abstracts/v13i1n10.html
9. Byrne, T., Fekete, S.P., Kalcsics, J., Kleist, L.: Competitive location problems: balanced facility location and the one-round Manhattan Voronoi game. In: Uehara, R., Hong, S.-H., Nandy, S.C. (eds.) WALCOM 2021. LNCS, vol. 12635, pp. 103–115. Springer, Cham (2021). https://doi.org/10.1007/978-3-030-68211-8_9
10. Chan, T.M.: An optimal randomized algorithm for maximum Tukey depth. In: Munro, J.I. (ed.) Proceedings 15th Annual ACM-SIAM Symposium on Discrete Algorithms (SODA 2004), pp. 430–436 (2004). http://dl.acm.org/citation.cfm?id=982792.982853

11. Chawla, S., Rajan, U., Ravi, R., Sinha, A.: Min-max payoffs in a two-player location game. Oper. Res. Lett. **34**(5), 499–507 (2006). https://doi.org/10.1016/j.orl.2005.10.002
12. Cheong, O., Har-Peled, S., Linial, N., Matousek, J.: The one-round Voronoi game. Discret. Comput. Geom. **31**(1), 125–138 (2003). https://doi.org/10.1007/s00454-003-2951-4
13. Dürr, C., Thang, N.K.: Nash equilibria in Voronoi games on graphs. In: Arge, L., Hoffmann, M., Welzl, E. (eds.) ESA 2007. LNCS, vol. 4698, pp. 17–28. Springer, Heidelberg (2007). https://doi.org/10.1007/978-3-540-75520-3_4
14. McKelvey, R.D., Wendell, R.E.: Voting equilibria in multidimensional choice spaces. Math. Oper. Res. **1**(2), 144–158 (1976)
15. Mustafa, N.H., Ray, S.: An optimal extension of the centerpoint theorem. Comput. Geom. **42**, 505–510 (2009). https://doi.org/10.1016/j.comgeo.2007.10.004
16. Teramoto, S., Demaine, E.D., Uehara, R.: Voronoi game on graphs and its complexity. In: 2006 IEEE Symposium on Computational Intelligence and Games, pp. 265–271 (2006). https://doi.org/10.1109/CIG.2006.311711
17. Wu, Y.-W., Lin, W.-Y., Wang, H.-L., Chao, K.-M.: Computing plurality points and Condorcet points in Euclidean space. In: Cai, L., Cheng, S.-W., Lam, T.-W. (eds.) ISAAC 2013. LNCS, vol. 8283, pp. 688–698. Springer, Heidelberg (2013). https://doi.org/10.1007/978-3-642-45030-3_64

General Space-Time Tradeoffs via Relational Queries

Shaleen Deep[1](✉), Xiao Hu[2], and Paraschos Koutris[3]

[1] Microsoft GSL, Redmond, USA
shaleen.deep@microsoft.com
[2] University of Waterloo, Waterloo, Canada
xiaohu@uwaterloo.ca
[3] University of Wisconsin-Madison, Madison, USA
paris@cs.wisc.edu

Abstract. In this paper, we investigate space-time tradeoffs for answering Boolean conjunctive queries. The goal is to create a data structure in an initial preprocessing phase and use it for answering (multiple) queries. Previous work has developed data structures that trade off space usage for answering time and has proved conditional space lower bounds for queries of practical interest such as the path and triangle query. However, most of these results cater to only those queries, lack a comprehensive framework, and are not generalizable. The isolated treatment of these queries also fails to utilize the connections with extensive research on related problems within the database community. The key insight in this work is to exploit the formalism of relational algebra by casting the problems as answering join queries over a relational database. Using the notion of boolean *adorned queries* and *access patterns*, we propose a unified framework that captures several widely studied algorithmic problems. Our main contribution is three-fold. First, we present an algorithm that recovers existing space-time tradeoffs for several problems. The algorithm is based on an application of the *join size bound* to capture the space usage of our data structure. We combine our data structure with *query decomposition* techniques to further improve the tradeoffs and show that it is readily extensible to queries with negation. Second, we falsify two proposed conjectures in the existing literature related to the space-time lower bound for path queries and triangle detection for which we show unexpectedly better algorithms. This result opens a new avenue for improving several algorithmic results that have so far been assumed to be (conditionally) optimal. Finally, we prove new conditional space-time lower bounds for star and path queries.

1 Introduction

Recent work has made remarkable progress in developing data structures and algorithms for answering set intersection problems [12], reachability oracles and directed reachability [3,4,9], histogram indexing [7,18], and problems related to document retrieval [2,20]. This class of problems splits an algorithmic task

P. Morin and S. Suri (Eds.): WADS 2023, LNCS 14079, pp. 309–325, 2023.
https://doi.org/10.1007/978-3-031-38906-1_21

into two phases: the *preprocessing phase*, which computes a space-efficient data structure, and the *answering phase*, which uses the data structure to answer the requests to minimize the answering time. A fundamental algorithmic question related to these problems is the tradeoff between the space S necessary for data structures and the answering time T for requests.

For example, consider the 2-Set Disjointness problem: given a universe of elements U and a collection of m sets $C_1, \ldots, C_m \subseteq U$, we want to create a data structure such that for any pair of integers $1 \leq i, j \leq m$, we can efficiently decide whether $C_i \cap C_j$ is empty or not. Previous work [9,12] has shown that the space-time tradeoff for 2-Set Disjointness is captured by the equation $S \cdot T^2 = N^2$, where N is the total size of all sets. The data structure obtained is conjectured to be optimal [12], and its optimality was used to develop conditional lower bounds for other problems, such as approximate distance oracles [3,4]. Similar tradeoffs have been independently established for other data structure problems as well. In the k-Reachability problem [8,12] we are given as an input a directed graph $G = (V, E)$, an arbitrary pair of vertices u, v, and the goal is to decide whether there exists a path of length k between u and v. In the edge triangle detection problem [12], we are given an input undirected graph $G = (V, E)$, and the goal is to develop a data structure that takes space S and can answer in time T whether a given edge $e \in E$ participates in a triangle or not. Each of these problems has been studied in isolation and, as a result, the algorithmic solutions are not generalizable.

In this paper, we cast many of the above problems into answering *Conjunctive Queries (CQs)* over a relational database. CQs are a powerful class of relational queries with widespread applications in data analytics and graph exploration [11,30,31]. For example, by using the relation $R(x, y)$ to encode that element x belongs to set y, 2-Set Disjointness can be captured by the following CQ: $\varphi(y_1, y_2) = R(x, y_1) \wedge R(x, y_2)$. The insight of casting data structure problems into CQs over a database allows for a unified treatment for developing algorithms within the same framework. In particular, we can leverage the techniques developed by the data management community through a long line of research on efficient join evaluation [22,23,32], including worst-case optimal join algorithms [22] and tree decompositions [13,26]. Building upon these techniques, we achieve the following:

- We obtain in a simple way general space-time tradeoffs for any Boolean CQ (a Boolean CQ is one that outputs only true or false). As a consequence, we recover state-of-the-art tradeoffs for several existing problems (e.g., 2-Set Disjointness as well as its generalization k-Set Disjointness and k-Reachability) as special cases of the general tradeoff. We can even obtain improved tradeoffs for some specific problems, such as edge triangles detection, thus falsifying existing conjectures. This also gives us a way to construct data structures for any new problem that can be cast as a Boolean CQ (e.g., finding any subgraph pattern in a graph).
- Space-time tradeoffs for enumerating (non-Boolean) query results under static and dynamic settings have been a subject of previous work [1,11,14,16,17,24].

The space-time tradeoffs from [11] can be applied to the setting of this paper by stopping the enumeration after the first result is observed. We improve upon this result by (*i*) showing a much simpler data structure construction and proofs, and (*ii*) shaving off a polylogarithmic factor from the tradeoff.

We next summarize our three main technical contributions.

1. We propose a unified framework that captures several widely-studied data structure problems. More specifically, we use the formalism of CQs and the notion of *Boolean adorned queries*, where the values of some variables in the query are fixed by the user (denoted as an *access pattern*), and aim to evaluate the Boolean query. We then show how this framework captures the 2-Set Disjointness and k-Reachability problems. Our first main result (Theorem 1) is an algorithm that builds a data structure to answer any Boolean CQ under a specific access pattern. We show how to recover existing and new tradeoffs using this general framework. The first main result may sometimes lead to suboptimal tradeoffs since it does not take into account the structural properties of the query. Our second main result (Theorem 2) combines tree decompositions of the query structure with access patterns to improve space efficiency. We then show how this algorithm can handle Boolean CQs with negation.
2. We explicitly improve the best-known space-time tradeoff for the k-Reachability problem for $k \geq 4$. For any $k \geq 2$, the tradeoff of $S \cdot T^{2/(k-1)} = O(|E|^2)$ was conjectured to be optimal by [12], where $|E|$ is the number of edges in the graph, and was used to conditionally prove other lower bounds on space-time tradeoffs. We show that for a regime of answer time T, it can be improved to $S \cdot T^{2/(k-2)} = O(|E|^2)$, thus breaking the conjecture. To the best of our knowledge, this is the first non-trivial improvement for the k-Reachability problem. We also refute a lower bound conjecture for the edge triangles detection problem established by [12] that appeared at WADS'17.
3. Our third main contribution applies our framework to CQs with negation. This allows us to construct space-time tradeoffs for tasks such as detecting open triangles in a graph. We also show a reduction between lower bounds for the problem of k-Set Disjointness for $k \geq 2$, which generalizes the 2-Set Disjointness to computing the intersection between k given sets.

2 Notation and Preliminaries

Data Model. A *schema* is defined as a collection of relation names, where each relation name R is associated with an arity n. Assuming a (countably infinite) domain $\mathbf{dom}$, a tuple t of relation R is an element of $\mathbf{dom}^n$. An instance of relation R with arity n is a finite set of tuples of R; the size of the instance will be denoted as $|R|$. An input database D is a set of relation instances over the schema. The size of the database $|D|$ is the sum of sizes of all its instances.

Conjunctive Queries. A *Conjunctive Query* (CQ) is an expression of the form $\varphi(\mathbf{y}) = R_1(\mathbf{x}_1) \wedge R_2(\mathbf{x}_2) \wedge \ldots \wedge R_n(\mathbf{x}_n)$. The expressions $\varphi(\mathbf{y}), R_1(\mathbf{x}_1), R_2(\mathbf{x}_2), \ldots, R_n(\mathbf{x}_n)$ are called *atoms*. The atom $\varphi(\mathbf{y})$ is the *head* of the query, while the atoms $R_i(\mathbf{x}_i)$ form the *body*. Here, $\mathbf{y}, \mathbf{x}_1, \ldots, \mathbf{x}_n$ are vectors where each position is a variable (typically denoted as $x, y, z, \ldots$) or a constant from **dom** (typically denoted $a, b, c, \ldots$). Each $\mathbf{x}_i$ must match the arity of the relation R_i, and the variables in $\mathbf{y}$ must occur in the body of the query. We use $\mathsf{vars}(\varphi)$ to denote the set of all variables occurring in φ, and $\mathsf{vars}(R_i)$ to denote the set of variables in atom $R_i(\mathbf{x}_i)$. A CQ is *full* if every variable in the body appears also in the head, and *Boolean* if the head contains no variables. Given variables $x_1, \ldots, x_k$ from $\mathsf{vars}(\varphi)$ and constants $a_1, \ldots, a_k$ from **dom**, we define $\varphi[a_1/x_1, \ldots, a_k/x_k]$ to be the CQ where every occurrence of a variable x_i, $i = 1, \ldots, k$, is replaced by the constant a_i. Given an input database D and a CQ φ, we define the query result $\varphi(D)$ as follows. A *valuation* v is a mapping from $\mathbf{dom} \cup \mathsf{vars}(\varphi)$ to **dom** such that $v(a) = a$ whenever a is a constant. Then, $\varphi(D)$ is the set of all tuples t such that there exists a valuation v for which $t = v(\mathbf{y})$ and for every atom $R_i(\mathbf{x}_i)$, we have $R_i(v(\mathbf{x}_i)) \in D$.[1]

Example 1. Suppose that we have a directed graph G that is represented through a binary relation $R(x, y)$: this means that there exists an edge from node x to node y. We can compute the pairs of nodes that are connected by a directed path of length k using the following CQ, which we call a *path query*: $P_k(x_1, x_{k+1}) = R(x_1, x_2) \wedge R(x_2, x_3) \wedge \cdots \wedge R(x_k, x_{k+1})$.

Output Size Bounds. Let $\varphi(\mathbf{y}) = R_1(\mathbf{x}_1) \wedge R_2(\mathbf{x}_2) \wedge \ldots \wedge R_n(\mathbf{x}_n)$ be a CQ. A weight assignment $\mathbf{u} = (u_i)_{i=1,\ldots,n}$ is called a *fractional edge cover* of $S \subseteq \mathsf{vars}(\varphi)$ if (i) for every atom R_i, $u_i \geq 0$ and (ii) for every $x \in S$, $\sum_{i: x \in \mathsf{vars}(R_i)} u_i \geq 1$. The *fractional edge cover number* of S, denoted by $\rho^*(S)$ is the minimum of $\sum_{i=1}^n u_i$ over all fractional edge covers of S. Whenever $S = \mathsf{vars}(\varphi)$, we call this a fractional edge cover of φ and simply use ρ^*. In a celebrated result, Atserias, Grohe and Marx [5] proved that for every fractional edge cover $\mathbf{u}$ of φ, the size of the output is bounded by the *AGM inequality*: $|\varphi(D)| \leq \prod_{i=1}^n |R_i|^{u_i}$. The above bound is constructive [22,23]: there exists an algorithm that computes the result $\varphi(D)$ in $O(\prod_i |R_i|^{u_i})$ time for every fractional edge cover $\mathbf{u}$.

Tree Decompositions. Let $\varphi(\mathbf{y}) = R_1(\mathbf{x}_1) \wedge R_2(\mathbf{x}_2) \wedge \ldots \wedge R_n(\mathbf{x}_n)$ be a CQ. A *tree decomposition* of φ is a tuple $(\mathcal{T}, (\mathcal{B}_t)_{t \in V(\mathcal{T})})$ where $\mathcal{T}$ is a tree, and every $\mathcal{B}_t$ is a subset of $\mathsf{vars}(\varphi)$, called the *bag* of t, such that

- For every atom R_i, the set $\mathsf{vars}(R_i)$ is contained in some bag; and
- For each variable $x \in \mathsf{vars}(\varphi)$, the set of nodes $\{t \mid x \in \mathcal{B}_t\}$ form a connected subtree of $\mathcal{T}$.

The *fractional hypertree width* of a decomposition is defined as $\max_{t \in V(\mathcal{T})} \rho^*(\mathcal{B}_t)$, where $\rho^*(\mathcal{B}_t)$ is the minimum fractional edge cover of the

[1] Here we extend the valuation to mean $v((a_1, \ldots, a_n)) = (v(a_1), \ldots, v(a_n))$.

vertices in $\mathcal{B}_t$. The fractional hypertree width of a query φ, denoted $\mathsf{fhw}(\varphi)$, is the minimum fractional hypertree width among all tree decompositions. We say that a query is *acyclic* if $\mathsf{fhw}(\varphi) = 1$.

Computational Model. To measure the running time of our algorithms, we will use the uniform-cost RAM model [15], where data values and pointers to databases are of constant size. Throughout the paper, all complexity results are with respect to data complexity, where the query is assumed fixed.

3 Framework

3.1 Adorned Queries

In order to model different access patterns, we will use the concept of *adorned queries* introduced by [28]. Let $\varphi(x_1, \dots, x_k)$ be the head of a CQ φ. In an adorned query, each variable in the head is associated with a binding type, which can be either *bound* (b) or *free* (f). We denote this as φ^η, where $\eta \in \{\mathsf{b}, \mathsf{f}\}^k$ is called the *access pattern*. The access pattern tells us for which variables the user must provide a value as input. Concretely, let $x_1, x_2, \dots, x_\ell$ be the bound variables. An *access request* is sequence of constants $a_1, \dots, a_\ell$, and it asks to return the result of the query $\varphi^\eta[a_1/x_1, \dots, a_\ell/x_\ell]$ on the input database. We next demonstrate how to capture several data structure problems in this way.

Example 2 (Set Disjointness and Set Intersection). In the set disjointness problem, we are given m sets $S_1, \dots, S_m$ drawn from the same universe U. Let $N = \sum_{i=1}^{m} |S_i|$ be the total size of input sets. Each access request is a pair of indexes $(i, j), 1 \leq i, j, \leq m$, for which we need to decide whether $S_i \cap S_j$ is empty or not. To cast this problem as an adorned query, we encode the family of sets as a binary relation $R(x, y)$, such that element x belongs to set y. Note that the relation will have size N. Then, the set disjointness problem corresponds to: $\varphi^{\mathsf{bb}}(y, z) = R(x, y) \wedge R(x, z)$. An access request in this case specifies two sets $y = S_i, z = S_j$, and issues the (Boolean) query $\varphi(S_i, S_j) = R(x, S_i) \wedge R(x, S_j)$. In the related set intersection problem, given a pair of indexes (i, j) for $1 \leq i, j, \leq m$, we instead want to enumerate the elements in the intersection $S_i \cap S_j$, which can be captured by the following adorned query: $\varphi^{\mathsf{bbf}}(y, z, x) = R(x, y) \wedge R(x, z)$.

Example 3 (k-Set Disjointness). The k-set disjointness problem is a generalization of 2-set disjointness problem, where each request asks whether the intersection between k sets is empty or not. Again, we can cast this problem into the following adorned query: $\varphi^{\mathsf{b} \dots \mathsf{b}}(y_1, \dots, y_k) = R(x, y_1) \wedge R(x, y_2) \wedge \dots \wedge R(x, y_k)$

Example 4 (k-Reachability). Given a direct graph G , the k-reachability problem asks, given a pair vertices (u, v), to check whether they are connected by a path of length k. Representing the graph as a binary relation $R(x, y)$ (which means that there is an edge from x to y), we can model this problem through the following adorned query: $\varphi^{\mathsf{bb}}(x_1, x_{k+1}) = R(x_1, x_2) \wedge R(x_2, x_3) \wedge \dots \wedge R(x_k, x_{k+1})$ Observe that we can also check whether there is a path of length at most k by combining the results of k such queries (one for each length $1, \dots, k$).

Example 5 (Edge Triangles Detection). Given a graph $G = (V, E)$, this problem asks, given an edge (u, v) as the request, whether (u, v) participates in a triangle or not. This task can be expressed as the following adorned query $\varphi_{\triangle}^{\mathsf{bb}}(x, z) = R(x, y) \wedge R(y, z) \wedge R(x, z)$ In the reporting version, the goal is to enumerate all triangles participated by edge (x, z), which can also be expressed by the following adorned query $\varphi_{\triangle}^{\mathsf{bbf}}(x, z, y) = R(x, y) \wedge R(y, z) \wedge R(x, z)$.

We say that an adorned query is *Boolean* if every head variable is bound. In this case, the answer for every access request is also Boolean, i.e., true or false.

3.2 Problem Statement

Given an adorned query φ^η and an input database D, our goal is to construct a data structure, such that we can answer any access request that conforms to the access pattern η as fast as possible. In other words, an algorithm can be split into two phases:

- **Preprocessing phase:** we compute a data structure using space S.
- **Answering phase:** given an access request, we compute the answer using the data structure built in the preprocessing phase, within time T.

In this work, our goal is to study the relationship between the space of the data structure S and the answering time T for a given adorned query φ^η. We will focus on Boolean adorned queries, where the output is just true or false.

4 Space-Time Tradeoffs via Worst-Case Optimal Algorithms

Let φ^η be an adorned query and $\mathcal{V}_{\mathsf{b}}$ denote its bound variables. For any fractional edge cover $\mathbf{u}$, we define the *slack* of $\mathbf{u}$ [11] as:

$$\alpha(\mathbf{u}) := \min_{x \in \mathsf{vars}(\varphi) \setminus \mathcal{V}_{\mathsf{b}}} \left(\sum_{i: x \in \mathsf{vars}(R_i)} u_i \right).$$

In other words, the slack is the maximum factor by which we can scale down the fractional cover $\mathbf{u}$ so that it remains a valid edge cover of the non-bound variables in the query[2]. Hence $\{u_i/\alpha(\mathbf{u})\}_i$ is a fractional edge cover of the nodes in $\mathsf{vars}(\varphi) \setminus \mathcal{V}_{\mathsf{b}}$. We always have $\alpha(\mathbf{u}) \geq 1$.

Example 6. Consider $\varphi^{\mathsf{b} \cdots \mathsf{b}}(y_1, \ldots, y_k) = R_1(x, y_1) \wedge R_2(x, y_2) \wedge \ldots R_k(x, y_k)$ with the optimal fractional edge cover $\mathbf{u}$, where $u_i = 1$ for $i \in \{1, \ldots, k\}$. The slack is $\alpha(\mathbf{u}) = k$, since the fractional edge cover $\hat{\mathbf{u}}$, where $\hat{u}_i = u_i/k = 1/k$ covers the only non-bound variable x.

[2] We will omit the parameter $\mathbf{u}$ from the notation of α whenever it is clear from the context.

Theorem 1. *Let φ^η be a Boolean adorned query. Let* $\mathbf{u}$ *be any fractional edge cover of* φ. *Then, for any input database* D, *we can construct a data structure that answers any access request in time* $O(T)$ *and takes space*

$$S = O\left(|D| + \prod_{i=1}^{n} |R_i|^{u_i}/T^\alpha\right)$$

We should note that Theorem 1 applies even when the relation sizes are different; this gives us sharper upper bounds compared to the case where each relation is bounded by the total size of the input. Indeed, if using $|D|$ as an upper bound on each relation, we obtain a space requirement of $O(|D|^{\rho^*}/T^\alpha)$ for achieving answering time $O(T)$, where ρ^* is the fractional edge cover number. Since $\alpha \geq 1$, this gives us at worst a linear tradeoff between space and time, i.e., $S \cdot T = O(|D|^{\rho^*})$. For cases where $\alpha \geq 1$, we can obtain better tradeoffs. The full proofs for all results in this paper can be found in [10].

Example 7. Continuing the example in this section $\varphi^{\mathsf{b}\ldots\mathsf{b}}(y_1, \ldots, y_k) = R_1(x, y_1) \wedge R_2(x, y_2) \wedge \cdots \wedge R_k(x, y_k)$. We obtain an improved tradeoff: $S \cdot T^k = O(|D|^k)$[3]. Note that this result matches the best-known space-time tradeoff for the k-Set Disjointness problem [12]. (Note that all atoms use the same relation symbol R, so $|R_i| = |D|$ for every $i = 1, \ldots, k$.)

Example 8 (Edge Triangles Detection). For the Boolean version, it was shown in [12] that - conditioned on the strong set disjointness conjecture - any data structure that achieves answering time T needs space $S = \Omega(|E|^2/T^2)$. A matching upper bound can be constructed by using a fractional edge cover $\mathbf{u} = (1, 1, 0)$ with slack $\alpha = 2$. Thus, Theorem 1 can be applied to achieve answering time T using space $S = O(|E|^2/T^2)$. Careful inspection reveals that a different fractional edge cover $\mathbf{u} = (1/2, 1/2, 1/2)$ with slack $\alpha = 1$, achieves a better tradeoff. Thus, Theorem 1 can be applied to obtain the following corollary.

Corollary 1. *For a graph* $G = (V, E)$, *there exists a data structure of size* $S = O(|E|^{3/2}/T)$ *that can answer the edge triangles detection problem in* $O(T)$.

The data structure implied by Theorem 1 is always better when $T \leq \sqrt{|E|}$[4], thus refuting the conditional lower bound in [12]. We should note that this does not imply that the strong set disjointness conjecture is false, as we have observed an error in the reduction used by [12].

Example 9 (Square Detection). Beyond triangles, we consider the edge square detection problem, which checks whether a given edge belongs in a square pattern in a graph $G = (V, E)$, $\varphi_\square^{\mathsf{bb}}(x_1, x_2) = R_1(x_1, x_2) \wedge R_2(x_2, x_3) \wedge R_3(x_3, x_4) \wedge R_4(x_4, x_1)$. Using the fractional edge cover $\mathbf{u} = (1/2, 1/2, 1/2, 1/2)$ with slack $\alpha = 1$, we obtain a tradeoff $S = O(|E|^2/T)$.

[3] For all results in this paper, S includes the space requirement of the input as well. If we are interested in only the space requirement of the constructed data structure, then the $|D|$ term in the space requirement of Theorem 1 can be removed.

[4] All answering times $T > \sqrt{|E|}$ are trivial to achieve using linear space by using the data structure for $T' = \sqrt{E}$ and holding the result back until time T has passed.

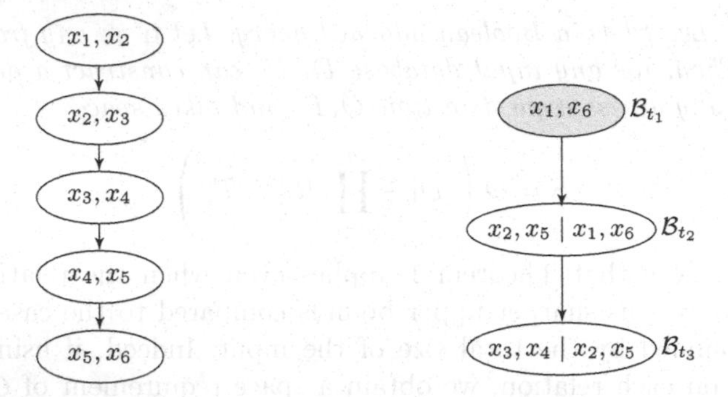

Fig. 1. Two tree decompositions for the length-5 path query: the left is unconstrained, while the right is a C-connex decomposition with $C = \{x_1, x_6\}$. The bound variables are colored red. The nodes in A are colored grey.

5 Space-Time Tradeoffs via Tree Decompositions

Theorem 1 does not always give us the optimal tradeoff. For the k-reachability problem with the adorned query $\varphi^{\mathsf{bb}}(x_1, x_{k+1}) = R_1(x_1, x_2) \wedge \cdots \wedge R_k(x_k, x_{k+1})$, Theorem 1 gives a tradeoff $S \cdot T = |D|^{\lceil (k+1)/2 \rceil}$, by taking the optimal fractional edge covering number $\rho^* = \lceil (k+1)/2 \rceil$ and slack $\alpha = 1$, which is far from efficient. In this section, we will show how to leverage tree decompositions to further improve the space-time tradeoff in Theorem 1.

Again, let φ^{η} be an adorned query. Given a set of nodes $C \subseteq \mathcal{V}$, a C-connex tree decomposition of φ is a pair $(\mathcal{T}, A)$, where (i) $\mathcal{T}$ is a tree decomposition of φ, and (ii) A is a connected subset of the tree nodes such that the union of their variables is exactly C. For our purposes, we choose $C = \mathcal{V}_{\mathsf{b}}$. Given a $\mathcal{V}_{\mathsf{b}}$-connex tree decomposition, we orient the tree from some node in A. We then define the bound variables for the bag t, $\mathcal{V}_{\mathsf{b}}^t$ as the variables in $\mathcal{B}_t$ that also appear in the bag of some ancestor of t. The free variables for the bag t are the remaining variables in the bag, $\mathcal{V}_{\mathsf{f}}^t = \mathcal{B}_t \setminus \mathcal{V}_{\mathsf{b}}^t$.

Example 10. Consider the 5-path query $\varphi^{\mathsf{bb}}(x_1, x_6) = R_1(x_1, x_2) \wedge \cdots \wedge R_5(x_5, x_6)$. Here, x_1 and x_6 are the bound variables. Figure 1 shows the unconstrained decomposition as well as the C-connex decomposition for $\varphi^{\mathsf{bb}}(x_1, x_6)$, where $C = \{x_1, x_6\}$. The root bag contains the bound variables x_1, x_6. Bag $\mathcal{B}_{t_2}$ contains x_1, x_6 as bound variables and x_2, x_5 as the free variables. Bag $\mathcal{B}_{t_3}$ contains x_2, x_5 as bound variables for $\mathcal{B}_{t_3}$ and x_3, x_4 as free variables.

Next, we use a parameterized notion of width for the $\mathcal{V}_{\mathsf{b}}$-connex tree decomposition that was introduced in [11]. The width is parameterized by a function δ that maps each node t in the tree to a non-negative number, such that $\delta(t) = 0$ whenever $t \in A$. The intuition here is that we will spend $O(|D|^{\delta(t)})$ in the node t while answering the access request. The parameterized width of a bag $\mathcal{B}_t$ is now defined as: $\rho_t(\delta) = \min_{\mathbf{u}} (\sum_F u_F - \delta(t) \cdot \alpha)$ where $\mathbf{u}$ is a fractional edge cover of the bag $\mathcal{B}_t$, and α is the slack (on the bound variables of the bag). The

δ-width of the decomposition is then defined as $\max_{t \notin A} \rho_t(\delta)$. Finally, we define the δ-height as the maximum-weight path from the root to any leaf, where the weight of a path P is $\sum_{t \in P} \delta(t)$. We now have all the necessary machinery to state our second main theorem.

Theorem 2. *Let φ^η be a Boolean adorned query. Consider any $\mathcal{V}_\mathsf{b}$-connex tree decomposition of φ. For some parametrization δ of the decomposition, let f be its δ-width, and h be its δ-height. Then, for any input database D, we can construct a data structure that answers any access request in time $T = O(|D|^h)$ with space $S = O(|D| + |D|^f)$.*

The function δ allows us to trade off between time and space. If we set $\delta(t) = 0$ for every node t in the tree, then the δ-height becomes $O(1)$, while the δ-width equals to the fractional hypetree width of the decomposition. As we increase the values of δ in each bag, the δ-height increases while the δ-width decreases, i.e., the answer time T increases while the space decreases. Additionally, we note that the tradeoff from Theorem 2 is at least as good as the one from Theorem 1. Indeed, we can always construct a tree decomposition where all variables reside in a single node of the tree. In this case, we recover exactly the tradeoff from Theorem 1.

Example 11. We continue with the 5-path query. Since $\mathcal{B}_{t_1} = \{x_1, x_6\} \in A$, we assign $\delta(t_1) = 0$. For $\mathcal{B}_{t_2} = \{x_1, x_2, x_5, x_6\}$, the only valid fractional edge cover assigns weight 1 to both R_1, R_5 and has slack 1. Hence, if we assign $\delta(t_2) = \tau$ for some parameter τ, the width is $2 - \tau$. For $\mathcal{B}_{t_3} = \{x_2, x_3, x_4, x_5\}$, the only fractional cover also assigns weight 1 to both R_2, R_4, with slack 1 again. Assigning $\delta(t_3) = \tau$, the width becomes $2 - \tau$ for t_3 as well. Hence, the δ-width of the tree decomposition is $2 - \tau$, while the δ-height is 2τ. Plugging this to Theorem 2, it gives us a tradeoff with answering time $T = O(|E|^{2\tau})$ and space usage $S = O(|E| + |E|^{2-\tau})$, which matches the state-of-the-art result in [12].

For the k-reachability problem, a general tradeoff $S \times T^{2/(k-1)} = O(|D|^2)$ was also shown by [12] using a careful recursive argument. The data structure generated using Theorem 2 is able to recover the tradeoff. In particular, we obtain the answering time as $T = O(|E|^{(k-1)\tau/2})$ using space $S = O(|E| + |E|^{2-\tau})$.

Example 12. Consider a variant of the square detection problem: given two vertices, the goal is to decide whether they occur in two opposites corners of a square, which can be captured by the following adorned query:

$$\varphi^{\mathsf{bb}}(x_1, x_3) = R_1(x_1, x_2) \wedge R_1(x_2, x_3) \wedge R_3(x_3, x_4) \wedge R_4(x_4, x_1).$$

Theorem 1 gives a tradeoff with answering time $O(T)$ and space $O(|E|^2/T)$. But we can obtain a better tradeoff using Theorem 2. Indeed, consider the tree decomposition where we have a root bag t_1 with $\mathcal{B}_{t_1} = \{x_1, x_3\}$, and two children of t_1 with Boolean $\mathcal{B}_{t_2} = \{x_1, x_2, x_3\}$ and $\mathcal{B}_{t_3} = \{x_1, x_3, x_4\}$. For $\mathcal{B}_{t_2}$, we can see that if assigning a weight of 1 to both hyperedges, we get a slack of 2. Hence, if $\delta(t_2) = \tau$, the δ-width is $2-2\tau$. Similarly for t_3, we assign $\delta(t_3) = \tau$, for a δ-width with $2 - 2\tau$. Applying Theorem 2, we obtain a tradeoff with time $T = O(|E|^{\tau})$ (since both root-leaf paths have only one node), and space $S = O(|E| + |E|^{2-2\tau})$. So the space usage can be improved from $O(|E|^2/T)$ to $O(|E|^2/T^2)$.

6 CQs with Negation

In this section, we present a simple but powerful extension of our result to adorned Boolean CQs with negation. A CQ with negation, denoted as $CQ^{\neg}$, is a CQ where some of the atoms can be negative, i.e., $\neg R_i(\mathbf{x}_i)$ is allowed. For $\varphi \in CQ^{\neg}$, we denote by φ^+ the conjunction of the positive atoms in φ and φ^- the conjunction of all negated atoms. A $CQ^{\neg}$ is said to be *safe* if every variable appears in at least some positive atom. In this paper, we restrict our scope to the class of safe $CQ^{\neg}$, a standard assumption [21,29] ensuring that query results are well-defined and do not depend on domains.

Given a query $\varphi \in CQ^{\neg}$, we build the data structure from Theorem 2 for φ^+ but impose two constraints on the decomposition: (*i*) no leaf node(s) contains any free variables, (*ii*) for every negated atom R^-, all variables of R^- must appear together as bound variables in some leaf node(s). In other words, there exists a leaf node such that $\mathsf{vars}(R^-)$ is present in it. It is easy to see that such a decomposition always exists. Indeed, we can fix the root bag to be $C = \mathcal{V}_{\mathsf{b}}$, its child bag with free variables as $\mathsf{vars}(\varphi^+) \setminus C$ and bound variables as C, and the leaf bag, which is connected to the child of the root, with bound variables as $\mathsf{vars}(\varphi^-)$ without free variables. Observe that the bag containing $\mathsf{vars}(\varphi^+)$ free variables can be covered by only using the positive atoms since φ is safe. The intuition is the following: during the query answering phase, we wish to find the join result over all variables $\mathcal{V}_{\mathsf{f}}$ before reaching the leaf nodes; and then, we can check whether there the tuples satisfy the negated atoms or not, in $O(1)$ time. The next example shows the application of the algorithm to adorned path queries containing negation.

Example 13. Consider the query $Q^{\mathsf{bb}}(x_1, x_6) = R(x_1, x_2) \wedge \neg S(x_2, x_3) \wedge T(x_3, x_4) \wedge \neg U(x_4, x_5) \wedge V(x_5, x_6)$. Using the decomposition in Fig. 2, we can now apply Theroem 2 to obtain the tradeoff $S = O(|D|^3/\tau)$ and $T = O(\tau)$. Both leaf nodes only require linear space since a single atom covers the variables. Given an access request, we check whether the answer for this request has been materialized or not. If not, we proceed to the query answering phase and find at most $O(\tau)$ answers after evaluating the join in the middle bag. For each of these answers, we can now check in constant time whether the tuples formed by values for x_2, x_3 and x_4, x_5 are not present in relations S and U respectively.

For adorned queries where $\mathcal{V}_{\mathsf{b}} \subseteq \mathsf{vars}(\varphi^-)$, we can further simplify the algorithm. In this case, we no longer need to create a constrained decomposition since the check to see if the negated relations are satisfied or not can be done in constant time at the root bag itself. Thus, we can directly build the data structure from Theorem 2 using the query φ^+.

Example 14 (Open Triangle Detection). Consider the query $\varphi^{\mathsf{bb}}(x_2, x_3) = R_1(x_1, x_2) \wedge \neg R_2(x_2, x_3) \wedge R_3(x_1, x_3)$, where φ^- is $\neg R_2(x_2, x_3)$ and φ^+ is $R_1(x_1, x_2) \wedge R_3(x_1, x_3)$ with the adorned view as $\varphi^{+\mathsf{bb}}(x_2, x_3) = R_1(x_1, x_2) \wedge R_3(x_1, x_3)$. Observe that $\{x_2, x_3\} \subseteq \mathsf{vars}(\varphi^-)$. We apply Theroem 2 to obtain the tradeoff $S = O(|E|^2/\tau^2)$ and $T = O(\tau)$ with root bag $C = \{x_2, x_3\}$, its child bag with $\mathcal{V}_\mathsf{b} = C$ and $\mathcal{V}_\mathsf{f} = \{x_1\}$, and the leaf bag to be $\mathcal{V}_\mathsf{b} = C$ and $\mathcal{V}_\mathsf{f} = \emptyset$. Given an access request (a, b), we check whether the answer for this request has been materialized or not. If not, we traverse the decomposition and evaluating the join to find if there exists a connecting value for x_1. For the last bag, we simply check whether (a, b) exists in R_2 or not in $O(1)$ time.

A Note on Optimality. It is easy to see that the algorithm obtained for Boolean CQs with negation is conditionally optimal assuming the optimality of Theroem 2. Indeed, if all negated relations are empty, the join query is equivalent to φ^+ and the algorithm now simply applies Theroem 2 to φ^+. In Example 14, assuming relation R_2 is empty, the query is equivalent to set intersection whose tradeoffs are conjectured to be optimal.

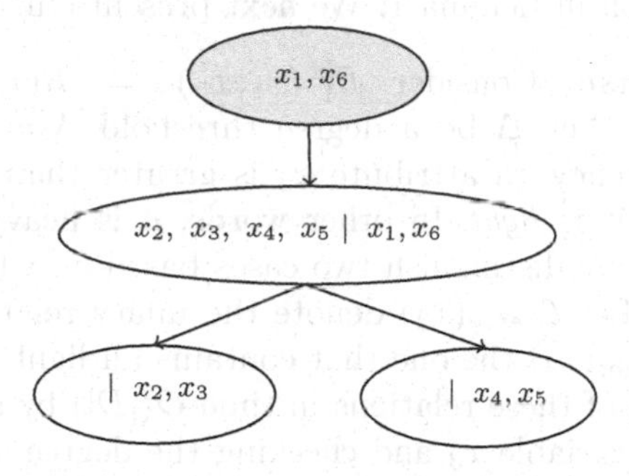

Fig. 2. C-connex decomposition for Example 13.

7 Path Queries

In this section, we present an algorithm for the adorned query $P_k^{\mathsf{bb}}(x_1, x_{k+1}) = R_1(x_1, x_2) \wedge \cdots \wedge R_k(x_k, x_{k+1})$ that improves upon the conjectured optimal solution. Before diving into the details, we first state the upper bound on the tradeoff between space and query time.

Theorem 3 (due to [12]). *There exists a data structure for solving $P_k^{\mathsf{bb}}(x_1, x_{k+1})$ with space S and answering time T such that $S \cdot T^{2/(k-1)} = O(|D|^2)$.*

Note that for $k = 2$, the problem is equivalent to SetDisjointness with the space/time tradeoff as $S \cdot T^2 = O(N^2)$. [12] also conjectured that the tradeoff is essentially optimal.

Conjecture 1 (due to [12]). Any data structure for $P_k^{\mathsf{bb}}(x_1, x_{k+1})$ with answering time T must use space $S = \tilde{\Omega}(|D|^2/T^{2/(k-1)})$.

Building upon Conjecture 1, [12] also showed a result on the optimality of approximate distance oracles. Our result implies that Theroem 3 can be improved further, thus refuting Theorem 1. The first observation is that the tradeoff in Theroem 3 is only useful when $T \leq |D|$. Indeed, we can always answer any Boolean path query in linear time using breadth-first search. Surprisingly, it is also possible to improve Theroem 3 for the regime of small answering time as well. In what follows, we will show the improvement for paths of length 4; we will generalize the algorithm for any length later.

7.1 Length-4 Path

Lemma 1. *There exists a parameterized data structure for solving $P_4^{\mathsf{bb}}(x_1, x_5)$ that uses space S and answering time $T \leq \sqrt{|D|}$ that satisfies the tradeoff $S \cdot T = O(|D|^2)$.*

For $k = 4$, Theorem 3 gives us the tradeoff $S \cdot T^{2/3} = O(|D|^2)$ which is always worse than the tradeoff in Lemma 1. We next present our algorithm in detail.

Preprocessing Phase. Consider $P_4^{\mathsf{bb}}(x_1, x_5) = R(x_1, x_2) \wedge S(x_2, x_3) \wedge T(x_3, x_4) \wedge U(x_4, x_5)$. Let Δ be a degree threshold. We say that a constant a is *heavy* if its frequency on attribute x_3 is greater than Δ in both relations S and T; otherwise, it is *light*. In other words, a is heavy if $|\sigma_{x_3=a}(S)| > \Delta$ and $|\sigma_{x_3=a}(T)| > \Delta$. We distinguish two cases based on whether a constant for x_3 is heavy or light. Let $\mathcal{L}_{\mathsf{heavy}}(x_3)$ denote the unary relation that contains all heavy values, and $\mathcal{L}_{\mathsf{light}}(x_3)$ the one that contains all light values. Observe that we can compute both of these relations in time $O(|D|)$ by simply iterating over the active domain of variable x_3 and checking the degree in relations S and T. We compute two views:

$$V_1(x_1, x_3) = R(x_1, x_2) \wedge S(x_2, x_3) \wedge \mathcal{L}_{\mathsf{heavy}}(x_3)$$
$$V_2(x_3, x_5) = \mathcal{L}_{\mathsf{heavy}}(x_3) \wedge T(x_3, x_4) \wedge U(x_4, x_5)$$

We store the views as a hash index that, given a value of x_1 (or x_5), returns all matching values of x_3. Both views take space $O(|D|^2/\Delta)$. Indeed, $|\mathcal{L}_{\mathsf{heavy}}| \leq |D|/\Delta$. Since we can construct a fractional edge cover for V_1 by assigning a weight of 1 to R and $\mathcal{L}_{\mathsf{heavy}}$, this gives us an upper bound of $|D| \cdot (|D|/\Delta)$ for the query output. The same argument holds for V_2. We also compute the following view for light values: $V_3(x_2, x_4) = S(x_2, x_3) \wedge \mathcal{L}_{\mathsf{light}}(x_3) \wedge T(x_3, x_4)$. This view requires space $O(|D| \cdot \Delta)$, since the degree of the light constants is at most Δ (i.e. $\sum_{x \in \mathcal{L}_{\mathsf{light}}(x_3)} |S(x_2, x) \wedge T(x, x_4)| \leq \sum_{x \in \mathcal{L}_{\mathsf{light}}(x_3)} |S(x_2, x)| \cdot |T(x, x_4)| \leq \sum_{x \in \mathcal{L}_{\mathsf{light}}(x_3)} |S(x_2, x)| \cdot \Delta \leq |D| \cdot \Delta$). We can now rewrite the original query as $P_4^{\mathsf{bb}}(x_1, x_5) = R(x_1, x_2) \wedge V_3(x_2, x_4) \wedge U(x_4, x_5)$.

The rewritten query is a three path query. Hence, we can apply Theorem 1 to create a data structure with answering time $T = O(|D|/\Delta)$ and space $S = O(|D|^2/(|D|/\Delta)) = O(|D| \cdot \Delta)$.

Query Answering. Given an access request, we first check whether there exists a 4-path that goes through some heavy value in $\mathcal{L}_{\mathsf{heavy}}(x_3)$. This can be done in time $O(|D|/\Delta)$ using the views V_1 and V_2. Indeed, we obtain at most $O(|D|/\Delta)$ values for x_3 using the index for V_1, and $O(|D|/\Delta)$ values for x_3 using the index for V_3. We then intersect the results in time $O(|D|/\Delta)$ by iterating over the $O(|D|/\Delta)$ values for x_3 and checking if the bound values for x_1 and x_5 from a tuple in V_1 and V_2 respectively. If we find no such 4-path, we check for a 4-path that uses a light value for x_3. From the data structure we have constructed in the preprocessing phase, we can do this in time $O(|D|/\Delta)$.

Tradeoff Analysis. From the above, we can compute the answer in time $T = O(|D|/\Delta)$. From the analysis in the preprocessing phase, the space needed is $S = O(|D|^2/\Delta + |D| \cdot \Delta)$. Thus, whenever $\Delta \geq \sqrt{|D|}$, the space becomes $S = O(|D| \cdot \Delta)$, completing our analysis.

7.2 General Path Queries

We can now use the algorithm for the 4-path query to improve the space-time tradeoff for general path queries of length greater than four.

Theorem 4. *Let D be an input instance. For $k \geq 4$, there is a data structure for $P_k^{\mathsf{bb}}(x_1, x_{k+1})$ with space $S = O(|D| \cdot \Delta)$ and answer time $T = O\left((\frac{|D|}{\Delta})^{\frac{k-2}{2}}\right)$ for $\Delta \geq \sqrt{|D|}$.*

The space-time tradeoff obtained from Theroem 4 is $S \cdot T^{2/(k-2)} = O(|D|^2)$, but only for $T \leq |D|^{(k-2)/4}$. To compare it with the tradeoff of $S \cdot T^{2/(k-1)} = O(|D|^2)$ obtained from Theroem 3, it is instructive to look at Figures Fig. 3a and 3b, which plot the space-time tradeoffs for $k = 4$ and $k = 6$ respectively. In general, as k grows, the new tradeoff line (labeled as ρ_1) becomes flatter and approaches Theorem 3.

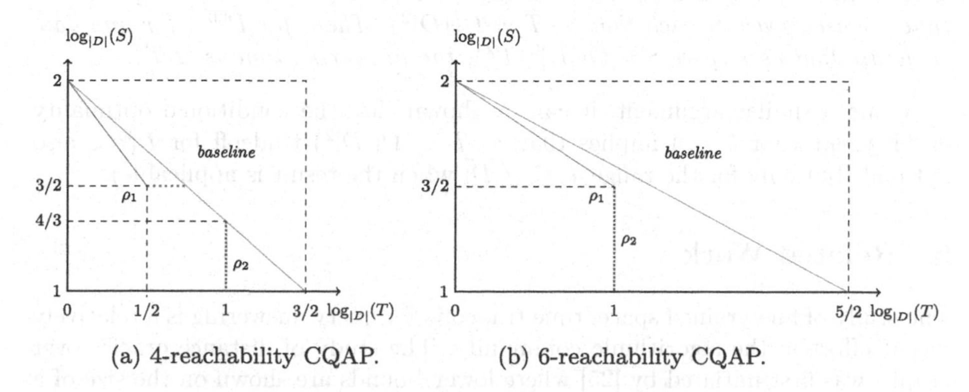

(a) 4-reachability CQAP. (b) 6-reachability CQAP.

Fig. 3. Space/time tradeoffs for path query of length $k \in \{4, 6\}$. The line in brown (baseline) shows the tradeoff obtained from Theroem 3. The red curve (ρ_1) is the new tradeoff obtained using Theorem 4 and ρ_2 shows the transition to when BFS takes over as the best algorithm.

8 Lower Bounds

In this section, we study the lower bounds for adorned star and path queries. We first present conditional lower bounds for the k-Set Disjointness problem using the conditional optimality of ℓ-Set Disjointness where $\ell < k$. First, we review the known results from [12] starting with the conjecture for k-Set Disjointness .

Conjecture 2. (due to [12]*).* Any data structure for k-Set Disjointness problem that answers queries in time T must use space $S = \Omega(|D|^k/T^k)$.

Conjecture 2 was shown to be conditionally optimal based on conjectured lower bound for the $(k+1)$-Sum Indexing problem, however, it was subsequently showed to be false [19], which implies that Conjecture 2 is still an open problem. 2 can be further generalized to the case when input relations are of unequal sizes as follows.

Conjecture 3. Any data structure for $\varphi_*^{\mathsf{b}\ldots\mathsf{b}}(y_1,\ldots,y_k) = R_1(x,y_1) \wedge \cdots \wedge R_k(x,y_k)$ that answers queries in time T must use space $S = \Omega(\Pi_{i=1}^k|R_i|/T^k)$.

We now state the main result for star queries.

Theorem 5. *Suppose that any data structure for* $\varphi_*^{\mathsf{b}\ldots\mathsf{b}}(y_1,\ldots,y_k)$ *with answering time* T *must use space* $S = \Omega(\Pi_{i=1}^k|R_i|/T^k)$*. Then, any data structure for* $Q_*^{\mathsf{b}\ldots\mathsf{b}}(y_1,\ldots,y_\ell)$ *with answering time* T *must use space* $S = \Omega(\Pi_{i=1}^\ell|R_i|/T^\ell)$*, for* $2 \leq \ell < k$.

Theorem 5 creates a hierarchy for k-Set Disjointness , where the optimality of smaller set disjointness instances depends on larger set disjointness instances. Next, we show conditional lower bounds on the space requirement of path queries. We begin by proving a simple result for optimality of P_2^{bb} (equivalent to 2-Set Disjointness) assuming the optimality of P_3^{bb} query.

Theorem 6. *Suppose that any data structure for* P_3^{bb} *that answers queries in time* T*, uses space* S *such that* $S \cdot T = \Omega(|D|^2)$*. Then, for* P_2^{bb} *, for any data structure that uses space* $S = O(|D|^2/T^2)$*, the answering time is* $\Omega(T)$.

Using a similar argument, it can be shown that the conditional optimality of Theroem 4 for $k = 4$ implies that $S \cdot T = \Omega(|D|^2)$ tradeoff for P_3^{bb} is also optimal (but only for the range $T \leq \sqrt{|D|}$ when the result is applicable).

9 Related Work

The study of fine-grained space/time tradeoffs for query answering is a relatively recent effort in the algorithmic community. The study of distance oracles over graphs was first initiated by [25] where lower bounds are shown on the size of a distance oracle for sparse graphs based on a conjecture about the best possible data structure for a set intersection problem. [9] also considered the problem of set intersection and presented a data structure that can answer boolean set intersection queries which is conditionally optimal [12]. There also exist another line of work that looks at the problem of approximate distance oracles. Agarwal et al. [3,4] showed that for stretch-2 and stretch-3 oracles, we can achieve $S \times T = O(|D|^2)$ and $S \times T^2 = O(|D|^2)$. They also showed that for any integer k, a stretch-$(1 + 1/k)$ oracle exhibits $S \times T^{1/k} = O(|D|^2)$ tradeoff. Unfortunately, no lower bounds are known for non-constant query time. The authors in [12] conjectured that the tradeoff $S \times T^{2/(k-1)} = O(|D|^2)$ for k-reachability is optimal which would also imply that stretch-$(1+1/k)$ oracle tradeoff is also optimal. A different line of work has considered the problem of enumerating query results [27] of a non-boolean query. [9] presented a data structure to enumerate the intersection of two sets with guarantees on the total answering time. This result was generalized to incorporate *full* adorned views over CQs [11]. Our work extends the results to the setting where the join variables are projected away from the query result (i.e. the adorned views are *non-full*) and makes the connection between several different algorithmic problems that have been studied independently. Further, we also consider boolean CQs that may contain negations. In the non-static setting, [6] initiated the study of answering conjunctive query results under updates. More recently, [16] presented an algorithm for counting the number of triangles under updates. There have also been some exciting developments in the space of enumerating query results with delay for a proper subset of CQs known as *hierarchical queries*. [17] presented a tradeoff between preprocessing time and delay for enumerating the results of any (not necessarily full) hierarchical queries under static and dynamic settings. It remains an interesting problem to find improved algorithms for more restricted set of CQs such as hierarchical queries.

10 Conclusion

In this paper, we investigated the tradeoffs between answering time and space required by the data structure to answer boolean queries. Our main contribution is a unified algorithm that recovers the best known results for several boolean queries of practical interests. We then apply our main result to improve upon the state-of-the-art algorithms to answer boolean queries over the four path query which is subsequently used to improve the tradeoff for all path queries of length greater than four and show conditional lower bounds. There are several questions that remain open. We describe the problems that are particularly engaging.

Unconditional Lower Bounds. It remains an open problem to prove unconditional lower bounds on the space requirement for answering boolean star and path queries in the RAM model. For instance, 2-Set Disjointness can be answered in constant time by materializing all answers using $\Theta(|D|^2)$ space but there is no lower bound to rule out if this can be achieved using sub-quadratic space.

Improved Approximate Distance Oracles. It would be interesting to investigate whether our ideas can be applied to existing algorithms for constructing distance oracles to improve their space requirement. [12] conjectured that the k-reachability tradeoff is optimal and used it to prove the conditional optimality of distance oracles. We believe our framework can be used to improve upon the bounds for k-reachability in conjunction with other techniques used to prove bounds for join query processing in the database theory community.

References

1. Abo Khamis, M., Kolaitis, P.G., Ngo, H.Q., Suciu, D.: Decision problems in information theory. In: ICALP (2020)
2. Afshani, P., Nielsen, J.A.S.: Data structure lower bounds for document indexing problems. In: ICALP (2016)
3. Agarwal, R.: The space-stretch-time tradeoff in distance oracles. In: Schulz, A.S., Wagner, D. (eds.) ESA 2014. LNCS, vol. 8737, pp. 49–60. Springer, Heidelberg (2014). https://doi.org/10.1007/978-3-662-44777-2_5
4. Agarwal, R., Godfrey, P.B., Har-Peled, S.: Approximate distance queries and compact routing in sparse graphs. In: INFOCOM, pp. 1754–1762. IEEE (2011)
5. Atserias, A., Grohe, M., Marx, D.: Size bounds and query plans for relational joins. SIAM J. Comput. **42**(4), 1737–1767 (2013)
6. Berkholz, C., Keppeler, J., Schweikardt, N.: Answering conjunctive queries under updates. In: PODS, pp. 303–318. ACM (2017)
7. Chan, T.M., Lewenstein, M.: Clustered integer 3SUM via additive combinatorics. In: STOC, pp. 31–40 (2015)
8. Cohen, H., Porat, E.: Fast set intersection and two-patterns matching. Theoret. Comput. Sci. **411**(40–42), 3795–3800 (2010)
9. Cohen, H., Porat, E.: On the hardness of distance oracle for sparse graph. arXiv preprint arXiv:1006.1117 (2010)
10. Deep, S., Hu, X., Koutris, P.: General space-time tradeoffs via relational queries. arXiv preprint arXiv:2109.10889 (2021)
11. Deep, S., Koutris, P.: Compressed representations of conjunctive query results. In: PODS, pp. 307–322. ACM (2018)
12. Goldstein, I., Kopelowitz, T., Lewenstein, M., Porat, E.: Conditional lower bounds for space/time tradeoffs. In: WADS 2017. LNCS, vol. 10389, pp. 421–436. Springer, Cham (2017). https://doi.org/10.1007/978-3-319-62127-2_36
13. Gottlob, G., Greco, G., Scarcello, F.: Treewidth and hypertree width. In: Tractability: Practical Approaches to Hard Problems, vol. 1 (2014)
14. Greco, G., Scarcello, F.: Structural tractability of enumerating CSP solutions. Constraints **18**(1), 38–74 (2013)
15. Hopcroft, J.E., Ullman, J.D., Aho, A.: The Design and Analysis of Computer Algorithms (1975)
16. Kara, A., Ngo, H.Q., Nikolic, M., Olteanu, D., Zhang, H.: Counting triangles under updates in worst-case optimal time. In: ICDT (2019)
17. Kara, A., Nikolic, M., Olteanu, D., Zhang, H.: Trade-offs in static and dynamic evaluation of hierarchical queries. In: PODS, pp. 375–392 (2020)
18. Kociumaka, T., Radoszewski, J., Rytter, W.: Efficient indexes for jumbled pattern matching with constant-sized alphabet. In: Bodlaender, H.L., Italiano, G.F. (eds.) ESA 2013. LNCS, vol. 8125, pp. 625–636. Springer, Heidelberg (2013). https://doi.org/10.1007/978-3-642-40450-4_53
19. Kopelowitz, T., Porat, E.: The strong 3SUM-indexing conjecture is false. arXiv preprint arXiv:1907.11206 (2019)

20. Larsen, K.G., Munro, J.I., Nielsen, J.S., Thankachan, S.V.: On hardness of several string indexing problems. Theoret. Comput. Sci. **582**, 74–82 (2015)
21. Nash, A., Ludäscher, B.: Processing unions of conjunctive queries with negation under limited access patterns. In: Bertino, E., et al. (eds.) EDBT 2004. LNCS, vol. 2992, pp. 422–440. Springer, Heidelberg (2004). https://doi.org/10.1007/978-3-540-24741-8_25
22. Ngo, H.Q., Porat, E., Ré, C., Rudra, A.: Worst-case optimal join algorithms. In: PODS, pp. 37–48. ACM (2012)
23. Ngo, H.Q., Ré, C., Rudra, A.: Skew strikes back: new developments in the theory of join algorithms. SIGMOD Rec. **42**(4), 5–16 (2013)
24. Olteanu, D., Schleich, M.: Factorized databases. ACM SIGMOD Rec. **45**(2), 5–16 (2016)
25. Patrascu, M., Roditty, L.: Distance oracles beyond the Thorup-zwick bound. In: FOCS, pp. 815–823. IEEE (2010)
26. Robertson, N., Seymour, P.D.: Graph minors. II. Algorithmic aspects of tree-width. J. Algorithms **7**(3), 309–322 (1986)
27. Segoufin, L.: Enumerating with constant delay the answers to a query. In: Proceedings of the 16th ICDT, pp. 10–20. ACM (2013)
28. Ullman, J.D.: An approach to processing queries in a logic-based query language. In: Brodie, M.L., Mylopoulos, J. (eds.) On knowledge base management systems. Topics in Information Systems, pp. 147–164. Springer, New York (1986). https://doi.org/10.1007/978-1-4612-4980-1_16
29. Wei, F., Lausen, G.: Containment of conjunctive queries with safe negation. In: Calvanese, D., Lenzerini, M., Motwani, R. (eds.) ICDT 2003. LNCS, vol. 2572, pp. 346–360. Springer, Heidelberg (2003). https://doi.org/10.1007/3-540-36285-1_23
30. Xirogiannopoulos, K., Deshpande, A.: Extracting and analyzing hidden graphs from relational databases. In: SIGMOD, pp. 897–912. ACM (2017)
31. Xirogiannopoulos, K., Khurana, U., Deshpande, A.: GraphGen: exploring interesting graphs in relational data. Proc. VLDB Endowment **8**(12), 2032–2035 (2015)
32. Yannakakis, M.: Algorithms for acyclic database schemes. In: VLDB, pp. 82–94 (1981)

Approximate Minimum Sum Colorings and Maximum k-Colorable Subgraphs of Chordal Graphs

Ian DeHaan[1](✉) and Zachary Friggstad[2]

[1] Department of Combinatorics and Optimization, University of Waterloo, Waterloo, Canada
ijdehaan@uwaterloo.ca

[2] Department of Computing Science, University of Alberta, Edmonton, Canada
zacharyf@ualberta.ca

Abstract. We give a $(1.796 + \epsilon)$-approximation for the minimum sum coloring problem on chordal graphs, improving over the previous 3.591-approximation by Gandhi et al. [2005]. To do so, we also design the first PTAS for the maximum k-colorable subgraph problem in chordal graphs.

1 Introduction

We consider a coloring/scheduling problem introduced by Kubicka in 1989 [14].

Definition 1. *In the* Minimum Sum Coloring *(*MSC*) problem, we are given an undirected graph $G = (V, E)$. The goal is to find a proper coloring $\phi : V \to \{1, 2, 3, \dots\}$ of vertices with positive integers which minimizes $\sum_{v \in V} \phi(v)$. In weighted* MSC*, each vertex $v \in V$ additionally has a weight $w_v \geq 0$ and the goal is then to minimize $\sum_{v \in V} w_v \cdot \phi(v)$.*

Naturally, in saying ϕ is a proper coloring, we mean $\phi(u) \neq \phi(v)$ for any edge $uv \in E$. MSC is often used to model the scheduling of unit-length dependent jobs that utilize shared resources. Jobs that conflict for resources cannot be scheduled at the same time. The goal in MSC is then to minimize the average time it takes to complete a job.

In contrast with the standard graph coloring problem, where we are asked to minimize the number of colors used, sum coloring is NP-hard on many simple graph types. Even on bipartite and interval graphs, where there are linear time algorithms for graph coloring, MSC remains APX-hard [3,10].

In [2], it was shown that if one can compute a maximum independent set in any induced subgraph of G in polynomial time, then iteratively coloring G by greedily choosing a maximum independent set of the uncolored nodes each

I. DeHaan—Supported by an NSERC Undergraduate Student Research Award held at the University of Alberta.
Z. Friggstad—Supported by an NSERC Discovery Grant and Accelerator Supplement.

P. Morin and S. Suri (Eds.): WADS 2023, LNCS 14079, pp. 326–339, 2023.
https://doi.org/10.1007/978-3-031-38906-1_22

step yields a 4-approximation for MSC. A series of improved approximations for other graph classes followed, these are summarized in Table 1. Of particular relevance for this paper are results for perfect graphs and interval graphs. For MSC in perfect graphs, the best approximation is $\mu^\star \approx 3.591$, the solution to $\mu \ln \mu = \mu + 1$. For MSC in interval graphs, the best approximation is $\frac{\mu^\star}{2} \approx 1.796$.

Table 1. Known results for sum coloring. The $O^\star$-notation hides poly($\log \log n$) factors. Our work appears in bold.

	u.b.	l.b.
General graphs	$O^\star(n/\log^3 n)$ [2,6]	$O(n^{1-\epsilon})$ [2,7]
Perfect graphs	3.591 [8]	APX-HARD [3]
Chordal graphs	**1.796**+ϵ	APX-HARD [10]
Interval graphs	1.796 [11]	APX-HARD [10]
Bipartite graphs	27/26 [15]	APX-HARD [3]
Planar graphs	PTAS [12]	NP-HARD [12]
Line graphs	1.8298 [13]	APX-HARD [16]

In this paper, we study MSC in chordal graphs. A graph is chordal if it does not contain a cycle of length at least 4 as an induced subgraph. Equivalently, every cycle of length at least 4 has a chord - an edge connecting two non-consecutive nodes on the cycle. Chordal graphs form a subclass of perfect graphs, so we can color them optimally in polynomial time. But MSC itself remains APX-HARD in chordal graphs [10], as they generalize interval graphs. The class of chordal graphs is well studied; linear-time algorithms have been designed to recognize them, to compute maximum independent sets, and to find minimum colorings, among other things. A comprehensive summary of many famous results pertaining to chordal graphs can be found in the excellent book by Golumbic [9]. Chordal graphs also appear often in practice; for example Pereira and Palsberg study register allocation problems (which can be viewed as a sort of graph coloring problem) and observe that the interference graphs for about 95% of the methods in the Java 1.5 library are chordal when compiled with a particular compiler [17].

Our main result is an improved approximation algorithm for MSC in chordal graphs.

Theorem 1. *For any constant $\epsilon > 0$, there is a polynomial-time $\frac{\mu^\star}{2} + \epsilon \approx 1.796 + \epsilon$ approximation for weighted* MSC *on chordal graphs.*

That is, we can approximate MSC in chordal graphs essentially within the same guarantee as for interval graphs. Prior to our work, the best approximation in chordal graphs was the same as in perfect graphs: a 3.591-approximation by Gandhi et al. [8].

To attain this, we study yet another variant of the coloring problem.

Definition 2. *In the weighted* Maximum k-Colorable Subgraph *(MkCS) problem, we are given a graph $G = (V, E)$, vertex weights $w_v \geq 0$, and a positive integer k. The goal is to find a maximum-weight subset of nodes $S \subseteq V$ such that the induced subgraph $G[S]$ is k-colorable.*

We also design a polynomial-time approximation scheme (PTAS) for weighted MkCS in chordal graphs.

Theorem 2. *For any $\epsilon > 0$, there is a $(1-\epsilon)$-approximation for weighted* MkCS *in chordal graphs.*

Prior to our work, the best approximation recorded in literature was a 1/2-approximation by Chakaravarthy and Roy [4]. Although one could also get a $(1 - 1/e)$-approximation by greedily finding and removing a maximum-weight independent set of nodes for k iterations, i.e., the maximum coverage algorithm.

Organization
We begin with a high-level discussion of our techniques. Then, Sect. 2 presents the proof of Theorem 1 assuming one has a PTAS for MkCS in chordal graphs. Theorem 2 is proven in Sect. 3.

1.1 Our Techniques

Our work is inspired by the 1.796-approximation for MSC in interval graphs by Halldórsson, Kortsarz, and Shachnai [11]. They show that if one has an exact algorithm for MkCS, then by applying it to values of k from a carefully selected geometric sequence and "concatenating" these colorings, one gets a 1.796-approximation. In interval graphs, MkCS can be solved in polynomial time using a greedy algorithm. We show that a similar result holds: we show Theorem 1 holds in any family of graphs that admit a PTAS for MkCS and a polynomial time algorithm for the standard minimum coloring problem. However, we need to use linear programming techniques instead of a greedy algorithm since their approach seems to heavily rely on getting *exact* algorithms for MkCS.

MkCS in Chordal Graphs
In chordal graphs, MkCS is NP-complete, but it can be solved in $n^{O(k)}$ time [19]. We rely on this algorithm for constant values of k, so we briefly summarize how it works to give the reader a complete picture of our PTAS.

Their algorithm starts with the fact that chordal graphs have the following representation. For each chordal graph $G = (V, E)$ there is a tree T with $O(n)$ nodes of maximum degree 3 plus a collection of subtrees $\mathcal{T} = \{T_v : v \in V\}$, one for each $v \in V$. These subtrees satisfy the condition that $uv \in E$ if and only if subtrees T_u and T_v have at least one node in common. For a subset $S \subseteq V$, we have $G[S]$ is k-colorable if and only if each node in T lies in at most k subtrees from $\{T_v : v \in S\}$. The tree T and subtrees $\mathcal{T}$ are computed in polynomial time and then a straightforward dynamic programming procedure is used to find the maximum k-colorable subgraph. The states of the DP algorithm are characterized by a node a of T and subtrees $\mathcal{S} \subseteq \mathcal{T}$ with $|\mathcal{S}| \leq k$ where each subtree in $\mathcal{S}$ includes a.

Our contribution is an approximation for large values of k. It is known that a graph G is chordal if and only if its vertices can be ordered as $v_1, v_2, \ldots, v_n$ such that for every $1 \leq i \leq n$, the set $N^{left}(v_i) := \{v_j : v_i v_j \in E \text{ and } j < i\}$ is a clique. Such an ordering is called a **perfect elimination ordering**. We consider the following LP relaxation based on a perfect elimination ordering. We have a variable x_v for every $v \in V$ indicating if we should include v in the subgraph.

$$\textbf{maximize} \left\{ \sum_{v \in V} w_v \cdot x_v : x_v + x(N^{left}(v)) \leq k \; \forall \, v \in V, x \in [0,1]^V \right\}.$$

The natural $\{0,1\}$ solution corresponding to a k-colorable induced subgraph $G[S]$ is feasible, so the optimum LP solution has value at least the size of the largest k-colorable subgraph of G.

We give an LP-rounding algorithm with the following guarantee.

Lemma 1. *Let x be a feasible LP solution. In $n^{O(1)}$ time, we can find a subset $S \subseteq V$ such that $G[S]$ is k-colorable and $\sum_{v \in S} w_v \geq \left(1 - \frac{2}{k^{1/3}}\right) \cdot \sum_{v \in V} w_v \cdot x_v$.*

Theorem 2 then follows easily. If $k \leq 8/\epsilon^3$, we use the algorithm from [19] which runs in polynomial time since k is bounded by a constant. Otherwise, we run our LP rounding procedure.

Linear Programming Techniques for MSC

We give a general framework for turning approximations for weighted MkCS into approximations for MSC.

Definition 3. *We say that an algorithm for weighted* MkCS *is a (ρ, γ) approximation if it always returns a $\gamma \cdot k$ colorable subgraph with vertex weight at least $\rho \cdot OPT$, where OPT is the maximum vertex weight of any k-colorable subgraph.*

For Theorem 1, we only need to consider the case $\rho = 1 - \epsilon$ and $\gamma = 1$. Still, we consider this more general concept since it is not any harder to describe and may have other applications.

We prove the following, where e denotes the base of the natural logarithm.

Lemma 2. *Suppose there is a (ρ, γ) approximation for weighted* MkCS *on graphs in a class of a graphs where minimum colorings can be found in polynomial time. Then, for any $1 < c < \min(e^2, \frac{1}{1-\rho})$, there is a $\frac{\rho \cdot \gamma \cdot (c+1)}{2 \cdot (1-(1-\rho) \cdot c) \cdot \ln c}$-approximation for* MSC *for graphs in the same graph class.*

Our main result follows by taking $\gamma = 1$ and $\rho = 1 - \epsilon$. For small enough ϵ, we then choose $c^* \approx 3.591$ to minimize the expression, resulting in an approximation guarantee of at most 1.796.

Roughly speaking, we prove Lemma 2 by considering a time-indexed configuration LP relaxation for latency-style problems. Configuration LPs have been considered for MSC in other graph classes, such as line graphs [13]. The configurations used in previous work have variables for each independent set. We use a stronger LP that has variables for each k-colorable subgraph for each $1 \leq k \leq n$.

Our configuration LP was inspired by one introduced by Chakrabarty and Swamy for the Minimum Latency Problem (a variant of the Travelling

SALESPERSON PROBLEM) [5], but is tailored for our setting. For each "time" $k \geq 1$ we have a family of variables, one for each k-colorable subgraph, indicating if this is the set of nodes that should be colored with integers $\leq k$. This LP can be solved approximately using the (ρ, γ)-approximation for MkCS, and it can be rounded in a manner inspired by [5,18].

Note that Theorem 2 describes a $(1-\epsilon, 1)$-approximation for MkCS for any constant $\epsilon > 0$. If we had a $(1, 1+\epsilon)$-approximation then the techniques in [11] could be easily adapted to prove Theorem 1. But these techniques don't seem to apply when given MkCS approximations that are inexact on the number of nodes included in the solution.

2 An LP-Based Approximation Algorithm for MSC

As mentioned earlier, our approach is inspired by a time-indexed LP relaxation for latency problems introduced by Chakrabarty and Swamy [5]. Our analysis follows ideas presented by Post and Swamy who, among other things, give a 3.591-approximation for the MINIMUM LATENCY PROBLEM [18] using a configuration LP.

2.1 The Configuration LP

For a value $k \geq 0$ (perhaps non-integer), $\mathcal{C}_k$ denotes the vertex subsets $S \subseteq V$ such that $G[S]$ can be colored using at most k colors. For integers $1 \leq k \leq n$ and each $C \in \mathcal{C}_k$, we introduce a variable $z_{C,k}$ that indicates if C is the set of nodes colored with the first k integers. We also use variables $x_{v,k}$ to indicate vertex v should receive color k. We only need to consider n different colors since no color will be "skipped" in an optimal solution.

$$\textbf{minimize:} \quad \sum_{v \in V} \sum_{k=1}^{n} w_v \cdot k \cdot x_{v,k} \qquad \textbf{(LP-MSC)}$$

$$\textbf{subject to:} \quad \sum_{k=1}^{n} x_{v,k} = 1 \qquad \forall\, v \in V \qquad (1)$$

$$\sum_{C \in \mathcal{C}_k} z_{C,k} \leq 1 \qquad \forall\, 1 \leq k \leq n \qquad (2)$$

$$\sum_{C \in \mathcal{C}_k : v \in C} z_{C,k} \geq \sum_{k' \leq k} x_{v,k'} \qquad \forall\, v \in V, 1 \leq k \leq n \qquad (3)$$

$$x, z \geq 0$$

Constraint (1) says each vertex should receive one color, constraint (2) ensures we pick just one subset of vertices to use the first k colors on, and constraint (3) enforces that each vertex colored by a value less than or equal to k must be in the set we use the first k colors on.

Recall that this work is not the first time a configuration LP has been used for MSC. In [13], the authors consider one that has a variable $x_{C,k}$ for every independent set C, where the variable models that C is the independent set used at time t. Our approach allows us to prove better bounds via LP rounding, but it has the stronger requirement that in order to (approximately) solve our LP, one needs to (approximately) solve the MkCS problem, rather than just the maximum independent set problem.

Let OPT denote the optimal MSC cost of the given graph and OPT_{LP} denote the optimal cost of (**LP-MSC**). Then $OPT_{LP} \leq OPT$ simply because the natural $\{0,1\}$ solution corresponding to OPT is feasible for this LP.

At a high level, we give a method to solve this LP approximately by using the approximation for MkCS to approximately separate the constraints of the dual LP, which is given as follows.

$$\textbf{maximize:} \quad \sum_{v \in V} \alpha_v - \sum_{k=1}^{n} \beta_k \qquad \text{(DUAL-MSC)}$$

$$\textbf{subject to:} \quad \alpha_v \leq w_v \cdot k + \sum_{\hat{k}=k}^{n} \theta_{v,\hat{k}} \qquad \forall\, v \in V, 1 \leq k \leq n \qquad (4)$$

$$\sum_{v \in C} \theta_{v,k} \leq \beta_k \qquad \forall\, 1 \leq k \leq n, C \in \mathcal{C}_k \qquad (5)$$

$$\beta, \theta \geq 0 \qquad (6)$$

Note (DUAL-MSC) has polynomially-many variables. We approximately separate the constraints in the following way. For values $\nu \geq 0, \rho \leq 1, \gamma \geq 1$, let $\mathcal{D}(\nu; \rho, \gamma)$ denote the following polytope:

$$\left\{(\alpha, \beta, \theta) : (4), (6), \sum_{v \in C} \theta_{v,k} \leq \beta_k \ \forall\, 1 \leq k \leq n \ \forall\, C \in \mathcal{C}_{\gamma \cdot k}, \sum_{v} \alpha_v - \frac{1}{\rho} \cdot \sum_{k} \beta_k \geq \nu \right\}$$

Lemma 3. *If there is a* (ρ, γ)*-approximation for* MkCS*, there is also a polynomial-time algorithm* $\mathcal{A}$ *that takes a single value* ν *plus values* (α, β, θ) *for the variables of (DUAL-MSC) and always returns one of two things:*

- *A (correct) declaration that* $(\alpha, \beta/\rho, \theta) \in \mathcal{D}(\nu; 1, 1)$.
- *A hyperplane separating* (α, β, θ) *from* $\mathcal{D}(\nu; \rho, \gamma)$.

Proof. First, check that (4), (6), and $\sum_v \alpha_v - \frac{1}{\rho} \cdot \sum_k \beta_k \geq \nu$ hold. If not, the violated constraint gives a hyperplane separating (α, β, θ) from $\mathcal{D}(\nu; \rho, \gamma)$. Then, for each k, we run the MkCS (ρ, γ)-approximation on the instance with vertex weights $\theta_{v,k}, v \in V$. If this finds a solution with weight exceeding β_k, we return the corresponding constraint as a separating hyperplane. Otherwise, we know that the maximum possible weight of a k-colorable subgraph is at most β_k/ρ. If the latter holds for all k, then $(\alpha, \beta/\rho, \theta) \in \mathcal{D}(\nu; 1, 1)$.

Lemma 3.3 from [5] takes such a routine and turns it into an approximate LP solver. The following is proven in the exact same manner where we let $\mathbf{LP}^{(\rho,\gamma)}$ be the same as (**LP-MSC**), except $\mathcal{C}_k$ is replaced by $\mathcal{C}_{\gamma\cdot k}$ in both (2) and (3) and the right-hand side of (2) is replaced by $1/\rho$. For the sake of space, we provide only a quick overview of the proof below the statement.

Lemma 4. *Given a (ρ,γ)-approximation for MkCS, we can find a feasible solution (x,z) to $\mathbf{LP}^{(\rho,\gamma)}$ with cost at most OPT_{LP} in polynomial time.*

As a reminder, here OPT_{LP} is the optimum value of (**LP-MSC**) itself.

Roughly speaking, Lemma 4 is obtained as follows. First, for any ν, we can run the ellipsoid method using the approximate separation oracle. It will either generate enough constraints to certify $\mathcal{D}(\nu;\rho,\gamma)=\emptyset$ or it will eventually find a point $(\alpha,\beta,\theta)\in\mathcal{D}(\nu;1,1)$. A binary search can be used to find the largest ν for which $\mathcal{D}(\nu;\rho,\gamma)$ is not certified to be empty. The constraints produced by the ellipsoid method can be used to determine a polynomial-size set of variables that need to be considered in $\mathbf{LP}^{(\rho,\gamma)}$ to get a solution with value $\leq OPT$. See [5] for details.

2.2 The Rounding Algorithm and Analysis

The rounding algorithm is much like that in [11] in that it samples k-colorable subgraphs for various values of k in a geometric sequence and concatenates these colorings to get a coloring of all nodes. For convenience, let $z_{C,k}=z_{C,\lfloor k\rfloor}$ for any real value $k\geq 0$.

Algorithm 1. MSCRound(G)

find a solution (x,z) to $\mathbf{LP}^{(\rho,\gamma)}$ with value $\leq OPT$ using Lemma 4
if necessary, increase $z_{\emptyset,k}$ until $\sum_{C\in\mathcal{C}_{\gamma\cdot k}} z_{C,k}=1/\rho$ for each k
let $1<c<\min(e^2,1/(1-\rho))$ be a constant we will optimize later
let $h=c^{\Gamma}$ be a random offset where Γ is sampled uniformly from $[0,1)$
$j\leftarrow 0$
$k\leftarrow 0$ ▷ The next color to use
while $G\neq\emptyset$ **do**
  $k_j\leftarrow h\cdot c^j$
  $k'_j\leftarrow\min\{n,\lfloor k_j\rfloor\}$
  choose C randomly from $\mathcal{C}_{\gamma\cdot k'_j}$ with probability according to the LP values $z_{C',k'_j}\cdot\rho$ for $C'\in\mathcal{C}_{\gamma\cdot k'_j}$
  color C with $\lfloor\gamma\cdot k'_j\rfloor$ colors, call the color classes $C_1,C_2,\ldots,C_{\lfloor\gamma\cdot k'_j\rfloor}$
  randomly permute the color classes, let $C'_1,C'_2,\ldots,C'_{\lfloor\gamma\cdot k'_j\rfloor}$ be the reordering
  finally, assign nodes in C'_i color $k+i$ for each $1\leq i\leq\lfloor\gamma\cdot k'_j\rfloor$ in the final solution
  $k\leftarrow k+\lfloor\gamma\cdot k'_j\rfloor$
  $G\leftarrow G-C$
  $j\leftarrow j+1$
end while

Note that nodes colored during iteration j get assigned colors at most $\gamma \cdot (k_0 + k_1 + \ldots + k_j)$ and the expected color of such a node is at most $\gamma \cdot (k_0 + k_1 + \ldots + k_{j-1} + (k_j + 1)/2)$. The number of iterations is $O(\log n)$ because each vertex will appear in each $\gamma \cdot n$ coloring, as this is the largest color considered in $\mathbf{LP}^{(\rho,\gamma)}$.

We note that despite our approach following the main ideas of the algorithm and analysis for minimum latency given in [18], there are some key details that change. In [18], each iteration of the algorithm produces a tree, which is then doubled and shortcutted to produce a cycle with cost at most double the tree. While we randomly permute the colors in our coloring, they randomly choose which direction to walk along the cycle. For a tree with cost k, this gives an expected distance of k for each node. We save a factor of 2 because we do not have a doubling step, but our average color is $\frac{k+1}{2}$ as opposed to $\frac{k}{2}$. Some extra work is required in our analysis to account for the extra $\frac{1}{2}$ on each vertex.

Let $p_{v,j}$ be the probability that vertex v is not colored by the end of iteration j. For $j < 0$, we use $p_{v,j} = 1$ and $k_j = 0$. Finally, for $v \in V$, let $\phi(v)$ denote the color assigned to v in the algorithm.

The following is essentially Claim 5.2 in [18], with some changes based on the differences in our setting as outlined above.

Lemma 5. *For a vertex v,*

$$\mathbf{E}[\phi(v)|h] \leq \frac{\gamma}{2} \cdot \frac{c+1}{c-1} \cdot \sum_{j \geq 0} p_{v,j-1} \cdot (k_j - k_{j-1}) + \gamma(\frac{1}{2} - \frac{h}{c-1})$$

Proof. There are at most $\gamma \cdot k_j$ colors introduced in iteration j. They are permuted randomly, so any vertex colored in iteration j has color, in expectation, at most $\gamma \cdot (k_j + 1)/2$ more than all colors used in previous iterations. That is, the expected color of v if colored in iteration j is at most

$$\begin{aligned}\gamma\left(k_0 + k_1 + \ldots + k_{j-1} + \frac{k_j + 1}{2}\right) &\leq \gamma\left(h \cdot \left(\frac{c^j - 1}{c-1} + \frac{c^j}{2}\right) + \frac{1}{2}\right) \\ &= \gamma\left(\frac{k_j}{2} \cdot \frac{c+1}{c-1} + \frac{1}{2} - \frac{h}{c-1}\right),\end{aligned}$$

where we have used $k_i = h \cdot c^i$ and summed a geometric sequence.

The probability v is colored in iteration j is $p_{v,j-1} - p_{v,j}$, so the expected color of v is bounded by

$$\frac{\gamma}{2} \cdot \frac{c+1}{c-1} \cdot \left(\sum_{j \geq 0} (p_{v,j-1} - p_{v,j}) \cdot k_j\right) + \gamma\left(\frac{1}{2} - \frac{h}{c-1}\right).$$

By rearranging, this is what we wanted to show.

For brevity, let $y_{v,j} = \sum_{k \le k_j} x_{v,k}$ denote the LP coverage for v up to color k_j. The next lemma is essentially Claim 5.3 from [18], but the dependence on ρ is better in our context[1].

Lemma 6. *For any $v \in V$ and $j \ge 0$, we have $p_{v,j} \le (1-y_{v,j})\cdot\rho + (1-\rho)\cdot p_{v,j-1}$.*

Proof. If v is not covered by iteration j, then it is not covered in iteration j itself and it is not covered by iteration $j-1$, which happens with probability

$$p_{v,j-1} \cdot \left(1 - \sum_{C \in \mathcal{C}_{\gamma \cdot k_j} : v \in C} \rho \cdot z_{C,k_j}\right) \le p_{v,j-1} \cdot (1 - \rho \cdot y_{v,j})$$
$$= p_{v,j-1} \cdot \rho \cdot (1 - y_{v,j}) + p_{v,j-1} \cdot (1 - \rho).$$

Note that the first inequality follows from constraint (3) and the definition of $y_{v,j}$. The lemma then follows by using $p_{v,j-1} \le 1$ and $y_{v,j} \le 1$ to justify dropping $p_{v,j-1}$ from the first term.

From these lemmas, we can complete our analysis. Here, for $v \in V$, we let $col_v = \sum_{k=1}^{n} k \cdot x_{v,k}$ denote the fractional color of v, so the cost of (x, z) is $\sum_{v \in V} w_v \cdot col_v$. The following lemma is essentially Lemma 5.4 in [18] but with our specific calculations from the previous lemmas.

Lemma 7. *For any $v \in V$, we have $\mathbf{E}[\phi(v)] \le \frac{\rho \cdot \gamma \cdot (c+1)}{2 \cdot (1-(1-\rho)\cdot c) \cdot \ln c} \cdot col_v$.*

Proof. For brevity, let $\Delta_j = k_j - k_{j-1}$. We first consider a fixed offset h. Let $A = \sum_{j \ge 0} p_{v,j-1} \cdot \Delta_j$ and recall, by Lemma 5, that the expected color of v for a given h is at most $\frac{\gamma}{2} \cdot \frac{c+1}{c-1} \cdot A + \gamma(\frac{1}{2} - \frac{h}{c-1})$.

Note $\Delta_j = c \cdot \Delta_{j-1}$ for $j \ge 2$ and $\Delta_0 + \Delta_1 = c \cdot \Delta_0$. So from Lemma 6,

$$A \le \sum_{j \ge 0} \rho \cdot (1 - y_{v,j}) \cdot \Delta_j + (1-\rho) \sum_{j \ge 0} p_{v,j-2} \cdot \Delta_j$$
$$= \sum_{j \ge 0} \rho \cdot (1 - y_{v,j}) \cdot \Delta_j + c \cdot (1-\rho) \cdot A.$$

Rearranging and using $c < 1/(1-\rho)$, we have that

$$A \le \frac{\rho}{1 - c \cdot (1-\rho)} \cdot \sum_{j \ge 0} (1 - y_{v,j}) \cdot \Delta_j.$$

For $1 \le k \le n$, let $\sigma(k)$ be k_j for the smallest integer j such that $k_j \ge k$. Simple manipulation and recalling $y_{v,j} = \sum_{k \le k_j} x_{v,j}$ shows $\sum_{j \ge 0}(1-y_{v,j}) \cdot \Delta_j = \sum_{k=1}^{n} \sigma(k) \cdot x_{v,k}$.

[1] We note [18] does have a similar calculation in a single-vehicle setting of their problem whose dependence is more like that in Lemma 6. They just don't have a specific claim summarizing this calculation that we can reference.

The expected value of $\sigma(k)$ over the random choice of h, which is really over the random choice of $\Gamma \in [0, 1)$, can be directly calculated as follows where j is the integer such that $k \in [c^j, c^{j+1})$.

$$\begin{aligned}\mathbf{E}_h[\sigma(k)] &= \int_0^{\log_c k - j} c^{\Gamma + j + 1} d\Gamma + \int_{\log_c k - j}^{1} c^{\Gamma + j} d\Gamma \\ &= \frac{1}{\ln c}(c^{\log_c k + 1} - c^{j+1} + c^{j+1} - c^{\log_c k}) = \frac{c-1}{\ln c} \cdot k.\end{aligned}$$

We have just shown $\mathbf{E}_h[\sum_{j \geq 0}(1 - y_{v,j}) \cdot \Delta_j] = \frac{c-1}{\ln c} \sum_{k \geq 0} k \cdot x_{v,k} = \frac{c-1}{\ln c} \cdot col_v$. So, we can now bound the unconditional color $\mathbf{E}_h[\phi(v)]$ using our previous lemmas.

$$\begin{aligned}\mathbf{E}_h[\phi(v)] &= \frac{\gamma}{2} \cdot \frac{c+1}{c-1} \cdot \mathbf{E}_h[A] + \gamma(\frac{1}{2} - \mathbf{E}_h[h]/(c-1)) \\ &\leq \frac{\rho \cdot \gamma \cdot (c+1)}{2 \cdot (1 - (1-\rho) \cdot c) \cdot \ln c} \cdot col_v + \gamma\left(\frac{1}{2} - \mathbf{E}_h[h]/(c-1)\right) \\ &= \frac{\rho \cdot \gamma \cdot (c+1)}{2 \cdot (1 - (1-\rho) \cdot c) \cdot \ln c} \cdot col_v + \gamma\left(\frac{1}{2} - \frac{1}{\ln c}\right) \\ &\leq \frac{\rho \cdot \gamma \cdot (c+1)}{2 \cdot (1 - (1-\rho) \cdot c) \cdot \ln c} \cdot col_v\end{aligned}$$

The first equality and inequality follow from linearity of expectation and known bounds on $\mathbf{E}[\phi(v)|h]$ and A. The second equality follows from the fact that $\mathbf{E}_h[h] = \int_0^1 c^\Gamma d\Gamma = \frac{c-1}{\ln c}$, and the last inequality is due to the fact that $c < e^2$ by assumption.

To finish the proof of Lemma 2, observe the expected vertex-weighted sum of colors of all nodes is then at most

$$\frac{\rho \cdot \gamma \cdot (c+1)}{2 \cdot (1 - (1-\rho) \cdot c) \cdot \ln c} \cdot \sum_{v \in V} w_v \cdot col_v \leq \frac{\rho \cdot \gamma \cdot (c+1)}{2 \cdot (1 - (1-\rho) \cdot c) \cdot \ln c} \cdot OPT.$$

Theorem 1 then follows by combing the $(1-\epsilon, 1)$ MkCS approximation (described in the next section) with this MSC approximation, choosing $c \approx 3.591$, and ensuring ϵ is small enough so $c < 1/\epsilon$.

We note Algorithm 1 can be efficiently derandomized. First, there are only polynomially-many offsets of h that need to be tried. That is, for each k_j, we can determine the values of h that would cause $\lfloor \gamma \cdot k_j \rfloor$ to change and try all such h over all j. Second, instead of randomly permuting the color classes in a $\gamma \cdot k_j$-coloring, we can order them greedily in non-increasing order of total vertex weight.

3 A PTAS for MkCS in Chordal Graphs

We first find a perfect elimination ordering of the vertices $v_1, v_2, \ldots, v_n$. This can be done in linear time, e.g., using lexicographical breadth-first search [9].

Let $N^{left}(v) \subseteq V$ be the set of neighbors of v that come before v in the ordering, so $N^{left}(v) \cup \{v\}$ is a clique. Recall that we are working with the following LP. The constraints we use exploit the fact that a chordal graph is k-colorable if and only if all left neighbourhoods of its nodes in a perfect elimination ordering have size at most $k - 1$.

$$\text{maximize:} \quad \sum_{v \in V} w_v \cdot x_v \qquad \textbf{(K-COLOR-LP)}$$

$$\text{subject to:} \quad x_v + x(N^{left}(v)) \leq k \qquad \forall\, v \in V \qquad (7)$$
$$x \in [0,1]^V$$

Let OPT_{LP} denote the optimal LP value and OPT denote the optimal solution to the problem instance. Of course, $OPT_{LP} \geq OPT$ since the natural $\{0,1\}$ integer solution corresponding to a k-colorable subgraph of G is a feasible solution.

We can now give a rounding algorithm as follows.

Algorithm 2. MCSRound(G, k)

let $0 \leq f(k) \leq 1$ be a value we will optimize later
find a perfect elimination ordering $v_1, v_2, \ldots, v_n$ for G
let x be an optimal feasible solution to **(K-COLOR-LP)**
form S' by adding each $v \in V$ to S' independently with probability $(1 - f(k)) \cdot x_v$.
$S \leftarrow \emptyset$
for $v \in \{v_1, v_2, \ldots, v_n\}$ **do**
  if $v \in S'$ and $S \cup \{v\}$ is k-colorable, add v to S
end for
return S

3.1 Analysis

Observe that when we consider adding some $v \in S'$ to S, $S \cup \{v\}$ is k-colorable if and only if $|S \cap N^{left}(v)| \leq k - 1$. This is easy to prove by noting that the restriction of a perfect elimination ordering of G to a subset S yields a perfect elimination ordering of $G[S]$. Because we consider the nodes v according to a perfect elimination ordering of G, by adding v the only possible left-neighbourhood of a node that could have size $\geq k$ is $N^{left}(v)$ itself.

We bound the probability that we select at least k vertices from $N^{left}(v)$. The second moment method is used so that derandomization is easy. Let Y_u indicate the event that $u \in S'$. Then $\mathbf{E}[Y_u^2] = \mathbf{E}[Y_u] = (1 - f(k)) \cdot x_u$. Fix some vertex v. Let $Y = \sum_{u \in N^{left}(v)} Y_u$. By constraint (7), we have

$$\mathbf{E}[Y] = \sum_{u \in N^{left}(v)} (1 - f(k)) \cdot x_u \leq (1 - f(k)) \cdot k.$$

And since each Y_u is independent, we have again by constraint (7) that

$$\mathbf{Var}[Y] = \sum_{u \in N^{left}(v)} \mathbf{Var}[Y_u] = \sum_{u \in N^{left}(v)} (\mathbf{E}[Y_u^2] - \mathbf{E}[Y_u]^2) \leq \sum_{u \in N^{left}(v)} \mathbf{E}[Y_u^2] \leq k.$$

We are interested in

$$\mathbf{Pr}[Y \geq k] \leq \mathbf{Pr}[|Y - E[Y]| \geq f(k) \cdot k].$$

By Chebyshev's inequality,

$$\mathbf{Pr}[|Y - E[Y]| \geq f(k) \cdot k] \leq \frac{\mathbf{Var}[Y]}{f(k)^2 \cdot k^2} \leq \frac{k}{f(k)^2 \cdot k^2} = \frac{1}{f(k)^2 \cdot k}.$$

From this, we find that the probability we actually select vertex v is at least

$$\mathbf{Pr}[Y_v \wedge (Y \leq k-1)] = \mathbf{Pr}[Y_v] \cdot \mathbf{Pr}[Y \leq k-1] \geq (1 - f(k)) \cdot x_v \cdot \left(1 - \frac{1}{f(k)^2 \cdot k}\right).$$

The first equality is justified because Y only depends on Y_u for $u \neq v$, so these two events are independent.

Choosing $f(k) = k^{-1/3}$ results in $v \in S$ with probability at least $x_v \cdot (1 - 2 \cdot k^{-1/3})$. By linearity of expectation, the expected value of S is at least $(1 - 2 \cdot k^{-1/3}) \cdot \sum_{v \in V} w_v \cdot x_v$.

The PTAS for MkCS in chordal graphs is now immediate. For any constant $\epsilon > 0$, if $k \geq 8/\epsilon^3$, then we run our LP rounding algorithm to get a k-colorable subgraph with weight at least $(1 - \epsilon) \cdot OPT_{LP}$. Otherwise, we run the exact algorithm in [19], which runs in polynomial time since k is bounded by a constant.

It is desirable to derandomize this algorithm so it always finds a solution with the stated guarantee. This is because we use it numerous times in the approximate separation oracle for (DUAL-MSC). Knowing it works all the time does not burden us with providing concentration around the probability we successfully approximately solve $\mathbf{LP}^{(\rho,\gamma)}$ as in Lemma 4. We can derandomize Algorithm 2 using standard techniques, since it only requires that the variables $Y_u, u \in V$ be pairwise-independent (in order to bound $\mathbf{Var}[Y]$).

4 Conclusion

It is natural to wonder if MSC admits a better approximation in perfect graphs. Unfortunately, our techniques do not extend immediately. In [1], Addario-Berry et al. showed MkCS is NP-HARD in a different subclass of perfect graphs than chordal graphs. Their proof reduces from the maximum independent set problem and it is easy to see it shows MkCS is APX-HARD in the same graph class if one reduces from bounded-degree instances of maximum independent set.

However, our approach, or a refinement of it, may succeed if one has good constant approximations for MkCS in perfect graphs. Note that MkCS can be approximated within $1 - 1/e$ in perfect graphs simply by using the *maximum coverage* approach. That is, for k iterations, we greedily compute a maximum independent set of nodes that are not yet covered. This is not sufficient to get an improved MSC approximation in perfect graphs using Lemma 2. Lemma 2 can be used if we get a sufficiently-good (≈ 0.704) approximation for MkCS. As a starting point, we ask if there is a ρ-approximation for MkCS in perfect graphs for some constant $\rho > 1 - 1/e$.

References

1. Addario-Berry, L., Kennedy, W., King, A., Li, Z., Reed, B.: Finding a maximum-weight induced k-partite subgraph of an i-triangulated graph. Discret. Appl. Math. **158**(7), 765–770 (2010)
2. Bar-Noy, A., Bellare, M., Halldórsson, M.M., Shachnai, H., Tamir, T.: On chromatic sums and distributed resource allocation. Inf. Comput. **140**(2), 183–202 (1998)
3. Bar-Noy, A., Kortsarz, G.: The minimum color-sum of bipartite graphs. J. Algorithms **28**(2), 339–365 (1998)
4. Chakaravarthy, V., Roy, S.: Approximating maximum weight k-colorable subgraphs in chordal graphs. Inf. Process. Lett. **109**(7), 365–368 (2009)
5. Chakrabarty, D., Swamy, C.: Facility location with client latencies: linear programming based techniques for minimum latency problems. In: Günlük, O., Woeginger, G.J. (eds.) IPCO 2011. LNCS, vol. 6655, pp. 92–103. Springer, Heidelberg (2011). https://doi.org/10.1007/978-3-642-20807-2_8
6. Feige, U.: Approximating maximum clique by removing subgraphs. SIAM J. Discret. Math. **18**(2), 219–225 (2004)
7. Feige, U., Kilian, J.: Zero knowledge and the chromatic number. J. Comput. Syst. Sci. **57**(2), 187–199 (1998)
8. Gandhi, R., Halldórsson, M.M., Kortsarz, G., Shachnai, H.: Improved bounds for sum multicoloring and scheduling dependent jobs with minsum criteria. In: Persiano, G., Solis-Oba, R. (eds.) WAOA 2004. LNCS, vol. 3351, pp. 68–82. Springer, Heidelberg (2005). https://doi.org/10.1007/978-3-540-31833-0_8
9. Golumbic, M.C.: Algorithmic Graph Theory and Perfect Graphs. Academic Press (1980)
10. Gonen, M.: Coloring problems on interval graphs and trees. Master's thesis, School of Computer Science, The Open University, Tel-Aviv (2001)
11. Halldórsson, M.M., Kortsarz, G., Shachnai, H.: Sum coloring interval and k-claw free graphs with application to scheduling dependent jobs. Algorithmica **37**, 187–209 (2003)
12. Halldórsson, M., Kortsarz, G.: Tools for multicoloring with applications to planar graphs and partial k-trees. J. Algorithms **42**(2), 334–366 (2002)
13. Halldórsson, M., Kortsarz, G., Sviridenko, M.: Sum edge coloring of multigraphs via configuration LP. ACM Trans. Algorithms **7**(2), 1–21 (2011)
14. Kubicka, E.: The chromatic sum and efficient tree algorithms. Ph.D. thesis, Western Michigan University (1989)
15. Malafiejski, M., Giaro, K., Janczewski, R., Kubale, M.: Sum coloring of bipartite graphs with bounded degree. Algorithmica **40**, 235–244 (2004)
16. Marx, D.: Complexity results for minimum sum edge coloring. Discret. Appl. Math. **157**(5), 1034–1045 (2009)

17. Pereira, F.M.Q., Palsberg, J.: Register allocation via coloring of chordal graphs. In: Yi, K. (ed.) APLAS 2005. LNCS, vol. 3780, pp. 315–329. Springer, Heidelberg (2005). https://doi.org/10.1007/11575467_21
18. Post, I., Swamy, C.: Linear programming-based approximation algorithms for multi-vehicle minimum latency problems. In: Proceedings of the Twenty-Sixth Annual ACM-SIAM Symposium on Discrete Algorithms, pp. 512–531 (2014)
19. Yannakakis, M., Gavril, F.: The maximum k-colorable subgraph problem for chordal graphs. Inf. Process. Lett. **24**(2), 133–137 (1987)

Differentially Private Range Query on Shortest Paths

Chengyuan Deng, Jie Gao(✉), Jalaj Upadhyay(✉), and Chen Wang(✉)

Rutgers University, Piscataway, USA
{cd751,jg1555,upadhyay,wc497}@rutgers.edu

Abstract. We consider range queries on a graph under the constraints of differential privacy and query ranges are defined as the set of edges on the shortest path of the graph. Edges in the graph carry sensitive attributes and the goal is to report the sum of these attributes on the shortest path for *counting query* or the minimum of the attributes in a *bottleneck query*. We use differential privacy to ensure that answering these queries does not violate the privacy of the sensitive edge attributes. Our goal is to design mechanisms that minimize the additive error of the output with the given privacy budget.

For this, we develop the first set of non-trivial results for private range queries on shortest paths. For counting range queries we can achieve an additive error of $\widetilde{O}(n^{1/3})$ for ε-DP and $\widetilde{O}(n^{1/4})$ for (ε, δ)-DP. We present two algorithms where we control the final error by carefully balancing perturbation added to the edge attributes directly versus perturbation added to (a subset of) range query answers. Bottleneck range queries are easier and can be answered with polylogarithmic additive errors.

Keywords: Range query · Differential privacy · Shortest path

1 Introduction

Range counting has been extensively studied in the literature, particularly for geometric ranges. In the typical setting, there is a set of points X in $\mathbb{R}^d$. A range query is often formulated by a geometric shape, and range counting reports the number of points inside the range [34]. The points can be weighted, in which case the goal is to return the weighted sum inside the query range. Compared to the huge literature on geometric range queries [48], there has been much less work on the study of range queries with non-geometric ranges.

In this paper, we study private range counting when the ranges are defined as paths on a graph. This setting becomes interesting with the exploding amount of graph data. Graphs are used as a natural mathematical structure to model pairwise relations between objects. Often, the pairwise relations or attributes can represent private and confidential information. As such, performing statistics on such a graph without any robust privacy guarantee can be problematic. We consider the scenario where both the graph topology and

P. Morin and S. Suri (Eds.): WADS 2023, LNCS 14079, pp. 340–370, 2023.
https://doi.org/10.1007/978-3-031-38906-1_23

the query ranges (paths on the graph) are public information, but attributes on the edges of the graph, that may come from private sources, are sensitive and protected. Our goal is to return (approximate) range queries while protecting data privacy.

The above model is applicable in many real-world scenarios. In financial analysis, graph-based techniques have been adopted to combat fraud [39]. One can consider a graph where edges represent transactions between two financial entities with attributes such as the total amount being transferred. Forensic analysis researchers may want to issue queries along certain paths that involve multiple financial entities to detect anomalies. In supply chain networks, vertices represent participants such as producers, transporters or retailers, and edges represent their relationships. Resilience is a critical factor in supply chains and metrics on edges such as Time-to-Stockout (TTS) [29] have been used for estimating end-to-end resilience of certain paths. Response time or cost are also important edge attributes. In these settings, privacy and security issues of the attributes are natural and crucial (e.g., as trade secrets) [38]. In road networks, ranges can be naturally defined as paths that users take and queries are about collective statistics of traffic along the path. Privacy is also crucial in healthcare information systems [46].

1.1 Our Setting and Results

We consider the setting when query ranges are taken as shortest paths based on *public* edge weights, and the query answer is a function of *private* attributes on the edges involved in a query range/path. Using shortest paths between two vertices is natural in many of the application settings discussed above. Further, if the range query is applied on arbitrary paths in a graph, the additive query error needed to ensure privacy can be as large as $\Omega(n)$, where n is the number of vertices in the graph. We give a proof of this in Appendix A.

We consider two types of query function f on a path P:

- *Counting query*: return the sum of the attribute values on edges of P;
- *Bottleneck query*: return the minimum of the attribute values on edges of P.

Since the attribute values are private and sensitive, the reported range query answers are perturbed to ensure differential privacy guarantees. Specifically, we consider two neighboring attribute value sets w and w' on the same graph G, which differ by a ℓ_1 norm of 1. A mechanism $\mathcal{A}$ is called (ε, δ)-differentially private if the probability of obtaining query outputs on input attributes w or w' is relatively bounded by a multiplicative error of e^{ε} and an additive error of δ. When $\delta = 0$, we call $\mathcal{A}$ ε-DP or pure-DP. The objective is to achieve the specified privacy requirement with noise perturbation as small as possible.

In this paper, we study the private range query (both counting and bottleneck) on the shortest paths. As standard in the literature of differential privacy, our aim is to understand the trade-off between privacy and additive error in the final query answer, i.e., for a given privacy budget, minimize the additive error. One can additionally consider the query time and space required for the data structure. We leave designing a differentially private data structure with a better query time-space trade-off as a direction of future research.

For counting queries, we present two algorithms with privacy guarantees of pure-DP and approximate-DP respectively (in Sect. 3 and Sect. 4), returning the counts with relatively small worst-case additive errors. Our main results are captured by the following theorem:

Result 1 **(ε-DP algorithm for counting query, informal version of Theorem 1).** *There exists an ε-differentially private algorithm that outputs counting queries along all pairs shortest paths with additive error at most $\widetilde{O}(\frac{n^{1/3}}{\varepsilon})$ with high probability.*

Result 2 **((ε, δ)-DP algorithm for counting query, informal version of Theroem 2).** *There exists an (ε, δ)-differentially private algorithm that outputs counting queries along all pairs shortest paths with additive error at most $\widetilde{O}(\frac{n^{1/4}}{\varepsilon} \log^{1/2} \frac{1}{\delta})$ with high probability.*

The above results are the first known upper bounds for this specific problem. Meanwhile, we establish a lower bound of $\Omega(n^{1/6})$ adapted from the construction of the lower bound for private all pairs shortest distances [10] (with details in Appendix D). The gap between the best-known upper and lower bounds provokes an interesting perspective of private range queries: we do not yet have optimal bounds for specific ranges, despite the results by [36] presenting optimal bounds for generic range query problems. Closing the gap for counting queries would also be an interesting open question. Our next result, however, shows that the bottleneck query yields simple algorithms using existing techniques to achieve logarithm additive error:

Result 3 **(DP algorithms for bottleneck query, informal version of Theroem 3).** *There exists an ε-differentially private algorithm and an (ε, δ)-differentially private algorithm, such that with high probability, outputs bottleneck queries along all pairs shortest paths with additive error at most $\widetilde{O}(\frac{\log n}{\varepsilon})$ and $\widetilde{O}(\frac{\sqrt{\log n \log \frac{1}{\delta}}}{\varepsilon})$ respectively.*

Collectively, our results give the first set of non-trivial bounds for privately releasing queries for shortest paths on range query systems. We further show that it is possible to use the VC-dimension of shortest paths queries to obtain a bound similar to Result 2, albeit with a much more complicated algorithm for generic range query applications from [36].

1.2 Main Techniques

In general, differentially private mechanisms add perturbation to data samples. There are two standard primitives, namely *output perturbation*, where random noises are added to the final data output, and *input perturbation*, where random noises are added to each data element.

We first explain the challenges in improving these two mechanisms. To guarantee privacy, the noise in the output perturbation should take a magnitude of the sensitivity of the range query function. If the edge attribute changes by 1 in the ℓ_1 norm, there can be up to $\Theta(n^2)$ query pairs being impacted – e.g., when $\Theta(n^2)$ shortest paths share one edge. As such, if we apply a crude output perturbation, the noise for each query should be $\widetilde{O}(n^2)$ for ε-differential privacy and $\widetilde{O}(n)$ for (ε, δ)-differential privacy. On the other hand, with input perturbation, one can add a Laplace noise of magnitude proportional to $1/\varepsilon$ to each edge attribute. This satisfies ε-privacy, but the shortest path may have up to order n edges, and the noises on edges are accumulated with a total error of $\widetilde{O}(n)$.

To improve the error bound, we actually need to combine input and output perturbations. In general, the error due to output perturbation is defined by the *sensitivity* of the function – how many entries will be changed when we have neighboring attributes. The error for input perturbation depends on the *graph hop diameter*, i.e., the maximum number of edge attributes that we need to sum up as the output of counting queries. Therefore, one natural idea is to introduce 'shortcuts' (to replace a selective set of shortest paths) to the graph such that the network diameter is reduced. We then apply output perturbation on the shortcuts and use input perturbation on the graph with shortcuts. Of course, when the shortcuts are introduced, we need to be mindful of their sensitivity. The natural question is, can we reduce the network diameter with no or limited increment to the edge sensitivity with the introduction of the shortcuts?

Pure-DP Algorithm. The main idea in our first solution is to choose shortcuts with small sensitivity. By the assumption of unique shortest path, any two shortest paths would either be completely disjoint or intersect at *exactly one* common sub-path. For every intersecting shortest path between vertices (u_1, u_2), we name u_1, u_2 as the *cut vertices*. Since there are $\binom{s}{2}$ shortest paths for all pairs in $\mathcal{S}$, there are at most $O(s^2)$ cut vertices on any shortest path $P(u, v)$ with $(u, v) \in \mathcal{S} \times \mathcal{S}$. For every $(u, v) \in \mathcal{S} \times \mathcal{S}$, we cut the path $P(u, v)$ along these cut vertices into $O(s^2)$ *canonical segments* and pre-compute their length using output perturbation. The good thing is that the maximum sensitivity for the length of a canonical segment is one – since no two canonical segments can share any common edge. Reducing sensitivity by a multiplicative factor of s^2 at the cost of increasing the hop diameter by an additive value of s^2 turns out to be beneficial when we calculate the final additive error, which is $\widetilde{O}(\sqrt{n/s + s^2})$, for our ε-DP algorithm. Plugging in $s = n^{1/3}$, we can get an error of $\widetilde{O}(n^{1/3})$ and an ε-DP algorithm.

Approximate-DP Algorithm. Our solution for (ε, δ)-DP exploits properties of strong composition [17], which allows us to massage k (ε, δ)-DP mechanisms into an (ε', δ')-DP mechanism, where $\varepsilon' \approx \varepsilon\sqrt{k}$ and $\delta' \approx k\delta$. Our strategy to leverage strong decomposition is to build a shortest path tree rooted at each vertex in the sampled set $\mathcal{S}$. Tree graphs admit much better differentially private mechanisms – one can get polylogarithmic additive error for running queries on a tree graph [18,45]. Now for any two vertices u, v in $\mathcal{G}$, if the shortest path $P(u,v)$ has more than $\widetilde{O}(n/s)$ vertices, $P(u,v)$ has at least one vertex w in $\mathcal{S}$ with high probability. Thus the length of $P(u,v)$ is taken as the sum of length $P(u,w)$ and $P(w,v)$, which, can be obtained by using pre-computed query values between (u,w) and (v,w) in the shortest path tree rooted at w. The sensitivity of an edge in this case goes up – an edge can appear in possibly all the s trees. Thus, on the trees we take $(O(\varepsilon/\sqrt{s}), \delta/2s)$-differentially private mechanisms. The composition of s of them gives (ε,δ)-DP. The final error bound is $\widetilde{O}\left(\sqrt{n/s}+\sqrt{s}\right)$. Optimizing the error by setting $s = \widetilde{O}(\sqrt{n})$ gives an (ε,δ)-DP mechanism with an additive error of $\widetilde{O}(n^{1/4})$.

Remark 1. Our scheme for the approximate-DP algorithm can also be applied to the pure-DP regime to obtain the same upper bound of $\widetilde{O}(n^{1/3})$, using the basic composition theorem (Proposition 4) and replacing Gaussian mechanism with Laplace mechanism. However, there will be an extra $\log^2 n$ on the additive error over the pure-DP algorithm described above.

Remark 2. The algorithm using canonical segments works only for undirected graphs, while the algorithm using shortest path trees can be extended for directed graphs. In particular, we can build two shortest path trees at each sampled vertex w, one $T_{in}(w)$ with edges pointing towards w and one tree $T_{out}(w)$ with edges pointing away from w. Any shortest path $P(u,v)$ that visits a vertex $w \in S$ is composed of the shortest path from u to w (captured in the tree $T_{in}(w)$) and then a path from w to v (captured in tree $T_{out}(w)$). With this in mind, throughout the paper we assume an undirected graph.

1.3 Related Work

Geometric Range Queries. Geometric range queries typically consider halfplane ranges, axis-parallel rectangles (orthogonal range query), or simplices (simplex range query). The majority of work on range counting considers upper and lower bounds on the running time for answering a query, with different data storage requirements [48]. Designing geometric data structures while preserving differential privacy has also gained attention in the recent past. For example, Biemel et al. [3,30] looked at the problem of the center point of a convex hull. They instantiated exponential mechanism with *Tukey depth* [49] as the score function. Since then, several works have looked at various geometric problems, like learning axis-aligned rectangles [4,44], where one can achieve optimal error bound under pure differential privacy using exponential mechanism; however, the case for approximate differential privacy is still open. There has been some recent work that studied differentially private geometric range queries (e.g., orthogonal range queries) under both the *central model* and *local model* of privacy [11,12,21,36,40,51,53].

Differentially Private Linear Queries. A fundamental class of queries studied in the literature of differential privacy are linear queries on a dataset [2,5–8,15,23–28,31,32,37,41,42,52]. Here, given a dataset from a data universe $\mathcal{U}$ of size d (usually represented in a form of a histogram $D \in \mathbb{R}^d$) and a query $q \in \mathbb{R}^d$, the goal is to estimate $q^\top D$. One can replace the query vector with a predicate $\phi : \mathcal{U}^n \rightarrow \{0,1\}$, where n is the size of the database, $D = \{d_1, \cdots, d_n\} \in \mathcal{U}^n$. The counting query is then simply $\sum_{i=1}^{n} \phi(d_i)$. Range queries can be seen as a special case of linear queries with a properly defined set of predicates.

The most relevant work to this paper is the work by Muthukrishnan and Nikolov [36], who proposed a differentially private mechanism for answering (generic) range queries when the ranges have bounded VC-dimension [36]. We can apply their techniques to get results for our setting of using shortest paths as ranges. Our algorithm can be easily extended to guarantee ε-differentially private with a slight change of parameters, while this substitution is non-trivial for the algorithm of Muthukrishnan and Nikolov [36], and to the best of our understanding, yields sub-optimal error bound. More discussion of this is in Sect. 6.

Private Release of Graph Data. Private release of graph data has been studied in recent years on many graph properties; see the survey [33]. There has been recent work on differentially private release of all pairs shortest path length [10,18,19,45]. Here, the edge weights w is considered sensitive, and the goal is to produce an approximate distance matrix for all pairs shortest paths length with differential privacy guarantees. In other words, the edge weights w *are* the sensitive attributes a. This is a harder problem than the problem considered in this paper. Specifically, the topology of the shortest paths are public information in our setting, but the knowledge of which edges are on the shortest path may reveal knowledge of the sensitive edge length w. It has been shown in [45] that when one releases the set of edges on an approximate shortest path in a differentially private manner, the additive error in the distance report has to be as large as $\Omega(n)$. The best known results for private release of all pairs shortest distance have an additive error of $\widetilde{O}(n^{2/3})$ for pure-DP and $\widetilde{O}(\sqrt{n})$ for approximate-DP [10,18,19] for general graphs. There is a lower bound of $\Omega(n^{1/6})$ for approximate-DP [10]. For trees the two problems are the same since for any two nodes the shortest path is unique regardless of edge length.

Differentially private range query on shortest paths has been done on a planar graph in [22], where they provide mechanisms with polylogarithmic additive error. But this problem has not been studied for the general graph setting.

2 Preliminaries

Notation. We use $\mathcal{G} = (\mathcal{V}, \mathcal{E})$ to denote a graph on vertex set $\mathcal{V}$ and edges $\mathcal{E}$. An edge $e \in \mathcal{E}$ is also denoted by the tuple (u, v) if u and v are its endpoints. For a pair of vertices (u, v), we denote $P(u, v)$ as their shortest path, and $d(u, v)$

as the shortest distance. We can define the attribute function $w : \mathcal{E} \to \mathbb{R}^m$ over all the edges *independent* of the shortest paths. On a path $P(u,v)$, we let $\gamma(u,v) := \min_e\{w(e)|e \in P(u,v)\}$ as the minimum attribute value along the shortest path $P(u,v)$. We use $\mathcal{R} = (X, \mathcal{S})$ to denote a set system, where $\mathcal{S}$ is a collection of sets with elements from X.

2.1 The Models for Range Query and Privacy

Shortest Paths as Ranges. Let $\mathcal{R} = (X, \mathcal{S})$ be a set system, where X is a set of elements and $\mathcal{S}$ is a collection of subsets $S_i \subseteq X$ called *ranges*. In a graph $\mathcal{G}$ when shortest paths are unique[1], we can define shortest paths as ranges. We take X to be the set of m edges in $\mathcal{G}$, and each set of $\mathcal{S}$ corresponds to a set of edges on a (u,v) shortest path. In particular, for an undirected graph $\mathcal{G}$, its corresponding $\mathcal{S}$ has $\binom{n}{2}$ order sets; and for a directed graph $\mathcal{G}$, $\mathcal{S}$ may have up to n^2 ordered sets.

Based on the set system $\mathcal{R} = (X, \mathcal{S})$, we can define *range queries* on $\mathcal{R}$ as $(\mathcal{R}, f)$ with a *query function* $f : \mathcal{S} \to \mathbb{R}$ as $\{f(S)\}_{S\in\mathcal{S}}$ for every set in $\mathcal{S}$. We can further extend this notion of range queries on shortest distances with *attribute* functions $w : X \to \mathbb{R}^{\geq 0}$, and the queries on each set S become $f(w(S))$, where $w(S)$ means to apply attribute function to each element in S. Note that the attribute function should *not* be considered as edge weights as it does not affect the shortest paths. Our goal is to release the statistics of *all* sets with small additive errors and privacy guarantees following the definitions in Definition 2.

We now formally define the privacy model for range queries on shortest paths.

Definition 1 (Range Queries with Neighboring Attributes). *Let $(\mathcal{R} = (X, \mathcal{S}), f)$ be a system of range queries, and let $w, w' : X \to \mathbb{R}^{\geq 0}$ be attribute functions that map each element in X to a non-negative real number. We say the attributes are* neighboring

$$\sum_{x\in X} |w(x) - w'(x)| \leq 1.$$

We emphasize that the attributes do not *change the shortest paths, i.e., the graphs operate on the same set system $\mathcal{R} = (X, \mathcal{S})$. When it is clear from context, we abuse the notation and denote the above by $\|w - w'\|_1 \leq 1$.*

We shall define the pure- and approximate DP with the notions of the neighboring attributes on range queries as follows.

Definition 2 (Differentially Private Range Queries). *Let $(\mathcal{R} = (X, \mathcal{S}), f)$ be a system of range queries and $w, w' : X \to \mathbb{R}^{\geq 0}$ be attribute functions as prescribed in Definition 1. Furthermore, let $\mathcal{A}$ be an algorithm that takes $(\mathcal{R}, f, w)$ as input. Then*

[1] One can use symbolic perturbation of edge distances to produce unique shortest paths.

$\mathcal{A}$ is (ε, δ)-differentially private on $\mathcal{G}$ if, for all pairs of neighboring attribute functions w, w' and all sets of possible outputs C, we have that

$$\Pr[\mathcal{A}(\mathcal{R}, f, w) \in C] \leq e^{\varepsilon} \cdot \Pr[\mathcal{A}(\mathcal{R}, f, w') \in C] + \delta.$$

If $\delta = 0$, we say $\mathcal{A}$ is ε-differentially private on $\mathcal{G}$.

We now define the notion that characterizes the *utility* of the algorithm. In the range query model, we say an algorithm $\mathcal{A}$ provides (α, β)-approximation to *all sets range queries* (ASRQ) if, given a range query system $(\mathcal{R} = (X, \mathcal{S}), f)$ and a attribute function w, with probability at least $1 - \beta$, algorithm $\mathcal{A}$ outputs an answer within an α additive error for the original query value on every set.

Definition 3 (Approximate-ASRQ). *We say a randomized algorithm $\mathcal{A}$ is an (α, β)-approximation for* all sets range queries (ASRQ) *on a range query system $(\mathcal{R} = (X, \mathcal{S}), f)$ with attribute function w if for any $S \in \mathcal{S}$,*

$$\Pr\left[|f(w(S)) - \mathcal{A}(w(S))| \leq \alpha\right] \geq 1 - \beta.$$

Since S contains the ranges of all-pairs shortest paths, the approximation in Definition 3 naturally corresponds to the additive approximation of shortest distances when f is the *counting query*. Trivially, if we output the range queries simply based on the elements and the attribute function w, we have $\alpha = \beta = 0$. However, such an output will *not* be private – and to guarantee both privacy and approximation is the main focus of this paper.

Remark 3. Our model of Definition 1 is closely related to the all-pair shortest distances release studied in [10,18,19,45]. In particular, in the model of private all-pair shortest distances, the neighboring graphs are also defined as the norm of attributes differing by at most 1. However, there is a subtle difference: in the shortest distances model, the shortest paths are private and subject to protection; while in the range query model, the shortest paths are known, and we do *not* have to protect their privacy. This allows us to bypass the $\Omega(n)$ additive error lower bound in [45] for any algorithm that privately reveal the shortest paths, and obtain much stronger results.

2.2 Standard Technical Tools

Tools from Probability Theory. We first introduce some well-known results from probability theory. We refer interested readers to the standard textbooks on this subject for more details [50].

Definition 4 (Laplace distribution) . *We say a zero-mean random variable X follows the Laplace distribution with parameter b (denoted by $X \sim \mathsf{Lap}(b)$) if the probability density function of X follows*

$$p(x) = \mathsf{Lap}(b)\,(x) = \frac{1}{2b} \cdot \exp\left(-\frac{|x|}{b}\right).$$

Definition 5 (Gaussian distribution). *We say a zero-mean random variable X follows the Gaussian distribution with variance σ^2 (denoted by $X \sim \mathcal{N}(0, \sigma^2)$) if the probability density function of X follows*

$$p(x) = \frac{1}{\sqrt{2\pi\sigma^2}} \cdot \exp\left(-\frac{x^2}{2\sigma^2}\right).$$

Both Laplace and Gaussian random variables have nice concentration properties. Furthermore, we can get stronger concentration results by the summation of both random variables [50].

Lemma 1 (Sum of Laplace random variables, [9,50]). *Let $\{X_i\}_{i=1}^m$ be a collection of independent random variables such that $X_i \sim \mathsf{Lap}(b_i)$ for all $1 \le i \le m$. Then, for $\nu \ge \sqrt{\sum_i b_i^2}$ and $0 < \lambda < \frac{2\sqrt{2}\nu^2}{b}$ for $b = \max_i \{b_i\}$,*

$$\Pr\left[\left|\sum_i X_i\right| \ge \lambda\right] \le 2 \cdot \exp\left(-\frac{\lambda^2}{8\nu^2}\right).$$

Lemma 2 (Sum of Gaussian random variables, [50]). *Let $\{X_i\}_{i=1}^m$ be a collection of independent random variables such that $X_i \sim \mathcal{N}(\mu, \delta^2)$ for all $1 \le i \le m$. Then,*

$$\Pr\left[\left|\frac{\sum_i X_i}{m} - \mu\right| \ge \lambda\right] \le 2 \cdot \exp\left(-\frac{m\lambda^2}{2\delta^2}\right).$$

Tools in Differential Privacy. We proceed to existing tools used frequently in differential privacy:

Definition 6 (Sensitivity). *Let $p \ge 1$. For any function $f : \mathcal{X} \to \mathbb{R}^k$ defined over a domain space $\mathcal{X}$, the ℓ_p-sensitivity of the function f is defined as*

$$\Delta_{f,p} = \max_{\substack{w,w' \in \mathcal{X} \\ w \sim w'}} \|f(w) - f(w')\|_p,$$

Here, $\|\mathbf{x}\|_p := \left(\sum_{i=1}^d |\mathbf{x}[i]|^p\right)^{1/p}$ is the ℓ_p-norm of the vector $\mathbf{x} \in \mathbb{R}^d$ and $\mathbf{x}[i]$ denote the i-th coordinate.

Based on Laplace distribution, we can now define Laplace mechanism – a standard DP mechanism that adds noise sampled from Laplace distribution with scale dependent on the ℓ_1-sensitivity of the function. The formal definition is as follows.

Definition 7 (Laplace mechanism). *For any function $f : \mathcal{X} \to \mathbb{R}^k$, the Laplace mechanism on input $w \in \mathcal{X}$ samples $Y_1, \ldots, Y_k$ independently from $\mathsf{Lap}(\frac{\Delta_{f,1}}{\varepsilon})$ and outputs*

$$M_\varepsilon(f) = f(w) + (Y_1, \ldots, Y_k).$$

The following privacy property of Laplace mechanism is known.

Proposition 1 (Laplace mechanism [15]). *The Laplace mechanism $M_\varepsilon(f)$ is ε-differentially private.*

Similar to Laplace mechanism, we can define the Gaussian mechanism:

Definition 8 (Gaussian mechanism). *For any function $f : \mathcal{X} \to \mathbb{R}^k$, the Gaussian mechanism on input $w \in \mathcal{X}$ samples $Y_1, \ldots, Y_k$ independently from the Gaussian distribution $\mathcal{N}\left(0, \frac{2\Delta_{f,2}^2 \log(1.25/\delta)}{\varepsilon^2}\right)$ and outputs*

$$M_\varepsilon(f) = f(w) + (Y_1, \ldots, Y_k).$$

The following privacy property of Gaussian mechanism is known.

Proposition 2 (Gaussian mechanism [14]). *For $\varepsilon \in (0,1)$, the Gaussian mechanism $M_{\varepsilon;\delta}(f)$ is (ε, δ)-differentially private.*

It is well-known that if a mechanism M provides (ε, δ)-DP output, any function g that takes the output of M as input is also (ε, δ)-DP. This is known as the *post-processing theorem*, formalized as follows.

Proposition 3 (Post-processing theorem [16]). *Let $M : \mathbb{R}^{d_1} \to \mathbb{R}^{d_2}$ be an (ε, δ)-differentially private mechanism and let $g : \mathbb{R}^{d_2} \to \mathbb{R}^{d_3}$ be an arbitrary function. Then, the function $g \circ M : \mathbb{R}^{d_1} \to \mathbb{R}^{d_3}$ is also (ε, δ)-differentially private.*

Finally, we introduce another useful property of differential privacy: privacy is preserved when combining multiple differentially private mechanisms even against adaptive adversary.

Proposition 4 (Composition theorem [15]). *For any $\varepsilon > 0$, the adaptive composition of k ε-differentially private algorithms is $k\varepsilon$-differentially private.*

Proposition 5 (Strong composition theorem [17]). *For any $\varepsilon, \delta \geq 0$ and $\delta' > 0$, the adaptive composition of k (ε, δ)-differentially private algorithms is $(\varepsilon', k\delta + \delta')$-differentially private for*

$$\varepsilon' = \sqrt{2k \ln(1/\delta')} \cdot \varepsilon + k\varepsilon(e^\varepsilon - 1).$$

Furthermore, if $\varepsilon' \in (0,1)$ and $\delta' > 0$, the composition of k ε-differentially private mechanism is (ε', δ')-differentially private for

$$\varepsilon' = \varepsilon \cdot \sqrt{8k \log(\frac{1}{\delta'})}.$$

The following proposition follows from strong composition theorem.

Proposition 6 (Corollary 3.21 in [16]). *Let $\mathcal{A}_1, \cdots, \mathcal{A}_k$ be k (ε', δ')-differentially private algorithm for*

$$\varepsilon' = \frac{\varepsilon}{\sqrt{8k \log(1/\delta)}}.$$

Then an algorithm $\mathcal{A}$ formed by adaptive composition of $\mathcal{A}_1, \cdots, \mathcal{A}_k$ is $(\varepsilon, k\delta' + \delta)$-differentially private.

3 An ε-DP Algorithm for Counting Queries

In the current and following section, we focus on private algorithms for the counting query function. As clarified in Remark 1, the algorithms using single-source shortest-path tree scheme can achieve ε and (ε, δ)-DP regime using only different parameters. However, we propose a different algorithmic idea for pure-DP algorithm, which shaves off a $\log^2 n$ factor. We formally state the results on ε-DP as follows.

Theorem 1. *For any $\varepsilon \geq 0$, there exists an ε-differentially private efficient algorithm that given a graph $\mathcal{G} = (\mathcal{V}, \mathcal{E}, w)$ as a range query system $(\mathcal{R} = (X, \mathcal{S}), f, w)$ such that $\mathcal{S}$ is the set of the shortest paths and f is the counting query, with high probability, outputs all pairs counting queries with additive error $O(\frac{n^{1/3}\log^{5/6} n}{\varepsilon})$. That is, the algorithm outputs an estimate $\widehat{f}(\cdot,\cdot)$ such that*

$$\Pr\left(\max_{u,v\in\mathcal{V}} |\widehat{f}(u,v) - f(u,v)| = O\left(\frac{n^{1/3}\log^{5/6} n}{\varepsilon}\right)\right) \geq 1 - \frac{1}{n}.$$

We start with some high-level intuitions. Our algorithm leverages both input-perturbation and output-perturbation, as mentioned in Sect. 1.2. A naive solution would be applying output-perturbation to the pair-wise counting queries for vertices in $\mathcal{S}$. However, the change of a single edge attribute may trigger the change of potentially all counting queries for vertices in $\mathcal{S}$. As such, by the composition theorem, we need to boost the privacy parameter by a factor of $|\mathcal{S}|^2$ since *each* counting query can change by 1. On the other hand, note that the ranges are shortest paths, which have special structures. With the standard assumption that all shortest paths are unique, two shortest paths only overlap by one common shortest path segment. Therefore instead of using output perturbation directly among vertices in $\mathcal{S}$, we will be better off by decomposing the shortest paths by how they overlap and privatize the decomposed segments. As will become evident, the size of decomposed segments is less than $|\mathcal{S}|^2$, hence the cumulative error is reduced.

To formalize the above intuition, we introduce the notion of *cut vertices* and *canonical segments*. Both notions are defined w.r.t a subset of vertices $\mathcal{S} \subseteq \mathcal{V}$. Informally, a vertex w becomes a cut vertex if it is a vertex of $\mathcal{S}$, or if it witnesses the branching – either 'merging' or 'splitting' – of two shortest paths between different pairs of vertices in $\mathcal{S}$. The formal definition is as follows.

Definition 9 (Cut Vertices). *Let $\mathcal{S} \subseteq \mathcal{V}$ be an arbitrary subset of vertices. For any pair of vertices $(u, v) \in \mathcal{S}$ and their shortest path $P(u, v)$, we say $w \in P(u, v)$ is a* cut vertex *for (u, v) if it satisfies* one *of the following two conditions:*

1. *$w \in \{u, v\}$;*
2. *$w \notin \{u, v\}$ and*
   (a) $w \in P(x, z)$ for some $(x, z) \in \mathcal{S}$ and $(x, z) \neq (u, v)$;

(b) *Without any loss of generality, suppose the path is from* x. *Let* pred(w) *be the vertex before* w *on* $P(x,z)$ *and* succ(w) *be the vertex after* w *on* $P(x,z)$. *Then either* pred$(w) \notin P(u,v)$ *or* succ$(w) \notin P(u,v)$.

See Fig. 1 (i) for an illustration of cut vertices. Based on Definition 9, we can now define the canonical segments as the path between two adjacent cut vertices along shortest paths of vertices in $\mathcal{S}$.

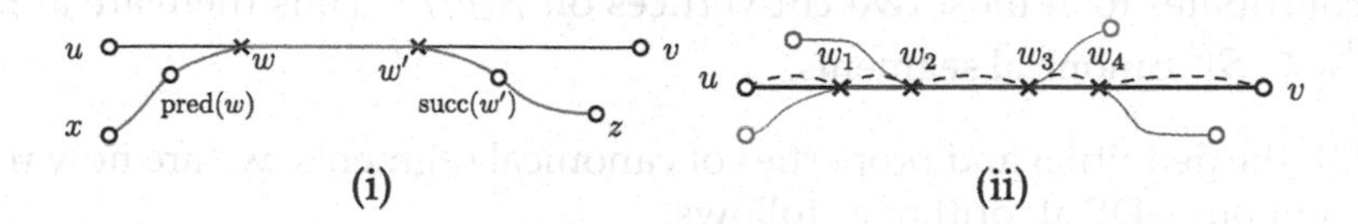

Fig. 1. (i) Two shortest paths $P(u,v)$ and $P(x,z)$, $u,v,x,z \in \mathcal{S}$, intersect at a common subpath as the shortest path between two cut vertices w, w'. (ii) The shortest path $P(u,v)$ is partitioned into canonical segments $P(u,w_1), P(w_1,w_2), \cdots, P(w_\ell, w_{\ell+1}), P(w_{\ell+1}, v)$, where $w_1, w_2, \cdots, w_{\ell+1}$ are (ordered) cut vertices along path $P(u,v)$.

Definition 10 (Canonical Segments). *Let* $\mathcal{S} \subseteq \mathcal{V}$ *be an arbitrary subset of vertices. For any pair of vertices* $(u,v) \in \mathcal{S}$ *and their shortest path* $P(u,v)$, *a subpath* $P(w,w')$ *of* $P(u,v)$ *is a* canonical segment *if*

1. w *is a cut vertex for some* $(x,z) \in \mathcal{S}$;
2. w' *is a cut vertex for some* $(x',z') \in \mathcal{S}$;
3. *None of the vertices between* w *and* w' *on* $P(u,v)$ *is a cut vertex for any* $(x'',z'') \in \mathcal{S}$.

Note that (u,v), (x,z), and (x',z') may or may not be the same in the above definition. One can think of the cut vertices as *all* vertices that witnesses the shortest path branching between *all* pairs of vertices in $\mathcal{S}$, and the canonical segments are exactly the collection of segments between adjacent cut vertices along shortest paths. See Fig. 1 (ii) for an example: $\{u, v, w_1, w_2, w_3, w_4\}$ are all cut vertices, which define 5 canonical segments.

For a fixed vertex pair $(u,v) \in \mathcal{S}$, we define $\mathsf{Canon}(\mathcal{S},u,v)$ as the set of canonical segments on the shortest path of (u,v). Note that the canonical segments need not to be among the edges between the vertices in $\mathcal{S}$: the shortest path between $(u,v) \in \mathcal{S}$ may well be outside of $\mathcal{S}$. We provide some observations about the basic properties of canonical segments.

Observation 3. *Canonical segments defined as in Definition 10 satisfy the following properties:*

1. *Any two canonical segments are* disjoint.
2. *The segments in Canon*$(\mathcal{S},u,v)$ *covers all edges in* $P(u,v)$, *i.e.*

$$P(u,v) = \cup_{P(x,z) \in \mathsf{Canon}(\mathcal{S},u,v)} P(x,z).$$

3. *For any pair of vertices* $(u, v) \in \mathcal{S}$, *there are at most* $|\mathcal{S}|^2$ *canonical segments in* $Canon(\mathcal{S}, u, v)$ *for* $|\mathcal{S}| \geq 2$.

Proof. Observation 1 is by definition. Concretely, if two canonical segments overlap, there must be one cut vertex inside another canonical segment, which is not possible by definition. Observation 2 follows from the fact that u and v themselves are cut vertices, and any other cut vertices on $P(u, v)$ only further divides the path. Finally, observation 3 holds since every pair of vertices in $\mathcal{S}$ contributes to at most two cut vertices on $P(u, v)$. Thus there are at most $2 \cdot \binom{|\mathcal{S}|}{2} \leq |\mathcal{S}|^2$ canonical segments.

With the definition and properties of canonical segments, we are now ready to present our ε-DP algorithm as follows.

CANON-APSD: **An ε-DP algorithm to release all pairs counting queries**
Input: An n vertices graph, $\mathcal{G} = (\mathcal{V}, \mathcal{E}, w)$ and privacy parameter $\varepsilon > 0$.

1. Sample a set $\mathcal{S}$ of $s = 100\zeta \cdot \log n$ vertices uniformly at random, where $\zeta = O(n^{1/3} \log^{-2/3} n)$
2. Compute all-pair shortest path for every vertex pair $(x, z) \in \mathcal{S}$ in $\mathcal{G}$, and let $P_{\mathcal{S}}$ be the set of the paths.
3. Compute $\mathsf{Canon}(\mathcal{S})$ based on the sampled vertices $\mathcal{S}$ and their shortest paths $P_{\mathcal{S}}$.
4. $\mathcal{S}$ **Perturbation:** For each canonical segment $P \in \mathsf{Canon}(\mathcal{S}, u, v)$, add an independent Laplace noise $\mathsf{Lap}(2/\varepsilon)$ to its shortest path length. Compute a function $f_{\mathcal{S}}(\cdot, \cdot)$ for counting queries between any vertices $(u, v) \in \mathcal{S}$, by summing up the noisy attributes of the canonical segments in $\mathsf{Canon}(\mathcal{S}, u, v)$.
5. **Non-$\mathcal{S}$ Perturbation:** For each edge in $\mathcal{G}$, add independent Laplace noise $\mathsf{Lap}(2/\varepsilon)$ to the edge attribute. For any vertices $u, v \in \mathcal{V}$, let $P(u, v)$ be the shortest path in $\mathcal{G}$ and $f'(u, v)$ be the sum of the noisy attributes of the edges along $P(u, v)$.
6. For each pair of vertices (u, v),
   - If there are at least two vertices in $P(u, v)$ that are in $\mathcal{S}$, let vertex x be the first one along $P(u, v)$ and z be the last one such that $x, z \in \mathcal{S}$, release $\widehat{f}(u, v) = f'(u, x) + f_{\mathcal{S}}(x, z) + f'(z, v)$.
   - Otherwise, release $\widehat{f}(u, v) = f'(u, v)$.

We now give the formal analysis of the privacy guarantee and bounds for the additive error.

3.1 Proof of Thorem 1

We start with an observation of the sensitivity of canonical segments. Since canonical segments do not overlap, the weight change of a single edge can only trigger changes of the shortest path distances of at most one canonical segment.

Claim. Fix any $\mathcal{S} \subseteq \mathcal{V}$, and let $g : (2^{\mathcal{V}}, 2^{\mathcal{E}}) \rightarrow \mathbb{R}^{|\mathsf{Canon}(\mathcal{S})|}$ be the function that computes the distances for canonical segments. Then, the ℓ_1 sensitivities for g is at most 1.

Proof. The claim follows from the fact that the canonical segments are disjoint (statement 1 of Lemma 3). Concretely, recall that for two neighboring graphs $\mathcal{G} \sim \mathcal{G}' \in \mathcal{X}$, we have

$$\sum_{e \in \mathcal{E}} |w(e) - w'(e)| \leq 1.$$

As such,

$$\Delta_{g,1} = \max_{\substack{w,w' \in \mathcal{X} \\ w \sim w'}} \left\| g(w) - g(w') \right\|_1 \leq \max_{\substack{w,w' \in \mathcal{X} \\ w \sim w'}} \left\| w - w' \right\|_1 \leq 1,$$

where the first inequality follows from the disjointness of canonical segments and the second inequality is by the neighboring graphs.

Notably, Sect. 3.1 is already sufficient for us to prove the *privacy* of the algorithm.

Lemma 4. *The* CANON-APSD *algorithm is* ε*-differentially private.*

Proof. We can simply use the (basic) composition theorem (Proposition 4) to obtained the desired privacy guarantee. Note that one can view $\mathcal{S}$ Perturbation and Non-$\mathcal{S}$ perturbation as two Laplace mechanisms as defined in Definition 7. As such, we only need to prove that both perturbation mechanisms are $O(\varepsilon)$-DP.

By Sect. 3.1, the functions in steps 4. is of ℓ_1 sensitivity at most 1. As such, by Propostion 1, its output is $\frac{\varepsilon}{2}$-DP. For the input perturbation, we are directly operating on the edge attributes. As such, we have $\|w - w'\|_1 \leq 1$. Therefore, by Proposition 1, the $\mathsf{Lap}(\frac{2}{\varepsilon})$ noise gives an $\varepsilon/2$-DP algorithm.

We now proceed to bounding the additive error, which follows a simple idea: we decompose the noise into different parts, and use the concentration of Laplace distribution to get the tight bound.

Lemma 5. *With high probability, for any vertex pair* $(u, v) \in V$*, the difference between* $f(u, v)$ *and* $\widehat{f}(u, v)$ *by* CANON-APSD *is* $O\left(\frac{1}{\varepsilon}\sqrt{\left(\frac{n}{\zeta} + \zeta^2 \log^2 n\right) \log n}\right)$*. More precisely,*

$$\left| f(u,v) - \widehat{f}(u,v) \right| \leq \frac{900}{\varepsilon} \cdot \sqrt{\left(\frac{n}{\zeta} + \zeta^2 \log^2 n\right) \cdot \log n}$$

for any $n \geq C \cdot \zeta \log n$ *where* C *is a sufficiently large absolute constant.*

Proof. We start with proving a structural lemma, which powers the algorithm to decompose the error into different parts to apply the concentration inequality of Laplace noise. The following lemma will be extensively used in the paper:

Lemma 6. *For any pair of vertices (u,v), if the number of edges on the shortest path $P(u,v)$, denoted by $|P(u,v)|$, is at least $\frac{n}{\zeta}$, then, with high probability, there exist at least two vertices $(x,z) \in P(u,v)$ such that*

1. *$x \in S$ and $z \in S$.*
2. *Suppose without any loss of generality, $|P(u,x)| \leq |P(u,z)|$, then the numbers of edges from u to x and from z to v are at most $\frac{n}{\zeta}$, i.e. $|P(u,x)| \leq \frac{n}{\zeta}$ and $|P(z,v)| \leq \frac{n}{\zeta}$.*

We defer the proof Lemma 6 to Appendix C. Now, coming back to the analysis on separate parts of additive error, fix a pair of vertices $(u,v) \in \mathcal{V}$ and their shortest path $P(u,v)$, the additive noises are:

1. At most $\frac{2n}{\zeta}$ independent noises sampled from $\mathsf{Lap}(\frac{2}{\varepsilon})$.
2. At most $s^2 = 100^2 \cdot \zeta^2 \cdot \log^2 n$ independent noises sampled from $\mathsf{Lap}(\frac{2}{\varepsilon})$ for the canonical segments.

The second line is obtained from statements 2 and 3 of Lemma 3: to compute the all-pairs shortest distances between pair in $\mathcal{S}$, it suffices to estimate the canonical segments, and there are at most s^2 many of them. As such, in the CANON-APSD algorithm, we let each Laplace noise be with variance $b_i = 2/\varepsilon$ for all i, we again pick $v = \sqrt{\sum_i b_i^2}$ and $\lambda = 30v\sqrt{\log n} = \frac{60}{\varepsilon} \cdot \sqrt{n \log n}$. Recall that $s = 100 \log n \cdot n^{1/3}$ (since $\zeta = n^{1/3}$), which implies $\frac{2\sqrt{2}v}{\max_i b_i} \geq 30\sqrt{\log n}$ (this only needs $n \geq C \cdot \zeta \log n$ for some constant C). Therefore, we can apply the concentration of Laplace tail in Lemma 1, which gives us

$$\Pr\left[\left|f(u,v) - \widehat{f}(u,v)\right| \geq 30\sqrt{\log n}\, v\right] \leq 2\exp\left(-\frac{900 \log n}{8}\right) \leq \frac{1}{n^3}.$$

Therefore, with probability $1 - \frac{1}{n^3}$,

$$\left|f(u,v) - \widehat{f}(u,v)\right| \leq 30\sqrt{\log n} \cdot v \leq \frac{90}{\varepsilon}\sqrt{\left(\frac{n}{\zeta} + 100^2 \cdot \zeta^2 \cdot \log^2 n\right) \cdot \log n}.$$

A union bound over the above event and the high probability event in Lemma 6 gives us the desired statement.

In fact, Lemma 5 holds for any $\zeta = n^{1-\Omega(1)}$ for sufficiently large n (as long as $n^{\Omega(1)} > 900 \log(n)$). We can now finalize the analysis of the additive error of the CANON-APSD algorithm.

Lemma 7. *With high probability, the* CANON-APSD *algorithm has an additive error of at most $O\left(\frac{n^{1/3}}{\varepsilon} \cdot \log^{5/6} n\right)$.*

Proof. We use Lemma 5 by setting the parameter $\zeta = \frac{1}{C} \cdot n^{1/3} \log^{-2/3} n$ with the C in Lemma 5. As such, the total additive error becomes

$$O\left(\frac{1}{\varepsilon} \cdot \sqrt{\left(\frac{n}{n^{1/3}\log^{-2/3} n} + (n^{1/3}\log^{-2/3} n)^2 \cdot \log^2 n\right) \cdot \log n}\right)$$
$$= O\left(\frac{n^{1/3}}{\varepsilon} \cdot \log^{5/6} n\right),$$

as claimed.

This concludes the proof of Theorem 1.

4 A Simple (ε, δ)-DP Algorithm for Counting Queries

Proceeding to the (ε, δ)-DP setting, we show that with the relaxation of approximate-DP, the worst case additive error can be reduced from $\widetilde{O}(n^{1/3})$ to $\widetilde{O}(n^{1/4})$, formally stated as follows.

Theorem 2. *For privacy parameters, $\varepsilon, \delta \in (0,1)$, there exists an (ε, δ)-differentially private efficient algorithm that given a graph $\mathcal{G} = (\mathcal{V}, \mathcal{E}, w)$ as a range query system $(\mathcal{R} = (X, \mathcal{S}), f, w)$ such that $\mathcal{S}$ is the set of the shortest paths and f is the counting query, with high probability, outputs all pairs counting queries with additive error $O\left(\frac{n^{1/4}\log^{2/3} n \log^{1/4} \frac{1}{\delta}}{\varepsilon}\right)$. That is, the algorithm outputs an estimate $\widehat{f}(\cdot,\cdot)$ such that*

$$\Pr\left(\max_{u,v\in\mathcal{V}} |\widehat{f}(u,v) - ft(u,v)| = O\left(\frac{n^{1/4}\log^{5/4} n \log^{1/2} \frac{1}{\delta}}{\varepsilon}\right)\right) \geq 1 - \frac{1}{n}.$$

At the high level, our algorithm builds single-source shortest path trees (see formal definition in Definition 11) for each vertex sampled uniformly at random, then employs an (ε, δ)-DP algorithm for distances release in the tree graph. Notice that the construction of single-source shortest-path trees follows from folklore algorithms based on Dijkstra's algorithm, which takes $O(m + n\log(n))$ time with the classical Fibonacci heap implementation. Further, our algorithm can be easily extended to guarantee ε-differentially private with slight change of parameters, while this substitution is non-trivial for the algorithm of Muthukrishnan and Nikolov [36], and to the best of our understanding, yields suboptimal error bound.

Definition 11 (Single-source shortest-path tree). *Given a graph $\mathcal{G} = (\mathcal{V}, \mathcal{E})$ and a vertex $s \in \mathcal{V}$, the single-source shortest-path tree rooted at s is a spanning tree $\mathcal{G}'$ such that the unique path from s to v in $\mathcal{G}'$ is the shortest path from s to v in $\mathcal{G}$.*

We will use the following result of the (ε, δ)-DP algorithm for tree graphs (see Appendix B).

Lemma 8 **((ε,δ)-DP for tree graph).** *Given a tree graph $\mathcal{G}=(\mathcal{V},\mathcal{E},w)$ and privacy parameter $\varepsilon,\delta\in(0,1)$, there exists an (ε,δ)-DP algorithm releasing shortest distances from the root vertex to the rest such that, with high probability, induce additive error at most $O(\frac{1}{\varepsilon}\log^{1.5} n\sqrt{\log(\frac{1}{\delta})})$.*

We have three remarks for Lemma 8. First for tree graphs, our problem and the private release of all pairs shortest distances are the same – since there is a unique path between any two vertices in a tree graph. Therefore private release of all pairs shortest distances in a tree graph can be used here. Prior work for this problem ([18,45]) focused on ε-DP. Between [18,45], Fan and Li's algorithm [18] uses heavy-light decomposition of the tree, with a better error bound only when the tree is shallow. Thus we present the version of (ε,δ)-DP based on Sealfon's algorithm [45]. Second, Seafon's algorithm exploits Laplace mechanism, which is replaced by Gaussian mechanism with $\sigma^2 := 1/\varepsilon^2\cdot\ln(1.25/\delta)\log n$ in Lemma 8. Third, the additive error bound for ε-DP on tree graph is $O(\frac{1}{\varepsilon}\log^{2.5} n)$ with high probability for single-source distance. Lemma 8 implies that the (ε,δ)-DP algorithm can shave off a $\log n$ factor, end up with a quadratic improvement on the logarithm term in the final algorithm for private all pairs shortest distances.

For simplicity, call the algorithm in Lemma 8 as $\mathsf{PrivateTree}(G)$ with an input tree graph G. Also we use $\mathsf{SSSP}(v)$ for the single-source shortest path tree algorithm, which takes any $v\in V$ as input and outputs a shortest path tree with v as the root. The (ε,δ)-DP algorithm is presented above.

SSSP-ASRQ: **An (ε,δ)-DP algorithm to release all pairs counting queries**
Input: An n vertices graph, $\mathcal{G}=(\mathcal{V},\mathcal{E},w)$ and privacy parameter $\varepsilon,\delta>0$.

1. Sample a set $\mathcal{S}$ of $s=\zeta\cdot\log n$ vertices uniformly at random, where $\zeta = O(\sqrt{n}\log^{-2.5} n)$.
2. For each vertex $v\in\mathcal{S}$, compute $T(v)=\mathsf{SSSP}(v)$. Call the set of all trees T.
3. $\mathcal{S}$ **Perturbation:** For each tree $T\in\mathcal{T}$, privatize it by running $\mathsf{PrivateTree}(T)$ with the Gaussian noise $\mathcal{N}(\mu=0,\sigma^2:=\frac{1}{\varepsilon_0^2}\ln(1.25/\delta_0)\log n)$, ε_0,δ_0 will be specified later, let the output of count query be $f_{\mathrm{T}}(u,v)$.
4. **Non-$\mathcal{S}$ Perturbation:** For each edge in $\mathcal{G}$ add independent Gaussian noise $\mathcal{N}(\mu=0,\sigma^2:=\frac{4}{\varepsilon^2}\ln(2.5/\delta)\log n)$. For any vertices $u,v\in\mathcal{V}$, let $P(u,v)$ be the shortest path in $\mathcal{G}$ and $f'(u,v)$ be the sum of the noisy attributes of the edges along $P(u,v)$.
5. For each pair of vertices (u,v)
   - If at least one of u,v is in $\mathcal{S}$, release $\widehat{f}(u,v)=f_{\mathrm{T}}(u,v)$.
   - If $u,v\notin\mathcal{S}$ and the path $P(u,v)$ has one vertex $x\in\mathcal{S}$, release $\widehat{f}(u,v)=f_{\mathrm{T}}(u,x)+f_{\mathrm{T}}(x,v)$.
   - Otherwise, release $\widehat{f}(u,v)=f'(u,v)$.

4.1 Proof of Theorem 2

Our analysis mainly hinges on the concentration of Laplace random variables (Lemma 1), a corollary (Proposition 6) of strong composition theorem (Proposition 5) and the observation that any shortest path with length larger than $\frac{n}{\zeta}$ goes through at least one vertex in the sampled set $\mathcal{S}$ with high probability (Lemma 6).

Lemma 9. *The* SSSP-ASRQ *algorithm is* (ε, δ)*-differentially private.*

Proof. First observe that any edge in $\mathcal{G}$ can only appear in at most s trees ($s = |\mathcal{S}|$), since we only build one single-source shortest path tree for each vertex in $\mathcal{S}$. Therefore, the PrivateTree algorithm (Lemma 8) is applied at most s times to any edge. In $\mathcal{S}$ perturbation, the Gaussian mechanism achieves $(\varepsilon_0, \delta_0)$-DP for each tree. Pick ε_0, δ_0 such that $\varepsilon_0 = \frac{\varepsilon}{4\sqrt{2s\ln(4/\delta)}}$ and $\delta_0 = \frac{\delta}{4s}$, using a corollary of strong composition theorem (Proposition 6) on s number of PrivateTree algorithms, we have that the $\mathcal{S}$ perturbation is $(\varepsilon/2, \delta/2)$-differentially private.

Combining with the Non-$\mathcal{S}$ perturbation, which is $(\varepsilon/2, \delta/2)$-differentially private, it is straightforward to see that the SSSP-ASRQ algorithm is (ε, δ)-differentially private.

The analysis of the additive error is again, similar as in Theorem 1 and Lemma 5. The only difference is that s takes various values to balance the contribution from output perturbation and the input perturbation, leading to different additive errors.

Lemma 10. *With high probability, the* SSSP-ASRQ *algorithm has additive error at most* $O(\frac{n^{1/4}}{\varepsilon} \cdot \log^{1.25} n\sqrt{\log \frac{1}{\delta}})$

Proof. We first show that with high probability, for any vertex pair $(u, v) \in V$, $\left|f(u,v) - \widehat{f}(u,v)\right|$ released by SSSP-ASRQ is at most

$$\max\left\{O(\sqrt{(n/\zeta)\log\frac{1}{\delta}}/\varepsilon), O\left(\sqrt{s\log\frac{2s}{\delta}}\cdot\log^{1.5} n\log\frac{1}{\delta}/\varepsilon\right)\right\}$$

.

Notice that the additive error is once again decomposed into noises from 'output perturbation' ($\mathcal{S}$ perturbation) and 'input perturbation' (Non-$\mathcal{S}$ perturbation). Fix a pair of vertices $(u, v) \in \mathcal{V}$ and denote their shortest path as $P(u, v)$. By Lemma 6 and Lemma 9, the additive noises must be either of the following two cases:

1. At most $\frac{2n}{\zeta}$ independent noises sampled from $\mathcal{N}(\mu, \sigma^2)$, with $\mu = 0, \sigma^2 := \frac{4}{\varepsilon^2}\ln(2.5/\delta)\log n$.
2. At most two independent noises induced by the PrivateTree algorithm, which is upper bounded by $O(\frac{2}{\varepsilon_0}\log^{1.5} n\sqrt{\log\frac{1}{\delta_0}})$.

The first case considers the third bullet point in Step 5. of the SSSP-ASRQ algorithm. From Lemma 6, we know that the additive error is the summation of at most $\frac{2n}{\zeta}$ independent Gaussian noises. The second case considers the first and second points in Step 5. of the SSSP-ASRQ algorithm, where $\widehat{f}(u,v)$ is decomposed into two distances output by the PrivateTree algorithm. Notice that only one of the two cases can happen, hence the additive error bound is the maximum of the two. This is different from the analysis in Lemma 5, where the two cases are combined together to construct the shortest paths. In the following, we give detailed upper bounds of the additive error of two terms.

We now apply the concentration of Gaussian tail (Lemma 2) for the first case,

$$\Pr\left[\left|f(u,v) - \widehat{f}(u,v)\right| \geq t\right] \leq 2\exp\left(-\frac{t^2}{2n/\zeta \cdot \delta^2}\right),$$

Let $t = (n/\zeta)^{1/4}\log^{0.5} n \cdot \delta$, the above probability is smaller than $\frac{1}{n^4}$. Apply union bound on all vertex pairs, then with high probability, $\left|\mathrm{d}(u,v) - \widehat{\mathrm{d}}(u,v)\right|$ for the first case is at most

$$t = \frac{(n/\zeta)^{1/2}\log^{0.5} n \cdot \delta}{s} = O\left(\frac{1}{\varepsilon}(n/\zeta)^{1/2}\sqrt{\log\frac{1}{\delta}}\right)$$

Next, we show the additive error in the second case. In the $\mathcal{S}$ perturbation that we pick the privacy parameter $(\varepsilon_0, \delta_0)$ for the Gaussian mechanism where $\varepsilon_0 = \frac{\varepsilon}{4\sqrt{2s\ln(2\delta)}}$ and $\delta_0 = \frac{\delta}{2s}$.

Recall Lemma 8, the additive error is at most

$$\frac{1}{\varepsilon_0}\log^{1.5} n\sqrt{\log\frac{1}{\delta_0}} = O\left(\frac{1}{\varepsilon}\sqrt{s\log(\frac{1}{\delta})}\cdot\log^{1.5} n\sqrt{\log\frac{2s}{\delta}}\right)$$

It only remains to balance the two terms to obtain the maximum additive error. Recall that $s = O(\zeta \cdot \log n)$, we pick $\zeta = C\sqrt{n}\log^{-2.5} n$, where C is a fixed constant, leading to the following additive error:

$$O\left(\frac{1}{\varepsilon}\sqrt{\frac{2n}{\zeta}}\cdot\log\frac{1}{\delta}\right) = O\left(n^{1/4}\log^{1.25} n\cdot\sqrt{\log\frac{1}{\delta}}\right)$$

5 Private Algorithms for the Bottleneck Edge Queries

We investigate the problem of private *bottleneck edge* queries under the range query model in this section. The problem has natural motivations in a bulk of applications where the resilience on the shortest path is quantified by a *bottleneck* attribute. For instance, in the Time-to-Stockout problem we discussed in

Sect. 1, the quantity of interest is usually the edge with the *minimum* value of the attribute among the shortest path. We show that we can release such information privately by simply applying the input perturbation technique. More formally, we have:

Theorem 3. *For privacy parameters $\varepsilon, \delta \in (0,1)$, there exist*

- *an ε-differentially private efficient algorithm that given a graph $\mathcal{G} = (\mathcal{V}, \mathcal{E}, w)$ as a range query system $(\mathcal{R} = (X, \mathcal{S}), f, w)$ such that $\mathcal{S}$ is the set of the shortest paths and f is the bottleneck query, with high probability, outputs all pairs bottleneck queries with additive error $O(\frac{\log n}{\varepsilon})$.*
- *an (ε, δ)-differentially private efficient algorithm that given a graph $\mathcal{G} = (\mathcal{V}, \mathcal{E}, w)$ as a range query system $(\mathcal{R} = (X, \mathcal{S}), f, w)$ such that $\mathcal{S}$ is the set of the shortest paths and f is the bottleneck query, with high probability, outputs all pairs bottleneck queries with additive error $O\left(\frac{\sqrt{\log n \log \frac{1}{\delta}}}{\varepsilon}\right)$.*

Remark 4. We remark that the bottleneck edge task cannot be trivially solved by the *top-k* selection problem in differential privacy (e.g. [13,35,43] and references therein). Note that although it is possible to directly apply top-1 selection to privately release the bottleneck edge on *a single* shortest path, the $O(n^2)$-many shortest paths may incur significant privacy loss if we simply use composition.

We now present the ε-DP and (ε, δ)-DP algorithms with the input perturbation technique first developed by [45]. Recall that $\gamma(u, v) = \min_{e \in P(u,v)} w(e)$ is the minimum edge weight on the shortest path between u and v. Both algorithms can be presented with only differences on a subroutine as follows.

Algorithms for minimum attribute edge on the shortest path

Input: A range system $\mathcal{R} = (X, \mathcal{S})$ and attribute function w, where X and w specifies a graph $\mathcal{G} = (\mathcal{V}, \mathcal{E}, w)$, and the ranges $\mathcal{S}$ specifies shortest paths; a privacy budget $\varepsilon, \delta \in (0, 1)$.

1. Perform the input perturbation depending on the application:
   - For ε-DP, use the Lap-perturb procedure: add Laplace noise $\mathsf{Lap}(\frac{1}{\varepsilon})$ to every output of w, and obtain $\widetilde{w}$.
   - For (ε, δ)-DP, use the Gaussian-perturb procedure: add Gaussian noise with $\sigma = \frac{\sqrt{2 \log(1.25/\delta)}}{\varepsilon}$ to every output of w, and obtain $\widetilde{w}$.
2. For each shortest path $S \in \mathcal{S}$, find $e_S^\star$ the edge with the minimum attribute on each shortest path $S \in \mathcal{S}$ with the *original* attribute function w, i.e. $e_S^\star = \text{argmin}_e \{w(e) \mid e \in S\}$.
3. Report $\widetilde{\gamma} = \widetilde{w}(e_S^\star)$ as the attribute of the bottleneck edge on each shortest path $S = P(u, v)$.

In other words, the whole algorithm can be framed as adding input noise to the attributes (Laplace noise for ε-DP and Gaussian noise for (ε, δ)-DP), identifying the bottleneck edge with the *original* attributes, and release the noisy attribute of that bottleneck edge. We now show that the algorithms are differentially private under their respective setting, and the additive error is small.

The Analysis of ε-DP Bottleneck Edge

The privacy guarantee follows from the input perturbation guarantee and the post-processing theorem (Proposition 3). More formally, we can show the following lemma.

Lemma 11. *The algorithm with* Lap-perturb *procedure is ε-differentially private.*

Proof. Let $f : \mathcal{E} \rightarrow \mathbb{R}^m$ be the attribute function. By the properties of neighboring attributes (Definition 1), it follows that the ℓ_1 sensitivity $\Delta_{f,1}$ is at most 1 since the total change of bottleneck edges can be at most 1. As such, by Propositio 1, the output of $\widetilde{w}$ is ε-DP. Since we only release information as post-processing of $\widetilde{w}$, by Proposition 3, the algorithm is ε-DP.

We now show that the additive error is bounded by $O(\frac{\log n}{\varepsilon})$ with high probability. The argument follows by using the concentration of Laplace distribution and union bound over poly(n) scenarios.

Lemma 12. *If the* Lap-perturb *procedure is applied, with high probability, for each pair of vertices (u, v), the difference between the output of $\widetilde{\gamma}(u, v)$ and the true bottleneck edge attribute $\gamma(u, v)$ is at most $O(\frac{\log n}{\varepsilon})$, i.e.*

$$\Pr\left(\max_{u,v\in\mathcal{V}} |\gamma(u,v) - \widetilde{\gamma}(u,v)| = O\left(\frac{\log n}{\varepsilon}\right)\right) \geq 1 - \frac{1}{n}.$$

Proof. For a fixed vertex pair (u, v), we need to take care of at most $n - 1$ edges on a shortest path. Note that for each edge (x, y) on the path $P(u, v)$, by the tail bound of Laplace distribution, the error induced by a single Laplace noise is at most $5\frac{\log n}{\varepsilon}$ with probability at least $1 - \frac{1}{n^5}$. As such, we have

$$\Pr\left(\max_{e\in P(u,v)} |w(e) - \widetilde{w}(e)| > 5 \cdot \frac{\log n}{\varepsilon}\right) \leq \frac{1}{n^4}.$$

Therefore, the additive error on the bottleneck edge is also at most $5 \cdot \frac{\log n}{\varepsilon}$ with probability at least $1 - \frac{1}{n^4}$. Applying a union bound over $\binom{n}{2}$ pairs gives us the desired statement.

The Analysis of (ε, δ)-DP Bottleneck Edge

We now turn to the algorithm for (ε, δ)-DP. Similar to the case in the ε-DP, we show that the approximate-DP [property holds by the Gaussian noise property and the post-processing theorem. The formal lemma can be shown as follows.

Lemma 13. *The algorithm with* Gaussian-perturb *procedure is* (ε, δ)*-differentially private.*

Proof. Similar to the proof of Lemma 11, we let attribute function $w : \mathcal{E} \rightarrow \mathbb{R}^m$ be the function of Definition 6. We can then bound the ℓ_2 sensitivity $\Delta_{f,2}$ of the attribute function by 1, again using the properties of neighboring attributes (Defintion 1). As such, by Proposition 2 and Proposition 3 and the right choice of σ, the algorithm is (ε, δ)-DP.

The benefit of allowing approximate-DP is a quadratic improvement on the additive error – conceptually, this follows straightforwardly by considering the lighter tail of the Gaussian distribution. We formalize the result as follows.

Lemma 14. *If the* Gaussian-perturb *procedure is applied, with high probability, for each pair of vertices* (u, v)*, the difference between the output of* $\widetilde{\gamma}(u, v)$ *and the true bottleneck edge attribute* $\gamma(u, v)$ *is at most* $O(\frac{\sqrt{\log n \log 1/\delta}}{\varepsilon})$*, i.e.*

$$\Pr\left(\max_{u,v \in \mathcal{V}} |\gamma(u,v) - \widetilde{\gamma}(u,v)| = O\left(\frac{\sqrt{\log n \log \frac{1}{\delta}}}{\varepsilon}\right)\right) \geq 1 - \frac{1}{n}.$$

Proof. Again, for a fixed vertex pair (u, v), there are at most $n - 1$ edges among a shortest path. Note that for each edge (x, y) on the path $P(u, v)$, by the tail bound of Gaussian distribution (Lemma 2), there is

$$\Pr\left(|\widetilde{w}((x,y)) - w((x,y))| > 5\sqrt{\log n}\sigma\right) \leq \exp(-10 \log n) \leq \frac{1}{n^5}.$$

As such, with probability at least $1 - \frac{1}{n^5}$, the attribute of a single edge is only different from the original with an additive error of $5\sqrt{\log n}\sigma$. Therefore, we have

$$\Pr\left(\max_{e \in P(u,v)} |w(e) - \widetilde{w}(e)| > 5\sqrt{\log n}\sigma\right) \leq \frac{1}{n^4}.$$

By the choice of σ, we have $5\sqrt{\log n}\sigma = O(\frac{\sqrt{\log n \log 1/\delta}}{\varepsilon})$. Applying a union bound over $\binom{n}{2}$ pairs gives us the desired statement.

6 VC-Dimension of Shortest Paths Ranges and Generic Algorithms

Under the range query context, it is possible to study the VC-dimension of shortest paths in a graph using a range system. The benefit of such a perspective is that one can apply generic algorithms for private range queries, most notably by the work of Muthukrishnan and Nikolov [36]. We discuss the problem from this perspective in this section.

Recall that we say a subset $A \subseteq X$ to be shattered by $\mathcal{S}$ if each of the subsets of A can be obtained as the intersection of some $S \in \mathcal{S}$ with A, i.e., if $\mathcal{S}|_A = 2^A$. The Vapnik-Chervonenkis (VC) d of a set system $(X, \mathcal{S})$ is defined as the size of the largest subset of X that can be shattered. Formally, the definition can be described as follows.

Definition 12 (Vapnik-Chervonenkis (VC) dimension). *Let $\mathcal{R} = (X, \mathcal{S})$ be a set system and let $A \subseteq X$ be a set. We say A is shattered by $\mathcal{S}$ if $\{S \cap A \mid S \in \mathcal{S}\} = 2^A$, i.e. the union of intersections between sets in S and A covers all subsets of A. The Vapnik-Chervonenkis (VC) dimension d of $\mathcal{R}$ is defined as the size of the largest $A \subseteq X$ that can be shattered by $\mathcal{S}$.*

In an undirected graph G, the VC-dimension of (unique) shortest paths[2] is 2 [1,47]. In a directed graph, the VC-dimension of (unique) shortest paths[3] is 3 [20].

A closely-related notion is the (primal) shatter function of a set system $\mathcal{R} = (X, \mathcal{S})$ (with parameter s), which is defined as the maximum number of distinct sets in $\{A \cap S \mid S \in \mathcal{S}\}$ for some $A \subseteq X$ such that $|A| = S$. More formally, the notion can be defined as follows.

Definition 13 (Primal Shatter Function). *Let $\mathcal{R} = (X, \mathcal{S})$ be a set system, and s be a positive integer. The primal shatter function of $\mathcal{R}$, denoted as $\pi_{\mathcal{R}}(s)$, is defined as* $\max_{A:\,|A|=s} |\{A \cap S \mid S \in \mathcal{S}\}|$

It is well known that if the VC-dimension of a range space is d, then $\pi_{\mathcal{R}}(s) = O(s^d)$ [34]. This immediately gives a bound of $O(s^2)$ for shortest paths in undirected graphs. We now show that shortest paths in directed graphs enjoys the same bound as well despite having a higher VC-dimension.

Lemma 15. *For a range query system $\mathcal{R} = (X, \mathcal{S})$ defined by shortest paths in (both directed and undirected) graphs, the primal shatter function is $\pi_{\mathcal{R}}(s) = O(s^2)$ for any positive integer s.*

---

[2] Any set of three vertices $\{u, v, w\}$ cannot be shattered: if one vertex w stays on the shortest path of the other two vertices u, v, then one cannot obtain the subset u, v; if none of them stays on the shortest path of the other two, then one cannot obtain the subset u, v, w .

[3] In a directed graph, a directed cycle of u, v, w can be shattered.

Proof. Take any set A of size s, any shortest path either does not contain any vertex in A, or contains a first vertex $x \in A$ and the last vertex $y \in A$ along the path. Notice that x, y might be the same vertex. Thus $\mathcal{S}|_A$ contains the subset of A as $A \cup S(x,y), \forall x, y \in A$ where $S(x,y)$ is the set of vertices on the shortest path from x to y. Therefore $\mathcal{S}|_A$ has at most $O(s^2)$ elements.

The benefit of understanding the VC-dimension and the primal shatter function for shortest system is that we can use generic algorithms for private range queries. In particular, Muthukrishnan and Nikolov [36] have developed a differentially private mechanism for answering range queries of bounded VC-dimension. The guarantee of the algorithm is as follows.

Proposition 7 (Muthukrishnan-Nikolov algorithm [36], rephrased). *Let $(\mathcal{R} = (X, \mathcal{S}), f)$ be a range query system, where f is the counting query and the primal shatter function of $\mathcal{R}$ is $\pi_{\mathcal{R}}(s) = O(s^d)$ for any s. There exists an algorithm that outputs* all queries *with*

- *Expected average squared error of* $O\left(\frac{n^{1-1/d}\log\frac{1}{\delta}}{\varepsilon^2}\right)$;
- *With probability at least* $1-\beta$, *worst case squared error of* $O\left(\frac{n^{1-1/d}\log\frac{1}{\delta}\log\frac{n}{\beta}}{\varepsilon^2}\right)$.

The algorithm is (ε, δ)-differentially private.

Using the algorithm of Proposition 7, the bound on the primal shatter functions of Lemma 15, and the fact that counting query sums up the attributes on the shortest paths, we can obtain an (ε, δ)-DP result with additive error $O(n^{1/4}\log^{1/2}(n)\log^{1/2}(1/\delta)/\varepsilon)$ with high constant probability. This matches our (ε, δ)-DP result in Theorem 2 up to lower order terms.

Remark 5. Although it is possible to recover the bound of Theroem 2 using Proposition 7 as a black-box, our constructions still enjoy multiple advantages. In particular, the construction of Proposition 7 does *not* give any non-trivial bound for ε-DP, and it is not trivial to adapt it to pure-DP within the framework. Furthermore, the algorithm of Proposition 7 requires to find a maximal set of ranges with the minimum symmetric differences on different levels, and by the packing lemma bound in [36], it appears that a straightforward implementation could take $\Theta(n^4)$ time in the worst case. On the other hand, our constructions for both Theroem 1 and Theroem 2 can be implemented in $\widetilde{O}(n^2)$ time. Finally, the algorithm of Proposition 7 is much more complicated and counter-intuitive, and our algorithm enjoys much better simplicity.

7 Conclusion and Future Work

We study the private release of shortest path queries under the range query context in this paper, where the graph topology and the shortest paths are public, and the attributes on the graphs (which do *not* affect shortest paths) are subject to privacy protection. Our upper bounds cannot be applied to the (harder) problem of private release of all pairs shortest distances [45]. Thus improving the bounds of private range query problem (with upper bound $\widetilde{O}(n^{1/3})$ for ε-DP and $\widetilde{O}(n^{1/4})$ for (ε, δ)-DP) and all pairs shortest distances release (with upper bound $\widetilde{O}(n^{2/3})$ for ε-DP and $\widetilde{O}(n^{1/2})$ for (ε, δ)-DP), where both have a lower bound of $\Omega(n^{1/6})$, remains an interesting open problem. Furthermore, since our algorithms are simple to implement, the empirical performances of our algorithms could be another future research direction.

Acknowledgements. We would like to thank Adam Sealfon, Shyam Narayanan, Justin Chen, Badih Ghazi, Ravi Kumar, Pasin Manurangsi, Jelani Nelson and Yinzhan Xu for useful discussion and suggestions. Deng and Gao have been partially supported by NSF through CCF-2118953, CNS-2137245, CCF-2208663, and CRCNS-2207440.

A Range Query on All Paths

When we allow queries along any path in a graph and require differential privacy guarantees, the following result provides a lower bound of $\Omega(n)$ on the additive error. To show the lower bound, we first consider a range query formulated by the incidence matrix A, with m columns corresponding to the m edges in the graph G and rows corresponding to all queries. A query along path P is represented by a row in the matrix with an element of 1 corresponding to edge e if e is on P and 0 otherwise. We will then talk about the discrepancy of matrix A.

The classical notion of discrepancy of a matrix A is the minimum value of $||Ax||_\infty$, where x is a vector with elements taking values $+1$ or -1. And the hereditary discrepancy of A is the maximum discrepancy of A limited on any subset of columns. As shown in [36], both discrepancy and hereditary discrepancy of A provides a lower bound on the additive error of differentially private range query using incidence matrix A.

Theorem 4. *A (ε, δ)-differential privacy mechanism that answers range queries where ranges are defined on any path of an input graph has to incur additive error of $\Omega(n)$.*

Proof. Consider a graph of $n+1$ vertices $v_1, v_2, \cdots, v_{n+1}$ and $2n$ edges. Between vertices v_i and v_{i+1} there are two parallel edges e_i and e_i'. On this graph there are 2^n paths from v_1 to v_{n+1}. We consider only queries along these paths and the incidence matrix is a tall matrix A of 2^n rows and $2n$ columns, corresponding to the $2n$ edges in the graph. Now we take a submatrix of A with only the columns corresponding to edges e_i. This gives a matrix A' of $2^n \times n$, with the rows corresponding to all subsets of $[n]$. A' has discrepancy of $\Omega(n)$. To see that, consider the specific vector x that minimizes $||A'x||_\infty$. Suppose x has k entries of $+1$ and $n-k$ entries of -1. Without loss of generality, we assume $k \geq n/2$, The row of A that has value 1 corresponding to the positive entries of x and value 0 corresponding to the negative entries of x, gives a value of $k \geq n/2$.

Thus $||Ax||_\infty$ is at least $n/2$. This means that the hereditary discrepancy of A is at least $\Omega(n)$.

By the same argument and use Corollary 1 in [36], we conclude that any (ε, δ)-differentially private mechanism has to have error of $\Omega(n)$.

B Proof of Lemma 8 – (ε, δ) Algorithm for Tree Graphs

Proof. We first claim that we can answer all pairs shortest distance on a tree with (α, β)-accuracy for

$$\alpha = O\left(\frac{1}{\varepsilon}\log n\sqrt{\log\left(\frac{n}{\beta}\right)\log\left(\frac{1}{\delta}\right)}\right)$$

showing the utility guarantee of Lemma 8. Specifically, if we wish to have high probability bounds for the shortest path distance errors, i.e., $\beta = O(1/n)$, the error is upper bounded by $O\left(\frac{1}{\varepsilon}\log^{1.5} n\sqrt{\log\left(\frac{1}{\delta}\right)}\right)$.

In Sealfon's algorithm [45], a tree rooted at v_0 is partitioned into subtrees each of at most $n/2$ vertices. Specifically, define v^* to be the vertex with at least $n/2$ descendants but none of v^*'s children has more than $n/2$ descendants. The tree is partitioned into the subtrees rooted at the children of v^*, and a subtree of the remaining vertices rooted at v_0. In Sealfon's algorithm a Laplace noise of $\mathsf{Lap}(\log n/\varepsilon)$ is added to the shortest path distance from v_0 to v^* and the edges from v^* to each of its children. The algorithm then repeatedly privatizes each of the subtrees recursively. Using Sealfon's algorithm, we know that for a given root node v_0, computing the single source (with the root being the source) shortest path distance requires adding at most $O(\log n)$ privatized edges. Further, their algorithm ensures that any edge can be in at most $\log n$ levels of recursion and hence can be used to compute $O(\log n)$ noisy answers. In other words, the number of adaptive compositions we need is $O(\log n)$.

We use the Gaussian mechanism to privatize the edges. Since we are concerned with approximate-DP guarantee, the variance of the noise required to preserve (ε, δ)-differential privacy is $\sigma^2 := O\left(\frac{1}{\varepsilon^2}\log(1/\delta)\log n\right)$.

Fix a node u. Let $\widehat{d}(u, v_0)$ be the distance estimated by using Sealfon's algorithm instantiated with the Gaussian mechanism instead of the Laplace mechanism. Now the noise added are zero mean. Therefore,

$$\mathbb{E}[\widehat{d}(u, v_0)] = d(u, v_0).$$

Using the standard concentration of Gaussian distribution [50] implies that

$$\mathsf{Pr}\left(\left|\widehat{d}(u, v_0) - \mathbb{E}\left[\widehat{d}(u, v_0)\right]\right| > a\right) \leq 2e^{-a^2/(2\sigma^2 \log n)}.$$

Setting $a = \frac{C}{\varepsilon} \log n \sqrt{\log\left(\frac{2n}{\beta}\right) \log\left(\frac{1}{\delta}\right)}$ for some constant $C > 0$, we have

$$\Pr\left(\left|\widehat{d}(u,v_0) - \mathbb{E}\left[\widehat{d}(u,v_0)\right]\right| > \frac{C}{\varepsilon} \log n \sqrt{\log\left(\frac{2n}{\beta}\right) \log\left(\frac{1}{\delta}\right)}\right)$$
$$\leq 2e^{-C\log(2n/\beta)} \leq \frac{\beta}{n}.$$

Now union bound gives that

$$\Pr\left(\max_{u \in \mathcal{V}} \left|\widehat{d}(u,v_0) - d(u,v_0)\right| \leq \frac{C}{\varepsilon} \log n \sqrt{\log\left(\frac{n}{\beta}\right) \log\left(\frac{1}{\delta}\right)}\right) \geq 1 - \beta.$$

We can now use the above result to answer all pair shortest paths by fixing a node v^* to be the root note and compute a single source shortest distance with the root node being the source node. Once we have all these estimates, to compute all pair shortest distance, for any two vertices, $(u, v) \in \mathcal{V} \times \mathcal{V}$, we first compute the least common ancestor z of u and v. We then compute the distance as follows:

$$\widehat{d}(u,v) = \widehat{d}(u,v^*) + \widehat{d}(v,v*) - 2\widehat{d}(z,v^*).$$

Since each of these estimates can be computed with an absolute error:

$$O\left(\frac{1}{\varepsilon} \log n \sqrt{\log\left(\frac{n}{\beta}\right) \log\left(\frac{1}{\delta}\right)}\right),$$

we get the final additive error bound. That is,

$$\Pr\left(\max_{u,v \in \mathcal{V}} |\widehat{\mathsf{d}}(u,v) - \mathsf{d}(u,v)| = O\left(\frac{1}{\varepsilon} \log n \sqrt{\log\left(\frac{n}{\beta}\right) \log\left(\frac{1}{\delta}\right)}\right)\right) \geq 1 - \beta$$

completing the proof of the claim.

C Proof of Lemma 6

Proof. (Proof of Lemma 6). The lemma is proved by a simple application of the Chernoff bound. For each path $P(u, v)$ with more than $\frac{n}{\zeta}$ edges, let v' be the $\left(\frac{n}{\zeta} + 1\right)$-th vertices on the path $P(u, v)$ from u. Similarly, let u' be the $\left(\frac{n}{\zeta} + 1\right)$-th vertices on the path $P(u, v)$ from v (traversing backward). We show that there must be two vertices sampled in S on both $P(u, v')$ and $P(u', v)$, which is sufficient to prove the lemma statement.

Define $X_{u,v'}$ as the random variable for the number of vertices on $P(u, v')$ that are sampled in S, and define X_z for each $z \in P(u, v')$ as the indicator random variable for z to be sampled in S. It is straightforward to see that

$X_{u,v'} = \sum_{z \in P(u,v')} X_z$. Since $P(u,v')$ has at least $\frac{n}{\zeta}$ vertices, and we are sampling $s = 100 \log n \cdot \zeta$ vertices uniformly at random as S, the expected number of vertices on $P(u,v')$ that are sampled is at least $100 \log n$. Formally, we have

$$\mathbb{E}\left[X_{u,v'}\right] \geq 100 \log n \cdot \frac{\zeta}{n} \cdot \frac{n}{\zeta} = 100 \log n.$$

As such, by applying the multiplicative Chernoff bound, we have

$$\Pr\left[X_{u,v'} \leq 2\right] \leq \exp\left(-\frac{0.8^2 \cdot 100 \log n}{3}\right) \leq \frac{1}{n^{10}}.$$

The same argument can be applied to $P(u',v)$ by defining $X_{u',v}$ as the total number of vertices that are sampled in S. We omit the repetitive details for simplicity. Finally, although the random variables for different (u,v) pairs are dependent, we can still apply a union bound regardless the dependence, and get the desired statement.

D A Remark on Range Query Shortest Path Lower Bound

For counting range queries with (ε,δ)-DP guarantee, there is a lower bound of $\Omega(n^{1/6})$ on the additive error, adapted from the construction of the lower bound for private all pairs shortest distances [10]. Specifically, the construction uses a graph where vertices are points in the plane and edges map to line segments between two points that do not contain other vertices. The edge length is the Euclidean length and therefore the shortest path between two vertices is the path corresponding to a straight line. The range query problem can be now formulated as a (special case) of linear queries, as in Sect. 6 and Sect. A, where the matrix A corresponds to the incidence matrix of the shortest paths and the edges in the graph. It is known that this matrix has a discrepancy lower bound of $\Omega(n^{1/6})$ [34]. By the connection of the discrepancy and linear query lower bounds [36], this is a lower bound for our problem.

References

1. Abraham, I., Delling, D., Fiat, A., Goldberg, A.V., Werneck, R.F.: VC-dimension and shortest path algorithms. In: Aceto, L., Henzinger, M., Sgall, J. (eds.) ICALP 2011. LNCS, vol. 6755, pp. 690–699. Springer, Heidelberg (2011). https://doi.org/10.1007/978-3-642-22006-7_58
2. Acs, G., Castelluccia, C., Chen, R.: Differentially private histogram publishing through lossy compression. In: 2012 IEEE 12th International Conference on Data Mining, pp. 1–10. IEEE (2012)
3. Beimel, A., Moran, S., Nissim, K., Stemmer, U.: Private center points and learning of halfspaces. In: Conference on Learning Theory, pp. 269–282. PMLR (2019)

4. Beimel, A., Nissim, K., Stemmer, U.: Private learning and sanitization: pure vs. approximate differential privacy. In: Raghavendra, P., Raskhodnikova, S., Jansen, K., Rolim, J.D.P. (eds.) APPROX/RANDOM -2013. LNCS, vol. 8096, pp. 363–378. Springer, Heidelberg (2013). https://doi.org/10.1007/978-3-642-40328-6_26
5. Bhaskara, A., Dadush, D., Krishnaswamy, R., Talwar, K.: Unconditional differentially private mechanisms for linear queries. In: Proceedings of the forty-fourth annual ACM Symposium on Theory of computing, pp. 1269–1284 (2012)
6. Blum, A., Dwork, C., McSherry, F., Nissim, K.: Practical privacy: the SuLQ framework. In: Proceedings of the Twenty-Fourth ACM SIGMOD-SIGACT-SIGART Symposium on Principles of Database Systems, pp. 128–138 (2005)
7. Blum, A., Ligett, K., Roth, A.: A learning theory approach to noninteractive database privacy. J. ACM **60**(2), 12 (2013)
8. Bun, M., Ullman, J., Vadhan, S.: Fingerprinting codes and the price of approximate differential privacy. SIAM J. Comput. **47**(5), 1888–1938 (2018)
9. Chan, T.H.H., Shi, E., Song, D.: Private and continual release of statistics. ACM Trans. Inf. Syst. Secur. (TISSEC) **14**(3), 1–24 (2011)
10. Chen, J.Y., et al.: Differentially private all-pairs shortest path distances: improved algorithms and lower bounds. In: 2023 Symposium on Discrete Algorithm (SODA 2023) (2023)
11. Cormode, G., Kulkarni, T., Srivastava, D.: Answering range queries under local differential privacy. Proc. VLDB Endowment **12**(10), 1126–1138 (2019)
12. Cormode, G., Procopiuc, C., Srivastava, D., Shen, E., Yu, T.: Differentially private spatial decompositions. In: 2012 IEEE 28th International Conference on Data Engineering, pp. 20–31. IEEE (2012)
13. Durfee, D., Rogers, R.M.: Practical differentially private top-k selection with pay-what-you-get composition. In: Wallach, H.M., Larochelle, H., Beygelzimer, A., d'Alché-Buc, F., Fox, E.B., Garnett, R. (eds.) Advances in Neural Information Processing Systems, vol. 32: Annual Conference on Neural Information Processing Systems 2019, NeurIPS 2019 (December), pp. 8–14. Vancouver, BC, Canada, pp. 3527–3537 (2019). https://proceedings.neurips.cc/paper/2019/hash/b139e104214a08ae3f2ebcce149cdf6e-Abstract.html
14. Dwork, C., Kenthapadi, K., McSherry, F., Mironov, I., Naor, M.: Our data, ourselves: privacy via distributed noise generation. In: Vaudenay, S. (ed.) EUROCRYPT 2006. LNCS, vol. 4004, pp. 486–503. Springer, Heidelberg (2006). https://doi.org/10.1007/11761679_29
15. Dwork, C., McSherry, F., Nissim, K., Smith, A.: Calibrating noise to sensitivity in private data analysis. J. Priv. Confidentiality **7**(3), 17–51 (2016)
16. Dwork, C., Roth, A.: The algorithmic foundations of differential privacy. Found. Trends® Theor. Comput. Sci. **9**(3–4), 211–407 (2014)
17. Dwork, C., Rothblum, G.N., Vadhan, S.: Boosting and differential privacy. In: 2010 IEEE 51st Annual Symposium on Foundations of Computer Science, pp. 51–60. IEEE (2010)
18. Fan, C., Li, P.: Distances release with differential privacy in tree and grid graph. arXiv preprint arXiv:2204.12488 (2022)
19. Fan, C., Li, P., Li, X.: Breaking the linear error barrier in differentially private graph distance release. arXiv preprint arXiv:2204.14247 (2022)
20. Funke, S., Nusser, A., Storandt, S.: On k-path covers and their applications. Proc. VLDB Endowment **7**(10), 893–902 (2014)
21. Ghane, S., Kulik, L., Ramamoharao, K.: A differentially private algorithm for range queries on trajectories. Knowl. Inf. Syst. **63**(2), 277–303 (2021)

22. Ghosh, A., Ding, J., Sarkar, R., Gao, J.: Differentially private range counting in planar graphs for spatial sensing. In: Proceedings of the 39th Annual IEEE International Conference on Computer Communications (INFOCOM 2020), pp. 2233–2242 (2020)
23. Gupta, A., Roth, A., Ullman, J.: Iterative constructions and private data release. In: Cramer, R. (ed.) TCC 2012. LNCS, vol. 7194, pp. 339–356. Springer, Heidelberg (2012). https://doi.org/10.1007/978-3-642-28914-9_19
24. Hardt, M., Ligett, K., McSherry, F.: A simple and practical algorithm for differentially private data release. In: Advances in Neural Information Processing Systems, vol. 25 (2012)
25. Hardt, M., Rothblum, G.N.: A multiplicative weights mechanism for privacy-preserving data analysis. In: 2010 51st Annual IEEE Symposium on Foundations of Computer Science (FOCS), pp. 61–70. IEEE (2010)
26. Hardt, M., Talwar, K.: On the geometry of differential privacy. In: Proceedings of the Forty-Second ACM Symposium on Theory of Computing, pp. 705–714. ACM (2010)
27. Hay, M., Li, C., Miklau, G., Jensen, D.: Accurate estimation of the degree distribution of private networks. In: 2009 Ninth IEEE International Conference on Data Mining, pp. 169–178. IEEE (2009)
28. Hay, M., Rastogi, V., Miklau, G., Suciu, D.: Boosting the accuracy of differentially-private histograms through consistency. arXiv preprint arXiv:0904.0942 (2009)
29. Hong, Y.C., Chen, J.: Graph database to enhance supply chain resilience for industry 4.0. IJISSCM **15**(1), 1–19 (2022)
30. Kaplan, H., Mansour, Y., Stemmer, U., Tsfadia, E.: Private learning of halfspaces: simplifying the construction and reducing the sample complexity. Adv. Neural. Inf. Process. Syst. **33**, 13976–13985 (2020)
31. Li, C., Hay, M., Rastogi, V., Miklau, G., McGregor, A.: Optimizing linear counting queries under differential privacy. In: Proceedings of the Twenty-Ninth ACM SIGMOD-SIGACT-SIGART Symposium on Principles of Database Systems, pp. 123–134. ACM (2010)
32. Li, C., Miklau, G.: Optimal error of query sets under the differentially-private matrix mechanism. In: Proceedings of the 16th International Conference on Database Theory, pp. 272–283 (2013)
33. Li, Y., Purcell, M., Rakotoarivelo, T., Smith, D., Ranbaduge, T., Ng, K.S.: Private graph data release: a survey. ACM Comput. Surv. **55**(11), 1–39 (2023). https://doi.org/10.1145/3569085
34. Matoušek, J.: Geometric Discrepancy. Springer, Berlin Heidelberg (1999). https://doi.org/10.1007/978-3-642-03942-3
35. McSherry, F., Talwar, K.: Mechanism design via differential privacy. In: 48th Annual IEEE Symposium on Foundations of Computer Science (FOCS 2007), pp. 94–103. IEEE (2007)
36. Muthukrishnan, S., Nikolov, A.: Optimal private halfspace counting via discrepancy. In: Proceedings of the Forty-Fourth Annual ACM Symposium on Theory of Computing, pp. 1285–1292 (2012)
37. Nikolov, A., Talwar, K., Zhang, L.: The geometry of differential privacy: the sparse and approximate cases. In: Proceedings of the Forty-Fifth Annual ACM Symposium on Theory of Computing, pp. 351–360 (2013)
38. Ogbuke, N.J., Yusuf, Y.Y., Dharma, K., Mercangoz, B.A.: Big data supply chain analytics: ethical, privacy and security challenges posed to business, industries and society. Prod. Plan. Control **33**(2–3), 123–137 (2022)
39. Pourhabibi, T., Ong, K.L., Kam, B.H., Boo, Y.L.: Fraud detection: a systematic literature review of graph-based anomaly detection approaches. Decis. Support Syst. **133**, 113303 (2020)

40. Qardaji, W., Yang, W., Li, N.: Differentially private grids for geospatial data. In: 2013 IEEE 29th International Conference on Data Engineering (ICDE), pp. 757–768. IEEE (2013)
41. Qardaji, W., Yang, W., Li, N.: Understanding hierarchical methods for differentially private histograms. Proc. VLDB Endowment **6**(14), 1954–1965 (2013)
42. Qardaji, W., Yang, W., Li, N.: Priview: practical differentially private release of marginal contingency tables. In: Proceedings of the 2014 ACM SIGMOD International Conference on Management of Data, pp. 1435–1446 (2014)
43. Qiao, G., Su, W.J., Zhang, L.: Oneshot differentially private top-k selection. In: Meila, M., Zhang, T. (eds.) Proceedings of the 38th International Conference on Machine Learning, ICML 2021, 18–24 July 2021, Virtual Event. Proceedings of Machine Learning Research, vol. 139, pp. 8672–8681. PMLR (2021). http://proceedings.mlr.press/v139/qiao21b.html
44. Sadigurschi, M., Stemmer, U.: On the sample complexity of privately learning axis-aligned rectangles. Adv. Neural. Inf. Process. Syst. **34**, 28286–28297 (2021)
45. Sealfon, A.: Shortest paths and distances with differential privacy. In: Proceedings of the 35th ACM SIGMOD-SIGACT-SIGAI Symposium on Principles of Database Systems, pp. 29–41 (2016)
46. Sharma, S., Chen, K., Sheth, A.: Toward practical privacy-preserving analytics for IoT and cloud-based healthcare systems. IEEE Internet Comput. **22**(2), 42–51 (2018)
47. Tao, Y., Sheng, C., Pei, J.: On k-skip shortest paths. In: Proceedings of the 2011 ACM SIGMOD International Conference on Management of data, SIGMOD 2011, pp. 421–432. Association for Computing Machinery, New York (2011)
48. Toth, C.D., O'Rourke, J., Goodman, J.E.: Handbook of Discrete and Computational Geometry (2017)
49. Tukey, J.W.: Mathematics and the picturing of data. In: Proceedings of the International Congress of Mathematicians, Vancouver, 1975, vol. 2, pp. 523–531 (1975)
50. Wainwright, M.J.: High-Dimensional Statistics: A Non-asymptotic Viewpoint, vol. 48. Cambridge University Press, Cambridge (2019)
51. Xiao, X., Wang, G., Gehrke, J.: Differential privacy via wavelet transforms. IEEE Trans. Knowl. Data Eng. **23**(8), 1200–1214 (2010)
52. Xiao, Y., Xiong, L., Fan, L., Goryczka, S.: DPCube: Differentially private histogram release through multidimensional partitioning. arXiv preprint arXiv:1202.5358 (2012)
53. Zhang, J., Xiao, X., Xie, X.: Privtree: A differentially private algorithm for hierarchical decompositions. In: Proceedings of the 2016 International Conference on Management of Data, pp. 155–170 (2016)

Revisiting Graph Persistence for Updates and Efficiency

Tamal K. Dey[1(✉)], Tao Hou[2(✉)], and Salman Parsa[2(✉)]

[1] Purdue University, West Lafayette, IN 47907, USA
tamaldey@purdue.edu
[2] DePaul University, Chicago, IL 60604, USA
{thou1,s.parsa}@depaul.edu

Abstract. It is well known that ordinary persistence on graphs can be computed more efficiently than the general persistence. Recently, it has been shown that zigzag persistence on graphs also exhibits similar behavior. Motivated by these results, we revisit graph persistence and propose efficient algorithms especially for local updates on filtrations, similar to what is done in ordinary persistence for computing the *vineyard*. We show that, for a filtration of length m, (i) switches (transpositions) in ordinary graph persistence can be done in $O(\log m)$ time; (ii) zigzag persistence on graphs can be computed in $O(m \log m)$ time, which improves a recent $O(m \log^4 n)$ time algorithm assuming n, the size of the union of all graphs in the filtration, satisfies $n \in \Omega(m^{\varepsilon})$ for any fixed $0 < \varepsilon < 1$; (iii) open-closed, closed-open, and closed-closed bars in dimension 0 for graph zigzag persistence can be updated in $O(\log m)$ time, whereas the open-open bars in dimension 0 and closed-closed bars in dimension 1 can be done in $O(\sqrt{m} \log m)$ time.

Keywords: Zigzag persistence · graph persistence · dynamic update · vines and vineyard

1 Introduction

Computing persistence for graphs has been a special focus within topological data analysis (TDA) [9,10] because graphs are abundant in applications and they admit more efficient algorithms than general simplicial complexes. It is well known that the persistence algorithm on a graph filtration with m additions can be implemented with a simple Union-Find data structure in $O(m\,\alpha(m))$ time, where $\alpha(m)$ is the inverse Ackermann's function (see e.g. [9]). On the other hand, the general-purpose persistence algorithm on a simplicial filtration comprising m simplices runs in $O(m^{\omega})$ time [18], where $\omega < 2.373$ is the exponent for matrix multiplication. In a similar vein, Yan et al. [21] have recently shown that extended persistence [4] for graphs can also be computed more efficiently in $O(m^2)$ time. The zigzag version [2] of the problem also exhibits similar behavior; see e.g. the survey [1]. Even though the general-purpose zigzag persistence algorithm runs

This work is partially supported by NSF grant CCF 2049010.

P. Morin and S. Suri (Eds.): WADS 2023, LNCS 14079, pp. 371–385, 2023.
https://doi.org/10.1007/978-3-031-38906-1_24

in $O(m^\omega)$ time on a zigzag filtration with m additions and deletions [3,8,17,18], a recent result in [6] shows that graph zigzag persistence can be computed in $O(m \log^4 n)$ time using some appropriate dynamic data structures [14,16] (n is the size of the union of all graphs in the filtration).

Motivated by the above developments, we embark on revisiting the graph persistence and find more efficient algorithms using appropriate dynamic data structures, especially in the dynamic settings [5,7]. In a dynamic setting, the graph filtration changes, and we are required to update the barcode (persistence diagram) accordingly. For general simplicial complexes as input, the *vineyard* algorithm of [5] updates the barcode in $O(m)$ time for a *switch* of two consecutive simplices (originally called a *transposition* in [5]). So, we ask if a similar update can be done more efficiently for a graph filtration. We show that, using some appropriate dynamic data structures, indeed we can execute such updates more efficiently. Specifically, we show the following:

1. In a standard (non-zigzag) graph filtration comprising m additions, a switch can be implemented in $O(\log m)$ time with a preprocessing time of $O(m \log m)$. See Sect. 3. As a subroutine of the update algorithm for switches on graph filtrations, we propose an update on the merge trees (termed as *merge forest* in this paper) of the filtrations, whose complexity is also $O(\log m)$.
2. The barcode of a graph zigzag filtration comprising m additions and deletions can be computed in $O(m \log m)$ time. Assuming $n \in \Omega(m^\varepsilon)$ for any fixed positive $\varepsilon < 1$, where n is the size of the union of all graphs in the filtration, this is an improvement over the $O(m \log^4 n)$ complexity of the algorithm in [6]. See Sect. 4. Also, our current algorithm using Link-Cut tree [20] is much easier to implement than the algorithm in [6] using the Dynamic Minimum Spanning Forest [16].
3. For switches [7] on graph zigzag persistence, the *closed-closed* intervals in dimension 0 can be maintained in $O(1)$ time; the *closed-open* and *open-closed* intervals, which appear only in dimension 0, can be maintained in $O(\log m)$ time; the *open-open* intervals in dimension 0 and *closed-closed* intervals in dimension 1 can be maintained in $O(\sqrt{m} \log m)$ time. All these can be done with an $O(m^{1.5} \log m)$ preprocessing time. See Sect. 5.

2 Preliminaries

Graph Zigzag Persistence. A *graph zigzag filtration* is a sequence of graphs

$$\mathcal{F} : G_0 \leftrightarrow G_1 \leftrightarrow \cdots \leftrightarrow G_m, \tag{1}$$

in which each $G_i \leftrightarrow G_{i+1}$ is either a forward inclusion $G_i \hookrightarrow G_{i+1}$ or a backward inclusion $G_i \hookleftarrow G_{i+1}$. For computation, we only consider *simplex-wise* filtrations starting and ending with *empty* graphs in this paper, i.e., $G_0 = G_m = \varnothing$ and each inclusion $G_i \leftrightarrow G_{i+1}$ is an addition or deletion of a single vertex or edge (both called a *simplex*). Such an inclusion is sometimes denoted as $G_i \xleftrightarrow{\sigma} G_{i+1}$ with σ indicating the vertex or edge being added or deleted. The p-th homology functor ($p = 0, 1$) applied on $\mathcal{F}$ induces a *zigzag module*:

$$\mathsf{H}_p(\mathcal{F}) : \mathsf{H}_p(G_0) \leftrightarrow \mathsf{H}_p(G_1) \leftrightarrow \cdots \leftrightarrow \mathsf{H}_p(G_m),$$

in which each $\mathsf{H}_p(G_i) \leftrightarrow \mathsf{H}_p(G_{i+1})$ is a linear map induced by inclusion. It is known [2,13] that $\mathsf{H}_p(\mathcal{F})$ has a decomposition of the form $\mathsf{H}_p(\mathcal{F}) \simeq \bigoplus_{k\in\Lambda} \mathcal{I}^{[b_k,d_k]}$, in which each $\mathcal{I}^{[b_k,d_k]}$ is an *interval module* over the interval $[b_k, d_k]$. The multiset of intervals $\mathsf{Pers}_p(\mathcal{F}) := \{[b_k, d_k] \mid k \in \Lambda\}$ is an invariant of $\mathcal{F}$ and is called the *p-th barcode* of $\mathcal{F}$. Each interval in $\mathsf{Pers}_p(\mathcal{F})$ is called a *p-th persistence interval* and is also said to be in dimension p. Frequently in this paper, we consider the barcode of $\mathcal{F}$ in all dimensions $\mathsf{Pers}_*(\mathcal{F}) := \bigsqcup_{p=0,1} \mathsf{Pers}_p(\mathcal{F})$.

Standard Persistence and Simplex Pairing. If all inclusions in Eq. (1) are forward, we have a *standard* (*non-zigzag*) graph filtration. We also only consider standard graph filtrations that are simplex-wise and start with empty graphs. Let $\mathcal{F}$ be such a filtration. It is well-known [11] that $\mathsf{Pers}_*(\mathcal{F})$ is generated from a *pairing* of simplices in $\mathcal{F}$ s.t. for each pair (σ, τ) generating a $[b, d) \in \mathsf{Pers}_*(\mathcal{F})$, the simplex σ creating $[b, d)$ is called *positive* and τ destroying $[b, d)$ is called *negative*. Notice that d may equal ∞ for a $[b, d) \in \mathsf{Pers}_*(\mathcal{F})$, in which case $[b, d)$ is generated by an *unpaired* positive simplex. For a simplex σ added from G_i to G_{i+1} in $\mathcal{F}$, we let its *index* be i and denote it as $\mathrm{idx}_{\mathcal{F}}(\sigma) := i$. For another simplex τ added in $\mathcal{F}$, if $\mathrm{idx}_{\mathcal{F}}(\sigma) < \mathrm{idx}_{\mathcal{F}}(\tau)$, we say that σ is *older* than τ and τ is *younger* than σ.

Merge Forest. Merge forests (more commonly called merge trees) encode the evolution of connected components in a standard graph filtration [10,19]. We rephrase its definition below:

Definition 1 (Merge forest). *For a simplex-wise standard graph filtration*

$$\mathcal{F} : \varnothing = G_0 \xhookrightarrow{\sigma_0} G_1 \xhookrightarrow{\sigma_1} \cdots\cdots \xhookrightarrow{\sigma_{m-1}} G_m,$$

its ***merge forest*** $\mathsf{MF}(\mathcal{F})$ *is a forest (acyclic undirected graph) where the leaves correspond to vertices in* $\mathcal{F}$ *and the internal nodes correspond to negative edges in* $\mathcal{F}$*. Moreover, each node in* $\mathsf{MF}(\mathcal{F})$ *is associated with a* **level** *which is the index of its corresponding simplex in* $\mathcal{F}$*. Let* $\mathsf{MF}^i(\mathcal{F})$ *be the subgraph of* $\mathsf{MF}(\mathcal{F})$ *induced by nodes at levels less than* i*. Notice that trees in* $\mathsf{MF}^i(\mathcal{F})$ *bijectively correspond to connected components in* G_i*. We then constructively define* $\mathsf{MF}^{i+1}(\mathcal{F})$ *from* $\mathsf{MF}^i(\mathcal{F})$*, starting with* $\mathsf{MF}^0(\mathcal{F}) = \varnothing$ *and ending with* $\mathsf{MF}^m(\mathcal{F}) = \mathsf{MF}(\mathcal{F})$*. Specifically, for each* $i = 0, 1, \ldots, m-1$*, do the following:*

- σ_i **is a vertex:** $\mathsf{MF}^{i+1}(\mathcal{F})$ *equals* $\mathsf{MF}^i(\mathcal{F})$ *union an isolated leaf at level* i *corresponding to* σ_i*.*
- σ_i **is a positive edge:** *Set* $\mathsf{MF}^{i+1}(\mathcal{F}) = \mathsf{MF}^i(\mathcal{F})$*.*
- σ_i **is a negative edge:** *Let* $\sigma_i = (u, v)$*. Since* u *and* v *are in different connected components* C_1 *and* C_2 *in* G_i*, let* T_1, T_2 *be the trees in* $\mathsf{MF}^i(\mathcal{F})$ *corresponding to* C_1, C_2 *respectively. To form* $\mathsf{MF}^{i+1}(\mathcal{F})$*, we add an internal node at level* i *(corresponding to* σ_i*) to* $\mathsf{MF}^i(\mathcal{F})$ *whose children are the roots of* T_1 *and* T_2*.*

In this paper, we do not differentiate a vertex or edge in $\mathcal{F}$ and its corresponding node in $\mathsf{MF}(\mathcal{F})$.

3 Updating Standard Persistence on Graphs

The switch operation originally proposed in [5] for general filtrations looks as follows on standard graph filtrations:

$$\begin{array}{l} \mathcal{F} : \varnothing = G_0 \hookrightarrow \cdots \hookrightarrow G_{i-1} \overset{\sigma}{\hookrightarrow} G_i \overset{\tau}{\hookrightarrow} G_{i+1} \hookrightarrow \cdots \hookrightarrow G_m \\ \mathcal{F}' : \varnothing = G_0 \hookrightarrow \cdots \hookrightarrow G_{i-1} \overset{\tau}{\hookrightarrow} G'_i \overset{\sigma}{\hookrightarrow} G_{i+1} \hookrightarrow \cdots \hookrightarrow G_m \end{array} \qquad (2)$$

In the above operation, the addition of two simplices σ and τ are switched from $\mathcal{F}$ to $\mathcal{F}'$. We also require that $\sigma \not\subseteq \tau$ [5] (σ is not a vertex of the edge τ) because otherwise G'_i is not a valid graph.

For a better presentation, we provide the idea at a high level for the updates in Algorithm 1; the full details are presented in the full version[1].

We also notice the following fact about the change on pairing caused by the switch in Eq. (2) when σ, τ are both positive or both negative. Let σ be paired with σ' and τ be paired with τ' in $\mathcal{F}$. (If σ or τ are *unpaired*, then let σ' or τ' be *null*.) By the update algorithm for general complexes [5], either (i) the pairing for $\mathcal{F}$ and $\mathcal{F}'$ stays the same, or (ii) the only difference on the pairing is that σ is paired with τ' and τ is paired with σ' in $\mathcal{F}'$.

Algorithm 1 (Update for switch on standard graph filtrations). For the switch operation in Eq. (2), the algorithm maintains a merge forest $\mathbb{T}$ (which initially represents $\mathsf{MF}(\mathcal{F})$) and a pairing of simplices Π (which initially corresponds to $\mathcal{F}$). The algorithm makes changes to $\mathbb{T}$ and Π so that they correspond to $\mathcal{F}'$ after the processing. For an overview of the algorithm, we describe the processing only for the cases (or sub-cases) where we need to make changes to $\mathbb{T}$ or Π:

A. Switch is a vertex-vertex switch: First let $v_1 := \sigma$ and $v_2 := \tau$. As illustrated in Fig. 1, the only situation where the pairing Π changes in this case is that v_1, v_2 are in the same tree in $\mathbb{T}$ and are both unpaired when e is added in $\mathcal{F}$, where e is the edge corresponding to the *nearest common ancestor* of v_1, v_2 in $\mathbb{T}$. In this case, v_1,v_2 are leaves at the *lowest* levels in the subtrees T_1, T_2 respectively (see Fig. 1), so that v_2 is paired with e and v_1 is the *representative* (the only unpaired vertex) in the merged connected component due to the addition of e. After the switch, v_1 is paired with e due to being younger and v_2 becomes the representative of the merged component. Notice that the structure of $\mathbb{T}$ stays the same.

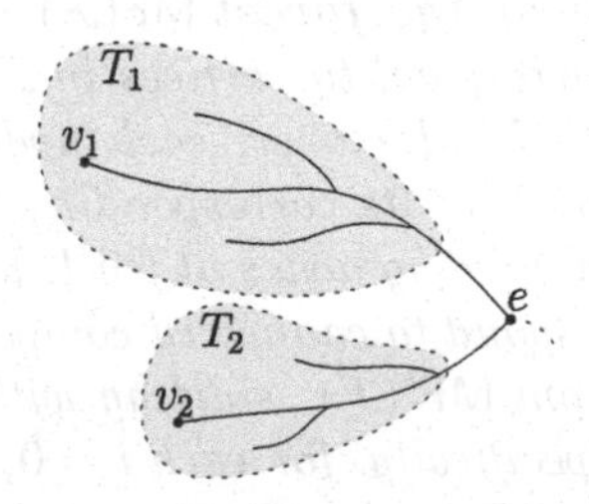

Fig. 1. Vertex-vertex switch.

[1] https://arxiv.org/pdf/2302.12796.pdf.

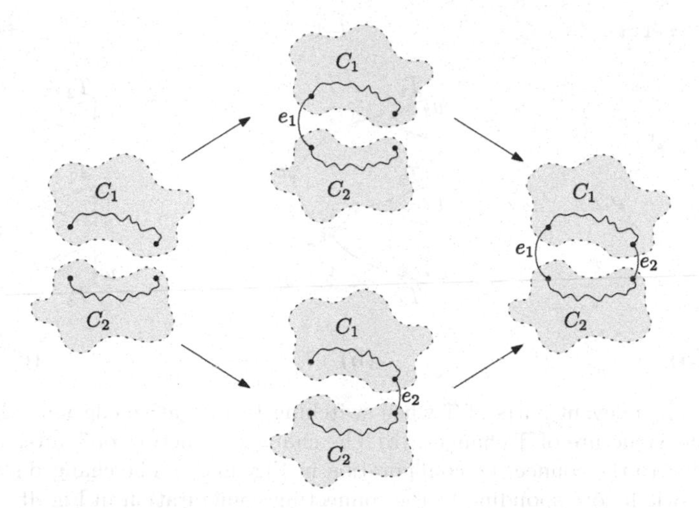

Fig. 2. The edges e_1, e_2 connect to the same two connected components causing the change in an edge-edge switch where e_1 is negative and e_2 is positive.

B. Switch is an edge-edge switch: Let $e_1 := \sigma$ and $e_2 := \tau$. We have the following sub-cases:

B.1. e_1**is negative and** e_2**is positive:** We need to make changes when e_1 is in a 1-cycle in G_{i+1} (see Fig. 2), which is equivalent to saying that e_1, e_2 connect to the same two connected components in G_{i-1}. In this case, e_1 becomes positive and e_2 becomes negative after the switch, for which we pair e_2 with the vertex that e_1 previously pairs with. The node in $\mathbb{T}$ corresponding to e_1 should now correspond to e_2 after the switch.

B.2. e_1**and** e_2**are both negative:** We need to make changes when the corresponding node of e_1 is a child of the corresponding node of e_2 in $\mathbb{T}$ (see Fig. 3a). To further illustrate the situation, let T_1, T_2 be the subtrees rooted at the two children of e_1 in $\mathbb{T}$, and let T_3 be the subtree rooted at the other child of e_2 that is not e_1 (as in Fig. 3a). Moreover, let u, v, w be the leaves at the lowest levels in T_1, T_2, T_3 respectively. Without loss of generality, assume that $\mathrm{idx}_{\mathcal{F}}(v) < \mathrm{idx}_{\mathcal{F}}(u)$. Since T_1, T_2, T_3 can be considered as trees in $\mathsf{MF}^{i-1}(\mathcal{F})$, let C_1, C_2, C_3 be the connected components of G_{i-1} corresponding to T_1, T_2, T_3 respectively (see Definition 1).
We have that C_1, C_2, C_3 are connected by e_1, e_2 in G_{i+1} in the two different ways illustrated in Fig. 4. For the two different connecting configurations, the structure of $\mathbb{T}$ after the switch is different, which is shown in Figs. 3b and 3c. Furthermore, if $\mathrm{idx}_{\mathcal{F}}(w) < \mathrm{idx}_{\mathcal{F}}(u)$ and e_2 directly connects C_1, C_3 as in Fig. 4b, then we swap the paired vertices of e_1, e_2 in Π. (See the full version of the paper for further details and justifications.)

In all cases, the algorithm also updates the levels of the leaves in $\mathbb{T}$ corresponding to σ and τ (if such leaves exist) due to the change of indices for the vertices. Notice that the positivity/negativity of simplices can be easily read off from the simplex pairing Π.

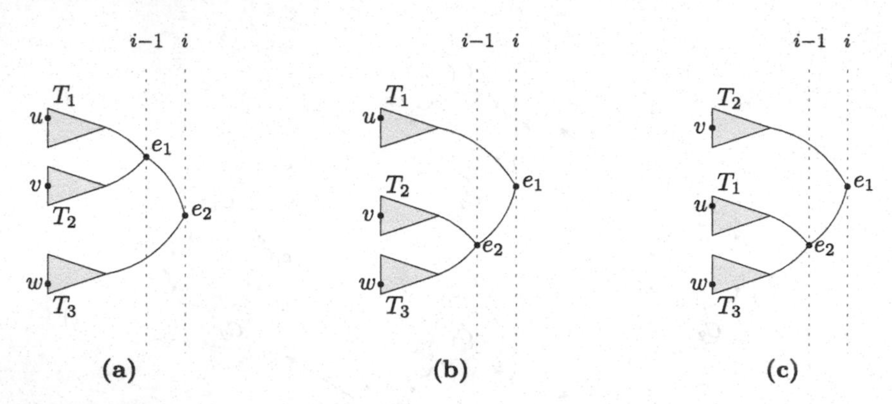

Fig. 3. (a) The relevant parts of $\mathbb{T}$ when switching two negative edges in Algorithm 1 for which the structure of $\mathbb{T}$ changes. (b) The changed structure of $\mathbb{T}$ after the switch corresponding to the connecting configuration in Fig. 4a. (c) The changed structure of $\mathbb{T}$ after the switch corresponding to the connecting configuration in Fig. 4b.

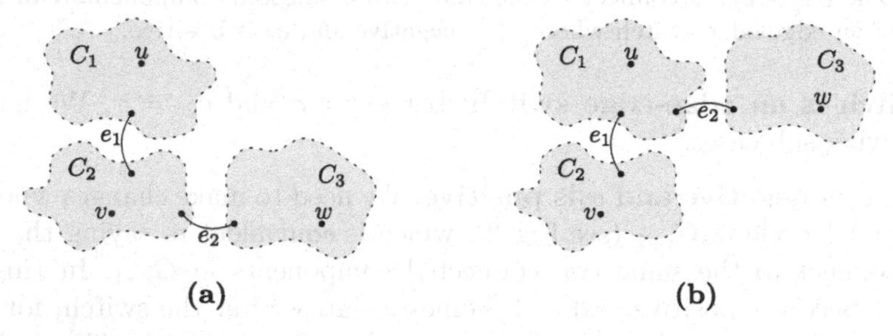

Fig. 4. Two different ways in which the edges connect the three components C_1, C_2, C_3 in G_{i+1} when switching two negative edges e_1, e_2. While e_1 always connects C_1 and C_2, e_2 could either directly connect C_2, C_3 (a) or C_1, C_3 (b).

Implementation. We use the DFT-Tree proposed by Farina and Laura [12] and Link-Cut tree proposed by Sleator and Tarjan [20] for implementing the updates in Algorithm 1, achieving the claimed $O(\log m)$ complexity. See the full version of the paper for details of the implementation and the preprocessing step.

Justification. We justify the correctness of Algorithm 1 for the different cases. Due to space restrictions, we only provide justifications for the vertex-vertex switch here (see Propositions 1 and 2) and justifications for the other cases are provided in the full version.

Proposition 1. *For the switch operation in Eq. (2) where $v_1 := \sigma$ and $v_2 := \tau$ are vertices (v_1, v_2 are thus both positive), the pairings for $\mathcal{F}$ and $\mathcal{F}'$ change if and only if the following two conditions hold:*

1. *v_1 and v_2 are in the same tree in $\mathsf{MF}(\mathcal{F})$;*
2. *v_1 and v_2 are both unpaired when e is added in $\mathcal{F}$, where e is the edge corresponding to the nearest common ancestor x of v_1, v_2 in $\mathsf{MF}(\mathcal{F})$.*

Proof. Suppose that the two conditions hold. Let $j = \mathrm{idx}_{\mathcal{F}}(e)$, i.e., e is added to G_j to form G_{j+1} and x is at level j. Based on the definition of nearest common ancestors, we have that v_1, v_2 are descendants of different children of x. Let T_1, T_2 be the two trees rooted at the two children of x in $\mathsf{MF}(\mathcal{F})$ respectively. WLOG, we can assume that v_1 is in T_1 and v_2 is in T_2 (see Fig. 1). Since x is at level j, we can view T_1, T_2 as trees in $\mathsf{MF}^j(\mathcal{F})$. Let C_1, C_2 be the connected components in G_j corresponding to T_1, T_2 respectively (see Definition 1). We have that $v_1 \in C_1$ and $v_2 \in C_2$. We then observe that as the simplices are added in a graph filtration, each connected component contains only one unpaired vertex which is the oldest one [11]. Since v_1, v_2 are both unpaired when e is added to $\mathcal{F}$, we must have that v_1 is the oldest vertex of C_1 and v_2 is the oldest vertex of a C_2. Then, when e is added in $\mathcal{F}$, v_2 must be paired with e because v_2 is younger than v_1 (see the pairing in the persistence algorithm [11]). However, after the switch, v_1 must be paired with e in $\mathcal{F}'$ because v_1 is now younger than v_2. Therefore, the pairing changes after the switch and we have finished the proof of the 'if' part of the proposition.

We now prove the 'only if' part of the proposition. Suppose that the pairing changes after the switch. First, if v_1, v_2 are in different trees in $\mathsf{MF}(\mathcal{F})$, then the two vertices are in different connected components in $G := G_m$. The pairings for v_1 and v_2 are completely independent in the filtrations and therefore cannot change due to the switch. Then, let T be the subtree of $\mathsf{MF}(\mathcal{F})$ rooted at x and let $j = \mathrm{idx}_{\mathcal{F}}(e)$. Similarly as before, we have that v_1 is in a connected component C_1 of G_j and v_2 is in a connected component C_2 of G_j for $C_1 \neq C_2$. For contradiction, suppose instead that at least one of v_1, v_2 is paired when e is added to G_j in $\mathcal{F}$. If v_1 is paired when e is added in $\mathcal{F}$, then let u be the oldest vertex of C_1. Notice that $u \neq v_1$ because the oldest vertex of C_1 must be unpaired when e is added in $\mathcal{F}$. We notice that the pairing for vertices in $C_1 \setminus \{u\}$ only depends on the index order of these vertices in the filtrations, before and after the switch. Since the indices for simplices in a filtration are unique and $v_2 \notin C_1$, changing the index of v_1 from $i-1$ to i does not change the index order of vertices in $C_1 \setminus \{u\}$. Therefore, the pairing for vertices in $C_1 \setminus \{u\}$ stays the same after the switch, which means that the pairing of v_1 does not change. This contradicts the assumption that the pairing for v_1, v_2 changes due to the switch. If v_2 is paired when e is added in $\mathcal{F}$, we can reach a similar contradiction, and the proof is done.

Proposition 2. *For the switch operation in Eq. (2) where $v_1 := \sigma$ and $v_2 := \tau$ are both vertices, $\mathsf{MF}(\mathcal{F})$ and $\mathsf{MF}(\mathcal{F}')$ have the same structure, with the only difference being on the levels of v_1, v_2 due to the index change of simplices.*

Proof. The structure of the merge forest for a graph filtration only depends on how the edges merge different connected components for the filtration. Thus, switching two vertices does not alter the structure of the merge forest.

We conclude the following:

Theorem 1. *Algorithm 1 correctly updates the pairing and the merge forest for a switch operation in $O(\log m)$ time.*

4 Computing Graph Zigzag Persistence

In this section, we show how the usage of the Link-Cut Tree [20], can improve the computation of graph zigzag persistence. For this purpose, we combine two recent results:

- We show in [8] that a given zigzag filtration can be converted into a standard filtration for a fast computation of zigzag barcode.
- Yan et al. [21] show that the extended persistence of a given graph filtration can be computed by using operations only on trees.

Building on the work of [8], we first convert a given simplex-wise graph zigzag filtration into a *cell*-wise *up-down* filtration, where all insertions occur before deletions. Then, using the extended persistence algorithm of Yan et al. [21] on the up-down filtration with the Link-Cut Tree [20] data structure, we obtain an improved $O(m \log m)$ algorithm for computing graph zigzag persistence.

4.1 Converting to Up-Down Filtration

First, we recall the necessary set up from [8] relevant to our purpose. The algorithm, called FastZigzag [8], builds filtrations on extensions of simplicial complexes called *Δ-complexes* [15], whose building blocks are called *cells* or *Δ-cells*. Notice that 1-dimensional Δ-complexes are nothing but graphs with *parallel edges* (also termed as *multi-edges*) [8].

Assume a *simplex*-wise graph zigzag filtration

$$\mathcal{F} : \varnothing = G_0 \xleftrightarrow{\sigma_0} G_1 \xleftrightarrow{\sigma_1} \cdots \xleftrightarrow{\sigma_{m-1}} G_m = \varnothing$$

consisting of simple graphs as input. We convert $\mathcal{F}$ into the following *cell*-wise up-down [3] filtration consisting of graphs with parallel edges:

$$\mathcal{U} : \varnothing = \hat{G}_0 \xhookrightarrow{\hat{\sigma}_0} \hat{G}_1 \xhookrightarrow{\hat{\sigma}_1} \cdots \xhookrightarrow{\hat{\sigma}_{k-1}} \hat{G}_k \xhookleftarrow{\hat{\sigma}_k} \hat{G}_{k+1} \xhookleftarrow{\hat{\sigma}_{k+1}} \cdots \xhookleftarrow{\hat{\sigma}_{m-1}} \hat{G}_m = \varnothing. \tag{3}$$

Cells $\hat{\sigma}_0, \hat{\sigma}_1, \ldots, \hat{\sigma}_{k-1}$ are *uniquely identified* copies of vertices and edges added in $\mathcal{F}$ with the addition order preserved (notice that the same vertex or edge could be repeatedly added and deleted in $\mathcal{F}$). Cells $\hat{\sigma}_k, \hat{\sigma}_{k+1}, \ldots, \hat{\sigma}_{m-1}$ are *uniquely identified* copies of vertices and edges deleted in $\mathcal{F}$, with the order also preserved. Notice that $m = 2k$ because an added simplex must be eventually deleted in $\mathcal{F}$.

Definition 2. *In $\mathcal{F}$ or $\mathcal{U}$, let each addition or deletion be uniquely identified by its index in the filtration, e.g., index of $G_i \xleftrightarrow{\sigma_i} G_{i+1}$ in $\mathcal{F}$ is i. Then, the* **creator** *of an interval $[b, d] \in \mathsf{Pers}_*(\mathcal{F})$ or $\mathsf{Pers}_*(\mathcal{U})$ is an addition/deletion indexed at $b - 1$, and the* **destroyer** *of $[b, d]$ is an addition/deletion indexed at d.*

As stated previously, each $\hat{\sigma}_i$ in $\mathcal{U}$ for $0 \leq i < k$ corresponds to an addition in $\mathcal{F}$, and each $\hat{\sigma}_i$ for $k \leq i < m$ corresponds to a deletion in $\mathcal{F}$. This naturally defines a bijection ϕ from the additions and deletions in $\mathcal{U}$ to the additions and deletions in $\mathcal{F}$. Moreover, for simplicity, we let the domain and codomain of ϕ be the sets of indices for the additions and deletions. The interval mapping in [8] (which uses the *Mayer-Vietoris Diamond* [2,3]) can be summarized as follows:

Theorem 2. *Given* $\mathsf{Pers}_*(\mathcal{U})$*, one can retrieve* $\mathsf{Pers}_*(\mathcal{F})$ *using the following bijective mapping from* $\mathsf{Pers}_*(\mathcal{U})$ *to* $\mathsf{Pers}_*(\mathcal{F})$*: an interval* $[b, d] \in \mathsf{Pers}_p(\mathcal{U})$ *with a creator indexed at* $b-1$ *and a destroyer indexed at* d *is mapped to an interval* $I \in \mathsf{Pers}_*(\mathcal{F})$ *with the same creator and destroyer indexed at* $\phi(b-1)$ *and* $\phi(d)$ *respectively. Specifically,*

- *If* $\phi(b-1) < \phi(d)$*, then* $I = [\phi(b-1)+1, \phi(d)] \in \mathsf{Pers}_p(\mathcal{F})$*, where* $\phi(b-1)$ *indexes the creator and* $\phi(d)$ *indexes the destroyer.*
- *Otherwise,* $I = [\phi(d)+1, \phi(b-1)] \in \mathsf{Pers}_{p-1}(\mathcal{F})$*, where* $\phi(d)$ *indexes the creator and* $\phi(b-1)$ *indexes the destroyer.*

Notice the decrease in the dimension of the mapped interval in $\mathsf{Pers}_*(\mathcal{F})$ when $\phi(d) < \phi(b-1)$ (indicating a swap on the roles of the creator and destroyer).

While Theorem 2 suggests a simple mapping rule for $\mathsf{Pers}_*(\mathcal{U})$ and $\mathsf{Pers}_*(\mathcal{F})$, we further interpret the mapping in terms of the different types of intervals in zigzag persistence. We define the following:

Definition 3. *Let* $\mathcal{L} : \varnothing = K_0 \leftrightarrow K_1 \leftrightarrow \cdots \leftrightarrow K_\ell = \varnothing$ *be a zigzag filtration. For any* $[b, d] \in \mathsf{Pers}_*(\mathcal{L})$*, the birth index* b *is* ***closed*** *if* $K_{b-1} \hookrightarrow K_b$ *is a forward inclusion; otherwise,* b *is* ***open****. Symmetrically, the death index* d *is* ***closed*** *if* $K_d \hookleftarrow K_{d+1}$ *is a backward inclusion; otherwise,* d *is* ***open****. The types of the birth/death ends classify intervals in* $\mathsf{Pers}_*(\mathcal{L})$ *into four types:* ***closed-closed, closed-open, open-closed,*** *and* ***open-open****.*

Table 1 breaks down the bijection between $\mathsf{Pers}_*(\mathcal{U})$ and $\mathsf{Pers}_*(\mathcal{F})$ into mappings for the different types, where $\mathsf{Pers}_0^{\mathrm{co}}(\mathcal{U})$ denotes the set of closed-open intervals in $\mathsf{Pers}_0(\mathcal{U})$ (meanings of other symbols can be derived similarly).

We notice the following:

- $\mathsf{Pers}_*(\mathcal{U})$ has no open-open intervals because there are no additions after deletions in $\mathcal{U}$.
- $\mathsf{Pers}_1(\mathcal{U})$ and $\mathsf{Pers}_1(\mathcal{F})$ contain only closed-closed intervals because graph filtrations have no triangles.
- $[b, d] \in \mathsf{Pers}_1^{\mathrm{cc}}(\mathcal{U})$ is mapped to an interval in $\mathsf{Pers}_0^{\mathrm{oo}}(\mathcal{F})$ when $\phi(d) < \phi(b-1)$.

Table 1. Mapping of different types of intervals for $\mathsf{Pers}_*(\mathcal{U})$ and $\mathsf{Pers}_*(\mathcal{F})$

$\mathcal{U}$		$\mathcal{F}$
$\mathsf{Pers}_0^{\mathrm{co}}(\mathcal{U})$	$\leftrightarrow$	$\mathsf{Pers}_0^{\mathrm{co}}(\mathcal{F})$
$\mathsf{Pers}_0^{\mathrm{oc}}(\mathcal{U})$	$\leftrightarrow$	$\mathsf{Pers}_0^{\mathrm{oc}}(\mathcal{F})$
$\mathsf{Pers}_0^{\mathrm{cc}}(\mathcal{U})$	$\leftrightarrow$	$\mathsf{Pers}_0^{\mathrm{cc}}(\mathcal{F})$
$\mathsf{Pers}_1^{\mathrm{cc}}(\mathcal{U})$	$\leftrightarrow$	$\mathsf{Pers}_0^{\mathrm{oo}}(\mathcal{F}) \cup \mathsf{Pers}_1^{\mathrm{cc}}(\mathcal{F})$

4.2 Extended Persistence Algorithm for Graphs

Yan et al. [21] present an extended persistence algorithm for graphs in a neural network setting (see also [22]) which runs in quadratic time. We adapt it to computing up-down zigzag persistence while improving its time complexity with a Link-Cut tree [20] data structure.

Definition 2 indicates that for the up-down filtration in Eq. (3), $\mathsf{Pers}_*(\mathcal{U})$ can be considered as generated from the *cell pairs* similar to the simplex pairs in standard persistence [11]. Specifically, an interval $[b, d] \in \mathsf{Pers}_*(\mathcal{U})$ is generated from the pair $(\hat{\sigma}_{b-1}, \hat{\sigma}_d)$. While each cell appears twice in $\mathcal{U}$ (once added and once deleted), we notice that it should be clear from the context whether a cell in a pair refers to its addition or deletion. We then have the following:

Remark 1. Every vertex-edge pair for a closed-open interval in $\mathsf{Pers}_0(\mathcal{U})$ comes from the *ascending* part $\mathcal{U}_u$ of the filtration $\mathcal{U}$, and every edge-vertex pair for an open-closed interval in $\mathsf{Pers}_0(\mathcal{U})$ comes from the *descending* part $\mathcal{U}_d$. These ascending and descending parts are as shown below:

$$\begin{aligned} \mathcal{U}_u &: \varnothing = \hat{G}_0 \xhookrightarrow{\hat{\sigma}_0} \hat{G}_1 \xhookrightarrow{\hat{\sigma}_1} \cdots \xhookrightarrow{\hat{\sigma}_{k-1}} \hat{G}_k, \\ \mathcal{U}_d &: \varnothing = \hat{G}_m \xhookrightarrow{\hat{\sigma}_{m-1}} \hat{G}_{m-1} \xhookrightarrow{\hat{\sigma}_{m-2}} \cdots \xhookrightarrow{\hat{\sigma}_k} \hat{G}_k. \end{aligned} \tag{4}$$

We first run the standard persistence algorithm with the Union-Find data structure on $\mathcal{U}_u$ and $\mathcal{U}_d$ to obtain all pairs between vertices and edges in $O(k\,\alpha(k))$ time, retrieving closed-open and open-closed intervals in $\mathsf{Pers}_0(\mathcal{U})$. We also have the following:

Remark 2. Each closed-closed interval in $\mathsf{Pers}_0(\mathcal{U})$ is given by pairing the first vertex in $\mathcal{U}_u$, that comes from a connected component C of $\hat{G}_k$, and the first vertex in $\mathcal{U}_d$ coming from C. There is no extra computation necessary for this type of pairing.

Remark 3. Each closed-closed interval in $\mathsf{Pers}_1(\mathcal{U})$ is given by an edge-edge pair in $\mathcal{U}$, in which one edge is a positive edge from the ascending filtration $\mathcal{U}_u$ and the other is a positive edge from the descending filtration $\mathcal{U}_d$.

To compute the edge-edge pairs, the algorithm scans $\mathcal{U}_d$ and keeps track of whether an edge is positive or negative. For every positive edge e in $\mathcal{U}_d$, it finds the cycle c that is created the earliest in $\mathcal{U}_u$ containing e and then pairs e with the youngest edge e' of c added in $\mathcal{U}_u$, which creates c in $\mathcal{U}_u$. To determine c and e', we use the following procedure from [21]:

Algorithm 2.

1. Maintain a spanning forest T of $\hat{G}_k$ while processing $\mathcal{U}_d$. Initially, T consists of all vertices of $\hat{G}_k$ and all negative edges in $\mathcal{U}_d$.
2. For every positive edge e in $\mathcal{U}_d$ (in the order of the filtration):
   (a) Add e to T and check the *unique* cycle c formed by e in T.
   (b) Determine the edge e' which is the youngest edge of c with respect to the filtration $\mathcal{U}_u$. The edge e' has to be positive in $\mathcal{U}_u$.
   (c) Delete e' from T. This maintains T to be a tree all along.
   (d) Pair the positive edge e from $\mathcal{U}_d$ with the positive edge e' from $\mathcal{U}_u$.

We propose to implement the above algorithm by maintaining T as a Link-Cut Tree [20], which is a dynamic data structure allowing the following operations in $O(\log N)$ time (N is the number of nodes in the trees): (i) insert or delete a node or an edge from the Link-Cut Trees; (ii) find the maximum-weight edge on a path in the trees. Notice that for the edges in T, we let their weights equal to their indices in $\mathcal{U}_u$.

We build T by first inserting all vertices of $\hat{G}_k$ and all negative edges in $\mathcal{U}_d$ into T in $O(m \log m)$ time. Then, for every positive edge e in $\mathcal{U}_d$, find the maximum-weight edge ϵ in the unique path in T connecting the two endpoints of e in $O(\log m)$ time. Let e' be the edge in $\{e, \epsilon\}$ whose index in $\mathcal{U}_u$ is greater (i.e., e' is the younger one in $\mathcal{U}_u$). Pair e' with e to form a closed-closed interval in $\mathsf{Pers}_1(\mathcal{U})$. After this, delete e' from T and insert ϵ into T, which takes $O(\log m)$ time. Therefore, processing the entire filtration $\mathcal{U}_d$ and getting all closed-closed intervals in $\mathsf{Pers}_1(\mathcal{U})$ takes $O(m \log m)$ time in total.

Theorem 3. *For a simplex-wise graph zigzag filtration $\mathcal{F}$ with m additions and deletions, $\mathsf{Pers}_*(\mathcal{F})$ can be computed in $O(m \log m)$ time.*

Proof. We first convert $\mathcal{F}$ into the up-down filtration $\mathcal{U}$ in $O(m)$ time [8]. We then compute $\mathsf{Pers}_*(\mathcal{U})$ in $O(m \log m)$ time using the algorithm described in this section. Finally, we convert $\mathsf{Pers}_*(\mathcal{U})$ to $\mathsf{Pers}_*(\mathcal{F})$ using the process in Theorem 2, which takes $O(m)$ time. Therefore, computing $\mathsf{Pers}_*(\mathcal{F})$ takes $O(m \log m)$ time.

5 Updating Graph Zigzag Persistence

In this section, we describe the update of persistence for switches on graph zigzag filtrations. In [7], we considered the updates in zigzag filtration for general simplicial complexes. Here, we focus on the special case of graphs, for which we find more efficient algorithms for switches. In a similar vein to the switch operation on standard filtrations [5] (see also Sect. 3), a switch on a zigzag filtration swaps two consecutive simplex-wise inclusions. Based on the directions of the inclusions, we have the following four types of switches (as defined in [7]), where $\mathcal{F}, \mathcal{F}'$ are both simplex-wise graph zigzag filtrations starting and ending with empty graphs:

- *Forward* switch is the counterpart of the switch on standard filtrations, which swaps two forward inclusions (i.e., additions) and also requires $\sigma \nsubseteq \tau$:

$$
\begin{aligned}
\mathcal{F} &: G_0 \leftrightarrow \cdots \leftrightarrow G_{i-1} \xhookrightarrow{\sigma} G_i \xhookrightarrow{\tau} G_{i+1} \leftrightarrow \cdots \leftrightarrow G_m \\
\mathcal{F}' &: G_0 \leftrightarrow \cdots \leftrightarrow G_{i-1} \xhookrightarrow{\tau} G'_i \xhookrightarrow{\sigma} G_{i+1} \leftrightarrow \cdots \leftrightarrow G_m
\end{aligned}
\tag{5}
$$

- *Backward* switch is the symmetric version of forward switch, requiring $\tau \nsubseteq \sigma$:

$$
\begin{aligned}
\mathcal{F} &: G_0 \leftrightarrow \cdots \leftrightarrow G_{i-1} \xhookleftarrow{\sigma} G_i \xhookleftarrow{\tau} G_{i+1} \leftrightarrow \cdots \leftrightarrow G_m \\
\mathcal{F}' &: G_0 \leftrightarrow \cdots \leftrightarrow G_{i-1} \xhookleftarrow{\tau} G'_i \xhookleftarrow{\sigma} G_{i+1} \leftrightarrow \cdots \leftrightarrow G_m
\end{aligned}
\tag{6}
$$

- The remaining switches swap two inclusions of opposite directions:

$$
\begin{aligned}
\mathcal{F} &: G_0 \leftrightarrow \cdots \leftrightarrow G_{i-1} \xhookrightarrow{\sigma} G_i \xhookleftarrow{\tau} G_{i+1} \leftrightarrow \cdots \leftrightarrow G_m \\
\mathcal{F}' &: G_0 \leftrightarrow \cdots \leftrightarrow G_{i-1} \xhookleftarrow{\tau} G'_i \xhookrightarrow{\sigma} G_{i+1} \leftrightarrow \cdots \leftrightarrow G_m
\end{aligned}
\tag{7}
$$

The switch from $\mathcal{F}$ to $\mathcal{F}'$ is called an *outward* switch and the switch from $\mathcal{F}'$ to $\mathcal{F}$ is called an *inward* switch. We also require $\sigma \neq \tau$ because if $\sigma = \tau$, e.g., for outward switch, we cannot delete τ from G_{i-1} in $\mathcal{F}'$ because $\tau \notin G_{i-1}$.

5.1 Update Algorithms

Instead of performing the updates in Eqs. (5 – 7) directly on the graph zigzag filtrations, our algorithms work *on the corresponding up-down filtrations* for $\mathcal{F}$ and $\mathcal{F}'$, with the conversion described in Sect. 4.1. Specifically, we maintain a pairing of cells for the corresponding up-down filtration, and the pairing for the original graph zigzag filtration can be derived from the bijection ϕ as defined in Sect. 4.1.

For outward and inward switches, the corresponding up-down filtration before and after the switch is the same and hence the update takes $O(1)$ time. Moreover, the backward switch is a symmetric version of the forward switch and the algorithm is also symmetric. Hence, in this section we focus on how to perform the forward switch. The symmetric behavior for backward switch is mentioned only when necessary.

For the forward switch in Eq. (5), let $\mathcal{U}$ and $\mathcal{U}'$ be the corresponding up-down filtrations for $\mathcal{F}$ and $\mathcal{F}'$ respectively. By the conversion in Sect. 4.1, there is also a forward switch (on the ascending part) from $\mathcal{U}$ to $\mathcal{U}'$, where $\hat{\sigma}, \hat{\tau}$ are Δ-cells corresponding to σ, τ respectively:

$$
\begin{aligned}
\mathcal{U} &: \hat{G}_0 \hookrightarrow \cdots \hookrightarrow \hat{G}_{j-1} \xhookrightarrow{\hat{\sigma}} \hat{G}_j \xhookrightarrow{\hat{\tau}} \hat{G}_{j+1} \hookrightarrow \cdots \hookrightarrow \hat{G}_k \hookleftarrow \cdots \hookleftarrow \hat{G}_m \\
\mathcal{U}' &: \hat{G}_0 \hookrightarrow \cdots \hookrightarrow \hat{G}_{j-1} \xhookrightarrow{\hat{\tau}} \hat{G}'_j \xhookrightarrow{\hat{\sigma}} \hat{G}_{j+1} \hookrightarrow \cdots \hookrightarrow \hat{G}_k \hookleftarrow \cdots \hookleftarrow \hat{G}_m
\end{aligned}
\tag{8}
$$

We observe that the update of the different types of intervals for up-down filtrations (see Table 1) can be done *independently*:

- To update the *closed-open* intervals for Eq. (8) (which updates the closed-open intervals for Eq. (5)), we run Algorithm 1 in Sect. 3 on the ascending part of the up-down filtration. This is based on descriptions in Sect. 4 (Table 1 and Remark 1).
- A backward switch in Eq. (6) causes a backward switch on the descending parts of the up-down filtrations, which may change the *open-closed* intervals for the filtrations. For this, we run Algorithm 1 on the descending part of the up-down filtration. Our update algorithm hence maintains two sets of data structures needed by Algorithm 1, for the ascending and descending parts separately.
- Following Remark 2 in Sect. 4, to update the *closed-closed* intervals in dimension 0 for the switches, we only need to keep track of the oldest vertices in the ascending and descending parts for each connected component of $\hat{G} := \hat{G}_k$ (defined in Eq. (8)). Since indices of no more than two vertices can change in a switch, this can be done in constant time by a simple bookkeeping.

We are now left with the update of the closed-closed intervals in dimension 1 for the switch on up-down filtrations, which are generated from the edge-edge pairs (see Remark 3 in Sect. 4). Due to space limitations, we only present a high-level algorithm (Algorithm 3) here by explicitly maintaining representatives (see Definition 4) for the edge-edge pairs. The implementation details of Algorithm 3 to achieve the claimed $O(\sqrt{m}\log m)$ complexity, its justification, and the required preprocessing step are described in the full version.

Definition 4. *For a cell-wise up-down filtration of graphs with parallel edges*

$$\mathcal{L} : \varnothing = H_0 \xhookrightarrow{\varsigma_0} H_1 \xhookrightarrow{\varsigma_1} \cdots \xhookrightarrow{\varsigma_{\ell-1}} H_\ell \xleftrightarrow{\varsigma_\ell} H_{\ell+1} \xhookleftarrow{\varsigma_{\ell+1}} \cdots \xhookleftarrow{\varsigma_{2\ell-1}} H_{2\ell} = \varnothing,$$

a ***representative cycle*** *(or simply* ***representative****) for an edge-edge pair* $(\varsigma_b, \varsigma_d)$ *is a 1-cycle* z *s.t.* $\varsigma_b \in z \subseteq H_{b+1}$ *and* $\varsigma_d \in z \subseteq H_d$.

Algorithm 3. We describe the algorithm for the forward switch in Eq. (8). The procedure for a backward switch on an up-down filtration is symmetric. The algorithm maintains a set of edge-edge pairs Π initially for $\mathcal{U}$. It also maintains a representative cycle for each edge-edge pair in Π. After the processing, edge-edge pairs in Π and their representatives correspond to $\mathcal{U}'$. As mentioned, a switch containing a vertex makes no changes to the edge-edge pairs. So, suppose that the switch is an edge-edge switch and let $e_1 := \hat{\sigma}$, $e_2 := \hat{\tau}$. Moreover, let $\mathcal{U}_u$ be the ascending part of $\mathcal{U}$. We have the following cases:

A. e_1 **and** e_2 **are both negative in** $\mathcal{U}_u$**:** Do nothing (negative edges are not in edge-edge pairs).
B. e_1 **is positive and** e_2 **is negative in** $\mathcal{U}_u$**:** No pairing changes by this switch.
C. e_1 **is negative and** e_2 **is positive in** $\mathcal{U}_u$**:** Let z be the representative cycle for the pair $(e_2, \epsilon) \in \Pi$. If $e_1 \in z$, pair e_1 with ϵ in Π with the same representative z (notice that e_2 becomes unpaired).

D. $e_1$and $e_2$are both positive in $\mathcal{U}_u$: Let z, z' be the representative cycles for the pairs $(e_1, \epsilon), (e_2, \epsilon') \in \Pi$ respectively. Do the following according to different cases:

- If $e_1 \in z'$ and the deletion of ϵ' is before the deletion of ϵ in $\mathcal{U}$: Let the representative for (e_2, ϵ') be $z + z'$. The pairing does not change.
- If $e_1 \in z'$ and the deletion of ϵ' is after the deletion of ϵ in $\mathcal{U}$: Pair e_1 and ϵ' in Π with the representative z'; pair e_2 and ϵ in Π with the representative $z + z'$.

We conclude with the following (see the full version for justification):

Theorem 4. *For the switches on graph zigzag filtrations, the* closed-closed *intervals in dimension 0 can be maintained in* $O(1)$ *time; the* closed-open *and* open-closed *intervals, which appear only in dimension* 0*, can be maintained in* $O(\log m)$ *time; the* open-open *intervals in dimension 0 and* closed-closed *intervals in dimension 1 can be maintained in* $O(\sqrt{m} \log m)$ *time.*

References

1. Berkouk, N., Nyckees, L.: One diamond to rule them all: old and new topics about zigzag, levelsets and extended persistence (2022). https://arxiv.org/abs/2210.00916, https://doi.org/10.48550/ARXIV.2210.00916
2. Carlsson, G., De Silva, V.: Zigzag persistence. Found. Comput. Math. **10**(4), 367–405 (2010)
3. Carlsson, G., De Silva, V., Morozov, D.: Zigzag persistent homology and real-valued functions. In: Proceedings of the Twenty-Fifth Annual Symposium on Computational Geometry, pp. 247–256 (2009)
4. Cohen-Steiner, D., Edelsbrunner, H., Harer, J.: Extending persistence using Poincaré and Lefschetz duality. Found. Comput. Math. **9**(1), 79–103 (2009)
5. Cohen-Steiner, D., Edelsbrunner, H., Morozov, D.: Vines and vineyards by updating persistence in linear time. In: Proceedings of the Twenty-Second Annual Symposium on Computational Geometry, pp. 119–126 (2006)
6. Dey, T.K., Hou, T.: Computing zigzag persistence on graphs in near-linear time. In: 37th International Symposium on Computational Geometry (2021)
7. Dey, T.K., Hou, T.: Updating barcodes and representatives for zigzag persistence. arXiv preprint arXiv:2112.02352 (2021)
8. Dey, T.K., Hou, T.: Fast computation of zigzag persistence. In: 30th Annual European Symposium on Algorithms, ESA 2022, 5–9 September 2022, Berlin/Potsdam, Germany, volume 244 of LIPIcs, pp. 43:1–43:15. Schloss Dagstuhl - Leibniz-Zentrum für Informatik (2022)
9. Dey, T.K., Wang, Y.: Computational Topology for Data Analysis. Cambridge University Press, Cambridge (2022)
10. Edelsbrunner, H., Harer, J.: Computational Topology: An Introduction. American Mathematical Soc, Providence (2010)
11. Edelsbrunner, H., Letscher, D., Zomorodian, A.: Topological persistence and simplification. In: Proceedings 41st Annual Symposium on Foundations of Computer Science, pp. 454–463. IEEE (2000)

12. Farina, G., Laura, L.: Dynamic subtrees queries revisited: the depth first tour tree. In: Lipták, Z., Smyth, W.F. (eds.) IWOCA 2015. LNCS, vol. 9538, pp. 148–160. Springer, Cham (2016). https://doi.org/10.1007/978-3-319-29516-9_13
13. Gabriel, P.: Unzerlegbare Darstellungen I. Manuscripta Math. **6**(1), 71–103 (1972)
14. Georgiadis, L., Kaplan, H., Shafrir, N., Tarjan, R.E., Werneck, R.F.: Data structures for mergeable trees. ACM Trans. Algorithms (TALG) **7**(2), 1–30 (2011)
15. Hatcher, A.: Algebraic Topology. Cambridge University Press, Cambridge (2002)
16. Holm, J., De Lichtenberg, K., Thorup, M.: Poly-logarithmic deterministic fully-dynamic algorithms for connectivity, minimum spanning tree, 2-edge, and biconnectivity. J. ACM (JACM) **48**(4), 723–760 (2001)
17. Maria, C., Oudot, S.Y.: Zigzag persistence via reflections and transpositions. In Proceedings of the Twenty-Sixth Annual ACM-SIAM Symposium on Discrete Algorithms, pp. 181–199. SIAM (2014)
18. Milosavljević, N., Morozov, D., Skraba, P.: Zigzag persistent homology in matrix multiplication time. In: Proceedings of the Twenty-Seventh Annual Symposium on Computational Geometry, pp. 216–225 (2011)
19. Parsa, S.: A deterministic $O(m \log m)$ time algorithm for the reeb graph. Discrete Comput. Geom. **49**(4), 864–878 (2013)
20. Sleator, D.D., Tarjan, R.E.: A data structure for dynamic trees. In: Proceedings of the Thirteenth Annual ACM Symposium on Theory of Computing, pp. 114–122 (1981)
21. Yan, Z., Ma, T., Gao, L., Tang, Z., Chen, C.: Link prediction with persistent homology: an interactive view. In: International Conference on Machine Learning, pp. 11659–11669. PMLR (2021)
22. Zhang, S., Mukherjee, S., Dey, T.K.: GEFL: extended filtration learning for graph classification. In: Rieck, B., Pascanu, R. (eds.), Learning on Graphs Conference, LoG 2022, volume 198 of Proceedings of Machine Learning Research, p. 16. PMLR (2022). https://proceedings.mlr.press/v198/zhang22b.html

Block Crossings in One-Sided Tanglegrams

Alexander Dobler(✉) and Martin Nöllenburg

Algorithms and Complexity Group, TU Wien, Vienna, Austria
{adobler,noellenburg}@ac.tuwien.ac.at

Abstract. Tanglegrams are drawings of two rooted binary phylogenetic trees and a matching between their leaf sets. The trees are drawn crossing-free on opposite sides with their leaf sets facing each other on two vertical lines. Instead of minimizing the number of pairwise edge crossings, we consider the problem of minimizing the number of *block crossings*, that is, two bundles of lines crossing each other locally.

With one tree fixed, the leaves of the second tree can be permuted according to its tree structure. We give a complete picture of the algorithmic complexity of minimizing block crossings in one-sided tanglegrams by showing NP-completeness, constant-factor approximations, and a fixed-parameter algorithm. We also state first results for non-binary trees.

1 Introduction

Tanglegrams [25] are drawings of two rooted n-leaf trees and a matching between their leaf sets drawn as straight edges. The trees are drawn such that one tree is on the left and the other is on the right with their leaf sets facing each other on two vertical lines (see Fig. 1a). An important application of tanglegrams is the comparison of two phylogenetic trees with the same leaf set [26,28], which can be used to study co-speciation or for comparison of hypothetical phylogenetic trees computed by different algorithms. Other applications are comparisons of dendrograms in hierarchical clustering or software hierarchies [21]. The readability of tanglegrams heavily depends on the order of the two leaf sets on the vertical lines, as this determines the number of edge crossings between the matching edges. The possible orders depend on the tree structure of both trees, so finding appropriate orders that minimize the number of pairwise crossings is a nontrivial problem known as the Tanglegram Layout Problem (TLP) [7,13,16]. In this paper we focus on minimizing block crossings instead of pairwise edge crossings; see Fig. 1 for an example. This is done by relaxing that the matching edges must be drawn as straight lines, but rather drawing them as x-monotone curves, which allows for shifting and grouping crossings more flexibly. A *block crossing* [11] is then defined as a crossing between two disjoint sets of edges in a confined region R, each of which forms a bundle of locally parallel curves in

This work has been supported by the Vienna Science and Technology Fund (WWTF) [10.47379/ICT19035].

P. Morin and S. Suri (Eds.): WADS 2023, LNCS 14079, pp. 386–400, 2023.
https://doi.org/10.1007/978-3-031-38906-1_25

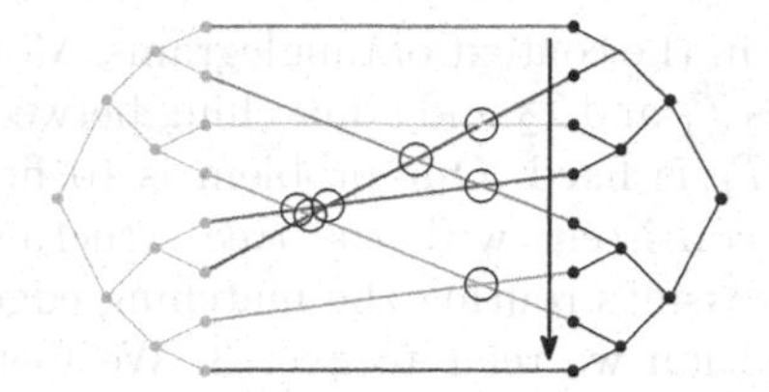

(a) Optimal solution with 7 pairwise crossings.

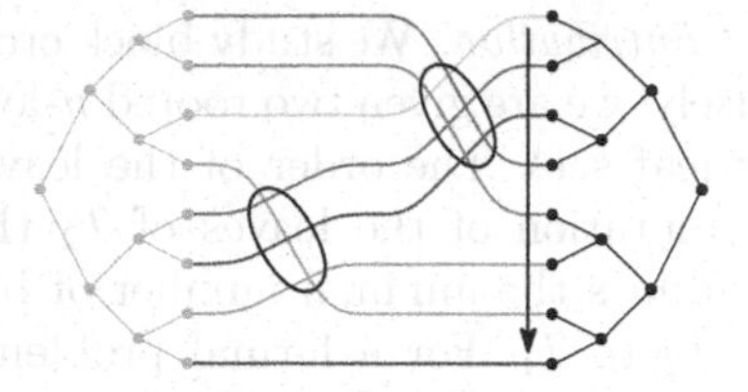

(b) Optimal solution with 2 block crossings (but 9 pairwise crossings).

Fig. 1. Two tanglegrams for the same input. The leaf order of the left tree is fixed, the right tree can be permuted to minimize pairwise (a) or block crossings (b).

R; no further edge may intersect with R. Block crossings provide the ability to group multiple crossings together, instead of having them scattered throughout the drawing. Furthermore, this mostly leads to fewer block crossings than the required number of pairwise crossings and thus reduces visual clutter [17]. We initiate the work on block crossings for tanglegrams and focus on tanglegrams for binary trees where the leaf order of one tree is fixed.

Related Work. The TLP has been studied mainly for binary trees, as phylogenetic trees are mostly binary: If both leaf sets can be ordered then the problem is known to be NP-complete [16], even if the trees are complete [7]. However, there exist approximation results [7], fixed-parameter algorithms [5,7,16], integer linear programming formulations [4], and heuristics [24]. If the leaf order of one tree is fixed, then the problem is solvable in polynomial time [16]; if the trees are not binary, however, then even this problem is NP-complete [9].

Edge bundling is a technique in network visualization that groups multiple edges together to reduce visual clutter. If two edge bundles locally cross each other this is called a *bundled crossing* [1,17]. For a collection of results on bundled crossings in general graphs we refer to [23]. Essentially, a *block crossing* is the same as a bundled crossing. But the term block crossing has been used mainly in contexts, where bundled crossings are determined in a purely combinatorial way using permutations, and no topology is required. Our work is similar, so we adopt that term, and give an overview of previous work on block crossings: Fink et al. [18] worked on minimizing block crossings amongst metro lines in a pre-specified metro network where multiple metro lines might be routed along the same edge. They considered general metro networks but mostly focused on special cases such as paths, trees, and upward trees. Their results mostly include approximation and fixed-parameter algorithms, as even the most restricted variant of their problem is NP-complete. Van Dijk et al. [11] studied block crossings in the context of storylines. They showed NP-completeness, fixed-parameter tractability, and an approximation algorithm. Both of the above works on block crossings pointed out the connection between block crossings and a problem stemming from computational biology called SORTING BY TRANSPOSITIONS [2], where a permutation has to be transformed into the identity permutation by exchanging adjacent blocks of elements, calling this operation a *transposition*.

Our Contribution. We study block crossings in the context of tanglegrams. More precisely, we are given two rooted n-leaf trees T_1 and T_2 and a matching between their leaf sets. The order of the leaves in T_1 is fixed. Our problem is to find a permutation of the leaves of T_2 that is consistent with the tree structure, and admits the minimal number of block crossings routing the matching edges from T_2 to T_1. For a formal problem definition we refer to Sect. 3. We focus mostly on binary trees T_2: In Sect. 4 we show that the problem is NP-complete even for complete binary trees. In Sect. 5 we give 2.25-approximation algorithms, the first for general binary trees, and a faster second one for complete binary trees. In Sect. 6 we show that the problem is fixed-parameter tractable (FPT) parameterized by the number of block crossings. In Sect. 7 we show that the techniques used in Sect. 5 to find a polynomial-time approximation algorithm do not extend to non-binary trees. We start by giving some preliminaries in the following section.

Some proofs are in the full version of this paper [12] *and the corresponding statements are marked* ($\star$).

2 Preliminaries

Let $\delta_{i,j}$ be the Kronecker delta function that is 1 if $i = j$ and 0 otherwise. Let $[n] = \{1, \dots, n\}$.

Permutations. A *permutation* $\pi : [k] \to X$ is a bijective function from $[k]$ to a set X, in particular, $\pi : [k] \to [k]$ is a permutation of $[k]$, and $\mathsf{inv}(\pi)$ is its inverse. We write π_i for $\pi(i)$ and use superscript if we want to tell apart multiple permutations. Sometimes we write permutations π as sequences of elements $(\pi_1, \dots, \pi_k)$. We denote by Π_n the set of all permutations from $[n]$ to $[n]$. For a permutation $\pi \in \Pi_n$ and $i \in [n]$, let $\pi \ominus \pi_i$ be the permutation of $[n-1]$ obtained by first removing π_i from π, and then decreasing all elements of π greater than π_i by one. For two permutations $\pi = (x_1, \dots, x_n)$ and $\pi' = (y_1, \dots, y_m)$ of disjoint elements, we denote by $\pi \star \pi'$ their *concatenation* $(x_1, \dots x_n, y_1, \dots, y_m)$. For two sets Π and Π' of permutations, we define $\Pi \star \Pi' = \{\pi \star \pi' \mid \pi \in \Pi, \pi' \in \Pi'\}$.

Transpositions. A *transposition* $\tau = \tau(i, j, k) \in \Pi_n$ with $1 \le i < j < k \le n+1$ is the permutation

$$(1, \dots, i-1, j, \dots, k-1, i, i+1, \dots, j-2, j-1, k, \dots, n).$$

This definition is different from the classic transpositions in discrete mathematics as it stems from computational biology [2]. Assume $\pi \in \Pi_n$. The permutation $\pi \circ \tau(i, j, k)$ has the effect of swapping the contiguous subsequences $\pi_i, \dots, \pi_{j-1}$ and $\pi_j, \dots, \pi_{k-1}$. A *block* in a permutation $\pi \in \Pi_n$ is a maximal contiguous subsequence $\pi_i, \pi_{i+1}, \dots, \pi_j$ that is also a contiguous subsequence of the identity permutation id_n. The number of blocks in π is denoted by $\mathsf{blocks}(\pi)$. An index $i \in [n] \cup \{0\}$ is a *breakpoint* if

- $i = 0$ and $\pi_1 \neq 1$,
- $1 \leq i \leq n-1$ and $\pi_i + 1 \neq \pi_{i+1}$, or
- $i = n$ and $\pi_n \neq n$.

Essentially, a breakpoint in π corresponds to a pair (x, y) of adjacent elements in the extended permutation $\pi^e = (0) \star \pi \star (n+1)$ such that $x + 1 \neq y$. Each breakpoint i in π has a corresponding breakpoint element π^e_{i+1} (this can include 0). Conversely, we say that π^e_{i+1} corresponds to breakpoint i. Let $\mathsf{bp}(\pi)$ be the number of breakpoints in π. The transposition distance $d_t(\pi)$ of π is the minimum number $k \in \mathbb{N}_0$ such that there exist transpositions $\tau^1, \ldots, \tau^k$ with $\pi \circ \tau^1 \circ \cdots \circ \tau^k = \mathsf{id}_n$. In this case we call $\tau^1, \ldots, \tau^k$ an id-transposition sequence for π. Note that there always exists an id-transposition sequence, as every permutation can be transformed to the identity-permutation by adjacent swaps (cf. Bubblesort). Let π be a permutation with $r > 0$ breakpoints. Then $\mathsf{gl}(\pi) \in \Pi_{r-1}$ is formed by "gluing" each block together into a single element. Furthermore, if π starts with 1 then the first block is removed, and if π ends with n then the block at the end is removed (see [10]). For instance, if $\pi = (3, 1, 2, 8, 9, 4, 5, 6, 7, 10)$, then $\mathsf{gl}(\pi) = (2, 1, 4, 3)$. Two important lemmata that will be used throughout the paper are given below.

Lemma 1 ([2]). *For $\pi \in \Pi_n$ we have $d_t(\pi) \geq \lceil \frac{\mathsf{blocks}(\pi)-1}{3} \rceil$ and $d_t(\pi) \geq \lceil \frac{\mathsf{bp}(\pi)}{3} \rceil$.*

Lemma 2 ([10]). *For $\pi \in \Pi_n$, $d_t(\pi) = d_t(\mathsf{gl}(\pi))$.*

A well-studied problem in genome rearrangement is Sorting by Transpositions. It asks for a permutation π and an integer k, whether $d_t(\pi) \leq k$. It is known that this problem is NP-complete [8], and the authors even showed the following result which will be used in our paper.

Lemma 3 ([8]). *For $\pi \in \Pi_n$ it is NP-hard to decide whether $d_t(\pi) = \frac{\mathsf{bp}(\pi)}{3}$.*

But there is a simple fixed-parameter algorithm outlined by Mahajan et al. [22]: First, if $\mathsf{bp}(\pi) > 3k$, we can immediately report that (π, k) is a no-instance by Lemma 1. Otherwise, we search for an id-transposition sequence of length at most k for $\mathsf{gl}(\pi)$ (see Lemma 2) using a simple search tree approach. As $\mathsf{gl}(\pi)$ has at most $3k$ elements, there are only $\mathcal{O}((3k)^3)$ possible transpositions. The search-tree depth is at most k, as we can perform at most k transpositions. Thus, we can determine in time $\mathcal{O}(n(3k)^{3k})$ if an id-transposition sequence for $\mathsf{gl}(\pi)$ exists, and also report it in the positive case. This transposition sequence can be easily transformed into an id-transposition sequence for π by transposing the blocks of π corresponding to the elements in $\mathsf{gl}(\pi)$ for each transposition.

Trees. We only consider ordered rooted trees T. Let $\mathsf{root}(T)$ be the *root* of T. Let $\mathsf{leaf}(T)$ be the set of *leaves* of T. For $v \in V(T)$ let $\mathsf{depth}(v)$ be the length of the shortest path between $\mathsf{root}(T)$ and v in T. For $v \in V(T)$, let $T(v)$ be the subtree of T rooted at v. For an internal node $v \in V(T)$ let $\mathsf{ch}(v)$ be the set of children of v and let $\mathsf{par}(v)$ be the parent of v. Further, let $\mathsf{anc}_T(v)$ be the set of *ancestors* of v in T. For two distinct vertices $v, w \in V(T)$ let $\mathsf{lca}(v, w)$ be the *lowest common ancestor* of v and w. Two vertices $v, w \in V(T)$ are *siblings* if they are children of the same vertex. If T is a binary tree, then we denote the two children by $\mathsf{lc}(v)$ and $\mathsf{rc}(v)$.

A rooted tree T encodes a set of permutations $\Pi(T)$ of its leaves which can be obtained by permuting children of an inner node: Namely, let $\Pi(T(v)) = \{(v)\}$ if $v \in \mathsf{leaf}(T)$. If $v \notin \mathsf{leaf}(T)$, let $\mathsf{ch}(v) = \{w_1, \ldots, w_k\}$ and we define

$$\Pi(T(v)) = \bigcup_{\psi \in \Pi_k} \Pi(T(w_{\psi_1})) \star \Pi(T(w_{\psi_2})) \star \cdots \star \Pi(T(w_{\psi_k})).$$

If $\pi \in \Pi(T)$, we say that π is *consistent* with T.

3 Block Crossings in Tanglegrams

In this section we want to properly define the problem we are dealing with. To reiterate, we are given two trees T_1 and T_2 with n leaves, a matching between their leaf sets, and a fixed leaf order of T_1. Notice, that w.l.o.g. we can assume that the leaf sets of both trees are labelled with $[n]$, and that the matching edges are between leaves labelled with the same integer. Further, by relabelling we can assume that the leaf order of T_1 is the identity permutation id_n. Now we want to find (1) a permutation π of the leaves of T_2 that conforms to the structure of T_2, and (2) a sequence of block crossings that route the matching edges from T_1 to T_2. (1) means that we are looking for a permutation $\pi \in \Pi(T_2)$. As a block crossing only changes the vertical order of two blocks of matching edges, (2) asks for a sequence of swaps of adjacent blocks of edges—which is a purely combinatorial procedure. Further, notice that each block crossing can be modelled by a transposition τ on the vertical order of edges. Hence, instead of looking for a sequence of block crossings, we equivalently look for a sequence of transposition transforming π into the identity permutation. This leads to the following decision variant of our problem, where $T := T_2$.

One-Tree Block Crossing Minimization (OTBCM)
Instance: A rooted tree T with $\mathsf{leaf}(T) = [n]$ and a positive integer k.
Question: Is there a permutation $\pi \in \Pi(T)$ such that $d_t(\pi) \leq k$?

Our algorithms will produce a witness in case of a YES-instance, that is, a permutation $\pi \in \Pi(T)$, and a id-transposition sequence $\tau^1, \ldots, \tau^\ell$ of π with $\ell \leq k$. In the following sections we will investigate the algorithmic complexity of this problem. We start with results that assume that T is a binary tree.

4 NP-Hardness

OTBCM implicitly contains as a subproblem to sort a permutation by a sequence of few transpositions. As Sorting by Transpositions is NP-complete [8], this suggests that OTBCM is also NP-complete. We show this for the restricted case, where the input tree is complete and binary. The proof, however, is not as straight-forward as the relation between OTBCM and SORTING BY TRANSPOSITIONS might suggest. The main idea is to construct for an input permutation π a tree T such that there exists $\pi' \in \Pi(T)$ with $\mathsf{gl}(\pi) = \mathsf{gl}(\pi')$, and π' is the only permutation in $\Pi(T)$ that minimizes the number breakpoints. As it is already NP-hard to decide whether $d_t(\pi)$ equals the lower bound $\mathsf{bp}(\pi)/3$ (Lemma 3), this shows NP-hardness for OTBCM.

Theorem 1 (⋆). *OTBCM is* NP*-complete for complete binary trees.*

5 Approximation

We present two approximation algorithms for OTBCM on binary trees. These rely on a polynomial algorithm by Walter and Dias [29] that sorts a permutation π of length n by at most $\frac{3}{4}\mathsf{bp}(\pi)$ transpositions in time $\mathcal{O}(n^2)$. Together with the lower bound of Lemma 1 this implies a 2.25-approximation algorithm for SORTING BY TRANSPOSITIONS[1]. Hence, our algorithms need to find the permutation $\pi \in \Pi(T)$ that minimizes the number of breakpoints. Applying the algorithm of Walter and Dias to this permutation gives a 2.25-approximation for OTBCM. We present two algorithms for minimizing the number of breakpoints. In the case of binary trees we present an $\mathcal{O}(n^3)$ algorithm, in the case of complete binary trees we can improve this to $\mathcal{O}(n^2)$. In both algorithms we start by discussing how to minimize the number of blocks instead of breakpoints, then we show how the algorithms can be adapted for breakpoints. Note that the number of blocks and breakpoints differ by at most one for any permutation, hence, minimizing blocks also leads to similar approximations.

Binary Trees. We start with an algorithm for binary trees. Essentially, finding $\pi \in \Pi(T)$ with minimal blocks is closely to the following previously studied problem: Given a complete binary tree T with n leaves and an arbitrary distance function $d : \mathsf{leaf}(T) \times \mathsf{leaf}(T) \to \mathbb{R}$ find $\pi \in \Pi(T)$ that minimizes $\sum_{i=1}^{n-1} d(\pi_i, \pi_{i+1})$. Defining $d(i,j) = (1 - \delta_{i+1,j})$ for $i, j \in \mathsf{leaf}(T)$ exactly captures our problem (note that d is not symmetric). Bar-Joseph et al. [3] gave an algorithm in the case where T is a binary tree and the distance function d is symmetric. We cannot directly apply their algorithm, so we show an adaptation to our problem, large parts of the algorithm are the same. The main idea is to do a bottom up dynamic program that computes the optimal ordering for the leaves of a subtree when the leftmost and rightmost leaf is fixed.

Theorem 2 (⋆). *Let T be a rooted binary tree with $\mathsf{leaf}(T) = [n]$. Then a permutation $\pi \in \Pi(T)$ minimizing the number of blocks (breakpoints) can be computed in $\mathcal{O}(n^3)$ time and $\mathcal{O}(n^2)$ space.*

[1] There exists a better approximation algorithm based on breakpoints leading to a 2-approximation, but the authors do not state any runtime [15].

Complete Binary Trees. If we are dealing with complete binary trees we can give a faster algorithm for minimizing blocks (breakpoints). This algorithm is based on an algorithm by Brandes [6] for the problem of finding $\pi \in \Pi(T)$ that minimizes $\sum_{i=1}^{n-1} d(\pi_i, \pi_{i+1})$ as described above. The algorithm of Brandes solves the problem for complete binary trees in $\mathcal{O}(n^2 \log n)$ time and $\mathcal{O}(n)$ space. This already gives an algorithm for our problem on complete binary trees T by setting $d(i,j) = (1-\delta_{i+1,j})$ for $i, j \in \mathsf{leaf}(T)$. But as we are not dealing with an arbitrary distance function, we can find an even faster algorithm.

Theorem 3. *Let T be a complete rooted binary tree with $\mathsf{leaf}(T) = [n]$ s.t. $n = 2^k$. Then a permutation $\pi \in \Pi(T)$ minimizing the number of blocks (breakpoints) can be computed in $\mathcal{O}(n^2)$ time and $\mathcal{O}(n \log n)$ space.*

We start by explaining the algorithm for blocks. The main idea is that, instead of applying dynamic programming bottom-up, we can apply dynamic programming building the permutation from left to right.

Brandes [6] pointed out that fixing a leaf of T to be at a specific position of a permutation $\pi \in \Pi(T)$ determines a partition into preceding and succeeding leaves. For $p \in [n]$ let $b^p = (b_k^p \dots b_1^p)$ be the k-bit string corresponding to the number $p-1$. For $i \in \mathsf{leaf}(T)$ we inductively define the jth parent $\mathsf{par}(i,j)$ of i as $\mathsf{par}(i,0) = i$ and $\mathsf{par}(i,j) = \mathsf{par}(\mathsf{par}(i,j-1))$. Furthermore, for $i \in \mathsf{leaf}(T)$ and $1 \le j \le k$ we define a precede-function $\mathsf{prec}(i,j)$ as

$$\mathsf{prec}(i,j) = \begin{cases} \mathsf{lc}(\mathsf{par}(i,j)), & \text{if } i \in \mathsf{leaf}(T(\mathsf{rc}(\mathsf{par}(i,j)))) \\ \mathsf{rc}(\mathsf{par}(i,j)), & \text{otherwise.} \end{cases}$$

The values $\mathsf{par}(i,j)$ and $\mathsf{prec}(i,j)$ can be precomputed in $\mathcal{O}(n \log n)$ time and stored using $\mathcal{O}(n \log n)$ space. The value $\mathsf{rob}(p) = \min\{i : 1 \le i \le k, b_i^p = 1\}$ is the *rightmost one bit* of the binary number corresponding to $p-1$ that is 1.

The key insight given by Brandes is stated in the following lemma.

Lemma 4 ([6]). *If leaf $i \in [n]$ is fixed at position $p \in [n] \setminus \{1\}$ then exactly $\mathsf{leaf}(T(\mathsf{prec}(i, \mathsf{rob}(p))))$ can precede i.*

Essentially, if a leaf i is fixed at a specific position p of $\pi \in \Pi(T)$, we know exactly which leaves $j \in \mathsf{leaf}(T)$ can precede it in π. Further, these leaves j exactly correspond to $\mathsf{leaf}(T(\mathsf{prec}(i, \mathsf{rob}(p))))$.This leads to the algorithm for Theorem 3.

Proof of Theorem 3. As in the algorithm of Brandes [6] we compute the values $\mathsf{opt}(i,p)$ that correspond to the minimal number of blocks of a prefix of length p ending with leaf i. When computing $\mathsf{opt}(i,p)$ we simply have to look at all values $\mathsf{opt}(j, p-1)$ for leaves j that can precede i when i is fixed at position p. But in our case, we know that $d(j,i)$ is 1 for all j with the only exception of $j = i-1$. Thus, we can store $\min_j \mathsf{opt}(j, p-1)$ at the internal vertex that corresponds to the lowest common ancestor of all these leaves j. Additionally, we have to check for the case where $i-1$ can precede i. Formally, we compute the following. For internal nodes v let $\mathsf{opt}(v,p) = \min(\mathsf{opt}(\mathsf{lc}(v),p), \mathsf{opt}(\mathsf{rc}(v),p))$. Clearly $\mathsf{opt}(i,1) = 1$ for all $i \in [n]$. For $i \in \mathsf{leaf}(T)$ and $p > 1$ we can compute $\mathsf{opt}(i,p)$ as

$$\mathsf{opt}(i,p) = \begin{cases} \mathsf{opt}(\mathsf{prec}(i,\mathsf{rob}(p)), p-1) + 1, & \text{if } i-1 \notin \mathsf{leaf}(T(\mathsf{prec}(i,\mathsf{rob}(p)))) \\ \min(\mathsf{opt}(\mathsf{prec}(i,\mathsf{rob}(p)), p-1) + 1, \mathsf{opt}(i-1,p-1)) & \text{otherwise.} \end{cases}$$

The case $i-1 \in \mathsf{leaf}(T(\mathsf{prec}(i,\mathsf{rob}(p))))$ exactly corresponds to the possibility of $i-1$ preceding i. This condition can be checked by precomputing $\mathsf{lca}(i-1,i)$ for all $2 \leq i \leq n$. Clearly, all values can be computed in $\mathcal{O}(n^2)$ time and

$$\min_{\pi' \in \Pi(T)} \mathsf{blocks}(\pi') = \min_{i \in [n]} \mathsf{opt}(i,n).$$

With the values in the opt-array, it is immediate how to also compute the permutation $\pi \in \Pi(T)$ with $\mathsf{blocks}(\pi) = \min_{i \in [n]} \mathsf{opt}(i,n)$ in $\mathcal{O}(n^2)$ space and time. But the space complexity can even be reduced to $\mathcal{O}(n \log n)$ as shown by Brandes [6]. We will only give the high-level idea here, the full description can be found in [6]. First, when computing the value $\min_{\pi' \in \Pi(T)} \mathsf{blocks}(\pi')$ we do not need to store the values $\mathsf{opt}(i,p)$ and $\mathsf{opt}(v,p)$ for all $p \in [n]$. We can simply iterate p from 1 to n and only keep opt-values for p and $p-1$. The main idea is then to only store the element in the middle of the optimal permutation and recursively determine the optimal permutation under this boundary condition in the first and the second half. This somewhat resembles a single-pivot quicksort approach. The time complexity is $T(n) = 2T(n/2) + n^2$ which solves to $T(n) = \mathcal{O}(n^2)$. This approach still works for our adaptation as the recursive procedure splits the search space into two equal parts, both still forming a complete binary tree. Thus, the opt-values can again be stored at internal nodes. The space complexity is dominated by $\mathcal{O}(n \log n)$ for storing $\mathsf{par}(i,j)$ and $\mathsf{prec}(i,j)$.

Lastly the algorithm can be adapted for breakpoints instead of blocks by setting $\mathsf{opt}(1,1) = 0$ in the beginning and setting $\mathsf{opt}(n,n) = \mathsf{opt}(n,n) - 1$ at the end of the algorithm. □

6 FPT-Algorithm

In this section we will show an FPT-algorithm for OTBCM for arbitrary binary trees parameterized by the number of transpositions. The proposed algorithm will be able to produce a witness for OTBCM in the same time if such a witness exists.

First, let us state two lemmata that are used for reduction rules. The first lemma shows that by removing an element from a permutation we never increase the transposition distance.

Lemma 5 (⋆). *Let* $\pi \in \Pi_n$ *and* $i \in [n]$. *Then* $d_t(\pi \ominus \pi_i) \leq d_t(\pi)$.

This can be used to show that, w.r.t. transposition distance, it is in some sense always "safe" to shift an element $x \in \pi$ after $x-1$ or before $x+1$ in a permutation. Let us give the formal definition and lemma.

Definition 1. *Let $\pi \in \Pi_n$, $i \in [n]$, and $\pi_i = x$. Define* $\mathsf{adjbef}(\pi, x)$ *as the permutation obtained from π by shifting x to the beginning if $x = 1$, otherwise to the position after $x - 1$. Formally,*

$$\mathsf{adjbef}(\pi, x) = \begin{cases} (x, \pi_1, \ldots, \pi_{i-1}, \pi_{i+1}, \ldots, \pi_n) & \textit{if } x = 1 \\ (\pi_1, \ldots, \pi_{j-1}, x - 1, x, \pi_{j+1}, \ldots, \pi_{i-1}, \pi_{i+1}, \ldots, \pi_n) & \\ \qquad \textit{if } x \neq 1 \textit{ and } i > j = \mathsf{inv}(\pi)(x - 1) & \\ (\pi_1, \ldots, \pi_{i-1}, \pi_{i+1}, \ldots \pi_{j-1}, x - 1, x, \pi_{j+1}, \ldots, , \pi_n) & \\ \qquad \textit{if } x \neq 1 \textit{ and } i < j = \mathsf{inv}(\pi)(x - 1). & \end{cases}$$

Similarly, define $\mathsf{adjaft}(\pi, x)$ *as the permutation obtained from π by shifting x to the end if $x = n$, otherwise to the position before $x + 1$. Formally,*

$$\mathsf{adjaft}(\pi, x) = \begin{cases} (\pi_1, \ldots, \pi_{i-1}, \pi_{i+1}, \ldots, \pi_n, x) & \textit{if } x = n \\ (\pi_1, \ldots, \pi_{j-1}, x, x + 1, \pi_{j+1}, \ldots, \pi_{i-1}, \pi_{i+1}, \ldots, \pi_n) & \\ \qquad \textit{if } x \neq n \textit{ and } i > j = \mathsf{inv}(\pi)(x + 1) & \\ (\pi_1, \ldots, \pi_{i-1}, \pi_{i+1}, \ldots \pi_{j-1}, x, x + 1, \pi_{j+1}, \ldots, , \pi_n) & \\ \qquad \textit{if } x \neq n \textit{ and } i < j = \mathsf{inv}(\pi)(x + 1). & \end{cases}$$

Lemma 6 (⋆). *Let $\pi \in \Pi_n$, $i \in [n]$, and $\pi_i = x$. Then $d_t(\mathsf{adjbef}(\pi, x)) \leq d_t(\pi)$ and $d_t(\mathsf{adjaft}(\pi, x)) \leq d_t(\pi)$.*

Intuitively, the algorithm in Theorem 4 at the end of this section first exhaustively applies a reduction rule based on Lemma 6 and removes leaves for which we know how they will be ordered in the tree in an optimal solution. Then, it applies a branching scheme that branches into the two orders of the children of an inner node in a bottom-up fashion, such that we can bound the depth of the resulting search tree. During this branching scheme, another reduction rule similar to the first one is applied, and in the leaf nodes of the search tree an FPT-algorithm for Sorting by Transpositions is applied.

We start with the first reduction rule (see Fig. 2a for an example) based on Lemma 6 that allows us to delete a pair of children from the input tree T which are siblings and whose difference is one.

Reduction Rule 1. *If there are two leaves $x, y \in \mathsf{leaf}(T)$ which are siblings such that $x + 1 = y$, (1) delete the leaf nodes corresponding to x and y from T, (2) replace their parent node, which is now a leaf, with x, and (3) for each leaf $z \in \mathsf{leaf}(T)$ with $z > y$, set $z = z - 1$.*

If $\mathsf{leaf}(T) = [n]$, then after the application of Reduction Rule 1 we have that $\mathsf{leaf}(T') = [n - 1]$ for the new tree T'. The safeness of Reduction Rule 1 follows directly from Lemma 6 and Lemma 2. Further, if we obtain a witness permutation $\pi' \in \Pi(T')$ with $d_t(\pi') \leq k$, this permutation can easily be transformed into a permutation $\pi \in \Pi(T)$ with $d_t(\pi) \leq k$ by increasing elements in the

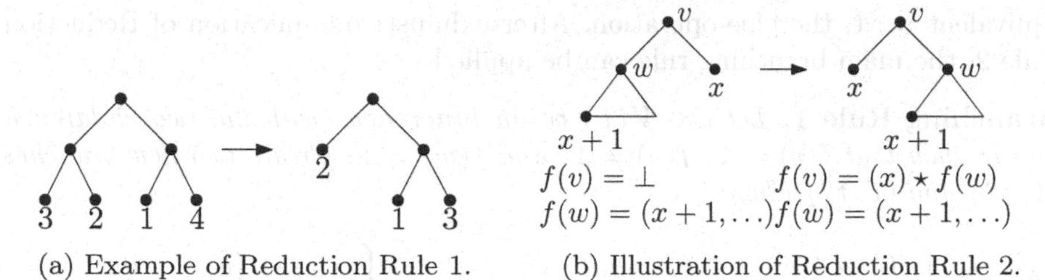

(a) Example of Reduction Rule 1. (b) Illustration of Reduction Rule 2.

Fig. 2. The reduction rules for the FPT-algorithm.

permutation that are larger than x by one and replacing the element x with the pair $(x, x+1)$. An id-transposition sequence for π' can then be mimicked in π.

After exhaustively applying Reduction Rule 1 we employ a search-tree algorithm such that in each search tree node some part of the ordering corresponding to sets $\mathsf{leaf}(T(v))$ for internal nodes v is already fixed. Formally, in each search tree node we are given a function f whose domain is $V(T)$ such that either $f(v) = \bot$, or $f(v) \in \Pi(T(v))$. Further, if $f(v) \in \Pi(T(v))$ and v is not a leaf node, then we have for the children w, x of v that $f(w) \neq \bot$ and $f(x) \neq \bot$, and $f(v)$ is *consistent with* $f(w)$ and $f(x)$, that is, $f(v) = f(w) \star f(x)$ or $f(v) = f(x) \star f(w)$. A permutation $\pi \in \Pi(T)$ is *consistent with* f if for all $v \in V(T)$ with $f(v) \neq \bot$, $\pi = \phi \star f(v) \star \psi$ for some ϕ, ψ which can also be the empty permutation. We call the problems (T, f) associated with the search tree nodes OTBCM-Extension. The question is if there is a permutation $\pi \in \Pi(T)$ consistent with f such that $d_t(\pi) \leq k$. In the root of our search tree and in the initial problem equivalent to OTBCM, $f(x) = (x)$ for all leaves $x \in \mathsf{leaf}(T)$ and $f(v) = \bot$ for all inner nodes.

A key observation is that f already tells us something about the number of breakpoint in any permutation $\pi \in \Pi(T)$ consistent with f. We thus let $\mathsf{bp}(f)$ be the number of pairs $x, y \in \mathsf{leaf}(T)$ that will form a breakpoint in any permutation $\pi \in \Pi(T)$ consistent with f. Notice that $\mathsf{bp}(f)$ can be determined in polynomial time as the breakpoint pairs (x, y) s.t. $x + 1 \neq y$ correspond to pairs of adjacent elements in some permutations $f(v) \neq \bot$.

Let us now present the next reduction rule that is applied during the search tree algorithm (see Fig. 2b for an illustration), and the branching rule that is employed once the reduction rule is not applicable.

Reduction Rule 2. *Let $v \in V(T)$ be an inner node with the two children x and w, such that $f(v) = \bot$, x is a leaf, $f(w) \neq \bot$, and $f(w)(1) = x + 1$ or $f(w)(|T(w)|) = x - 1$. If $f(w)(1) = x + 1$, set $f(v) = (x) \star f(w)$. Otherwise, set $f(v) = f(w) \star (x)$.*

The safeness of Reduction Rule 2 again follows from Lemma 6, as it essentially shifts x before $x + 1$ or after $x - 1$ in any permutation consistent with f. Notice that if $f(w)(1) = x+1$ and $f(w)(|T(w)|) = x-1$ it does not matter how we order the children of v, as resulting permutations $\pi \in \Pi(T)$ consistent with f will be equivalent w.r.t. the glue-operation. After exhaustive application of Reduction Rule 2, the main branching rule can be applied.

Branching Rule 1. *Let $v \in V(T)$ be an inner node with the two children u and w such that $f(v) = \bot$, $f(u) \neq \bot$, and $f(w) \neq \bot$. Create two new branches (T, f^α) and (T, f^β) where*

$$f^\alpha(x) = \begin{cases} f(u) \star f(w) & \text{if } x = v, \\ f(x) & \text{otherwise,} \end{cases} \qquad f^\beta(x) = \begin{cases} f(w) \star f(u) & \text{if } x = v, \\ f(x) & \text{otherwise.} \end{cases}$$

Branching Rule 1 essentially tries ordering the two children of an inner node in the two possible ways, if the corresponding orderings of subtrees rooted at the children are already determined by f. If u and w are both leaves, then

$$\mathsf{bp}(f^\alpha) = \mathsf{bp}(f^\beta) = \mathsf{bp}(f) + 1, \tag{1}$$

as the pair (u, w) or (w, u) will contribute one breakpoint. This is the case as Reduction Rule 1 was already applied exhaustively. Also, if only one of u and w is a leaf, and Reduction Rule 2 was already applied exhaustively, then Eq. (1) also holds: If, e.g., u is a leaf, then $(u, f(w)(1))$ or $(f(w)(|\mathsf{leaf}(T(w))|), u)$ will be a breakpoint, depending on the chosen order of u and w. This insight will allow us to bound the depth of the search tree.

With Branching Rule 1 we are now ready to give the theorem that captures the algorithm.

Theorem 4 ($\star$)**.** OTBCM *is solvable in time $\mathcal{O}(2^{6k} \cdot (3k)^{3k} \cdot n^{\mathcal{O}(1)})$ for rooted binary trees, i.e.,* OTBCM *is FPT for rooted binary trees when parameterized by the number of transpositions (= block crossings) k.*

While the full algorithm and proof are given in the full version [12], the intuitive idea is to first exhaustively apply Reduction Rule 1. Then a search tree is employed such that each search tree node corresponds to an instance of OTBCM-Extension. In each node, Reduction Rule 2 is first applied exhaustively. Then Branching Rule 1 creates two new child instances examining the two orders associated with the children of an inner node. Once $\mathsf{bp}(f) > 3k$ the search tree node can report failure because of Lemma 1. Otherwise, if $f(\mathsf{root}(T)) \neq \bot$, a total order on the leaves is fixed and the $\mathcal{O}(n(3k)^{3k})$-time algorithm sketched in Sect. 2 for Sorting by Transpositions can be applied. The correctness follows from the safeness of Reduction Rules 1 and 2, and because Branching Rule 1 essentially tries out all possible leaf orders. The search tree depth can be bounded because, essentially, if Branching Rule 1 is applied to an inner node without leaf-children, there are already breakpoints in the left and right subtree rooted at its children.

While we are giving an FPT algorithm here, we think that the asymptotic running time can be further optimized. Nonetheless, the bottleneck of the algorithm is still an FPT subprocedure for Sorting by Transpositions. So any better algorithm for Sorting by Transpositions will also improve Theorem 4.

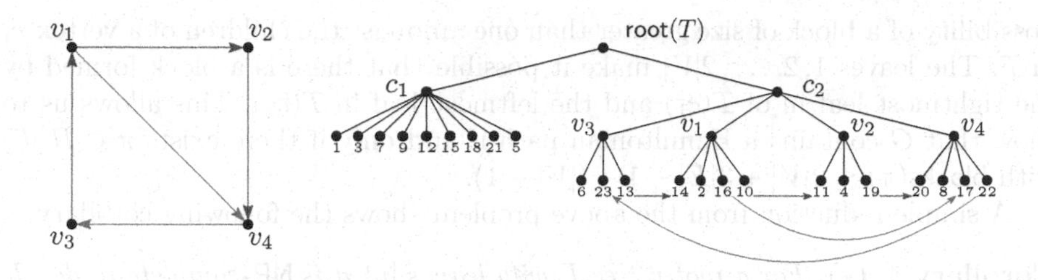

Fig. 3. Example of reduction in Theorem 5.

We also believe that a similar algorithm can be constructed for non-binary trees. In that case, the algorithm should have running time $\mathcal{O}(f(k, \Delta) \cdot n^c)$, where Δ is the maximum degree of the input tree T, c is a constant, and f is some function that only depends on k and Δ.

7 Beyond Binary Trees

In Sect. 5 we have given two approximation algorithms for OTBCM on binary trees. The key step was to find in polynomial time a permutation consistent with the input tree that minimizes the number of breakpoints. In the following we show that this is not possible for non-binary trees. We start by showing that finding a permutation with less than or equal k blocks is NP-complete.

Theorem 5 ($\star$). *For a rooted tree T with leaf set $[n]$ and an integer k it is* NP*-complete to decide whether there exists $\pi \in \Pi(T)$ s.t.* $\mathsf{blocks}(\pi) \leq k$.

We only outline the reduction from the Hamiltonian Path problem on directed graphs with at most one arc between a pair of vertices, which is NP-complete [19]. Let $G = (V, E)$ be an instance of Hamiltonian Path with directed arcs E. The problem is to find a path P in G that visits every vertex exactly once, called *Hamiltonian path.* Let $V = \{v_1, v_2, \dots, v_{|V|}\}$ and $E = \{e_1, e_2, \dots, e_{|E|}\}$. Without loss of generality, we assume that every vertex in G is incident to at least one edge. We construct a rooted tree T such that $\mathsf{leaf}(T) = [3|E| + 2|V|]$. An illustration is given in Fig. 3. The root of T has two children c_1 and c_2. Vertex c_1 contains $|V| + |E|$ children

$$\{1, 3, \dots, 2|V| - 1\} \cup \{2|V| + 1, 2|V| + 4, 2|V| + 7, \dots, 2|V| + 3|E| - 2\}.$$

Vertex c_2 contains $|V|$ children corresponding to the vertex set V. Let $\delta_G(v)$ be the *degree* of a vertex v in G. In T, vertex v_i has $1 + \delta_G(v_i)$ children, which are the following leaves: one child is $2i$; for each edge e_j incident to v_i one child is $2|V| + 3(j-1) + 2$ if v_j is the source of e_j, and $2|V| + 3(j-1) + 3$ otherwise. Notice that T contains $2|V| + 3|E|$ leaves and that $\mathsf{leaf}(T) = [2|V| + 3|E|]$, as intended. The intuition is that each edge $e \in E(G)$ corresponds to a triple $(\ell, \ell+1, \ell+2)$ of leaves in T such that ℓ is the child of c_1, and $\ell+1$, $\ell+2$ are children of vertices corresponding to the source and target of e. The leaf ℓ rules out the possibility of a block of size greater than one amongst the children of a vertex v_i in T. The leaves $1, 2, \ldots, 2|V|$ make it possible that there is a block formed by the rightmost leaf in of $T(c_1)$ and the leftmost leaf in $T(c_2)$. This allows us to show that G contains a Hamiltonian path if and only if there exists $\pi \in \Pi(T)$ with $\mathsf{blocks}(\pi) \leq 2|V| + 3|E| - 1 - (|V| - 1)$.

A simple reduction from the above problem shows the following corollary.

Corollary 1 **($\star$).** *For a rooted tree T with leaves $[n]$ it is* NP*-complete to decide whether there exists $\pi \in \Pi(T)$ s.t. $\mathsf{bp}(\pi) \leq k$.*

We believe though, that both problems become tractable if we fix an upper bound on the maximum degree of the tree or the number of blocks/breakpoints.

Hence, the techniques applied in Sect. 5 to obtain a polynomial time approximation algorithm do not extend to non-binary trees. This does not imply, however, that there is no polynomial time approximation for OTBCM.

8 Summary and Open Problems

We have analyzed the complexity of minimizing block crossings in one-sided (binary) tanglegrams from different theoretical perspectives. A number of open problems and interesting research directions remain:

- We have considered one tree to be fixed. What happens if we can permute the leaves of both trees?
- Sorting by Transpositions admits better approximations than those we have presented here. However, these algorithms rely on a more complicated structure called *breakpoint graph* (see, e.g., [2,14,20,27]). Can these approximations be utilized for tanglegrams?
- A pair of edges could cross multiple times in the tanglegrams produced by our algorithms. Can this be prevented using a similar notion such as *monotone block crossings* (see [18])?
- How do our algorithms perform in practice? What is the relation between number of pairwise crossings and number block crossings in practice?

References

1. Alam, M.J., Fink, M., Pupyrev, S.: The bundled crossing number. In: Hu, Y., Nöllenburg, M. (eds.) GD 2016. LNCS, vol. 9801, pp. 399–412. Springer, Cham (2016). https://doi.org/10.1007/978-3-319-50106-2_31
2. Bafna, V., Pevzner, P.A.: Sorting by transpositions. SIAM J. Discret. Math. **11**(2), 224–240 (1998). https://doi.org/10.1137/S089548019528280X
3. Bar-Joseph, Z., Demaine, E.D., Gifford, D.K., Srebro, N., Hamel, A.M., Jaakkola, T.S.: K-ary clustering with optimal leaf ordering for gene expression data. Bioinformatics **19**(9), 1070–1078 (2003). https://doi.org/10.1093/bioinformatics/btg030
4. Baumann, F., Buchheim, C., Liers, F.: Exact bipartite crossing minimization under tree constraints. In: Festa, P. (ed.) SEA 2010. LNCS, vol. 6049, pp. 118–128. Springer, Heidelberg (2010). https://doi.org/10.1007/978-3-642-13193-6_11

5. Böcker, S., Hüffner, F., Truss, A., Wahlström, M.: A faster fixed-parameter approach to drawing binary tanglegrams. In: Chen, J., Fomin, F.V. (eds.) IWPEC 2009. LNCS, vol. 5917, pp. 38–49. Springer, Heidelberg (2009). https://doi.org/10.1007/978-3-642-11269-0_3
6. Brandes, U.: Optimal leaf ordering of complete binary trees. J. Discrete Algorithms **5**(3), 546–552 (2007). https://doi.org/10.1016/j.jda.2006.09.003
7. Buchin, K., et al.: Drawing (complete) binary tanglegrams - hardness, approximation, fixed-parameter tractability. Algorithmica **62**(1–2), 309–332 (2012). https://doi.org/10.1007/s00453-010-9456-3
8. Bulteau, L., Fertin, G., Rusu, I.: Sorting by transpositions is difficult. SIAM J. Discret. Math. **26**(3), 1148–1180 (2012). https://doi.org/10.1137/110851390
9. Bulteau, L., Gambette, P., Seminck, O.: Reordering a tree according to an order on its leaves. In: Bannai, H., Holub, J. (eds.) Proceedings of 33rd Symposium on Combinatorial Pattern Matching (CPM). LIPIcs, vol. 223, pp. 24:1–24:15 (2022). https://doi.org/10.4230/LIPIcs.CPM.2022.24
10. Christie, D.A.: Genome Rearrangement Problems. Ph.D. thesis, University of Glasgow (1998). https://theses.gla.ac.uk/74685/
11. van Dijik, T.C., et al.: Block crossings in storyline visualizations. J. Graph Algorithms Appl. **21**(5), 873–913 (2017). https://doi.org/10.7155/jgaa.00443
12. Dobler, A., Nöllenburg, M.: Block crossings in one-sided tanglegrams. CoRR abs/2305.04682 (2023). https://doi.org/10.48550/arXiv.2305.04682
13. Dwyer, T., Schreiber, F.: Optimal leaf ordering for two and a half dimensional phylogenetic tree visualisation. In: Churcher, N., Churcher, C. (eds.) Proceedings of Australasian Symposium on Information Visualisation (InVis.au). CRPIT, vol. 35, pp. 109–115. Australian Computer Society (2004). https://crpit.scem.westernsydney.edu.au/abstracts/CRPITV35Dwyer.html
14. Elias, I., Hartman, T.: A 1.375-approximation algorithm for sorting by transpositions. IEEE ACM Trans. Comput. Biol. Bioinform. **3**(4), 369–379 (2006). https://doi.org/10.1109/TCBB.2006.44
15. Eriksson, H., Eriksson, K., Karlander, J., Svensson, L.J., Wästlund, J.: Sorting a bridge hand. Discret. Math. **241**(1–3), 289–300 (2001). https://doi.org/10.1016/S0012-365X(01)00150-9
16. Fernau, H., Kaufmann, M., Poths, M.: Comparing trees via crossing minimization. J. Comput. Syst. Sci. **76**(7), 593–608 (2010). https://doi.org/10.1016/j.jcss.2009.10.014
17. Fink, M., Hershberger, J., Suri, S., Verbeek, K.: Bundled crossings in embedded graphs. In: Kranakis, E., Navarro, G., Chávez, E. (eds.) LATIN 2016. LNCS, vol. 9644, pp. 454–468. Springer, Heidelberg (2016). https://doi.org/10.1007/978-3-662-49529-2_34
18. Fink, M., Pupyrev, S., Wolff, A.: Ordering metro lines by block crossings. J. Graph Algorithms Appl. **19**(1), 111–153 (2015). https://doi.org/10.7155/jgaa.00351
19. Garey, M.R., Johnson, D.S.: Computers and Intractability: A Guide to the Theory of NP-Completeness. Freeman, W. H (1979)
20. Hartman, T., Shamir, R.: A simpler and faster 1.5-approximation algorithm for sorting by transpositions. Inf. Comput. **204**(2), 275–290 (2006). https://doi.org/10.1016/j.ic.2005.09.002
21. Holten, D., van Wijk, J.J.: Visual comparison of hierarchically organized data. Comput. Graph. Forum **27**(3), 759–766 (2008). https://doi.org/10.1111/j.1467-8659.2008.01205.x

22. Mahajan, M., Rama, R., Raman, V., Vijaykumar, S.: Approximate block sorting. Int. J. Found. Comput. Sci. **17**(2), 337–356 (2006). https://doi.org/10.1142/S0129054106003863
23. Nöllenburg, M.: Crossing layout in non-planar graph drawings. In: Hong, S.-H., Tokuyama, T. (eds.) Beyond Planar Graphs, pp. 187–209. Springer, Singapore (2020). https://doi.org/10.1007/978-981-15-6533-5_11
24. Nöllenburg, M., Völker, M., Wolff, A., Holten, D.: Drawing binary tanglegrams: an experimental evaluation. In: Finocchi, I., Hershberger, J. (eds.) Proceedings of 11th Workshop on Algorithm Engineering and Experiments (ALENEX), pp. 106–119. SIAM (2009). https://doi.org/10.1137/1.9781611972894.11
25. Page, R.D.M.: Tangled Trees: Phylogeny, Cospeciation, and Coevolution. University of Chicago Press, Chicago (2003)
26. Scornavacca, C., Zickmann, F., Huson, D.H.: Tanglegrams for rooted phylogenetic trees and networks. Bioinformatics **27**(13), 248–256 (2011). https://doi.org/10.1093/bioinformatics/btr210
27. Silva, L.A.G., Kowada, L.A.B., Rocco, N.R., Walter, M.E.M.T.: A new 1.375-approximation algorithm for sorting by transpositions. Algorithms Mol. Biol. **17**(1), 1 (2022). https://doi.org/10.1186/s13015-022-00205-z
28. Venkatachalam, B., Apple, J., John, K.S., Gusfield, D.: Untangling tanglegrams: comparing trees by their drawings. IEEE ACM Trans. Comput. Biol. Bioinform. **7**(4), 588–597 (2010). https://doi.org/10.1109/TCBB.2010.57
29. Walter, M.E.T., Dias, Z., Meidanis, J.: A new approach for approximating the tranposition distance. In: de la Fuente, P. (ed.) Proceedings of 7th Symposium on String Processing and Information Retrieval (SPIRE), pp. 199–208. IEEE Computer Society (2000). https://doi.org/10.1109/SPIRE.2000.878196

Observation Routes and External Watchman Routes

Adrian Dumitrescu[1] and Csaba D. Tóth[2,3](✉)

[1] Algoresearch L.L.C., Milwaukee, WI, USA
ad.dumitrescu@algoresearch.org
[2] California State University Northridge, Los Angeles, CA, USA
csaba.toth@csun.edu
[3] Tufts University, Medford, MA, USA

Abstract. We introduce the Observation Route Problem (ORP) defined as follows: Given a set of n pairwise disjoint compact regions in the plane, find a shortest tour (route) such that an observer walking along this tour can see (observe) some point in each region from some point of the tour. The observer does *not* need to see the entire boundary of an object. The tour is *not* allowed to intersect the interior of any region (i.e., the regions are obstacles and therefore out of bounds). The problem exhibits similarity to both the Traveling Salesman Problem with Neighborhoods (TSPN) and the External Watchman Route Problem (EWRP). We distinguish two variants: the range of visibility is either limited to a bounding rectangle, or unlimited. We obtain the following results:

(I) Given a family of n disjoint convex bodies in the plane, computing a shortest observation route does not admit a $(c \log n)$-approximation unless $\mathsf{P} = \mathsf{NP}$ for an absolute constant $c > 0$. (This holds for both limited and unlimited vision.)

(II) Given a family of disjoint convex bodies in the plane, computing a shortest external watchman route is NP-hard. (This holds for both limited and unlimited vision; and even for families of axis-aligned squares.)

(III) Given a family of n disjoint fat convex polygons, an observation tour whose length is at most $O(\log n)$ times the optimal can be computed in polynomial time. (This holds for limited vision.)

(IV) For every $n \geq 5$, there exists a convex polygon with n sides and all angles obtuse such that its perimeter is *not* a shortest external watchman route. This refutes a conjecture by Absar and Whitesides (2006).

1 Introduction

Path planning and visibility are two central areas in computational geometry and robotics. In path planning, a short collision-free path between two specified points is desired, and the robot has to see or detect obstacles in order to avoid them in its path. Hence there is a close relation between short paths and visibility. Moreover, visibility of an object (say, an obstacle) can be accomplished at various

P. Morin and S. Suri (Eds.): WADS 2023, LNCS 14079, pp. 401–415, 2023.
https://doi.org/10.1007/978-3-031-38906-1_26

degrees; for instance, sometimes it may suffice to simply detect the presence of an obstacle, and other times the robot may need to map or recognize (e.g., see) the entire boundary of an obstacle in order to select a meaningful action.

In the Traveling Salesman with Neighborhoods problem (TSPN), given a set of regions (neighborhoods) in the plane, one is to compute a shortest closed route (tour) that visits each neighborhood; whereas in the External Watchman Route Problem (EWRP), given a set of disjoint regions in the plane, one is to compute a shortest closed route (tour) so that every point in the exterior of a region (i.e., in the *free space*), is visible from some point of the tour. The problems were posed about three decades ago by Arkin and Hassin [3] and by Ntafos and Gewali [36], respectively. A small example that illustrates ORP and EWRP appears in Fig. 1. Here we introduce the following related problem we call the Observation Route Problem (ORP):

ORP: Given a set of n pairwise disjoint compact regions in the plane, find a shortest route (tour) such that an observer going along this tour can see (observe) each of the regions from at least one point of the tour. The route cannot enter the interior of any region.

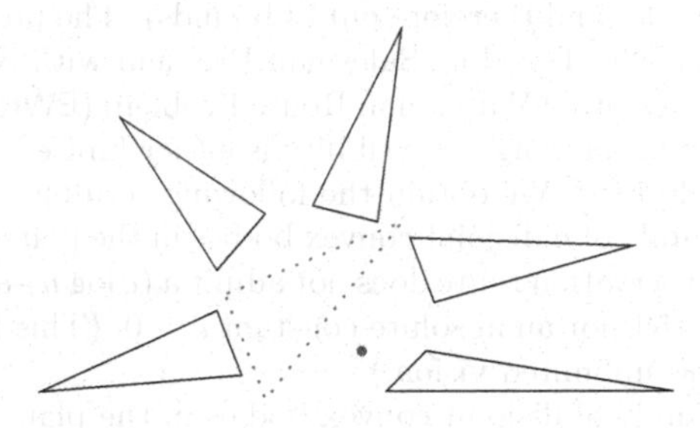

Fig. 1. An observation route (the blue point) and an external watchman route (dotted, in red) for a set of five triangles. (Color figure online)

Related Work. In the Euclidean Traveling Salesman problem (ETSP), given a set of points in the plane (or in the Euclidean space $\mathbb{R}^d$, $d \geq 3$), one seeks a shortest tour (closed curve) that visits each point. In the *TSP with neighborhoods* (TSPN), each point is replaced by a (possibly disconnected) region [3]. The tour must visit at least one point in each of the given regions (i.e., it must intersect each region). Since ETSP is NP-hard in $\mathbb{R}^d$ for every $d \geq 2$ [20,21,39], TSPN is also NP-hard for every $d \geq 2$.

At about the same time, Arora [4] and Mitchell [28] independently showed that ETSP in $\mathbb{R}^d$, for constant d, admits a polynomial-time approximation scheme (PTAS). In contrast, TSPN is harder to approximate and certain instances are known to be APX-hard. However, better approximations can be obtained for neighborhoods with "nice" geometric properties: connected, pairwise disjoint, or fat, or of comparable sizes, etc. Arkin and Hassin [3] gave constant-factor approximations for translates of a connected region; Dumitrescu and Mitchell [15] extended the above result to connected neighborhoods of comparable diameters.

For n connected (possibly overlapping) neighborhoods in the plane, TSPN can be approximated with ratio $O(\log n)$ by the algorithms of (i) Mata and Mitchell [27], (ii) Gudmundsson and Levcopoulos [23], and (iii) Elbassioni, Fishkin, and Sitters [18]. The $O(\log n)$-approximation stems from the following early result by Levcopoulos and Lingas [25]: Every (simple) rectilinear polygon P with n vertices, r of which are reflex, can be partitioned in $O(n \log n)$ time into rectangles whose total perimeter is $\log r$ times the perimeter of P. We will use any of the three algorithms mentioned above as a subroutine in our approximation algorithm for ORP in Sect. 3.

In the Watchman Route Problem (WRP), given a polygonal domain P, the goal is to find a shortest closed curve within P such that every point of P is seen from some point along the curve. Thus WRP is dual to EWRP in the sense that the former deals with the interior of a polygonal domain whereas the latter deals with the exterior of one or more polygons. The watchman route problem in a simple polygon P was first considered by Chin and Ntafos as early as 1986 [7–9]. After more than one decade of being in a tangle [6,42,43], its polynomial status appears to have been settled by Tan et al. [44]. The current fastest algorithm, running in $O(n^4 \log n)$ time, is due to Dror et al. [12]. A linear-time 2-approximation algorithm is due to Tan [43]. Ntafos and Gewali [36] showed that a shortest external watchman route for an n-vertex convex polygon can be found in $O(n)$ time. They also studied the case of two convex polygons [22]. The first polynomial-time approximation algorithm for the watchman route problem in n-vertex polygons with holes was given by Mitchell [30]; its approximation ratio is $O(\log^2 n)$. Among recent work on guarding tours in polygons with holes, including inapproximability, we mention that of Nilsson and Żyliński [34].

Regarding the degree of approximation achievable, TSPN for arbitrary neighborhoods is APX-hard [10,40], and approximating TSPN for connected regions in the plane within a factor smaller than 2 is intractable (NP-hard) [40]. The problem is also APX-hard for disconnected regions [40], the simplest case being point-pair regions [13]. It is conjectured that approximating TSPN for disconnected regions in the plane within a $O(\log^{1/2} n)$ factor is intractable [40]. Computing the minimum number of point guards, vertex guards, or edge guards, are all APX-hard [17] and finding the minimum number of point guards is $\exists\mathbb{R}$-complete [1], even for simple polygons (without holes). See [5] and [24] for recent approximations and parameterized hardness results. Historically, watchman routes under limited visibility have been considered by Ntafos [35]; other variants are discussed in [29,31]. The problem of computing shortest *external* watchman routes for collections of disjoint polygons was suggested by Ntafos and Gewali [36].

Definitions and Notations. A curve is called *simple* if it has no self-intersections. A *simple polygon* P is a polygon without holes, that is, the interior of the polygon is topologically equivalent to a disk. A *polygon with holes* is obtained by removing a set of nonoverlapping, strictly interior, simple subpolygons from P [38]. The Euclidean length of a curve γ is denoted by $\text{len}(\gamma)$, or just $|\gamma|$ when there is no danger of confusion. Similarly, the total (Euclidean) length of the edges of a geometric graph G or a polygon P is denoted by $\mathrm{len}(G)$ and $\mathrm{per}(P)$, respectively.

A TSP tour for a set $\mathcal{F}$ of regions (neighborhoods) in $\mathbb{R}^d$, $d \geq 2$, is a closed curve in the ambient space that intersects $\mathcal{F}$ (i.e., γ intersects each region in $\mathcal{F}$). For $\alpha \geq 1$, an approximation algorithm (for ORP, EWRP, or TSPN) has ratio α if its output tour ALG satisfies $\mathrm{len}(\mathsf{ALG}) \leq \alpha \, \mathrm{len}(\mathsf{OPT})$, where OPT is an optimal tour for the respective problem.

A *convex body* $C \subset \mathbb{R}^d$ is a compact convex set with nonempty interior. Its *boundary* is denoted by ∂C and its *interior* by $\mathring{C}$. The *width* of a convex body C is the minimum width of a strip of parallel lines enclosing C. Informally, a convex body is *fat* if its width is comparable with its diameter. More precisely, for $0 \leq \lambda \leq 1$, a convex body C is λ-*fat* if its width w is at least λ times the diameter: $w \geq \lambda \cdot \mathrm{diam}(C)$, and C is *fat* if the inequality holds for a constant λ. For instance, a square is $\frac{1}{\sqrt{2}}$-fat, a 3×1 rectangle is $\frac{1}{\sqrt{10}}$-fat and a segment is 0-fat. Let γ be a closed curve. The *geometric dilation* of γ is $\delta(\gamma) := \sup_{p,q \in \gamma} \frac{d_\gamma(p,q)}{|pq|}$, where $d_\gamma(p,q)$ is the shortest distance along γ between p and q. For example, the geometric dilation of a square is 2 and that of a 3×1 rectangle is 4.

Points p and q are mutually *visible* if the segment pq does not intersect the interior of any region in $\mathcal{F}$ [37]. An object $O \in \mathcal{F}$ can be *seen* (or *observed*) from a point p if there is a point $q \in \partial O$ such that p and q are mutually visible. The convex hull of a set $A \subset \mathbb{R}^d$ is denoted by $\mathrm{conv}(A)$.

1.1 Our Results

In Sect. 2, as a preliminary result, we show that given a set of convex polygons $\mathcal{F}$, determining whether $\mathcal{F}$ can be observed from a single point can be done by a polynomial time algorithm. In Sect. 3 we show that given a set of n pairwise disjoint fat convex polygons, an observation tour whose length is at most $O(\log n)$ times the optimal can be computed in polynomial time. The algorithm reduces the ORP problem to TSPN for polygons with holes and then executes additional local transformations of the tour that only increase the total length by at most a constant factor.

Theorem 1. *Given a family of n pairwise disjoint fat convex polygons, an observation tour whose length is at most $O(\log n)$ times the optimal can be computed in polynomial time.*

In the full version of the paper, we prove the NP-hardness of both ORP and EWRP for both limited and unlimited vision. Throughout this paper, the term *limited vision* refers to unrestricted vision in a given bounding box of the family.

Theorem 2. *Given a family of disjoint convex bodies in the plane, computing a shortest observation route is* NP*-hard. (This holds for both limited and unlimited vision.) The problem remains so even for families of axis-aligned squares.*

Theorem 3. *Given a family of disjoint convex bodies in the plane, computing a shortest external watchman route is* NP*-hard. (This holds for both limited and unlimited vision.) The problem remains so even for families of axis-aligned squares.*

In Sect. 4, we prove that one cannot approximate the minimum length of an observation route for n convex bodies in the plane within a factor of $c \log n$, for some $c > 0$, unless P = NP. The inapproximabilty is proved via a reduction from Set Cover.

Theorem 4. *Given a family of n disjoint convex bodies in the plane, the length of a shortest observation route cannot be approximated within a factor of $c \log n$ unless* P = NP*, where $c > 0$ is an absolute constant. (This holds for both limited and unlimited vision.)*

In Sect. 5 we study the structure of shortest external watchman routes for a convex polygon (i.e., $|\mathcal{F}| = 1$). In 2006, Absar and Whitesides conjectured that all convex polygons with all angles *obtuse* have convex-hull routes as their shortest external watchman routes [2]. Theorem 5 below refutes this conjecture for every $n \geq 5$.

Theorem 5. *For every $n \geq 5$ there exists a convex polygon with n sides and all angles obtuse such that its perimeter is not a shortest external watchman route.*

2 Preliminaries

Throughout the paper we consider families of disjoint compact convex sets in the plane; and are only concerned with *external* visibility. For a set $\mathcal{F}$ of n disjoint polygons in a rectangle R, the *visibility region* of a polygon $P \in \mathcal{F}$, denoted $V(P)$, is the set of all point $p \in R$ such that there exists a point $q \in \partial P$ such that the line segment pq is disjoint from the interior of P' for all $P' \in \mathcal{F}$. See Fig. 2 for an example.

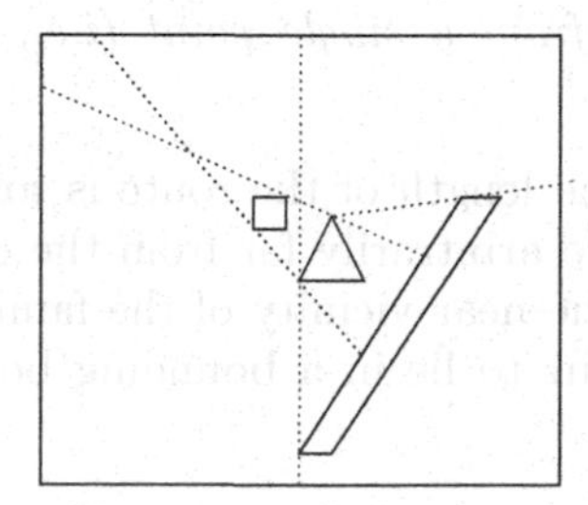
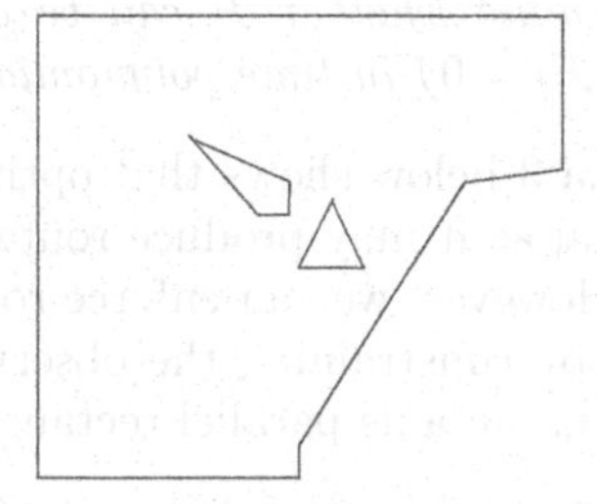

Fig. 2. Left: three convex polygons in a bounding box. Right: the visibility region V of the triangle is a polygon with two holes.

Lemma 1. *Given a set $\mathcal{F}$ of n disjoint convex polygons with a total of m vertices in a rectangle R, for every $C \in \mathcal{F}$, the visibility region $V(C)$ is a polygon with $O(m+n^2)$ vertices and $O(n^2)$ holes.*

Proof. Note that C is always a hole of $V(C)$. The boundary of $V(C)$ is contained in the union of the boundary of the free space $R \setminus \bigcup_{C' \in \mathcal{F}} C'$ and all common inner and outer tangents between C and other polygons in $\mathcal{F}$. Consequently, $V(C)$ is a polygon with $O(m+n^2)$ vertices. Since the polygons in $\mathcal{F}$ are disjoint, every hole of $V(C)$ has a vertex at the intersection of two tangents or one tangent and the boundary $\partial C'$, $C' \in \mathcal{F} \setminus \{C\}$. Now $O(n)$ tangents and $O(n)$ convex curves yield $O(n^2)$ intersections, and so the number of holes in $V(C)$ is also $O(n^2)$.

It remains to prove that $V(C)$ is connected. Let p and q be any two points in $V(C)$. By definition, there exist $p', q' \in \partial C$ such that p' is visible from p, and q' is visible from q. Let $\rho(p', q')$ denote the shortest path connecting p' and q' on the boundary of C. Then one can reach q from p via the 3-leg path that connects p to p' via a straight-line segment, follows $\rho(p', q')$ on C's boundary and connects q' to q via a straight-line segment.

Next we show that given a set $\mathcal{F}$ of convex polygons, one can determine in polynomial-time whether all polygons in $\mathcal{F}$ can be observed from a single point. Note that for the variant with unlimited visibility, visibility regions may be unbounded. We need a simple lemma regarding zero-instances of TSPN.

Lemma 2. *Given a family $\mathcal{F}$ of (possibly unbounded) polygonal regions with a total of m vertices, one can determine whether there exists a point contained in all polygons in $\mathcal{F}$ (i.e., whether $\mathsf{OPT}_{\mathsf{TSPN}}(\mathcal{F}) = 0$) in time polynomial in m.*

Proof. This is equivalent to determining whether the intersection $\bigcap_{P \in \mathcal{F}} P$ is empty. Since the total complexity of the visibility regions is polynomial in m, the complexity of the intersection is also polynomial in m. Consequently, the resulting algorithm takes time polynomial in m.

Applying Lemma 2 to visibility regions immediately yields the following.

Corollary 1. *Given a set $\mathcal{F}$ of convex polygons with a total of m vertices, one can determine whether $\mathcal{F}$ can be observed from a single point (i.e., whether $\mathsf{OPT}_{\mathsf{ORP}}(\mathcal{F}) = 0$) in time polynomial in m.*

Lemma 3 below shows that optimizing the length of the route is sometimes impractical as it may produce routes that are arbitrarily far from the observed objects. However, we can enforce routes in the near vicinity of the family to be observed by constraining the observation tour to lie in a bounding box of the family (e.g., an axis-parallel rectangle).

Lemma 3. *For every $\Delta > 0$, there exists a configuration $\mathcal{F} = \mathcal{F}(\Delta)$ of $O(1)$ axis-parallel unit squares such that:* (i) $\mathrm{diam}(\mathrm{conv}(\mathcal{F})) = O(1)$, (ii) *$\mathsf{OPT}_{\mathsf{ORP}}(\mathcal{F}) = 0$, i.e., the configuration can be observed from a single point,* and (iii) *every single observation point is at a distance at least Δ from* $\mathrm{conv}(\mathcal{F})$. *Alternatively, $\mathcal{F}$ can be realized from unit disks.*

Proof. We exhibit and analyze a configuration (family $\mathcal{F}$) of six axis-parallel unit squares; refer to Fig. 3. An analogous unit disk configuration (with six elements) can be derived from a piece of the hexagonal disk packing by slightly shrinking each disk from its center; its analysis is left to the reader.

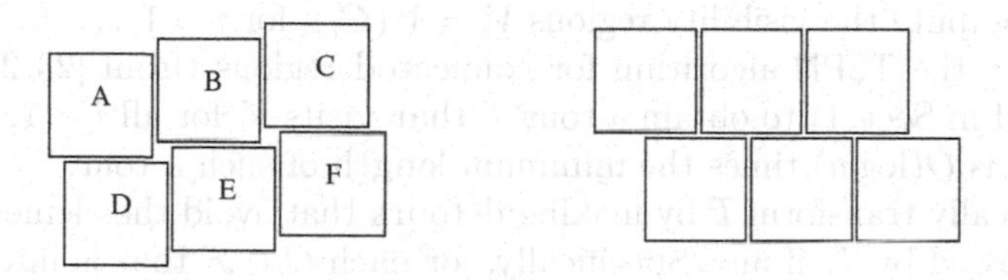

Fig. 3. Left: This family can be only observed from single points far away up or down. Right: This family can be observed from any single point on a horizontal line that separates the upper chain of squares from the lower one.

Let $\varepsilon > 0$ be small. Successive squares (from left to right) are horizontally separated by ε and shifted vertically by 2ε. Let ℓ be a horizontal line separating the squares B and E. Without loss of generality let p be an observation point on or above ℓ. If p lies inside $\text{conv}(\mathcal{F})$, depending on its position, either C or D is not observable from p (if $\varepsilon > 0$ is sufficiently small). Suppose that p lies outside $\text{conv}(\mathcal{F})$, and on or to the right of the vertical axis of symmetry of B. Then D is not observable from p unless p is above the common internal tangent to A and B of positive slope and above the common internal tangent to B and C of negative slope. These tangents continuously depend on ε and are almost vertical as ε tends to zero, hence the lowest point in the intersection of the corresponding halfplanes can be arbitrarily high, as claimed. The case when p lies outside $\text{conv}(\mathcal{F})$ on or to the left of the vertical axis of symmetry of B is similar. It is clear that one can always find a suitable $\varepsilon = \varepsilon(\Delta)$, as required. □

The following lemma relates fatness to geometric dilation for closed curves:

Lemma 4. *Let C be a λ-fat convex curve. Then $\delta(C) \leq \min(\pi\lambda^{-1}, 2(\lambda^{-1}+1))$.*

Proof. If C is a convex curve, it is known [16, Lemma 11] that $\delta(C) = \frac{|C|}{2h}$. It is also known [14, Theorem 8] that $h \geq w/2$, where $h = h(C)$ is the *minimum halving distance* of C (i.e., the minimum distance between two points on C that divide the length of C in two equal parts), and $w = w(C)$ is the width of C. Putting these together one deduces that $\delta(C) \leq \frac{|C|}{w}$. Let D denote the diameter of C. The isoperimetric inequality $|C| \leq D\pi$ and the obvious inequality $|C| \leq 2D+2w$ lead to the following dilation bounds $\delta(C) \leq \pi\frac{D}{w}$ and $\delta(C) \leq 2\left(\frac{D}{w}+1\right)$, see also [14,41]. Since C is λ-fat, direct substitution yields the two bounds given in the lemma. Note that the latter bound is better for small λ. □

3 Fat Convex Polygons

In this section we prove Theorem 1. The following algorithm computes a tour for a family $\mathcal{F} = \{C_1, \ldots, C_n\}$ of convex polygons in a rectangle R.

Algorithm 1.
STEP 1: Compute the visibility regions $V_i = V(C_i)$, for $i = 1, \ldots, n$.
STEP 2: Use the TSPN algorithm for connected regions (from [23,27], or [18], as explained in Sect. 1) to obtain a tour T that visits V_i for all $i = 1, \ldots, n$ such that $\text{len}(T)$ is $O(\log n)$ times the minimum length of such a tour.
STEP 3: Locally transform T by making detours that avoid the elements $C_i \in \mathcal{F}$ that are crossed by T, if any. Specifically, for each $C \in \mathcal{F}$ that is intersected by T, replace the subpath $\varrho = C \cap T$ by the shortest path along ∂C connecting the start and end points of ϱ; as shown in Fig. 4. Output the resulting tour T'.

Fig. 4. Local replacements to obtain T' from T.

Algorithm Analysis. By Lemma 1, all visibility regions $V_i = V(C_i)$ are connected, and so $\mathcal{F}$ represents a valid input for the TSPN algorithm. Recall that the tour T returned by the TSPN algorithm visits all visibility regions; this means that each C_i is seen from some point in $T \setminus (\bigcup_{i=1}^n C_i)$. The local replacements in STEP 3 ensure that the resulting tour T' does not intersect (the interior of) any obstacle, and maintain the property that the tour visits all visibility regions; i.e., each C_i is seen from some point in $T' \setminus (\bigcup_{i=1}^n C_i)$. Consequently, T' is an observation route for $\mathcal{F}$. Since each region V_i has polynomial complexity by Lemma 1 (i.e., polynomial in the total number of vertices of the polygons in $\mathcal{F}$), Algorithm 1 runs in polynomial time.

It remains to bound $\text{len}(T')$ from above. Let $\mathcal{V} = \{V_1, \ldots, V_n\}$. An observation route for $\mathcal{F}$ must visit the visibility regions V_i for all $i = 1, \ldots, n$, and so $\text{OPT}_{\text{TSPN}}(\mathcal{V}) \leq \text{OPT}_{\text{ORP}}(\mathcal{F})$, which implies

$$\text{len}(T) \leq O(\text{OPT}_{\text{TSPN}}(\mathcal{V}) \log n) \leq O(\text{OPT}_{\text{ORP}}(\mathcal{F}) \log n).$$

Recall that all C_i are fat and thus by Lemma 4, the local replacements in STEP 3 increase the length by at most a constant factor (that depends on $\lambda = \Omega(1)$), that is, $\text{len}(T') \leq O(\text{len}(T))$. This concludes the proof of Theorem 1.

4 Inapproximability Results

We deduce the inapproximability of ORP from that of Set Cover. A *set system* is a pair $(U, \mathcal{S})$, where U is a set and $\mathcal{S}$ is a collection of subsets of U. Given a set system $(U, \mathcal{S})$, the Set Cover problem asks for the minimum number of sets in $\mathcal{S}$ whose union is U. Set Cover cannot be approximated within a factor of $(1-o(1)) \ln n$ unless P = NP [11], where $n = |U|$. Furthermore, for any $c \in (0,1)$, Set Cover cannot be approximated within a factor of $c \ln n$ over instances where $m \leq O(n^{f(c)})$ for some function $f : (0,1) \to \mathbb{R}$, unless P = NP [11,33]; see also [19,26,32].

Given a set system $(U, \mathcal{S})$ with $|U| = n$ and $|\mathcal{S}| = m$, we construct a family $\mathcal{F}$ of disjoint convex polygons in four stages. We first construct an arrangement of lines $\mathcal{L}$ in $\mathbb{R}^2$, and then "thicken" the lines into narrow corridors. The family $\mathcal{F}$ will consist of the convex faces of this arrangement and $n+3$ additional axis-parallel rectangles inserted in the corridors at strategic locations. We continue with the details; see Fig. 5.

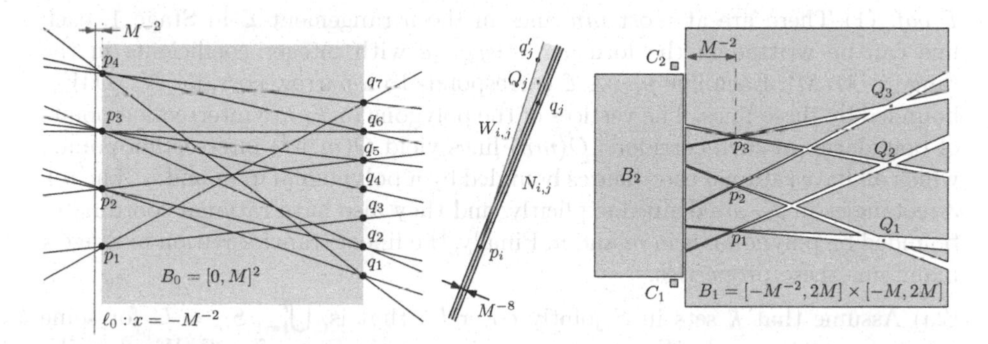

Fig. 5. Left: A line arrangement $\mathcal{L}$ constructed in Stage 1; here $n = 7$ and $m = 4$. Middle: A slab $N_{i,j}$ and a wedge $W_{i,j}$ for a line $p_i q_j \in \mathcal{L}$. Right: A schematic picture of the family $\mathcal{F}$ (not to scale).

Stage 1. We are given a set system $(U, \mathcal{S})$ with $U = \{1, \ldots, n\}$ and $\mathcal{S} = \{S_1, \ldots, S_m\}$. Let $M = \max\{12m, n+1\}$, and consider the axis-aligned square $B_0 = [0, M]^2$. Each set S_i, $i \in \{1, \ldots, m\}$, is represented by the point $p_i = (0, i)$ at the left side of B_0, and each element $j \in U$ corresponds to the point $q_j = (M, j)$ at the right side of B_0. Let $\mathcal{L}$ be the set of lines $p_i q_j$ such that $j \in S_i$.

Stage 2. For each line $p_i q_j \in \mathcal{L}$, let $N_{i,j}$ be the $\frac{1}{2} M^{-8}$-neighborhood of ℓ, which is a slab of width M^{-8}. For each line $p_i q_j \in \mathcal{L}$, we also create a cone $W_{i,j}$ bounded by the ray $\overrightarrow{p_i q_j}$ from below and the ray $\overrightarrow{p_i q'_j}$ from above, where $q'_j = (M, j + M^{-4})$ is a point at distance M^{-4} above q_j, and in particular $q'_j \notin N_{i,j}$. Note that $N_{i,j} \cup W_{i,j}$ is a simply connected region; see Fig. 5(middle).

Consider the axis-aligned rectangle $B_1 = [-M^{-2}, 2M] \times [-M, 2M] \supset B_0$. Let $\mathcal{F}_1$ be the bounded components of $B_1 \setminus \bigcup_{i,j} (N_{i,j} \cup W_{i,j})$, that is, we create a convex polygon for each bounded face of the arrangement $\mathcal{L}$ within B_1.

Stage 3. Create a family $\mathcal{F}_2$ of the following $n+3$ disjoint axis-aligned rectangles: a small square Q_j of side length M^{-8} centered at the midpoint of $q_j q'_j$ for all

$j \in U$; two unit squares, denoted C_1 and C_2, resp., centered at $(-1,-1)$ and $(-1, M+1)$; and a large rectangle $B_2 = [-M, -M^{-2} - M^{-4}] \times [0, M]$.

Stage 4. Apply the linear transformation $T : \mathbb{R}^2 \to \mathbb{R}^2$, $T(x,y) = \left(\frac{M^2 x}{2}, \frac{M^{-1} y}{24}\right)$, and let $\mathcal{F} = \{T(C) : C \in \mathcal{F}_1 \cup \mathcal{F}_2\}$.

Lemma 5. *For an instance $(U, \mathcal{S})$ of Set Cover with $|U| = n$ and $|\mathcal{S}| = m$, let $\mathcal{F}$ be the family of disjoint convex polygons constructed above.*

1. *There is a polynomial $f(m,n)$ such that the total number of vertices of the polygons in $\mathcal{F}$ is at most $f(m,n)$, each vertex has rational coordinates where both numerators and denominators are bounded by $f(m,n)$.*
2. *For every integer k, $1 \le k \le m$, the union of k sets in $\mathcal{S}$ covers U if and only if $\mathcal{F}$ admits an observation tour of length at most $k + \frac{1}{2}$.*

Proof. (1) There are at most mn lines in the arrangement $\mathcal{L}$ in Stage 1, each line can be written in the form $ax + by = c$ with integer coefficients in the range $[-M, M]$. Each line $p_i q_j \in \mathcal{L}$ corresponds to a narrow corridor $N_{i,j} \cup W_{i,j}$ bounded by three lines. The vertices of the polygons in $\mathcal{F}_1$ are intersection points of boundaries of such corridors: $O(mn)$ lines yield $O(m^2 n^2)$ intersection points, which all have rational coordinates bounded by a polynomial in m and n. The $n+3$ rectangles in $\mathcal{F}_2$ are defined explicitly, and they also have rational coordinates bounded by polynomials in m and n. Finally, the linear transformation in Stage 4 maintains these properties.

(2a) Assume that k sets in $\mathcal{S}$ jointly cover U, that is, $\bigcup_{t=1}^{k} S_{i_t} = U$ for some $i_1, \ldots, i_k \in \{1, \ldots m\}$. We construct an observation tour for $\mathcal{F}$. We describe the tour in terms of the polygons before Stage 4, since a linear transformation maintains visibility (but distorts distances). Let the initial tour γ_0 traverse the left side of the rectangle B_1 twice. The upper-left and lower-left corners of B_1 see the squares C_1 and C_2, and every point in γ_0 can see B_2. The tour γ_0 intersects every line $p_i q_j \in \mathcal{L}$. The point $\gamma_0 \cap p_i q_j$ can see all convex bodies in $\mathcal{F}_1$ whose boundaries touch the slab $N_{i,j}$. Since the upper arc of every polygon in $\mathcal{F}_1$, other than the polygon containing the top side of B_1, is formed by the bottom sides of the slabs $N_{i,j}$, and so γ_0 can see every polygon in $\mathcal{F}_1$.

We expand γ_0 to a tour of $\mathcal{F}_1 \cup \mathcal{F}_2$ as follows: For $t = 1, \ldots, k$, choose an arbitrary point $q_{j_t} \in S_{i_t}$, and add a loop from the point $p_{i_t} q_{j_t} \cap \gamma_0$ to the point p_{i_t} in the corridor $N(p_{i_t} q_{j_t})$. The point p_{i_t} can see the small squares representing all $j \in S_{i_t}$. Since $\bigcup_{t=1}^{k} S_{i_t} = U$, every small square in $\mathcal{F}_2$ is visible from some point in the tour. After the linear transformation in Stage 4, we obtain an observation tour γ for $\mathcal{F}$. We bound the length of γ using the L_1-norm of its edges. The linear transformation in Stage 4 compresses the y-extents of each edge, so the L_1 norm is dominated by the x-extents: The x-extent of an edge between the left side of B_1 and p_i is exactly M^{-2}, thus the sum of x-extents is $2kM^{-2}$. Accounting for the y-extents of these $2k$ edges and γ_0, and applying the linear transformation in Stage 4, we obtain $|\gamma| \le k + \frac{1}{2}$, as required.

(2b) Now assume that $\mathcal{F}$ admits an observation tour γ with $|\gamma| \le k + \frac{1}{2}$. We analyze the construction before Stage 4, hence the sum of x-extents of all edges of the tour is at most $(2k+1)M^{-2}$. The observation tour intersects the visibility region $V(Q_j)$ for all $j \in U$. A square Q_j lies in the corridor $N_{i,j} \cup W_{i,j}$ iff $j \in S_i$, and Q_j is only visible from such corridors. Each line $p_i q_j \in \mathcal{L}$ is incident to two points on opposite vertical sides of a square, and so its slope is in the range $[-1,1]$. By construction, $Q_j \subset W_{i,j}$ and the cone $W_{i,j}$ is convex, consequently $W_{i,j} \subset V(Q_j)$. However, Q_j is above the slab $N_{i,j}$, every vertical segment between them has length at least $\frac{1}{3}M^{-4}$, and so the x-coordinate of the leftmost point in $V(Q_j) \cap N_{i,j}$ is at least $-M \cdot M^{-8}/(\frac{1}{3}M^{-4}) = -3M^{-3}$. Overall, the x-coordinate of the leftmost point in $V(Q_j)$ is at least $-3M^{-3}$.

Note that the visibility region $V(C_1)$ is disjoint from $V(Q_j)$ for all $j \in U$. Since $|\gamma| \leq (2k+1)M^{-2} < M^{-1}$, γ must be contained in the vertical slab $\{(x,y) \in \mathbb{R}^2 : |x| < M^{-1}\}$. Consequently, γ visits each visibility region $V(Q_j)$ in the vertical slab $\{(x,y) \in \mathbb{R}^2 : -3M^{-3} < x < M^{-1}\}$.

Let I be the set of indices $i \in \{1,\ldots,m\}$ such that γ visits a point $g_i \in (N_{i,j} \cup W_{i,j}) \cap V(Q_j)$ for some $j \in U$. Since γ is an observation tour, we have $U = \cup_{i \in I} S_i$. It remains to show that $|I| \leq k$.

The tour γ determines a cyclic order on the points $G = \{g_i : i \in I\}$. We claim that the x-extent of the arc of γ between two consecutive points in G is at least $2M^{-2} - 6M^{-3}$. To prove the claim, note that every point $p_i = (0, 2i)$ has integer coordinates, and the slope of every line $p_i q_j \in \mathcal{L}$ is in the range $[-1,1]$. This implies that if two lines in $\mathcal{L}$ cross in the slab $\{(x,y) \in \mathbb{R}^2 : |x| < M^{-1}\}$, then they cross at p_i for some $i \in U$. Consequently, if visibility corridors, say $N_{i,j} \cup W_{i,j}$ and $N_{i',j'} \cup W_{i',j'}$ intersect, then the intersection is in a M^{-8}-neighborhood of p_i for some $i \in U$. It follows that the arc of γ between any two points in G must reach the line $x = -M^{-2}$ on the left or the line $x = 1$ on the right. In both cases, its arclength is at least $2M^{-2} - 6M^{-3}$, as claimed. We can bound the length of γ by the summation of the arcs between consecutive points in G:

$$|\gamma| \geq |I|(2M^{-2} - 6M^{-3}) \geq (2|I|)M^{-2} - 6mM^{-3} \geq \left(2|I| - \frac{1}{2}\right)M^{-2}.$$

In combination with $|\gamma| \leq (2k+1)M^{-2}$ and $M \geq 12m$, we obtain that $|I| \leq k + \frac{3}{4}$. As both k and $|I|$ are integers, $|I| \leq k$ follows. □

Proof of Theorem 4. As noted above [11], there exists a constant $\kappa > 0$ such that Set Cover cannot be approximated within a factor of $\kappa \log n$ on instances $(U, \mathcal{S})$ with $n = |U|$, $m = |\mathcal{S}|$, and $m \leq O(n^\alpha)$ for a constant $\alpha > 0$ unless P = NP. By Lemma 5, for every such instance $(U, \mathcal{S})$ of Set Cover, there is a family $\mathcal{F}$ of N disjoint convex polygons in the plane such that (i) for every $k \leq n$, there is a set cover of size k iff $\mathcal{F}$ admits a observation route of length at most $k + \frac{1}{2}$, and (ii) $N \leq \binom{mn}{2} + n + 3 \leq m^2 n^2 = O(n^{2\alpha+2})$.

Suppose that for $\delta = \kappa(2\alpha+2)^{-1} > 0$, there exists a polynomial-time $(\delta \log N)$-factor approximation algorithm for ORP with N convex bodies. Since $\delta \log N \leq \delta(2\alpha+2)\log n = \kappa \log n$, this yields a $(\kappa \log n)$-approximation for Set Cover, which is a contradiction unless P = NP. □

5 External Watchman Tours for a Convex Polygon

Given a polygon, the External Watchman Route problem (EWRP) is that of finding a shortest route such that each point in the exterior of the polygon is visible from some point along the route. For a convex polygon, this requirement is tantamount to requiring that each point on the boundary of the polygon is visible from some point along the route.

Let P be a convex polygon. Ntafos and Gewali [36] distinguished between two types of external watchman tours: those that wrap around the perimeter and those that do not. They also showed that the second type of route can be obtained by doubling a simple open curve that wraps around a part of P's boundary and is extended at both ends until the vision encompasses the entire boundary of P; see Fig. 6 for an example. They referred to the second type as "a 2-leg watchman route", see [36].

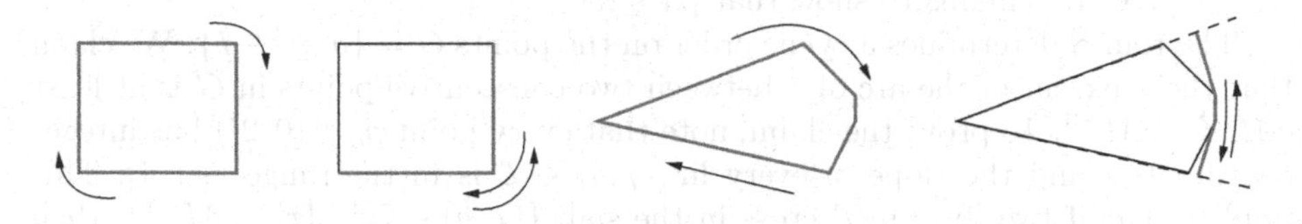

Fig. 6. The two types of external watchman tours.

Proof of Theorem 5. Let P be a convex pentagon whose angles listed in clockwise order from the top are 120°, 120°, 90°, 90°, 120°; refer to Fig. 7. Its horizontal base has length 2 and each of its vertical sides is of length ε, for a small $\varepsilon > 0$. Note that the length of each slanted top side is $a = 2/\sqrt{3}$. Observe that all angles are at least 90°. By slightly shortening the base to $2(1-\varepsilon^2)$ the pentagon becomes one with all angles obtuse, say P', whose angles are 120°, $(120-\delta)°$, $(90+\delta)°$, $(90+\delta)°$, $(120-\delta)°$, for some small $\delta > 0$. Its side lengths (listed in the same order) are a, $\varepsilon(1+\varepsilon^2)^{1/2}$, $2(1-\varepsilon^2)$, $\varepsilon(1+\varepsilon^2)^{1/2}$, a. Thus

$$\operatorname{per}(P') = 2(a + (1-\varepsilon^2) + \varepsilon(1+\varepsilon^2)^{1/2}) > 2(1+a-\varepsilon).$$

Observe that the red curve in the figure is a 3-polygonal chain that makes a valid watchman path for P'. Its length is $< 2(1+\varepsilon)$ and thus doubling it yields an external watchman route of length $< 4(1+\varepsilon)$. It remains to show that $4(1+\varepsilon) < 2(1+a-\varepsilon)$ or $\varepsilon < (a-1)/3 = \frac{2-\sqrt{3}}{3\sqrt{3}}$, which clearly holds for small $\varepsilon > 0$. Note that the ratio between the length of the double red curve and the perimeter is in fact smaller than some absolute constant < 1, e.g., $0.93 < 1$ in the range $\varepsilon \leq 0.001$.

Alternatively, by cutting two small right-angled isosceles triangles, one from each side of the base of the pentagon one gets a convex heptagon all whose angles are at least 120°. A similar calculation as above shows that this heptagon

Fig. 7. Left: a pentagon with all angles obtuse can be obtained by slightly perturbing this one. Right: a second type of watchman route can be obtained by doubling the red 3-chain polygonal curve; its two legs are perpendicular to the extensions of the two top sides. The figure is not to scale. (Color figure online)

provides yet another counterexample in which all angles are even larger, here at least 120°.

By slightly 'shaving' the pentagon in a repeated manner, one can increase the number of vertices while maintaining all angles obtuse and thereby obtain counterexamples to Conjecture 1 in [2] for every $n \geq 5$. Note, however, that since every triangle or convex quadrilateral has at least one nonobtuse angle, the range for n in Theorem 5 cannot be improved. □

6 Concluding Remarks

We conclude with a few open problems regarding the remaining gaps and the quality of approximation.

1. Is there a constant-ratio approximation algorithm for the shortest observation tour problem for families of disjoint axis-aligned unit squares (or unit disks, or translates of a convex polygon)?
2. What approximations can be computed for the shortest observation tour for families of disjoint convex polygons?
3. Is there a constant-ratio approximation algorithm for the shortest external watchman route for families of disjoint axis-aligned unit squares (or unit disks or translates of a convex polygon)?
4. What approximations can be computed for the shortest external watchman route for families of disjoint convex polygons? Is the problem APX-hard?

Acknowledgment. Research on this paper was partially supported by the NSF awards DMS 1800734 and DMS 2154347.

References

1. Abrahamsen, M., Adamaszek, A., Miltzow, T.: The art gallery problem is ∃ℝ-complete. J. ACM **69**(1), 4:1–4:70 (2022)
2. Absar, R., Whitesides, S.: On computing shortest external watchman routes for convex polygons. In: Proceedings 18th Canadian Conference on Computational Geometry (CCCG), Kingston, ON (2006)
3. Arkin, E.M., Hassin, R.: Approximation algorithms for the geometric covering salesman problem. Discret. Appl. Math. **55**(3), 197–218 (1994)

4. Arora, S.: Polynomial time approximation schemes for Euclidean traveling salesman and other geometric problems. J. ACM **45**(5), 753–782 (1998)
5. Bonnet, É., Miltzow, T.: Parameterized hardness of art gallery problems. ACM Trans. Algorithms **16**(4), 42:1-42:23 (2020)
6. Carlsson, S., Jonsson, H., Nilsson, B.J.: Finding the shortest watchman route in a simple polygon. Discret. Comput. Geom. **22**(3), 377–402 (1999)
7. Chin, W.-P., Ntafos, S.: Optimum watchman routes. In: Proceedings of the 2nd ACM Symposium on Computational Geometry (SoCG), pp. 24–33 (1986)
8. Chin, W.-P., Ntafos, S.: Optimum watchman routes. Inf. Process. Lett. **28**(1), 39–44 (1988)
9. Chin, W.-P., Ntafos, S.: Shortest watchman routes in simple polygons. Discrete Comput. Geom. **6**(1), 9–31 (1991)
10. de Berg, M., Gudmundsson, J., Katz, M., Levcopoulos, C., Overmars, M., van der Stappen, A.: TSP with neighborhoods of varying size. J. Algorithms **57**, 22–36 (2005)
11. Dinur, I., Steurer, D.: Analytical approach to parallel repetition. In: Proceedings of the 46th STOC, pp. 624–633 (2014)
12. Dror, M., Efrat, A., Lubiw, A., Mitchell, J.S.B.: Touring a sequence of polygons. In: Proceedings of the 35th ACM Symposium on Theory of Computing (STOC), pp. 473–482 (2003)
13. Dror, M., Orlin, J.B.: Combinatorial optimization with explicit delineation of the ground set by a collection of subsets. SIAM J. Disc. Math. **21**(4), 1019–1034 (2008)
14. Dumitrescu, A., Ebbers-Baumann, A., Grüne, A., Klein, R., Rote, G.: On the geometric dilation of closed curves, graphs, and point sets. Comput. Geom. **36**(1), 16–38 (2007)
15. Dumitrescu, A., Mitchell, J.S.B.: Approximation algorithms for TSP with neighborhoods in the plane. J. Algorithms **48**(1), 135–159 (2003)
16. Ebbers-Baumann, A., Grüne, A., Klein, R.: Geometric dilation of closed planar curves: New lower bounds. Comput. Geom. **37**(3), 188–208 (2007)
17. Eidenbenz, S.J., Stamm, C., Widmayer, P.: Inapproximability results for guarding polygons and terrains. Algorithmica **31**(1), 79–113 (2001)
18. Elbassioni, K.M., Fishkin, A.V., Sitters, R.: Approximation algorithms for the Euclidean traveling salesman problem with discrete and continuous neighborhoods. Int. J. Comput. Geom. Appl. **19**(2), 173–193 (2009)
19. Feige, U.: A threshold of $\ln n$ for approximating set cover. J. ACM **45**, 634–652 (1998)
20. Garey, M.R., Graham, R.L., Johnson, D.S.: Some NP-complete geometric problems. In: Proceedings of the 8th ACM Symposium on Theory of Computing (STOC), pp. 10–22 (1976)
21. Garey, M.R., Johnson, D.S.: Computers and Intractability: A Guide to the Theory of NP-Completeness. W. H. Freeman and Company, New York (1979)
22. Gewali, L.P., Ntafos, S.C.: Watchman routes in the presence of a pair of convex polygons. J. Inf. Sci. **105**(1–4), 123–149 (1998)
23. Gudmundsson, J., Levcopoulos, C.: A fast approximation algorithm for TSP with neighborhoods. Nord. J. Comput. **6**(4), 469 (1999)
24. Kirkpatrick, D.G.: An $o(\lg \lg)$-approximation algorithm for multi-guarding galleries. Discret. Comput. Geom. **53**(2), 327–343 (2015)
25. Levcopoulos, C., Lingas, A.: Bounds on the length of convex partitions of polygons. In: Joseph, M., Shyamasundar, R. (eds.) FSTTCS. LNCS, vol. 181, pp. 279–295. Springer, Heidelberg (1984). https://doi.org/10.1007/3-540-13883-8_78

26. Lund, C., Yannakakis, M.: On the hardness of approximating minimization problems. J. ACM **41**(5), 960–981 (1994)
27. Mata, C.S., Mitchell, J.S.B.: Approximation algorithms for geometric tour and network design problems. In: Proceedings of the 11th ACM Symposium on Computational Geometry (SoCG), pp. 360–369 (1995)
28. Mitchell, J.S.B.: Guillotine subdivisions approximate polygonal subdivisions: a simple polynomial-time approximation scheme for geometric TSP, k-MST, and related problems. SIAM J. Comput. **28**(4), 1298–1309 (1999)
29. Mitchell, J.S.B.: Geometric shortest paths and network optimization. In: Handbook of Computational Geometry, pp. 633–701. Elsevier (2000)
30. Mitchell, J.S.B.: Approximating watchman routes. In: Proceedings of the 24th ACM-SIAM Symposium on Discrete Algorithms (SODA), pp. 844–855 (2013)
31. Mitchell, J.S.B.: Shortest paths and networks. In: Handbook of Discrete and Computational Geometry, 3rd (edn.), vol. 31, pp. 811–848. CRC Press (2017)
32. Moshkovitz, D.: The projection games conjecture and the NP-hardness of $\ln n$-approximating set-cover. Theory Comput. **11**, 221–235 (2015)
33. Nelson, J.: A note on set cover inapproximability independent of universe size. In: Electronic Colloquium Computational Complexity TR07-105 (2007)
34. Nilsson, B.J., Żyliński, P.: How to keep an eye on small things. Int. J. Comput. Geom. Appl. **30**(2), 97–120 (2020)
35. Ntafos, S.C.: Watchman routes under limited visibility. Comput. Geom. **1**, 149–170 (1992)
36. Ntafos, S.C., Gewali, L.P.: External watchman routes. Vis. Comput. **10**(8), 474–483 (1994)
37. O'Rourke, J.: Visibility. In: Handbook of Discrete and Computational Geometry, 3rd (edn.) vol. 33, pp. 875–896. CRC Press (2017)
38. O'Rourke, J., Suri, S., Tóth, C.D.: Polygons. In: Handbook of Discrete and Computational Geometry, 3rd (edn.), vol. 30, pp. 787–810. CRC Press (2017)
39. Papadimitriou, C.H.: The Euclidean traveling salesman problem is NP-complete. Theor. Comput. Sci. **4**(3), 237–244 (1977)
40. Safra, S., Schwartz, O.: On the complexity of approximating tsp with neighborhoods and related problems. Comput. Complex. **14**(4), 281–307 (2006)
41. Scott, P.R., Awyong, P.W.: Inequalities for convex sets. J. Inequalities Pure Appl. Math. **1**, 6 (2000)
42. Tan, X.: Fast computation of shortest watchman routes in simple polygons. Inf. Process. Lett. **77**(1), 27–33 (2001)
43. Tan, X.: A linear-time 2-approximation algorithm for the watchman route problem for simple polygons. Theor. Comput. Sci. **384**(1), 92–103 (2007)
44. Tan, X., Hirata, T., Inagaki, Y.: Corrigendum to "An incremental algorithm for constructing shortest watchman routes". Int. J. Comput. Geom. Appl. **9**(3), 319–323 (1999)

Lower Bounds for Non-adaptive Shortest Path Relaxation

David Eppstein(✉)

Computer Science Department, University of California, Irvine,
Irvine, CA 92697, USA
eppstein@uci.edu

Abstract. We consider single-source shortest path algorithms that perform a sequence of relaxation steps whose ordering depends only on the input graph structure and not on its weights or the results of prior steps. Each step examines one edge of the graph, and replaces the tentative distance to the endpoint of the edge by its minimum with the tentative distance to the start of the edge, plus the edge length. As we prove, among such algorithms, the Bellman-Ford algorithm has optimal complexity for dense graphs and near-optimal complexity for sparse graphs, as a function of the number of edges and vertices in the given graph. Our analysis holds both for deterministic algorithms and for randomized algorithms that find shortest path distances with high probability.

1 Introduction

Dijkstra's algorithm finds shortest paths in directed graphs when all edge weights are non-negative, but the problem becomes more difficult when negative edge weights (but not negative cycles) are allowed. In this case, despite recent breakthroughs on near-linear time bounds for graphs with small integer edge weights [5], the best strongly-polynomial time bound for single-source shortest paths remains that of the Bellman-Ford algorithm [4,10,18], which takes time $O(mn)$ on graphs with m edges and n vertices, or $O(n^3)$ on dense graphs.

Both Dijkstra's algorithm and the Bellman-Ford algorithm (as well as an unnamed linear-time algorithm for single-source shortest paths in directed acyclic graphs) can be unified under the framework of *relaxation algorithms*, also called *label-correcting algorithms* [8]. These algorithms initialize tentative distances $D[v]$ from the source vertex to each other vertex v, by setting $D[s] = 0$ and $D[v] = +\infty$ for $v \neq s$. Then, they repeatedly *relax* the edges of the graph. This means, that for a given edge $u \to v$, the algorithm updates $D[v]$ to $D[u] + \text{length}(u \to v)$. In Dijkstra's algorithm, each edge $u \to v$ is relaxed once, in sorted order by the tentative distance $D[u]$. In the Bellman-Ford algorithm, an edge can be relaxed many times. The algorithm starts with the tentative distance equal to the correct distance for s, but not for the other vertices. Whenever the algorithm relaxes an edge $u \to v$ in the shortest path tree, at a time when u already has the correct distance, the tentative distance to v becomes correct as well. Thus, the goal in designing the algorithm is to perform these

P. Morin and S. Suri (Eds.): WADS 2023, LNCS 14079, pp. 416–429, 2023.
https://doi.org/10.1007/978-3-031-38906-1_27

distance-correcting relaxations while wasting as little effort as possible on other relaxations that do not correct any distance, and on the overhead in selecting which relaxation to perform.

We would like to prove or disprove the optimality of the Bellman–Ford algorithm among a general class of strongly-polynomial shortest path algorithms, without restricting the types of computation such an algorithm can perform, but such a result appears to remain far out of reach. Instead, in this work we focus only on relaxation algorithms, asking: how few relaxation steps are needed? Note that, without further assumptions, a shortest path algorithm could "cheat", computing a shortest path tree in some other way and then performing only $n-1$ relaxation steps in a top-down traversal of a shortest path tree. To focus purely on relaxation, and prevent such cheating, we consider *non-adaptive relaxation algorithms*, in which the sequence of relaxation steps is determined only by the structure of the given graph, and not on its weights nor on the outcome of earlier relaxation steps. Dijkstra's algorithm is adaptive, but the linear-time DAG algorithm is non-adaptive. Another example of a non-adaptive algorithm comes from past work on the graphs in which, like DAGs, it is possible to relax every edge once in a fixed order and guarantee that all tentative distances are correct [12]. As usually described, the Bellman-Ford algorithm is adaptive. Its typical optimizations include adaptive rules that disallow repeatedly relaxing any edge $u \to v$ unless the tentative distance to u has decreased since the previous relaxation, and that stop the entire algorithm when no more allowed relaxations can be found. However, its same asymptotic time bounds can be achieved by a non-adaptive version of the Bellman-Ford algorithm, with a *round-robin* relaxation sequence, one that merely repeats $n-1$ rounds of relaxing all edges in the same order per round. A non-adaptive asynchronous distributed form of the Bellman-Ford algorithm is widely used in *distance vector routing* of internet traffic, to maintain paths of minimum hop count between major internet gateways [13].

1.1 Known Upper Bounds

We do not require non-adaptive relaxation algorithms to be round-robin, but we are unaware of any way to take advantage of this extra flexibility. Nevertheless, among round-robin algorithms, there is still freedom to choose the ordering of edges within each round, and this freedom can lead to improved constant factors in the number of relaxation steps performed by the Bellman-Ford algorithm.

Yen [21] described a method based on the following idea. Choose an arbitrary linear ordering for the vertices, and partition the edges into two subsets: the edges that are directed from an earlier vertex to a later vertex in the ordering, and the edges that are directed from a later vertex to an earlier vertex. Both of these two edge subsets define directed acyclic subgraphs of the given graph, with the chosen linear ordering or its reverse as a topological ordering. Use a round-robin edge ordering that first relaxes all of the edges of the first subgraph, in its topological order, and then relaxes all of the edges of the second subgraph, in its topological order. If any shortest path is divided into contiguous subpaths that lie within one of these two DAGs, then each two consecutive subpaths from the first and second DAG will be relaxed in order by each round of the algorithm. In the worst case, there is a single shortest path of $n-1$ edges, alternating between the two DAGs, requiring $\lceil n/2 \rceil$ rounds of relaxation. For complete directed graphs, this method uses $\left(\frac{1}{2}+o(1)\right)n^3$ relaxation steps, instead of the $\left(1+o(1)\right)n^3$ that might be used by a less-careful round-robin method.

As we showed in earlier work [2], an additional constant factor savings can be obtained by a randomized algorithm that selects from a random distribution of non-adaptive relaxation sequences, and that obtains a correct output with high probability rather than with certainty. To do so, use Yen's method, but choose the vertex ordering as a uniformly random permutation of the vertices, rather than arbitrarily. In any shortest path tree, each vertex with more than one child reduces the number of steps from the source to the deepest leaf by one, reducing the number of alternations between the two DAGs. For each remaining vertex with one child in the tree, the probability that it lies between its parent and child in the randomly selected ordering is $\frac{1}{3}$, and when this happens, it does not contribute to the bound on the number of alternations. With high probability, the number of these non-contributing vertices is close to one third of the single-child vertices. Therefore, with high probability, the maximum number of alternations between the two DAGs among paths on the shortest path tree is $\left(\frac{2}{3}+o(1)\right)n$, and an algorithm that uses this method to perform $\left(\frac{1}{3}+o(1)\right)n^3$ relaxation steps will find the correct shortest paths with high probability.

The worst-case asymptotic time of these methods remains $O(n^3)$ for complete graphs, and $O(mn)$ for arbitrary graphs with m vertices and n edges. Both Yen's method and the randomized permutation method can also be used in adaptive versions of the Bellman-Ford algorithm, with better constant factors and in the randomized case leading to a Las Vegas algorithm rather than a Monte Carlo algorithm, but it is their non-adaptive variants that concern us here.

1.2 New Lower Bounds

We provide the following results:

- Any deterministic non-adaptive relaxation algorithm for single-source shortest paths on a complete directed graph with n vertices must use $\left(\frac{1}{6}-o(1)\right)n^3$ relaxation steps.
- Any randomized non-adaptive relaxation algorithm for shortest paths on a complete directed graph with n vertices, that with high probability sets all distances correctly, must use $\left(\frac{1}{12}-o(1)\right)n^3$ relaxation steps.
- For any m and n with $n \leq m \leq 2\binom{n}{2}$, there exists a directed graph on m edges and n vertices on which any deterministic or high-probability randomized non-adaptive relaxation algorithm for shortest paths must use $\Omega(mn/\log n)$ relaxation steps. When $m = \Omega(n^{1+\varepsilon})$ for some $\varepsilon > 0$, the lower bound improves to $\Omega(mn)$.

These lower bounds hold even on graphs for which all edges weights are zero and one, for which an adaptive algorithm, Dial's algorithm, can find shortest paths in linear time [9].

1.3 Related Work

Although we are not aware of prior work in the precise model of computation that we use, variants of the Bellman-Ford algorithm have been studied and shown optimal for some other related problems:

- The k-walk problem asks for a sequence of exactly k edges, starting and one vertex and ending at the other, allowing repeated edges. The Bellman-Ford algorithm can be modified to find the shortest k-walk between two vertices in time $O(kn^2)$, non-adaptively. In any non-adaptive relaxation algorithm, the only arithmetic operations on path lengths and edge weights are addition and minimization, and these operations are performed in a fixed order. Therefore, the sequence of these operations can be expanded into a circuit, with two kinds of gates: minimization and addition. The resulting (min, +)-circuit model of computation is somewhat more general than the class of relaxation algorithms, because the sequence of operations performed in this model does not need to come from a sequence of relaxation steps. The k-walk version of the Bellman-Ford algorithm is nearly optimal in the (min, +)-circuit model: circuit size $\Omega(k(n-k)n)$ is necessary [14]. However, this k-walk problem is different from the shortest path problem, so this bound does not directly apply to shortest paths.
- Under conditional hypotheses that are standard in fine-grained complexity analysis, the $O(km)$ time of Bellman-Ford for finding paths of at most k steps, for graphs of m edges, is again nearly optimal: neither the exponent of k nor the exponent of m can be reduced to a constant less than one. For large-enough k, the shortest path of at most k steps is just the usual shortest path, but this lower bound applies only for choices of k that are small enough to allow the result to differ from the shortest path [15].
- Another related problem is the all hops shortest path problem, which asks to simultaneously compute k paths, having distinct numbers of edges from one to a given parameter k. Again, this can be done in time $O(km)$ by a variant of the Bellman-Ford algorithm, and it has an unconditional $\Omega(km)$ lower bound for algorithms that access the edge weights only by path length comparisons, as Bellman-Ford does [6,11]. Because it demands multiple paths as output, this lower bound does not apply to algorithms that compute only a single shortest path.
- Meyer et al. [17] study a version of the Bellman-Ford algorithm, in which edges are relaxed in a specific (adaptive) order. They construct sparse graphs, with $O(n)$ edges, on which this algorithm takes $\Omega(n^2)$ time, even in the average case for edge weights uniformly drawn from a unit interval. This bound applies only to this algorithm and not to other relaxation orders.

2 Deterministic Lower Bound for Complete Graphs

The simplest of our results, and the prototype for our other results, is a lower bound on the number of relaxations needed by a deterministic non-adaptive relaxation algorithm, in the worst case, on a complete directed graph with n vertices.

Theorem 1. *Any deterministic non-adaptive relaxation algorithm for single-source shortest paths on a complete directed graph with n vertices must use at least $\left(\frac{1}{6} - o(1)\right)n^3$ relaxation steps.*

Proof. Fix the sequence σ of relaxation steps chosen by any such algorithm. We will find an assignment of weights for the complete directed graph, such that the distances obtained by the relaxation algorithm are not all correct until $\left(\frac{1}{6} - o(1)\right)n^3$ relaxation steps have taken place. Therefore, in order for the algorithm to be correct, it must make this many steps. For the weights we choose, the shortest path tree will form a single directed path, of $n-1$ edges, starting at the source vertex. In order for the relaxation algorithm to achieve correct distances to all vertices, its sequence of relaxations must include a subsequence consisting of all path edges in order. The weights of these edges are unimportant (because we are considering only non-adaptive algorithms) so we may set all path edges to have weight zero and all other edges to have weight one.

To determine this path, we choose one at a time its edges in even positions: its second, fourth, sixth, etc., edge. These chosen edges include every vertex in the path, so choosing them will also determine the edges in odd positions. When choosing the ith edge (for an even number i), we make the choice greedily, to maximize the position in σ of the step that relaxes this edge and makes its endpoint have the correct distance. Let s_i denote this position, with $s_0 = 0$ as a base case recording the fact that, before we have relaxed any edges, the source vertex already has the correct distance. Then the length of σ is at least equal to the telescoping sum

$$(s_2 - s_0) + (s_4 - s_2) + (s_6 - s_4) + \cdots .$$

When choosing edge i, for an even position i, there are $i-1$ earlier vertices, whose position in the shortest path is already determined, and $n-i+1$ remaining vertices. Between step s_{i-2} and step s_i of the relaxation sequence σ, it must relax all $n-i+1$ edges from the last endpoint of edge $i-2$ to one of these remaining vertices, and all $2\binom{n-i+1}{2}$ edges between pairs of the vertices that remain to be corrected. For, if it did not do so, there would be an edge that it had not relaxed, and choosing this edge next would cause s_i to be greater; but this would violate the greedy choice of edge i to make s_i as large as possible. Therefore,

$$s_i - s_{i-2} \geq (n-i+1) + 2\binom{n-i+1}{2} = (n-i+1)^2.$$

Summing over all $\lfloor (n-1)/2 \rfloor$ choices of edges in even positions gives, as a lower bound on the total number of relaxation steps,

$$\sum_{i=2,4,6,\ldots} s_i - s_{i-2} \geq \sum_{i=2,4,6,\ldots} (n-i+1)^2 = \frac{n^3-n}{6},$$

where the closed form for the summation follows easily by induction.

3 Randomized Lower Bound for Complete Graphs

It does not make much sense to consider expected time analysis for non-adaptive algorithms, because these algorithms have a fixed stopping time (determined as a function of the given graph), and we want their output to be correct with high probability rather than in any expected sense. Nevertheless, it is often easier to lower-bound the expected behavior of randomized algorithms, by using Yao's principle [20], according to which the expected cost of a randomized algorithm on its worst-case input can be lower bounded by the cost of the best deterministic algorithm against any random distribution of inputs.

In order to convert high-probability time bounds into expectations, we consider randomized non-adaptive algorithms that are guaranteed to produce the correct distances, and we define the *reduced cost* of such an algorithm to be the number of relaxations that it performs until all distances are correct, ignoring any remaining relaxations after that point.

Lemma 1. *If a randomized non-adaptive relaxation algorithm $\mathcal{A}$ takes $s(G)$ steps on any weighted input graph G and computes all distances from the source vertex correctly with probability $1 - o(1)$, then there exists a randomized non-adaptive relaxation algorithm $\mathcal{B}$ that is guaranteed to produce correct distances and whose expected reduced cost, on weighted graphs G with n vertices and m edges, is at most $s(G) + o(mn)$.*

Proof. Construct algorithm $\mathcal{B}$ by using the relaxation sequence from algorithm $\mathcal{A}$, appending onto it the sequence of relaxations from a conventional non-adaptive deterministic Bellman-Ford algorithm. Then with probability $1 - o(1)$ the relaxed cost of $\mathcal{B}$ counts only the relaxation sequence from algorithm $\mathcal{A}$, of length $s(G)$. With probability $o(1)$ the relaxed cost extends into the deterministic Bellman-Ford part of the sequence, of length $O(mn)$. Because this happens with low probability, its contribution to the expected reduced cost is $o(mn)$.

Corollary 1. *Any lower bound on expected reduced cost is also a valid lower bound, up to an additive $o(mn)$ term, on the number of relaxation steps for a randomized non-adaptive relaxation algorithm that produces correct distances with high probability.*

With this conversion to expected values in hand, we may now formulate Yao's principle as it applies to our problem. We need the following notation:

Definition 1. *For any graph G, with a specified source vertex, let W_G be the family of assignments of real weights to edges of G. Let $\mathcal{D}_G$ be the family of probability distributions of weights in W_G, and let Σ_G be the class of relaxation sequences on G that are guaranteed to produce correct distances from the specified source vertex. For any randomized non-adaptive relaxation algorithm $\mathcal{A}$ and weight vector $w \in W_G$, let $r_G(\mathcal{A}, w)$ denote the expected reduced cost of running algorithm $\mathcal{A}$ on G with edges weighted by w. For $\sigma \in \Sigma_G$ and $D \in \mathcal{D}_G$ let $\rho_G(\sigma, D)$ be the expected reduced cost of sequence σ on weight vectors drawn from D.*

Lemma 2 (Yao's principle). *For any graph G with specified source vertex, and any randomized non-adaptive relaxation algorithm $\mathcal{A}$,*

$$\min_{\mathcal{A}} \max_{w \in W_G} r_G(\mathcal{A}, w) = \max_{D \in \mathcal{D}_G} \min_{\sigma \in \Sigma_G} \rho_G(\sigma, D).$$

Proof. This is just the minimax principle for zero-sum games, applied to a game in which one player chooses a relaxation sequence $\sigma \in \Sigma_G$, the other player chooses a weight vector $w \in W_G$, and the outcome of the game is the reduced cost for σ on w. According to that principle, the value of the best mixed strategy for the sequence player, against its worst-case pure strategy (the left hand side of the equality in the lemma) equals the value of the best mixed strategy for the weight player, against its worst-case pure strategy (the right hand side).

Corollary 2. *For any weight distribution $D \in \mathcal{D}_G$, $\min_{\sigma \in \Sigma_G} \rho_G(\sigma, D)$ is a valid lower bound on the expected reduced cost of any randomized non-adaptive relaxation algorithm that is guaranteed to produce correct distances.*

Proof. An arbitrary algorithm $\mathcal{A}$ can only have a greater or equal value to the left hand side of Lemma 2, and an arbitrary weight distribution D can only have a smaller or equal value to the right hand side. So the expected reduced cost of the algorithm, on a worst-case input, can only be greater than or equal to the value given for D in the statement of the corollary.

Theorem 2. *Any randomized non-adaptive relaxation algorithm for shortest paths on a complete directed graph with n vertices, that with high probability sets all distances correctly, must use at least $\left(\frac{1}{12} - o(1)\right)n^3$ relaxation steps.*

Proof. We apply Corollary 2 to a weight distribution D defined as follows: we choose a random permutation of the vertices of the given complete graph, starting with the source vertex, we make the weight of edges connecting consecutive vertices in order along this permutation zero, and we make all other weights one. Thus, each weighting of the complete graph drawn from this distribution will have a unique shortest path tree in the form of a single path, with all paths from the source vertex equally likely. For any weight vector w drawn from D, let π_w be this path.

Let σ be any relaxation sequence in Σ_D. As in the proof of Theorem 1, we define s_i (for a weight vector w to be determined) to be the step at which the second endpoint of the ith edge of π_w has its shortest path distance set correctly.

Let C_i denote the conditional probability distribution obtained from D by fixing the choice of the first i edges of π_w. Under condition C_i, the remaining $n-i-1$ vertices remain equally likely to be permuted in any order. There are $2\binom{n-i-1}{2}$ choices for edge $i+2$, each of which is equally likely. Therefore, the expected value of $s_{i+2} - s_i$ is greater than or equal to the average, among these edges, of their distance along sequence σ from position s_i. (It is greater than or equal, rather than equal, because this analysis does not take into account the requirement that edge $i+1$ must be relaxed first, before we relax edge $i+2$.) Sequence σ can minimize this average if, in σ, the next $2\binom{n-i-1}{2}$ relaxation

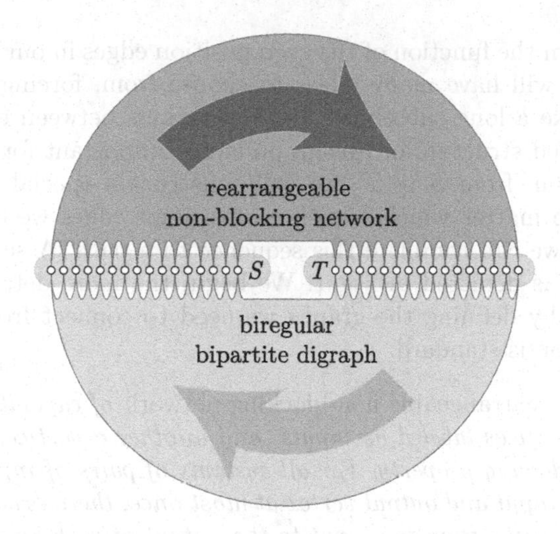

Fig. 1. Schematic view of the graphs used for our lower bound construction

steps after s_i are exactly these distinct edges. When σ packs the edges in this minimizing way, the average is $2\binom{n-i-1}{2}/2$; for other sequences it can only be greater. Therefore,

$$E[s_{i+2} - s_i \mid C_i] \geq \binom{n-i-1}{2}.$$

Summing these expected differences, over the sequence of values s_i for even i, and applying Corollary 1 and Corollary 2, gives the result.

4 Lower Bounds for Incomplete Graphs

In our lower bounds for complete graphs, the edges in even and odd positions of the shortest paths perform very different functions. The edges in even positions are the ones that, at each step in the shortest path, force the relaxation sequence to have a large subsequence of relaxation steps. Intuitively, this is because there are many possible choices for the edge at the next step and all of these possibilities (in the deterministic bound) or many of these possibilities (in the randomized bound) must be relaxed before reaching the edge that is actually chosen. The edges in odd positions, on the other hand, do not contribute much directly to the length of the sequence of relaxation steps. Instead, they are used to connect the edges in the even positions into a single shortest path.

To construct graphs that are not complete, for which we can prove analogous lower bounds, we make this dichotomy more explicit. For a chosen "capacity" parameter c, we will construct graphs that have two designated subsets of c vertices, S and T (with the source vertex contained in subset S). We will connect the vertices in T to the vertices in S by a biregular bipartite directed graph of some degree $d \approx m/2c$, a graph in which each vertex in T has exactly d outgoing neighbors and each vertex in S has exactly d incoming neighbors. This biregular graph will perform the function of the even position edges in our complete graph lower bounds: it will have many edges to choose from, forcing any relaxation algorithm to make a long subsequence of relaxations between each two chosen edges. The detailed structure of this graph is not important for our bounds. In the other direction, from S to T, we will construct a special graph with the property that, no matter which sequence of disjoint edges we choose from the biregular graph, we can complete this sequence to a path. A schematic view of this construction is depicted in Fig. 1. We begin the more detailed description of this structure by defining the graphs we need to connect from S to T. The following definition is standard:

Definition 2. *A* rearrangeable non-blocking network *of capacity c is a directed graph G with c vertices labeled as inputs, and another c vertices labeled as outputs, with the following property. For all systems of pairs of inputs and outputs that include each input and output vertex at most once, there exists in G a system of vertex-disjoint paths from the input to the output of each pair.*

Observation 3. *A complete bipartite graph $K_{c,c}$, with its edges directed from c input vertices to c output vertices, is a rearrangeable non-blocking network of capacity c, with $2c$ vertices and c^2 edges. In this case, the disjoint paths realizing any system of disjoint input-output pairs is just a matching, formed by the edges from the input to the output in each pair.*

Lemma 4. *For any capacity c, there exist rearrangeable non-blocking network of capacity c with $O(c \log c)$ vertices and edges.*

Pippenger [19] credits the proof of Lemma 4 to Beizer [3], who used a recursive construction. A more recent construction of Alon and Capalbo [1] is based on blowing up an expander graph, producing enough copies of each vertex that a system of edge-disjoint paths in the expander can be transformed into a system of vertex-disjoint paths in the non-blocking network. Their networks are non-blocking in a stronger sense (the vertex-disjoint paths can be found incrementally and efficiently), but we do not need that additional property. A simple counting argument shows that $o(c \log c)$ edges is not possible: to have enough subsets of edges to connect $c!$ possible systems of pairs, the number of edges must be at least $\log_2 c!$. For non-blocking networks with fewer vertices and more edges we turn to an older construction of Clos [7]:

Lemma 5 (Clos [7]). *Suppose that there exists a rearrangeable non-blocking network G_c of capacity c with n vertices and m edges. Then there exists a rearrangeable non-blocking network of capacity c^2 with $3cn - 2c^2$ vertices and $3cm$ edges.*

Proof. Construct $3c$ copies of G_c, identified as c input subunits, c internal subunits, and c output subunits. The input subunits have together c^2 input vertices, which will be the inputs of the whole network. Similarly, the output subunits have together c^2 output vertices, which will be the outputs of the whole network. Identify each output vertex of an input subunit with an input vertex of an

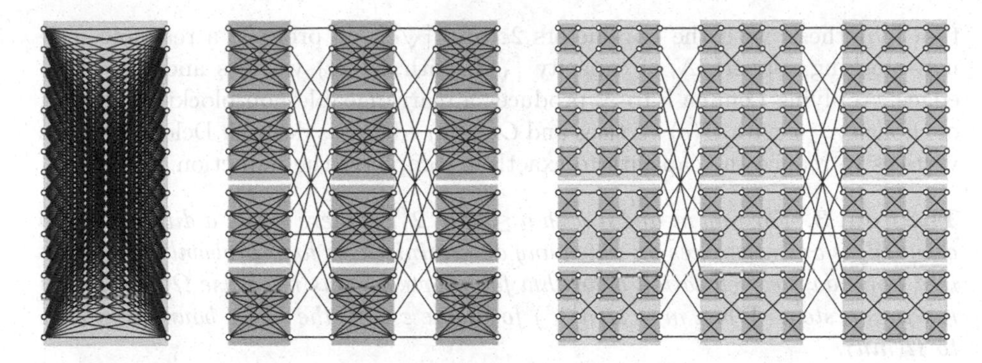

Fig. 2. Three rearrangeable non-blocking networks of capacity 16. Each network's input vertices are in its left column and its output vertices are in the right column. Left: Complete bipartite graph. Center: Three-stage Clos network, with pairs of input and output vertices in consecutive stages connected by edges rather than being identified as single vertices. Right: Nine-stage network obtained by expanding each subunit of the center network into a three-stage network.

internal subunit, in such a way that each pair of these subunits has exactly one identified vertex. Similarly, identify each output vertex of an internal subunit with an input vertex of an output subunit, in such a way that each pair of these subunits has exactly one identified vertex.

An example of this network, for $c = 4$ and $G_c = K_{4,4}$, can be seen in an expanded form as the middle network of Fig. 2. For greater legibility of the figure, instead of identifying pairs of vertices between subunits, these pairs have been connected by added edges. Contracting these edges would produce the network described above.

To produce vertex-disjoint paths connecting any system of disjoint pairs of inputs and outputs, consider these pairs as defining a multigraph connecting the input subunits to the output subunits of the overall network. This multigraph has maximum degree c (each input or output subunit participates in at most c pairs), and we may apply a theorem of Dénes Kőnig according to which every bipartite multigraph with maximum degree c has an edge coloring using c colors [16]. These colors may be associated with the c internal subunits, and used to designate which internal subunit each path should pass through. Once this designation is made, each subunit has its own system of disjoint pairs of inputs and outputs through which its paths should go, and the paths through each subunit can be completed using the assumption that it is rearrangeable non-blocking.

Corollary 3. *For any constant $\varepsilon > 0$ and any integer $c \geq 1$, there exist rearrangeable non-blocking networks of capacity c with $O(c)$ vertices and $O(c^{1+\varepsilon})$ edges.*

Proof. We prove the result by induction on the integer $i = \lceil \log_2 1/\varepsilon \rceil$. As a base case this is true for $\varepsilon = 1$ (for which $i = 0$) and for arbitrary c, using the complete bipartite graph as the network. For smaller values of ε, apply the induction hypothesis with the parameters 2ε and $\lceil\sqrt{c}\rceil$, to produce a rearrangeable non-blocking network N of capacity $\lceil\sqrt{c}\rceil$ with $O(\sqrt{c})$ vertices and $O(c^{1/2+\varepsilon})$ edges. Applying Lemma 5 to N produces a rearrangeable non-blocking network of capacity $\geq c$ with $O(c)$ vertices and $O(c^{1+\varepsilon})$ edges, as desired. Deleting excess vertices to reduce the capacity to exactly c completes the induction.

Theorem 3. *For any m and n with $n \leq m \leq 2\binom{n}{2}$, there exists a directed graph on m edges and n vertices on which any deterministic or high-probability randomized non-adaptive relaxation algorithm for shortest paths must use $\Omega(mn/\log n)$ relaxation steps. When $m = \Omega(n^{1+\varepsilon})$ for some $\varepsilon > 0$, the lower bound improves to $\Omega(mn)$.*

Proof. We construct a graph according to the construction outlined above, in which we choose a capacity c, set up two disjoint sets S and T of c vertices, connect T to S by a biregular bipartite digraph of some degree d, and connect S to T by a rearrangeable non-blocking network of capacity c. We allocate at least $m/2$ edges to the biregular graph, and the rest to the non-blocking network, giving $d \approx m/2c$. For the $\Omega(mn/\log n)$ bound, we use the non-blocking network of Lemma 4, with $c = \Theta(n/\log n)$. For the $\Omega(mn)$ bound, we use the non-blocking network of Corollary 3, with $c = \Theta(n)$. In both cases, we can choose the parameters of these networks to achieve these asymptotic bounds without exceeding the given numbers n and m of vertices and edges. We pad the resulting graph with additional vertices and edges in order to make the numbers of vertices and edges be exactly n and m, and set the weights of these padding edges to be high enough that they do not interfere with the remaining construction.

Next, we choose a random distribution on weights for the resulting network so that, for every relaxation sequence σ, the expected reduced cost of σ, for weights from this distribution, matches the lower bound in the statement of the lemma. For deterministic non-adaptive relaxation algorithms, this will give the desired lower bound directly, via the simple fact that the worst case of any distribution is always at least its expectation. For randomized algorithms, the lower bound will follow using Corollary 1 and Corollary 2 to convert the lower bound on expected reduced cost into a high-probability lower bound.

As in Theorem 2, the random distribution on weights that we use is determined from a random distribution on paths from the source, such that the shortest path tree for the weighted graph will contain the chosen path. We can accomplish this by setting the lengths of the path edges to zero and all other edge lengths to one. Unlike in Theorem 2, these paths will not necessarily include all vertices in the graph and the shortest path tree may contain other branches. To choose a random path, we simply choose a sequence of edges in the biregular graph, one at a time, in order along the path. In each step, we choose uniformly at random among the subset of edges in the biregular graph that are disjoint from already-chosen edges. Because of the biregularity of the biregular part of our graph, each chosen edge is incident to at most $2(d-1)$ other edges, and eliminates these other edges from being chosen later. At least $c/2$ choices are possible before there are no more disjoint edges, and throughout the first $c/4$

choices there will remain at least $m/4$ edges to choose from, disjoint from all previous edges. The sequence ends when there are no more such edges to choose. Once we have chosen this sequence of edges from the biregular graph, we construct a set of vertex-disjoint paths in the rearrangeable nonblocking network that connects them in sequence into a single path.

For any given relaxation sequence σ, as in the proof of Theorem 2, let τ be the subsequence of edges in σ that belong to the biregular part of the graph, and consider a modified relaxation algorithm that, after relaxing each edge in τ, immediately relaxes all edges of the non-blocking network. Define the reduced cost for τ to be the number of relaxation steps made from τ before all distances are correct, not counting the relaxation steps in the non-blocking network. Clearly, this is at most equal to the reduced cost for σ, because σ might fail to relax a path in the non-blocking network when τ succeeds, causing the computation of shortest path distances using σ to fall behind that for τ. Define t_i to be the step in the relaxation sequence for τ that relaxes the ith chosen edge from the biregular graph, making the distance to its ending vertex correct. Then the expectation of $t_i - t_{i-1}$ (conditioned on the choice of the first $i-1$ edges is at least the average, over all edges that were available to be chosen as the ith edge, of the number of steps along τ from t_{i-1} to the next occurrence of that edge. This expectation is minimized when the edges occurring immediately following position t_{i-1} in τ are exactly the next available edges, and is equal to half the number of available edges; for other possibilities for τ, the expectation can only be even larger. The expected reduced cost for τ equals the sum of these differences $t_i - t_{i-1}$. Since there are $\Omega(c)$ steps in which the number of available edges is $\Omega(m)$, the expected reduced cost for τ is $\Omega(cm)$. The expected reduced cost for σ can only be larger, and plugging in the value of c (coming from our choice of which type of non-blocking network to use) gives the result.

5 Conclusions and Open Problems

We have shown that, for a wide range of choices for m and n, the Bellman-Ford algorithm is asymptotically optimal among non-adaptive relaxation algorithms. Adaptive versions of the Bellman-Ford algorithm are faster, but only by constant factors. Is it possible to prove that, among adaptive relaxation algorithms, Bellman-Ford is optimal? Doing so would require a careful specification of what information about the results of relaxation steps can be used in choosing how to adapt the relaxation sequence.

The constant factors of $\frac{1}{6}$ and $\frac{1}{12}$ in our deterministic and randomized lower bounds for complete graphs are far from the constant factors of $\frac{1}{2}$ and $\frac{1}{3}$ in the corresponding upper bounds. Can these gaps be tightened? Is it possible to make them tight enough to distinguish deterministic and randomized complexity? Alternatively, is it possible to improve the deterministic methods to match the known randomized upper bound?

For sparse graphs ($m = O(n)$), our lower bound falls short of the Bellman-Ford upper bound by a logarithmic factor. Can the lower bound in this range be improved, or can the Bellman-Ford algorithm for sparse graphs be improved?

In this work, we considered the worst-case number of relaxation steps used by non-adaptive relaxation algorithms for the parameters m and n. But it is also natural to look at this complexity for individual graphs, with unknown weights. For any given graph, there is some relaxation sequence that is guaranteed to find shortest path distances for all weightings of that graph, with as few relaxation steps as possible. An algorithm of Haddad and Schäffer [12] can find such a sequence for the special case of graphs for which it is as short as possible, one relaxation per edge. What is the complexity of finding or approximating it more generally?

Acknowledgements. This research was supported in part by NSF grant CCF-2212129.

References

1. Alon, N., Capalbo, M.R.: Finding disjoint paths in expanders deterministically and online. In: 48th Annual IEEE Symposium on Foundations of Computer Science (FOCS 2007), October 20–23, 2007, Providence, RI, USA, Proceedings, pp. 518–524. IEEE Computer Society (2007). https://doi.org/10.1109/FOCS.2007.19
2. Bannister, M.J., Eppstein, D.: Randomized speedup of the Bellman-Ford algorithm. In: Martínez, C., Hwang, H-K (eds.), Proceedings of the 9th Meeting on Analytic Algorithmics and Combinatorics, ANALCO 2012, Kyoto, Japan, January 16, 2012, pp. 41–47. SIAM (2012). https://doi.org/10.1137/1.9781611973020.6
3. Beizer, B.: The analysis and synthesis of signal switching networks. In: Proceedings of the Symposium on Mathematical Theory of Automata, New York, April 1962, pp. 563–572. Polytechnic Press of the Polytechnic Institute of Brooklyn (1962)
4. Bellman, R.: On a routing problem. Q. Appl. Math. **16**, 87–90 (1958). https://doi.org/10.1090/qam/102435
5. Bernstein, A., Nanongkai, D., Wulff-Nilsen, C.: Negative-weight single-source shortest paths in near-linear time. In: 63rd IEEE Annual Symposium on Foundations of Computer Science, FOCS 2022, Denver, CO, USA, October 31 - November 3, 2022, pp. 600–611. IEEE (2022). https://doi.org/10.1109/FOCS54457.2022.00063
6. Cheng, G., Ansari, N.: Finding all hops shortest paths. IEEE Commun. Lett. **8**(2), 122–124 (2004). https://doi.org/10.1109/LCOMM.2004.823365
7. Clos, C.: A study of non-blocking switching networks. Bell Syst. Tech. J. **32**(2), 406–424 (1953). https://doi.org/10.1002/j.1538-7305.1953.tb01433.x
8. Narsingh Deo and Chi Yin Pang: Shortest-path algorithms: taxonomy and annotation. Networks **14**(2), 275–323 (1984). https://doi.org/10.1002/net.3230140208
9. Dial, R.B.: Algorithm 360: Shortest-path forest with topological ordering [H]. Commun. ACM **12**(11), 632–633 (1969). https://doi.org/10.1145/363269.363610
10. Ford, L.R., Jr. Fulkerson, D.R.: A shortest chain algorithm. In: Flows in Networks, pp. 130–134. Princeton University Press (1962)
11. Guérin, R., Orda, A.: Computing shortest paths for any number of hops. IEEE/ACM Trans. Networking **10**(5), 613–620 (2002). https://doi.org/10.1109/TNET.2002.803917
12. Haddad, R.W., Schäffer, A.A.: Recognizing Bellman-Ford-orderable graphs. SIAM J. Discret. Math. **1**(4), 447–471 (1988). https://doi.org/10.1137/0401045

13. Hedrick, C.: Routing Information Protocol. Request for Comments, RFC 1058. Network Working Group (1988). https://www.rfc-editor.org/rfc/rfc1058
14. Jukna, S., Schnitger, G.: On the optimality of Bellman-Ford-Moore shortest path algorithm. Theoret. Comput. Sci. **628**, 101–109 (2016). https://doi.org/10.1016/j.tcs.2016.03.014
15. Kociumaka, T., Polak, A.: Bellman-Ford is optimal for shortest hop-bounded paths. Electronic preprint arxiv:2211.07325 (2023)
16. Kőnig, D.: Über Graphen und ihre Anwendung auf Determinantentheorie und Mengenlehre. Math. Ann. **77**, 453–465 (1916). https://doi.org/10.1007/BF01456961
17. Meyer, U., Negoescu, A., Weichert, V.: New bounds for old algorithms: on the average-case behavior of classic single-source shortest-paths approaches. In: Marchetti-Spaccamela, A., Segal, M. (eds.) TAPAS 2011. LNCS, vol. 6595, pp. 217–228. Springer, Heidelberg (2011). https://doi.org/10.1007/978-3-642-19754-3_22
18. Moore, E.F.: The shortest path through a maze. In: Proceedings of International Symposium on Switching Theory 1957, Part II, pp. 285–292. Harvard University Press, Cambridge, Massachusetts (1959)
19. Pippenger, N.: On rearrangeable and non-blocking switching networks. J. Comput. Syst. Sci. **17**(2), 145–162 (1978). https://doi.org/10.1016/0022-0000(78)90001-6
20. Yao, A.C.C.: Probabilistic computations: toward a unified measure of complexity. In: 18th Annual Symposium on Foundations of Computer Science, Providence, Rhode Island, USA, 31 October - 1 November 1977, pp. 222–227. IEEE Computer Society (1977). https://doi.org/10.1109/SFCS.1977.24
21. Yen, J.Y.: *Shortest Path Network Problems*, volume 18 of *Mathematical Systems in Economics*. Verlag Anton Hain, Meisenheim am Glan (1975)

Shortest Coordinated Motion for Square Robots

Guillermo Esteban[1,2(✉)], Dan Halperin[3], Víctor Ruíz[4], Vera Sacristán[4], and Rodrigo I. Silveira[4]

[1] Departamento de Física y Matemáticas, Universidad de Alcalá, Alcalá de Henares, Spain
g.esteban@uah.es
[2] School of Computer Science, Carleton University, Ottawa, Canada
[3] School of Computer Science, Tel Aviv University, Tel Aviv, Israel
[4] Departament de Matemàtiques, Universitat Politècnica de Catalunya, Barcelona, Spain

Abstract. We study the problem of determining minimum-length coordinated motions for two axis-aligned square robots translating in an obstacle-free plane: Given feasible start and goal configurations, find a continuous motion for the two squares from start to goal, comprising only robot-robot collision-free configurations, such that the total Euclidean distance traveled by the two squares is minimal among all possible such motions. We present an adaptation of the tools developed for the case of discs by Kirkpatrick and Liu [*Characterizing minimum-length coordinated motions for two discs.* Proceedings 28th CCCG, 252-259, 2016; CoRR abs/1607.04005, 2016.] to the case of squares. Certain aspects of the case of squares are more complicated, requiring additional and more involved arguments over the case of discs. Our contribution can serve as a basic component in optimizing the coordinated motion of two squares among obstacles, as well as for *local planning* in sampling-based algorithms, which are often used in practice, in the same setting.

Keywords: Motion planning · Coordinated motions · Geometric algorithms

1 Introduction

The basic motion planning problem is, given start and goal placements for moving objects (robots), to decide whether the objects can move from start to goal without colliding with obstacles in the environment nor with one another, and

Partially supported by project PID2019-104129GB-I00/MCIN/AEI/10.13039/501100011033 of the Spanish Ministry of Science and Innovation. Work by D.H. has been supported in part by the Israel Science Foundation (grant no. 1736/19), by NSF/US-Israel-BSF (grant no. 2019754), by the Israel Ministry of Science and Technology (grant no. 103129), by the Blavatnik Computer Science Research Fund, and by the Yandex Machine Learning Initiative for Machine Learning at Tel Aviv University.

P. Morin and S. Suri (Eds.): WADS 2023, LNCS 14079, pp. 430–443, 2023.
https://doi.org/10.1007/978-3-031-38906-1_28

if so, to plan such a motion. This problem has been intensively investigated for almost five decades now; see, e.g., several books and surveys [1,4,5,9,11,15]. The basic problem is relatively well understood, has general theoretical solutions as well as an arsenal of more practical approaches used by practitioners in robotics, molecular biology, animation, computer games, and additional domains where one needs to automatically plan or simulate feasible collision-free motions; see, e.g., [10].

Finding a feasible collision-free motion is typically insufficient in practice, and we aim to find paths of high quality, such as short paths, paths with high clearance from obstacles, paths requiring minimum energy and so on. Optimizing motion plans is in general significantly harder than finding a feasible solution. The topic is discussed in the books and surveys mentioned above; the specific topic of geometric shortest path problems is reviewed in [13].

1.1 Optimal Motion in the Absence of Obstacles

Let $\mathbb{A}$ and $\mathbb{B}$ be two axis-aligned square robots in the plane. The position of robot $\mathbb{A}$ (resp., $\mathbb{B}$) at a given moment is denoted by A (resp., B), and refers to the coordinates of its center. We define the *radius* of a square as the length of its apothem (half an edge, in the case of a square). Let $r_\mathbb{A}$ and $r_\mathbb{B}$ be the radii of robot $\mathbb{A}$ and robot $\mathbb{B}$, respectively, then we define $r = r_\mathbb{A} + r_\mathbb{B}$. The movement of robot $\mathbb{A}$ in the presence of robot $\mathbb{B}$ can be described as the movement of the point A in the presence of robot $\mathbb{B}'$ with radius r. Thus, from this point on, for convenience of exposition, we will assume that one robot is shrunk to a point and the other robot has the expanded radius r.

Given a point X in the plane, we denote by $sq(X)$ the open axis-aligned square centered at X, with radius r.

We say that a pair of positions (A, B) is *feasible* if $A \notin sq(B)$. Notice that this implies that $B \notin sq(A)$. An *instance* of our problem consists of a feasible pair of initial and final positions (A_0, B_0) and (A_1, B_1), respectively. A *trajectory* from a point X_0 to a point X_1 in the plane is any continuous rectifiable curve $m_X : [0, 1] \to \mathbb{R}^2$ such that $m_X(0) = X_0$ and $m_X(1) = X_1$. Given an instance of our problem, a *coordinated motion* m is a pair of trajectories $m = (m_A, m_B)$. Throughout this paper we refer to *coordinated motions* simply as *motions*. A motion is *feasible* if for all $t \in [0, 1]$ the pair of positions $m(t) = (m_A(t), m_B(t))$ is feasible. We denote the Euclidean arc-length of a trajectory m_X by $\ell(m_X)$. We define the *length* of a motion $m = (m_A, m_B)$ as $\ell(m) = \ell(m_A) + \ell(m_B)$. We focus in this paper on minimum-length coordinated translational motion for two squares. Our goal is to find a description of a minimum-length feasible motion (m_A, m_B), for each instance of the problem. An example of an optimal coordinated motion for two square robots is shown in Fig. 1. We note that, in the figures of this paper, squares representing robots are drawn with filled color, while squares $sq(X)$ are depicted with a white filling.

Feasible motion for two squares translating among polygonal obstacles with n vertices can be found in $O(n^2)$ time, if one exists [16]. For an arbitrary number of unit squares in the same setting, the problem is known to be PSPACE-

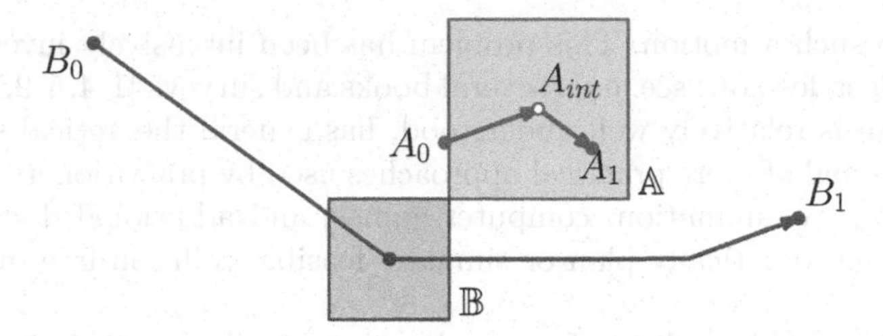

Fig. 1. Example of an optimal motion: first, $\mathbb{A}$ is translated from A_0 to A_{int}, then $\mathbb{B}$ moves from B_0 to B_1 avoiding $\mathbb{A}$, finally, $\mathbb{A}$ is translated from A_{int} to A_1. The squares show the positions of the two robots when $\mathbb{A}$ is at A_{int} and $\mathbb{B}$ is starting to slide around $\mathbb{A}$.

hard [17]. Recently, Kirkpatrick and Liu [7,8] solved the minimum-length coordinated motion for two discs, as we discuss below in Sect. 1.2.

We remark that optimal motion in the absence of obstacles was also investigated for a single wheeled vehicle; see, for example, the work by Dubins [2], or by Reeds and Shepp [14]. The major complication in deriving optimal paths for such systems even in the absence of obstacles stems from their being non-holonomic. The resulting paths bear some similarity to the results obtained in [7] and in the current paper in that the optimal motion of a single robot/vehicle comprises a small set of simple primitive paths. For more details and additional references, see [11, Section 15.3].

1.2 The Kirkpatrick-Liu Analysis for Two Discs

Kirkpatrick and Liu [7,8] describe, for any pair of initial and final positions of the discs, two motions that involve at most six (straight or circular-arc) segments. Then, either (i) a single motion is feasible and optimal, or (ii) among the two motions, one is optimal among all clockwise[1] motions and the other is optimal among all counterclockwise motions. The proof of the optimality of the motions involves an extensive case analysis, which depends on the relative initial and final positions of the discs. However, all motions have a simple structure:

1. Move robot $\mathbb{A}$ from its initial position to an intermediate position A_{int}.
2. Move robot $\mathbb{B}$ from its initial position to its final position.
3. Move robot $\mathbb{A}$ from the intermediate position A_{int} to its final position.

The main mathematical tool employed in [7,8] is *Cauchy's surface area formula.* Its use in the context of optimal motion planning was introduced by Icking et al. [6] for a line segment translating and rotating in the plane. The study of the full rigid motion of a segment involves rotation, which raises the question how to measure the distance between two configurations of the moving object.

[1] Formally defined in Sect. 2; roughly, clockwise here refers to the direction of rotation of a vector from the center of one robot to the center of the other robot throughout the motion, from start to goal.

Icking et al. [6] focus on what they call the d_2-distance, which measures the length of the motion of a segment $\overline{pq}$ by averaging the distance travelled by its two endpoints p and q. We remark that the case of a segment has attracted much attention—the interesting history of the problem, as well as other distance measures, are reviewed in [6].

There is a close relation between minimizing the d_2-distance traveled by a segment and the minimum-length coordinated motion of two discs. Assume the sum of the radii of the two discs equals the length $|pq|$ of the segment. Then, if the two discs remain in contact throughout the whole motion, the two problems are equivalent.

In this paper, we adapt the techniques from [7,8] to square robots. Our work for squares makes several contributions beyond the work for discs. Primarily, as we explain below, there are certain complications for squares that do not arise with discs, making the analysis more complex. Secondly, we complete details that are similar to both systems but have not been dealt with previously. For square robots the shape and relative position matters. The square corners create discontinuities in tangency points, which impacts the definition of the cases that need to be analyzed. The relative position of the two squares (more precisely, the slope of the line passing through the centers of the two squares) also has to be taken into account. For example, the possible shapes of optimal motions can vary considerably between horizontally-aligned and non-horizontally-aligned squares, giving rise to situations that do not exist for disc robots. Additionally, when a disc slides along the boundary of another disc, the distance between their centers remains constant, as opposed to the case for squares, where the distance varies with the angle of contact. This forces us to derive new conditions that guarantee the optimality of motions.

Optimal motion planning for two squares is a fundamental problem. As mentioned above, in the presence of obstacles, we do not even know if the problem has a polynomial-time solution. Understanding the obstacle-free case is a first significant step toward devising solutions to the more complex case with obstacles. Also, these results could serve as good *local motion plans* [11], for more practicable solutions based on sampling.

2 The General Approach

In this section we present the general framework to prove that a motion is optimal, based on the work by Kirkpatrick and Liu [7,8] and Icking et al. [6].

The *trace* of a trajectory m_X is defined as the image of m_X in the plane. For the remaining of this paper, we will use the notation m_X to refer to a trajectory m_X and its trace. Let $\overline{m}_X$ be the boundary of the convex hull of m_X. The trajectory m_X is said to be *convex* if $\overline{m}_X = m_X \cup \overline{X_0X_1}$, where $\overline{X_0X_1}$ is the segment whose endpoints are $m_X(0) = X_0$ and $m_X(1) = X_1$.

A motion $m = (m_A, m_B)$ from (A_0, B_0) to (A_1, B_1) is optimal if both m_A and m_B are convex, and $\ell(\overline{m}_A) + \ell(\overline{m}_B)$ is minimized over all motions from (A_0, B_0) to (A_1, B_1), see [8, Lemma 3.1]. The convexity of m_A and m_B is easy to verify, but the minimality of $\ell(\overline{m}_A) + \ell(\overline{m}_B)$ is not. In order to facilitate proving optimality,

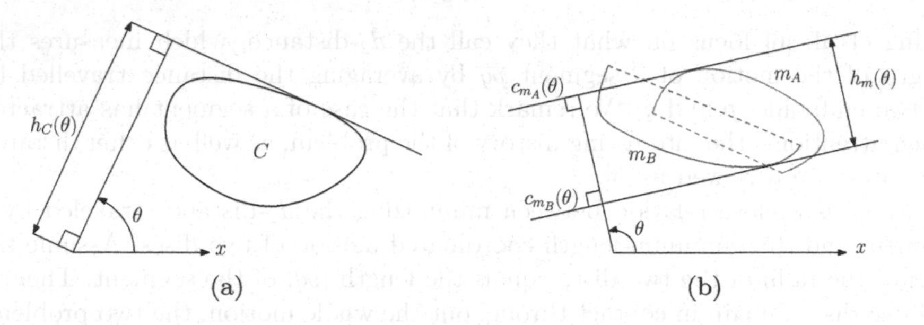

Fig. 2. Left: support function $h_C(\theta)$ of a closed curve C. Right: support function $h_m(\theta)$ of a motion $m = (m_A, m_B)$.

the problem of measuring $\ell(\overline{m}_A) + \ell(\overline{m}_B)$ can be translated into the problem of computing the width of a strip defined by a pair of *supporting lines*, one for $\overline{m}_A$ and one for $\overline{m}_B$ [6]. Given a closed curve C, its *support function* $h_C(\theta)$ can be seen as the (signed) distance from the origin to the extremal supporting line of C in the direction $\theta + \pi/2$. See Fig. 2a for an illustration. As noted in [6], Cauchy's surface area formula [3] implies that if C_1 and C_2 are two closed convex curves, then $\ell(C_1) + \ell(C_2) = \int_0^{2\pi} (h_{C_1}(\theta) + h_{C_2}(\theta + \pi))d\theta$.

We define $h_{m_A}(\theta)$ (resp., $h_{m_B}(\theta)$) to be the support function of $\overline{m}_A$ (resp., $\overline{m}_B$). Then $h_m(\theta) = h_{m_A}(\theta) + h_{m_B}(\theta+\pi)$ can be interpreted as follows. For each $\theta \in S^1$, let $c_{m_A}(\theta)$ be the supporting line for $\overline{m}_A$ in direction $\theta + \pi/2$ and $c_{m_B}(\theta)$ the supporting line for $\overline{m}_B$ in direction $\theta + \pi + \pi/2$. Then $h_m(\theta)$ is the width of the strip formed by the two supporting lines at the given angle, as illustrated in Fig. 2b.

Given two points X, Y, we define $\angle(X, Y)$ as the angle that vector $\overrightarrow{YX}$ forms with the positive x-axis. Let $\theta_0 = \angle(A_0, B_0)$ and $\theta_1 = \angle(A_1, B_1)$. If I is the range of angles $\angle(m_A(t), m_B(t))$ for all $t \in [0, 1]$, then either $[\theta_0, \theta_1] \subseteq I$ or $S^1 \backslash [\theta_0, \theta_1] \subseteq I$ (or both), due to the continuity of the trajectories. In the first case, we call m a *counterclockwise* (CCW) motion. In the second case, we call it a *clockwise* (CW). All motions fall in at least one of the two categories, counterclockwise or clockwise. Therefore, our strategy consists in finding a feasible minimum-length motion in each of the two categories, and taking the best of both.

Given $r = r_{\mathbb{A}} + r_{\mathbb{B}}$ and two angles θ and θ', we define $d_r(\theta, \theta')$ as the width of the minimum strip containing any pair of points X and Y such that i) the bounding lines of the strip have slope $\theta + \pi/2$, ii) $\angle(X, Y) = \theta'$ and iii) Y lies on the boundary of $sq(X)$. Refer to Fig. 3.

Given an instance of our problem with $\theta_0 = \angle(A_0, B_0)$ and $\theta_1 = \angle(A_1, B_1)$, we define the function $s : [\theta_0, \theta_1] \to \mathbb{R}$ as $s(\theta) = \max_{\theta' \in [\theta_0, \theta_1]} d_r(\theta, \theta')$. $s(\theta)$ is a point-wise lower bound for $h_m(\theta)$ for any counterclockwise motion m. Intuitively, the idea is that any counterclockwise motion has to cover the angles from $\angle(A_0, B_0)$ to $\angle(A_1, B_1)$, and, for each of them, $s(\theta)$ gives the minimum valid distance between the centers of the two robots at that angle; hence, it is a lower bound for $h_m(\theta)$.

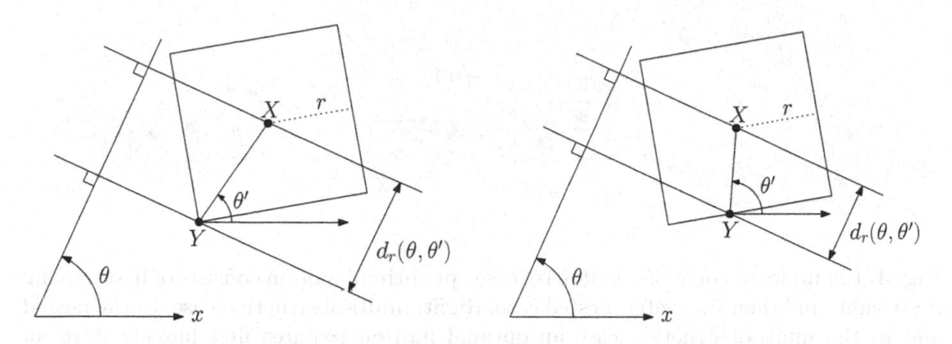

Fig. 3. Illustration of $d_r(\theta, \theta')$ for the same value of θ and two values of θ'.

Let $1_{[\theta_0,\theta_1]}$ be the indicator function of the interval $[\theta_0, \theta_1]$, and let $\overline{h}_m(\theta) = \overline{h}_{m_A}(\theta) + \overline{h}_{m_B}(\theta + \pi), \theta \in S^1$, where $\overline{h}_{m_A}(\theta)$ is the support function of the segment $\overline{A_0A_1}$, and $\overline{h}_{m_B}(\theta)$ is that of $\overline{B_0B_1}$. We define the lower bound function $LB : S^1 \to \mathbb{R}$ as $LB(\theta) = \max\{\overline{h}_m(\theta), s(\theta) \cdot 1_{[\theta_0,\theta_1]}\}$. Since for every angle θ the support function $h_{m_A}(\theta)$ (resp. $h_{m_B}(\theta)$) is lower-bounded by the support function $\overline{h}_{m_A}(\theta)$ (resp. $\overline{h}_{m_B}(\theta)$), we can then observe the following.

Observation 1. *If $m = (m_A, m_B)$ is a feasible counterclockwise motion from (A_0, B_0) to (A_1, B_1), then $h_m(\theta) \geq LB(\theta)$ for all $\theta \in S^1$.*

An analogous result holds for clockwise motions by replacing $1_{[\theta_0,\theta_1]}$ by $1_{S^1\setminus[\theta_0,\theta_1]}$ in the definition of function $LB(\theta)$. Finally, we present two conditions guaranteeing that the support function $h_m(\theta)$ coincides with $LB(\theta)$, hence implying that m has minimum length.

Lemma 1. *Let $m = (m_A, m_B)$ be a motion from (A_0, B_0) to (A_1, B_1). For any angle $\theta \in S^1$, if the support points for $h_m(\theta)$ are A_i and B_j, for $i, j \in \{0, 1\}$, then $h_m(\theta) = LB(\theta)$.*

Lemma 2. *Let $m = (m_A, m_B)$ be a motion from (A_0, B_0) to (A_1, B_1). For any angle $\theta \in [\theta_0, \theta_1]$, if the support points for $h_m(\theta)$ are one point X of m_A or m_B and a boundary point of its square $sq(X)$, then $h_m(\theta) = LB(\theta)$.*

Intuitively, what these conditions say is the following. Recall that by Cauchy's surface area formula, the length of a (convex) motion equals the integral of $h_m(\theta)$ over all angles θ. The first condition says that for a particular θ where the supporting lines are at an endpoint of motion m_A for robot $\mathbb{A}$ and at an endpoint of motion m_B for robot $\mathbb{B}$, the distance between the supporting lines has to be at least $\overline{h}_m(\theta)$. The second condition applies to the angles where the supporting lines are at a point X for one robot and at the boundary of $sq(X)$ for the other robot. In this case, the distance between the supporting lines has to be at least $s(\theta)$, achieved when one robot is sliding around the other one. Any smaller distance at that angle would imply a collision.

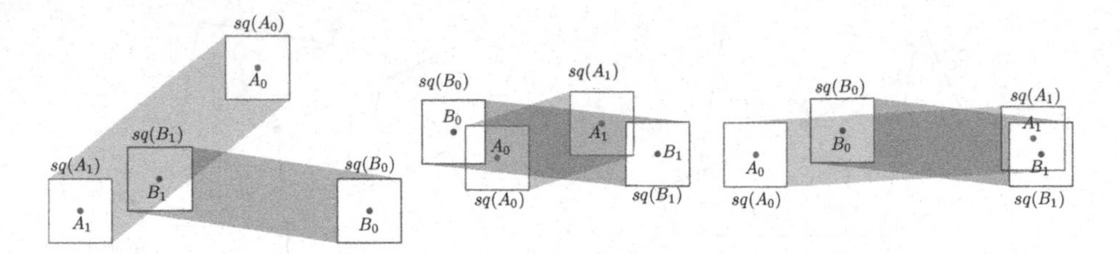

Fig. 4. Examples of corridors. Left: easy case, an optimal motion consists of first moving $\mathbb{A}$ straight, and then $\mathbb{B}$. Center: nested case. Right: multi-obstruction case. In the nested and in the multi-obstruction case an optimal motion requires first moving $\mathbb{A}$ to an intermediate position, then moving $\mathbb{B}$, and then again $\mathbb{A}$.

3 The Minimum-Length Motions at a Glance

In this section we make the minimum-length motions more precise. In particular, we claim that a minimum-length motion can always be obtained by sequentially and alternatively moving one robot at a time, in at most three movements.[2]

The *corridor* $corr_{\mathbb{A}}$ is the Minkowski sum of the closed line-segment $\overline{A_0A_1}$ and a square $sq(X)$, where X is the origin. The definition of $corr_{\mathbb{B}}$ is analogous. See Fig. 4, where $corr_{\mathbb{A}}$ is depicted in red and $corr_{\mathbb{B}}$ in blue.

The relative positions of A_0, A_1, B_0, B_1 with respect to $corr_{\mathbb{A}}$ and $corr_{\mathbb{B}}$ completely determine the shape of an optimal motion. Up to symmetry, exchanging the roles of $\mathbb{A}$ and $\mathbb{B}$, or exchanging the roles of A_0 by A_1, and B_0 by B_1, we can classify these relative positions in three types; see Table 1 and Fig. 4.

- *Easy*: $A_0 \notin corr_{\mathbb{B}}$ and $B_1 \notin corr_{\mathbb{A}}$.
- *Nested*: $A_0 \in corr_{\mathbb{B}}$, $A_1 \in corr_{\mathbb{B}}$, $B_0 \notin corr_{\mathbb{A}}$ and $B_1 \notin corr_{\mathbb{A}}$.
- *Multi-obstruction* (multi): $A_0 \in corr_{\mathbb{B}}$, $B_0 \in corr_{\mathbb{A}}$.

Our main result is summarized by the following theorem.

Theorem 2. *Let $\mathbb{A}$ and $\mathbb{B}$ be two axis-aligned square robots. Let their initial positions be A_0 and B_0, and their final positions be A_1 and B_1, respectively. Up to exchanging the roles of $\mathbb{A}$ and $\mathbb{B}$, there exists a position $A_{int} \notin sq(B_0) \cup sq(B_1)$ (possibly equal to A_0 or A_1) such that the following is a minimum-length feasible motion:*

1. *Move robot $\mathbb{A}$ along the shortest path from A_0 to A_{int} avoiding robot $\mathbb{B}$ (currently located at B_0).*
2. *Move robot $\mathbb{B}$ along the shortest path from B_0 to B_1 avoiding robot $\mathbb{A}$ (currently located at A_{int}).*
3. *Move robot $\mathbb{A}$ along the shortest path from A_{int} to A_1 avoiding robot $\mathbb{B}$ (currently located at B_1).*

[2] One movement is translating a robot from one point in the plane to another location where it does not intersect any other robot by minimizing the distance traveled.

Table 1. Exhaustive description of the cases.

	$A_0 \notin corr_{\mathbb{B}}$ $A_1 \notin corr_{\mathbb{B}}$	$A_0 \notin corr_{\mathbb{B}}$ $A_1 \in corr_{\mathbb{B}}$	$A_0 \in corr_{\mathbb{B}}$ $A_1 \notin corr_{\mathbb{B}}$	$A_0 \in corr_{\mathbb{B}}$ $A_1 \in corr_{\mathbb{B}}$
$B_0 \notin corr_{\mathbb{A}}$ $B_1 \notin corr_{\mathbb{A}}$	easy	easy	easy	nested
$B_0 \notin corr_{\mathbb{A}}$ $B_1 \in corr_{\mathbb{A}}$	easy	multi	easy	multi
$B_0 \in corr_{\mathbb{A}}$ $B_1 \notin corr_{\mathbb{A}}$	easy	easy	multi	multi
$B_0 \in corr_{\mathbb{A}}$ $B_1 \in corr_{\mathbb{A}}$	nested	multi	multi	multi

For each case (easy, nested, multi), the main challenges are i) defining an intermediate position A_{int} and ii) proving that the motion defined in Theorem 2 with that value of A_{int} is optimal.

We start by noting that in the easy case, an optimal motion consists of translating each of the robots directly from its initial to its final position along a straight-line segment, but the order might be relevant. In the remaining cases, a straight-line motion is not possible, so we need a finer analysis.

The full proof, covering the rest of the cases, requires the analysis of a large number of situations. Due to space limitations, in Sect. 4 we only explain in some detail the nested case. The remaining details can be found in the full version.

4 Details on Minimum-Length Motions for the Nested Case

In the following, it will be more convenient to assume, without loss of generality, that B_0 and B_1 are horizontally aligned; this can be achieved with a suitable rotation (note that in the figures that follow, squares have been rotated accordingly). Moreover, we will also assume that B_0 lies to the left of B_1, since otherwise a 180° rotation would take us to this situation. These assumptions are for ease of presentation, allowing us to properly use expressions like "upper tangent" as well as the notions "above", "below", "left" and "right".

Since we are in the nested case, without loss of generality, we can assume that $A_0, A_1 \in corr_{\mathbb{B}}$ and $B_0, B_1 \notin corr_{\mathbb{A}}$. Observe that $B_0, B_1 \notin corr_{\mathbb{A}}$ implies that $A_i \notin sq(B_j)$ for $i, j \in \{0, 1\}$.

Let t_{ij} denote the upper tangent line from a point A_i to $sq(B_j)$, for $i, j = 0, 1$. Let p_{ij} be the support point of line t_{ij} in $sq(B_j)$. We will consider t_{ij} as a directed line, directed from A_i to p_{ij}. We say that two tangents t_{i0} and t_{i1} are *twisted* if p_{i1} is to the left of t_{i0}, for some $i \in \{0, 1\}$. See Fig. 8, where tangents t_{00} and t_{01} are non-twisted, and Fig. 5, where t_{00} and t_{01}, and t_{10} and t_{11} are twisted.

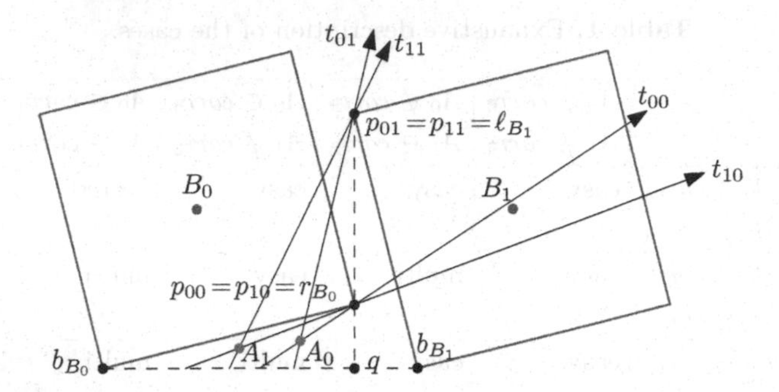

Fig. 5. t_{00} and t_{01}, and t_{10} and t_{11} are twisted. Note that A_1 and A_0 are below the polyline $b_{B_0} r_{B_0} q$.

We will use the notation $\overrightarrow{pq}$ to denote the line through p and q in the direction of the vector $q - p$. Note that t_{i0}, t_{i1} are twisted if A_i is to the right of the line $\overrightarrow{b_{B_0} r_{B_0}}$ and to the left of the line $\overrightarrow{p_{i0} p_{i1}}$.

For the nested case, we need to differentiate three cases, depending on whether the motion is *fully* CCW (i.e., it does not contain CW (sub)motion parts) or not, and, in the former case, depending on whether at least one of the pairs of tangents t_{i0}, t_{i1} is twisted or not.

In the following we only examine motions that are counterclockwise; a clockwise optimal motion can be obtained by reflecting the initial and final placements across the x-axis, and then computing an optimal counterclockwise motion.

4.1 Motion is Fully CCW, t_{00} and t_{01}, or t_{10} and t_{11} are Twisted

If the motion is fully CCW, but a pair of tangents t_{i0}, t_{i1} is twisted, the motion described in Theorem 2 might not be optimal. In this case, we prove that there exists a CW motion that is feasible and globally optimal, therefore we can ignore instances with twisted tangents.

If t_{i0} and t_{i1} are twisted, the sides of the squares are not axis-aligned. So, let $r_{B_i}, b_{B_i}, \ell_{B_i}$, and t_{B_i} be, respectively, the right, bottom, left, and top corner of $sq(B_i)$, for $i \in \{0, 1\}$. Let $r_{A_i}, b_{A_i}, \ell_{A_i}$, and t_{A_i} be, respectively, the right, bottom, left, and top corner of $sq(A_i)$, for $i \in \{0, 1, int\}$. Let q be the intersection between the lines $\overrightarrow{\ell_{B_1} r_{B_0}}$ and $\overrightarrow{b_{B_0} b_{B_1}}$; see Fig. 5. Then, we know the following:

Observation 3. *If t_{i0} and t_{i1} are twisted, A_i is below or on the polyline $b_{B_0} r_{B_0} q$ and above or on the line $\overrightarrow{b_{B_0} b_{B_1}}$. Hence, the highest point A_i could be is r_{B_0}, and the highest point ℓ_{A_i} and r_{A_i} could be is on the line $\overrightarrow{B_0 B_1}$. Also, since we are in the nested case, A_0 is visible from A_1.*

We will prove that if t_{00} and t_{01}, or t_{10} and t_{11} are twisted, there is a clockwise motion as short or shorter than any counterclockwise motion, for any possible position of the intermediate point A_{int} in the counterclockwise motion.

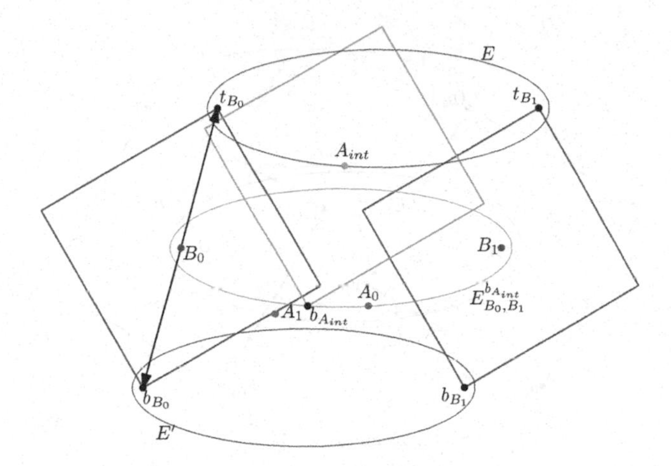

Fig. 6. Ellipse $E^{b_{A_{int}}}_{B_0,B_1}$ translated by the vectors $\overrightarrow{B_0 t_{B_0}}$ and $\overrightarrow{B_0 b_{B_0}}$.

Theorem 4. *In a fully counterclockwise motion, if t_{i0} and t_{i1} are twisted, for some $i \in \{0, 1\}$, then the minimum-length motion is clockwise.*

Proof. We will show that there is an intermediate position A'_{int} for the motion of $\mathbb{A}$ such that the length of motion m_B around A'_{int} in the CCW and in the CW motion is the same. In addition, we will need that the distance traveled by $\mathbb{A}$ in the CW motion is not larger than the distance traveled in the CCW motion.

Let $E^z_{x,y}$ denote the ellipse through z, whose two foci are x and y. We distinguish the case where $b_{A_{int}}$ belongs to the interior of both $E^{t_{A_0}}_{B_0,B_1}$ and $E^{t_{A_1}}_{B_0,B_1}$, and a second case where $b_{A_{int}}$ does not belong to the interior of one ellipse $E^{t_{A_j}}_{B_0,B_1}$, for some $j \in \{0, 1\}$.

Let E and E' be the ellipses $E^{b_{A_{int}}}_{B_0,B_1}$ translated by, respectively, the vectors $\overrightarrow{B_0 t_{B_0}}$, and $\overrightarrow{B_0 b_{B_0}}$; see the dark-green ellipses in Fig. 6. First, if $b_{A_{int}}$ belongs to the interior of both $E^{t_{A_0}}_{B_0,B_1}$ and $E^{t_{A_1}}_{B_0,B_1}$, by the construction of E' and the two robots having the same orientation, we know that E' does not contain A_0 nor A_1. Similarly, using Observation 3, and the construction of E, we know that it does not contain A_0 nor A_1.

Consider the point A'_{int} in E', which is the reflection of A_{int} about the segment $\overline{B_0 B_1}$ and then about its bisector; see Fig. 7. Using the symmetry of E and E', we get that $|A_i A'_{int}| + |A'_{int} A_{1-i}| \leq |A_{1-i} A_{int}| + |A_{int} A_i|$.

This means that the distance from A_0 to A_1 through A'_{int} is not larger than the distance through A_{int}. Thus, we found a position A'_{int} such that the distance traveled by $\mathbb{B}$ in the CCW motion around A_{int} is the same than the distance in the CW motion around A'_{int}. However, the distance traveled by $\mathbb{A}$ in the CW motion is not larger than the distance traveled in the CCW motion.

Similarly, let $j \in \{0, 1\}$ be such that $b_{A_{int}}$ does not belong to the interior of $E^{t_{A_j}}_{B_0,B_1}$. This case can be proved in a similar fashion, using Observation 3, and the properties of the ellipses. See the full version for more details. □

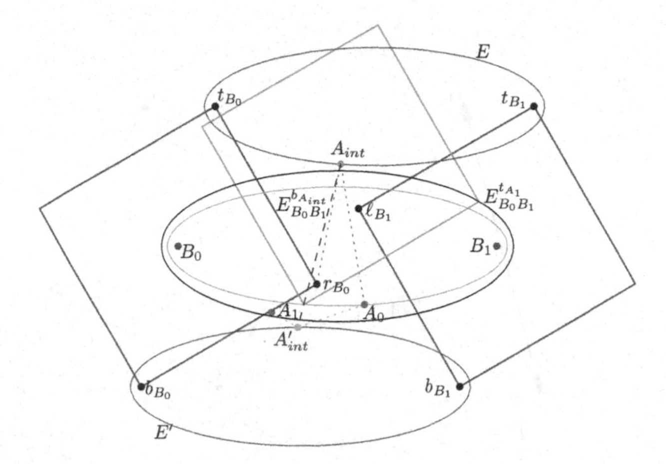

Fig. 7. A'_{int} belongs to the ellipse E', and the distance $|A_0A'_{int}| + |A'_{int}A_1|$ is shorter than the distance $|A_0A_{int}| + |A_{int}r_{B_0}| + |r_{B_0}A_1|$.

Theorem 4 allows us to ignore the case defined in this section, and the optimal motion will be obtained when analyzing the CW motion. It is worth noting that, since A_i is below the polyline $b_{B_0}r_{B_0}q$, and A_0 is visible from A_1, t_{i0} and t_{i1} cannot be twisted in the CW case if they are twisted in the CCW case.

4.2 Motion is Fully CCW, and no Tangents t_{i0} and t_{i1} are Twisted

If the motion is fully CCW, and no tangents are twisted, we prove that the sufficient conditions from Lemmas 1 and 2 are fulfilled, and that the motion is always feasible. To that end, we analyze the position of the two supporting lines for each possible angle $\theta \in [\theta_0, \theta_1]$, and argue that for each value of θ, either both supporting lines are touching A_i and B_j, for $i, j \in \{0, 1\}$, or one is touching a point X of one of the motions, and the other is touching its square $sq(X)$. This implies that the sufficient conditions hold, thus the motion is optimal. In addition, the motion is feasible since $A_{int} \notin (sq(B_0) \cup sq(B_1))$ and A_{int} is visible from both A_0 and A_1.

Let $vis(A_0)$ be the region of $corr_{\mathbb{B}} \backslash (sq(B_0) \cup sq(B_1))$ that is visible from A_0. $vis(A_0)$ is decomposed into four *zones*, defined by the tangents t_{ij}; see Fig. 8. For each zone we specify a different location for the intermediate point A_{int}. In this section we explain the case where A_1 belongs to Zone I, the locus of all points to the right of t_{00} and to the left of t_{01}, and defer to the full version the rest of the cases.

If $A_1 \in$ Zone I, $A_{int} = A_1$. Recall that $c_{m_A}(\theta)$ is the supporting line for $\overline{m}_A$ in direction $\theta + \pi/2$ and $c_{m_B}(\theta)$ is the supporting line for $\overline{m}_B$ in direction $\theta + \pi + \pi/2$. So, when the supporting point of $c_{m_B}(\theta)$ is B_j, the supporting point of $c_{m_A}(\theta)$ is A_i, for $i, j \in \{0, 1\}$ and the conditions from Lemma 1 are fulfilled. Hence, we just need to prove that when $c_{m_B}(\theta)$ is touching a corner of $sq(A_{int})$, $c_{m_A}(\theta)$ is touching $A_{int} = A_1$.

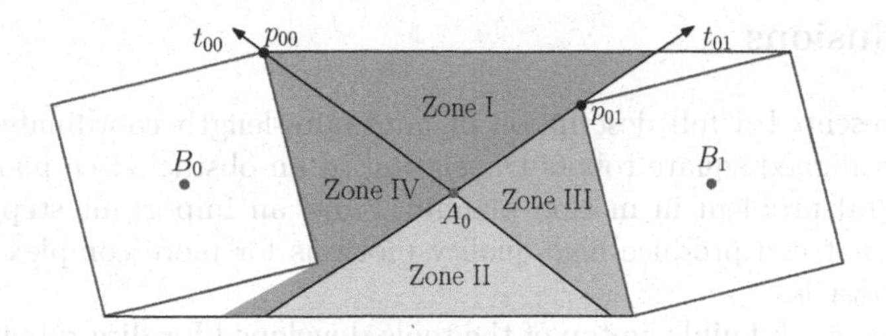

Fig. 8. Possible regions for A_1 in the nested case.

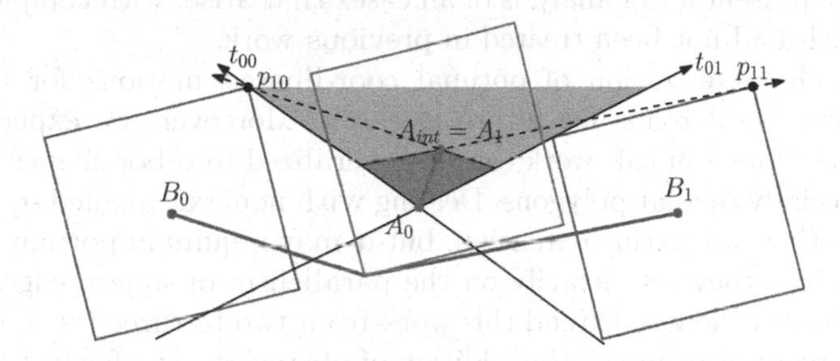

Fig. 9. A_0 has to be in the shaded region when $A_1 \in$ Zone I.

The first segment of the motion m_B is parallel to the line $\overrightarrow{A_1 p_{10}}$, and the last segment is parallel to the line $\overrightarrow{A_1 p_{11}}$; see Fig. 9. The motion of the supporting lines is counterclockwise, so we have to argue that from the moment $c_{m_B}(\theta)$ starts touching a corner of $sq(A_{int})$ until it stops touching a corner of $sq(A_{int})$, $c_{m_A}(\theta)$ is touching A_1. This is analogous to showing that from the moment $c_{m_A}(\theta)$ touches p_{10} until it touches p_{11}, $c_{m_A}(\theta)$ does not touch A_0, i.e., that A_0 is on or below the *lower envelope*[3] of the lines $\overrightarrow{A_1 p_{10}}$ and $\overrightarrow{A_1 p_{11}}$. One can prove this property by using the fact that both supporting lines are parallel, and the definition of Zone I.

4.3 Motion Contains CW Parts

If the motion contains clockwise (sub)motion parts, there can be some angles for which the motion does not satisfy any of the two sufficient conditions from Lemmas 1 and 2. This means that the techniques used previously to prove the optimality of the motions cannot be directly applied. However, we show in the full version that in such a case, there exists an alternative motion m', which is different in the coordination scheme from m, but that is fully counterclockwise and has exactly the same trace for $\mathbb{A}$ and $\mathbb{B}$ (thus also the same length). Since m' is fully counterclockwise, we can apply the results in Sects. 4.1 and 4.2 to prove that m' has minimum length. Therefore, since m has the same length as m', we can conclude that m is optimal as well.

[3] If we think of a finite set of lines as graphs of (partially defined) functions, the lower envelope is the graph of the pointwise minimum [12].

5 Conclusions

We have presented a full description of minimum-length coordinated motions for two axis-aligned square robots translating in an obstacle-free plane. This is a fundamental problem in motion planning, and an important step to design algorithms that can produce high-quality motions for more complex situations involving obstacles.

While our work builds on top of the tools developed for disc robots [7,8], we show that squares present several additional difficulties over the case of discs. Moreover, we present a full analysis of all cases that arise, with complete details, many of which had not been treated in previous work.

Our full characterization of optimal coordinated motions for two square robots opens several roads for future research. Moreover, we expect that the techniques developed in this work can be generalized to robot shapes other than squares, especially regular polygons. Dealing with non-axis aligned square robots would be another interesting extension, but it may require important changes to our proofs, since they rely heavily on the parallelism of square edges. Another natural question is how to extend this work from two to three (or more) robots. Finally, as already mentioned, the addition of obstacles, even for just two robots, is probably the most relevant next step.

References

1. Choset, H., et al.: Principles of Robot Motion: Theory, Algorithms, and Implementation. MIT Press, Cambridge (2005)
2. Dubins, L.E.: On curves of minimal length with a constraint on average curvature, and with prescribed initial and terminal positions and tangents. Am. J. Math. **79**, 497–516 (1957)
3. Eggleston, H.G.: Convexity (1966)
4. Halperin, D., Kavraki, L., Solovey, K.: Robotics. In: Goodman, J.E., O'Rourke, J., Tóth, C., (eds.), Handbook of Discrete and Computational Geometry, chapter 51, pp. 1343–1376. 3rd edn Chapman & Hall/CRC (2018)
5. Halperin, D., Salzman, O., Sharir, M.: Algorithmic motion planning. In: Goodman, J.E., O'Rourke, J., Tóth, C., (eds.) Handbook of Discrete and Computational Geometry, chapter 50, pp. 1311–1342. 3rd edn Chapman & Hall/CRC (2018)
6. Icking, C., Rote, G., Welzl, E., Yap, C.: Shortest paths for line segments. Algorithmica **10**(2), 182–200 (1993)
7. Kirkpatrick, D.G., Liu, P.: Characterizing minimum-length coordinated motions for two discs. In Proceedings of 28th CCCG, pp. 252–259 (2016)
8. Kirkpatrick, D.G., Liu, P.: Characterizing minimum-length coordinated motions for two discs. CoRR, abs/1607.04005 (2016). http://arxiv.org/1607.04005
9. Latombe, J.-C.: Robot Motion Planning. Kluwer, Boston (1991)
10. Latombe, J.-C.: Motion planning: a journey of robots, molecules, digital actors, and other artifacts. Int. J. Robot. Res. **18**(11), 1119–1128 (1999)
11. Steven, M.: LaValle. Cambridge University Press, Planning Algorithms (2006)
12. Matousek, J.: Lectures on Discrete Geometry, vol. 212. Springer, Cham (2013)

13. Mitchell, J.S.B.: Shortest paths and networks. In: Goodman, J.E., O'Rourke, J., Tóth, C., (eds.), Handbook of Discrete and Computational Geometry, chapter 31, 3rd edn., pp. 811–848. Chapman & Hall/CRC (2018)
14. Reeds, J.A., Shepp, L.A.: Optimal paths for a car that goes both forwards and backwards. Pac. J. Math. **145**(2), 367–393 (1990)
15. Salzman, O.: Sampling-based robot motion planning. Commun. ACM **62**(10), 54–63 (2019). https://doi.org/10.1145/3318164
16. Sharir, M., Sifrony, S.: Coordinated motion planning for two independent robots. Ann. Math. Artif. Intell. **3**(1), 107–130 (1991). https://doi.org/10.1007/BF01530889
17. Solovey, K., Halperin, D.: On the hardness of unlabeled multi-robot motion planning. Int. J. Robot. Res. **35**(14), 1750–1759 (2016). https://doi.org/10.1177/0278364916672311

Linear Layouts of Bipartite Planar Graphs

Henry Förster[1], Michael Kaufmann[1], Laura Merker[2(✉)], Sergey Pupyrev[3], and Chrysanthi Raftopoulou[4]

[1] Department of Computer Science, University of Tübingen, Tübingen, Germany
{henry.foerster,michael.kaufmann}@uni-tuebingen.de
[2] Institute of Theoretical Informatics, Karlsruhe Institute of Technology, Karlsruhe, Germany
laura.merker2@kit.edu
[3] Meta Platforms Inc., Menlo Park, CA, USA
[4] National Technical University of Athens, Athens, Greece
crisraft@mail.ntua.gr

Abstract. A linear layout of a graph G consists of a linear order, $\prec$, of the vertices and a partition of the edges. A part is called a *queue* (*stack*) if no two edges nest (cross), that is, two edges (v, w) and (x, y) with $v \prec x \prec y \prec w$ ($v \prec x \prec w \prec y$) may not be in the same queue (stack). The best known lower and upper bounds for the number of queues needed for planar graphs are 4 [Alam et al., Algorithmica 2020] and 42 [Bekos et al., Algorithmica 2022], respectively. While queue layouts of special classes of planar graphs have received increased attention following the breakthrough result of [Dujmović et al., J. ACM 2020], the meaningful class of bipartite planar graphs has remained elusive so far, explicitly asked for by Bekos et al. In this paper we investigate bipartite planar graphs and give an improved upper bound of 28 by refining existing techniques. In contrast, we show that two queues or one queue together with one stack do not suffice; the latter answers an open question by Pupyrev [GD 2018]. We further investigate subclasses of bipartite planar graphs and give improved upper bounds; in particular we construct 5-queue layouts for 2-degenerate quadrangulations.

Keywords: bipartite planar graphs · queue number · mixed linear layouts · graph product structure

1 Introduction

Since the 1980s, linear graph layouts have been a central combinatorial problem in topological graph theory, with a wealth of publications [7,8,12,18,21–23,28,

This research was initiated at the GNV workshop in Heiligkreuztal, Germany, June 26 – July 1, 2022, organized by Michalis Bekos and Michael Kaufmann. Thanks to the organizers and other participants for creating a productive environment.

P. Morin and S. Suri (Eds.): WADS 2023, LNCS 14079, pp. 444–459, 2023.
https://doi.org/10.1007/978-3-031-38906-1_29

29,38]. A *linear layout* of a graph consists of a linear order $\prec$ of the vertices and a partition of the edges into *stacks* and *queues*. A part is called a *queue* (*stack*) if no two edges of this part nest (cross), that is, two edges (v, w) and (x, y) with $v \prec x \prec y \prec w$ ($v \prec x \prec w \prec y$) may not be in the same queue (stack). Most notably, research has focused on so-called *stack layouts* (also known as book-embeddings) and *queue layouts* where either all parts are *stacks* or all parts are *queues*, respectively. While these kinds of graph layouts appear quite restrictive on first sight, they are in fact quite important in practice. For instance, stack layouts are used as a model for chip design [18], while queue layouts find applications in three-dimensional network visualization [14,23,25]. For these applications, it is important that the edges are partitioned into as few stacks or queues as possible. This notion is captured by the *stack number*, $\text{sn}(G)$, and *queue number*, $\text{qn}(G)$, of a graph G, which denote how many stacks or queues are required in a stack and a queue layout of G, respectively. Similarly, *mixed linear layouts*, where both stacks and queues are allowed, have emerged as a research direction in the past few years [3,19,33].

Recently, queue layouts have received much attention as several breakthroughs were made which pushed the field further. Introduced in 1992 [29], it was conjectured in the same year [28], that all planar graphs have a bounded queue number. Despite various attempts at settling the conjecture [5,6,10,21], it remained unanswered for almost 30 years. In 2019, the conjecture was finally affirmed by Dujmović, Joret, Micek, Morin, Ueckerdt and Wood [22]. Their proof relies on three ingredients: First, it was already known that graphs of bounded treewidth have bounded queue number [36]. Second, they showed that the *strong product* of a graph of bounded queue number and a path has bounded queue number. Finally, and most importantly, they proved that every planar graph is a subgraph of the *strong product* of a path and a graph of treewidth at most 8. In the few years following the result, both queue layouts [1,7,13,20,30] and graph product structure [9,16,17,31,35,37] have become important research directions. Yet, after all recent developments, the best known upper bound for the queue number of planar graphs is 42 [7] whereas the best known corresponding lower bound is 4 [2]. This stands in contrast to a tight bound of 4 for the stack number of planar graphs [11,38].

It is noteworthy that better upper bounds of the queue number are known only for certain subclasses of planar graphs, such as planar 3-trees [2] and posets [1], or for relaxed variants of the queue number [30]. It remains elusive how other properties of a graph, such as a bounded degree or bipartiteness, can be used to reduce the gap between the lower and the upper bounds on the queue number; see also the open problems raised in [7] which contains the currently best upper bound. This is partially due to the fact that it is not well understood how these properties translate into the product structure of the associated graph classes. In fact, the product structure theorem has been improved for general planar graphs [35] while, to the best of our knowledge, there are very few results that yield stronger properties for subclasses thereof. Here, we contribute to this line of research by studying bipartite planar graphs.

Results. Our paper focuses on the queue number of bipartite graphs and subclasses thereof. All necessary definitions are given in Sect. 2. We start by revisiting results from the existing literature in Sect. 2. In particular, we discuss techniques that are used to bound the queue number of general planar graphs by 42 [7,22] and refine them to obtain an upper bound of 28 on the queue number of bipartite planar graphs (Theorem 1). We then improve this bound for interesting subfamilies of bipartite planar graphs. In particular, in Sect. 3, we obtain an upper bound of 21 for the queue number of stacked quadrangulations which is a family of graphs that may be regarded as a bipartite variant of planar 3-trees (Theorem 5). For this we first prove a product structure theorem which states that every stacked quadrangulation is a subgraph of $H \boxtimes C_4$, where H is a planar 3-tree (Theorem 4). We remark that we avoid the path factor that is present in most known product structure theorems.

Complementing our upper bounds, we provide lower bounds on the queue number and the mixed page number of bipartite planar graphs in Sect. 4. Namely, we show a bipartite planar graph whose queue number is at least 3 (Theorem 6) and additionally construct a bipartite planar graph that does not admit a 1-queue 1-stack layout (Theorem 7), answering an open question discussed in [3,19, 33]. For this purpose, we use a family of 2-degenerate quadrangulations. Finally, inspired by the lower bound construction, we prove that every 2-degenerate quadrangulation has queue number at most 5 (Theorem 8) in Sect. 5.

2 Preliminaries

In this section, we introduce basic definitions and tools that we use to analyze the queue number of bipartite planar graphs.

Classes of Bipartite Planar Graphs. We focus on the *maximal bipartite planar graphs* which are exactly the quadrangulations of the plane, that is, every face is a quadrangle. In the following, we introduce interesting families of quadrangulations.

One such family are the *stacked quadrangulations* that can be constructed as follows. First, a square is a stacked quadrangulation. Second, if G is a stacked quadrangulation and f is a face of G, then inserting a plane square S into f and connecting the four vertices of S with a planar matching to the four vertices of f again yields a stacked quadrangulation. Note that every face has four vertices, that is, the constructed graph is indeed a quadrangulation. We are particularly interested in this family of quadrangulations as stacked quadrangulations can be regarded as the bipartite variant of the well-known graph class *planar 3-trees* which are also known as *stacked triangulations.* This class again can be recursively defined as follows: A *planar 3-tree* is either a triangle or a graph that can be obtained from a planar 3-tree by adding a vertex v into some face f and connecting v to the three vertices of f. This class is particularly interesting in the context of queue layouts as it provides the currently best lower bound on the queue number of planar graphs [2].

The notions of stacked triangulations and stacked quadrangulations can be generalized as follows using once again a recursive definition. For $t \geqslant 3$ and $s \geqslant 1$, any connected planar graph of order at most s is called a (t, s)-stacked graph. Moreover, for a (t, s)-stacked graph G and a connected planar graph G' with at most s vertices, we obtain another (t, s)-stacked graph G'' by connecting G' to the vertices of a face f of G in a planar way such that each face of G'' has at most t vertices. Note that we do not require that the initial graph G and the connected stacked graph G' of order at most s in the recursive definition have only faces of size at most t as this will not be required by our results in Sect. 3. If in each recursive step the edges between G' and the vertices of f form a matching, then the resulting graph is called an *(t, s)-matching-stacked graph.* Now, in particular $(3, 1)$-stacked graphs are the planar 3-trees while the stacked quadrangulations are a subclass of the $(4, 4)$-matching-stacked graphs. In addition, $(3, 3)$-stacked graphs are the *stacked octahedrons*, which were successfully used to construct planar graphs that require four stacks [11,39].

In addition to graphs obtained by recursive stacking operations, we will also study graphs that are restricted by *degeneracy*. Namely, we call a graph $G = (V, E)$ d-degenerate if there exists a total order $(v_1, \ldots, v_n)$ of V, so that for $1 \leqslant i \leqslant n$, v_i has degree at most d in the subgraph induced by vertices $(v_1, \ldots, v_i)$. Of particular interest will be a family of 2-degenerate quadrangulations discussed in Sect. 4. It is worth pointing out that there are recursive constructions for triconnected and simple quadrangulations that use the insertion of degree-2 vertices and $(4, 4)$-matching stacking in their iterative steps [26].

Linear Layouts. A *linear layout* of a graph $G = (V, E)$ consists of an order $\prec$ of V and a partition $\mathcal{P}$ of E. Consider two disjoint edges $(v, w), (x, y) \in E$. We say that (v, w) *nests* (x, y) if $v \prec x \prec y \prec w$, and we say that (v, w) and (x, y) *cross* if $v \prec x \prec w \prec y$. For each part $P \in \mathcal{P}$ we require that either no two edges of P nest or that no two edges of P cross. We call P a *queue* in the former and a *stack* in the latter case. Let $\mathcal{Q} \subseteq \mathcal{P}$ denote the set of queues and let $\mathcal{S} \subseteq \mathcal{P}$ denote the set of stacks; such a linear layout is referred to as a *$|\mathcal{Q}|$-queue $|\mathcal{S}|$-stack* layout. If $\mathcal{Q} \neq \emptyset$ and $\mathcal{S} \neq \emptyset$, we say that the linear layout is *mixed*, while when $\mathcal{S} = \emptyset$, it is called a *$|\mathcal{Q}|$-queue* layout. The *queue number*, $\mathrm{qn}(G)$, of a graph G is the minimum value q such that G admits a q-queue layout. Heath and Rosenberg [29] characterize graphs admitting a 1-queue layout in terms of *arched-leveled* layouts. In particular, this implies that each *leveled planar graph* admits a 1-queue layout. Leveled planar graphs are an important family of bipartite graphs whose vertices can be assigned a *layer value* $\lambda : V \to \mathbb{N}$ such that for each edge (u, v) it holds that $\lambda(u) = \lambda(v) \pm 1$. Indeed a bipartite graph has queue number 1 if and only if it is leveled planar [4].

Important Tools. The *strong product* $G_1 \boxtimes G_2$ of two graphs G_1 and G_2 is a graph with vertex set $V(G_1) \times V(G_2)$ and an edge between two vertices (v_1, v_2) and (w_1, w_2) if (i) $v_1 = w_1$ and $(v_2, w_2) \in E(G_2)$, (ii) $v_2 = w_2$ and $(v_1, w_1) \in E(G_1)$, or (iii) $(v_1, w_1) \in E(G_1)$ and $(v_2, w_2) \in E(G_2)$.

Given a graph G, an *H-partition* of G is a pair $\big(H, \{V_x : x \in V(H)\}\big)$ consisting of a graph H and a partition of V into sets $\{V_x : x \in V(H)\}$ called *bags* such that for every edge $(u, v) \in E$ one of the following holds: (i) $u, v \in V_x$ for some $x \in V(H)$, or (ii) there is an edge (x, y) of H with $u \in V_x$ and $v \in V_y$. In Case (i), we call (u, v) an *intra-bag* edge, while we call it *inter-bag* in Case (ii). To avoid confusion with the vertices of G, the vertices of H are called *nodes*. The *width* of an H-partition is defined as the maximum size of a bag. It is easy to see that a graph G has an H-partition of width w if and only if it is a subgraph of $H \boxtimes K_w$ (compare Observation 35 of [22]). In the case where H is a tree, we call the H-partition a *tree-partition*. We refer to Fig. 4 for an example of a tree-partition. Another well-known tool we use are *tree decompositions* [15], for which we assume familiarity but provide a brief introduction in [27].

2.1 General Bipartite Planar Graphs

Following existing works on queue layouts of planar graphs, we prove:

Theorem 1. *The queue number of bipartite planar graphs is at most* 28.

Proof (Sketch). We closely follow the arguments in [7,22]. Given a bipartite planar graph G, we compute an H-partition $(H, \{A_x : x \in V(H)\})$ of G and a *BFS-layering* $\mathcal{L} = (V_0, V_1, \ldots)$ such that (i) V_i contains exactly the vertices with graph-theoretic distance i from a specified vertex $r \in V(G)$, (ii) H has treewidth at most 3, and (iii) the H-partition has layered-width at most 3, that is, for each pair i, x we have $|A_x \cap V_i| \leqslant 3$. While the arguments in [7,22] assume the input graph to be triangulated, we show that a bipartite planar graph G can be triangulated in such a way, that every layer V_i contains only vertices of one part of $V(G)$. We note that the modifications of Bekos et al. [7] on the algorithm by Dujmović et al. [22] respect the layering, that is, they can also be applied for bipartite planar graphs. Now, given an H-partition and a BFS-layering as defined above, by [7] we can compute a queue layout where the *intra-layer* edges (that is, edges between vertices in the same layer $V_i \in \mathcal{L}$) are partitioned into 14 queues while the *inter-layer* edges (that is, edges between vertices in two consecutive layers $V_i, V_{i+1} \in \mathcal{L}$) are partitioned into 28 distinct queues. Since the BFS-layering respects the bipartition of $V(G)$, there are no intra-layer edges. Hence, we use the 28 queues for inter-layer edges. See [27] for details. □

3 Structure of Stacked Quadrangulations

In this section we investigate the structure of stacked quadrangulations and then deduce an upper bound on the queue number. In particular we show that every stacked quadrangulation is a subgraph of the strong product $H \boxtimes C_4$, where H is a planar 3-tree. Recall that stacked quadrangulations are $(4,4)$-matching-stacked graphs. We first prove a product structure theorem for general (t, s)-stacked graphs as it prepares the proof for the matching variant.

Theorem 2. *For* $t \geqslant 3, s \geqslant 1$, *every* (t, s)*-stacked graph* G *is a subgraph of* $H \boxtimes K_s$ *for some planar graph* H *of treewidth at most* t.

In order to prove the theorem, we need to find an H-partition of G and a tree decomposition T of H with treewidth at most t, where H is a planar graph. Given a (t, s)-stacked graph G for $t \geqslant 3$ and $s \geqslant 1$ and its construction sequence, we define the H-partition $(H, \{V_x : x \in V(H)\})$ of width at most s as follows. In the base case we have a graph of size at most s whose vertices all are assigned to a single bag. Then, in each construction step we add a new bag, say $V_{v'}$, that contains all new vertices (those of the inserted graph G'). Thus each bag contains at most s vertices, that is, the width of the H-partition is at most s. For $u \in V_x$ and $v \in V_y$, if G contains edge (u, v), we connect x and y in H.

We give a full proof of Theorem 2 in [27], where we find a tree decomposition of width at most t for the constructed graph H. We now extend our ideas to the following product structure theorem, shown in [27]. Note that we use the same H-partition and show that the treewidth is at most $t - 1$.

Theorem 3. *For $t \geqslant 3, s \geqslant 2$, every (t, s)-matching-stacked graph G is a subgraph of $H \boxtimes K_s$ for some planar graph H of treewidth at most $t - 1$.*

Theorem 2 shows that every stacked quadrangulation is a subgraph of $H \boxtimes K_4$, where H is a planar graph with treewidth at most 4. However, stacked quadrangulations are also $(4, 4)$-matching-stacked graphs and, hence, by Theorem 3 they are subgraphs of $H \boxtimes K_4$, where H is a planar 3-tree. In the following we improve this result by replacing K_4 by C_4.

Theorem 4. *Every stacked quadrangulation is a subgraph of $H \boxtimes C_4$, where H is a planar 3-tree.*

Proof. Given a stacked quadrangulation G and its construction sequence, we compute its H-partition as described above. Note that in Theorem 3, we show that H is a planar 3-tree. It remains to show that H does not only certify that $G \subseteq H \boxtimes K_4$ (Theorem 3) but also the stronger statement that $G \subseteq H \boxtimes C_4$. That is, we need to show that in order to find G in the product, we only need edges that show up in the product with C_4. In each step of the construction of G we insert a 4-cycle into some face f. Hence, for each node x of H the bag V_x consists of a 4-cycle denoted by $(v_0^x, v_1^x, v_2^x, v_3^x)$. We label the four vertices with $0, 1, 2, 3$ such that for $i = 0, 1, 2, 3$ vertex v_{i+k}^x is labeled i for some offset k (all indices and labels taken modulo 4).

As the labels appear consecutively along the 4-cycle, the strong product allows for edges between two vertices of distinct bags if and only if their labels differ by at most 1 (mod 4). That is we aim to label the vertices of each inserted 4-cycle so that each inter-bag edge connects two vertices whose labels differ by at most 1. We even prove a slightly stronger statement, namely that offset k can be chosen for a newly inserted bag V_x such that the labels of any two vertices v and v' in G connected by an inter-bag edge differ by exactly 1. Indeed, assuming this for now, we see that if (v, v') is an inter-bag edge with $v \in V_x$ and $v' \in V_y$, then $(x, y) \in E(H)$. For this recall that V_x and V_y induce two 4-cycles. Assume that $v = v_i^x$ with label i and $v' = v_j^y$ is labeled j, where $0 \leqslant i, j \leqslant 3$. As the two labels differ by exactly 1, we have $j = i + 1 \pmod 4$ or $j = i - 1 \pmod 4$. Now in the strong product $H \boxtimes C_4$, vertex v_i^x is connected to all of $v_{i-1}^y, v_i^y, v_{i+1}^y$, that is, the edge (v, v') exists in $H \boxtimes C_4$.

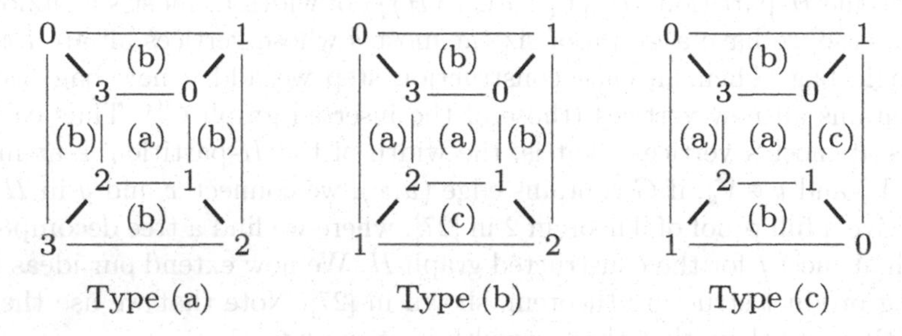

Fig. 1. The three types of faces defined in the proof of Theorem 4, each with a 4-cycle stacked inside, where the labels represent the position of the vertex in the 4-cycle. For better readability, vertices are represented by their labels, where a vertex with label $i + j$ (same $i \in \{0, 1, 2, 3\}$ for all eight vertices) is written as j (labels taken mod 4). That is, each type represents four situations that can occur in a face, e.g., the other three cases of Type (b) have labels 1, 2, 3, 2 ($i = 1$), resp. 2, 3, 0, 3 ($i = 2$), resp. 3, 0, 1, 0 ($i = 3$) for the outer vertices and also plus i for the labels of the stacked 4-cycle. The inter-bag edges are drawn thick, and indeed the labels of their endpoints differ exactly by 1 (mod 4).

It is left to prove that the labels can indeed by chosen as claimed. As shown in Fig. 1, starting with a 4-cycle with labels $0, 1, 2, 3$ from the initial 4-cycle, we obtain three types of faces that differ in the labeling of their vertices along their boundary: Type **(a)** has labels $i, i + 1, i + 2, i + 3$, $(0 \leqslant i \leqslant 3)$, Type **(b)** has labels $i, i + 1, i + 2, i + 1$, $(0 \leqslant i \leqslant 3)$ and Type **(c)** has labels $i, i + 1, i, i + 1$ $(0 \leqslant i \leqslant 3)$. In each of the three cases we are able to label the new 4-cycle so that the labels of the endpoints of each inter-bag edge differ by exactly 1, see Fig. 1 and also Fig. 2 for an example. This concludes the proof. □

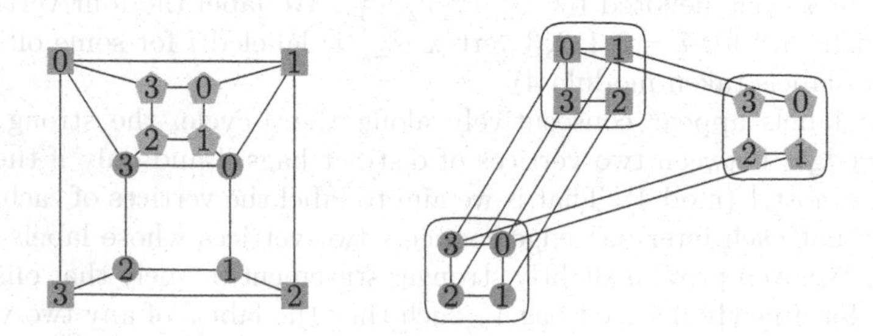

Fig. 2. A stacked quadrangulation G with its H-partition showing $G \subseteq H \boxtimes C_4$, where H is a triangle in this case. The labels used in the proof of Theorem 4 are written inside the vertices.

Theorem 4 gives an upper bound of 21 on the queue number of stacked quadrangulations, compared to the upper bound of 5 on the queue number of planar 3-trees (stacked triangulations) [2].

Theorem 5. *The queue number of stacked quadrangulations is at most* 21.

Proof. In general, we have that $\text{qn}(H_1 \boxtimes H_2) \leqslant |V(H_2)| \cdot \text{qn}(H_1) + \text{qn}(H_2)$ by taking a queue layout of H_2 and replacing each vertex with a queue layout of H_1 [22, Lemma 9]. In particular, we conclude that $\text{qn}(H \boxtimes C_4) \leqslant 4 \cdot \text{qn}(H) + 1$. As the queue number of planar 3-trees is at most 5 [2], Theorem 4 gives an upper bound of $4 \cdot 5 + 1 = 21$. □

4 Lower Bounds

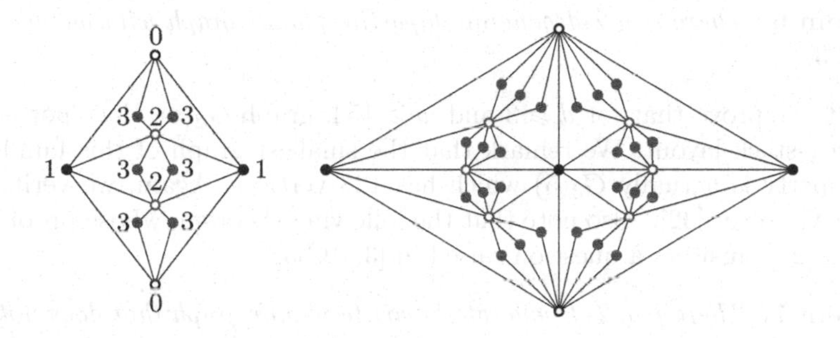

Fig. 3. The graphs $G_3(2)$ and $G_3(3)$, where the numbers and style of the vertices indicate their depth.

This section is devoted to lower bounds on the queue number and mixed page number of bipartite planar graphs. We use the same family of 2-degenerate quadrangulations $G_d(w)$ for both lower bounds. The graph $G_d(w)$ is defined as follows, where we call d the *depth* and w the *width* of $G_d(w)$; see Fig. 3. Let $G_0(w)$ consist of two vertices, which we call *depth-0 vertices.* For $i \geqslant 0$, the graph $G_{i+1}(w)$ is obtained from $G_i(w)$ by adding w vertices into each inner face (except $i = 0$, where we use the unique face) and connecting each of them to the two depth-i vertices of this face (if two exist). If the face has only one depth-i vertex v on the boundary, then we connect the new vertices to v and the vertex opposite of v with respect to the face, that is the vertex that is not adjacent to v. The new vertices are then called *depth-*$(i+1)$ *vertices.* Observe that the resulting graph is indeed a quadrangulation and each inner face is incident to at least one and at most two depth-$(i+1)$ vertices. The two neighbors that a depth-$(i+1)$ vertex u has when it is added are called its *parents*, and u is called a *child* of its parents. If two vertices u and v have the same two parents, they are called *siblings.* We call two vertices of the same depth a *pair* if they have a common child.

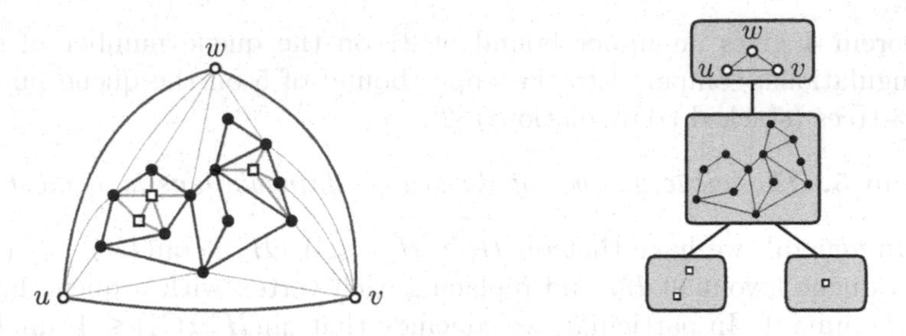

Fig. 4. A planar graph and its tree-partition of shadow width 4.

We prove combinatorially that $G_d(w)$ does not admit a 2-queue layout for $d \geqslant 3, w \geqslant 24$ in [27]. We also verified with a SAT-solver [32] that the smallest graph in the family of queue number 3 is $G_4(4)$ containing 259 vertices.

Theorem 6. *There is a 2-degenerate bipartite planar graph with queue number at least* 3.

Next, we prove that for $d \geqslant 3$ and $w \geqslant 154$, graph $G_d(w)$ does not admit a 1-queue 1-stack layout. We remark that the smallest graph of this family with this property is actually $G_3(5)$ which has 128 vertices. Again, we verified this with a SAT-solver [32]. Also note that the following theorem, whose proof can be found in [27], answers a question raised in [3,19,33].

Theorem 7. *There is a 2-degenerate bipartite planar graph that does not admit a* 1*-queue* 1*-stack layout.*

5 2-Degenerate Quadrangulations

Note that the graph $G_d(w)$ defined in Sect. 4 is a 2-degenerate quadrangulation. Recall that it can be constructed from a 4-cycle by repeatedly adding a degree-2 vertex and keeping all faces of length 4. Hence, every 2-degenerate quadrangulation is a subgraph of a 4-tree. This can also be observed by seeing $G_d(w)$ as a (4, 1)-stack graph, together with Theorem 2. Thus, by the result of Wiechert [36], it admits a layout on $2^4-1 = 15$ queues. In this section, we improve this bound by showing that 2-degenerate quadrangulations admit 5-queue layouts. Full proofs of all lemmas and claims in this section are in [27].

Our proof is constructive and uses a special type of tree-partition. Let T be a tree-partition of a given graph G, that is, an H-partition where H is a rooted tree. For every node x of T, if y is the parent node of x in T, the set of vertices in T_y having a neighbor in T_x is called the *shadow* of x; we say that the shadow is *contained* in node y. The *shadow width* of a tree-partition is the maximum size of a shadow contained in a node of T; see Fig. 4.

Let $\mathcal{S} = \{C_i \subseteq V \ : \ 1 \leqslant i \leqslant |\mathcal{S}|\}$ be a collection of vertex subsets for a graph $G = (V, E)$, and let π be an order of V. Consider two elements from $\mathcal{S}$, $C_x = [x_1, x_2, \ldots, x_{|C_x|}]$ and $C_y = [y_1, y_2, \ldots, y_{|C_y|}]$, where the vertices are ordered according to π. We say that C_x *precedes* C_y with respect to π if $x_i \leqslant y_i$, for all $1 \leqslant i \leqslant \min(|C_x|, |C_y|)$; we denote this relation by $C_1 \prec C_2$. We say that $\mathcal{S}$ is *nicely ordered* if $\prec$ is a total order on $\mathcal{S}$, that is, $C_i \prec C_j$ for all $1 \leqslant i < j \leqslant |\mathcal{S}|$. A similar concept of clique orders has been considered in [23] and [34].

Lemma 1. *Let $G = (V, E)$ be a graph with a tree-partition $\left(T, \{T_x : x \in V(T)\}\right)$ of shadow width k. Assume that for every node x of T, the following holds: (i) there exists a q-queue layout of T_x with vertex order π, and (ii) all the shadows contained in x are nicely ordered with respect to π. Then* $\text{qn}(G) \leqslant q + k$.

Before applying Lemma 1 to 2-degenerate quadrangulations, we remark that the result provides a 5-queue layout of planar 3-trees, as shown by Alam et al. [2]. Indeed, a breadth-first search (starting from an arbitrary vertex) on 3-trees yields a tree-partition in which every bag is an outerplanar graph. The shadow width of the tree-partition is 3 (the length of each face), and it is easy to construct a nicely ordered 2-queue layout for every outerplanar graph [24]. Thus, Lemma 1 yields a 5-queue layout for planar 3-trees.

Lemma 2. *Every 2-degenerate quadrangulation admits a tree-partition of shadow width 4 such that every bag induces a leveled planar graph.*

Proof (Sketch). We assign a *layer value* $\lambda(v)$ to each vertex v of G such that (i) the subgraph G_λ induced by vertices of layer λ ($\lambda \geqslant 0$) is a leveled planar graph, and (ii) the connected components of all subgraphs G_λ define the bags of a tree-partition of shadow width 4. Note that the assigned layer values do not necessarily correspond to a BFS-layering of G.

In order to compute the layer value of the vertices of G, we use its construction sequence. The four vertices of the starting 4-cycle have layer value equal to 0. Consider a set of siblings $u_1, \ldots, u_k$ with parents v and v', that are placed inside a 4-face $f = (v, w, v', w')$ of the constructed subgraph. We compute the layer value of vertices $u_1, \ldots, u_k$ as follows. Assume without loss of generality that $\lambda(v) \leqslant \lambda(v')$ and $\lambda(w) \leqslant \lambda(w')$ and that $u_1, \ldots, u_k$ are such that the cyclic order of edges incident to v are $(v, w), (v, u_1), (v, u_2), \ldots, (v, u_k), (v, w')$ whereas the cyclic order of edges incident to v' are $(v', w'), (v', u_k), (v', u_{k-1}), \ldots, (v', u_1), (v', w)$; see Fig. 5. We will insert these vertices in the order $u_1, u_k, u_2 \ldots, u_{k-1}$, that is, u_1 is inserted inside $f_1 = f$, u_k inside $f_k = (v, u_1, v', w')$ and every subsequent u_i inside $f_i = (v, u_{i+1}, u_k, w')$. Then the layer value of vertex u_i is defined as $\lambda(u_i) = 1 + \min\{\lambda(x), x \in f_i\}$, for $1 \leqslant i \leqslant k$. In the following, we associate f with the layer values of its vertices, that is, we call f a $[\lambda(v), , \lambda(w), \lambda(v')]\lambda(w')$*-face*, where vertices v and v' form a pair whose children are placed inside f.

Note that, by the definition of the layer values, if a vertex u that is placed inside a face f has layer λ, then all vertices of f have layer value $\lambda - 1$ or λ. This implies that connected components of layer λ are adjacent to at most four vertices of layer value $\lambda - 1$ (that form a 4-cycle). In the following, we survey the structure of a subgraph of layer value λ placed inside a $[\lambda - 1, \lambda - 1, \lambda - 1, \lambda - 1]$-face f_0. For simplification, we assume $\lambda = 1$.

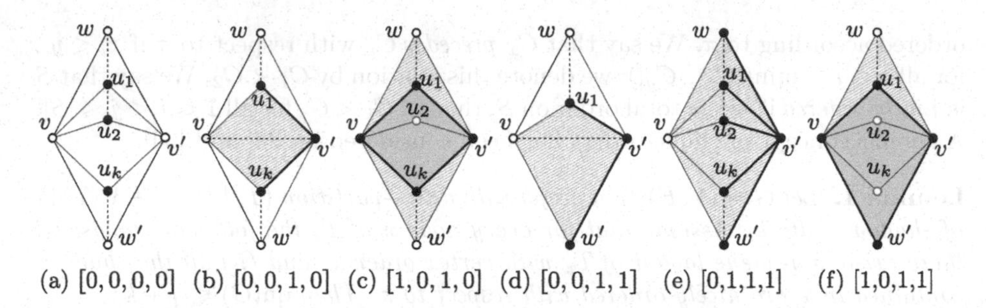

Fig. 5. Cases for faces that contain a set of children in their interior. Vertices at layer values 0, 1 and 2 are drawn as black circles, black disks and red circles resp. Edges that connect vertices of the same (different) layer value are drawn black (red resp.). The dotted diagonals inside a face connect the parents of the children placed inside. Faces of type [0, 1, 1, 1] are shaded in green, and faces that contain vertices of layer value at least 2 are shaded in gray. (Color figure online)

[0, 0, 0, 0]-Face: In this case, all vertices $u_1, \ldots, u_k$ will have layer value 1 (see Fig. 5a). The newly created faces (w, v, u_1, v') and (w', v, u_k, v') are [0, 0, 1, 0]-faces, while all other faces (u_i, v, u_{i+1}, v') for $i = 1, \ldots, k-1$ are [1, 0, 1, 0]-faces.

[0, 0, 1, 0]-Facev: In a [0, 0, 1, 0]-face, we add only vertices u_1 and u_k, which creates the [0, 0, 1, 1]-faces (w, v, u_1, v') and $(w,' v, u_k, v')$, and one [0, 1, 1, 1]-face (v, u_1, v', u_k) (see Fig. 5b). Note that the remaining siblings $u_2, \ldots, u_{k-1}$ will be added as children of v and v' inside face (v, u_1, v', u_k).

[1, 0, 1, 0]-Face: In a [1, 0, 1, 0]-face, vertices u_1 and u_k have layer value 1, while all other siblings u_i, $i = 2, \ldots, k-1$ have layer value 2. Note that in a BFS-layering, vertex u_1 would have layer value 2 instead of 1. In this case, (w, v, u_1, v') and (w', v, u_k, v') are [0, 1, 1, 1]-faces, while the other faces cannot contain vertices of layer value 1 (see Fig. 5c).

We next consider the two new types of faces, [0, 0, 1, 1] and [0, 1, 1, 1], introduced in the previous two cases.

[0, 0, 1, 1]-Face: In a [0, 0, 1, 1]-face, we add only vertex u_1, which creates two faces, namely the [0, 0, 1, 1]-face (w, v, u_1, v'), and the [0, 1, 1, 1]-face (v, w', v', u_1). Note that the remaining vertices $u_2, \ldots, u_k$ will be added as children of v and v' inside face (v, w', v', u_1); see Fig. 5d.

[0, 1, 1, 1]-Face: A [0, 1, 1, 1]-face is split into $k+1$ faces of type [1, 0, 1, 1] (Fig. 5e).

[1, 0, 1, 1]-Face: Finally, in a [1, 0, 1, 1]-face, only vertex u_1 has layer value 1, and face (w, v, u_1, v') is of type [0, 1, 1, 1] (see Fig. 5f).

In order to create a layered planar drawing, we first consider faces of type [0, 1, 1, 1], since they are contained in almost all other types of faces. For a face f, let G_f be the subgraph induced by all vertices of layer value 1 inside f (including its boundary vertices).

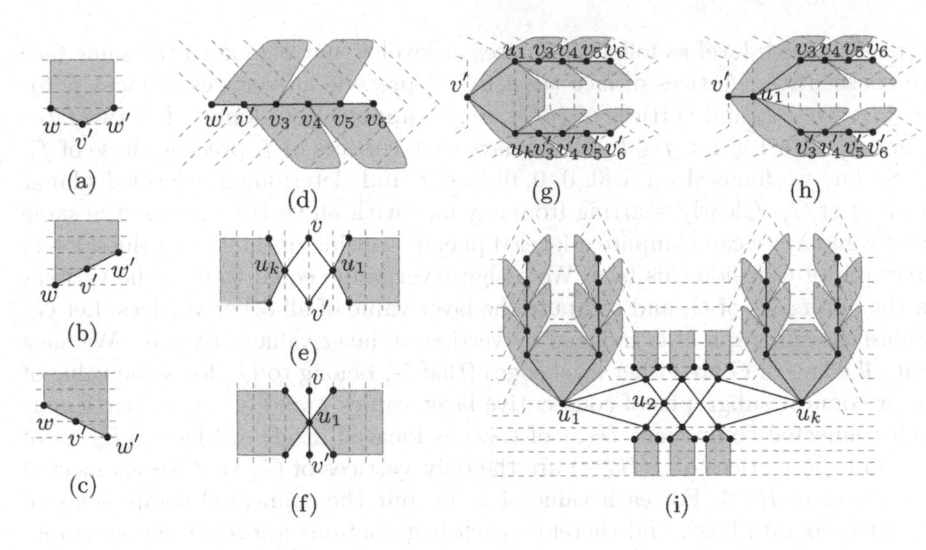

Fig. 6. Illustrations of Γ_f when f is (a)–(c) a [0, 1, 1, 1]-face, (d) a [0, 0, 1, 1]-face, (e)–(f) a [1, 0, 1, 0]-face, (g)–(h) a [0, 0, 1, 0]-face, and (i) a [0, 0, 0, 0]-face.

Claim 1. *The subgraph G_f of a [0, 1, 1, 1]-face $f = (v, w, v', w')$ has a leveled planar drawing such that level 0 contains one vertex among w, v' and w' (and no other vertex); see Figs. 6a–6c.*

For faces of type [0, 0, 1, 1] we can prove that they can be decomposed into an empty [0, 0, 1, 1]-face and a series of [0, 1, 1, 1]-faces (refer to [27] for details). By appropriately combining leveled planar drawings of the contained [0, 1, 1, 1]-faces, we can construct a leveled planar drawing of the [0, 0, 1, 1]-face. We follow a similar approach for [0, 0, 1, 1], [1, 0, 1, 0] and [0, 0, 1, 0]-faces.

Claim 2. *The subgraph G_f of a [0, 0, 1, 1]-face $f = (v, w, v', w')$ has a leveled planar drawing such that level 0 contains only vertex w'; see Fig. 6d.*

Claim 3. *The subgraph G_f of a [1, 0, 1, 0]-face $f = (v, w, v', w')$ has a leveled planar drawing such that level 0 contains only vertices v and v', vertices of the [0, 1, 1, 1]-face (w, v, u_1, v') are drawn on levels above level 0, while vertices of the [0, 1, 1, 1]-face (w', v, u_k, v') are drawn on levels below level 0; see Fig. 6e. In the special case where $u_1 = u_k$, then also u_1 is on level 0; see Fig. 6f.*

Claim 4. *The subgraph G_f of a [0, 0, 1, 0]-face $f = (v, w, v', w')$ has a leveled planar drawing such that level 0 contains only vertex v'; see Figs. 6g and 6h.*

Next consider [0, 0, 0, 0]-faces. We combine leveled planar drawings for faces (w, v, u_1, v'), (w', v, u_k, v') and faces $f_i = (u_i, v, u_{i+1}, v')$ for $i = 1, \ldots, k-1$. We create drawings for the [0, 0, 1, 0]-faces (w, v, u_1, v') and (w', v, u_k, v') using Claim 4, such that u_1 and u_k are the only vertices placed at level 0. Then for every face f_i ($i = 1, \ldots, k-1$), we use Claim 3; vertices u_i, for $i = 1, \ldots, k$ are placed at level 0. We combine the drawings as in Fig. 6i, by ordering the vertices on each level as follows. Vertices at level ℓ that belong to the same face are consecutive. Vertices of face (w, v, u_1, v') precede all vertices of faces f_i for $i = 1, \ldots, k-1$, and vertices of (w', v, u_k, v') appear last along ℓ. For two faces f_i and f_j with $1 \leqslant i < j \leqslant k-1$, we have that vertices of f_i precede those of f_j.

So far, we focused on a [0, 0, 0, 0]-face f and determined a leveled planar drawing of G_f. Clearly, starting from any face with all vertices having the same layer value λ, we can compute a leveled planar drawing of the layer value-$(\lambda+1)$ subgraph of G inside this face. We assign layer value equal to 0 to the vertices on the outer face of G, and compute the layer value of all other vertices. Let G_λ denote the subgraph of G induced by vertices of layer value λ $(\lambda \geqslant 0)$. We have that all edges of G are either level edges (that is, belong to G_λ for some value of λ), or connect subgraphs of consecutive layer values λ and $\lambda + 1$. In particular, each connected component $H_{\lambda+1}$ of $G_{\lambda+1}$ is located inside a 4-face $f(H_{\lambda+1})$ of G_λ, and the vertices of $f(H_{\lambda+1})$ are the only vertices of G_λ that are connected to vertices of $H_{\lambda+1}$. For each value of λ, we put the connected components of G_λ into separate bags, and therefore each bag contains a leveled planar graph. For a bag that contains connected component H_λ, we define its parent to be the bag that contains the component where $f(H_\lambda)$ belongs to. The defined bags create a tree T with root-bag consisting of the outer vertices of G (with layer value 0). Additionally, the shadow of each bag consists of at most four vertices, and therefore the shadow width of T is at most 4. The lemma follows. □

Theorem 8. *Every 2-degenerate quadrangulation admits a 5-queue layout.*

Proof. Combine Lemma 2 with Lemma 1. Observe that *every* leveled planar graph admits a 1-queue layout in which the faces are nicely ordered with respect to the layout. That implies that we can apply Lemma 1 with $q = 1$ (the queue number of the bags) and $k = 4$ (the shadow width of the tree-partition). □

6 Open Questions

In this work, we focused on the queue number of bipartite planar graphs and related subfamilies. Next we highlight a few questions for future work.

- First, there is still a significant gap between our lower and upper bounds for the queue number of bipartite planar graphs.
- Second, although 2-degenerate quadrangulations always admit 5-queue layouts, the question of determining their exact queue number remains open.
- Third, for stacked quadrangulations, our upper bound relies on the strong product theorem. We believe that a similar approach as for 2-degenerate quadrangulations could lead to a significant improvement.
- Perhaps the most intriguing questions are related to mixed linear layouts of bipartite planar graphs: One may ask what is the minimum q so that each bipartite planar graph admits a 1-stack q-queue layout.
- Finally, the recognition of 1-stack 1-queue graphs remains an important open problem even for bipartite planar graphs.

References

1. Alam, J.M., Bekos, M.A., Gronemann, M., Kaufmann, M., Pupyrev, S.: Lazy queue layouts of posets. In: GD 2020. LNCS, vol. 12590, pp. 55–68. Springer, Cham (2020). https://doi.org/10.1007/978-3-030-68766-3_5
2. Alam, J.M., Bekos, M.A., Gronemann, M., Kaufmann, M., Pupyrev, S.: Queue layouts of planar 3-trees. Algorithmica **82**(9), 2564–2585 (2020). https://doi.org/10.1007/s00453-020-00697-4
3. Angelini, P., Bekos, M.A., Kindermann, P., Mchedlidze, T.: On mixed linear layouts of series-parallel graphs. Theor. Comput. Sci. **936**, 129–138 (2022). https://doi.org/10.1016/j.tcs.2022.09.019
4. Auer, C., Gleißner, A.: Characterizations of deque and queue graphs. In: Kolman, P., Kratochvíl, J. (eds.) WG 2011. LNCS, vol. 6986, pp. 35–46. Springer, Heidelberg (2011). https://doi.org/10.1007/978-3-642-25870-1_5
5. Bannister, M.J., Devanny, W.E., Dujmović, V., Eppstein, D., Wood, D.R.: Track layouts, layered path decompositions, and leveled planarity. Algorithmica **81**(4), 1561–1583 (2018). https://doi.org/10.1007/s00453-018-0487-5
6. Battista, G.D., Frati, F., Pach, J.: On the queue number of planar graphs. SIAM J. Comput. **42**(6), 2243–2285 (2013). https://doi.org/10.1137/130908051
7. Bekos, M., Gronemann, M., Raftopoulou, C.N.: An improved upper bound on the queue number of planar graphs. Algorithmica 1–19 (2022). https://doi.org/10.1007/s00453-022-01037-4
8. Bekos, M.A., Bruckdorfer, T., Kaufmann, M., Raftopoulou, C.N.: The book thickness of 1-planar graphs is constant. Algorithmica **79**(2), 444–465 (2016). https://doi.org/10.1007/s00453-016-0203-2
9. Bekos, M.A., Da Lozzo, G., Hliněný, P., Kaufmann, M.: Graph product structure for h-framed graphs. CoRR abs/2204.11495 (2022). https://doi.org/10.48550/arXiv.2204.11495
10. Bekos, M.A., Förster, H., Gronemann, M., Mchedlidze, T., Montecchiani, F., Raftopoulou, C.N., Ueckerdt, T.: Planar graphs of bounded degree have bounded queue number. SIAM J. Comput. **48**(5), 1487–1502 (2019). https://doi.org/10.1137/19M125340X
11. Bekos, M.A., Kaufmann, M., Klute, F., Pupyrev, S., Raftopoulou, C.N., Ueckerdt, T.: Four pages are indeed necessary for planar graphs. J. Comput. Geom. **11**(1), 332–353 (2020). https://doi.org/10.20382/jocg.v11i1a12
12. Bernhart, F., Kainen, P.C.: The book thickness of a graph. J. Comb. Theory Ser. B **27**(3), 320–331 (1979). https://doi.org/10.1016/0095-8956(79)90021-2
13. Bhore, S., Ganian, R., Montecchiani, F., Nöllenburg, M.: Parameterized algorithms for queue layouts. J. Graph Algorithms Appl. **26**(3), 335–352 (2022). https://doi.org/10.7155/jgaa.00597
14. Biedl, T.C., Shermer, T.C., Whitesides, S., Wismath, S.K.: Bounds for orthogonal 3D graph drawing. J. Graph Algorithms Appl. **3**(4), 63–79 (1999). https://doi.org/10.7155/jgaa.00018
15. Bodlaender, H.L.: A linear-time algorithm for finding tree-decompositions of small treewidth. SIAM J. Comput. **25**(6), 1305–1317 (1996). https://doi.org/10.1137/S0097539793251219
16. Bose, P., Morin, P., Odak, S.: An optimal algorithm for product structure in planar graphs. In: Czumaj, A., Xin, Q. (eds.) SWAT 2022. LIPIcs, vol. 227, pp. 19:1–19:14. Schloss Dagstuhl - Leibniz-Zentrum für Informatik (2022). https://doi.org/10.4230/LIPIcs.SWAT.2022.19

17. Campbell, R., et al.: Product structure of graph classes with bounded treewidth. CoRR abs/2206.02395 (2022). https://doi.org/10.48550/arXiv.2206.02395
18. Chung, F.R.K., Leighton, F.T., Rosenberg, A.L.: Embedding graphs in books: a layout problem with applications to VLSI design. SIAM J. Algebraic Discrete Methods **8**(1), 33–58 (1987)
19. de Col, P., Klute, F., Nöllenburg, M.: Mixed linear layouts: complexity, heuristics, and experiments. In: Archambault, D., Tóth, C.D. (eds.) GD 2019. LNCS, vol. 11904, pp. 460–467. Springer, Cham (2019). https://doi.org/10.1007/978-3-030-35802-0_35
20. Dujmović, V., Eppstein, D., Hickingbotham, R., Morin, P., Wood, D.R.: Stack-number is not bounded by queue-number. Combinatorica (2), 1–14 (2021). https://doi.org/10.1007/s00493-021-4585-7
21. Dujmović, V., Frati, F.: Stack and queue layouts via layered separators. J. Graph Algorithms Appl. **22**(1), 89–99 (2018). https://doi.org/10.7155/jgaa.00454
22. Dujmović, V., Joret, G., Micek, P., Morin, P., Ueckerdt, T., Wood, D.R.: Planar graphs have bounded queue-number. J. ACM (JACM) **67**(4), 1–38 (2020). https://doi.org/10.1145/3385731
23. Dujmović, V., Morin, P., Wood, D.R.: Layout of graphs with bounded tree-width. SIAM J. Comput. **34**(3), 553–579 (2005). https://doi.org/10.1137/S0097539702416141
24. Dujmovic, V., Pór, A., Wood, D.R.: Track layouts of graphs. Discret. Math. Theor. Comput. Sci. **6**(2), 497–522 (2004)
25. Dujmović, V., Wood, D.R.: Three-dimensional grid drawings with sub-quadratic volume. In: Liotta, G. (ed.) GD 2003. LNCS, vol. 2912, pp. 190–201. Springer, Heidelberg (2004). https://doi.org/10.1007/978-3-540-24595-7_18
26. Felsner, S., Huemer, C., Kappes, S., Orden, D.: Binary labelings for plane quadrangulations and their relatives. Discret. Math. Theor. Comput. Sci. **12**(3), 115–138 (2010)
27. Förster, H., Kaufmann, M., Merker, L., Pupyrev, S., Raftopoulou, C.: Linear layouts of bipartite planar graphs. arXiv abs/2305.16087 (2023). https://doi.org/10.48550/arXiv.2305.16087
28. Heath, L.S., Leighton, F.T., Rosenberg, A.L.: Comparing queues and stacks as mechanisms for laying out graphs. SIAM J. Discret. Math. **5**(3), 398–412 (1992). https://doi.org/10.1137/0405031
29. Heath, L.S., Rosenberg, A.L.: Laying out graphs using queues. SIAM J. Comput. **21**(5), 927–958 (1992). https://doi.org/10.1137/0221055
30. Merker, L., Ueckerdt, T.: The local queue number of graphs with bounded treewidth. In: GD 2020. LNCS, vol. 12590, pp. 26–39. Springer, Cham (2020). https://doi.org/10.1007/978-3-030-68766-3_3
31. Morin, P.: A fast algorithm for the product structure of planar graphs. Algorithmica **83**(5), 1544–1558 (2021). https://doi.org/10.1007/s00453-020-00793-5
32. Pupyrev, S.: A SAT-based solver for constructing optimal linear layouts of graphs. https://github.com/spupyrev/bob
33. Pupyrev, S.: Mixed linear layouts of planar graphs. In: Frati, F., Ma, K.-L. (eds.) GD 2017. LNCS, vol. 10692, pp. 197–209. Springer, Cham (2018). https://doi.org/10.1007/978-3-319-73915-1_17
34. Pupyrev, S.: Improved bounds for track numbers of planar graphs. J. Graph Algorithms Appl. **24**(3), 323–341 (2020). https://doi.org/10.7155/jgaa.00536
35. Ueckerdt, T., Wood, D.R., Yi, W.: An improved planar graph product structure theorem. Electron. J. Comb. **29**(2) (2022). https://doi.org/10.37236/10614

36. Wiechert, V.: On the queue-number of graphs with bounded tree-width. Electr. J. Comb. **24**(1), P1.65 (2017). https://doi.org/10.37236/6429
37. Wood, D.R.: Product structure of graph classes with strongly sublinear separators. CoRR abs/2208.10074 (2022). https://doi.org/10.48550/arXiv.2208.10074
38. Yannakakis, M.: Embedding planar graphs in four pages. J. Comput. Syst. Sci. **38**(1), 36–67 (1989). https://doi.org/10.1016/0022-0000(89)90032-9
39. Yannakakis, M.: Planar graphs that need four pages. J. Comb. Theory Ser. B **145**, 241–263 (2020). https://doi.org/10.1016/j.jctb.2020.05.008

Adaptive Data Structures for 2D Dominance Colored Range Counting

Younan Gao(✉)

Faculty of Computer Science, Dalhousie University, Halifax, Canada
gaoyounan@dal.ca

Abstract. We propose an optimal adaptive data structure for 2D dominance colored range counting in the word RAM model; namely, for n colored points in two-dimensional rank space, we present a linear space data structure that supports each query in $O(1 + \log_w k)$ time, where w denotes the number of bits in a word and k denotes the number of distinct colors in the query range. To this end, we design adaptive data structures for 2D 3-sided stabbing counting. Previously, in the orthogonal colored counting problems, the adaptive data structure is only known for 1D [TODS'2014].

Keywords: Computational geometry · Adaptive data structures · Colored range counting · Dominance range counting · Stabbing queries

1 Introduction

Studies on the retrieval of color information over points associated with colors have received considerable attention [1,3,6,8–14,17–20]. In this paper, we study one of the most fundamental among them, *2D dominance colored range counting.* In this problem, we preprocess a set of n points in the plane, each assigned a color from $\{0, 1, \cdots, C-1\}$, where $C \leq n$, into a data structure such that given a query point $Q = (q.x, q.y)$, the number, k, of distinct colors in the range $(-\infty, q.x] \times (-\infty, q.y]$ can be computed efficiently. This problem emerges in database applications, and it is also used as a basis of other orthogonal colored range counting problems [12].

On one hand, it is known that 2D dominance colored range counting can be reduced to 2D 3-sided stabbing counting queries [17], and 2D 3-sided stabbing counting can be reduced to 2D dominance range counting [5]. Using the solution by JáJá et al. [15] to 2D dominance range counting, one can solve 2D dominance colored range counting in $O(n)$ word space and $O(\log_w n)$ query time in the word RAM model. On the other hand, Nekrich [19] show that 2D 3-sided colored range reporting can be answered in $O(1 + k)$ time using a linear space data structure, provided that the point coordinates are in 2D rank space. In the reporting problem, we want to output the distinct colors in the query range.

This work was supported by NSERC of Canada.

P. Morin and S. Suri (Eds.): WADS 2023, LNCS 14079, pp. 460–473, 2023.
https://doi.org/10.1007/978-3-031-38906-1_30

In 2D rank space, all x- and y-coordinates of n points are drawn from $[n]$, and no two points in the set share the same x- or y-coordinates. Obviously, the data structure for 2D 3-sided colored range reporting can be used for 2D dominance colored range counting as well. Hence, when the output number k is small, i.e., $k = o(\log_w n)$, k can be found more efficiently using the reporting data structure.

In this paper, we aim at designing under the word RAM model a data structure for 2D dominance colored range counting that achieves better query time when the output number k is small. In the literature, the data structures that is sensitive to k are also called *adaptive data structures* [4,16,19].

Related Work. Chan and Wilkinsson [4] proposed the first adaptive data structure for 2D 4-sided orthogonal range counting. In their solution, they first solve the K-capped version of the problem, in which a fixed value K is provided before the preprocessing, and the data structure returns $\min(K, k)$ as the result. Then by constructing K-capped data structures for $O(\lg\lg n)$ of different values of K, they achieve $O(n \lg\lg n)$-word data structure that solves 4-sided range counting in $O(\lg\lg n + \log_w k)$, where w denotes the number of bits in a word. Note that their result holds under the assumption that the data structure is built for points in 2D rank space; otherwise, an extra term $O(\lg\lg U)$ is required in the query bound, given that the point coordinates are drawn from universe $\{0, 1, \cdots, U-1\}$. Whether an $O(n)$-word data structure with $O(\lg^\epsilon n + \log_w k)$ query time exists for 2D orthogonal range counting is still an open problem.

The optimal adaptive data structure for 2D 3-sided orthogonal range counting has been found later by Nekrich [19]. The data structure requires $O(n)$ words of space and $O(1 + \log_w k)$ query time provided that the point coordinates are bounded by $O(n)$. To achieve that, Nekrich proposed a new data structure, called *path-range tree*, which combines the properties of *range trees* and *priority trees*. The 2D 3-sided orthogonal range counting can be used to solve 1D orthogonal colored range counting, following the reduction by Gupta et al. [11]. Therefore, in 1D, colored counting can be solved in linear space and $O(1+\log_w k)$ time as well. However, the reduction by Gupta et al. does not work for higher dimensional space, even for 2D space [12].

Obviously, 2D dominance colored range counting is a generalized version of 1D colored range counting, since the solution to the prior problem can be used for solving the latter one, following a simple transformation: For each point $(p.x)$ in 1D, we turn it into $(p.x, -p.x)$ in 2D. And after transformation, the number of original points in the range $[a, b]$ in 1D space is exactly the same as the number of transformed points in the range $[a, +\infty) \times [-b, +\infty)$ in 2D space.

As mentioned before, 2D dominance colored range counting can be reduced to 2D dominance range counting, following the reduction from the prior problem to 2D 3-sided stabbing counting and the reduction from 2D 3-sided stabbing counting to the latter problem. However, this method might not give an adaptive data structure, although the adaptive data structure for 2D 3-sided orthogonal range counting by Nekrich is sufficient for solving 2D dominance range counting. Due to the second reduction, one has to compute the number, $\overline{k}$, of distinct colors that are not the in the query range and then return $C - \overline{k}$ as the result, where C denotes the total number of distinct colors in the input set. As a result, the query time of the solution is bounded by $O(\log_w \overline{k})$. When k is as small as a constant, $\overline{k}$ approaches closely to C, while C could be as large as n.

Overview of Our Solution. We design an adaptive data structure for stabbing counting queries over 2D 3-sided rectangles, which results in a solution to adaptive 2D dominance colored range counting, following the reduction from dominance colored counting to stabbing counting. The result is shown as follows:

Theorem 1. *Let P denote a set of n colored points in 2D rank space. There is a linear space data structure constructed over P to support 2D dominance colored range counting in $O(1 + \log_w k)$ time, where k denotes the number of distinct colors in the query range.*

Inspired by the work of Chan and Wilkinsson [4], we first present a data structure of $O(n \lg\lg n)$ words (in Sect. 3). The supper-linear space cost is due to the data structure that solves the K-capped version of the problem for $O(\lg\lg n)$ of different values of K, and the data structure for each K requires $O(n)$ words. One of the technique that is used in our solution to the K-version of this problem is the *K-level shallow cutting* [20] built for 2D 3-sided rectangles. Essentially, a K-level shallow cutting is a set of $O(n/K)$ cells, each intersecting at most $2K$ rectangles of the input set.

To improve the overall space cost from $O(n \lg\lg n)$ to $O(n)$ words, we make use of the ideas from the realm of *index data structures*. Precisely, our goal is to improve the space cost of the data structure for each K, apart from storing the representation of the rectangles in the input set. For example, storing no more than $2K$ rectangles for each cell, after reducing them into rank space within each cell, saves space cost from $O(K \lg n)$ to $O(K \lg K)$ bits. However, applying rank space reduction over a query point might still require a data structure of $O(K \lg n)$ bits. The *ball inheritance* structure [2] that is used in the solution of Chan and Wilkinsson [4] for the rank space reduction and that is shared across all of $O(\lg\lg n)$ of K-versions is no longer applicable here: The linear space tradeoff of the ball inheritance structure requires $O(\lg^{\epsilon} n)$ query time, which is not affordable in our solution, when the output number k is small.

To circumvent this difficulty, we divide the rectangles that each cell intersects into two types, each with at most K rectangles, and we handle two types of rectangles separately. In particular, a type-1 rectangle of a cell has at least one vertical edge lying in the cell. By making use of these vertical edges lying in the cell, we manage to construct a global data structure that is shared across all K-versions, to support finding x- and y-rank of a query point within a cell in $O(1 + \log_w K)$ time (in Sect. 4.1). A type-2 rectangle of a cell completely spans the slabs of a cell. To cope with type-2 rectangles, we apply the nested property that holds across different levels of shallow cuttings: A cell in a low level shallow cutting is strictly contained in a cell in a higher shallow cutting. As a result, a type-2 rectangle of a high level cell C' is a type-2 rectangle of a lower level cell

C'' as well, as long as C'' is strictly contained in C'. This property allows us to store only $\sqrt{K}$ rectangles of type-2 in the data structure built for each cell, instead of K, thereby saving the space cost (Sect. 4.2). Finally, we manage to improve the space cost for each K from $O(n \lg n)$ to $O(n(\lg K + \lg n / \log_w K + \lg n/\sqrt{K}))$ additional bits (Sect. 4.3). Although these techniques involved in our solution, such as shallow cutting and Fusion trees, are not new, combining all these techniques in a correct way, in order to achieve an optimal solution, is the challenge.

2 Preliminaries

Notation. All our results are in the standard word RAM model with word size $w = \Theta(\lg n)$ bits, where n denotes the input size. W.l.o.g, we assume that n is always a power of 2. We denote as $[x]$ the set $\{0, 1, 2, \cdots, x-1\}$ for an integer $x > 0$, and we use $\lg x$ to denote $\log_2 x$ and $\lg^{(3)} x$ to denote $\lg \lg \lg x$ for short. We call a point q dominates p if $p.x \leq q.x$ and $p.y \leq q.y$.

Predecessor and Successor Queries. Let A be a set of integers. A predecessor (resp. successor) query for an integer x returns the largest (resp. smallest) element of S that is no more (resp. less) than x as well as the number of elements of S that are no more (resp. less) than x. We apply the well known data structure, *Fusion Tree*, for predecessor/successor queries.

Lemma 1 (Fusion Tree [7]**).** *Let $A[0..n'-1]$ be a sequence of w-bit integers. The Fusion tree constructed over A requires $O(n'w)$ bits of space and supports each predecessor/successor query in $O(1 + \log_w n')$ time.*

Solving 2D Dominance Range Counting. In 2D dominance range counting problem for a set P of points in the plane, we are asked to compute the number of all the points of P that are dominated by a query point Q. We apply two different solutions. In the first one, the query time is decided by the input size.

Lemma 2 (JáJá et al. [15]**).** *Let P be a set of n' points in the plane. There is a data structure of $O(n')$ words of space, constructed over P, to support 2D dominance range counting queries in $O(1 + \log_w n')$ time.*

The second solution might uses less space cost when the universe that the y-coordinates of the points are drawn from is small.

Lemma 3 (Lemma 3.2 [4]**).** *Let $A[0..n'-1]$ be a sequence drawn from universe $[U]$ such that each entry $A[i]$ represents a point $(i, A[i])$ in the plane. There is a data structure of $O(n' \lg U)$ bits, constructed over n' points represented by A, to support 2D dominance range counting queries in $O(1 + \log_w U)$ time.*

Solving 2D 3-Sided Stabbing Counting. The problem, 2D 3-sided stabbing counting, can be regarded as a dual version of 2D 3-sided range counting. In stabbing counting, the input is a set S of rectangles in the plane, each bounded by 3 sides. Given a query point $Q = (q.x, q.y)$, we are asked to compute the number, $|S \cap Q|$, of all the rectangles that contain Q. For example, a rectangle $r = [x_1, x_2] \times (-\infty, y_2]$ contains Q if and only if (iff) $x_1 \leq Q.x \leq x_2$ and $Q.y \leq y_2$. In general, Edelsbrunner and Overmars [5] show that stabbing counting in d-dimensional space can be reduced to d-dimensional dominance range counting. Taking 2D 3-sided stabbing counting as an example, we present the reduction to 2D dominance range counting in Appendix A. Following the reduction and Lemma 3 for 2D dominance range counting, we solve stabbing counting.

Lemma 4. *Given a set S of n' 2D 3-sided rectangles in rank space*[1]*, there is a data structure of $O(n' \lg n')$ bits of space constructed over S to support each stabbing counting query in $O(1 + \log_w n')$ time.*

Reducing 2D Dominance Colored Range Counting To 2D 3-Sided Stabbing Counting. Kaplan et al. [17] show that in low dimensional space, e.g. 2D or 3D, dominance colored range counting over n colored points can be reduced to stabbing counting queries over $O(n)$ rectangles. Here, we introduce their reduction from 2D dominance colored range counting to 2D 3-sided stabbing counting.

Let P denote a set of n' colored points in 2D rank space. Let P_c denote the subset of P that contains all points colored in c. We call a point p of P_c a *skyline point* if p does not dominate any point in P_c except itself, and we store all the skyline points of P_c in array $S_c[0..|S_c| - 1]$ sorted by their x-coordinates. Then we traverse S_c backwards, from the last entry to the first entry, and operate the following operations for each entry: Shoot a vertical ray upwards and a horizontal ray rightwards, both from the skyline point $S_c[i]$. The vertical ray does not stop until it hits $+\infty$, while the horizontal ray stops as soon as it hits the vertical ray shot from $S_c[i+1]$ provided that $i+1 < |S_c|$. Let $U(P_c)$ denote the region $\bigcup_{p \in P_c} [p.x, +\infty) \times [p.y, +\infty)$. All the vertical and horizontal rays divide the region $U(P_c)$ into $O(|S_c|)$ pairwise rectangles, each bounded by at most 3 sides. Precisely, the rectangles include $[S_c[i].x, S_c[i+1].x) \times [S_c[i].y, +\infty)$ for each $0 \leq i < |S_c| - 1$, as well as the two-sided rectangle, $[S_c[t].x, +\infty) \times [S_c[t].y, +\infty)$, where $t = |S_c| - 1$. It turns out that the query point Q dominates any point that is colored in c iff Q is contained in one of these $O(|S_c|)$ rectangles.

We construct rectangles for the point set P_c for each different color in a similar way. Overall, this transformation generates $\sum_c O(|S_c|) = \sum_c O(|P_c|) = O(n')$ rectangles in total. Since the rectangles built for the same set P_c are pairwise disjoint, at most one of them contains Q. It follows that the number of rectangles that contain Q is equal to the number of distinct colors in the range dominated by Q. As a result, we reduce 2D dominance colored range counting to

[1] In rank space, all n' rectangles lie on a $2n' \times n'$ grid, and there is no two rectangles that share the same x_1-, x_2-, or y_2-coordinates.

2D 3-sided stabbing counting. Following Lemma 4 and the reduction introduced above, we achieve an $O(n)$-word data structure that supports each 2D dominance colored range counting query in $O(1 + \log_w n)$ time. However, this solution is non-adaptive.

In the rest of this paper, we focus on designing adaptive data structures for 2D 3-sided stabbing counting queries, thereby solving 2D dominance colored range counting by applying the reduction. W.l.o.g, we assume that the input set S contains n rectangles in 2D rank space and each rectangle is of the form $[x_1, x_2] \times [y_1, +\infty)$.

Shallow Cutting Data Structure. Rahul [20] first discovers the shallow cutting data structure that is applicable to rectangles and uses it to solve the approximate version of colored range counting. Since his shallow cutting will be extensively used in our solution, we give a brief overview of the data structure.

Let S denote a set of n rectangles of the form $[x_1, x_2] \times [y_1, +\infty)$ in 2D rank space. Our goal is to build a *t-level shallow cutting* for S. To this end, we divide the vertical edges of the rectangles into slabs of size t; namely, the grid is partitioned into vertical slabs $[i \cdot t, (i+1) \cdot t - 1] \times (-\infty, +\infty)$ for each $0 \le i \le 2n/t - 1$, and each slab contains exactly t vertical edges of the rectangles. For the i-th slab, we create a cell $[i \cdot t, (i+1) \cdot t - 1] \times (-\infty, \hat{y}_i]$, in which $\hat{y}_i$ is chosen as follows: Let Y_i denote the set that contains y-coordinates of the rectangles that span the interval $[i \cdot t, (i+1) \cdot t - 1]$. If $|Y_i|$ is less than t, then we set $\hat{y}_i$ to be $+\infty$; otherwise, $\hat{y}_i$ is set to be the t-th smallest one in Y_i.

The t-level shallow cutting is essentially the set of $2n/t$ cells. Conceptually, we assign rectangles of S to the cells as follows: If a rectangle r has at least one vertical edge between the slabs of a cell, then r is assigned to the cell as a *type-1* rectangle of the cell; if r completely spans the slabs of a cell, then r is assigned to the cell as a *type-2* rectangle of the cell. Note that the same rectangle might be assigned to multiple cells. It turns out that each cell can have at most t rectangles of type-1 and at most t rectangles of type-2. The properties of the shallow cutting are summarized in Lemma 5.

Lemma 5 (Rahul *[20]***).** *Let S be a set of n rectangles of the form $[x_1, x_2] \times [y_1, +\infty]$ in 2D rank space. A t-level shallow cutting of S is a set of interior-disjoint cells of the form $[x_1, x_2] \times (-\infty, y_2]$ such that*

- *the shallow cutting contains exactly $2n/t$ cells,*
- *if a query point Q is not contained in any of these cells, then $|S \cap Q| > t$,*
- *and each cell intersects at most $2t$ rectangles in S.*

The concept, *nested shallow cuttings*, involves a hierarchy of cuttings. Basically, it says that cells of shallow cuttings in lower levels are contained in the cells of shallow cutting in higher levels:

Lemma 6 (Observation 1 *[20]***).** *Let t_1 and t_2 be any two integers such that $t_1 = 2^i$ and $t_2 = 2^j$, where $0 \le i \le j \le \lg n$. A cell of a t_2-level shallow cutting* *contains exactly 2^{j-i} cells of a t_1-level shallow cutting. And any cell of a t_1-level shallow cutting is completely contained by a cell of t_2-level shallow cutting.*

3 An $O(n \lg\lg n)$ -Word Space Solution

We begin with a solution to the K-capped version of 2D 3-sided stabbing counting problem in Sect. 3.1. Then we present the first adaptive data structure that uses $O(n \lg\lg n)$ words of space in Sect. 3.2.

3.1 K-Capped 2D 3-Sided Stabbing Counting

In the K-capped version of this problem, we are given a fixed value K before building the data structure. Our data structure returns failure if $|S \cap Q| > K$; otherwise, it computes and returns $|S \cap Q|$.

We construct a K-level shallow cutting, denoted by Γ, of S, and thus Γ contains $2n/K$ cells of the form $[x_1, x_2] \times (-\infty, y_2]$, where y_2 could be $+\infty$. Following Lemma 5, each cell of Γ intersects at most $2K$ rectangles of S. In each cell of the shallow cutting, we construct the non-adaptive data structure over these rectangles for 2D 3-sided stabbing counting. As shown in Appendix A, the 2D 3-sided stabbing counting queries can be reduced to 2D dominance range counting, for which Lemma 2 is applied in our solution. The data structure for each cell requires $O(K)$ words. Since the shallow cutting contains $O(n/K)$ cells, the total cost of the space is $O(n)$ words.

Given a query point $Q = (q.x, q.y)$, we first decide whether or not there is a cell that contains Q. If no cell contains Q, then there are more than K rectangles that contain Q following Lemma 5, and we simply return failure. The only cell that possibly contains Q is the $\lfloor q.x/K \rfloor$-th one from the left, which can be determined in constant time. We denote that cell by γ if it exists. Observe that every rectangle in $S \cap Q$ intersects γ as well. Hence, $|S \cap Q|$ can be computed by querying over the stabbing counting data structure built for γ, using Q as the query point. Since there are at most $2K$ rectangles that intersect γ, the query time is bounded by $O(1 + \log_w K)$.

Lemma 7. *There is a data structure for K-capped 2D 3-sided stabbing counting requiring $O(n)$ words of space and $O(1 + \log_w K)$ query time.*

3.2 Adaptive 2D 3-Sided Stabbing Counting

In this section, we present the adaptive data structure for stabbing counting queries. The idea behind it is to build the data structure for K-capped versions as shown in Lemma 7 for different values of K. Let $\delta(i)$ to be 2^{2^i}. For each $\lg^{(3)} n \leq i \leq \lg\lg n$, we build the $\delta(i)$-capped data structure of Lemma 7. The total space cost is $O(n \lg\lg n)$ words.

Let k denote $|S \cap Q|$ for a query point Q. Note that if $k \leq \lg n$, we can simply use the $\delta(\lg^{(3)} n)$-capped data structure of Lemma 7 to answer the query in constant time. Starting from $i = \lg^{(3)} n$, we query over $\delta(i)$-capped data structure using Q as the query point and keep incrementing i until no failure is reported. Let i' denote the last value that is set to i. Then, we have $2^{2^{i'-1}} \le k < 2^{2^{i'}}$. As a result, by querying over $\delta(i')$-capped data structure, we find the correct answer. The query time is bounded by $\sum_{i=\lg^{(3)} n}^{i'} O(\log_w \delta(i)) = O(\lg \lg k + \log_w k)$.

Improvement in the Query Time. In the previous solution, to find i' such that $2^{2^{i'-1}} \le k < 2^{2^{i'}}$, we iterate through index $i \in [\lg^{(3)} n, \lg \lg n]$. As a result, the previous solution has an extra $\lg \lg k$ term in the query time. Next, we improve the query time by getting rid of this extra term.

Let cell_i denote a cell of $\delta(i)$-level shallow cutting. We refer to each $\text{cell}_{\lg^{(3)} n}$ as a base cell. Following Lemma 6, each cell_{i+1} contains 2^{2^i} cell_i's, and each cell_i is contained by one cell_{i+1}, since $\delta(i+1) = \delta(i) \cdot 2^{2^i}$. We refer to cell_{i+1} that contains cell_i as the parent of cell_i, where $i \le \lg \lg n - 1$. Clearly, each cell_i has a single parent. The ancestors of cell_i are defined as follows: cell_i is an ancestor of itself; the ancestors of the parent of cell_i are ancestors of cell_i as well.

Finding the index i' such that $2^{2^{i'-1}} \le k < 2^{2^{i'}}$ is equivalent to finding the smallest cell $\text{cell}_{i'}$ that contains Q, among all the cells of $O(\lg \lg n)$ of shallow cuttings. In the new solution, we will show that finding $\text{cell}_{i'}$ only requires $O(1)$ time, thereby achieving the optimal query time.

Following Lemma 6, from a base cell to its furthest ancestor cell, their heights are in a non-decreasing order. For each base cell, we store the heights of its ancestor cells in a Fusion tree as shown in Lemma 1. Note that if a cell is vertically unbounded, then we set its height to be n, and thus the height of a cell in the shallow cutting is drawn from universe $[n+1]$. Since a base cell has $O(\lg \lg n)$ ancestors, including itself, the Fusion tree uses $O(\lg n \cdot \lg \lg n)$ bits of space and supports predecessor/successor queries over these heights in constant time. Meanwhile, for each height stored in the Fusion tree, we store a pointer that points to the cell that the height is from. The data structure contains $n/\delta(\lg^{(3)} n) = O(n/\lg n)$ base cells, and each of them requires $O(\lg n \cdot \lg \lg n)$ bits for storing the Fusion tree, for a total of $O(n \lg \lg n)$ bits.

Given a query point $Q = (q.x, q.y)$, we first find the base cell that contains Q along x-axis in constant time. The base cell is the $\lfloor q.x/\lg n \rfloor$-th one from the left. Then we query over the Fusion tree that stores for the base cell to search for the successor y' of $q.y$. Following the pointer that is stored for y', we find the smallest cell $\text{cell}_{i'}$ that contains Q. Overall, finding $\text{cell}_{i'}$ takes $O(1)$ time.

After finding $\text{cell}_{i'}$, we compute the number of rectangles of S that contains Q by querying over the stabbing counting data structure built for $\text{cell}_{i'}$. Hence,

Lemma 8. *There is an $O(n \lg \lg n)$-word data structure that supports 2D 3-sided stabbing counting queries in $O(1 + \log_w k)$ time, where k denotes the number of rectangles that contains the query point.*

4 A Linear Space Solution

In this section, we improve the solution of Sect. 3 and present a linear space data structure. In the previous solution, we construct K-level shallow cuttings for different K's. During a query, we look for the smallest cell cell$_{i'}$ of the shallow cuttings that contains the query point and then query over the data structure built for the rectangles of S that intersect cell$_{i'}$ to find the answer. Let cell$_{i'+1}$ denote the parent of cell$_{i'}$. Following Lemma 6, these at most $2^{2^{i'}}$ type-1 rectangles of cell$_{i'}$ consist of a subset of type-1 rectangles of cell$_{i'+1}$. Moreover, the type-2 rectangles of cell$_{i'}$ are either type-1 or type-2 rectangles of cell$_{i'+1}$. By performing queries over the rectangles that intersect cell$_{i'+1}$, we can correctly compute the number of rectangles that contain the query point as well. Recall that the total number of rectangles that intersect cell$_{i'+1}$ is bounded by $2 \cdot 2^{2^{i'+1}}$, while the total number of type-1 and type-2 rectangles of cell$_{i'}$ is no more than $2 \cdot 2^{2^{i'}}$. Since $\log_w(2 \cdot 2^{2^{i'+1}}) = \Theta(\log_w(2 \cdot 2^{2^{i'}}))$, the shift from cell$_{i'}$ to its parent, during a query, does not affect the upper bound of the query time. Meanwhile, due to this shift, we are allowed to build more space-efficient data structure, thereby improving the overall space cost. In the new solution, we handle type-1 and type-2 rectangles of each cell differently. Before showing the solution, we introduce the notation used throughout this section. We use K to represent $\delta(i)$ for each $\lg^{(3)} n \leq i \leq \lg\lg n$. Let Γ denote the K-level shallow cutting constructed over S, and thus Γ contains $2n/K$ cells, each of the form $[x_1, x_2] \times (-\infty, y_2]$.

4.1 Data Structures for Type-1 Rectangles

Let S_t denote the set of the type-1 rectangles of the t-th cell of Γ for each $0 \leq t < 2n/K$. Recall that each rectangle of S_t has at least one vertical edge between the slabs of the cell. Let B_t denote a subset of S_t, in which each rectangle r has one edge lying on the left of the t-th cell, namely, $r.x_1 < t \cdot K$. First, we construct an array $V[0..2n/K - 1]$ such that entry $V[t]$ stores the size of B_t. Obviously, each entry in V uses $O(\lg K)$ bits. Array V occupies $O(n \lg K/K)$ bits and will be used for finding the x- and y-rank of Q within a cell. We apply rank reduction over the endpoints of the rectangles of each S_t. After rank reduction, the rectangles of S_t lie on a $2|S_t| \times |S_t|$ grid, where $K/2 \leq |S_t| \leq K$. Then we build the data structure over S_t for stabbing counting queries by applying Lemma 4. Since there are $2n/K$ sets of rectangles in total and the data structure built for each set requires $O(|S_t| \lg |S_t|) = O(K \lg K)$ bits of space, and the total space cost is $2n/K \times O(K \lg K) = O(n \lg K)$ bits.

The data structure is built over the rectangles on a $2|S_t| \times |S_t|$ grid, while the query point still lies on a $2n \times n$ grid. Observe that to query point Q, the index of the cell that matters is $\lfloor q.x/K \rfloor$: If Q is not contained in the $\lfloor q.x/K \rfloor$-th cell, then there are more than K rectangles in S that contains Q and we simply return failure; otherwise, each rectangle that contains Q intersects this cell. Within this cell, the x-rank of Q is simply $V[\lfloor q.x/K \rfloor] + (q.x \mod K)$. Next, we consider the data structure that supports finding the y-rank of Q.

Lemma 9. *Let* $r = [x_1, x_2] \times [y_1, +\infty)$ *be any rectangle of* S_t *in rank space, where* $x_1, x_2 \in [2|S_t|]$, *and* $y_1 \in [|S_t|]$, *where* $K/2 \leq |S_t| \leq K$ *and* $K \in [\lg n, n]$. *Assume that given* $r.x_1$ *and* $r.x_2$, *the original coordinate in* $[n]$ *of* $r.y_1$ *can be found in* $O(1)$ *time. There is a data structure that uses* $O(K(\lg n / \log_w K + \lg K))$ *additional bits of space and finds the* y*-rank of* $Q.y$, *for any* $Q.y \in [n]$, *in* $O(\log_w K)$ *time.*

Proof. We sort rectangles of S_t by their y_1's in an increasing order and store x_1's and x_2's of the rectangles in array $\check{X}_1[0..|S_t|-1]$ and $\check{X}_2[0..|S_t|-1]$ such that entry $\check{X}_1[S_t[\ell].y_1] = S_t[\ell].x_1$ and $\check{X}_2[S_t[\ell].y_1] = S_t[\ell].x_2$ for each $0 \leq \ell < |S_t|$. Since each entry of $\check{X}_1$ and $\check{X}_2$ can be encoded with $\lg |S_t|$ bits, $\check{X}_1$ and $\check{X}_2$ together use $O(|S_t| \lg |S_t|) = O(K \lg K)$ bits. Let Δ denote $\lceil \log_w K \rceil$. We partition both $\check{X}_1$ and $\check{X}_2$ into blocks of size Δ. Let $\check{S}_t$ denote the set of the rectangles at positions $j\Delta - 1$ for each $1 \leq j \leq \lfloor |S_t|/\Delta \rfloor$ in the sorted list. We construct a fusion tree $\check{T}$ over the original y_1-coordinates of the rectangles of $\check{S}_t$ by Lemma 1. Since $|\check{S}_t| \leq \lfloor K/\Delta \rfloor$ and each original y_1-coordinate is encoded with $\lg n$ bits, tree $\check{T}$ uses $O(K \lg n / \log_w K)$ bits of space and supports predecessor and successor queries for any $Q.y \in [n]$ in $O(\log_w K)$ time. Overall, the space cost of our data structure is $O(K \lg K + K \lg n / \log_w K)$ bits.

To find the y-rank of $Q.y$ in set S_t, we find in $O(\log_w K)$ time the block that the predecessor of $Q.y$ stays using the fusion tree. Then we iterate through entries $\check{X}_1[\ell]$ and $\check{X}_2[\ell]$ from that block, compute in constant time the original y_1-coordinate of the rectangle whose y-rank is ℓ and compare this original y_1-coordinate with $Q.y$, until we find the predecessor of $Q.y$. Since a block contains Δ entries of $\check{X}_1$ and $\check{X}_2$ and each comparison takes $O(1)$ time, finding the predecessor of $Q.y$ within a block requires $O(\log_w K)$ time. Overall, the y-rank of $Q.y$ can be computed in $O(\log_w K)$ time. □

In fact, the assumption in Lemma 9 can be naturally supported by the shallow cutting structure. Recall that when building the shallow cutting structure, the plane is divided into $2n/K$ vertical slabs such that each K consecutive vertical edges of the rectangles of S lie in the same slab. To support the assumption, we construct a global array $A[0..2n-1]$ such that entry $A[j]$, for each $0 \leq j < 2n$, stores the the original y_1-coordinate of the rectangle whose x_1- or x_2-coordinate is ranked by j along x-axis. Note that each entry in A is encoded with $\lg n$ bits. The global array uses $O(n \lg n)$ bits of space.

Let r denote any rectangle in S_t, and let ℓ_1 and ℓ_2 denote the x-rank of $r.x_1$ and $r.x_2$, respectively. Since r has at least one edge lying in the t-th cell of Γ, between ℓ_1 and ℓ_2, at least one of them is between $V[t]+1$ and $V[t]+K$. W.l.o.g, assume that $V[t] + 1 \leq \ell_1 \leq V[t] + K$. Then, the original y_1-coordinate of r is $A[t \cdot K + \ell_1 - V[t] - 1]$, which can be computed in $O(1)$ time. Although we will construct $O(\lg \lg n)$ shallow cutting data structures over set S, the global array A can be constructed once and shared across all of these data structures.

4.2 Data Structures for Type-2 Rectangles

Each rectangle of S is of the form $[x_1, x_2] \times [y_1, +\infty)$; henceforth, we call y_1, i.e., the y-coordinate of the lower horizontal edge, the height of the rectangle. Let $\hat{S}_t$ denote the set of the type-2 rectangles of the t-th cell of C for each

$0 \le t < 2n/K$. Recall that each rectangle of $\hat{S}_t$ completely spans the slabs of the cell. We select the lowest $\sqrt{K}$ rectangles in $\hat{S}_t$. If $|\hat{S}_t| < \sqrt{K}$, then we select all the rectangles of $\hat{S}_t$. We store the heights of the selected rectangles in a Fusion tree as shown in Lemma 1. Note that the height of each rectangle is drawn from universe $[n]$, encoded with $O(\lg n)$ bits. The K-level shallow cutting contains $O(2n/K)$ cells, and the Fusion tree for each cell requires $O(\sqrt{K} \lg n)$ bits, for a total of $O(n \lg n/\sqrt{K})$ bits of space. Given a query point $Q = (q.x, q.y)$, the Fusion tree supports finding the number of selected rectangles whose height is no more than $q.y$ in $O(\log_w \sqrt{K})$ time. We defer to the next section the reason that we choose $\sqrt{K}$ rectangles from each cell, instead of K rectangles.

4.3 Wrap-Up

The Data Structure. We construct the $\delta(i)$-level shallow cutting for each $\lg^{(3)} n \le i \le \lg\lg n$. For the type-1 rectangles of each cell$_i$, we apply rank reduction over these rectangles and construct the stabbing counting data structure over the rectangles in rank space by Lemma 4 and the data structure that supports finding the y-rank of Q within cell$_i$ by Lemma 9. As shown in Sect. 4.1, the space cost for all the cells of the $\delta(i)$-level shallow cutting is $O(2n/\delta(i) \cdot (\delta(i) \lg \delta(i) + \delta(i) \lg n/ \log_w \delta(i))) = O(n(\lg \delta(i) + \lg n/ \log_w \delta(i)))$ bits. The global array A mentioned in the end of Sect. 4.1 is also necessary, which requires extra $O(n \lg n)$ bits of space. For the type-2 rectangles of each cell$_i$, we build the Fusion tree over the height of the lowest $\sqrt{\delta(i)}$ rectangles as shown in Sect. 4.2. The Fusion trees for all the cells of the $\delta(i)$-level shallow cutting require $O(n \lg n/\sqrt{\delta(i)})$ bits.

To find in constant time the smallest cell cell$_{i'}$ that contains query point Q, thereby finding its parent cell$_{i'+1}$, we build Fusion trees mentioned in the end of Sect. 3, for a total $O(n \lg\lg\lg n)$ bits of space. Finally, we construct $\lg n$-capped and n-capped stabbing counting data structures by Lemma 7, respectively, using extra $O(n)$ words of space.

For all $\lg^{(3)} n \le i \le \lg\lg n$, the overall space cost is $O(n \lg\lg\lg n) + O(n \lg n) + \sum_{i=\lg^{(3)} n}^{\lg\lg n} O(n \cdot (\lg \delta(i) + \lg n/ \log_w \delta(i) + \lg n/\sqrt{\delta(i)})) = O(n \lg n)$ bits of space.

The Query Algorithm. Let k denote $|S \cap Q|$. Given a query point $Q = (q.x, q.y)$, we first find in $O(1)$ time the smallest cell cell$_{i'}$ that contains Q using the same algorithm shown in Sect. 3, thereby finding the parent cell, cell$_{i'+1}$, of cell$_{i'}$. If cell$_{i'}$ happens to be a base cell, then $k \le \lg n$, and we use the $\lg n$-capped data structure of Lemma 7 to find k in $O(1)$ time. Otherwise, if cell$_{i'+1}$ does not exist, then i' is equal to $\lg\lg n$; namely, $k \ge 2^{2^{i'-1}} = \sqrt{n}$. So, we can simply query over the n-capped data structure of Lemma 7 to compute $|S \cap Q|$ in $O(\log_w n) = O(\log_w k)$.

If cell$_{i'+1}$ does exist, we apply the rank reduction over Q using the data structure shown in Lemma 9 and compute the number, k_1, of the type-1 rectangles of cell$_{i'+1}$ that contains Q. As shown in Sect. 4.1, k_1 can be computed in $O(\log_w 2^{2^{i'+1}}) = O(\log_w k)$ time, since $2^{i'+1} \le 4 \lg k$.

Then we query over the Fusion tree constructed for $\text{cell}_{i'+1}$ to find the number, k_2, of the rectangles whose height is no more than $Q.y$, among the lowest $2^{2^{i'}}$ type-2 rectangles of $\text{cell}_{i'+1}$. Since the Fusion tree contains at most $2^{2^{i'}}$ entries and $2^{i'} \leq 2\lg k$, k_2 can be computed in $O(\log_w 2^{2^{i'}}) = O(\log_w k)$ time.

Finally, we return $k_1 + k_2$ as the answer, and the query time is $O(1+\log_w k)$.

Proofs of Correctness. Next we prove that the sum of k_1+k_2 that is returned is equal to k. The rectangle set $S \cap Q$ can be divided into three pairwise subsets, $\tilde{A}$, $\tilde{B}$, and $\tilde{C}$, in which $\tilde{A}$ represents the subset of the rectangles that has at least one vertical edge between the horizontal span of $\text{cell}_{i'}$, $\tilde{B}$ represents the subset of rectangles that completely span the slabs of $\text{cell}_{i'}$ and have at least one vertical edge between the horizontal span of $\text{cell}_{i'+1}$, and $\tilde{C}$ represents the subset of rectangles that completely span the slabs of $\text{cell}_{i'+1}$. Lemma 6 tells that $\text{cell}_{i'}$ is contained by its parent $\text{cell}_{i'+1}$. Therefore, both $\tilde{A}$ and $\tilde{B}$ are a subset of type-1 rectangles of $\text{cell}_{i'+1}$. By querying over the stabbing counting data structure built for $\text{cell}_{i'+1}$, we get $|\tilde{A}| + |\tilde{B}|$, which is k_1.

It remains to show that k_2 is equal to $|\tilde{C}|$. Cell $\text{cell}_{i'+1}$, as a cell of a $\delta(i'+1)$-level shallow cutting, can have as many as $2^{2^{i'+1}}$ type-2 rectangles. However, the Fusion tree constructed for $\text{cell}_{i'+1}$ only handles the lowest $2^{2^{i'}}$ ones of them. Is querying over these lowest $2^{2^{i'}}$ ones sufficient? The answer is yes. Observe that a rectangle that spans the slabs of $\text{cell}_{i'+1}$ spans the slabs of $\text{cell}_{i'}$ as well. We can read $|\tilde{C}|$ as the number of type-2 rectangles of $\text{cell}_{i'}$ that contain Q and span the slabs of $\text{cell}_{i'+1}$. These k_2 rectangles whose height is no higher than Q also belong to the type-2 rectangles of $\text{cell}_{i'}$; therefore, we have $|\tilde{C}| \geq k_2$. Assume that $|\tilde{C}| > k_2$. So, there is at least one type-2 rectangle, r, of $\text{cell}_{i'}$ that contains Q, spans the slabs of $\text{cell}_{i'+1}$, but does not appear in the set of the lowest $2^{2^{i'}}$ type-2 rectangles of $\text{cell}_{i'+1}$. Including r, we find at least $2^{2^{i'}}+1$ type-2 rectangles for $\text{cell}_{i'}$, which generates a contradiction. Therefore, we prove that $|\tilde{C}|$ equals k_2. And we have $k \triangleq |\tilde{A}| + |\tilde{B}| + |\tilde{C}| \triangleq k_1 + k_2$. Hence, we achieve Theorem 2.

Theorem 2. *There is a linear space data structure that supports in $O(\log_w k)$ time stabbing counting queries over n of 3-sided rectangles in 2D rank space, where k denotes the number of rectangles that contains the query point.*

Following the reduction from 2D dominance colored range counting to 2D 3-sided stabbing counting queries, we achieve Theorem 1.

Acknowledgments. The author is thankful to Dr. Meng He for fruitful discussions and suggestions on this project. The author would like to thank the anonymous reviewers for their valuable comments and suggestions.

A Stabbing Counting to Dominance Range Counting

Let S denote a set of n' rectangles of the form $[x_1, x_2] \times [-\infty, y_2]$ in rank space. Let $\hat{A}$, $\hat{B}$, and $\hat{C}$ denote two-dimensional point set such that $\hat{A}$, $\hat{B}$, and $\hat{C}$ contains

$(r.x_1, 0)$, $(r.x_2, 0)$, and $(r.y_2, 0)$ for each rectangle $r \in S$, respectively. Then we build 2D dominance range counting data structure over $\hat{A}$, $\hat{B}$, and $\hat{C}$ applying Lemma 3. Note that the x-coordinates of points in $\hat{A}$, $\hat{B}$, or $\hat{C}$ might not be consecutive from 0 to $2n-1$. To apply Lemma 3, we can simply add some dump points in both sets as follows: For each $0 \le i \le 2n-1$, if i is not a x-coordinate of any point in the set, we append a dump point (i, n'). Let $\hat{D}$ and $\hat{E}$ denote the point sets such that $\hat{D}$ contains point $(-r.x_1, r.y_2)$ and $\hat{E}$ contains point $(r.x_2, r.y_2)$ for each $r \in S$, respectively. And we construct data structures over $\hat{D}$ and $\hat{E}$ for 2D dominance range counting apply Lemma 3. Again, to apply Lemma 3, we need to add some dump points in sets $\hat{D}$ and $\hat{E}$ as well: For each $0 \le i \le 2n-1$, if i is not a x-coordinate of any point in the set (either $\hat{D}$ or $\hat{E}$), we append a dump point (i, n'). By searching for the number of points in $\hat{A}$ that are dominated by $(q.x, 1)$, we can find the number, k_1, of rectangles that lie completely on the right side of Q, taking $O(\log_w n')$ time. In a similar way, we can find the number, k_2, of rectangles that lie completely on the left side of Q, and the number, k_3, of rectangles that lie completely below Q. Since each rectangle is unbounded downwards, no rectangles can completely lie above Q without containing it. The number, k_4, of rectangles that lie completely on the southeast side of Q and the number, k_5, of rectangles that lie completely on the southwest side of Q can be computed by querying over the dominance range counting data structures built for $\hat{D}$ and $\hat{E}$ using $(-Q.x, Q.y)$ and $(Q.x, Q.y)$ as the query point, respectively. Both k_4 and k_5 can be found in $O(\log_w n')$ time applying Lemma 3. Note that a rectangle on either southeast or southwest side of Q is counted twice in the sum of k_1, k_2, and k_3. Therefore, the number of rectangles that do not contain Q is $k_1 + k_2 + k_3 - k_4 - k_5$, and we return $n' - (k_1 + k_2 + k_3) + k_4 + k_5$ as the result.

References

1. Chan, T.M., He, Q., Nekrich, Y.: Further results on colored range searching. In: 36th International Symposium on Computational Geometry, SoCG 2020, vol. 164, pp. 1–15. Schloss Dagstuhl - Leibniz-Zentrum für Informatik (2020). https://doi.org/10.4230/LIPICS.SOCG.2020.28
2. Chan, T.M., Larsen, K.G., Pătraşcu, M.: Orthogonal range searching on the RAM, revisited. In: 27th International Symposium on Computational Geometry, SoCG 2011, pp. 1–10. ACM (2011). https://doi.org/10.1145/1998196.1998198
3. Chan, T.M., Nekrich, Y.: Better data structures for colored orthogonal range reporting. In: Proceedings of the Fourteenth Annual ACM-SIAM Symposium on Discrete Algorithms, pp. 627–636. Society for Industrial and Applied Mathematics (2020). https://doi.org/10.1137/1.9781611975994.38
4. Chan, T.M., Wilkinson, B.T.: Adaptive and approximate orthogonal range counting. ACM Trans. Algorithms **12**(4), 1–15 (2016). https://doi.org/10.1145/2830567
5. Edelsbrunner, H., Overmars, M.H.: On the equivalence of some rectangle problems. Inf. Process. Lett. **14**(3), 124–127 (1982). https://doi.org/10.1016/0020-0190(82)90068-0

6. El-Zein, H., Munro, J.I., Nekrich, Y.: Succinct color searching in one dimension. In: 28th International Symposium on Algorithms and Computation (ISAAC 2017), vol. 92, pp. 1–11. Schloss Dagstuhl - Leibniz-Zentrum für Informatik (2017). https://drops.dagstuhl.de/opus/volltexte/2017/8209/
7. Fredman, M.L., Willard, D.E.: BLASTING through the information theoretic barrier with FUSION TREES. In: Proceedings of the twenty-second annual ACM symposium on Theory of computing - STOC '90. pp. 1–7. ACM Press (1990). DOI: https://doi.org/10.1145/100216.100217
8. Gao, Y., He, M.: Space efficient two-dimensional orthogonal colored range counting. In: 29th Annual European Symposium on Algorithms (ESA 2021), vol. 204, pp. 1–17. Schloss Dagstuhl - Leibniz-Zentrum für Informatik (2021). https://doi.org/10.4230/LIPICS.ESA.2021.46
9. Gao, Y., He, M.: Faster path queries in colored trees via sparse matrix multiplication and min-plus product. In: 30th Annual European Symposium on Algorithms (ESA 2022), vol. 244, pp. 1–15. Schloss Dagstuhl - Leibniz-Zentrum für Informatik (2022). https://doi.org/10.4230/LIPICS.ESA.2022.59
10. Grossi, R., Vind, S.: Colored range searching in linear space. In: Ravi, R., Gørtz, I.L. (eds.) SWAT 2014. LNCS, vol. 8503, pp. 229–240. Springer, Cham (2014). https://doi.org/10.1007/978-3-319-08404-6_20
11. Gupta, P., Janardan, R., Smid, M.: Further results on generalized intersection searching problems: counting, reporting, and dynamization. J. Algorithms **19**(2), 282–317 (1995). https://doi.org/10.1006/jagm.1995.1038
12. Gupta, P., Janardan, R., Rahul, S., Smid, M.: Computational geometry: generalized (or colored) intersection searching. In: Handbook of Data Structures and Applications, pp. 1043–1058. Chapman and Hall/CRC (2018)
13. Gupta, P., Janardan, R., Smid, M.: Algorithms for generalized halfspace range searching and other intersection searching problems. Comput. Geom. **6**(1), 1–19 (1996). https://doi.org/10.1016/0925-7721(95)00012-7
14. He, M., Kazi, S.: Data structures for categorical path counting queries. Theoret. Comput. Sci. **938**, 97–111 (2022). https://doi.org/10.1016/j.tcs.2022.10.011
15. JaJa, J., Mortensen, C.W., Shi, Q.: Space-efficient and fast algorithms for multidimensional dominance reporting and counting. In: Fleischer, R., Trippen, G. (eds.) ISAAC 2004. LNCS, vol. 3341, pp. 558–568. Springer, Heidelberg (2004). https://doi.org/10.1007/978-3-540-30551-4_49
16. Jørgensen, A.G., Larsen, K.G.: Range selection and median: tight cell probe lower bounds and adaptive data structures. In: Proceedings of the Twenty-Second Annual ACM-SIAM Symposium on Discrete Algorithms. Society for Industrial and Applied Mathematics (2011). https://doi.org/10.1137/1.9781611973082.63
17. Kaplan, H., Rubin, N., Sharir, M., Verbin, E.: Efficient colored orthogonal range counting. SIAM J. Comput. **38**(3), 982–1011 (2008). https://doi.org/10.1137/070684483
18. Kaplan, H., Sharir, M., Verbin, E.: Colored intersection searching via sparse rectangular matrix multiplication. In: Proceedings of the Twenty-Second Annual Symposium on Computational Geometry. ACM (2006). https://doi.org/10.1145/1137856.1137866
19. Nekrich, Y.: Efficient range searching for categorical and plain data. ACM Trans. Database Syst. **39**(1), 1–21 (2014). https://doi.org/10.1145/2543924
20. Rahul, S.: Approximate range counting revisited. J. Comput. Geom. **12**, 40–69 (2021). https://doi.org/10.20382/JOCG.V12I1A3

Zip-Zip Trees: Making Zip Trees More Balanced, Biased, Compact, or Persistent

Ofek Gila[1(✉)], Michael T. Goodrich[1], and Robert E. Tarjan[2]

[1] University of California, Irvine, CA 92697, USA
{ogila,goodrich}@uci.edu
[2] Princeton University, Princeton, NJ 08544, USA
ret@cs.princeton.edu

Abstract. We define simple variants of zip trees, called ***zip-zip trees***, which provide several advantages over zip trees, including overcoming a bias that favors smaller keys over larger ones. We analyze zip-zip trees theoretically and empirically, showing, e.g., that the expected depth of a node in an n-node zip-zip tree is at most $1.3863 \log n - 1 + o(1)$, which matches the expected depth of treaps and binary search trees built by uniformly random insertions. Unlike these other data structures, however, zip-zip trees achieve their bounds using only $O(\log \log n)$ bits of metadata per node, w.h.p., as compared to the $\Theta(\log n)$ bits per node required by treaps. In fact, we even describe a "just-in-time" zip-zip tree variant, which needs just an expected $O(1)$ number of bits of metadata per node. Moreover, we can define zip-zip trees to be strongly history independent, whereas treaps are generally only weakly history independent. We also introduce ***biased zip-zip trees***, which have an explicit bias based on key weights, so the expected depth of a key, k, with weight, w_k, is $O(\log(W/w_k))$, where W is the weight of all keys in the weighted zip-zip tree. Finally, we show that one can easily make zip-zip trees partially persistent with only $O(n)$ space overhead w.h.p.

1 Introduction

A ***zip tree*** is a randomized binary search tree introduced by Tarjan, Levy, and Timmel [26]. Each node contains a specified key and a small randomly generated ***rank***. Nodes are in symmetric order by key, smaller to larger, and in max-heap order by rank. At a high level, zip trees are similar to other random search structures, such as the ***treap*** data structure of Seidel and Aragon [23], the ***skip list*** data structure of Pugh [20], and the ***randomized binary search tree*** (RBST) data structure of Martínez and Roura [16], but with two advantages:

1. Insertions and deletions in zip trees are described in terms of simple "zip" and "unzip" operations rather than sequences of rotations as in treaps and RBSTs, which are arguably more complicated; and

Research at Princeton Univ. was partially supported by a gift from Microsoft. Research at Univ. of California, Irvine was supported by NSF Grant 2212129.

P. Morin and S. Suri (Eds.): WADS 2023, LNCS 14079, pp. 474–492, 2023.
https://doi.org/10.1007/978-3-031-38906-1_31

2. Like treaps, zip trees organize keys using random ranks, but the ranks used by zip trees use $\Theta(\log\log n)$ bits each, whereas the key labels used by treaps and RBSTs use $\Theta(\log n)$ bits each. Also, as we review and expand upon, zip trees are topologically isomorphic to skip lists, but use less space.

In addition, zip trees have a desirable privacy-preservation property with respect to their ***history independence*** [15]. A data structure is ***weakly history independent*** if, for any two sequences of operations X and Y that take the data structure from initialization to state A, the distribution over memory after X is performed is identical to the distribution after Y. Thus, if an adversary observes the final state of the data structure, the adversary cannot determine the sequence of operations that led to that state. A data structure is ***strongly history independent***, on the other hand, if, for any two (possibly empty) sequences of operations X and Y that take a data structure in state A to state B, the distribution over representations of B after X is performed on a representation, r, is identical to the distribution after Y is performed on r. Thus, if an adversary observes the states of the data structure at different times, the adversary cannot determine the sequence of operations that lead to the second state beyond just what can be inferred from the states themselves. For example, it is easy to show that skip lists and zip trees are strongly history independent, and that treaps and RBSTs are weakly history independent.[1]

Indeed, zip trees and skip lists are strongly history independent for exactly the same reason, since Tarjan, Levy, and Timmel [26] define zip trees using a tie-breaking rule for ranks that makes zip trees isomorphic to skip lists, so that, for instance, a search in a zip tree would encounter the same keys as would be encountered in a search in an isomorphic skip list. This isomorphism between zip trees and skip lists has a potentially undesirable property, however, in that there is an inherent bias in a zip tree that favors smaller keys over larger keys. For example, as we discuss, the analysis from Tarjan, Levy, and Timmel [26] implies that the expected depth of the smallest key in an (original) zip tree is $0.5\log n$ whereas the expected depth of the largest key is $\log n$. Moreover, this same analysis implies that the expected depth for any node in a zip tree is at most $1.5\log n + O(1)$, whereas Seidel and Aragon [23] show that the expected depth of any node in a treap is at most $1.3863\log n + 1$, and Martínez and Roura [16] prove a similar result for RBSTs.

As mentioned above, the inventors of zip trees chose their tie-breaking rule to provide an isomorphism between zip trees and skip lists. But one may ask if there is a (hopefully simple) modification to the tie-breaking rule for zip trees that makes them more balanced for all keys, ideally while still maintaining the property that they are strongly history independent and that the metadata for keys in a zip tree requires only $O(\log\log n)$ bits per key w.h.p.

In this paper, we show how to improve the balance of nodes in zip trees by a remarkably simple change to its tie-breaking rule for ranks. Specifically, we

[1] If the random priorities used in a treap are distinct and unchanging for all keys and all time (which occurs only probabilistically), then the treap is strongly history independent.

describe and analyze a zip-tree variant we call ***zip-zip trees***, where we give each key a rank pair, $r = (r_1, r_2)$, such that r_1 is chosen from a geometric distribution as in the original definition of zip trees, and r_2 is an integer chosen uniformly at random, e.g., in the range $[1, \log^c n]$, for $c \geq 3$. We build a zip-zip tree just like an original zip tree, but with these rank pairs as its ranks, ordered and compared lexicographically. We also consider a just-in-time (JIT) variant of zip-zip trees, where we build the secondary r_2 ranks bit by bit as needed to break ties. Just like an original zip tree, zip-zip trees (with static secondary ranks) are strongly history independent, and, in any variant, each rank in a zip-zip tree requires only $O(\log \log n)$ bits w.h.p. Nevertheless, as we show (and verify experimentally), the expected depth of any node in a zip-zip tree storing n keys is at most $1.3863 \log n - 1 + o(1)$, whereas the expected depth of a node in an original zip tree is $1.5 \log n + O(1)$, as mentioned above. We also show (and verify experimentally) that the expected depths of the smallest and largest keys in a zip-zip tree are the same—namely, they both are at most $0.6932 \log n + \gamma + o(1)$, where $\gamma = 0.577721566\ldots$ is the Euler-Mascheroni constant.

In addition to showing how to make zip trees more balanced, by using the zip-zip tree tie-breaking rule, we also describe how to make them more biased for weighted keys. Specifically, we study how to store weighted keys in a zip-zip tree, to define the following variant (which can also be implemented for the original zip-tree tie-breaking rule):

- ***biased zip-zip trees***: These are a biased version of zip-zip trees, which support searches with expected performance bounds that are logarithmic in W/w_k, where W is the total weight of all keys in the tree and w_k is the weight of the search key, k.

Biased zip-zip trees can be used in simplified versions of the link-cut tree data structure of Sleator and Tarjan [25] for dynamically maintaining arbitrary trees, which has many applications, e.g., see Acar [1].

Zip-zip trees and biased zip-zip trees utilize only $O(\log \log n)$ bits of metadata per key w.h.p. (assuming polynomial weights in the weighted case) and are strongly history independent. The just-in-time (JIT) variant utilizes only $O(1)$ bits of metadata per operation w.h.p. but lacks history independence. Moreover, if zip-zip trees are implemented using the tiny pointers technique of Bender, Conway, Farach-Colton, Kuszmaul, and Tagliavini [5], then all of the non-key data used to implement such a tree requires just $O(n \log \log n)$ bits overall w.h.p.

Additional Prior Work. Before we provide our results, let us briefly review some additional related prior work. Although this analysis doesn't apply to treaps or RBSTs, Devroye [7,8] shows that the expected height of a randomly-constructed binary search tree tends to $4.311 \log n$ in the limit, which tightened a similar earlier result of Flajolet and Odlyzko [11]. Reed [21] tightened this bound even further, showing that the variance of the height of a randomly-constructed binary search tree is $O(1)$. Eberl, Haslbeck, and Nipkow [10] show that this analysis also applies to treaps and RBSTs, with respect to their expected height. Papadakis, Munro, and Poblete [19] provide an analysis for the expected search cost in a skip list, showing the expected cost is roughly $2 \log n$.

With respect to weighted keys, Bent, Sleator, and Tarjan [6] introduce a ***biased search tree*** data structure, for storing a set, $\mathcal{K}$, of n weighted keys, with a search time of $O(\log(W/w_k))$, where w_k is the weight of the search key, k, and $W = \sum_{k \in \mathcal{K}} w_k$. Their data structure is not history independent, however. Seidel and Aragon [23] provide a weighted version of treaps, which are weakly history independent and have expected $O(\log(W/w_k))$ access times, but weighted treaps have weight-dependent key labels that use exponentially more bits than are needed for weighted zip-zip trees. Afek, Kaplan, Korenfeld, Morrison, and Tarjan [2] provide a fast concurrent self-adjusting biased search tree when the weights are access frequencies. Zip trees and by extension zip-zip trees would similarly work well in a concurrent setting as most updates only affect the bottom of the tree, although such an implementation is not explored in this paper. Bagchi, Buchsbaum, and Goodrich [4] introduce randomized ***biased skip lists***, which are strongly history independent and where the expected time to access a key, k, is likewise $O(\log(W/w_k))$. Our weighted zip-zip trees are dual to biased skip lists, but use less space.

2 A Review of Zip Trees

In this section, we review the (original) zip tree data structure of Tarjan, Levy, and Timmel [26].

A Brief Review of Skip Lists. We begin by reviewing a related structure, namely, the ***skip list*** structure of Pugh [20]. A skip list is a hierarchical, linked collection of sorted lists that is constructed using randomization. All keys are stored in level 0, and, for each key, k, in level $i \geq 0$, we include k in the list in level $i + 1$ if a random coin flip (i.e., a random bit) is "heads" (i.e., 1), which occurs with probability 1/2 and independent of all other coin flips. Thus, we expect half of the elements from level i to also appear in level $i + 1$. In addition, every level includes a node that stores a key, $-\infty$, that is less than every other key, and a node that stores a key, $+\infty$, that is greater than every other key. The highest level of a skip list is the smallest i such that the list at level i only stores $-\infty$ and $+\infty$. (See Fig. 1). The following theorem follows from well-known properties of skip lists.

Theorem 1. *Let S be a skip list built from n distinct keys. The probability that the height of S is more than $\log n + f(n)$ is at most $2^{-f(n)}$, for any monotonically increasing function $f(n) > 0$.*

Proof. Note that the highest level in S is determined by the random variable $X = \max\{X_1, X_2, \ldots, X_n\}$, where each X_i is an independent geometric random variable with success probability 1/2. Thus, for any $i = 1, 2, \ldots, n$,

$$\Pr(X_i > \log n + f(n)) < 2^{-(\log n + f(n))} = 2^{-f(n)}/n;$$

hence, by a union bound, $\Pr(X > \log n + f(n)) < 2^{-f(n)}$. □

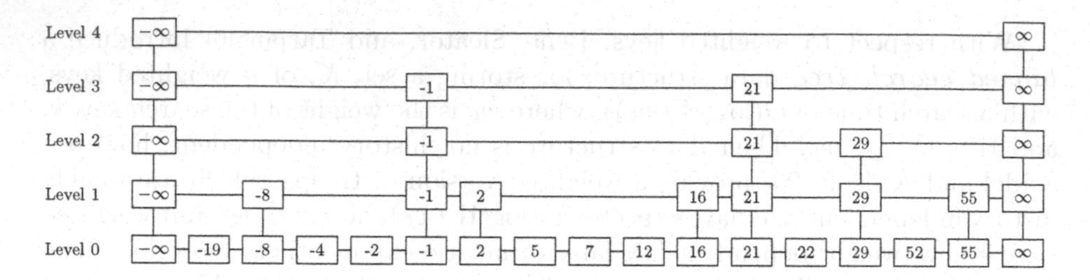

Fig. 1. An example skip list.

Zip Trees and Their Isomorphism to Skip Lists. Let us next review the definition of the (original) zip tree data structure [26]. A zip tree is a binary search tree where nodes are max-heap ordered according to random ***ranks***, with ties broken in favor of smaller keys, so that the parent of a node has rank greater than that of its left child and no less than that of its right child [26]. The rank of a node is drawn from a geometric distribution with success probability 1/2, starting from a rank 0, so that a node has rank k with probability $1/2^{k+1}$.

As noted by Tarjan, Levy, and Timmel [26], there is a natural isomorphism between a skip-list, L, and a zip tree, T, where L contains a key k in its level-i list if and only if k has rank at least i in T. That is, the rank of a key, k, in T equals the highest level in L that contains k. See Fig. 2. As we review in an appendix, insertion and deletion in a zip tree are done by simple "unzip" and "zip" operations, and these same algorithms also apply to the variants we discuss in this paper, with the only difference being the way we define ranks.

An advantage of a zip tree, T, over its isomorphic skip list, L, is that T's space usage is roughly half of that of L, and T's search times are also better. Nevertheless, there is a potential undesirable property of zip trees, in that an original zip tree is biased towards smaller keys, as we show in the following.

Theorem 2. *Let T be an (original) zip tree storing n distinct keys. Then the expected depth of the smallest key is $0.5 \log n + O(1)$ whereas the expected depth of the largest key is $\log n + O(1)$.*

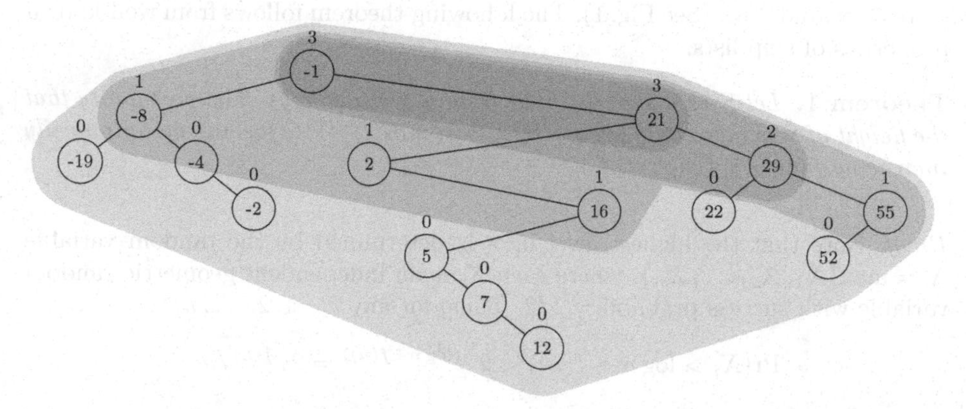

Fig. 2. An example zip tree, corresponding to the skip list in Fig. 1.

Proof. The bound for the largest (resp., smallest) key follows immediately from Lemma 3.3 (resp., Lemma 3.4) from Tarjan, Levy, and Timmel [26] and the fact that the expect largest rank in T is at most $\log n + O(1)$. □

That is, the expected depth of the largest key in an original zip tree is twice that of the smallest key. This bias also carries over, unfortunately, into the characterization of Tarjan, Levy, and Timmel [26] for the expected depth of a node in an original zip tree, which they show is at most $1.5 \log n + O(1)$. In contrast, the expected depth of a node in a treap or randomized binary search tree can be shown to be at most $1.39 \log n + O(1)$ [16,23].

3 Zip-Zip Trees

In this section, we define and analyze the zip-zip tree data structure.

Uniform Zip Trees As a warm-up, let us first define a variant to an original zip tree, called a ***uniform zip tree***, which is a zip tree where we define the rank of each key to be a random integer drawn independently from a uniform distribution over a suitable range. We perform insertions and deletions in a uniform zip tree exactly as in an original zip tree, except that rank comparisons are done using these uniform ranks rather than using ranks drawn from a geometric distribution. Thus, if there are no rank ties that occur during its construction, then a uniform zip tree is a treap [23]. But if a rank tie occurs, we resolve it using the tie-breaking rule for a zip tree, rather than doing a complete tree rebuild, as is done for a treap [23]. Still, we introduce uniform zip trees only as a stepping stone to our definition of zip-zip trees, which we give next.

Zip-Zip Trees. A ***zip-zip tree*** is a zip tree where we define the rank of each key to be the pair, $r = (r_1, r_2)$, where r_1 is drawn independently from a geometric distribution with success probability $1/2$ (as in the original zip tree) and r_2 is an integer drawn independently from a uniform distribution on the interval $[1, \log^c n]$, for $c \geq 3$. We perform insertions and deletions in a zip-zip tree exactly as in an original zip tree, except that rank comparisons are done lexicographically based on the (r_1, r_2) pairs. That is, we perform an update operation focused primarily on the r_1 ranks, as in the original zip tree, but we break ties by reverting to r_2 ranks. And if we still get a rank tie for two pairs of ranks, then we break these ties as in original zip tree approach, biasing in favor of smaller keys. Fortunately, as we show, such ties occur with such low probability that they don't significantly impact the expected depth of any node in a zip-zip tree, and this also implies that the expected depth of the smallest key in a zip-zip tree is the same as for the largest key.

Let x_i be a node in a zip-zip tree, T. Define the ***r_1-rank group*** of x_i as the connected subtree of T comprising all nodes with the same r_1-rank as x_i. That is, each node in x_i's r_1-rank group has a rank tie with x_i when comparing ranks with just the first rank coordinate, r_1.

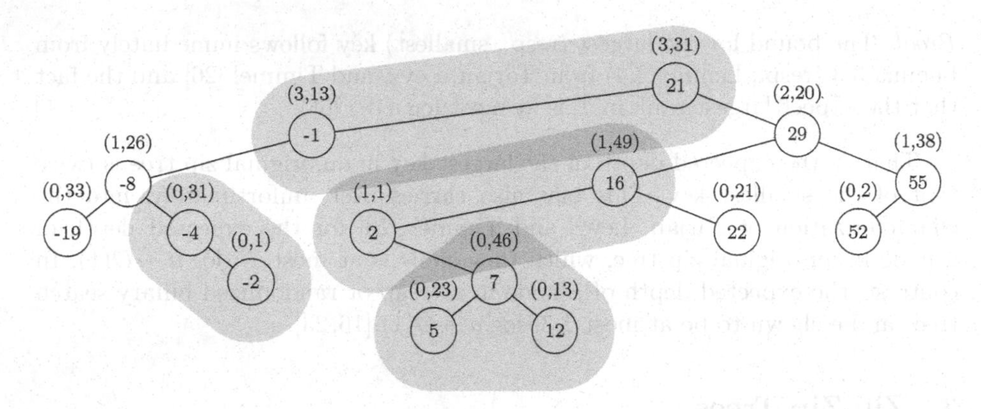

Fig. 3. A zip-zip tree, with each node labeled with its (r_1, r_2) rank. Each shaded subtree is an r_1-rank group defining a uniform zip tree based on r_2 ranks.

Lemma 1. *The r_1-rank group for any node, x_i, in a zip-zip tree is a uniform zip tree defined using r_2-ranks.*

Proof. The proof follows immediately from the definitions. □

Incidentally, Lemma 1 is the motivation for the name "zip-zip tree," since a zip-zip tree can be viewed as a zip tree comprised of little zip trees. Moreover, this lemma immediately implies that a zip-zip tree is strongly history independent, since both zip trees and uniform zip trees are strongly history independent. See Fig. 3.

Lemma 2. *The number of nodes in an r_1-rank group in a zip-zip tree, T storing n keys has expected value 2 and is at most $2 \log n$ w.h.p.*

Proof. By the isomorphism between zip trees and skip lists, the set of nodes in an r_1-rank group in T is dual to a sequence of consecutive nodes in a level-r_1 list in the skip list but not in the level-$(r_1 + 1)$ list. Thus, the number of nodes, X, in an r_1-rank group is a random variable drawn from a geometric distribution with success probability $1/2$; hence, $E[X] = 2$ and X is at most $2 \log n$ with probability at least $1 - 1/n^2$. Moreover, by a union bound, all the r_1-rank groups in T have size at most $2 \log n$ with probability at least $1 - 1/n$. □

We can also define a variant of a zip-zip tree that is not history independent but which uses only $O(1)$ bits of metadata per key in expectation.

Just-in-Time Zip-Zip Trees. In a ***just-in-time (JIT) zip-zip tree***, we define the (r_1, r_2) rank pair for a key, x_i, so that r_1 is (as always) drawn independently from a geometric distribution with success probability $1/2$, but where r_2 is an initially empty string of random bits. If at any time during an update in a JIT zip-zip tree, there is a tie between two rank pairs, $(r_{1,i}, r_{2,i})$ and $(r_{1,j}, r_{2,j})$, for two keys, x_i and x_j, respectively, then we independently add unbiased random bits, one bit at a time, to $r_{2,i}$ and $r_{2,j}$ until x_i and x_j no longer have a tie in their rank pairs, where r_2-rank comparisons are done by viewing the binary strings as binary fractions after a decimal point.

Note that the definition of an r_1-rank group is the same for a JIT zip-zip tree as a (standard) zip-zip tree. Rather than store r_1-ranks explicitly, however, we store them as a difference between the r_1-rank of a node and the r_1-rank of its parent (except for the root). Moreover, by construction, each r_1-rank group in a JIT zip-zip tree is a treap; hence, a JIT zip-zip tree is topologically isomorphic to a treap. We prove the following theorem in an appendix.

Theorem 3. *Let T be a JIT zip-zip tree resulting from n update operations starting from an initially empty tree. The expected number of bits for rank metadata in any non-root node in T is $O(1)$ and the number of bits required for all the rank metadata in T is $O(n)$ w.h.p.*

Depth Analysis. The main theoretical result of this paper is the following.

Theorem 4. *The expected depth, δ_j, of the j-th smallest key in a zip-zip tree, T, storing n keys is equal to $H_j + H_{n-j+1} - 1 + o(1)$, where $H_n = \sum_{i=1}^{n}(1/i)$ is the n-th harmonic number.*

Proof. Let us denote the ordered list of (distinct) keys stored in T as $L = (x_1, x_2, \ldots, x_n)$, where we use "$x_j$" to denote both the node in T and the key that is stored there. Let X be a random variable equal to the depth of the j-th smallest key, x_j, in T, and note that

$$X = \sum_{i=1,\ldots,j-1,j+1,\ldots,n} X_i,$$

where X_i is an indicator random variable that is 1 iff x_i is an ancestor of x_j. Let A denote the event where the r_1-rank of the root, z, of T is more than $3\log n$, or the total size of all the r_1-rank groups of x_j's ancestors is more than $d\log n$, for a suitable constant, d, chosen so that, by Lemma 3 (in an appendix), $\Pr(A) \le 2/n^2$. Let B denote the event, conditioned on A not occurring, where the r_1-rank group of an ancestor of x_j contains two keys with the same rank, i.e., their ranks are tied even after doing a lexicographic rank comparison. Note that, conditioned on A not occurring, and assuming $c \ge 4$ (for the sake of a $o(1)$ additive term[2]), the probability that any two keys in any of the r_1-rank groups of x_j's ancestors have a tie among their r_2-ranks is at most $d^2\log^2 n/\log^4 n$; hence, $\Pr(B) \le d^2/\log^2 n$. Finally, let C denote the complement event to both A and B, that is, the r_1-rank of z is less than $3\log n$ and each r_1-rank group for an ancestor of x_j has keys with unique (r_1, r_2) rank pairs. Thus, by the definition of conditional expectation,

[2] Taking $c = 3$ would only cause an $O(1)$ additive term.

$$\begin{aligned}\delta_j = E[X] &= E[X|A] \cdot \Pr(A) + E[X|B] \cdot \Pr(B) + E[X|C] \cdot \Pr(C) \\ &\leq \frac{2n}{n^2} + \frac{d^3 \log n}{\log^2 n} + E[X|C] \\ &\leq E[X|C] + o(1).\end{aligned}$$

So, for the sake of deriving an expectation for X, let us assume that the condition C holds. Thus, for any x_i, where $i \neq j$, x_i is an ancestor of x_j iff x_i's rank pair, $r = (r_1, r_2)$, is the unique maximum such rank pair for the keys from x_i to x_j, inclusive, in L (allowing for either case of $x_i < x_j$ or $x_j < x_i$, and doing rank comparisons lexicographically). Since each key in this range has equal probability of being assigned the unique maximum rank pair among the keys in this range,

$$\Pr(X_i = 1) = \frac{1}{|i-j|+1}.$$

Thus, by the linearity of expectation,

$$E[X|C] = H_j + H_{n+1-j} - 1.$$

Therefore, $\delta_j = H_j + H_{n+1-j} - 1 + o(1)$. □

This immediately gives us the following:

Corollary 1. *The expected depth, δ_j, of the j-th smallest key in a zip-zip tree, T, storing n keys can be bounded as follows:*

1. *If $j = 1$ or $j = n$, then $\delta_j < \ln n + \gamma + o(1) < 0.6932 \log n + \gamma + o(1)$, where $\gamma = 0.57721566\ldots$ is the Euler-Mascheroni constant.*
2. *For any $1 \leq j \leq n$, $\delta_j < 2 \ln n - 1 + o(1) < 1.3863 \log n - 1 + o(1)$.*

Proof. The bounds all follow from Theorem 4, the fact that $\ln 2 = 0.69314718\ldots$, and Franel's inequality (see, e.g., Guo and Qi [13]):

$$H_n < \ln n + \gamma + \frac{1}{2n}.$$

Thus, for (1), if $j = 1$ or $j = n$, $\delta_j = H_n < \ln n + \gamma + o(1)$.
For (2), if $1 \leq j \leq n$,

$$\begin{aligned}\delta_j &= H_j + H_{n-j+1} - 1 \\ &< \ln j + \ln(n-j+1) + 2\gamma - 1 + o(1) \\ &\leq 2 \ln n - 1 + o(1),\end{aligned}$$

since $\ln 2 > \gamma$ and $j(n-j+1)$ is maximized at $j = n/2$ or $j = (n+1)/2$. □

Incidentally, these are actually tighter bounds than those derived by Seidel and Aragon for treaps [23], but similar bounds can be shown to hold for treaps.

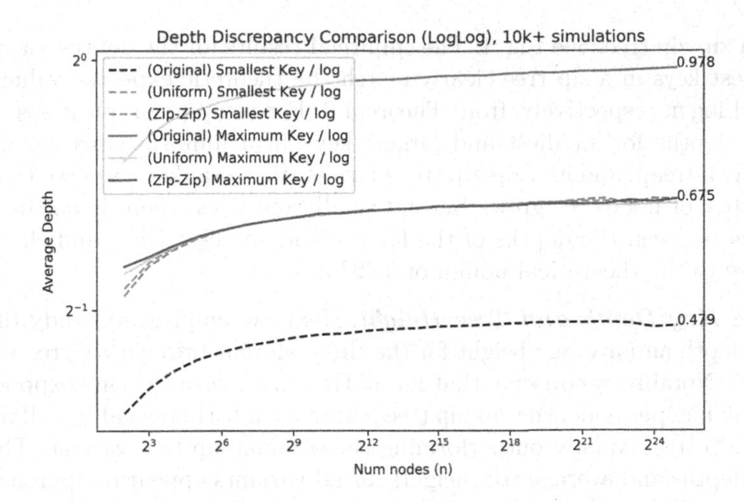

Fig. 4. Experimental results for the depth discrepancy between the smallest and largest keys in the original, uniform (treap), and zip-zip variants of the zip tree. Each data point is scaled down by a factor of $\log n$ (base 2).

Making Zip-Zip Trees Partially Persistent. A data structure that can be updated in a current version while also allowing for queries in past versions is said to be ***partially persistent***, and Driscoll, Sarnak, Sleator, and Tarjan [9] show how to make any bounded-degree linked structure, like a binary search tree, T, into a partially persistent data structure by utilizing techniques employing "fat nodes" and "node splitting." They show that if a sequence of n updates on T only modifies $O(n)$ data fields and pointers, then T can be made partially persistent with only an constant-factor increase in time and space for processing the sequence of updates, and allows for queries in any past instance of T. We show in an appendix that zip-zip trees have this property, w.h.p., thereby proving the following theorem.

Theorem 5. *One can transform an initially empty zip-zip tree, T, to be partially persistent, over the course of n insert and delete operations, so as to support, w.h.p., $O(\log n)$ amortized-time updates in the current version and $O(\log n)$-time queries in the current or past versions, using $O(n)$ space.*

4 Experiments

We augment our theoretical findings with experimental results, where we repeatedly constructed search trees with keys, $\{0, 1, \ldots, n-1\}$, inserted in order (since insertion order doesn't matter). For both uniform zip trees and zip-zip trees with static r_2-ranks, we draw integers independently for the uniform ranks from the intervals $[1, n^c]$, and $[1, \log^c n]$, respectively, choosing $c = 3$.

Depth Discrepancy. First, we consider the respective depths of the smallest and the largest keys in an original zip tree, compared with the depths of these keys in a zip-zip tree. See Fig. 4. The empirical results for the depths for smallest and largest keys in a zip tree clearly match the theoretic expected values of 0.5 $\log n$ and $\log n$, respectively, from Theorem 2. For comparison purposes, we also plot the depths for smallest and largest keys in a uniform zip tree, which is essentially a treap, and in a zip-zip tree (with static r_2-ranks). Observe that, after the number of nodes, n, grows beyond small tree sizes, there is no discernible difference between the depths of the largest and smallest keys, and that this is very close to the theoretical bound of $0.69 \log n$.

Average Key Depth and Tree Height. Next, we empirically study the average key depth and average height for the three aforementioned zip tree variants. See Fig. 5. Notably, we observe that for all tree sizes, despite using exponentially fewer rank bits per node, the zip-zip tree performs indistinguishably well from the uniform zip tree, equally outperforming the original zip tree variant. The average key depths and average tree heights for all variants appear to approach some constant multiple of $\log n$. For example, the average depth of a key in an original zip tree, uniform zip tree, and zip-zip tree reached $1.373 \log n$, $1.267 \log n$, $1.267 \log n$, respectively. Interestingly, these values are roughly 8.5% less than the original zip tree and treap theoretical average key depths of $1.5 \log n$ [26] and $1.39 \log n$ [23], respectively, suggesting that both variants approach their limits at a similar rate. Also, we note that our empirical average height bounds for uniform zip trees and zip-zip trees get as high as $2.542 \log n$. It is an open problem to bound these expectations theoretically, but we show in an appendix that the height of a zip-zip tree is at most $3.82 \log n$ with probability $1 - o(1)$, which clearly beats the $4.31107 \log n$ expected height for a randomly-constructed binary search tree [7,8,11,21].

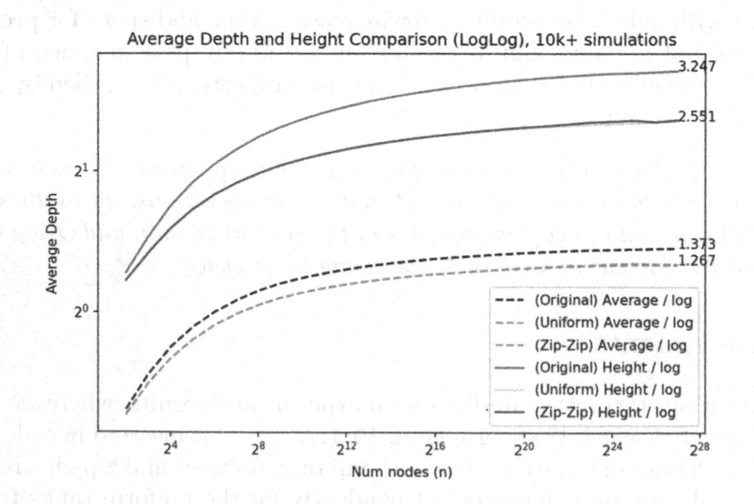

Fig. 5. Experimental results for the average node depth and tree height, comparing the original, uniform (treap-like), and zip-zip variants of the zip tree. Each data point is scaled down by a factor of $\log n$ (base 2).

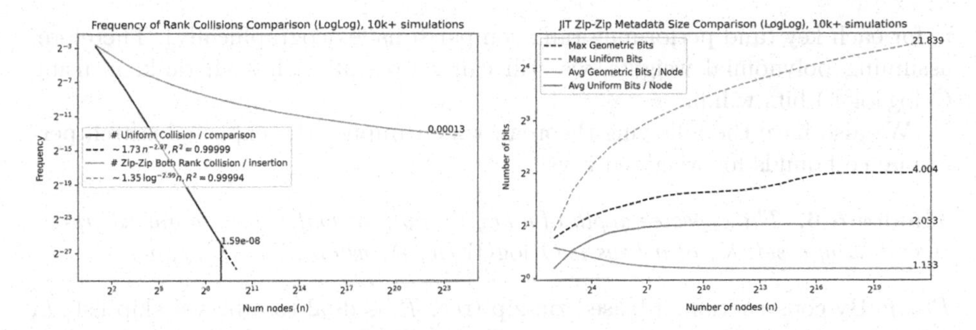

Fig. 6. (Left) The frequency of encountered rank ties per rank comparison for the uniform variant and per element insertion for the zip-zip variant. (Right) The metadata size for the just-in-time implementation of the zip-zip tree.

Rank Comparisons. Next, we experimentally determine the frequency of complete rank ties (collisions) for the uniform and zip-zip variants. See Fig. 6 (left). The experiments show how the frequencies of rank collisions decrease polynomially in n for the uniform zip tree and in $\log n$ for the second rank of the zip-zip variant. This reflects how these rank values were drawn uniformly from a range of n^c and $\log^c n$, respectively. Specifically, we observe the decrease to be polynomial to $n^{-2.97}$ and $\log^{-2.99} n$, matching our chosen value of c being 3.

Just-in-Time Zip-Zip Trees. Finally, we show how the just-in-time zip-zip tree variant uses an expected constant number of bits per node. See Fig. 6 (right). We observe a results of only 1.133 bits per node for storing the geometric (r_1) rank differences, and only 2.033 bits per node for storing the uniform (r_2) ranks, leading to a remarkable total of 3.166 expected bits per node of rank metadata to achieve ideal treap properties.

5 Biased Zip-Zip Trees

In this section, we describe how to make zip-zip trees biased for weighted keys. In this case, we assume each key, k, has an associated weight, w_k, such as an access frequency. Without loss of generality, we assume that weights don't change, since we can simulate a weight change by deleting and reinserting a key with its new weight.

Our method for modifying zip-zip trees to accommodate weighted keys is simple—when we insert a key, k, with weight, w_k, we now assign k a rank pair, $r = (r_1, r_2)$, such that r_1 is $\lfloor \log w_k \rfloor + X_k$, where X_k is drawn independently from a geometric distribution with success probability 1/2, and r_2 is an integer independently chosen uniformly in the range from 1 to $\lceil \log^c n \rceil$, where $c \geq 3$. Thus, the only modification to our zip-zip tree construction to define a biased zip-zip tree is that the r_1 component is now a sum of a logarithmic rank and a value drawn from a geometric distribution. As with our zip-zip tree definition for unweighted keys, all the update and search operations for biased zip-zip trees are the same as for the original zip trees, except for this modification to the rank,

r, for each key (and performing rank comparisons lexicographically). Therefore, assuming polynomial weights, we still can represent each such rank, r, using $O(\log \log n)$ bits w.h.p.

We also have the following theorem, which implies the expected search performance bounds for weighted keys.

Theorem 6. *The expected depth of a key, k, with weight, w_k, in a biased zip-zip tree storing a set, $\mathcal{K}$, of n keys is $O(\log(W/w_k))$, where $W = \sum_{k \in \mathcal{K}} w_k$.*

Proof. By construction, a biased zip-zip tree, T, is dual to a biased skip list, L, defined on $\mathcal{K}$ with the same r_1 ranks as for the keys in $\mathcal{K}$ as assigned during their insertions into T. Bagchi, Buchsbaum, and Goodrich [4] show that the expected depth of a key, k, in L is $O(\log(W/w_k))$. Therefore, by Theorem 1, and the linearity of expectation, the expected depth of k in T is $O(\log(W/w_k))$, where, as mentioned above, W is the sum of the weights of the keys in T and w_k is the weight of the key, k. □

Thus, a biased zip-zip tree has similar expected search and update performance as a biased skip list, but with reduced space, since a biased zip-zip tree has exactly n nodes, whereas, assuming a standard skip-list representation where we use a linked-list node for each instance of a key, k, on a level in the skip list (from level-0 to the highest level where k appears) a biased skip list has an expected number of nodes equal to $2n + 2\sum_{k \in \mathcal{K}} \log w_k$. For example, if there are n^ε keys with weight n^ε, then such a biased skip list would require $\Omega(n \log n)$ nodes, whereas a dual biased zip-zip tree would have just n nodes.

Further, due to their simplicity and weight biasing, we can utilize biased zip-zip trees as the biased auxiliary data structures in the link-cut dynamic tree data structure of Sleator and Tarjan [25], thereby providing a simple implementation of link-cut trees.

A Insertion and Deletion in Zip Trees and Zip-Zip Trees

The insertion and deletion algorithms for zip-zip trees are the same as those for zip trees, except that in a zip-zip tree a node's rank is a pair (r_1, r_2), as we explain above, and rank comparisons are done lexicographically on these pairs.

To insert a new node x into a zip tree, we search for x in the tree until reaching the node y that x will replace, namely the node y such that $y.rank \leq x.rank$, with strict inequality if $y.key < x.key$. From y, we follow the rest of the search path for x, unzipping it by splitting it into a path, P, containing each node with key less than $x.key$ and a path, Q, containing each node with key greater than $x.key$ (recall that we assume keys are distinct) [26]. To delete a node x, we perform the inverse operation, where we do a search to find x and let P and Q be the right spine of the left subtree of x and the left spine of the right subtree of x, respectively. Then we zip P and Q to form a single path R by merging them from top to bottom in non-increasing rank order, breaking a tie in favor of the smaller key [26]. See Fig. 7.

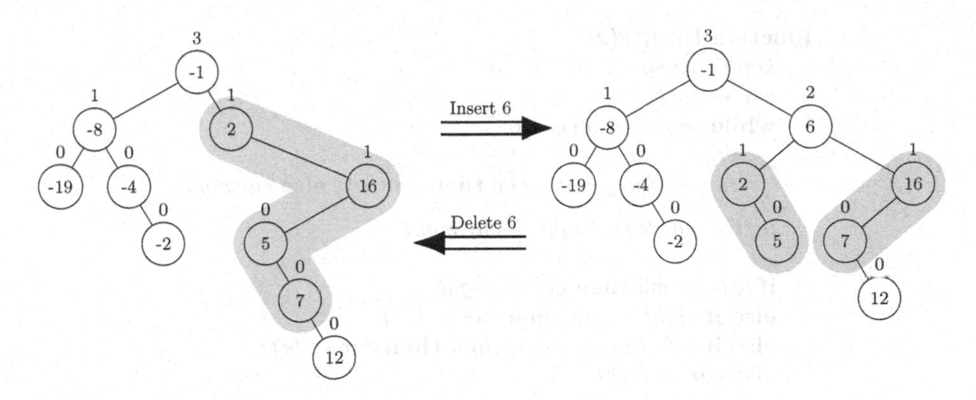

Fig. 7. How insertion in a zip tree is done via unzipping and deletion is done via zipping.

For completeness, we give the pseudo-code for the insert and delete operations, from Tarjan, Levy, and Timmel [26], in Figs. 8 and 9.

```
function Insert(x)
    rank ← x.rank ← RandomRank
    key ← x.key
    cur ← root
    while cur ≠ null and (rank < cur.rank or (rank = cur.rank and key > cur.key)) do
        prev ← cur
        cur ← if key < cur.key then cur.left else cur.right

    if cur = root then root ← x
    else if key < prev.key then prev.left ← x
    else prev.right ← x

    if cur = null then { x.left ← x.right ← null; return }
    if key < cur.key then x.right ← cur else x.left ← cur
    prev ← x

    while cur ≠ null do
        fix ← prev

        if cur.key < key then
            repeat { prev ← cur; cur ← cur.right }
            until cur = null or cur.key < key
        else
            repeat { prev ← cur; cur ← cur.left }
            until cur = null or cur.key > key

        if fix.key > key or (fix = x and prev.key > key) then
            fix.left ← cur
        else
            fix.right ← cur
```

Fig. 8. Insertion in a zip tree (or zip-zip tree), from [26].

```
function DELETE(x)
    key ← x.key
    cur ← root
    while key ≠ cur.key do
        prev ← cur
        cur ← if key < cur.key then cur.left else cur.right

    left ← cur.left; right ← cur.right

    if left = null then cur ← right
    else if right = null then cur ← left
    else if left.rank ≥ right.rank then cur ← left
    else cur ← right

    if root = x then root ← cur
    else if key < prev.key then prev.left ← cur
    else prev.right ← cur

    while left ≠ null and right ≠ null do
        if left.rank ≥ right.rank then
            repeat { prev ← left; left ← left.right }
            until left = null or left.rank < right.rank
            prev.right ← right
        else
            repeat { prev ← right; right ← right.left }
            until right = null or left.rank ≥ right.rank
            prev.left ← left
```

Fig. 9. Deletion in a zip tree (or zip-zip tree), from [26].

B Omitted Proofs

In this appendix, we provide proofs that were omitted in the body of this paper. We start with a simple lemma that our omitted proofs use.

Lemma 3. *Let X be the sum of n independent geometric random variables with success probability $1/2$. Then, for $t \geq 2$,*

$$\Pr(X > (2+t)n) \leq e^{-tn/10}.$$

Proof. The proof follows immediately by a Chernoff bound for a sum of n independent geometric random variables (see, e.g., Goodrich and Tamassia [12, pp. 555–556]). □

Compacting a JIT Zip-Zip Tree. We prove the following theorem in this appendix.

Theorem 7 (Same as Theorem 3). *Let T be a JIT zip-zip tree resulting from n update operations starting from an initially-empty tree. The expected number of bits for rank metadata in any non-root node in T is $O(1)$ and the number of bits required for all the rank metadata in T is $O(n)$ w.h.p.*

Proof. By the duality between zip trees and skip lists, the set of nodes in an r_1-rank group in T is dual to a sequence, L, of consecutive nodes in a level-r_1 list in the skip list but not in the level-(r_1+1) list. Thus, since v is not the root, there is a node, u, that is the immediate predecessor of the first node in L in the level-r_1 list in the skip list, and there is a node, w, that is the immediate successor of the last node in L in the level-r_1 list in the skip list. Moreover, both u and v are in the level-(r_1+1) list in the skip list, and (since v is not the root) it cannot be the case that u stores the key $-\infty$ and w stores the key $+\infty$. As an over-estimate and to avoid dealing with dependencies, we will consider the r_1-rank differences determined by predecessor nodes (like u) separate from the r_1-rank differences determined by successor nodes (like w). Let us focus on predecessors, u, and suppose u does not store $-\infty$. Let $r_1' > r_1$ be the highest level in the skip list where u appears. Then the difference between the r_1-rank of v and its parent is at most $r_1' - r_1$. That is, this rank difference is at most a random variable that is drawn from a geometric distribution with success probability $1/2$ (starting at level r_1+1); hence, its expected value is at most 2. Further, for similar reasons, the sum of all the r_1-rank differences for all nodes in T that are determined because of a predecessor node (like u) can be bounded by the sum, X, of n independent geometric random variables with success probability $1/2$. (Indeed, this is also an over-estimate, since a r_1-rank difference for a parent in the same r_1-rank group is 0, and some r_1-rank differences may be determined by a successor node that has a lower highest level in the dual skip list that a predecessor node.) By Lemma 3, X is $O(n)$ with (very) high probability, and a similar argument applies to the sum of r_1-rank differences determined by successor nodes. Thus, with (very) high probability, the sum of all r_1-rank differences between children and parents in T is $O(n)$.

Let us next consider all the r_2-ranks in a JIT zip-zip tree. Recall that each time there is a rank tie when using existing (r_1, r_2) ranks, during a given update, we augment the two r_2 ranks bit by bit until they are different. That is, the length of each such augmentation is a geometric random variable with success probability $1/2$. Further, by the way that the zip and unzip operations work, the number of such encounters that could possibly have a rank tie is upper bounded by the sum of the r_1-ranks of the keys involved, i.e., by the sum of n geometric random variables with success probability $1/2$. Thus, by Lemma 3, the number of such encounters is at most $N = 12n$ and the number of added bits that occur during these encounters is at most $12N$, with (very) high probability. □

Partially-Persistent Zip-Zip Trees. We show in this appendix that one can efficiently make a zip-zip tree partially persistent.

Theorem 8 (Same as Theorem 5). *One can transform an initially-empty zip-zip tree, T, to be partially persistent, over the course of n insert and delete operations, so as to support, w.h.p., $O(\log n)$ amortized-time updates in the current version and $O(\log n)$-time queries in the current or past versions, using $O(n)$ space.*

Proof. By the way that the zip and unzip operations work, the total number of data or pointer changes in T over the course of n insert and delete operations can be upper bounded by the sum of r_1-ranks for all the keys involved, i.e., by the sum of n geometric random variables with success probability 1/2. Thus, by Lemma 3, the number of data or pointer changes in T is at most $N = 12n$ with (very) high probability. Driscoll, Sarnak, Sleator, and Tarjan [9] show how to make any bounded-degree linked structure, like a binary search tree, T, into a partially persistent data structure by utilizing techniques employing "fat nodes" and "node splitting," so that if a sequence of n updates on T only modifies $O(n)$ data fields and pointers, then T can be made partially persistent with only an constant-factor increase in time and space for processing the sequence of updates, and this allows for queries in any past instance of T in the same asymptotic time as in the ephemeral version of T plus the time to locate the appropriate prior version. Alternatively, Sarnak and Tarjan [22] provide a simpler set of techniques that apply to binary search trees without parent pointers. Combining these facts establishes the theorem. □

For example, we can apply this theorem with respect to a sequence of n updates of a zip-zip tree that can be performed in $O(n \log n)$ time and $O(n)$ space w.h.p., e.g., to provide a simple construction of an $O(n)$-space planar point-location data structure that supports $O(\log n)$-time queries. A similar construction was provided by Sarnak and Tarjan [22], based on the more-complicated red-black tree data structure; hence, our construction can be viewed as simplifying their construction.

Theorem 9. *The height of a zip-zip tree, T, holding a set, S, of n keys is at most $3.82 \log n$ with probability $1 - o(1)$.*

Proof. As in the proof of Theorem 4, we note that the depth, X, in T of the i-th smallest key, x_i, can be characterized as follows. Let

$$L_i = \sum_{1 \le j < i} X_j, \quad \text{and} \quad R_i = \sum_{i < j \le n} X_j,$$

where X_j is a 0–1 random variable that is 1 if and only if x_j is an ancestor of x_i, where x_i is the i-th smallest key in S and x_j is the j-th smallest key. Then $X = 1 + L_i + R_i$. Further, note that the random variables that are summed in L_i (or, respectively, R_i) are independent, and, focusing on $E[X|C]$, as in the proof of Theorem 4, $E[L_i] = H_i - 1$ and $E[R_i] = H_{n-i+1} - 1$, where $H_m = \sum_{k=1}^{m} 1/k$ is the m-th Harmonic number; hence, $E[X|C] = H_i + H_{n-i+1} - 1 < 2 \ln n - 1$. Thus, we can apply a Chernoff bound to characterize X by bounding L_i and

R_i separately (w.l.o.g., we focus on L_i), conditioned on C holding. For example, for the high-probability bound for the proof, it is sufficient that, for some small constant, $\varepsilon > 0$, there is a reasonably small $\delta > 0$ such that

$$P(L_i > (1+\delta)\ln n) < 2^{-((1+\varepsilon)/\ln 2)(\ln 2)\log n} = 2^{-(1+\varepsilon)\log n} = 1/n^{1+\varepsilon},$$

which would establish the theorem by a union bound. In particular, we choose $\delta = 1.75$ and let $\mu = E[L_i]$. Then by a Chernoff bound, e.g., see [3,14,17,18,24], for $\mu = \ln n$, we have the following:

$$\begin{aligned}
\Pr(L_i > 2.75\ln n) &= \Pr(L_i > (1+\delta)\mu) \\
&< \left(\frac{e^{\delta}}{(1+\delta)^{1+\delta}}\right)^{\mu} \\
&= \left(\frac{e^{1.75}}{2.75^{2.75}}\right)^{\ln n} \\
&\le 2.8^{-(\ln 2)\log n} \\
&\le 2.04^{-\log n} \\
&= \frac{1}{n^{\log 2.04}},
\end{aligned}$$

which establishes the above bound for $\varepsilon = \log_2 2.04 - 1 > 0$. Combining this with a similar bound for R_i, and the derived from Markov's inequality with respect to $E[X|A]$ and $E[X|B]$, given in the proof of Theorem 4 for the conditional events A and B, we get that the height of a zip-zip tree is at most

$$2(2.75)(\ln 2)\log n \le 3.82\log n,$$

with probability $1 - o(1)$. □

References

1. Acar, U.A.: Self-adjusting computation. Ph.D. thesis, Carnegie Mellon University (2005)
2. Afek, Y., Kaplan, H., Korenfeld, B., Morrison, A., Tarjan, R.E.: The CB tree: a practical concurrent self-adjusting search tree. Distrib. Comput. **27**(6), 393–417 (2014). https://doi.org/10.1007/s00446-014-0229-0
3. Alon, N., Spencer, J.H.: The Probabilistic Method, 4th edn. Wiley, Hoboken (2016)
4. Bagchi, A., Buchsbaum, A.L., Goodrich, M.T.: Biased skip lists. Algorithmica **42**, 31–48 (2005)
5. Bender, M.A., Conway, A., Farach-Colton, M., Kuszmaul, W., Tagliavini, G.: Tiny pointers. In: ACM-SIAM Symposium on Discrete Algorithms (SODA), pp. 477–508 (2023). https://doi.org/10.1137/1.9781611977554.ch21, https://epubs.siam.org/doi/abs/10.1137/1.9781611977554.ch21
6. Bent, S.W., Sleator, D.D., Tarjan, R.E.: Biased search trees. SIAM J. Comput. **14**(3), 545–568 (1985)
7. Devroye, L.: A note on the height of binary search trees. J. ACM **33**(3), 489–498 (1986)

8. Devroye, L.: Branching processes in the analysis of the heights of trees. Acta Inform. **24**(3), 277–298 (1987)
9. Driscoll, J.R., Sarnak, N., Sleator, D.D., Tarjan, R.E.: Making data structures persistent. J. Comput. Syst. Sci. **38**(1), 86–124 (1989). https://doi.org/10.1016/0022-0000(89)90034-2, https://www.sciencedirect.com/science/article/pii/0022000089900342
10. Eberl, M., Haslbeck, M.W., Nipkow, T.: Verified analysis of random binary tree structures. In: Avigad, J., Mahboubi, A. (eds.) ITP 2018. LNCS, vol. 10895, pp. 196–214. Springer, Cham (2018). https://doi.org/10.1007/978-3-319-94821-8_12
11. Flajolet, P., Odlyzko, A.: The average height of binary trees and other simple trees. J. Comput. Syst. Sci. **25**(2), 171–213 (1982)
12. Goodrich, M.T., Tamassia, R.: Algorithm Design and Applications. Wiley, Hoboken (2015)
13. Guo, B.N., Qi, F.: Sharp bounds for harmonic numbers. Appl. Math. Comput. **218**(3), 991–995 (2011). https://doi.org/10.1016/j.amc.2011.01.089, https://www.sciencedirect.com/science/article/pii/S009630031100124X
14. Hagerup, T., Rüb, C.: A guided tour of Chernoff bounds. Inf. Process. Lett. **33**(6), 305–308 (1990)
15. Hartline, J.D., Hong, E.S., Mohr, A.E., Pentney, W.R., Rocke, E.C.: Characterizing history independent data structures. Algorithmica **42**, 57–74 (2005)
16. Martínez, C., Roura, S.: Randomized binary search trees. J. ACM **45**(2), 288–323 (1998). https://doi.org/10.1145/274787.274812
17. Mitzenmacher, M., Upfal, E.: Probability and Computing: Randomization and Probabilistic Techniques in Algorithms and Data Analysis, 2nd edn. Cambridge University Press, Cambridge (2017)
18. Motwani, R., Raghavan, P.: Randomized Algorithms. Cambridge University Press, Cambridge (1995)
19. Papadakis, T., Ian Munro, J., Poblete, P.V.: Average search and update costs in skip lists. BIT Numer. Math. **32**(2), 316–332 (1992)
20. Pugh, W.: Skip lists: a probabilistic alternative to balanced trees. Commun. ACM **33**(6), 668–676 (1990). https://doi.org/10.1145/78973.78977
21. Reed, B.: The height of a random binary search tree. J. ACM **50**(3), 306–332 (2003)
22. Sarnak, N., Tarjan, R.E.: Planar point location using persistent search trees. Commun. ACM **29**(7), 669–679 (1986)
23. Seidel, R., Aragon, C.R.: Randomized search trees. Algorithmica **16**(4–5), 464–497 (1996)
24. Shiu, D.: Efficient computation of tight approximations to Chernoff bounds. Comput. Stat. **38**, 1–15 (2022)
25. Sleator, D.D., Tarjan, R.E.: A data structure for dynamic trees. In: 13th ACM Symposium on Theory of Computing (STOC), pp. 114–122 (1981)
26. Tarjan, R.E., Levy, C., Timmel, S.: Zip trees. ACM Trans. Algorithms **17**(4), 34:1–34:12 (2021). https://doi.org/10.1145/3476830

External-Memory Sorting with Comparison Errors

Michael T. Goodrich and Evrim Ozel[(✉)]

University of California, Irvine, USA
{goodrich,eozel}@uci.edu

Abstract. We provide several algorithms for sorting an array of n comparable distinct elements subject to probabilistic comparison errors in external memory. In this model, which has been extensively studied in internal-memory settings, the comparison of two elements returns the wrong answer according to a fixed probability, $p_e < 1/2$, and otherwise returns the correct answer. The ***dislocation*** of an element is the distance between its position in a given (current or output) array and its position in a sorted array. There are various existing algorithms that can be utilized for sorting or near-sorting elements subject to probabilistic comparison errors, but these algorithms do not translate into efficient external-memory algorithms, because they all make heavy use of noisy binary searching. In this paper, we provide new efficient methods that are in the external-memory model for sorting with comparison errors. Our algorithms achieve an optimal number of I/Os, in both cache-aware and cache-oblivious settings.

1 Introduction

Given n distinct comparable elements, we study the problem of efficiently sorting them subject to noisy probabilistic comparisons. In this framework, which has been extensively studied in internal-memory settings [2,5,7–9,11–16,21], the comparison of two elements, x and y, results in a true and accurate result independently according to a fixed probability, $p < 1/2$, and otherwise returns the opposite (false) result. In the case of ***persistent*** errors [2,7–9,13], the result of a comparison of two given elements, x and y, always returns the same result. In the case of ***non-persistent*** errors [5,11,12,14,16,21], however, the probabilistic determination of correctness is determined independently for each comparison, even if it is for a pair of elements, (x, y), that were previously compared.

Motivation for sorting with comparison errors comes from multiple sources, including ranking objects online via A/B testing [22], which evaluates the impact of a new technology or technological choice by executing a system in a real production environment and testing two instances of its performance (an "A" and "B") on a random subset of the users of the platform. Such systems can involve many users and choices to compare via A/B testing, e.g., see [10,20]; hence, we feel that managing such implementations could benefit from external-memory solutions.

P. Morin and S. Suri (Eds.): WADS 2023, LNCS 14079, pp. 493–506, 2023.
https://doi.org/10.1007/978-3-031-38906-1_32

Since one cannot always correctly sort an array, A, subject to persistent comparison errors, we follow Geissmann *et al.* [7–9], and define the ***dislocation*** of an element, x, in an array, A, as the absolute value of the difference between x's index in A and its index in the correctly sorted permutation of A. Further, define the ***maximum dislocation*** of A as the maximum dislocation for the elements in A, and define the ***total dislocation*** of A is the sum of the dislocations of the elements in A. By known lower bounds [7–9], the best a sorting algorithm can achieve under persistent comparison errors is a maximum dislocation of $O(\log n)$ and a total dislocation of $O(n)$.

In this paper, we are interested in sorting algorithms that are in the ***external-memory*** model. Unfortunately, the existing algorithms for sorting with noisy comparisons are not easily converted into efficient external-memory algorithms, because they all make use of noisy binary search, which involves a random walk in a binary search tree [5,8,14]. Instead, we desire efficient sorting algorithms that tolerate noisy comparisons and have an efficient number of input/output operations, primarily for the persistent model, since we can sort an array with maximum dislocation of $O(\log n)$ in the non-persistent model by a single scan where we repeat each comparison in internal memory $O(\log n)$ times.

Intuitively, the main disadvantage of relying on noisy binary search is that it is cache inefficient, in that it requires performing memory accesses for widely-distributed storage locations. Large-scale applications need to minimize the number of input/output (I/O) operations to external memory. Thus, we also desire sorting algorithms that tolerate noisy comparisons and minimize the number of I/Os. In external-memory applications, I/Os occur in terms of memory blocks. In this context, we use M to denote the size of internal memory and B to denote the size of a block of memory, and we note that the best I/O bound that is possible for sorting an array of size N in external-memory is $\Theta((N/B)\log_{M/B}(N/B))$, see, e.g., [19]. Thus, we also desire sorting algorithm that tolerate noisy comparisons and have this bound on their number of I/Os. Moreover, we desire solutions that are either cache-aware (taking advantage of knowledge of the parameters M and B) or cache-oblivious (which don't know the parameters M and B).

Related Prior Results. The non-persistent error model traces back to a classic problem by Rényi [18] of playing a game involving posing questions to someone who lies with a given probability; see, e.g., the survey by Pelc [17]. Braverman and Mossel [2] introduced the persistent-error model, where comparison errors are persistently wrong with a fixed probability, $p < 1/2 - \varepsilon$, and they achieved a running time of $O(n^{3+f(p)})$ time with maximum expected dislocation $O(\log n)$ and total dislocation $O(n)$. Klein, Penninger, Sohler, and Woodruff [13] improve the running time to $O(n^2)$, but with $O(n \log n)$ total dislocation w.h.p. The internal-memory running time for sorting in the persistent-error model optimally with respect to maximum and total dislocation was subsequently improved to $O(n^2)$, $O(n^{3/2})$, and ultimately to $O(n \log n)$, in a sequence of papers by Geissmann, Leucci, Liu, and Penna [7–9].

Feige, Raghavan, Peleg, and Upfal [5] provide a parallel algorithm for sorting with non-persistent errors that, with high probability, runs in $O(\log n)$ time and

$O(n \log n)$ work in the CRCW PRAM model, and Leighton, Ma, and Plaxton [14] show how to achieve these bounds in the EREW PRAM model.

None of these prior algorithms translate into efficient external-memory algorithms, however, where we focus on optimizing the number of input/output (I/O) operations. The main reason is that they all use noisy binary searching, which is a random walk in a binary search tree, where each step involves a noisy comparison. As an external-memory algorithm, this search algorithm unfortunately involves far-flung comparisons; hence, it causes a lot of I/Os.

Frigo, Leiserson, Prokop and Ramachandran [6] introduced the notion of cache-oblivious algorithms, which are algorithms that do not have any variables dependent on hardware parameters such as cache or block size that need to be tuned for it to perform optimally. The authors also introduced the ***(M,B) ideal-cache model*** to analyze cache oblivious algorithms, and defined the ***cache complexity*** $Q(n)$ and ***work complexity*** $W(n)$ of an algorithm with input size n, which respectively measure the number of cache misses the algorithm incurs in the ideal-cache model, and the conventional running time of the algorithm in a RAM model. The authors then introduced a cache-oblivious sorting algorithm, Funnelsort, and showed, assuming $M = \Omega(B^2)$ (also known as the ***tall-cache assumption***), that Funnelsort is cache-oblivious, has work complexity $O(n \log n)$ and cache complexity $O(1 + \frac{n}{B}(1 + \log_M n))$, which matches the $\Omega(\frac{n}{B} \log_{M/B} \frac{n}{B})$ lower bound for sorting in the external-memory model.

Our Results. In this paper, we provide efficient sorting algorithms in the external-memory model that tolerate noisy comparisons. All our algorithms utilize an optimal number of I/Os. In particular, we provide solutions for either the persistent or non-persistent error models, and for the cache-aware and cache-oblivious external-memory models. Our algorithms avoid using noisy binary searching by instead utilizing a generalized subroutine that is an external-memory version of window-sort. This allows us to then design windowed versions of external-memory merge-sort and funnel-sort. Both algorithms run in time $O(n \log^2 n)$ in internal memory, or in external memory with an optimal $O(n/B) \log_{M/B}(n/B)$ I/O's, subject to comparison errors with probability $p_e < 1/2$ so as to have a maximum dislocation of $O(\log n)$ w.h.p. For both algorithms, we assume that the block size is at least logarithmic in the problem size, i.e., $B = \Omega(\log n)$. Our windowed version of funnel-sort will also use the tall-cache assumption, i.e., $M = \Omega(B^2)$. In the sections that follow, we describe our algorithms for sorting with comparison errors.

2 Window-Sort

We begin with a version of window-sort [8], which will be useful as a subroutine in our algorithms. We provide the pseudo-code at a high level in Algorithm 1, for approximately sorting an array of size n that has maximum dislocation at most $d_1 \leq n$ so that it will have maximum dislocation at most $d_2 = d_1/2^k$, for some integer $k \geq 1$, with high probability as a function of d_2.

Algorithm 1: Window-Sort($A = \{a_0, a_1, \ldots, a_{n-1}\}, d_1, d_2$)

1 **for** $w \leftarrow 2d_1, d_1, d_1/2, \ldots, 2d_2$ **do**
2   **foreach** $i \leftarrow 0, 1, 2, \ldots, n-1$ **do**
3         $r_i \leftarrow \max\{0, i-w\} + |\{a_j < a_i \;:\; |j-i| \leq w\}|$
4   Sort A (deterministically) by nondecreasing r_i values (i.e., using r_i as the comparison key for a_i)
5 **return** A

We note that determining the r_i values can be done by scans; hence, that step is I/O efficient in either cache-aware or cache-oblivious settings. Moreover, the sorting step can be done with an I/O efficient algorithm, in either the cache-aware or cache-oblivious settings, e.g., see [3,4,6]. For completeness, we provide below an analysis of window-sort. We note that to simplify our presentation, we assume $p_e \leq 1/16$, however this constraint can be relaxed to any $p_e < 1/2$ to obtain the same asymptotic results.

Lemma 1. *Suppose the comparison error probability, p_e, is at most* $1/16$. *If an array, A, has maximum dislocation at most d' prior to an iteration of window-sort for $w = 2d'$ (line 1 of Algorithm 1), then after this iteration, A will have maximum dislocation at most $d'/2$ with probability at least* $1 - n2^{-d'/8}$.

Proof. Let a_i be an element in A. Let W denote the window of elements in A for which we perform comparisons with a_i in this iteration; hence, $2d' \leq |W| \leq 4d'$. Because A has maximum dislocation d', by assumption, there are no elements to the left (resp., right) of W that are greater than a_i (resp., less than a_i). Thus, a_i's dislocation after this iteration depends only on the comparisons between a_i and elements in its window. Let X be a random variable that represents a_i's dislocation after this iteration, and note that $X \leq Y$, where Y is the number of incorrect comparisons with a_i performed in this iteration. Note further that we can write Y as the sum of $|W|$ independent indicator random variables and that $\mu = E[Y] = p_e|W| \leq d'/4$. Thus, if we let $R = d'/2$, then $R \geq 2\mu$; hence, we can use a Chernoff bound as follows:

$$\Pr(X > d'/2) \leq \Pr(Y > d'/2) = \Pr(Y > R) \leq 2^{-R/4} = 2^{-d'/8}.$$

Thus, with the claimed probability, the maximum dislocation for all of A will be at most $d'/2$, by a union bound. □

This allows us to implement window-sort in external memory, as follows.

Theorem 1. *Suppose the comparison error probability, p_e, is at most* $1/16$. *If an array, A, of size n has maximum dislocation at most $d_1 \geq \log n$, then executing Window-Sort(A, d_1, d_2) runs in $O(d_1 n)$ time in internal memory. It can be implemented in external memory with $O(n/B)$ I/Os if $n \leq M$; otherwise, it* *can be implemented with* $O((nd_1/B) + (\log(d_1/d_2))(n/B)\log_{M/B}(n/B))$ *I/Os. Executing Window-Sort*(A, d_1, d_2) *results in* A *having maximum dislocation of* $d_2/2$ *with probability at least* $1 - 2n2^{-d_2/8}$, *where* $d_2 = d_1/2^k$, *for some integer* $k \geq 1$.

Proof. For the internal-memory running time, note that we can perform the deterministic sorting step using any efficient sorting algorithm in $O(n \log n)$ time. The running times for the windowed comparison steps (step 3 of Algorithm 1) form a geometric sum adding up to $O(d_1 n)$ and the total time for all the deterministic sorting steps (step 4 of Algorithm 1) is $O((\log(d_1/d_2))n \log n)$, which is at most $O(d_1 n)$ for $d_1 \geq \log n$. For the external-memory model in both the cache-aware and cache-oblivious settings, a cache-efficient sorting algorithm can be used, requiring at most $O(\log(d_1/d_2))(n/B)\log_{M/B}(n/B))$ I/Os for all the sorting steps. The scanning step can also be done in an cache efficient way, requiring at most $O(nd_1/B)$ I/Os.

For the maximum dislocation bound, note once $w - 2d_2$ and the array A prior to this iteration has maximum dislocation at most d_2, then it will result in having maximum dislocation at most $d_2/2$ with probability at least $1 - n2^{-d_2/8}$, by Lemma 1. Thus, by a union bound, the overall failure probability is at most

$$n\left(2^{-d_2/8} + 2^{-2d_2/8} + 2^{-4d_2/8} + \cdots + 2^{-d_1/8}\right) < n2^{-d_2/8}\sum_{i=0}^{\infty} 2^{-i} = 2n2^{-d_2/8}.$$

□

3 Window-Merge-Sort

In this section, we describe a simple external-memory algorithm for sorting with noisy comparisons, which achieves a maximum dislocation of $O(\log n)$. The number of I/Os for this algorithm is optimal. As is common (see, e.g., [1]), we assume that the block size is at least logarithmic in the problem size, i.e., $B \geq \log n$.

Our window-merge-sort method is a windowed version of merge sort; hence, it is deterministic. Suppose we are given an array, A, of n elements (to keep track of the original input size, we use n to denote the original size of A, and N to denote the size of the subproblem we are currently working on recursively). We take as input another parameter d, which determines the resulting maximum dislocation after running the algorithm.

For expository reasons, we first describe an internal-memory method that runs in $O(n \log^2 n)$ time and then we show how to generalize this method to an efficient external-memory method that uses an optimal number of I/Os. We give the pseudo-code for this method in Algorithm 2, with $d = c \log n$ for a constant $c \geq 1$ set in the analysis.

Algorithm 2: Window-Merge-Sort($A = \{a_0, a_1, \dots, a_{N-1}\}, n, d$)

```
1  if N ≤ 6d then
2      return Window-Sort(A, 4d, d)
3  Divide A into two subarrays, A1 and A2, of roughly equal size
4  Window-Merge-Sort(A1, n, d)
5  Window-Merge-Sort(A2, n, d)
6  Let B be an initially empty output list
7  while |A1| + |A2| > 6d do
8      Let S1 be the first min{3d, |A1|} elements of A1
9      Let S2 be the first min{3d, |A2|} elements of A2
10     Let S ← S1 ∪ S2
11     Window-Sort(S, 4d, d)
12     Let B' be the first d elements of (the near-sorted) S
13     Add B' to the end of B and remove the elements of B' from A1 and A2
14 Call Window-Sort(A1 ∪ A2, 4d, d) and add the output to the end of B
15 return B
```

Our method begins by checking if the current problem size, N, satisfies $N \leq 6d$, in which case we're done. Otherwise, if $N > 6d$, then we divide A into 2 subarrays, A_1 and A_2, of roughly equal size and recursively approximately sort each one. For the merge of the two sublists, A_1 and A_2, we inductively assume that A_1 and A_2 have maximum dislocation at most $3d/2 = (3c/2)\log n$. We then copy the first $3d$ elements of A_1 and the first $3d$ elements of A_2 into a temporary array, S, and we note that, by our induction hypothesis, S contains the smallest $3d/2$ elements currently in A_1 and the smallest $3d/2$ elements currently in A_2. We then call Window-Sort($S, 4d, d$), and copy the first d elements from the output of this window-sort to the output of the merge, removing these same elements from A_1 and A_2. Then we repeat this merging process until we have at most $6d$ elements left in $A_1 \cup A_2$, in which case we call window-sort on the remaining elements and copy the result to the output of the merge. The following lemma establishes the correctness of this algorithm.

Lemma 2. *If A_1 and A_2 each have maximum dislocation at most $3d/2$, then the result of the merge of A_1 and A_2 has maximum dislocation at most $3d/2$ with probability at least $1 - 12N2^{-d/8}$.*

Proof. By Lemma 1 and a union bound, each of the calls to window-sort performed during the merge of A_1 and A_2 will result in an output with maximum dislocation at most $d/2$, with at least the claimed probability. So, let us assume each of the calls to window-sort performed during the merge of A_1 and A_2 will result in an output with maximum dislocation at most $d/2$. Consider, then, merge step i, involving the i-th call to Window-Sort($S, 4d, d$), where S consists of the current first $3d$ elements in A_1 and the current first $3d$ elements in A_2, which, by assumption, contain the current smallest $3d/2$ elements in A_1 and current smallest $3d/2$ elements in A_2. Thus, since this call to window-sort results in an array with maximum dislocation at most $d/2$, the subarray, B_i, of the d elements moved to the output in step i includes the $d/2$ current smallest elements in $A_1 \cup A_2$. Moreover, the first $d/2$ elements in B_i have no smaller elements that remain in S. In addition, for the $d/2$ elements in the second half of B_i, let S' denote the set of elements that remain in S that are smaller than at least one of these $d/2$ elements. Since the output of Window-Sort$(S, 4d, d)$ has maximum dislocation at most $d/2$, we know that $|S'| \leq d/2$. Moreover, the elements in S' are a subset of the smallest $d/2$ elements that remain in S and there are no elements in $(A_1 \cup A_2) - S$ smaller than the elements in S' (since S includes the $3d/2$ smallest elements in A_1 and A_2, respectively. Thus, all the elements in S' will be included in the subarray, B_{i+1}, of d elements output in merge step $i+1$. In addition, a symmetric argument applies to the first $d/2$ elements with respect to the d elements in B_{i-1}. Therefore, the output of the merge of A_1 and A_2 will have maximum dislocation at most $3d/2$ with the claimed probability. □

As an internal-memory algorithm, window-merge-sort runs in $O(n \log^2 n)$ time. To convert this algorithm to an external-memory one, we just need to make a few changes. First, rather than divide A into 2 subarrays for the recursive calls, we divide A into $m = \Theta(M/B) \geq 2$ subarrays, $A_1, A_2, \ldots, A_m$, each of roughly equal size, and recursively sort each one. For the merge step, we bring the first $\max\{3d, |A_i|\}$ elements from each A_i, group them together into a list, S, and call Window-Sort$(S, 4md, d)$ on this list, performing this computation entirely in internal memory (so it does not require any additional I/Os). Then we output the first d elements from this window-sort, and continue as in Algorithm 2. This implies the following.

Lemma 3. *If $A_1, A_2, \ldots, A_m$ each have maximum dislocation at most $3d/2$, then the result of the their merge has maximum dislocation at most $3d/2$ with probability at least $1 - 6mN2^{-d/8}$.*

Proof. The proof follows by similar arguments used in the proof of Lemma 2. □

This gives us the following.

Theorem 2. *Given an array, A, of n distinct comparable elements, one can deterministically sort A in internal memory in $O(n \log^2 n)$ time or in external memory with $O((n/B) \log_{M/B}(n/B))$ I/Os subject to comparison errors with probability $p_e \leq 1/16$, so as to have maximum dislocation of $O(\log n)$ w.h.p., assuming $B \geq \log n$.*

4 Window Funnelsort

In this section we describe WindowFunnelSort (see Algorithm 3), a noise-tolerant version of the Funnelsort algorithm that sorts n distinct comparable elements so as to have at most $O(\log n)$ maximum dislocation, with $W(n) = O(n \log^2 n)$ work complexity and $Q(n) = O(1 + (n/B)(1 + \log_M n))$ cache complexity, which matches the lower bound of $\Omega(\frac{n}{B} \log_{M/B} \frac{n}{B})$ for sorting in the external-memory model. In our pseudocode, n denotes the original input size, while N denotes the size of the input array given to each function call, which can be less than n during recursive calls.

We require a stronger assumption on the cache size for our algorithm: in addition to the tall-cache assumption $M = \Omega(B^2)$, we also require that the block size B be at least logarithmic in the problem size, i.e. $B \geq \gamma \log n$ for some constant $\gamma > 0$ that will be determined in the analysis. In our analysis, we follow the same general proof structure used in [6] with the ideal cache model. For the remainder of this section, we assume that the maximum dislocation bound we would like to obtain, d, is $(3c/2)\log n$ for some constant $c > 0$ that will be determined later. The main difference in our analysis compared to [6] is that in the recursive definition of a k-merger, we define the base cases differently such that each base case k-merger will now use a similar merging method to the one used in Algorithm 2, and our base cases are defined over multiple values of k, instead of just $k = 2$ as done in [6].

Algorithm 3: Window-Funnel-Sort($A = \{a_0, a_1, \ldots, a_{N-1}\}, n$)

1 **if** $N \leq (c \log n)^{3/2}$ **then**
2   **return WindowSort**($A, N, c \log n$)
3 Divide A into $N^{1/3}$ subarrays, $A_1 \ldots, A_{N^{1/3}}$, each of size $N^{2/3}$
4 **for** $i = 1, \ldots, N^{1/3}$ **do**
5   A_i = WINDOW-FUNNEL-SORT(A_i)
6 $A \leftarrow$ output of merging $A_1 \ldots, A_{N^{1/3}}$ using a $N^{1/3}$-merger, as described in Section 4
7 **return** A

We first describe how to construct a k-merger, which is defined recursively in terms of smaller mergers. We follow the same general structure for constructing a k-merger in the original Funnelsort algorithm [6], however in the recursive definition of a k-merger, instead of having $k = 2$ as the base case, we view k-mergers with $\sqrt{c \log n} \leq k < c \log n$ as base cases. As an invariant, each k-merger outputs the next k^3 elements of the approximately sorted sequence obtained by merging its k input sequences.

Our base case k-merger works similarly to the merging procedure in Algorithm 2. We read in $3c \log n$ elements from each of the k inputs into an array S, call WINDOWSORT($S, 4kc \log n, c \log n$), then output the last $c \log n$ elements from this call. Then, we replace the $c \log n$ elements in the k-merger that were just written to the output as follows: for each element e written to the output, we read into the k-merger a new element from the input queue that e belonged to. We then call WINDOWSORT again on this updated set of elements, and repeat this process until the k-merger has outputted k^3 elements.

For all other values of $k \geq c \log n$, a k-merger will work the same way as in [6], which we describe here for completeness. A (non-base case) k-merger is built recursively out of $\sqrt{k}$-mergers by first partitioning the k inputs into $\sqrt{k}$ sets of $\sqrt{k}$ elements, which forms the input to $\sqrt{k}$ left mergers $L_1, L_2, \ldots, L_{\sqrt{k}}$, each of which is a $\sqrt{k}$-merger. Each L_i is connected to an output buffer i, implemented as a circular queue that can hold up to $2k^{3/2}$ elements. Each buffer is then connected as input to R, which is another $\sqrt{k}$-merger. The output of R then becomes the output of the whole k-merger. Following our invariant, in order to output k^3 elements, the k-merger will invoke R $k^{3/2}$ times. Since the input queues connected to R might become empty, the k-merger first fills all buffers that have less than $k^{3/2}$ elements before each invocation of R, which is done by invoking the corresponding left merger L_i that connects to buffer i. Since each left merger invocation will output $k^{3/2}$ elements to the corresponding buffer, each L_i will only need to be invoked at most once before each invocation of R.

Let us first consider the cache complexity of WINDOWFUNNELSORT. Following the proof in [6], we first consider how much space a k-merger requires.

Lemma 4. *A k-merger requires at most $O(k^2)$ contiguous memory locations when $k \geq c\log n$.*

Proof. A k-merger with $k \geq c\log n$ requires $O(k^2)$ memory locations for the buffers, and it also requires space for its $\sqrt{k}+1$ $\sqrt{k}$-mergers. Thus, the space $S(k)$ required by a k-merger satisfies the recurrence relation

$$S(k) \leq (\sqrt{k}+1)S(\sqrt{k}) + \beta k^2,$$

for some constant $\beta > 0$. We prove inductively that $S(k) \leq Zk^2$ for some constant Z. For k-mergers with $\sqrt{c\log n} \leq k < c\log n$, we will read in $c\log n$ elements from k input queues, then perform WINDOWSORT on them, requiring $S(k) = O(k\log n)$ space for some constant $\beta > 0$. Thus for $c\log n \leq k < (c\log n)^2$, we have $S(k) \leq (\sqrt{k}+1)O(\sqrt{k}\log n) + \beta k^2 \leq Zk^2$ for sufficiently large Z.

For $k \geq (c\log n)^2$, we inductively have

$$\begin{aligned} S(k) &\leq (\sqrt{k}+1)S(\sqrt{k}) + \beta k^2 \\ &\leq (\sqrt{k}+1)Zk + \beta k^2 \leq Zk^2 \end{aligned}$$

for sufficiently large Z. Thus we have $S(k) = O(k^2)$ for any $k \geq c\log n$. □

Any k-merger with $\sqrt{c\log n} \leq k < c\log n$ reads in $3c\log n$ elements from less than $c\log n$ inputs, and will require $O(\log^2 n)$ space in total. Therefore we require that the block size B is at least $\gamma\log n$ for an appropriate constant $\gamma > 0$ such that after applying the tall-cache assumption $M = \Omega(B^2)$, any k-merger with $\sqrt{c\log n} \leq k < c\log n$ will fit inside the cache. Therefore, more generally, through Lemma 4, any k-merger with $\sqrt{c\log n} \leq k \leq c\log n \leq \alpha\sqrt{M}$, where α is a sufficiently small constant, will also fit inside the cache and run without any additional cache misses.

The following lemma, which is proved in [6], shows that the $\sqrt{k}$ buffers used in a k-merger can be managed cache-efficiently.

Lemma 5 (Lemma 4.2. in [6]). Performing r insert and remove operations on a circular queue causes $O(1+r/B)$ cache misses if two cache lines are available for the buffer.

We now bound the cache complexity Q_k of one invocation of a k-merger.

Lemma 6 *One invocation of a k-merger incurs*

$$Q_k = O(k + k^3/B + k^3 \log_M k/B)$$

cache misses.

Proof We first consider the case $\sqrt{c \log n} \leq k \leq \alpha\sqrt{M}$. From Lemma 4 and our assumption on the cache size, we know that any k-merger with $\sqrt{c \log n} \leq k \leq \alpha\sqrt{M}$ will fit inside the cache and run with no additional cache misses. Each k-merger has k input queues, and loads a total of $O(k^3)$ elements. Let r_i denote the number of elements extracted from the ith queue. Since $k \leq \alpha\sqrt{M}$ and $B = O(\sqrt{M})$, there are at least $M/B = \Omega(k)$ cache lines available for the input buffers. Thus, through Lemma 5, the total number of cache misses for accessing the input queues is

$$\sum_{i=1}^{k} O(1 + r_i/B) = O(k + k^3/B).$$

Similarly, the cache complexity of writing to the output queue is $O(1 + k^3/B)$. The k-merger incurs an additional $O(k^2/B)$ cache misses through using its internal data structures, for a total of $Q_k = O(k + k^3/B)$ cache misses.

We then consider the case $k > \alpha\sqrt{M}$. We prove by induction that $Q(k) \leq (Zk^3 \log_M k)/B - A(k)$ for some constant $Z > 0$, where $A(k) = o(k^3)$. We first verify that values of $\alpha M^{1/4} < k \leq \alpha\sqrt{M}$ also satisfy this inequality: from the first case, we have $Q(k) = O(k + k^3/B) = O(k^3/B)$ since $B = O(\sqrt{M}) = O(k^2)$ and $k = \Omega(1)$.

For $k > \alpha\sqrt{M}$, for a k-merger to output k^3 elements, the number of times the left mergers are invoked is bounded by $k^{3/2} + 2\sqrt{k}$. The right merger R is also invoked $k^{3/2}$ times. The k-merger also has to check before each invocation of R whether any of the buffers are empty. This requires at most $\sqrt{k}$ cache misses and is repeated exactly $k^{3/2}$ times, for a total of at most k^2 cache misses. Therefore the cache complexity Q_k of a k-merger satisfies the following recurrence relation:

$$\begin{aligned} Q_k &\leq (2k^{3/2} + 2\sqrt{k})Q_{\sqrt{k}} + k^2 \\ &\leq (2k^{3/2} + 2\sqrt{k})\left(\frac{Zk^{3/2} \log_M k}{2B} - A(\sqrt{k})\right) + k^2 \\ &\leq \frac{Z}{B} k^3 \log_M k + k^2\left(1 + \frac{Z}{B} \log_M k\right) - (2k^{3/2} + 2\sqrt{k})A(\sqrt{k}), \end{aligned}$$

which is at most $(Zk^3 \log_M k)/B - A(k)$ if $A(k) = k(1 + (2Z \log_M k)/B)$. $\square$

Theorem 3 WindowFunnelSort *incurs at most $Q(n)$ cache misses, where*

$$Q(n) = O(\frac{n}{B}\log_{M/B}\frac{n}{B}).$$

Proof If $n \leq \alpha M$ for a sufficiently small constant α, the algorithm will incur at most $O(1 + n/B)$ cache misses, since only one k-merger will be active at any time, and the largest possible k-merger will require $O(n^{2/3}) < O(n)$ space. This case also covers the base case in Line 1 of Algorithm 3 through our assumption on the cache size.

If $n > \alpha M$, have the recurrence

$$Q(n) = n^{1/3}Q(n^{2/3}) + Q_{n^{1/3}}.$$

From Lemma 6, we have $Q_{n^{1/3}} = O(n^{1/3} + n/B + (n \log_M n)/B)$. Therefore the recurrence simplifies to

$$Q(n) = n^{1/3}Q(n^{2/3}) + O((n \log_M n)/B),$$

which has solution $Q(n) = O(1+(n/B)(1+\log_M n))$ by induction, which matches the $\Omega(\frac{n}{B}\log_{M/B}\frac{n}{B})$ lower bound for sorting in the external-memory model. □

We now prove that windowFunnelSort is tolerant to persistent comparison errors.

Lemma 7 *Given k input queues with maximum dislocation at most $\frac{3}{2}c\log n$ for some constant $c > 0$, one invocation a k-merger outputs k^3 elements with maximum dislocation at most $\frac{3}{2}c\log n$ with probability at least $1 - 2Zk^3(c\log n)^5 2^{-(c\log n)/8}$ for some constant $Z > 0$.*

Proof We first consider k-mergers with $\sqrt{c\log n} \leq k < c\log n$. Each such k-merger will call WindowSort $\frac{k^3}{c\log n} < (c\log n)^2$ times, with each call working on at most $(c\log n)^2$ elements. Therefore, using a similar argument to Lemmas 2 and 3 and a union bound, the resulting sequence after $(c\log n)^2$ calls to windowSort will have maximum dislocation at most $(3c/2)\log n$ with probability at least $1 - 2(c\log n)^4 2^{-(c\log n)/8}$.

We then consider the case $k \geq c\log n$. We have $\sqrt{k}$ left $\sqrt{k}$-mergers, along with a $\sqrt{k}$-merger R. Each left merger inductively outputs $k^{3/2}$ elements with dislocation at most $\frac{3}{2}c\log n$, which is used as the input to the $\sqrt{k}$-merger R that also inductively outputs $k^{3/2}$ elements with dislocation at most $\frac{3}{2}c\log n$. Using a similar argument to Lemma 2, the output queue of the k-merger will also have dislocation at most $\frac{3}{2}c\log n$. To find the success probability, we consider the number of times windowSort is called. Since the number of invocations of smaller k-mergers is bounded by $2k^{3/2} + 2\sqrt{k}$, the number of invocations of windowSort, $I(k)$, satisfies the recurrence relation

$$I(k) = \begin{cases} (2k^{3/2} + 2\sqrt{k})I(\sqrt{k}) & k \geq c\log n \\ 1 & \sqrt{c\log n} \leq k < c\log n, \end{cases}$$

which has solution $I(k) = Zk^3 \log k$ for some constant $Z > 0$ using a similar derivation to the one in Lemma 6. Therefore, using a union bound, the probability that a k-merger outputs k^3 elements with maximum dislocation at most $\frac{3}{2}c \log n$ is at least $1 - 2Zk^3(c \log n)^5 2^{-(c \log n)/8}$. □

Theorem 4 *Given an array A of n distinct comparable elements, and assuming $B = \Omega(\log n)$, one can deterministically sort A subject to comparison errors with probability $p_e \leq 1/16$, so as to have maximum dislocation of at most $\frac{c}{2} \log n$ for some constant $c > 0$ w.h.p., with at most $O(\frac{n}{B} \log_{M/B} \frac{n}{B})$ cache misses in the cache-oblivious model, and taking $O(n \log^2 n)$ time in a RAM model.*

Proof By induction, each of the $n^{1/3}$ input sequences given to the $n^{1/3}$-merger has maximum dislocation at most $\frac{3c}{2} \log n$ w.h.p. From Lemma 7, we have that a $n^{1/3}$ merger outputs n elements with maximum dislocation at most $\frac{3c}{2} \log n$ with probability at least $1 - 2Zn(c \log n)^5 2^{-(c \log n)/8}$ for some constant $Z > 0$. Choosing an appropriate value for c establishes this theorem. □

We now bound the work complexity of WINDOWFUNNELSORT, by first bounding the work complexity W_k of a k-merger.

Lemma 8 *The work complexity W_k of one invocation of a k-merger is $O(k^3 \log^2 n)$.*

Proof We first consider k-mergers with $\sqrt{c \log n} \leq k < c \log n$. The k-merger reads $3c \log n$ elements from k input queues, each of which have maximum dislocation at most $O(\log n)$ from Theorem 4, for a total of $3kc \log n$ elements, then performs window-sort on these elements, which takes $O(k \log^2 n)$ time. To output k^3 elements, the k-merger needs to repeat this $O(\frac{k^3}{\log n})$ times, taking a total of $O(k^4 \log n)$ time, which is bounded by $O(k^3 \log^2 n)$ since $k < c \log n$.

For k-mergers where $k \geq c \log n$, to output k^3 elements, the left mergers and right merger are invoked at most $k^{3/2} + 2\sqrt{k}$ and $k^{3/2}$ times respectively. The k-merger also has to check before each invocation of R whether any of the buffers are empty. This takes $O(\sqrt{k})$ time and is repeated exactly $k^{3/2}$ times, for a total of $O(k^2)$ time. Therefore the total work complexity $W(k)$ of a k-merger satisfies the following recurrence relation:

$$W_k \leq (2k^{3/2} + 2\sqrt{k})W_{\sqrt{k}} + O(k^2).$$

Using a derivation similar to the one in Lemma 6, we can show that $W_k = O(k^3 \log^2 n)$ by induction. □

Theorem 5 *The work complexity $W(n)$ of* WINDOWFUNNELSORT *is $O(n \log^2 n)$ for any input sequence of n elements.*

Proof We have the recurrence

$$W(n) = n^{1/3} W(n^{2/3}) + W_{n^{1/3}}.$$

From Lemma 8, we have $W_{n^{1/3}} = O(n \log^2 n)$. Therefore the recurrence simplifies to

$$W(n) = n^{1/3} W(n^{2/3}) + O(n \log^2 n),$$

which has solution $W(n) = O(n \log^2 n)$ by induction. □

5 Conclusions and Future Work

We provided efficient sorting algorithms that tolerate noisy comparisons and are cache efficient in both cache-aware and cache-oblivious external memory models. In [6], the authors introduced another cache-oblivious sorting algorithm based on distribution-sort, that has the same work and cache complexities as funnel-sort. One direction for future work could be to design and analyze a windowed version of the cache-oblivious distribution sort algorithm that has similar bounds on the work and cache complexities.

Acknowledgements. We would like to graciously thank Riko Jacob and Ulrich Meyer for several helpful discussions regarding the topics of this paper. This work was partially support by NSF Grant 2212129.

References

1. Bender, M., Demaine, E., Farach-Colton, M.: Cache-oblivious B-trees. In: 41st IEEE Symposium on Foundations of Computer Science (FOCS), pp. 399–409 (2000)
2. Braverman, M., Mossel, E.: Noisy sorting without resampling. In: 19th ACM-SIAM Symposium on Discrete Algorithms (SODA), pp. 268–276 (2008)
3. Brodal, G.S., Fagerberg, R.: Cache oblivious distribution sweeping. In: Widmayer, P., Eidenbenz, S., Triguero, F., Morales, R., Conejo, R., Hennessy, M. (eds.) ICALP 2002. LNCS, vol. 2380, pp. 426–438. Springer, Heidelberg (2002). https://doi.org/10.1007/3-540-45465-9_37
4. Brodal, G.S., Fagerberg, R., Vinther, K.: Engineering a cache-oblivious sorting algorithm. ACM J. Exp. Algorithmics **12**, 1–23 (2008). https://doi.org/10.1145/1227161.1227164
5. Feige, U., Raghavan, P., Peleg, D., Upfal, E.: Computing with noisy information. SIAM J. Comput. **23**(5), 1001–1018 (1994)
6. Frigo, M., Leiserson, C.E., Prokop, H., Ramachandran, S.: Cache-oblivious algorithms. ACM Trans. Algorithms **8**(1), 1–22 (2012). https://doi.org/10.1145/2071379.2071383
7. Geissmann, B., Leucci, S., Liu, C.H., Penna, P.: Sorting with recurrent comparison errors. In: Okamoto, Y., Tokuyama, T. (eds.) 28th International Symposium on Algorithms and Computation (ISAAC). LIPIcs, vol. 92, pp. 38:1–38:12 (2017)
8. Geissmann, B., Leucci, S., Liu, C.H., Penna, P.: Optimal sorting with persistent comparison errors. In: Bender, M.A., Svensson, O., Herman, G. (eds.) 27th European Symposium on Algorithms (ESA). LIPIcs, vol. 144, pp. 49:1–49:14 (2019)

9. Geissmann, B., Leucci, S., Liu, C.H., Penna, P.: Optimal dislocation with persistent errors in subquadratic time. Theory Comput. Syst. **64**(3), 508–521 (2020). This work appeared in preliminary form in STACS 2018
10. Gilotte, A., Calauzènes, C., Nedelec, T., Abraham, A., Dollé, S.: Offline A/B testing for recommender systems. In: Eleventh ACM International Conference on Web Search and Data Mining (WSDM), pp. 198–206. New York (2018). https://doi.org/10.1145/3159652.3159687
11. Karp, R.M., Kleinberg, R.: Noisy binary search and its applications. In: 18th ACM-SIAM Symposium on Discrete Algorithms (SODA), pp. 881–890 (2007)
12. Khadiev, K., Ilikaev, A., Vihrovs, J.: Quantum algorithms for some strings problems based on quantum string comparator. Mathematics **10**(3), 377 (2022)
13. Klein, R., Penninger, R., Sohler, C., Woodruff, D.P.: Tolerant algorithms. In: Demetrescu, C., Halldórsson, M.M. (eds.) ESA 2011. LNCS, vol. 6942, pp. 736–747. Springer, Heidelberg (2011). https://doi.org/10.1007/978-3-642-23719-5_62
14. Leighton, T., Ma, Y., Plaxton, C.G.: Breaking the $\Theta(n \log^2 n)$ barrier for sorting with faults. J. Comput. Syst. Sci. **54**(2), 265–304 (1997)
15. Mao, C., Weed, J., Rigollet, P.: Minimax rates and efficient algorithms for noisy sorting. In: Janoos, F., Mohri, M., Sridharan, K. (eds.) Proceedings of Algorithmic Learning Theory. Proceedings of Machine Learning Research, vol. 83, pp. 821–847 (2018)
16. Pelc, A.: Searching with known error probability. Theor. Comput. Sci. **63**(2), 185–202 (1989)
17. Pelc, A.: Searching games with errors–fifty years of coping with liars. Theor. Comput. Sci. **270**(1), 71–109 (2002)
18. Rényi, A.: On a problem in information theory. Magyar Tud. Akad. Mat. Kutató Int. Közl. **6**, 505–516 (1961). https://mathscinet.ams.org/mathscinet-getitem?mr=0143666
19. Vitter, J.S.: External memory algorithms and data structures: dealing with massive data. ACM Comput. Surv. **33**(2), 209–271 (2001)
20. Wang, J., Huang, P., Zhao, H., Zhang, Z., Zhao, B., Lee, D.L.: Billion-scale commodity embedding for e-commerce recommendation in Alibaba. In: 24th ACM SIGKDD International Conference on Knowledge Discovery and Data Mining (KDD), pp. 839–848. Association for Computing Machinery, New York (2018). https://doi.org/10.1145/3219819.3219869
21. Wang, Z., Ghaddar, N., Wang, L.: Noisy sorting capacity. arXiv abs/2202.01446 (2022)
22. Xu, Y., Chen, N., Fernandez, A., Sinno, O., Bhasin, A.: From infrastructure to culture: A/B testing challenges in large scale social networks. In: 21th ACM SIGKDD International Conference on Knowledge Discovery and Data Mining (KDD), pp. 2227–2236 (2015)

Verifying the Product of Generalized Boolean Matrix Multiplication and Its Applications to Detect Small Subgraphs

Wing-Kai Hon[1(✉)], Meng-Tsung Tsai[2(✉)], and Hung-Lung Wang[3(✉)]

[1] Computer Science, National Tsing Hua University, Hsinchu, Taiwan
wkhon@cs.nthu.edu.tw

[2] Information Science, Academia Sinica, Taipei City, Taiwan
mttsai@iis.sinica.edu.tw

[3] Computer Science and Information Engineering, National Taiwan Normal University, Taipei City, Taiwan
hlwang@gapps.ntnu.edu.tw

Abstract. Given three n by n integer matrices A, B, and P, determining whether the product AB equals P can be done in randomized $O(n^2)$ time by Freivalds' algorithm. In this paper, we consider some generalized Boolean matrix multiplication $AB = P_f$, which is defined to be setting the entry p_{ij} of P_f for $i, j \in [n]$ as the value of a given function f of the entries on the ith row of A and jth column of B. We show that, for a family of functions f, it takes deterministic $O(n^2)$ time to verify whether the generalized product P_f contains only FALSE entries. Then, we present how to apply such a result to detect small subgraphs efficiently, including:

- Detect any designated colored 4-cycle in randomized $O(n^2)$ time for n-node k-edge-colored (for $k = O(1)$) complete graphs with maximum color degree at most 2 (including 1-edge-colored general graphs), which unifies several independently discovered algorithms each corresponding to a customized Ramsey-type theorem. As a complementary result, we show that if the maximum color degree is at least 3, any combinatorial algorithm that solves this problem requires $\Omega(n^{3-\varepsilon})$ time for any constant $\varepsilon > 0$, assuming the hardness of triangle detecting.
- Detect any designated 4-node induced subgraph in randomized $O(n^2)$ time for n-node triangle-free graphs. In contrast, the known best algorithm for this problem on general graphs needs triangle time.

Keywords: Colored 4-cycles · Small Induced Subgraphs · Ramsey-type Theorems

This research was supported in part by the National Science and Technology Council under contract NSTC grants 109-2221-E-001-025-MY3, 110-2221-E-003-003-MY3, and 110-2221-E-007-043-MY3.

P. Morin and S. Suri (Eds.): WADS 2023, LNCS 14079, pp. 507–520, 2023.
https://doi.org/10.1007/978-3-031-38906-1_33

1 Introduction

Freivalds' algorithm [10] is a randomized algorithm that verifies with high probability whether the product XY of two n by n integer matrices X and Y equals a given matrix Z. The running time is $O(n^2)$, despite that the known fastest algorithm for matrix multiplication takes $O(n^{2.3729})$ time [2,5,26]. Since then, Freivalds' algorithm has become one of the main building blocks that verifies the correctness of large-scale matrix multiplications and several computational linear algebra problems, whose error sources may include software bugs, hardware logical errors, faulty communication, etc. See [9,11,19,25] for example.

It is still unknown whether Freivalds' algorithm can be efficiently derandomized [11], though the number of random bits required can be reduced from linear to logarithmic [4,20,23] and the failure probability can be greatly reduced at no cost of running time [18]. However, a recent work [11] shows that, if the matrix Z has at most k erroneous entries, then these entries can be identified and also corrected in deterministic $\tilde{O}(kn^2)$[1] time. The running time can be further reduced to $\tilde{O}(\sqrt{k}n^2 + k^2)$ as shown in [21].

We extend the product verification problem to the following *generalized Boolean matrix multiplication.* Let A, B be a pair of n by n Boolean matrices. For any matrix C, we denote by c_{ij} the entry of C at the intersection of the ith row and jth column. When C is Boolean, let $\bar{C}$ be the complement matrix of C; that is, $\bar{c}_{ij}$ is TRUE if and only if c_{ij} is FALSE. Let $[n]$ denote the set $\{1, 2, \ldots, n\}$. Let A_i be the set of indices of the TRUE entries in the ith row of A; that is, $A_i = \{k \in [n] : a_{ik} \text{ is TRUE}\}$. Let B_j be the set of indices of the TRUE entries in the jth column of B; that is, $B_j = \{k \in [n] : b_{kj} = \text{TRUE}\}$. The standard Boolan matrix multiplication $C = AB$ is defined to be setting the entries c_{ij} as the indicator whether the intersection of A_i and B_j is nonempty. Let Γ be any nonempty subset of the collection of the products with the first matrix from $\{A, \bar{A}\}$ and the second from $\{B, \bar{B}\}$; that is,

$$\Gamma \subseteq \{AB, A\bar{B}, \bar{A}B, \bar{A}\bar{B}\} \text{ and } \Gamma \neq \varnothing.$$

We define the generalized Boolean matrix multiplication $C = (A, B)_\Gamma$, with parameter Γ, to be setting

$$c_{ij} = \bigwedge_{D \in \Gamma} d_{ij} \text{ for } i, j \in [n].$$

In particular for $\Gamma = \{AB\}$, $(A, B)_\Gamma$ denotes the standard Boolean matrix multiplication.

Our main result is that:

Theorem 1. *For any two n by n Boolean matrices A and B, for any nonempty subset Γ of $\{AB, A\bar{B}, \bar{A}B, \bar{A}\bar{B}\}$, verifying whether the product $P = (A, B)_\Gamma$ contains only* FALSE *entries can be done in $O(n^2)$ time deterministically. Moreover, the algorithm knows a product in Γ that makes p_{ij}* FALSE *for each $i, j \in [n]$.*

Here we present some applications of our main theorem.

[1] $\tilde{O}(f(n))$ suppresses polylogarithmic terms in $f(n)$.

Detecting Any Designated Colored 4-Cycle

An application of our main result, i.e. Theorem 1, is detecting all distinct colored 4-cycles, up to isomorphism, for k-edge-colored (for $k = O(1)$) complete graphs with maximum color degree at most 2. See Fig. 1 for an example. Here is some notation for the description of our results. We use $\{u, v\}$ to denote an undirected edge that connects nodes u and v, and reserve (u, v) to denote an ordered tuple. We use $\ell : E \to \mathcal{L}$ to denote an edge label function. If $\ell(e) = r$ for $e \in E$, $r \in \mathcal{L}$, then we say that e is an r-edge. Let C_k bc a cycle of length k that consists of nodes $v_1, v_2, \ldots, v_k$ and edges $\{v_i, v_{i+1}\}$ for $i \in [k]$, assuming that $v_{k+1} \equiv v_1$. We say that C is a $c_1c_2 \ldots c_k$-cycle if $\ell(\{v_i, v_{i+1}\}) = c_i$ for every $i \in [k]$.

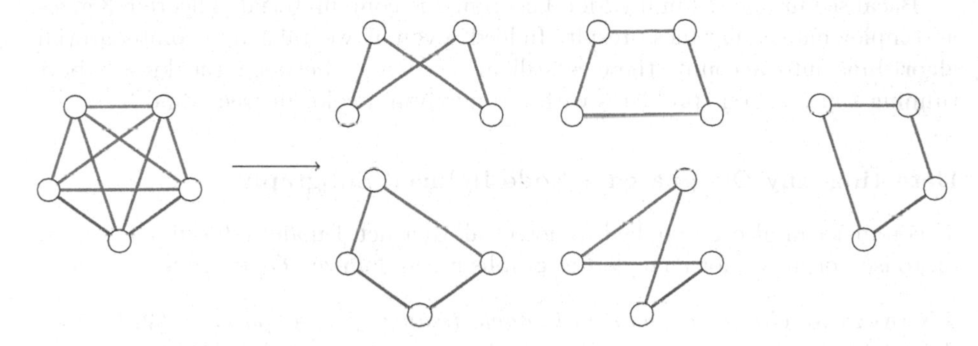

Fig. 1. All kinds of colored 4-cycles for a 3-edge-colored K_5 with maximum color degree 2. Note that in this example there are more than one 4-cycle for each kind. Our algorithm reports one for each.

Our result is:

Theorem 2. *Given an n-node complete graph $G = (V, E)$ and an edge label function $\ell : E \to \mathcal{L}$, for any $q_1, q_2, q_3, q_4 \in \mathcal{L}$, outputting a $q_1q_2q_3q_4$-cycle in G or reporting that G contains no such cycles can be done in deterministic $O(n^2 \log n)$ time or randomized $O(tn^2)$ time with failure probability 2^{-t} if for each node v there are at most two distinct labels on the edges incident with v.*

Theorem 2 unifies a number of deterministic $O(n^2)$-time algorithms that are independently discovered for detecting colored 4-cycles for n-node 2-edge-colored complete graphs [12,13,22,24,29,30]. For each colored 4-cycle except the monochromatic one, its detecting algorithm uses a customized Ramsey-type theorem. More generally, our result can be applied for any k-edge-colored (for $k = O(1)$) n-node complete graphs if for every node v the number of colors of the edges incident with v is at most 2. We say that such graphs have *maximum color degree* at most 2. Surprisingly, if we relax the maximum color degree of the input graph from 2 to 3, then detecting all distinct colored 4-cycles cannot be done by any $O(n^{3-\delta})$-time combinatorial algorithm (formally stated in Theorem 3), assuming a standard hardness conjecture (Conjecture 4).

Theorem 3. *Given an n-node complete graph $G = (V, E)$ and an edge label function $\ell : E \rightarrow \mathcal{L}$ with $|\mathcal{L}| = 3$, for some $q_1, q_2, q_3, q_4 \in \mathcal{L}$, outputting a $q_1q_2q_3q_4$-cycle in G or reporting that G contains no such cycles cannot be done by an $O(n^{3-\delta})$-time combinatorial algorithm for any constant $\delta > 0$ unless detecting a triangle for an n-node undirected simple graph can be done by an $O(n^{3-\delta'})$-time combinatorial algorithm for some constant $\delta' > 0$.*

Conjecture 4. *(Hardness of Triangle Detecting* [28]*).* Given an n-node undirected simple graph G, any combinatorial algorithm (i.e. excluding the algorithms that use fast matrix multiplication) that either outputs a triangle in G or reports the triangle-freeness of G requires $\Omega(n^{3-\varepsilon})$ time for any constant $\varepsilon > 0$.

Because our algorithm used for Theorem 2 is combinatorial, Theorem 3 gives a complementary hardness result. Indeed, even if we take non-combinatorial algorithms into account, there is still an $n^{\Omega(1)}$-gap between the known best running times for our problems with the maximum color degree 2 and 3.

Detecting Any Designated 4-Node Induced Subgraph

Theorem 1 can also be applied to detect all distinct 4-node induced subgraphs, up to isomorphism, for triangle-free graphs in randomized $O(n^2)$ time. Formally,

Theorem 5. *Given an undirected simple triangle-free graph $G = (V, E)$ with $|V| = n$ and any 4-node graph H, outputting an induced subgraph in G isomorphic to H or reporting that G does not contain H as an induced subgraph can be done in randomized $O(n^2)$ time with constant failure probability < 1.*

In contrast, the known best algorithm for this problem on general graphs needs triangle time [27]. For geometric intersection graphs, subquadratic algorithms have been proposed [3].

For $H = 2K_2$ ($2K_2$ is a 4-node 2-edge graph in which the two edges have no end-nodes in common), to our knowledge, Theorem 5 gives the first quadratic-time algorithm; for other Hs, Theorem 5 either simplifies the existing results listed below or gives an alternative way to view the computation.

- $H = 2K_2$: A graph that contains none of C_3, C_5, and $2K_2$ as an induced subgraph is called a difference graph [15] or a bipartite chain [16]. We say such a graph is $\{C_3, C_5, 2K_2\}$-free. It is known that the class of $\{C_3, C_5, 2K_2\}$-free graphs is exactly the class of $2K_2$-free bipartite graphs, which can be recognized in linear time [16]. The known best algorithm to test the $2K_2$-freeness for general graphs needs $O(n^\omega)$ time where ω denotes the exponent of fast matrix multiplications [27]. It was unknown whether $2K_2$-freeness can be decided in quadratic time for triangle-free graphs, and Theorem 5 gives the first quadratic-time algorithm. Because the bipartiteness is easy to test and bipartite graphs contain no triangles, the algorithm to recognize difference graphs can be unified by our algorithm.

- $H = P_4$ (P_4 is a path of four nodes): A graph containing no induced P_4 is a cograph, whose recognition draws considerable attention in the literature [6–8, 14]. Most of these recognition algorithms are developed based on the construction of *cotree*, which decomposes the input graph using *union* and *join* operations. Although the construction can be done in linear time, complicated data structures have to be maintained. Habib et al. [14] proposed a simple algorithm without the construction of a cotree. Our algorithm provides a simple alternative that detects whether a triangle-free graph is a cograph.

Other Variants

In [1], the authors showed that for any n-node m-edge undirected unweighted graph G distinguishing whether G has diameter 2 or 4 takes $O(m\sqrt{n \log n})$ time. They mentioned also that it is still unknown whether there exists a combinatorial algorithm that can distinguish graphs with diameter 2 or 3 in $O(n^{3-\delta})$ time for some constant $\delta > 0$. This implies that it is unknown how to verify by an $O(n^{3-\delta})$-time combinatorial algorithm for some constant $\delta > 0$ whether the product of a pair of n by n Boolean matrices contains only TRUE entries. This implies further that, for any nonempty subset $\Gamma \subseteq \{AB, A\bar{B}, \bar{A}B, \bar{A}\bar{B}\}$, it is unknown how to verify by an $O(n^{3-\delta})$-time combinatorial algorithm for any constant $\delta > 0$ whether $(A, B)_\Gamma$ contains only TRUE entries. Here is why. We assume w.l.o.g. that $AB \in \Gamma$. For each $i, j \in [n]$ we replace the universe of each A_i, B_j from $[n]$ to $[n + 3]$ and let $A_i \cap \{n + 1, n + 2, n + 3\} = \{n + 1\}$ and $B_j \cap \{n+1, n+2, n+3\} = \{n+2\}$. Hence, for every $i, j \in [n]$, $A_i \cap \bar{B}_j$, $\bar{A}_i \cap B_j$, and $\bar{A}_i \cap \bar{B}_j$ are not empty. As a result, verifying whether $(A, B)_\Gamma$ contains only TRUE entries is as hard as that for the standard Boolean matrix multiplication $(A, B)_{\{AB\}}$.

Though there is no known subcubic-time combinatorial algorithm to verify whether the product of the generalized Boolean matrix multiplication contains only TRUE entries, there are some special cases that can be computed in quadractic time. For example, if A is *row-unimodal* or B is *column-unimodal*, then computing all entries in $(A, B)_\Gamma$ (not just verifying whether all entries are identical) for any $\Gamma \subseteq \{AB, A\bar{B}, \bar{A}B, \bar{A}\bar{B}\}$ can be easily done in $O(n^2)$ time. We say that A is row-unimodal if there exists a sequence of column interchanges so that the TRUE entries in each row of the resulting A are consecutive. We say that B is column-unimodal if there exists a sequence of row interchanges so that the TRUE entries in each column of the resulting B are consecutive. In this way, if B is column-unimodal, testing whether $A_i \cap B_j = \varnothing$ is equivalent to asking whether some TRUE entry in A_i is located in a consecutive range. Hence, the computation can be done in $O(1)$ time by precomputing the standard prefix sum of each row in A. We observe that the adjacency matrix of any $2K_2$-free bipartite graph is both row-unimodal and column-unimodal.

Our Contributions

We present an $O(n^2)$-time deterministic algorithm for each of the 15 cases (classified into 5 representative cases in Table 1, detailed in Sect. 2) of the generalized Boolean matrix multiplications.

Table 1. A list of all representative Γs.

	Γ
$\lvert\Gamma\rvert = 1$	$\{AB\}$
$\lvert\Gamma\rvert = 2$	$\{AB, A\bar{B}\}, \{A\bar{B}, \bar{A}B\}$
$\lvert\Gamma\rvert = 3$	$\{AB, A\bar{B}, \bar{A}B\}$
$\lvert\Gamma\rvert = 4$	$\{AB, A\bar{B}, \bar{A}B, \bar{A}\bar{B}\}$

We apply the two representative cases with $|\Gamma| = 2$ to detect in quadratic time any designated colored 4-cycles for $O(1)$-edge-colored complete graphs with the maximum color degree at most 2. This result unifies two groups of algorithms: (i) the algorithm to detect 4-cycles for 1-edge-colored general graphs by the pigeonhole principle [24,30], and (ii) the algorithms to detect any designated colored 4-cycles for 2-edge-colored complete graphs [12,13,22,29], each corresponding to a customized Ramsey-type theorem. Our algorithm can handle more problem instances than the previous ones because, even when the maximum color degree is at most 2, the number of edge colors can be greater than 2. We conclude also that the constraint on the maximum color degree is essential by showing that detecting some colored 4-cycles for complete graph with the maximum color degree 3 requires cubic running time given some standard hardness assumption.

Then we apply the three representative cases with $|\Gamma| \leq 2$ to detect in quadratic time any designated 4-node induced subgraphs for triangle-free graphs. This result not only simplifies the known algorithms to recognize difference graphs and triangle-free cographs, but also gives the first quadratic-time algorithm, to our knowledge, to detect the induced $2K_2$ for triangle-free graphs. The latter is essential to conclude that all 4-node induced subgraphs can be detected in quadratic time for triangle-free graphs.

Paper Organization

In Sect. 2, for each Γ listed in Table 1 we devise a deterministic $O(n^2)$-time algorithm to compute the generalized Boolean matrix multiplication $(A, B)_\Gamma$; that is, a proof of Theorem 1. The rest of the claimed results are placed in the full version of this paper [17] due to the space constraint. Then, in Section 3 of [17], we apply the generalized Boolean matrix multiplication to detecting all kinds of colored 4-cycles, up to isomorphism, for n-node k-edge-colored (for $k = O(1)$) complete graphs in either deterministic $O(n^2 \log n)$ time or randomized

$O(tn^2)$ time with failure probability 2^{-t}, thereby proving Theorem 2. In Section 4 of [17], we show how to detect all 4-node induced subgraphs for n-node triangle-free graphs in randomized $O(n^2)$ time using the generalized Boolean matrix multiplication. We establish, in Section 5 of [17], that our algorithm for detecting colored 4-cycles is sharp by proving Theorem 3. Finally, in Sect. 3, we make a conclusion on our results.

2 Generalized Boolean Matrix Multiplication

In this section, we will prove Theorem 1. That is, we will devise a deterministic $O(n^2)$-time algorithm for each nonempty $\Gamma \subseteq \{AB, A\bar{B}, \bar{A}B, \bar{A}\bar{B}\}$ to compute the generalized product $(A, B)_\Gamma$ for any pair of n by n Boolean matrices A and B. Let $\Gamma_0 = \{AB\}$. Verifying whether $(A, B)_{\Gamma_0}$ contains only FALSE entries is equivalent to deciding whether

$$\left(\bigcup_{i\in[n]} A_i\right) \cap \left(\bigcup_{j\in[n]} B_j\right) = \varnothing,$$

which can be done in deterministic $O(n^2)$ time. Let $\Gamma_1 = \{A\bar{B}\}$. Computing $(A, B)_{\Gamma_1}$ is equivalent to computing $(A, \bar{B})_{\Gamma_0}$, so we can avoid a redundant discussion for the computation of $(A, B)_{\Gamma_1}$. Let $\Gamma_2 = \{AB, A\bar{B}\}$ and $\Gamma_3 = \{A\bar{B}, \bar{A}\bar{B}\}$. Computing $(A, B)_{\Gamma_3}$ is equivalent to computing $((\bar{B}^T, A^T)_{\Gamma_2})^T$, where C^T denotes the transpose matrix of C. Again, we can avoid a redundant discussion for the computation of $(A, B)_{\Gamma_3}$. By a similar argument, we reduce the 15 cases for all nonempty $\Gamma \subseteq \{AB, A\bar{B}, \bar{A}B, \bar{A}\,\bar{B}\}$ to the representative Γs listed in Table 1.

Though the randomized algorithms to verify the product for the representative Γs listed in Table 1 are alike, our derandomization for each case requires a different technique. The algorithm in Sect. 2.1 is based on an efficient computation of components in an auxiliary graph, that in Sect. 2.2 is based on an observation on the cardinalities of A_is and B_js, that in Sect. 2.3 is a mixture of the previous two but needs to compute components dynamically, and finally in Sect. 2.4 we employ the symmetry of such a Γ.

2.1 Computing $(A, B)_\Gamma$ with $\Gamma = \{AB, A\bar{B}\}$

Let $\Gamma = \{AB, A\bar{B}\}$ throughout this subsection. To compute the generalized product $(A, B)_\Gamma$, we reduce it to verifying the product of some integer matrix multiplication defined below.

Let X be a 0-1 n by n integer matrix whose entry x_{ij} for every $i, j \in [n]$ equals 1 iff $a_{ij} =$ TRUE. Let Y be a 0-1 n by n integer matrix whose entry y_{ij} for every $i, j \in [n]$ equals 1 iff $b_{ij} =$ TRUE. Let Z be the product of the integer matrix multiplication XY. We observe that:

Observation 6. *Let $P = (A, B)_\Gamma$. P contains only* FALSE *entries if and only if $z_{ij} \in \{0, |A_i|\}$ for all $i, j \in [n]$.*

Proof. Let $C^{(\alpha)} = AB$ and $C^{(\beta)} = A\bar{B}$. By definition, we have that for every $i, j \in [n]$

$$p_{ij} = \text{FALSE iff } c_{ij}^{(\alpha)} = \text{FALSE or } c_{ij}^{(\beta)} = \text{FALSE}. \tag{1}$$

On the RHS of (1), the former condition corresponds to $z_{ij} = 0$ and the latter corresponds to $z_{ij} = |A_i|$. □

Given Observation 6, $(A, B)_\Gamma$ contains only FALSE entries if and only if the product Z is equal to one of the 2^{n^2} matrices. Note that we do not have Z initially. However, we will show how to pick a guess $Z^\dagger$ of Z in $O(n^2)$ time from the 2^{n^2} possible matrices by excluding one of the two choices in $\{0, |A_i|\}$ for all $i, j \in [n]$ so that $(A, B)_\Gamma$ contains only FALSE entries if and only if $AB = Z^\dagger$. By Freivalds' algorithm, we can verify whether $XY = Z^\dagger$ in randomized $O(n^2)$ time with success probability at least 1/2. This answers whether $(A, B)_\Gamma$ contains only FALSE entries.

In what follows, we will show how to obtain $Z^\dagger$ and verify whether $(A, B)_\Gamma$ contains only FALSE entries in deterministic $O(n^2)$ time.

Observation 7. *There exists some matrix $Z^\dagger$ where $z_{ij}^\dagger \in \{0, |A_i|\}$ for each $i, j \in [n]$ so that $(A, B)_\Gamma$ contains only* FALSE *entries if and only if $XY = Z^\dagger$. Such a matrix $Z^\dagger$ can be determined in deterministic $O(n^2)$ time.*

Proof. If A_i is an empty set, then clearly $z_{ij} = 0$, so $z_{ij}^\dagger = 0$. Hence, w.l.o.g. let e be an element in the set A_i. If e is contained in the set B_j, then z_{ij} cannot be 0, so set $z_{ij}^\dagger = |A_i|$. Otherwise $e \notin B_j$, then $z_{ij} \neq |A_i|$, so set $z_{ij}^\dagger = 0$. Hence, for all $i, j \in [n]$, we can exclude one of the two choices (as mentioned in Observation 6) of z_{ij}. Thus, we obtain a matrix $Z^\dagger$ in $O(n^2)$ time as desired. □

For general matrices, it is not known whether Freivalds' algorithm has a deterministic alternative that runs in (nearly) $O(n^2)$ time. Surprisingly, as we show below, for such three matrices X, Y, and $Z^\dagger$, deciding whether $XY = Z^\dagger$ can be done in deterministic $O(n^2)$ time; that is, Freivalds' algorithm can be derandomized for such a case.

To have a deterministic alternative of Freivalds' algorithm for this particular case, we need to identify all components in an auxiliary graph G_{aux} in deterministic $O(n^2)$ time. Define an undirected graph $G_{\text{aux}} = (V_{\text{aux}}, E_{\text{aux}})$ so that V_{aux} consists of n nodes $v_1, v_2, \ldots, v_n$ and, for every $i, j \in [n]$, v_i, v_j is joined by an edge if i, j both are elements in A_k for some $k \in [n]$. It is not known how to construct such a G_{aux} by an $O(n^{3-\delta})$-time combinatorial algorithm for any constant $\delta > 0$ because DIAMETER2OR3 can be reduced to the construction of G_{aux}. We refer readers to the discussion on the problem of DIAMETER2OR3 in Sect. 1. However, all components in G_{aux} can be identified in $O(n^2)$ time, as shown in:

Observation 8. *All components of the auxiliary graph G_{aux} can be identified (by identifying a component we mean to compute its node set) in deterministic $O(n^2)$ time.*

Proof. For every $k \in [n]$, the number of pairs in A_k that induce an edge in G_{aux} is $\binom{|A_k|}{2}$. We do not enumerate all of these pairs in A_k for all $k \in [n]$ because the running time can be as large as $\Omega(n^3)$. Instead, we construct a "sparse" subgraph H of G_{aux} so that the components in H are the same as those in G_{aux}. Initially, H is an empty graph. For each $k \in [n]$, where $A_k = \{\alpha_1, \alpha_2, \ldots, \alpha_{|A_k|}\}$, for each $i \in [|A_k| - 1]$ we add an edge $\{v_{\alpha_i}, v_{\alpha_{i+1}}\}$ to H. Clearly the components in G_{aux} are the same as those in H. Such an H can be constructed in $O(n^2)$ time, and its components can be identified by any $O(n^2)$-time graph traversal, say BFS. □

Given the above observations, we are ready to prove the following lemma.

Lemma 9. *Let $\Gamma = \{AB, A\bar{B}\}$. Verifying whether the product $(A, B)_\Gamma$ contains only* FALSE *entries can be done in deterministic $O(n^2)$ time.*

Proof. For any $i, i', j \in [n]$ with $A_i \cap A_{i'} \neq \varnothing$, to have $(|A_i \cap B_j|, |A_{i'} \cap B_j|) \in \{0, |A_i|\} \times \{0, |A_{i'}|\}$, the only possibilities are

$$(|A_i \cap B_j|, |A_{i'} \cap B_j|) \in \{(0,0), (|A_i|, |A_{i'}|)\}\,. \tag{2}$$

Here is why. Let e be an element in $A_i \cap A_{i'}$. If $e \in B_j$, then $|A_i \cap B_j| \neq 0$ and $|A_{i'} \cap B_j| \neq 0$; otherwise $e \notin B_j$, then $|A_i \cap B_j| \neq |A_i|$ and $|A_{i'} \cap B_j| \neq |A_{i'}|$. By applying (2) in a serial manner, we have that for any $A_{i_1}, A_{i_2}, \ldots, A_{i_t}$ with $t \geq 2$ and $A_{i_k} \cap A_{i_{k+1}} \neq \varnothing$ for $k \in [t-1]$, to have $(|A_{i_1} \cap B_j|, \ldots, |A_{i_t} \cap B_j|) \in \{0, |A_{i_1}|\} \times \cdots \times \{0, |A_{i_t}|\}$, the only possibilities are

$$(|A_{i_1} \cap B_j|, \ldots, |A_{i_t} \cap B_j|) \in \{(0, \ldots, 0), (|A_{i_1}|, \ldots, |A_{i_t}|)\}\,. \tag{3}$$

We partition the collection $S{:=}\{A_i : i \in [n]\}$ into subcollections $S_{r_1}, \ldots, S_{r_\ell}$ where

- ℓ equals the number of components in G_{aux},
- r_i denotes an arbitrary node in a unique component in G_{aux} for $i \in [\ell]$, and
- $S_{r_i} = \{A \in S : \exists e \in A$ so that v_e, r_i are connected in $G_{\text{aux}}\}$ for $i \in [\ell]$.

Let U_k for each $k \in [\ell]$ be $\bigcup_{A \in S_{r_k}} A$. Note that $\{U_k : k \in [\ell]\}$ is a partition of $[n]$ because the components of the auxiliary graph G_{aux} is a partition of its node set.

By (3), for each $k \in [\ell]$, the only possibilities that $|A \cap B_j| \in \{0, |A|\}$ for each $A \in S_{r_k}$ are

$$\text{either } |A \cap B_j| = 0 \text{ for all } A \in S_{r_k} \text{ or } |A \cap B_j| = |A| \text{ for all } A \in S_{r_k}, \tag{4}$$

or equivalently,

$$\text{either } B_j \cap \left(\bigcup_{A_i \in S_{r_k}} A_i \right) = \varnothing \text{ or } \bar{B}_j \cap \left(\bigcup_{A_i \in S_{r_k}} A_i \right) = \varnothing. \tag{5}$$

Given the components in G_{aux}, we can compute U_k for all $k \in [\ell]$ in a total of $O(n^2)$ time. Because U_k and $U_{k'}$ are disjoint for every $k \neq k' \in [\ell]$, for each

$j \in [n]$ verifying whether $U_k \cap B_j = \varnothing$ and whether $U_k \cap \bar{B}_j = \varnothing$ for all $k \in [\ell]$ can be done in

$$O(n) + O\left(\sum_{k\in[\ell]} |U_k|\right) = O(n) \text{ time.}$$

Because of the equivalence of (4) and (5), the computation of $|B_j \cap U_k|$ for $j \in [n], k \in [\ell]$ suffices to decide whether $|A_i \cap B_j| \in \{0, |A_i|\}$ for all $i, j \in [n]$. The total running time of the above is deterministic $O(n^2)$. By Observation 6, this is equivalent to verifying whether $(A, B)_\Gamma$ contains only FALSE entries. □

2.2 Computing $(A, B)_\Gamma$ with $\Gamma = \{A\bar{B}, \bar{A}B\}$

Let $\Gamma = \{A\bar{B}, \bar{A}B\}$ throughout this subsection. We reuse the definition of X, Y, Z defined in Sect. 2.1.

Observation 10. *Let $P = (A, B)_\Gamma$. P contains only* FALSE *entries if and only if $z_{ij} \in \{|A_i|, |B_j|\}$ for all $i, j \in [n]$.*

Proof. Let $C^{(\alpha)} = A\bar{B}$ and $C^{(\beta)} = \bar{A}B$. By definition, we have that for every $i, j \in [n]$

$$p_{ij} = \text{FALSE iff } c_{ij}^{(\alpha)} = \text{FALSE or } c_{ij}^{(\beta)} = \text{FALSE}. \tag{6}$$

On the RHS of (6), the former condition corresponds to $z_{ij} = |A_i|$ and the latter corresponds to $z_{ij} = |B_j|$. □

One can exclude one of the two choices of z_{ij} for each $i, j \in [n]$, as stated in Observation 10, by comparing the cardinalities of A_i and B_j. A detailed explanation is included in the proof of Lemma 11.

Lemma 11. *Let $\Gamma = \{A\bar{B}, \bar{A}B\}$. Verifying whether the product $(A, B)_\Gamma$ contains only* FALSE *entries can be done in deterministic $O(n^2)$ time.*

Proof. For any $i, j \in [n]$, $z_{ij} \in \{|A_i|, |B_j|\}$ iff $A_i \subseteq B_j$ or $B_j \subseteq A_i$. $A_i \subseteq B_j$ holds only if $|A_i| \leq |B_j|$, and $B_j \subseteq A_i$ holds only if $|A_i| \geq |B_j|$.

For each $j \in [n]$, let $\mathcal{S}_j$ (resp. $\mathcal{L}_j$) denote the union of A_is for $i \in [n]$ whose $|A_i| \leq |B_j|$ (resp. $|A_i| \geq |B_j|$). Preparing $\mathcal{S}_j$ and $\mathcal{L}_j$ for all $j \in [n]$ can be done in deterministic $O(n^2)$ time by sorting A_is for $i \in [n]$ by their cardinalities and computing the prefix union and the suffix union of these ordered A_is. If $z_{ij} = |A_i|$ or $|B_j|$ for all $i, j \in [n]$, then for every B_j with $j \in [n]$, both $B_j \supseteq \mathcal{S}_j$ and $B_j \subseteq \mathcal{L}_j$ hold, which can be checked in $O(n)$ time given $\mathcal{S}_j$ and $\mathcal{L}_j$. Summing the running time over all $j \in [n]$ gives the bound. □

2.3 Computing $(A, B)_\Gamma$ with $\Gamma = \{AB, A\bar{B}, \bar{A}B\}$

Let $\Gamma = \{AB, A\bar{B}, \bar{A}B\}$ throughout this subsection. We reuse the definition of X, Y, Z defined in Sect. 2.1.

Observation 12. *Let $P = (A, B)_\Gamma$. P contains only* FALSE *entries if and only if $z_{ij} \in \{0, |A_i|, |B_j|\}$ for all $i, j \in [n]$.*

Proof. Let $C^{(\alpha)} = AB$, $C^{(\beta)} = A\bar{B}$, and $C^{(\gamma)} = \bar{A}B$. By definition, we have that for every $i, j \in [n]$

$$p_{ij} = \text{FALSE iff } c_{ij}^{(\alpha)} = \text{FALSE}, c_{ij}^{(\beta)} = \text{FALSE}, \text{ or } c_{ij}^{(\gamma)} = \text{FALSE}. \quad (7)$$

On the RHS of (7), the first condition corresponds to $z_{ij} = 0$, the second corresponds to $z_{ij} = |A_i|$, and the third corresponds to $z_{ij} = |B_j|$. □

In the proof of Lemma 13, we combine the ideas used to exclude one of the two choices for each z_{ij} in Observation 6 and Observation 10 to exclude two of the three choices for each z_{ij} in Observation 12. However, an efficient implementation of the combination may need to identify all components in an auxiliary graph dynamically.

Lemma 13. *Let $\Gamma = \{AB, A\bar{B}, \bar{A}B\}$. Verifying whether the product $(A, B)_\Gamma$ contains only* FALSE *entries can be done in deterministic $O(n^2)$ time.*

Proof. Let $\mathcal{S}$ be a sequence of the sets in $\{A_i : i \in [n]\} \cup \{B_j : j \in [n]\}$ ordered by their cardinalities in non-decreasing order. For each B_j, for each A_i that precedes B_j in $\mathcal{S}$, p_{ij} = FALSE iff $z_{ij} = 0$ or $|A_i|$. The case $z_{ij} = |B_j|$ can be ignored because $z_{ij} = |B_j|$ only if $|A_i| = |B_j|$. Thus we reuse the trick in the proof of Lemma 9. For any $i, i', j \in [n]$ with $A_i \cap A_{i'} \neq \varnothing$, $(|A_i \cap B_j|, |A_{i'} \cap B_j|) \in \{0, |A_i|\} \times \{0, |A_{i'}|\}$ iff $|A_{i'} \cap B_j|)$ is either $(|A_i|, |A_{i'}|)$ or $(0, 0)$. Then we extend the idea of the auxiliary graph G_{aux} as follows. Let $G_{\text{aux},j}$ denote the auxiliary graph induced by all the A_is that precede B_j in $\mathcal{S}$ for $i \in [n]$. The components in $G_{\text{aux},j}$ for all $j \in [n]$ can be computed in $O(n^2)$ time in total by keeping a spanning forest of $G_{\text{aux},j-1}$ and running a graph traversal, say BFS, on the union of the spanning forest of $G_{\text{aux},j-1}$ and the sparse representation of the A_is (as that in Observation 8) that are in-between $G_{\text{aux},j-1}$ and $G_{\text{aux},j}$. Note that both the sparse representation of A_i and the spanning forest of $G_{\text{aux},j-1}$ have $O(n)$ edges, so BFS takes $O(n)$ time. Thus, fix B_j, determining whether p_{ij} = FALSE for all i, j with that A_i precedes B_j in $\mathcal{S}$ can be done in $O(n)$ time by checking whether the component containing A_i in $G_{\text{aux},j}$ is contained in B_j. Summing the running time over all $j \in [n]$ yields $O(n^2)$.

Analogously, for each A_i, for each B_j that precedes A_i in $\mathcal{S}$, p_{ij} = FALSE iff $z_{ij} = 0$ or $|B_j|$. We repeat the above on another auxiliary graph $H_{\text{aux},i}$ induced by all B_js that precede A_i in $\mathcal{S}$. The rest goes analogously, so the total running time is bounded by $O(n^2)$. □

2.4 Computing $(A, B)_\Gamma$ with $\Gamma = \{AB, A\bar{B}, \bar{A}B, \bar{A}\bar{B}\}$

Let $\Gamma = \{AB, A\bar{B}, \bar{A}B, \bar{A}\bar{B}\}$ throughout this subsection. We reuse the definition of X, Y, Z defined in Sect. 2.1.

Observation 14. *Let $P = (A, B)_\Gamma$. P contains only* FALSE *entries if and only if $z_{ij} \in \{0, |A_i|, |B_j|, |A_i| + |B_j| - n\}$ for all $i, j \in [n]$.*

Proof. Let $C^{(\alpha)} = AB$, $C^{(\beta)} = A\bar{B}$, $C^{(\gamma)} = \bar{A}B$, and $C^{(\delta)} = \bar{A}\bar{B}$. By definition, we have that for every $i, j \in [n]$

$$p_{ij} = \text{FALSE iff } c_{ij}^{(\alpha)} = \text{FALSE}, c_{ij}^{(\beta)} = \text{FALSE}, c_{ij}^{(\gamma)} = \text{FALSE, or } c_{ij}^{(\delta)} = \text{FALSE}. \tag{8}$$

On the RHS of (8), the first condition corresponds to $z_{ij} = 0$, the second corresponds to $z_{ij} = |A_i|$, and the third corresponds to $z_{ij} = |B_j|$. The last corresponds to $z_{ij} = |A_i| + |B_j| - n$ because $c_{ij}^{(\delta)} =$ FALSE yields that $\bar{A}_i \subseteq B_j$ and hence $|A_i \cap B_j| = |B_j| - |\bar{A}_i| = |B_j| + |A_i| - n$. □

In the proof of Lemma 15, we employ the symmetry of Γ to exclude the last choice of $z_{ij} = |A_i| + |B_j| - n$ and perform a reduction from this reduced case to the case discussed in Sect. 2.3.

Lemma 15. *Let $\Gamma = \{AB, A\bar{B}, \bar{A}B, \bar{A}\bar{B}\}$. Verifying whether the product $(A, B)_\Gamma$ contains only* FALSE *entries can be done in deterministic $O(n^2)$ time.*

Proof. The computation of $(A, B)_\Gamma$ for this Γ satisfies a symmetric property, detailed below. Define A' to be a matrix so that for each $i \in [n]$ either $A'_i = A_i$ or $A'_i = \bar{A}_i$. Define B' analogously that for each $j \in [n]$ either $B'_j = B_j$ or $B'_j = \bar{B}_j$. Then $(A, B)_\Gamma = (A', B')_\Gamma$. This equality holds because of the symmetry of Γ. Hence, we assume w.l.o.g. that, for every $i, j \in [n]$, $|A_i| \le |\bar{A}_i|$ and $|B_j| \le |\bar{B}_j|$. This yields that $|A_i| \le n/2$ and $|B_j| \le n/2$ and $|A_i| + |B_j| - n \le 0$. Thus the last choice of z_{ij} defined in Observation 14 can be excluded, so Lemma 13 can be applied to verify the product of $(A, B)_\Gamma$ in deterministic $O(n^2)$ time. □

To determine for each $i, j \in [n]$ which product in Γ witnesses that $p_{ij} =$ FALSE, it suffices to know which of $\{0, |A_i|, |B_j|, |A_i|+|B_j|-n\}$ corresponds to the evaluation of $z_{ij}^\dagger$. It is possible that some evaluations of $\{0, |A_i|, |B_j|, |A_i|+|B_j|-n\}$ have the same value, but only one of them is left after the exclusion and is used for the evaluation of $z_{ij}^\dagger$. Note that $|A_i|+|B_j|-n$ is always excluded because we transform A and B into a certain form. A postprocessing is needed to perform the inverse of the transformation, to identify which entries $z_{ij}^\dagger$ correspond to an evaluation of $|A_i|+|B_j|-n$. As a result, Lemmas 9, 11, 13 and 15 together prove Theorem 1.

3 Conclusion

We present a generalization of the standard Boolean matrix multiplication and devise a deterministic quadratic-time algorithm to verify whether the generalized product contains only FALSE entries. In the full version of this work [17] we present how to apply such an efficient matrix multiplication verifier to detect all distinct colored 4-cycles, up to isomorphism, for $O(1)$-edge-colored complete graphs with the maximum color degree at most 2, which unifies a number of algorithms studied in the previous work, each corresponding to a customized Ramsey-type theorem. In addition, we show how to detect all distinct four-node induced subgraphs for triangle-free graphs in quadratic running time and the key step is by our generalized Boolean matrix multiplication.

Our results are seemingly sharp in the sense that (1) verifying whether the product contains only TRUE entries is no easier than some diameter computation, and (2) our problem is triangle-hard if the input graph has the maximum color degree at least 3. We note also that finding one type of colored 4-cycle can be a different computational problem from finding another, so they were solved by different Ramsey-type theorems. A concrete example is that finding an $rrbb$-cycle in the streaming model can be done using $O(n)$ space for n-node graphs given the Ramsey-type theorem [17, Lemma 16], but finding an $rbgb$-cycle in the same model requires $\Omega(n^2)$ space for n-node graphs since it is triangle-hard as shown in [17, Corollary 26], so they cannot be problems with the same complexities.

Acknowledgments. We sincerely thank the anonymous reviewers for their helpful comments.

References

1. Aingworth, D., Chekuri, C., Indyk, P., Motwani, R.: Fast estimation of diameter and shortest paths (without matrix multiplication). SIAM J. Comput. **28**(4), 1167–1181 (1999)
2. Alman, J., Williams, V.V.: A refined laser method and faster matrix multiplication. In: Proceedings of the 32th ACM-SIAM Symposium on Discrete Algorithms (SODA), pp. 522–539 (2021)
3. Chan, T.M.: Finding triangles and other small subgraphs in geometric intersection graphs. In: Proceedings of the 34th Annual ACM-SIAM Symposium on Discrete Algorithms (SODA), pp. 1777–1805 (2023)
4. Chen, Z., Kao, M.: Reducing randomness via irrational numbers. In: Proceedings of the 29th Annual ACM Symposium on the Theory of Computing (STOC), pp. 200–209 (1997)
5. Coppersmith, D., Winograd, S.: Matrix multiplication via arithmetic progressions. J. Symb. Comput. **9**(3), 251–280 (1990)
6. Corneil, D.G., Perl, Y., Stewart, L.K.: A linear recognition algorithm for cographs. SIAM J. Comput. **14**(4), 926–934 (1985)
7. Dahlhaus, E.: Efficient parallel recognition algorithms of cographs and distance hereditary graphs. Discret. Appl. Math. **57**(1), 29–44 (1995)
8. Dahlhaus, E., Gustedt, J., McConnell, R.M.: Efficient and practical algorithms for sequential modular decomposition. J. Algorithm **41**(2), 360–387 (2001)
9. Dumas, J.G., van der Hoeven, J., Pernet, C., Roche, D.S.: LU factorization with errors. In: Proceedings of the 32th International Symposium on Symbolic and Algebraic Computation (ISSAC), pp. 131–138 (2019)
10. Freivalds, R.: Probabilistic machines can use less running time. In: Proceedings of the 7th IFIP Congress Information Processing, pp. 839–842. North-Holland (1977)
11. Gąsieniec, L., Levcopoulos, C., Lingas, A., Pagh, R., Tokuyama, T.: Efficiently correcting matrix products. Algorithmica **79**(2), 428–443 (2017)

12. Geerdes, H.F., Szabó, J.: A unified proof for Karzanov's exact matching theorem. Technical report. QP-2011-02, Egerváry Research Group, Budapest (2011)
13. Gurjar, R., Korwar, A., Messner, J., Thierauf, T.: Exact perfect matching in complete graphs. ACM Trans. Comput. Theory **9**(2), Article no. 8 (2017)
14. Habib, M., Paul, C.: A simple linear time algorithm for cograph recognition. Discret. Appl. Math. **145**(2), 183–197 (2005)
15. Hammer, P.L., Peled, U.N., Sun, X.: Difference graphs. Discret. Appl. Math. **28**(1), 35–44 (1990)
16. Heggernes, P., Kratsch, D.: Linear-time certifying recognition algorithms and forbidden induced subgraphs. Nordic J. Comput. **14**(1), 87–108 (2007)
17. Hon, W., Tsai, M., Wang, H.: Verifying the product of generalized Boolean matrix multiplication and its applications to detect small subgraphs. CoRR (2023)
18. Ji, H., Mascagni, M., Li, Y.: Gaussian variant of Freivalds' algorithm for efficient and reliable matrix product verification. Monte Carlo Methods Appl. **26**(4), 273–284 (2020)
19. Kaltofen, E.L., Nehring, M., Saunders, B.D.: Quadratic-time certificates in linear algebra. In: Proceedings of the 24th International Symposium on Symbolic and Algebraic Computation (ISSAC), pp. 171–176 (2011)
20. Kimbrel, T., Sinha, R.K.: A probabilistic algorithm for verifying matrix products using $O(n^2)$ time and $\log_2 n+O(1)$ random bits. Inform. Process. Lett. **45**(2), 107–110 (1993)
21. Künnemann, M.: On nondeterministic derandomization of Freivalds' algorithm: consequences, avenues and algorithmic progress. In: 26th Annual European Symposium on Algorithms (ESA), pp. 56:1–56:16 (2018)
22. Li, R., Broersma, H., Yokota, M., Yoshimoto, K.: Edge-colored complete graphs without properly colored even cycles: a full characterization. J. Graph Theor. **98**(1), 110–124 (2021)
23. Naor, J., Naor, M.: Small-bias probability spaces: efficient constructions and applications. SIAM J. Comput. **22**(4), 838–856 (1993)
24. Richards, D., Liestman, A.L.: Finding cycles of a given length. In: Cycles in Graphs, North-Holland Mathematics Studies, vol. 115, pp. 249–255. North-Holland (1985)
25. Roche, D.S.: Error correction in fast matrix multiplication and inverse. In: Proceedings of the 31th ACM International Symposium on Symbolic and Algebraic Computation (ISSAC), pp. 343–350 (2018)
26. Strassen, V.: Gaussian elimination is not optimal. Numer. Math. **13**(4), 354–356 (1969)
27. Williams, V.V., Wang, J.R., Williams, R.R., Yu, H.: Finding four-node subgraphs in triangle time. In: Proceedings of the 26th Annual ACM-SIAM Symposium on Discrete Algorithms (SODA), pp. 1671–1680 (2015)
28. Williams, V.V., Williams, R.R.: Subcubic equivalences between path, matrix, and triangle problems. J. ACM **65**(5), 27:1–27:38 (2018)
29. Yeo, A.: A note on alternating cycles in edge-coloured graphs. J. Comb. Theory B **69**(2), 222–225 (1997)
30. Yuster, R., Zwick, U.: Finding even cycles even faster. SIAM J. Discret. Math. **10**(2), 209–222 (1997)

Reconfiguration of Time-Respecting Arborescences

Takehiro Ito[1], Yuni Iwamasa[2], Naoyuki Kamiyama[3](✉), Yasuaki Kobayashi[4], Yusuke Kobayashi[5], Shun-ichi Maezawa[6], and Akira Suzuki[1]

[1] Graduate School of Information Sciences, Tohoku University, Sendai, Japan
{takehiro,akira}@tohoku.ac.jp

[2] Graduate School of Informatics, Kyoto University, Kyoto, Japan
iwamasa@i.kyoto-u.ac.jp

[3] Institute of Mathematics for Industry, Kyushu University, Fukuoka, Japan
kamiyama@imi.kyushu-u.ac.jp

[4] Graduate School of Information Science and Technology, Hokkaido University, Sapporo, Japan
koba@ist.hokudai.ac.jp

[5] Research Institute for Mathematical Sciences, Kyoto University, Kyoto, Japan
yusuke@kurims.kyoto-u.ac.jp

[6] Department of Mathematics, Tokyo University of Science, Tokyo, Japan
maezawa@rs.tus.ac.jp

Abstract. An arborescence, which is a directed analogue of a spanning tree in an undirected graph, is one of the most fundamental combinatorial objects in a digraph. In this paper, we study arborescences in digraphs from the viewpoint of combinatorial reconfiguration, which is the field where we study reachability between two configurations of some combinatorial objects via some specified operations. Especially, we consider reconfiguration problems for time-respecting arborescences, which were introduced by Kempe, Kleinberg, and Kumar. We first prove that if the roots of the initial and target time-respecting arborescences are the same, then the target arborescence is always reachable from the initial one and we can find a shortest reconfiguration sequence in polynomial time. Furthermore, we show if the roots are not the same, then the target arborescence may not be reachable from the initial one. On the other hand, we show that we can determine whether the target arborescence is reachable form the initial one in polynomial time. Finally, we prove that it is NP-hard to find a shortest reconfiguration sequence in the case where the roots are not the same. Our results show an interesting contrast to the previous results for (ordinary) arborescences reconfiguration problems.

Keywords: Arborescence · Temporal network · Combinatorial reconfiguration

This work was partially supported by JSPS KAKENHI Grant Numbers JP18H04091, JP19K11814, JP20H05793, JP20H05794, JP20H05795, JP20K11666, JP20K11692, JP20K19742, JP20K23323, JP22K17854, JP22K13956.

P. Morin and S. Suri (Eds.): WADS 2023, LNCS 14079, pp. 521–532, 2023.
https://doi.org/10.1007/978-3-031-38906-1_34

1 Introduction

An arborescence, which is a directed analogue of a spanning tree in an undirected graph, is one of the most fundamental combinatorial objects in a digraph. For example, the problem of finding a minimum-cost arborescence in a digraph with a specified root vertex has been extensively studied (see, e.g., [4,7,10]). Furthermore, the theorem on arc-disjoint arborescences proved by Edmonds [6] is one of the most important results in graph theory.

Motivated by a variety of settings, such as communication in distributed networks, epidemiology and scheduled transportation networks, Kempe, Kleinberg and Kumar [16] introduced the concept of temporal networks, which can be used to analyze relationships involving time over networks. More formally, a temporal network consists of a digraph $D = (V, A)$ and a time label function $\lambda\colon A \to \mathbb{R}_+$, where $\mathbb{R}_+$ denotes the set of non-negative real numbers. For each arc $a \in A$, the value $\lambda(a)$ specifies the time at which the two end vertices of a can communicate. Thus, in an arborescence T with a root r, in order that the root r can send information to every vertex, for every vertex v, the time labels of the arcs of the directed path from r to v must be non-decreasing. An arborescence T in D satisfying this property is called a *time-respecting* arborescence. For example, Fig. 1 illustrates four different time-respecting arborescences in a digraph.

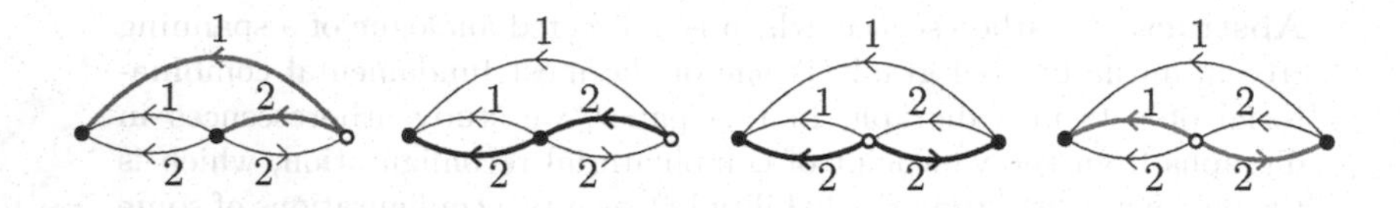

Fig. 1. Examples of time-respecting arborescences.

In this paper, we study time-respecting arborescences in digraphs from the viewpoint of combinatorial reconfiguration. Combinatorial reconfiguration [13,17] analyzes the reachability (and its related questions) of the solution space formed by combinatorial objects under a prescribed adjacency relation. The algorithmic studies of combinatorial reconfiguration were initiated by Ito et al. [13], and have been actively studied for this decade. (See, e.g., a survey [17]).

1.1 Our Problem and Related Work

In this paper, we introduce the Time-Respecting Arborescence Reconfiguration problem, as follows: Given two time-respecting arborescences in a digraph D, we are asked to determine whether or not we can transform one into the other by exchanging a single arc in the current arborescence at a time, so that all intermediate results remain time-respecting arborescences in D. (We call this sequence of arborescences a *reconfiguration sequence*.) For example, Fig. 1 shows such a transformation between the blue and red arborescences, and hence it is a yes-instance.

This is the first paper, as far as we know, which deals with the TIME-RESPECTING ARBORESCENCE RECONFIGURATION problem. However, reconfiguration problems have been studied for spanning trees and arborescences without time-respecting condition. For undirected graphs without time-respecting condition, it is well-known that every two spanning trees can be transformed into each other by exchanging a single edge at a time, because the set of spanning trees forms the family of bases of a matroid [8,13]. For digraphs without time-respecting condition, Ito et al. [15] proved that every two arborescences can be transformed into each other by exchanging a single arc at a time.[1] Interestingly, we note that this property does not hold when considering the time-respecting condition (See Fig. 2). Furthermore, in both undirected [13] and directed [15] cases, shortest transformations can be found in polynomial time, because there is a transformation that exchanges only edges (or arcs) in the symmetric difference of two given spanning trees (resp. arborescences).

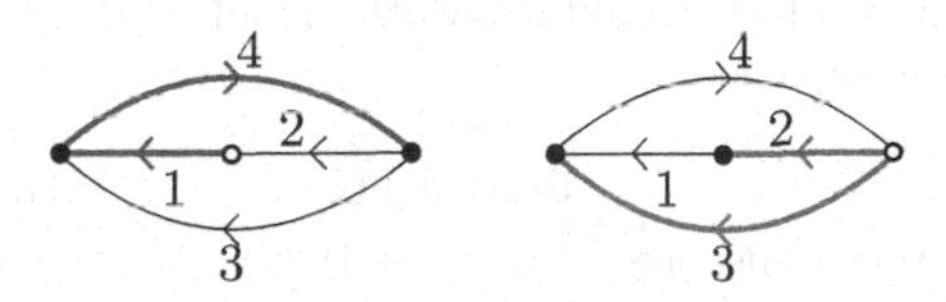

Fig. 2. There is no desired transformation from the time-respecting arborescence induced by blue arcs to the time-respecting arborescence induced by red arcs. (Color figure online)

We here review some further previous work. Reconfiguration problems have been studied for various kinds of combinatorial objects, mainly in undirected graphs. Reconfiguration of spanning trees with additional constrains was studied in [2,3]. Hanaka et al. [12] introduced the framework of subgraph reconfiguration problems, and studied the problem for several combinatorial objects, including trees: they showed that every two trees (that are not necessarily spanning) in an undirected graph can be transformed into each other by exchanging a single edge at a time unless two input trees have different numbers of edges. Motivated by applications in motion planning, Biasi and Ophelders [1], Demaine et al. [5], and Gupta et al. [11] studied some variants of reconfiguring undirected paths. These variants are shown to be PSPACE-complete in general, while they are fixed-parameter tractable when parameterized by the length of input paths.

1.2 Our Contribution

Our contributions are summarized as follows. We first prove that if the roots of the initial and target time-respecting arborescences are the same, then the

[1] Note that, in the paper [15], an arborescence is not necessarily a spanning subgraph. Arborescence in our paper corresponds to spanning arborescence in [15].

target arborescence is always reachable from the initial one and we can find a shortest reconfiguration sequence in polynomial time. Furthermore, we show if the roots of the initial and target time-respecting arborescences are not the same, then the target arborescence may not be reachable from the initial one. On the other hand, we show that we can determine whether the target arborescence is reachable from the initial one in polynomial time. Finally, we prove that it is NP-hard to find a shortest reconfiguration sequence in the case where the roots of the initial and target time-respecting arborescences are not the same. Our results show an interesting contrast to the results for (ordinary) arborescence reconfiguration problems [15]. See Table 1 for the summary of our results.

Table 1. The column Reachability shows the answer for the question whether every arborescence is reachable from any other arborescence. The column Reachability shows the results for the problem of determining whether the target arborescence is reachable from the initial one. The symbol P means that the problem can be solved in polynomial time. The column Shortest variant shows the results for the problem of finding a shortest reconfiguration sequence.

	Reachability	Shortest sequence
Arborescences without time-respecting	always yes [15]	P [15]
Identical roots with time-respecting	always yes [Theorem 1]	P [Theorem 1]
Non-identical roots with time-respecting	P [Theorem 2]	NP-complete [Theorem 3]

2 Preliminaries

Let $D = (V, A)$ be a digraph with possibly multiple arcs. We write $V(D)$ and $A(D)$ to denote the vertex set and arc set of D, respectively. For an arc $e = (u, v) \in A$, we call v the *head* of e, denoted head(e), and u the *tail* of e, denoted tail(e). For $v \in V$, we denote by $\delta_D^-(v)$ the set of arcs incoming to v in D (i.e. $\delta_D^-(v) = \{e \mid e = (u, v) \in A\}$) and by $\delta_D^+(v)$ the set of outgoing arcs from v (i.e. $\delta_D^+(v) = \{e \mid e = (v, u) \in A\}$). We extend these notations to sets: For $X \subseteq V$, we define $\delta_D^-(X) = \{e = (u, v) \in E \mid u \in V \setminus X, v \in X\}$ and $\delta_D^+(X) = \{e = (u, v) \in E \mid u \in X, v \in V \setminus X\}$. We may omit the subscript when no confusion arises. For $e \in A$ (resp. $f \in V \times V$), we denote by $D - e$ (resp. $D + f$) the digraph obtained from D by removing e (resp. adding f).

Let $r \in V$. An *r-arborescence* in D is a spanning acyclic subgraph of D in which there is exactly one (directed) path to any vertex from r. An *arborescence* (without specifying r) in D is an r-arborescence for some $r \in V$. Let $\lambda \colon A \to \mathbb{R}_+$. For a directed path P that traverses arcs $e_1, e_2, \ldots, e_k$ in this order, we say that P is *time-respecting for* λ if $\lambda(e_i) \leq \lambda(e_{i+1})$ for $1 \leq i < k$. An (r-)arborescence T is time-respecting for λ if every directed path in it is time-respecting. When λ is clear from the context, we may just say P (or T) is time-respecting.

For two arborescences T_1 and T_2 in D, a *reconfiguration sequence* between T_1 and T_2 is a sequence of arborescences $(T^0, T^1, \ldots, T^\ell)$ in D with $T^0 = T_1$ and $T^\ell = T_2$ such that for $0 \leq i < \ell$, $|A(T^i) \setminus A(T^{i+1})| = |A(T^{i+1}) \setminus A(T^i)| = 1$. In other words, T^{i+1} is obtained from T^i by removing an arc $e \in A(T^i)$ and adding an arc $f \notin A(T^i)$ (i.e., $T^{i+1} = T^i - e + f$). The *length* of the reconfiguration sequence is defined as ℓ.

3 Minimal Time-Respecting r-Arborescences

In this section, we give a polynomial-time algorithm for computing a *minimal* time-respecting r-arborescence of a digraph $D = (V, A)$ for λ. This arborescence plays a vital role in the subsequent section.

Let $r \in V$. We assume that D has at least one time-respecting r-arborescence. A function $d\colon V \to \mathbb{R}_+$ is defined as

$$d(v) = \min\{\lambda(e) \mid e \in \delta^-(v) \text{ and } \exists \text{ time-respecting path from } r \text{ containing } e\},$$

where we define $d(r) = 0$. Since every directed path from r in the r-arborescence is time-respecting, this function d is well-defined (under the assumption that D has at least one time-respecting r-arborescence). A time-respecting arborescence T is said to be *minimal* if the unique arc e of T incoming to v satisfies $\lambda(e) = d(v)$ for $v \in V \setminus \{r\}$. Now, we claim that (under the assumption) D has a minimal time-respecting r-arborescence, which can be found by the following algorithm.

1. Set $R := \{r\}$, $T := \emptyset$, and $d'(r) := 0$.
2. Repeat the following two steps until $R = V$.
3. Let $e \in \delta^+_D(R)$ minimizing $\lambda(e)$ subject to $\lambda(e) \geq d'(\text{tail}(e))$. If there is no such an arc $e \in \delta^+_D(R)$, the algorithm halts.
4. $R := R \cup \{\text{head}(e)\}$, $T := T \cup \{e\}$, and $d'(\text{head}(e)) := \lambda(e)$.

Note that T is an arc set in the algorithm, which is sometimes identified with the subgraph induced by T. In the following, we use R, T, and d' to denote the values of R, T, and d' after the execution of the above algorithm, respectively.

Lemma 1. *Suppose that D has a time-respecting r-arborescence. Then, T forms a minimal time-respecting r-arborescence of D.*

Proof. By the construction of T, every (directed) path from r in T is time-respecting, and hence we have $d'(v) \geq d(v)$ for $v \in R$. For $1 \leq i \leq |T|$, we let e_i be the arc selected at Step 3 in the ith iteration of the execution. Let $v_0 = r$ and let $v_i = \text{head}(e_i)$. In the following, we first show, by induction, that $d'(v_i) = d(v_i)$ for all $0 \leq i \leq |T|$. The base case $i = 0$ is clear from the definition (i.e., $d'(r) = d(r)$).

Let $i \geq 1$ and let $R_i = \{v_j \mid 0 \leq j < i\}$ be the set of vertices that are "reached" from r before the ith iteration. Suppose for contradiction that $d'(v_i) > d(v_i)$. Let P be a time-respecting path from r to v_i in which the unique arc e incoming to v_i satisfies $\lambda(e) = d(v_i)$. Since $r \in R_i$ and $v_i \notin R_i$, there is an arc

e' on P such that tail$(e') \in R_i$ and head$(e') \notin R_i$, i.e., $e' \in \delta^+_D(R_i)$. For such an arc e', we have $\lambda(e') \leq \lambda(e) = d(v_i) < d'(v_i) = \lambda(e_i)$. This implies that e_i cannot be selected at Step 3 in the ith iteration, because $e' \in \delta^+(R_i)$ and $\lambda(e') \geq d(\text{tail}(e')) = d'(\text{tail}(e'))$ by the induction hypothesis.

We next show that $R = V$, which implies that T forms a minimal time-respecting r-arborescence of D. As we have shown above, $d'(v) = d(v)$ for all $v \in R$. Suppose for contradiction that $R \neq V$. We note that every arc $e \in \delta^+_D(R)$ satisfies that $\lambda(e) < d'(\text{tail}(e))$. Let $v \in V \setminus R$ and let P be an arbitrary time-respecting path from r to v in D. We can choose such a vertex v in such a way that all vertices of P except for v belong to R, as $r \in R$ and any subpath of P starting from r is also time-respecting. Let e be the unique arc of P incoming to v. Since P is time-respecting and tail$(e) \in R$, $\lambda(e) \geq d(\text{tail}(e)) = d'(\text{tail}(e))$. This contradicts the fact that every arc $e' \in \delta^+_D(R)$ satisfies $\lambda(e') < d'(\text{tail}(e'))$. Therefore, the lemma follows. □

As a consequence of this lemma, we have the following corollary.

Corollary 1. *In polynomial time, we can compute a minimal time-respecting r-arborescence of D or conclude that D has no time-respecting r-arborescence.*

4 Time-Respecting r-Arborescence Reconfiguration

In this section, we give a polynomial-time algorithm for finding a shortest reconfiguration sequence between given two time-respecting r-arborescences of a graph D such that all intermediates are also time-respecting r-arborescences of D (if any). We show, in fact, that such a (shortest) reconfiguration sequence between T_1 and T_2 always exists, in contrast to the fact that, for some digraph, there is no reconfiguration sequence between time-respecting arborescences of distinct roots (see Fig. 2).

Let $D = (V, A)$ be a digraph with $\lambda\colon A \to \mathbb{R}_+$ and let $r \in V$. Let T_1 and T_2 be time-respecting r-arborescences of D. We construct a digraph $D^* = (V, A^*)$, where $A^* = A(T_1) \cup A(T_2)$. As there is a time-respecting r-arborescence of D^* (say, T_1), a minimal time-respecting r-arborescence T^* of D^* can be computed in polynomial time with the algorithm in Corollary 1.

To show the existence of a reconfiguration sequence between T_1 and T_2 in D, it suffices to give a reconfiguration sequence between T_1 and T^* in D^* as T_2 is symmetric and D^* is a subgraph of D.

We transform T_1 into the minimal time-respecting r-arborescence T^* as follows. Let $k = |A(T^*)|$. Let $e_1, e_2, \ldots, e_k$ be the arcs of T^* such that e_i is selected at Step 3 in the ith iteration of the algorithm in Corollary 1. We set $T^0 = T_1$. For $1 \leq i \leq k$ in increasing order, we define $T^i = T^{i-1} - f_i + e_i$ (possibly $e_i = f_i$), where f_i is the unique arc of T_1 incoming to head(e_i). Clearly, $T^k = T^*$. Since the update operation preserves the indegree of each vertex and the reachability of each vertex from r, every T^i is always an r-arborescence of D^*. Moreover, the following lemma ensures that T^i is time-respecting.

Lemma 2. *For* $0 \leq i \leq k$, T^i *is a time-respecting* r*-arborescence of* D^*.

Proof. It suffices to show that for $1 \leq i \leq k$, T^i is time-respecting, assuming that T^{i-1} is time-respecting. It suffices to consider the case when $f_i \neq e_i$. Since $\text{head}(e_i) = \text{head}(f_i)$, $T^{i-1}+e_i$ has exactly two paths P and P' from r to $\text{head}(e_i)$, where P (resp. P') is the one that contains e_i (resp. f_i). Since T^{i-1} is time-respecting, P' is indeed time-respecting. Similarly, since $P(\subseteq \{e_1, e_2, \ldots, e_i\})$ is a path in T^*, it is also time-respecting. As T^* is minimal, we have $d(\text{head}(f_i)) = \lambda(e_i)$, which implies that $\lambda(f_i) \geq \lambda(e_i)$. Therefore, T^i is time-respecting. □

As $T^k = T^*$, this lemma shows that there is a reconfiguration sequence between T_1 and T^*. Since we update $T \leftarrow T - e_i + f_i$ only when $e_i \neq f_i$, the obtained sequence has length $|A(T^*) \setminus A(T_1)|$. Similarly, there is a sequence between T_2 and T^* of length $|A(T^*) \setminus A(T_2)|$. By combining them, we obtain a reconfiguration sequence from T_1 to T_2 of length $|A(T^*)\setminus A(T_1)|+|A(T^*)\setminus A(T_2)|$.

We now show that this length is equal to $|A(T_1) \setminus A(T_2)|$, which implies that the sequence is shortest among all reconfiguration sequences between T_1 and T_2. For any $e \in A(T_1) \cap A(T_2)$, since e is a unique arc entering $\text{head}(e)$ in D^*, we obtain $e \in A(T^*)$. This means that $A(T_1) \cap A(T_2) \subseteq A(T^*)$. Then, we obtain

$$\begin{aligned}|A(T^*) \setminus A(T_1)| + |A(T^*) \setminus A(T_2)| &= |A(T^*) \setminus (A(T_1) \cap A(T_2))| \\ &= |A(T^*)| - |A(T_1) \cap A(T_2)| \\ &= |A(T_1) \setminus A(T_2)|,\end{aligned}$$

where we use $A(T^*) \subseteq A(T_1) \cup A(T_2)$ in the first equality, use $A(T_1) \cap A(T_2) \subseteq A(T^*)$ in the second equality, and use $|A(T^*)| = |A(T_1)|$ in the last equality.

Theorem 1. *There is a reconfiguration sequence between two time-respecting* r*-arborescence* T_1 *and* T_2 *of* D *with length* $|A(T_1) \setminus A(T_2)|$. *Moreover, such a reconfiguration sequence can be found in polynomial time.*

5 Time-Respecting Arborescence Reconfiguration

For two time-respecting arborescences T_1 and T_2 in a digraph $D = (V, A)$ with $\lambda : A \to \mathbb{R}_+$, we consider the problem of determining whether there exists a reconfiguration sequence from T_1 to T_2, which we call TIME-RESPECTING ARBORESCENCE RECONFIGURATION. By Theorem 1, for any two r-arborescences T_1 and T_2, there exists a reconfiguration sequence from T_1 to T_2, that is, the answer of TIME-RESPECTING r-ARBORESCENCE RECONFIGURATION is always yes. However, TIME-RESPECTING ARBORESCENCE RECONFIGURATION does not have the property, that is, there is a no-instance in the problem (see Fig. 2). In this section, we show that the problem can be solved in polynomial time.

Theorem 2. *We can solve* TIME-RESPECTING ARBORESCENCE RECONFIGURATION *in polynomial time.*

To prove the theorem, we introduce some concepts. For $t \in \mathbb{R}_+$, we say that a subgraph $H = (V(H), A(H))$ of D is *t-labeled extendible* if

1. $\lambda(e) = t$ for all $e \in A(H)$ and
2. the digraph D' obtained from D by contracting H into a vertex r_H has a time-respecting r_H-arborescence T such that $\lambda(e) \geq t$ for $e \in A(T)$.

We can see that if a t-labeled extendible subgraph H contains an arborescence T_H, then $A(T_H) \cup A(T)$ induces a time-respecting arborescence in D.

Let $D = (V, A)$ be a digraph and define the *reconfiguration graph* $\mathcal{G}(D)$ as follows: the vertex set consists of all the time-respecting arborescences of D and two time-respecting arborescences are joined by an (undirected) edge in the reconfiguration graph if and only if one is obtained from the other by exchanging a single arc. For a vertex $r \in V$, let $\mathcal{G}_r$ be the subgraph of $\mathcal{G}(D)$ induced by the time-respecting r-arborescences. By Theorem 1, $\mathcal{G}_r$ is connected for $r \in V(D)$. Let $\mathcal{G}'(D)$ be the graph obtained from $\mathcal{G}(D)$ by contracting $\mathcal{G}_r$ into a vertex v_r for each $r \in V(D)$. We show the following necessary and sufficient condition for two vertices in $\mathcal{G}'(D)$ to be adjacent to each other.

Lemma 3. *Let r_1 and r_2 be two distinct vertices in a digraph $D = (V, A)$ with $\lambda : A \to \mathbb{R}_+$. Then v_{r_1} and v_{r_2} are adjacent in $\mathcal{G}'(D)$ if and only if one of the following holds:*

(i) *there exist an arc $f = (r_2, r_1)$ and a time-respecting r_1-arborescence T_1 such that $\lambda(f) \leq \lambda(e')$ for each $e' \in \delta^+_{T_1}(r_1) \setminus \delta^-_{T_1}(r_2)$,*
(ii) *there exist an arc $e = (r_1, r_2)$ and a time-respecting r_2-arborescence T_2 such that $\lambda(e) \leq \lambda(e')$ for each $e' \in \delta^+_{T_2}(r_2) \setminus \delta^-_{T_2}(r_1)$,*
(iii) *for some $t \in \mathbb{R}_+$, D has a t-labeled extendible directed cycle C that contains both r_1 and r_2.*

Proof. **[Necessity ("only if" part)]** Suppose that v_{r_1} and v_{r_2} are adjacent in $\mathcal{G}'(D)$. Then there exist a time-respecting r_1-arborescence T_1 and two arcs e and f in $A(D)$ such that $T_1 - e + f$ is a time-respecting r_2-arborescence. Let $T_2 := T_1 - e + f$. Since T_1 is an r_1-arborescence and T_2 is an r_2-arborescence, $\text{head}(e) = r_2$ and $\text{head}(f) = r_1$. Then, $T_1 + f$ contains a directed cycle C, which contains e. (Otherwise, T_2 contains a directed cycle, which contradicts the fact that T_2 is an arborescence.) Let ℓ denote the length of C. Suppose that C traverses arcs $e_1, e_2, \ldots, e_\ell$ in this order when starting from r_1, that is, $e_\ell = f$ and $\text{head}(e_i) = \text{tail}(e_{i+1})$ for each i, where $e_{\ell+1} = e_1$. We can easily see that (i) holds if $f = (r_2, r_1)$ and (ii) holds if $e = (r_1, r_2)$.

Hence it suffices to consider the case when $f \neq (r_2, r_1)$ and $e \neq (r_1, r_2)$. Let j be the index such that $e_j = e$. Note that $j \neq 1$ by $e \neq (r_1, r_2)$, $j \neq \ell - 1$ by $f \neq (r_2, r_1)$, and $j \neq \ell$ by $e \neq f$. Since T_1 is a time-respecting r_1-arborescence, we obtain the following inequalities.

$$\lambda(e_1) \leq \lambda(e_2) \leq \cdots \leq \lambda(e_{j+1}) \leq \cdots \leq \lambda(e_{\ell-1}) \tag{1}$$

Since T_2 is a time-respecting r_2-arborescence, we obtain the following inequalities.

$$\lambda(e_{j+1}) \leq \lambda(e_{j+2}) \leq \cdots \leq \lambda(e_\ell) \leq \lambda(e_1) \tag{2}$$

By (1) and (2), we obtain $\lambda(e_1) = \lambda(e_2) = \cdots = \lambda(e_\ell)$. Let D' be the digraph obtained from D by contracting C into a vertex r_C. Then the digraph induced by the arcs in $A(T_1) \setminus A(C)$ is a time-respecting r_C-arborescence in D'. Hence (iii) holds.

[Sufficiency ("if" part)] Suppose that (i) holds. Let e be the arc in T_1 such that head$(e) = r_2$. Then $T_1 - e + f$ is a time-respecting r_2-arborescence in D. If (ii) holds, then we obtain a time-respecting r_1-arborescence in D from T_2 by one step by a similar argument above.

Suppose that (iii) holds. Let e and f be the arcs in C such that head$(e) = r_2$ and head$(f) = r_1$, respectively. Let D' be the digraph obtained from D by contracting C into r_C, and let T_C be a time-respecting r_C-arborescence in D'. Then the digraph T_1 induced by the arcs in $A(T_C) \cup A(C) \setminus \{f\}$ is a time-respecting r_1-arborescence in D and the digraph T_2 induced by the arcs in $A(T) \cup A(C) \setminus \{e\}$ is a time-respecting r_2-arborescence in D. Since $T_2 = T_1 - e + f$, v_{r_1} and v_{r_2} are adjacent in $\mathcal{G}'(D)$.

This completes the proof of Lemma 3. □

Let r_1 and r_2 be two distinct vertices in D. To check condition (i) in Lemma 3, it suffices to give a polynomial-time algorithm for finding a time-respecting r_1-arborescence T_1 such that $\lambda(e') \geq \lambda(f)$ for each $e' \in \delta^+_{T_1}(r_1) \setminus \delta^-_{T_1}(r_2)$, where $f = (r_2, r_1)$. This can be done in polynomial time by removing all the arcs $e \in \delta^+_D(r_1) \setminus \delta^-_D(r_2)$ with $\lambda(e) < \lambda(f)$ from D and by applying Corollary 1 to find a time-respecting r_1-arborescence in the obtained digraph. Similarly, condition (ii) in Lemma 3 can be checked in polynomial time.

We consider to check condition (iii) in Lemma 3. However, it is NP-hard to find a directed cycle in a digraph containing two specified vertices [9]. To overcome this difficulty, we consider a supergraph of $\mathcal{G}'(D)$, which is a key ingredient in our algorithm.

For $t \in \mathbb{R}_+$, let D_t denote the subgraph of D induced by the edges of label t. We consider the following condition instead of (iii):

(iii)' for some $t \in \mathbb{R}_+$, r_1 and r_2 are contained in the same strongly connected component in D_t, which is t-labeled extendible.

We can see that (iii)' is a relaxation of (iii) as follows. If r_1 and r_2 are contained in a t-labeled extendible directed cycle C, then they are contained in the same connected component H of D_t. Furthermore, since C is t-labeled extendible, so is H, which means that (iii)' holds.

Define $\hat{\mathcal{G}}(D)$ as the graph whose vertex set is the same as $\mathcal{G}'(D)$, and v_{r_1} and v_{r_2} are adjacent in $\mathcal{G}'(D)$ if and only if r_1 and r_2 satisfy (i), (ii), or (iii)'. Since (iii)' is a relaxation of (iii), Lemma 3 shows that $\hat{\mathcal{G}}(D)$ is a supergraph of $\mathcal{G}'(D)$. We now show the following lemma.

Lemma 4. *Let r_1 and r_2 be distinct vertices in D that satisfy condition (iii)'. Then, $\mathcal{G}'(D)$ contains a path between v_{r_1} and v_{r_2}.*

Proof. Let H be the t-labeled extendible strongly connected component in D_t that contains r_1 and r_2, where $t \in \mathbb{R}_+$. Since H is strongly connected, it contains a directed path from r_1 to r_2 that traverses vertices $p_0, p_1, \ldots, p_k$ in this order, where $p_0 = r_1$ and $p_k = r_2$. Then, for $0 \leq i \leq k-1$, H has a directed cycle C_i containing arc (p_i, p_{i+1}) as H is strongly connected. Since H is t-labeled extendible and strongly connected, we see that C_i is also t-labeled extendible. Therefore, condition (iii) in Lemma 3 shows that v_{p_i} and $v_{p_{i+1}}$ are adjacent in $\mathcal{G}'(D)$, which implies that $\mathcal{G}'(D)$ contains a path connecting $v_{r_1} = v_{p_0}$ and $v_{r_2} = v_{p_k}$. □

By this lemma and by the fact that $\hat{\mathcal{G}}(D)$ is a supergraph of $\mathcal{G}'(D)$, we obtain the following lemma.

Lemma 5. *For any distinct vertices r_1 and r_2 in D, $\hat{\mathcal{G}}(D)$ has a v_{r_1}-v_{r_2} path if and only if $\mathcal{G}'(D)$ has one.* □

This lemma shows that it suffices to check the reachability in $\hat{\mathcal{G}}(D)$ to solve Time-respecting Arborescence Reconfiguration. For distinct vertices r_1 and r_2 in D, (i) and (ii) can be checked in polynomial time as described above. We can check condition (iii)' by applying the following algorithm for each $t \in \{\lambda(e) \mid e \in A(D)\}$.

1. Construct the subgraph D_t of D induced by the edges of label t.
2. If r_1 and r_2 are contained in the same strongly connected component H in D_t, then go to Step 3. Otherwise, (iii)' does not hold for the current t.
3. Contract H into a vertex r_H to obtain a digraph D'. Remove all the arcs $e \in A(D')$ with $\lambda(e) < t$ from D' and find a time-respecting r_H-arborescence in this digraph by Corollary 1.
4. If a time-respecting r_H-arborescence is found, then r_1 and r_2 satisfy condition (iii)'. Otherwise, (iii)' does not hold for the current t.

Since we can decompose a digraph into strongly connected components in polynomial time, this algorithm runs in polynomial time. Therefore, by checking (i), (ii), and (iii)' for every pair of r_1 and r_2, we can construct $\hat{\mathcal{G}}(D)$ in polynomial time.

By Theorem 1 and Lemma 5, a time-respecting r_1-arborescence T_1 can be reconfigured to a time-respecting r_2-arborescence T_2 if and only if $\hat{\mathcal{G}}(D)$ contains a v_{r_1}-v_{r_2} path, which can be checked in polynomial time. This completes the proof of Theorem 2. □

6 NP-Completeness of Shortest Reconfiguration

For two time-respecting arborescences T_1 and T_2 in a digraph $D = (V, A)$ with $\lambda \colon A \to \mathbb{R}_+$ and for a positive integer ℓ, we consider the problem of determining whether there exists a reconfiguration sequence from T_1 to T_2 of length at most ℓ, which we call TIME-RESPECTING ARBORESCENCE SHORTEST RECONFIGURATION. Note that the length is defined as the number of swap operations, which is equal to the number of time-respecting arborescences in the sequence minus one. In this section, we prove the NP-completeness of this problem.

Theorem 3. TIME-RESPECTING ARBORESCENCE SHORTEST RECONFIGURATION *is NP-complete.*

Proof. The proof of Theorem 2 shows that if T_1 is reconfigurable to T_2, then there exists a reconfiguration sequence whose length is bounded by a polynomial in $|V|$. This implies that TIME-RESPECTING ARBORESCENCE SHORTEST RECONFIGURATION is in NP.

To show the NP-hardness, we reduce VERTEX COVER to TIME-RESPECTING ARBORESCENCE SHORTEST RECONFIGURATION. Recall that, in VERTEX COVER, we are given a graph $G = (V, E)$ and a positive integer k, and the task is to determine whether G contains a vertex cover of size at most k or not.

Suppose that $G = (V, E)$ and k form an instance of VERTEX COVER. We construct a digraph $D = (W, A)$ with multiple arcs as follows:

$$\begin{aligned}
W &= \{r_1, r_2\} \cup \{w_v \mid v \in V\} \cup \{w_e \mid e \in E\},\\
A_1 &= \{(r_1, w_e), (r_2, w_e) \mid e \in E\}, \quad A_2 = \{(r_1, r_2), (r_2, r_1)\},\\
A_3 &= \{a_v = (r_2, w_v) \mid v \in V\}, \quad A_4 = \{(w_v, w_e) \mid e \in \delta_G(v)\},\\
A_5 &= \{a'_v = (r_2, w_v) \mid v \in V\}, \quad A = A_1 \cup A_2 \cup A_3 \cup A_4 \cup A_5.
\end{aligned}$$

Here, w_v and w_e are newly introduced vertices associated with $v \in V$ and $e \in E$, respectively. Note that $a_v \in A_3$ and $a'_v \in A_5$ are distinct, that is, they form multiple arcs. For $i \in \{1, 2, 3, 4, 5\}$ and for $a \in A_i$, we define $\lambda(a) = i$. Let $\ell = 2|E| + 2k + 1$. Let

$$\begin{aligned}
T_1 &= \{(r_1, r_2)\} \cup \{(r_1, w_e) \mid e \in E\} \cup A_5,\\
T_2 &= \{(r_2, r_1)\} \cup \{(r_2, w_e) \mid e \in E\} \cup A_5.
\end{aligned}$$

One can easily see that T_1 and T_2 are time-respecting arborescences in D. This completes the construction of an instance of TIME-RESPECTING ARBORESCENCE SHORTEST RECONFIGURATION.

The following lemma ensures that by this construction we can reduce VERTEX COVER to TIME-RESPECTING ARBORESCENCE SHORTEST RECONFIGURATION. In the current version, the proof of the lemma is omitted, which can be found in the full version [14].

Lemma 6. *G contains a vertex cover of size at most k if and only if D admits a reconfiguration sequence from T_1 to T_2 of length at most ℓ.*

Hence TIME-RESPECTING ARBORESCENCE SHORTEST RECONFIGURATION is NP-hard. □

References

1. Biasi, M.D., Ophelders, T.: The complexity of snake and undirected NCL variants. Theor. Comput. Sci. **748**, 55–65 (2018). https://doi.org/10.1016/j.tcs.2017.10.031
2. Bousquet, N., et al.: Reconfiguration of spanning trees with many or few leaves. In: Grandoni, F., Herman, G., Sanders, P. (eds.) Proceedings of the 28th Annual European Symposium on Algorithms (ESA 2020). Leibniz International Proceedings in Informatics, Wadern, Germany, vol. 173, pp. 24:1–24:15. Schloss Dagstuhl - Leibniz-Zentrum für Informatik (2020). https://doi.org/10.4230/LIPIcs.ESA.2020.24
3. Bousquet, N., et al.: Reconfiguration of spanning trees with degree constraint or diameter constraint. In: Berenbrink, P., Monmege, B. (eds.) Proceedings of the 39th International Symposium on Theoretical Aspects of Computer Science (STACS 2022). Leibniz International Proceedings in Informatics, Wadern, Germany, vol. 219, pp. 15:1–15:21. Schloss Dagstuhl - Leibniz-Zentrum für Informatik (2022). https://doi.org/10.4230/LIPIcs.STACS.2022.15
4. Chu, Y.J.: On the shortest arborescence of a directed graph. Sci. Sinica **14**, 1396–1400 (1965)
5. Demaine, E.D., et al.: Reconfiguring undirected paths. In: Friggstad, Z., Sack, J.-R., Salavatipour, M.R. (eds.) WADS 2019. LNCS, vol. 11646, pp. 353–365. Springer, Cham (2019). https://doi.org/10.1007/978-3-030-24766-9_26
6. Edmonds, J.: Edge-disjoint branchings. In: Rustin, R. (ed.) Combinatorial Algorithms, pp. 91–96. Academic Press, New York (1973)
7. Edmonds, J.: Optimum branchings. J. Res. Natl. Bureau Stand. B **71**(4), 233–240 (1967). https://doi.org/10.6028/jres.071b.032
8. Edmonds, J.: Matroids and the greedy algorithm. Math. Program. **1**(1), 127–136 (1971). https://doi.org/10.1007/BF01584082
9. Fortune, S., Hopcroft, J., Wyllie, J.: The directed subgraph homeomorphism problem. Theor. Comput. Sci. **10**(2), 111–121 (1980). https://doi.org/10.1016/0304-3975(80)90009-2
10. Georgiadis, L.: Arborescence optimization problems solvable by Edmonds' algorithm. Theor. Comput. Sci. **301**(1), 427–437 (2003). https://doi.org/10.1016/S0304-3975(02)00888-5
11. Gupta, S., Sa'ar, G., Zehavi, M.: The parameterized complexity of motion planning for snake-like robots. J. Artif. Intell. Res. **69**, 191–229 (2020). https://doi.org/10.1613/jair.1.11864
12. Hanaka, T., et al.: Reconfiguring spanning and induced subgraphs. Theor. Comput. Sci. **806**, 553–566 (2020). https://doi.org/10.1016/j.tcs.2019.09.018
13. Ito, T., et al.: On the complexity of reconfiguration problems. Theor. Comput. Sci. **412**(12–14), 1054–1065 (2011). https://doi.org/10.1016/j.tcs.2010.12.005
14. Ito, T., et al.: Reconfiguration of time-respecting arborescences. CoRR arXiv:2305.07262 (2023)
15. Ito, T., Iwamasa, Y., Kobayashi, Y., Nakahata, Y., Otachi, Y., Wasa, K.: Reconfiguring (non-spanning) arborescences. Theor. Comput. Sci. **943**, 131–141 (2023). https://doi.org/10.1016/j.tcs.2022.12.007
16. Kempe, D., Kleinberg, J.M., Kumar, A.: Connectivity and inference problems for temporal networks. J. Comput. Syst. Sci. **64**(4), 820–842 (2002). https://doi.org/10.1006/jcss.2002.1829
17. Nishimura, N.: Introduction to reconfiguration. Algorithms **11**(4), 52 (2018). https://doi.org/10.3390/a11040052

Algorithmic Theory of Qubit Routing

Takehiro Ito[1(✉)], Naonori Kakimura[2(✉)], Naoyuki Kamiyama[3(✉)], Yusuke Kobayashi[4(✉)], and Yoshio Okamoto[5(✉)]

[1] Graduate School of Information Sciences, Tohoku University, Sendai, Japan
takehiro@tohoku.ac.jp

[2] Faculty of Science and Technology, Keio University, Yokohama, Japan
kakimura@math.keio.ac.jp

[3] Institute of Mathematics for Industry, Kyushu University, Fukuoka, Japan
kamiyama@imi.kyushu-u.ac.jp

[4] Research Institute for Mathematical Sciences, Kyoto University, Kyoto, Japan
yusuke@kurims.kyoto-u.ac.jp

[5] Graduate School of Informatics and Engineering, The University of Electro-Communications, Chofu, Japan
okamotoy@uec.ac.jp

Abstract. The qubit routing problem, also known as the swap minimization problem, is a (classical) combinatorial optimization problem that arises in the design of compilers of quantum programs. We study the qubit routing problem from the viewpoint of theoretical computer science, while most of the existing studies investigated the practical aspects. We concentrate on the linear nearest neighbor (LNN) architectures of quantum computers, in which the graph topology is a path. Our results are three-fold. (1) We prove that the qubit routing problem is NP-hard. (2) We give a fixed-parameter algorithm when the number of two-qubit gates is a parameter. (3) We give a polynomial-time algorithm when each qubit is involved in at most one two-qubit gate.

Keywords: Qubit routing · Qubit allocation · Fixed-parameter tractability

1 Introduction

The qubit routing problem captures a (classical) combinatorial problem in designing compilers of quantum programs. We rely on the formulation introduced by Siraichi et al. [12]. In their setting, a quantum program is designed as a quantum circuit. In a quantum circuit, there are wires and quantum gates such as Hadamard gates and controlled NOT gates. Each wire corresponds to one quantum bit (or a qubit for short) and gates operate on one or more qubits

This work was partially supported by JSPS KAKENHI Grant Numbers JP19K11814, JP20H05793, JP20H05795, JP20K11670, JP20K11692, JP21H03397, JP22H05001, JP23K10982. A full version is available at https://arxiv.org/abs/2305.02059.

P. Morin and S. Suri (Eds.): WADS 2023, LNCS 14079, pp. 533–546, 2023.
https://doi.org/10.1007/978-3-031-38906-1_35

at a time. A quantum circuit is designed at the logic level and needs to be implemented at the physical level. This requires a mapping of logical qubits to physical qubits in such a way that all the gate operations can be performed. However, due to physical restrictions, some sets of qubits could be mapped to places where an operation on those qubits cannot be physically performed. This problem is essential for some of the superconducting quantum computers such as IBM Quantum systems. Figure 1 shows the graph topology of such computers.

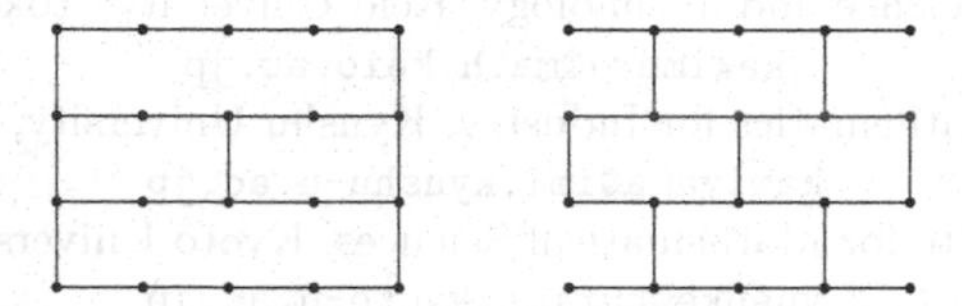

Fig. 1. The graph topology of IBM quantum devices called "Johannesburg" (left) and "Almaden" (right). Each vertex represents a physical qubit, and we may perform a two-input gate operation only for a pair of adjacent qubits. In our problem formulation, the graph topology is taken into account as the graph G. Source: https://www.ibm.com/blogs/research/2019/09/quantum-computation-center/.

To overcome this issue, we insert so-called swap gates into the quantum circuit. A swap gate can swap two qubits on different wires, and can be implemented by a sequence of three controlled NOT gates. With swap operations, we will be able to perform all the necessary gate operations that were designed in the original circuit. The process is divided into two phases. In the first phase, we find an initial mapping of logical qubits to physical qubits. In the second phase, we determine how swap operations are performed. Since a swap operation incurs a cost, we want to minimize the number of swap operations. The quantum routing problem focuses on the second phase of the process.

Several methods have been proposed to solve the quantum routing problem. The existing research has mainly focused on the practical aspects of the problem. On the other hand, the theoretical investigation has been scarce. Some authors have claimed that the problem is NP-hard, but no formal proof was given to date. Until this paper, it was not known whether the problem can be solved in polynomial time for the simplest case where the physical placement of qubits forms a path and the program has no two gate operations that can be performed at the same time. A path corresponds to the case of the linear nearest neighbor architectures that have been extensively studied in the literature of quantum computing (e.g. [11]).

This paper focuses on the theoretical aspects of the quantum routing problem and gives a better understanding of the problem from the viewpoint of theoretical computer science. We are mainly concerned with the case where the physical placement of qubits forms a path. Under this restriction, we obtain the following results.

1. We prove that the qubit routing problem is NP-hard even when the program has no two gate operations that can be performed at the same time (Theorem 1).
2. We give a fixed-parameter algorithm when the number of gates in a given quantum circuit is a parameter (Theorem 4).
3. We give a polynomial-time algorithm when each qubit is involved in at most one two-qubit operation (Theorem 5).

As a side result, we also prove that the problem is NP-hard when the physical placement of qubits forms a star and any set of gate operations can be performed at the same time. See the full version for the details.

Problem Formulation. We formulate the problem Qubit Routing in a purely combinatorial way as follows. As input, we are given an undirected graph $G = (V, E)$, a partially ordered set (a poset for short) $P = (S, \preceq)$, a set T of tokens, a map $\varphi\colon S \to \binom{T}{2}$, where $\binom{T}{2}$ denotes the set of unordered pairs of T, and an initial token placement $f_0\colon V \to T$, which is defined as a bijection. The undirected graph G corresponds to the physical architecture of a quantum computer in which each vertex corresponds to a physical qubit and an edge corresponds to a pair of qubits on which a gate operation can be performed. The poset P corresponds to a quantum circuit that we want to run on the quantum computer. Each token in T corresponds to a logical qubit. The token placement f_0 corresponds to an initial mapping of the logical qubits in T to the physical qubits in V (e.g., as a result of the qubit allocation, see "Related Work" below). For the notational convenience, we regard f_0 as a mapping from V to T; this is not mathematically relevant since f_0 is bijective and to construct a mapping from T to V, one uses the inverse f_0^{-1}. The bijectivity of a token placement is irrelevant since if there are more logic qubits than physical qubits, then we may introduce dummy logical qubits so that their numbers can be equal. Each element in S corresponds to the pair of logical qubits on which a gate operation is performed. The correspondence is given by φ. We note that φ does not have to be injective. The poset P determines the order along which the gate operations determined by φ are performed. The order is partial since some pairs of operations may be performed independently: in that case, the corresponding elements of S are incomparable in P. An example is given in Fig. 2.

A *swap* $f_i \rightsquigarrow f_{i+1}$ is defined as an operation that transforms one token placement (i.e., a bijection) $f_i\colon V \to T$ to another token placement $f_{i+1}\colon V \to T$ such that there exists an edge $\{u, v\} \in E$ such that $f_{i+1}(u) = f_i(v)$, $f_{i+1}(v) = f_i(u)$ and $f_{i+1}(x) = f_i(x)$ for all $x \in V \setminus \{u, v\}$. This corresponds to a swap operation of two logical qubits $f_i(u)$ and $f_i(v)$ assigned to two different physical qubits u and v.

As an output of the qubit routing problem, we want a swap sequence $f_0 \rightsquigarrow f_1 \rightsquigarrow \cdots \rightsquigarrow f_\ell$ that satisfies the following condition: there exists a map $i\colon S \to \{0, 1, 2, \ldots, \ell\}$ such that $f_{i(s)}^{-1}(\varphi(s)) \in E$ for every $s \in S$ and if $s \preceq s'$, then $i(s) \leq i(s')$. A swap sequence with this condition is said to be *feasible*. The objective is to minimize the length ℓ of a swap sequence. The condition states

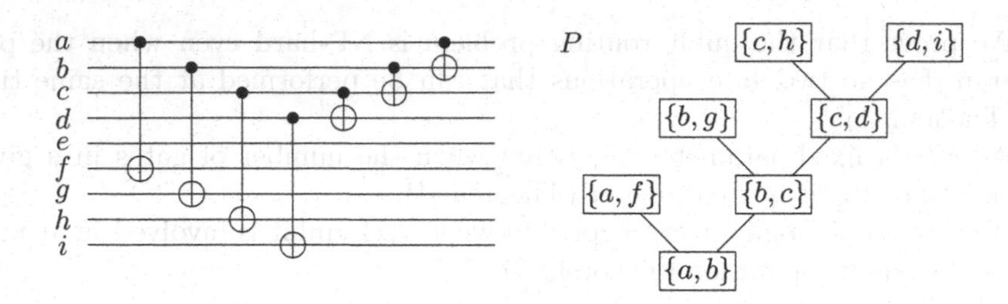

Fig. 2. The left figure shows an instance of a quantum circuit, which is a part of the circuits given in [2]. The right figure shows the Hasse diagram of the corresponding poset P in our problem formulation.

that all the operations in the quantum circuit that corresponds to the poset P are performed as they follow the order P.

In summary, our problem is described as follows.

Qubit Routing
Input. A graph $G = (V, E)$, a poset $P = (S, \preceq)$, a set T of tokens, a map $\varphi\colon S \to \binom{T}{2}$, and an initial token placement $f_0\colon V \to T$.
Question. Find a feasible swap sequence of minimum length.

Note that the minimum length of a feasible swap sequence is at most $|V||S|$ (if G is connected), which implies that determining the existence of a feasible swap sequence of length bounded by a given number is in NP.

Related Work. The qubit allocation problem was introduced by Siraichi et al. [12]. Following Nannicini et al. [10], we divide the task in the qubit allocation problem into two phases. The first phase involves qubit assignment that gives an initial placement of the logical qubits on the physical qubits, and the second phase involves qubit routing that inserts swap operations at appropriate positions so that all the gate operations in a given circuit can be performed. The problem formulation in this paper solves the second-phase problem. The qubit routing problem is often called the swap minimization, too.

For qubit routing, several heuristic algorithms have been given [8,12,13], and integer-programming formulations have been given [6,10] that attempt to solve problem instances to optimality. Siraichi et al. [12,13] wrote that the qubit routing problem is NP-hard since it is more general than the so-called token swapping problem [14] that is known to be NP-hard [3,7,9] even for trees [1]. However, their argument was informal and no formal proof was not found.

A similar problem was studied by Botea, Kishimoto, and Marinescu [4]. In their problem, a mapping of logical qubits to physical qubits is injective but not bijective, and a logical qubit can only be moved to a physical qubit that is not occupied by another logical qubit. They proved that the makespan minimization in their setting is NP-hard.

2 Hardness: Paths and Chains

In this section, we show that the problem is NP-hard even when G is a path and P is a chain (i.e., a totally ordered set).

Theorem 1. QUBIT ROUTING *is NP-hard even when G is a path and P is a chain.*

To show this theorem, we first introduce notation. For a swap sequence $f_0 \rightsquigarrow f_1 \rightsquigarrow \cdots \rightsquigarrow f_\ell$, which is denoted by $\mathbf{f}$, we say that f_0 and f_ℓ are the *initial* and *target* token placements of $\mathbf{f}$, respectively. We also say that $\mathbf{f}$ is from f_0 to f_ℓ. The length ℓ of $\mathbf{f}$ is denoted by $|\mathbf{f}|$. For two swap sequences $\mathbf{f}_1$ and $\mathbf{f}_2$, if the target token placement of $\mathbf{f}_1$ is equal to the initial token placement of $\mathbf{f}_2$, then its concatenation is denoted by $\mathbf{f}_1 \circ \mathbf{f}_2$. Note that in the concatenation of swap sequences, the target token placement of $\mathbf{f}_1$ and the initial token placement of $\mathbf{f}_2$ are identified, and thus $|\mathbf{f}_1 \circ \mathbf{f}_2| = |\mathbf{f}_1| + |\mathbf{f}_2|$. When the initial and the target token placements of $\mathbf{f}$ coincide, for a positive integer h, we denote $\mathbf{f}^h = \mathbf{f} \circ \mathbf{f} \circ \cdots \circ \mathbf{f}$, where $\mathbf{f}$ appears h times.

Throughout this section, we only consider the case where P is a chain. For a chain $P = (S, \preceq)$, let $s_1, s_2, \ldots, s_{|S|}$ be distinct elements in S such that $s_1 \prec s_2 \prec \cdots \prec s_{|S|}$. Then, the information of P and φ can be represented as a sequence $Q = (q_1, q_2, \ldots, q_{|S|})$, where $q_i := \varphi(s_i) \in \binom{T}{2}$ for each i. We say that a swap sequence $f_0 \rightsquigarrow f_1 \rightsquigarrow \cdots \rightsquigarrow f_\ell$ *realizes* Q if there exist $0 \leq i_1 \leq i_2 \leq \cdots \leq i_{|S|} \leq \ell$ such that $f_{i_j}^{-1}(q_j) \in E$ for $j = 1, \ldots, |S|$. In particular, if the swap sequence consisting of a single token placement f realizes Q, then we say that f realizes Q. With this terminology, when P is a chain, QUBIT ROUTING is to find a shortest swap sequence that realizes a given sequence of token pairs. For two sequences Q_1 and Q_2 of token pairs, its concatenation is denoted by $Q_1 \circ Q_2$. For a sequence Q of token pairs and a positive integer h, we denote $Q^h = Q \circ Q \circ \cdots \circ Q$, where Q appears h times.

To show the NP-hardness of QUBIT ROUTING, we reduce OPTIMAL LINEAR ARRANGEMENT, which is known to be NP-hard [5].

> OPTIMAL LINEAR ARRANGEMENT
> **Input.** A graph $H = (V(H), E(H))$ and a positive integer k.
> **Question.** Is there a bijection $g\colon V(H) \to \{1, 2, \ldots, |V(H)|\}$ that satisfies $\sum_{\{u,v\} \in E(H)} |g(u) - g(v)| \leq k$?

Suppose that we are given an instance of OPTIMAL LINEAR ARRANGEMENT that consists of a graph $H = (V(H), E(H))$ and a positive integer k. Denote $V(H) = \{v_1, v_2, \ldots, v_n\}$ and $E(H) = \{e_1, e_2, \ldots, e_m\}$, where $n = |V(H)|$ and $m = |E(H)|$. We may assume that $k < nm$ since otherwise, any bijection g is a solution to OPTIMAL LINEAR ARRANGEMENT. Let $\alpha = 2nm + 1$, which is an odd integer, and let β and γ be sufficiently large integers (e.g., $\beta = n^2\alpha$ and $\gamma = 4k\alpha$).

We now construct an instance of QUBIT ROUTING as follows. Define a set of tokens as $T = \{t_{v,i} \mid v \in V(H),\ i \in \{1, 2, \ldots, \alpha\}\}$. Let $G = (V, E)$ be a path with $n\alpha$ vertices. We define

$$\begin{aligned} Q_v &:= (\{t_{v,1}, t_{v,2}\}, \{t_{v,2}, t_{v,3}\}, \dots, \{t_{v,\alpha-1}, t_{v,\alpha}\}) && (v \in V(H)), \\ Q &:= Q_{v_1} \circ Q_{v_2} \circ \dots \circ Q_{v_n}, \\ \psi(e) &:= (\{t_{u,nm+1}, t_{v,nm+1}\}) && (e = \{u, v\} \in E(H)), \\ R &:= Q^\gamma \circ \psi(e_1) \circ Q^\gamma \circ \psi(e_2) \circ \dots \circ \psi(e_m) \circ Q^\gamma, \end{aligned}$$

and R^β is the sequence of token pairs that has to be realized. The initial token placement f_0 is defined arbitrarily. This gives an instance (G, T, R^β, f_0) of QUBIT ROUTING.

To show the validity of the reduction, it suffices to show the following, whose proof is given in the full version.

Proposition 1. *The above instance* (G, T, R^β, f_0) *of* QUBIT ROUTING *has a solution of length less than* $2\beta(k\alpha + \alpha - m)$ *if and only if the original instance of* OPTIMAL LINEAR ARRANGEMENT *has a solution of objective value at most* k.

This completes the proof of Theorem 1. □

3 Algorithm Parameterized by the Number of Gates

In this section, we assume that $G = (V, E)$ is a path. Let $k = |S|$ be the size of a poset $P = (S, \preceq)$. The purpose of this section is to design a fixed-parameter algorithm for the problem parameterized by k. Since G is a path, we suppose for simplicity that $V = \{1, 2, \dots, n\}$ and $E = \{\{i, i+1\} \mid i = 1, 2, \dots, n-1\}$.

We first observe that we may assume that a poset forms a chain. Indeed, suppose that P is not a chain. Then, if we have a fixed-parameter algorithm for a chain, we can apply the algorithm for all the linear extensions of P. Since the number of the linear extensions is at most $k!$, it is a fixed-parameter algorithm for P. Thus, we may assume that $S = \{s_1, s_2, \dots, s_k\}$ such that $s_1 \prec s_2 \prec \dots \prec s_k$. Let $\tilde{T} = \bigcup_{s \in S} \varphi(s)$ and let $\tilde{k} = |\tilde{T}|$. Then, we have $\tilde{k} \leq 2k$. We denote $[\tilde{k}] = \{1, 2, \dots, \tilde{k}\}$.

Let f be a token placement. Let $\tilde{V}$ be the positions where tokens in $\tilde{T}$ are placed, i.e., $\tilde{V} = \{f^{-1}(t) \mid t \in \tilde{T}\}$. We denote $\tilde{V} = \{v_1, v_2, \dots, v_{\tilde{k}}\}$ where $v_1 < v_2 < \dots < v_{\tilde{k}}$. Define a vector $x \in \mathbb{Z}^{\tilde{k}}$ so that $x_i = v_{i+1} - v_i$, where $v_0 = 0$, for every index $i = 0, 1, \dots, \tilde{k}-1$. We note that $\sum_{i=0}^{\tilde{k}-1} x_i = v_{\tilde{k}} \leq n$ holds. We further define a bijection $\sigma \colon [\tilde{k}] \to \tilde{T}$ as $\sigma(i) = f(v_i)$. Let Σ denote the set of all bijections from $[\tilde{k}]$ to $\tilde{T}$. We call the pair (x, σ) the *signature* of the token placement f, which is denoted by $\mathsf{sig}(f)$. The signature maintains the information only on the tokens in $\tilde{T}$, which suffices for finding a shortest feasible swap sequence since swapping two tokens not in $\tilde{T}$ is redundant in the swap sequence.

We first present a polynomial-time algorithm when k is a fixed constant in Sect. 3.1, and then a fixed-parameter algorithm in Sect. 3.2.

3.1 Polynomial-Time Algorithm for a Fixed Constant k

Define $\mathcal{R} = \{x \in \mathbb{Z}^{\tilde{k}} \mid \sum x_j \leq n, x_j \geq 1\ (j = 0, 1, \ldots, \tilde{k} - 1)\}$ and $\mathcal{S} = \{(x, \sigma) \mid x \in \mathcal{R}, \sigma \in \Sigma\}$. We see that, for any $(x, \sigma) \in \mathcal{S}$, there exists a token placement f such that $\mathsf{sig}(f) = (x, \sigma)$, and such a placement f can be found in polynomial time even when k is not a constant. It holds that $|\mathcal{S}| = O(\tilde{k}! n^{\tilde{k}})$.

For two signatures (x^0, σ^0) and (x^t, σ^t), a *swap sequence from* (x^0, σ^0) *to* (x^t, σ^t) means a swap sequence $f_0 \rightsquigarrow f_1 \rightsquigarrow \cdots \rightsquigarrow f_\ell$ for some token placements f_0 and f_ℓ such that $\mathsf{sig}(f_0) = (x^0, \sigma^0)$ and $\mathsf{sig}(f_\ell) = (x^t, \sigma^t)$. Its length is defined to be ℓ.

We first show that we can find a shortest swap sequence between two signatures in polynomial time when k is a fixed constant.

Lemma 1. *For two signatures (x^0, σ^0) and (x^t, σ^t), we can find a shortest swap sequence from (x^0, σ^0) to (x^t, σ^t) in $O(\mathrm{poly}(|\mathcal{S}|))$ time.*

Proof. We construct a graph on the vertex set $\mathcal{S}$ such that (x, σ) and (x', σ') are adjacent if and only if there exist two token placements f, f' such that $\mathsf{sig}(f) = (x, \sigma)$, $\mathsf{sig}(f') = (x', \sigma')$, and $f \rightsquigarrow f'$. Then a path from (x^0, σ^0) to (x^t, σ^t) in the graph corresponds to a swap sequence from (x^0, σ^0) to (x^t, σ^t). Therefore, we can find a shortest swap sequence by finding a shortest path from (x^0, σ^0) to (x^t, σ^t) in the graph. Since the number of vertices of the constructed graph is $|\mathcal{S}|$, it can be done in $O(\mathrm{poly}(|\mathcal{S}|))$ time. □

The above lemma allows us to design a polynomial-time algorithm for a fixed constant k.

Theorem 2. *Let $P = (S, \preceq)$ be a chain of k elements, where k is a fixed constant. For a token placement f_0, we can find a shortest feasible swap sequence from f_0 in polynomial time.*

Proof. Let $\mathcal{S}_i = \{(x, \sigma) \in \mathcal{S} \mid \exists f \text{ s.t. } \mathsf{sig}(f) = (x, \sigma) \text{ and } f^{-1}(\varphi(s_i)) \in E\}$, which is the set of signatures (x, σ) that correspond to token placements in which $\varphi(s_i)$ are adjacent. Note that if $(x, \sigma) \in \mathcal{S}_i$, then $\mathsf{sig}(f) = (x, \sigma)$ implies $f^{-1}(\varphi(s_i)) \in E$ for any f, because $\varphi(s_i) \subseteq \tilde{T}$. Moreover, let $\mathcal{S}_0 = \{\mathsf{sig}(f_0)\}$.

Define a digraph $\mathcal{G} = (\bigcup_{i=0}^{k} \mathcal{S}_i, \bigcup_{i=0}^{k-1} E_i)$, where $E_i = \{((x, \sigma), (x', \sigma')) \mid (x, \sigma) \in \mathcal{S}_i, (x', \sigma') \in \mathcal{S}_{i+1}\}$ for $i = 0, 1, \ldots, k-1$. We suppose that $e = ((x, \sigma), (x', \sigma'))$ in E_i has a length equal to the shortest length of a swap sequence from (x, σ) to (x', σ'), which can be computed in polynomial time by Lemma 1.

We see that the shortest path from the vertex in $\mathcal{S}_0$ to some vertex in $\mathcal{S}_k$ corresponds to a shortest feasible swap sequence from f_0. The number of vertices of the graph is bounded by $O(k|\mathcal{S}|)$, which is polynomial when k is a constant. Thus, the theorem holds. □

3.2 Fixed-Parameter Algorithm

In this section, we present a fixed-parameter algorithm parameterized by k by dynamic programming.

We first observe that we can compute the shortest length of a swap sequence between two signatures with the same bijection σ.

Lemma 2. *Suppose that we are given two signatures (x, σ) and (y, σ) with the same bijection σ. Then, the shortest length of a swap sequence from (x, σ) to (y, σ) is equal to $\sum_{i=1}^{\tilde{k}} |v_i - w_i|$, where $v_i = \sum_{j=0}^{i-1} x_j$ and $w_i = \sum_{j=0}^{i-1} y_j$ for $i = 1, \ldots, \tilde{k}$. Moreover, there exists a shortest swap sequence such that all the token placements in the sequence have the same bijection σ in their signatures.*

Proof. Since the initial and target token placements have the same σ in their signature, we need $|w_i - v_i|$ swaps to move the token $\sigma(i)$ to w_i for any $i = 1, \ldots, \tilde{k}$. Thus the shortest swap sequence has length at least $\sum_{i=1}^{\tilde{k}} |v_i - w_i|$. To see that they are equal, we show the existence of a swap sequence of length $\sum_{i=1}^{\tilde{k}} |v_i - w_i|$ by induction on this value. If $\sum_{i=1}^{\tilde{k}} |v_i - w_i| = 0$, then the claim is obvious. Otherwise, let p be the minimum index such that $v_p \neq w_p$. By changing the roles of x and y if necessary, we may assume that $v_p > w_p$. Then, starting from (x, σ), we apply swap operations $v_p - w_p$ times to obtain a new signature (x', σ) such that $x'_{p-1} = x_{p-1} - (v_p - w_p)$ and $x'_p = x_p + (v_p - w_p)$. That is, $v'_i = v_i$ for $i \in [\tilde{k}] \setminus \{p\}$ and $v'_p = w_p$, where v'_i is defined as $v'_i = \sum_{j=0}^{i-1} x'_j$. Note that this operation is possible without changing the bijection σ, because $v'_p = w_p > w_{p-1} = v_{p-1}$ by the minimality of p. By the induction hypothesis, there exists a swap sequence of length $\sum_{i=1}^{\tilde{k}} |v'_i - w_i|$ between (x', σ) and (y, σ). Therefore, we obtain a swap sequence between (x, σ) and (y, σ) whose length is $|v_p - w_p| + \sum_{i=1}^{\tilde{k}} |v'_i - w_i| = \sum_{i=1}^{\tilde{k}} |v_i - w_i|$. Moreover, each token placement in the obtained swap sequence has the same bijection σ. □

By the lemma, the shortest length of a swap sequence from (x, σ) to (y, σ) does not depend on σ. Thus, we denote it by $d(x, y)$ for $x, y \in \mathcal{R}$.

Let $\mathbf{f}$ be a feasible swap sequence $f_0 \rightsquigarrow f_1 \rightsquigarrow \cdots \rightsquigarrow f_\ell$. Suppose that $\varphi(s_j)$ is realized at token placement f_{i_j}, that is, $f_{i_j}^{-1}(\varphi(s_j)) \in E$ for $j = 1, 2, \ldots, k$ and $i_1 \leq i_2 \leq \cdots \leq i_k$. We define $i_0 = 0$. Note that the number of distinct values in $\{i_1, \ldots, i_k\}$, denoted by α, is at most k. Also, let β be the number of times that σ in the signature changes in the swap sequence. We call $\alpha + \beta$ the *signature length* of $\mathbf{f}$.

The following lemma says that the signature length is bounded by a function of k, which we denote by $\ell_{\max}$.

Lemma 3. *For a shortest feasible swap sequence $f_0 \rightsquigarrow f_1 \rightsquigarrow \cdots \rightsquigarrow f_\ell$, the signature length is bounded by $((2k)! + 1)k$ from above.*

Proof. Consider the partial swap sequence $f_{i_{j-1}} \rightsquigarrow \cdots \rightsquigarrow f_{i_j}$ for $j = 1, 2, \ldots, k$. We observe that, in the partial swap sequence, the number of times that bijections change is at most $\tilde{k}!$. Indeed, in this partial sequence, token placements with the same bijection appear sequentially, as otherwise, we can short-cut between them by Lemma 2. Therefore, the total signature length is bounded by $k(\tilde{k}!) + k \leq k((2k)!) + k$. □

Let $g^\ell(x,\sigma,P)$ be the shortest length of a swap sequence to realize P from some token placement f_0 with $\mathsf{sig}(f_0)=(x,\sigma)$ such that it has signature length at most ℓ. In what follows, we derive a recursive equation on $g^\ell(x,\sigma,P)$ for dynamic programming.

We first give notation. For a bijection $\sigma \in \Sigma$ and a non-negative integer j with $1 \leq j \leq \tilde{k}-1$, let $\sigma_j \in \Sigma$ be the bijection obtained from σ by swapping the j-th and $(j+1)$-st tokens. We define $\mathcal{R}_j = \{x \in \mathcal{R} \mid x_j = 1\}$, which is the set of signatures such that the j-th token and the $(j+1)$-st token are adjacent.

To derive a recursive equation on $g^\ell(x,\sigma,P)$, consider the following two cases in which the signature length is decreased at least by one, separately.

The first case is when the bijection σ is changed, i.e., β decreases. Suppose that we change σ to σ_j for some $1 \leq j \leq \tilde{k}-1$. In this case, we first move to (x',σ) for some $x' \in \mathcal{R}_j$, and then change (x',σ) to (x',σ_j). By Lemma 2, the number of swaps is $d(x,x')+1$. After moving to (x',σ_j), we can recursively consider finding a shortest swap sequence to realize P from (x',σ_j) with the signature length at most $\ell-1$. Therefore, the total length in this case is $g^{\ell-1}(x',\sigma_j,P)+d(x,x')+1$, and hence the shortest length when we change σ is equal to

$$\min_{1\leq j\leq \tilde{k}-1}\ \min_{x'\in\mathcal{R}_j} \left\{g^{\ell-1}(x',\sigma_j,P)+d(x,x')+1\right\}.$$

The other case is when s_1 is realized without changing σ, i.e., α decreases. Then, it is necessary that $s_1 = (\sigma(h),\sigma(h+1))$ for some $1 \leq h \leq \tilde{k}-1$. To realize s_1, we move (x,σ) to (x',σ) for some $x' \in \mathcal{R}_h$. By recursion, the total length in this case is $g^{\ell-1}(x',\sigma,P')+d(x,x')$ by Lemma 2, where P' is the poset obtained from P by removing the first element s_1, that is, P' forms the chain $s_2 \prec s_3 \prec \cdots \prec s_k$. Thus, the shortest length in this case is

$$\min_{x'\in\mathcal{R}_h} \left\{g^{\ell-1}(x',\sigma,P')+d(x,x')\right\}.$$

In summary, we have that, for any $x \in \mathcal{R}$, $\sigma \in \Sigma$, and $1 \leq \ell \leq \ell_{\max}$,

$$g^\ell(x,\sigma,P) = \min\left\{ \min_{1\leq j\leq \tilde{k}-1}\ \min_{x'\in\mathcal{R}_j} \left\{g^{\ell-1}(x',\sigma_j,P)+d(x,x')+1\right\},\right.$$
$$\left.\min_{x'\in\mathcal{R}_h} \left\{g^{\ell-1}(x',\sigma,P')+d(x,x')\right\}\right\}, \tag{1}$$

where $s_1 = (\sigma(h),\sigma(h+1))$ for some h. If such h does not exist, the second term is defined to be $+\infty$.

It follows from (1) that we can design a dynamic programming algorithm. However, the running time would become $O(k\cdot k!\ell_{\max}|\mathcal{R}|)$, and this does not give a fixed-parameter algorithm since $|\mathcal{R}| = O(n^{\tilde{k}})$. In what follows, we will reduce the running time by showing that the minimum is achieved at an extreme point.

For $i = 0,1,\ldots \tilde{k}-1$, let e_i denote the unit vector whose i-th entry is one and the other entries are zeros. For a vector $x \in \mathcal{R}$ and $1 \leq j \leq \tilde{k}-1$, define $N_j(x)$ as

$$N_j(x) = \{x + ae_{j-1} - (x_j-1)e_j + be_{j+1} \mid a+b = x_j - 1,\ a,b \in \mathbb{Z}_+\},$$

where we regard $e_{\tilde{k}}$ as the zero vector to simplify the notation. Then, $x' \in N_j(x)$ satisfies that

$$x'_j = 1,$$
$$x'_{j-1} + x'_{j+1} = x_{j-1} + x_j + x_{j+1} - 1,$$
$$x'_i = x_i \text{ for } i \notin \{j-1, j, j+1\},$$

where $x_{\tilde{k}} = n - \sum_{i=0}^{\tilde{k}-1} x_i$ and $x'_{\tilde{k}} = n - \sum_{i=0}^{\tilde{k}-1} x'_i$. The signature (x', σ) with $x' \in N_j(x)$ means that it is obtained from (x, σ) by only moving two tokens $\sigma(j)$ and $\sigma(j+1)$ so that the two tokens are adjacent. Moreover, define y^j and y'^j to be vectors in $N_j(x)$ in which $(a, b) = (0, x_j - 1)$ and $(a, b) = (x_j - 1, 0)$, respectively. Then,

$$(y^j)_{j-1} = x_{j-1}, \qquad (y^j)_{j+1} = x_j + x_{j+1} - 1,$$
$$(y'^j)_{j-1} = x_{j-1} + x_j - 1, \qquad (y'^j)_{j+1} = x_{j+1}.$$

Thus, (y^j, σ) ((y'^j, σ), resp.,) is obtained from (x, σ) by only moving one token $\sigma(j+1)$ ($\sigma(j)$, resp.,) so that $\sigma(j)$ and $\sigma(j+1)$ are adjacent.

The following theorem asserts that the minimum of (1) is achieved at either y^j or y'^j.

Theorem 3. *For any $x \in \mathcal{R}$, $\sigma \in \Sigma$, and $1 \le \ell \le \ell_{\max}$, it holds that*

$$g^\ell(x, \sigma, P) = \min \left\{ \min_{1 \le j \le \tilde{k}-1, y \in \{y^j, y'^j\}} \{g^{\ell-1}(y, \sigma_j, P) + d(x, y) + 1\}, \right.$$
$$\left. \min_{y \in \{y^h, y'^h\}} \{g^{\ell-1}(y, \sigma, P') + d(x, y)\} \right\}.$$

To prove the theorem, we show the following two lemmas; their proofs can be found in the full version. We first show that the minimum of (1) is achieved at some point in $N_j(x)$.

Lemma 4. *It holds that, for any $x \in \mathcal{R}$, $\sigma \in \Sigma$, $1 \le \ell \le \ell_{\max}$ and $1 \le j \le \tilde{k}-1$,*

$$\min_{x' \in \mathcal{R}_j} \{g^{\ell-1}(x', \sigma, P) + d(x, x')\} = \min_{y \in N_j(x)} \{g^{\ell-1}(y, \sigma, P) + d(x, y)\}.$$

□

By Lemma 4, it holds that

$$g^\ell(x, \sigma, P) = \min \left\{ \min_{1 \le j \le \tilde{k}-1} \min_{y \in N_j(x)} \{g^{\ell-1}(y, \sigma_j, P) + x_j\}, \right.$$
$$\left. \min_{y \in N_h(x)} \{g^{\ell-1}(y, \sigma, P') + x_h - 1\} \right\}, \tag{2}$$

in which we note $d(x, y) = x_j - 1$ for $y \in N_j(x)$.

Lemma 5. *The function $g^\ell(x, \sigma, P)$ can be expressed as the minimum of linear functions on x. That is, for any $\sigma \in \Sigma$ and $1 \leq \ell \leq \ell_{\max}$, there exist vectors $c_1, \ldots, c_p$ and real numbers $\delta_1, \ldots, \delta_p$ such that*

$$g^\ell(x, \sigma, P) = \min_{1 \leq i \leq p} \{c_i x + \delta_i\}$$

for any $x \in \mathcal{R}$. □

Proof. (Proof of Theorem 3). By (2), it suffices to show that, for any $\sigma \in \Sigma$ and $1 \leq j \leq \tilde{k} - 1$, either y^j or y'^j achieves the minimum of

$$\min_{y \in N_j(x)} \left\{g^{\ell-1}(y, \sigma, P) + x_j\right\}.$$

Since $g^{\ell-1}$ is in the form of the minimum of linear functions by Lemma 5, the above can be expressed as

$$\min_{1 \leq i \leq p'} \min_{a, b \in \mathbb{Z}_+ : a+b = x_j - 1} \{c'_i(x + ae_{j-1} - (x_j - 1)e_j + be_{j+1}) + x_j + \delta'_i\}$$

for some linear functions $c'_i y + \delta'_i$ for $i = 1, \ldots, p'$. For given x and j, the minimum is attained either at $(a, b) = (x_j - 1, 0)$ or at $(a, b) = (0, x_j - 1)$. They correspond to y^j and y'^j, respectively. Thus, the theorem holds. □

Let f_0 be the initial placement and $\mathsf{sig}(f_0) = (x^0, \sigma^0)$. Let $v_i^0 = \sum_{j=0}^{i-1} x_j^0$ for each i. For $i = 0, 1, \ldots, \tilde{k} - 1$, define $B_i = \{f_0(v_i^0 + 1), f_0(v_i^0 + 2), \ldots, f_0(v_{i+1}^0 - 1)\}$, which are tokens not in $\tilde{T}$. Then, $|B_i| = x_i^0 - 1$. Theorem 3 shows that, when we compute $g^\ell(x, \sigma, P)$ by using the formula, it suffices to consider signatures (y, σ') such that all tokens in B_i appear consecutively along the path. That is, we only consider vectors y such that, for each i, $y_i - 1 = \sum_{h=i'}^{j'} |B_h|$ for some $0 \leq i', j' \leq \tilde{k} - 1$ (possibly, $y_i - 1 = 0$). This shows that each y_i can take one of $O(\tilde{k}^2)$ values, and hence y has $O(\tilde{k}^{2\tilde{k}})$ choices. Therefore, the size of the DP table is $O(k \cdot k! \ell_{\max} \tilde{k}^{2\tilde{k}})$, and each step can be processed in a fixed-parameter time. Thus, we have the following theorem.

Theorem 4. *There exists a fixed-parameter algorithm for* Qubit Routing *when G is a path and $k = |S|$ is a parameter.* □

4 Polynomial-Time Algorithm for Disjoint Two-Qubit Operations

We say that an instance $(G, P = (S, \preceq), T, \varphi, f_0)$ of Qubit Routing has *disjoint pairs* if $\varphi(s) \cap \varphi(s') = \emptyset$ for any pair of distinct elements $s, s' \in S$. The objective of this section is to give a polynomial-time algorithm for instances with disjoint pairs when the graph G is a path.

Theorem 5. Qubit Routing *can be solved in polynomial time when a given graph is a path and the instance has disjoint pairs.*

Let $(G, P = (S, \preceq), T, \varphi, f_0)$ be an instance with disjoint pairs. Since G is a path, we suppose for simplicity that $V = \{1, 2, \ldots, n\}$ and $E = \{\{i, i+1\} \mid i = 1, 2, \ldots, n-1\}$. Let f be a token placement. For an element $s \in S$ with $f^{-1}(\varphi(s)) = \{\alpha_1, \alpha_2\}$, $\alpha_1 < \alpha_2$, we define $\mathsf{gap}_f(s) := \alpha_2 - \alpha_1 - 1$. Then $\mathsf{gap}_f(s)$ is a lower bound on the number of swaps to realize $\varphi(s)$ if the initial token placement is f. For distinct elements $s, s' \in S$ such that $f^{-1}(\varphi(s)) = \{\alpha_1, \alpha_2\}$, $\alpha_1 < \alpha_2$, and $f^{-1}(\varphi(s')) = \{\beta_1, \beta_2\}$, $\beta_1 < \beta_2$, we say that s and s' *cross* if $\alpha_1 < \beta_1 < \alpha_2 < \beta_2$ or $\beta_1 < \alpha_1 < \beta_2 < \alpha_2$. One can observe that, if $S = \{s, s'\}$ such that s and s' cross, we can realize both $\varphi(s)$ and $\varphi(s')$ by $\mathsf{gap}_f(s) + \mathsf{gap}_f(s') - 1$ swaps. Note that $\alpha_1, \alpha_2, \beta_1$, and β_2 are always distinct since the instance has disjoint pairs. We define

$$\mathsf{gap}(f) := \sum_{s \in S} \mathsf{gap}_f(s),$$

$$\mathsf{cross}(f) := \left| \left\{ \{s, s'\} \in \binom{S}{2} \;\middle|\; s \text{ and } s' \text{ cross} \right\} \right|,$$

$$\mathsf{value}(f) := \mathsf{gap}(f) - \mathsf{cross}(f),$$

where gap, cross, and value are regarded as functions only in f by fixing G, P, T, and φ.

We show the following proposition, whose proof is given in the full version.

Proposition 2. *The optimal value for the instance* $(G, P = (S, \preceq), T, \varphi, f_0)$ *is* $\mathsf{value}(f_0)$. *Such a sequence can be constructed in polynomial time.*

This completes the proof of Theorem 5. □

5 Concluding Remarks

We initiated algorithmic studies on the quantum routing problem, also known as the swap minimization problem, from the viewpoint of theoretical computer science. The problem is of central importance in compiler design for quantum programs when they are implemented in some of the superconducting quantum computers such as IBM Quantum systems.

Most notably, we proved the quantum routing problem is NP-hard even when the graph topology of a quantum computer is a path, which corresponds to the so-called linear nearest neighbor architecture. In our proof, the initial token placement can be chosen arbitrarily. This implies that the combined optimization of the quantum assignment and the quantum routing is also NP-hard for the same architecture.

We also gave some algorithmic results, but they were restricted to the case of the linear nearest neighbor architectures. Possible future work is to give algorithmic results with theoretical guarantees for other graph topologies.

Acknowledgment. We thank Toshinari Itoko at IBM Research Tokyo for bringing the qubit allocation problem to our attention and for helpful comments.

References

1. Aichholzer, O., et al.: Hardness of token swapping on trees. In: Chechik, S., Navarro, G., Rotenberg, E., Herman, G. (eds.) 30th Annual European Symposium on Algorithms, ESA 2022, Berlin/Potsdam, Germany, 5–9 September 2022. LIPIcs, vol. 244, pp. 3:1–3:15. Schloss Dagstuhl - Leibniz-Zentrum für Informatik (2022). https://doi.org/10.4230/LIPIcs.ESA.2022.3
2. Asaka, R., Sakai, K., Yahagi, R.: Quantum circuit for the fast Fourier transform. Quantum Inf. Process. **19**(8), 277 (2020). https://doi.org/10.1007/s11128-020-02776-5
3. Bonnet, É., Miltzow, T., Rzazewski, P.: Complexity of token swapping and its variants. Algorithmica **80**(9), 2656–2682 (2018). https://doi.org/10.1007/s00453-017-0387-0
4. Botea, A., Kishimoto, A., Marinescu, R.: On the complexity of quantum circuit compilation. In: Bulitko, V., Storandt, S. (eds.) Proceedings of the Eleventh International Symposium on Combinatorial Search, SOCS 2018, Stockholm, Sweden, 14–15 July 2018, pp. 138–142. AAAI Press (2018)
5. Garey, M., Johnson, D., Stockmeyer, L.: Some simplified NP-complete graph problems. Theor. Comput. Sci. **1**(3), 237–267 (1976). https://doi.org/10.1016/0304-3975(76)90059-1
6. van Houte, R., Mulderij, J., Attema, T., Chiscop, I., Phillipson, F.: Mathematical formulation of quantum circuit design problems in networks of quantum computers. Quantum Inf. Process. **19**(5), 141 (2020). https://doi.org/10.1007/s11128-020-02630-8
7. Kawahara, J., Saitoh, T., Yoshinaka, R.: The time complexity of permutation routing via matching, token swapping and a variant. J. Graph Algorithms Appl. **23**(1), 29–70 (2019). https://doi.org/10.7155/jgaa.00483
8. Li, G., Ding, Y., Xie, Y.: Tackling the qubit mapping problem for NISQ-era quantum devices. In: Bahar, I., Herlihy, M., Witchel, E., Lebeck, A.R. (eds.) Proceedings of the Twenty-Fourth International Conference on Architectural Support for Programming Languages and Operating Systems, ASPLOS 2019, Providence, RI, USA, 13–17 April 2019, pp. 1001–1014. ACM (2019). https://doi.org/10.1145/3297858.3304023
9. Miltzow, T., Narins, L., Okamoto, Y., Rote, G., Thomas, A., Uno, T.: Approximation and hardness of token swapping. In: Sankowski, P., Zaroliagis, C.D. (eds.) 24th Annual European Symposium on Algorithms, ESA 2016, Aarhus, Denmark, 22–24 August 2016. LIPIcs, vol. 57, pp. 66:1–66:15. Schloss Dagstuhl - Leibniz-Zentrum für Informatik (2016). https://doi.org/10.4230/LIPIcs.ESA.2016.66
10. Nannicini, G., Bishop, L.S., Günlük, O., Jurcevic, P.: Optimal qubit assignment and routing via integer programming. ACM Trans. Quantum Comput. **4**(1) (2022). https://doi.org/10.1145/3544563
11. Saeedi, M., Wille, R., Drechsler, R.: Synthesis of quantum circuits for linear nearest neighbor architectures. Quantum Inf. Process. **10**(3), 355–377 (2011). https://doi.org/10.1007/s11128-010-0201-2
12. Siraichi, M.Y., dos Santos, V.F., Collange, C., Pereira, F.M.Q.: Qubit allocation. In: Knoop, J., Schordan, M., Johnson, T., O'Boyle, M.F.P. (eds.) Proceedings of the 2018 International Symposium on Code Generation and Optimization, CGO 2018, Vösendorf/Vienna, Austria, 24–28 February 2018, pp. 113–125. ACM (2018). https://doi.org/10.1145/3168822

13. Siraichi, M.Y., dos Santos, V.F., Collange, C., Pereira, F.M.Q.: Qubit allocation as a combination of subgraph isomorphism and token swapping. Proc. ACM Program. Lang. **3**(OOPSLA), 120:1–120:29 (2019). https://doi.org/10.1145/3360546
14. Yamanaka, K., et al.: Swapping labeled tokens on graphs. Theor. Comput. Sci. **586**, 81–94 (2015). https://doi.org/10.1016/j.tcs.2015.01.052

3-Coloring C_4 or C_3-Free Diameter Two Graphs

Tereza Klimošová[1(✉)] and Vibha Sahlot[2(✉)]

[1] Charles University in Prague, Prague, Czechia
tereza@kam.mff.cuni.cz

[2] University of Cologne, Cologne, Germany
sahlotvibha@gmail.com

Abstract. The question whether 3-COLORING can be solved in polynomial time for the diameter two graphs is a well-known open problem in the area of algorithmic graph theory. We study the problem restricted to graph classes that avoid cycles of given lengths as induced subgraphs. Martin et al. [CIAC 2021] showed that the problem is solvable in polynomial time for C_5-free or C_6-free graphs, and, (C_4, C_s)-free graphs where $s \in \{3, 7, 8, 9\}$. We extend their result proving that it is polynomial-time solvable for (C_4, C_s)-free graphs, for any constant $s \geq 5$, and for (C_3, C_7)-free graphs. Our results also hold for the more general problem List 3-Colouring.

Keywords: 3-coloring · List 3-Coloring · Diameter 2 Graphs · Induced C_4 free Graphs · Induced C_3 free Graphs

1 Introduction

In graph theory, k-COLORING is one of the most extensively studied problems in theoretical computer science. Here, given a graph $G(V, E)$, we ask if there is a function $c : V(G) \rightarrow \{1, 2, \ldots, k\}$ coloring all the vertices of the graph with k colors such that adjacent vertices get different colors. If such a function exists, then we call the graph G *k-colorable.* The k-COLORING is one of Karp's 21 NP-complete problems and is NP-complete for $k \geq 3$ [11].

The 3-COLORING is NP-hard even on planar graphs [9]. It motivates study of 3-COLORING under various graph constraints. For example, lots of research has been done on hereditary classes of graphs, i.e., classes that are closed under vertex deletion [2,4,6,10,12]. It has also led to the development of many powerful algorithmic techniques.

However, many natural classes of graphs are not hereditary, for example, graphs with bounded diameter, where the *diameter* of a graph is the maximum distance between any two vertices in the graph. These graph classes are not

Tereza Klimošová is supported by the Center for Foundations of Modern Computer Science (Charles Univ. project UNCE/SCI/004) and by GACR grant 22-19073S.

P. Morin and S. Suri (Eds.): WADS 2023, LNCS 14079, pp. 547–560, 2023.
https://doi.org/10.1007/978-3-031-38906-1_36

hereditary as the deletion of a vertex may increase the diameter of the graph. Graphs with bounded diameter are interesting as they come up in many real-life scenarios, for example, real-world graphs like Facebook. In this paper, we restrict our attention to 3-COLORING on graphs with diameter two.

The structure of the diameter two graphs is not simple, as adding a vertex to any graph G such that it is adjacent to all other vertices, makes the diameter of the graph at most two. Hence, the fact that 3-Coloring is NP-complete for general graph class implies that 4-Coloring is NP-complete for diameter two graphs.

Mertzios and Spiraki [15] gave a very non-trivial NP-hardness construction proving that 3-Coloring is NP-complete for the class of graphs with diameter three, even for triangle-free graphs. Furthermore, they presented a subexponential algorithm for 3-COLORING for n-vertex graphs with diameter two with runtime $2^{\mathcal{O}(\sqrt{n \log n})}$. Dębski et al. provided a further improved algorithm for 3-COLORING on n-vertex graphs with diameter two with runtime $2^{\mathcal{O}(n^{\frac{1}{3}} \log^2 n)}$. 3-COLORING on graphs with diameter two has been posed as an open problem in several papers [1,3,5,13,15,16].

The problem has been studied for various subclasses and is known to be solvable in polynomial time for:

- graphs that have at least one articulation neighborhood [15].
- (C_3, C_4)-free graphs [13].
- C_5-free or C_6-free graphs, (C_4, C_s)-free graphs where $s \in \{3, 7, 8, 9\}$ [14].
- $K_{1,r}^2$-free or $S_{1,2,2}$-free graphs, where $r \geq 1$ [13].

Continuing this line of research, we further investigate 3-COLORING for C_4-free and C_3-free diameter 2 graphs. In particular, we consider the following two problems:

1. 3-COLORING (C_4, C_s)-FREE DIAMETER TWO, where given an undirected induced (C_4, C_s)-free diameter two graph G for constant natural number s, we ask if there exists a 3-coloring of G.
2. 3-COLORING (C_3, C_7)-FREE DIAMETER TWO, where given an undirected induced (C_3, C_7)-free diameter two graph G, we ask if there exists a 3-coloring of G.

In fact, we consider a slightly more general problem of LIST 3-COLORING. A *list assignment* on G is a function L which assigns to every vertex $u \in V(G)$ a list of admissible colours. A list assignment is a *list k-assignment* if each list is a subset of a given k-element set. The problem of LIST 3-COLORING is then to decide whether there is a coloring c of G that *respects* a given list 3-assignment L, that is, for each vertex $u \in V(G)$, $c(u) \in L(u)$. This problem has also been considered in many of the previously mentioned works, in particular the aforementioned results from [14], some of which we use as subroutines in our algorithms, hold for LIST 3-COLORING as well.

The following two theorems summarize the main results of our paper.

Theorem 1. *The problem of* LIST 3-COLORING *is solvable in polynomial time on* (C_4, C_s)*-free graphs with diameter two for any constant* $s \geq 5$.

Theorem 2. *The problem of* LIST 3-COLORING *is solvable in polynomial time on* (C_3, C_7)*-free graphs with diameter two.*

It follows that 3-COLORING (C_4, C_s)-FREE DIAMETER TWO for any constant $s \geq 5$ and 3-COLORING (C_3, C_7)-FREE DIAMETER TWO are solvable in polynomial time.

The paper is organized as follows. We define the terminology and notation used in this paper in Sect. 2. We give some preprocessing rules in Sect. 3. We prove Theorem 1 in Sect. 4 and Theorem 2 in Sect. 5.

2 Preliminaries

In this section, we state the graph theoretic terminology and notation used in this paper. The set of consecutive integers from 1 to n is denoted by $[n]$. The vertex set and the edge set of a graph G are denoted by $V(G)$ and $E(G)$, respectively (or simply V and E when the underlying graph G is clear from the context). By $|G|$, we denote the order of G, that is $\max\{|V(G)|, |E(G)|\}$. An edge between vertices u and v is denoted by (u, v). For an unweighted and undirected graph $G(V, E)$, we define *distance* $d(u, v)$ between two vertices $u, v \in V(G)$ to be the length of a shortest path between u, v, if u is reachable from v, else it is defined as $+\infty$. The length of a path is defined as the number of edges in the path.

Let $f : A \rightarrow B$ be a function. Then, for any non-empty set $A' \subseteq A$, by $f(A')$, we denote the set $\{f(a) | a \in A'\}$.

For a vertex $v \in V(G)$, its *neighborhood* $N(v)$ is the set of all vertices adjacent to it and its *closed neighborhood* $N[v]$ is the set $N(v) \cup \{v\}$. Moreover, for a set $A \subseteq V$, $N_A(v) = N(v) \cap A$, similarly, $N_A[v] = N_A(v) \cup \{v\}$. We define $N_G[S] = N[S] = \bigcup_{v \in S} N_G[v]$ and $N_G(S) = N(S) = N_G[S] \backslash S$ where $S \subseteq V(G)$. The *degree* of a vertex $v \in V(G)$, denoted by $deg_G(v)$ or simply $deg(v)$, is the size of $N_G(v)$. A complete graph on q vertices is denoted by K_q.

A graph G' is a *subgraph* of G if $V(G') \subseteq V(G)$ and $E(G') \subseteq E(G)$. A graph G' is an *induced subgraph* of G if for all $x, y \in V(G')$ such that $(x, y) \in E(G)$, then $(x, y) \in E(G')$. For further details on graphs, refer to [7].

We say that list assignment L is a k*-list assignment* if $|L(v)| \leq k$ for each vertex $v \in V(G)$.

In k-LIST COLORING, given a graph G and a k-list assignment L, we ask if G has a coloring that respects L.

3 Preprocessing Rules for LIST 3-COLORING on Graphs with Diameter Two

The main idea of our algorithms is to reduce the problem to polynomially instances of 2-LIST COLORING. Due to the following result, this leads to polynomial-time algorithms.

Theorem 3 ([8]). *The problem of* 2-LIST COLORING *is linear-time solvable.*

For this purpose, we introduce several preprocessing rules which either reduce the size of some lists or resolve the instance completely.

More formally, a *preprocessing rule* is a rule which we apply to the given instance to produce another instance or an answer YES or NO. It is said to be *safe* if applying it to the given instance produces an equivalent instance. We say that a preprocessing rule is *applicable* on an instance if the output is different from the input instance. Now we list the preprocessing rules that we will use in later sections.

Consider a diameter two graph G with a list 3-assignment such that each vertex $v \in V(G)$ is assigned a list (a set) $L(v)$ of colors from the set $\{a, b, c\}$. When $|L(v)| = 1$ for some vertex v, we say that v is *colored* and let $c(v)$ be the only element of $L(v)$.

Preprocessing Rule 1. *If* $|L(u)| = 1$, *for every neighbor* u *of* v, *then assign* $L(v) \setminus \sum_{u \in N(v)} L(u)$ *to* $L(v)$.

Preprocessing Rule 2. *If* $L(v) = \emptyset$ *for any vertex* $v \in V(G)$, *then* G *is not list* k*-colorable.*

Preprocessing Rule 3. *If* $0 < |L(v)| \leq 2$ *for all vertices* $v \in V(G)$, *then resolve a* 2-LIST COLORING *problem instance (takes linear time in* $|G|$ *by Theorem 3).*

We call K_4 minus an edge a *diamond.*

Preprocessing Rule 4. *If* G *contains a diamond* $\{v, w, x, y\}$ *such that it does not contain the edge* (v, x), *then assign* $L(v) \cap L(x)$ *to both* $L(x)$ *and* $L(v)$.

Preprocessing Rule 5. *If* G *contains a triangle* $\{v, w, x\}$ *such that* $|L(v)| = 2$ *and* $L(v) = L(w)$, *then assign* $L(x) \setminus L(v)$ *to* $L(x)$.

Preprocessing Rule 6. *If* G *contains an induced* C_4 $\{v, w, x, y\}$ *such that* $|L(v)| = |L(w)| = |L(x)| = 2$ *and* $L(v), L(w), L(x)$ *are pairwise different, then assign* $L(y) \setminus (L(v) \cap L(x))$ *to* $L(v) \cap L(x)$.

Proposition 1. *The Preprocessing Rules 1, 2, 3, 4, 5 and 6 are safe.*

Proof. The Preprocessing Rules 1, 2, 3, and 5 are easy to see.

Safety of Preprocessing Rule 4 follows from the fact that any 3-colouring assigns v and x the same colour. Similarly, safety of Preprocessing Rule 5 follows from the fact that any 3-colouring assigns x a color not in $L(v)$, since both colors in $L(v) = L(w)$ are used to color v and w.

Now we consider Preprocessing Rule 6. Without loss of generality, assume $L(v) = \{b, c\}, L(w) = \{a, c\}, L(x) = \{a, b\}$. If w is colored a, then x will be colored b. Else if w is colored c, then v is colored b. Thus, one of v, x is always colored b. Hence, y cannot be colored b. □

The Preprocessing Rules are to be applied exhaustively in the given order to reduce and simplify the instance. Observe that this takes polynomial time as every application of a rule either yields an answer YES or NO or reduces size of list of at least one vertex.

The following proposition asserts that if a neighborhood of some vertex is fully colored, application of Preprocessing Rules completely resolves the instance. We use this repeatedly in the rest of the paper, sometimes without explicit mention.

Proposition 2. *If the given diameter two graph G has a vertex that has its neighborhood colored, application of Preprocessing Rules yields an answer YES or NO.*

Proof. Suppose $v \in V(G)$ such that its neighborhood is completely colored. Observe that since G has diameter two, every vertex which is not colored is in $N(N(v))$ and thus, it has a colored neighbor. Hence, applying Preprocessing Rule 1 yields an instance where all vertices have lists of size at most two. Then, either there is a vertex with empty list and applying Preprocessing Rule 2 yield NO or applying Preprocessing Rule 3 yields an answer YES or NO. □

This proposition is particularly useful for branching arguments when a neighborhood of some vertex admits only a few possible colorings. For instance, it can be used to show that LIST 3-COLORING is solvable in polynomial time on diameter two graphs with minimum degree bounded by a constant.

4 Polynomial-Time Algorithm for LIST 3-COLORING on (C_4, C_s)-Free Graphs with Diameter Two

In this section, we prove Theorem 1, which we restate below.

Theorem 1. *The problem of* LIST 3-COLORING *is solvable in polynomial time on* (C_4, C_s)*-free graphs with diameter two for any constant* $s \geq 5$.

Consider graph G with a list 3-assignment L, where $L(v) = \{a, b, c\}$ for all $v \in V(G)$ initially (notice that in case of LIST 3-COLORING, we can initialise with any given list 3-assignment L).

We may assume that G contains an induced C_5, otherwise, the problem is solvable in polynomial time as 3-COLORING on C_5-free diameter two graphs is polynomial time solvable [14]. Consider a C_5 as $C_5^1 = (1, 2, 3, 4, 5, 1)$ in G. Note that all colorings of C_5 are equivalent up to renaming and cyclic ordering of the colors. Without loss of generality, assume $c(1) = a$, $c(2) = b$, $c(3) = a$, $c(4) = b$ and $c(5) = c$.

Let the open neighborhood of vertices in C_5^1, that is, $N(C_5^1) = N_1$ and the remaining vertices except for the vertices in C_5^1 are $N(N_1)\backslash C_5^1 = N_2$. G has diameter 2, hence $V - (C_5^1 \cup N_1 \cup N_2) = \emptyset$. Let Col_1 be the set of vertices in N_1 that has list size one (i.e. their color is directly determined by the coloring of C_5^1 or during the algorithm). Similarly, Col_2 are the vertices of N_2 that have list size one.

As vertices of C_5^1 are already colored, the size of the list for the vertices in N_1 is at most two. Consider $N_1(i) = N(i) \backslash (C_5^1 \cup Col_1)$ for all $i \in [5]$. For example, $N_1(1)$ is the open neighborhood of 1 except for the neighbors in C_5^1 and Col_1. Let $L_2 \subseteq N_2$ be the set of vertices with list size two and $L_3 \subseteq N_2$ be the set of vertices with list size three.

In the lemmas below, we assume that G is a (C_4, C_s)-free diameter two graph with a list 3-assignment L on which Preprocessing Rules have been exhaustively applied. By Proposition 2, we may assume that no vertex has its neighborhood fully colored.

Lemma 4.1. *If there are at most $k \in \mathbb{N}$ connected components in some $N_1(i)$ for $i \in [5]$, then list 3-coloring can be resolved by solving at most 2^k instances of* 2-List Coloring.

Proof. Without loss of generality, assume that $i = 1$. Notice that for any valid 3-coloring of graph G, each connected component in $N_1(1)$ should be a bipartite graph. For contradiction, assume that there is an odd cycle in a connected component of $N_1(1)$. It requires at least three colors to color any odd cycle. But all the vertices in the odd cycle are adjacent to 1. Hence, we require a fourth color to color 1. This is a contradiction for any valid list 3-coloring of G.

Now for each connected component in $N_1(1)$, arbitrarily choose a vertex and consider both possibilities of colors in its list. Propagate the coloring to the rest of the vertices in that connected component. As there are at most k connected components in $N_1(1)$ by the assumption, we have 2^k possibilities of coloring all the vertices in $N_1(1)$. By Proposition 2, each of the possible color assignments to $N(1)$ leads to an instance which is resolved by application of Preprocessing Rules (in polynomial time). Thus, the given instance can be resolved by solving at most 2^k instances obtained by assigning all possible colorings to $N(1)$. □

Lemma 4.2. *Each vertex in N_1 that has a list of size two is adjacent exactly to one vertex in C_5^1.*

Proof. Consider a vertex v in $N_1 \backslash Col_1$, thus $|L(v)| = 2$. Then there are two possibilities. Either it is adjacent to more than one differently colored vertices in C_5^1 or it is adjacent to more than one same colored vertices in C_5^1, say i and $i+2$ for $i \in \{1, 2\}$. The first case implies that $|L(v)| = 1$, which is a contradiction. In the second case, as G is C_4-free, i, $i+1$, $i+2$ and v either form a K_4 (which implies G is not list 3-colorable) or a diamond, where $i+1$ is a common neighbor of i and $i+2$ in C_5^1. But by the Preprocessing Rule 4, the diamond will imply that $v \in Col_1$, which is a contradiction. Hence, the second case is not possible and v is adjacent to exactly one vertex in C_5^1. □

Lemma 4.3. *There are no edges between vertices in $N_1(i)$ and in $N_1(i+1)$ for all $i \in [4]$ and between $N_1(1)$ and $N_1(5)$.*

Proof. Consider a vertex $v \in N_1(1)$ and a vertex $u \in N_1(2)$. Let $(u, v) \in E(G)$. As G is C_4-free, therefore, the cycle $(1, v, u, 2, 1)$ has an edge, $(1, u)$ or $(2, v)$. This will reduce the size of the list of u or v, respectively, to one by the Preprocessing

Rule 4. But this is a contradiction to the fact that $u, v \notin Col_1$. Similar arguments work for the remaining cases. □

Lemma 4.4. *Every vertex in $N_1(i)$ has at most one neighbor in $N_1(j)$ for all $i, j \in [5]$ and $i \neq j$.*

Proof. Suppose not. Assume a vertex $v \in N_1(1)$ is adjacent to two vertices $x, y \in N_1(j)$ where j can only be 3, 4 from the Lemma 4.3. As G is C_4 free, the cycle (v, x, j, y, v) should have a chord. By Lemma 4.2, $(v, j) \notin E(G)$. Thus, $(x, y) \in E(G)$, which is a contradiction as it implies $c(v) = c(j)$ by the Preprocessing Rule 4, but $v \notin Col_1$. Similar arguments hold for the remaining cases. □

Lemma 4.5. *We have $|N_1(1)| = |N_1(3)|$ and $|N_1(2)| = |N_1(4)|$. Also, $G[N_1(1), N_1(3)]$, $G[N_1(2), N_1(4)]$ are perfect matchings.*

Proof. Consider a vertex $v \in N_1(1)$ and a vertex $u \in N_1(3)$. As, v is not adjacent to 2 or 4, using Lemma 4.2, v should be adjacent to some neighbor of 3 in N_1 to keep the distance between v and 3 as at most two. Also, by Lemma 4.4, it can have at most one neighbor in $N_1(3)$. This holds for all vertices in $N_1(1)$. Hence the graph induced on $N_1(1)$ and $N_1(3)$, i.e., $G[N_1(1), N_1(3)]$ is a perfect matching and $|N_1(1)| = |N_1(3)|$. Another case can be proven similarly. □

Lemma 4.6. *Every vertex in $N_2 \backslash Col_2$ has at most one neighbor in $N_1(i)$, $\forall i \in [5]$ and every vertex in L_3 has exactly one neighbor in $N_1(i)$, $\forall i \in [5]$.*

Proof. Consider a vertex $v \in N_2$ that is adjacent to two vertices $z_1, z_2 \in N_1(1)$. Then $(z_1, z_2) \in E(G)$ as otherwise $(1, z_1, v, z_2, 1)$ forms a C_4 but G is C_4-free. This implies that the color of v is the same as the color of the vertex 1 by the Preprocessing Rule 4, which is a contradiction as $v \notin Col_2$. Analogous arguments can be extended for the remaining cases. Hence, any vertex in $N_2 \backslash Col_2$ can be adjacent to at most one neighbor in $N_1(i)$, $\forall i \in [5]$.

Suppose $v \in L_3$ and v is not adjacent to any vertex in $N_1(1)$. As the diameter of the graph is two, the distance between v and 1 is at most two. This implies there is a vertex $y \in Col_1$ such that $(v, y), (y, 1) \in E(G)$ by Lemma 4.2. But this reduces the list size of $L(v)$ to at most two as now v is adjacent to a colored vertex. This is a contradiction as $|L(v)|$ is three. Hence, v is adjacent to a vertex in $N_1(1)$. We can argue similarly for the remaining cases. Thus every vertex in L_3 has exactly one neighbor in $N_1(i)$, $\forall i \in [5]$. □

Lemma 4.7. *Any pair of vertices $x \in N_1(1)$ and $y \in N_1(3)$ such that $(x, y) \in E(G)$, do not share a common neighbour in L_2 or L_3. Similarly, any pair of vertices $w \in N_1(2)$ and $z \in N_1(4)$ such that $(w, z) \in E(G)$, don't share a common neighbour in L_2 or L_3.*

Proof. Suppose not and there is a vertex in $v \in L_2 \cup L_3$ that is adjacent to both x and y, for any two vertices $x \in N_1(1)$ and $y \in N_1(3)$ such that $(x, y) \in E(G)$. Both x, y have list $\{b, c\}$. Hence v is colored a, which is a contradiction as $v \in L_2 \cup L_3$. The other case can be proved using similar arguments. □

Lemma 4.8. *Let $z \in L_3$ and $u \in N_1(i)$ for some $i \in [5]$ such that $uz \notin E(G)$. Then there is at most one vertex $z' \in N_2 \backslash \{z\}$ such that $(z, z'), (u, z') \in E(G)$.*

Proof. For contradiction, assume that there are at least two common neighbors $z', z'' \in N_2 \backslash \{z\}$ of u and z. Then, (u, z', z, z'', u) forms a diamond (there exists the edge (z', z'')) as G is C_4-free and $uz \notin E(G)$ by assumption). This implies that the size of the list of z is two as the size of list of u is two by the Preprocessing Rule 4. This is a contradiction to the assumption that $z \in L_3$.

□

Lemma 4.9. *Either $G[L_2 \cup L_3]$ contains an induced path $P_\ell*$ of length $\ell - 1$ for some $\ell \in \mathbb{N}$, or whether G is list 3-colorable can be decided by solving at most $\mathcal{O}(3^{6\ell})$ 2-LIST COLORING instances. Here $P_\ell* = (p_1, p_2, \ldots, p_\ell)$ is such that the neighborhood of p_1 and p_ℓ in N_1 is disjoint from neighborhood of vertices $p_2, p_3, \ldots, p_{\ell-1}$.*

Proof. Pick any vertex $p_1 \in L_3$. Let $j = 0$. Repeat the following, until $P_\ell*$ is constructed or step 2 fails. In the later case, we claim that whether G is list 3-colorable can be decided by solving at most $\mathcal{O}(3^{6\ell})$ 2-LIST COLORING instances. Note that during the following, we only modify the lists, not the sets L_3, N_1, etc.

For $i = 2j + 1$:

1. Color p_i and its five neighbors in N_1 and apply Preprocessing Rules exhaustively.
2. If there are vertices $x \in L_3$ and $y \in N_2$ satisfying the following:
   (i) x has a list of size three,
   (ii) y is a common neighbor of x and p_i and $|L(y)| = 2$, and
   (iii) neighbors of y in N_1 are not adjacent to p_1,
   set $p_{i+1} = y$, $p_{i+2} = x$, color p_{i+1} and its (at most five) neighbors in N_1, increase j by one, apply Preprocessing Rules exhaustively and proceed to the next iteration.

Note that if step 2 fails because there is no x satisfying (i), we have a 2-LIST COLORING instance. If there is such x, since G has diameter two, p_i and x have a common neighbor y. We next argue that every such y satisfies (ii). As x has a list of size three, it has no colored neighbors and since neighbors of p_i in N_1 are colored, it follows that $y \in N_2$ and moreover, $L(y) \geq 2$. On the other hand, $L(y) \leq 2$ as it does not contain $c(p_i)$.

Before discussing the case when step 2 fails because there is no pair of x and y satisfying (iii), we make a few observations about adjacencies in G.

First, observe that from the fact that p_{2j+1}, $j \geq 1$, was chosen as a vertex with list of size three, it follows that it is adjacent to none of the already colored vertices, in particular, to none of $p_1, \ldots, p_{2j}$ and their neighbors in N_1.

Claim 1. In the above procedure for p_i, where $i \neq 1$ and i is odd, every neighbor of p_1 is adjacent to at most one neighbor of p_i.

Proof. Suppose there are at least two such common neighbors s and t of p_1 and p_i. Hence, (p_1, s, p_i, t, p_1) forms a diamond with the edge (s, t) (since G has no induced C_4 and (p_1, p_i) is not an edge). By the Preprocessing Rule 4, $L(p_i) := L(p_1)$ which is a contradiction with the fact that p_i has a list of size three after coloring p_1 and applying Preprocessing Rules. □

Claim 2. In the above procedure for p_i, where $i \neq 1$ and i is odd, p_i has at most five neighbors in N_2 adjacent to neighbors of p_1 in N_1.

Proof. Assume $q \in N_1$ is a neighbor of p_1 adjacent to two neighbors $v, w \in N_2$ of p_i. Hence, (q, v, p_i, w, q) forms a diamond with the edge (v, w). Again, application of Preprocessing Rule 4 after coloring p_1 implies $|L(p_i)| = |L(p_1)| = 1$, contradicting the choice of p_i.

□

So if in any iteration step 2 fails because of (iii), from Claim 2 it follows that all vertices with list of size three are adjacent to one of at most five neighbors of p_i in N_2 which are adjacent to neighbors of p_1 in N_1. Thus, any coloring of these at most five vertices yields an instance of 2-List Coloring.

In total, if the process stops before constructing $P_\ell*$, at most 6ℓ vertices are colored before reaching a 2-List Coloring instance (including vertices colored if step 2 fails because of (iii)). For each such vertex, we have at most three possible choices of color. So, we can decide whether the instance is list 3-colorable by solving at most $\mathcal{O}(3^{6\ell})$ 2-List Coloring instances or we construct $P_\ell*$. □

Proof for Theorem 1. Assume that G contains $P_\ell* = (p_1, p_2, \dots, p_\ell)$ as in Lemma 4.9 (i.e. the neighborhood of p_1 and p_ℓ in N_1 is disjoint from neighborhood of vertices $p_2, p_3, \dots, p_{\ell-1}$) for $\ell = s - 4$. Then, we can construct a $C_s = (p_1, p_2, \dots, p_\ell, c_3, 3, 2, c_2, p_1)$, where $c_2 \in N_{N_1(2)}(p_1)$ and $c_3 \in N_{N_1(3)}(p_\ell)$. This contradicts (C_4, C_s)-freeness of G. Hence, by Lemma 4.9, we can decide whether the instance is 3-colorable by solving at most $\mathcal{O}(3^{6s})$ 2-List Coloring instances. For a constant s, the runtime is polynomial using Theorem 3. □

5 Polynomial-Time Algorithm for List 3-Coloring on (C_3, C_7)-Free Graphs with Diameter Two

In this section, we prove Theorem 2, which we restate below.

Theorem 2. The problem of List 3-Coloring *is solvable in polynomial time on (C_3, C_7)-free graphs with diameter two.*

Consider graph G with a list 3-assignment L, where $L(v) = \{a, b, c\}$ for all $v \in V(G)$ initially (notice that in case of List 3-Coloring, we can initialise with any given list 3-assignment L). Similar to the previous section, in our algorithm, we try to reduce the size of list of vertices to get 2-List Coloring instance.

G has a C_5, otherwise, we can solve the problem in polynomial time as 3-Coloring on C_5-free diameter two graphs is polynomial time solvable [14].

Consider a C_5 as $C_5^1 = (1, 2, 3, 4, 5, 1)$ in G and assume $c(1) = a$, $c(2) = b$, $c(3) = a$, $c(4) = b$ and $c(5) = c$.

Let the open neighborhood of vertices in C_5^1, that is, $N(C_5^1) = N_1$ and the remaining vertices except for the vertices in C_5^1 are $N(N_1)\backslash C_5^1 = N_2$. Let Col_1 and Col_2 be the set of vertices in N_1 and N_2, respectively, that have list size one.

As the vertices of C_5^1 are already colored, the size of the list for the vertices in N_1 is at most two. Consider $A = (N(1) \cup N(3))\backslash(C_5^1 \cup Col_1)$, $B = (N(2) \cup N(4))\backslash(C_5^1 \cup Col_1)$ and $C = N(5)\backslash(C_5^1 \cup Col_1)$. We further partition A into A_1, A_3 and A_{13}, where the vertices in A_1 are adjacent to 1 but not 3, the vertices in A_3 are adjacent to 3 but not 1 and the vertices in A_{13} are adjacent to both 1 and 3. Similarly, we partition B into B_2, B_4 and B_{24}, where the vertices in B_2 are adjacent to 2 but not 4, the vertices in B_4 are adjacent to 4 but not 2 and the vertices in B_{24} are adjacent to both 2 and 4. We partition $N_2\backslash Col_2$ into L_3 and L_2. The set L_3 contains the vertices that have list size three and L_2 contains the vertices that have list size two.

By Proposition 2, throughout the algorithm (or in lemmas below) we may assume that no vertex has its neighborhood fully colored.

Lemma 5.1. *1. The vertices in A are not adjacent to* $2, 4, 5$. *Similarly, vertices in B are not adjacent to* $1, 3, 5$ *and vertices in C are not adjacent to* $1, 2, 3, 4$.
2. Each of the vertex $v \in L_3$ has at least one neighbor in each of A, B and C.
3. The sets A_1, A_3, A_{13}, B_2, B_4, B_{24} and C are independent. Also, there is no edge between the vertices in A_1 and A_{13}, A_3 and A_{13}, B_2 and B_{24}, B_2 and B_{24}.

Proof. Consider the first part of the lemma. The vertices in A have list of size two, thus they cannot be adjacent to $2, 4, 5$. Similar arguments can be extended for the remaining cases.

Consider the second part of the lemma. Let $v \in L_3$. As the diameter of G is two, there should be a common neighbor of v and 1 (or 3) in N_1. But as $|L(v)| = 3$, it cannot be adjacent to any vertex in Col_1. Thus v has at least one neighbor in A. More precisely, v has at least one neighbor in each of A_1 and A_3, or v has at least one neighbor in A_{13}. Similar arguments can be extended for the remaining cases.

Consider the third part of the lemma. As G is C_3-free, there cannot be an edge between neighbors of any vertex. Hence, A_1, A_3, A_{13}, B_2, B_4, B_{24} and C are independent sets. Similarly, the vertices in A_1 and A_{13} are adjacent to 1. Thus, there is no edge between the vertices in A_1 and A_{13}. Similar arguments can be extended for the rest of the cases. □

Lemma 5.2. *The vertices in A_3 and B_2 do not have neighbors in $N_2\backslash Col_2$ that sees C. Similarly, any vertex in $N_2\backslash Col_2$ doesn't have neighbors in both A_1 and B_4.*

Proof. Suppose there exist vertices $z \in N_2\backslash Col_2$, $u_b \in B_2$ and $u_c \in C$ such that z is adjacent to both u_b and u_c, then there is an induced C_7 $(u_b, z, u_c, 5, 4, 3, 2, u_b)$.

To see this, notice that u_b and u_c are both neighbors of z, hence $(u_b, u_c) \notin E(G)$. As per Lemma 5.1, u_b is not adjacent to 3, 4, 5. Similarly, u_c is not adjacent to 2, 3, 4. As $z \in L_3$, it is not adjacent to 2, 3, 4, 5 by construction. This is a contradiction as G is C_7-free.

Similarly, if there are vertices $u_a \in A_3$, $u'_c \in C$ and $z' \in N_2 \backslash Col_2$ such that $(z', u_a), (z', u'_c) \in E(G)$, then there is an induced C_7 $(u_a, z', u'_c, 5, 1, 2, 3, u_a)$. This can be verified using similar arguments as in the previous case. It is a contradiction as G is C_7-free.

Likewise, if there are vertices $u_a \in A_1$, $u_b \in B_4$ and $z \in N_2 \backslash Col_2$ such that z is is adjacent to both u_a and u_b, then there is an induced C_7 $(u_a, z, u_b, 4, 3, 2, 1, u_a)$ based on similar arguments as in the previous cases. This is a contradiction as G is C_7-free. Hence, the claim is true. □

Lemma 5.3. *Any vertex $z \in N_2 \backslash Col_2$ that has a neighbor in C, neither sees any vertex in A_3, nor in B_2. Hence, any vertex $v \in L_3$ has neighbors both in A_{13} and B_{24}.*

Proof. Suppose that z has a neighbor $u_c \in C$. Assume that $u_a \in N_{A_3}(z)$. Then we have an induced C_7 $(z, u_c, 5, 1, 2, 3, u_a, z)$. To see this, notice that $(u_a, u_c) \notin E(G)$ as both u_a and u_c are neighbors of z and G is C_3-free. As per Lemma 5.1, u_a is not adjacent to 1, 2, 5 and u_c is not adjacent to 1, 2, 3. This is a contradiction as G is C_7-free. Thus, z does not have any neighbor in A_3.

Now assume that $u_b \in N_{B_2}(z)$. Then we have an induced C_7 $(z, u_b, 2, 3, 4, 5, u_c, z)$ based on similar arguments. But it a contradiction as G is C_7-free. Thus, z does not have any neighbor in B_2.

By Lemma 5.1, any vertex $v \in L_3$ has a neighbor in C. Hence, neighborhood of v in A_3 and B_2 is empty. As G has diameter two and v has list size three, v should have neighbor both in A_{13} and B_{24}. □

Lemma 5.4. *Vertices in L_3 are isolated in $G[N_2]$.*

Proof. Consider a vertex $z \in L_3$. For contradiction, assume it has a neighbor $z' \in N_2$. Note that $z' \notin Col_2$.

We first argue that z' has no neighbors in A_{13}, B_{24}, and C. Assume that z' has a neighbor $u_c \in C$. By Lemma 5.3, z has neighbors both in A_{13} and B_{24}. Let $u_a \in N_{A_{13}}(z)$ and $u_b \in N_{B_{24}}(z)$. Then, $(u_a, u_c) \in E(G)$, else we have an induced C_7 $(u_a, 3, 4, 5, u_c, z', z, u_a)$. Similarly, $(u_b, u_c) \in E(G)$, otherwise we have an induced C_7 $(u_b, 2, 1, 5, u_c, z', z, u_b)$. Consider the 4-cycle (z, u_a, u_c, u_b, z). As, u_a,u_b and u_c have all different lists with list size two,by the Preprocessing Rule 6, z should have a list of size at most 2, which contradicts that $z \in L_3$. The cases when z' has a neighbor in A_{13} or B_{24} are analogous.

By Lemma 5.2 z' does not have neighbors both in A_1 and B_4. Assume z' does not have any neighbor in B_4, the other case is analogous. Since G has diameter two and z' has no neighbor in B_{24}, B_4 and C, it has a neighbor in Col_1 adjacent to 5 and a neighbor in Col_1 adjacent to 4. Observe that since $|L(z')| \geq 2$ and $L(z')$ does not contain colors of colored neighbors of z', all neighbors of z' in Col_1 have the same color, namely the color a, as it must be different from $c(4) = b$

and $c(5) = c$. It follows that z' has no neighbor in Col_1 adjacent 1 or 3, since $c(1) = c(3) = a$.

Thus, since z' has no neighbor in A_{13}, it has neighbors in both A_1 and A_3 (as diameter of G is two). Let $a_1 \in N_{A_1}(z')$ and $a_3 \in N_{A_3}(z')$. Then G contains an induced C_7 $(z', a_1, 1, 5, 4, 3, a_3, z')$. This is a contradiction as G is C_7-free. Thus, any vertex $z \in L_3$ has no neighbor in N_2. □

Proof of Theorem 2. We argue that coloring any vertex $z_1 \in L_3$ by color c, applying Preprocessing Rules, then coloring any vertex z_2, which still has a list of size three (if it exists—otherwise, we have a 2-List coloring instance) by color b and applying Preprocessing Rules again, yields a 2-List coloring instance. This leads to the following algorithm which requires resolving $O(|V^2|)$ instances of 2-List coloring on G:

- resolve the 2-List coloring instance obtained by setting $L(z) = \{a, b\}$ for all $z \in L_3$, if it is a YES-instance, return YES
- else: for all $z_1 \in L_3$:
  - color z_1 by color c and apply Preprocessing Rules exhaustively
  - if the resulting instance is a 2-List coloring instance, resolve it and if it is a YES-instance, return YES
  - else:
    - resolve the 2-List coloring instance obtained by setting $L(z') = \{a, c\}$ for all z' with lists of size three, if it is a YES-instance, return YES
    - else: for all z_2 with list of size three
      - color z_2 by b and apply Preprocessing Rules exhaustively
      - resolve the resulting 2-List coloring instance, if it is a YES-instance, return YES
- return NO

Note that in the following, the sets of vertices L_3, Col_2, A, B, B_{24}, etc., are not modified, coloring and application of Preprocessing Rules change only the lists of colors available for the vertices.

Consider a vertex $z_1 \in L_3$ and color it c, (if no such vertex exists, then G is a 2-List Coloring instance).

Apply Preprocessing Rules and assume that it does not yield a 2-List Coloring instance. Observe that all neighbors of z_1 in A and B are colored.

Consider a vertex z_2 with list of size three. It has a common neighbor u_c with z_1 in C, as z_2 is not adjacent to any other neighbor of z_1 and G has diameter two. Color z_2 by color b. Applying Preprocessing Rules colors all neighbors of z_2 in C and B in particular, u_c is colored a.

We claim that there is no vertex with list of size three in the resulting instance. For contradiction, assume there is such a vertex z_3. It has a common neighbor u_b with z_2 in B, as z_3 is not adjacent to any other neighbor of z_2 and G has diameter two.

By Lemma 5.3 z_3 has a neighbor $u_a \in A_{13}$. Moreover, z_3 has a common neighbor with z_1 in $C \backslash \{u_c\}$, say v_c, as z_3 is not adjacent to any other neighbor of z_1, and G has diameter two.

By Lemma 5.3, z_1 has a neighbor v_b in B_{24}. Notice that $(u_a, v_b) \in E(G)$, otherwise, we have an induced C_7 $(z_3, u_a, 3, 4, v_b, z_1, v_c, z_3)$ which is a contradiction as G is C_7-free. But now, we have an induced C_7 $(z_1, u_c, z_2, u_b, z_3, u_a, v_b, z_1)$ which is a contradiction as G is C_7-free. Hence, we do not have such z_3. Thus, we must have reduced our given initial instance to some 2-List Coloring instance (or a polynomial number of instances). Hence, List 3-Coloring on (C_3, C_7)-free diameter two graphs is solvable in polynomial time. □

6 Conclusions

We have proved that 3-coloring on diameter two graphs is polytime solvable for (C_4, C_s)-free graphs where $s \geq 5$ is a constant, and (C_3, C_7)-free graphs. In the first case, we give an FPT on parameter s. Further, our algorithms also work for List 3-Coloring on the same graph classes. This opens avenues for further research on this problem for general C_4-free or C_3-free graphs. A less ambitious question is to extend similar FPT results to (C_3, C_s)-free with parameter s.

Acknowledgements. The authors would like to thank Kamak 2022 organised by Charles University for providing a platform for several fruitful discussions.

References

1. Bodirsky, M., Kára, J., Martin, B.: The complexity of surjective homomorphism problems–a survey. Discrete Appl. Math. **160**(12), 1680–1690 (2012)
2. Bonomo, F., Chudnovsky, M., Maceli, P., Schaudt, O., Stein, M., Zhong, M.: Three-coloring and list three-coloring of graphs without induced paths on seven vertices. Combinatorica **38**(4), 779–801 (2018). https://doi.org/10.1007/s00493-017-3553-8
3. Broersma, H., Fomin, F.V., Golovach, P.A., Paulusma, D.: Three complexity results on coloring P_k-free graphs. Eur. J. Comb. **34**(3), 609–619 (2013)
4. Chudnovsky, M., Robertson, N., Seymour, P., Thomas, R.: The strong perfect graph theorem. Ann. Math. **164**(1), 51–229 (2006)
5. Dębski, M., Piecyk, M., Rzaźewski, P.: Faster 3-coloring of small-diameter graphs. SIAM J. Discrete Math. **36**(3), 2205–2224 (2022)
6. Demaine, E.D., Hajiaghayi, M.T., Kawarabayashi, K.I.: Algorithmic graph minor theory: decomposition, approximation, and coloring. In: 46th Annual IEEE Symposium on Foundations of Computer Science (FOCS 2005), pp. 637–646. IEEE (2005)
7. Diestel, R.: Graph Theory. Springer, Heidelberg (2005)
8. Edwards, K.: The complexity of colouring problems on dense graphs. Theor. Comput. Sci. **43**, 337–343 (1986)
9. Garey, M.R., Johnson, D.S., Stockmeyer, L.J.: Some simplified NP-complete graph problems. Theor. Comput. Sci. **1**(3), 237–267 (1976)
10. Golovach, P.A., Johnson, M., Paulusma, D., Song, J.: A survey on the computational complexity of coloring graphs with forbidden subgraphs. J. Graph Theory **84**(4), 331–363 (2017)

11. Karp, R.M.: Reducibility among combinatorial problems. In: Miller, R.E., Thatcher, J.W. (eds.) Proceedings of a Symposium on the Complexity of Computer Computations. The IBM Research Symposia Series, 20–22 March 1972, pp. 85–103. Plenum Press, New York (1972)
12. Kratochvíl, J.: Can they cross? And how? (The hitchhiker's guide to the universe of geometric intersection graphs). In: Proceedings of the Twenty-Seventh Annual Symposium on Computational Geometry, pp. 75–76 (2011)
13. Martin, B., Paulusma, D., Smith, S.: Colouring H-free graphs of bounded diameter. Schloss Dagstuhl-Leibniz-Zentrum für Informatik (2019)
14. Martin, B., Paulusma, D., Smith, S.: Colouring graphs of bounded diameter in the absence of small cycles. Discrete Appl. Math. **314**, 150–161 (2022)
15. Mertzios, G.B., Spirakis, P.G.: Algorithms and almost tight results for 3-colorability of small diameter graphs. Algorithmica **74**(1), 385–414 (2016). https://doi.org/10.1007/s00453-014-9949-6
16. Paulusma, D.: Open problems on graph coloring for special graph classes. In: Mayr, E.W. (ed.) WG 2015. LNCS, vol. 9224, pp. 16–30. Springer, Heidelberg (2016). https://doi.org/10.1007/978-3-662-53174-7_2

Colored Constrained Spanning Tree on Directed Graphs

Hung-Yeh Lee, Hsuan-Yu Liao, and Wing-Kai Hon[(✉)]

Department of CS, National Tsing Hua University, Hsinchu, Taiwan
{s110062520,s110062581}@m110.nthu.edu.tw, wkhon@cs.nthu.edu.tw

Abstract. We study the κ-Colored Constrained Spanning Tree (κ-CCST) and the κ-Colored Out-Constrained Spanning Tree (κ-COCST) problems on edge-colored directed graphs. These problems target to find a spanning tree such that for each vertex, the number of incident edges (or, out-going edges) that share any specific color is bounded by some constant κ. We show that both problems are in general NP-hard, but for special cases, polynomial-time algorithms exist. In particular, when the input graph is a directed acyclic graph (DAG), κ-COCST can be solved via a maximum-flow-based algorithm. Furthermore, if the edges in the DAG are limited to use 2 colors and κ is restricted to 1, κ-CCST can be solved via dynamic programming.

Keywords: edge-colored graph · spanning tree · colored constraint · NP-hardness

1 Introduction

Consider a network where some pairs of stations are communicating through optical fibers, where signals are transmitted through lights of varying colors. However, due to certain constraints such as bandwidth and power, a pair of such stations may only be communicating through a limited choices of colors, and for a particular station, it may only communicate to a limited number of neighbors by optical fibers with a specific color. Suppose now a certain source station s wants to broadcast a message to every station in the network. How should we configure the network, preferably using the fewest number of communication links, to achieve the task?

One can model the network as an edge-colored graph, where the vertices are the stations, and the colored-edges denote the fibers between stations with their acceptable colors. The problem then becomes finding a spanning tree that satisfies the following constraint: For each vertex, the number of its incident edges of any specific color does not exceed a given number κ. We call this the κ-Colored Constrained Spanning Tree (κ-CCST) problem.

A series of works on edge-colored graphs examine the sufficient conditions for the existence of properly colored (i.e., $\kappa = 1$) and heterochromatic

H.-Y. Lee and H.-Y. Liao—These authors contribute equally.

P. Morin and S. Suri (Eds.): WADS 2023, LNCS 14079, pp. 561–573, 2023.
https://doi.org/10.1007/978-3-031-38906-1_37

subgraphs, where a heterochromatic subgraph requires all edges to have distinct colors. Grossman and Häggkvist [10] gave sufficient conditions of properly colored cycles on undirected graphs with two colors, while Yeo [14] generalized it to undirected graphs with arbitrary colors. Suzuki [13] proposed a necessary and sufficient condition for the existence of a heterochromatic spanning tree on edge-colored graphs. Cheng et al. [5] and Kano et al. [11] proposed different sufficient conditions for the existence of a properly colored spanning tree (i.e., a 1-CCST) based on the number of different edge colors incident to each vertex in the input graph. Kano et al. [12] further sharpened the sufficient conditions on edge-colored bipartite graphs for existence of either a heterochromatic or properly colored spanning tree.

Less attention has been given to algorithms that find a subgraph with constraints on edge-colored graphs. Borozan et al. [3] proposed an algorithm to find a properly colored spanning tree (or a maximum tree if no spanning tree exists) in an undirected edge-colored graph without any properly colored cycle. However, the results are not applicable on directed graphs.

The κ-CCST problem combines both degree bounds and edge colors constraints on subgraphs, which in general has not been studied before. Nevertheless, special cases of the κ-CCST problem have been identified to be NP-hard. Let λ denote the number of distinct colors in the input edge-colored graphs. If $\lambda = 1$ (i.e., the graph is uncolored), the undirected version of the problem is the Degree Constrained Spanning Tree problem [9]; as for the directed case, the problem is known as the Arborescence problem [2]. Additionally, if $\lambda = 2$ and $\kappa = 1$, the problem is the Properly Colored Hamiltonian Path problem, which is also NP-hard [3].

It is natural to ask for what classes of graphs the κ-CCST problem admits an efficient solution. Directed acyclic graphs (DAG) caught our attention, as a number of NP-hard problems on general graphs can be solved efficiently when restricted to DAGs (e.g., the longest path problem). We show that when $\lambda = 2$ and $\kappa = 1$, the problem can be solved efficiently via dynamic programming, using linear time and linear space. Surprisingly, when the input graph is a general directed graph instead, this problem becomes NP-hard with the same λ and κ settings. Also, when the input graph remains to be a DAG, but λ is increased slightly to 3, the problem is again NP-hard for $\kappa = 1$ (based on a reduction from Exact Cover by 3-Sets (X3C) [9]).

We also consider a slightly different constraint called *out-constraint*, which only limits the number of outgoing edges of the same color. The problem is the same as before, except that we are finding a rooted spanning tree satisfying this new constraint instead. We call this the κ-Colored Out-Constrained Spanning Tree (κ-COCST) problem. We show that this problem is NP-hard on directed graphs (based on a reduction from κ-CCST), but admits a polynomial-time (maximum-flow-based) algorithm on rooted DAGs. Figure 1 provides an example to the κ-CCST problem and the κ-COCST problem.

This paper provides a comprehensive analysis of the computational complexity for nearly all parameter settings in both the κ-CCST and κ-COCST problems

Fig. 1. T is a spanning tree of the edge-colored directed graph G. T is a 3-CCST but not a 2-CCST, and T is a 2-COCST but not a 1-COCST. (Color figure online)

on directed edge-colored graphs. The results are organized in Table 1. We assume that the input directed graph is *rooted*, meaning that the graph contains at least one "root" vertex capable of reaching all other vertices. Note that a rooted DAG has exactly one root vertex. Additionally, we do not require the graph to be simple. It is possible that the graph contains edges that share the same endpoints but have different colors.

Table 1. Summary of our results.

Graph Class	λ	κ	Problem	Hardness	Reference
Directed	≥ 2	$= 1$	κ-CCST	NPH	Borozan et al. [3]
	≥ 2	≥ 1	κ-CCST	NPH	Theorem 3
	≥ 2	≥ 1	κ-COCST	NPH	Theorem 4
DAG	≥ 3	≥ 1	κ-CCST	NPH	Theorem 1
	$= 2$	$= 1$	κ-CCST	P	Theorem 2
	$= 2$	≥ 2	κ-CCST	Open	
	≥ 2	≥ 1	κ-COCST	P	Theorem 5

1.1 Paper Organization

We prove that the κ-CCST problem is NP-hard on general DAGs in Sect. 2, and give a linear time and space algorithm when $\lambda = 2$ and $\kappa = 1$ in Sect. 3. We show that the κ-COCST problem is NP-hard on directed graphs in Sect. 4, and present an efficient algorithm when the input graph is a DAG in Sect. 5. We conclude the paper in Sect. 6, with an application of our problem to automaton compression. Due to space limitations, we omit some of the proofs. The proofs will be given in the full version of this paper.

2 κ-CCST on DAGs is NP-Hard

Theorem 1. *For any $\lambda \geq 3$ and $\kappa \geq 1$, κ-CCST on DAGs is NP-hard.*

Proof. To show the above theorem, it is sufficient to show κ-CCST on DAGs is NP-hard with $\lambda = 3$ and $\kappa = 1$. This is because we can reduce the κ-CCST problem on DAGs with $\lambda = 3$ and $\kappa = 1$ to the κ-CCST problem on DAGs for any $\lambda \geq 3$ and $\kappa \geq 1$ using the following method: Given an edge-colored graph G with three colors, we construct an edge-colored graph G' with λ colors from G such that G has a 1-CCST if and only if G' has a κ-CCST. In the construction process, for each vertex v in G and for each color c among the three colors present in G, we first create $(\kappa - 1)$ dummy vertices. Then, we connect v to each of the dummy vertices using edges colored with c. Lastly, we create $(\lambda - 3)$ dummy vertices and connect the root vertex to each of them using $(\lambda - 3)$ distinct new colors. Consequently, G has a 1-CCST if and only if G' has a κ-CCST, since the dummy vertices added in the first step restrict the quota of the colors appearing in G for each vertex in G'.

It remains to show κ-CCST problem on DAGs is NP-hard with $\lambda = 3$ and $\kappa = 1$. The proof is based on a reduction from the NP-complete problem X3C [9]:

	EXACT COVER BY 3-SETS (X3C)
Input:	A ground set $X = \{x_1, x_2, \ldots, x_{3n}\}$ with $3n$ items;
	A collection $C = \{c_1, c_2, \ldots, c_m\}$ of m 3-subsets of X;
Question:	Are there n elements of C whose union is X?

Let (X, C) be an instance of X3C, with X and C defined as above. For each 3-subset c_i, we label its items by c_i^1, c_i^2, c_i^3. We construct an edge-colored DAG $G = (V, E)$ such that G has a 1-CCST if and only if there exists a subcollection $C' \subseteq C$ such that every item of X occurs in exactly one member of C'.

The construction is made up of several components, including two binary subtrees for communication, m truth-setting components, m offset components, and one exact-cover-testing component. The components are augmented by additional edges to communicate the various components. The edges are colored by exactly three colors ℓ_1, ℓ_2, and ℓ_3. We use (u, v, c) to denote a directed edge from u to v with color c.

Before moving towards the construction details, we introduce the intuition of our gadget design. The m truth-setting components have 1–1 correspondence to the m elements of C. There are $3n$ vertices in the exact-cover-testing component, which have 1–1 correspondence to the $3n$ items of X. We design the structure of truth-setting components such that the following statement holds: A truth-setting component corresponds to c_i being *chosen* if and only if such truth-setting component connects to all the three vertices c_i^1, c_i^2, and c_i^3 in the exact-cover-testing component. Thus, there exists a solution to X3C if and only if each vertex of the exact-cover-testing component is connected by exactly one edge from all the chosen truth-setting components. Figure 2 illustrates the reduction with a sample instance.

Binary Tree B. To construct the first binary tree $B = (V_B, E_B)$, we first create a vertex r. The vertex r will be the root of the desired spanning tree on DAG G if it exists. From vertex r, we create children b_1 and b_2, then connect them with two edges (r, b_1, ℓ_1) and (r, b_2, ℓ_2). We repeat the following steps until we have exactly m leaves on binary tree B: We choose an arbitrary leaf, create two

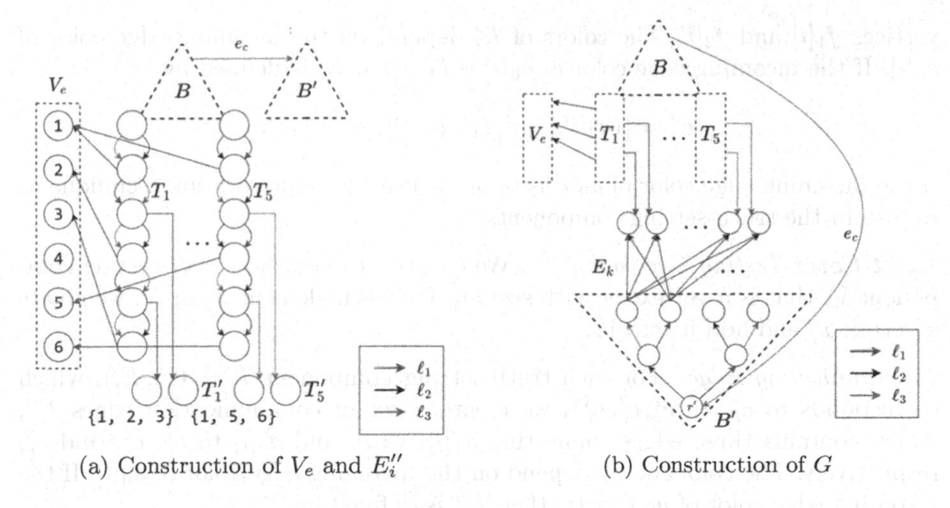

(a) Construction of V_e and E_i'' (b) Construction of G

Fig. 2. Reducing an X3C instance with $X = \{1, 2, 3, 4, 5, 6\}$ and $C = \{\ \{1, 2, 3\}, \{2, 3, 4\}, \{1, 2, 5\}, \{2, 5, 6\}, \{1, 5, 6\}\ \}$ to a κ-CCST instance. The edge colors ℓ_1, ℓ_2, and ℓ_3 are represented by blue, black, and red, respectively. (Color figure online)

children, then create two edges to connect them. The two edges are colored by the two colors other than the incoming edge color of the chosen leaf.

Binary Tree B'. To construct the second binary tree $B' = (V_B', E_B')$, we create a vertex r', and apply the same strategy as we construct B: From vertex r', we repeatedly grow the tree until B' has exactly $2n$ leaves. Finally, we use a communicating edge $e_c = (r, r', \ell_3)$ to connect the two binary trees B and B'.

Truth-Setting Components T_i. We construct m truth-setting components. We label the m leaves of binary tree B with $a_0[1]$, $a_0[2]$, ..., and $a_0[m]$. For each 3-subset $c_i = \{c_i^1, c_i^2, c_i^3\}$ in C, we create a truth-setting component $T_i = (V_i, E_i)$ emanating from $a_0[i]$. V_i contains five vertices $a_1[i]$, $a_2[i]$, $a_3[i]$, $a_4[i]$, and $a_5[i]$. E_i is the union of two sets of five edges L_i and R_i, which we call the "left route" and the "right route." The edges of the two routes are designed so that each route form a path connecting the five vertices in V_i with alternating colors. Furthermore, the colors of the edges depend on the incoming edge color of $a_0[i]$. If the incoming edge color of $a_0[i]$ is ℓ_1, then L_i and R_i are defined by

$$\begin{aligned}L_i = \{&(a_0[i], a_1[i], \ell_3), (a_1[i], a_2[i], \ell_1),\\ &(a_2[i], a_3[i], \ell_3), (a_3[i], a_4[i], \ell_1), (a_4[i], a_5[i], \ell_3)\},\\ R_i = \{&(a_0[i], a_1[i], \ell_2), (a_1[i], a_2[i], \ell_3), (a_2[i], a_3[i], \ell_2),\\ &(a_3[i], a_4[i], \ell_3), (a_4[i], a_5[i], \ell_2)\}.\end{aligned}$$

If the incoming edge color of $a_0[i]$ is ℓ_2 (or ℓ_3), we rotate the colors once (or twice): Replace ℓ_1 with ℓ_2, ℓ_2 with ℓ_3, and ℓ_3 with ℓ_1.

Offset Components T_i'. For each truth-setting component $T_i = (V_i, E_i)$, we create an offset component $T_i' = (V_i', E_i')$ emanating from T_i, where V_i' contains two vertices $f_1[i]$ and $f_2[i]$. The colors of E'_i depend on the incoming edge color of $a_0[i]$. If the incoming edge color of $a_0[i]$ is ℓ_1, then E'_i is defined by

$$E'_i = \{(a_2[i], f_2[i], \ell_1), (a_4[i], f_1[i], \ell_1)\}.$$

If the incoming edge color of $a_0[i]$ is ℓ_2 or ℓ_3, use the same rotation technique as we use in the truth-setting components.

Exact-Cover-Testing Component V_e. We construct one exact-cover-testing component V_e that is a vertex set with size $3n$. For each element x_j in X, we create a vertex x_j and add it into V_e.

Communicating Edges. For each truth-setting component $T_i = (V_i, E_i)$, which corresponds to $c_i = \{c_i^1, c_i^2, c_i^3\}$, we create a set of communication edges E''_i, which contains three edges connecting $a_1[i]$, $a_3[i]$, and $a_5[i]$ to c_i^1, c_i^2, and c_i^3, respectively. The colors of E''_i depend on the incoming edge color of $a_0[i]$. If the incoming edge color of $a_0[i]$ is ℓ_1, then E''_i is defined by

$$E''_i = \{(a_1[i], c_i^1, \ell_2), (a_3[i], c_i^2, \ell_2), (a_5[i], c_i^3, \ell_2)\}.$$

If the incoming edge color of $a_0[i]$ is ℓ_2 or ℓ_3, use the same rotation technique as we use in the truth-setting components.

For each leaf g_k in the second binary tree B', we create a set of communication edges E_k, which contains $2m$ edges that connect g_k to each vertex in the m offset components. The colors of E_k depend on the incoming edge color of g_k. If the incoming edge color of g_k is ℓ_1, then E_k is defined by

$$E_k = \left\{ (g_k, v, \ell_2) \;\middle|\; v \in \bigcup_{i=1}^{m} V'_i \right\}.$$

If the incoming edge color of g_k is ℓ_2 or ℓ_3, use the same rotation technique as we use in the truth-setting components.

One can show the correctness of the reduction by proving the following statement: There exists a subcollection $C' \subseteq C$ such that every item of X occurs in exactly one member of C' if and only if there exists a 1-CCST of G. Combining with the fact that G can be constructed in polynomial $O(nm)$ time, the proof of the theorem completes. In the full version of this paper, we provide a more detailed step-by-step reduction example, and present the complete correctness proof of the reduction. □

3 Solving 1-CCST on DAGs with $\lambda = 2$ in Linear Time

Theorem 2. *Let $G = (V, E)$ be an edge-colored DAG with $\lambda = 2$. Then, there exists an algorithm that runs in $O(|V| + |E|)$ time and*

1. *determines if G contains a 1-CCST or not;*
2. *outputs a desired spanning tree if G contains one.*

The remainder of this section describes a corresponding algorithm to prove the above theorem.

Given an edge-colored DAG $G = (V, E)$ with $\lambda = 2$ (we simply call the two colors red and blue), where the vertices $v_1, v_2, \ldots, v_{|V|}$ are topologically sorted. In this section we define two shapes of tree. An *alternating path* refers to a tree consisting of a single directed path in which each vertex is incident with edges of contrasting colors. A *V-shaped tree* is characterized by two alternating paths, where the root serves as the only shared starting vertex of these paths. In order for a 1-CCST to exist on G, the tree must take the form of either an alternating path or a V-shaped tree, as determined by the parameters $\kappa = 1$ and $\lambda = 2$. Specifically, the root vertex is capable of being incident with a maximum of two outgoing edges with contrast colors, while the other vertices are limited to having exactly one incoming edge and at most one outgoing edge with contrast colors. Therefore, it is possible to build a 1-CCST on G through dynamic programming by keeping track of the terminating vertices and their incoming edge colors in the two alternating paths. In the following, the terminating vertex of a path and its incoming edge color shall be referred to as the *ending*, which is represented by a vertex bearing a superscript indicating the incoming edge color. For example, v_i^{red} signifies v_i having a red incoming edge. In the case that the tree is an alternating path, it is treated as a V-shaped tree with one path terminating at a special ending *NIL*.

In the dynamic programming, we successively maintain, for each i, all feasible trees with $\kappa = 1$ that span $\{v_1, v_2, \ldots, v_i\}$. Among these trees, either v_i^{red} or v_i^{blue} must be the ending of an alternating path, as v_i has the largest topological order within the tree. Thus, we maintain a table DP of size $2 \cdot |V|$ indexed by endings v_i^{red} and v_i^{blue} for i from 1 to $|V|$. Each table entry $DP[v_i^{\text{red}}]$ (and similarly for $DP[v_i^{\text{blue}}]$) is a set that stores the possible endings of an alternating path given that the other alternating path terminates at v_i^{red} on the V-shaped trees spanning $\{v_1, v_2, \ldots, v_i\}$.

To compute DP, we discuss the following cases. When $i = 1$ (i.e., the tree that contains only the root v_1), we consider it as a special case and initialize $DP[v_1^{\text{red}}]$ as $\{v_1^{\text{blue}}, NIL\}$ and $DP[v_1^{\text{blue}}]$ as $\{v_1^{\text{red}}, NIL\}$. When $i > 1$, we will obtain $DP[v_i^{\text{red}}]$ and $DP[v_i^{\text{blue}}]$ from $DP[v_{i-1}^{\text{red}}]$ and $DP[v_{i-1}^{\text{blue}}]$. We concentrate on obtaining $DP[v_i^{\text{red}}]$, noting that the approach for $DP[v_i^{\text{blue}}]$ is symmetrical. Based on how trees spanning $\{v_1, v_2, \ldots, v_{i-1}\}$ can be expanded by adding v_i with a red incoming edge, we discuss the two cases (as depicted in Fig. 3). Firstly, we try to append v_i with a red edge subsequent to an ending v_{i-1}^{blue}. If an edge $(v_{i-1}, v_i, \text{red})$ is present in E, then all trees featuring an ending v_{i-1}^{blue} can be expanded by appending v_i after v_{i-1} with a red edge. As a result, $DP[v_i^{\text{red}}]$ is inclusive of all endings in $DP[v_{i-1}^{\text{blue}}]$. Secondly, we try to append v_i with a red edge to the path that does not terminate at v_{i-1}. For any incoming edge of v_i with red color (v_j, v_i, red), if the ending v_j^{blue} is present in $DP[v_{i-1}^{\text{red}}]$, then the ending v_{i-1}^{red} will be included in $DP[v_i^{\text{red}}]$. Similarly, if the ending v_j^{blue} is present in $DP[v_{i-1}^{\text{blue}}]$, then the ending v_{i-1}^{blue} will be included in $DP[v_i^{\text{red}}]$.

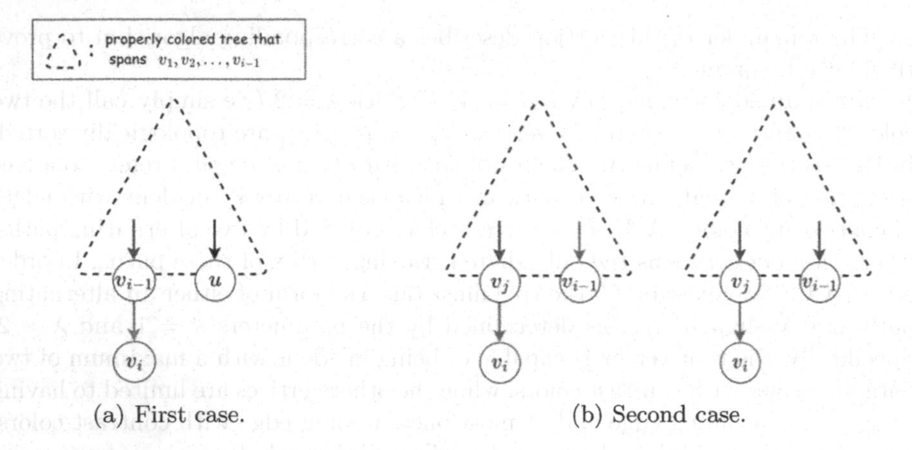

(a) First case.

(b) Second case.

Fig. 3. The two cases in adding v_i^{red}. (a) The right ending denotes an arbitrary ending in $DP[v_{i-1}^{\text{blue}}]$. (Color figure online)

Consequently, G has a 1-CCST if and only if any of $DP[v_{|V|}^{\text{red}}]$ or $DP[v_{|V|}^{\text{blue}}]$ is non-empty. Furthermore, a 1-CCST can be reconstructed using entries in DP. Together with the fact that this algorithm runs in linear time and linear space (the corresponding analysis is given in the full version of the paper), this completes the proof of Theorem 2.

4 κ-COCST on Directed Graphs is NP-Hard

Borozan et al. [3] showed that 1-CCST on *undirected* graphs with $\lambda = 2$ is NP-hard. In this section, we extend their results to show that for any $\lambda \geq 2$ and $\kappa \geq 1$, κ-CCST on *directed* graphs is NP-hard. Finally, we reduce κ-CCST to κ-COCST, so as to establish the hardness result of this section.

Given an edge-colored *undirected* graph G with two colors, we construct an edge-colored *undirected* graph G' with λ colors from G such that G has a 1-CCST if and only if G' has a κ-CCST. In the construction process, for each vertex v in G and for each color c among the two colors present in G, we first create $(\kappa - 1)$ dummy vertices. Then, we connect v to each of the dummy vertices using edges colored with c. Lastly, we create $(\lambda - 2)$ dummy vertices and connect the root vertex to each of them using $(\lambda - 2)$ distinct new colors. Consequently, G has a 1-CCST if and only if G' has a κ-CCST, since the dummy vertices added in the first step restrict the quota of the colors appearing in G for each vertex in G'. Finally, we transform G' into a *directed* graph G'', by replacing each color-c edge $\{u, v\}$ with two directed edges (u, v) and (v, u) with the same color c. Then, the *undirected* G' has a κ-CCST if and only if the *directed* G'' has a κ-CCST rooted at any vertex. This gives the following theorem.

Theorem 3. *For any $\lambda \geq 2$ and $\kappa \geq 1$, κ-CCST on directed graphs is NP-hard.*

In the following, we present the reduction from κ-CCST to κ-COCST. Let $G = (V, E)$ be an input directed graph for κ-CCST. We will construct a directed graph G' from G, such that a κ-CCST on G exists if and only if a κ-COCST on G' exists.

The construction of G' contains a single step: Replacing every edge $e = (u, v, c)$ in G with a corresponding gadget G_e, as shown in Fig. 4. Formally, a gadget $G_e = (V_e, E_e)$ corresponding to $e = (u, v, c)$ is defined by

$$V_e = \{ \, d_e^i \mid 0 \le i \le \kappa \, \}$$

and

$$E_e = \{ \, (u, d_e^0, c), (d_e^0, v, c), (v, d_e^0, \#_e) \, \} \cup \{ \, (d_e^0, d_e^i, c) \mid 1 \le i \le \kappa \, \} \cup \{ \, (v, d_e^i, c) \mid 1 \le i \le \kappa \, \},$$

where $\#_e$ is a special color distinct from any other existing color in G'.

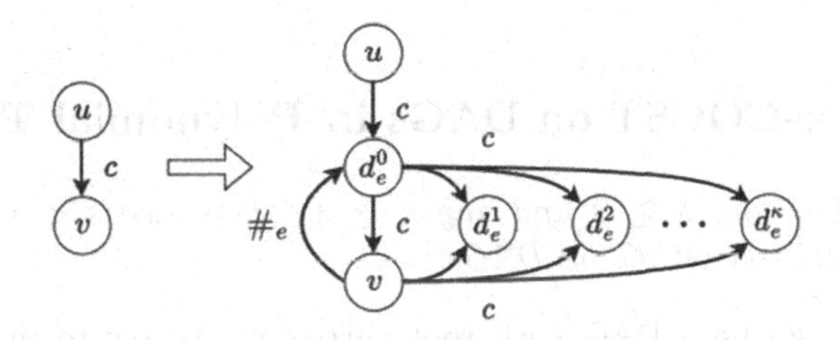

Fig. 4. The gadget G_e to replace an edge $e = (u, v, c)$. The edge colors are represented by labels beside the edges.

To simplify the discussion, we give special names to the vertices and the edges in the gadget: d_e^0 is the midpoint vertex, while d_e^i for $1 \le i \le \kappa$ are the offset vertices; edge (u, d_e^0, c) is the starting edge, and edge (d_e^0, v, c) is the ending edge; edges (d_e^0, d_e^i, c) for $1 \le i \le \kappa$ are the forward communicating edges, while edges (v, d_e^i, c) for $1 \le i \le \kappa$ are the backward communicating edges; edge $(v, d_e^0, \#_e)$ is the securing edge, where $\#_e$ is a special color distinct from any other existing color in G'. Additionally, for vertices in G' that are neither midpoint vertices nor offset vertices, we call them the backbone vertices.

The design of the gadget G_e enforces the following two key properties:

Property 1. If the ending edge of G_e is in a κ-COCST of G', then at least one backward communicating edge of G_e is also in the spanning tree of G'. (That is, if we add the ending edge of G_e with $e = (u, v, c)$ into a κ-COCST of G', vertex v will have one more outgoing edge with color c correspondingly.)

Proof. Otherwise, all κ offset vertices are connected by the forward communicating edges with color c in the spanning tree. Then d_e^0 has $\kappa + 1$ outgoing edges with color c, which violates the κ-colored out-constraint. □

Property 2. If the ending edge of G_e is in a κ-COCST of G' and the midpoint vertex of G_e is not the root of the spanning tree, then the starting edge of G_e is also in the spanning tree of G'. (That is, if we add the ending edge of G_e with $e = (u, v, c)$ into a κ-COCST of G', vertex u will have one more outgoing edge with color c correspondingly.)

Proof. Only the starting edge and the securing edge can connect to the midpoint vertex. However, we cannot add the securing edge into the spanning tree; otherwise, it generates a cycle. □

One can show the correctness of the reduction by proving the following statement: There exists a κ-CCST on G if and only if there exists a κ-COCST on G'. (See the full version of this paper for the complete proof.) Together with the fact that G' can be constructed in $O(\kappa \cdot |E|)$ time, we have the following theorem.

Theorem 4. *For any $\lambda \geq 2$ and $\kappa \geq 1$, κ-COCST on directed graphs is NP-hard.*

5 Solving κ-COCST on DAGs in Polynomial Time

Theorem 5. *For any $\lambda \geq 2$ and any $\kappa \geq 1$, there exists a polynomial-time algorithm to solve κ-COCST on DAGs.*

Let $G = (V, E)$ be a DAG with root vertex r. We try to match each vertex except r to one of its incoming edges, such that when considering only the matched edges, these edges form a rooted spanning tree, and every vertex obeys the κ-colored out-constraint. To do so, we will reduce the problem to the MAXIMUM FLOW problem.

We first construct a flow network $G' = (V', E')$. We let V' be the union of the source vertex s, the sink vertex t, and two vertex sets V_1 and V_2. To construct V_1, for each vertex u in G and each of its outgoing edge color ℓ, we create a vertex u_ℓ and add it into V_1. To construct V_2, for each vertex v except r in G, we create a vertex v and add it into V_2. We let E' be the union of three edge sets E_1, E_2, and E_3. We use $c(u, v)$ to denote the capacity of the edge (u, v) in G'. To construct E_1, we create $|V_1|$ edges that connect s to each vertex in V_1, where each of the edges has capacity κ. To construct E_2, for each edge $e = (u, v, \ell)$ in G, we create an edge (u_ℓ, v) with capacity 1 and add it into E_2. That is, there is a 1–1 correspondence between the edges in G and the edges in E_2. To construct E_3, we create $|V_2|$ edges that connect each vertex in V_2 to t, where each of the edges has capacity 1. Figure 5 gives an example of the above flow network construction.

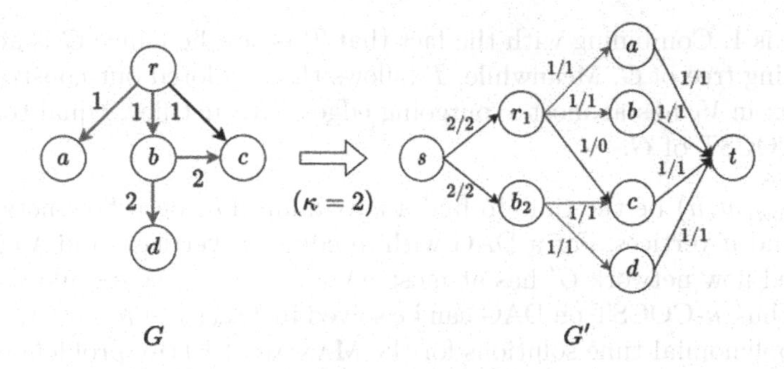

Fig. 5. Construct a flow network G' from a DAG G. We use a number to denote the color of an edge in G. We use a pair of numbers to denote the capacity and the net flow of an edge in G'. The red edges show how we construct the desired spanning tree according to the net flow of edges in E_2. (Color figure online)

Based on the above reduction, we have the following lemma:

Lemma 1. *G has a κ-COCST if and only if the maximum flow of G' has value $|V|-1$.*

Proof. According to the Integrality Theorem [6], if the capacities of edges in a flow network all have integral values, then there exists a maximum flow such that the values of net flows on every edge is an integer. Let f^* be such a maximum flow of G' with value $|f^*|$. For each edge (u, v) in E_2 and E_3, the value of its net flow $f^*(u, v)$ is binary. Therefore, each vertex in V_1 has at most κ outgoing edges with net flow equal to 1, as the total flow entering it is bounded by κ.

We first prove that if there exists a κ-COCST of G, then $|f^*| = |V| - 1$. Suppose T is a κ-COCST of G. We construct a flow f as follows: For each edge (v, t) in E_3, we set $f(v, t)$ to be 1. Then, for each edge (u, v, ℓ) in T, we set $f(u_\ell, v)$ to be 1. Since T is a spanning tree, each vertex in V_2 has exactly one incoming edge with net flow equal to 1. Thus, all vertices in V_2 obey the flow conservation. Also, for each edge (s, u_ℓ), we set $f(s, u_\ell)$ to be the sum of positive net flows leaving u_ℓ. Since T is κ-colored out-constrained, $f(s, u_\ell)$ is bounded by κ, and thus it will not exceed $c(s, u_\ell)$. So, f is a valid flow, and it is easy to verify that $|f| = |V| - 1$. On the other hand, the value of any flow on G' does not exceed $|V| - 1$, as the capacity sum of t's incoming edges is $|V| - 1$. Thus, f is also a maximum flow, so that $|f^*| = |f| = |V| - 1$.

It remains to prove that if $|f^*| = |V| - 1$, then there exists a κ-COCST of G. Initially, we let $T = (V_T, E_T)$ be an empty graph with $V_T = \varnothing$ and $E_T = \varnothing$. We will add edges and their endpoints from G to T in order to construct a κ-COCST. The construction of T contains a single step: For each edge (u_ℓ, v) in E_2 with $f(u_\ell, v) = 1$, we add (u, v, ℓ) into E_T. (If the endpoints are not in V_T, add them into V_T.) Because $|f^*| = |V| - 1$ and there are $|V| - 1$ edges with capacity 1 in E_3, we have $f(v, t) = 1$ for each edge (v, t) in E_3. Thus, each vertex in V_2 has at most one incoming edge with net flow equal to 1. Recall that each edge in E_2 corresponds to an edge in G, we can see that the indegree of each vertex in T

(except r) is 1. Combining with the fact that T is acyclic (since G is acyclic), T is a spanning tree of G. Meanwhile, T follows the κ-colored out-constraint since each vertex in V_1 has at most κ outgoing edges with net flow equal to 1. Thus, T is a κ-COCST of G. □

Let $T_{\text{flow}}(m, n)$ be the time to find a maximum flow on a flow network with m edges and n vertices. For a DAG with m edges, n vertices, and λ colors, the constructed flow network G' has at most $n\lambda + m + n - 1$ edges and $n\lambda + n + 1$ vertices. Thus, κ-COCST on DAG can be solved in $T_{\text{flow}}(n\lambda+m+n-1, n\lambda+n+1)$ time. As polynomial-time solutions for the MAXIMUM FLOW problem exist, this completes the proof of Theorem 5.

Remark. In particular, the MAXIMUM FLOW problem can be solved with Gao-Liu-Peng algorithm [8] in $\widetilde{O}(m^{\frac{3}{2}-\frac{1}{128}} \log U)$ time, where m and U are the number of edges and the maximum capacity on the edges. One can further reduce the κ-COCST problem on DAGs to BIPARTITE b-MATCHING problem, which can be solved in randomized $\widetilde{O}(m + n^{1.5})$ time by Brand et al.'s algorithm [4].

6 Conclusion

We have studied the κ-CCST and κ-COCST problems on directed graphs and DAGs. Some of our results are relatively tight, in a sense that:

- For κ-CCST on DAGs and $\kappa = 1$, if $\lambda = 2$, we show that polynomial-time solutions exist, while if $\lambda \geq 3$, the problem is NP-hard.
- For κ-COCST, for any $\lambda \geq 2$ and $\kappa \geq 1$, if the input graph is a DAG, polynomial-time solutions exist, while if the input graph is a directed graph, the problem is NP-hard.

Yet, for κ-CCST on DAGs with $\lambda = 2$, though we have a polynomial-time dynamic programming algorithm for $\kappa = 1$, it is not readily extendable for $\kappa > 1$. We leave this as an open problem for further studies.

Finally, we remark that the concept of COCST has an application in automaton compression. Gagie et al. [7] studied a special class of automaton called *Wheeler graph*, and showed that such automaton can be stored compactly in succinct space (whose space usage is within a factor of $1+o(1)$ of the information-theoretic minimum), and at the same time strings can be processed on the automaton efficiently. Alanko et al. [1] showed that for some special form of NFA called 2-NFA, whether an automaton is a Wheeler graph can be determined efficiently. Indeed, when we examine the work of Alanko et al. more closely, we find that if the state diagram of an NFA (not necessarily a 2-NFA) has a 2-COCST, then determining whether such an NFA is a Wheeler graph can also be done efficiently. This gives the following corollary based on Alanko et al.'s result:

Corollary 1. *Let $\mathcal{N}$ be an NFA whose state diagram is a DAG. We can determine (i) if $\mathcal{N}$ contains a 2-COCST, and (ii) if so, further determine whether $\mathcal{N}$ is a Wheeler graph in polynomial time.*

Acknowledgments. We thank the anonymous reviewers for their careful reading and insightful comments. This research is supported in part by Taiwan NSTC Grant 110-2221-E-007-043.

References

1. Alanko, J., D'Agostino, G., Policriti, A., Prezza, N.: Regular languages meet prefix sorting. In: ACM-SIAM Symposium on Discrete Algorithms (SODA), pp. 911–930 (2020)
2. Bansal, N., Khandekar, R., Nagarajan, V.: Additive guarantees for degree-bounded directed network design. SIAM J. Comput. **39**(4), 1413–1431 (2010)
3. Borozan, V., et al.: Maximum colored trees in edge-colored graphs. Eur. J. Comb. **80**, 296–310 (2019)
4. van den Brand, J., et al.: Bipartite matching in nearly-linear time on moderately dense graphs. In: IEEE Symposium on Foundations of Computer Science (FOCS), pp. 919–930 (2020)
5. Cheng, Y., Kano, M., Wang, G.: Properly colored spanning trees in edge-colored graphs. Discrete Math. **343**(1), 111629 (2020)
6. Cormen, T.H., Leiserson, C.E., Rivest, R.L., Stein, C.: Introduction to Algorithms, 3rd edn. MIT Press, Cambridge (2009)
7. Gagie, T., Manzini, G., Sirén, J.: Wheeler graphs: a framework for BWT-based data structures. Theor. Comput. Sci. **698**, 67–78 (2017)
8. Gao, Y., Liu, Y.P., Peng, R.: Fully dynamic electrical flows: sparse maxflow faster than Goldberg-Rao. In: IEEE Symposium on Foundations of Computer Science (FOCS), pp. 516–527 (2022)
9. Garey, M.R., Johnson, D.S.: Computers and Intractability: A Guide to the Theory of NP-Completeness. W. H. Freeman (1979)
10. Grossman, J.W., Häggkvist, R.: Alternating cycles in edge-partitioned graphs. J. Comb. Theory Ser. B **34**(1), 77–81 (1983)
11. Kano, M., Maezawa, S.I., Ota, K., Tsugaki, M., Yashima, T.: Color degree sum conditions for properly colored spanning trees in edge-colored graphs. Discrete Math. **343**(11), 112042 (2020)
12. Kano, M., Tsugaki, M.: Rainbow and properly colored spanning trees in edge-colored bipartite graphs. Graphs Comb. **37**(5), 1913–1921 (2021). https://doi.org/10.1007/s00373-021-02334-5
13. Suzuki, K.: A necessary and sufficient condition for the existence of a heterochromatic spanning tree in a graph. Graphs Comb. **22**(2), 261–269 (2006). https://doi.org/10.1007/s00373-006-0662-3
14. Yeo, A.: A note on alternating cycles in edge-coloured graphs. J. Comb. Theory Ser. B **69**(2), 222–225 (1997)

Geometric Hitting Set for Line-Constrained Disks

Gang Liu(✉) and Haitao Wang

University of Utah, Salt Lake City, UT 84112, USA
{u0866264,haitao.wang}@utah.edu

Abstract. Given a set P of n weighted points and a set S of m disks in the plane, the hitting set problem is to compute a subset P' of points of P such that each disk contains at least one point of P' and the total weight of all points of P' is minimized. The problem is known to be NP-hard. In this paper, we consider a line-constrained version of the problem in which all disks are centered on a line ℓ. We present an $O((m+n)\log(m+n)+\kappa\log m)$ time algorithm for the problem, where κ is the number of pairs of disks that intersect. For the unit-disk case where all disks have the same radius, the running time can be reduced to $O((n+m)\log(m+n))$. In addition, we solve the problem in $O((m+n)\log(m+n))$ time in the L_∞ and L_1 metrics, in which a disk is a square and a diamond, respectively.

Keywords: Hitting set · Line-constrained · Disks · Coverage

1 Introduction

Let S be a set of m disks and P a set of n points in the plane such that each point of P has a weight. The *hitting set problem* is to find a subset $P_{opt} \subseteq P$ of minimum total weight so that each disk of S contains a least one point of P_{opt} (i.e., each disk is *hit* by a point of P_{opt}). The problem is NP-hard even if all disks have the same radius and all point weights are the same [8,14,17].

In this paper, we consider the *line-constrained* version of the problem in which centers of all disks of S are on a line ℓ (e.g., the x-axis). To the best of our knowledge, this line-constrained problem was not particularly studied before. We give an algorithm of $O((m+n)\log(m+n)+\kappa\log m)$ time, where κ is the number of pairs of disks that intersect. We also present an alternative algorithm of $O(nm\log(m+n))$ time. For the *unit-disk case* where all disks have the same radius, we give a better algorithm of $O((n+m)\log(m+n))$ time. We also consider the problem in L_∞ and L_1 metrics (the original problem is in the L_2 metric), where a disk becomes a square and a diamond, respectively; we solve the problem in $O((m+n)\log(m+n))$ time in both metrics. The 1D case where all disks are line segments can also be solved in $O((m+n)\log(m+n))$ time.

This research was supported in part by NSF under Grants CCF-2005323 and CCF-2300356. A full version of this paper is available at http://arxiv.org/abs/2305.09045.

P. Morin and S. Suri (Eds.): WADS 2023, LNCS 14079, pp. 574–587, 2023.
https://doi.org/10.1007/978-3-031-38906-1_38

In addition, by a reduction from the element uniqueness problem, we prove an $\Omega((m+n)\log(m+n))$ time lower bound in the algebraic decision tree model even for the 1D case (even if all segments have the same length and all points of P have the same weight). The lower bound implies that our algorithms for the unit-disk, L_∞, L_1, and 1D cases are all optimal.

Related Work. The hitting set and many of its variations are fundamental and have been studied extensively; the problem is usually hard to solve, even approximately [15]. Hitting set problems in geometric settings have also attracted much attention and most problems are NP-hard, e.g., [4,5,10,12,16], and some approximation algorithms are known [10,16].

A "dual" problem is the coverage problem. For our problem, we can define its *dual coverage problem* as follows. Given a set P^* of n weighted disks and a set S^* of m points, the problem is to find a subset $P^*_{opt} \subseteq P^*$ of minimum total weight so that each point of S^* is covered by at least one disk of P^*_{opt}. This problem is also NP-hard [11]. The line-constrained problem was studied before and polynomial time algorithms were proposed [19]. The time complexities of the algorithms of [19] match our results in this paper. Specifically, an algorithm of $O((m+n)\log(m+n)+\kappa^*\log n)$ time was given in [19] for the L_2 metric, where κ^* is the number of pairs of disks that intersect [19]; the unit-disk, L_∞, L_1, and 1D cases were all solved in $O((n+m)\log(m+n))$ time [19]. Other variations of line-constrained coverage have also been studied, e.g., [1,3,18].

Our Approach. We propose a novel and interesting method, dubbed *dual transformation*, by reducing our hitting set problem to the 1D dual coverage problem and consequently solve it by applying the 1D dual coverage algorithm of [19]. Indeed, to the best of our knowledge, we are not aware of such a dual transformation in the literature. Two issues arise for this approach: The first one is to prove a good upper bound on the number of segments in the 1D dual coverage problem and the second is to compute these segments efficiently. These difficulties are relatively easy to overcome for the 1D, unit-disk, and L_1 cases. The challenge, however, is in the L_∞ and L_2 cases. Based on many interesting observations and techniques, we prove an $O(n+m)$ upper bound and present an $O((n+m)\log(n+m))$ time algorithm to compute these segments for the L_∞ case; for the L_2 case, we prove an $O(m+\kappa)$ upper bound and derive an $O((n+m)\log(n+m)+\kappa\log m)$ time algorithm.

Outline. In Sect. 2, we define notation and some concepts. Section 3 introduces the dual transformation and solves the 1D, unit-disk, and L_1 cases. Algorithms for the L_∞ and L_2 cases are presented in Sects. 4 and 5, respectively. The lower bound proof can be bound in Sect. 6. Due to the space limit, many details and proofs are ommited but can be found in the full paper.

2 Preliminaries

We follow the notation defined in Sect. 1, e.g., P, S, P_{opt}, κ, ℓ, etc. In this section, unless otherwise stated, all statements, notation, and concepts are applicable for all three metrics, i.e., L_1, L_2, and L_∞, as well as the 1D case. Recall that we assume ℓ is the x-axis, which does not lose generality for the L_2 case but is special for the L_1 and L_∞ cases.

We assume that all points of P are above or on ℓ since if a point $p \in P$ is below ℓ, we could replace p by its symmetric point with respect to ℓ and this would not affect the solution as all disks are centered at ℓ. For ease of exposition, we make a general position assumption that no two points of P have the same x-coordinate and no point of P lies on the boundary of a disk of S (these cases can be handled by standard perturbation techniques [9]). We also assume that each disk of S is hit by at least one point of P since otherwise there would be no solution (we could check whether this is the case by slightly modifying our algorithms).

For any point p in the plane, we use $x(p)$ and $y(p)$ to refer to its x- and y-coordinates, respectively. We sort all points of P in ascending order of their x-coordinates; let $\{p_1, p_2, \cdots, p_n\}$ be the sorted list.

For any point $p \in P$, we use $w(p)$ to denote its weight. We assume that $w(p) > 0$ for each $p \in P$ since otherwise one could always include p in the solution.

We sort all disks of S by their centers from left to right; let $s_1, s_2, \cdots, s_m$ be the sorted list. For each disk $s_j \in S$, let l_j and r_j denote its leftmost and rightmost points on ℓ, respectively. Note that l_j is the leftmost point of s_j and r_j is the rightmost point of s_j. More specifically, l_j (resp., r_j) is the only leftmost (resp., rightmost) point of s_j in the 1D, L_1, and L_2 cases. For each of exposition, we make a general position assumption that no two points of $\{l_i, r_i \mid 1 \leq i \leq m\}$ are coincident.

For $1 \leq j_1 \leq j_2 \leq m$, let $S[j_1, j_2]$ denote the subset of disks $s_j \in S$ for all $j \in [j_1, j_2]$.

We often talk about the relative positions of two geometric objects O_1 and O_2 (e.g., two points, or a point and a line). We say that O_1 is to the *left* of O_2 if $x(p) \leq x(p')$ holds for any point $p \in O_1$ and any point $p' \in O_2$, and *strictly left* means $x(p) < x(p')$. Similarly, we can define *right, above, below,* etc.

Non-containment Property. We observe that to solve the problem it suffices to consider only a subset of S with certain property, called the *Non-Containment subset*, defined as follows. We say that a disk of S is *redundant* if it contains another disk of S. The Non-Containment subset, denoted by $\widehat{S}$, is defined as the subset of S excluding all redundant disks. We have the following observation, called the *Non-Containment property.*

Observation 1. (Non-Containment Property) *For any two disks $s_i, s_j \in \widehat{S}$, $x(l_i) < x(l_j)$ if and only if $x(r_i) < x(r_j)$.*

Observe that it suffices to work on $\widehat{S}$ instead of S. Indeed, suppose P_{opt} is an optimal solution for $\widehat{S}$. Then, for any disk $s \in S \backslash \widehat{S}$, there must be a disk $s' \in \widehat{S}$ such that s contains s'. Hence, any point of P_{opt} hitting s' must hit s as well.

We can easily compute $\widehat{S}$ in $O(m \log m)$ time in any metric. Indeed, because all disks of S are centered at ℓ, a disk s_k contains another disk s_j if and only the segment $s_k \cap \ell$ contains the segment $s_j \cap \ell$. Hence, it suffices to identify all redundant segments from $\{s_j \cap \ell \mid s_j \in S\}$. This can be easily done in $O(m \log m)$ time, e.g., by sweeping endpoints of disks on ℓ; we omit the details.

In what follows, to simplify the notation, we assume $S = \widehat{S}$, i.e., S does not have any redundant disk. As such, S has the Non-Containment property in Observation 1. As will be seen later, the Non-Containment property is very helpful in designing algorithms.

3 Dual Transformation and the 1D, Unit-Disk, and L_1 Problems

By making use of the Non-Containment property of S, we propose a *dual transformation* that can reduce our hitting set problem on S and P to an instance of the 1D dual coverage problem. More specifically, we will construct a set S^* of points and a set P^* of weighted segments on the x-axis such that an optimal solution for the coverage problem on S^* and P^* corresponds to an optimal solution for our original hitting set problem. We refer to it as the *1D dual coverage problem*. To differentiate from the original hitting set problem on P and S, we refer to the points of S^* as *dual points* and the segments of P^* as *dual segments*.

As will be seen later, $|S^*| = m$, but $|P^*|$ varies depending on the specific problem. Specifically, $|P^*| \leq n$ for the 1D, unit-disk, and L_1 cases, $|P^*| = O(n+m)$ for the L_∞ case, and $|P^*| = O(m+\kappa)$ for the L_2 case. In what follows, we present the details of the dual transformation by defining S^* and P^*.

For each disk $s_j \in S$, we define a dual point s_j^* on the x-axis with x-coordinate equal to j. Define S^* as the set of all m points $s_1^*, s_2^*, \ldots, s_m^*$. As such, $|S^*| = m$.

We next define the set P^* of dual segments. For each point $p_i \in P$, let I_i be the set of indices of the disks of S that are hit by p_i. We partition the indices of I_i into maximal intervals of consecutive indices and let $\mathcal{I}_i$ be the set of all these intervals. By definition, for each interval $[j_1, j_2] \in \mathcal{I}_i$, p_i hits all disks s_j with $j_1 \leq j \leq j_2$ but does not hit either s_{j_1-1} or s_{j_2+1}; we define a dual segment on the x-axis whose left (resp., right) endpoint has x-coordinate equal to j_1 (resp., j_2) and whose weight is equal to $w(p_i)$ (for convenience, we sometimes also use the interval $[j_1, j_2]$ to represent the dual segment and refer to dual segments as intervals). We say that the dual segment is *defined* or *generated* by p_i. Let P_i^* be the set of dual segments defined by the intervals of $\mathcal{I}_i$. We define $P^* = \bigcup_{i=1}^n P_i^*$. The following observation follows the definition of dual segments.

Observation 2. *p_i hits a disk s_j if and only if a dual segment of P_i^* covers the dual point s_j^*.*

Suppose we have an optimal solution P_{opt}^* for the 1D dual coverage problem on P^* and S^*, we obtain an optimal solution P_{opt} for the original hitting set problem on P and S as follow: for each segment of P_{opt}^*, if it is from P_i^* for some i, then we include p_i into P_{opt}.

Clearly, $|S^*| = m$. We will prove later in this section that $|P_i^*| \leq 1$ for all $1 \leq i \leq n$ in the 1D problem, the unit-disk case, and the L_1 metric, and thus $|P^*| \leq n$ for all these cases. Since $|P_i^*| \leq 1$ for all $1 \leq i \leq n$, in light of Observation 2, P_{opt} constructed above is an optimal solution of the original hitting set problem. Therefore, one can solve the original hitting set problem for the above cases with the following three main steps: (1) Compute S^* and P^*; (2) apply the algorithm for the 1D dual coverage problem in [19] to compute P^*_{opt}, which takes $O((|S^*|+|P^*|)\log(|S^*|+|P^*|))$ time [19]; (3) derive P_{opt} from P^*_{opt}. For the first step, computing S^* is straightforward. For P^*, we will show later that for all above three cases (1D, unit-disk, L_1), P^* can be computed in $O((n+m)\log(n+m))$ time. As $|S^*| = m$ and $|P^*| \leq n$, the second step can be done in $O((n+m)\log(n+m))$ time [19]. As such, the hitting set problem of the above three cases can be solved in $O((n+m)\log(n+m))$ time.

For the L_∞ metric, we will prove in Sect. 4 that $|P^*| = O(n+m)$ but each P_i^* may have multiple segments. If P_i^* has multiple segments, a potential issue is the following: If two segments of P_i^* are in P^*_{opt}, then the weights of both segments will be counted in the optimal solution value (i.e., the total weight of all segments of P^*_{opt}), which corresponds to counting the weight of p_i twice in P_{opt}. To resolve the issue, we prove in Sect. 4 that even if $|P_i^*| \geq 2$, at most one dual segment of P_i^* will appear in any optimal solution P^*_{opt}. As such, P_{opt} constructed above is an optimal solution for the original hitting set problem. Besides proving the upper bound $|P^*| = O(n+m)$, another challenge of the L_∞ problem is to compute P^* efficiently, for which we propose an $O((n+m)\log(n+m))$ time algorithm. Consequently, the L_∞ hitting set problem can be solved in $O((n+m)\log(n+m))$ time.

For the L_2 metric, we will show in Sect. 5 that $|P^*| = O(m+\kappa)$. Like the L_∞ case, each P_i^* may have multiple segments but we can also prove that P_i^* can contribute at most one segment to any optimal solution P^*_{opt}. Hence, P_{opt} constructed above is an optimal solution for the original hitting set problem. We present an algorithm that can compute P^* in $O((n+m)\log(n+m)+\kappa\log m)$ time. As such, the L_2 hitting set problem can be solved in $O((m+n)\log(m+n)+\kappa\log m)$ time. Alternatively, a straightforward approach can prove $|P^*| = O(nm)$ and compute P^* in $O(nm)$ time; hence, we can also solve the problem in $O(nm\log(n+m))$ time.

In the rest of this section, following the above framework, we solve the unit-disk case. Due to the space limit, the 1D and the L_1 cases are omitted but can be found in the full paper.

3.1 The Unit-Disk Case

In the unit-disk case, all disks of S have the same radius. We follow the dual transformation and have the following lemma.

Lemma 1. *In the unit-disk case, $|P_i^*| \leq 1$ for any $1 \leq i \leq n$. In addition, P_i^* for all $1 \leq i \leq n$ can be computed in $O((n+m)\log(n+m))$ time.*

Proof. Consider a point $p_i \in P$. Observe that p_i hits a disk s_j if and only if the segment $D(p_i) \cap \ell$ covers the center of s_j, where $D(p_i)$ is the unit disk centered at p_i. By definition, the indices of the disks whose centers are covered by the segment $D(p_i) \cap \ell$ must be consecutive. Hence, $|P_i^*| \leq 1$ must hold.

To compute P_i^*, it suffices to determine the disks whose centers are covered by $D(p_i) \cap \ell$. This can be easily done in $O((n+m)\log(n+m))$ time for all $p_i \in P$ (e.g., first sort all disk centers and then do binary search on the sorted list with the two endpoints of $D(p_i) \cap \ell$ for each $p_i \in P$). □

In light of Lemma 1, using the dual transformation, we have the following result.

Theorem 1. *The line-constrained unit-disk hitting set problem is solvable in $O((n+m)\log(n+m))$ time.*

Note that in both the 1D and the L_1 cases we can prove results similar to Lemma 1 and thus solve both cases in $O((n+m)\log(n+m))$ time. The details are omitted but can be found in the full paper.

4 The L_∞ Metric

In this section, following the dual transformation, we present an $O((m+n)\log(m+n))$ time algorithm for L_∞ case.

In the L_∞ metric, each disk is a square whose edges are axis-parallel. For a disk $s_j \in S$ and a point $p_i \in P$, we say that p is *vertically above* s_j if p_i is outside s_j and $x(l_j) \leq x(p_i) \leq x(r_j)$.

In the L_∞ metric, using the dual transformation, it is easy to come up with an example in which $|P_i^*| \geq 2$. Observe that $|P_i^*| \leq \lceil m/2 \rceil$ as the indices of S can be partitioned into at most $\lceil m/2 \rceil$ disjoint maximal intervals. Despite $|P_i^*| \geq 2$, the following critical lemma shows that each P_i^* can contribute at most one segment to any optimal solution of the 1D dual coverage problem on P^* and S^*. The proof of the lemma can be found in the full paper.

Lemma 2. *In the L_∞ metric, for any optimal solution P^*_{opt} of the 1D dual coverage problem on P^* and S^*, P^*_{opt} contains at most one segment from P_i^* for any $1 \leq i \leq n$.*

Lemma 2 implies that an optimal solution to the 1D dual coverage problem on P^* and S^* still corresponds to an optimal solution of the original hitting set problem on P and S. As such, it remains to compute the set P^* of dual segments. In what follows, we first prove an upper bound for $|P^*|$.

4.1 Upper Bound for $|P^*|$

As $|P_i^*| \leq \lceil m/2 \rceil$, an obvious upper bound for $|P^*|$ is $O(mn)$. Below we reduce it to $O(m+n)$.

Our first observation is that if the same dual segment of P^* is defined by more than one point of P, then we only need to keep the one whose weight is minimum. In this way, all segments of P^* are distinct (i.e., P^* is not a multi-set).

We sort all points of P from top to bottom as $q_1, q_2, \ldots, q_n$. For ease of exposition, we assume that no point of P has the same y-coordinate as the upper edge of any disk of S. For each $2 \leq i \leq n$, let S_i denote the subset of disks whose upper edges are between q_{i-1} and q_i. Let S_1 denote the subset of disks whose upper edges are above q_1. For each $1 \leq i \leq n$, let $m_i = |S_i|$.

We partition the indices of disks of S_1 into a set $\mathcal{I}_1$ of maximal intervals. Clearly, $|\mathcal{I}_1| \leq m_1$. The next lemma shows that other than the dual segments corresponding to the intervals in $\mathcal{I}_1$, q_1 can generate at most two dual segments in P^*.

Lemma 3. *The number of dual segments of $P^* \setminus \mathcal{I}_1$ defined by q_1 is at most 2.*

Proof. Assume to the contrary that q_1 defines three intervals $[j_1, j_1']$, $[j_2, j_2']$, and $[j_3, j_3']$ in $P^* \setminus \mathcal{I}_1$, with $j_1' < j_2$ and $j_2' < j_3$. By definition, $\mathcal{I}_1$ must have an interval, denoted by I_k, that strictly contains $[j_k, j_k']$ (i.e., $[j_k, j_k'] \subset I_k$), for each $1 \leq k \leq 3$. Then, I_2 must contain an index j that is not in $[j_1, j_1'] \cup [j_2, j_2'] \cup [j_3, j_3']$ with $j_1' < j < j_3$ (e.g., see Fig. 1). As such, q_1 does not hit s_j. Also, since $j \in I_2$, s_j is in S_1.

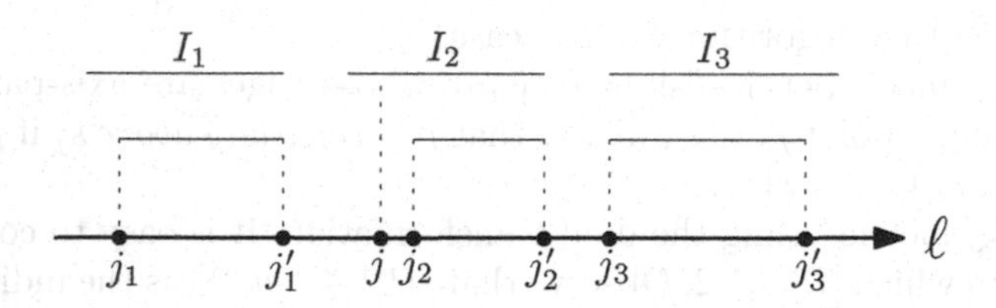

Fig. 1. Illustrating a schematic view of the intervals $[j_k, j_k']$ and I_k for $1 \leq k \leq 3$.

Since $j_1' < j < j_3$, due to the Non-Containment property of S, $x(l_j) \leq x(l_{j_3})$ and $x(r_{j_1'}) \leq x(r_j)$. As q_1 hits both $s_{j_1'}$ and s_{j_3}, we have $x(l_{j_3}) \leq x(p_1) \leq x(r_{j_1'})$. Hence, we obtain $x(l_j) \leq x(q_1) \leq x(r_j)$. Since q_1 does not hit s_j, the upper edge of s_j must be below q_1. But this implies that s_j is not in S_1, which incurs contradiction. □

Now we consider the disks of S_2 and the dual segments defined by q_2. For each disk s_j of S_2, we update the intervals of $\mathcal{I}_1$ by adding the index j, as follows. Note that by definition, intervals of $\mathcal{I}_1$ are pairwise disjoint and no interval contains j.

1. If neither $j+1$ nor $j-1$ is in any interval of $\mathcal{I}_1$, then we add $[j, j]$ as a new interval to $\mathcal{I}_1$.
2. If $j+1$ is contained in an interval $I \in \mathcal{I}_1$ while $j-1$ is not, then $j+1$ must be the left endpoint of I. In this case, we add j to I to obtain a new interval I' (which has j as its left endpoint) and add I' to $\mathcal{I}_1$; but we still keep I in $\mathcal{I}_1$.

3. Symmetrically, if $j-1$ is contained in an interval $I \in \mathcal{I}_1$ while $j+1$ is not, then we add j to I to obtain a new interval I' and add I' to $\mathcal{I}_1$; we still keep I in $\mathcal{I}_1$.
4. If both $j+1$ and $j-1$ are contained in intervals of $\mathcal{I}_1$, they must be contained in two intervals, respectively; we merge these two intervals into a new interval by padding j in between and adding the new interval to $\mathcal{I}_1$. We still keep the two original intervals in $\mathcal{I}_1$.

Let $\mathcal{I}_1'$ denote the updated set $\mathcal{I}_1$ after the above operation. Clearly, $|\mathcal{I}_1'| \leq |\mathcal{I}_1| + 1$.

We process all disks $s_j \in S_2$ as above; let $\mathcal{I}_2$ be the resulting set of intervals. It holds that $|\mathcal{I}_2| \leq |\mathcal{I}_1| + |S_2| \leq m_1 + m_2$. Also observe that for any interval I of indices of disks of $S_1 \cup S_2$ such that I is not in $\mathcal{I}_2$, $\mathcal{I}_2$ must have an interval I' such that $I \subset I'$ (i.e., $I \subseteq I'$ but $I \neq I'$). Using this property, by exactly the same analysis as Lemma 3, we can show that other than the intervals in $\mathcal{I}_2$, q_2 can generate at most two intervals in P^*. Since $\mathcal{I}_1 \subset \mathcal{I}_2$, combining Lemma 3, we obtain that other than the intervals of $\mathcal{I}_2$, the number of intervals of P^* generated by q_1 and q_2 is at most 4.

We process disks of S_i and point q_i in the same way as above for all $i = 3, 4, \ldots, n$. Following the same argument, we can show that for each i, we obtain an interval set $\mathcal{I}_i$ with $\mathcal{I}_{i-1} \subseteq \mathcal{I}_i$ and $|\mathcal{I}_i| \leq \sum_{k=1}^{i} m_k$, and other than the intervals of $\mathcal{I}_i$, the number of intervals of P^* generated by $\{q_1, q_2, \ldots, q_i\}$ is at most $2i$. In particular, $|\mathcal{I}_n| \leq \sum_{k=1}^{n} m_k \leq m$, and other than the intervals of $\mathcal{I}_n$, the number of intervals of P^* generated by $P = \{q_1, q_2, \ldots, q_n\}$ is at most $2n$. We thus achieve the following conclusion.

Lemma 4. *In the L_∞ metric, $|P^*| \leq 2n + m$.*

4.2 Computing P^*

Using Lemma 4, we present an algorithm that computes P^* in $O((n+m)\log(n+m))$ time.

For each segment $I \in P^*$, let $w(I)$ denote its weight. We say that a segment I of P^* is *redundant* if there is another segment I' such that $I \subset I'$ and $w(I) \geq w(I')$. Clearly, any redundant segment of P^* cannot be used in any optimal solution for the 1D dual coverage problem on S^* and P^*. A segment of P^* is *non-redundant* if it is not redundant.

In the following algorithm, we will compute a subset P_0^* of P^* such that segments of $P^* \backslash P_0^*$ are all redundant (i.e., the segments of P^* that are not computed by the algorithm are all redundant and thus are useless). We will show that each segment reported by the algorithm belongs to P^* and thus the total number of reported segments is at most $2n+m$ by Lemma 4. We will show that the algorithm spends $O(\log(n+m))$ time reporting one segment and each segment is reported only once; this guarantees the $O((n+m)\log(n+m))$ upper bound of the runtime of the algorithm.

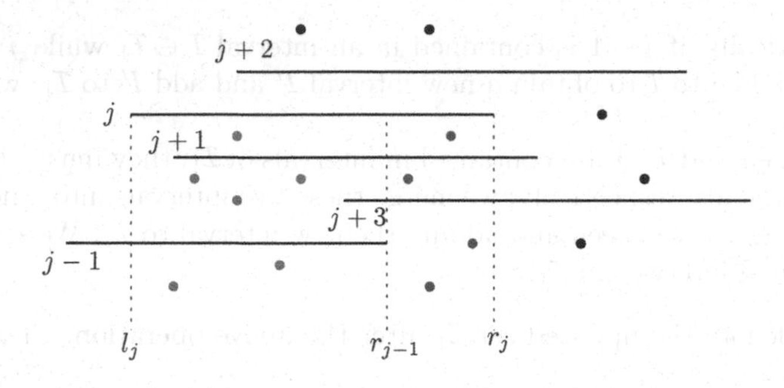

Fig. 2. Illustrating P_j^1 (the red points) and P_j^2 (the blue points). Only the upper edges of disks are shown. The numbers are the indices of disks. (Color figure online)

For each disk $s_j \in S$, we use $y(s_j)$ to denote the y-coordinate of the upper edge of s_j.

Our algorithm has m iterations. In the j-th iteration, it computes all segments in P_j^*, where P_j^* is the set of all non-redundant segments of P^* whose starting indices are j, although it is possible that some redundant segments with starting index j may also be computed. Points of P defining these segments must be inside s_j; let P_j denote the set of points of P inside s_j. We partition P_j into two subsets (e.g., see Fig. 2): P_j^1 consists of points of P_j to the left of r_{j-1} and P_j^2 consists of points of P_j to the right of r_{j-1}. We will compute dual segments of P_j^* defined by P_j^1 and P_j^2 separately; one reason for doing so is that when computing dual segments defined by a point of P_j^1, we need to additionally check whether this point also hits s_{j-1} (if yes, such a dual segment does not exist in P^* and thus will not be reported). In the following, we first describe the algorithm for P_j^1 since the algorithm for P_j^2 is basically the same but simpler. Note that our algorithm does not need to explicitly determine the points of P_j^1 or P_j^2; rather we will build some data structures that can implicitly determine them during certain queries.

If the upper edge of s_{j-1} is higher than that of s_j, then all points of P_j^1 are in s_{j-1} and thus no point of P_j^1 defines any dual segment of P^* starting from j. Indeed, assume to the contrary that a point $p_i \in P_j^1$ defines such a dual segment $[j, j']$. Then, since p_i is in s_{j-1}, $[j, j']$ cannot be a maximal interval of indices of disks hit by p_i and thus cannot be a dual segment defined by p_i. In what follows, we assume that the upper edge of s_{j-1} is lower than that of s_j. In this case, it suffices to only consider points of P_j^1 above s_{j-1} since points below the upper edge of s_{j-1} (and thus are inside s_{j-1}) cannot define any dual segments due to the same reason as above. Nevertheless, our algorithm does not need to explicitly determine these points.

We start with performing the following *rightward segment dragging query*: Drag the vertical segment $x(l_j) \times [y(s_{j-1}), y(s_j)]$ rightwards until a point $p \in P$ and return p (e.g., see Fig. 3). Such a segment dragging query can be answered in $O(\log n)$ time after $O(n \log n)$ time preprocessing on P (e.g., using Chazelle's result [6] one can build a data structure of $O(n)$ space in $O(n \log n)$ time such that

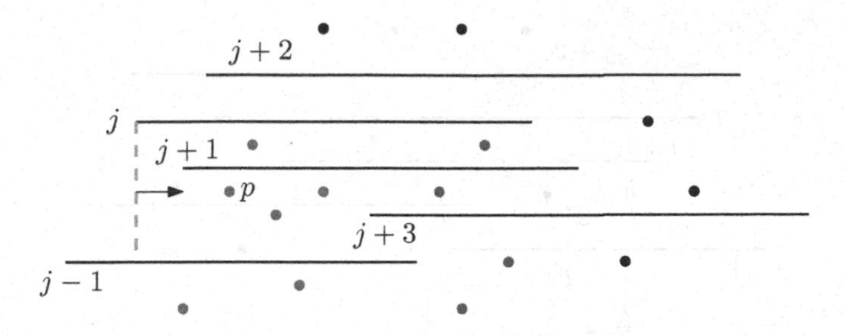

Fig. 3. Illustrating the rightward segment dragging query: the green dashed segment is the dragged segment $x(l_j) \times [y(s_{j-1}), y(s_j)]$. (Color figure online)

each query can be answered in $O(\log n)$ time; alternatively, if one is satisfied with an $O(n \log n)$ space data structure, then an easier solution is to use fractional cascading [7] and one can build a data structure in $O(n \log n)$ time and space with $O(\log n)$ query time). If the query does not return any point or if the query returns a point p with $x(p) > x(r_{j-1})$, then P_j^1 does not have any point above s_{j-1} and we are done with the algorithm for P_j^1. Otherwise, suppose the query returns a point p with $x(p) \leq x(r_{j-1})$; we proceed as follows.

We perform the following *max-range query* on p: Compute the largest index k such that all disks in $S[j, k]$ are hit by p (e.g., in Fig. 3, $k = j + 2$). We will show later in Lemma 5 that after $O(m \log m)$ time and $O(m)$ space processing, each such query can be answered in $O(\log m)$ time. Such an index k must exist as s_j is hit by p. Observe that $[j, k]$ is a dual segment in P^* defined by p. However, the weight of $[j, k]$ may not be equal to $w(p)$, because it is possible that a point with smaller weight also defines $[j, k]$. Our next step is to determine the minimum-weight point that defines $[j, k]$.

We perform a *range-minima query* on $[j, k]$: Find the lowest disk among all disks in $S[j, k]$ (e.g., in Fig. 3, s_{j+1} is the answer to the query). This can be easily done in $O(\log m)$ time with $O(m)$ space and $O(m \log m)$ time preprocessing. Indeed, we can build a binary search tree on the upper edges of all disks of S with their y-coordinates as keys and have each node storing the lowest disk among all leaves in the subtree rooted at the node. A better but more complicated solution is to build a range-minima data structure on the y-coordinates of the upper edges of all disks in $O(m)$ time and each query can be answered in $O(1)$ time [2,13]. However, the binary search tree solution is sufficient for our purpose. Let y^* be the y-coordinate of the upper edge of the disk returned by the query.

We next perform the following *downward min-weight point query* for the horizontal segment $[x(l_k), x(r_{j-1})] \times y^*$: Find the minimum weight point of P below the segment (e.g., see Fig. 4). We will show later in Lemma 6 that after $O(n \log n)$ time and space preprocessing, each query can be answered in $O(\log n)$ time. Let p' be the point returned by the query. If $p' = p$, then we report $[j, k]$ as a dual segment with weight equal to $w(p)$. Otherwise, if p' is inside s_{j-1} or s_{k+1}, then $[j, k]$ is a redundant dual segment (because a dual segment defined by p' strictly contains $[j, k]$ and $w(p') \leq w(p)$) and thus we do not need to report it. In any case, we proceed as follows.

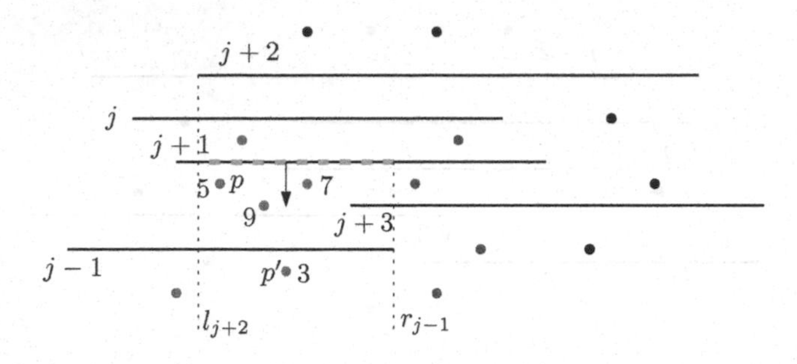

Fig. 4. Illustrating the downward min-weight point query (with $k = j + 2$): the green dashed segment is the dragged segment $[x(l_k), x(r_{j-1})] \times y^*$. The numbers besides the points are their weights. The answer to the query is p', whose weight is 3. (Color figure online)

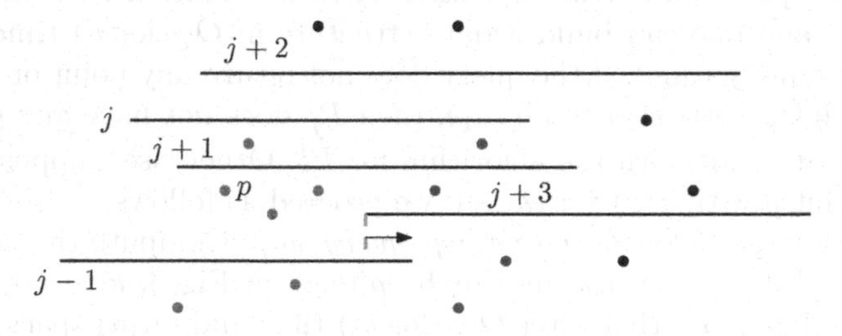

Fig. 5. Illustrating the rightwards segment dragging query: the green dashed segment is the dragged segment $x(l_{k+1}) \times [y(s_{j-1}), y']$. (Color figure online)

The above basically determines that $[j, k]$ is a dual segment in P^*. Next, we determine those dual segments $[j, k']$ with $k' > k$. If such a segment exists, the interval $[j, k']$ must contain index $k + 1$. Hence, we next consider s_{k+1}. If $y(s_{k+1}) > y(s_{j-1})$, then let $y' = \min\{y^*, y(s_{k+1})\}$; we perform a rightward segment dragging query with the vertical segment $x(l_{k+1}) \times [y(s_{j-1}), y']$ (e.g., see Fig. 5) and then repeat the above algorithm. If $y(s_{k+1}) \leq y(s_{j-1})$, then points of P_j^1 above s_{j-1} are also above s_{k+1} and thus no point of P_j^1 can generate any dual segment $[j, k']$ with $k' > k$ and thus we are done with the algorithm on P_j^1.

For time analysis, we charge the time of the above five queries to the interval $[j, k]$, which is in P^*. Note that $[j, k]$ will not be charged again in future because future queries in the j-th iteration will be charged to $[j, k']$ for some $k' > k$ and future queries in the j'-th iteration for any $j' > j$ will be charged to $[j', k'']$. As such, each dual segment of P^* is charged $O(1)$ times in the entire algorithm. As each query takes $O(\log(n + m))$ time, the total time of all queries in the entire algorithm is $O(|P^*| \log(n + m))$, which is $((n + m) \log(n + m))$ by Lemma 4.

Proofs of Lemmas 5 and 6 are in the full paper.

Lemma 5. *With $O(m \log m)$ time and $O(m)$ space preprocessing on S, each max-range query can be answered in $O(\log m)$ time.*

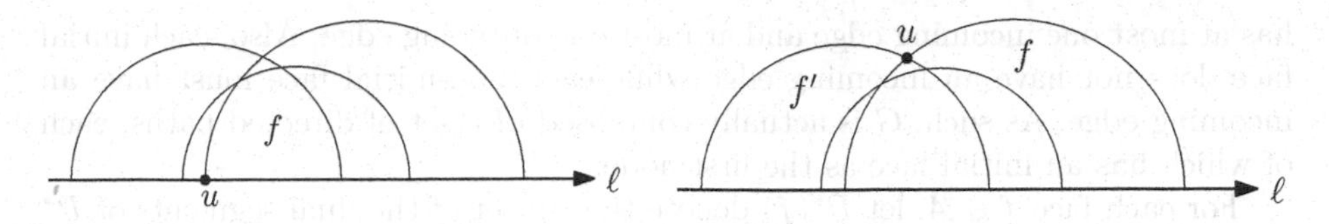

Fig. 6. Illustrating an initial face f with leftmost vertex u.

Fig. 7. A non-initial face f with leftmost vertex u and its opposite face f'.

Lemma 6. *With $O(n \log n)$ time and space preprocessing on P, each downward min-weight point query can be answered in $O(\log n)$ time.*

This finishes the description of the algorithm for P_j^1. The algorithm for P_j^2 is similar with the following minor changes. First, when doing each rightward segment dragging query, the lower endpoint of the query vertical segment is at $-\infty$ instead of $y(s_{j-1})$. Second, when the downward min-weight point query returns a point, we do not have to check whether it is in s_{j-1} anymore. The rest of the algorithm is the same. In this way, all non-redundant intervals of P^* starting at index j can be computed. As analyzed above, the runtime of the entire algorithm is bounded by $O((n+m)\log(n+m))$.

As such, using the dual transformation, we have the following result.

Theorem 2. *The line-constrained L_∞ hitting set can be solved in $O((n+m)\log(n+m))$ time.*

5 The L_2 Case – A Sketch

Due to the space limit, we only sketch our result for the L_2 case; the full details can be found in the full paper.

As in the L_∞ case, $|P_i^*| \geq 2$ is possible and $|P_i^*| \leq \lceil m/2 \rceil$. We first prove a lemma similar to Lemma 2, following a similar proof scheme. As such, it suffices to find an optimal solution to the 1D dual coverage problem on P^* and S^*.

Upper bound for $|P^*|$. We then prove the upper bound $|P^*| = O(m+\kappa)$, where κ is the number of pairs of disks of S that intersect. To this end, we consider the arrangement $\mathcal{A}$ of the boundaries of all disks of S in the half-plane above ℓ. An easy but critical observation is that points of P located in the same face of $\mathcal{A}$ define the same subset of dual segments of P^*. As such, it suffices to consider the dual segments defined by all faces of $\mathcal{A}$.

We define *initial faces* of $\mathcal{A}$. Roughly speaking, a face is an *initial face* if its leftmost vertex is on ℓ (e.g., see Fig. 6).

We define a directed graph G as follows. The faces of $\mathcal{A}$ form the node set of G. There is an edge from a node f' to another node f if the face f is a non-initial face and f' is the *opposite* face of f (i.e., the rightmost vertex of f' is the leftmost vertex of f; e.g., in Fig. 7, there is a directed edge from f' to f). Since each face of $\mathcal{A}$ has only one leftmost vertex and only one rightmost vertex, each node G

has at most one incoming edge and at most one outgoing edge. Also, each initial face does not have an incoming edge while each non-initial face must have an incoming edge. As such, G is actually composed of a set of directed paths, each of which has an initial face as the first node.

For each face $f \in \mathcal{A}$, let $P^*(f)$ denote the subset of the dual segments of P^* generated by f (i.e., generated by any point in f). Our goal is to obtain an upper bound for $|\bigcup_{f \in \mathcal{A}} P^*(f)|$, which is an upper bound for $|P^*|$ as $P^* \subseteq \bigcup_{f \in \mathcal{A}} P^*(f)$. To this end, we first show that $|P^*(f)| = 1$ for each initial face f and we then show that for any two adjacent faces f' and f in any path of G, the symmetric difference of $P^*(f)$ and $P^*(f')$ is $O(1)$. As such, we can obtain $|P^*| = O(m + \kappa)$ as $\mathcal{A}$ has $O(m + \kappa)$ faces.

Computing P^*. To compute the set P^*, following the above idea, it suffices to compute the dual segments generated by all faces of $\mathcal{A}$ (or equivalently, generated by all nodes of the graph G). The main idea is to directly compute for each path $\pi \in G$ the dual segments defined by the initial face of π and then for each non-initial face $f \in \pi$, determine $P^*(f)$ indirectly based on $P^*(f')$, where f' is the predecessor face of f in π. To this end, after constructing $\mathcal{A}$ and G and other preprocessing, we first show that $P^*(f)$ for each initial face f can be computed in $O(\log m)$ time, and we then show that $P^*(f)$ can be determined in $O(\log m)$ time based on $P^*(f')$, where f' is the predecessor face of f in π, for each path π of G. As such, P^* can be computed in $O(n \log(n + m) + (m + \kappa) \log m)$ time. Consequently, using the dual transformation, the L_2 hitting set problem on P and S can be solved in $O((n + m) \log(n + m) + \kappa \log m)$ time.

6 Lower Bound

We can prove an $\Omega((n + m) \log(n + m))$ time lower bound for the problem even for the 1D unit-disk case (i.e., all segments have the same length), by a simple reduction from the element uniqueness problem (Pedersen and Wang [19] used a similar approach to prove the same lower bound for the 1D coverage problem). Indeed, the element uniqueness problem is to decide whether a set $X = \{x_1, x_2, \ldots, x_N\}$ of N numbers are distinct. We construct an instance of the 1D unit-disk hitting set problem with a point set P and a segment set S on the x-axis ℓ as follows. For each $x_i \in X$, we create a point p_i on ℓ with x-coordinate equal to x_i and create a segment on ℓ that is the point p_i itself. Let $P = \{p_i \mid 1 \leq i \leq N\}$ and S the set of segments defined above (and thus all segments have the same length); then $|P| = |S| = N$. We set the weights of all points of P to 1. Observe that the elements of X are distinct if and only if the total weight of points in an optimal solution to the 1D unit disk hitting set problem on P and S is n. As the element uniqueness problem has an $\Omega(N \log N)$ time lower bound under the algebraic decision tree model, $\Omega((n+m) \log(n+m))$ is a lower bound for our 1D unit disk hitting set problem.

References

1. Alt, H., et al.: Minimum-cost coverage of point sets by disks. In: Proceedings of the 22nd Annual Symposium on Computational Geometry (SoCG), pp. 449–458 (2006)
2. Bender, M.A., Farach-Colton, M.: The LCA problem revisited. In: Gonnet, G.H., Viola, A. (eds.) LATIN 2000. LNCS, vol. 1776, pp. 88–94. Springer, Heidelberg (2000). https://doi.org/10.1007/10719839_9
3. Bilò, V., Caragiannis, I., Kaklamanis, C., Kanellopoulos, P.: Geometric clustering to minimize the sum of cluster sizes. In: Brodal, G.S., Leonardi, S. (eds.) ESA 2005. LNCS, vol. 3669, pp. 460–471. Springer, Heidelberg (2005). https://doi.org/10.1007/11561071_42
4. Bus, N., Mustafa, N.H., Ray, S.: Practical and efficient algorithms for the geometric hitting set problem. Discrete Appl. Math. **240**, 25–32 (2018)
5. Chan, T.M., Grant, E.: Exact algorithms and APX-hardness results for geometric packing and covering problems. Comput. Geom.: Theory Appl. **47**, 112–124 (2014)
6. Chazelle, B.: An algorithm for segment-dragging and its implementation. Algorithmica **3**(1–4), 205–221 (1988)
7. Chazelle, B., Guibas, L.J.: Fractional cascading: I. A data structuring technique. Algorithmica **1**(1), 133–162 (1986)
8. Durocher, S., Fraser, R.: Duality for geometric set cover and geometric hitting set problems on pseudodisks. In: Proceedings of the 27th Canadian Conference on Computational Geometry (CCCG) (2015)
9. Edelsbrunner, H., Mücke, E.P.: Simulation of simplicity: a technique to cope with degenerate cases in geometric algorithms. ACM Trans. Graph. **9**, 66–104 (1990)
10. Even, G., Rawitz, D., Shahar, S.: Hitting sets when the VC-dimension is small. Inf. Process. Lett. **95**, 358–362 (2005)
11. Feder, T., Greene, D.H.: Optimal algorithms for approximate clustering. In: Proceedings of the 20th Annual ACM Symposium on Theory of Computing (STOC), pp. 434–444 (1988)
12. Ganjugunte, S.K.: Geometric hitting sets and their variants. Ph.D. thesis, Duke University (2011)
13. Harel, D., Tarjan, R.E.: Fast algorithms for finding nearest common ancestors. SIAM J. Comput. **13**, 338–355 (1984)
14. Karp, R.M.: Reducibility among combinatorial problems. Complexity of Computer Computations, pp. 85–103 (1972)
15. Moreno-Centeno, E., Karp, R.M.: The implicit hitting set approach to solve combinatorial optimization problems with an application to multigenome alignment. Oper. Res. **61**, 453–468 (2013)
16. Mustafa, N.H., Ray, S.: PTAS for geometric hitting set problems via local search. In: Proceedings of the 25th Annual Symposium on Computational Geometry (SoCG), pp. 17–22 (2009)
17. Mustafa, N.H., Ray, S.: Improved results on geometric hitting set problems. Discrete Comput. Geom. **44**, 883–895 (2010)
18. Pedersen, L., Wang, H.: On the coverage of points in the plane by disks centered at a line. In: Proceedings of the 30th Canadian Conference on Computational Geometry (CCCG), pp. 158–164 (2018)
19. Pedersen, L., Wang, H.: Algorithms for the line-constrained disk coverage and related problems. Comput. Geom.: Theory Appl. **105–106**(101883), 1–18 (2022)

An ETH-Tight Algorithm for Bidirected Steiner Connectivity

Daniel Lokshtanov[1], Pranabendu Misra[2], Fahad Panolan[3](✉), Saket Saurabh[4,5], and Meirav Zehavi[6]

[1] University of California, Santa Barbara, USA
[2] Chennai Mathematical Institute, Chennai, India
[3] Indian Institute of Technology, Hyderbad, India
fahad@cse.iith.ac.in
[4] The Institute of Mathematical Sciences, HBNI, Chennai, India
[5] University of Bergen, Bergen, Norway
[6] Ben-Gurion University, Beersheba, Israel

Abstract. In the Strongly Connected Steiner Subgraph problem, we are given an n-vertex digraph D, a weight function $w\colon A(D) \mapsto \mathbb{R}^+$ on the arc set of D, and a set of k terminals $Q \subseteq V(D)$, and our objective is to find a strongly connected subgraph of D containing Q with minimum total weight. The problem is known to be W[1]-hard on general digraphs. However on bi-directed graphs (digraphs where, if uv is an arc then so is vu) with symmetric weight function $w\colon A(D) \mapsto \mathbb{R}^+$ (i.e., $w(uv) = w(vu)$ for any $uv \in A(D)$), Chitnis, Feldmann and Manurangsi [TALG 2021] showed that the problem is fixed parameter tractable (FPT) with running time $2^{\mathcal{O}(k^2)}n^{\mathcal{O}(1)}$, where n is the input length. They also show that, unless the Exponential Time Hypothesis (ETH) fails, there is no algorithm for the problem on bi-directed graphs with running time $2^{o(k)}n^{\mathcal{O}(1)}$. They left the existence of a single-exponential in k time algorithm as an open problem. We resolve this question, by designing an algorithm for the problem running in time $2^{\mathcal{O}(k)}n^{\mathcal{O}(1)}$ that is asymptotically tight under ETH, thereby closing the gap between the upper and lower-bounds for this problem.

Chitnis, Feldmann and Manurangsi [TALG 2021] showed that an optimum solution to this problem can always be described by a collection of trees, that are mapped to the input graph via homomorphisms, and glued together at the terminal vertices. This structural result allows us to design an algorithm via the combination of a Dreyfus-Wagner style dynamic programming algorithm and the notion of representative sets over linear matroids.

Keywords: FPT · Graph Connectivity · Representative Family · Matroids

Supported by NSF award CCF-2008838, Swarnajayanti Fellowship grant DST/SJF/MSA-01/2017-18, and ERC grants LOPRE (grant agreement No. 819416) and PARAPATH.

P. Morin and S. Suri (Eds.): WADS 2023, LNCS 14079, pp. 588–604, 2023.
https://doi.org/10.1007/978-3-031-38906-1_39

1 Introduction

A central family of problems in network design addresses the objective of connecting a given set of terminals in a given network in a particular way, as to ensure that the terminals can transmit information from one to the other in that particular way. For instance, such problems naturally arise when modelling radio or ad-hoc wireless networks. Here, a very general setting is when the input consists of an arc-weighted digraph D as well as a collection $\mathcal{Q}$ of pairs of *terminals* (designated vertices) in D, while the objective is to find a subgraph of D with minimum cost (in terms of arc weights) such that for each terminal pair $(s,t) \in \mathcal{Q}$, there is a directed path from s to t in the subgraph. Unfortunately, this problem, called Directed Steiner Network (DSN) (also known as Directed Steiner Forest, although the sought solution may not be a forest), is particularly hard. On the one hand, it is W[1]-hard with respect to $k = |\mathcal{Q}|$ [27], which means that it is unlikely to be *fixed-parameter tractable (FPT)*—that is, solvable in time $f(k) \cdot n^{\mathcal{O}(1)}$ for some computable function f of k. In fact, unless Exponential Time Hypothesis (ETH) fails, it is not even solvable in time $f(k) \cdot n^{o(k)}$ for any computable function f of k [10]. On the other hand, DSN cannot be approximated within factor $O(2^{\log^{1-\epsilon} n})$ in polynomial time unless NP $\subseteq$ DTIME$(2^{\text{polylog}(n)})$ [16]. Furthermore, even combining both FPT and approximation also does not seem to be of much help. Indeed, Dinur and Manurangsi [15] have shown that the approximation of DSN within factor $k^{\frac{1}{4}-o(1)}$ is already not FPT unless Gap-ETH fails. Recently, Manurangsi et al. [31] proved that there is no FPT approximation algorithm for DSN with a factor of $o(k^{\frac{1}{2}})$ assuming Strongish Planted Clique Hypothesis. On the positive side, DSN is solvable in time $n^{\mathcal{O}(k)}$ [19], and can be approximated within factors $\mathcal{O}(n^{\frac{2}{3}+\epsilon})$ and $\mathcal{O}(k^{\frac{1}{2}+\epsilon})$ in polynomial time [2,6,20]. When the arc costs are uniform, DSN can be approximated within a factor $\mathcal{O}(n^{\frac{3}{5}+\epsilon})$ [11].

When dealing with undirected graphs, the problem is *significantly* easier at both parameterized complexity and approximation fronts. Here, we are given an undirected graph G and a set of terminals Q, and the objective is to find a subtree of G with minimum number of edges containing Q. This problem is called Steiner Tree (ST), and a more general version called Steiner Forest is also well studied. Steiner Tree is one of the 21 NP-hard problems in the seminal paper of Karp [28]. Already in 1971, it was shown to be FPT with respect to $k = |Q|$—specifically, it was shown to be solvable in time $3^k \cdot n^{\mathcal{O}(1)}$ [18]. The running time was improved to $(2+\epsilon)^k \cdot n^{f(\frac{1}{\epsilon})}$ and later to $2^k \cdot n^{\mathcal{O}(1)}$ for some computable function f of $\frac{1}{\epsilon}$ and $\epsilon > 0$ [4,23]. The algorithm in [4] works for the weighted version of the problem with edge weights from $\{1, 2, \ldots, M\}$, and the running time will have an additional multiplicative factor of M. On the other hand, unless ETH fails, Steiner Tree cannot be solved in time $2^{o(n)}$ and hence neither in time $2^{o(k)} \cdot n^{\mathcal{O}(1)}$. (In fact, unless the so called Set Cover Conjecture (SeCoCo) fails, it is not even solvable in time $(2-\epsilon)^k \cdot n^{\mathcal{O}(1)}$ [13].) From the approximation perspective, ST can be approximated within a constant factor in polynomial-time: for over several decades, there is an ongoing quest to find the best possible approximation factor (see, e.g., [5,25,33,37]); to the best of our knowledge, the current best approximation factor is of $\ln(4)+\epsilon < 1.39$ [5,25]. On the negative side, unless P=NP, ST is inapproximable within factor $\frac{96}{95}$ [3,12]. It is known that unless PH collapses, ST does not admit a polynomial kernel [17].

Between these two extremes, lies the STRONGLY CONNECTED STEINER SUBGRAPH (SCSS) problem. Here, given an arc weighted digraph D and a collection of terminals Q, the objective is to find a strongly connected subgraph of D with minimum number of arcs containing Q. Notice that the SCSS problem is precisely the special case of DSN where $\mathcal{Q} = \{(u,v) : u, v \in Q\}$. Further, the flavor of SCSS is precisely that of ST in the sense that every terminal can transmit information to every other terminal in the sought subnetwork, which is a realistic demand in network design. Unlike DSN, SCSS is known to be approximable within factor 2 in $3^k \cdot n^{\mathcal{O}(1)}$ time [9]. Unfortunately, like DSN, SCSS is W[1]-hard, and in fact it cannot even be approximated within factor $(2-\epsilon)$ in time $f(k) \cdot n^{\mathcal{O}(1)}$ unless Gap-ETH fails [8]. We remark that the special case of SCSS where $Q = V(D)$, known as MINIMUM STRONGLY CONNECTED SPANNING SUBRAPH, can be approximated in polynomial time within factor 2 [22] and when all the edge weights are equal to 1, it has a factor $\frac{3}{2}$ approximation in polynomial time [35]. Also, it can be solved in single-exponential time in n (specifically, $2^{4\omega n} n^{\mathcal{O}(1)}$ where ω is the matrix multiplication exponent) [21].

Knowing that SCSS is unlikely to be FPT, how can we make its flavor more similar to ST so that it will be tractable (FPT) yet will still deal naturally with digraphs? Recently, Chitnis, Feldmann and Manurangsi [8] initiated a study where they restricted DSN to *bi-directed graphs* called BI-DSN. Formally, a digraph D is bi-directed if for any pair of vertices $\{u, v\}$ in D, either both arcs uv and vu belong to D, or none of them does. Moreover, if $uv, vu \in A(D)$, then their weights are also equal. As noted in [8], bi-directed graphs model some realistic settings [7,30,34,36]—for instance, when nodes have the same transmitter model (e.g., in some wireless networks), thus if a node u can transmit information to a node v, the node v can transmit information to the node u as well. Critically, unlike undirected graphs which also model this property, bi-directed graphs capture the property that if we want to transmit information in both directions, then we need to pay *twice*, once for each direction. Thus, similarly to general digraphs, orientation plays a role. In particular, it can completely change the set of optimal solutions; for example, see Fig. 1, taken from [8]. Chitnis et al. proved that BI-DSN is W[1]-hard, and it admits a polynomial time 4-approximation algorithm and a 2-approximation algorithm running in time $2^{\mathcal{O}(k)} n^{\mathcal{O}(1)}$. Formally, BI-STRONGLY CONNECTED STEINER SUBGRAPH (BI-SCSS) is defined as follows.

BI-SCSS **Parameter:** $k = \|Q\|$
Input: A bi-directed graph D, a set of terminals $Q \subseteq V(D)$ and a symmetric weight function $w : A(D) \to \mathbb{R}^+$ such that $w(uv) = w(vu)$ for any $uv \in A(D)$.
Output: A strongly connected subgraph of D containing Q of minimum total weight.

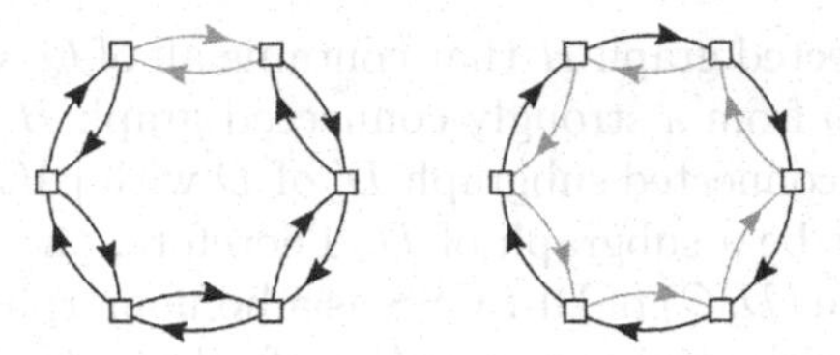

Fig. 1. An example in [8] of a BI-SCSS instance where all vertices are terminals. Left: Black edges show a solution which takes an undirected optimum twice. Right: The actual optimum solution is shown in black.

In [8], it was proved that BI-SCSS is solvable in time $2^{\mathcal{O}(k^2)} \cdot n^{\mathcal{O}(1)}$, as well as that it is NP-hard and that unless ETH fails, it cannot be solved in time $2^{o(k)} \cdot n^{\mathcal{O}(1)}$. It was noted in [8] that bidirected inputs are the first example where SCSS remains NP-hard but turns out to be FPT parameterized by k. The results of [8] leave a gap between the known upper and conditional lower bounds and they posed the following open question.

> Can we obtain a single-exponential FPT algorithm for BI-SCSS?

In this paper, we answer the above question in the affirmative.

Theorem 1. *The* BI-SCSS *problem is solvable in time* $2^{\mathcal{O}(k)} \cdot n^{\mathcal{O}(1)}$.

In particular, due to the aforementioned lower bound, *our algorithm is optimal under the ETH.* Our algorithm builds on a structural result which states that a minimal solution can always be described as a sum of trees mapped by homomorphisms into the graph and glued together at the terminal vertices. This result follows from [8]. The structural result allows us to find an optimal solution by making use of representative sets combined with a Dreyfus-Wagner style dynamic programming algorithm. We note the technique of representative sets has been used to improve the running times of FPT and exact algorithms of various problems and design efficient kernelization algorithms [21,26,29]. We describe our algorithms for the unweighted version of the problem, which easily extends to the symmetric weighted version via weighted representative sets. A short technical overview of our result is given below.

Our Methods. We start with the definition of homomorphism on digraphs. A *homomorphism* from a digraph H to a digraph D is a function ϕ from $V(H)$ to $V(D)$ such that for any $uv \in A(H)$, $\phi(u)\phi(v) \in A(D)$. Our first step is a structural analysis of solutions to an instance of BI-SCSS. This result shows that optimal solutions can be decomposed into a sum of trees mapped by a homomorphism from a strongly connected digraph and glued together at the terminal vertices. Towards that we make the following observation. Any solution D' to an instance of BI-SCSS (D, Q) can be viewed as the image of a homomorphism from a strongly connected graph H that contains all of Q. Observe that, if there is a homomorphism ϕ from a strongly connected graph H to D then the image of ϕ forms a strongly connected subgraph D' of D with $|A(D')| \leq |A(H)|$. Let us note that H need not be a subgraph of D. Therefore, one can view a minimum solution to an instance (D, Q) of Bi-SCSS as a homomorphism ϕ from a strongly connected graph H with minimum number of edges such that $Q \subseteq \phi(V(H))$. Formally, consider an instance (D, Q) of Bi-SCSS, and let H be a strongly connected graph and ϕ be a homomorphism from H to D such that $Q \subseteq \phi(V(H))$. Then $\phi(H)$ is a solution to (D, Q) corresponding to the pair (H, ϕ).

We consider both D and H as k-boundaried graphs. In a *k-boundaried digraph* D, a certain subset of k vertices, denoted by $\beta(G)$, are tagged as *boundary vertices* and are labeled with $1, \ldots, k$. For an instance (D, Q) of Bi-SCSS, the boundary vertices in D are exactly the vertices in Q (i.e., $Q = \beta(D)$). Then we say that (H, ϕ), where H is a strongly connected k-boundaried digraph and ϕ is a homomorphism that preserves labels (i.e., the vertex with label i in H will be mapped to the vertex in D with label i), is a *solution* to the k-boundaried bi-directed graph D. The *elements* of a k-boundaried digraph H are defined as follows. Let C be a connected component in $\overline{H} - \beta(H)$, where $\overline{H}$ is the underlying undirected graph of H. Here, $\overline{H}$ is a simple graph that has an edge between u and v if and only if $uv \in A(H)$ or $vu \in A(H)$. Then the subgraph of H induced on the closed neighborhood of $V(C)$, excluding the arcs incident only on $\beta(H)$, i.e., $H[N[V(C)]] - A(\beta(H))$, is called an element of H. One can show that there is an optimal solution (H, ϕ) such that for every element J of H, $\overline{J}$ is a tree. This result follows directly from [8, Lemma 5.2]. In other words, we can view H as a collection of trees that are glued together at boundary vertices union the set of edges within the boundary (see Fig. 2 for an illustration).

Having established the above structural result, we proceed to an algorithm for finding an optimum solution to the given instance. We design a Dreyfus-Wagner style dynamic programming (DP) algorithm via *representative families* over graphic matroids. Our algorithm builds upon a well known characterization of strongly connected digraph: a digraph is strongly connected if and only if for any vertex u there is a spanning *in-tree* and a spanning *out-tree* rooted at u. This characterization can be captured using a collection of linear matroids [24]. Then, we present a dynamic programming algorithm, that gradually builds an optimum solution to the given instance. Here, we crucially use the fact that this solution can be decomposed into a collection of trees, and these trees can be constructed via a Dreyfus-Wagner style dynamic programming [18]. To ensure that the algorithm runs in $2^{\mathcal{O}(k)}n^{\mathcal{O}(1)}$ time, we need an additional tool. At each step of the dynamic programming algorithm, we must prune the collection of partial solutions using the notion of representative sets [21] over linear matroids. Otherwise, the set of partial solution cannot be bounded by a function of k alone.

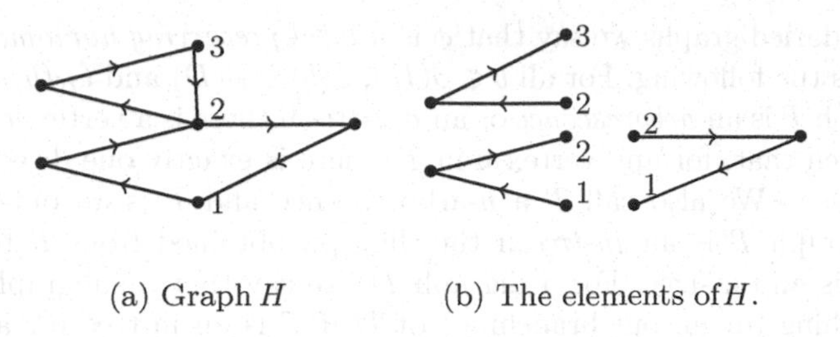

(a) Graph H (b) The elements of H.

Fig. 2. A strongly connected graph H with boundary vertices colored red is drawn in the left side. Its elements are drawn in the right side. (Color figure online)

2 Preliminaries

We refer the reader to [14] for standard graph theoretic notions that are not explicitly stated here. For a positive integer $t \in \mathbb{N}$, $[t]$ denotes the set $\{1, \ldots, t\}$. For a function $f\colon A \to B$ and $X \subseteq A$, $f(X) = \{f(x) : x \in X\}$. We use the term graph to denote an undirected graph without self-loops, but with parallel edges. However, the digraphs considered in this paper are without self-loops and parallel arcs. For a graph G, we use $V(G)$ and $E(G)$ to denote its vertex set and edge set, respectively. Let G be a graph, $u \in V(G)$, and $U \subseteq V(G)$. We use $d_G(u)$ to denote the number of edges in G that are incident to u. We use $N[U]$ to denote the closed neighborhood of U and $G[U]$ to denote the subgraph of G induced on U. We use $E(U)$ to denote the set $\{uv \in E(G) : u, v \in U\}$.

For a digraph D, we use $V(D)$ and $A(D)$ to denote its vertex set and arc set, respectively. For a digraph D, we use $\overline{D}$ to denote the underlying undirected graph of D. Moreover, if there exist two vertices $u, v \in V(D)$ such that $uv, vu \in A(D)$, then there will be two edges between u and v in $\overline{D}$. For a (di)graph D and an arc/edge subset F of D, $D[F]$ denotes the graph induced on F, i.e., $D[F]$ has vertex set $V(D)$ and arc/edge set F. A connected component of a digraph D is a maximal subgraph C of D such that $\overline{C}$ is a connected graph.

A *k-boundaried (di)graph* is a (di)graph G with a set $W \subseteq V(G)$ of cardinality at most k such that each vertex $w \in W$ has a unique label $\ell_G(w) \in \{1, \ldots, k\}$. The vertex subset W is referred to as the *boundary* of G. For a k-boundaried (di)graph G, the function $\beta(G)$ returns the boundary of G. For two k-boundaried (di)graphs G_1 and G_2, $G_1 \oplus G_2$ denotes the k-boundaried graph obtained by "gluing together the same labelled boundary vertices". That is, the gluing operation takes the disjoint union of G_1 and G_2 and identifies the vertices of $\beta(G_1)$ and $\beta(G_2)$ with the same label. If there are vertices $u_1, v_1 \in \beta(G_1)$ and $u_2, v_2 \in \beta(G_2)$ such that $\ell_{G_1}(u_1) = \ell_{G_2}(u_2)$ and $\ell_{G_1}(v_1) = \ell_{G_2}(v_2)$, then G has vertices u and v formed by unifying u_1 and u_2, and v_1 and v_2, respectively. Moreover, $uv \in E(G_1 \oplus G_2)$ if and only if $u_1v_1 \in E(G_1)$ or $u_2v_2 \in E(G_2)$. Also, u and v are boundary vertices in $G_1 \oplus G_2$ such that $\ell_{G_1 \oplus G_2}(u) = \ell_{G_1}(u_1)$ and $\ell_{G_1 \oplus G_2}(v) = \ell_{G_1}(v_1)$.

A *homomorphism* from a digraph H to a digraph D is a function ϕ from $V(H)$ to $V(D)$ such that for any $uv \in A(H)$, $\phi(u)\phi(v) \in A(D)$. Also, if D and

H are boundaried graphs we say that ϕ is a *label preserving homomorphism* if ϕ also satisfies the following: For all $b \in \beta(H)$, $\phi(b) \in \beta(D)$ and $\ell_H(b) = \ell_D(\phi(b))$.

A digraph T is an *arborescence* or an *out-tree* if there is a vertex u in T, called the root, such that, for any vertex v in T, there is exactly one directed path in T from u to v. We also call T a u-arborescence and T is an out-tree rooted at u. A digraph T is an *in-tree* if the digraph obtained from T by reversing all its arcs is an out-tree. For a digraph D, we say that a subgraph T of D is an in-branching (or an out-branching) of D if T is an in-tree (or an out-tree) and $V(D) = V(T)$. The following observation is an alternate characterization of in-trees and out-trees.

Observation 1. *Let T be a digraph and $u \in V(T)$. The digraph T is an in-tree rooted at u if and only if $\overline{T}$ is a tree, the out-degree of u is 0, and for each $v \in V(T) \setminus \{u\}$, there is exactly one arc vw in T for some $w \in V(T)$. Similarly, T is an out-tree if and only if $\overline{T}$ is a tree, in-degree of u is 0, and for each $v \in V(T) \setminus \{u\}$, there is exactly one arc wv in T for some $w \in V(T)$.*

3 Algorithm

Recall that, the input to Bi-SCSS is a pair (D, Q) where D is bi-directed graph D, and $Q \subseteq V(D)$ is the set of terminal vertices. Also recall that we present an algorithm for the unweighted version of the problem. First we state that a certain class of solutions for this problem admit a decomposition into trees. For notational convenience, instead of using a pair (D, Q) to denote an instance of Bi-SCSS, we consider it as a k-boundaried bi-directed graph D with boundary $\beta(D)$, where the boundary vertices $\beta(D)$ play the role of terminals. Recall that we have a labeling $\ell_D : \beta(D) \to [k]$ where $k = |\beta(D)|$. Let us begin with the following lemma that allows us to view a solution to an instance of Bi-SCSS as a homomorphism between graphs. Similar methods, that is, viewing solutions as homomorphisms were also considered before, for example [32].

Lemma 1. *Let D be an instance of* Bi-SCSS, *and H be a strongly connected digraph with boundary $\beta(H)$ such that $|\beta(H)| = |\beta(D)|$. If there is a label preserving homomorphism ϕ from H to D, then there is a strongly connected subgraph D' of D containing $\beta(D)$ such that $|A(D')| \leq |A(H)|$.*

Proof. We define the graph D' as follows. The vertex set of D' is $V(D') = \phi(V(H)) \subseteq V(D)$. For any arc $uv \in A(H)$, $\phi(u)\phi(v)$ belongs to $A(D')$. This completes the construction of D'.

Since H is strongly connected, between any two vertices x and y in H, there is a path P_1 from x to y, and there is a path from y to x in H. Each of these paths corresponds to a walk in D'. This implies that there is a path from $\phi(x)$ to $\phi(y)$, and there is a path from $\phi(y)$ to $\phi(x)$ in D'. Thus, D' is a strongly connected graph. Moreover, $\beta(D) \subseteq V(D')$ and $|A(D')| \leq |A(H)|$. □

Observe that, for a solution D' to the instance D, there is a trivial label preserving homomorphism from $H = D'$ to D. Thus, because of Lemma 1, we can view the solutions of a BI-SCSS instance D as homomorphisms from strongly connected graphs H to D that induce a one-to-one map from $\beta(H)$ to $\beta(D)$ that also preserves the labels. That is, a solution to an instance G of BI-SCSS is a pair (H, ϕ) where H is a strongly connected k-boundaried digraph with $|\beta(H)| = |\beta(D)|$, and ϕ is a label preserving homomorphism from H to D.

Next we define a decomposition of a solution into its *elements*, and then prove that the underlying undirected graphs of elements of an optimum solution, are trees (see Fig. 2).

Definition 1 (Elements of a solution). *Let H be a k-boundaried graph and $Q = \beta(H)$. Let $C_1, \ldots, C_r$ be the connected components of $\overline{H} - Q$. Then, the k-boundaried graphs $H[N[V(C_1)]] - E(Q), \ldots, H[N[V(C_r)]] - E(Q)$ are called the* elements *of H. Here, for each element J, we have $\beta(J) = Q \cap V(J)$ and $\ell_J(b) = \ell_H(b)$ for all $b \in \beta(J)$.*

The following lemma (Lemma 2) directly follows from [8, Lemma 5.2]. As [8, Lemma 5.2] holds in the case of symmetric weight functions, Lemma 2 can be extended to the case of symmetric weight functions as well.

Lemma 2. *Let D be an instance of* BI-SCSS. *Then, there exist a strongly connected subgraph D' of D containing all the boundary vertices with the least number of edges and a solution (H, ϕ) to D such that*

- $D' = (\phi(V(H)), \{\phi(u)\phi(v) \ : \ uv \in A(H)\})$, *and*
- *for every element J of H, $\overline{J}$ is a tree.*

We say that (H, ϕ) is a *homomorphism minimal* solution to D, if for every element J of H, $\overline{J}$ is a tree. Lemma 2 implies that there is an optimum homomorphism minimal solution.

Our algorithm is a Dreyfus-Wagner [18] style dynamic programming algorithm combined with representative families [21] on disjoint union of graphic matroids to prune the sizes of DP table entries. Let D be an instance of BI-SCSS, where D is a k-boundaried graph. Let (H, ϕ) be an optimal homomorphism minimal solution to the instance D. Lemma 2 implies that for every element J of H, $\overline{J}$ is a tree. The tree structure of the elements of H allows us to do a DP algorithm in a manner that is designed for STEINER TREE by Dreyfus and Wagner [18]. But, both the sizes of elements and the number of potential choices for elements that need to be considered as partial solutions are not bounded by a function of k. To overcome this hurdle, we use the notion of representative families on graphic matroid to prune the list of partial solutions in our dynamic programming table.

The following proposition will allow us to see solutions as a union of in-branchings and out-branchings. This was used in previous results (for example [10]). The underlying tree structures in in-branchings and out-branchings are useful for applying representative families on graphic matroids.

Proposition 1. ([1]). *Let H be a digraph and $r \in V(H)$. Then, H is strongly connected if and only if there is an in-branching rooted at r and an out-branching rooted at r in H.*

Observation 2 ($\star$).[1] *Let D be an instance of* Bi-SCSS *and (H, ϕ) be a homomorphism minimal solution to D. Then, for any element J of H and $u \in V(J)$ with $d_{\overline{J}}(u) \leq 1$, $u \in \beta(H)$.*

From now on, throughout the section we fix a k-boundaried digraph D as the input of Bi-SCSS *and $n = |V(D)|$.* Since we have to explicitly keep track of parts of in-branching and out-branching in a "partial solution" while doing DP, we refine the notion of solution from a pair to a quadruple.

Definition 2. *A* homomorphism minimal solution *to D is a quadruple $(H, \phi, T_{in}, T_{out})$, where H is a k-boundaried digraph, $|\beta(H)| = |\beta(D)|$, and $A(T_{in}) \cup A(T_{out}) = A(H)$ such that the following holds. Let $y \in \beta(H)$ be the vertex in H such that $\ell_H(y) = 1$.*

(i) The function ϕ is a label preserving homomorphism from H to D.
(ii) T_{in} is an in-branching of H rooted at y and T_{out} is an out-branching of H rooted at y.
(iii) For any element J of H and leaf u in J, $\overline{J}$ is a tree and $u \in \beta(H)$.

Observation 3. *Let $(H, \phi, T_{in}, T_{out})$ be a homomorphism minimal solution to D and J be an element of H. For any $v \in V(J) \setminus \beta(J)$, all the arcs in $A(H)$ that are incident to v, are also present in J.*

Observation 3 follows from the definition of elements. By Observation 3, for a homomorphism minimal solution $(H, \phi, T_{in}, T_{out})$, each element is a subgraph J of H such that $\overline{J}$ is a tree and the arcs in $A(H)$ that are incident to a non-boundary vertex $v \in V(J) \setminus \beta(J)$ are also present in the graph J. Thus, to have complete information about the elements we define an element tuple as follows. For a homomorphism minimal solution $(H, \phi, T_{in}, T_{out})$ and an element J of H, the *element quadruple* associated with J is a quadruple $(J, \psi, A_{in}, A_{out})$, where $\psi = \phi|_{V(J)}$, $A_{in} = A(J) \cap A(T_{in})$ and $A_{out} = A(J) \cap A(T_{out})$. We remark that the underlying undirected graphs of $J[A_{in}]$ and $J[A_{out}]$, are connected. The proof of the following observation follows from Observation 3.

Observation 4. *Let $(H, \phi, T_{in}, T_{out})$ be a homomorphism minimal solution to D and J be an element of H. Let $(J, \psi, A_{in}, A_{out})$ be the element quadruple associated with J. Let $u \in V(J)$ and A_u be the set of arcs in J that are incident with u. Let $F \subseteq A_u$ and let J' be the connected component of $J - F$ containing u. Then, for any $w \in V(J') \setminus (\beta(J) \cup \{u\})$, all the arcs of A_{in} and A_{out} that are incident with w, are also present in J'.*

In the DP algorithm, the first step is to compute a "representative family" of element quadruples and in the second step we compose element quadruples and the edges between terminals in an iterative manner using representative families again, to form a solution.

[1] The proofs of the results marked with $\star$ are deferred to the full version of the paper.

3.1 Step 1: Representative Family of Element Quadruples

To compute a representative family of element quadruples using a DP algorithm, we first define *element partial solutions.*

Definition 3. *An elemental partial solution is a 6-tuple* $(R, \phi, A_{in}, A_{out}, h, z)$*, where* R *is a* k*-boundaried digraph,* $A_{in} \subseteq A(R)$*,* $A_{out} \subseteq A(R)$*,* $A(R) = A_{in} \cup A_{out}$*,* $h \in V(R)$*, and* $z \in V(D)$ *such that the following holds.*

(1) $\overline{R}$ *is a tree.*
(2) If there is a vertex $x \in V(R)$ *with label* $k+1$*, then* $x = h$*.*
(3) ϕ *is a label preserving homomorphism from* R *to* D *such that* $\phi(h) = z$[2]*.*
(4) The set of vertices $\{x \in V(R) : \ell_R(x) \in [k]\}$ *forms an independent set in* $\overline{R}$*.*
(5) For any $u \in V(R) \setminus \beta(R)$*, there is exactly one arc of the form* uw *in* A_{in} *and there is exactly one arc of the form* $w'u$ *in* A_{out}*.*

Here, notice that we have used one more label for R. This is similar to the idea of having one more terminal vertex in the Dreyfus-Wagner algorithm. That is, this extra boundary vertex is used to join two partial solutions. Recall that, in the Dreyfus-Wagner algorithm, we compute a minimum spanning tree using a dynamic programming algorithm. Here, we use a similar kind of dynamic programming algorithm, but instead of computing one solution, we compute a set of elemental partial solutions such that a solution to D can be composed of the elemental partial solutions from the set we computed and arcs between boundary vertices.

We define the *size* of an elemental partial solution $(R, \phi, A_{in}, A_{out}, h, z)$ to be $|A(R)|$. Suppose we have computed the set $\mathcal{S}$ of all possible elemental partial solutions of size at most $2n-2$, then a solution can be constructed by composing the element quadruples associated with it and the arcs between the terminal vertices. But the cardinality of $\mathcal{S}$ is not bounded by a function of k. Thus, instead of computing $\mathcal{S}$, we compute only a "representative of $\mathcal{S}$". In fact, in our DP algorithm we compute subsets of elemental partial solutions in the increasing order of their size that preserve some candidates which can be extended to an optimum solution. In short, we use ideas similar to the Dreyfus-Wagner algorithm to construct partial solutions of larger size from partial solutions of smaller size and we prune the partial solutions using representative sets on matroids to reduce its cardinality.

Definition 4. *Let* $Q = (R, \phi, A_{in}, A_{out}, h, z)$ *be a tuple satisfying the conditions in Definition 3. Let* $Z = (L, \psi, B_{in} \subseteq A(L), B_{out} \subseteq A(L), h^\star \in V(L), z)$ *be a tuple where* L *is a* $(k+1)$*-boundaried digraph,* $h^\star \in \beta(L)$*, such that*

(a) if there is a vertex x *such that* $\ell_L(x) = k+1$*, then* $x = h^\star$ *(i.e.,* $\ell_L^{-1}(k+1) \subseteq \{h^\star\}$*), and*
(b) ψ *is a label preserving homomorphism from* L *to* D *such that* $\psi(h^\star) = z$*.*

[2] Here, we slightly abuse the notation and consider z to be a labelled vertex in D with label $k+1$ if $\ell_R(h) = k+1$.

Let y be the vertex in $R \oplus L$ labelled with 1. We say that Q and Z **can be merged**, *if the following holds.*

(i) $(V(R \oplus L), A_{in} \cup B_{in})$ and $(V(R \oplus L), A_{out} \cup B_{out})$ are an in-branching and an out-branching rooted at y of $R \oplus L$, respectively, and
(ii) there is a vertex with label j in $R \oplus L$ for all $j \in \{1, \ldots, k\}$.

That is, $(R \oplus L, \psi^\star)$ forms a solution to D, where $\psi^\star(x) = \phi(x)$ for all $x \in V(R)$, and $\psi^\star(x) = \psi(x)$ for all $x \in V(L)$.

It is easy to prove that the function $\psi^\star$ in the above definition is well defined. Intuitively, in Definition 4, Q is a partial solution and it can be extended to a solution using the tuple Z which is yet to be computed by our DP algorithm. Now we define the notion of representative set of elemental partial solutions and later we will prove that a *small* set which represents any given set of elemental partial solutions can be computed *efficiently*.

Definition 5. *Let $\mathcal{F}$ be a set of elemental partial solutions. We say that a subset $\mathcal{F}' \subseteq \mathcal{F}$* represents *$\mathcal{F}$ if the following holds. Suppose there is a tuple $Z = (L, \psi, B_{in} \subseteq A(L), B_{out} \subseteq A(L), h^\star \in V(L), z \in V(D))$ and there is a tuple $Q \in \mathcal{F}$, such that Q and Z can be merged, then there is a tuple $Q' \in \mathcal{F}'$ such that Q' and Z can be merged and the size of Q' is at most the size of Q.*

Recall that $\mathcal{S}$ is the set of all possible elemental partial solutions. Next, we explain how to compute a subfamily $\mathcal{S}'$ that represents $\mathcal{S}$, of cardinality $2^{\mathcal{O}(k)}n$ using a Dreyfus-Wagner style dynamic programming algorithm along with representative families on the disjoint sum of two graphic matroids. In what follows we explain how to compute $\mathcal{S}'$ assuming Lemma 3 which can be proved using the notion of representative families on linear matroids and the proof is omitted here.

Lemma 3. *There is an algorithm that given an instance D (a k-boundaried graph) of* BI-SCSS, *and a set of elemental partial solutions $\mathcal{F}$ of D, runs in time $\mathcal{O}(|\mathcal{F}| \cdot n)$ and outputs a subfamily $\mathcal{F}'$ of $\mathcal{F}$ of cardinality at most $20^{k+1}n$ that represents $\mathcal{F}$.*

Now, we are ready to explain a DP algorithm to compute a subfamily $\mathcal{S}'$ of $\mathcal{S}$, of cardinality $2^{\mathcal{O}(k)}n$, that represents $\mathcal{S}$. Let $\mathcal{S}_i$ be the set of all the elemental partial solutions of size at most i. At stage i, our algorithm will compute a subfamily $\mathcal{S}_i'$ of $\mathcal{S}_i$ that represents $\mathcal{S}_i$ in time $2^{\mathcal{O}(k)}n^3$ using already computed sets of partial solutions $\mathcal{S}_1', \ldots, \mathcal{S}_{i-1}'$. Then, at the end it is straightforward to prove that the family $\widehat{\mathcal{S}} = \mathcal{S}_1' \cup \ldots \cup \mathcal{S}_{2n-2}'$ is a representative for $\mathcal{S}$. Then, in the final step using similar pruning technique we compute an optimum solution again by a DP algorithm using $\widehat{\mathcal{S}}$.

Now, we get back to the computation of $\mathcal{S}_i'$ that represents $\mathcal{S}_i$, for all $i \in \{1, \ldots, 2n-2\}$. At the first step, we enumerate $\mathcal{S}_1$ (its cardinality will be bounded by n^2) and use Lemma 3 to compute a subfamily $\mathcal{S}_1'$ of size at most $20^{k+1}n$.

Computation of $\mathcal{S}'_i$ for $i \in \{2, \dots, 2n-2\}$: Inductively, assume that we have computed $\mathcal{S}'_j$ of cardinality at most $20^{k+1}n$, for all $j \in \{1, \dots, i-1\}$. First we set $\mathcal{S}^\star_i = \emptyset$. Then, we add the following tuples to $\mathcal{S}^\star_i$.

(i) For all $j, r \in [i-1]$ such that $i = j + r$, and for every pair of tuples $(R_1, \phi_1, A'_{in}, A'_{out}, h_1, z) \in \mathcal{S}'_j$ and $(R_2, \phi_2, A''_{in}, A''_{out}, h_2, z) \in \mathcal{S}'_r$ such that $\ell_{R_1}(h_1) = \ell_{R_2}(h_2)$, we add a tuple $Q = (R = R_1 \oplus R_2, \phi, A'_{in} \cup A''_{in}, A'_{out} \cup A''_{out}, h, z)$ to $\mathcal{S}^\star_i$, if Q is an elemental partial solution, where h is the vertex in R such that $\ell_R(h) = \ell_{R_1}(h_1) = \ell_{R_2}(h_2)$, and ϕ are defined as follows. For any $x \in V(R_q)$, $\phi(x) = \phi_q(x)$ for all $q \in \{1, 2\}$. Since $\phi_1|_{\beta(R_1) \cap \beta(R_2)}$ is same as $\phi_2|_{\beta(R_1) \cap \beta(R_2)}$, the function ϕ is well defined. Here we compute elemental partial solutions from already computed partial solutions by gluing at the vertices labelled $k+1$.

(ii) For each $(R', \phi', A'_{in}, A'_{out}, h', z') \in \mathcal{S}'_{i-1}$ and *arc* $z'z \in A(D)$ let $Q_1 = (R, \phi, A_{in} \cup \{h'h\}, A_{out}, h, z)$, $Q_2 = (R, \phi, A_{in}, A_{out} \cup \{h'h\}, h, z)$, and $Q_3 = (R, \phi, A_{in} \cup \{h'h\}, A_{out} \cup \{h'h\}, h, z)$, where R is the graph $(V(R') \cup \{h\}, A(R') \cup \{h'h\})$, $\beta(R) = (\beta(R') \cup \{h\}) \setminus \{h'\}$, and ϕ is defined as follows: for any $y \in V(R')$, $\phi(y) = \phi'(y)$, and $\phi(h) = z$. Here, $\ell_R(h) = k+1$ and $\ell_R(x) = \ell_{R'}(x)$ for all $x \in \beta(R) \setminus \{h\}$. Then, for any $j \in \{1, 2, 3\}$, we add Q_j to $\mathcal{S}^\star_i$ if Q_j is an elemental partial solution. Here we compute partial solutions from already computed partial solutions by gluing an arc where the head of the arc is a "new vertex".

(iii) We have three more cases which are omitted here, as they are identical to Case (ii). In those cases, we compute partial solutions from already computed partial solutions by gluing an arc where, (a) the tail of the arc is a "new vertex", (b) head of the arc is a "new vertex" and both the endpoints of the arc are boundary vertices, (c) tail of the arc is a "new vertex" and both the endpoints of the arc are boundary vertices, respectively.

Clearly, the cardinality of $\mathcal{S}^\star_i$ is bounded by $2^{\mathcal{O}(k)}n^2$, because $|\mathcal{S}'_j| \le 20^{k+1}n$, for all $j \in \{1, \dots, i-1\}$. Now, we use Lemma 3 and compute a representative $\mathcal{S}'_i$ of $\mathcal{S}^\star_i$. The construction of $\mathcal{S}'_i$ takes time $2^{\mathcal{O}(k)}n^3$ and the cardinality of $\mathcal{S}'_i$ is at most $20^{k+1}n$. Thus, the total running time to compute $\mathcal{S}'_1, \dots, \mathcal{S}'_{2n-2}$ is $2^{\mathcal{O}(k)}n^4$.

Now, let $\widehat{\mathcal{S}} = \mathcal{S}'_1 \cup \ldots \cup \mathcal{S}'_{2n-2}$. Next, we prove that $\mathcal{S}'_i$ is a representative of $\mathcal{S}_i$, and $\widehat{\mathcal{S}}$ represents $\mathcal{S}$.

Lemma 4 ($\star$). *For all $i \in \{1, \dots, 2n-2\}$, $\mathcal{S}'_i$ represents $\mathcal{S}_i$.*

Lemma 5. *$\widehat{\mathcal{S}}$ represents $\mathcal{S}$ and $|\widehat{\mathcal{S}}|$ is upper bounded by $\mathcal{O}(20^k n^2)$.*

Proof. Let $Q \in \mathcal{S}$ and Z be a tuple such that Q and Z can be merged. Let the size of Q be i. Then, we have that there exists $Q' \in \mathcal{S}'_i \subseteq \widehat{\mathcal{S}}$ such that Q' and Z can be merged. Because of Lemma 3, we have that $|\widehat{\mathcal{S}}| \le \mathcal{O}(20^k n^2)$. □

3.2 Step 2: Composition of Element Quadruples

The next step is to compose the element quadruples to form a solution. In this step as well we use the notion of representative families and do a dynamic programming.

Definition 6. *A partial solution is a quadruple* $(Y, \phi, A_{in}, A_{out})$*, where* Y *is a* k*-boundaried digraph,* $A_{in} \subseteq A(Y)$*,* $A_{out} \subseteq A(Y)$*, and* $A(Y) = A_{in} \cup A_{out}$ *such that the following holds.*

(1) ϕ *is a label preserving homomorphism from* Y *to* D*.*
(2) For any $u \in V(Y) \setminus \beta(Y)$*, there is exactly one arc of the form* uw *in* A_{in} *and there is exactly one arc of the form* $w'u$ *in* A_{out}*.*

The *size* of a partial solution $(Y, \phi, A_{in}, A_{out})$ is $|A(Y)|$. For any two partial solutions $Q_1 = (Y_1, \phi_1, F_1, F_1')$ and $Q_2 = (Y_2, \phi_2, F_2, F_2')$, we define $Q_1 \circ Q_2$ to be the quadruple $(Y = Y_1 \oplus Y_2, \phi, F_1 \cup F_2, F_1' \cup F_2')$, where $\phi(x) = \phi_1(x)$ for all $x \in V(Y_1)$ and $\phi(x) = \phi_2(x)$ otherwise. Similar to the case of elemental partial solution, we define the notion of a representative for partial solutions and we prove that small representative families can be computed efficiently.

Definition 7. *Let* $\mathcal{F}$ *be a set of partial solutions. We say that a subset* $\mathcal{F}' \subseteq \mathcal{F}$ represents $\mathcal{F}$ *if the following holds. Suppose there is a partial solution* $Q \in \mathcal{F}$ *and there is another partial solution* Z *(not necessarily in* $\mathcal{F}$*), such that* $Q \circ Z$ *is a solution to* G*, then there exists* $Q' \in \mathcal{F}'$ *such that* $Q' \circ Z$ *is a solution to* D *and the size of* Q' *is at most the size of* Q*.*

Lemma 6 (⋆). *There is an algorithm that given an instance* D *of* Bi-SCSS*, where* D *is a* k*-boundaried bi-directed graph and a set of partial solutions* $\mathcal{F}$ *of* D*, runs in time* $2^{\mathcal{O}(k)}|\mathcal{F}| \cdot n$ *and outputs a subfamily* $\mathcal{F}'$ *of* $\mathcal{F}$ *of cardinality at most* 20^k *that represents* $\mathcal{F}$*.*

For any solution $(H, \phi, T_{in}, T_{out})$ of D, we know that the elements of H do not contain arcs between the boundary vertices. Let $\mathcal{B}$ be the set of all partial solutions $(Y, \phi, A_{in}, A_{out})$ of size 1 with $\beta(Y) = V(Y)$ and $|V(Y)| = 2$. Let $\widehat{\mathcal{R}}$ be the set of partial solutions extracted from $\widehat{\mathcal{S}}$ which is constructed in the previous subsection. That is,

$$\widehat{\mathcal{R}} = \{Q = (Y, \phi, A_{in}, A_{out}) : Q \text{ is a partial solution and } \exists h, z \text{ s.t. } (Y, \phi, A_{in}, A_{out}, h, z) \in \widehat{\mathcal{S}}\}.$$

Now compute sets $\mathcal{P}_0, \mathcal{P}_1, \ldots, \mathcal{P}_{2n-2}$ in this order as follows. We set $\mathcal{P}_0 = \{(\emptyset, \emptyset, \emptyset, \emptyset)\}$. $\mathcal{P}_1 = \widehat{\mathcal{R}}$. For any $i \in [2n-2]$, first we set $\mathcal{Q}_i = \{Q \circ R : Q \in \mathcal{P}_{i-1} \text{ and } R \in \widehat{\mathcal{R}}\}$. Then, let $\mathcal{P}_i$ be the representative of $\mathcal{Q}_i$, computed using Lemma 6. That is, $\mathcal{P}_i$ will be a representative subfamily of partial solutions which are composed of i partial solutions from $\widehat{\mathcal{R}}$ (that corresponds to elemental partial solutions). Now, for each $i \in \{0, 1, 2n-1\}$ and $j \in \{0, \ldots, k^2\}$, we construct a set $\mathcal{P}_{i,j}$ as follows in the increasing order of j. We set $\mathcal{P}_{i,0} = \mathcal{P}_i$ for all $i \in \{0, \ldots, 2n-2\}$. Now, for any $j \in [k^2]$, let $\mathcal{Q}_{i,j} = \{P \circ B : P \in \mathcal{P}_{i,j-1} \text{ and } B \in \mathcal{B}\}$. Then, by using Lemma 6, we compute $\mathcal{P}_{i,j}$ which is a representative of $\mathcal{Q}_{i,j}$. Now, among all the tuples in $\bigcup_{i,j} \mathcal{Q}_{i,j}$ that are solutions of D, we output the solution with minimum size. This completes the algorithm.

Now, we prove the correctness of our algorithm. For each optimum solution $(H, \phi, T_{in}, T_{out})$, H is a union of elements and arcs between boundary vertices of H. That is, $(H, \phi, T_{in}, T_{out})$ is composed of tuples from $\mathcal{S}$ and $\mathcal{B}$. That is, if H is composed of i elements and j arcs between boundary vertices, then there exist $P_1, \ldots, P_i \in \mathcal{R}$ (where $\mathcal{R}$ is defined below) and $B_1, \ldots B_j \in \mathcal{B}$ such that

$$(H, \phi, T_{in}, T_{out}) = ((((((P_1 \circ P_2) \circ \ldots) \quad \circ P_i) \circ B_1) \circ B_2) \circ \ldots) \circ B_j, \text{ where}$$
$$\mathcal{R} = \{Q = (Y, \phi, A_{in}, A_{out}) : Q \text{ is a partial solution and } \exists h, z \text{ s.t. } (Y, \phi, A_{in}, A_{out}, h, z) \in \mathcal{S}\}$$

For each $i \in \{0, \ldots, 2n-2\}$ and $j \in \{0, \ldots, k^2\}$, we define $\mathcal{Z}_{i,j}$ as follows. We define $\mathcal{Z}_{0,0} = \{(\emptyset, \emptyset, \emptyset, \emptyset)\}$. For any $i \in [2n-2]$, $\mathcal{Z}_{i,0} = \{Z \circ P : Z \in \mathcal{Z}_{i-1,0}$ and $P \in \mathcal{R}\}$. Now, for any $j \in [k^2]$, we define $\mathcal{Z}_{i,j} = \{Z \circ P : Z \in \mathcal{Z}_{i,j-1}$ and $P \in \mathcal{B}\}$. That is, for an optimum solution $Z = (H, \phi, T_{in}, T_{out})$, if H is composed of i elements and j arcs between boundary vertices, then $Z \in \mathcal{Z}_{i,j}$. The correctness of our algorithm follows from the lemma below.

Lemma 7. *For any $i \in \{0, \ldots, 2n-2\}$ and $j \in \{0, \ldots, k^2\}$, $\mathcal{P}_{i,j}$ is a representative for $\mathcal{Z}_{i,j}$.*

Proof (Proof sketch). First using induction on i, we prove that $\mathcal{P}_{i,0}$ is a representative for $\mathcal{Z}_{i,0}$. The base case is when $i = 0$ and it is true because $\mathcal{P}_{0,0} = \mathcal{Z}_{0,0}$. Now, consider the induction step when $i > 0$. Suppose there is a partial solution $Z \in \mathcal{Z}_{i,0}$ and another partial solution F such that $Z \circ F$ is a solution to D. Then, there exists $Z_1 \in \mathcal{Z}_{i-1,0}$ and $Z_2 \in \mathcal{R}$ such that $Z = Z_1 \circ Z_2$. Let $F_1 = Z_2 \circ F$. Notice that $Z \circ F = (Z_1 \circ Z_2) \circ F = Z_1 \circ (Z_2 \circ F) = Z_1 \circ F_1$. Then, by induction hypothesis, there exists $\widehat{Z}_1 \in \mathcal{P}_{i-1,0}$ such that $\widehat{Z}_1 \circ F_1$ is a solution to D. Notice that $Z_2 \circ (\widehat{Z}_1 \circ F) = \widehat{Z}_1 \circ F_1$. Since $\widehat{\mathcal{S}}$ is a representative of $\mathcal{S}$ and from the construction of $\widehat{\mathcal{R}}$, there exists $\widehat{Z}_2 \in \widehat{\mathcal{R}}$ such that $\widehat{Z}_2 \circ (\widehat{Z}_1 \circ F)$ is a solution D. Notice that $\widehat{Z}_2 \circ (\widehat{Z}_1 \circ F)$ is equal to $(\widehat{Z}_1 \circ \widehat{Z}_2) \circ F$, where $\widehat{Z}_1 \in \mathcal{P}_{i-1,0}$ and $\widehat{Z}_2 \in \widehat{\mathcal{R}}$. By the definition of $\mathcal{Q}_{i,0}$, we get that $\widehat{Z}_1 \circ \widehat{Z}_2 \in \mathcal{Q}_{i,0}$. Then, since $\mathcal{P}_{i,0}$ is a representative of $\mathcal{Q}_{i,0}$, there is a partial solution $\widehat{Z} \in \mathcal{P}_{i,0}$ such that $\widehat{Z} \circ F$ is a solution to D and size of $\widehat{Z}$ is at most the size of $\widehat{Z}_1 \circ \widehat{Z}_2$ which is at most the size of Z. Thus, we have proved that for any $i \in \{0, \ldots, 2n-2\}$, $\mathcal{P}_{i,0}$ is a representative for $\mathcal{Z}_{i,0}$.

Using similar arguments, one can prove by induction on j that for any $j \in \{0, \ldots, k^2\}$, $\mathcal{P}_{i,j}$ is a representative for $\mathcal{Q}_{i,j}$. □

Running Time Analysis. We have already mentioned that the computation of $\mathcal{S}'_1, \ldots, \mathcal{S}'_{2n-2}$ takes time $2^{\mathcal{O}(k)}n^4$ and by Lemma 5 the cardinality of $\widehat{\mathcal{S}}$ is at most $\mathcal{O}(20^k n^2)$. Hence the cardinality of $\widehat{\mathcal{R}}$ is upper bounded by $\mathcal{O}(20^k n^2)$. By Lemma 6, the computation of $\mathcal{P}_0, \ldots, \mathcal{P}_{2n-2}$, takes time $2^{\mathcal{O}(k)}n^2$, and $|\mathcal{P}_i| \leq 20^k$, for all $i \in \{0, \ldots, 2n-2\}$. Because of Lemma 6 the computation of $\mathcal{P}_{i,j}$ for all $i \in \{0, \ldots, 2n-2\}$ and $j \in \{0, \ldots, k^2\}$ together takes time $2^{\mathcal{O}(k)}n^2$ and the cardinality of each $\mathcal{Q}_{i,j}$ is upper bounded by 20^k. Thus, the total running time of the algorithm is $2^{\mathcal{O}(k)}n^4$.

References

1. Bang-Jensen, J., Gutin, G.Z.: Digraphs: Theory, Algorithms and Applications, 2nd edn. Springer, Berlin (2008). Incorporated
2. Berman, P., Bhattacharyya, A., Makarychev, K., Raskhodnikova, S., Yaroslavtsev, G.: Approximation algorithms for spanner problems and directed Steiner forest. Inf. Comput. **222**, 93–107 (2013)
3. Bern, M.W., Plassmann, P.E.: The Steiner problem with edge lengths 1 and 2. Inf. Process. Lett. **32**(4), 171–176 (1989)
4. Björklund, A., Husfeldt, T., Kaski, P., Koivisto, M.: Fourier meets möbius: fast subset convolution. In: Proceedings of the 39th Annual ACM Symposium on Theory of Computing, San Diego, California, USA, 11–13 June 2007, pp. 67–74 (2007)
5. Byrka, J., Grandoni, F., Rothvoß, T., Sanità, L.: Steiner tree approximation via iterative randomized rounding. J. ACM **60**(1), 6:1–6:33 (2013)
6. Chekuri, C., Even, G., Gupta, A., Segev, D.: Set connectivity problems in undirected graphs and the directed Steiner network problem. ACM Trans. Algorithms **7**(2), 18:1–18:17 (2011)
7. Chen, W., Huang, N.: The strongly connecting problem on multihop packet radio networks. IEEE Trans. Commun. **37**(3), 293–295 (1989)
8. Chitnis, R., Feldmann, A.E., Manurangsi, P.: Parameterized approximation algorithms for bidirected steiner network problems. ACM Trans. Algorithms **17**(2), 12:1–12:68 (2021). https://doi.org/10.1145/3447584
9. Chitnis, R.H., Hajiaghayi, M., Kortsarz, G.: Fixed-parameter and approximation algorithms: a new look. In: Parameterized and Exact Computation - 8th International Symposium, IPEC 2013, Sophia Antipolis, France, 4–6 September 2013, Revised Selected Papers, pp. 110–122 (2013)
10. Chitnis, R.H., Hajiaghayi, M., Marx, D.: Tight bounds for planar strongly connected Steiner subgraph with fixed number of terminals (and extensions). In: Proceedings of the Twenty-Fifth Annual ACM-SIAM Symposium on Discrete Algorithms, SODA 2014, Portland, Oregon, USA, 5–7 January 2014, pp. 1782–1801 (2014)
11. Chlamtáč, E., Dinitz, M., Kortsarz, G., Laekhanukit, B.: Approximating spanners and directed Steiner forest: upper and lower bounds. ACM Trans. Algorithms (TALG) **16**(3), 1–31(2020). https://doi.org/10.1145/3381451
12. Chlebík, M., Chlebíková, J.: The Steiner tree problem on graphs: inapproximability results. Theor. Comput. Sci. **406**(3), 207–214 (2008)
13. Cygan, M., et al.: On problems as hard as CNF-SAT. ACM Trans. Algorithms **12**(3), 41:1–41:24 (2016)
14. Diestel, R.: Graph Theory, 4th Ed. Graduate texts in mathematics, vol. 173. Springer, Cham (2012)
15. Dinur, I., Manurangsi, P.: ETH-hardness of approximating 2-CSPs and directed Steiner network. In: 9th Innovations in Theoretical Computer Science Conference, ITCS 2018, 11–14 January 2018, Cambridge, MA, USA, pp. 36:1–36:20 (2018)
16. Dodis, Y., Khanna, S.: Design networks with bounded pairwise distance. In: Proceedings of the Thirty-First Annual ACM Symposium on Theory of Computing, 1–4 May 1999, Atlanta, Georgia, USA, pp. 750–759 (1999)
17. Dom, M., Lokshtanov, D., Saurabh, S.: Kernelization lower bounds through colors and IDs. ACM Trans. Algorithms **11**(2), 1–20 (2014). https://doi.org/10.1145/2650261

18. Dreyfus, S.E., Wagner, R.A.: The Steiner problem in graphs. Networks **1**(3), 195–207 (1971)
19. Feldman, J., Ruhl, M.: The directed Steiner network problem is tractable for a constant number of terminals. SIAM J. Comput. **36**(2), 543–561 (2006)
20. Feldman, M., Kortsarz, G., Nutov, Z.: Improved approximation algorithms for directed Steiner forest. J. Comput. Syst. Sci. **78**(1), 279–292 (2012)
21. Fomin, F.V., Lokshtanov, D., Panolan, F., Saurabh, S.: Efficient computation of representative families with applications in parameterized and exact algorithms. J. ACM **63**(4), 29:1–29:60 (2016)
22. Frederickson, G.N., JáJá, J.: Approximation algorithms for several graph augmentation problems. SIAM J. Comput. **10**(2), 270–283 (1981)
23. Fuchs, B., Kern, W., Mölle, D., Richter, S., Rossmanith, P., Wang, X.: Dynamic programming for minimum Steiner trees. Theory Comput. Syst. **41**(3), 493–500 (2007)
24. Gabow, H.N.: A matroid approach to finding edge connectivity and packing arborescences. J. Comput. Syst. Sci. **50**(2), 259–273 (1995). https://doi.org/10.1006/jcss.1995.1022
25. Goemans, M.X., Olver, N., Rothvoß, T., Zenklusen, R.: Matroids and integrality gaps for hypergraphic Steiner tree relaxations. In: Proceedings of the 44th Symposium on Theory of Computing Conference, STOC 2012, New York, NY, USA, 19–22 May 2012, pp. 1161–1176 (2012)
26. Goyal, P., Misra, P., Panolan, F., Philip, G., Saurabh, S.: Finding even subgraphs even faster. J. Comput. Syst. Sci. **97**, 1–13 (2018). https://doi.org/10.1016/j.jcss.2018.03.001
27. Guo, J., Niedermeier, R., Suchý, O.: Parameterized complexity of arc-weighted directed Steiner problems. SIAM J. Discrete Math. **25**(2), 583–599 (2011)
28. Karp, R.M.: Reducibility among combinatorial problems. In: Miller, R.E., Thatcher, J.W., Bohlinger, J.D. (eds.) Complexity of Computer Computations. The IBM Research Symposia Series, pp. 85–103. Springer, Boston (1972). https://doi.org/10.1007/978-1-4684-2001-2_9
29. Kratsch, S., Wahlström, M.: Representative sets and irrelevant vertices: new tools for kernelization. J. ACM **67**(3), 16:1–16:50 (2020). https://doi.org/10.1145/3390887
30. Lam, N.X., Nguyen, T.N., An, M.K., Huynh, D.T.: Dual power assignment optimization and fault tolerance in WSNS. J. Comb. Optim. **30**(1), 120–138 (2015)
31. Manurangsi, P., Rubinstein, A., Schramm, T.: The strongish planted clique hypothesis and its consequences. In: Lee, J.R. (ed.) 12th Innovations in Theoretical Computer Science Conference (ITCS 2021). Leibniz International Proceedings in Informatics (LIPIcs), vol. 185, pp. 10:1–10:21. Schloss Dagstuhl-Leibniz-Zentrum für Informatik, Dagstuhl, Germany (2021). https://doi.org/10.4230/LIPIcs.ITCS.2021.10
32. Nederlof, J.: Fast polynomial-space algorithms using inclusion-exclusion. Algorithmica **65**(4), 868–884 (2013). https://doi.org/10.1007/s00453-012-9630-x
33. Prömel, H.J., Steger, A.: A new approximation algorithm for the Steiner tree problem with performance ratio 5/3. J. Algorithms **36**(1), 89–101 (2000)
34. Ramanathan, R., Hain, R.: Topology control of multihop wireless networks using transmit power adjustment. In: Proceedings IEEE INFOCOM 2000, The Conference on Computer Communications, Nineteenth Annual Joint Conference of the IEEE Computer and Communications Societies, Reaching the Promised Land of Communications, Tel Aviv, Israel, 26–30 March 2000, pp. 404–413 (2000)

35. Vetta, A.: Approximating the minimum strongly connected subgraph via a matching lower bound. In: Proceedings of the Twelfth Annual Symposium on Discrete Algorithms, 7–9 January 2001, Washington, DC, USA, pp. 417–426 (2001)
36. Wang, C., Park, M.A., Willson, J., Cheng, Y., Farago, A., Wu, W.: On approximate optimal dual power assignment for biconnectivity and edge-biconnectivity. Theor. Comput. Sci. **396**(1–3), 180–190 (2008)
37. Zelikovsky, A.: An 11/6-approximation algorithm for the network Steiner problem. Algorithmica **9**(5), 463–470 (1993)

From Curves to Words and Back Again: Geometric Computation of Minimum-Area Homotopy

Hsien-Chih Chang[1], Brittany Terese Fasy[2,3], Bradley McCoy[3](✉), David L. Millman[5], and Carola Wenk[4]

[1] Department of Computer Science, Dartmouth College, Hanover, USA
[2] Department of Mathematical Sciences, Montana State University, Bozeman, USA
[3] School of Computing, Montana State University, Bozeman, USA
bradleymccoy@montana.edu
[4] Department of Computer Science, Tulane University, New Orleans, USA
[5] Hanover, USA

Abstract. Let γ be a generic closed curve in the plane. Samuel Blank, in his 1967 Ph.D. thesis, determined if γ is self-overlapping by *geometrically* constructing a combinatorial word from γ. More recently, Zipei Nie, in an unpublished manuscript, computed the minimum homotopy area of γ by constructing a combinatorial word *algebraically*. We provide a unified framework for working with both words and determine the settings under which Blank's word and Nie's word are equivalent. Using this equivalence, we give a new geometric proof for the correctness of Nie's algorithm. Unlike previous work, our proof is constructive which allows us to naturally compute the actual homotopy that realizes the minimum area. Furthermore, we contribute to the theory of self-overlapping curves by providing the first polynomial-time algorithm to compute a self-overlapping decomposition of any closed curve γ with minimum area.

Keywords: curve representation · crossing sequence · homotopy area · self-overlapping curve · fundamental group · Dehn twist · change of basis · cancellation norm

1 Introduction

A *closed curve* in the plane is a continuous map γ from the circle $\mathbb{S}^1$ to the plane $\mathbb{R}^2$. In the plane, any closed curve is homotopic to a point. A homotopy that sweeps out the minimum possible area is a *minimum homotopy*. Chambers and Wang [4] introduced the minimum homotopy area between two simple homotopic curves with common endpoints as a way to measure the similarity between the two curves. They suggest that homotopy area is more robust against noise than another popular similarity measure on curves called the *Fréchet distance*. However, their algorithm requires that each curve be simple, which is restrictive.

P. Morin and S. Suri (Eds.): WADS 2023, LNCS 14079, pp. 605–619, 2023.
https://doi.org/10.1007/978-3-031-38906-1_40

Fasy, Karakoç, and Wenk [11] proved that the problem of finding the minimum homotopy area is easy on a closed curve that is the boundary of an immersed disk. Such curves are called *self-overlapping* [10,14,16,20,21,23]. They also established a tight connection between minimum-area homotopy and self-overlapping curves by showing that any generic closed curve can be decomposed at some vertices into self-overlapping subcurves such that the combined homotopy from the subcurves is minimum. This structural result gives an exponential-time algorithm for the minimum homotopy area problem by testing each decomposition in a brute-force manner.

Nie, in an unpublished manuscript [17], described a polynomial-time algorithm to determine the minimum homotopy area of any closed curve in the plane. Nie's algorithm borrows tools from geometric group theory by representing the curve as a word in the fundamental group $\pi_1(\gamma)$, and connects minimum homotopy area to the *cancellation norms* [2,3,18] of the word, which can be computed using a dynamic program. However, the algorithm does not naturally compute an associated *minimum-area homotopy.*

Alternatively, one can interpret the words from the dynamic program *geometrically* as crossing sequences by traversing any subcurve cyclicly and recording the crossings along with their directions with a collection of nicely-drawn *cables* from each face to a point at infinity. Such geometric representation is known as the Blank words [1,19]. In fact, the first application of these combinatorial words given by Blank is an algorithm that determines if a curve is self-overlapping. Blank words are geometric in nature and thus the associated objects are polynomial in size. When attempting to interpret Nie's dynamic program from the geometric view, one encounters the question of how to extend Blank's definition of cables to *subcurves*, where the cables inherited from the original curve are no longer positioned well with respect to the subcurves. To our knowledge, no geometric interpretation of the dynamic program is known.

1.1 Our Contributions

We first show that Blank and Nie's word constructions are, in fact, equivalent under the right assumptions (Sect. 3). Next, we extend the definition of Blank's word to subcurves and arbitrary cable drawings (Sect. 4.1), and interpret the dynamic program by Nie geometrically (Sect. 4.2). Using the self-overlapping decomposition theorem by Fasy, Karakoç, and Wenk [11] we provide a correctness proof to the algorithm. Finally, we conclude with a new result that a minimum-area self-overlapping decomposition can be found in polynomial time. We emphasize that extending Blank words to allow arbitrary cables is in no way straightforward. In fact, many assumptions on the cables have to be made in order to connect self-overlapping curves and minimum-area homotopy; handling arbitrary cable systems, as seen in the dynamic program, requires further tools from geometric topology like *Dehn twists.*

2 Background

In this section, we introduce concepts and definitions that are used throughout the paper. We assume the readers are familiar with the basic terminology for curves and surfaces.

2.1 Curves and Graphs

A *closed curve* in the plane is a continuous map $\gamma : \mathbb{S}^1 \to \mathbb{R}^2$, and a *path* in the plane is a continuous map $\zeta : [0,1] \to \mathbb{R}^2$. A path ζ is *closed* when $\zeta(0) = \zeta(1)$. In this work, we are presented with a *generic* curve; that is, one where there are a finite number of self-intersections, each of which is transverse and no three strands cross at the same point. See Fig. 1 for an example. The image of a generic closed curve is naturally associated with a four-regular plane graph. The self-intersection points of a curve are *vertices*, the paths between vertices are *edges*, and the connected components of the complement of the curve are *faces*. Given a curve, choose an arbitrary starting point $\gamma(0) = \gamma(1)$ and orientation for γ.

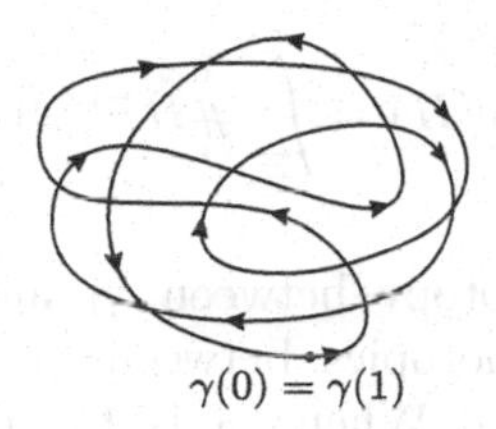

Fig. 1. A generic plane curve induces a four-regular graph.

The *dual graph* γ^* is another (multi-)graph, whose vertices represent the faces of γ, and two vertices in γ^* are joined by an edge if there is an edge between the two corresponding faces in γ. The dual graph is another plane graph with an inherited embedding from γ.

Let T be a spanning tree of γ. Let E denote the set of edges in γ, the tree T partitions E into two subsets, T and $T^*{:=}E \backslash T$. The edges in T^* define a spanning tree of γ^* called the *cotree*. The partition of the edges (T, T^*) is called the *tree-cotree pair*.

We call a rooted spanning cotree T^* of γ^* a *breadth-first search tree* (BFS-tree) if it can be generated from a breadth-first search rooted at the vertex in γ^* corresponding to the unbounded face in γ. Each bounded face f of γ is a vertex in a breadth-first search tree T^*, we associate f with the unique edge incident to f^* in the direction of the root. Thus, there is a correspondence between edges of T^* and faces of γ.

2.2 Homotopy and Isotopy

A *homotopy* between two closed curves γ_1 and γ_2 that share a point p_0 is a continuous map $H\colon [0,1] \times \mathbb{S}^1 \to \mathbb{R}^2$ such that $H(0,\cdot) = \gamma_1$, $H(1,\cdot) = \gamma_2$, and $H(s,0) = p_0 = H(s,1)$. We define a homotopy between two paths similarly, where the two endpoints are fixed throughout the continuous morph. Notice that homotopy between two closed curves as *closed curves* and the homotopy between them as *closed paths* with an identical starting points are different. A homotopy between two injective paths ζ_1 and ζ_2 is an *isotopy* if every intermediate path $H(s,\cdot)$ is injective for all s. The notion of isotopy naturally extends to a collection of paths.

We can think of γ as a topological space and consider the *fundamental group* $\pi_1(\gamma)$. Elements of the fundamental group are called *words*, whose letters correspond to equivalence classes of homotopic closed paths in γ. The fundamental group of γ is a free group with basis consisting of the classes corresponding to the cotree edges of any tree-cotree pair of γ.

Let H be a homotopy between curves γ_1 and γ_2. Let $\#H^{-1}(x)\colon \mathbb{R}^2 \to \mathbb{Z}$ be the function that assigns to each $x \in \mathbb{R}^2$ the number of times the intermediate curves H sweep over x. The homotopy area of H is

$$Area(H) := \int_{\mathbb{R}^2} \#H^{-1}(x)\, dx.$$

The minimum area homotopy between γ_1 and γ_2 is the infimum of the homotopy area over all homotopies between γ_1 and γ_2. We denote this by $Area_H(\gamma_1,\gamma_2) := \inf_H \mathrm{Area}(H)$. When γ_2 is the constant curve at a specific point p_0 on γ_1, define $Area_H(\gamma) := \mathrm{Area}_H(\gamma, p_0)$. See Fig. 2 for an example of a homotopy.

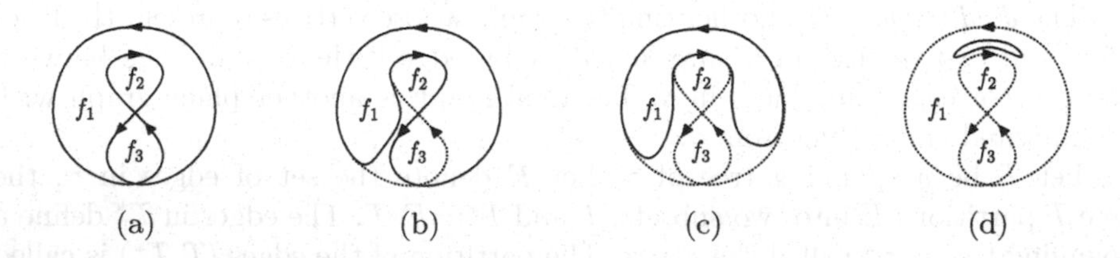

Fig. 2. (a) A generic closed curve in the plane. (b) We see a homotopy that sweeps over the face f_3. (c) The homotopy sweeps f_3 again. (d) The homotopy avoids sweeping over the face f_2. This is a minimum area homotopy for the curve, the area is $\mathrm{Area}(f_1) + 2 \cdot \mathrm{Area}(f_3)$.

For each $x \in \mathbb{R}\backslash\gamma$, the *winding number* of γ at x, denoted as $wind(x,\gamma)$, is the number of times γ "wraps around" x, with a *positive* sign if it is counterclockwise, and *negative* sign otherwise. The winding number is a constant on each face. The *winding area* of γ is defined to be the integral

$$Area_W(\gamma) := \int_{\mathbb{R}^2} |\text{wind}(x,\gamma)|\, dx = \sum_{\text{face } f} |\text{wind}(f,\gamma)| \cdot \text{Area}(f).$$

The *depth* of a face f is the minimal number of edges crossed by a path from f to the exterior face. The depth is a constant on each face. We say the depth of a curve is equal the maximum depth over all faces. We define the *depth area* to be

$$Area_D(\gamma) := \int_{\mathbb{R}^2} \text{depth}(x,\gamma)\, dx = \sum_{\text{face } f} \text{depth}(f) \cdot \text{Area}(f).$$

Chambers and Wang [4] showed that the winding area gives a lower bound for the minimum homotopy area. On the other hand, there is always a homotopy with area $\text{Area}_D(\gamma)$; one such homotopy can be constructed by smoothing the curve at each vertex into simple depth cycles [5], then contracting each simple cycle. Therefore we have

$$\text{Area}_W(\gamma) \leq \text{Area}_H(\gamma) \leq \text{Area}_D(\gamma). \tag{1}$$

2.3 Self-overlapping Curves

A generic curve γ is *self-overlapping* if there is an immersion of the two disk $F : \mathbb{D}^2 \to \mathbb{R}^2$ such that $\gamma = F|_{\partial \mathbb{D}^2}$. We say a map F *extends* γ. The image $F(\mathbb{D}^2)$ is the *interior* of γ. There are several equivalent ways to define self-overlapping curves [10,14,16,20,21]. Properties of self-overlapping curves are well-studied [9]; in particular, any self-overlapping curve has rotation number 1, where the *rotation number* of a curve γ is the winding number of the derivative γ' about the origin [23]. Also, the minimum homotopy area of any self-overlapping curve is equal to its winding area: $\text{Area}_W(\gamma) = \text{Area}_H(\gamma)$ [11].

The study of self-overlapping curves traces back to Whitney [23] and Titus [21]. Polynomial-time algorithms for determining if a curve is self-overlapping have been given [1,20], as well as NP-hardness result for extensions to surfaces and higher-dimensional spaces [7].

For any curve, the *intersection sequence*[1] $[\gamma]_V$ is a cyclic sequence of vertices $[v_0, v_1, \ldots, v_{n-1}]$ with $v_n = v_0$, where each v_i is an intersection point of γ. Each vertex appears exactly twice in γ_V. Two vertices x and y are *linked* if the two appearances of x and y in γ_V alternate in cyclic order: $\ldots x \ldots y \ldots x \ldots y \ldots$.

A pair of symbols of the same vertex x induces two natural subcurves generated by *smoothing* the vertex x; see Fig. 3 for an example. (In this work, every smoothing is done in the way that respects the orientation and splits the curve into two subcurves.) A *vertex pairing* is a collection of pairwise unlinked vertex pairs in $[\gamma]_V$.

[1] also known as the unsigned Gauss code [5,12].

A *self-overlapping decomposition* Γ of γ is a vertex pairing such that the induced subcurves are self-overlapping; see Fig. 3b and Fig. 3d for examples. The subcurves that result from a vertex pairing are not necessary self-overlapping; see Fig. 3c. For a self-overlapping decomposition Γ of γ, denote the set of induced subcurves by $\{\gamma_i\}_{i=1}^{\ell}$. Since each γ_i is self-overlapping, the minimum homotopy area is equal to its winding area. We define the *area of self-overlapping decomposition* to be

$$Area_\Gamma(\gamma) := \sum_{i=1}^{\ell} \mathrm{Area}_W(\gamma_i) = \sum_{i=1}^{\ell} \mathrm{Area}_H(\gamma_i).$$

Fasy, Karakoç, and Wenk [11,13] proved the following structural theorem.

Theorem 1 (Self-Overlapping Decomposition *[11,* **Theorem 20])** *Any curve γ has a self-overlapping decomposition whose area is minimum over all null-homotopies of γ.*

3 From Curves to Words

In order to work with plane curves, one must choose a *representation.* An important class of representations for plane curves are the various *combinatorial words.* One example is the *Gauss code* [12]. Determining whether a Gauss code corresponds to an actual plane curve is one of the earliest computational topology questions [8].

A plane curve (and its homotopic equivalents) can also be viewed as a word in the fundamental group $\pi_1(\gamma)$ of γ [1,17,19]. If we put a point p_i in each bounded face f_i, the curve γ is generated by the unique generators of each $\mathbb{R}^2 - \{p_i\}$. Nie [17] represents curves as words in the fundamental group to find the minimum area swept out by contracting a curve to a point. If the curve lies in a plane with punctures, one can define the *crossing sequence* of the curve with respect to a *system of arcs*, cutting the plane open into a simply-connected region. Blank [1] represents curves using a crossing sequences to determine if a curve is self-overlapping. While Blank constructed the words *geometrically* by drawing arcs and Nie defined the words *algebraically*, the dual view between the system of arcs and fundamental group suggests that the resemblance between Blank and Nie's constructions is not a coincidence.

In this section, we describe the construction by Blank; then, we interpret Blank's construction as a way of choosing the basis for the fundamental group under further restriction [19]. We prove that the Blank word is indeed unique when the restriction is enforced, providing clarification to Blank's original definition. In the full version of the paper, we give a complete description of Nie's word construction and prove that Nie's word and Blank's word are equivalent.

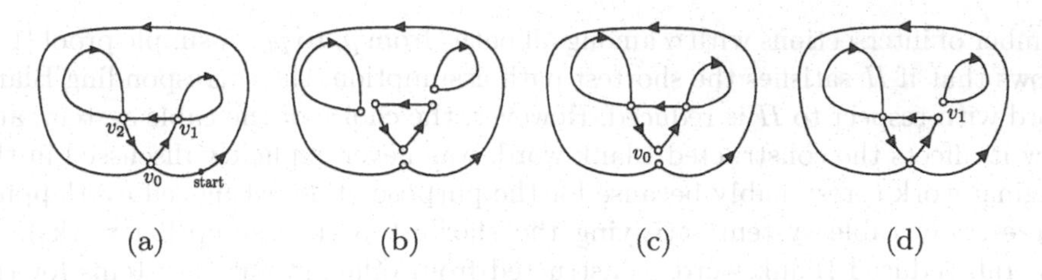

Fig. 3. (a) Curve γ with intersection sequence $\gamma_V = [v_0, v_1, v_1, v_2, v_2, v_0]$. (b) All vertices are paired. (c) One of the subcurves is not self-overlapping. (d) Both subcurves are self-overlapping.

3.1 Blank's Word Construction

We now describe Blank's word construction [1, page 5]. Let γ be a generic closed curve in the plane, pick a point in the unbounded face of γ, call it the *basepoint* p_0. From each bounded face f_i, pick a *representative point* p_i. Now connect each p_i to p_0 by a simple path in such a way that no two paths intersect each other. We call the collection of such simple paths a *cable system*, denoted as Π, and each individual path π_i from p_i to p_0 as a *cable*.

Orient each π_i from p_i to p_0. Now traverse γ from an arbitrary *starting point* of γ and construct a cyclic word by writing down the indices of γ crossing the cables π_i in the order they appear on γ; each index i has a *positive* sign if we cross π_i from right to left and a *negative* sign if from left to right. We denote negative crossing with an overline $\bar{\texttt{i}}$. We call the resulting combinatorial word over the faces a *Blank word* of γ with respect to Π, denoted as $[\gamma]_B(\Pi)$. Figure 4 provides an example of Blank's construction.

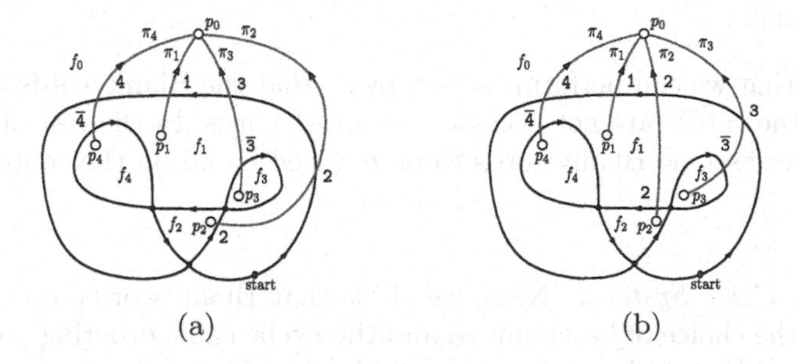

Fig. 4. (a) A curve γ with labeled faces and edges, Π_a is drawn in blue. The Blank word of γ corresponding to Π_a is $[\gamma]_B(\Pi_a) = [\texttt{2314}\overline{\texttt{2}}\overline{\texttt{3}}\overline{\texttt{4}}]$. (b) The same curve with a different choice of cables Π_b. The corresponding Blank word is $[\gamma]_B(\Pi_b) = [\texttt{3214}\overline{\texttt{3}}\texttt{2}\overline{\texttt{4}}]$. (Color figure online)

A word w is *reduced* if there are no two consecutive symbols in w that are identical and with opposite signs. We can enforce every Blank word to be reduced by imposing the following *shortest path assumption*: each cable has a minimum number of intersections with γ among all paths from p_i to p_0. A simple proof [1,6] shows that if Π satisfies the shortest path assumption, the corresponding Blank word with respect to Π is reduced. However, the choice of the cable system, and how it affects the constructed Blank word, was never explicitly discussed in the original work (presumably because for the purpose of detecting self-overlapping curves, any cable system satisfying the shortest path assumption works). In general, reduced Blank words constructed from different cable systems for the same curve are not identical, see Fig. 4a and Fig. 4b for an example. In this paper, we show that if the two cable systems have the same *cable ordering*–the (cyclic) order of cables around point p_0 in the unbounded face–then their corresponding (reduced) Blank words are the same, under proper assumptions on the cable system.

Our first observation is that the Blank words are invariant under cable isotopy; therefore the cable system can be specified up to isotopy.

Lemma 1 (Isotopy Invariance). *The reduced Blank word is invariant under cable isotopy.*

Proof. Let γ be a curve. Discretize the isotopy of the cables and consider all the possible *homotopy moves* [5] performed on γ and the cables involving up to two strands from γ and a cable, because isotopy disallows the crossing of two cables. No $11{\leftrightarrow}0$ move—the move that creates/destroys a self-loop—is possible as cables do not self-intersect. Any $2{\leftrightarrow}0$ move which creates/destroys a bigon is in between a cable and a strand from γ, which means the two intersections must have opposite signs, and therefore the reduced Blank word does not change. Any $3{\rightarrow}3$ move which moves a strand across another intersection does not change the signs of the intersections, so while the order of strands crossing the cable changes, the order of cables crossed by γ remains the same. Thus the reduced Blank word stays the same. □

We remark that we can perform an isotopy so that the Blank words are reduced even when the cables are not necessary shortest paths. In the rest of the paper, we sometimes assume Blank words to be reduced based on the context.

Manage the Cable Systems. Next, we show that Blank words are well-defined once we fix the choice of basepoint p_0 and the cyclic cable ordering around p_0, as long as the cables are drawn in a reasonable way. Fix a tree-cotree pair (T, T^*) of γ, where the root of the cotree is on p_0. We say that a cable system Π is *managed* with respect to the cotree T^* if each path π_i has to be a path on T^* from the root p_0 to the leaf p_i. Given such a collection of cotree paths, one can slightly perturb them to ensure that all paths are simple and disjoint.[2] See Fig. 5 for examples. Not every cable system can be managed with respect to T^*.

[2] In other words, the cables are *weakly-simple* [22].

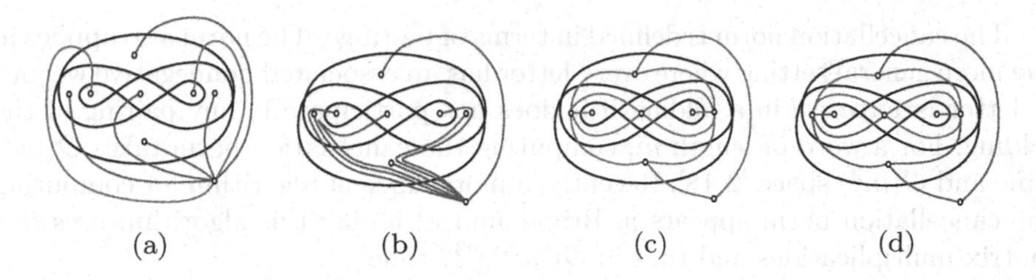

Fig. 5. (a) A cable system Π_1 on γ that is not managed. The red cables do not follow existing paths to the exterior face. (b) A managed cable system Π_2 on γ. (c) The dual γ^* in red. (d) The spanning tree T^* in γ^* generated by the managed cable system Π_2. (Color figure online)

In the full paper, we show that if two managed cable systems satisfying shortest path assumption with identical cable ordering around p_0, their corresponding Blank words are the same. Note that managed cable systems require a fixed tree-cotree pair. We emphasize that the shortest path assumption is necessary; one can construct two (not necessarily shortest) cable systems having the same cable ordering but different corresponding reduced Blank words.

Lemma 2 (Blank Word is Unique). *Given a curve γ, if the basepoint p_0 and the cable ordering of a managed cable system Π satisfying the shortest path assumption is fixed, then the Blank word of γ is unique.*

Therefore, given any plane curve γ, the Blank word is well-defined (if exists), independent of the cable system after specifying a cyclic permutation of all the bounded faces of γ.

4 Foldings and Self-overlapping Decompositions

In this section, we give a geometric proof of the correctness to Nie's dynamic program. To do so, we show that the minimum homotopy area of a curve can be computed from its Blank word using an algebraic quantity of the word called the *cancellation norm*, which is independent of the drawing of the cables. We then show a minimum-area self-overlapping decomposition can be found in polynomial time.

4.1 The Cancellation Norm and Blank Cuts

Given a (cyclic) word w, a *pairing* is a letter and its inverse $(\mathtt{f}, \overline{\mathtt{f}})$ in w. Two letter pairings, $(\mathtt{f}_1, \overline{\mathtt{f}}_1)$ and $(\mathtt{f}_2, \overline{\mathtt{f}}_2)$, are *linked* in a word if the letter pairs occur in alternating order in the word, $[\cdots \mathtt{f}_1 \cdots \mathtt{f}_2 \cdots \overline{\mathtt{f}}_1 \cdots \overline{\mathtt{f}}_2 \cdots]$. A *folding* of a word is a set of letter pairings such that no two pairings in the set are linked. For example, in the word $[2\overline{3}154\overline{6}\overline{5}4\overline{6}\overline{2}3]$ the set $\{(5, \overline{5}), (\overline{3}, 3)\}$ is a folding while $\{(5, \overline{5}), (6, \overline{6})\}$ is not.

The cancellation norm is defined in terms of pairings. The norm also applies in the more general setting where every letter has an associated nonnegative weight. A letter is *unpaired* in a folding if it does not participate in any pairing of the folding. For a word of length m, computing the cancellation norm takes $\mathcal{O}(m^3)$ time and $\mathcal{O}(m^2)$ space [2,18]. Recently, a more efficient algorithm for computing the cancellation norm appears in Bringmann *et al.* [3]; this algorithm uses fast matrix multiplications and runs in $\mathcal{O}(m^{2.8603})$ time.

The *weighted cancellation norm* of a word w is defined to be the minimum sum of weights of all the unpaired letters in w across all foldings of w [2,18]. If w is a word where each letter $\mathtt{f_i}$ corresponds to a face f_i of a curve, we define the weight of $\mathtt{f_i}$ to be Area(f_i). The *area of a folding* is the sum of weights of all the unpaired symbols in a folding. The weighted cancellation norm becomes $||w||{:=}\min_{\mathcal{F}} \sum_i \text{Area}(f_i)$ where $\mathcal{F}$ is the set of all foldings of w and i ranges over all unpaired letter in w.

A dynamic program, similar to the one for matrix chain multiplication, is applied on the word. Let $w = f_1 f_2 \cdots f_\ell$ where $\ell \geq 2$. Assume we have computed the cancellation norm of all subwords with length less than ℓ. Let $w' = f_1 f_2 \cdots f_{\ell-1}$. If f_ℓ is not the inverse of f_i for $1 \leq i \leq \ell - 1$, then f_ℓ is unpaired and $||w|| = ||w'|| + Area(f_\ell)$. Otherwise, f_ℓ participates in a folding and there exits at least one k where $1 \leq k \leq \ell - 1$ and $f_k = f_\ell^{-1}$. Let $w_1 = f_1 \cdots f_{k-1}$ and $w_2 = f_{k+1} \cdots f_{\ell-1}$. Then, we find the k that minimizes $||w_1||+||w_2||$. We have

$$||w|| = \min\{||w'|| + Area(f_\ell), \min_k\{||w_1|| + ||w_2||\}\}$$

Nie shows that the weighted cancellation norm whose weights correspond to face areas is equal to the minimum homotopy area using the triangle inequality and geometric group theory. Our proof that follows is more geometric and leads to a natural homotopy that achieves the minimum area.

We now show how to interpret the cancellation norm geometrically. Let $(\mathtt{f}, \bar{\mathtt{f}})$ be a face pairing in a folding of the word $[\gamma]_B(\Pi)$ for some cable system Π. Denote the cable in Π ending at face f as π_f. Cable π_f intersects γ at two points corresponding to the pairing $(\mathtt{f}, \bar{\mathtt{f}})$, which we denote as p and q respectively. Let π'_f be the simple subpath of π_f so that $\pi'_f(0) = q$ and $\pi'_f(1) = p$. We call π'_f a *Blank cut* [1,10,15] (see Fig. 6). Any face pairing defines a Blank cut, and the result of a Blank cut produces two curves each with fewer faces than the original curve: namely, γ_1 which is the restriction of γ from q to p following by the reverse of path π'_f, and γ_2 which is the restriction of γ from p to q followed by path π'_f.

In order to not partially cut any face, we require all Blank cuts to occur along the boundary of the face being cut. When cutting face f_i along path π_j, we reroute all cables crossing the interior of f_i, including π_j but excluding π_i, along the boundary of f_i through an *isotopy*, so that no cables intersect π_i. Lemma 1 ensures that the reduced Blank word remains unchanged. See Fig. 7 for an example. Notice that different cables crossing f_i might be routed around different sides of f_i in order to avoid intersecting cable π_i and puncture p_i. This way, we ensure the face areas of the subcurves are in one-to-one correspondence with the symbols in the subwords induced by a folding.

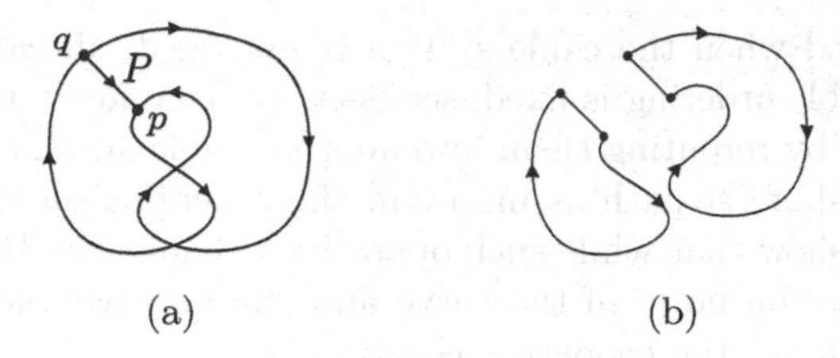

Fig. 6. (a) A curve with labeled path P. (b) The two induced subcurves from cutting along P.

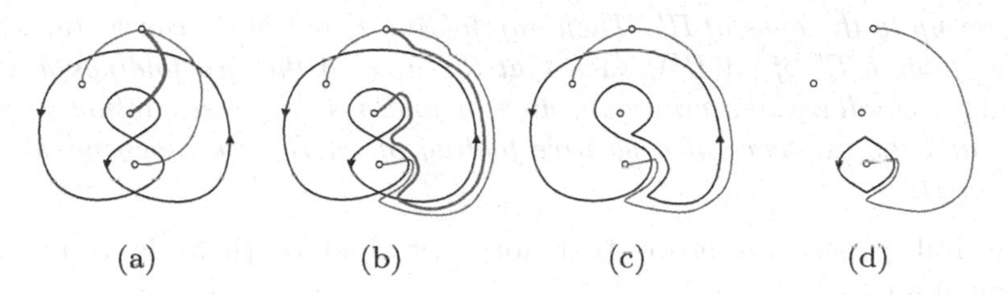

Fig. 7. (a) A curve with cables. (b) Isotopy the cables to not partially cut any faces. (c) One subcurve resulting from cutting along the middle cable. The curve is weakly simple and there are two cables in this face. (d) The other subcurve.

Using the concept of Blank cut we can determine if a curve is self-overlapping. A subword σ of w is *positive* if $\sigma = \mathtt{f_1 f_2 \ldots f_k}$, where each letter $\mathtt{f_i}$ is positive. A pairing $(\mathtt{f}, \bar{\mathtt{f}})$ is *positive* if one of the two subwords of the (cyclic) word w in between the two symbols $\mathtt{f}, \bar{\mathtt{f}}$ is positive; in other words, $w = [\mathtt{f}p\bar{\mathtt{f}}w']$ for some positive word p and some word w'. A folding of w is called a *positive folding*[3] if all pairings in w are positive, and the word constructed by replacing each positive pairing (including the positive word in-between) $\mathtt{f}p\bar{\mathtt{f}}$ in the folding with the empty string is still positive. Words that have positive foldings are called *positively foldable.* Blank established the characterization of self-overlapping curves through Blank cuts.

Theorem 2 (Self-Overlapping Detection [1]**).** *Curve γ is self-overlapping if and only if γ has rotation number 1 and $[\gamma]_B(\Pi)$ is positively foldable for any shortest Π.*

However, we face a difficulty when interpreting Nie's dynamic program geometrically. In our proof we have to work with *subcurves* (and their extensions) of the original curve and the induced cable system. For example, after a Blank cut or a vertex decomposition, there might be multiple cables connecting to the same face creating multiple punctures per face, and cables might not be managed or follow shortest paths to the unbounded face (see Fig. 7c and Fig. 9b). In other words, the subword corresponding to a subcurve with respect to the induced cable system might not be a regular Blank word (remember that Blank word

[3] Blank called these pairings *groupings*.

is only well-defined when the cable system is managed, all cables are shortest paths, and the cable ordering is fixed; see Sect. 2). To remedy this, we tame the cable system first by rerouting them into another cable system that is managed and satisfies the shortest path assumption, then merging all the cables ending at each face. We show that while such operations change the Blank word of the curve, the cancellation norm of the curve and the positive foldability does not change. We summarize the property needed below.

Lemma 3 (Cable Independence). *Let γ be any curve with two cable systems Π and Π' such that the weights of the cables in Π ending at any fixed face sum up to the ones of Π'. Then any folding F of $[\gamma](\Pi)$ can be turned into another folding F' of $[\gamma](\Pi')$, such that the area of the two foldings are identical. As a corollary, the minimum area of foldings (the cancellation norm) of $[\gamma](\Pi)$ and the existence of a positive folding of $[\gamma](\Pi)$ are independent of the choice of Π.*

In the full paper, we prove that for each folding there is a homotopy with equal area.

Lemma 4 (Folding to Homotopy). *Let γ be a curve and Π be a managed cable system satisfying the shortest path assumption, and let F be a folding of $[\gamma](\Pi)$. There exists a null-homotopy of γ with area equal to the area of F.*

4.2 Compute Min-Area Homotopy from Self-overlapping Decomp

A self-overlapping decomposition is a vertex decomposition where each subcurve is self-overlapping [11]. By Theorem 1, there exists a self-overlapping decomposition and an associated homotopy whose area is equal to the minimum homotopy area of the original curve.

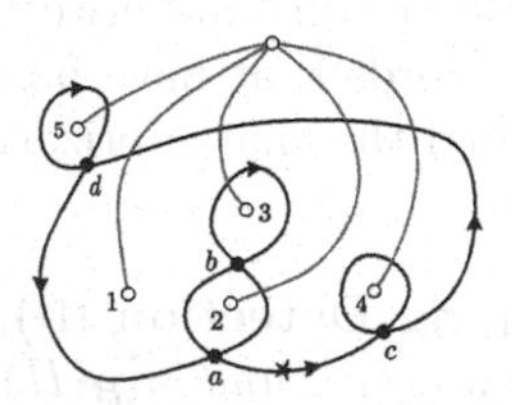

Fig. 8. A curve with combined word [c4c4231d$\bar{5}$da2b$\bar{3}$ba].

In order to relate vertex decompositions and face decompositions, we define a word that includes both the faces and vertices. Given any curve γ and cable system Π, traverse γ and record both self-crossings and (signed) cable intersections; we call the resulting sequence of vertices and faces the *combined word* $[[\gamma]](\Pi)$. See Fig. 8 for an example.

We now show that every self-overlapping decomposition (with respect to the vertex word of γ) determines a folding (of the face word of γ) using the combined word.

Theorem 3 (S-O Decomp. to Folding). *Given a self-overlapping decomposition Γ and a cable system Π of γ, there exists a folding F of $[\gamma](\Pi)$ whose area is $Area_\Gamma(\gamma)$.*

Corollary 1 (Geometric Correctness). *The dynamic programming algorithm computes the minimum-area homotopy for any curve γ.*

Proof. By Theorem 1, there exists a self-overlapping decomposition with minimum homotopy area. By Theorem 3, some folding achieves a minimum area. Using Lemma 4, the minimum-area folding produces a minimum-area homotopy. □

4.3 Min-Area Self-Overlapping Decomposition in Polynomial Time

Finally, we show that any *maximal* folding—where adding any extra pairs are linked—can be used to construct a self-overlapping decomposition.

Theorem 4 (Folding to S.O.D.). *Let γ be a curve and Π be a cable system. Given a maximal folding F of $[\gamma](\Pi)$, there is a self-overlapping decomposition of γ whose area is equal the area induced by the folding F.*

Proof. Begin with the combined word $[[\gamma]](\Pi)$. Decompose $[[\gamma]](\Pi)$ at the vertices given by the self-overlapping decomposition. Let $\Gamma = \{\gamma_1, \gamma_2, \cdots, \gamma_s\}$ be the self-overlapping subcurves and $[[\gamma]](\Pi)_i$ be the corresponding subwords of $[[\gamma]](\Pi)$. If we remove the vertex symbols and turn each $[[\gamma]](\Pi)_i$ into a face word $[\gamma_i]'$, such word may not correspond to Blank words of the subcurves; indeed, when decomposing γ into subcurves by Γ, the subcurve along with the relevant cables may contain multiple cables per face and cables might not be managed or follow shortest paths. See Fig. 9 for an example. However, we can first tame the cable system by choosing a new managed cable system Π^* where the cables follow shortest paths and has one cable per face (as in Sect. 3.1). Lemma 3 ensures that the cancellation norm and positive foldability of the subcurve remain unchanged. Denote the new face word of γ_i with respect to Π^* as $[\gamma_i] = [\gamma_i](\Pi^*)$.

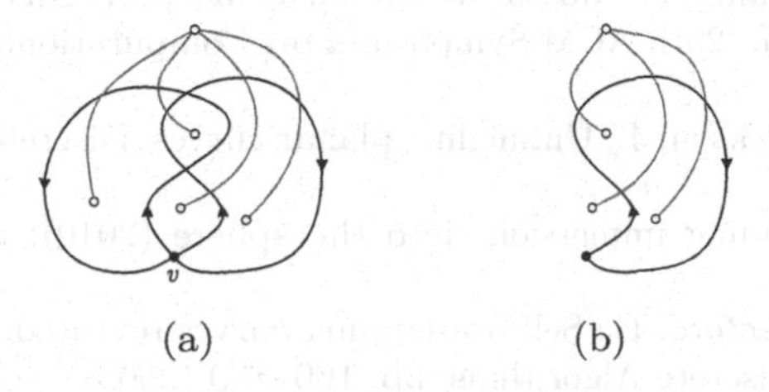

Fig. 9. We decompose the curve in (a) at vertex v into self-overlapping subcurves, the cable system on the induced subcurve in (b) has more than one marked point in a face and cables do not follow shortest paths.

Since each γ_i is a self-overlapping subcurve in Γ, we can find a positive folding F_i of $[\gamma_i]$ by Theorem 2, and the minimum homotopy area of γ_i is equal to the area of folding F_i. Now Lemma 3 implies that the subword $[\gamma_i]'$ from the original combined word also has a positive folding F_i' whose area is equal to the minimum homotopy area of γ_i. By combining all foldings F_i' of each face subword $[\gamma_i]'$, we create a folding F for $[\gamma](\Pi)$ (no pairings between different F_i's can be linked). The area of folding F is equal to the sum of areas of foldings F_i', which in turns is equal to $\sum_i \text{Area}_H(\gamma_i)$, that is, the homotopy area of self-overlapping decomposition $\text{Area}_\Gamma(\gamma)$. This proves the theorem. □

The above theorem implies a polynomial-time algorithm to compute a self-overlapping decomposition with minimum area.

Corollary 2 (Polynomial Optimal Self-Overlapping Decomposition). *Let γ be a curve. A self-overlapping decomposition of γ with area equal to minimum homotopy area of γ can be found in polynomial time.*

Proof. Apply the dynamic programming algorithm to compute the minimum-area folding F for $[\gamma](\Pi)$ with respect to some cable system Π. By Theorem 3 the area of F is equal to the minimum homotopy area of γ, and so does the corresponding self-overlapping decomposition given by Theorem 4. □

Acknowledgements. Brittany Terese Fasy and Bradley McCoy are supported by NSF grant DMS 1664858 and CCF 2046730. Carola Wenk is supported by NSF grant CCF 2107434.

References

1. Blank, S.J.: Extending Immersions and Regular Homotopies in Codimension 1. PhD thesis, Brandeis University (1967)
2. Brandenbursky, M., Gal, Ś, Kędra, J., Marcinkowski, M.: The cancelation norm and the geometry of bi-invariant word metrics. Glasg. Math. J. **58**, 153–176 (2015)
3. Bringmann, K., Grandoni, F., Saha, B., Williams, V.V.: Truly subcubic algorithms for language edit distance and RNA folding via fast bounded-difference min-plus product. SIAM J. Comput. **48**(2), 481–512 (2019)
4. Chambers, E., Wang, Y.: Measuring similarity between curves on 2-manifolds via homotopy area. In: 29th ACM Symposium on Computational Geometry, pp. 425–434 (2013)
5. Chang, H.-C., Erickson, J.: Untangling planar curves. Discrete Comput. Geom. **58**, 889 (2017)
6. Frisch, D.: Extending immersions into the sphere (2010). http://arxiv.org/abs/1012.4923
7. Eppstein, D., Mumford, E.: Self-overlapping curves revisited. In: 20th ACM-SIAM Symposium on Discrete Algorithms, pp. 160–169 (2009)
8. Erickson, J.: One-dimensional computational topology lecture notes. Lecture 7 (2020). Dhttps://mediaspace.illinois.edu/channel/CS+598+JGE+-%C2%A0Fall+2020/177766461/

9. Evans, P.: On Self-Overlapping Curves, Interior Boundaries, and Minimum Area Homotopies. Bachelor's thesis, Tulane University (2018)
10. Evans, P., Fasy, B.T., Wenk, C.: Combinatorial properties of self-overlapping curves and interior boundaries. In: 36th ACM Symposium on Computer Geometry (2020)
11. Fasy, B.T., Karakoç, S., Wenk, C.: On minimum area homotopies of normal curves in the plane (2017). http://arxiv.org/abs/1707.02251
12. Gauss, C.F.: Nachlass. I. Zur geometria situs. Werke **8**, 271–281 (1900)
13. Karakoç, S.: On Minimum Homotopy Areas. PhD thesis, Tulane University (2017)
14. Li, Y., Barbič, J.: Immersion of self-intersecting solids and surfaces. ACM Trans. on Graph. **45**, 1–14 (2018)
15. Marx, M.L.: Extensions of normal immersions of S^1 into R. Trans. Amer. Math. Soc. **187**, 309–326 (1974)
16. Mukherjee, U.: Self-overlapping curves: analysis and applications. Comput. Aided Des. **40**, 227–232 (2014)
17. Nie, Z.: On the minimum area of null homotopies of curves traced twice (2014). http://arxiv.org/abs/1412.0101
18. Nussinov, R., Jacobson, A.B.: Fast algorithm for predicting the secondary structure of single-stranded RNA. Proc. Natl. Acad. Sci. U.S.A. **77**(11), 6309–6313 (1980)
19. Poénaru, V.: Extension des immersions en codimension 1 (d'après Samuel Blank). Séminaire N. Bourbaki **1966–1968**(10), 473–505 (1968)
20. Shor, P., Van Wyk, C.: Detecting and decomposing self-overlapping curves. Comput. Geom. Theory Appl. **2**, 31–50 (1992)
21. Titus, C.J.: The combinatorial topology of analytic functions on the boundary of a disk. Acta Math. **106**(1–2), 45–64 (1961)
22. Toussaint, G.: On separating two simple polygons by a single translation. Discrete Comput. Geom. **4**(3), 265–278 (1989)
23. Whitney, H.: On regular closed curves in the plane. Compos. Math. **4**, 276–284 (1937)

Fully Dynamic Clustering and Diversity Maximization in Doubling Metrics

Paolo Pellizzoni[1,2], Andrea Pietracaprina[2(✉)], and Geppino Pucci[2]

[1] Max Planck Institute of Biochemistry, Martinsried, Germany
pellizzoni@biochem.mpg.de

[2] Department of Information Engineering, University of Padova, Padova, Italy
{andrea.pietracaprina,geppino.pucci}@unipd.it

Abstract. We present deterministic approximation algorithms for some variants of center-based clustering and related problems in the fully dynamic setting, where the pointset evolves through an arbitrary sequence of insertions and deletions. Specifically, we target the following problems: k-center (with and without outliers), matroid-center, and diversity maximization. All algorithms employ a coreset-based strategy and rely on the same data structure, a suitably augmented cover tree, which can thus serve queries for different problems at the same time. For all of the aforementioned problems, our algorithms yield approximations comparable to those achievable in the off-line setting. For spaces of bounded doubling dimension, the running times are dramatically smaller than those that would be required to compute solutions on the entire pointset from scratch after each update. To the best of our knowledge, ours are the first solutions for the matroid-center and diversity maximization problems in the fully dynamic setting.

Keywords: dynamic algorithms · k-center · outliers · matroid center · diversity maximization · cover tree

1 Introduction

Clustering, that is the task of partitioning a set of points according to some similarity metric, is one of the fundamental primitives in unsupervised learning and data mining, with applications in several fields [27]. A commonly used formulation is the *k-center* problem, where, given a set of points from a metric space and a parameter k, one seeks to select k points as cluster centers so that the maximum distance from any point to its closest center is minimized. Since finding the optimal solution is known to be NP-hard [20], in practice one has to settle for approximate solutions.

Several important variants of the k-center problem have been intensely studied over the years. Since the k-center's objective function involves a maximum, the optimal solution may be heavily affected by *outlier* points, which are markedly distant from all the other ones. A robust formulation of the problem, referred to as *k-center with z outliers*, has been introduced in [15], where

P. Morin and S. Suri (Eds.): WADS 2023, LNCS 14079, pp. 620–636, 2023.
https://doi.org/10.1007/978-3-031-38906-1_41

the objective function disregards the z most distant points from the centers. Another popular variant of k-center is the *matroid center* problem [16], which, given a set of points and a matroid defined on it, aims at finding a set of centers forming an independent set of the matroid, minimizing the maximum distance of any point from the closest center. The matroid can be employed to model a wide range of constraints on the solution (most notably, *fairness* [17]). As for k-center, matroid center admits a robust formulation with z outliers [16]. Finally, a problem closely related to k-center is *diversity maximization* [1], whose goal is somehow dual with respect to the one of k-center, as the k points to be selected should maximize a suitable notion of diversity among them. As for k-center, a variant of diversity maximization exists where the diverse point selection is subject to a further matroid constraint [5,7].

Conceivably, in practical scenarios the inputs to the above problems are not static, but evolve with time, hence it is important to develop algorithms that can maintain good solutions under insertion and deletion of arbitrary points. Indeed, in recent years there has been a surge in research efforts in the development of *fully dynamic* clustering algorithms, whose main focus is on the efficient handling of arbitrary insertions and deletions [2,12,13,21].

In this paper, we tackle all of the aforementioned problems in the fully dynamic setting, targeting solutions for general metric spaces of bounded doubling dimension, which is a widely used notion of dimensionality for general metrics.

1.1 Related Work

In the offline setting, k-center can be approximated within a factor 2 in polynomial time, but there is no polynomial-time $(2-\epsilon)$-approximation unless P = NP [20]. For the robust formulation with z outliers, there exists a simple, combinatorial 3-approximation algorithm [15], which has been extended to weighted pointsets in [6]. More complex LP-based 2-approximate algorithms have been developed [11,23], but it is not clear whether they can be employed in the weighted case. For what concerns the matroid-center problem, [16] provides a 3-approximate algorithm for the standard formulation and a 7-approximate algorithm for the robust formulation, which has been improved to a 3-approximation in [23]. The diversity maximization problem admits several instantiations, depending on the specific diversity function embodied in its objective, which are all NP-hard but admit polynomial-time $O(1)$-approximation algorithms (e.g., see [1,7] and references therein).

In [12], the authors developed the first fully dynamic k-center algorithm, which is able to return $(2+\epsilon)$-approximate solutions under arbitrary insertions and deletions of a non-adaptive adversary in general metrics. The algorithm is randomized and has an amortized update time of $O\left(k^2\epsilon^{-1}\log\Delta\right)$, where Δ denotes the *aspect ratio* of the input pointset, namely the ratio between the largest and the smallest distance between any two points. This approach has been recently improved in [2], where an algorithm with expected amortized

$O\left((k+\log n)\epsilon^{-1}\log\Delta\log n\right)$ update time is presented, where n is the maximum number of points at any time. We remark that the data structures used in these works for storing the dynamically changing pointset, must be statically configured to deal with *fixed* values of k and ϵ, and answering queries for different clustering granularities and/or accuracies would, in principle, require building the data structures from scratch. Also, both works use data structures whose size is superlinear in the pointset size by at least a factor $O\left(\log\Delta/\epsilon\right)$. Recently, [13] presented a randomized fully dynamic algorithm for k-center with z outliers that returns a bicriteria $(14+\epsilon)$-approximation solution when discarding at most $(1+\lambda)z$ outliers. Again, their data structure works for fixed k and ϵ and requires superlinear size by a factor $O\left(\log\Delta/\epsilon\right)$.

In [21], the authors propose a fully dynamic k-center algorithm for metric spaces of bounded doubling dimension D based on the *navigating net* data structure [26], which affords insertions and deletions in $O\left((1/\epsilon)^{O(D)}\log\Delta\log\log\Delta\log\epsilon^{-1}\right)$ time. While their approach allows clustering queries for arbitrary values of k, the data structure is built for a specific value of the accuracy parameter ϵ, and must be rebuilt from scratch if a different accuracy ϵ' is sought for. Moreover, the data structure requires superlinear space, by a factor $O\left((1/\epsilon)^{O(D)}\log\Delta\log\log\Delta\log\epsilon^{-1}\right)$. Finally, k-center clustering has been studied in semi-dynamic frameworks, such as the insertion-only [6,10,14,24] or the sliding window [4,18,29,30] settings.

1.2 Our Contributions

We present approximation algorithms for the problems introduced before, in the fully dynamic setting. All algorithms are deterministic and crucially rely on a core data structure aimed at maintaining the necessary information for solving the problem at hand. At any point in time, when a solution to any of the problems is required with a given accuracy, a small subset of representative points, dubbed *coreset*, is extracted from the data structure, and a sequential algorithm is run on such a coreset to compute the solution. This allows for fast execution times, which are independent of the current number of points stored in the tree. Our specific contributions are the following.

1. We define the *augmented cover tree* data structure to maintain the current set of points, by extending and improving the cover tree of [3] in several ways: (i) we store richer information at each node regarding the subset of points stored in its subtree, such as its cardinality and, for matroid center, a maximal independent set of the submatroid induced by the subset; (ii) we devise iterative algorithms to handle insertion and deletions maintaining the aforementioned richer information, which also afford simpler correctness proofs and simpler implementations compared to the original recursive ones [3]; (iii) we make the complexity analysis of the update operations parametric in the doubling dimension D and the aspect ratio Δ (Theorems 1 and 2), fixing some flaws in the original analysis, as pointed out in [19].

2. We devise a fully dynamic $(2+\epsilon)$-approximate algorithm for k-center. Unlike all previous works, our data structure allows to query for solutions for *arbitrary values* of k and ϵ. Compared to [21], our data structure has smaller (linear) size and can handle insertions and deletions at a lower asymptotic cost. Also, our time to compute the solution has only a linear dependency on k (Theorem 3).
3. We devise a fully dynamic $(3+\epsilon)$ approximation algorithm for k-center with z outliers, which allows to choose k, ϵ, and z at query time. We remark that the only previously available algorithm for general metrics can only return a $(14+\epsilon)$ bi-criteria solution (i.e. with an additional slackness on the number of outliers), and requires to fix k, z and ϵ beforehand (Theorem 4).
4. We present the first fully dynamic algorithm for matroid center. Our algorithm returns a $(3+\epsilon)$ approximate solution (Theorem 5).
5. We present the first fully dynamic algorithms for diversity maximization. Our algorithms return $(\alpha_{\text{div}}+\epsilon)$ approximate solutions, where α_{div} is the best sequential approximation factor (Theorem 6).

An important feature of our algorithms is that they are oblivious to D, in the sense that the value D only influences the analysis, but must not be known by the algorithms. This is a very desirable feature, since, in practice, this value is difficult to estimate. Moreover, our algorithms are completely deterministic, while most previously available algorithms make use of randomization.

The rest of the paper is organized as follows. Section 2 provides the formal definitions of the problems and some technical facts used in the analysis. Section 3 is dedicated to describing the novel augmented cover tree, while Sect. 4 details how to maintain such data structure efficiently. In Sect. 5, we present the fully dynamic algorithms for the problems introduced before. Section 6 concludes the paper with some final remarks.

Due to space constraints, pseudocodes and some proofs are omitted in this extended abstract but can be found in [31].

2 Preliminaries

This section formally defines the problems studied in this paper, and states some important technical facts.

(Robust) k-Center Problem. Consider a metric space (U, dist) and a set $S \subseteq U$ of n points. For any $p \in U$ and any subset $C \subseteq S$, we use the notation $\text{dist}(p, C) = \min_{q \in C} \text{dist}(p, q)$, and define the *radius of C with respect to S* as

$$r_C(S) = \max_{p \in S} \text{dist}(p, C).$$

For a positive integer $k < n$, the *k-center* problem requires to find a subset $C \subseteq S$ of size at most k which minimizes $r_C(S)$. The points of the solution C are referred to as *centers*. Note that C induces a partition of S into $|C|$ clusters, by assigning each point to its closest center (with ties broken arbitrarily). We denote the radius of the optimal solution by $r_k^*(S)$. The seminal work by Gonzalez [20] presents a greedy algorithm that returns a 2-approximate solutions to k center in $O(nk)$-time.

The algorithms presented in this paper crucially rely on confining the computation of the solution on a succinct *coreset* T, efficiently extracted from the (possibly large) input S, which contains a close enough "representative" for each point in S. The quality of a coreset T is captured by the following definition.

Definition 1. *Given a pointset S and a value $\epsilon > 0$, a subset $T \subseteq S$ is an (ϵ, k)-coreset for S (w.r.t. the k-center problem) if $r_T(S) \leq \epsilon r_k^*(S)$.*

In real world applications, large datasets often include noisy points which, if very distant from all other points, may severely distort the optimal center selection. To handle these scenarios, the following robust formulation of the k-center problem has been introduced [15]. For positive $k, z < n$, the *k-center problem with z outliers* (a.k.a. *(k, z)-center*) requires to find a subset $C \subseteq S$ of size k minimizing $r_C(S - Z_C)$, where Z_C is the set of z points in S with the largest distances from C, which are regarded as outliers. We denote the radius of the optimal solution of this problem by $r_{k,z}^*(S)$. Observe that the (k, z)-center problem reduces to the k-center problem for $z = 0$. Also, it is straightforward to argue that

$$r_{k+z}^*(S) \leq r_{k,z}^*(S). \tag{1}$$

A well known sequential 3-approximation algorithm for the k-center problem with z outliers, which runs in $O(n^2 \log n)$ time was devised in [15]. In this work, we will make use of a more general formulation of the problem, referred to as *weighted (k, z)-center*, where each point $p \in S$ carries a positive integer weight $w(p)$, and the desired set C of k centers must minimize $r_C(S - Z_C)$, where Z_C is the set of points with the largest distances from C, of maximum cardinality and aggregate weight at most z.

Matroid Center Problem. Another variant of the k-center problem requires the solution C to satisfy an additional constraint, specified through a matroid. A *matroid* [28] on a pointset S is a pair $M = (S, I)$, where I is a family of subsets of S, called *independent sets*, satisfying the following properties: (i) the empty set is independent; (ii) every subset of an independent set is independent (*hereditary property*); and (iii) if $A, B \in I$ and $|A| > |B|$, then there exists $x \in A \backslash B$ such that $B \cup \{x\} \in I$ (*augmentation property*). An independent set is *maximal* if it is not properly contained in another independent set. Given a matroid $M = (S, I)$, the *matroid center problem* on M requires to determine an independent set $C \in I$ minimizing the radius $r_C(S)$. We let $r^*(M) = \min_{C \in I} r_C(S)$ to denote the radius of the optimal solution. The augmentation property ensures that all maximal independent sets of M have the same size, which is referred to as *rank* of M, denoted by rank(M). It is easy to argue that $r_{\text{rank}(M)}^*(S) \leq r^*(M)$, since each solution to the matroid center problem is also a solution to the k-center problem with $k = \text{rank}(M)$. The definition of matroid implies that any subset $S' \subseteq S$, induces a matroid $M' = (S', I')$, where $I' \subseteq I$ is the restriction of I

to subsets of S'. The following fact, which will be exploited in the derivation of our results, is a consequence of the *extended augmentation property* proved in [8, Lemma 2.1] (see also [24]).

Fact 1. *Let $M = (S, I)$ be a matroid, and let $S_1, \ldots, S_h$ be a partition of S into h disjoint subsets. If $A_1 \subseteq S_1, \ldots, A_h \subseteq S_h$ are maximal independent sets of the submatroids $M_1 = (S_1, I_1), \ldots, M_h = (S_h, I_h)$, then $\cup_{I=1}^{h} A_i$ contains a maximal independent set of M.*

As in previous works [1,7,24], in the paper we will assume that a constant-time oracle is available to check independence of any subset of S. A combinatorial 3-approximation algorithm for the matroid center problem in general metrics is presented in [16], which runs in time polynomial in $|S|$ and $\text{rank}(M)$.

An important instantiation of the problem is based on the *partition matroid* $M_P = (S, I_P)$, where each point in S is associated to one of $m \leq k$ of categories, and I_P consists of all subsets with at most k_i points of the i-th category, with $\sum_{i=0}^{m} k_i = k$. For instance, this matroid can be employed to model fairness constraints [17,25].

Diversity Maximization. Let $\text{div} : 2^S \rightarrow \mathbb{R}$ be a *diversity function* that maps any subset $X \subset S$ to some non-negative real number. For a specific diversity function div and a positive integer $k \leq n$, the goal of the *diversity maximization problem* is to find a set $C \subseteq S$ of size k that maximizes $\text{div}(C)$. We denote the optimal value of the objective function as $\text{div}_k^*(S) = \max_{C \subseteq S, |C|=k} \text{div}(C)$. In this paper, we will focus on several variants of the problem based on different diversity functions amply studied in the previous literature, which are reported in Table 1. All variants are NP-hard, and Table 1 lists the best known approximation ratios attainable in polynomial time (see [1,10] and references therein).

Table 1. Diversity functions considered in this paper. $w(\text{MST}(X))$ (resp., $w(\text{TSP}(X))$) denotes the minimum weight of a spanning tree (resp., Hamiltonian cycle) of the complete graph induced by the points of X and their pairwise distances.

Name	Diversity measure div(X)	Sequential approx. α_{div}
remote-edge	$\min_{p,q \in X} d(p,q)$	2
remote-clique	$\sum_{p,q \in X} d(p,q)$	2
remote-star	$\min_{c \in X} \sum_{q \in X \setminus \{c\}} d(c,q)$	2
remote-bipartition	$\min_{Q \subset X, \lvert Q\rvert = \lfloor \lvert X\rvert/2 \rfloor} \sum_{q \in Q, z \in X \setminus Q} d(q,z)$	3
remote-tree	$w(\text{MST}(X))$	4
remote-cycle	$w(\text{TSP}(X))$	3

Doubling Dimension. We will relate the performance of our algorithms to the dimensionality of the data which, for a general metric space (U, dist), can be captured by the notion of doubling dimension, reviewed below. For any $p \in U$

and $r > 0$, the *ball of radius r centered at p*, denoted as $B(p, r)$, is the subset of all points of U at distance at most r from p. The *doubling dimension* of U is the minimum value D such that, for all $p \in U$, any ball $B(p, r)$ is contained in the union of at most 2^D balls of radius $r/2$ centered at points of U. The notion of doubling dimension has been used extensively for a variety of applications (e.g., see [9,10,22,29] and references therein). The following fact is a simple consequence of the definition.

Fact 2. *Let X be a set of points from a metric space of doubling dimension D, and let $Y \subseteq X$ be such that any two distinct points $a, b \in Y$ have pairwise distance $dist(a, b) > r$. Then for every $R \geq r$ and any point $p \in X$, we have $|B(p, R) \cap Y| \leq 2^{\lceil \log_2(2R/r) \rceil \cdot D} \leq (4R/r)^D$. If $R/r = 2^i$, then the bound can be lowered to $|B(p, R) \cap Y| \leq (2R/r)^D$.*

3 Augmented Cover Trees

Let S be a set of n points from a metric space (U, dist) of doubling dimension D. Our algorithms employ an augmented version of the cover tree data structure of [3]. For completeness, we first review the original data structure and then discuss the augmentation, providing simpler update algorithms and a new correct analysis. Conceptually, a *cover tree* T for S is an infinite tree, where each node corresponds to a single point of S, while each point of S is associated to one or more nodes. The levels of the tree are indexed by integers, decreasing from the root towards the leaves. For $\ell \in (-\infty, +\infty)$, we let T_ℓ to be the set of nodes of level ℓ, and let $\text{pts}(T_\ell)$ be the points associated with the nodes of T_ℓ, which are required to be distinct. For each node $u \in T$ we maintain: its associated point (u.point); a pointer to its parent (u.parent); the list of pointers to its children (u.children); and its level in T (u.level). For brevity, in what follows, for any two nodes $u, v \in T$ (resp., any point $p \in U$ and node $u \in T$), we will use $\text{dist}(u, v)$ (resp., $\text{dist}(p, u)$) to denote $\text{dist}(u.\text{point}, v.\text{point})$ (resp., $\text{dist}(p, u.\text{point})$). For each level ℓ, the set T_ℓ must satisfy the following three properties:

1. $\text{pts}(T_\ell) \subseteq \text{pts}(T_{\ell-1})$;
2. for each $u \in T_\ell$, $\text{dist}(u, u.\text{parent}) \leq 2^{\ell+1}$;
3. for all $u, v \in T_\ell$, $\text{dist}(u, v) > 2^\ell$.

For every $p \in S$, let $\ell(p)$ denote the largest index such that $p \in \text{pts}(T_{\ell(p)})$. The definition implies that for every $\ell < \ell(p)$, $p \in \text{pts}(T_\ell)$, and the node u in T_ℓ with $u.\text{point} = p$ is a *self-child* of itself, in the sense that $u.\text{parent}.\text{point} = p$, since for every other $v \in T_{\ell+1}$, with $v.\text{point} \neq p$, $\text{dist}(u, v) > 2^{\ell+1}$. Let $d_{\min}$ and $d_{\max}$ denote, respectively, the minimum and maximum distances between two points of S, and define $\Delta = d_{\max}/d_{\min}$ as the *aspect ratio* of S. It can be easily seen that for every $\ell < \log_2 d_{\min}$, $\text{pts}(T_\ell) = S$, and for every $\ell \geq \log_2 d_{\max}$, $|\text{pts}(T_\ell)| = 1$. We define $\ell_{\min}$ (resp., $\ell_{\max}$) as the largest (resp., smallest) index such that $\text{pts}(T_\ell) = S$ (resp., $|\text{pts}(T_\ell)| = 1$), and note that every node u in a level

T_j with $j > \ell_{\max}$ or $j \leq \ell_{\min}$ has only the self-child in u.children. Therefore, we will consider only the portion of the tree constituted by the levels T_ℓ with $\ell \in [\ell_{\min}, \ell_{\max}]$ and will regard the unique node $r \in T_{\ell_{\max}}$ as the root of the tree. The above observations imply that the number of levels in this portion of the tree is $O(\log \Delta)$.

Update operations on a cover tree crucially rely on the notion of *cover set* [3]. For any point $p \in U$, and for every index $\ell \leq \ell_{\max}$ the *cover set (for p at level ℓ)* $Q_\ell^p \subseteq T_\ell$ is defined inductively as follows:

$$\begin{aligned} Q_{\ell_{\max}}^p &= \{r\} \\ Q_\ell^p &= \{u \in T_\ell \;:\; u.\text{parent} \in Q_{\ell+1}^p \wedge \text{dist}(u,p) \leq 2^{\ell+1}\} \quad \text{for } \ell < \ell_{\max} \end{aligned} \tag{2}$$

A simple inductive argument proves that for any node $u \in T_\ell - Q_\ell^p$, $\text{dist}(p,u) > 2^{\ell+1}$. Thus, intuitively, Q_ℓ^p contains all points of T_ℓ somewhat "close" to p, at the scale prescribed by the level.

We now relate the size of the cover sets to the doubling dimension D of (U, dist). We wish to remark that, when the doubling dimension D_S of the subspace (S, dist) is lower than D, e.g. if the current points belong to a lower-dimensional manifold of U, D_S can replace D in the subsequent analyses, yielding tighter results.

Lemma 1. *For every point $p \in U$ and $\ell \leq \ell_{\max}$, we have that $|Q_\ell^p| \leq 4^D$ and $\sum_{u \in Q_\ell^p} |u.\text{children}| \leq 12^D$.*

Proof. For each $u_1, u_2 \in T_\ell$, we have that $\text{dist}(u_1,u_2) > 2^\ell$. Also, $\text{pts}(Q_\ell^p) \subseteq B(p, 2^{\ell+1}) \cap T_\ell$. Then, by Fact 2, it follows that $|Q_\ell^p| \leq (2 \cdot 2^{\ell+1}/2^\ell)^D = 4^D$. Moreover, for each $u \in Q_\ell^p$ and each $u' \in u$.children, we have that $\text{dist}(u',p) \leq \text{dist}(u',u)+\text{dist}(u,p) \leq 2^\ell+2^{\ell+1} = 3 \cdot 2^\ell$, whence $u'.\text{point} \in B(p, 3 \cdot 2^\ell) \cap \text{pts}(T_{\ell-1})$. Then, again by Fact 2, it follows that $|\{u' \in u.\text{children s.t. } u \in Q_\ell^p\}| \leq (2 \cdot 3 \cdot 2^\ell/2^{\ell-1})^D = 12^D$.

We maintain a pointer to the root r of T, and recall that $r.\text{level} = \ell_{\max}$. Note that the naive representation of T, referred to as *implicit representation* in [3], would require $O(n \log \Delta)$ space. In order to save space, a more compact representation of T, referred to as *explicit representation* in [3], is used, where chains of 1-child nodes, which correspond to instances of the same point, are represented solely by the first node of the chain, which inherits the children of the last node of the chain. It is easy to argue that this compact representation takes only $O(n)$ space. Given a point $p \in U$, the cover sets Q_ℓ^p for $\ell \leq \ell_{\max}$ can be constructed in a top-down fashion from the explicit representation of T by simply recreating the contracted chains of implicit nodes. By Lemma 1, Q_ℓ^p can be constructed from $Q_{\ell+1}^p$ in $O(12^D)$ time.

Augmenting the Basic Structure. We augment the cover tree data structure so to maintain at its nodes two additional data fields, which will be exploited in the target applications presented later. Suppose that a matroid $M = (U, I)$ is defined over the universe U. An *augmented cover tree* T for S (with respect to

M) is a cover tree such that each node $u \in U$ stores the following two additional fields: a positive weight $u.weight = |S_u|$, where S_u is the subset of points of S associated with nodes in the subtree rooted at u, and a set of points $u.mis$, which is a maximal independent set of the submatroid $M_u = (S_u, I_u)$ spanned by the subsets of S_u. The size of T becomes $O(n \cdot \text{rank}(M))$, where $\text{rank}(M)$ denotes the size of a maximum independent set of S. For the applications where the matroid information is not needed, the fields u.mis will be always set to null (as if $M = (U, \emptyset)$), and, in this case, the size of T will be $O(n)$.

4 Dynamic Maintenance of Augmented Cover Trees

Let T be an augmented cover tree for a set S of n points. In this section, we show how to update T efficiently when a point p is added to or deleted from S.

Insertion. Let p be a new point to be inserted in T. The insertion of p is accomplished as follows. First, if p is very far from the root r, namely $\text{dist}(p, r) > 2^{\ell_{\max}}$, then both $\ell_{\max}$ and r.level are increased to $\lfloor \log_2 \text{dist}(p, r) \rfloor$. Then, an explicit node u is created with u.point $= p$, u.weight $= 1$ and u.mis $= \{p\}$. In order to determine the level $\ell(p)$ where u must be placed, all cover sets Q^p_ℓ are computed, as described above, for every $\ell \in [\bar{\ell}, \ell_{\max}]$, where $\bar{\ell}$ is the largest index in $(-\infty, \ell_{\max}]$ such that $Q^p_{\bar{\ell}} = \emptyset$. Note that such empty cover set must exist and it is easy to see that $\bar{\ell} \geq \lceil \log_2 \text{dist}(p, S) \rceil - 2$. Then, u.level is set to the smallest index $\ell(p) \geq \bar{\ell}$ such that $\text{dist}(p, \text{pts}(Q^p_{\ell(p)})) > 2^{\ell(p)}$ and $\text{dist}(p, \text{pts}(Q^p_{\ell(p)+1})) \leq 2^{\ell(p)+1}$. At this point, an arbitrary node $v \in Q^p_{\ell(p)+1}$ such that $\text{dist}(p, v) \leq 2^{\ell(p)+1}$ (which must exists for sure) is determined. Let $q = v$.point. If v has no explicit self-child at level $\ell(p)$, a new node w with w.point $= q$, w.level $= \ell(p)$, and w.children $= v$.children, is created, and v.children is set to $\{u, w\}$. If instead such an explicit self-child w of v exists at level $\ell(p)$, then u is simply added as a further child of v. Finally, the path from the newly added node u to r is traversed, and for every ancestor v of u, v.weight is increased by 1 and p is added to the independent set v.mis, if $v.\text{mis} \cup \{p\}$ is still an independent set.

In [31], we provide a fully detailed pseudocode for the insertion procedure.

Theorem 1. *Let T be an augmented cover tree for a set S of n points, with respect to a matroid $M = (U, I)$. The insertion algorithm described above yields an augmented cover tree for $S \cup \{p\}$ in time $O\left(12^D \log \Delta\right)$ where D is the doubling dimension of the metric space and Δ is the aspect ratio of S.*

Proof. It is easy to see that the insertion algorithm enforces, for every level ℓ, Properties 1, 2, 3 of the definition of cover tree, restricted to the nodes in Q^p_ℓ plus the new node created for p (for $\ell = \ell(p)$). These properties immediately extend to the entire level ℓ, since, as observed before, $\text{dist}(p, u) > 2^{\ell+1}$, for every $u \in T_\ell - Q^p_\ell$. For what concerns the update of the .weight and .mis fields, correctness is trivially argued for the .weight fields, while Fact 1 ensures correctness of the updates of the .mis fields. can be argued as for the insertion algorithm. The complexity bounds follow by observing that there are $O(\log \Delta)$ levels in the explicit representation of T and that, at each such level ℓ, the algorithm performs work linear in the number of children of Q_ℓ^p, which are at most 12^D, by Lemma 1.

Deletion. Let $p \in S$ be the point to be removed. We assume that p is not the only point in S, otherwise the removal is trivial. The deletion of p is accomplished as follows. In the first, top-down phase, all cover sets Q_ℓ^p are computed, for every $\ell \in [\bar{\ell}, \ell_{\max}]$, where $\bar{\ell} \leq \ell_{\max}$ is the level of the leaf node corresponding to p in the explicit tree. Also, a list $R_{\bar{\ell}}$ of explicit nodes at level $\bar{\ell}$ to be relocated is created and initialized to the empty list. In the second, bottom-up phase, the following operations are performed iteratively, for every $\ell = \bar{\ell}, \bar{\ell}+1, \ldots, \ell_{\max}-1$.

- If Q_ℓ^p contains a node u with u.point $= p$ and u.level $= \ell$, the following additional operations are performed. Let v be the parent of u, and observe that all children of v must also be explicit nodes at level ℓ. If u is the self child of v (i.e., v.point $= u$.point $= p$) u is removed from T, u's siblings are detached from v and added to R_ℓ (v will be later removed at iteration v.level). If instead u is not the self child of v, but it is the only child of v besides the self-child, u is removed from T and v and its self-child are merged together in the explicit tree.
- An empty list $R_{\ell+1}$ is created. Then, R_ℓ is scanned sequentially, and, for every $w \in R_\ell$, a node $w' \in Q_{\ell+1}^p \cup R_{\ell+1}$ is searched for such that $d(w, w') \leq 2^{\ell+1}$. If no such node exists, then w is added to $R_{\ell+1}$, raising w.level to $\ell + 1$. Otherwise, if w' is found, it becomes parent of w as follows. If w' is internal and its children are at level ℓ, w is simply added as a further child. Otherwise, a new explicit node z, is created with z.point $= w'$.point, z.level $= \ell$, and z.children $= w'$.children, and w'.children is set to $\{z, w\}$.
- For all nodes in $w \in Q_{\ell+1}^p \cup R_{\ell+1}$ with w.level $= \ell + 1$, their .weight and .mis fields are updated based on the values of the corresponding fields of their children. The update of the .weight fields is straightforward, while, based on Fact 1, the update of the .mis field of one such node w can be accomplished by computing a maximal independent set in the union of the elements of the .mis fields of w's children.

Once the above operations are performed, a slightly more complex iteration is required for $\ell = \ell_{\max}$, since it might be necessary to modify $\ell_{\max}$ and/or create a new root, in case $p = r$.point. A fully detailed pseudocode for the deletion procedure is provided in [31].

Theorem 2. *Let T be an augmented cover tree for a set S of n points, with respect to a matroid $M = (U, I)$. The deletion algorithm described above yields an augmented cover tree for $S - \{p\}$ in time $O\left((16^D + 12^D \mathit{rank}(M)) \log \Delta\right)$ where D is the doubling dimension of the metric space and Δ is the aspect ratio of S.*

Proof. It is easy to see that the bottom-up phase of the deletion algorithm enforces, for every level ℓ, Properties 1, 2, 3 of the definition of cover tree, restricted to the nodes in $Q_\ell^p \cup R_\ell$. As argued before, these properties immediately extend to the entire level ℓ. Finally, correctness of the update of the .weight and .mis fields can be argued as for the insertion algorithm. For what concerns the complexity bound, first observe that for every level ℓ, the nodes in $Q_\ell^p \cup R_\ell$ represent the new coversets $\hat{Q}_\ell^p$ associated to p after its deletion from T, thus Lemma 1 holds. As a consequence, the work needed to process the nodes in R_ℓ is $\Theta(|R_\ell|(|Q_{\ell+1}^p \cup R_{\ell+1}|) = O(16^D)$, while, by Fact 1, recreating the .weight and .mis fields for all nodes in $Q_{\ell+1}^p \cup R_{\ell+1}$ is upper bounded by the number of their children multiplied by $(1+\text{rank}(M))$. The final bound follows by observing that there are $O(\log \Delta)$ levels in the explicit representation of T.

5 Extracting Solutions from the Augmented Cover Tree

We show how to employ the augmented cover tree presented before to extract accurate solutions to the various problems introduced in Sect. 2. For all these problems, we rely on the extraction from the cover tree of a small (ϵ, k)-coreset (see Definition 1), for suitable values of ϵ and k.

Let T be an augmented cover tree for a set S of n points from a metric space of doubling dimension D. Given ϵ and k, an (ϵ, k)-coreset for S can be constructed as follows. Let $T_{\ell(k)}$ be the level of largest index (in the implicit representation of T) such that $|T_{\ell(k)}| \leq k$ and $|T_{\ell(k)-1}| > k$. Then, define

$$\ell^*(\epsilon, k) = \max\{\ell_{\min}, \ell(k) - \lceil \log_2(8/\epsilon) \rceil\}. \tag{3}$$

(For ease of notation, in what follows we shorthand $\ell^*(\epsilon, k)$ with ℓ^* whenever the parameters are clear from the context.) We have:

Lemma 2. *The set of points pts(T_{ℓ^*}) is an (ϵ, k)-coreset for S of size at most $k(64/\epsilon)^D$ and can be constructed in time $O\left(k((64/\epsilon)^D + \log \Delta)\right)$.*

Proof. We first show that pts(T_{ℓ^*}) is an (ϵ, k)-coreset for S. If $\ell^* = \ell_{\min}$, we have pts$(T_{\ell^*}) = S$, so the property is trivially true. Suppose that $\ell^* = \ell(k) - \lceil \log_2(8/\epsilon) \rceil > \ell_{\min}$ and consider an arbitrary point $p \in S$. There must exist some node $u \in T_{\ell_{\min}}$ such that $p = u$.point. Letting v be the ancestor of u in T_{ℓ^*}, by the cover tree properties, we know that dist$(v.\text{point}, p) \leq 2^{\ell^*+1} \leq (\epsilon/4)2^{\ell(k)}$. Also, since all pairwise distances among points of pts$(T_{\ell(k)-1})$ are greater than $2^{\ell(k)-1}$ and $|T_{\ell(k)-1}| \geq k+1$, there must be two points $q, q' \in$ pts$(T_{\ell(k)-1})$ which belong to the same cluster in the optimal k-center clustering of S. Therefore, $2^{\ell(k)-1} < \text{dist}(q, q') \leq 2r_k^*(S)$, which implies that $2^{\ell(k)} \leq 4r_k^*(S)$. Putting it all together, we have that for any $p \in S$, dist$(p, \text{pts}(T_{\ell^*})) \leq \epsilon r_k^*(S)$. Let us now bound the size of T_{ℓ^*}. By construction $|T_{\ell(k)}| \leq k$, and we observe that T_{ℓ^*} can be partitioned into $|T_{\ell(k)}|$ subsets $T_{\ell^*}^u$, for every $u \in T_{\ell(k)}$, where $T_{\ell^*}^u$ is the set of descendants of u in T_{ℓ^*}. The definition of cover tree implies that for each $u \in T_{\ell(k)}$ and $v \in T_{\ell^*}^u$, dist$(u, v) \leq 2^{\ell(k)+1}$. Moreover, since the pairwise distance between points of pts$(T_{\ell^*}^u)$ is greater than 2^{ℓ^*}, by applying Fact 2 with $Y =$ pts$(T_{\ell^*}^u)$, $R = 2^{\ell(k)+1}$ and $r = 2^{\ell^*}$, we obtain that $|T_{\ell^*}^u| \leq 2^{(\lceil \log_2(8/\epsilon) \rceil + 2) \cdot D} \leq (64/\epsilon)^D$, and the bound on $|T_{\ell^*}|$ follows. T_{ℓ^*} can be constructed on the explicit tree through a simple level-by-level visit up to level ℓ^*, which can be easily determined from $\ell(k)$ and the fact that $\ell_{\min}$ is the largest level for which all nodes in the explicit representation of T are leaves. The construction time is linear in $\sum_{\ell=\ell^*}^{\ell_{\max}} |T_\ell| = \sum_{\ell=\ell^*}^{\ell(k)-1} |T_\ell| + \sum_{\ell=\ell(k)}^{\ell_{\max}} |T_\ell|$. The second summation is clearly upper bounded by $k \log \Delta$, while, using again Fact 2 it is easy to argue that $|T_\ell| \leq k \cdot 2^{(\ell(k)+2-\ell)\cdot D}$, for every $\ell^* \leq \ell \leq \ell(k) - 1$, whence the first sum is $O\left(k(64/\epsilon)^D\right)$.

5.1 Solving k-Center

Suppose that an augmented cover tree T for S is available. We can compute an $O(2 + O(\epsilon))$-approximate solution C to k-center on S as follows. First, we extract the coreset $Q = \text{pts}(T_{\ell^*})$, where $\ell^* = \ell^*(\epsilon, k)$ is the index defined in Eq. 3, and then run a sequential algorithm for k-center on Q. To do so, we could use Gonzalez's 2-approximation algorithm. However, this would contribute an $O(k|Q|)$ term to the running time, which, based on the size bound stated in Lemma 2, would yield a quadratic dependency on k. The asymptotic dependency on k can be lowered by computing the solution through an adaptation of the techniques presented in [21], as explained below. Let us define a generalization of the cover tree data structure, dubbed (α, β)-*cover tree*, where the three properties that each level ℓ must satisfy are rephrased as follows:

1. $\text{pts}(T_\ell) \subseteq \text{pts}(T_{\ell-1})$;
2. for each $u \in T_\ell$, $\text{dist}(u, u.\text{parent}) \leq \beta \cdot \alpha^{\ell+1}$;
3. for all $u, v \in T_\ell$, $\text{dist}(u, v) > \beta \cdot \alpha^\ell$.

By adapting the insertion procedure and its analysis, it is easily seen that the insertion of a new point in the data structure can be supported in $O\left(12^D \cdot \log_\alpha \Delta\right)$ time. For a given integer parameter m, we construct m generalized cover trees for the coreset Q, namely an $(\alpha, \alpha^{i/m})$-cover tree $T^{(i)}$ for every $1 \leq i \leq m$. Each cover tree is constructed by inserting one point of Q at a time. Let ℓ_i be the smallest index such that level $T^{(i)}_{\ell_i}$ in $T^{(i)}$ has at most k nodes. The returned solution C is the set $\text{pts}(T^{(i)}_{\ell_i})$ such that $T^{(i)}_{\ell_i}$ minimizes $\alpha^{\ell_i+i/m}$. By selecting $\alpha = 2/\epsilon$ and $m = O\left(\epsilon^{-1} \ln \epsilon^{-1}\right)$, and by using the argument of [21], it can be shown that C is a $(2 + O(\epsilon))$-approximation for k-center on Q. The following theorem is a consequence of the above discussion, Lemma 2, and the definition of (ϵ, k)-coreset.

Theorem 3. *Given an augmented cover tree T for S, the above procedure returns a $(2+O(\epsilon))$-approximation C to the k-center problem for S, and can be implemented in time $O\left((k/\epsilon)(768/\epsilon)^D \log \Delta\right)$.*

Proof. By Lemma 2, Q is an (ϵ, k)-coreset for S. Let $C^* = \{c_1, c_2, \ldots, c_k\}$ be an optimal solution for k-center on S. The coreset property of Q ensures that for each c_i there is a point $c'_i \in Q$ such that $d(c_i, c'_i) \leq \epsilon r^*_k(S)$. This implies that the set $C' = \{c'_1, c'_2, \ldots, c'_k\}$ is a solution to k-center on Q with $r_{C'}(Q) \leq (1+\epsilon) r^*_k(S)$, hence $r_k^*(Q) \leq (1+\epsilon)r_k^*(S)$. Suppose that the $(2+O(\epsilon))$-approximation algorithm outlined above is used in Phase 2 to compute the solution C on Q. Then, $r_C(Q) \leq (2+O(\epsilon))r_k^*(Q) \leq (2+O(\epsilon))(1+\epsilon)r_k^*(S) = (2+O(\epsilon))r_k^*(S)$. By the coreset property and the triangle inequality, it follows that $r_C(S) \leq (2+O(\epsilon))r_k^*$. For what concerns the running time, we have that the construction of the (ϵ, k)-coreset Q requires $O\left(k((64/\epsilon)^D + \log \Delta)\right)$ time (see Lemma 2), while the running time of Phase 2 is dominated by the construction of the $m = O\left(\epsilon^{-1} \ln \epsilon^{-1}\right)$ $(\alpha, \alpha^{p/m})$-cover trees $T^{(p)}$, for $1 \leq p \leq m$, by successive insertions of the elements of Q. As observed above, an insertion takes $O\left(12^D \log_\alpha \Delta\right)$ time, hence the cost for constructing each $T^{(p)}$ is $O\left(|Q| 12^D \log_\alpha \Delta\right)$. Since $\alpha = 2/\epsilon$ and $|Q| \leq k(64/\epsilon)^D$, the total cost is thus $O\left((k/\epsilon)(768/\epsilon)^D \log \Delta\right)$, which dominates over the cost of Phase 1.

5.2 Solving k-Center with z Outliers

For the k-center problem with z outliers, an approach similar to the one adopted for k-center can be employed. Let T be an augmented cover tree for S. We can compute a $(3+O(\epsilon))$-approximation to k-center with z outliers on S, by proceeding as follows. First, we extract the coreset $Q = \text{pts}(T_{\ell^*})$, where $\ell^* = \ell^*(k+z)$ is the index defined in Eq. 3. Each point $q \in Q$ is associated to the weight $w_q = u.\text{weight}$, where $u \in T_{\ell^*}$ is such that $u.\text{point} = q$. Then, we extract the solution C from this weighted coreset Q using the techniques from [6]. We have:

Theorem 4. *Given an augmented cover tree T for S, the above procedure returns a $(3+O(\epsilon))$-approximation C to the k-center problem with z outliers for S, and can be implemented in time $O\left((k+z)^2(64/\epsilon)^{2D}(1/\epsilon)\log \Delta\right)$.*

Proof. By Lemma 2, $Q = \text{pts}(T_{\ell^*})$ is an $(\epsilon, k+z)$-coreset for S. Consider a point $q \in Q$ and let $u \in T_{\ell^*}$ be such that $u.\text{point} = q$. It is easy to see that each of the $u.\text{weight} = w_q$ points associated with the descendants of u in T (forming set $S_u \subseteq S$) is at distance at most $\epsilon r_{k+z}^*(S)$ from q. Since the S_u's form a partition of S, we have that each point of S can be associated to a point of Q at distance at most $\epsilon r_{k+z}^*(S) \leq r_{k,z}^*(S)$, referred to as its *proxy*. Suppose that algorithm OutliersCluster described in [6] is run on the weighted coreset Q with parameters k, r, and ϵ, where r is a guess of the optimal radius. The analysis in [6] shows that the algorithm returns two subsets $X, Q' \subseteq Q$ such that

- $|X| \leq k$
- For every $p \in S$ whose proxy is in $Q - Q'$, $\text{dist}(p, X) \leq \epsilon r_{k,z}^*(S) + (3+4\epsilon)r$;
- if $r \geq r_{k,z}^*(S)$, then $\sum_{q \in Q'} w_q \leq z$.

Then, we can repeatedly run OutliersCluster for $r = 2^{\ell_{\max}}/(1+\epsilon)^i$, for $i = 0, 1, \ldots$, stopping at the smallest guess r which returns a pair (X, Q') where Q' has aggregate weight at most z and returning $C = X$ as the final solution. Therefore, the bound on the approximation ratio immediately follows from the properties of OutliersCluster. The running time is dominated by the repeated executions of OUTLIERSCLUSTER. Each execution of OUTLIERSCLUSTER can be performed in $O\left(|Q|^2 + k|Q|\right) = O\left((k+z)^2(64/\epsilon)^{2D}\right)$ time. The bound on the running time follows by observing that $2^{\ell_{\max}}/r^*_{k,z}(S) = O(\Delta)$, whence $O\left(\log_{1+\epsilon}\Delta\right) = O\left((1/\epsilon)\log\Delta\right)$ executions suffice.

5.3 Solving Matroid Center

Consider a matroid $M = (S, I)$ defined on a set S, and suppose that an augmented cover tree T for S w.r.t. M is available. We can compute an $O\left(3 + O\left(\epsilon\right)\right)$-approximate solution C to the matroid center problem on M as follows. First we determine a coreset Q as the union of the independent sets associated with the nodes of level T_{ℓ^*} where $\ell^* = \ell^*(\epsilon, \text{rank}(M))$ is the index defined in Eq. 3. Namely, $Q = \bigcup_{u \in T_{\ell^*}} u.\text{mis}$. (Note that rank$(M)$ is easily obtained as the size of r.mis, where r is the root of T.) Then, solution C is computed by running the 3-approximation algorithm by [16] on Q. The resulting performance is summarized by following theorem, whose proof can be found in [31].

Theorem 5. *Given an augmented cover tree T for S w.r.t. $M = (S, I)$, the above procedure returns a $(3 + O(\epsilon))$-approximation C to the matroid center problem on M, and can be implemented in time $O\left(poly(rank(M)), (64/\epsilon)^D\right) + rank(M)\log\Delta\right)$.*

5.4 Solving Diversity Maximization

In [10], the authors present a coreset-based approach to yield approximate solutions to all variants of the diversity maximization problem listed in Table 1. Specifically, for a pointset S and a diversity function div$(\cdot)$, the polynomial-time sequential algorithm A_{div} yielding the α_{div} approximation mentioned in the table, is run on a coreset $Q' \subseteq S$ which is derived from an (ϵ, k)-coreset Q for k-center on S, as follows. For the remote edge and the remote cycle variants, simply setting $Q' = Q$ suffices, while for all the other variants, Q' is obtained by selecting, for each $p \in Q$, $\min\{k, |S_p|\}$ points from the subset S_p of a partition $\{S_p : p \in Q\}$ of S into disjoint subsets, where each S_p contains points $q \in S$ with dist$(p, q) \leq \epsilon r^*_k$. It is shown in [10] that running A_{div} on Q' yields an $(\alpha_{\text{div}} + O(\epsilon))$-approximate solution for S. Now, given an (augmented) cover tree T for S, Q' can be easily constructed as follows. For the remote edge and remote clique variants, Q' is set to pts(T_{ℓ^*}), where $\ell^* = \ell^*(\epsilon, k)$ is the index defined in Eq. 3. The other variants require the use of an augmented cover tree w.r.t. to the simple *k-bounded cardinality matroid* $M_{k,S}$, whose independent sets are all subsets of S of at most k points, so that each node of T will store a set of (at most) k points from its subtree. Then, $Q' = \cup_{u \in T_{\ell^*}} u.\text{mis}$. The resulting performance is summarized by following theorem, where $t_{A_{\text{div}}}(\cdot)$ denotes the running time of A_{div}. The proof of the theorem can be found in [31].

Theorem 6. *Consider a cover tree T for S (augmented w.r.t. the k-bounded cardinality matroid $M_{k,S}$, when necessary). For each diversity variant in*

Table 1, running A_{div} on the coreset Q' extracted from T returns an $(\alpha_{\text{div}} + O(\epsilon))$-approximate solution to the diversity maximization problem in time $O\left(t_{A_{\text{div}}}(k(64/\epsilon)^D) + k \log \Delta\right)$ for the remote edge and cycle variants, and time $O\left(t_{A_{\text{div}}}(k^2(64/\epsilon)^D) + k \log \Delta\right)$ for the other variants.

6 Conclusions

In this paper, we provided novel coreset-based fully dynamic algorithms for (robust) k-center, matroid center and diversity maximization. All algorithms are deterministic and run on the same augmented cover tree data structure, which can thus serve queries for different problems at the same time. It is important to remark that for all problems, when the input S is large and both the spread Δ and the doubling dimension D of the metric are small, the dynamic maintenance of the data structure and the extraction of solutions can be accomplished in time dramatically smaller than the time that would be required to compute solutions on the entire pointset from scratch.

We remark that the coreset-based approaches developed in [8] for robust matroid center, and in [7] for diversity maximization under matroid constraints, can be integrated with the approach presented in this paper, to yield fully dynamic approximations for these more general versions of the problems, with similar accuracy-performance tradeoffs.

Acknowledgments. This work was supported, in part, by MUR of Italy, under Projects PRIN 20174LF3T8 (AHeAD: Efficient Algorithms for HArnessing Networked Data), and PNRR CN00000013 (National Centre for HPC, Big Data an d Quantum Computing), and by the University of Padova under Project SID 2020 (RATED-X: Resource-Allocation TradEoffs for Dynamic and eXtreme data).

References

1. Abbassi, Z., Mirrokni, V.S., Thakur, M.: Diversity maximization under matroid constraints. In: Proceedings of ACM KDD, pp. 32–40 (2013)
2. Bateni, M., et al.: Optimal fully dynamic k-center clustering for adaptive and oblivious adversaries. In: Proceedings of ACM-SIAM SODA, pp. 2677–2727 (2023)
3. Beygelzimer, A., Kakade, S., Langford, J.: Cover trees for nearest neighbor. In: Proceedings of ICML, pp. 97–104 (2006)
4. Borassi, M., Epasto, A., Lattanzi, A., Vassilvitskii, S., Zadimoghaddam, M.: Better sliding window algorithms to maximize subadditive and diversity objectives. In: Proceedings of ACM PODS, pp. 254–268 (2019)
5. Ceccarello, M., Pietracaprina, A., Pucci, G.: Fast coreset-based diversity maximization under matroid constraints. In: Proceedings of ACM WSDM, pp. 81–89 (2018)
6. Ceccarello, M., Pietracaprina, A., Pucci, G.: Solving k-center clustering (with outliers) in MapReduce and streaming, almost as accurately as sequentially. PVLDB **12**(7), 766–778 (2019)

7. Ceccarello, M., Pietracaprina, A., Pucci, G.: A general coreset-based approach to diversity maximization under matroid constraints. ACM Trans. Knowl. Discov. Data **14**(5), 60:1–60:27 (2020)
8. Ceccarello, M., Pietracaprina, A., Pucci, G., Soldà, F.: Scalable and space-efficient robust matroid center algorithms. J. Big Data **10**, 49 (2023). https://doi.org/10.1186/s40537-023-00717-4
9. Ceccarello, M., Pietracaprina, A., Pucci, G., Upfal, E.: A practical parallel algorithm for diameter approximation of massive weighted graphs. In: Proceedings of IEEE IPDPS, pp. 12–21 (2016)
10. Ceccarello, M., Pietracaprina, A., Pucci, G., Upfal, E.: MapReduce and streaming algorithms for diversity maximization in metric spaces of bounded doubling dimension. PVLDB **10**(5), 469–480 (2017)
11. Chakrabarty, D., Goyal, P., Krishnaswamy, R.: The non-uniform k-center problem. ACM Trans. Algorithms **16**(4), 46:1–46:19 (2020)
12. Chan, T.H., Guerqin, A., Sozio, M.: Fully dynamic k-center clustering. In: Proceedings of WWW, pp. 579–587 (2018)
13. Chan, T.H.H., Lattanzi, S., Sozio, M., Wang, B.: Fully dynamic k-center clustering with outliers. In: Proceedings of COCOON, LNCS, vol. 13595, pp. 150–161 (2023).
14. Charikar, M., Chekuri, C., Feder, T., Motwani, R.: Incremental clustering and dynamic information retrieval. In: Proceedings of ACM STOC, pp. 626–635 (1997)
15. Charikar, M., Khuller, S., Mount, D., Narasimhan, G.: Algorithms for facility location problems with outliers. In: Proceedings of ACM-SIAM SODA, pp. 642–651 (2001)
16. Chen, D.Z., Li, J., Liang, H., Wang, H.: Matroid and knapsack center problems. Algorithmica **75**(1), 27–52 (2015). https://doi.org/10.1007/s00453-015-0010-1
17. Chiplunkar, A., Kale, S., Ramamoorthy, S.: How to solve fair k-center in massive data models. In: Proceedings of ICML, pp. 1877–1886 (2020)
18. Cohen-Addad, V., Schwiegelshohn, C., Sohler, C.: Diameter and k-center in sliding windows. In: Proceedings of ICALP (2016)
19. Elkin, Y., Kurlin, V.: Counterexamples expose gaps in the proof of time complexity for cover trees introduced in 2006. In: Proceedings of IEEE TopoInVis, pp. 9–17 (2022)
20. Gonzalez, T.: Clustering to minimize the maximum intercluster distance. Theor. Comput. Sci. **38**, 293–306 (1985)
21. Goranci, G., Henzinger, M., Leniowski, D., Schulz, C., Svozil, A.: Fully dynamic k-center clustering in low dimensional metrics. In: Proceedings of ALENEX 2021, pp. 143–153 (2021)
22. Gottlieb, L.A., Kontorovich, A., Krauthgamer, R.: Efficient classification for metric data. IEEE Trans. Inf. Theory **60**(9), 5750–5759 (2014)
23. Harris, D., Pensyl, T., Srinivasan, A., Trinh, K.: A lottery model for center-type problems with outliers. ACM Trans. Algorithms **15**(3), 36:1–36:25 (2019)
24. Kale, S.: Small space stream summary for matroid center. In: Proceedings of APPROX/RANDOM, pp. 20:1–20:22 (2019)
25. Kleindessner, M., Awasthi, P., Morgenstern, J.: Fair k-center clustering for data summarization. In: Proceedings of ICML, pp. 3448–3457 (2019)
26. Krauthgamer, R., Lee, J.: Navigating nets: simple algorithms for proximity search. In: Proceedings of ACM-SIAM SODA, pp. 798–807 (2004)
27. Leskovec, J., Rajaraman, A., Ullman, J.: Mining of Massive Data Sets. Cambridge University Press, Cambridge (2014)
28. Oxley, J.: Matroid Theory. Oxford University Press, Oxford (2006)

29. Pellizzoni, P., Pietracaprina, A., Pucci, G.: Dimensionality-adaptive k-center in sliding windows. In: Proceedings of DSAA, pp. 197–206 (2020)
30. Pellizzoni, P., Pietracaprina, A., Pucci, G.: k-center clustering with outliers in sliding windows. Algorithms **15**(2), 52 (2022)
31. Pellizzoni, P., Pietracaprina, A., Pucci, G.: Fully dynamic clustering and diversity maximization in doubling metrics. arXiv preprint arXiv:2302.07771 (2023)

Quick Minimization of Tardy Processing Time on a Single Machine

Baruch Schieber[1(✉)] and Pranav Sitaraman[2(✉)]

[1] New Jersey Institute of Technology, Newark, NJ 07102, USA
sbar@njit.edu
[2] Edison Academy Magnet School, Edison, NJ 08837, USA
sitaraman.pranav@gmail.com

Abstract. We consider the problem of minimizing the total processing time of tardy jobs on a single machine. This is a classical scheduling problem, first considered by [Lawler and Moore 1969], that also generalizes the Subset Sum problem. Recently, it was shown that this problem can be solved efficiently by computing (max, min)-skewed-convolutions. The running time of the resulting algorithm is the same, up to logarithmic factors, as the time it takes to compute a (max, min)-skewed-convolution of two vectors of integers whose sum is $O(P)$, where P is the sum of the jobs' processing times. We further improve the running time of the minimum tardy processing time computation by introducing a job "bundling" technique and achieve a $\tilde{O}\left(P^{2-1/\alpha}\right)$ running time, where $\tilde{O}(P^{\alpha})$ is the running time of a (max, min)-skewed-convolution of vectors of size P. This results in a $\tilde{O}\left(P^{7/5}\right)$ time algorithm for tardy processing time minimization, an improvement over the previously known $\tilde{O}\left(P^{5/3}\right)$ time algorithm.

Keywords: scheduling · convolution · tardy processing time

1 Introduction

The input to the Minimum Tardy Processing Time (MTPT) Problem consists of n jobs each of which is associated with a due date and processing time $p_i \in \mathbb{N}$. Consider a (nonpreemptive) schedule of these jobs on a single machine that can execute only one job at a time. A job is *tardy* if it terminates after its due date. The MTPT Problem is to find a schedule of the jobs that minimizes the total processing time of the tardy jobs. In the standard scheduling notation the MTPT problem is denoted $1||\sum p_j U_j$.

Consider an instance of MTPT in which all the jobs have the same due date d. Let $P = \sum_{j=1}^{n} p_j > d$. The decision whether the total processing time of the tardy jobs is exactly $P - d$ (which is optimal in this case) is equivalent to finding whether there exists a subset of the jobs whose processing time sums to d. This is equivalent to the Subset Sum problem. It follows that MTPT is NP-hard. MTPT is weakly NP-hard and Lawler and Moore [6] gave an $O(P \cdot n)$ time algorithm for this problem.

P. Morin and S. Suri (Eds.): WADS 2023, LNCS 14079, pp. 637–643, 2023.
https://doi.org/10.1007/978-3-031-38906-1_42

Bringmann *et al.* [2] introduced a new convolution variant called a (max, min)-skewed-convolution. They gave an algorithm for MTPT that uses (max, min)-skewed-convolutions, and proved that up to logarithmic factors, the running time of this algorithm is equivalent to the time it takes to compute a (max, min)-skewed-convolution of integers that sum to $O(P)$. They also gave an $\tilde{O}(P^{7/4})$ time algorithm[1] for computing a (max, min)-skewed-convolution of integers that sum to $O(P)$, which results in an $\tilde{O}(P^{7/4})$ time algorithm for the MTPT problem. Klein *et al.* [3] further improved the algorithm for computing a (max, min)-skewed-convolution and achieved an $\tilde{O}(P^{5/3})$ running time, and thus an $\tilde{O}(P^{5/3})$ time algorithm for the MTPT problem.

A natural approach to further improve the MTPT algorithm is by improving the running time of a (max, min)-skewed computation. However, obtaining an $\tilde{o}(P^{3/2})$ time algorithm for computing a (max, min)-skewed-convolution seems difficult as this would imply an improvement to the best known (and decades old) algorithm for computing a (max, min)-convolution [5]. We were able to "break" the $\tilde{O}(P^{3/2})$ barrier by introducing a job "bundling" technique. Applying this technique in conjunction with the best known algorithm for computing a (max, min)-skewed-convolution yields an $\tilde{O}(P^{7/5})$ time algorithm for the MTPT problem. This algorithm outperforms Lawler and Moore's algorithm [6] in instances where $n = \tilde{\omega}(P^{2/5})$. In general, applying our technique in conjunction with an $\tilde{O}(P^{\alpha})$ time algorithm for computing a (max, min)-skewed-convolution yields an $\tilde{O}(P^{2-1/\alpha})$ time for the MTPT problem.

The rest of the paper is organized as follows. In Sect. 2 we introduce our notations and describe the prior work that we apply in our algorithm. Section 3 specifies our algorithm, and Sect. 4 concludes with a summary and open problems.

2 Preliminaries

Our notations follow the notations in [2]. The input to the MTPT problem is a set of n jobs $\mathcal{J} = \{J_1, J_2, \dots, J_n\}$. Each job $J_j \in \mathcal{J}$ has due date $e_j \in \mathbb{N}$ and processing time $p_j \in \mathbb{N}$. Let $D_\#$ denote the number of distinct due dates, and denote the monotone sequence of distinct due dates by $d_1 < d_2 < d_3 < \cdots < d_{D_\#}$, with $d_0 = 0$. Let $\mathcal{J}_k \subseteq \mathcal{J}$ be the set of jobs with due date d_k. Let $D = \sum_{i=1}^{D_\#} d_i$ and $P = \sum_{i=1}^{n} p_i$. For a consecutive subset of indices $I = \{i_0, \dots, i_1\}$, where $1 \le i_0 \le i_1 \le D_\#$, let $\mathcal{J}_I = \bigcup_{i \in I} \mathcal{J}_i$ and $P_I = \sum_{J_i \in \mathcal{J}_I} p_i$.

Recall that the goal is to schedule the jobs in $\mathcal{J}$ so that the total processing time of tardy jobs is minimized. Since we only consider non-preemptive schedules, any schedule S corresponds to a permutation $\sigma_S : \{1, \dots, n\} \to \{1, \dots, n\}$ of the job indices. The completion time of job $J_j \in \mathcal{J}$ in schedule S is $C_j = \sum_{\sigma_S(i) \le \sigma_S(j)} p_i$, and j is tardy in S if $C_j > e_j$. Therefore, we can consider that our algorithm seeks to minimize $\sum_{J_j \in \mathcal{J}, C_j > e_j} p_j$.

[1] The notation $\tilde{O}(\cdot)$ hides the poly-logarithmic factors.

Next, we recall the definition of convolutions and describe the techniques developed in [2] and used by our algorithm. Given two vectors A and B of dimension $n+1$ and two binary operations $\circ$ and $\bullet$, the $(\circ, \bullet)$-convolution applied on A and B results in a $2n+1$ dimensional vector C, defined as:

$$C[k] = \bigcirc_{i=\max\{0,k-n\}}^{\min\{k,n\}} A[i] \bullet B[k-i],\ \forall\ k \in \{0, \dots, 2n\}.$$

A $(\max, \min)$-skewed-convolution applied on A and B results in a $2n+1$ dimensional vector C, defined as:

$$C[k] = \max_{i=\max\{0,k-n\}}^{\min\{k,n\}} \min\{A[i], B[k-i]+k\},\ \forall\ k \in \{0, \dots, 2n\}.$$

Bringmann *et al.* [2] apply an equivalent form of $(\max, \min)$-skewed-convolution defined as

$$C[k] = \max_{i=\max\{0,k-n\}}^{\min\{k,n\}} \min\{A[i], B[k-i]-i\},\ \forall\ k \in \{0, \dots, 2n\}.$$

Below, we use this equivalent form as well.

Let X and Y be two integral vectors. Define the *sumset* $X \oplus Y = \{x+y : x \in X, y \in Y\}$. It is not difficult to see that the sumset can be inferred from a $(+, \cdot)$-convolution of X_1 and X_2 which can be computed in $\tilde{O}(P)$ time for $X, Y \subseteq \{0, \dots, P\}$ as in [1].

The set of all subset sums of entries of X, denoted $\mathcal{S}(X)$, is defined as $\mathcal{S}(X) = \{\sum_{x \in Z} x : Z \subseteq X\}$. These subset sums can be calculated in $\tilde{O}(\sum_{x \in X} x)$ time by successive computations of sumsets [4]. We note that we always have $0 \in \mathcal{S}(X)$. Define the *t-prefix* and *t-suffix* of $\mathcal{S}(X)$ as $\mathsf{pref}(\mathcal{S}, t) = \{x \in \mathcal{S}(X) \wedge x \leq t\}$ and $\mathsf{suff}(\mathcal{S}, t) = \{x \in \mathcal{S}(X) \wedge x > t\}$.

We say that a subset of jobs $\mathcal{J}' \subseteq \mathcal{J}$ can be scheduled *feasibly* starting at time t if there exists a schedule of these jobs starting at time t such that all jobs are executed by their due date. Note that it is enough to check whether all jobs in $\mathcal{J}'$ are executed by their due date in the *earliest due date first* (EDD) schedule of these jobs starting at t.

For a consecutive subset of indices $I = \{i_0, \dots, i_1\}$, where $1 \leq i_0 \leq i_1 \leq D_{\#}$, define an integral vector $M(I)$ as follows. The entry $M(I)[x]$ equals $-\infty$ if none of the subsets of jobs in $\mathcal{J}_I$ with total processing time exactly x can be scheduled feasibly. Otherwise, $M(I)[x]$ equals the latest time t starting at which a subset of jobs in $\mathcal{J}_I$ with total processing time x can be scheduled feasibly. Applying the algorithm for $(\max, \min)$-skewed-convolutions given in [3], we get an $\tilde{O}\left({P_I}^{5/3}\right)$ time algorithm for computing $M(I)$, where $P_I = \sum_{J_i \in \mathcal{J}_I} p_i$. This implies an $\tilde{O}(P^{5/3})$ time algorithm for the MTPT problem.

In addition to the algorithm that uses $(\max, \min)$-skewed-convolutions, Bringmann *et al.* [2] gave a second algorithm for the MTPT problem. The running time of this algorithm is $\tilde{O}(P \cdot D_{\#})$. We use a version of this algorithm in our algorithm and for completeness we describe it in Algorithm 1.

Algorithm 1 The $\tilde{O}(P \cdot D_{\#})$ time algorithm

1: Let $d_1 < \cdots < d_{D_{\#}}$ denote the different due dates of jobs in $\mathcal{J}$.
2: **for** $i = 1, \ldots, D_{\#}$ **do**
3: Compute $X_i = \{p_j : J_j \in \mathcal{J}_i\}$
4: Compute $\mathcal{S}(X_i)$
5: Let $S_0 = \emptyset$.
6: **for** $i = 1, \ldots, D_{\#}$ **do** ▷ *compute the sumsets and exclude infeasible sums*
7: Compute $S_i = S_{i-1} \oplus \mathcal{S}(X_i)$.
8: Remove any $x \in S_i$ with $x > d_i$.
9: Return $P - x$, where x is the maximum value in $S_{D_{\#}}$.

3 The Algorithm

We define job bundles by coloring due dates in *red* and *blue*. The blue due dates are the bundled ones.

Initially, all due dates are uncolored. Choose some $\delta \in (0, 1)$. For each $k = 1, 2, \ldots D_{\#}$, color the due date d_k red if $\sum_{J_i \in \mathcal{J}_k} p_i > P^{1-\delta}$. To determine the bundles we repeat the following procedure until all due dates are colored.

Let m be the largest index for which due date d_m is not yet colored. Find the smallest $k \leq m$ that satisfies the following conditions.
Condition 1: None of the due dates $d_k, \ldots, d_m$ are colored red.
Condition 2: $\sum_{i=k}^{m} \sum_{J_j \in \mathcal{J}_i} p_j \leq P^{1-\delta}$.

Color all due dates $d_k, \ldots, d_m$ blue and "bundle" them into one group, denoted $B(k, m)$. We say that due date d_k is the *start* of the bundle and d_m is the *end* of the bundle.

Lemma 1. *The number of red due dates is* $O(P^\delta)$ *and the number of bundles is* $O(P^\delta)$.

Proof. Clearly, there can be at most P^δ due dates with $\sum_{J_i \in \mathcal{J}_k} p_i > P^{1-\delta}$. Consider a bundle $B(k, m)$. Due date d_1 can be the start of at most one bundle. Assume that $k > 1$. Since $k \leq m$ is the smallest index that satisfies the two conditions above, it is either true that d_{k-1} is red or $\sum_{i=k-1}^{m} \sum_{J_j \in \mathcal{J}_i} p_j > P^{1-\delta}$.

(i) Since there are at most P^δ red due dates, there can be at most only P^δ bundles $B(k, m)$ for which d_{k-1} is red.
(ii) Consider the sum $\sum_{i=k-1}^{m} \sum_{J_j \in \mathcal{J}_i} p_j$, for a bundle $B(k, m)$. Note that p_j of a job $J_j \in \mathcal{J}_{k-1} \cup \mathcal{J}_m$ may appear in at most one more sum that corresponds to a different bundle, while p_j of a job $J_j \in \bigcup_{i=k}^{m-1} \mathcal{J}_i$ cannot appear in any other such sum. Thus, the total of all sums cannot exceed $2P$. It follows that there are at most $2P^\delta$ bundles $B(k, m)$ for which $\sum_{i=k-1}^{m} \sum_{J_j \in \mathcal{J}_i} p_j > P^{1-\delta}$. ∎

Algorithm 2 called SOLVE($\mathcal{J}$), given below, follows the structure of Algorithm 1 with additional processing of entire bundles that avoids processing each due date in the bundles individually. Note that coloring the due dates takes $O(P)$ time.

We prove later that processing a bundle takes $\tilde{O}(P^{(1-\delta)\cdot\alpha}+P)$ time, where $\tilde{O}(P^{\alpha})$ is the running time of the algorithm needed for computing a (max, min)-skewed-convolution. Processing each red due date takes $\tilde{O}(P)$ time. Substituting $\delta = 1 - \frac{1}{\alpha}$ yields a total running time of $\tilde{O}(P \cdot P^{1-1/\alpha}) = \tilde{O}(P^{2-1/\alpha})$.

Algorithm 2 SOLVE($\mathcal{J}$)

1: Let $T = \{0\}$
2: For each red due date d_i, compute $X_i = \{p_j : e_j = d_i\}$ and $\mathcal{S}(X_i)$
3: **for** $i = 1, \ldots, D_{\#}$ **do**
4: **if** d_i is a red due date **then**
5: Compute $T = T \oplus \mathcal{S}(X_i)$
6: Remove any $x \in T$ with $x > d_i$
7: **else if** d_i is the end of some bundle $B(k, i)$ **then**
8: Let $I = \{k, \ldots, i\}$
9: Compute the vector $M(I)$
10: Let $S_i = \{x \in \{0, \ldots, P_I\} : M(I)[x] \neq -\infty\}$
11: **if** $d_k - P_I \geq 0$ **then**
12: Let $T = T \cup (\mathsf{pref}(T, d_k - P_I) \oplus S_i)$
13: Let M' be an integral vector of dimension d_k and initialize $M' = -\infty$
14: For each $x \in \mathsf{suff}(T, d_k - P_I)$, let $M'[x] = 0$
15: **for** $y = 0, \ldots, d_k - 1 + P_I$ **do**
16: Let $C[y] = \max_{x=0}^{y} \min\{M'[x], M(I)[y-x] - x\}$
17: Let $T_i = \{y \in \{0, \ldots, d_k - 1 + P_I\} : C[y] = 0\}$
18: Let $T = T \cup T_i$
19: Remove any $x \in T$ with $x > d_i$
20: Return $P - x$, where x is the maximum value in T

Theorem 1. *Algorithm* SOLVE*($\mathcal{J}$) returns the longest feasible schedule that starts at d_0.*

Proof. Consider iteration i of Algorithm SOLVE($\mathcal{J}$), for an index i such that either d_i is a red due date or d_i is the end of some bundle $B(k, i)$. To prove the theorem it suffices to prove that at the end of any such iteration i the set T consists of the processing times of all feasible schedules of jobs in $\bigcup_{j=1}^{i} \mathcal{J}_j$ that start at d_0. The proof is by induction. The basis is trivial since T is initialized to $\{0\}$. Consider such an iteration i and suppose that the claim holds for all iterations $i' < i$ such that either $d_{i'}$ is a red due date or $d_{i'}$ is the end of some bundle $B(k', i')$. We distinguish two cases.

Case 1: d_i is a red due date. By our induction hypothesis, at the start of iteration i the set T consists of the processing times of all feasible schedules of subsets of jobs in $\bigcup_{j=1}^{i-1} \mathcal{J}_j$ that start at d_0. Since iteration i sets $T = T \oplus \mathcal{S}(X_i)$ (Line 5), the claim follows.

Case 2: d_i is the end of some bundle $B(k, i)$. By our induction hypothesis, at the start of iteration i the set T consists of the processing times of all feasible schedules of subsets of jobs in $\bigcup_{j=1}^{k-1} \mathcal{J}_j$ that start at d_0. Let $I = \{k, \ldots, i\}$. The maximum length of any feasible schedule of subsets of jobs in $\mathcal{J}_I$ is P_I. Since the earliest due date of these jobs is d_k we are guaranteed that any such feasible schedule can start at any time up to (and including) $d_k - P_I$ (assuming that $d_k - P_I \geq 0$). By the definition of $M(I)$ the set $S_i = \{x \in \{0, \ldots, P_I\} : M(I)[x] \neq -\infty\}$ consists of the processing times of all feasible schedules of subsets of jobs in $\mathcal{J}_I$ (Line 10). $\mathsf{pref}(T, d_k - P_I)$ consists of the processing times of all feasible schedules of subsets of jobs in $\bigcup_{j=1}^{k-1} \mathcal{J}_j$ that start at d_0 and end at any time up to (and including) $d_k - P_I$. Since iteration i sets $T = T \cup (\mathsf{pref}(T, d_k - P_I) \oplus S_i)$ (Line 12), after this line T consists of all the feasible schedules of subsets of jobs in $\bigcup_{j=1}^{i} \mathcal{J}_j$ that start at d_0 and also satisfy the condition that the sum of the lengths of the jobs in $\bigcup_{j=1}^{k-1} \mathcal{J}_j$ that are scheduled is at most $d_k - P_I$.

The set T is still missing the lengths of all the feasible schedules of subsets of jobs in $\bigcup_{j=1}^{i} \mathcal{J}_j$ that start at d_0 in which the sum of the lengths of the jobs in $\bigcup_{j=1}^{k-1} \mathcal{J}_j$ exceeds $d_k - P_I$. These schedules are added to T in Lines 13–18 of Solve($\mathcal{J}$). Consider such a feasible schedule of length y in which the length of the jobs in $\bigcup_{j=1}^{k-1} \mathcal{J}_j$ is some $x > d_k - P_I$, which implies that $M'[x] = 0$ (Line 14). To complement the prefix of this schedule by a feasible schedule of a subset of jobs in J_I that starts at x and is of length $y - x$ we must have $M(I)[y - x] \geq x$ or $\min\{M'[x], M(I)[y - x] - x\} = 0$. Lines 15–16 of Solve($\mathcal{J}$) check if such a feasible schedule exists. ∎

Lemma 2. *The running time of algorithm* Solve($\mathcal{J}$) *is* $\tilde{O}(P^{(1-\delta)\cdot\alpha} + P) \cdot P^{\delta}$.

Proof. By Lemma 1 the number of iterations that are not vacuous is P^{δ}. It is not difficult to see that all operations other than the computation of the vectors $M(I)$, C, and T_i take $\tilde{O}(P)$ time. (Note that the initialization of the first $d_k - P_I$ entries of vector M' can be done "implicitly".) The vector $M(I)$ is computed as in [2] in $\tilde{O}({P_I}^{\alpha})$ time. The vector C is also computed via a (max, min)-skewed-convolution and thus its computation time is proportional to the sum of lengths of the vectors $M(I)$ and M' (up to logarithmic factors). Naively, this sum of lengths is $d_k + P_I$. However, since $M'[x] = -\infty$ for all $x \leq d_k - P_I$, we can ignore these entries and implement the convolution in $\tilde{O}({P_I}^{\alpha})$ time. Since $M'[x] = -\infty$, for all $x \leq d_k - P_I$, we have also $C[x] = -\infty$, and thus T_i can be computed in $O(P_I)$ time (Line 17). Recall that by the definition of bundles $P_I \leq P^{1-\delta}$. Thus, the lemma is proved. ∎

4 Conclusions

We have shown a $\tilde{O}(P^{7/5})$ time algorithm for tardy processing time minimization, an improvement over the previously known $\tilde{O}(P^{5/3})$ time algorithm. Improving this bound further is an interesting open problem. In general, by applying our job "bundling" technique we can achieve a $\tilde{O}(P^{2-1/\alpha})$ running time, where $\tilde{O}(P^{\alpha})$ is the running time of a (max, min)-skewed-convolution of vectors of size P. Since it is reasonable to assume that computing a (max, min)-skewed-convolution requires $\tilde{\Omega}(P^{3/2})$ time, our technique is unlikely to yield a $\tilde{o}(P^{4/3})$ running time. It will be interesting to see whether this running time barrier can be broken, and whether the MTPT problem can be solved without computing a (max, min)-skewed-convolution.

References

1. Aho, A.V., Hopcroft, J.E., Ullman, J.D.: The Design and Analysis of Computer Algorithms. Addison-Wesley, Boston (1974)
2. Bringmann, K., Fischer, N., Hermelin, D., Shabtay, D., Wellnitz, P.: Faster minimization of tardy processing time on a single machine. Algorithmica **84**, 1341–1356 (2022)
3. Klein, K.M., Polak, A., Rohwedder, L.: On minimizing tardy processing time, max-min skewed convolution, and triangular structured ILPs. In: Proceedings of the 2023 Annual ACM-SIAM Symposium on Discrete Algorithms (SODA), pp. 2947–2960. Society for Industrial and Applied Mathematics, USA (2023)
4. Koiliaris, K., Xu, C.: Faster pseudopolynomial time algorithms for subset sum. ACM Trans. Algorithms **15**(3), 1–20 (2019)
5. Kosaraju, S.R.: Efficient tree pattern matching. In: Proceedings of the 30th Annual Symposium on Foundations of Computer Science (FOCS), pp. 178–183. IEEE Computer Society, USA (1989)
6. Lawler, E.L., Moore, J.M.: A functional equation and its application to resource allocation and sequencing problems. Manage. Sci. **16**, 77–84 (1969)

Space-Efficient Functional Offline-Partially-Persistent Trees with Applications to Planar Point Location

Gerth Stølting Brodal(✉), Casper Moldrup Rysgaard, Jens Kristian Refsgaard Schou, and Rolf Svenning

Department of Computer Science, Aarhus University, Aarhus, Denmark
{gerth,rysgaard,jkrs,rolfsvenning}@cs.au.dk

Abstract. In 1989 Driscoll, Sarnak, Sleator, and Tarjan presented general space-efficient transformations for making ephemeral data structures persistent. The main contribution of this paper is to adapt this transformation to the functional model. We present a general transformation of an ephemeral, linked data structure into an offline, partially persistent, purely functional data structure with additive $\mathcal{O}(n \log n)$ construction time and $\mathcal{O}(n)$ space overhead; with n denoting the number of ephemeral updates. An application of our transformation allows the elegant slab-based algorithm for planar point location by Sarnak and Tarjan 1986 to be implemented space efficiently in the functional model using linear space.

Keywords: Data structures · Functional · Persistence · Point location

1 Introduction

The functional model has many well-known advantages such as modulation, shared resources, no side effects, and easier formal verification [4,14]. These advantages are given by restricting the model to only use functions and immutable data. As all data are immutable, side effects in functions are not possible, allowing modules to work independently of the context they are used in and reducing the complexity of formal verification [14].

In 1999 Okasaki [19] gave a seminal work on techniques for designing efficient (purely) functional data structures, and our result follows this line of research. Adapting existing data structures to the functional model is non-trivial since modifications are prohibited. However, it also means that updates do not destroy earlier versions of the data structure making functional data structures inherently persistent, but not necessarily space efficient. The focus of this paper is to adapt existing imperative techniques for persistence to the functional model in a space-efficient manner.

We introduce a purely functional framework that adapts classical tree structures to support offline partial persistence with an additive overhead. By offline partial persistence, we mean that all updates are made before queries. This

Work supported by Independent Research Fund Denmark, grant 9131-00113B.

P. Morin and S. Suri (Eds.): WADS 2023, LNCS 14079, pp. 644–659, 2023.
https://doi.org/10.1007/978-3-031-38906-1_43

Table 1. Previous and new results for planar point location, where † are expected bounds and * are results based on persistent data structures.

Reference	Construction	Query	Space	Model
David Kirkpatrick [17]	$\mathcal{O}(n \log n)$	$\mathcal{O}(\log n)$	$\mathcal{O}(n)$	Imperative
Seidel [26]	$\mathcal{O}(n \log n)$	$\mathcal{O}(\log n)$	$\mathcal{O}(n)$	Imperative†
Dobkin and Munro [11]	$\mathcal{O}(n \log n)$	$\mathcal{O}(\log^2 n)$	$\mathcal{O}(n \log n)$	Imperative*
Richard Cole [8]	$\mathcal{O}(n^2)$	$\mathcal{O}(\log n)$	$\mathcal{O}(n)$	Imperative*
Sarnak and Tarjan [24]	$\mathcal{O}(n \log n)$	$\mathcal{O}(\log n)$	$\mathcal{O}(n)$	Imperative*
Sarnak and Tarjan [24]	$\mathcal{O}(n \log n)$	$\mathcal{O}(\log n)$	$\mathcal{O}(n \log n)$	Functional*
New	$\mathcal{O}(n \log n)$	$\mathcal{O}(\log n)$	$\mathcal{O}(n)$	Functional*

restriction allows us to store update information without immediately being able to handle queries efficiently. A data structure that is not persistent, i.e., does not support queries and updates in previous versions, is said to be *ephemeral*. We show that by recording when an ephemeral tree structure is updated, we can build a query structure that can efficiently answer queries to previous versions of the structure. In the imperative paradigm, it is possible to efficiently interweave updates and queries [12,24], but in the functional paradigm, it incurs a multiplicative logarithmic space overhead, which we show how to circumvent, in the offline setting.

Planar point location is a classic computational geometry problem [13,17,24]. Given a planar straight-line graph with n edges (interchangeably line segments) and report the region containing a query point q. The task is to create a data structure that supports these queries while minimizing the construction time, query time, and space of the data structure. Sarnak and Tarjan [24] showed that the planar point location problem can be solved elegantly using *partially-persistent* sorted sets, that in the functional setting can be solved with balanced search trees using path copying resulting in a space usage of $\mathcal{O}(n \log n)$. As an application of our technique, we show how the algorithm of Sarnak and Tarjan can be implemented in the functional model to only use space $\mathcal{O}(n)$. For an overview of results for the planer point location problem see Table 1.

Below we state sufficient conditions for a functional or imperative data structure to be augmented with a functional support structure that supports offline partial persistence queries.

Definition 1 (`TUNA` conditions).

T: The data structure forms a rooted tree of constant degree d.

U: Updates create $\mathcal{O}(1)$ new edges and nodes.

N: No cycles are created by updates when considering the edges that have been created across all versions.

A: Attribute values of nodes are static, i.e., the information stored in a node is not changed after its insertion. This does not include fields pointing to the children.

This definition is not too restrictive, we show that binary search trees (BST), Treaps [3], Red/Black Trees [5], and Functional Random Access Arrays [18] all can be modified to satisfy Definition 1, without asymptotically significant overhead. In Sect. 5 we discuss how some of these requirements can be relaxed or generalized further.

Theorem 1. *For any ephemeral, linked data structure that satisfies the TUNA conditions, an equivalent, functional, offline-partially-persistent data structure preserving the asymptotic update and query times, can be created, with an additive construction overhead of time $\mathcal{O}(n \log n)$ and space $\mathcal{O}(n)$ for a series of n updates.*

The main idea behind Theorem 1 is to store the update information in a list, including edge insertion and deletion timestamps. After all the updates have been applied, a bottom-up topological sort produces a directed acyclic graph (DAG) that supports queries to any previous version of the data structure. This is essentially implementing the node copying approach of [24], while carefully avoiding the creation of cycles.

Building upon [24], Theorem 1 immediately implies a state-of-the-art functional planar point location solution, summarised in the following corollary.

Corollary 1. *There exists a purely functional solution to the planar point location problem with construction time $\mathcal{O}(n \log n)$, query time $\mathcal{O}(\log n)$, and space $\mathcal{O}(n)$.*

We implemented unbalanced binary search trees in the purely functional programming language Haskell and report on some experiments in Sect. 6.

1.1 Persistence

A data structure is said to be *persistent* if it is possible to query previous versions of it and *ephemeral* if it is only the current version that is available. A *partially-persistent* data structure is *persistent* and allows updates only to the latest version. The stronger notion of *full persistence* implies that any version can be both queried and updated. An update to a persistent data structure never changes an existing version, but instead creates a new version derived from the version the update is applied to. In this way, the different versions of the partial persistent structure form a version list, whereas full persistence forms a version tree. General transformations to make data structures persistent were studied by Driscoll et al. [12] and Overmars [21,22]. In this paper, we focus on *offline* partial persistence for linked data structures, where all updates are performed before all queries, which was also explored in [11].

A naive idea to achieve partial persistence is to store a copy of every previous version. If the underlying structure is a list, then this approach generates an overhead of $\Omega(n^2)$ space for n insertions into an initially empty list. To improve upon this, the crucial observation is that when structures have a large overlap between updates, it is possible to reuse large parts of the previous versions and greatly reduce space usage. To achieve linear space, the notions of *fat nodes* and *timestamps* were introduced by Driscoll et al. [12], where each pointer field in a node is replaced by a list of pointers and each pointer in the list has an associated timestamp, denoting when the pointer was updated. Specifically, for a binary search tree, each node will, instead of containing a pointer to the `left` and `right` subtree, contain a list of timestamped `left` pointers and a list of timestamped `right` pointers. An update to a given pointer now adds a new pointer to the pointer list with the current timestamp, resulting in $\mathcal{O}(1)$ space overhead per update. As the pointer in the version is the last in the list, there is no overhead in finding the active pointer in the current version. To locate the correct pointer at a given older timestamp, a binary search on the list of pointers can be performed, which then imposes a multiplicative $\mathcal{O}(\log n)$ overhead on queries.

In [24] the notion of *path copying* was introduced for BSTs, where all nodes on the path to the node being updated are copied. Any existing pointer along the path can then point freely to parts of the old structure. For BSTs, this has a space overhead of the length of the path, i.e., for balanced BSTs an $\mathcal{O}(\log n)$ space overhead per update. It does however impose no overhead on the query time, apart from initially finding the correct root to query.

Thus, the fat node technique has a query time overhead, whereas path copying has a space overhead. By combining these two techniques, it is possible to have no overhead on query time and space. In [12,24] the authors achieve this by introducing the *node copying* technique for partially-persistent general pointer-based data structures. Here nodes are allowed to hold a constant number of additional time-stamped pointers. When some operation requires a node to update a pointer, the timestamp of the old pointer is updated to end at the current time. We denote a pointer that has ended as *expired*. If the pointer is replaced by a new pointer, then the new pointer is placed in one of the free extra pointer slots of the node, with a timestamp starting at the current time. If there is no free pointer slot *node copying* is performed, where the non-expired pointers and the new pointer are copied to a new node. Any pointer in another node to the copied node at the current time must be split, which may cause node copying to cascade up the structure. However, an amortization argument [24] shows that this technique only has an additive $\mathcal{O}(n)$ overhead in space, when the indegree of every node in the underlying structure is constant. Finally, as the number of pointers in each node is constant the overhead on the query time is also constant.

1.2 Persistence and the Functional Model

The functional programming paradigm is well suited for persistence as stated by Okasaki: "*A distinctive property of functional data structures is that they are always persistent*" [19]. In purely functional programming, there are no side effects and variables are immutable, meaning that any modifications to a structure S are obtained by creating a new structure S' without altering S. In this new structure S', substructures may be references to (immutable) old substructures from S.

Purely functional data structures are consequently particularly interesting when persistence is critical, and it is natural that they are less efficient than imperative data structures, since they inherently solve a more general problem. We note that there has been some work in describing to what degree mutability increases the capabilities (efficiency) of a language, and Pippenger [23] gave an example of a problem with a logarithmic-factor separation under certain conditions. However, for many data structure problems, purely functional solutions have been developed that match their imperative counterparts with only constant overhead. Examples include optimal *confluently* persistent deques which were developed over a number of papers [7,9,16] and optimal priority queues [6].

1.3 Functional vs. Space Efficient Imperative Persistence

The fat node and node copying persistence techniques mentioned in Sect. 1.1 rely upon the imperative paradigm's ability to modify pointers in the nodes in a graph structure where nodes can have multiple ingoing edges. As the functional model cannot mutate pointers, directly translating these solutions leads to significant overhead in the update time, as, even with path copying, all ancestors of an updated node must be remade to point to the newly created node(s). For this reason, we focus our attention on data structures with a tree structure where ancestors appear on a single path to the root.

1.4 Planar Point Location

Dobkin and Lipton [10] solved the planar point location problem by drawing vertical lines through every node resulting in vertical slabs. For every slab, the line segments spanning the slab are stored in a BST. This method allows efficient queries by performing a binary search horizontally for the slab, and then a binary search vertically in the slab for the region. This gives an overall query time of $\mathcal{O}(\log n)$. The drawback of this method is that each line could potentially be stored in almost every slab, resulting in $\Theta(n^2)$ space. A number of different results [8,13,17,24], show that the space can be reduced to $\mathcal{O}(n)$ (non-functional) without affecting the query time. The solution by Cole [8] is particularly interesting as it exploits that neighboring slabs are very similar, meaning that the problem can be reduced to creating persistent, sorted sets. This observation is vital to the work on persistent search trees by Sarnak and Tarjan [24]. On the other hand, the approach by Kirkpatrick [17] is completely different and is based on repeatedly triangulating the graph and removing a constant fraction of the nodes with degree at most 11. Then a DAG is created bottom-up based on the overlap between two consecutive triangulations. Since the DAG is created bottom-up, similarly to our approach described in Sect. 3.2, we conjecture that this approach could be adapted to the purely functional model and leave it as an open problem.

2 Initial Analysis of Binary Search Trees

In this section, we introduce ephemeral (i.e., non-persistent) unbalanced binary search trees (BSTs) and illustrate how the fat node technique can be adapted to the functional paradigm. The main obstacle in making the adaptation is to avoid cycles between nodes. We solve this by making new copies of nodes that should be moved above their parent. In Sect. 5 these techniques are extended to cover balanced search trees.

2.1 Ephemeral Binary Search Trees

Any BST node is either an empty leaf or a node containing an element and two pointers to a left and right sub-tree, which are in turn also BSTs. Forthwith, these pointers will be denoted as *edges*. Likewise, let T and T' denote BSTs over a totally ordered set of elements X and let $x \in X$ denote an element. BSTs are ordered such that all elements in the left subtree are smaller than the element in the node, and all elements in the right subtree are larger. BSTs support many operations; we focus on the following three basic operations:

- `Insert`(T, x): Insert x into T and return the resulting tree T'.
- `Delete`(T, x): Delete x from T and return the resulting tree T'.
- `Search`(T, x): Return the smallest element x' in T, such that $x \leq x'$.

All operations can be implemented in time linear in the height of the tree. We call operations that modify the data structure *updates* (`Insert` and `Delete`). We call operations that query the data structure without changing it *queries* (`Search`). When constructing the data structure, we consider a sequence of n updates $(u_1, \ldots, u_n)$ and for version $0 \leq t \leq n$ updates $u_1, \ldots, u_t$ have been applied.

2.2 Fat Node Binary Search Trees

In this subsection, we describe how to adapt imperative unbalanced BSTs to adhere to the `TUNA` conditions (Definition 1), using fat nodes. Most importantly, the `Delete` update is changed slightly from the classic behavior since it can cause nodes to be reordered in the tree.

When performing `Insert`, assuming that x is not present in T, a path to the correct leaf position is found, and a new node, containing x is created at the position of that leaf. The difference between the old and the new tree is a single edge at the bottom of the tree. By adding *creation timestamps* on the edges to represent creation time, queries can detect if a particular edge should be considered to exist in a specific version or not.

When performing `Delete`, if the deleted element x is at the bottom of the tree, then, in effect, the edge e from its parent p to the deleted node ceases to exist. We record this by adding an *ending timestamp* to the edge. If the deleted element x is in some internal node, see Fig. 1 (Left), then a predecessor

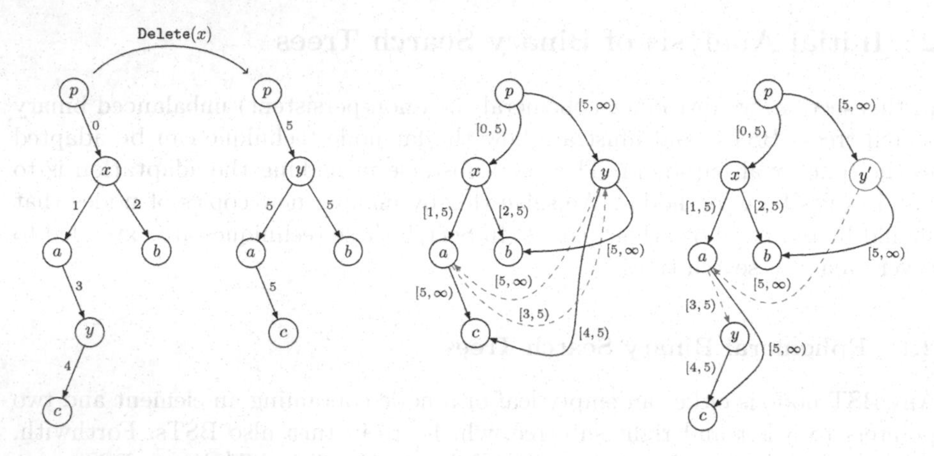

Fig. 1. We let a single number above an edge denote its creation time. In the graphs, the interval above an edge is the living span of the edge. Recall that a persistent search tree query is a timestamp/version and a value. Left: a deletion in a BST. Middle: the resulting graph contains the dashed cycle between y and a. Right: by creating a new copy y' of y the resulting graph remains a DAG and is TUNA compliant.

or successor y (depending on implementation) in the subtree rooted at x is found and moved to the deleted spot. This introduces up to four new edges.

When transforming the BST into the functional model, the predecessor cannot be directly moved and reused, as it might be moved above its parent, which would create a cycle in the structure when also considering expired edges, as seen in Fig. 1 (Middle). We avoid creating cycles by replacing x with a copy of its predecessor which breaks the cycle as seen in Fig. 1 (Right). However, it remains a crucial issue that b (and c) still have two parents, when also considering the expired edges, which cannot be updated efficiently in the functional model (see Sect. 1.3).

3 Freezer and Query Structure Construction

In this section we present our main contribution: the concept of a *Freezer*, which stores update information as it comes in, and how to use it to build a query DAG after all updates have occurred, using fat nodes and node copying. We give an amortized analysis argument that the total number of node copies is linear in the number of updates and from the proof we deduce that the TUNA conditions 1 are sufficient to prove Theorem 1.

3.1 The Freezer

The underlying ephemeral structure forms a tree, but when performing updates using the TUNA-compliant fat node method (see Fig. 3) to get persistence, then the data structure forms a DAG. This is problematic since in the functional model a DAG cannot be maintained efficiently using path copying, as any node in the data structure which can reach an updated node would also have to be copied to reflect the change. Thus, we store all edges not present in the most recent version of the data structure separately in a list which we call the *freezer*. That is, only the most recent version of the data structure is explicitly maintained. This is sufficient for partial persistence since updates are only allowed in the most recent version. Finally, after all the updates have been applied, the DAG is built bottom-up. For this reason, persistence is restricted to offline.

To store the edges in the freezer, a unique *id* is assigned to each node and each edge in the freezer is stored as a 5-tuple $(id_{\text{from}}, id_{\text{to}}, \mathit{field}, t_{\text{start}}, t_{\text{end}})$. The ids are used to identify the nodes the edge connects, the field denotes which field of the node with id id_{from} the edge originates from (in the BST this is either `left` or `right`), and the time stamps t_{start} and t_{end} denote the *living span* of the edge as the half-open interval $[t_{\text{start}}, t_{\text{end}})$.

The freezer in addition stores which node is the root at any given time, and a map from ids to the value contained in the node with the given id. Further, note that storing the deletion time of an edge can be omitted and instead be read from the starting time of the next edge in the corresponding field of the same node, with the modification that empty fields have an edge to a special `Nil` leaf.

3.2 Offline Construction of the Fat Node Query DAG

In this section, we describe how to construct the fat node DAG structure from the expired edges in the freezer and the final non-expired structure.

Any edge that ends in the freezer must have been created in some version of the tree as the result of an update operation. An `Insert` operation creates one new edge, and a `Delete` operation creates at most four new edges. Thus, after all n updates, the freezer contains $\mathcal{O}(n)$ edges. Similarly, each `Insert` and `Delete` creates at most one new node, so the number of nodes is $\mathcal{O}(n)$.

Using Kahn's algorithm [15], which topologically sorts the nodes by repeatedly extracting nodes with outdegree zero, we build the DAG bottom-up. The while loop of the imperative algorithm is replaced by a recursive function in the functional algorithm, for each iteration calling with the new values needed. As mentioned, by copying nodes when they were moved around by the updates we avoid introducing cycles (see Figs. 1 and 3), which is required for Kahn's algorithm. By having a map from node ids to the values contained in the nodes, it is possible to explicitly build the DAG over the edges of the freezer. This creates the DAG of fat nodes.

The construction time of, a non-functional implementation of, Kahn's algorithm is linear in the number of nodes and edges. This however relies on being able to effectively fetch the ingoing and outgoing edges of nodes, and reduce the outdegree of nodes in time $\mathcal{O}(1)$. The construction relies on efficient maps from ids to nodes. As random access is not part of the functional model, maps with $\mathcal{O}(1)$ lookup time and update times do not exist. We instead use balanced trees which introduces an overhead of $\mathcal{O}(\log n)$ for each of the operations, yielding a DAG construction time of $\mathcal{O}(n \log n)$. The space usage remains $\mathcal{O}(n)$.

3.3 Bounding Node Outdegree of the Query DAG

Having arbitrary outdegree of fat nodes affects the query time, as stated in Sect. 1.1. However, by limiting the number of extra pointers in each node, limited node copying similar to [24] can be implemented to remove the query overhead, while still maintaining linear space. The difference here is that in [24] the copy is performed during the update phase as soon as there are too many pointers in a node. We restrict ourselves to the offline setting and as such do not need to be able to handle queries before all updates have been applied, this means that we can allow nodes to become arbitrarily fat in the update phase.

Let d be the degree of the underlying ephemeral structure. After the update phase, we handle the high degree by recursively splitting the fat nodes into multiple nodes, each with degrees $\mathcal{O}(d)$ by interleaving the node splitting idea with Kahn's algorithm. As discussed in Sect. 3.2 we visit the nodes of the freezer bottom-up in topological order. Recall that the freezer contains edges, so nodes are inferred. When visiting a node v, that has an outdegree larger than $d + e$, for a parameter $e = \mathcal{O}(d)$ indicating the extra edges we allow every node to hold, we split it into a left node v_l to represent "the past" and a right node v_r to represent "the future", essentially performing node copying. Note that some edges will be active in both of these nodes and thus are duplicated, similar to regular node copying. We perform the split such that the outdegree of the left node is at most $d + e$, and that the left and right nodes in total represent the original node. If the outdegree of v_r is larger than $d + e$ we recursively split it until v has been split into a number of nodes each of outdegree at most $d + e$. By requiring $1 \leq e \leq cd$, for some constant c, we ensure that the query time is proportional to that of the underlying ephemeral structure since the new nodes will contain $\Theta(d)$ edges. We call this procedure *node copying* and the following lemma shows that the space remains linear in the number of updates n.

Lemma 1. *The number of nodes introduced by* node copying *is* $\mathcal{O}(n)$ *when* $1 \leq e \leq cd$ *for some constant* c.

Proof. We define the potential function

$$\Phi = \sum_v \max\{0, O_v - (d + e)\},$$

where O_v is the number of outgoing edges from the node v, and the terms in the sum indicate the number of outgoing edges above the threshold $d + e$. Observe that the potential is nonnegative and that $\Phi = \mathcal{O}(n)$ when initializing *node copying* due to property U.

We now analyze how the potential changes when we split a node v, which must have $O_v > d + e$, into a left node v_l and a right node v_r. We consider the edges (consisting of a start and end timestamp) in nondecreasing order by

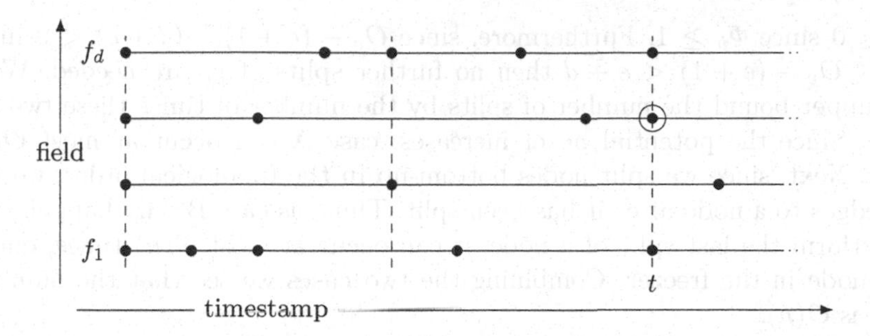

Fig. 2. Illustration of the fields of a node over time. The number of fields is $d = 4$ and $e = 4$. Each horizontal line represents one of the fields, with the dots denoting times when the value of the field change. In other words, the segments between dots represent time intervals where the field is unaltered. The vertical dashed lines are the times when the node is split. The dot with the circle around it is the start of the $(d+e+1)$th edge (when counting only edges after the previous split) introducing the split at time t. This results in $s_1 = 1$, $s_2 = 2$ and $a = 1$.

their start timestamp. The split is performed at the $(d + e + 1)$th smallest start timestamp t. We include all edges with start timestamps strictly less than t in v_l. In v_r we include all edges with an end timestamp strictly larger than t.

Expanding on this, at time t we let s_1 denote the number of edges with start timestamp t among the first $d+e$ edges, and let s_2 be the number of edges with start timestamp t not among the first $d+e$ edges. Together s_1+s_2 is the number of edges changing at time t, or equivalently the number of fields updated at time t. Furthermore, we call an edge *active* if its living span contains the splitting time, that is $t_{start} < t < t_{end}$, and let a denote the number of active edges at time t. See Fig. 2 for an example. Since all nodes have degree d exactly[1], we have $d = s_1+s_2+a$, where $s_2 \geq 1$ from the edge where we perform the split. The number of outgoing edges in v_l is $O_{v_l} = d+e-s_1 \leq d+e$. Likewise, the number of outgoing edges in v_r is $O_{v_r} = O_v - O_{v_l} + a = O_v - d - e + s_1 + d - s_1 - s_2 \leq O_v - (e+1)$, where the inequality follows from $s_2 \geq 1$.

Finally, at time t there is at most one incoming edge to v from a parent v_p, since the underlying structure forms a tree. This edge must potentially be split in two which increases the outdegree of the parent by one. The difference in potential before and after this split is then at most:

$$\begin{aligned}\Delta\Phi &= \Delta\Phi_{v_p} + \Phi_{v_l} + \Phi_{v_r} - \Phi_v \\ &\leq 1 + 0 + \max\{0, O_v - (e+1) - (d+e)\} - (O_v - (d+e)).\end{aligned}$$

Now there are two cases depending on which term is larger in the maximum. We first look at the case when $0 \leq O_v - (e+1) - (d+e)$. We call this case A and here we get $\Delta\Phi \leq -e$. The other case we call case B and here we get

[1] Each node has exactly d fields and each field always holds an edge to either another node or to `Nil`.

$\Delta\Phi \leq 0$ since $\Phi_v \geq 1$. Furthermore, since $O_v - (e+1) - (d+e) < 0$ implies $O_{v_r} \leq O_v - (e+1) < e+d$ then no further splits of v_r are needed. We can now upper bound the number of splits by the number of times these two cases occur. Since the potential never increases, case A can occur at most $\mathcal{O}(n/e)$ times. Next, since we split nodes bottom-up in the topological order, we never add edges to a node after it has been split. Thus, as case B only happens when we perform the last split of a node, it can occur at most $\mathcal{O}(n)$ times, once for each node in the freezer. Combining the two cases we see that the number of splits is $\mathcal{O}(n)$. □

4 Proof of Theorem 1

Let D be a data structure that satisfies the TUNA conditions (Definition 1). We identify each node by a unique id and store timestamps with each edge denoting their living span. The freezer, see Sect. 3.1, stores edges on the form $(id_{\text{from}}, id_{\text{to}}, \mathit{field}, t_{\text{start}}, t_{\text{end}})$, where field denotes the outgoing field of the edge from the node with id id_{from}, and the edge was live in the version interval $[t_{\text{start}}, t_{\text{end}})$. The freezer further records what id the root of the data structure has for each time step, as well as what static value is associated with each id.

Applying an update to the structure can be done without saving the old structure, as condition A ensures that all nodes still present contain the same value. Furthermore, condition T ensures that the outgoing number of edges is d, leading to a constant number of updates to the freezer for each node updated, resulting in updates having unaltered asymptotic running time.

After applying all updates, recording the relevant information in the freezer, and obtaining tree T, condition U ensures that the freezer contains $\mathcal{O}(n)$ edges. For ease of argument, we then enter all edges from T into the freezer. Note that the number of elements in the freezer remains $\mathcal{O}(n)$.

To build the query DAG, apply a modified version of Kahn's algorithm [15] as described in Sect. 3.2, to perform a bottom-up topological sort of the graph induced by the edges in the freezer in time $\mathcal{O}(n \log n)$. Kahn's algorithm requires that the graph is acyclic, which condition N ensures. First, give each node, defined by id, the value stored in the freezer and the edges with matching id_{from}. Second, we employ the fat node technique [12] by allowing nodes to have $d+e$ edges. A fat node can overflow, if it gets more than $d+e$ edges. When an overflow is encountered the node is split into two nodes with the same value and the same parent but only a subset of the edges. Lemma 1 guarantees that splits only cause limited cascading while keeping nodes of degree $\mathcal{O}(d)$ and thus within a constant factor of their degree in the ephemeral structure as promised by condition T.

We now have a list of roots, sorted by time, that can be used to access previous versions of D. For queries, we assume the search starts from the relevant root. Otherwise, the relevant root can be found in $\mathcal{O}(\log r)$ time, where r is the total number of roots. Timestamps on edges represent the versions in which the edge was present, so it is easy to adapt queries to the DAG to only take into account the relevant edges among the $\Theta(d)$ present edges in the node, leading to the same asymptotic running time for queries. □

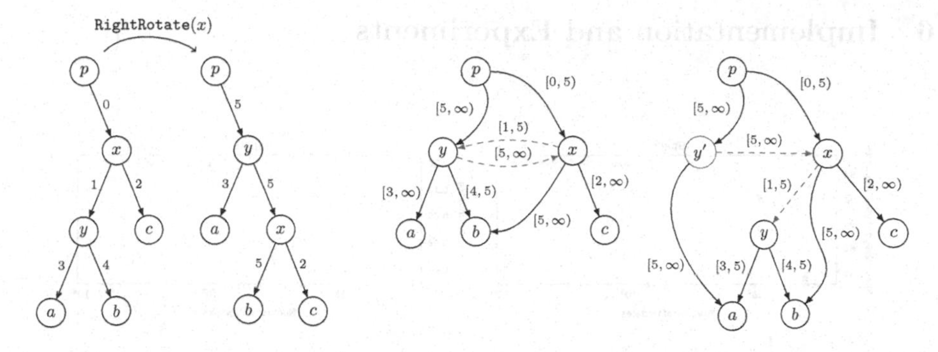

Fig. 3. Left: a rotation in a BST. Middle: the resulting graph contains the dashed cycle between x and y. Right: by creating a new copy y' of y the resulting graph remains a DAG and is TUNA compliant. We use the same notation as in Fig. 1.

5 Further Applications

Okasaki [18] introduces functional *random access arrays*, achieving, for an array of size n, worst-case lookup and update time $\mathcal{O}(\min\{i, \log n\})$, where i is the index of the queried element. This data structure easily satisfies the TUNA conditions, and can therefore be made offline partially persistent with only a constant factor space overhead.

Condition U of Definition 1 ensures that each of the n updates makes $\mathcal{O}(1)$ edges, which then totals to $\mathcal{O}(n)$ edges in the final structure. In the analysis of the space complexity, it is however not important exactly how many edges each update adds, and in fact, the more general property holds that the space usage is linear in the number of edges created by the ephemeral structure. Condition U can therefore be relaxed to each update producing for example amortized or expected $\mathcal{O}(1)$ new edges. This allows for balanced *Red-Black trees* [5] to fulfill the TUNA conditions even if the colors should be stored for persistent queries, as they make amortized $\mathcal{O}(1)$ color changes for each update [27]. The simpler functional implementation of Red-Black trees by Okasaki [20] makes amortized $\mathcal{O}(1)$ changes per insertion by a similar argument. See Fig. 3 on how rotations can be handled. Similarly, *Treaps* [3] fulfill the relaxed TUNA conditions, as each update makes expected $\mathcal{O}(1)$ rotations and therefore $\mathcal{O}(1)$ new edges.

Condition A ensures that all nodes can be reused and that storing edges is sufficient to produce the query DAG. The underlying tree is always represented explicitly, and as updates always operate on the current structure, it is possible to relax condition A, to allow for dynamic update information in each node. This information can be altered during the update phase and is not needed to produce the query DAG. Recall that our structure imposes additive $\mathcal{O}(n \log n)$ construction time independent of original update complexity. This allows for *AVL-trees* [1] to fulfill the TUNA conditions, as the balance value (the height of the subtree rooted at the node) is only used for balancing during the updates, and updates only perform amortized $\mathcal{O}(1)$ rotations [2].

6 Implementation and Experiments

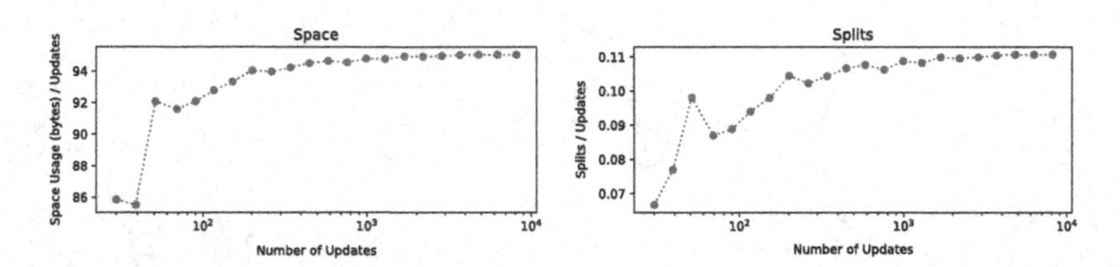

Fig. 4. Space usage experiments. Elements 1 to n are inserted into a persistent BST in order to create a path. Element $n+1$ was then inserted and deleted n times, to introduce cascading node splits up the path, for a total of $3n$ updates.

The construction described in this chapter has been implemented and tested in Haskell[2]. Experiments were performed on WSL Ubuntu 20.04.6 on Windows 10.0.19044, with Processor 11th Gen Intel(R) Core(TM) i7-1165G7 @ 2.80 GHz, 4 Core(s), 8 Logical Processor(s), 16 GB RAM, running `ghc` haskell compiler version 8.6.5, without any compiler flags. Each runtime measurement is the average of a constant number of runs on the same input. A random input is generated by providing the algorithm with a seed to a pseudo-random number generator.

For the following, let the version of a data structure made in our offline-partially-persistent framework be denoted as *persistent* and the version made in regular Haskell without necessarily saving the root of the structure be denoted *ephemeral*. The implementation uses time intervals for the edges in the freezer, and not only a start timestamp. Edges in the current structure are therefore not recorded in the freezer before they are removed from the live structure. Therefore, if any larger part of the structure is to be removed, it must be done so recursively, to correctly enter all of the edges into the freezer. Note that this does not alter the amortized running time of updates.

To test the *correctness* of the implementation, we apply various deterministic and random sequences of updates to both an ephemeral and a persistent unbalanced binary search tree, saving the roots of the ephemeral versions. We then construct the query DAG as described in Sect. 3.2 based on the persistent version. Then, for each timestamp, we tested if both versions produced the same tree. The test did not reveal any errors.

To measure space usage of a data structure, Haskell provides the function `recursiveSizeNF` which recursively measures the size of the object in bytes, i.e., traverses the whole pointer structure on the heap. Note that parts of the structure reachable from multiple places only are counted once. The space of the persistent data structure is measured after building the query DAG. Note that due to [25], a sequence of uniformly random insertions in an unbalanced binary search tree produces expected depth $\Theta(\log n)$.

[2] Available at https://github.com/Crowton/Persistent-Functional-Trees.

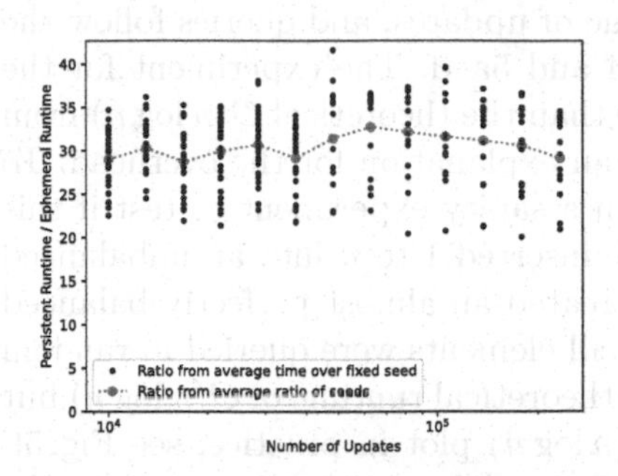

(a) Inserting elements 1 to n in random order.

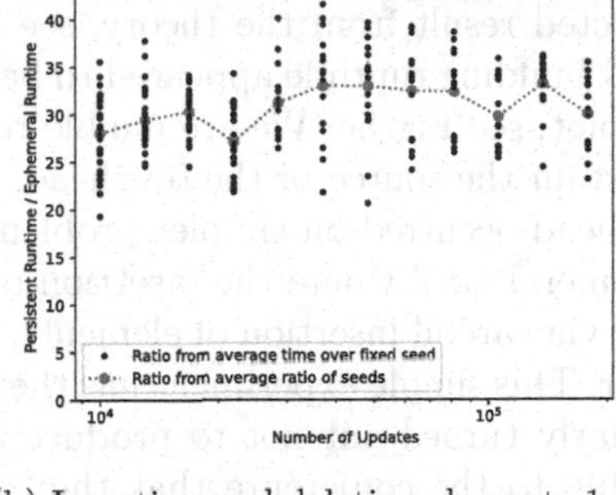

(b) Inserting and deleting elements 1 to n in random order.

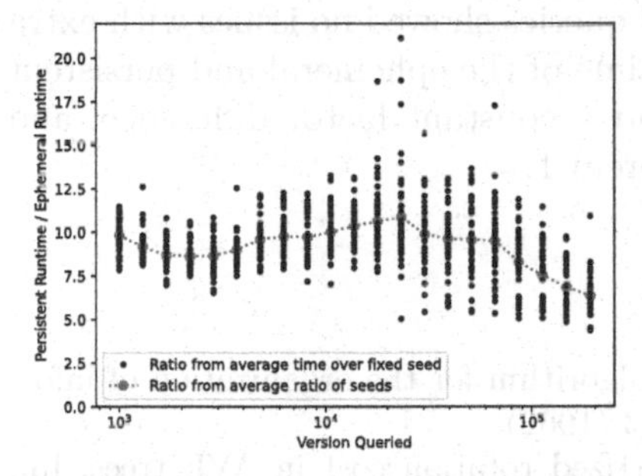

(c) Relative running time over querying all elements of a persistent BST in order at different timestamps. The tree is created by inserting elements 1 to 200000 in random order.

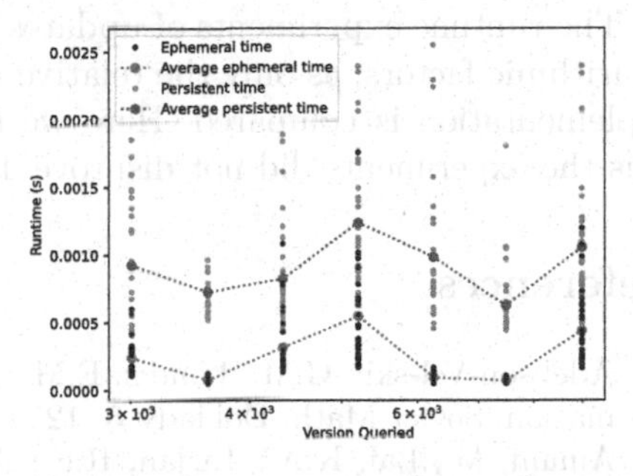

(d) Elements 1 to n are inserted to make a path. Element $n+1$ is inserted and deleted n times, for $n = 3000$. Time measured for querying element $n+1$ at different timestamps.

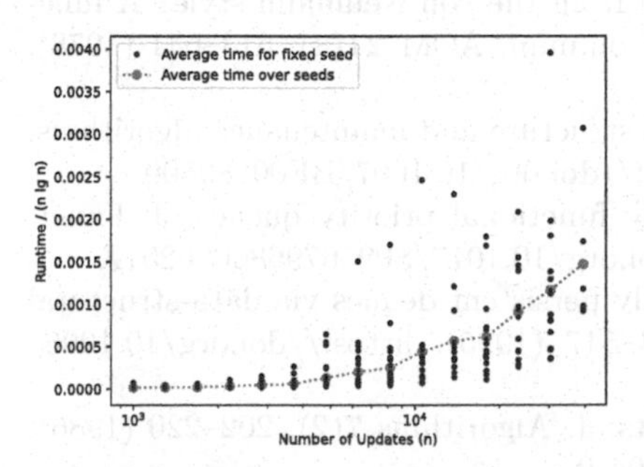

(e) DAG building running time experiment. Elements 1 to n were inserted and deleted from a persistent BST in random order. Time was measured on the DAG building alone.

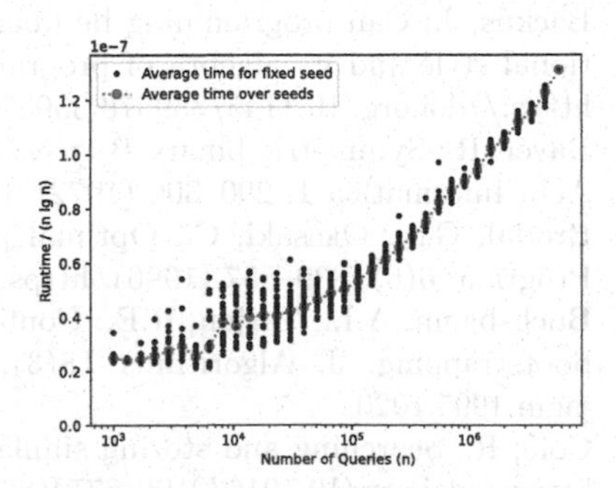

(f) Sanity experiment. Querying all elements of an emphemeral perfectly balanced BST of size n in random order.

Fig. 5. Running time experiments.

The experiments for space usage, runtime of updates, and queries follow the expected result from the theory, see Figs. 4 and 5a–d. The experiment for the DAG building runtime appeared to be more than the theoretical $\mathcal{O}(n \log n)$ from the plot, see Fig. 5e. We are unable to find an explanation for the overhead. To ascertain the source of the overhead, we ran a sanity experiment to test if this overhead occurred on simpler problems. We inserted 1 to n into an unbalanced ephemeral BST where the insertion order created an almost perfectly balanced tree, via careful insertion of elements. Then all elements were queried in random order. This simple experiment has the clean theoretical runtime of $\mathcal{O}(n \log n)$ but similarly turned out not to produce an $\mathcal{O}(n \log n)$ plot in practice, see Fig. 5f, leading to the conjecture that the extra overhead in runtime is not from the program itself, but the Haskell compiler, specific implementations of underlying structures, and/or the environment the code is executed in.

The runtime experiments of updates and queries showed no issues with extra logarithmic factors, as only the relative runtime of the ephemeral and persistent implementation is compared. Here we found a constant factor difference, and thus the experiments did not disprove Theorem 1.

References

1. Adel'son-Vel-skii, G.M., Landis, E.M.: An algorithm for the organization of information. Soviet Math. Doklady **3**, 1259–1263 (1962)
2. Amani, M., Lai, K.A., Tarjan, R.E.: Amortized rotation cost in AVL trees. Inf. Process. Lett. **116**(5), 327–330 (2016). https://doi.org/10.1016/j.ipl.2015.12.009
3. Aragon, C.R., Seidel, R.: Randomized search trees. In: 30th Annual Symposium on Foundations of Computer Science, Research Triangle Park, North Carolina, USA, 30 October–1 November 1989, pp. 540–545. IEEE Computer Society (1989). https://doi.org/10.1109/SFCS.1989.63531
4. Backus, J.: Can programming be liberated from the von Neumann style? A functional style and its algebra of programs. Commun. ACM **21**(8), 613–641 (1978). https://doi.org/10.1145/359576.359579
5. Bayer, R.: Symmetric binary B-trees: data structure and maintenance algorithms. Acta Informatica **1**, 290–306 (1972). https://doi.org/10.1007/BF00289509
6. Brodal, G.S., Okasaki, C.: Optimal purely functional priority queues. J. Funct. Program. **6**(6), 839–857 (1996). https://doi.org/10.1017/S095679680000201X
7. Buchsbaum, A.L., Tarjan, R.E.: Confluently persistent deques via data-structural bootstrapping. J. Algorithms **18**(3), 513–547 (1995). https://doi.org/10.1006/jagm.1995.1020
8. Cole, R.: Searching and storing similar lists. J. Algorithms **7**(2), 202–220 (1986). https://doi.org/10.1016/0196-6774(86)90004-0
9. Demaine, E.D., Langerman, S., Price, E.: Confluently persistent tries for efficient version control. In: Gudmundsson, J. (ed.) SWAT 2008. LNCS, vol. 5124, pp. 160–172. Springer, Heidelberg (2008). https://doi.org/10.1007/978-3-540-69903-3_16
10. Dobkin, D., Lipton, R.J.: Multidimensional searching problems. SIAM J. Comput. **5**(2), 181–186 (1976). https://doi.org/10.1137/0205015
11. Dobkin, D.P., Munro, J.I.: Efficient uses of the past. In: Proceedings of the 21st Annual Symposium on Foundations of Computer Science, SFCS 1980, pp. 200–206. IEEE Computer Society, USA (1980). https://doi.org/10.1109/SFCS.1980.18

12. Driscoll, J.R., Sarnak, N., Sleator, D.D., Tarjan, R.E.: Making data structures persistent. In: Proceedings of the Eighteenth Annual ACM Symposium on Theory of Computing, STOC 1986, pp. 109–121. Association for Computing Machinery, New York (1986). https://doi.org/10.1145/12130.12142
13. Edelsbrunner, H., Guibas, L.J., Stolfi, J.: Optimal point location in a monotone subdivision. SIAM J. Comput. **15**(2), 317–340 (1986). https://doi.org/10.1137/0215023
14. Hughes, J.: Why functional programming matters. In: Turner, D.A. (ed.) Research Topics in Functional Programming. The UT Year of Programming Series, pp. 17–42. Addison-Wesley (1990). https://doi.org/10.1093/comjnl/32.2.98
15. Kahn, A.B.: Topological sorting of large networks. Commun. ACM **5**(11), 558–562 (1962). https://doi.org/10.1145/368996.369025
16. Kaplan, H., Tarjan, R.E.: Persistent lists with catenation via recursive slow-down. In: Proceedings of the Twenty-Seventh Annual ACM Symposium on Theory of Computing, pp. 93–102 (1995). https://doi.org/10.1145/225058.225090
17. Kirkpatrick, D.: Optimal search in planar subdivisions. SIAM J. Comput. **12**(1), 28–35 (1983). https://doi.org/10.1137/0212002
18. Okasaki, C.: Purely functional random-access lists. In: Proceedings of the Seventh International Conference on Functional Programming Languages and Computer Architecture, FPCA 1995, pp. 86–95. Association for Computing Machinery, New York (1995). https://doi.org/10.1145/224164.224187
19. Okasaki, C.: Purely Functional Data Structures. Cambridge University Press, Cambridge (1999)
20. Okasaki, C.: Red-black trees in a functional setting. J. Funct. Program. **9**(4), 471–477 (1999). https://doi.org/10.1017/s0956796899003494
21. Overmars, M.H.: Searching in the past II: general transforms. Technical report. RUU (1981)
22. Overmars, M.H.: The Design of Dynamic Data Structures. LNCS, vol. 156. Springer, Heidelberg (1983). https://doi.org/10.1007/BFb0014927
23. Pippenger, N.: Pure versus impure lisp. In: Proceedings of the 23rd ACM SIGPLAN-SIGACT Symposium on Principles of Programming Languages, POPL 1996, pp. 104–109. Association for Computing Machinery, New York (1996). https://doi.org/10.1145/237721.237741
24. Sarnak, N., Tarjan, R.E.: Planar point location using persistent search trees. Commun. ACM **29**(7), 669–679 (1986). https://doi.org/10.1145/6138.6151
25. Sedgewick, R., Flajolet, P.: An Introduction to the Analysis of Algorithms. Addison-Wesley Longman Publishing Co., Inc., Boston (1996)
26. Seidel, R.: A simple and fast incremental randomized algorithm for computing trapezoidal decompositions and for triangulating polygons. Comput. Geom. **1**(1), 51–64 (1991). https://doi.org/10.1016/0925-7721(91)90012-4
27. Tarjan, R.E.: Amortized computational complexity. SIAM J. Algebraic Discret. Methods **6**(2), 306–318 (1985). https://doi.org/10.1137/0606031

Approximating the Discrete Center Line Segment in Linear Time

Joachim Gudmundsson and Yuan Sha(✉)

The University of Sydney, Camperdown, NSW 2006, Australia
{joachim.gudmundsson,ysha3185}@sydney.edu.au

Abstract. Let P be a set of n points in the plane. The discrete center line segment of P is the line segment with both its endpoints in P such that the maximum distance from any point in P to it is minimized. In 2021, Daescu and Teo [6] gave a quadratic time algorithm using quadratic space. In this paper, we give a $(1+\varepsilon)$-approximation algorithm for the problem which runs in $O(n + \frac{1}{\varepsilon^4} \log \frac{1}{\varepsilon})$ time using linear space.

1 Introduction

Given a set P of n points in $\mathbb{R}^2$, the *discrete center line segment problem* is to find a line segment with both its endpoints in P such that the maximum distance from a point in P to the segment is minimized.

The general discrete p-center problem is a fundamental problem in clustering and facility location. The problem was shown to be NP-complete by Fowler et al. [8] in 1981. Megiddo and Supowit [10] then showed that the problem has no polynomial time 1.154-approximation algorithm unless P = NP, which was later improved to 1.822 by Feder and Greene [7].

Due to the hardness of the general discrete p-center problem, the problem for constant p, where $p = 1$ or 2, has been considered in the literature. The discrete 1-center problem can easily be solved in $O(n \log n)$ time using a farthest-neighbour Voronoi diagram, while Hershberger and Suri [9] gave a near-quadratic time algorithm for the discrete 2-center problem. This was later improved to $O(n^{4/3} \log^5 n)$ time by Agarwal et al. [1]. They also noticed that the discrete 2-center problem is harder to efficiently solve than the continuous 2-center problem, since determining whether the intersection of a set of congruent disks contains points is harder than determining whether the intersection is nonempty.

The discrete center line segment problem is closely related to the discrete 2-center problem in that it searches for two points in P that decides a discrete center segment which has a minimum maximum distance from a point to it, while the discrete 2-center problem searches for two points in P that minimize the maximum distance from a point to them.

The discrete center line segment problem was proposed by Daescu and Teo [6], who gave a quadratic time algorithm for the problem, using $O(n^2)$ space. In this paper, we give a strong linear-time approximation scheme (LTAS) [4] for

P. Morin and S. Suri (Eds.): WADS 2023, LNCS 14079, pp. 660–674, 2023.
https://doi.org/10.1007/978-3-031-38906-1_44

the problem, i.e. an algorithm that given a set P of n points and some $\varepsilon > 0$ computes in $O(n + \frac{1}{\varepsilon^4} \log \frac{1}{\varepsilon})$ time a $(1 + \varepsilon)$-approximate discrete center line segment.

Our algorithm consists of two main steps. The first step is to construct an approximate point set $\widetilde{P}$ such that we can get a $(1 + \varepsilon)$-approximate center segment of P by computing a $(1 + \varepsilon/6)$-approximate center segment of $\widetilde{P}$. This step takes $O(n + 1/\varepsilon)$ time and is described in Sect. 3.1.

The second step (Sect. 3.3) efficiently computes a $(1+\varepsilon/6)$-approximate center segment of $\widetilde{P}$. This can be done in $O(n + 1/\varepsilon^7)$ time. In Sects. 3.4 and 3.5 we show improvements of the second step which gives the final running time of $O(n + \frac{1}{\varepsilon^4} \log \frac{1}{\varepsilon})$.

2 Preliminaries

Let P be a set of n points in $\mathbb{R}^2$. A segment of P is a segment joining two points of P. For any two points a and b in $\mathbb{R}^2$, let ab denote the line segment joining a and b and let $\overline{ab}$ denote the line going through a and b. For any point p in $\mathbb{R}^2$, let $d(p, ab)$ denote the distance from p to segment ab and let $d_\perp(p, ab)$ denote the distance from p to line $\overline{ab}$. Let $|ab|$ denote the length of segment ab.

The *diameter* of P is the maximum distance between two points of P. A diametral point pair of P is a pair of points in P that realize the diameter of P. Let d^* denote the maximum distance from a point in P to a center segment of P.

3 The Algorithm

The algorithm consists of several parts. The first part computes an approximate point set of the input point set and is presented in Sect. 3.1. The second part concerns with computing a $(1 + \varepsilon)$-approximate center segment of the approximate point set and is presented in Sect. 3.3. The third part contains further optimizations for the second part and is presented in Sects. 3.4, 3.5 and 3.6.

3.1 Compute an Approximate Point Set of P

In this section, we start by finding an orientation of P such that the width of P along it is some constant times d^*. We can find such an orientation by computing a $(1 + \varepsilon)$-approximate diametral point pair of P, as shown by Lemma 1 below. We first state two simple but useful facts.

Fact 1. *Let p, q be any two points in $\mathbb{R}^2$ and let ab be a segment in the plane. Then $d(p, ab) \leqslant |pq| + d(q, ab)$ and $d_\perp(p, ab) \leqslant |pq| + d_\perp(q, ab)$.*

Let $\mathcal{D}$ denote the diameter of P.

Fact 2. *Let u and v be any two points inside $CH(P)$, then $|uv| \leqslant \mathcal{D}$.*

The following lemma shows that the width of P along the orientation perpendicular to the line through a $(1+\varepsilon)$-approximate diametral point pair of P is at most 10 times d^*.

Lemma 1. *The width of P along the orientation perpendicular to the line through a $(1+\varepsilon)$-approximate diametral point pair (a,b) of P is at most 10 times d^*, where d^* is the maximum distance from a point in P to a center segment of P.*

Proof. Let xy be a center segment of P and d^* be the maximum distance from a point in P to it. $\mathcal{D} \leqslant |ab| \cdot (1+\varepsilon)$ by definition. We will prove the lemma by showing that the maximum distance from a point in P to line $\overline{ab}$ is at most $5d^*$. We prove in a case-by-case manner. We assume that P is rotated so that xy is horizontal.

The first case is where ab and xy intersect. Let o be the intersection point and θ be the smaller intersection angle, as shown in Fig. 1(a). Let p be any point in P. Let q be p's nearest point on xy. By Fact 1, we have

$$\begin{aligned} d_\perp(p, ab) &\leqslant |pq| + d_\perp(q, ab) \\ &\leqslant d^* + d_\perp(q, ab) \end{aligned}$$

Since ab and xy intersect at o, either x or y is the farthest point on xy to line $\overline{ab}$. We have

$$\begin{aligned} d_\perp(p, ab) &\leqslant d^* + \max\{d_\perp(x, ab), d_\perp(y, ab)\} \\ &= d^* + \max\{|xo| \cdot \sin\theta, |yo| \cdot \sin\theta\} \\ &\leqslant d^* + |xy| \cdot \sin\theta \\ &\leqslant d^* + (1+\varepsilon)|ab| \cdot \sin\theta \\ &\leqslant d^* + (1+\varepsilon) \cdot 2\max\{|ao|, |bo|\} \cdot \sin\theta \\ &\leqslant d^* + (1+\varepsilon) \cdot 2\max\{d_\perp(a, xy), d_\perp(b, xy)\} \\ &\leqslant d^* + (1+\varepsilon) \cdot 2d^* \leqslant 5d^*. \end{aligned}$$

The second case is where ab and xy don't intersect. We only consider the case where ab has negative or infinity slope. The case where ab has positive slope is symmetric . When ab has negative or infinity slope, we consider cases where ab lies above line $\overline{xy}$ or where ab intersects $\overline{xy}$. The case where ab lies below $\overline{xy}$ is symmetric. We have the following cases.

b) Segment ab lies above $\overline{xy}$, xy lies below $\overline{ab}$ and b lies to the right of y. This configuration is shown in Fig. 1(b). Since xy lies below $\overline{ab}$ and ab has negative slope, x is the farthest point on xy from line $\overline{ab}$. Thus by Fact 1, for any point p in P,

$$\begin{aligned} d_\perp(p, ab) &\leqslant d(p, xy) + d_\perp(x, ab) \\ &\leqslant d^* + d_\perp(x, ab). \end{aligned}$$

Consider the horizontal line ℓ_b through b as shown in Fig. 1(b). Let x' be the point on ℓ_b such that $|x'b| = |xy|$. Let x'' be the point on $\overline{ab}$ closest to x'. We then have

$$\begin{aligned} d_\perp(p, ab) &\leqslant d^* + |xx'| + |x'x''| \\ &= d^* + |by| + |bx'| \cdot \sin\theta \\ &= d^* + d(b, xy) + |xy| \cdot \sin\theta \\ &\leqslant d^* + d^* + (1+\varepsilon)|ab| \cdot \sin\theta \\ &\leqslant 2d^* + (1+\varepsilon) d_\perp(a, xy) \\ &\leqslant 2d^* + (1+\varepsilon)d^* \leqslant 4d^*. \end{aligned}$$

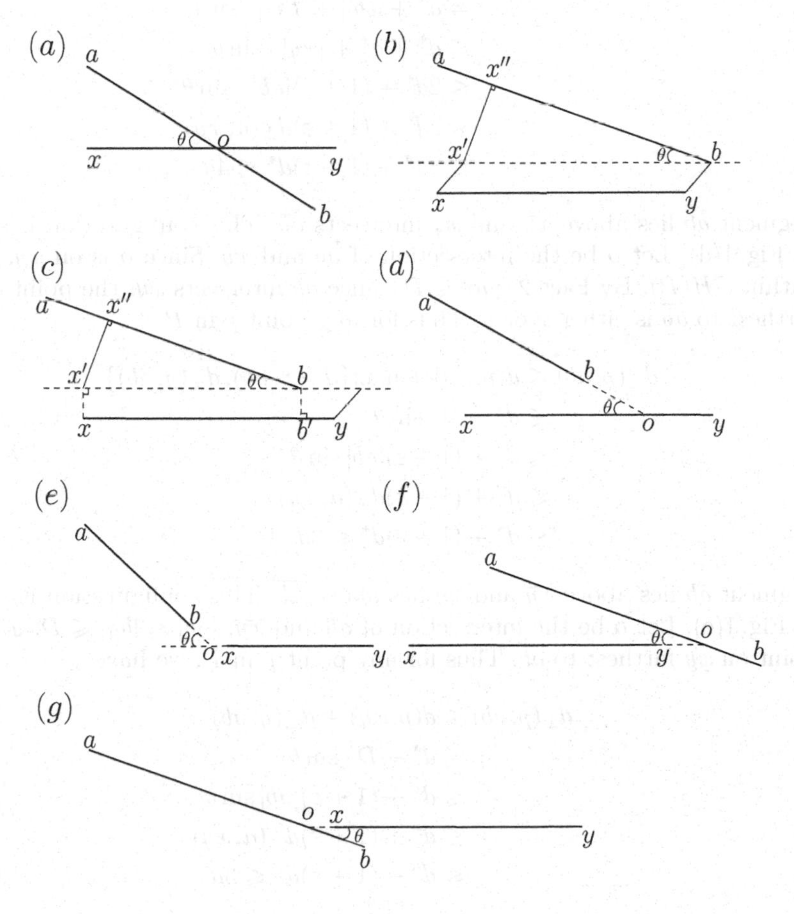

Fig. 1. Possible configurations.

c) Segment ab lies above $\overline{xy}$, xy lies below $\overline{ab}$ and b lies to the right of x and to the left of y. This configuration is shown in Fig. 1(c). As above, x is the point on xy farthest from $\overline{ab}$. Thus by Fact 1, for any point p in P,

$$\begin{aligned} d_\perp(p, ab) &\leqslant d(p, xy) + d_\perp(x, ab) \\ &\leqslant d^* + d_\perp(x, ab). \end{aligned}$$

Let ℓ_b be the line through b as shown in Fig. 1(c). Let b' be the point on xy closest to b and let x' be the point on ℓ_b closest to x. Finally let x'' be the point on line $\overline{ab}$ closest to x'. Thus

$$\begin{aligned} d_\perp(p, ab) &\leqslant d^* + |xx'| + |x'x''| \\ &= d^* + |bb'| + |bx'| \cdot \sin\theta \\ &\leqslant d^* + d^* + |xy| \cdot \sin\theta \\ &\leqslant 2d^* + (1+\varepsilon)|ab| \cdot \sin\theta \\ &\leqslant 2d^* + (1+\varepsilon)d_\perp(a, xy) \\ &\leqslant 2d^* + (1+\varepsilon)d^* \leqslant 4d^*. \end{aligned}$$

d) Segment ab lies above $\overline{xy}$ and xy intersects $\overline{ab}$. This configuration is shown in Fig. 1(d). Let o be the intersection of $\overline{ab}$ and xy. Since o is on xy, o lies within $CH(P)$. By Fact 2, $|ao| \leqslant \mathcal{D}$. Since $\overline{ab}$ intersects xy, the point on xy farthest to $\overline{ab}$ is either x or y. Thus for any point p in P,

$$\begin{aligned} d_\perp(p, ab) &\leqslant d(p, xy) + \max\{d_\perp(x, ab), d_\perp(y, ab)\} \\ &\leqslant d^* + \mathcal{D} \cdot \sin\theta \\ &\leqslant d^* + (1+\varepsilon)|ab| \sin\theta \\ &\leqslant d^* + (1+\varepsilon)d_\perp(a, xy) \\ &\leqslant d^* + (1+\varepsilon)d^* \leqslant 3d^*. \end{aligned}$$

e) Segment ab lies above $\overline{xy}$ and xy lies above $\overline{ab}$. This configuration is shown in Fig. 1(e). Let o be the intersection of $\overline{ab}$ and $\overline{xy}$. $|yo| \leqslant |by| \leqslant \mathcal{D}$. y is the point on xy farthest to $\overline{ab}$. Thus for any point p in P, we have

$$\begin{aligned} d_\perp(p, ab) &\leqslant d(p, xy) + d_\perp(y, ab) \\ &\leqslant d^* + \mathcal{D} \cdot \sin\theta \\ &\leqslant d^* + (1+\varepsilon)|ab| \sin\theta \\ &\leqslant d^* + (1+\varepsilon)d_\perp(a, xy) \\ &\leqslant d^* + (1+\varepsilon)d^* \leqslant 3d^*. \end{aligned}$$

f) Segment ab intersects $\overline{xy}$ and xy lies below $\overline{ab}$. Let o be the intersection of $\overline{xy}$ and ab. o lies to the right of y. This configuration is shown in Fig. 1(f). Since o is on ab, o lies within $CH(P)$. By Fact 2, $|xo| \leqslant \mathcal{D}$. $\overline{xy}$ intersects ab at o, so

$$|ab| \cdot \sin\theta \leqslant 2\max\{|ao| \cdot \sin\theta, |bo| \cdot \sin\theta\} \leqslant 2d^*.$$

Since x is the point on xy farthest to $\overline{ab}$, for any point p in P we have

$$\begin{aligned} d_{\perp}(p, ab) &\leqslant d(p, xy) + d_{\perp}(x, ab) \\ &\leqslant d^* + |xo| \cdot \sin\theta \\ &\leqslant d^* + \mathcal{D} \cdot \sin\theta \\ &\leqslant d^* + (1+\varepsilon)|ab| \sin\theta \\ &\leqslant d^* + (1+\varepsilon) \cdot 2d^* \leqslant 5d^*. \end{aligned}$$

g) Segment ab intersects $\overline{xy}$ and xy lies above $\overline{ab}$. Let o be the intersection of $\overline{xy}$ and ab. This configuration is shown in Fig. 1(g). Similar to case f), we can prove that for any point p in P, $d_{\perp}(p, ab) \leqslant 5d^*$.

The above cases are exhaustive. This finished the proof of the lemma. □

Now we are ready to compute the approximate point set. We use the approximate diameter algorithm of [3] to get a $(1+\varepsilon)$-approximate diametral point pair of P. Let (a, b) be the computed pair. Rotate P around the origin of the coordinate system so that segment ab is parallel to the y-axis after rotation. With a slight abuse of notation, we denote the rotated point set by P. We compute an approximate convex hull of P along the x-axis by using the algorithm in [3]. The approximate convex hull algorithm of [3] uses a strip method and works as follows. Let x_{min}, x_{max} be the extreme x-coordinates of the points in P and ε_1 be the precision parameter of the algorithm. Divide the plane into strips of width $\varepsilon_1/10 \cdot (x_{max} - x_{min})$. Index the points in P into the strips. For each strip find the two points of P in the strip with extreme y-coordinate. Let S' be the set of extreme y-coordinate points in the strips together with the two extreme x-coordinate points. Finally compute and return the convex hull of S', which is the approximate convex hull of P, denoted as $ACH(P)$.

For every point in P lying outside $ACH(P)$, we shift it along the x-axis to the closest boundary of $ACH(P)$. See Fig. 2. The approximate point set $\widetilde{P}$ is the union of all points in P that are inside $ACH(P)$ (thus not shifted) and all the shifted points. Note that we can get $\widetilde{P}$ from P by sweeping the stripped P and $ACH(P)$ along the x-axis. We have

Lemma 2. *The approximate point set* $\widetilde{P}$ *can be computed in* $O(n + \frac{1}{\varepsilon_1})$ *time.*

Proof. The $(1+\varepsilon)$-approximate diametral point pair (a, b) is computed in $O(n + \frac{1}{\varepsilon})$ time [3]. The approximate convex hull of P is computed in $O(n + \frac{1}{\varepsilon_1})$ time [3]. When sweeping P and $ACH(P)$ along the x-axis strip by strip and making the shifts, a shift takes constant time and $\widetilde{P}$ can be computed in $O(n + \frac{1}{\varepsilon_1})$ time. □

3.2 $\widetilde{P}$ is almost as Good as P

In this section, we show that we can get a $(1+\varepsilon)$-approximate center segment of P by computing a $(1 + \frac{\varepsilon}{6})$-approximate center segment of $\widetilde{P}$. The approximate

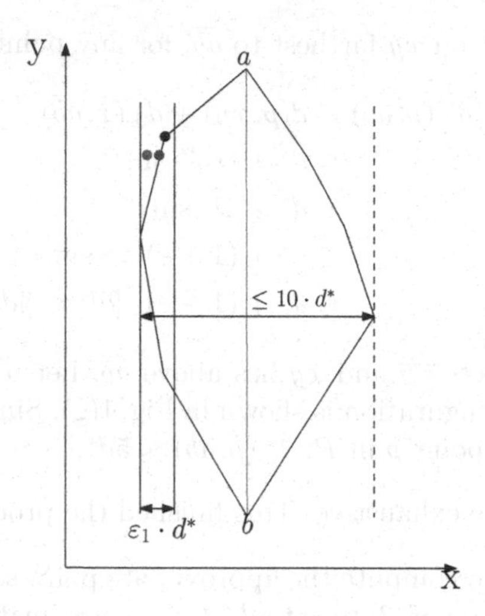

Fig. 2. The blue point is in P and lies outside $ACH(P)$. We shift it along the positive x-axis direction by at most $\varepsilon_1 d^*$ until it lies on the boundary of $ACH(P)$. The red point is the shifted blue point.

convex hull of P has $O(\frac{1}{\varepsilon_1})$ vertices. The width of the strips is at most $\varepsilon_1 \cdot d^*$, by Lemma 1. For any point in P lying outside $ACH(P)$, the shifted distance is at most $\varepsilon_1 d^*$. See Fig. 2 for an illustration.

For any point in $\widetilde{P}$ that is a shifted point, we keep track of its original point in P. Thus each segment of $\widetilde{P}$ corresponds to an original segment of P. Let $\tilde{x}\tilde{y}$ be a center segment of $\widetilde{P}$ and let $\tilde{d}$ be the maximum distance from a point in $\widetilde{P}$ to $\tilde{x}\tilde{y}$. Assume that the original points of $\tilde{x}$ and $\tilde{y}$ are x and y, respectively. The following lemma shows that xy is a $(1+4\varepsilon_1)$-approximate center segment of P.

Lemma 3. *The maximum distance from a point in P to xy is at most $(1+4\varepsilon_1) \cdot d^*$.*

By Lemma 3, the original segment of a $(1+\varepsilon_2)$-approximate center segment of $\widetilde{P}$ is a $(1+\varepsilon)$-approximate center segment of P. In the proof of Lemma 3, we have shown that $\tilde{d} \leqslant (1+2\varepsilon_1)d^*$, where $\tilde{d}$ is the maximum distance from a point in $\widetilde{P}$ to a center segment of $\widetilde{P}$. For any point p' in P, its distance to the original segment of a $(1+\varepsilon_2)$-approximate center segment of $\widetilde{P}$ is at most $(1+\varepsilon_2)\tilde{d}+2\varepsilon_1 d^* \leqslant (1+\varepsilon_2)(1+2\varepsilon_1)d^*+2\varepsilon_1 d^*$. By setting $\varepsilon_1 = \varepsilon_2 = \frac{\varepsilon}{6}$, we have $(1+\varepsilon_2)\tilde{d}+2\varepsilon_1 d^* \leqslant (1+\varepsilon)d^*$. Thus

Lemma 4. *In $O(n+\frac{1}{\varepsilon})$ time, we can compute an approximate point set $\widetilde{P}$ of P. The original segment of a $(1+\varepsilon/6)$-approximate center segment of $\widetilde{P}$ is a $(1+\varepsilon)$-approximate center segment of P.*

3.3 Compute a $(1+\varepsilon/6)$-Approximate Center Segment of $\widetilde{P}$

To compute a $(1+\varepsilon/6)$-approximate center segment of $\widetilde{P}$, we need a good estimation of $\tilde{d}$. We can get one by examining the diagonals of $CH(\widetilde{P})$. We call a diagonal of $CH(\widetilde{P})$ with the minimum maximum distance to a point in $\widetilde{P}$ among all the diagonals a *center diagonal* of $CH(\widetilde{P})$. The following lemma shows that the maximum distance from a point in $\widetilde{P}$ to a center diagonal is at most $2\tilde{d}$. Since we will not refer to the input point set P below, with a slight abuse of notation, from now on we will use p instead of $\tilde{p}$ to represent a point in $\widetilde{P}$.

Lemma 5. *Let $\mathcal{G}$ be a center diagonal of $CH(\widetilde{P})$. The maximum distance from a point in $\widetilde{P}$ to $\mathcal{G}$ is at most $2\tilde{d}$.*

There are $O(\frac{1}{\varepsilon_1^2})$ diagonals of $CH(\widetilde{P})$, however we will show next that we only need to consider $O(\frac{1}{\varepsilon_1})$ diagonals to find a center diagonal. Let $m = |CH(\widetilde{P})|$ and let $v_1, v_2, \ldots, v_m$ be a counterclockwise ordering of the hull vertices along $CH(\widetilde{P})$. Consider all diagonals having vertex $v_i (1 \leq i \leq m)$ as one endpoint. Let v_j, $i < j < i+m$, be the other endpoint of the diagonal. Let $ccw(i,j)$ denote the counterclockwise chain from v_i to v_j on the boundary of $CH(\widetilde{P})$ and $cw(i,j)$ denote the clockwise chain from v_i to v_j on the boundary of $CH(\widetilde{P})$. We prove the following monotonicity properties.

Lemma 6. *Let $f(i,j)$ be the maximum distance from a vertex on $ccw(i,j)$ to diagonal v_iv_j and let $g(i,j)$ be the maximum distance from a vertex on $cw(i,j)$ to v_iv_j. Then*[1]

(a) $f(i,j) \leqslant f(i,j+1)$.
(b) $g(i,j) \geqslant g(i,j+1)$.
(c) $f(i+1,j) \leqslant f(i,j)$.
(d) $g(i+1,j) \geqslant g(i,j)$

Let $f(i)$ be the minimum j such that $f(i,j)$ is greater than $g(i,j)$. Then

(e) $f(i+1) \geqslant f(i)$.

Proof. We first prove (a). Let ℓ_i be the line through v_i and perpendicular to v_iv_j and let H_i be the half-plane delimited by ℓ_i and not including v_j. Define ℓ_j and H_j similarly. Let $\bar{H}_i$ and $\bar{H}_j$ be the complement of H_i and H_j, respectively. Let v_k be any vertex on $ccw(i,j)$. We can show that its distance to v_iv_{j+1} is at least its distance to v_iv_j.

If v_k lies in H_i, its distance to v_iv_j equals $|v_kv_i|$. Since $\angle v_kv_iv_{j+1}$ is obtuse, its distance to v_iv_{j+1} also equals $|v_kv_i|$. See v_{i+1} in Fig. 3(a) for an illustration. If v_k lies in $\bar{H}_i \cap \bar{H}_j$, its distance to v_iv_j is less than its distance to v_iv_{j+1}. If v_k lies in H_j, the segment joining v_k and its nearest point on v_iv_{j+1} must intersect v_iv_j. Thus its distance to v_iv_{j+1} is greater than its distance to v_iv_j. This finishes the proof of (a).

We can use the same arguments for the proof of (b), (c) and (d). To prove (e), we know $f(i,f(i)-1) \leqslant g(i,f(i)-1)$ by the definition of $f(i)$, and $f(i+1,f(i)-1) \leqslant g(i+1,f(i)-1)$ by (c) and (d). Thus $f(i+1) \geqslant f(i)$. □

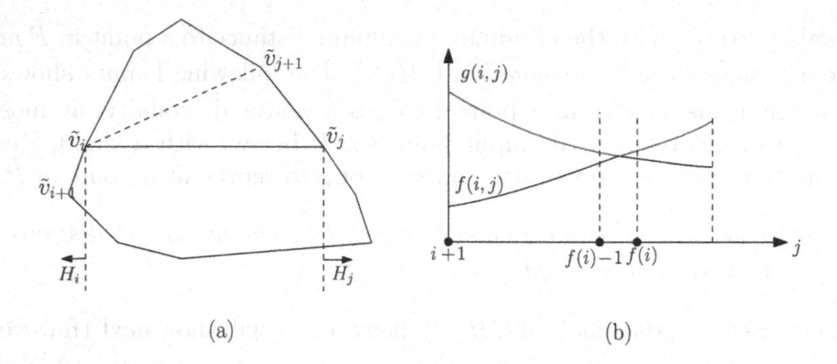

Fig. 3. Illustrating the monotone properties in Lemma 6.

Recall that $\max\{f(i,j), g(i,j)\}$ is the maximum distance from a hull vertex to diagonal $v_i v_j$. For a fixed i, $\max\{f(i,j), g(i,j)\}$ is a unimodal function of j and its minimum takes place at either $j = f(i)$ or $j = f(i) - 1$. See Fig. 3(b). After we get $f(i)$, we can search for $f(i+1)$ from $j = f(i)$ rather than from $j = i+2$, by Lemma 6(e). In this way, we only consider $O(\frac{1}{\varepsilon_1})$ candidate diagonals in search of a center diagonal.

We now discuss how to compute the farthest point in $\widetilde{P}$ and its distance to a segment pq of $\widetilde{P}$. We have the following lemma.

Lemma 7. *The farthest point in $\widetilde{P}$ to a segment pq of $\widetilde{P}$ is a vertex of $CH(\widetilde{P})$.*

Thus we only consider vertices of $CH(\widetilde{P})$ in search of the farthest point in $\widetilde{P}$ to pq. For any vertex of $CH(\widetilde{P})$ in H_p, the point on pq closest to it is p. For vertices of $CH(\widetilde{P})$ in H_p or H_q, we can make a *half plane farthest neighbour query* to find the farthest vertex to p in H_p or the farthest vertex to pq in H_q. Given a point set, a query point and a line through the query point, the *half plane farthest neighbor* of the query point is the farthest point in the point set that lies in the half-plane delimited by the line through the query point. Let $\delta(\widetilde{P}, p, pq)$ denote the maximum distance from a $CH(\widetilde{P})$ vertex in H_p to p and $\delta(\widetilde{P}, q, pq)$ denote the maximum distance from a $CH(\widetilde{P})$ vertex in H_q to q. We can compute $\delta(\widetilde{P}, p, pq)$ or $\delta(\widetilde{P}, q, pq)$ by using the data structure in [5].

For any hull vertex in $\bar{H}_p \cap \bar{H}_q$, its distance to pq equals its distance to the line through p and q. For any hull vertex in H_p or H_q, its distance to pq is at least its distance to the line through p and q. Let $d_\perp(\widetilde{P}, pq)$ denote the maximum distance from a hull vertex to the line through p and q. The maximum distance from a hull vertex to pq equals the maximum of $\delta(\widetilde{P}, p, pq)$, $\delta(\widetilde{P}, q, pq)$, and $d_\perp(\widetilde{P}, pq)$. Since $CH(\widetilde{P})$ has already been computed, $d_\perp(\widetilde{P}, pq)$ can be computed in $O(\log \frac{1}{\varepsilon_1})$ time by binary search the vertices of $CH(\widetilde{P})$.

Let uv be a diagonal of $CH(\widetilde{P})$. After spending $O(\frac{1}{\varepsilon_1} \log \frac{1}{\varepsilon_1})$ time in building the data structure [5], we can compute $\delta(\widetilde{P}, v, uv)$ or $\delta(\widetilde{P}, u, uv)$ in $O(\log^2 \frac{1}{\varepsilon_1})$

[1] $i+1$, $j+1$ should take modulo m, we omit it for brevity.

time. Thus we can compute the maximum distance from a $CH(\widetilde{P})$ vertex to uv in $O(\log^2 \frac{1}{\varepsilon_1})$ time. There are $O(\frac{1}{\varepsilon_1})$ diagonals to consider, so we can compute a center diagonal and $\bar{d}$ in $O(\frac{1}{\varepsilon_1} \log^2 \frac{1}{\varepsilon_1})$ time, where $\bar{d}$ is the maximum distance from a point in $\widetilde{P}$ to the center diagonal of $CH(\widetilde{P})$.

Lemma 8. *Given $CH(\widetilde{P})$, a center diagonal of $CH(\widetilde{P})$ and $\bar{d}$ can be computed in $O(\frac{1}{\varepsilon_1} \log^2 \frac{1}{\varepsilon_1})$ time.*

By Lemma 5, $\bar{d}$ is a 2-approximation of $\tilde{d}$. If we place squares with side length $2\bar{d}$ such that the centers of the squares are vertices of $CH(\widetilde{P})$, the center segment of $\widetilde{P}$ must have both ends inside the squares, as suggested by the following observation.

Observation 1. *The center segment of $\widetilde{P}$ must have both ends inside the squares which have side length $2\bar{d}$ and the centers of which are vertices of $CH(\widetilde{P})$.*

Now we can compute a $(1+\varepsilon/6)$-approximate center segment of $\widetilde{P}$ as follows. Place an axis-aligned grid of side length $\frac{\varepsilon \bar{d}}{48\sqrt{2}}$ over the plane. Index all points in $\widetilde{P}$ into cells of the grid. An axis-aligned square which has side length $2\bar{d}$ and the center of which is a vertex v of $CH(\widetilde{P})$ is called *v's $\bar{d}$-square.* Call the intersection of the grid and a vertex's $\bar{d}$-square a *vertex grid.* An example vertex grid is shown in Fig. 4. For each cell of a vertex grid, randomly pick one point indexed into the cell (if any) and the point is called the *representative* of the cell. By Observation 1, the center segment of $\widetilde{P}$ must have each of its endpoints inside a cell of some vertex grid. Let these two cells be A and B. The segment joining the representatives of A and B is a $(1+\varepsilon/6)$-approximate center segment of $\widetilde{P}$. We can consider all pairs of vertex grids and for each pair consider all candidate segments joining two representatives from the pair, to get a $(1+\varepsilon/6)$-approximate center segment of $\widetilde{P}$.

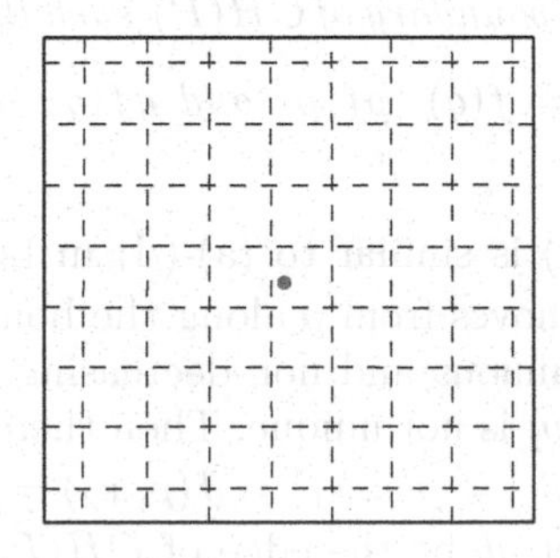

Fig. 4. A vertex grid.

A vertex grid has $O(\frac{1}{\varepsilon^2})$ cells. Since $CH(\widetilde{P})$ has $O(\frac{1}{\varepsilon_1})$ vertices and there are $O(\frac{1}{\varepsilon^2})$ representatives of cells in a vertex grid, and $O(\frac{1}{\varepsilon_1^2} \cdot \frac{1}{\varepsilon^4})$ candidate segments in total. By Lemma 7, we can compute the maximum distance from a point in $\widetilde{P}$ to a candidate segment trivially in $O(\frac{1}{\varepsilon_1})$ time. Thus given $CH(\widetilde{P})$ and $\bar{d}$, we can compute a $(1+\varepsilon/6)$-approximate center segment of $\widetilde{P}$ in $O(\frac{1}{\varepsilon_1^3} \cdot \frac{1}{\varepsilon^4})$ time. Combined with Lemma 4 and Lemma 8, we have

Theorem 1. *We can compute a* $(1+\varepsilon)$*-approximate center segment of* P *in* $O(n + \frac{1}{\varepsilon^7})$ *time.*

The following sections describe how to further improve the running time of the algorithm given in this section. Section 3.4 notes that we only need to consider $O(\frac{1}{\varepsilon_1})$ pairs of vertex grids, instead of $O(\frac{1}{\varepsilon_1^2})$. Section 3.5 shows that for a pair of vertex grids, we only need to consider $O(\frac{1}{\varepsilon^3})$ candidate segments, instead of $O(\frac{1}{\varepsilon^4})$. Finally, Sect. 3.6 gives details on how to compute the farthest point to a candidate segment in $O(\log \frac{1}{\varepsilon_1})$ time.

3.4 Only Consider $O(\frac{1}{\varepsilon_1})$ Pairs of Vertex Grids

Let p, r, q, s be any four points on the boundary of the convex hull of $\widetilde{P}$, in counterclockwise order. Let $ccw(p,q)$ denote the counterclockwise chain from p to q on the boundary of $CH(\widetilde{P})$ and let $cw(p,q)$ denote the clockwise chain from p to q on the boundary of $CH(\widetilde{P})$. The main lemma of this section depends on the following monotone properties.

Lemma 9. *Let* $f(p,q)$ *be the maximum distance from a vertex on* $ccw(p,q)$ *to* pq *and* $g(p,q)$ *be the maximum distance from a vertex on* $cw(p,q)$ *to* pq*. Then*

(a) $f(p,q) \leqslant f(p,s)$.
(b) $g(p,q) \geqslant g(p,s)$.
(c) $f(r,q) \leqslant f(p,q)$.
(d) $g(r,q) \geqslant g(p,q)$

Let $f(p)$ *be the point on the boundary of* $CH(\widetilde{P})$ *such that* $f(p,f(p)) = g(p,f(p))$.

(e) $f(p)$ *is unique.* $f(p) \neq f(q)$. $pf(p)$ *and* $qf(q)$ *intersect at a point inside* $CH(\widetilde{P})$.

Proof. The proof of (a)-(d) is similar to (a)-(d) in Lemma 6. We prove (e) by contradiction. As point x moves from p along the boundary of $CH(\widetilde{P})$ counterclockwisely, $f(p,x)$ is continuous and non-decreasing, $g(p,x)$ is continuous and non-increasing. Assume $f(p)$ is not unique. Then there are two points x_1 and x_2 on $CH(\widetilde{P})$ such that $f(p,x_1) = g(p,x_1) = f(p,x_2) = g(p,x_2)$. Assume x_2 is on $cw(p,x_1)$. See Fig. 5(a). Let ab be the edge of $CH(\widetilde{P})$ where $f(p)$ lies. Assume without loss of generality that $\angle bpx_1 \leq \pi/2$. If $\angle bpx_1 < \pi/2$, $f(p,x_1) < f(p,x_2)$, a contradiction. If $\angle bpx_1 = \pi/2$, $g(p,x_2) < g(p,x_1)$, a contradiction. Thus $f(p)$ is unique.

Assume $f(p) = f(q)$. Then $f(q,f(q)) \neq g(q,f(q))$, a contradiction. Similarly, we can show that $pf(p)$ and $qf(q)$ intersect at a point inside $CH(\widetilde{P})$. □

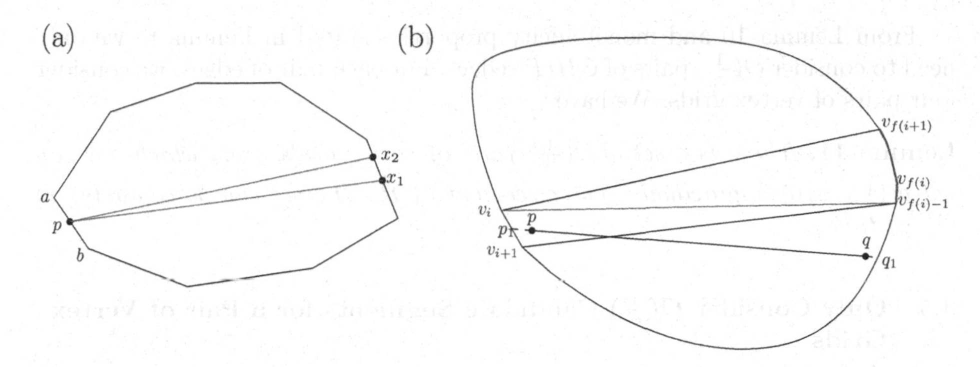

Fig. 5. (a) Illustration for Lemma 9(e). (b) For v_iv_{i+1}, only consider $v_{(f(i)-1}v_{f(i)}, \ldots, v_{f(i+1)-1}v_{f(i+1)}$.

As described in the last section, we can exhaustively consider all pairs of vertex grids to get a $(1 + \varepsilon/6)$-approximate center segment of $\widetilde{P}$. A candidate segment joins one representative from one vertex grid with one representative from another vertex grid.

As for Lemma 6, let $v_1, v_2, \ldots, v_m$ be a counterclockwise ordering of the hull vertices on $CH(\widetilde{P})$. Consider all candidate segments whose extensions have one endpoint on edge v_iv_{i+1} of $CH(\widetilde{P})$. Let $f(i)$ be as defined in Lemma 6. We have

Lemma 10. *Among all candidate segments whose extensions intersect edge v_iv_{i+1} of $CH(\widetilde{P})$, we only need to consider those segments whose extensions also intersect one of the edges $v_{f(i)-1}v_{f(i)}, v_{f(i)}v_{f(i)+1}, \ldots, v_{f(i+1)-1}v_{f(i+1)}$ (the thick blue chain in Fig. 5(b)).*

Proof. Among all candidate segments whose extensions have an endpoint on v_iv_{i+1}, consider any candidate segment pq whose extension has the other endpoint q_1 on $ccw(v_{i+1}, v_{f(i)-1})$. Let p_1 be the endpoint of pq's extension on v_iv_{i+1}. See Fig. 5(b). By Lemma 9,

$$f(p_1, q_1) \leqslant f(v_i, v_{f(i)-1}) \leqslant g(v_i, v_{f(i)-1}) \leqslant g(p_1, q_1).$$

Thus the maximum distance from a point in $\widetilde{P}$ to p_1q_1 is at least the maximum distance from a point in $\widetilde{P}$ to $v_iv_{f(i)-1}$. The maximum distance from a point in $\widetilde{P}$ to pq is at least the maximum distance from a point in $\widetilde{P}$ to $v_iv_{f(i)-1}$. Thus we do not need to consider any candidate segment whose extension has the other endpoint on $ccw(v_{i+1}, v_{f(i)-1})$.

By a similar argument, the maximum distance from a point in $\widetilde{P}$ to a candidate segment whose extension has the other endpoint on $cw(v_i, v_{f(i+1)})$ is at least the maximum distance from a point in $\widetilde{P}$ to $v_{i+1}v_{f(i+1)}$. Thus we do not need to consider any candidate segment whose extension has the other endpoint on $cw(v_i, v_{f(i+1)})$. This completes the proof of the lemma. □

From Lemma 10 and monotonicity properties stated in Lemma 6, we only need to consider $O(\frac{1}{\varepsilon_1})$ pairs of $CH(\widetilde{P})$ edges. For each pair of edges, we consider four pairs of vertex grids. We have

Lemma 11. *There is a set of $O(\frac{1}{\varepsilon_1})$ pairs of vertex grids, from which one can get a $(1+\varepsilon/6)$-approximate center segment of $\widetilde{P}$. The set can be computed in $O(\frac{1}{\varepsilon_1})$ time.*

3.5 Only Consider $O(\frac{1}{\varepsilon^3})$ Candidate Segments for a Pair of Vertex Grids

Consider the $\bar{d}$-square of a vertex v of $CH(\widetilde{P})$, as shown in Fig. 6(a), together with the cells of the $\bar{d}$-square. From a point outside the $\bar{d}$-square, the edges of the $\bar{d}$-square are divided into two groups. In Fig. 6(a), for example, from point y the two blue edges are called the *through* edges and the two black edges are called the *non-through* edges. From y we can draw a sequence of segments ending at cell corners which are on a through edge from y, as the blue dashed segments shown in Fig. 6(a). Every cell is intersected by one of the segments. No cell can fit between two neighbouring segments without being intersected. Let yt be one such segment where t is a corner of a cell on a through edge from y. Note that yt intersects $O(\frac{1}{\varepsilon})$ cells. See Fig. 6(b). Let $\mathcal{C}$ be the set of representatives of the cells that are intersected by yt. Let q be the representative in $\mathcal{C}$ such that the maximum distance from a point in $\widetilde{P}$ to yq is the minimum among all the segments between y and a point in $\mathcal{C}$.

Let o be yt's intersection with the non-through edge from y and let r_i, r_{i+1} be the two cell corners on the edge immediately above and below o. Let q_i be the representative whose nearest point on $r_i t$ is farthest from r_i among all representatives of the cells that are intersected by $r_i t$. Let q_{i+1} be the representative whose nearest point on $r_{i+1}t$ is farthest from r_{i+1} among all representatives of the cells that are intersected by $r_{i+1}t$. The segment yq can be approximated by two segments, as shown by the following lemma.

Lemma 12. *Either the maximum distance from a point in $\widetilde{P}$ to yq_i or the maximum distance from a point in $\widetilde{P}$ to yq_{i+1} is at most $\varepsilon\tilde{d}/6$ greater than the maximum distance from a point in $\widetilde{P}$ to yq.*

Proof. Each cell intersected by yt must be intersected by either $r_i t$ or $r_{i+1}t$. Without loss of generality, assume the cell containing q is intersected by $r_i t$. Let s be the point on $r_i t$ closest to q and let s_i be the point on $r_i t$ closest to q_i. Note that q_i may not be q. See Fig. 6(c). By definition, s_i is farther to r_i than s. Let the extension of qs and the extension of $q_i s_i$ intersect yt at s' and s'_i, respectively. s'_i is farther to y than s'. For any point x in $\widetilde{P}$, its distance to yq_i is at most $\frac{\varepsilon\bar{d}}{24}$ more than its distance to ys'_i, and its distance to ys' is at most $\frac{\varepsilon\bar{d}}{48}$ more than its distance to yq. Since the distance from x to ys' is at least the distance from x to ys'_i, the distance from x to yq_i is at most $\varepsilon\bar{d}/16$ more than its distance to yq. Thus the distance from x to yq_i is at most $\varepsilon\tilde{d}/6$ more than its distance to yq. When the cell containing yq is intersected by $r_{i+1}t$, we can use the same argument to show that the distance from any point in $\widetilde{P}$ to yq_{i+1} is at most $\varepsilon\tilde{d}/6$ more than its distance to yq. □

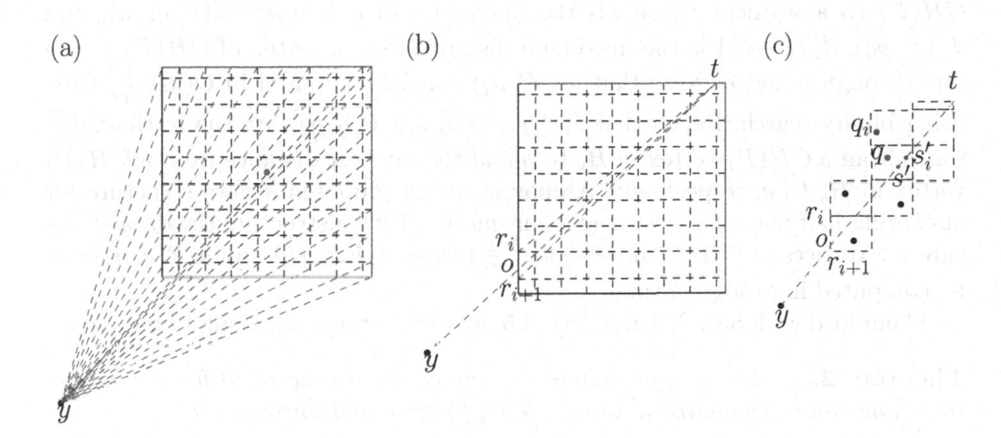

Fig. 6. (a) A vertex's $\bar{d}$-square and its grid. The dashed segments from y intersect every grid cell. (b) yt intersects the non-through edge at o. (c) The cell containing q is intersected by $r_i t$.

For a vertex grid, we can consider every pair of cell corners (r_i, t) where r_i is a cell corner on a non-through edge and t is a cell corner on a through edge, and compute the representative q_i for (r_i, t). We then store the representatives for all such pairs in a table. The representative q_i for a pair (r_i, t) can be trivially computed in $O(\frac{1}{\varepsilon})$ time. A vertex grid has $O(\frac{1}{\varepsilon^2})$ such pairs, thus the table can be built in $O(\frac{1}{\varepsilon^3})$ time. There are $O(\frac{1}{\varepsilon_1})$ vertex grids, so we build tables for them in $O(\frac{1}{\varepsilon^3\varepsilon_1})$ time.

Now consider a pair of vertex grids $(\mathcal{B}_1, \mathcal{B}_2)$. For a representative p_1 in $\mathcal{B}_1$, we can draw a sequence of segments from p_1 to cell corners on the through edges of $\mathcal{B}_2$, as in Fig. 6(a). For a segment p_1t in the sequence that intersects a non-through edge at o, let r_i and r_{i+1} be the two cell corners on the edge immediately above and below o. We can find q_i for (r_i, t) and q_{i+1} for (r_{i+1}, t) by looking up $\mathcal{B}_2$'s table. For all candidate segments with y at p_1 and the other endpoint at a representative of the cells intersected by p_1t, we consider p_1q_i and p_1q_{i+1} instead. By Lemma 12, this leads to an additive error of at most $\varepsilon\tilde{d}/6$. There are $O(\frac{1}{\varepsilon})$ segments in the sequence, we have

Lemma 13. *After $O(\frac{1}{\varepsilon^4})$ time preprocessing, one can compute a set of $O(\frac{1}{\varepsilon^3})$ candidate segments for any pair of vertex grids $(\mathcal{B}_1, \mathcal{B}_2)$ in $O(\frac{1}{\varepsilon^3})$ time. The optimal candidate segment in the set is a $(1+\varepsilon/6)$-approximation of the optimal candidate segment for $(\mathcal{B}_1, \mathcal{B}_2)$.*

3.6 Compute the Farthest Point in $\widetilde{P}$ to a Segment in $O(\log \frac{1}{\varepsilon_1})$ Time

We already know that the farthest point in $\widetilde{P}$ to a segment is a vertex of $CH(\widetilde{P})$. As mentioned earlier in the paper, the maximum distance from a vertex of $CH(\widetilde{P})$ to a segment pq equals the maximum of $\delta(\widetilde{P}, p, pq)$, $\delta(\widetilde{P}, q, pq)$, and $d_{\perp}(\widetilde{P}, pq)$. $d_{\perp}(\widetilde{P}, pq)\}$ is the maximum distance from a vertex of $CH(\widetilde{P})$ to the line through p and q. Note that $d_{\perp}(\widetilde{P}, pq)$ can be computed in $O(\log \frac{1}{\varepsilon_1})$ time using binary search. Recall that $\delta(\widetilde{P}, p, pq)$ and $\delta(\widetilde{P}, q, pq)$ are the farthest distance from a $CH(\widetilde{P})$ vertex in H_p to p and the farthest distance from a $CH(\widetilde{P})$ vertex in H_q to q, respectively. Aronov et al. [2] proposed a data structure for answering half plane farthest neighbour queries. By constructing the data structure for vertices of $CH(\widetilde{P})$ in $O(\frac{1}{\varepsilon_1} \log^3 \frac{1}{\varepsilon_1})$ time, $\delta(\widetilde{P}, p, pq)$ and $\delta(\widetilde{P}, q, pq)$ can be computed in $O(\log \frac{1}{\varepsilon_1})$ time.

Combined with Sect. 3.4 and Sect. 3.5, we obtain the main result of the paper.

Theorem 2. *A $(1+\varepsilon)$-approximate discrete center line segment for n points in the plane can be computed in $O(n + \frac{1}{\varepsilon^4} \log \frac{1}{\varepsilon})$ time and linear space.*

References

1. Agarwal, P.K., Sharir, M., Welzl, E.: The discrete 2-center problem. Discret. Comput. Geom. **20**(3), 287–305 (1998)
2. Aronov, B., et al.: Data structures for halfplane proximity queries and incremental voronoi diagrams. Algorithmica **80**(11), 3316–3334 (2018)
3. Bentley, J.L., Faust, M.G., Preparata, F.P.: Approximation algorithms for convex hulls. Commun. ACM **25**(1), 64–68 (1982)
4. Chan, T.M.: Approximating the diameter, width, smallest enclosing cylinder, and minimum-width annulus. Int. J. Comput. Geom. Appl. **12**(1–2), 67–85 (2002)
5. Daescu, O., Mi, N., Shin, C., Wolff, A.: Farthest-point queries with geometric and combinatorial constraints. Comput. Geom. **33**(3), 174–185 (2006)
6. Daescu, O., Teo, K.Y.: The discrete median and center line segment problems in the plane. In: He, M., Sheehy, D. (eds.) Proceedings of the 33rd Canadian Conference on Computational Geometry, CCCG 2021, 10–12 August 2021, Dalhousie University, Halifax, Nova Scotia, Canada, pp. 312–319 (2021)
7. Feder, T., Greene, D.H.: Optimal algorithms for approximate clustering. In: Simon, J. (ed.) Proceedings of the 20th Annual ACM Symposium on Theory of Computing, 2–4 May 1988, Chicago, Illinois, USA, pp. 434–444. ACM (1988)
8. Fowler, R.J., Paterson, M., Tanimoto, S.L.: Optimal packing and covering in the plane are NP-complete. Inf. Process. Lett. **12**(3), 133–137 (1981)
9. Hershberger, J., Suri, S.: Finding tailored partitions. J. Algorithms **12**(3), 431–463 (1991)
10. Megiddo, N., Supowit, K.J.: On the complexity of some common geometric location problems. SIAM J. Comput. **13**(1), 182–196 (1984)

Density Approximation for Moving Groups

Max van Mulken(✉), Bettina Speckmann, and Kevin Verbeek

TU Eindhoven, Eindhoven, The Netherlands
{m.j.m.v.mulken,b.speckmann,k.a.b.verbeek}@tue.nl

Abstract. Sets of moving entities can form groups which travel together for significant amounts of time. Tracking such groups is an important analysis task in a variety of areas, such as wildlife ecology, urban transport, or sports analysis. Correspondingly, recent years have seen a multitude of algorithms to identify and track meaningful groups in sets of moving entities. However, not only the mere existence of one or more groups is an important fact to discover; in many application areas the actual shape of the group carries meaning as well. In this paper we initiate the algorithmic study of the shape of a moving group. We use kernel density estimation to model the density within a group and show how to efficiently maintain an approximation of this density description over time. Furthermore, we track persistent maxima which give a meaningful first idea of the time-varying shape of the group. By combining several approximation techniques, we obtain a kinetic data structure that can approximately track persistent maxima efficiently.

Keywords: Group density · Quadtrees · Topological persistence

1 Introduction

Devices that track the movement of humans, animals, or inanimate objects are ubiquitous and produce significant amounts of data. Naturally this wealth of data has given rise to a large number of analysis tools and techniques which aim to extract patterns from said data. One of the most important patterns formed by both humans and animals are groups: sets of moving entities which travel together for a significant amount of time. Identifying and tracking groups is an important task in a variety of research areas, such as wildlife ecology, urban transport, or sports analysis. Consequently, in recent years various definitions and corresponding detection and tracking algorithms have been proposed, such as herds [16], mobile groups [17], clusters [2,18], and flocks [5,26].

In computational geometry, there is a sequence of papers on variants of the *trajectory grouping structure* which allows a compact representation of all groups within a set of moving entities [8,12,19,20,28]. The grouping structure uses a

A full version of this paper can be found at https://arxiv.org/abs/2212.03685.

P. Morin and S. Suri (Eds.): WADS 2023, LNCS 14079, pp. 675–688, 2023.
https://doi.org/10.1007/978-3-031-38906-1_45

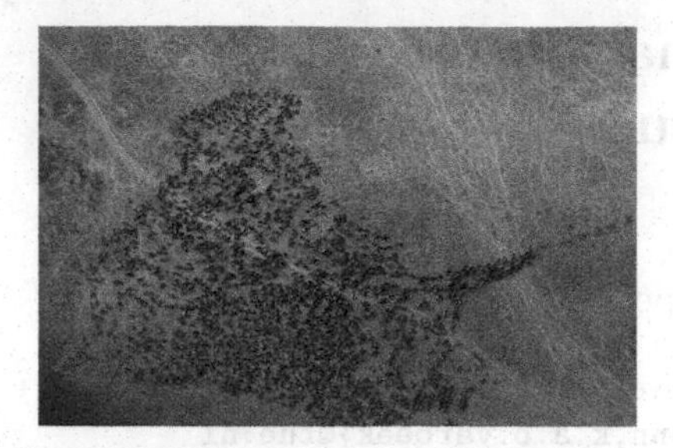

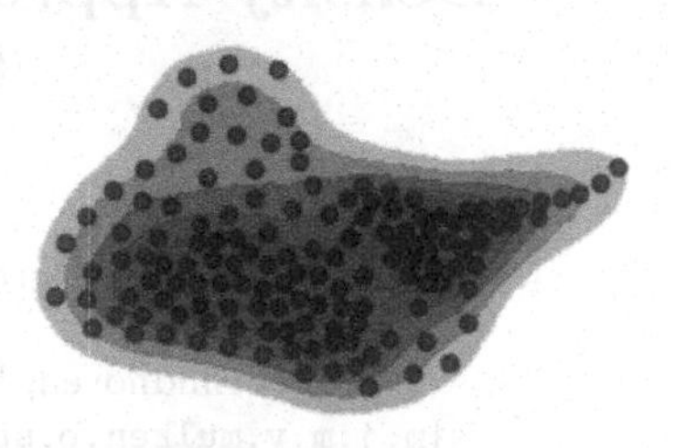

Fig. 1. A herd of wildebeest, the shape of the group shows migratory behavior (https://commons.wikimedia.org/wiki/File:Wbeest_Mara.jpg on 25/11/2022.).

size parameter m, a temporal parameter δ, and a spatial parameter σ. A set of entities forms a group during time interval I if it consists of at least m entities, I is of length at least δ, and the union of the discs of radius σ centered around all entities forms a single connected component.

However, not only the mere existence of these groups is important to discover; in many applications areas it is equally important to detect how individuals within a group or the group as a whole are moving. Several papers hence focus on defining [29], detecting [14], and categorizing [10,29] movement patterns of complete groups. These patterns are based on coordinated behavior of the individuals within the group, such as foraging behavior, flying in a V-shape, or following a leader [3,21]. Beyond animal ecology, papers focus on detecting roles in sports teams [22,23] and identifying formations in football teams [6,7,13,15].

Apart from the actions of the individual moving entities in a group, the actual shape of the group and the density distribution also carry meaning. Consider, for example, the herd of wildebeest in Fig. 1. Both its global shape and the distribution of dense areas indicate that this herd is migrating. Research in wildlife ecology [11,24] has established that animals often stay close together when not under threat and respond to immediate danger by spreading out. Hence from the density and the extent of a herd we can infer fear levels and external disturbances. The density distribution and general shape of a group are not only relevant in wildlife ecology, but they can also provide useful insights when monitoring, for example, visitors of a festival.

In this paper we initiate the algorithmic study of the shape of a moving group. Specifically, we identify and track particularly dense areas which provide a meaningful first idea of the time-varying shape of the group. It is our goal to develop a solid theoretical foundation which will eventually form the basis for a software system that can track group shapes in real time. We believe that algorithm engineering efforts are best rooted in as complete as possible an understanding of the theoretical tractability of the problem at hand. To develop an efficient algorithmic pipeline, we are hence making several simplifying assumptions on the trajectories of the moving objects (known ahead of time, piecewise linear, within a bounding box). In the discussion in Sect. 6 we (briefly) explain how we intend to build further on our theoretical results to lift these restrictions, trading theoretical guarantees for efficiency.

Problem Statement. Our input consists of a set P of n moving points in $\mathbb{R}^2$; we assume that the points follow piecewise linear motion and are contained in a bounding box $\mathcal{D} = [0, D] \times [0, D]$ with size parameter D. The position of a point $p \in P$ over time is described by a function $p(t)$, where $t \in [0, T]$ is the time parameter. We omit the dependence on t when it is clear from the context, and simply denote the position of a point as p. We refer to the x- and y-coordinates of a point p as $x(p)$ and $y(p)$, respectively.

We assume that the set P continuously forms a single group; we aim to monitor the density of P over time. We measure the density of P at position (x, y) using the well-known concept of *kernel density estimation* (KDE) [25]. KDE uses *kernels* around each point $p \in P$ to construct a function $\text{KDE}_P \colon \mathbb{R}^2 \to \mathbb{R}^+$ such that $\text{KDE}_P(x, y)$ estimates the density of P at position (x, y). We are mainly interested in how the density peaks of P, that is, the local maxima of the function KDE_P, change over time. Not all local maxima are equally relevant (some are minor "bumps" caused by noise), hence we focus on tracking *topologically persistent* local maxima through time.

Approach and Organization. We could maintain the entire function KDE_P over time and keep track of its local maxima. However, doing so would be computationally expensive. Furthermore, there are good reasons to approximate KDE_P: (1) KDE is itself also an approximation of the density, and (2) approximating KDE_P may eliminate local maxima that are not relevant. For a simple approach, consider building a quadtree on P. Observe that in areas where P is dense, the quadtree cells will be relatively small. Hence, we can associate the size of a quadtree cell with the density of P. This approach has the added advantage that the spatial resolution near the density peaks is higher. However, there are two main drawbacks: (1) the approximation does not depend on the kernel size for the KDE, which is an important parameter for analysis, and (2) we cannot guarantee that all significant local maxima of KDE_P are preserved.

The approach we present in this paper builds on the simple approach described above, but eliminates its drawbacks. We first introduce how to approximate any function f using a quadtree based on the volume underneath f (see Fig. 2), which is described in more detail in Sect. 3. Next, in Sect. 4, we shift our focus to the function KDE_P, and describe how to construct a point set to esti-

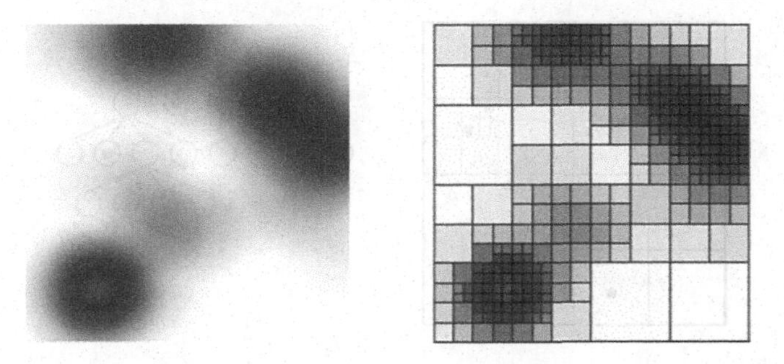

Fig. 2. A 2-dimensional function (left) approximated as a step function (right).

mate the volume underneath KDE_P. This will allow us to efficiently maintain the approximation as the underlying points move. In Sect. 5 we describe how to maintain the approximation of KDE_P, as well as its maxima, using a kinetic data structure. Lastly, in Sect. 6 we reflect upon our approach and sketch how our theoretical results can serve as a guide towards a solution that is efficient in practice. Omitted proofs can be found in the full paper.

2 Preliminaries

Kernel Density Estimation. Let P be a set of points in $\mathbb{R}^2$. To obtain a kernel density estimation (KDE) [25,27] KDE_P of P, we need to choose a kernel function $K\colon \mathbb{R}^2 \to \mathbb{R}^+$ that captures the influence of a single point on the density. Let σ denote the *kernel width* and $\|(x, y)\|$ the Euclidean norm of (x, y). An example of a common kernel is the cone kernel:

$$K(x,y) = \begin{cases} 1 - \|(x,y)\|/\sigma, & \text{for } \|(x,y)\| < \sigma \\ 0, & \text{otherwise} \end{cases}$$

In Sect. 3 we show that our approach works with any kernel that has bounded slope. Furthermore, for simplicity we assume that the range in which the kernel function produces nonzero values is bounded by the width of the kernel σ. We scale the input, without loss of generality, such that $\sigma = 1$, and the kernel function such that the volume under the kernel function is always 1.

Given a kernel function K and a point set P, we compute the KDE as follows:

$$\text{KDE}_P(x,y) = \frac{1}{n} \sum_{p \in P} K(x - x(p), y - y(p)) \qquad \text{for } (x,y) \in \mathbb{R}^2. \tag{1}$$

Quadtrees. Consider a bounded domain of size D in $\mathbb{R}^2$ which is defined by the square $\mathcal{D} = [0, D] \times [0, D]$. We build a *quadtree* T on this domain, where each node $v \in T$ represents a region $R(v) \in \mathcal{D}$. We refer to a leaf in a quadtree

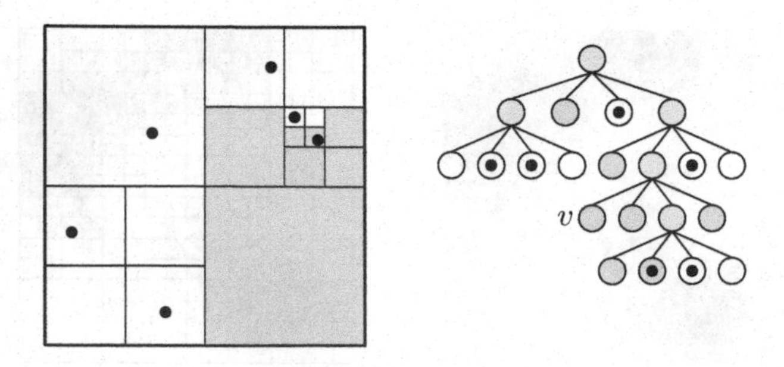

Fig. 3. The spatial neighborhood $\mathcal{N}(v)$ of the red node v is shown in blue.

T also as a *cell*. For a node $v \in T$ we further denote the side length of $R(v)$ as $s(v)$. Note that, if w is a child of v in T, then $s(w) = s(v)/2$. We denote the spatial neighborhood of a leaf $v \in T$ by $\mathcal{N}(v)$; a leaf $u \in T$ is in $\mathcal{N}(v)$ if and only if the closed regions $R(u)$ and $R(v)$ share a (piece of a) boundary, that is, $R(u) \cap R(v) \neq \emptyset$ (see Fig. 3).

To use a quadtree T to represent a function, we augment the quadtree by adding a value $h(v) \in \mathbb{R}^+$ to each leaf node $v \in T$. The corresponding function $f_T : \mathcal{D} \to \mathbb{R}^+$ is then defined as the function for which $f_T(x, y) = h(v)$ if $(x, y) \in R(v)$ for some leaf $v \in T$. While f_T is ill-defined on the boundary between cells, for our approach the function is allowed to take the value $h(v)$ of any of the adjacent cells. Hence, we will simply ignore the values at the boundaries of cells.

Coresets. In this paper, we define a *coreset* Q of a data set P with respect to an algorithm $\mathcal{A}$ as a subset $Q \subset P$ such that $\mathcal{A}$ applied to the coreset Q approximates the result of applying $\mathcal{A}$ to P. We consider a specific type of coreset that can estimate the density of geometric objects: *ε-approximations.* To define ε-approximations we need the concept of *range spaces.* A range space $(S, \mathcal{R})$ consists of a finite set of objects S and a set of ranges $\mathcal{R}$, where $\mathcal{R}$ is a set of subsets of S. Given a range space $(S, \mathcal{R})$, an *ε-approximation* for some $\varepsilon > 0$ is a subset $Q \subset S$ such that, for all ranges $R \in \mathcal{R}$ we have that $\left| \frac{|R \cap Q|}{|Q|} - \frac{|R|}{|S|} \right| \leq \varepsilon$. The *VC-dimension* d of a range space $(S, \mathcal{R})$ is the size of the largest subset $Y \subseteq S$ such that $\mathcal{R}$ restricted to Y contains all subsets of Y, that is, $2^Y \subseteq \mathcal{R}_{|Y}$, where $\mathcal{R}_{|Y} = \{R \cap Y \mid R \in \mathcal{R}\}$. We say that Y is *shattered* by $\mathcal{R}$.

Kinetic Data Structures. Kinetic Data Structures (KDS) were introduced by Basch [4] to efficiently track attributes of time-varying geometric objects, such as the convex hull of a set of moving points. A KDS uses *certificates* to ensure that the attribute in question is unchanged. Certificates are geometric expressions which are parameterized by the trajectories of the objects. In a classic KDS we assume that these trajectories, also called flight plans, are known at all times. We refer to the failure of a certificate as an event; all certificates are sorted by failure time and stored in an event queue. The KDS then proceeds to handle events one by one, updating its certificates and possibly also the tracked attribute.

Topological Persistence. Let $f : \mathbb{R}^2 \to \mathbb{R}$ be a (smooth) 2d function. A *critical point* of f is a point (x, y) such that the gradient of f at (x, y) is $(0, 0)$. Critical points capture the overall (topological) structure of the function f, and are hence important for analysis. Non-degenerate critical points of a 2d function come in three types: *local minima, saddle points,* and *local maxima.* While some critical points capture relevant features of the function f, other critical points may exist only due to noise. We can measure the relevance of a critical point via the concept of *topological persistence.*

Let $\hat{f}$ be an approximation of f, such that the high persistence local maxima of f are in some sense preserved as maxima in $\hat{f}$. In this paper we use the following lemma, which follows from results of Cohen-Steiner *et al.* [9].

Lemma 1. *Given two functions $f, \hat{f}\colon \mathbb{R}^2 \to \mathbb{R}$ such that $|f(x,y) - \hat{f}(x,y)| < \varepsilon$ for all $(x,y) \in \mathbb{R}^2$, there exists an injection from the local maxima of f with persistence $> 2\varepsilon$ to the local maxima of $\hat{f}$.*

3 Volume-Based Quadtree

In this section we analyze the approximation of a continuous, two-dimensional function f by a quadtree T. Specifically, we prove certain properties on the structure of T and how well the corresponding function f_T approximates f.

We are given a function $f\colon \mathcal{D} \to \mathbb{R}^+$ on a bounded square domain $\mathcal{D} = [0,D] \times [0,D]$ of size D. Without loss of generality, we assume that the total volume under the function f (over the whole domain $\mathcal{D}$) is exactly 1 (this can be achieved by scaling the domain/function). We construct the quadtree T using a threshold value $\rho > 0$. Specifically, starting from the root, we subdivide a node $v \in T$ when the volume under f restricted to the region $R(v)$ exceeds ρ, and we recursively apply this rule to all newly created child nodes. As a result, for every cell (leaf) $v \in T$, the volume under f restricted to $R(v)$ is at most ρ. Finally, for every cell $v \in T$, we set the value $h(v)$ to the average of f in $R(v)$, which implies that the volume under f and f_T is equal when restricted to $R(v)$. For convenience, we will denote this volume as $V(v) = s(v)^2 h(v)$. We will refer to the quadtree T constructed in this manner as the *volume-based quadtree* of f.

We want to choose a threshold value ρ such that f_T has small additive error with respect to f. This is not possible in general, but we prove that it is always possible for functions f which are *Lipschitz continuous*. A function is Lipschitz continuous if there is some Lipschitz constant λ such that λ is the maximum absolute slope of f in any direction. Thus, we also use the Lipschitz constant λ to express our bounds.

The following lemma captures some basic properties of T.

Lemma 2. *Let T be the volume-based quadtree of $f\colon [0,D]^2 \to \mathbb{R}^+$ with threshold ρ. Let $z^* = \max_{(x,y)\in[0,D]^2} f(x,y)$. Then, for any cell $v \in T$, we have that $s(v) > \frac{1}{2}\sqrt{\rho/z^*}$. Additionally, the depth of T is at most $\log\left(\frac{2D}{\sqrt{\rho/z^*}}\right)$ and T has $O\left(\frac{1}{\rho}\log\left(\frac{D}{\sqrt{\rho/z^*}}\right)\right)$ nodes in total.*

Lemma 3 relates the function values of a Lipschitz continuous function f to the volume under f.

Lemma 3. *Let $f\colon \mathcal{D} \to \mathbb{R}^+$ be a 2-dimensional function with Lipschitz constant λ, and let R be a square region in $\mathcal{D}$ with side length s. If $f(x,y) = z$ for some $(x,y) \in R$, then the volume V under f restricted to R satisfies:*

(a) If $z < \sqrt{2}\lambda s$, then $V \geq \frac{z^3}{6\lambda^2}$,
(b) If $z \geq \sqrt{2}\lambda s$, then $V \geq s^2(z - \frac{2\sqrt{2}}{3}\lambda s)$,
(c) $V \leq s^2(z + \frac{2\sqrt{2}}{3}\lambda s)$.

With the help of Lemma 3 we can now prove how well the function f_T implied by a volume-based quadtree T approximates a Lipschitz continuous function f.

Lemma 4. *Let T be the volume-based quadtree of $f: \mathcal{D} \to \mathbb{R}^+$ with threshold ρ. Then, for any cell $v \in T$, we have that $|f(x,y) - f_T(x,y)| \leq \min\left(\frac{2\sqrt{2}}{3}\lambda s(v), \sqrt[3]{6\lambda^2\rho}\right)$ for all $(x,y) \in R(v)$, where λ is the Lipschitz constant of f.*

Proof. We first show that $f_T(x,y) - f(x,y) \leq \frac{2\sqrt{2}}{3}\lambda s(v)$. Pick any $(x,y) \in R(v)$ and let $z = f(x,y)$. From Lemma 3(c) it follows that $V(v) \leq s(v)^2(z + \frac{2\sqrt{2}}{3}\lambda s(v))$. But that directly implies that $h(v) \leq z + \frac{2\sqrt{2}}{3}\lambda s(v)$, and hence $f_T(x,y) - f(x,y) \leq \frac{2\sqrt{2}}{3}\lambda s(v)$.

We now show that $f(x,y) - f_T(x,y) \leq \frac{2\sqrt{2}}{3}\lambda s(v)$. We choose coordinates $(x,y) \in R(v)$ such that $f(x,y)$ is maximized in $R(v)$, and let $z = f(x,y)$. If $z \geq \sqrt{2}\lambda s(v)$, then Lemma 3(b) states that $V(v) \geq s(v)^2(z - \frac{2\sqrt{2}}{3}\lambda s(v))$. But then $h(v) \geq z - \frac{2\sqrt{2}}{3}\lambda s(v)$ and hence $f(x,y) - f_T(x,y) \leq \frac{2\sqrt{2}}{3}\lambda s(v)$. Otherwise, let $z = c\sqrt{2}\lambda s(v)$ for some constant $c \in [0,1)$. From Lemma 3(a) it follows that $V(v) \geq \frac{z^3}{6\lambda^2} = \frac{\sqrt{2}}{3}c^3\lambda s(v)^3$. This also implies that $h(v) \geq \frac{\sqrt{2}}{3}c^3\lambda s(v)$. Finally note that $c - \frac{c^3}{3} \leq \frac{2}{3}$ for all $c \in [0,1)$, and hence $f(x,y) - f_T(x,y) \leq z - h(v) \leq \sqrt{2}\lambda s(v)(c - \frac{c^3}{3}) \leq \frac{2\sqrt{2}}{3}\lambda s(v)$.

In the remainder we can assume that $\sqrt[3]{6\lambda^2\rho} < \frac{2\sqrt{2}}{3}\lambda s(v) < \lambda s(v)$, for otherwise the stated bound already holds. We can rewrite this inequality (by cubing and eliminating some factors) as $6\rho < \lambda s(v)^3$ or $\frac{\rho}{s(v)^2} < \frac{1}{6}\lambda s(v)$. We then get that $h(v) = \frac{V(v)}{s(v)^2} \leq \frac{\rho}{s(v)^2} < \frac{1}{6}\lambda s(v)$. Together with the bounds already proven above, this implies that $f(x,y) < (\frac{1}{6} + \frac{2\sqrt{2}}{3})\lambda s(v) < \sqrt{2}\lambda s(v)$ for all $(x,y) \in R(v)$. Now let $(x,y) \in R(v)$ be the coordinates that maximize $f(x,y)$ in $R(v)$, and let $z = f(x,y)$. From Lemma 3(a) it follows that $\rho \geq V(v) \geq \frac{z^3}{6\lambda^2}$. But then $z^3 \leq 6\lambda^2\rho$, or $z \leq \sqrt[3]{6\lambda^2\rho}$. We thus obtain that $0 \leq f(x,y) \leq \sqrt[3]{6\lambda^2\rho}$ for all $(x,y) \in R(v)$, and since $h(v)$ must be the average of $f(x,y)$ over all $(x,y) \in R(v)$, we directly obtain that $|f(x,y) - f_T(x,y)| \leq \sqrt[3]{6\lambda^2\rho}$. □

Now that we have established that f_T approximates f well, we need to argue that we can find local maxima of f_T efficiently. A (weak) local maximum of f_T corresponds to a cell $v \in T$ such that $h(v) \geq h(w)$ for all $w \in \mathcal{N}(v)$, where $\mathcal{N}(v)$ is the (spatial) neighborhood of v. If the size $|\mathcal{N}(v)|$ of this spatial neighborhood is bounded, then we can efficiently determine if v is a local maximum. Lemma 5 below establishes that $|\mathcal{N}(v)|$ is indeed bounded for cells with a sufficiently small side length $s(v)$. A priori, (weak) maxima can of course also exist in cells with large side length. In the full paper, we further prove that cells with large side length cannot correspond to high persistence maxima in f.

Lemma 5. *Let T be the volume-based quadtree of $f: \mathcal{D} \to \mathbb{R}^+$ with threshold ρ, and let λ be the Lipschitz constant of f. If a cell $v \in T$ satisfies $\frac{\lambda s(v)^3}{\rho} \leq \frac{90}{\sqrt{2}}$, then for all $w \in \mathcal{N}(v)$ with $h(v) \geq h(w)$ it holds that $s(w) \geq \frac{1}{4}s(v)$.*

4 From Volume to Points

We aim to maintain a volume-based quadtree for $f = \text{KDE}_P$ over time. Most common kernels, with the exception of the uniform kernel, are Lipschitz continuous and hence the resulting KDE is also Lipschitz continuous. We scale KDE_P such that the volume underneath KDE_P is 1. Additionally, for ease of analysis, we assume the kernel width σ to be 1. As a result, for common kernels such as the cone kernel, $f = \text{KDE}_P$ is Lipschitz continuous with a small Lipschitz constant. Furthermore, the maximum value of f is bounded as well.

We want to approximate the volume under $f = \text{KDE}_P$, as the points in P are moving, via a (small) set of moving points Q. We use $V_R(f(t))$ to refer to the volume under a function f at time t restricted to a region R. For a chosen value $\varepsilon_{\text{cor}} > 0$, we require the following property on Q: for any square region $R \subseteq \mathcal{D}$ and time t, we have that $\left|\frac{|Q \cap R|}{|Q|} - V_R(\text{KDE}_P(t))\right| < \varepsilon_{\text{cor}}$. We plan to use ε-approximations to construct a suitable point set Q. An ε-approximation needs an initial point set from which to construct Q. We therefore first take a dense point sample S under each kernel K to represent its volume (see Sect. 4.1). Then we combine the samples for the individual kernels into a set $\mathcal{S}$ which serves as the input for the ε-approximation that will ultimately result in Q (see Sect. 4.2). In Sect. 4.3 we then show how to replace the actual volume under KDE_P with the points in Q when constructing the volume-based quadtree.

4.1 Approximating a Single Kernel

Let $K\colon [-1,1]^2 \to \mathbb{R}^+$ denote the kernel function. We aim to represent the volume under K using a set of points S. To obtain S, we consider a regular $r \times r$ grid G on the domain $[-1,1]^2$, for some value r to be chosen later. Note that the area of a single grid cell $c \in G$ is $\frac{4}{r^2}$. We construct a *grid-based sampling* $S(r)$ of K by arbitrarily placing $\lceil rz(c) \rceil$ points in every cell $c \in G$, where $z(c)$ is the average value of K in the corresponding grid cell c. See Fig. 4 for an example.

For ease of exposition we assume in the remainder of the paper that the maximum value of the kernel K is bounded by 1 (this holds for most common kernels). We can prove the following property on $S(r)$.

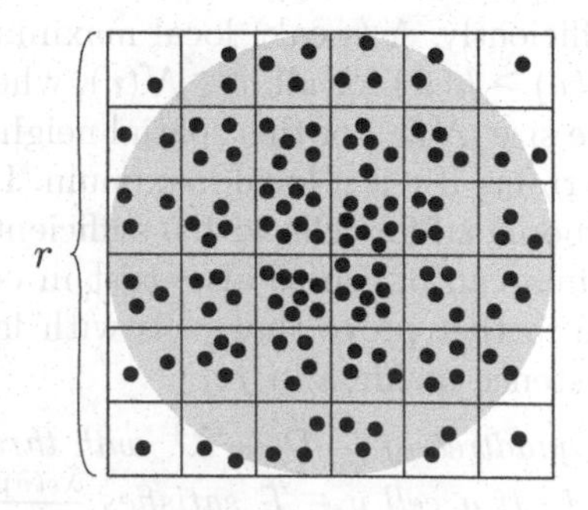

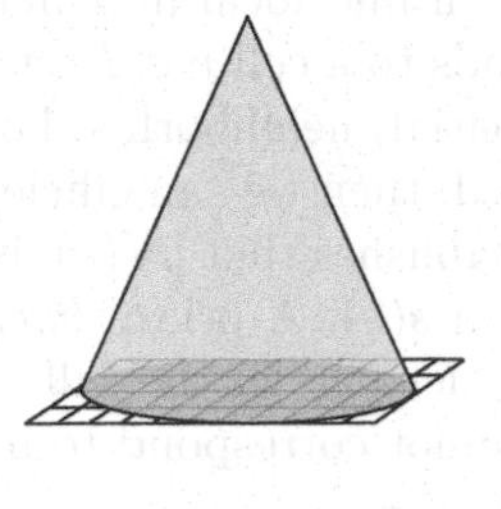

Fig. 4. Approximating a cone kernel: in each cell, $\lceil rz(c) \rceil$ points are placed arbitrarily.

Lemma 6. *Let $K\colon [-1,1]^2 \to \mathbb{R}^+$ be a function such that the total volume under K is 1 and $K(x,y) \leq 1$ for all $(x,y) \in [-1,1]^2$. If $S(r)$ is a grid-based sampling of K with parameter r, then for any square region R that overlaps with the domain of K we have that $\left|\frac{|S(r)\cap R|}{|S(r)|} - V_R(K)\right| \leq \frac{36}{r}$.*

Lemma 6 implies that, for a chosen error $\varepsilon_{\text{dsc}} > 0$, we can choose $r = \frac{36}{\varepsilon_{\text{dsc}}}$ to obtain a grid-based sampling S with $O(\frac{1}{\varepsilon_{\text{dsc}}^3})$ points that approximates the volume under K with error at most ε_{dsc}.

4.2 Coreset

We now construct an approximation for the volume under KDE_P as the points in P are moving. For a chosen error $\varepsilon_{\text{dsc}} > 0$, we construct a grid-based sampling S_p of $O(\frac{1}{\varepsilon_{\text{dsc}}^3})$ points around each point $p \in P$, resulting in $O(\frac{n}{\varepsilon_{\text{dsc}}^3})$ points in total. We let the points in S_p move in the same direction as the corresponding point $p \in P$; that is, each S_p becomes a set of linearly moving points. The complete set of moving points $\mathcal{S} = \bigcup_{p\in P} S_p$ provides an approximation for the volume under KDE_P with error at most ε_{dsc} for all times t.

We use an algorithm by Agarwal *et al.* [1] to construct an ε-approximation Q of $\mathcal{S}$. The details can be found in the full paper. For this algorithm to be applicable, we need to show that the range space $X = (\mathcal{S}, \mathcal{R})$ has bounded VC-dimension, where $\mathcal{R}$ contains all subsets of points in $\mathcal{S}$ that may appear in a square region R at some time t. Note that this is non-trivial, since the points in $\mathcal{S}$ are moving points (see Fig. 5). As already stated earlier, we assume that the points follow linear motion.

Lemma 7. *Let $X_2 = (S_2, \mathcal{R}_2)$ be a range space where S_2 contains a set of z-monotone lines in $\mathbb{R}^3$, and $\mathcal{R}_2$ contains all subsets of lines in S_2 that can be intersected by an axis-aligned square in $\mathbb{R}^3$ with constant z-coordinate. The VC-dimension of X_2 is at most* 38.

From Lemma 7 we can conclude that the range space $X = (\mathcal{S}, \mathcal{R})$ has VC-dimension 38. In the remainder of this paper we will simply refer to the set of

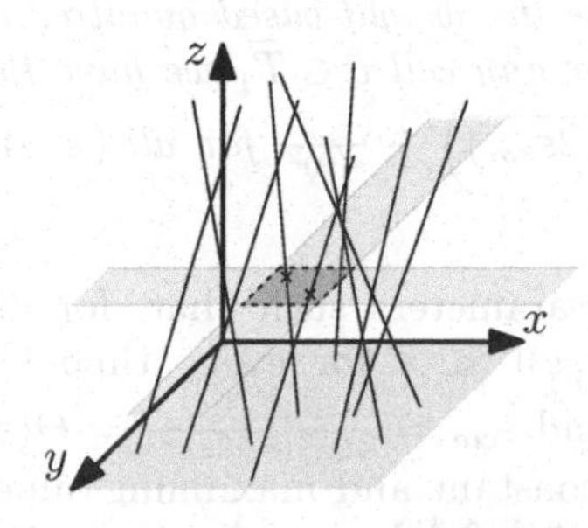

Fig. 5. An example of the range space $X = (\mathcal{S}, \mathcal{R})$ plotted over time. The lines denote set $\mathcal{S}$, and the blue square is an example of a range in $\mathcal{R}$.

linearly moving points Q as the coreset of KDE_P, where the additive error with respect to the volume is ε_{cor}. We summarize the result in the following lemma.

Lemma 8. *Let KDE_P be a KDE function on a set of n linearly moving points P. For any $\varepsilon_{cor} > 0$, we can construct a coreset Q of linearly moving points such that, for any time t and any square region R, we get that*

$$\left| \frac{|Q \cap R|}{|Q|} - V_R(KDE_P(t)) \right| < \varepsilon_{cor},$$

where $V_R(KDE_P(t))$ is the volume under KDE_P at time t restricted to R. Q has $O(\frac{1}{\varepsilon_{cor}^2} \log\left(\frac{1}{\varepsilon_{cor}}\right))$ points and can be constructed in $O(n \operatorname{poly}\left(\frac{\log n}{\varepsilon_{cor}}\right))$ time.

4.3 Weight-Based Quadtree

The coreset Q of KDE_P functions as a proxy for the volume under KDE_P restricted to some square region. We can therefore approximate the volume-based quadtree T of KDE_P with the *weight-based quadtree* $\widetilde{T}$ of Q, in which we subdivide a node $v \in \widetilde{T}$ when the fraction $\frac{|Q \cap R(v)|}{|Q|}$ exceeds ρ. However, we do not subdivide nodes with $s(v) \leq \sqrt{\rho}$, so that the lower bound on cell size in Lemma 2 is preserved in $\widetilde{T}$. We refer to $\frac{|Q \cap R(v)|}{|Q|}$ as the *weight* of cell $v \in \widetilde{T}$, denoted by $W(v)$. As a result, for every cell (leaf) $v \in \widetilde{T}$, the weight of v is at most $\rho + \varepsilon_{\text{cor}}$ when $s(v) \leq \sqrt{\rho}$, and at most ρ otherwise. Finally, for every cell $v \in \widetilde{T}$, we set the value $h(v)$ to $\frac{W(v)}{s(v)^2}$.

By construction, the minimum cell size and maximum depth in Lemma 2 are preserved by $\widetilde{T}$. Since the total weight of all cells in $\widetilde{T}$ is 1 by construction, the number of nodes in Lemma 2 also holds for $\widetilde{T}$. However, the error made by the volume-based quadtree in Lemma 4 does not directly hold for $\widetilde{T}$, as we need to incorporate the error on the volume under the function $f = \text{KDE}_P$. We therefore give a new bound on the error for $\widetilde{T}$.

Lemma 9. *Let $f = KDE_P$ be a KDE function on a set of linearly moving points P at a fixed time t, and let Q be a coreset for KDE_P with additive error ε_{cor}. Furthermore, let $\widetilde{T}$ be the weight-based quadtree on Q with threshold ρ at the same time t. Then, for any cell $v \in \widetilde{T}$, we have that $|f(x,y) - f_{\widetilde{T}}(x,y)| < \min\left(\frac{2\sqrt{2}}{3}\lambda s(v), \sqrt[3]{6\lambda^2(\rho + 2\varepsilon_{cor})}\right) + \frac{\varepsilon_{cor}}{s(v)^2}$ for all $(x,y) \in R(v)$, where λ is the Lipschitz constant of f.*

We may now choose parameters such that, for any error $\varepsilon > 0$, we get that $|\text{KDE}_P(x,y) - f_{\widetilde{T}}(x,y)| < \varepsilon$ for every time t. Specifically, we choose $\rho = \frac{\varepsilon^3}{6\lambda^2(8+2\varepsilon z^*)} = \Theta(\varepsilon^3)$ and $\varepsilon_{\text{cor}} = \frac{\varepsilon^4 z^*}{48\lambda^2(8+2\varepsilon z^*)} = \Theta(\varepsilon^4)$, where $\lambda = \Theta(1)$ and $z^* \leq 1$ are the Lipschitz constant and maximum value of KDE_P, respectively. Observe that $\rho + 2\varepsilon_{\text{cor}} = \frac{8\varepsilon^3 + 2\varepsilon^4 z^*}{48\lambda^2(8+2\varepsilon z^*)} = \frac{\varepsilon^3}{48\lambda^2}$, and hence $\sqrt[3]{6\lambda^2(\rho + 2\varepsilon_{\text{cor}})} = \frac{\varepsilon}{2}$. Furthermore, by Lemma 2 we know that $s(v)^2 \geq \frac{1}{4}\rho z^*$ for all $v \in \widetilde{T}$. As

$\frac{\varepsilon_{\text{cor}}}{\rho} = \frac{\varepsilon z^*}{8}$, we get that $\frac{\varepsilon_{\text{cor}}}{s(v)^2} \leq \frac{\varepsilon}{2}$ for all $v \in \widetilde{T}$. With these choices of ρ and ε_{cor}, it now follows from Lemma 9 that $|\text{KDE}_P(x,y) - f_{\widetilde{T}}(x,y)| < \varepsilon$ for every time t. This also implies that the weight-based quadtree $\widetilde{T}$ has at most $O(\frac{1}{\varepsilon^3}\log\left(\frac{n}{\varepsilon}\right))$ nodes (Lemma 2), and that the coreset Q contains $O(\frac{1}{\varepsilon^8}\log\left(\frac{1}{\varepsilon}\right))$ points in total (Lemma 8). In the remainder of this paper we assume that Q and $\widetilde{T}$ are constructed with the parameters ρ and ε_{cor} chosen above.

5 KDS for Density Approximation

In this section we describe a KDS to efficiently maintain the weight-based quadtree $\widetilde{T}$ on a set of linearly moving points Q over time. By the results of Sect. 4 and Lemma 1, keeping track of the local maxima of $f_{\widetilde{T}}$ is sufficient to track the local maxima of KDE_P with persistence at least 2ε. For every cell in the weight-based quadtree $\widetilde{T}$ we store a boolean value which indicates if this cell is currently a local maximum or not. To efficiently update this information as points move in and out of quadtree cells, we store a set of pointers $\mathcal{M}(v)$ in each node $v \in \widetilde{T}$ (including internal nodes) to all nodes $w \in \widetilde{T}$ with $\frac{1}{4}s(v) \leq s(w) \leq 4s(v)$ such that $R(v)$ and $R(w)$ share (a piece of) boundary. It is easy to see that $|\mathcal{M}(v)| = O(1)$ for all $v \in \widetilde{T}$. These pointers will be used to efficiently update the boolean values.

An event is triggered only when a point from Q crosses a boundary of a cell in $\widetilde{T}$. Assume that some point $q \in Q$ moved from a cell $v \in \widetilde{T}$ to a cell $u \in \mathcal{N}(v)$. We can determine u using a simple point query on $\widetilde{T}$. Next, we must update the weights $W(v)$ and $W(u)$ accordingly. Based on these new weights and the threshold value ρ, we may need to (possibly recursively) split u into four smaller cells and/or (possibly recursively) merge v with its siblings in $\widetilde{T}$.

Finally, we need to update the boolean values that indicate which cells are local maxima; these can change for each affected node $v \in \widetilde{T}$, as well as their neighborhoods $\mathcal{N}(v)$. We use a generalization of Lemma 5 to weight-based quadtrees to bound the number of neighbors we need to consider. Hence, we can update the boolean values efficiently via the pointers in $\mathcal{M}(v)$. Refer to the full paper for a more detailed description of the event handling, as well as a quality analysis for the KDS. We summarize the result in the following theorem.

Theorem 1. *Let $f = \text{KDE}_P$ be a KDE function on a set P of n linearly moving points in $[0,D]^2$. For any $\varepsilon > 0$, there exists a KDS that approximately maintains the local maxima of f with persistence at least 2ε. The KDS can be initialized in $O\left(n \operatorname{poly}\left(\frac{\log n}{\varepsilon}\right)\right)$ time, processes at most $O\left(D \operatorname{poly}\left(\frac{1}{\varepsilon}\right)\right)$ events, and can handle events and flight plan updates in $O\left(\log D + \operatorname{poly}\left(\frac{1}{\varepsilon}\right)\right)$ and $O\left(\operatorname{poly}\left(\frac{\log n}{\varepsilon}\right)\right)$ time, respectively.*

The coreset Q changes only when a point changes its trajectory. We can handle such a flight plan update efficiently using the data structure by Agarwal *et al.* [1]. We only prove results on maintaining persistent local maxima, our approach actually maintains an approximation of the KDE function, and hence could also be used to track other shape features based on the density of P.

6 Discussion

We presented a KDS that efficiently tracks persistent local maxima of a KDE on a set of linearly moving points P. To develop this KDS, we first showed how to approximate a density function via a volume-based quadtree. We then proved that we can compute a coreset of moving points which approximates the volume under a density function. A weight-based quadtree on this coreset in turn approximates the volume-based quadtree on the density function. For any $\varepsilon > 0$, we can compute this coreset of size $O(\text{poly}(\frac{1}{\varepsilon}))$ in $O\left(n \,\text{poly}\left((\log n)/\varepsilon\right)\right)$ time, where n is the number of points in P and ε is the error bound between the weight-based quadtree and the density function.

Various bounds on the quadtree complexity and the KDS quality measures depend on the size of the domain D. As we assume that our input points represent a single group, it makes sense to assume that the kernel functions of any point (its region of influence) must overlap with the kernel function of at least one other point. Since we scale the input such that the kernel width is $\sigma = 1$, this directly implies that $D = O(n)$ for a static set of points, although it is likely much smaller. However, when points move in a single direction for a long time, they may easily leave a domain of that size. To address this problem without blowing up the size of the domain, we can move the domain itself along a piecewise-linear trajectory. A change of direction of the domain directly changes the trajectories of all points, and all events in the KDS must be recomputed. The coreset, however, does not need to change during such an event. We can limit the number of domain flight plan changes by using a slightly larger domain than needed at any point in time.

We believe that approximating the density surface we want to maintain via a suitable coreset of moving points is a promising direction also in practice. Below we briefly sketch the necessary adaptations that we foresee. For the coreset to exist, the range space formed by the trajectories of the (samples around) the input objects and a set of square regions needs to have bounded VC-dimension. We proved an upper bound on this VC-dimension in the case that all trajectories are linear. Real-world animal trajectories are certainly not linear. However, since animals cannot move at arbitrary speeds and subgroups can often be observed to stay together, we still expect the corresponding range space to have bounded VC-dimension. A formal proof seems out of reach for more than very restrictive motion models, but bounds might be deduced from experimental data.

The actual computation of the coreset via the algorithm of Agarwal *et al.* [1] is impossible if the trajectories are not known ahead of time. However, random sampling (that is, sample a point $p \in P$, and then sample from its kernel) can be expected to result in a coreset of good size and quality in practice. Since we have bounds on the maximum speeds of the animals, a black-box KDS could be used to maintain such a random sampling coreset efficiently.

Our theoretical results inform the direction of our future engineering efforts in two ways. First of all, we now know that we can approximate well with a coreset whose size depends only on the desired approximation factor and not on the input size. Second, we know how to sample to find such a coreset, by essentially constructing randomly shifted copies of the input points.

References

1. Agarwal, P., de Berg, M., Gao, J., Guibas, L., Har-Peled, S.: Staying in the middle: exact and approximate medians in R1 and R2 for moving points. In: Proceedings of the 17th Canadian Conference on Computational Geometry, pp. 43–46 (2005)
2. Andrienko, G., Andrienko, N.: Interactive cluster analysis of diverse types of spatiotemporal data. ACM SIGKDD Explorations **11**(2), 19–28 (2010)
3. Andrienko, N., Andrienko, G., Barrett, L., Dostie, M., Henzi, P.: Space transformation for understanding group movement. IEEE Trans. Vis. Comput. Graph. **19**(12), 2169–2178 (2013)
4. Basch, J.: Kinetic data structures. Ph.D. thesis, Stanford University (1999)
5. Benkert, M., Gudmundsson, J., Hübner, F., Wolle, T.: Reporting flock patterns. Comput. Geom. **41**(3), 111–125 (2008)
6. Bialkowski, A., Lucey, P., Carr, P., Yue, Y., Sridharan, S., Matthews, I.: Identifying team style in soccer using formations learned from spatiotemporal tracking data. In: Proceedings of 2014 IEEE International Conference on Data Mining Workshop, pp. 9–14 (2014)
7. Bialkowski, A., Lucey, P., Carr, P., Yue, Y., Sridharan, S., Matthews, I.: Large-scale analysis of soccer matches using spatiotemporal tracking data. In: Proceedings of 2014 IEEE International Conference on Data Mining, pp. 725–730 (2014)
8. Buchin, K., Buchin, M., van Kreveld, M., Speckmann, B., Staals, F.: Trajectory grouping structure. J. Comput. Geom. **6**(1), 75–98 (2015)
9. Cohen-Steiner, D., Edelsbrunner, H., Harer, J.: Stability of persistence diagrams. Discret. Comput. Geom. **37**, 103–120 (2007)
10. Dodge, S., Weibel, R., Lautenschütz, A.K.: Towards a taxonomy of movement patterns. Inf. Vis. **7**, 240–252 (2008)
11. Eikelboom, J.: Sentinel animals: enriching artificial intelligence with wildlife ecology to guard rhinos. Ph.D. thesis, Wageningen University (2021)
12. van Goethem, A., van Kreveld, M., Löffler, M., Speckmann, B., Staals, F.: Grouping time-varying data for interactive exploration. In: Proceedings of the 32nd International Symposium on Computational Geometry, pp. 61:1–61:16 (2016)
13. Gudmundsson, J., Horton, M.: Spatio-temporal analysis of team sports. ACM Comput. Surv. **50**(2), 1–34 (2017)
14. Gudmundsson, J., Laube, P., Wolle, T.: Movement patterns in spatio-temporal data. Encycl. GIS **726**, 732 (2008)
15. Gudmundsson, J., Wolle, T.: Football analysis using spatio-temporal tools. Comput. Environ. Urban Syst. **47**, 16–27 (2014)
16. Huang, Y., Chen, C., Dong, P.: Modeling herds and their evolvements from trajectory data. In: Proceedings of the 5th International Conference on Geographic Information Science, pp. 90–105 (2008)
17. Hwang, S.Y., Liu, Y.H., Chiu, J.K., Lim, E.P.: Mining mobile group patterns: a trajectory-based approach. In: Proceedings of the 9th Conference on Advances in Knowledge Discovery and Data Mining, pp. 713–718 (2005)
18. Kalnis, P., Mamoulis, N., Bakiras, S.: On discovering moving clusters in spatio-temporal data. In: Proceedings of the 9th International Conference on Advances in Spatial and Temporal Databases, pp. 364–381 (2005)

19. Kostitsyna, I., Kreveld, M.V., Löffler, M., Speckmann, B., Staals, F.: Trajectory grouping structure under geodesic distance. In: Proceedings of the 31st International Symposium on Computational Geometry, pp. 674–688 (2015)
20. van Kreveld, M., Löffler, M., Staals, F., Wiratma, L.: A refined definition for groups of moving entities and its computation. Int. J. Comput. Geom. Appl. **28**(02), 181–196 (2018)
21. Laube, P., Imfeld, S., Weibel, R.: Discovering relative motion patterns in groups of moving point objects. Int. J. Geogr. Inf. Sci. **19**(6), 639–668 (2005)
22. Lucey, P., Bialkowski, A., Carr, P., Morgan, S., Matthews, I., Sheikh, Y.: Representing and discovering adversarial team behaviors using player roles. In: Proceedings of 2013 IEEE Conference on Computer Vision and Pattern Recognition, pp. 2706–2713 (2013)
23. Lucey, P., Bialkowski, A., Carr, P., Yue, Y., Matthews, I.: How to get an open shot: analyzing team movement in basketball using tracking data. In: Proceedings of the 8th Annual MIT SLOAN Sports Analytics Conference (2014)
24. Nilsson, A.: Predator behaviour and prey density: evaluating density-dependent intraspecific interactions on predator functional responses. J. Anim. Ecol. **70**(1), 14–19 (2001)
25. Parzen, E.: On estimation of a probability density function and mode. Ann. Math. Stat. **33**(3), 1065–1076 (1962)
26. Reynolds, C.: Flocks, herds and schools: a distributed behavioral model. In: Proceedings of the 14th Annual Conference on Computer Graphics and Interactive Techniques, pp. 25–34 (1987)
27. Rosenblatt, M.: Remarks on some nonparametric estimates of a density function. Ann. Math. Stat. **27**(3), 832–837 (1956)
28. Wiratma, L., van Kreveld, M., Löffler, M., Staals, F.: An experimental evaluation of grouping definitions for moving entities. In: Proceedings of the 27th ACM SIGSPATIAL International Conference on Advances in Geographic Information Systems, pp. 89–98 (2019)
29. Wood, Z.M.: Detecting and identifying collective phenomena within movement data. Ph.D. thesis, University of Exeter (2011)

Dynamic Convex Hulls Under Window-Sliding Updates

Haitao Wang(✉)

School of Computing, University of Utah, Salt Lake City, UT 84112, USA
haitao.wang@utah.edu

Abstract. We consider the problem of dynamically maintaining the convex hull of a set S of points in the plane under the following special sequence of insertions and deletions (called *window-sliding updates*): insert a point to the right of all points of S and delete the leftmost point of S. We propose an $O(|S|)$-space data structure that can handle each update in $O(1)$ amortized time, such that all standard binary-search-based queries on the convex hull of S can be answered in $O(\log |S|)$ time, and the convex hull itself can be output in time linear in its size.

Keywords: Convex hulls · Dynamic update · Window-sliding

1 Introduction

As a fundamental structure in computational geometry, the convex hull $CH(S)$ of a set S of points in the plane has been well studied in the literature. Several $O(n \log n)$ time algorithms are known for computing $CH(S)$, e.g., see [5,26], where $n = |S|$, and the time matches the $\Omega(n \log n)$ lower bound. Output-sensitive $O(n \log h)$ time algorithms have also been given [9,21], where h is the number of vertices of $CH(S)$. If the points of S are already sorted, e.g., by x-coordinate, then $CH(S)$ can be computed in $O(n)$ time by Graham's scan [14].

Due to a wide range of applications, the problem of dynamically maintaining $CH(S)$, where points can be inserted and/or deleted from S, has also been extensively studied. Overmars and van Leeuwen [24] proposed an $O(n)$-space data structure that can support each insertion and deletion in $O(\log^2 n)$ worst-case time; Preparata and Vitter [27] gave a simpler method with the same performance. If only insertions are involved, then the approach of Preparata [25] can support each insertion in $O(\log n)$ worst-case time. For deletions only, Hershberger and Suri's method [18] can support each update in $O(\log n)$ amortized time. If both insertions and deletions are allowed, a breakthrough was given by Chan [10], who developed a data structure of linear space that can support each update in $O(\log^{1+\epsilon} n)$ amortized time, for an arbitrarily small $\epsilon > 0$. Subsequently, Brodal and Jacob [7], and independently Kaplan et al. [20] reduced the

This research was supported in part by NSF under Grants CCF-2005323 and CCF-2300356. A full version is available at http://arxiv.org/abs/2305.08055.

P. Morin and S. Suri (Eds.): WADS 2023, LNCS 14079, pp. 689–703, 2023.
https://doi.org/10.1007/978-3-031-38906-1_46

update time to $O(\log n \log\log n)$. Finally, Brodal and Jacob [6] achieved $O(\log n)$ amortized time performance for each update, with $O(n)$ space.

Under certain special situations, better and simpler results are also known. If the insertions and deletions are given offline, the data structure of Hershberger and Suri [19] can support $O(\log n)$ amortized time update. Schwarzkopf [28] and Mulmuley [23] presented algorithms to support each update in $O(\log n)$ expected time if the sequence of updates is random in a certain sense. In addition, Friedman et al. [13] considered the problem of maintaining the convex hull of a simple path P such that vertices are allowed to be inserted and deleted from P at both ends of P, and they gave a linear space data structure that can support each update in $O(\log |P|)$ amortized time (more precisely, $O(1)$ amortized time for each deletion and $O(\log |P|)$ amortized time for each insertion). There are also other special dynamic settings for convex hulls, e.g., [12,17].

In most applications, the reason to maintaining $CH(S)$ is to perform queries on it efficiently. As discussed in Chan [11], there are usually two types of queries, depending on whether the query is *decomposable* [4], i.e., if S is partitioned into two subsets, then the answer to the query for S can be obtained in constant time from the answers of the query for the two subsets. For example, the following queries are decomposable: find the most extreme vertex of $CH(S)$ along a query direction; decide whether a query line intersects $CH(S)$; find the two common tangents to $CH(S)$ from a query point outside $CH(S)$, while the following queries are not decomposable: find the intersection of $CH(S)$ with a vertical query line or more generally an arbitrary query line. It seems that the decomposable queries are easier to deal with. Indeed, most of the above mentioned data structures can handle the decomposable queries in $O(\log n)$ time each. However, this is not the case for the non-decomposable queries. For example, none of the data structures of [6,7,10,13,20] can support $O(\log n)$-time non-decomposable queries. More specifically, Chan's data structure [10] can be modified to support each non-decomposable query in $O(\log^{3/2} n)$ time but the amortized update time also increases to $O(\log^{3/2} n)$. Later Chan [11] gave a randomized algorithm that can support each non-decomposable query in expected $O(\log^{1+\epsilon} n)$ time, for an arbitrarily small $\epsilon > 0$, and the amortized update time is also $O(\log^{1+\epsilon} n)$.

Another operation on $CH(S)$ is to output it explicitly, ideally in $O(h)$ time. To achieve this, one usually has to maintain $CH(S)$ explicitly in the data structure, e.g., in [18,24]. Unfortunately, most other data structures are not able to do so, e.g., [6,7,10,13,19,20,27], although they can output $CH(S)$ in a slightly slower $O(h \log n)$ time. In particular, Bus and Buzer [8] considered this operation for maintaining the convex hull of a simple path P such that vertices are allowed to be inserted and deleted from P at both ends of P, i.e., in the same problem setting as in [13]. Based on a modification of the algorithm in [22], they achieved $O(1)$ amortized update time such that $CH(S)$ can be output explicitly in $O(h)$ time [8]. However, no other queries on $CH(S)$ were considered in [8].

Our Results. We consider a special sequence of insertions and deletions: the point inserted by an insertion must be to the right of all points of the current set S, and a deletion always happens to the leftmost point of the current set S.

Equivalently, we may consider the points of S contained in a window bounded by two vertical lines that are moving rightwards (but the window width is not fixed), so we call them the *window-sliding updates.*

To solve the problem, one can apply any previous data structure for arbitrary point updates. For example, the method in [6] can handle each update in $O(\log n)$ amortized time and answer each decomposable query in $O(\log n)$ time. Alternatively, if we connect all points of S from left to right by line segments, then we can obtain a simple path whose ends are the leftmost and rightmost points of S, respectively. Therefore, the data structure of Friedman et al. [13] can be applied to handle each update in $O(\log n)$ amortized time and support each decomposable query in $O(\log n)$ time. In addition, although the data structure in [18] is particularly for deletions only, Hershberger and Suri [18] indicated that their method also works for the window-sliding updates, in which case each update (insertion and deletion) takes $O(\log n)$ amortized time. Further, as the data structure [18] explicitly stores the edges of $CH(S)$ in a balanced binary search tree, it can support both decomposable and non-decomposable queries each in $O(\log n)$ time as well as output $CH(S)$ in $O(h)$ time.

In this paper, we provide a new data structure for the window-sliding updates. Our data structure uses $O(n)$ space and can handle each update in $O(1)$ amortized time. All decomposable and non-decomposable queries on $CH(S)$ mentioned above can be answered in $O(\log n)$ time each. Further, after each update, the convex hull $CH(S)$ can be output explicitly in $O(h)$ time. Specifically, the following theorem summarizes our result.

Theorem 1. *We can dynamically maintain the convex hull $CH(S)$ of a set S of points in the plane to support each window-sliding update (i.e., either insert a point to the right of all points of S or delete the leftmost point of S) in $O(1)$ amortized time, such that the following operations on $CH(S)$ can be performed. Let $n = |S|$ and h be the number of vertices of $CH(S)$ right before each operation.*

1. *The convex hull $CH(S)$ can be explicitly output in $O(h)$ time.*
2. *Given two vertical lines, the vertices of $CH(S)$ between the vertical lines can be output in order along the boundary of $CH(S)$ in $O(k + \log n)$ time, where k is the number of vertices of $CH(S)$ between the two vertical lines.*
3. *Each of the following queries can be answered in $O(\log n)$ time.*
   - (a) *Given a query direction, find the most extreme point of S along the direction.*
   - (b) *Given a query line, decide whether the line intersects $CH(S)$.*
   - (c) *Given a query point outside $CH(S)$, find the two tangents from the point to $CH(S)$.*
   - (d) *Given a query line, find its intersection with $CH(S)$, or equivalently, find the edges of $CH(S)$ intersecting the line.*
   - (e) *Given a query point, decide whether the point is in $CH(S)$.*
   - (f) *Given a convex polygon (represented in any data structure that supports binary search), decide whether it intersects $CH(S)$, and if not, find their common tangents (both outer and inner).*

Comparing to all previous work, albeit on a very special sequence of updates, our result is particularly interesting due to the $O(1)$ amortized update time as well as its simplicity.

Applications. Although the updates in our problem are quite special, the problem still finds applications. For example, Becker et al. [2] considered the problem of finding two rectangles of minimum total area to enclose a set of n rectangles in the plane. They gave an algorithm of $O(n \log n)$ time and $O(n \log \log n)$ space. Their algorithm has a subproblem of processing a dynamic set of points to answer queries of Type 3a of Theorem 1 with respect to window-sliding updates (see Section 3.2 [2]). The subproblem is solved using subpath convex hull query data structure in [15], which costs $O(n \log \log n)$ space. Using Theorem 1, we can reduce the space of the algorithm to $O(n)$ while the runtime is still $O(n \log n)$. Note that Wang [29] recently improved the space of the result of [15] to $O(n)$, which also leads to an $O(n)$ space solution for the algorithm of [2]. However, the approach of Wang [29] is much more complicated.

Becker et al. [1] extended their work above and studied the problem of enclosing a set of simple polygons using two rectangles of minimum total area. They gave an algorithm of $O(n\alpha(n) \log n)$ time and $O(n \log \log n)$ space, where n is the total number of vertices of all polygons and $\alpha(n)$ is the inverse Ackermann function. The algorithm has a similar subproblem as above (see Section 4.2 [1]). Similarly, our result can reduce the space of their algorithm to $O(n)$ while the runtime is still $O(n\alpha(n) \log n)$.

Outline. After introducing notation in Sect. 2, we will prove Theorem 1 gradually as follows. First, in Sect. 3, we give a data structure for a special problem setting. Then we extend our techniques to the general problem in Sect. 4. The data structures in Sect. 3 and 4 can only perform the first operation in Theorem 1 (i.e., output $CH(S)$), we will enhance the data structure in Sect. 5 so that other operations can be handled. Due to the space limit, all lemma proofs are omitted but can be found in the full paper.

2 Preliminaries

Let $\mathcal{U}(S)$ denote the upper hull of $CH(S)$. We will focus on maintaining $\mathcal{U}(S)$, and the lower hull can be treated likewise. The data structure for both hulls together will achieve Theorem 1.

For any two points p and q in the plane, we say that p is to the *left* of q if the x-coordinate of p is smaller than or equal to that of q. Similarly, we can define "to the right of", "above", and "below". We add "strictly" in front of them to indicate that the tie case does not happen. For example, p is strictly below q if the y-coordinate of p is smaller than that of q.

For a line segment s and a point p, we say that p is *vertically below* s if the vertical line through p intersects s at a point above p ($p \in s$ is possible). For any two line segments s_1 and s_2, we say that s_1 is vertically below s_2 if both endpoints of s_1 are vertically below s_2.

Suppose $\mathcal{L}$ is a sequence of points and p and q are two points of $\mathcal{L}$. We follow the convention that a subsequence of $\mathcal{L}$ *between* p *and* q includes both p and q, but a subsequence of $\mathcal{L}$ *strictly* between p and q does not include either one.

For ease of exposition, we make a general position assumption that no two points of S have the same x-coordinate and no three points are collinear.

3 A Special Problem Setting with a Partition Line

In this section we consider a special problem setting. Specifically, let $L = \{p_1, p_2, \ldots, p_n\}$ (resp., $R = \{q_1, q_2, \ldots, q_n\}$) be a set of n points sorted by increasing x-coordinate, such that all points of L are strictly to the left of a known vertical line ℓ and all points of R are strictly to the right of ℓ. We want to maintain the upper convex hull $\mathcal{U}(S)$ of a consecutive subsequence S of $L \cup R = \{p_1, \ldots, p_n, q_1, \ldots, q_n\}$, i.e., $S = \{p_i, p_{i+1}, \ldots, p_n, q_1, q_2, \ldots, q_j\}$, with $S = L$ initially, such that a deletion will delete the leftmost point of S and an insertion will insert the point of R right after the last point of S. Further, deletions only happen to points of L, i.e., once p_n is deleted from S, no deletion will happen. Therefore, there are a total of n insertions and n deletions.

Our result is a data structure that supports each update in $O(1)$ amortized time, and after each update we can output $\mathcal{U}(S)$ in $O(|\mathcal{U}(S)|)$ time. The data structure can be built in $O(n)$ time on $S = L$ initially. Note that L is given offline because $S = L$ initially, but points of R are given online. We will extend the techniques to the general problem setting in Sect. 4, and the data structure will be enhanced in Sect. 5 so that other operations on $CH(S)$ can be handled.

3.1 Initialization

Initially, we construct the data structure on L, as follows. We run Graham's scan to process points of L leftwards from p_n to p_1. Each vertex $p_i \in L$ is associated with a stack $Q(p_i)$, which is empty initially. Each vertex p_i also has two pointers $l(p_i)$ and $r(p_i)$, pointing to its left and right neighbors respectively if p_i is a vertex of the current upper hull. Suppose we are processing a point p_i. Then, the upper hull of $p_{i+1}, p_{i+2}, \ldots, p_n$ has already been maintained by a doubly linked list with p_{i+1} as the head. To process p_i, we run Graham's scan to find a vertex p_j of the current upper hull such that $\overline{p_i p_j}$ is an edge of the new upper hull. Then, we push p_i into the stack $Q(p_j)$, and set $l(p_j) = p_i$ and $r(p_i) = p_j$. The algorithm is done after p_1 is processed.

The stacks essentially maintain the left neighbors of the vertices of the historical upper hulls so that when some points are deleted in future, we can traverse leftwards from any vertex on the current upper hull after those deletions. More specifically, if p_i is a vertex on the current upper hull, then the vertex at the top of $Q(p_i)$ is the left neighbor of p_i on the upper hull. In addition, notice that once the right neighbor pointer $r(p_i)$ is set during processing p_i, it will never be changed. Hence, in future if p_i becomes a vertex of the current upper hull after some deletions, $r(p_i)$ is the right neighbor of p_i on the current upper hull. Therefore, we do not need another stack to keep the right neighbor of p_i.

The above builds our data structure for $\mathcal{U}(S)$ initially when $S = L$. In what follows, we discuss the general situation when S contains both points of L and R. Let $S_1 = S \cap L$ and $S_2 = S \cap R$. The data structure described above is used for maintaining $\mathcal{U}(S_1)$. For S_2, we only use a doubly linked list to store its upper hull $\mathcal{U}(S_2)$, and the stacks are not needed. In addition, we explicitly maintain the common tangent $\overline{t_1 t_2}$ of the two upper hulls $\mathcal{U}(S_1)$ and $\mathcal{U}(S_2)$, where t_1 and t_2 are the tangent points on $\mathcal{U}(S_1)$ and $\mathcal{U}(S_2)$, respectively. We also maintain the leftmost and rightmost points of S. This completes the description of our data structure for S.

Using the data structure we can output $\mathcal{U}(S)$ in $O(|\mathcal{U}(S)|)$ time as follows. Starting from the leftmost vertex of S_1, we follow the right neighbor pointers until we reach t_1, and then we output $\overline{t_1 t_2}$. Finally, we traverse $\mathcal{U}(S_2)$ from t_2 rightwards until the rightmost vertex. In the following, we discuss how to handle insertions and deletions.

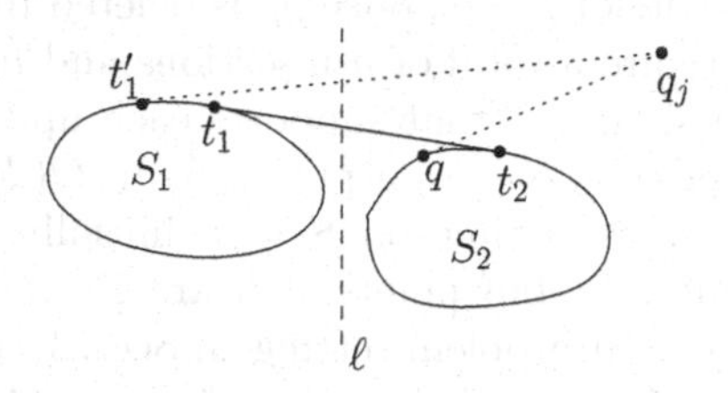

Fig. 1. Illustrating the insertion of q_j.

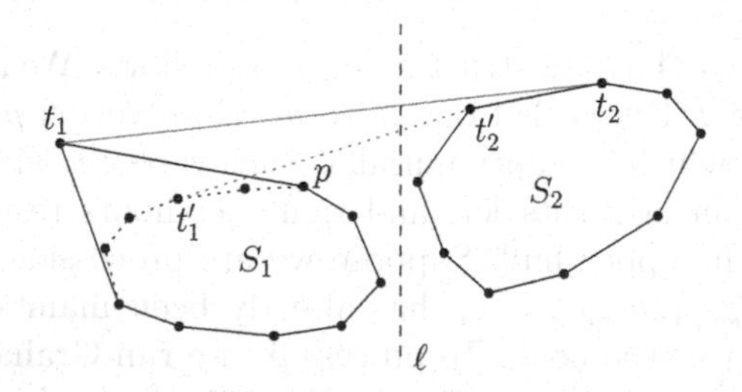

Fig. 2. Illustrating the deletion of p_i where $p_i = t_1$. $\overline{t_1' t_2'}$ are the new tangent of $\mathcal{U}(S_1)$ and $\mathcal{U}(S_2)$ after p_i is deleted.

3.2 Insertions

Suppose a point $q_j \in R$ is inserted into S. If $j = 1$, then this is the first insertion. We set $t_2 = q_1$ and find t_1 on $\mathcal{U}(S_1)$ by traversing it leftwards from p_n (i.e., by Graham's scan). This takes $O(n)$ time but happens only once in the entire algorithm (for processing all $2n$ insertions and deletions), so the amortized cost for the insertion of q_1 is $O(1)$. In the following we consider the general case $j > 1$.

We first update $\mathcal{U}(S_2)$ by Graham's scan. This procedure takes $O(n)$ time in total for all n insertions, and thus $O(1)$ amortized time per insertion. Let q be the vertex such that $\overline{qq_j}$ is the edge of the new hull $\mathcal{U}(S_2)$ (e.g., see Fig. 1). If q is strictly to the right of t_2, or if $q = t_2$ and $\overline{t_1t_2}$ and $\overline{t_2q_j}$ make a right turn at t_2, then $\overline{t_1t_2}$ is still the common tangent and we are done with the insertion. Otherwise, we update $t_2 = q_j$ and find the new t_1 by traversing $\mathcal{U}(S_1)$ leftwards from the current t_1, and we call it the *insertion-type tangent searching procedure*, which takes $O(1 + k)$ time, with k equal to the number of vertices on $\mathcal{U}(S_1)$ strictly between the original t_1 and the new t_1 (and we say that those vertices are *involved* in the procedure). The following lemma implies the amortized cost of the procedure is $O(1)$.

Lemma 1. *Each point $p \in L \cup R$ can be involved in the insertion-type tangent searching procedure at most once in the entire algorithm.*

3.3 Deletions

Suppose a point $p_i \in L$ is deleted from S_1. If $i = n$, then this is the last deletion. In this case, we only need to maintain $\mathcal{U}(S_2)$ for insertions only in future, which can be done by Graham's scan. In the following, we assume that $i < n$.

Note that p_i must be the leftmost vertex of the current hull $\mathcal{U}(S_1)$. Let $p = r(p_i)$ (i.e., p is the right neighbor of p_i on $\mathcal{U}(S_1)$). According to our data structure, p_i is at the top of the stack $Q(p)$. We pop p_i out of $Q(p)$. If $p_i \neq t_1$, then p_i is strictly to the left of t_1 and $\overline{t_1t_2}$ is still the common tangent of the new S_1 and S_2, and thus we are done with the deletion. Otherwise, we find the new tangent by simultaneously traversing on $\mathcal{U}(S_1)$ leftwards from p and traversing on $\mathcal{U}(S_2)$ leftwards from t_2 (e.g., see Fig. 2). Specifically, we first check whether $\overline{pt_2}$ is tangent to $\mathcal{U}(S_1)$ at p. If not, then we move p leftwards on the new $\mathcal{U}(S_1)$ until $\overline{pt_2}$ is tangent to $\mathcal{U}(S_1)$ at p. Then, we check whether $\overline{pt_2}$ is tangent to $\mathcal{U}(S_2)$ at t_2. If not, then we move t_2 leftwards on $\mathcal{U}(S_2)$ until $\overline{pt_2}$ is tangent to $\mathcal{U}(S_2)$ at t_2. If the new $\overline{pt_2}$ is not tangent to $\mathcal{U}(S_1)$ at p, then we move p leftwards again. We repeat the procedure until the updated $\overline{pt_2}$ is tangent to $\mathcal{U}(S_1)$ at p and also tangent to $\mathcal{U}(S_2)$ at t_2. Note that both p and t_2 are monotonically moving leftwards on $\mathcal{U}(S_1)$ and $\mathcal{U}(S_2)$, respectively. Note also that traversing leftwards on $\mathcal{U}(S_1)$ is achieved by using the stacks associated with the vertices while traversing on $\mathcal{U}(S_2)$ is done by using the doubly-linked list that stores $\mathcal{U}(S_2)$. We call the above the *deletion-type tangent searching procedure*, which takes $O(1 + k_1 + k_2)$ time, where k_1 is the number of points on $\mathcal{U}(S_1)$ strictly between p and the new tangent point t_1, i.e., the final position of p after the algorithm finishes (we say that these points are involved in the procedure), and k_2 is the number of points on $\mathcal{U}(S_2)$ strictly between the original t_2 and the new t_2 (we say that these points are involved in the procedure). The following lemma implies that the amortized cost of the procedure is $O(1)$.

Lemma 2. *Every point in $L \cup R$ can be involved in the deletion-type tangent searching procedure at most once in the entire algorithm.*

As a summary, in the special problem setting, we can perform each insertion and deletion in $O(1)$ amortized time, and after each update, the upper hull $\mathcal{U}(S)$ can be output in $|\mathcal{U}(S)|$ time.

4 The General Problem Setting

In this section, we extend our algorithm in Sect. 3 to the general problem setting without the restriction on the existence of the partition line ℓ. Specifically, we want to maintain the upper hull $\mathcal{U}(S)$ under window-sliding updates, with $S = \emptyset$ initially. We will show that each update can be handled in $O(1)$ amortized time and after each update $\mathcal{U}(S)$ can be output in $O(|\mathcal{U}(S)|)$ time. We will enhance the data structure in Sect. 5 so that it can handle other operations on $CH(S)$.

During the course of processing updates, we maintain a vertical line ℓ that will move rightwards. At any moment, ℓ plays the same role as in the problem setting in Sect. 3. In addition, ℓ always contains a point of S. Let S_1 be the subset of S to the left of ℓ (including the point on ℓ), and $S_2 = S \backslash S_1$. For S_1, we use the same data structure as before to maintain $\mathcal{U}(S_1)$, i.e., a doubly linked list for vertices of $\mathcal{U}(S_1)$ and a stack associated with each point of S_1, and we call it *the list-stack data structure*. For S_2, as before, we only use a doubly linked list to store the vertices of $\mathcal{U}(S_2)$. Note that $S_2 = \emptyset$ is possible. If $S_2 \neq \emptyset$, we also maintain the common tangent $\overline{t_1t_2}$ of $\mathcal{U}(S_1)$ and $\mathcal{U}(S_2)$, with $t_1 \in \mathcal{U}(S_1)$ and $t_2 \in \mathcal{U}(S_2)$. We can output the upper hull $\mathcal{U}(S)$ in $O(|\mathcal{U}(S)|)$ time as before.

For each $i \geq 1$, let p_i denote the i-th inserted point. Let U denote the universal set of all points p_i that will be inserted. Note that points of U are given online and we only use U for the reference purpose in our discussion (our algorithm has no information about U beforehand). We assume that S initially consists of two points p_1 and p_2. We let ℓ pass through p_1. According to the above definitions, we have $S_1 = \{p_1\}$, $S_2 = \{p_2\}$, $t_1 = p_1$, and $t_2 = p_2$. We assume that during the course of processing updates S always has at least two points, since otherwise we could restart the algorithm from this initial stage. Next, we discuss how to process updates.

Deletions. Suppose a point p_i is deleted. If p_i is not the only point of S_1, then we do the same processing as before in Sect. 3. We briefly discuss it here. If $p_i \neq t_1$, then we pop p_i out of the stack $Q(p)$ of p, where $p = r(p_i)$. In this case, we do not need to update $\overline{t_1t_2}$. Otherwise, we also need to update $\overline{t_1t_2}$, by the deletion-type tangent searching procedure as before. Lemma 2 is still applicable here (replacing $L \cup R$ with U), so the procedure takes $O(1)$ amortized time.

If p_i is the only point in S_1, then we do the following. We move ℓ to the rightmost point of S_2, and thus, the new set S_1 consists of all points in the old set S_2 while the new S_2 becomes $\emptyset$. Next, we build the list-stack data structure for S_1 by running Graham's scan leftwards from its rightmost point, which takes $|S_1|$ time. We call it the *left-hull construction procedure*. The following lemma implies that the amortized cost of the procedure is $O(1)$.

Lemma 3. *Every point of U is involved in the left-hull construction procedure at most once in the entire algorithm.*

Insertions. Suppose a point p_j is inserted. We first update $\mathcal{U}(S_2)$ using Graham's scan, and we call it the *right-hull updating procedure*. After that, p_j becomes the rightmost vertex of the new $\mathcal{U}(S_2)$. The procedure takes $O(1+k)$ time, where k is the number of vertices got removed from the old $\mathcal{U}(S_2)$ (we say that these points are *involved* in the procedure). By the standard Graham's scan, the amortized cost of the procedure is $O(1)$. Note that although the line ℓ may move rightwards, we can still use the same analysis as the standard Graham's scan. Indeed, according to our algorithm for processing deletions discussed above, once ℓ moves rightwards, all points in S_2 will be in the new set S_1 and thus will never be involved in the right-hull updating procedure again in future.

After $\mathcal{U}(S_2)$ is updated as above, we check whether the upper tangent $\overline{t_1t_2}$ needs to be updated, and if yes (in particular, if $S_2 = \emptyset$ before the insertion), we run an insertion-type tangent searching procedure to find the new tangent in the same way as before in Sect. 3. Lemma 1 still applies (replacing $L \cup R$ with U), and thus the procedure takes $O(1)$ amortized time. This finishes the processing of the insertion, whose amortized cost is $O(1)$.

As a summary, in the general problem setting, we can perform each insertion and deletion in $O(1)$ amortized time, and after each update, the upper hull $\mathcal{U}(S)$ can be output in $|\mathcal{U}(S)|$ time.

5 Convex Hull Queries

In this section, we enhance the data structure described in Sect. 4 to support logarithmic time convex hull queries as stated in Theorem 1. This is done by incorporating an interval tree into our data structure. Below, we first describe the interval tree in Sect. 5.1. We incorporate the interval tree into our data structure in Sect. 5.2. The data structure can support $O(\log |U|)$ time queries, where U is the universe of all points that will be inserted, under the assumption that the size $|U|$ is known initially when $S = \emptyset$. We finally lift the assumption in Sect. 5.3 and also reduce the query time to $O(\log n)$, with $n = |S|$.

5.1 The Interval Tree

We borrow an idea from Guibas et al. [15] and use interval trees. We build a complete binary search tree T whose leaves from left to right correspond to the indices from 1 to $|U|$. So the height of T is $O(\log |U|)$. For each leaf, if it corresponds to index i, then we assign i as the *index* of the leaf. For each internal node v, if i is the index of the rightmost leaf in its left subtree, then we assign $i + 1/2$ as the index of v, although it is not an integer. In this way, the sorted order of the indices of all nodes of T follows the symmetric order of the nodes. For a line segment $\overline{p_ip_j}$ connecting two points p_i and p_j of U, we say that the segment *spans* a node v, if the index of v is in the range $[i, j]$. Comparing to the interval tree in [15], which is defined with respect to the actual x-coordinates of the points, our tree is more abstract because it is defined on indices only.

Consider the set S maintained by our algorithm, which is a subset of U. We can store its upper hull $\mathcal{U}(S)$ in T as follows [15]. For each edge of $\mathcal{U}(S)$ connecting two vertices p_i and p_j, we store $\overline{p_i p_j}$ at the lowest common ancestor of leaves i and j in T (i.e., the highest node spanned by $\overline{p_i p_j}$; e.g., see Fig. 3). By also storing the lower hull of S in T as above, all queries on $CH(S)$ as specified in Theorem 1(3) can be answered in $O(\log |U|)$ time, by following a path from the root to a leaf [15] (the main idea is that the hull edge spanning a node v is stored either at v or at one of its ancestors, and only at most two ancestor edges closest to v need to be remembered during the search in T). In fact, our problem is slightly more complex because we need to store not only the edges of $\mathcal{U}(S)$ but also some historical hull edges. In the following, we incorporate this interval tree T into our data structure.

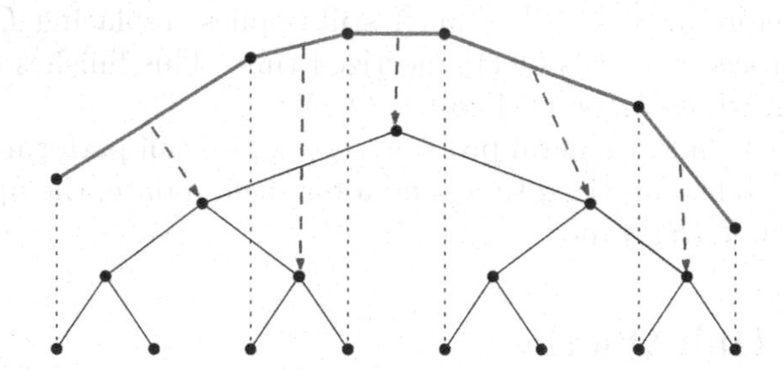

Fig. 3. Illustrating an upper hull and an interval tree that stores all hull edges: the (blue) dashed lines with arrows show where edges are stored. (Color figure online)

5.2 The Enhanced Data Structure

Unless otherwise stated, we follow the same notation as that in Sect. 4.

In addition to the data structure for storing S_1 and S_2 described in Sect. 4, we initially build the interval tree T. Then, we preprocess T in $O(|U|)$ time so that given any two nodes of the tree, their lowest common ancestor can be found in $O(1)$ time [3,16]. The amortized cost of this preprocessing is $O(1)$, for there will be $|U|$ insertions. In addition, we associate each node v of T with a stack, which is $\emptyset$ initially. For any two points p_i and p_j of U, we use $lca(p_i, p_j)$ to refer to the lowest common ancestor of the two leaves of T corresponding to the indices of the two points, respectively.

During the left-hull construction procedure for computing the list-stack data structure for S_1, we make the following changes. Whenever a vertex p_i is processed and an edge (p_i, p_j) is added as an edge to the current upper hull, in addition to setting $l(p_j) = p_i$, $r(p_i) = p_j$, and pushing p_i into the stack $Q(p_j)$ as before, we also push the edge (p_i, p_j) into the stack associated with the node $lca(p_i, p_j)$ of T. Thanks to the $O(1)$-time query performance of the lowest common ancestor query data structure [3,16], this change only adds constant time to each step in the original algorithm, and thus does not affect the time complexity asymptotically.

We make similar changes in the right-hull updating procedure for computing $\mathcal{U}(S_2)$. Specifically, assume that we are inserting a point p_j. We run Graham's scan on the vertices of $\mathcal{U}(S_2)$ from right to left. Suppose that we are scanning an edge $\overline{pp'}$ of $\mathcal{U}(S_2)$ and we find that it needs to be removed from the current upper hull. Then, we also pop $\overline{pp'}$ out of the stack associated with the node $lca(p, p')$ of T. After the Graham's scan is done, let p_k be the vertex that connects to p_j in the new hull (i.e., $\overline{p_k p_j}$ is the new edge). Then, we push $\overline{p_k p_j}$ into the stack associated with the node $lca(p_k, p_j)$ of T. Again, these changes do not change the time complexity of the algorithm asymptotically.

In addition, we store the common tangent $\overline{t_1 t_2}$ at the top of the stack associated with the node $lca(t_1, t_2)$ of T.

Deletions and Insertions. Consider the deletion of a point p_i. As before, depending on whether p_i is the only point of S_1, there are two cases.

1. If p_i is not the only point in S_1, then we do the same processing as before with the following changes. First, we pop the common tangent $\overline{t_1 t_2}$ out of the stack associated with the node $lca(t_1, t_2)$ of T. Second, when we pop p_i out of the stack $Q(p)$ with $p = r(p_i)$, we also pop the edge $\overline{p_i p}$ out of the stack at the node $lca(p_i, p)$ in T. Third, we push the new tangent (t_1, t_2) into the stack of the new $lca(t_1, t_2)$ of T. Note that the push and the pop operations of the common tangent $\overline{t_1 t_2}$ are always needed following the above order even if it does not change (otherwise, assume that we do not perform the pop operation in the first step, then in the second step when we attempt to pop $\overline{p_i p}$ out of the stack at the node $lca(p_i, p)$, $\overline{p_i p}$ may not be at the top of the stack because $\overline{t_1 t_2}$ may be at the top of the same stack).
2. If p_i is the only point in S_1, then according to our algorithm, we need to run the left-hull construction procedure on the points $\{p_{i+1}, p_{i+2}, \cdots, p_j\}$, where p_j is the rightmost point of S_2. Here we make the following changes. First, for each edge $\overline{p_k p_{k'}}$ of $\mathcal{U}(S_2)$, we pop $\overline{p_k p_{k'}}$ out of the stack associated with the node $lca(p_k, p_{k'})$ of T. Then, we run the left-hull construction procedure with the changes discussed above.

Consider the insertion of a point p_j. We first pop the tangent $\overline{t_1 t_2}$ out of the stack at the node $lca(t_1, t_2)$. Then, we run the right-hull updating procedure with the changes discussed above. Finally, we push the new tangent $\overline{t_1 t_2}$ (which may be the same as the original one) into the stack at the node $lca(t_1, t_2)$.

As discussed above, due to the $O(1)$ query time of the lowest common ancestor data structure, the amortized time of each insertion/deletion is still $O(1)$.

Queries. Next we discuss how to answer convex hull queries using the interval tree T. One difference between our interval tree T and that used in [15] is that there are stacks associated with the nodes of T, which may store edges not on $\mathcal{U}(S)$. Therefore, we cannot directly use the query algorithm in [15]. Rather, we need to make sure that non-hull edges in the stacks will not give us trouble. To this end, we first prove the following lemma (which further leads to Corollary 1).

Lemma 4. *Suppose a stack associated with a node of T contains more than one edge, and let e_1 and e_2 be any two edges in the stack such that e_1 is above e_2 in the stack (i.e., e_1 is stored closer to the top of the stack than e_2 is). Then, e_2 is vertically below e_1.*

Corollary 1. *The stack at any node of T can store at most one edge of $\mathcal{U}(S)$, and if it stores such an edge, then the edge must be at the top of the stack.*

Lemma 5 provides a foundation that guarantees the convex hull queries on $CH(S)$ can be answered in $O(\log|U|)$ time each. It resembles Lemma 4.1 in [15], but its proof relies on Lemma 4 to handle the non-hull edges in stacks.

Lemma 5. *We can walk in $O(\log|U|)$ time from the root to any leaf in T, at each node knowing which edge of $\mathcal{U}(S)$ spans the current node, or if none, to which side $\mathcal{U}(S)$ lies.*

With the algorithm in Lemma 5 as a "template", the convex hulls queries in Theorem 1(3) can all be answered in $O(\log|U|)$ time each. For example, consider the following query: Given a vertical line l, find the edge of $\mathcal{U}(S)$ that intersects l. If l is strictly to the left of the leftmost point of S or strictly to the right of the rightmost point of S, then the answer to the query is $\emptyset$. Otherwise, the query algorithm starts from the root of T. For each node v, we apply the algorithm in Lemma 5 to determine the edge e of $\mathcal{U}(S)$ that spans v (if there is no such a spanning edge, then the algorithm in Lemma 5 can determine to which side of v $\mathcal{U}(S)$ lies, and we proceed on the child of v on that side). If e intersects l, then we report e. Otherwise, if l is to the left (resp., right) of e, then we proceed on the left (resp., right) child of v. The time of the query algorithm is $O(\log|U|)$. Other queries can be handled similarly (see [15] for some details).

We can still output $\mathcal{U}(S)$ in $O(|\mathcal{U}(S)|)$ time as before. Hence, the performance of the first operation in Theorem 1 can be achieved. For the second operation in Theorem 1, we consider the upper hull first. Let l_1 and l_2 be the two query lines. Without loss of generality, we assume that both lines intersect $\mathcal{U}(S)$ and l_1 is to the left of l_2. Using T, we first find in $O(\log|U|)$ time the edge $\overline{p_ip_j}$ of $\mathcal{U}(S)$ that intersects l_1. Without loss of generality, we assume that p_j is to the right of p_i. Then, following the right neighbor pointer $r(p_j)$ of p_j, we can output the vertices of $\mathcal{U}(S)$ to the left of l_2 in $O(1)$ time per vertex in a way similar to that for outputting vertices of $\mathcal{U}(S)$. The lower hull can be treated likewise. Hence, the second operation of Theorem 1 can be performed in $O(k + \log|U|)$ time.

5.3 A Further Improvement

We further improve our data structure to remove the assumption that $|U|$ is known initially and reduce the query time from $O(\log|U|)$ to $O(\log|S|)$. The idea is to still use an interval tree T, but instead of building it initially, we periodically rebuild it during processing updates so that the number of leaves of T is always no more than $4|S|$, and thus the height of T is $O(\log|S|)$, which guarantees the $O(\log|S|)$ query time. As will be seen later, our algorithm maintains an invariant that whenever T is rebuilt, the number of its leaves is equal to $2|S|$, where S is the set when T is being rebuilt.

Initially when $S = \{p_1, p_2\}$, we build T with 4 leaves corresponding to indices $\{1, 2, 3, 4\}$. Let $T.max$ denote the index of the rightmost leaf of T. Initially $T.max = 4$. Let $|T|$ denote the number of leaves of T. During processing the updates, we keep track of the size of $|S|$ using a variable σ.

For each deletion, we decrease σ by one. If $\sigma = |T|/4$, we claim that at least σ deletions have happened since the current tree T was built. Indeed, let m be the size of S when T was just built. According to the algorithm invariant, $|T| = 2m$. Now that $\sigma = |T|/4$, at least $m - \sigma = |T|/4$ points have been deleted from S since T was built. The claim thus follows; let P denote the set of points in those deletions specified in the claim. We rebuild a tree T of $2 \cdot \sigma$ leaves corresponding to the indices $i, i+1, \ldots, i+2\cdot\sigma-1$, where i is the index of the leftmost point of S, and set $T.max = i + 2\cdot\sigma - 1$. Note that rebuilding T also includes adding the edges in the data structure for S to the stacks of the nodes of the new T, which can be done by running the left-hull construction procedure on S_1 and running the right-hull updating procedure on S_2. The total time for building the new tree is $O(\sigma)$. We charge the $O(\sigma)$ time to the deletions of P, whose size is at least σ by the above claim. In this way, each deletion is charged at most once, and thus the amortized cost for rebuilding T during all deletions is $O(1)$.

Consider an insertion of a point p_j (i.e., this is the j-th insertion). We first increment σ by one. If $j = T.max$, then we rebuild a new tree T with $2\cdot\sigma$ leaves corresponding to the indices $i, i+1, \ldots, i+2\cdot\sigma-1$, where i is the index of the leftmost point of S, and set $T.max = i+2\cdot\sigma-1$. The time for building the new tree is $O(\sigma)$. We charge the time to the insertions of the points of the second half of S (i.e., the rightmost $\sigma/2$ points of S; note that $\sigma = |S|$). Lemma 6 implies that each insertion will be charged at most once in the entire algorithm and thus the amortized cost for rebuilding T during all insertions is $O(1)$.

Lemma 6. *No point in the second half of S was charged before.*

References

1. Becker, B., Franciosa, P., Gschwind, S., Leonardi, S., Ohler, T., Widmayer, P.: Enclosing a set of objects by two minimum area rectangles. J. Algorithms **21**, 520–541 (1996)
2. Becker, B., Franciosa, P.G., Gschwind, S., Ohler, T., Thiem, G., Widmayer, P.: An optimal algorithm for approximating a set of rectangles by two minimum area rectangles. In: Bieri, H., Noltemeier, H. (eds.) CG 1991. LNCS, vol. 553, pp. 13–25. Springer, Heidelberg (1991). https://doi.org/10.1007/3-540-54891-2_2
3. Bender, M.A., Farach-Colton, M.: The LCA problem revisited. In: Gonnet, G.H., Viola, A. (eds.) LATIN 2000. LNCS, vol. 1776, pp. 88–94. Springer, Heidelberg (2000). https://doi.org/10.1007/10719839_9
4. Bentley, J.: Decomposable searching problems. Inf. Process. Lett. **8**, 244–251 (1979)
5. de Berg, M., Cheong, O., van Kreveld, M., Overmars, M.: Computational Geometry – Algorithms and Applications, 3rd edn. Springer, Berlin (2008). https://doi.org/10.1007/978-3-540-77974-2

6. Brodal, G., Jacob, R.: Dynamic planar convex hull. In: Proceedings of the 43rd IEEE Symposium on Foundations of Computer Science (FOCS), pp. 617–626 (2002). Full version available at arXiv:1902.11169
7. Brodal, G.S., Jacob, R.: Dynamic Planar Convex Hull with Optimal Query Time and O(log n · log log n) Update Time. In: SWAT 2000. LNCS, vol. 1851, pp. 57–70. Springer, Heidelberg (2000). https://doi.org/10.1007/3-540-44985-X_7
8. Bus, N., Buzer, L.: Dynamic convex hull for simple polygonal chains in constant amortized time per update. In: Proceedings of the 31st European Workshop on Computational Geometry (EuroCG) (2015)
9. Chan, T.: Optimal output-sensitive convex hull algorithms in two and three dimensions. Discrete Comput. Geom. **16**, 361–368 (1996). https://doi.org/10.1007/BF02712873
10. Chan, T.: Dynamic planar convex hull operations in near-logarithmic amortized time. J. ACM **48**, 1–12 (2001)
11. Chan, T.: Three problems about dynamic convex hulls. Int. J. Comput. Geom. Appl. **22**, 341–364 (2012)
12. Chan, T., Hershberger, J., Pratt, S.: Two approaches to building time-windowed geometric data structures. Algorithmica **81**, 3519–3533 (2019). https://doi.org/10.1007/s00453-019-00588-3
13. Friedman, J., Hershberger, J., Snoeyink, J.: Efficiently planning compliant motion in the plane. SIAM J. Comput. **25**, 562–599 (1996)
14. Graham, R.: An efficient algorithm for determining the convex hull of a finite planar set. Inf. Process. Lett. **1**, 132–133 (1972)
15. Guibas, L., Hershberger, J., Snoeyink, J.: Compact interval trees: a data structure for convex hulls. Int. J. Comput. Geom. Appl. **1**(1), 1–22 (1991)
16. Harel, D., Tarjan, R.: Fast algorithms for finding nearest common ancestors. SIAM J. Comput. **13**, 338–355 (1984)
17. Hershberger, J., Snoeyink, J.: Cartographic line simplification and polygon CSG formula in $O(n \log^* n)$ time. Comput. Geom. Theory Appl. **11**, 175–185 (1998)
18. Hershberger, J., Suri, S.: Applications of a semi-dynamic convex hull algorithm. BIT **32**, 249–267 (1992). https://doi.org/10.1007/BF01994880
19. Hershberger, J., Suri, S.: Offline maintenance of planar configurations. J. Algorithms **21**, 453–475 (1996)
20. Kaplan, H., Tarjan, R., Tsioutsiouliklis, K.: Faster kinetic heaps and their use in broadcast scheduling. In: Proceedings of the 20th Annual ACM-SIAM Symposium on Discrete Algorithms (SODA), pp. 836–844 (2001)
21. Kirkpatrick, D., Seidel, R.: The ultimate planar convex hull algorithm? SIAM J. Comput. **15**, 287–299 (1986)
22. Melkman, A.: On-line construction of the convex hull of a simple polygon. Inf. Process. Lett. **25**, 11–12 (1987)
23. Mulmuley, K.: Randomized multidimensional search trees: lazy balancing and dynamic shuffling. In: Proceedings of the 32nd Annual Symposium of Foundations of Computer Science (FOCS), pp. 180–196 (1991)
24. Overmars, M., van Leeuwen, J.: Maintenance of configurations in the plane. J. Comput. Syst. Sci. **23**(2), 166–204 (1981)
25. Preparata, F.: An optimal real-time algorithm for planar convex hulls. Commun. ACM **22**, 402–405 (1979)
26. Preparata, F., Shamos, M.: Computational Geometry. Springer, New York (1985). https://doi.org/10.1007/978-1-4612-1098-6

27. Preparata, F.P., Vitter, J.S.: A simplified technique for hidden-line elimination in terrains. In: Finkel, A., Jantzen, M. (eds.) STACS 1992. LNCS, vol. 577, pp. 133–146. Springer, Heidelberg (1992). https://doi.org/10.1007/3-540-55210-3_179
28. Schwarzkopf, O.: Dynamic maintenance of geometric structures made easy. In: Proceedings of the 32nd Annual Symposium of Foundations of Computer Science (FOCS), pp. 197–206 (1991)
29. Wang, H.: Algorithms for subpath convex hull queries and ray-shooting among segments. In: Proceedings of the 36th International Symposium on Computational Geometry (SoCG), pp. 69:1–69:14 (2020)

Realizability Makes A Difference: A Complexity Gap For Sink-Finding in USOs

Simon Weber[1(✉)] and Joel Widmer[2]

[1] Department of Computer Science, ETH Zurich, 8092 Zurich, Switzerland
simon.weber@inf.ethz.ch

[2] Department of Mathematics, ETH Zurich, 8092 Zurich, Switzerland

Abstract. Algorithms for finding the sink in Unique Sink Orientations (USOs) of the hypercube can be used to solve many algebraic, geometric, and combinatorial problems, most importantly including the P-Matrix Linear Complementarity Problem and Linear Programming. The *realizable* USOs are those that arise from the reductions of these problems to the USO sink-finding problem. Finding the sink of realizable USOs is thus highly relevant to various theoretical fields, yet it is unknown whether realizability can be exploited algorithmically to find the sink more quickly. However, the best known unconditional lower bounds for sink-finding make use of USOs that are provably not realizable. This indicates that the sink-finding problem might indeed be strictly easier on realizable USOs.

In this paper we show that this is true for a subclass of all USOs. We consider the class of Matoušek-type USOs, which are a translation of Matoušek's LP-type problems into the language of USOs. We show a query complexity gap between sink-finding in *all*, and sink-finding in only the *realizable* n-dimensional Matoušek-type USOs. We provide concrete deterministic algorithms and lower bounds for both cases, and show that in the realizable case $O(\log^2 n)$ vertex evaluation queries suffice, while in general exactly n queries are needed. The Matoušek-type USOs are the first USO class found to admit such a gap.

Keywords: Unique Sink Orientation · Realizability · Query Complexity

1 Introduction

A Unique Sink Orientation (USO) is an orientation of the hypercube graph with the property that each non-empty face has a unique sink. The most studied algorithmic problem related to USOs is that of finding the global sink: An algorithm has access to a *vertex evaluation oracle*, which can be queried with a vertex and

A version of this paper containing the omitted proofs and additional explanations can be accessed at https://arxiv.org/abs/2207.05985.

P. Morin and S. Suri (Eds.): WADS 2023, LNCS 14079, pp. 704–718, 2023.
https://doi.org/10.1007/978-3-031-38906-1_47

returns the orientation of all incident edges. The task is to determine the unique sink of the USO using as few such vertex evaluation queries as possible.

Progress on this problem has stalled for a long time. Since Szábo and Welzl introduced USOs in 2001 [19], their deterministic and randomized algorithms—both requiring an exponential number of queries in terms of the hypercube dimension—are still the best known in the general case. Only for special cases, such as for acyclic USOs, better algorithms are known [5].

Realizability. The study of USOs and sink-finding was originally motivated by a reduction of the P-Matrix Linear Complementarity Problem (P-LCP) to sink-finding in USOs [18]. In the P-LCP, one is given a so-called P-matrix[1] M and a vector q, and is tasked to find non-negative vectors w, z such that $w - Mz = q$ and $w^T \cdot z = 0$. Many widely studied optimization problems have been shown to be reducible either to the P-LCP or to sink-finding in USOs directly, the most notable example being Linear Programming (LP) [8], but also Convex Quadratic Programming [15] or the Smallest Enclosing Ball problem [9]. It has also been shown that various games on graphs reduce to the P-LCP, such as simple stochastic games, mean payoff games, and parity games [7], the unresolved complexity statuses of which have sparked considerable interest. To make progress on this wide array of problems, one would not need to find an algorithm which can find the sink quickly in *all* USOs, but only in the USOs which can arise from these reductions. One might think that these different problems reduce to sink-finding in wildly different classes of USOs, but in fact all these problems reduce to sink-finding in USOs that can also be generated by the reduction from P-LCP. We call USOs that can be obtained from this reduction *realizable.* In the literature, the realizable USOs are also referred to as *P-USOs* [3], *PLCP-orientations* [12], or *P-cubes* [14].

The number of n-dimensional realizable USOs is much smaller than the total number of USOs, namely $2^{\Theta(n^3)}$ in contrast to $2^{\Theta(2^n \log n)}$ [3]. Furthermore, a simple combinatorial property, the Holt-Klee condition, is known to hold for all realizable USOs [6,10]: In a realizable USO, there must be n vertex-disjoint directed paths from the source to the sink. Thus, one can for example conclude that the USO shown in Fig. 1 is not realizable. The best-known lower bound for the query complexity of the sink-finding problem (for deterministic algorithms) is $\Omega(n^2/\log n)$ [16]. This lower bound however relies on USOs which fail the Holt-Klee condition and are therefore not realizable [1]. For sink-finding on realizable USOs, only a lower bound of $\Omega(n)$ is known [1].

We thus see differences between general and realizable USOs in their number, structure, as well as the current knowledge of lower bounds. These three facts indicate that it may be possible to algorithmically exploit the features of realizable USOs to beat the algorithms of Szábo and Welzl on this important class of USOs. For P-LCP and certain cases of LP, the fastest deterministic combinatorial algorithms are already today based on sink-finding in USOs [2,8]. Improved sink-finding algorithms for realizable USOs would directly translate

[1] A P-matrix is a matrix whose determinants of all principal submatrices are positive.

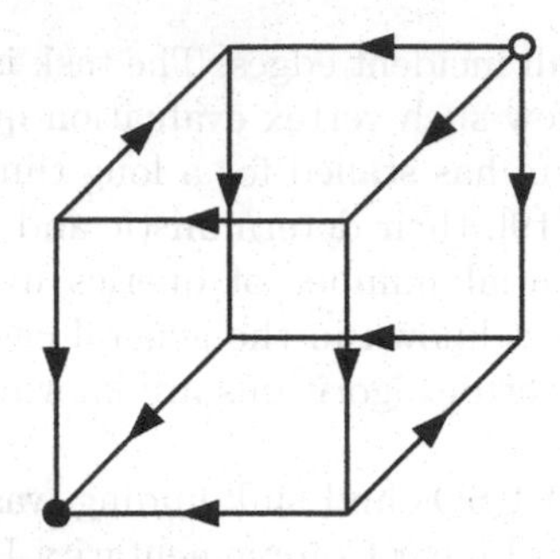

Fig. 1. A three-dimensional USO that is not realizable: there are no three vertex-disjoint paths from the source (top right) to the sink (bottom left). This USO thus does not satisfy the Holt-Klee condition.

to advances in P-LCP and LP algorithms. In particular, a sink-finding algorithm using polynomially many vertex evaluations on any realizable USO would imply the existence of a strongly polynomial-time algorithm for LP, answering a question from Smale's list of *mathematical problems for the next century* [17].

Matoušek(-type) USOs. In 1994, Matoušek introduced a family of LP-type problems (a combinatorial generalization of linear programs) to show a superpolynomial lower bound on the runtime of the Sharir-Welzl algorithm [13]. This result was later translated by Gärtner into the framework of USOs, where the *Matoušek USOs* provide a superpolynomial lower bound on the query complexity of the RANDOM FACET algorithm [5]. In the same paper it was also shown that the sink of all Matoušek USOs that fulfill the Holt-Klee property (thus including all realizable ones) is found by the RANDOM FACET algorithm in a quadratic number of queries. Therefore, RANDOM FACET is strictly (and substantially) faster on the realizable Matoušek USOs than on all Matoušek USOs. In this paper we aim to provide a similar result for the query complexity of the problem itself, instead of a concrete algorithm.

The Matoušek USOs all have the sink at the same vertex. This does not pose a problem when analyzing a fixed algorithm which does not exploit this fact (e.g., RANDOM FACET), but it does not allow us to derive algorithm-independent lower bounds. To circumvent this issue, we consider the class of *Matoušek-type USOs*, which simply contains all orientations isomorphic to classical Matoušek USOs.

It was recently discovered that all Matoušek-type USOs fulfilling the Holt-Klee property are also realizable [20], showing that the Holt-Klee property is not only necessary but also sufficient for realizability in Matoušek-type USOs. The proof of this result employed a novel view on Matoušek-type USOs, describing such a USO completely by (i) the location of the sink, and (ii) a directed graph with the n dimensions of the hypercube as vertices, called the *dimension influence graph*. We make heavy use of this view in the proofs of our results.

1.1 Results

We are now ready to state our results.

We show a query complexity gap between sink-finding on the realizable and sink-finding on all Matoušek-type USOs. We achieve this by proving the following two main theorems:

Theorem 1. *For every deterministic sink-finding algorithm $\mathcal{A}$ and any $n \geq 2$, there exists some n-dimensional Matoušek-type USO on which $\mathcal{A}$ requires at least n vertex evaluations to find the sink.*

Theorem 2. *There exists a deterministic algorithm finding the sink of any n-dimensional* realizable *Matoušek-type USO using $O(\log^2 n)$ vertex evaluations.*

In addition, we show that the result about general Matoušek-type USOs is tight. For the realizable case, we provide a lower bound of $\Omega(\log n)$ vertex evaluations.

1.2 Discussion

The Matoušek-type USOs form the first known USO class admitting such a complexity gap. We hope that this result motivates further research into tailored algorithms using the property of realizability for larger, more relevant classes of USOs.

Note that an artificial class of USOs exhibiting such a complexity gap could easily be constructed by combining a set R of easy-to-solve realizable USOs with a set N of difficult-to-solve non-realizable USOs. For R, one can take any set of realizable USOs which all have the same vertex as their sink. An algorithm to find the sink of USOs in R could then always output this vertex without needing to perform any vertex evaluations. For the set N, one can take the set of USOs constructed in the lower bound of Schurr and Szábo [16], and change each USO such that it becomes non-realizable. This can be achieved without destroying the lower bound. The resulting class $R \cup N$ then also exhibits a complexity gap.

The Matoušek-type USOs are not such an artificially constructed class of USOs. First off, they are well-studied due to their significance in proving the lower bound for the Random Facet algorithm [5,13]. Second, they can be considered a natural choice for proving unconditional lower bounds for realizable USOs: The only known unconditional lower bound for randomized algorithms on general USOs uses decomposable USOs [16], and the realizable Matoušek-type USOs are the only known class of realizable decomposable USOs.

Even on a natural USO class, a complexity gap could be trivial, for example if the class contains no (or only very few) realizable USOs. This is also *not* the case for the Matoušek-type USOs, as there are $2^{\Theta(n \log n)}$ realizable n-dimensional Matoušek-type USOs, while the overall number of Matoušek-type USOs is $2^{\Theta(n^2)}$. This is a much larger realizable fraction than one observes on the set of all USOs.

There is also an interesting connection between Matoušek-type USOs and *D-cubes*. The D-cubes are a subset of realizable USOs, obtained by the reduction to sink-finding from P-LCP instances where the P-matrix M is *symmetric*. They also include the USOs arising from the reduction of LP to sink-finding. Gao, Gärtner and Lamperski recently discovered that in a D-cube, the *L-graphs* at all vertices have to be acyclic [4]. The L-graphs are graphs that describe the local structure of the USO around a single vertex, and can be seen as a local version of the *dimension influence graph* encoding the structure of a Matoušek-type USO. In fact, a USO is a Matoušek-type USO if and only if the L-graphs are the same at every vertex. The techniques developed in this work to find the sink in the more rigid Matoušek-type USOs may be useful in developing algorithms for D-cubes, but this would first require a better understanding of the possible L-graphs in D-cubes. While all Matoušek-type USOs fulfill the necessary condition for being D-cubes, it remains open whether the realizable Matoušek-type USOs are in fact D-cubes.

1.3 Proof Techniques

For both the lower and the upper bounds, we need to view the Matoušek-type USOs by their dimension influence graphs. We first show an equivalence of finding a certain subset of the vertices in the dimension influence graph to finding the sink in the USO itself. Considering the adjacency matrix M of the dimension influence graph, finding the desired subset of vertices can be viewed as solving a linear system of equations $Mx = y$ over $\mathbb{Z}_2$, where M is only given by a matrix-vector product oracle, answering queries $q \in \{0,1\}^n$ with Mq.

For the lower bounds, we provide adversarial constructions to adapt the adjacency matrix of the dimension influence graph to the queries of the algorithm. Starting with the identity matrix, M is changed after some or all queries to ensure that the algorithm is not able to deduce the solution x. The main technical difficulties are to simultaneously ensure three conditions on a changing M: (i) that the previously given replies are consistent with the new matrix, (ii) that the algorithm still cannot deduce x after the change, and (iii) that the matrix describes a legal dimension influence graph of a (realizable) Matoušek-type USO.

For the upper bound in the realizable case, our algorithm does not find the sink of the USO directly, but instead recovers the whole dimension influence graph and thus the orientation of all USO edges. It can then compute the location of the sink without needing any more queries. The reply to each query contains only n bits of information, and the adjacency matrix M consists of n^2 bits. Any sublinear algorithm discovering the whole adjacency matrix must therefore leverage some additional structure of the graph. We use a previous result stating that the dimension influence graph of every realizable Matoušek-type USOs is the reflexive transitive closure of a branching [20]. Thanks to this rigid structure, we can split the problem into multiple subproblems, and we can combine queries used to solve different subproblems into a single query. This allows us to make progress on many subproblems in "parallel", leading to fewer queries needed overall.

2 Preliminaries

We begin with some basic notation. All vectors and matrices in this paper are defined over the field $\mathbb{Z}_2$. We write $\oplus$ for bit-wise addition ("xor") in $\mathbb{Z}_2$. By $\mathbf{0}$ (or $\mathbf{1}$) we denote the all-zero (or all-ones) n-dimensional vector. By e_i we denote the i-th standard basis vector. I denotes the $n \times n$ identity matrix. For a natural number x, we write $Bin(x)_i \in \{0,1\}$ for the i-th least significant bit of the binary representation of x, such that $\sum_{i=0}^{\infty} Bin(x)_i \cdot 2^i = x$.

2.1 Orientations and USOs

The n-dimensional hypercube Q_n is an undirected graph (V, E) consisting of the vertex set $V = \{0,1\}^n$, where two vertices are connected by an edge if they differ in exactly one coordinate. An orientation of the hypercube assigns a direction to each of the $n2^{n-1}$ edges.

Definition 3 (Hypercube Orientation). *An* orientation o *of* Q_n *is described by a function* $o : \{0,1\}^n \rightarrow \{0,1\}^n$ *assigning to each vertex its* outmap. *An edge* $(v, v \oplus e_i)$ *is directed away from* v *if* $o(v)_i = 1$. *To ensure consistent orientation of all edges,* o *has to fulfill* $o(v)_i \neq o(v \oplus e_i)_i$ *for all* $v \in V$ *and* $i \in [n]$.

Definition 4 (Unique Sink Orientation). *A* Unique Sink Orientation (USO) *is an orientation of* Q_n*, such that for each non-empty face* F *of the hypercube, the subgraph induced by* F *has a unique sink, i.e., a unique vertex which has no outgoing edges.*

A sink-finding algorithm has access to the orientation function o as an oracle. Given a vertex $v \in \{0,1\}^n$, the oracle returns its outmap $o(v)$. We are only interested in the number of such *vertex evaluation* queries made, and allow the algorithm to perform an unbounded amount of additional computation and put no restriction on the allowed memory consumption.

Definition 5 (Realizability). *A USO* o *is* realizable, *if there exists a non-degenerate P-Matrix Linear Complementarity Problem instance* (M, q) *such that the reduction of this instance to USO sink-finding produces* o.

We omit the formal definitions of the P-LCP and this reduction, as they are not needed to derive or understand our results. This reduction first appeared in the seminal paper of Stickney and Watson [18] in 1978. As their orientations were only named USOs much later [19], we point the interested reader to the comprehensive PhD thesis of Klaus [11], which uses more modern language.

2.2 (Realizable) Matoušek-Type USOs

A Matoušek USO, as defined by Gärtner [5], is an orientation o characterized by an invertible, upper-triangular matrix $A \in \{0,1\}^{n \times n}$ (thus all diagonal entries of A are 1). The matrix defines the orientation $o(v) = Av$. It is easy to see that each principal submatrix of A must also be invertible. This implies that there must be a unique sink in each face of the hypercube. In particular, the whole hypercube has a unique sink at the vertex $\mathbf{0}$.

To eliminate this commonality among Matoušek USOs, we define the more general Matoušek-type USOs, which are all orientations isomorphic to a Matoušek USO. Isomorphisms on the hypercube allow for mirroring of any subset of dimensions, and for relabeling the dimensions.

Definition 6 (Matoušek-type USO). *A* Matoušek-type USO *is an orientation o, with*

$$\forall v \in \{0,1\}^n: \quad o(v) = M(v \oplus s), \text{ where } M := PAP^T$$

for some permutation matrix $P \in \{0,1\}^{n\times n}$, some invertible, upper-triangular matrix $A \in \{0,1\}^{n\times n}$, and the desired location of the sink $s \in \{0,1\}^n$.

We can view the matrix M as the adjacency matrix of a directed graph $G = ([n], E_M)$ on the dimensions $[n]$, where $(i,j) \in E_M$ if $M_{j,i} = 1$. As A is invertible and upper-triangular, and as M is equal to A with rows and columns permuted in the same way, G is an acyclic graph with additional loop edges (i,i) at every vertex i. We also say that G is the *reflexive closure* of an acyclic graph.

We call this graph G the *dimension influence graph* of the Matoušek-type USO. The name is motivated by the following observation.

Remark 7. Let $\lambda, \varphi \in [n]$ be two distinct dimensions of a Matoušek-type USO o. For any vertex $v \in \{0,1\}^n$, it holds that

$$o(v)_\lambda \neq o(v \oplus e_\varphi)_\lambda \iff M_{\lambda,\varphi} = 1 \iff (\varphi, \lambda) \in E_M.$$

Intuitively, this means that in a Matoušek-type USO, any 2-dimensional face spanned by the same two dimensions λ and φ has the same structure: "Walking along" an edge in dimension φ either always changes the direction of the adjacent λ-edge, or never (see Fig. 2). If it always changes, we say that φ *influences* λ, and there is an edge from φ to λ in the dimension influence graph. See Fig. 3 for two example dimension influence graphs, their corresponding Matoušek-type USOs with the sink at vertex $\mathbf{0}$, and their adjacency matrices.

Weber and Gärtner showed that a Matoušek-type USO is realizable if and only if its dimension influence graph does not contain one of the two graphs in Fig. 3 as an induced subgraph [20]. This implies the following characterization.

Lemma 8 ([20, **Theorem 4.5**]). *A graph is the dimension influence graph of a realizable Matoušek-type USO if and only if it is the reflexive transitive closure of a branching*[2].

[2] Branching: a forest of rooted trees, with all edges directed away from the roots.

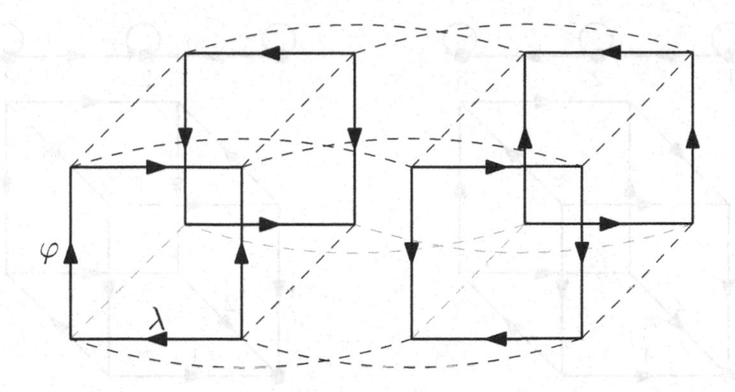

Fig. 2. All 2-faces spanned by λ and φ in this 4-dimensional Matoušek-type USO have the same structure: φ *influences* λ, but not the other way around.

2.3 Equivalence of Sink-Finding and Solving $Mx = y$

Based on Definition 6, we formulate an algebraic problem equivalent to sink-finding in Matoušek-type USOs.

Definition 9 ($Mx = y$ Problem). *For a matrix $M \in \{0,1\}^{n\times n}$ and a vector $y \in \{0,1\}^n$, the associated $Mx = y$ problem is to find the vector $x \in \{0,1\}^n$ fulfilling $Mx = y$. M is not explicitly included as part of the problem instance, but only an oracle is provided to the algorithm. The oracle answers* matrix-vector queries*: for any query $q \in \{0,1\}^n$, it returns Mq.*

We can show that up to a single additional query, the query complexities of the sink-finding problem and the $Mx = y$ problem are the same on Matoušek-type USOs and the matrices describing their dimension influence graphs. We formalize this in the following theorem, the proof of which we omit.

Theorem 10. *For any subclass $\mathcal{U}$ of n-dimensional Matoušek-type USOs closed under mirroring and dimension permutations, and for any $f : \mathbb{N} \to \mathbb{N}$, the following two statements are equivalent: (i) there exists a deterministic algorithm $\mathcal{A}$ to find the sink of any $o \in \mathcal{U}$ in $f(n)$ vertex evaluations, and (ii) there exists a deterministic algorithm $\mathcal{B}$ to find x fulfilling $Mx = y$ where M can be the adjacency matrix of the dimension influence graph of any $o \in \mathcal{U}$ in $f(n) - 1$ matrix-vector queries.*

As the mapping $q \mapsto Mq$ computed by the oracle is linear and invertible, the following two observations hold for any algorithm solving the $Mx = y$ problem:

Remark 11. If q is a linear combination of previously asked queries, the reply Mq can be computed without querying the oracle. Thus, every optimal algorithm only emits linearly independent queries.

Remark 12. If y is a linear combination of previously given replies, the algorithm can find the solution x with no additional queries.

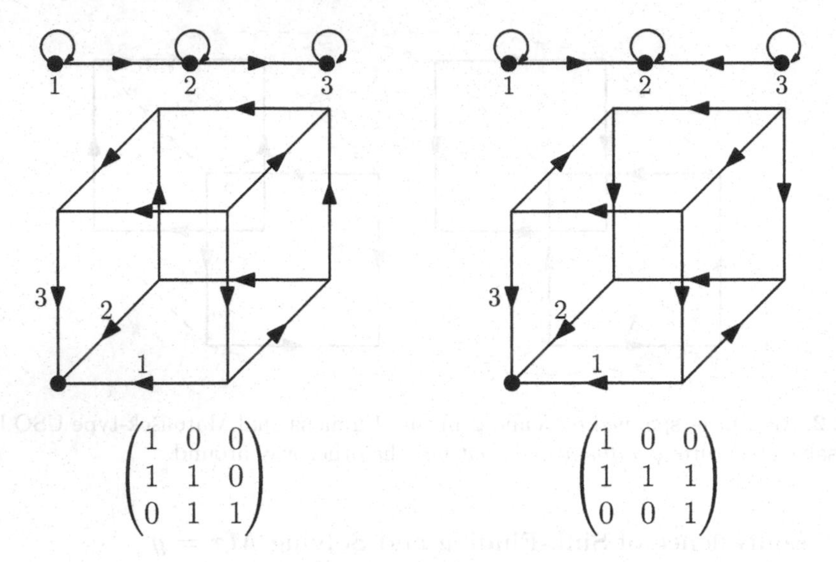

$$\begin{pmatrix} 1 & 0 & 0 \\ 1 & 1 & 0 \\ 0 & 1 & 1 \end{pmatrix} \qquad \begin{pmatrix} 1 & 0 & 0 \\ 1 & 1 & 1 \\ 0 & 0 & 1 \end{pmatrix}$$

Fig. 3. The two forbidden induced subgraphs (as in the text above Lemma 8) of realizable Matoušek-type USOs, their corresponding Matoušek-type USOs (with the sink at **0**), and the adjacency matrices of the forbidden graphs.

In some places it will be useful to interpret the matrix-vector queries in terms of vertex sets of the dimension influence graph.

Remark 13. For a query $q \in \{0,1\}^n$, the vertex set $\{i \in [n] : (Mq)_i = 1\}$ contains exactly the vertices which have an odd number of in-neighbors among the vertex set $\{i \in [n] : q_i = 1\}$ in G.

3 The General Case

It is not difficult to derive linear-time algorithms to find the sink in Matoušek-type USOs. The classical JumpAntipodal algorithm [15] can find the sink in n vertex evaluations on all *decomposable* USOs [16], which the Matoušek-type USOs are too. We are now going to prove Theorem 1, the matching lower bound to JumpAntipodal. The proof works in the framework of the $Mx = y$ problem. Algorithm 1 shows a strategy for an adversary to adaptively construct the matrix M in a way to force every deterministic algorithm to use at least $n - 1$ queries to find x. In terms of the dimension influence graph, this strategy can be seen as picking a sink j in every iteration, and adding/removing some edges towards j. This ensures that the graph represented by M always remains acyclic apart from the loops at each vertex, and thus a legal dimension influence graph.

Theorem 1. *For every deterministic sink-finding algorithm $\mathcal{A}$ and any $n \geq 2$, there exists some n-dimensional Matoušek-type USO on which $\mathcal{A}$ requires at least n vertex evaluations to find the sink.*

Algorithm 1. Adversarial Construction

1: $M^{(0)} \leftarrow I$
2: **for** $k \in \{1, \ldots, n-1\}$ **do**
3: $\quad x^{(k)} \leftarrow$ new linearly independent query from algorithm
4: $\quad$**if** $y \in \text{span}(M^{(k-1)}x^{(1)}, \ldots, M^{(k-1)}x^{(k)})$ **then**
5: $\quad\quad X \leftarrow \left(x^{(1)} \cdots x^{(k)}\right)^\top$
6: $\quad\quad$ *freevars* $\leftarrow$ non-pivot positions in the reduced row echelon form of X
7: $\quad\quad$ Pick $z^{(k)}$ such that $Xz^{(k)} = e_k$ and $z_i^{(k)} = 0$ for all $i \in$ *freevars*
8: $\quad\quad$ Pick $j \in$ *freevars* such that e_j is an eigenvector of $M^{(k-1)}$ $\quad \triangleright$ j is a sink
9: $\quad\quad M^{(k)} \leftarrow M^{(k-1)} + e_j z^{(k)T}$ $\quad \triangleright$ Add $z^{(k)}$ to j-th row of $M^{(k-1)}$
10: $\quad$**else**
11: $\quad\quad M^{(k)} \leftarrow M^{(k-1)}$ $\quad \triangleright$ No need to change M
12: $\quad$Answer query with $y^{(k)} := M^{(k)}x^{(k)}$

Proof (sketch). To prove this, we show three auxiliary properties of the instances constructed by Algorithm 1:

- Consistency: The replies to previous queries remain consistent.
- Legality: The graph defined by M remains a legal dimension influence graph, i.e., it is acyclic with added loops at every vertex.
- Uncertainty: After $k < n - 1$ queries, the algorithm cannot yet distinguish between some instances with different solutions.

We omit the proof that $z^{(k)}$ and j can be picked as described at least $n-1$ times.

Consistency: From the way $z^{(k)}$ is picked on line 7, it follows that we have $M^{(k)}x^{(k')} = M^{(k-1)}x^{(k')}$ for all $k' < k$, i.e., the replies to previous queries remain consistent.

Legality: We begin with a legal adjacency matrix. The loops at each vertex are never removed, as $z_j^{(k)} = 0$ due to lines 7–8. As e_j is an eigenvector, edges are only added/removed to sinks, preserving acyclicity.

Uncertainty: We can first show that after each iteration, the algorithm cannot deduce the solution through linear combination, i.e.,

$$\forall 0 \leq k \leq n-1: \;\; y \notin \text{span}(M^{(k)}x^{(1)}, \ldots, M^{(k)}x^{(k)}). \tag{1}$$

We can then show that after $k < n - 1$ queries, there exist two matrices which are both consistent with the given replies but have different solutions. The first such matrix is $M^{(k)}$. The second matrix is the matrix $M^{(k+1)}$ constructed by Algorithm 1 if it would be given the solution for $M^{(k)}$ as an additional query $x^{(k+1)} := {M^{(k)}}^{-1}y$. Due to Equation (1), $M^{(k+1)}x = y$ must have a different solution than $M^{(k)}x = y$.

Conclusion: Given any $n - 1$ queries, Algorithm 1 produces a series of legal matrices $M^{(k)}$ (Legality) which are always consistent with the previously given replies (Consistency). No algorithm can know the solution in fewer than $n - 1$

queries, as it cannot distinguish between matrices with different solutions (Uncertainty). By Theorem 10, no algorithm can find the sink of an n-dimensional Matoušek-type USO in fewer than n vertex evaluations in the worst case.

4 The Realizable Case

In this section we prove our second main result, the upper bound for the realizable case.

Theorem 2. *There exists a deterministic algorithm finding the sink of any n-dimensional* realizable *Matoušek-type USO using $O(\log^2 n)$ vertex evaluations.*

To prove Theorem 2, we provide a concrete algorithm in the matrix-vector query model to recover the matrix M. The algorithm makes heavy use of the structure of the graph G described by this matrix, which has to be the reflexive transitive closure of a branching (see Lemma 8). Recall that we can view the matrix-vector queries also as sets of vertices of the dimension influence graph (Remark 13). In a slight abuse of notation, we will sometimes use the name v of a vector $v \in \{0,1\}^n$ to also denote the set $\{i \in [n] : v_i = 1\}$.

We first take a closer look at the structure of G. The underlying branching (technically, the unique reflexive transitive reduction of G) can be decomposed into levels, where the roots are on level 0, and the children of a vertex on level ℓ are on level $\ell + 1$. In the reflexive transitive closure G, we can see that the in-degree of each vertex is equal to its level plus 1 (due to the loops).

Remark 14. A vertex v on level ℓ has exactly $\ell + 1$ incoming edges, and v has exactly one in-neighbor on each level $\ell' \in \{0, \dots, \ell\}$.

Definition 15. *For a vertex v on level ℓ and some level $\ell' < \ell$, the* ℓ'-ancestor *of v is the unique in-neighbor of v on level ℓ'. The* parent *of v is the $\ell - 1$-ancestor of v.*

Our proposed algorithm works in two main phases. In the first phase, the *levelling*, it determines the level of each vertex in $O(\log n)$ queries. In the second phase, we use a divide-and-conquer approach to partition the vertices and perform queries to find the edges within each partition simultaneously, requiring $O(\log^2 n)$ queries in total.

Lemma 16. *The level of each vertex can be computed using $O(\log n)$ queries.*

Proof (sketch). The strategy is laid out in detail in Algorithm 2, "Levelling", in the full version of this paper.

In essence, in each iteration i, we query the vector corresponding to the set of vertices on levels ℓ with $Bin(\ell)_{i'} = 1$ for all $i' < i$. From the reply we can determine which vertices lie on levels ℓ' with $Bin(\ell')_i = 1$. After $O(\log n)$ iterations, all bits of the levels of all vertices are known.

We now know how to compute the level of each vertex in $O(\log n)$ time. It remains to show that we can also determine the edges connecting each consecutive two levels, and thus recover the whole graph.

We first give the intuition for a simple strategy that given a level ℓ finds the ℓ-ancestor of all vertices on levels $\geq \ell + 1$. We can perform $\lceil \log_2 n \rceil$ queries, where the i-th query $q^{(i)}$ contains all vertices v on level ℓ with $Bin(v)_i = 1$. If a vertex is "hit", i.e., it is contained in the i-th reply $Mq^{(i)}$, we know that its ℓ-ancestor must be in $q^{(i)}$. As each binary representation uniquely determines an integer, after all $\lceil \log_2 n \rceil$ queries, the ℓ-ancestors of every vertex are found.

It would be too costly to use this procedure for all levels on their own, as there can be up to n levels. We thus make use of the following crucial observation, which follows directly from the transitivity and reflexivity of the dimension influence graph.

Remark 17. Let ℓ be some level, q some query, and Mq the corresponding reply. Furthermore, let w be some vertex on level $\ell' > \ell$, with the ℓ-ancestor of w being a. We form the alternative query q' by removing from q the vertex a as well as all vertices of levels strictly smaller than level ℓ. It holds that

$$(Mq')_w = (Mq)_w \oplus (Mq)_a.$$

This observation allows us to "filter out" the effect of querying vertices above a certain level ℓ on the vertices below ℓ, as long as their ℓ-ancestors are known. Using this crucial tool, we can now solve the $Mx = y$ problem with a divide-and-conquer approach. If we use the previously given strategy to find all ℓ-ancestors for a level ℓ roughly in the middle of the branching, we can then split the graph into two halves, the one above ℓ and the one below ℓ. Using Remark 17, we can proceed in each of the two subproblems simultaneously, as the effect of queries used to make progress in the upper half can be filtered out of the responses to the queries used to make progress in the lower half. We describe this process in detail in Algorithm 3, "Divide-And-Conquer", in the full version of this paper.

In essence, the algorithm keeps a list of disjoint subproblems, where each subproblem is described by an interval of levels $\{a_i, a_i + 1, \ldots, b_i\}$. In each iteration, a median level m_i of each subproblem is picked, and the previously mentioned procedure is performed to determine the m_i-ancestors of all vertices in the subproblem. Remark 17 is applied with $\ell := b_i$ to filter out the effect of the vertices queried for the subproblem $\{a_i, \ldots, b_i\}$ from the later subproblems. Finally, each subproblem is split into two parts, the one above m_i, and the one strictly below m_i.

Lemma 18. *Knowing the level of each vertex, Divide-And-Conquer can determine the parent of every vertex v in $O(\log^2 n)$ queries.*

Proof (sketch). In each iteration, Divide-And-Conquer finds a closer ancestor of each vertex, until the parent of all vertices is known. As the size of the subproblems is halved in each iteration, $O(\log n)$ iterations are needed. Each iteration takes $O(\log n)$ queries, resulting in $O(\log^2 n)$ total queries needed.

Proof (of theorem 2). Using Levelling to compute the level of each vertex in the dimension influence graph G and Divide-And-Conquer to find the parent of every vertex, G can be completely recovered in $O(\log^2 n)$ matrix-vector queries. Furthermore, knowing all of the parents, M^{-1} can be constructed efficiently, and $x = M^{-1}y$ can be computed with no additional queries. By Theorem 10, the sink of a realizable Matoušek-type USO can be found in $O(\log^2 n)$ vertex evaluations.

We believe that $\Theta(\log^2 n)$ is the best-possible number of queries to find the sink of a realizable Matoušek-type USO. Due to the rigid structure of the dimension influence graph, a large portion of the n bits of information in every reply from the oracle is redundant. We did not manage to prove this matching lower bound, but we can show a lower bound of $\Omega(\log n)$.

Theorem 19. *Every deterministic algorithm requires at least $\Omega(\log n)$ queries to find the sink of an n-dimensional realizable Matoušek-type USO in the worst case.*

Proof (sketch). The proof is similar to the proof of the lower bound for the general case (Theorem 1), but much simpler. The adversary sets $y = \mathbf{1}$, resulting in the algorithm needing to find the roots of the dimension influence graph. The graph is set to be the union of disjoint paths, and the adversary adaptively chains these paths together. This is again done in a manner that ensures queries to be consistent, and the algorithm to remain uncertain of the solution. The main insight is that a path can be added to the end of a path P with $|P \cap x^{(i)}|$ even for all i, without changing the response to these queries $x^{(i)}$.

5 Conclusion

We have determined the asymptotic query complexity of finding the sink in general Matoušek-type USOs exactly. For realizable Matoušek-type USOs, there remains an $O(\log n)$ gap between our lower and upper bound. While it would be interesting to close this gap, our main result—the gap between the realizable and the general case—is already well established by our bounds.

As the best-known sink-finding algorithms are randomized, it would be desirable to establish a complexity gap also for randomized algorithms. The upper bound naturally carries over, as all deterministic algorithms are also randomized algorithms. The most natural approach to establish a lower bound would be applying Yao's principle [21]. On the flip side, it might also be possible to improve upon our algorithms both for the realizable and the general case using randomness, but we did not observe any straightforward benefit of randomness.

The most important open question implied by our results is whether there are other (larger and more practically relevant) USO classes which also admit such a complexity gap. Ultimately, we hope for such a gap to exist for the class of *all* USOs. Considering the lack of a strong lower bound and good characterizations of realizability, this goal is still far away.

Finally, the connections between Matoušek-type USOs and D-cubes mentioned in Sect. 1 can be examined further. Are all or at least some of the realizable Matoušek-type USOs also D-cubes? Can our techniques used to find the sink of a Matoušek-type USO be adapted to work for the less rigid D-cubes?

Acknowledgments. We thank Bernd Gärtner for his valuable insights and feedback. This research was supported by the Swiss National Science Foundation under project no. 204320.

References

1. Borzechowski, M., Weber, S.: On degeneracy in the P-matroid oriented matroid complementarity problem (2023). https://doi.org/10.48550/ARXIV.2302.14585, appeared in Abstracts of the 39th European Workshop on Computational Geometry (EuroCG '23), Barcelona, Spain, 2023, 9:1–9:7
2. Fearnley, J., Gordon, S., Mehta, R., Savani, R.: Unique end of potential line. J. Comput. Syst. Sci. **114**, 1–35 (2020). https://doi.org/10.1016/j.jcss.2020.05.007
3. Foniok, J., Gärtner, B., Klaus, L., Sprecher, M.: Counting unique-sink orientations. Discret. Appl. Math. **163**, 155–164 (2014). https://doi.org/10.1016/j.dam.2013.07.017
4. Gao, Y., Gärtner, B., Lamperski, J.: A new combinatorial property of geometric unique sink orientations (2020). https://doi.org/10.48550/ARXIV.2008.08992
5. Gärtner, B.: The random-facet simplex algorithm on combinatorial cubes. Random Struct. Algorithms **20**(3), 353–381 (2002). https://doi.org/10.1002/rsa.10034
6. Gärtner, B., Morris jr., W.D., Rüst, L.: Unique sink orientations of grids. Algorithmica **51**(2), 200–235 (2008). https://doi.org/10.1007/s00453-007-9090-x
7. Gärtner, B., Rüst, L.: Simple stochastic games and P-Matrix generalized linear complementarity problems. In: Liśkiewicz, M., Reischuk, R. (eds.) FCT 2005. LNCS, vol. 3623, pp. 209–220. Springer, Heidelberg (2005). https://doi.org/10.1007/11537311_19
8. Gärtner, B., Schurr, I.: Linear programming and unique sink orientations. In: Proceedings of the 17th Annual ACM-SIAM Symposium on Discrete Algorithms (SODA), pp. 749–757 (2006). https://doi.org/10.5555/1109557.1109639
9. Gärtner, B., Welzl, E.: Explicit and implicit enforcing - randomized optimization. In: Alt, H. (ed.) Computational Discrete Mathematics. LNCS, vol. 2122, pp. 25–46. Springer, Heidelberg (2001). https://doi.org/10.1007/3-540-45506-X_3
10. Holt, F., Klee, V.: A proof of the strict monotone 4-step conjecture. Contemp. Math. **223**, 201–216 (1999)
11. Klaus, L.: A fresh look at the complexity of pivoting in linear complementarity. Ph.D. thesis, ETH Zürich (2012). https://doi.org/10.3929/ethz-a-007604201
12. Klaus, L., Miyata, H.: Enumeration of PLCP-orientations of the 4-cube. Eur. J. Comb. **50**, 138–151 (2015). https://doi.org/10.1016/j.ejc.2015.03.010, http://www.sciencedirect.com/science/article/pii/S0195669815000712, combinatorial Geometries: Matroids, Oriented Matroids and Applications. Special Issue in Memory of Michel Las Vergnas
13. Matoušek, J.: Lower bounds for a subexponential optimization algorithm. Random Struct. Algorithms **5**(4), 591–607 (1994). https://doi.org/10.1002/rsa.3240050408
14. Morris jr., W.D.: Randomized pivot algorithms for p-matrix linear complementarity problems. Math. Program. **92**(2), 285–296 (2002). https://doi.org/10.1007/s101070100268

15. Schurr, I.: Unique Sink Orientations of Cubes. Ph.D. thesis, ETH Zürich (2004). https://doi.org/10.3929/ethz-a-004844278
16. Schurr, I., Szabó, T.: Finding the sink takes some time: an almost quadratic lower bound for finding the sink of unique sink oriented cubes. Discret. Comput. Geom. **31**(4), 627–642 (2004). https://doi.org/10.1007/s00454-003-0813-8
17. Smale, S.: Mathematical problems for the next century. Math. Intell. **20**, 7–15 (1998). https://doi.org/10.1007/BF03025291
18. Stickney, A., Watson, L.: Digraph models of Bard-type algorithms for the linear complementarity problem. Math. Oper. Res. **3**(4), 322–333 (1978). https://www.jstor.org/stable/3689630
19. Szabó, T., Welzl, E.: Unique sink orientations of cubes. In: Foundations of Computer Science, 2001. Proceedings. 42nd IEEE Symposium on, pp. 547–555. IEEE (2001). https://doi.org/10.1109/SFCS.2001.959931
20. Weber, S., Gärtner, B.: A characterization of the realizable Matoušek unique sink orientations (2021). https://doi.org/10.48550/ARXIV.2109.03666
21. Yao, A.C.: Probabilistic computations: toward a unified measure of complexity. In: 18th Annual Symposium on Foundations of Computer Science (SFCS 1977), pp. 222–227 (1977). https://doi.org/10.1109/SFCS.1977.24

Author Index

P. Morin and S. Suri (Eds.): WADS 2023, LNCS 14079, pp. 719–721, 2023.
https://doi.org/10.1007/978-3-031-38906-1

Printed in the United States
by Baker & Taylor Publisher Services

Printed in the United States
by Baker & Taylor Publisher Services